SEVENTH EDITION

NUMERICAL MATHEMATICS AND COMPUTING

Ward Cheney & David Kincaid
The University of Texas at Austin

BROOKS/COLE
CENGAGE Learning

Australia • Brazil • Japan • Korea • Mexico • Singapore • Spain • United Kingdom • United States

BROOKS/COLE
CENGAGE Learning

Numerical Mathematics and Computing, Seventh Edition, International Edition
Ward Cheney & David Kincaid

Publisher: Richard Stratton

Acquisitions Editor: Molly Taylor

Assistant Editor: Shaylin Walsh Hogan

Editorial Assistant: Alex Gontar

Media Editor: Andrew Coppola

Marketing Manager: Jennifer P. Jones

Marketing Communications Manager: Mary Anne Payumo

Content Project Manager: Alison Eigel Zade

Senior Art Director: Linda May

Manufacturing Planner: Doug Bertke

Rights Acquisition Specialist: Shalice Shah-Caldwell

Production Service: MPS Limited

Interior Chapter Opener image: Roofoo/©istockphoto

Illustrator: MPS Limited

Cover Designer: KeDesign

Cover Image: Roofoo/©istockphoto

Compositor: MPS Limited

International Edition:

ISBN-13: 978-1-133-49181-1

ISBN-10: 1-133-49181-2

Cengage Learning International Offices

Asia
www.cengageasia.com
tel: (65) 6410 1200

Australia/New Zealand
www.cengage.com.au
tel: (61) 3 9685 4111

Brazil
www.cengage.com.br
tel: (55) 11 3665 9900

India
www.cengage.co.in
tel: (91) 11 4364 1111

Latin America
www.cengage.com.mx
tel: (52) 55 1500 6000

UK/Europe/Middle East/Africa
www.cengage.co.uk
tel: (44) 0 1264 332 424

Represented in Canada by Nelson Education, Ltd.
www.nelson.com
tel: (416) 752 9100 / (800) 668 0671

Cengage Learning is a leading provider of customized learning solutions with office locations around the globe, including Singapore, the United Kingdom, Australia, Mexico, Brazil, and Japan. Locate your local office at: **www.cengage.com/global**

For product information and free companion resources: **www.cengage.com/international**
Visit your local office: **www.cengage.com/global**

Printed in the United States of America
1 2 3 4 5 6 7 16 15 14 13 12

Contents

Contents

Preface

Our basic objective is to acquaint students of science and engineering with the potentialities of using computers for solving numerical problems that may arise in their professions. A secondary objective is to give students an opportunity to hone their skills in computer programming and problem solving. A final objective is to help students arrive at an understanding of the important subject of *errors* that inevitably arise in scientific computing as well as learning methods for detecting, predicting, and controlling them.

Much of science today involves complex computations built upon mathematical software systems. The users may have little knowledge of the underlying numerical algorithms used in these problem-solving environments. By studying numerical methods, one can become a more informed user and be better prepared to evaluate and judge the accuracy of the results. Students should study mathematical algorithms to learn not only how they work, but also how they can fail! Critical thinking and constant skepticism are traits we want students to acquire. An extensive numerical calculation, even when carried out by state-of-the-art software, may need to be subjected to independent verification, if possible.

We have tried to achieve an elementary style of presentation since we want this book to be accessible to students who may not be advanced in their formal study of mathematics and computer sciences. Toward this end, we have provided numerous examples and figures for illustrative purposes and fragments of pseudocode, which are informal descriptions of computer algorithms.

Believing that most students at this level need a *survey* of the subject of numerical mathematics and computing, we have presented a wide diversity of topics, including some rather advanced ones that play an important role in scientific computing. We recommend that the reader have at least a one-year study of calculus, plus some basic knowledge of matrices, vectors, and ordinary differential equations.

Seventh Edition Features

Following suggestions and comments from the reviewers and based on our experience in teaching this material, we have revised and enhanced the entire book.

- Some chapters have been combined and the order of others changed. Also, the Linear Systems chapter has been moved earlier because it is a topic that arises frequently and needs to be presented sooner.
- A new section on Fourier Series has been added because it arises in various engineering applications. The section on lower and upper sums in numerical integrations has been removed.
- Now displayed in double columns, the problems have been renamed as exercises to clarify that they are intended for practice and for further learning.
- In an effort to make the new edition more *student friendly*, margin notes have been added as well as additional figures, tables, examples, and exercises.

Suggestions for Use

Numerical Mathematics and Computing, Seventh Edition, can be used in a variety of ways, depending on the emphasis the instructor prefers and the inevitable time constraints. Exercises have been supplied in abundance to enhance the book's versatility. They are divided into two categories: *Exercises* and *Computer Exercises*. In the first category, there are more than 800 exercises in analysis that require pencil, paper, and possibly a calculator. In the second category, there are approximately 500 computer exercises involving programming and using a computer. Students are asked to solve some exercises using advanced software systems such as MATLAB, Mathematica, or Maple. While students may be asked to write their own computer codes, they can often follow a model or example to assist them, but in other cases they must proceed on their own from a mathematical description.

In some of the Computer Exercises, there is something to be learned beyond simply writing code—a *moral*, if you like. For example, this can happen if the exercise being solved and the pseudocode provided are somehow mismatched. Some Computing Exercises are designed to give the students experience in using, mathematical software.

The *Student Solution Manual* is sold as a separate publication. Also, teachers who adopt the book can obtain the *Instructor Solution Manual* from the publisher or via the Instructor Companion website. Sample programs based on the pseudocode have been coded in several programming languages and are on the textbook website.

To access course materials and companion resources, please visit the websites on p. iv; e.g., `www.cengagebrain.com`. At the `CengageBrain.com` home page, one can search using either the International Standard Book Number (ISBN), the title of the book, or the last name of one of the authors. This takes you to the product page, where free companion resources can be found. The authors have established the following website:

`www.ma.utexas.edu/CNA/NMC7/`

The arrangement of chapters reflects our own view of how the material might best unfold for a student new to the subject. In some cases, there may be little mutual dependence among the chapters, and the instructor can order the sequence of presentation in various ways. Certainly, instructor may omit some sections and chapters for want of time.

Our own recommendations for courses based on this text are as follows:

- A one-term course carefully covering Chapters 1 through 7 (possibly omitting Sections 4.2, 6.3, and 7.3–5, for example), followed by a selection of material from the remaining chapters as time permits.
- A one-term survey rapidly skimming over almost all of chapters and omitting some of the more advanced material.
- A two-term course carefully covering all chapters.

Student Research Projects

Throughout there are some Computer Exercises designated as *Student Research Projects*. These are opportunities for students to explore topics beyond the scope of the textbook. Many of these involve application areas for numerical methods. These projects usually include computer programming and numerical experiments. A favorable aspect of these assignments is to allow students to choose a topic of interest to them, possibly something that may arise in their future profession or their major study area. For example, any topic suggested may be delved into more deeply by consulting other texts and references. In preparing such a project, the students have to learn about the topic, locate the significant references (books, research papers, and websites), do the computing, and write a report that explains all this in a coherent way. Students can avail themselves of mathematical software systems such as MATLAB, Maple, or Mathematica, as well as doing their own computer programming in whatever programming language they prefer.

Acknowledgments

We have profited from advice and suggestions kindly offered by a large number of colleagues, students, and users of the previous editions.

We are grateful to have had the opportunity and privilege of teaching classes in scientific computing, numerical analysis, and various other topics in the Department of Mathematics and the Department of Computer Sciences of The University of Texas at Austin. Without their support and the use of their computing facilities, writing this book would not have been possible. In particular, we thank Maorong Zou and Margaret Combs for help with computer issues.

Recently, the second author taught classes in the Department of Petroleum Engineering and Geosystems Engineering of The University of Texas at Austin and in the Applied Mathematics and Computer Sciences Division of the King Abdullah Univeristy of Science and Technology in Saudi Arabia and wishes to thank them both.

Valuable comments and suggestions were made by our colleagues and friends. In particular, we miss our friend David Young who was always very generous with suggestions for improving the accuracy and clarity of the exposition in previous editions. Some parts of those editions were typed with great care and attention to detail by Sheri Brice, Katy Burrell, Kata Carbone, Margaret Combs, and Belinda Trevino. Aaron Naiman was particularly helpful in sharing material he prepared for his courses.

We wish to acknowledge the reviewers who have provided detailed critiques for this new edition: Eugino Aulisa, Texas Tech University; Erin Bach, University of Wisconsin, Madison; Marcin Bownik, University of Oregon; Olga Brezhneva, Miami University; George Grossman, Central Michigan University; Luke Olson, University of Illinois at Urbana-Champaign; Ronald Taylor, Wright State University.

Reviewers from previous editions were Krishan Agrawal, Eric Back, Neil Berger, Thomas Boger, Marcin Bownik, Olga Brezhneva, Jose E. Castillo, Charles Collins, Charles Cullen, Elias Y. Deeba, F. Emad, Gentil A. Estévez, Terry Feagin, Jose Flores, Leslie Foster, Bob Funderlic, Mahadevan Ganesh, William Gearhart, Juan Gil, John Gregory, George Grossman, Bruce P. Hillam, Patrick Lang, Ren Chi Li, Wu Li, Xiaofan Li, Vania Mascioni, Bernard Maxum, Edward Neuman, Roy Nicolaides, Luke Olson, Amar Raheja, J. N. Reddy, Daniel Reynolds, Asok Sen, Ching-Kuang Shene, William Slough, Ralph Smart, Thiab Taha, Ronald Taylor, Jin Wang, Stephen Wirkus, Marcus Wright, Quiang Ye, Tjalling Ypma, and Shangyou Zhan.

Many individuals took the time to write us with suggestions and criticisms. We are grateful to the following individuals and others who have send us e-mails concerning the textbook or solution manuals: A. Aawwal, Nabeel S.Abo-Ghander, Krishan Agrawal, Roger Alexander, Husain Ali Al-Mohssen, Kistone Anand, Keven Anderson, Vladimir Andrijevik, Jon Ashland, Hassan Basir, Steve Batterson, Neil Berger, Adarsh Beohar, Bernard Bialecki, Jason Brazile, Keith M. Briggs, Carl de Boor, Jose E. Castillo, Fatih Celiker, Debao Chen, Ellen Chen, Hwen Chin, Edmond Chow, Lloyd Clark, John Cook, Brad Copper, Roger Crawfis, Charles Cullen, Antonella Cupillari, Jonathan Dautrich, James Arthur Davis, Tim Davis, Elias Y. Deeba, Suhrit Dey, Alan Donoho, Jason Durheim, Wayne

Dymacek, John Eisenmenger, Fawzi P. Emad, Paul Enigenbury, Terry Feagin, Leslie Foster, Peter Fraser, Richard Gardner, Mohamad El Gharamti, John Gregory, Katherine Hua Guo, Scott Hagerup, Kent Harris, Scott Henry, Bruce P. Hillam, Tom Hogan, Jackie Hohnson, Christopher M. Hoss, Jason S. Howel, Kwang-il In, Victoria Interrante, Sadegh Jokar, Erni Jusuf, Jason Karns, Jacob Y. Kazakia, Grant Keady, Achim Kehrein, Jacek Kierzenka, Daniel Kopelove, S. A. (Seppo) Korpela, Andrew Knyazev, Gary Krenz, Jihoon Kwak, Kim Kyungjin, Minghorng Lai, Patrick Lang, Kevin Lee, Wu Li, Grace Liu, Wenguo Liu, Stacy Long, Mark C. Malburg, Igor Malkiman, P. W. Manual, Hamidreza Mashalyekh, Peter McNamara, Juan Meza, F. Milianazzo, Milan Miklavcic, Sue Minkoff, George Minty, Baharen Momken, Justin Montgomery, Ramon E. Moore, Harunrashid Muhammad, Aaron Naiman, Asha Nallana, Edward Neuman, Durene Ngo, Roy Nicolaides, Jeff Nunemacher, Valia Guerra Ones, David Parker, Tony Praseuth, Rolfe G. Petschek, Terri Prakash, Mihaela Quirk, Helia Niroomand Rad, Jeremy Rahe, Frank Roberts, Frank Rogers, Simen Rokaas, Hossein Roodi, Robert S. Raposo, Chris C. Seib, Granville Sewell, Keh-Ming Shyue, Daniel Somerville, Nathan Smith, Mandayam Srinivas, Alexander Stromberger, Xingping Sun, Thiab Taha, Hidajaty Thajeb, Joseph Traub, Phuoc Truong, Vincent Tsao, Bi Roubolo Vona, David Wallace, Charles Walters, Kegnag Wang, Layne T. Watson, Andre Weideman, Perry Wong, Richard Fa Wai, Yuan Xu, and Rick Zaccone.

It is our pleasure to thank those who helped with the task of preparing the new edition. The staff of Cengage Learning and associated individuals have been most understanding and patient in bringing this book to fruition. In particular, we thank Shaylin Walsh, Alison Eigel Zade, Charu Khanna, and Christine Sabooni for their efforts on behalf of this project. Some of those who were involved with previous editions were Seema Atwal, Craig Barth, Carol Benedict, Stacy Green, Jeremy Hayhurst, Janet Hill, Cheryll Linthicum, Gary Ostedt, Merrill Peterson, Bob Pirtle, Sara Planck, Elizabeth Rammel, Ragu Raghavan, Elizabeth Rodio, Anne Seitz, and Marlene Thom.

We offer our heartfelt gratitude to Victoria Cheney, Joyce Pfluger, and Martha Wells.

We would appreciate any comments, questions, suggestions, or corrections that readers may wish to communicate to us using the e-mail address below.

Ward Cheney & David Kincaid
kincaid@ices.utexas.edu

Dedication

In memory of
our friend and colleague

David M. Young, Jr.
(“Dr. SOR”)
(1923–2008)

1 Mathematical Preliminaries and Floating-Point Representation

The Taylor series for the natural logarithm $\ln(1 + x)$ is

$$\ln 2 = 1 - \frac{1}{2} + \frac{1}{3} - \frac{1}{4} + \frac{1}{5} - \frac{1}{6} + \frac{1}{7} - \frac{1}{8} + \cdots$$

Adding together the eight terms shown, we obtain $\ln 2 \approx 0.63452$,* which is a poor approximation to $\ln 2 = 0.69315\ldots$. On the other hand, the Taylor series for $\ln[(1 + x)/(1 - x)]$ gives us (with $x = \frac{1}{3}$)

$$\ln 2 = 2\left(3^{-1} + \frac{3^{-3}}{3} + \frac{3^{-5}}{5} + \frac{3^{-7}}{7} + \cdots\right)$$

By adding the four terms shown between the parentheses and multiplying by 2, we obtain $\ln 2 \approx 0.69313$. This illustrates the fact that rapid convergence of a Taylor series can be expected near the point of expansion but not at remote points. Evaluating the series $\ln[(1 + x)/(1 - x)]$ at $x = \frac{1}{3}$ is a mechanism for evaluating $\ln 2$ near the point of expansion. It also gives an example in which the properties of a function can be exploited to obtain a more rapidly convergent series. Taylor series and Taylor's Theorem are two of the principal topics we discuss in this chapter. They are ubiquitous features in much of numerical analysis.

Computers usually do *not* use base-10 arithmetic for storage or computation. Numbers that have a finite expression in one number system may have an infinite expression in another system. This phenomenon is illustrated when the familiar decimal number 1/10 is converted into the binary system:

$$(0.1)_{10} = (0.0\,0011\,0011\,0011\,0011\,0011\,0011\,0011\,0011\,\ldots)_2$$

We explain the floating-point number system and develop basic facts about roundoff errors. Another topic is loss of significance, which occurs when nearly equal numbers are subtracted. It is studied and shown to be avoidable by various programming techniques.

*The symbol $\approx$ means "approximately equal to."

1.1 Introduction

Objectives

The objective of this text is to help the reader to understand some of the many methods for solving scientific problems using computers. We intentionally limit ourselves to the typical problems that arise in science, engineering, and technology. Thus, we do *not* touch on problems of accounting, modeling in the social sciences, information retrieval, artificial intelligence, and so on.

Usually, our treatment of problems do *not* begin at the source, for that would take us far afield into such areas as physics, engineering, and chemistry. Instead, we consider problems after they have been cast into certain standard mathematical forms. The reader is therefore asked to accept on faith the assertion that the chosen topics are indeed important ones in **scientific computing**.

To survey many topics, we must treat some in a superficial way. But it is hoped that the reader acquires a good bird's-eye view of the subject and therefore is better prepared for a further, deeper study of **numerical analysis**.

For each principal topic, we list good current sources for more information. In any realistic computing situation, considerable thought should be given to the choice of method to be employed. Although most procedures presented here are useful and important, they may not be the optimum ones for a particular problem. In choosing among available methods for solving a problem, the analyst or programmer should consult recent references.

Limitations

Becoming familiar with basic numerical methods without realizing their limitations would be foolhardy. Numerical computations are almost invariably contaminated by errors, and it is important to understand the source, propagation, magnitude, and rate of growth of these errors. Numerical methods that provide approximations *and* error estimates are more valuable than those that provide only approximate answers. While we cannot help but be impressed by the speed and accuracy of the modern computer, we should temper our admiration with generous measures of skepticism. As the eminent numerical analyst Carl-Erik Fröberg once remarked:

> *Never in the history of mankind has it been possible to produce so many wrong answers so quickly!*

Thus, one of our goals is to help the reader arrive at this state of skepticism, armed with methods for detecting, estimating, and controlling errors.

Programming

The reader is expected to be familiar with the rudiments of programming. Algorithms are presented as **pseudocode**, and *no* particular programming language is adopted. The pseudocodes are an important intermediate step in translating the algorithms presented in the textbook before coding, running, and debugging them in a computer language and on a computer.

Some of the primary issues related to numerical methods are the nature of numerical errors, the propagation of errors, and the efficiency of the computations involved, as well as the number of operations and their possible reduction.

Many students have graphing calculators and access to mathematical software systems that can produce solutions to complicated numerical problems with minimal difficulty. The purpose of a numerical mathematics course is to examine the underlying algorithmic techniques so that students learn how the software or calculator found the answer. In this way, they would have a better understanding of the inherent limits on the accuracy that must be anticipated in working with such systems.

Strategies

One of the fundamental strategies behind many numerical methods is the replacement of a difficult problem with a string of simpler ones. By carrying out an iterative process, the solutions of the simpler problems can be put together to obtain the solution of the original, difficult problem. This strategy succeeds in solving linear systems (Chapters 2 and 8), finding zeros of functions (Chapter 3), interpolation (Chapter 4), numerical integration (Chapters 5), and more.

Students majoring in computer science and mathematics as well as those majoring in engineering and other sciences are usually well aware that numerical methods are needed to solve problems that they frequently encounter. It may *not* be as well recognized that scientific computing is quite important for solving problems that come from fields other than engineering and science, such as economics. For example, finding zeros of functions may arise in problems using the formulas for loans, interest, and payment schedules. Also, problems in areas such as those involving the stock market may require least-squares solutions (Chapter 9). In fact, the field of computational finance requires solving quite complex mathematical problems utilizing a great deal of computing power. Economic models routinely require the analysis of linear systems of equations with thousands of unknowns.

Significant Digits of Precision: Examples

Significant digits are digits beginning with the leftmost *nonzero* digit and ending with the rightmost *correct* digit, including final zeros that are exact.

EXAMPLE 1 Using an industrial laser cutting machine, a technician cuts a 2-meter by 3-meter rectangular sheet of steel into two equal triangular pieces.
What is the diagonal measurement of each triangle?
Can these pieces be slightly modified so the diagonals are exactly 3.6 meters?

Solution Since the piece is rectangular, the Pythagorean Theorem can be invoked. Thus, to compute the diagonal, d, in Figure 1.1 (p. 4), we write $2^2 + 3^2 = d^2$. It follows that

$$d = \sqrt{4+9} = \sqrt{13} = 3.60555\ 1275$$

This last number is obtained by using a handheld calculator. The accuracy of d as given can be verified by computing $(3.60555\ 1275) * (3.60555\ 1275) = 13$.

Is this numerical value for the diagonal to be taken seriously? Certainly not. To begin with, the given dimensions of the rectangle cannot be expected to be precisely 2 and 3. If the dimensions are accurate to 1 millimeter, the dimensions may be as large as 2.001 and 3.001.

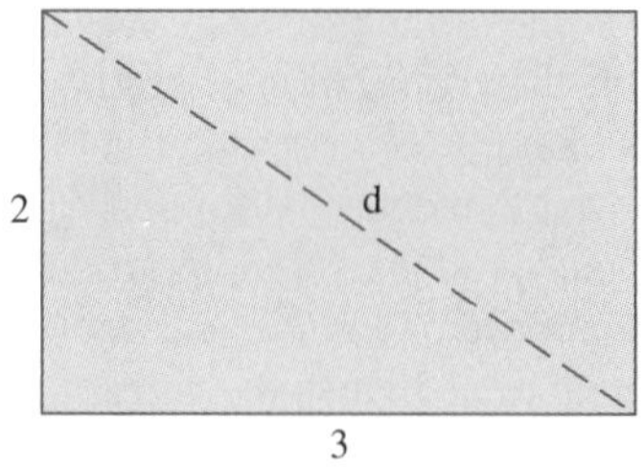

FIGURE 1.1 Rectangular sheet of steel

Using the Pythagorean Theorem again, one finds that the diagonal may be as large as

$$d = \sqrt{2.001^2 + 3.001^2} = \sqrt{4.004001 + 9.006001} = \sqrt{13.01002} \approx 3.6069$$

Similar reasoning indicates that d may be as small as 3.6042. These are both *worst cases*. We can conclude that

$$3.6042 \leqq d \leqq 3.6069$$

No greater accuracy can be claimed for the diagonal.

If we want the diagonal to be exactly 3.6, we require

$$(3 - c)^2 + (2 - c)^2 = 3.6^2$$

For simplicity, we reduce each side by the same amount. This leads to the equation

$$c^2 - 5c + 0.02 = 0$$

Using the quadratic formula, we find the smaller root to be

$$c = 2.5 - \sqrt{6.23} \approx 0.00400$$

By cutting off 4 millimeters from the two perpendicular sides, we have triangular pieces of sizes 1.996 by 2.996 meters. **Check:** $(1.996)^2 + (2.996)^2 \approx 3.6^2$. ■

To show the effect of the number of significant digits used in a calculation, we consider the problem of solving a linear system of equations.

EXAMPLE 2 Let's concentrate on solving for the variable y in this **linear system of equations** in two variables

$$\begin{cases} 0.1036\,x + 0.2122\,y = 0.7381 \\ 0.2081\,x + 0.4247\,y = 0.9327 \end{cases} \tag{1}$$

First, carry only three significant digits of precision in the calculations. Second, repeat with four significant digits throughout. Finally, use ten significant digits.

Solution **Step 1.** In the first task, we round all numbers in the original problem to three digits and round all the calculations, keeping only **three significant digits**. We take a multiple α of the first equation and subtract it from the second equation to eliminate the x-term in the second equation. The **multiplier** is $\alpha = 0.208/0.104 \approx 2.00$. Thus, in the second equation, the new coefficient of the x-term is $0.208 - (2.00)(0.104) \approx 0.208 - 0.208 = 0$ and the new y-term coefficient is $0.425 - (2.00)(0.212) \approx 0.425 - 0.424 = 0.001$. The right-hand side is $0.933 - (2.00)(0.738) = 0.933 - 1.48 = -0.547$. Hence, we find that

$$y = -0.547/(0.001) \approx -547$$

Step 2. We decide to keep **four significant digits** throughout and repeat the calculations. Now the multiplier is $\alpha = 0.2081/0.1036 \approx 2.009$. In the second equation, the new coefficient of the x-term is $0.2081 - (2.009)(0.1036) \approx 0.2081 - 0.2081 = 0$, the new coefficient of the y-term is $0.4247 - (2.009)(0.2122) \approx 0.4247 - 0.4263 = -0.001600$, and the new right-hand side is $0.9327 - (2.009)(0.7381) \approx 0.9327 - 1.483 \approx -0.5503$. Hence, we find

$$y = -0.5503/(-.001600) \approx 343.9$$

We are shocked to find that the *answer* has changed from -547 to 343.9, which is a huge difference!

Step 3. In fact, if we repeat this process and carry **ten significant digits**, we find that even 343.9 is *not* accurate, since we obtain $y \approx 356.2907199$.

Using a computer, we find

$$y \approx 3.56290\,74421\,51324 \times 10^2$$

The lesson learned in this example is that data thought to be accurate should be carried with full precision and *not* be rounded prior to each of the calculations. ■

Figure 1.2 shows a geometric illustration of what can happen in solving two equations in two unknowns. The point of intersection of the two lines is the exact solution. As is shown by the dotted lines, there may be a **degree of uncertainty** from errors in the measurements or roundoff errors. So instead of a sharply defined point, there may be a small trapezoidal area containing many possible solutions. However, if the two lines are nearly parallel, then this area of possible solutions can increase dramatically! This is related to well-conditioned and ill-conditioned systems of linear equations, which are discussed more in Chapter 8.

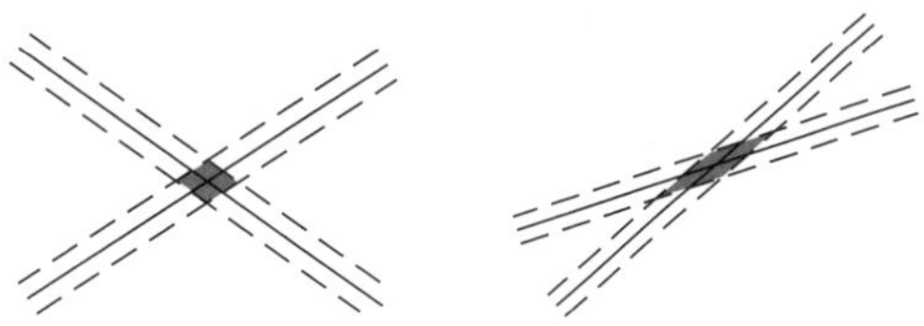

FIGURE 1.2 In 2D, well-conditioned and ill-conditioned linear systems

In most computers, the arithmetic operations are carried out in a double-length accumulator that has twice the precision of the stored quantities. However, even this may not avoid a loss of accuracy! Loss of accuracy can happen in various ways such as from roundoff errors and subtracting nearly equal numbers. We shall discuss loss of precision in Section 1.4, and the solving of linear systems in Chapter 2.

Errors: Absolute and Relative

Suppose that α and β are two numbers, of which one is regarded as an approximation to the other. The **error** of β as an approximation to α is $\alpha - \beta$; that is, the error equals the exact value minus the approximate value. The **absolute error** of β as an approximation to α is

$$|\alpha - \beta|$$

The **relative error** of β as an approximation to α is

$$\frac{|\alpha - \beta|}{|\alpha|}$$

Notice that in computing the absolute error, the roles of α and β are the same, whereas in computing the relative error, it is essential to distinguish one of the two numbers as *correct*. (Observe that the relative error is undefined in the case $\alpha = 0$.) For practical reasons, the relative error is usually more meaningful than the absolute error.

In summary, we have

Errors

$$\text{Absolute Error} = |\text{Exact Value} - \text{Approximate Value}|$$

$$\text{Relative Error} = \frac{|\text{Exact Value} - \text{Approximate Value}|}{|\text{Exact Value}|}$$

Here the exact value may be the *true* value or the *best* known value.

EXAMPLE 3 Let $\alpha_1 = 1.333,\ \beta_1 = 1.334$, and $\alpha_2 = 0.001,\ \beta_2 = 0.002$.

What are the absolute errors and relative errors of β_i as an approximation to α_i?

Solution The absolute error of β_i as an approximation to α_i is the same in both cases—namely, 10^{-3}. However, the relative errors are $\frac{3}{4} \times 10^{-3}$ and 1, respectively. The relative error clearly indicates that β_1 is a *good* approximation to α_1, but that β_2 is a *poor* approximation to α_2. ■

EXAMPLE 4 Consider $x = 0.00347$ rounded to $\widetilde{x} = 0.0035$ and $y = 30.158$ rounded to $\widehat{y} = 30.16$. In each case, what are the number of significant digits, absolute errors, and relative errors? Interpret the results.

Solution **Case 1.** $\widetilde{x} = 0.35 \times 10^{-2}$ has two significant digits, absolute error 0.3×10^{-4} and relative error 0.865×10^{-2}.

Case 2. $\widehat{y} = 0.3016 \times 10^{2}$ has four significant digits, absolute error 0.2×10^{-2} and relative error 0.66×10^{-4}.

Clearly, the relative error is a better indication of the number of significant digits than the absolute error. ■

Accuracy and Precision

Accurate to *n* decimal places means that you can trust n digits to the right of the decimal place. **Accurate to *n* significant digits** means that you can trust a total of n digits as being meaningful beginning with the leftmost nonzero digit.

Suppose you use a ruler graduated in millimeters to measure lengths. So the measurements are accurate to 1 millimeter, or 0.001m, which is three decimal places written in meters. A measurement such as 12.345m would be accurate to three decimal places. A measurement such as 12.34567 89m would be meaningless, since the ruler produces only three decimal places, and it should be 12.345m or 12.346m. If the measurement 12.345m has five dependable digits, then it is accurate to five significant figures. On the other hand, a measurement such as 0.076m has only two significant figures.

When using a calculator or computer in a laboratory experiment, one may get a *false sense* of having higher precision than is warranted by the data. For example, the result

$$(1.2) + (3.45) = 4.65$$

actually has only two significant digits of accuracy because the second digit in 1.2 may be the effect of rounding 1.24 down or rounding 1.16 up to two significant figures. Then the left-hand side could be as large as

$$(1.249) + (3.454) = (4.703)$$

or as small as

$$(1.16) + (3.449) = (4.609)$$

There are really only two significant decimal places in the answer! In adding and subtracting numbers, the result is accurate only to the *smallest* number of significant digits used in any step of the calculation. In the above example, the term 1.2 has two significant digits; therefore, the final calculation has an *uncertainty* in the third digit.

In multiplication and division of numbers, the results may be even more *misleading*. For instance, perform these computations on a calculator:

$$(1.23)(4.5) = 5.535, \qquad (1.23)/(4.5) = 0.27333\,3333$$

You think that there are four and nine significant digits in the results, but there are really only two! As a rule of thumb, one should keep as many significant digits in a sequence of calculations as there are in the least accurate number involved in the computations.

Rounding and Chopping

Rounding reduces the number of significant digits in a number. The result of rounding is a number similar in magnitude that is a *shorter* number having fewer nonzero digits. There are several slightly different rules for rounding. The **round-to-even** method is also known as *statistician's rounding* or *bankers' rounding*. It is discussed next. Over a large set of data, the round-to-even rule tends to reduce the total rounding error with (on average) an equal portion of numbers rounding up as well as rounding down.

We say that a number x is **chopped to n digits** or figures when all digits that follow the nth digit are discarded and none of the remaining n digits are changed. Conversely, x is **rounded to n digits** or figures when x is replaced by an n-digit number that approximates x with *minimum* error.

The question of whether to round up or down an $(n+1)$-digit decimal number that ends with a 5 is best handled by always selecting the rounded n-digit number with an *even* nth digit. This may seem strange at first, but remarkably, this is essentially what computers do in rounding decimal calculations when using the standard floating-point arithmetic! (This is a topic discussed in Section 1.3.)

EXAMPLE 5 Give some examples of rounding three-decimal numbers to two digits.

Solution The results of rounding are

$$0.217 \approx 0.22, \quad 0.365 \approx 0.36, \quad 0.475 \approx 0.48, \quad 0.592 \approx 0.59$$

while chopping them gives

$$0.217 \approx 0.21, \quad 0.365 \approx 0.36, \quad 0.475 \approx 0.47, \quad 0.592 \approx 0.59$$

■

On computers, the user sometimes has the option to have all arithmetic operations done with either chopping or rounding. The latter is usually preferable, of course!

Computation of π

History of Computing π

Computing π has a rich and colorful history. In 1650 BCE, the Rhind Papyrus of ancient Egypt contained numerical algorithms such as this approximation

$$\pi\left(\frac{9}{2}\right)^2 \approx 8^2 \Rightarrow \pi \approx \frac{256}{81} \approx 3.1605$$

In 287–212 BC, Archimedes determined that

$$3.1408 \approx \frac{223}{71} < \pi < \frac{22}{7} \approx 3.142$$

by noting that the value of π is between the length of the permieter of a polygon inscribing and a polygon circumscribing a circle of radius one-half. Around 1700, John Machin dicovered the identity

$$\pi = 16 \tan\frac{1}{5} - 4\tan\frac{1}{239}$$

and calculated the first hundred digits of π. In 1973, the first million digits of π were determined. Since then, many more sophisticated techniques have been developed for computing π. (See Moler [2011], for a recent article.)

Nested Multiplication and Horner's Algorithm

We begin with some remarks on evaluating a polynomial efficiently and on rounding and chopping real numbers. To evaluate the polynomial

$$p(x) = a_0 + a_1x + a_2x^2 + \cdots + a_{n-1}x^{n-1} + a_nx^n \tag{2}$$

we group the terms in a **nested multiplication**:

$$p(x) = a_0 + x(a_1 + x(a_2 + \cdots + x(a_{n-1} + x(a_n))\cdots))$$

The pseudocode[‡] that evaluates $p(x)$ starts with the innermost parentheses and works outward. It can be written as

Nested Multiplication Pseudocode

```
integer i, n; real p, x
real array (a_i)_{0:n}
p ← a_n
for i = n − 1 to 0
    p ← a_i + xp
end for
```

Here we assume that numerical values have been assigned to the integer variable n, the real variable x, as well as the coefficients $a_0, a_1, \ldots, a_n$, which are stored in a real linear array. (Throughout, we use semicolons between these declarative statements to save space.) The **left-pointing arrow** ($\leftarrow$) means that the value on the right is stored in the location

[‡]A **pseudocode** is a compact and informal description of an algorithm that uses the conventions of a programming language but omits the detailed syntax. When convenient, it may be augmented with natural language. Usually, writing the pseudocode is a good way of organizing the ideas in a mathematical algorithm before coding them in a particular programming language.

named on the left (i.e., **overwrites** from right to left). The for-loop index i runs backward, taking values $n-1, n-2, \ldots, 0$. The final value of p is the value of the polynomial at x. This nested multiplication procedure is also known as **Horner's algorithm** or **synthetic division**.

In the pseudocode (p. 8), there is exactly *one* addition and *one* multiplication each time the loop is traversed. Consequently, Horner's algorithm can evaluate a polynomial with only n additions and n multiplications. This is the minimum number of operations possible.

A *naive* method of evaluating a polynomial would require many more operations.

EXAMPLE 6 Show how $p(x) = 5 + 3x - 7x^2 + 2x^3$ should be computed.

Solution Let

$$p(x) = 5 + x(3 + x(-7 + x(2)))$$

for a given value of x. We have *avoided* all the exponentiation operations by using nested multiplication! ■

The polynomial in Equation (2) can be written in an alternative form by utilizing the mathematical symbols for **sum** $\sum$ and **product** $\prod$; namely,

Mathematical Notation

$$p(x) = \sum_{i=0}^{n} a_i x^i = \sum_{i=0}^{n} \left(a_i \prod_{j=1}^{i} x \right)$$

Recall that if $n \leqq m$, we write

$$\sum_{k=n}^{m} x_k = x_n + x_{n+1} + \cdots + x_m$$

and

$$\prod_{k=n}^{m} x_k = x_n x_{n+1} \cdots x_m$$

By convention, whenever $m < n$, we define

$$\sum_{k=n}^{m} x_k = 0 \quad \text{and} \quad \prod_{k=n}^{m} x_k = 1$$

Horner's algorithm can be used in the **deflation** of a polynomial. This is the process of removing a linear factor from a polynomial. If r is a **root** of the polynomial p, then $x - r$ is a factor of p. The remaining roots of p are the $n - 1$ roots of a polynomial q of degree 1 less than the degree of p such that

Deflation

$$p(x) = (x - r)q(x) + p(r) \tag{3}$$

where

$$q(x) = b_0 + b_1 x + b_2 x^2 + \cdots + b_{n-1} x^{n-1} \tag{4}$$

The pseudocode for **Horner's algorithm** can be written as follows:

Synthetic Division Pseudocode

integer i, n; **real** r
real array $(a_i)_{0:n}, (b_i)_{0:n-1}$
$b_{n-1} \leftarrow a_n$
for $i = n - 1$ **to** 0
  $b_{i-1} \leftarrow a_i + rb_i$
end for

Notice that $b_{-1} = p(r)$ in this pseudocode. If f is an exact root, then $b_{-1} = p(r) = 0$.

If the calculation in Horner's algorithm is to be carried out with pencil and paper, the following arrangement is often used:

$$\begin{array}{r|cccccc} & a_n & a_{n-1} & a_{n-2} & \cdots & a_1 & a_0 \\ r\,) & & rb_{n-1} & rb_{n-2} & \cdots & rb_1 & rb_0 \\ \hline & b_{n-1} & b_{n-2} & b_{n-3} & \cdots & b_0 & b_{-1} \end{array}$$

EXAMPLE 7 Use Horner's algorithm to evaluate $p(3)$, where p is the polynomial

$$p(x) = x^4 - 4x^3 + 7x^2 - 5x - 2$$

Solution We arrange the calculation as suggested:

$$\begin{array}{r|ccccc} & 1 & -4 & 7 & -5 & -2 \\ 3\,) & & 3 & -3 & 12 & 21 \\ \hline & 1 & -1 & 4 & 7 & \boxed{19} \end{array}$$

Thus, we obtain $p(3) = 19$, and we can write

$$p(x) = (x - 3)(x^3 - x^2 + 4x + 7) + 19$$ ■

In the **deflation process**, if r is a **zero** of the polynomial p, then $x - r$ is a factor of p, and conversely. The remaining zeros of p are the $n - 1$ zeros of $q(x)$.

EXAMPLE 8 Deflate the polynomial p of the preceding example, using the fact that 2 is one of its zeros.

Solution We use the same arrangement of computations as explained previously:

$$\begin{array}{r|ccccc} & 1 & -4 & 7 & -5 & -2 \\ 2\,) & & 2 & -4 & 6 & 2 \\ \hline & 1 & -2 & 3 & 1 & \boxed{0} \end{array}$$

Thus, we have $p(2) = 0$, and

$$x^4 - 4x^3 + 7x^2 - 5x - 2 = (x - 2)(x^3 - 2x^2 + 3x + 1)$$ ■

We can use Horner's algorithm for evaluating a derivative of a polynomial. By Equation (3), we can write polynomial p with root r as

$$p(x) = (x - r)q(x) + p(r)$$

where $q(x)$ is given by Equation (4). If we differentiate, we obtain

$$p'(x) = q(x) + (x - r)q'(x)$$

Clearly, we have

$$p'(r) = q(r)$$

The pseudocode for **Horner's algorithm** for determining $p(r)$ and $p'(r)$ can be written as follows:

Horner's Algorithm Pseudocode

> **integer** i, n; **real** p, r
> **real array** $(a_i)_{0:n}, (b_i)_{0:n-1}$
> $\alpha \leftarrow a_n$; $\beta \leftarrow 0$
> **for** $i = n - 1$ **to** 0
> $\beta \leftarrow \alpha + r\beta$
> $\alpha \leftarrow a_i + r\alpha$
> **end for**

Notice that the final values are $\alpha = p(r)$ and $\beta = p'(r)$.

If the calculation in Horner's algorithm (synthetic division) is to be carried out with pencil and paper, the following arrangement is often used:

$$\begin{array}{rcccccccl}
 & a_n & a_{n-1} & a_{n-2} & \cdots & a_2 & a_1 & a_0 & \\
r\,) & & rb_{n-1} & rb_{n-2} & \cdots & rb_2 & rb_1 & rb_0 & \\
\hline
 & b_{n-1} & b_{n-2} & b_{n-3} & \cdots & b_1 & b_0 & \boxed{b_{-1}} & = p(r) \\
 & & rc_{n-2} & rc_{n-3} & \cdots & c_2 & rc_1 & & \\
\hline
 & c_{n-2} & c_{n-3} & c_{n-4} & \cdots & c_0 & \boxed{c_{-1}} & = p'(r) &
\end{array}$$

EXAMPLE 9 Given the polynomial

$$p(x) = x^4 - 4x^3 + 7x^2 - 5x - 3$$

Use synthetic division to find $p(3)$ and $p'(3)$.

Solution We use this arrangement of computation as explained previously:

$$\begin{array}{rccccl}
 & 1 & -4 & 7 & -5 & -2 \\
3\,) & & 3 & -3 & 12 & 21 \\
\hline
 & 1 & -1 & 4 & 7 & \boxed{19} = p(3) \\
 & & 3 & 6 & 30 & \\
\hline
 & 1 & 2 & 10 & \boxed{37} = p'(3) &
\end{array}$$

Thus, we have

$$x^4 - 4x^3 + 7x^2 - 5x - 2 = (x - 3)(x^3 - x^2 + 4x + 7) + 19$$

So $p(x) = (x-3)q(x)+19$ and $p'(x) = q(x)+(x-3)q'(x)$ where $q(x) = x^3-x^2+4x+7$. Hence, we have $p(3) = 19$ and $p'(3) = q(3) = 37$. ■

Pairs of Easy/Hard Problems

Sometimes in scientific computing, we encounter a pair of problems, one of which is *easy* and the other *hard*, and they are inverses of each other. This is the main idea in **cryptology**, in which multiplying two numbers together is trivial, but the reverse problem (factoring a huge number) verges on the impossible!

The same phenomenon arises with **polynomials**. Given the roots, we can easily find the power form of the polynomial as in (2). Given the power form, it may be a hard problem to compute the roots (or it may be an *ill-conditioned problem*). Computer Exercise 1.1.24 calls for the writing of code to compute the coefficients in the power form of a polynomial from its roots. It is a do-loop with simple formulas. One adjoins one factor $(x - r)$ at a time. This theme arises again in linear algebra, in which computing $\boldsymbol{b} = \boldsymbol{A}\boldsymbol{x}$ is trivial, but finding $\boldsymbol{x}$ from $\boldsymbol{A}$ and $\boldsymbol{b}$ (the inverse problem) is hard. (See Section 2.1.)

Sample Problem Pairs

Easy/hard problems come up again in **two-point boundary value problems**. Finding Df, $f(0)$, and $f(1)$ when f is given and D is a differential operator is easy, but finding f from knowledge of Df, $f(0)$ and $f(1)$ is hard. (See Section 9.1.)

Likewise, computing the **eigenvalues/eigenvectors** of a matrix is a hard problem. Suppose we are given the eigenvalues $\lambda_1, \lambda_2, \dots, \lambda_n$ of an $n \times n$ matrix and corresponding eigenvectors $\boldsymbol{v}_1, \boldsymbol{v}_2, \dots, \boldsymbol{v}_n$ of an $n \times n$ matrix. We can get $\boldsymbol{A}$ by putting the eigenvalues on the diagonal of a diagonal matrix $\boldsymbol{D}$ and the eigenvectors as columns in a matrix $\boldsymbol{V}$. Then $\boldsymbol{A}\boldsymbol{V} = \boldsymbol{V}\boldsymbol{D}$, and we can get $\boldsymbol{A}$ from this by solving the equation for $\boldsymbol{A}$. But finding λ_i and $\boldsymbol{v}_i$ from $\boldsymbol{A}$ itself is hard. (See Section 8.2.)

The reader may think of other examples.

It needs to be pointed out that there may be a huge gap between the *complexity* of factoring integers and that of the other problems mentioned, which are probably all solvable in *polynomial time*. Also, there may be a cultural difference at work here. To a numerical analyst, *hard* usually means: "I can solve it in polynomial time, but the degree of the polynomial is too high," whereas to a complexity theorist, *hard* means: "It is not solvable in polynomial time, as far as I know." On the other hand, there are NP-hard problems with a numerical flavor such as the *traveling salesman problem*.

First Programming Experiment

We conclude this section with a short programming experiment involving numerical computations. Here we consider, from the computational point of view, a familiar operation in calculus—namely, taking the derivative of a function. Recall that the **derivative** of a function f at a point x is defined by the equation

Derivative Formula

$$f'(x) = \lim_{h \to 0} \frac{f(x+h) - f(x)}{h}$$

A computer has the capacity of imitating the limit operation by using a sequence of numbers h such as

$$h = 4^{-1}, 4^{-2}, 4^{-3}, \dots, 4^{-n}, \dots$$

This sequence certainly approaches zero rapidly! Of course, many other simple sequences are possible, such as $1/n$, $1/n^2$, and $1/10^n$. The sequence $1/4^n$ consists of machine numbers in a binary computer and, for this experiment on a 32-bit computer, it is sufficiently close to zero when n is 10.

If $f(x) = \sin x$, here is pseudocode for computing $f'(x)$ at the point $x = 0.5$:

***First* Pseudocode**

```
program First
integer i, imin, n ← 30
real error, y, x ← 0.5, h ← 1, emin ← 1
for i = 1 to n
    h ← 0.25h
    y ← [sin(x + h) − sin(x)]/h
```

(Continued)

```
        error ← |cos(x) − y|;
        output i, h, y, error
        if error < emin then
            emin ← error;   imin ← i
        end if
    end for
    output imin, emin
    end program First
```

We have neither explained the purpose of the experiment nor shown the output from this pseudocode. We invite the reader to discover this by coding and running it (or one like it) on a computer. (See Computer Exercises 1.1.1—1.1.3.)

Mathematical Software

The algorithms and programming problems in this book have been coded and tested in a variety of ways, and they are available on the website for this book as given in the Preface. Some are best done by using a scientific programming language such as C, C++, Fortran, or any other that allows for calculations with adequate precision. Sometimes it is instructive to utilize mathematical software systems such as MATLAB, Maple, Mathematica, or Octave, since they contain built-in problem-solving procedures. Alternatively, one could use a mathematical program library such as IMSL, NAG, or others when locally available. Some numerical libraries may have been specifically optimized for the processor, such as Intel and AMD. Software systems are particularly useful for obtaining graphical results as well as for experimenting with various numerical methods for solving a difficult problem. Mathematical software packages containing symbolic-manipulation capabilities, such as in Maple, Mathematica, and Macsyma, are particularly useful for obtaining exact as well as numerical solutions. In solving the computer problems, students should focus on gaining insights and better understanding of the numerical methods involved. Appendix A offers advice on computer programming for scientific computations. The suggestions are independent of the particular language being used.

Websites

With World Wide Web and the Internet, good mathematical software has become easy to locate and to use. Browsers, search engines, and websites may be used to find software that is applicable to a particular area of interest. Collections of mathematical software exist, ranging from large, comprehensive libraries to smaller versions of these libraries for personal computers; some of these may be interactive. Also, references to computer programs and collections of routines can be found in books and technical reports. The textbook website is given in the Preface. It contains an overview of mathematical software and other available supporting material.

Additional Reading

For additional study of topics found in this book, see the references in the Bibliography as well as an extensive list of items at the textbook website. Two interesting papers containing numerous examples of why numerical methods are critically important are Forsythe [1970] and McCartin [1998]. See Briggs [2004] and Friedman and Littman [1994] for some industrial and real-world problems.

Summary 1.1

- Absolute Error = |Exact Value − Approximate Value|

  $$\text{Relative Error} = \frac{|\text{Exact Value} - \text{Approximate Value}|}{|\text{Exact Value}|}$$

- Use **nested multiplication** to evaluate a polynomial efficiently:

  $$\begin{aligned} p(x) &= a_0 + a_1x + a_2x^2 + \cdots + a_{n-1}x^{n-1} + a_nx^n \\ &= a_0 + x(a_1 + x(a_2 + \cdots + x(a_{n-1} + x(a_n))\cdots)) \end{aligned}$$

  A segment of pseudocode for doing this is

```
p ← a_n
for k = 1 to n
    p ← xp + a_{n−k}
end for
```

- Deflation of the polynomial $p(x)$ is removing a linear factor:

  $$p(x) = (x - r)q(x) + p(r)$$

  where

  $$q(x) = b_0 + b_1x + b_2x^2 + \cdots + b_{n-1}x^{n-1}$$

  The pseudocode for **Horner's algorithm** for deflation of a polynomial is:

```
b_{n−1} ← a_n
for i = n − 1 to 0
    b_{i−1} ← a_i + rb_i
end for
```

  Hence, we obtain $b_{-1} = p(r)$.

Exercises 1.1

1. In high school, some students have been misled to believe that 22/7 is either the actual value of π or an acceptable approximation to π. Show that 355/113 is a better approximation in terms of both absolute and relative errors. Find some other simple rational fractions n/m that approximate π. For example, ones for which

   $$|\pi - n/m| < 10^{-9}$$

   Hint: See Exercise 1.1.4.

*Exercises marked with [a] have answers in the back of the book.

[a]**2.** A real number x is represented approximately by 0.6032, and we are told that the relative error is at most 0.1%. What is x?
Note: There are *two* answers.

[a]**3.** What is the relative error involved in rounding 4.9997 to 5.000?

[a]**4.** The value of π can be generated by the computer to nearly full machine precision by the assignment statement

```
pi ← 4.0 arctan(1.0)
```

Suggest at least four other ways to compute π using basic functions on your computer system.

5. A given doubly subscripted array $(a_{ij})_{n\times n}$ can be added in any order. Write the pseudocode segments for each of the following parts. Which is best?

[a]**a.** $\sum_{i=1}^{n}\sum_{j=1}^{n} a_{ij}$ **b.** $\sum_{j=1}^{n}\sum_{i=1}^{n} a_{ij}$

c. $\sum_{i=1}^{n}\left(\sum_{j=1}^{i} a_{ij} + \sum_{j=1}^{i-1} a_{ji}\right)$

[a]**d.** $\sum_{k=0}^{n-1}\sum_{|i-j|=k} a_{ij}$ **e.** $\sum_{k=2}^{2n}\sum_{i+j=k}^{n} a_{ij}$

[a]**6.** Count the number of operations involved in evaluating a polynomial using nested multiplication. Do *not* count subscript calculations.

7. For small x, show that $(1+x)^2$ can sometimes be more accurately computed from $(x+2)x+1$. Explain. What other expressions can be used to compute it?

8. Show how these polynomials can be efficiently evaluated:

[a]**a.** $p(x) = x^{32}$ **b.** $p(x) = 3(x-1)^5 + 7(x-1)^9$

[a]**c.** $p(x) = 6(x+2)^3 + 9(x+2)^7 + 3(x+2)^{15} - (x+2)^{31}$

d. $p(x) = x^{127} - 5x^{37} + 10x^{17} - 3x^7$

9. Using the exponential function $\exp(x)$, write an efficient pseudocode segment for the statement $y = 5e^{3x} + 7e^{2x} + 9e^x + 11$.

[a]**10.** Write a pseudocode segment to evaluate the expression

$$z = \sum_{i=1}^{n} b_i^{-1} \prod_{j=1}^{i} a_j$$

where $(a_1, a_2, \dots, a_n)$ and $(b_1, b_2, \dots, b_n)$ are linear arrays containing given values.

11. Write segments of pseudocode to evaluate the following expressions efficiently:

a. $p(x) = \sum_{k=0}^{n-1} kx^k$

[a]**b.** $z = \sum_{i=1}^{n}\prod_{j=1}^{i} x^{n-j+1}$

c. $z = \prod_{i=1}^{n}\sum_{j=1}^{i} x_j$

d. $p(t) = \sum_{i=1}^{n} a_i \prod_{j=1}^{i-1}(t - x_j)$

12. Using summation and product notation, write mathematical expressions for the following pseudocode segments:

a.
```
integer i, n; real v, x
real array (a_i)0:n
v ← a_0
for i = 1 to n
    v ← v + x a_i
end for
```

[a]**b.**
```
integer i, n; real v, x
real array (a_i)0:n
v ← a_n
for i = 1 to n
    v ← v x + a_{n−i}
end for
```

c.
```
integer i, n; real v, x
real array (a_i)0:n
v ← a_0
for i = 1 to n
    v ← v x + a_i
end for
```

d.
```
integer i, n; real v, x, z
real array (a_i)0:n
v ← a_0
z ← x
for i = 1 to n
    v ← v + z a_i
    z ← x z
end for
```

[a]**e.**
```
integer i, n; real v
real array (a_i)0:n
v ← a_n
for i = 1 to n
    v ← (v + a_{n−i}) x
end for
```

[a]**13.** Express in mathematical notation *without* parentheses the final value of z in the following pseudocode segment:

```
integer k, n; real z
real array (b_i)0:n
z ← b_n + 1
for k = 1 to n − 2
    z ← z b_{n−k} + 1
end for
```

[a]14. How many multiplications occur in executing the following pseudocode segment?

```
integer i, j, n; real x
real array (a_ij)_0:n×0:n, (b_ij)_0:n×0:n
x ← 0.0
for j = 1 to n
    for i = 1 to j
        x ← x + a_ij b_ij
    end for
end for
```

15. Criticize the following pseudocode segments and write improved versions:

a.
```
integer i, n; real x, z; real array (a_i)_0:n
for i = 1 to n
    x ← z^2 + 5.7
    a_i ← x/i
end for
```

[a]b.
```
integer i, j, n
real array (a_ij)_0:n×0:n
for i = 1 to n
    for j = 1 to n
        a_ij ← 1/(i + j − 1)
    end for
end for
```

c.
```
integer i, j, n; real array (a_ij)_0:n×0:n
for j = 1 to n
    for i = 1 to n
        a_ij ← 1/(i + j − 1)
    end for
end for
```

16. **a.** Solve Example 1.1.2 to full precision.

b. Repeat for this augmented matrix

$$\left[\begin{array}{cc|c} 3.5713 & 2.1426 & 7.2158 \\ 10.714 & 6.4280 & 1.3379 \end{array}\right]$$

for a system of two equations and two unknowns x and y.

c. Can small changes in the data lead to massive change in the each of these solutions?

17. A base 60 approximation (circa 1750 BCE) is

$$\sqrt{2} \approx 1 + \frac{24}{60} + \frac{51}{60^2} + \frac{10}{60^3}$$

Determine how accurate it is. See Sauer [2006] for additional details.

18. Use Horner's algorithm to evaluate each of these polynomials at the point indicated.

a. $2x^4 + 9x^2 - 16x + 12$ at -6

b. $2x^4 - 3x^3 - 5x^2 + 3x + 8$ at 2

c. $3x^5 - 38x^3 + 5x^2 - 1$ at 4

Computer Exercises 1.1

1. Write and run a computer program that corresponds to the pseudocode program *First* described on pp. 12–13 and interpret the results.

2. (Continuation) Select a function f and a point x and carry out a computer experiment like the one given in the text. Interpret the results. Do not select too simple a function. For example, you might consider $1/x$, $\log x$, e^x, $\tan x$, $\cosh x$, or $x^3 - 23x$.

3. (Continuation) As we saw in the computer experiment *First*, the accuracy of a formula for numerical differentiation may deteriorate as the step-size h decreases. Study the following **central difference formula**:

$$f'(x) \approx \frac{f(x+h) - f(x-h)}{2h}$$

as $h \to 0$. We learn in Section 4.3 that the **truncation error** for this formula is $-\frac{1}{6}h^2 f'''(\xi)$ for some ξ in the interval $(x - h, x + h)$.

Modify and run the code for the experiment *First* so that approximate values for the rounding error and truncation error are computed. On the same graph, plot the rounding error, the truncation error, and the total error (sum of these two errors) using a log-scale; that is, the axes in the plot should be $-\log_{10}|\text{error}|$ versus $\log_{10} h$. Analyze these results.

[a]4. The limit

$$e = \lim_{n\to\infty}\left(1 + \frac{1}{n}\right)^n$$

defines the number e in calculus. Estimate e by taking the value of this expression for $n = 8, 8^2, 8^3, \ldots, 8^{10}$.

Compare with e obtained from $e \leftarrow \exp(1.0)$. Interpret the results.

5. It is not difficult to see that the numbers

$$p_n = \int_0^1 x^n e^x \, dx$$

satisfy the inequalities $p_1 > p_2 > p_3 > \cdots > 0$. Establish this fact. Next, use integration by parts to show that

$$p_{n+1} = e - (n+1)p_n$$

and that $p_1 = 1$. In the computer, use the recurrence relation to generate the first 20 values of p_n and explain why the inequalities shown are violated. Do not use subscripted variables. (See Dorn and McCracken [1972], pp. 120–129.)

6. (Continuation) Let $p_{20} = \frac{1}{8}$ and use the formula in the preceding computer problem to compute $p_{19}, p_{18}, \ldots, p_2$, and p_1. Do the numbers generated obey the inequalities $1 = p_1 > p_2 > p_3 > \cdots > 0$? Explain the difference in the two procedures. Repeat with $p_{20} = 20$ or $p_{20} = 100$. Explain what happens.

7. Write an efficient routine that accepts as input a list of real numbers $a_1, a_2, \ldots, a_n$ and then computes the following:

Arithmetic mean $\quad m = \dfrac{1}{n}\displaystyle\sum_{k=1}^{n} a_k$

Variance $\quad v = \dfrac{1}{n-1}\displaystyle\sum_{k=1}^{n} (a_k - m)^2$

Standard deviation $\quad \sigma = \sqrt{v}$

Test the routine on a set of data of your choice.

8. (Continuation) Show that another formula is

Variance $\quad v = \dfrac{1}{n-1}\left[\displaystyle\sum_{k=1}^{n} a_k^2 - nm^2\right]$

Of the two given formulas for v, which is more accurate in the computer? Verify on the computer with a data set. *Hint*: Use a large set of real numbers that vary in magnitude from very small to very large.

[a]**9.** Let a_1 be given. Write a program to compute for $1 \leqq n \leqq 1000$ the numbers

$$b_n = na_{n-1}, \qquad a_n = b_n/n$$

Print the numbers $a_{100}, a_{200}, \ldots, a_{1000}$. *Do not* use subscripted variables. What should a_n be? Account for the deviation of fact from theory. Determine four values for a_1 so that the computation does deviate from theory on your computer.

Hint: Consider extremely small and large numbers and print to full machine precision.

[a]**10.** In a computer, it can happen that $a + x = a$ when $x \neq 0$. Explain why. Describe the set of n for which $1 + 2^{-n} = 1$ in your computer. Write and run appropriate programs to illustrate the phenomenon.

11. Write a program to test the programming suggestion concerning the roundoff error in the computation of $t \leftarrow t+h$ versus $t \leftarrow t_0 + ih$. For example, use $h = \frac{1}{10}$ and compute $t \leftarrow t+h$ in double precision for the correct single-precision value of t; print the absolute values of the differences between this calculation and the values of the two procedures. What is the result of the test when h is a machine number, such as $h = \frac{1}{128}$, on a binary computer (with more than seven bits per word)?

[a]**12.** The Russian mathematician P. L. **Chebyshev** (1821–1894) spelled his name Qebywev. Many transliterations from the Cyrillic to the Latin alphabet are possible. *Cheb* can alternatively be rendered as Ceb, Tscheb, or Tcheb. The *y* can be rendered as *i*. *Shev* can also be rendered as schef, cev, cheff, or scheff. Taking all combinations of these variants, program a computer to print all possible spellings.

13. Compute $n!$ using logarithms, integer arithmetic, and double-precision floating-point arithmetic. For each part, print a table of values for $0 \leqq n \leqq 30$, and determine the largest correct value.

14. Given two arrays, a real array $v = (v_1, v_2, \ldots, v_n)$ and an integer permutation array $p = (p_1, p_2, \ldots, p_n)$ of integers $1, 2, \ldots, n$, can we form a new permuted array $v = (v_{p_1}, v_{p_2}, \ldots, v_{p_n})$ by overwriting v and *not* involving another array in memory? If so, write and test the code for doing it. If *not*, use an additional array and test. Consider these cases:

Case 1. $v = (6.3, 4.2, 9.3, 6.7, 7.8, 2.4, 3.8, 9.7)$,
$p = (2, 3, 8, 7, 1, 4, 6, 5)$

Case 2. $v = (0.7, 0.6, 0.1, 0.3, 0.2, 0.5, 0.4)$,
$p = (3, 5, 4, 7, 6, 2, 1)$

15. Using a computer algebra system (e.g., Maple, Mathematica, MATLAB, etc.), print 200 decimal digits of $\sqrt{10}$.

16. **a.** Repeat the Example 1.1.1 on loss of significant digits of accuracy, but perform the calculations with twice the precision before rounding them. Does this help?

b. Use Maple or some other mathematical software system in which you can set the number of digits of precision.
Hint: In Maple, use `Digits`.

17. In 1706, Machin used the formula

$$\pi = 16 \arctan\left(\frac{1}{5}\right) - 4 \arctan\left(\frac{1}{239}\right)$$

to compute 100 digits of π. Derive this formula. Reproduce Machin's calculations by using suitable software. *Hint*: Let $\tan\theta = \frac{1}{5}$, and use standard trigonometric identities.

18. Using a symbol-manipulating program such as Maple, Mathematica, MATLAB, or Macsyma, carry out the following tasks. Record your work in some manner, for example, by using a `diary` or `script` command.

a. Find the Taylor series, up to and including the term x^{10}, for the function $(\tan x)^2$, using 0 as the point x_0.

b. Find the indefinite integral of $(\cos x)^{-4}$.

c. Find the definite integral $\int_0^1 \log|\log x|\,dx$.

d. Find the first prime number greater than 27448.

e. Obtain the numerical value of $\int_0^1 \sqrt{1 + \sin^3 x}\,dx$.

f. Find the solution of the differential equation $y' + y = (1 + e^x)^{-1}$.

g. Define the function $f(x, y) = 9x^4 - y^4 + 2y^2 - 1$. You want to know the value of $f(40545, 70226)$. Compute this in the straightforward way by direct substitution of $x = 40545$ and $y = 70226$ in the definition of $f(x, y)$, using first 6-decimal accuracy, then 7-, 8-, and so on up to 24-decimal digits of accuracy. Next, prove by means of elementary algebra that

$$f(x, y) = (3x^2 - y^2 + 1)(3x^2 + y^2 - 1)$$

Use this formula to compute the same value of $f(x, y)$, again using different precisions, from 6-decimal to 24-decimal. Describe what you have learned. To force the program to do floating-point operations instead of integer arithmetic, write your numbers in the form 9.0, 40545.0, and so forth.

19. Consider the following pseudocode segments:

```
a. integer i; real x, y, z
   for i = 1 to 20
       x ← 2 + 1.0/8^i
       y ← arctan(x) − arctan(2)
       z ← 8^i y
       output x, y, z
   end for
```

```
b. real epsi ← 1
   while 1 < 1 + epsi
       epsi ← epsi/2
       output epsi
   end while
```

What is the purpose of each program? Is it achieved? Explain. Code and run each one to verify your conclusions.

20. Consider some oversights involving assignment statements.

[a]**a.** What is the difference between the following two assignment statements? Write a code that contains them and illustrate with specific examples to show that sometimes $x = y$ and sometimes $x \neq y$.

```
integer m, n; real x, y
x ← real(m/n)
y ← real(m)/real(n)
output x, y
```

b. What value does n receive?

```
integer n;   real x, y
x ← 7.4
y ← 3.8
n ← x + y
output n
```

What happens when the last statement is replaced with the following?

```
n ← integer(x) + integer(y)
```

21. Write a computer code that contains the following assignment statements exactly as shown. Analyze the results.

a. Print these values first using the default format and then with an extremely large format field:

```
real p, q, u, v, w, x, y, z
x ← 0.1
y ← 0.01
z ← x − y
p ← 1.0/3.0
q ← 3.0p
u ← 7.6
v ← 2.9
w ← u − v
output x, y, z, p, q, u, v, w
```

b. What values would be computed for x, y, and z if this code is used?

```
integer n; real x, y, z
for n = 1 to 10
    x ← (n − 1)/2
    y ← n²/3.0
    z ← 1.0 + 1/n
    output x, y, z
end for
```

c. What values would the following assignment statements produce?

```
integer i, j; real c, f, x, half
x ← 10/3
i ← integer(x + 1/2)
half ← 1/2
j ← integer(half)
c ← (5/9)(f − 32)
f ← 9/5c + 32
output x, i, half, j, c, f
```

d. Discuss what is wrong with the following pseudocode segment:

```
real area, circum, radius
radius ← 1
area ← (22/7)(radius)²
circum ← 2(3.1416)radius
output area, circum
```

22. Criticize the following pseudocode for evaluating $\lim_{x\to 0} \arctan(|x|)/x$. Code and run it to see what happens.

```
integer i; real x, y
x ← 1
for i = 1 to 24
    x ← x/2.0
    y ← arctan(|x|)/x
    output x, y
end for
```

23. Carry out some computer experiments to illustrate or test the programming suggestions in Appendix A. Specific topics to include are these:

a. When to avoid arrays.

b. When to limit iterations.

c. Checking for floating-point equality.

d. Ways for taking equal floating-point steps.

e. Various ways to evaluate functions.

Hint: Comparing single and double precision results may be helpful.

24. (**Easy/Difficult Problem Pairs**) Write a computer program to obtain the power form of a polynomial from its roots. Let the roots be $r_1, r_2, \ldots, r_n$. Then (except for a scalar factor) the polynomial is the product

$$p(x) = (x - r_1)(x - r_2)\cdots(x - r_n).$$

Find the coefficients in the expression

$$p(x) = \sum_{j=0}^{n} a_j x^j$$

Test your code on the **Wilkinson polynomials** in Computer Exercises 3.1.10 and 3.3.9. Explain why this task of getting the power form of the polynomial is *trivial*, whereas the inverse problem of finding the roots from the power form is quite difficult.

25. A **prime number** is a positive integer that has no integer factors other than itself and 1. How many prime numbers are there in each of these open intervals: $(1, 40)$, $(1, 80)$, $(1, 160)$, and $(1, 2000)$? Make a guess as to the percentage of prime numbers among all numbers.

26. Mathematical software systems such as Maple, Mathematica, and MATLAB are able to do both numerical calculations and symbolic manipulations. Verify symbolically that a nested multiplication is correct for a general polynomial of degree ten.

27. In MATLAB, the `rat` function finds a rational fraction approximation (numerator and denominator) within a certain tolerance to a given floating-point number. For example, `[a,b]=rat(pi, 8000e-6)` return `a=22` and `b=7`. However, the relative error between $19/6$ and π is 0.007981306248670 in format `long`, which is less than the tolerance 0.008. What's going on here? In terms of absolute and relative errors, is $19/6$ or $22/7$ the better approximation to π?

28. Use mathematical software to reproduce the three solutions to Example 1.1.2.
Hint: In MATLAB, use commands `str2nun(num2str (x,4))` for rounding to four significant decimal digits as well as `format long`.

29. Explain the results from coding and executing the following pseudocode using mathematical software such as in MATLAB with `format long`:

```
integer k; real dt, s, t
t ← 0;   s ← 1
dt ← 0.1
for k = 1 to 10
t ← t + dt
s ← s * dt
output k, t, s
end
```

Hint: Print results with a very large number of decimal places.

30. By plotting $\ln x$ and $\ln[(1+x)/(1-x)]$, show that they both contain the point $\ln 2$. Are there other values that match up?

1.2 Mathematical Preliminaries

Most students have encountered infinite series (particularly Taylor series) in their study of calculus without necessarily having acquired a good understanding of this topic. Consequently, this section is particularly important for numerical analysis and deserves careful study.

Once students are well grounded with a basic understanding of Taylor series, the Mean-Value Theorem, and alternating series (all topics in this section) as well as computer number representation (Section 1.3), they can proceed to study the fundamentals of numerical methods with better comprehension. Well-prepared students may wish to skip over some of this material.

Taylor Series

Familiar (and useful) examples of Taylor series are the following:

Common Taylor series

$$e^x = 1 + x + \frac{x^2}{2!} + \frac{x^3}{3!} + \cdots = \sum_{k=0}^{\infty} \frac{x^k}{k!} \qquad (|x| < \infty) \tag{1}$$

$$\sin x = x - \frac{x^3}{3!} + \frac{x^5}{5!} - \cdots = \sum_{k=0}^{\infty} (-1)^k \frac{x^{2k+1}}{(2k+1)!} \qquad (|x| < \infty) \tag{2}$$

$$\cos x = 1 - \frac{x^2}{2!} + \frac{x^4}{4!} - \cdots = \sum_{k=0}^{\infty} (-1)^k \frac{x^{2k}}{(2k)!} \qquad (|x| < \infty) \tag{3}$$

$$\frac{1}{1-x} = 1 + x + x^2 + x^3 + \cdots = \sum_{k=0}^{\infty} x^k \qquad (|x| < 1) \tag{4}$$

$$\ln(1+x) = x - \frac{x^2}{2} + \frac{x^3}{3} - \cdots = \sum_{k=1}^{\infty} (-1)^{k-1} \frac{x^k}{k} \qquad (-1 < x \leqq 1) \tag{5}$$

For each case, the series represents the given function and converges in the interval specified. Series (1)–(5) are Taylor series expanded about $c = 0$.

A Taylor series expanded about $c = 1$ is

$$\ln(x) = (x-1) - \frac{(x-1)^2}{2} + \frac{(x-1)^3}{3} - \cdots = \sum_{k=1}^{\infty} (-1)^{k-1} \frac{(x-1)^k}{k}$$

where $0 < x \leqq 2$. The reader should recall the **factorial** notation

$$n! = 1 \cdot 2 \cdot 3 \cdot 4 \cdot \cdots \cdot n$$

for $n \geqq 1$ and the special definition of $0! = 1$.

Series of this type are often used to compute good approximate values of complicated functions at specific points.

EXAMPLE 1 Use five terms in the $\ln(1 + x)$ series (5) to approximate $\ln(1.1)$.

Solution Taking $x = 0.1$ in the first five terms of the series for $\ln(1 + x)$ gives us

$$\ln(1.1) \approx 0.1 - \frac{0.01}{2} + \frac{0.001}{3} - \frac{0.0001}{4} + \frac{0.00001}{5} = 0.09531\,03333\ldots$$

where $\approx$ means **approximately equal**. This value is correct to six decimal places of accuracy. ■

On the other hand, such good results are not always obtained in using series.

EXAMPLE 2 Try to compute e^8 by using the e^x series (1).

Solution The result is

$$e^8 = 1 + 8 + \frac{64}{2} + \frac{512}{6} + \frac{4096}{24} + \frac{32768}{120} + \cdots$$

It is apparent that many terms are needed to compute e^8 with reasonable precision. By repeated squaring, we find $e^2 = 7.38905\,6$, $e^4 = 54.59815\,00$, and $e^8 = 2980.95798\,7$. The first six terms given yield $570.06666\,5$. ■

These examples illustrate a general rule:

■ **Rule**

Rule of Thumb

A Taylor series converges rapidly near the point of expansion and slowly (or not at all) at more remote points.

A graphical depiction of the phenomenon can be obtained by graphing a few partial sums of a Taylor series. In Figure 1.3 (p. 22), we show the function

$$y = \sin x$$

and the partial-sum functions

Partial Summations for sin *x*

$$S_1 = x$$

$$S_3 = x - \frac{x^3}{6}$$

$$S_5 = x - \frac{x^3}{6} + \frac{x^5}{120}$$

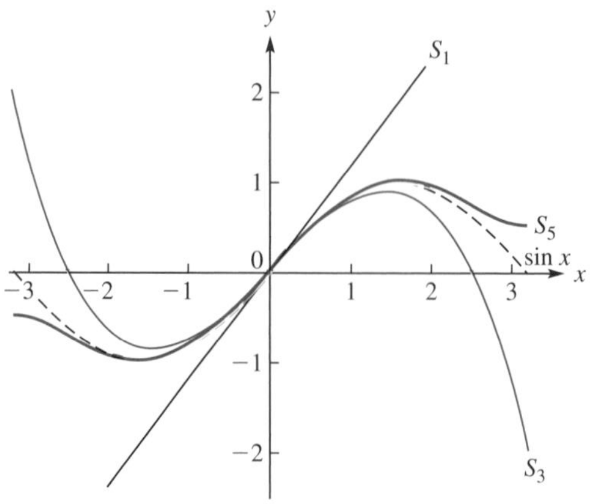

FIGURE 1.3 Approximations to sin x

which come from the $\sin x$ series (2). While S_1 may be an acceptable approximation to $\sin x$ when $x \approx 0$, the graphs for S_3 and S_5 match that of $\sin x$ on larger intervals about the origin.

All of the series illustrated here are examples of the following general series:

■ **Theorem 1** **Formal Taylor Series for f about c**

$$f(x) \sim f(c) + f'(c)(x-c) + \frac{f''(c)}{2!}(x-c)^2 + \frac{f'''(c)}{3!}(x-c)^3 + \cdots$$

$$f(x) \sim \sum_{k=0}^{\infty} \frac{f^{(k)}(c)}{k!}(x-c)^k \tag{6}$$

Here, rather than using $=$, we have written $\sim$ to indicate that we are *not* allowed to assume that $f(x)$ *equals* the series on the right. All we have at the moment is a formal series that can be written down provided that the successive derivatives f', f'', f''', ... exist at the point c. Series (6) is called the **Taylor series of f at the point c.**

In the special case $c = 0$, the Formal Taylor series (6) is also called a **Maclaurin series**:

$$f(x) \sim f(0) + f'(0)x + \frac{f''(0)}{2!}x^2 + \frac{f'''(0)}{3!}x^3 + \cdots$$

$$f(x) \sim \sum_{k=0}^{\infty} \frac{f^{(k)}(0)}{k!}x^k \tag{7}$$

The first term is $f(0)$ when $k = 0$.

EXAMPLE 3 What is the Taylor series of the function

$$f(x) = 3x^5 - 2x^4 + 15x^3 + 13x^2 - 12x - 5$$

at the point $c = 2$?

Solution To compute the coefficients in the series, we need the numerical values of $f^{(k)}(2)$ for $k \geqq 0$. Here are the details of the computation:

$$
\begin{aligned}
f(x) &= 3x^5 - 2x^4 + 15x^3 + 13x^2 - 12x - 5 &\Rightarrow\quad f(2) &= 207 \\
f'(x) &= 15x^4 - 8x^3 + 45x^2 + 26x - 12 &\Rightarrow\quad f'(2) &= 396 \\
f''(x) &= 60x^3 - 24x^2 + 90x + 26 &\Rightarrow\quad f''(2) &= 590 \\
f'''(x) &= 180x^2 - 48x + 90 &\Rightarrow\quad f'''(2) &= 714 \\
f^{(4)}(x) &= 360x - 48 &\Rightarrow\quad f^{(4)}(2) &= 672 \\
f^{(5)}(x) &= 360 &\Rightarrow\quad f^{(5)}(2) &= 360 \\
f^{(k)}(x) &= 0 &\Rightarrow\quad f^{(k)}(2) &= 0 \qquad (\text{k} \geqq 6)
\end{aligned}
$$

Therefore, we have

$$
\begin{aligned}
f(x) \sim 207 &+ 396(x-2) + 295(x-2)^2 \\
&+ 119(x-2)^3 + 28(x-2)^4 + 3(x-2)^5
\end{aligned}
$$

In this example, it is *not* difficult to see that $\sim$ may be replaced by $=$. Simply expand all the terms in the Taylor series and collect them to get the original form for f. Taylor's Theorem, discussed soon, allows us to draw this conclusion without doing any work! ■

It is interesting to show how well we can approximate a function $f(x)$ at a point $x = a$ by taking only a few terms of the Maclaurin series (7). We illustrated with three cases. With only the first term, the function is assumed to be a constant: $f(a) \approx f(0)$. With two terms, the slope at 0 is taken into account by way of the straight line from $f(0)$ to $f(a)$; namely, $f(a) \approx f(0) + f'(0)x$ when $x = a$. With three terms, the curivature due to $f''(0)$ comes into play and we obtain a parabolic curve: $f(a) \approx f(0) + f'(0)x + \frac{1}{2}f''(0)x^2$ when $x = a$. Each additional term improves the accuracy in the approximation for $f(a)$. In Figure 1.4, we show these partial sums in the Maclaurin series (7) when $f(x) = e^x$ and $x = a = 1$.

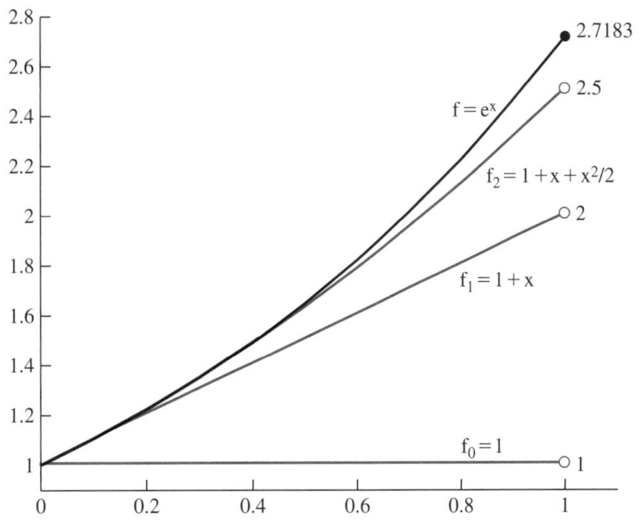

FIGURE 1.4
Approximations to e^x

Complete Horner's Algorithm

An application of Horner's algorithm is that of finding the Taylor expansion of a polynomial about any point. Let $p(x)$ be a given polynomial of degree n with coefficients a_k as in (2) in Section 1.1, and suppose that we desire the coefficients c_k in the equation

$$\begin{aligned} p(x) &= a_n x^n + a_{n-1}x^{n-1} + \cdots + a_0 \\ &= c_n(x-r)^n + c_{n-1}(x-r)^{n-1} + \cdots + c_1(x-r) + c_0 \end{aligned}$$

Of course, Taylor's Theorem asserts that $c_k = p^{(k)}(r)/k!$, but we seek a more efficient algorithm. Notice that $p(r) = c_0$, so this coefficient is obtained by applying Horner's algorithm to the polynomial p with the point r. The algorithm also yields the polynomial

$$q(x) = \frac{p(x) - p(r)}{x - r} = c_n(x-r)^{n-1} + c_{n-1}(x-r)^{n-2} + \cdots + c_1$$

This shows that the second coefficient, c_1, can be obtained by applying Horner's algorithm to the polynomial q with point r, because $c_1 = q(r)$. Notice that the first application of Horner's algorithm does not yield q in the form shown, but rather as a sum of powers of x. (See (3)–(4) in Section 1.1.) This process is repeated until all coefficients c_k are found.

We call the algorithm just described the **complete Horner's algorithm**. The pseudocode for executing it is arranged so that the coefficients c_k **overwrite** the input coefficients a_k.

Complete Horner's Algorithm Pseudocode

```
integer n, k, j;
real r; real array (a_i)_{0:n}
for k = 0 to n − 1
    for j = n − 1 to k
        a_j ← a_j + r a_{j+1}
    end for
end for
```

This procedure can be used in carrying out Newton's method for finding roots of a polynomial, which we discuss in Chapter 3. Moreover, it can be done in complex arithmetic to handle polynomials with complex roots or coefficients.

EXAMPLE 4 Using the complete Horner's algorithm, find the Taylor expansion of the polynomial

$$p(x) = x^4 - 4x^3 + 7x^2 - 5x + 2$$

about the point $r = 3$.

Solution The work can be arranged as follows:

$$\begin{array}{r|rrrrr}
 & 1 & -4 & 7 & -5 & 2 \\
3) & & 3 & -3 & 12 & 21 \\ \cline{2-6}
 & 1 & -1 & 4 & 7 & \boxed{23} \\
 & & 3 & 6 & 30 & \\ \cline{2-5}
 & 1 & 2 & 10 & \boxed{37} & \\
 & & 3 & 15 & & \\ \cline{2-4}
 & 1 & 5 & \boxed{25} & & \\
 & & 3 & & & \\ \cline{2-3}
 & \boxed{1} & \boxed{8} & & &
\end{array}$$

The calculation shows that

$$p(x) = (x-3)^4 + 8(x-3)^3 + 25(x-3)^2 + 37(x-3) + 23$$

■

Taylor's Theorem in Terms of $(x - c)$

■ **Theorem 2** **Taylor's Theorem for $f(x)$**

If the function f possesses continuous derivatives of orders $0, 1, 2, \ldots, (n+1)$ in a closed interval $I = [a, b]$, then for any c and x in I,

$$f(x) = \sum_{k=0}^{n} \frac{f^{(k)}(c)}{k!}(x-c)^k + E_{n+1} \tag{8}$$

where the error term E_{n+1} can be given in the form

$$E_{n+1} = \frac{f^{(n+1)}(\xi)}{(n+1)!}(x-c)^{n+1}$$

Here ξ is a point that lies between c and x and depends on both.

In practical computations with Taylor series, it is usually necessary to **truncate** the series because it is *not* possible to carry out an infinite number of additions. A series is said to be **truncated** if we ignore all terms after a certain point. Thus, if we truncate the exponential series (1) after seven terms, the result is

***e^x* Series**

$$e^x \approx 1 + x + \frac{x^2}{2!} + \frac{x^3}{3!} + \frac{x^4}{4!} + \frac{x^5}{5!} + \frac{x^6}{6!}$$

This no longer represents e^x except when $x = 0$. But the truncated series should *approximate* e^x. Here is where we need Taylor's Theorem. With its help, we can assess the difference between a function f and its truncated Taylor series.

The explicit assumption in this theorem is that $f(x), f'(x), f''(x), \ldots, f^{(n+1)}(x)$ are all continuous functions in the interval $I = [a, b]$. The final term E_{n+1} in (8) is the **remainder** or **error term**. The given formula for E_{n+1} is valid when we assume only that $f^{(n+1)}$ exists at each point of the open interval (a, b). The error term is similar to the terms preceding it, but notice that $f^{(n+1)}$ must be evaluated at a point other than c. This point ξ depends on x and is in the open interval (c, x) or (x, c). Other forms of the remainder are possible; the one given here is **Lagrange's form**. (We do *not* prove Taylor's Theorem here.)

EXAMPLE 5 Derive the Taylor series for e^x at $c = 0$, and prove that it converges to e^x by using Taylor's Theorem.

Solution If $f(x) = e^x$, then $f^{(k)}(x) = e^x$ for $k \geqq 0$. Therefore, $f^{(k)}(c) = f^{(k)}(0) = e^0 = 1$ for all k. From (8), we have

$$e^x = \sum_{k=0}^{n} \frac{x^k}{k!} + \frac{e^\xi}{(n+1)!}x^{n+1} \tag{9}$$

Now let us consider all the values of x in some symmetric interval around the origin, for example, $-s \leqq x \leqq s$. Then $|x| \leqq s$, $|\xi| \leqq s$, and $e^{\xi} \leqq e^{s}$. Hence, the remainder term satisfies this inequality:

$$\lim_{n\to\infty} \left| \frac{e^{\xi}}{(n+1)!} x^{n+1} \right| \leqq \lim_{n\to\infty} \frac{e^{s}}{(n+1)!} s^{n+1} = 0$$

Thus, if we take the limit as $n \to \infty$ on both sides of (9), we obtain

$$e^x = \lim_{n\to\infty} \sum_{k=0}^{n} \frac{x^k}{k!} = \sum_{k=0}^{\infty} \frac{x^k}{k!}$$

■

This example illustrates how we can establish, in specific cases, that a formal Taylor series (6) actually represents the function. Let's examine another example to see how the formal series can *fail* to represent the function.

EXAMPLE 6 Derive the formal Taylor series for $f(x) = \ln(1+x)$ at $c = 0$, and determine the range of positive x for which the series represents the function.

Solution We need $f^{(k)}(x)$ and $f^{(k)}(0)$ for $k \geqq 1$. Here is the work:

$$\begin{array}{lcl}
f(x) = \ln(1+x) & \Rightarrow & f(0) = 0 \\
f'(x) = (1+x)^{-1} & \Rightarrow & f'(0) = 1 \\
f''(x) = -(1+x)^{-2} & \Rightarrow & f''(0) = -1 \\
f'''(x) = 2(1+x)^{-3} & \Rightarrow & f'''(0) = 2 \\
f^{(4)}(x) = -6(1+x)^{-4} & \Rightarrow & f^{(4)}(0) = -6 \\
\vdots & \Rightarrow & \vdots \\
f^{(k)}(x) = (-1)^{k-1}(k-1)!(1+x)^{-k} & \Rightarrow & f^{(k)}(0) = (-1)^{k-1}(k-1)!
\end{array}$$

Hence by Taylor's Theorem, we obtain

$$\begin{aligned}
\ln(1+x) &= \sum_{k=1}^{n} (-1)^{k-1} \frac{(k-1)!}{k!} x^k + \frac{(-1)^n n!(1+\xi)^{-n-1}}{(n+1)!} x^{n+1} \\
&= \sum_{k=1}^{n} (-1)^{k-1} \frac{x^k}{k} + \frac{(-1)^n}{n+1} (1+\xi)^{-n-1} x^{n+1} \qquad (10)
\end{aligned}$$

For the *infinite* series to represent $\ln(1+x)$, it is necessary and sufficient that the error term converge to zero as $n \to \infty$. Assume that $0 \leqq x \leqq 1$. Then $0 \leqq \xi \leqq x$ (because *zero* is the point of expansion); thus, $0 \leqq x/(1+\xi) \leqq 1$. Hence, the error term converges to zero in this case. If $x > 1$, the terms in the series do not approach zero, and the series does not converge. Hence, the series represents $\ln(1+x)$ if $0 \leqq x \leqq 1$, but *not* if $x > 1$. (The series also represents $\ln(1+x)$ for $-1 < x < 0$, but not if $x \leqq -1$.) ■

Mean-Value Theorem

The special case $n = 0$ in Taylor's Theorem is known as the **Mean-Value Theorem**. It is usually stated, however, in a somewhat more precise form.

Theorem 3

Mean-Value Theorem

If f is a continuous function on the closed interval $[a, b]$ and possesses a derivative at each point of the open interval (a, b), then

$$f(b) = f(a) + (b - a)f'(\xi)$$

for some ξ in (a, b).

Hence, the ratio $[f(b) - f(a)]/(b - a)$ is equal to the derivative of f at some point ξ between a and b; that is, for some $\xi \in (a, b)$,

$$f'(\xi) = \frac{f(b) - f(a)}{b - a}$$

The right-hand side could be used as an *approximation* for $f'(x)$ at any x within the interval (a, b). The approximation of derivatives is discussed more fully in Section 4.3.

Taylor's Theorem in Terms of *h*

Other forms of Taylor's Theorem are often useful. These can be obtained from (8) by changing the variables.

Corollary 1

Taylor's Theorem for $f(x + h)$

If the function f possesses continuous derivatives of order $0, 1, 2, \ldots, (n + 1)$ in a closed interval $I = [a, b]$, then for any x in I,

$$f(x + h) = \sum_{k=0}^{n} \frac{f^{(k)}(x)}{k!} h^k + E_{n+1} \tag{11}$$

where h is any value such that $x + h$ is in I and where

$$E_{n+1} = \frac{f^{(n+1)}(\xi)}{(n + 1)!} h^{n+1}$$

for some ξ between x and $x + h$.

The form (11) is obtained from (8) by replacing x by $x + h$ and replacing c by x. Notice that because h can be positive or negative, the requirement on ξ means $x < \xi < x + h$ if $h > 0$ or $x + h < \xi < x$ if $h < 0$.

The **error term** E_{n+1} depends on h in two ways: First, h^{n+1} is explicitly present; second, the point ξ generally depends on h. As h converges to zero, E_{n+1} converges to zero with essentially the same rapidity with which h^{n+1} converges to zero. For large n, this is quite rapid. To express this qualitative fact, we write

Big O Notation

$$E_{n+1} = \mathcal{O}(h^{n+1})$$

as $h \to 0$. This is called **big O notation**, and it is shorthand for the inequality

$$|E_{n+1}| \leqq C|h|^{n+1}$$

where C is a constant. In the present circumstances, this constant could be any number for which $|f^{(n+1)}(t)|/(n+1)! \leq C$, for all t in the initially given interval, I. Roughly speaking, $E_{n+1} = \mathcal{O}(h^{n+1})$ means that the behavior of E_{n+1} is similar to the much simpler expression h^{n+1}.

It is important to realize that (11) corresponds to an entire sequence of theorems, one for each value of n. For example, we can write out the cases $n = 0, 1, 2$ as follows:

$$\begin{aligned} f(x+h) &= f(x) + f'(\xi_1)h \\ &= f(x) + \mathcal{O}(h) \\ f(x+h) &= f(x) + f'(x)h + \frac{1}{2!} f''(\xi_2)h^2 \\ &= f(x) + f'(x)h + \mathcal{O}(h^2) \\ f(x+h) &= f(x) + f'(x)h + \frac{1}{2!} f''(x)h^2 + \frac{1}{3!} f'''(\xi_3)h^3 \\ &= f(x) + f'(x)h + \frac{1}{2!} f''(x)h^2 + \mathcal{O}(h^3) \end{aligned}$$

The importance of the error term in Taylor's Theorem cannot be stressed too much. In later chapters, many situations require an estimate of errors in a numerical process by use of Taylor's Theorem. Here are some elementary examples.

EXAMPLE 7 Expand $\sqrt{1+h}$ in powers of h. Then compute $\sqrt{1.00001}$ and $\sqrt{0.99999}$.

Solution Let $f(x) = x^{1/2}$. Then $f'(x) = \frac{1}{2}x^{-1/2}$, $f''(x) = -\frac{1}{4}x^{-3/2}$, $f'''(x) = \frac{3}{8}x^{-5/2}$, and so on. Now use (11) with $x = 1$. Taking $n = 2$ for illustration, we have

$$\sqrt{1+h} = 1 + \frac{1}{2}h - \frac{1}{8}h^2 + \frac{1}{16}h^3\xi^{-5/2} \tag{12}$$

where ξ is an unknown number that satisfies $1 < \xi < 1 + h$, if $h > 0$. It is important to notice that the function $f(x) = \sqrt{x}$ possesses derivatives of all orders at any point $x > 0$.

In (12), let $h = 10^{-5}$. Then

$$\sqrt{1.00001} \approx 1 + 0.5 \times 10^{-5} - 0.125 \times 10^{-10} = 1.00000\,49999\,87500$$

By substituting $-h$ for h in the series, we obtain

$$\sqrt{1-h} = 1 - \frac{1}{2}h - \frac{1}{8}h^2 - \frac{1}{16}h^3\xi^{-5/2}$$

Hence, we have

$$\sqrt{0.99999} \approx 0.99999\,49999\,87500$$

Since $1 < \xi < 1 + h$, the absolute error does not exceed

$$\frac{1}{16}h^3\xi^{-5/2} < \frac{1}{16}10^{-15} = 0.00000\,00000\,00000\,0625$$

and both numerical values are correct to all 15 decimal places shown. ■

Alternating Series

Another theorem from calculus is often useful in establishing the convergence of a series and in estimating the error involved in truncation. From it, we have the following important principle for alternating series:

Alternative Series Principle

Principle. If the magnitudes of the terms in an alternating series converge monotonically to zero, then the error in truncating the series is no larger than the magnitude of the first omitted term.

This theorem applies only to **alternating series**—that is, series in which the successive terms are alternately positive and negative.

■ **Theorem 4** **Alternating Series Theorem**

If $a_1 \geqq a_2 \geqq \cdots \geqq a_n \geqq \cdots 0$ for all n and $\lim_{n\to\infty} a_n = 0$, then the alternating series

$$a_1 - a_2 + a_3 - a_4 + \cdots$$

converges; that is,

$$\sum_{k=1}^{\infty} (-1)^{k-1} a_k = \lim_{n\to\infty} \sum_{k=1}^{n} (-1)^{k-1} a_k = \lim_{n\to\infty} S_n = S$$

where S is its sum and S_n is the nth partial sum. Moreover, for all n,

$$|S - S_n| \leqq a_{n+1}$$

EXAMPLE 8 If the sine series is to be used in computing $\sin 1$ with an error less than $\frac{1}{2} \times 10^{-6}$, how many terms are needed?

Solution From Series (2), we have

$$\sin 1 = 1 - \frac{1}{3!} + \frac{1}{5!} - \frac{1}{7!} + \cdots$$

If we stop at $1/(2n-1)!$, the error does not exceed the first neglected term, which is $1/(2n+1)!$. Thus, we should select n so that

$$\frac{1}{(2n+1)!} < \frac{1}{2} \times 10^{-6}$$

Using logarithms to base 10, we obtain $\log(2n+1)! > \log 2 + 6 = 6.3$. With a calculator, we compute a table of values for $\log n!$ and find that $\log 10! \approx 6.6$. Hence, if $n \geqq 5$, the error is acceptable. ■

EXAMPLE 9 If the logarithmic series (5) is to be used for computing $\ln 2$ with an error of less than $\frac{1}{2} \times 10^{-6}$, how many terms are required?

Solution To compute $\ln 2$, we take $x = 1$ in the series, and using $\approx$ to mean approximate equality, we have

$$S = \ln 2 \approx 1 - \frac{1}{2} + \frac{1}{3} - \frac{1}{4} + \cdots + \frac{(-1)^{n-1}}{n} = S_n$$

By the Alternating Series Theorem, the error involved when the series is truncated with n terms is

$$|S - S_n| \leqq \frac{1}{n+1}$$

We select n so that

$$\frac{1}{n+1} < \frac{1}{2} \times 10^{-6}$$

Hence, more than two million terms would be needed! We conclude that this method of computing ln 2 is not practical. (See Exercises 1.2.10–1.2.12 for several good alternatives.) ■

Caution

A word of caution is needed about this technique of calculating the number of terms to be used in a series by just making the $(n+1)$st term less than some tolerance. This procedure is valid only for alternating series in which the terms decrease in magnitude to zero, although it is occasionally used to get rough estimates in other cases. For example, it can be used to identify a nonalternating series as one that converges slowly. When this technique cannot be used, a bound on the remaining terms of the series has to be established. Determining such a bound may be somewhat difficult.

EXAMPLE 10 It is known that

$$\frac{\pi^4}{90} = 1^{-4} + 2^{-4} + 3^{-4} + \cdots$$

How many terms should we take to compute $\pi^4/90$ with an error of at most $\frac{1}{2} \times 10^{-6}$?

Solution A naive approach is to take

$$1^{-4} + 2^{-4} + 3^{-4} + \cdots + n^{-4}$$

where n is chosen so that the next term, $(n+1)^{-4}$, is less that $\frac{1}{2} \times 10^{-6}$. This value of n is 37, but this is an erroneous answer because the partial sum

$$S_{37} = \sum_{k=1}^{37} k^{-4}$$

differs from $\pi^4/90$ by approximately 6×10^{-6}. What we should do, of course, is to select n so that *all* the omitted terms add up to less than $\frac{1}{2} \times 10^{-6}$; that is,

$$\sum_{k=n+1}^{\infty} k^{-4} < \frac{1}{2} \times 10^{-6}$$

By a technique familiar from calculus (see Figure 1.5), we have

$$\sum_{k=n+1}^{\infty} k^{-4} < \int_{n}^{\infty} x^{-4}\,dx = \left.\frac{x^{-3}}{-3}\right|_{n}^{\infty} = \frac{1}{3n^3}$$

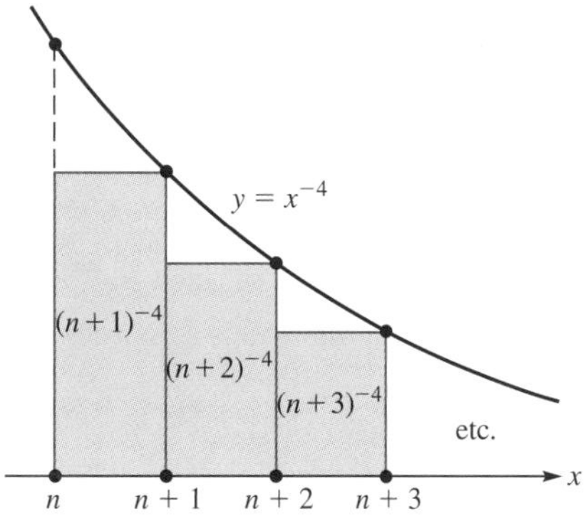

FIGURE 1.5 Illustrating Example 10

Thus, it suffices to select n so that $(3n^3)^{-1} < \frac{1}{2} \times 10^{-6}$, or $n \geqq 88$. (A more sophisticated analysis improves this considerably.) ■

Summary 1.2

- Complete Horner's Algorithm:

```
for k = 0 to n − 1
    for j = n − 1 to k
        a_j ← a_j + r a_{j+1}
    end for
end for
```

- The **Taylor series expansion** about c for $f(x)$ is

$$f(x) = \sum_{k=0}^{n} \frac{f^{(k)}(c)}{k!}(x-c)^k + E_{n+1}$$

with error term

$$E_{n+1} = \frac{f^{(n+1)}(\xi)}{(n+1)!}(x-c)^{n+1}$$

A more useful form for us is the **Taylor series expansion** for $f(x+h)$, which is

$$f(x+h) = \sum_{k=0}^{n} \frac{f^{(k)}(x)}{k!}h^k + E_{n+1}$$

with error term

$$E_{n+1} = \frac{f^{(n+1)}(\xi)}{(n+1)!}h^{n+1} = \mathcal{O}(h^{n+1})$$

- An **alternating series**

$$S = \sum_{k=1}^{\infty}(-1)^{k-1}a_k$$

converges when the terms a_k converge downward to zero. Furthermore, the partial sums S_n differ from S by an amount that is bounded by

$$|S - S_n| \leqq a_{n+1}$$

Exercises 1.2

1. The Maclaurin series for $(1+x)^n$ is also known as the **binomial series**. It states that

$$(1+x)^n = 1 + nx + \frac{n(n-1)}{2!}x^2 + \cdots \quad (x^2 < 1)$$

Derive this series. Then give its particular forms in summation notation by letting $n = 2$, $n = 3$, and $n = \frac{1}{2}$. Next use the last form to compute $\sqrt{1.0001}$ correct to 15 decimal places (rounded).

2. (Continuation) Use the series in the preceding problem to obtain series (4). How could this series be used on a computing machine to produce x/y if only addition and multiplication are built-in operations?

3. (Continuation) Use the previous problem to obtain a series for $(1 + x^2)^{-1}$.

4. Why do the following functions not possess Taylor series expansions at $x = 0$?

[a]**a.** $f(x) = \sqrt{x}$ [a]**b.** $f(x) = |x|$

c. $f(x) = \arcsin(x - 1)$ **d.** $f(x) = \cot x$

[a]**e.** $f(x) = \log x$ **f.** $f(x) = x^{\pi}$

[a]**5.** Determine the Taylor series for $\cosh x$ about zero. Evaluate $\cosh(0.7)$ by summing four terms. Compare with the actual value.

6. Determine the first two nonzero terms of the series expansion about zero for the following:

[a]**a.** $e^{\cos x}$ [a]**b.** $\sin(\cos x)$

c. $(\cos x)^2(\sin x)$

[a]**7.** Find the smallest nonnegative integer m such that the Taylor series about m for $(x - 1)^{1/2}$ exists. Determine the coefficients in the series.

[a]**8.** Determine how many terms are needed to compute e correctly to 15 decimal places (rounded) using e^x series (1) for e^x.

[a]**9.** (Continuation) If $x < 0$ in the preceding problem, what are the signs of the terms in the series? Loss of significant digits can be a serious problem in using the series. Will the formula $e^{-x} = 1/e^x$ be helpful in reducing the error? Explain. (See Section 1.5 for further discussion.) Try high-precision computer arithmetic to see how bad the floating-point errors can be.

10. Show how the simple equation $\ln 2 = \ln[e(2/e)]$ can be used to speed up the calculation of $\ln 2$ in series (10).

[a]**11.** What is the series for $\ln(1 - x)$? What is the series for $\ln[(1 + x)/(1 - x)]$?

[a]**12.** (Continuation) In the series for $\ln[(1 + x)/(1 - x)]$, determine what value of x to use if we wish to compute $\ln 2$. Estimate the number of terms needed for ten digits (rounded) of accuracy. Is this method practical?

13. Use the Alternating Series Theorem to determine the number of terms in series (5) needed for computing $\ln 1.1$ with error less than $\frac{1}{2} \times 10^{-8}$.

14. Write the Taylor series for the function $f(x) = x^3 - 2x^2 + 4x - 1$, using $x = 2$ as the point of expansion; that is, write a formula for $f(2 + h)$.

15. Determine the first four nonzero terms in the series expansion about zero for

[a]**a.** $f(x) = (\sin x) + (\cos x)$ and find an approximate value for $f(0.001)$.

[a]**b.** $g(x) = (\sin x)(\cos x)$ and find an approximate value for $g(0.0006)$.

Compare the accuracy of these approximations to those obtained from tables or via a calculator.

[a]**16.** Verify this Taylor series and prove that it converges on the interval $-e < x \leqq e$.

$$\ln(e + x) = 1 + \frac{x}{e} - \frac{x^2}{2e^2} + \frac{x^3}{3e^3} - \frac{x^4}{4e^4} + \cdots$$
$$= 1 + \sum_{k=1}^{\infty} \frac{(-1)^{k-1}}{k} \left(\frac{x}{e}\right)^k$$

[a]**17.** How many terms are needed in series (3) to compute $\cos x$ for $|x| < \frac{1}{2}$ accurate to 12 decimal places (rounded)?

[a]**18.** A function f is defined by the series

$$f(x) = \sum_{k=1}^{\infty} (-1)^k \left(\frac{x^k}{k^4}\right)$$

Determine the minimum number of terms needed to compute $f(1)$ with error less than 10^{-8}.

19. Verify that the partial sums $s_k = \sum_{i=0}^{k} x^i/i!$ in the series (1) for e^x can be written recursively as $s_k = s_{k-1} + t_k$, where $s_0 = 1$, $t_1 = x$, and $t_k = (x/k)t_{k-1}$.

[a]**20.** What is the fifth term in the Taylor series of $(1 - 2h)^{1/2}$?

21. Show that if $E = \mathcal{O}(h^n)$, then $E = \mathcal{O}(h^m)$ for any nonnegative integer $m \leqq n$. Here $h \to 0$.

22. Show how $p(x) = 6(x + 3) + 9(x + 3)^5 - 5(x + 3)^8 - (x + 3)^{11}$ can be efficiently evaluated.

[a]**23.** What is the second term in the Taylor series of $\sqrt[4]{4x - 1}$ about 4.25?

[a]**24.** How would you compute a table of $\log n!$ for $1 \leqq n \leqq 1000$?

25. For small x, the approximation $\sin x \approx x$ is often used. For what range of x is this good to a *relative* accuracy of $\frac{1}{2} \times 10^{-14}$?

26. In the Taylor series for the function $3x^2 - 7 + \cos x$ (expanded in powers of x), what is the coefficient of x^2?

27. In the Taylor series (about $\pi/4$) for the function $\sin x + \cos x$, find the third nonzero term.

[a]**28.** By using Taylor's Theorem, one can be sure that for all x that satisfy $|x| < \frac{1}{2}$, $|\cos x - (1 - x^2/2)|$ is less than or equal to what numerical value?

29. Find the value of ξ that serves in Taylor's Theorem when $f(x) = \sin x$, with $x = \pi/4$, $c = 0$, and $n = 4$.

30. Use Taylor's Theorem to find a linear function that approximates $\cos x$ best in the vicinity of $x = 5\pi/6$.

31. For the alternating series $S_n = \sum_{k=0}^{n}(-1)^k a_k$, with $a_0 > a_1 > \cdots > 0$, show by induction that $S_0 > S_2 > S_4 > \cdots$, that $S_1 < S_3 < S_5 < \cdots$, and that $0 < S_{2n} - S_{2n+1} = a_{2n+1}$.

[a]**32.** What is the Maclaurin series for the function $f(x) = 3 + 7x - 1.33x^2 + 19.2x^4$? What is the Taylor series for this function about $c = 2$?

33. In the text, it was asserted that $\sum_{k=0}^{6} x^k/k!$ represents e^x only at the point $x = 0$. Prove this.

34. Determine the first three terms in the Taylor series in terms of h for e^{x-h}. Using three terms, one obtains $e^{0.999} \approx Ce$, where C is a constant. Determine C.

[a]**35.** What is the least number of terms required to compute π as 3.14 (rounded) using the series

$$\pi = 4 - \frac{4}{3} + \frac{4}{5} - \frac{4}{7} + \cdots$$

36. Using the Taylor series expansion in terms of h, determine the first three terms in the series for $e^{\sin(x+h)}$. Evaluate $e^{\sin 90.01^\circ}$ accurately to ten decimal places as Ce for constant C.

37. Develop the first two terms and the error in the Taylor series in terms of h for $\ln(3 - 2h)$.

[a]**38.** Determine a Taylor series to represent $\cos(\pi/3+h)$. Evaluate $\cos(60.001^\circ)$ to eight decimal places (rounded). *Hint*: π radians equal 180 degrees.

[a]**39.** Determine a Taylor series to represent $\sin(\pi/4+h)$. Evaluate $\sin(45.0005^\circ)$ to nine decimal places (rounded).

40. Establish the first three terms in the Taylor series for $\csc(\pi/6 + h)$. Compute $\csc(30.00001^\circ)$ to the same accuracy as the given data.

41. Establish the Taylor series in terms of h for the following:

a. e^{x+2h} **b.** $\sin(x - 3h)$

c. $\ln[(x - h^2)/(x + h^2)]$

[a]**42.** Determine the first three terms in the Taylor series in terms of h for $(x - h)^m$, where m is an integer constant.

43. Given the series

$$-1 + 2^{-4} - 3^{-4} + 4^{-4} - \cdots$$

how many terms are needed to obtain four decimal places (chopped) of accuracy?

44. How many terms are needed in the series

$$\operatorname{arccot} x = \frac{\pi}{2} - x + \frac{x^3}{3} - \frac{x^5}{5} + \frac{x^7}{7} - \cdots$$

to compute $\operatorname{arccot} x$ for $x^2 < 1$ accurate to 12 decimal places (rounded)?

45. Determine the first three terms in the Taylor series to represent $\sinh(x + h)$. Evaluate $\sinh(0.0001)$ to 20 decimal places (rounded) using this series.

46. Determine a Taylor series to represent C^{x+h} for constant C. Use the series to find an approximate value of $10^{1.0001}$ to five decimal places (rounded).

[a]**47.** **Stirling's formula** states that $n!$ is greater than, and very close to, $\sqrt{2\pi n}\, n^n e^{-n}$. Use this to find an n for which $1/n! < \frac{1}{2} \times 10^{-14}$.

48. Develop the first two nonzero terms and the error term in the Taylor series in powers of h for $\ln[1 - (h/2)]$. Approximate $\ln(0.9998)$ using these two terms.

49. **L'Hôpital's rule** states that under suitable conditions,

$$\lim_{x\to a} \frac{f(x)}{g(x)} = \frac{f'(a)}{g'(a)}$$

It is true, for instance, when f and g have continuous derivatives in an open interval containing a, and $f(a) = g(a) = 0 \neq g'(a)$. Establish L'Hôpital's rule using the Mean-Value Theorem.

50. (Continuation) Evaluate the following numerically and use the previous problem to show that

a. $\lim_{x\to 0} \dfrac{\sin x}{x} = 1$ [a]**b.** $\lim_{x\to 0} \dfrac{\arctan x}{x} = 1$

[a]**c.** $\lim_{x\to \pi} \dfrac{\cos x + 1}{\sin x} = 0$

[a]**51.** Verify that if we take only the terms up to and including $x^{2n-1}/(2n-1)!$ in the series (2) for $\sin x$ and if $|x| < \sqrt{6}$, then the error involved does not exceed $|x|^{2n+1}/(2n+1)!$. How many terms are needed to compute $\sin(23)$ with an error of at most 10^{-8}? What problems do you foresee in using the series to compute $\sin(23)$? Show how to use periodicity to compute $\sin(23)$. Show that each term in the series can be obtained from the preceding one by a simple arithmetic operation.

[a]**52.** Expand the **error function**

$$\operatorname{erf}(x) = \frac{2}{\sqrt{\pi}} \int_0^x e^{-t^2}\, dt$$

in a series by using the exponential series and integrating. Obtain the Taylor series of $\operatorname{erf}(x)$ about zero directly. Are the two series the same? Evaluate $\operatorname{erf}(1)$ by adding four terms of the series and compare with the value

$\text{erf}(1) \approx 0.8427$, which is correct to four decimal places.
Hint: Recall from the **Fundamental Theorem of Calculus** that

$$\frac{d}{dx}\int_0^x f(t)\,dt = f(x)$$

[a]**53.** Establish the validity of the Taylor series

$$\arctan x = \sum_{k=1}^{\infty}(-1)^{k+1}\frac{x^{2k-1}}{2k-1} \qquad (-1 \leqq x \leqq 1)$$

Is it practical to use this series directly to compute $\arctan(1)$ if ten decimal places (rounded) of accuracy are required? How many terms of the series would be needed? Will loss of significance occur?
Hint: Start with the series for $1/(1+x^2)$ and integrate term by term. Note that this procedure is only formal; the convergence of the resulting series can be proved by appealing to certain theorems of advanced calculus.

[a]**54.** It is known that

$$\pi = 4 - 8\sum_{k=1}^{\infty}(16k^2 - 1)^{-1}$$

Discuss the numerical aspects of computing π by means of this formula. How many terms would be needed to yield ten decimal places (rounded) of accuracy?

55. Taylor's Theorem for $f(x)$ expanded about c involves this equation:

$$f(x) = f(c) + (x-c)f'(c) + \frac{1}{2}(x-c)^2 f''(c) + \cdots + \frac{1}{n!}(x-c)^n f^{(n)}(\xi)$$

Use this to determine how many terms in the series for e^x are needed to compute e with error at most 10^{-10}.
Hint: Use these approximate values of $n!$: $9! = 3.6\times 10^5$, $11! = 4.0\times 10^7$, $12! = 4.8\times 10^8$, $13! = 6.2\times 10^9$, $14! = 8.7\times 10^{10}$, and $15! = 1.3\times 10^{12}$.

56. To illustrate your understanding of the material, do the following:

a. Repeat Example 1.2.3 using the complete Horner's algorithm.

b. Repeat Example 1.2.6 using the Taylor series of the polynomial $p(x)$.

Computer Exercises 1.2

[a]**1.** Everyone knows the quadratic formula $(-b \pm \sqrt{b^2 - 4ac})/(2a)$ for the roots of the quadratic equation $ax^2 + bx + c = 0$. Using this formula, by hand and by computer, solve the equation $x^2 + 10^8 x + c = 0$ when $c = 1$ and 10^8. Interpret the results.

2. Use a computer algebra system to obtain graphs of the first five partial sums of the series

$$\arctan x = \sum_{k=1}^{\infty}(-1)^{k+1}\frac{x^{2k-1}}{2k-1}$$

3. Use a graphical computer package to reproduce the graphs in Figure 1.3 as well as the next two partial sums—that is, S_4 and S_5. Analyze the results.

4. Use a computer algebra system to obtain the Taylor series given in (1)–(5), obtaining the final form at once without displaying all the derivatives.

5. Use two or more computer algebra systems to carry out Example 1.2.6 to 50 decimal places. Are their answers the same and correct to all digits obtained? Repeat using $\sqrt{x}$ expanded about $x_0 = 1$.

6. Use a computer algebra system to verify the results in Examples 1.2.7 and 1.2.9.

7. Design and carry out an experiment to check the computation of x^y on your computer.
Hint: Compare the computations of some examples, such as $32^{2.5}$ and $81^{1.25}$, to their correct values. A more elaborate test can be made by comparing single-precision results to double-precision results in various cases.

8. Verify that $x^y = e^{y \ln x}$. Try to find values of x and y for which these two expressions differ in your computer. Interpret the results.

9. (Continuation) For

$$\cos(x-y) = (\cos x)(\cos y) + (\sin x)(\sin y)$$

repeat the preceding computer problem.

10. The number of combinations of n distinct items taken m at a time is given by the **binomial coefficient**

$$\binom{n}{m} = \frac{n!}{m!\,(n-m)!}$$

for integers m and n, with $0 \leqq m \leqq n$.
Recall that $\binom{n}{0} = \binom{n}{n} = 1$.

a. Write

> **integer function** *ibin*(n, m)

which uses the definition above to compute $\binom{n}{m}$.

b. Verify the formula

$$\binom{n}{m} = \prod_{k=1}^{\min(m,n-m)} \left[\frac{n-k+1}{k}\right]$$

for computing the binomial coefficients. Write

> **integer function** *jbin*(n, m)

that is based on this formula.

c. Verify the formulas (**Pascal's triangle**)

$$\begin{cases} a_{i0} = a_{ii} = 1 & (0 \leqq i \leqq n) \\ a_{ij} = a_{i-1,\,j-1} + a_{i-1,\,j} & (2 \leqq i \leqq n,\ 1 \leqq j \leqq i-1) \end{cases}$$

Using Pascal's triangle, compute the binomial coefficients

$$\binom{i}{j} = a_{i,\,j} \qquad (0 \leqq i,\, j \leqq n)$$

and store them in the lower triangular part of the array $(a_{ij})_{n \times n}$. Write

> **integer function** *kbin*(n, m)

that does an array look-up after first allocating and computing entries in the array.

11. The length of the curved part of a **unit semicircle** is π. We can approximate π by using triangles and elementary mathematics. Consider the semicircle with the arc bisected as in Figure (a). The hypotenuse of the right triangle is $\sqrt{2}$. Hence, a rough approximation to π is given by $2\sqrt{2} \approx 2.8284$. In Figure (b), we consider an angle θ that is a fraction $1/k$ of the semicircle. The secant shown has length $2\sin(\theta/2)$, and so an approximation to π is $2k\sin(\theta/2)$. From trigonometry, we have

$$\sin^2 \frac{1}{2}\theta = \frac{1}{2}(1 - \cos\theta) = \frac{1}{2}\left(1 - \sqrt{1 - \sin^2\theta}\right) = \frac{\sin^2\theta}{2 + 2\sqrt{1 - \sin^2\theta}}$$

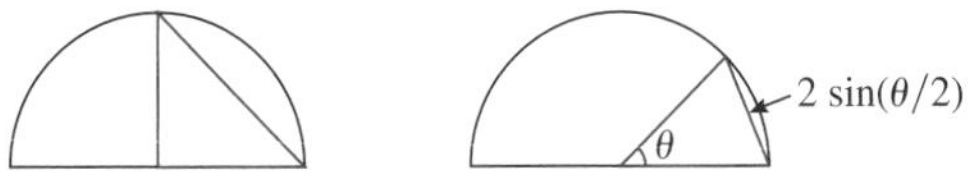

(a) (b)

Now let θ_n be the angle that results from division of the semicircular arc into 2^{n-1} pieces. Next let $S_n = \sin^2\theta_n$ and $P_n = 2^n\sqrt{S_{n+1}}$. Show that $S_{n+1} = S_n/(2 + 2\sqrt{1 - S_n})$ and P_n is an approximation to π. Starting with $S_2 = 1$ and $P_1 = 2$, compute S_{n+1} and P_n recursively for $2 \leqq n \leqq 20$.

12. The **irrational number** π can be computed by approximating the area of a unit circle as the limit of a sequence $p_1, p_2, \ldots$ described as follows. Divide the unit circle into 2^n sectors. (The figure shows the case $n = 3$.)

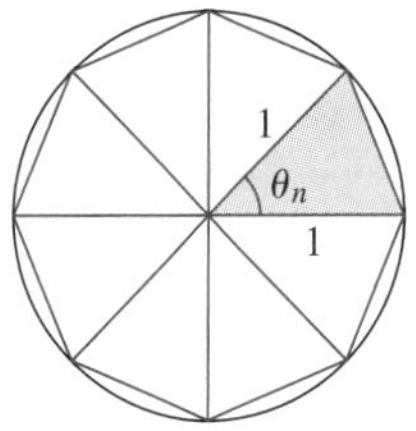

Approximate the area of the sector by the area of the isosceles triangle. The angle θ_n is $2\pi/2^n$. The area of the triangle is $\frac{1}{2}\sin\theta_n$. (Verify.) The nth approximation to π is then $p_n = 2^{n-1}\sin\theta_n$. Prove that $\sin\theta_n = \sin\theta_{n-1}/\{2[1+(1-\sin^2\theta_{n-1})^{1/2}]\}^{1/2}$ by means of well-known trigonometric identities. Use this recurrence relation to generate the sequences $\sin\theta_n$ and p_n $(3 \leqq n \leqq 20)$ starting with $\sin\theta_2 = 1$. Compare with the computation $4.0\arctan(1.0)$.

13. (Continuation) Calculate π by a method similar to that of the preceding computer problem, where the area of the unit circle is approximated by a sequence of trapezoids as illustrated by the figure.

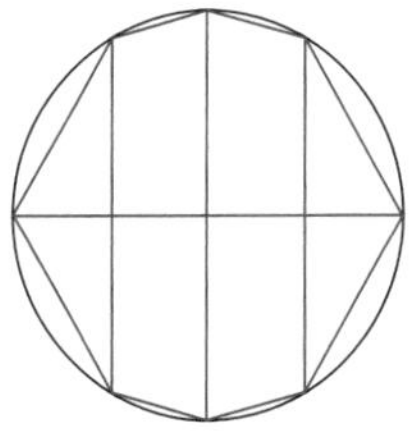

[a]**14.** Write a routine in double or long double precision to implement the following algorithm for computing π.

> **integer** k; **real** a, b, c, d, e, f, g
> $a \leftarrow 0$
> $b \leftarrow 1$
> $c \leftarrow 1/\sqrt{2}$
> $d \leftarrow 0.25$
> $e \leftarrow 1$

```
for k = 1 to 5
    a ← b
    b ← (b + c)/2
    c ← √(ca)
    d ← d − e(b − a)^2
    e ← 2e
    f ← b^2/d
    g ← (b + c)^2/(4d)
    output k, f, |f − π|, g, |g − π|
end for
```

Which converges faster, f or g? How accurate are the final values? Also compare with the double- or long-double-precision computation of 4.0 arctan(1.0).
Hint: The value of π correct to 36 digits is

3.14159 26535 89793 23846 26433 83279 50288

Note: A new formula for computing π was discovered in the early 1970s. This algorithm is based on that formula, which is a direct consequence of a method developed by Gauss for calculating elliptic integrals and of Legendre's elliptic integral relation, both known for over 150 years! The error analysis shows that rapid convergence occurs in the computation of π, and the number of significant digits doubles after each step. (The interested reader should consult Brent [1976], Borwein and Borwein [1987], and Salamin [1976].)

15. Another **quadratically convergent** scheme for computing π was discovered by Borwein and Borwein [1984] and can be written as

```
integer k; real a, b, t, x
a ← √2
b ← 0
x ← 2 + √2
for k = 1 to 5
    t ← √a
    b ← t(1 + b)/(a + b)
    a ← ½(t + 1/t)
    x ← xb(1 + a)/(1 + b)
    output k, x, |x − π|
end for
```

Numerically verify that $|x - \pi| \leqq 10^{-2k}$.
Note: Ludolf van Ceulen (1540–1610) was able to calculate π to 36 digits. With modern mathematical software packages such as MATLAB, Maple, and Mathematica, anyone can easily compute π to tens of thousands of digits in seconds!

[a]**16.** The **Fibonacci sequence** 1, 1, 2, 3, 5, 8, 13, 21, ... is defined by the linear recurrence relation

$$\begin{cases} \lambda_1 = 1, & \lambda_2 = 1 \\ \lambda_n = \lambda_{n-1} + \lambda_{n-2} & (n \geqq 3) \end{cases}$$

A formula for the nth **Fibonacci number** is

$$\lambda_n = \frac{1}{\sqrt{5}}\left\{\left[\frac{1}{2}\left(1+\sqrt{5}\right)\right]^n - \left[\frac{1}{2}\left(1-\sqrt{5}\right)\right]^n\right\}$$

Compute λ_n $(1 \leqq n \leqq 50)$, using both the recurrence relation and the formula. Write three programs that use integer, single-precision, and double-precision arithmetic, respectively. For each n, print the results using integer, single-precision, and double-precision formats, respectively.

[a]**17.** (Continuation) Repeat the experiment, using the sequence given by the recurrence relation

$$\begin{cases} \alpha_1 = 1, & \alpha_2 = \frac{1}{2}\left(1+\sqrt{5}\right) \\ \alpha_n = \alpha_{n-1} + \alpha_{n-2} & (n \geqq 3) \end{cases}$$

A closed-form formula is

$$\alpha_n = \left[\frac{1}{2}\left(1+\sqrt{5}\right)\right]^n$$

18. (Continuation) Change $+\sqrt{5}$ to $-\sqrt{5}$, and repeat the computation of α_n. Explain the results.

19. The **Bessel functions** J_n are defined by

$$J_n(x) = \frac{1}{\pi}\int_0^{\pi} \cos(x \sin\theta - n\theta)\, d\theta$$

Establish that $|J_n(x)| \leqq 1$.

a. It is known that

$$J_{n+1}(x) = 2nx^{-1}J_n(x) - J_{n-1}(x)$$

Use this recurrence relation to compute $J_0(1)$, $J_1(1), \ldots, J_{20}(1)$, starting from known values $J_0(1) \approx 0.76519\,76865$ and $J_1(1) \approx 0.44005\,05857$. Account for the fact that the inequality $|J_n(x)| \leqq 1$ is violated.

b. Another recursive relation is

$$J_{n-1}(x) = 2nx^{-1}J_n(x) - J_{n+1}(x)$$

Starting with the known values $J_{20}(1) \approx 3.87350\,3009 \times 10^{-25}$ and $J_{19}(1) \approx 1.54847\,8441 \times 10^{-23}$, use this equation to compute $J_{18}(1)$, $J_{17}(1), \ldots, J_1(1), J_0(1)$. Analyze the results.

20. A calculus student is asked to determine $\lim_{n\to\infty}(100^n/n!)$ and writes a program to evaluate $x_0, x_1, x_2, \ldots, x_n$ as follows:

```
integer parameter n ← 100
integer i; real x; x ← 1
for i = 1 to n
    x ← 100x/i
    output i, x
end for
```

The numbers printed become ever larger, and the student concludes that $\lim_{n\to\infty} x_n = \infty$. What is the moral here?

21. (**Maclaurin Series Function Approximations**) By using the truncated Maclaurin series, a function $f(x)$ with n continuous derivatives can be approximated by an nth-degree polynomial

$$f(x) \approx p_n(x) = \sum_{i=0}^{n} c_i x^i$$

where $c_i = f^{(i)}(0)/i!$.

a. Produce and compare computer plots for $f(x) = e^x$ and the polynomials $p_2(x)$, $p_3(x)$, $p_4(x)$, $p_5(x)$. Do the higher-order polynomials approximate the exponential function e^x satisfactorily on increasing intervals about zero?

b. Repeat for $g(x) = \ln(1 + x)$.

22. (Continuation) **Padé rational approximation** is the *best* approximation of a function by a rational function of a given order. Often it gives a better approximation of the function than truncating its Taylor series, and it may work even when the Taylor series does not converge! Consequently, the Padé rational approximations are frequently used in computer calculations such as for the basic function $\sin x$ as discussed in Computer Exercise 2.2.17. Rather than using high-order polynomials, we use ratios of low-order polynomials. These are called **rational approximations**. Let

$$f(x) \approx \frac{p_m(x)}{q_k(x)} = \frac{\sum_{i=0}^{m} a_i x^i}{\sum_{j=0}^{k} b_j x^j} = R_{m,k}(x)$$

where $b_0 = 1$. Here we have normalized with respect to $b_0 \neq 0$ and the values of m and k are modest. We choose the k coefficients b_j and the $m+1$ coefficients a_i in $R_{m,k}$ to match f and a specified number of its derivatives at the fixed point $x = 0$.

First, we construct the truncated Maclaurin series $\sum_{i=0}^{n} c_i x^i$ in which $c_i = f^{(i)}(0)/i!$ and $c_i = 0$ for $i < 0$. Next, we match the first $m + k + 1$ derivatives of $R_{m,k}$ with respect to x at $x = 0$ to the first $m+k+1$ coefficients c_i. This leads to the following displayed equations. Since $b_0 = 1$, we solve this $k \times k$ system of equations for $b_1, b_2, \ldots, b_k$

$$\begin{bmatrix} c_m & c_{m-1} & \cdots & c_{m-(k-2)} & c_{m-(k-1)} \\ c_{m+1} & c_m & \cdots & c_{m-(k-3)} & c_{m-(k-2)} \\ \vdots & \vdots & \ddots & \vdots & \vdots \\ c_{m+(k-2)} & c_{m+(k-3)} & \cdots & c_m & c_{m-1} \\ c_{m+(k-1)} & c_{m+(k-2)} & \cdots & c_{m+1} & c_m \end{bmatrix} \times \begin{bmatrix} b_1 \\ b_2 \\ \vdots \\ b_{k-1} \\ b_k \end{bmatrix} = \begin{bmatrix} -c_{m+1} \\ -c_{m+2} \\ \vdots \\ -c_{m+(k-1)} \\ -c_{m+k} \end{bmatrix}$$

(Solving systems of linear equations numerically is discussed in Chapter 2.) Finally, we evaluate these $m + 1$ equations for $a_0, a_1, \ldots, a_m$.

$$a_j = \sum_{\ell=0}^{j} c_{j-\ell} b_\ell \qquad (j = 0, 1, \ldots, m)$$

Note that $a_j = 0$ for $j > m$ and $b_j = 0$ for $j > k$. Also, if $k = 0$, then $R_{m,0}$ is a truncated Maclaurin series for f. Moreover, the Padé approximations may contain singularities.

a. Determine the **rational functions** $R_{1,1}(x)$ and $R_{2,2}(x)$. Produce and compare computer plots for $f(x) = e^x$, $R_{1,1}$, and $R_{2,2}$. Do these low-order rational functions approximate the exponential function e^x satisfactorily within $[-1, 1]$? How do they compare to the truncated Maclaurin polynomials of the preceding problem?

b. Repeat using $R_{2,2}(x)$ and $R_{3,1}(x)$ for the function $g(x) = \ln(1 + x)$.

Information on the life and work of the French mathematician **Herni Eugène Padé** (1863–1953) can be found in Wood [1999]. This reference also has examples and exercises similar to these. Further examples of Padé approximation can be seen.

23. (Continuation) Repeat for the **Bessel function** $J_0(2x)$, whose Maclaurin series is

$$1 - x^2 + \frac{x^4}{4} - \frac{x^6}{36} + \cdots = \sum_{i=0}^{\infty} (-1)^i \left(\frac{x^i}{i!}\right)^2$$

Then determine $R_{2,2}(x)$, $R_{4,3}(x)$, and $R_{2,4}(x)$ as well as comparing plots.

24. Carry out the details in the introductory example to this chapter by first deriving the Taylor series for $\ln(1 + x)$

and computing $\ln 2 \approx 0.63452$ using the first eight terms. Then establish the series $\ln[(1+x)/(1-x)]$ and calculate $\ln 2 \approx 0.69313$ using the terms shown. Determine the absolute error and relative errors for each of these answers.

25. Reproduce Figure 1.3 using your computer as well as adding the curve for S_4.

26. Use a mathematical software system that does symbolic manipulations such as Maple or Mathematica to carry out

a. Example 1.2.3 **b.** Example 1.2.6

27. Can you obtain the following numerical results?

$$\sqrt{1.00001} = 1.00000\,49999\,87500\,06249\,96093\,77734\,37500\,0000$$

$$\sqrt{0.99999} = 0.99999\,49999\,87499\,93749\,96093\,72265\,62500\,00000$$

Are these answers accurate to all digits shown?

28. Sometimes the values of a Taylor series cannot be easily reformulated. For 15 values of $x = 1, 0.1, 0.01, \ldots,$ compare the following.

a. $(-1+e^x)/x$ versus eight terms in the truncated Taylor series.

b. $(1-\cos x)/x^2$ versus $\sin^2 x/[x^2+\cos x]$.

1.3 Floating-Point Representation

The standard way to represent a nonnegative real number in decimal form is with an **integer part** and a **fractional part** with a **decimal point** between them such as

$$37.21829, \qquad 0.00227\,1828, \qquad 30\,00527.11059$$

(We group five digits together as shown.)

Another standard form, often called **normalized scientific notation**, is obtained by shifting the decimal point and supplying appropriate powers of 10. Thus, the preceding numbers have alternative representations as

$$\begin{aligned} 37.21829 &= 0.37218\,29 \times 10^2 \\ 0.00227\,1828 &= 0.22718\,28 \times 10^{-2} \\ 30\,00527.11059 &= 0.30005\,27110\,59 \times 10^7 \end{aligned}$$

In normalized scientific notation, the number is represented by a fraction multiplied by a power of 10, and the leading digit in the fraction is *not* zero (except when the number involved *is* zero). Thus, we write 79325 as 0.79325×10^5, *not* $0.07932\,5 \times 10^6$ or 7.9325×10^4 or some other way.

Normalized Floating-Point Representation

In the context of computer science, normalized scientific notation is also called **normalized floating-point representation**. In the decimal system, any real number x (other than zero) can be represented in normalized floating-point form as

$$x = \pm 0.d_1 d_2 d_3 \ldots \times 10^n$$

where $d_1 \neq 0$ and n is an integer (positive, negative, or zero). Each of the numbers d_1, d_2, $d_3, \ldots$ are the decimal digits 0, 1, 2, 3, 4, 5, 6, 7, 8, or 9.

Stated another way, the real number x, if different from zero, can be represented in normalized floating-point decimal form as

$$x = \pm r \times 10^n \qquad \left(\tfrac{1}{10} \leqq r < 1\right)$$

This representation consists of three parts: a sign that is either $+$ or $-$, a number r in the interval $\left[\frac{1}{10}, 1\right)$, and an integer power of 10. The number r is called the **normalized mantissa** and n the **exponent**.

The floating-point representation in the binary system is similar to that in the decimal system in several ways. If $x \neq 0$, it can be written as

$$x = \pm q \times 2^m \qquad \left(\tfrac{1}{2} \leqq q < 1\right)$$

The mantissa q would be expressed as a sequence of zeros or ones in the form $q = (0.b_1 b_2 b_3 \ldots)_2$, where $b_1 \neq 0$. Hence, $b_1 = 1$ and then necessarily $q \geqq \frac{1}{2}$ and $q < 1$.

A floating-point number system within a computer is similar to what we have just described, with one important difference: Every computer has only a finite word length and a finite total capacity, so only numbers with a finite number of digits can be represented. A number is allotted only one word of storage in the single-precision mode (two or more words in double or long-double precision). In either case, the degree of precision is strictly limited. Clearly, irrational numbers cannot be represented, nor can those rational numbers that do *not* fit the finite format imposed by the computer. Furthermore, numbers may be either too large or too small to be representable. The real numbers that are representable in a computer are called its **machine numbers**.

Since any number used in calculations with a computer must conform to the format of numbers in that computer system, it must have a **finite expansion**. Numbers that have a nonterminating expansion cannot be accommodated precisely. Moreover, a number that has a terminating expansion in one base may have a nonterminating expansion in another. A good example of this is the following simple fraction as given in the introductory example to this chapter:

$$\begin{aligned}\frac{1}{10} = (0.1)_{10} &= (0.06314\,6314\,6314\,6314\,\ldots)_8 \\ &= (0.0\,0011\,0011\,0011\,0011\,0011\,0011\,0011\,0011\,\ldots)_2\end{aligned}$$

The important point here is that most real numbers cannot be represented exactly in a computer. (See Appendix B for a discussion of representation of numbers in different bases.)

We illustrate that the effective number system for a computer is *not* a continuum, but a rather peculiar discrete set.

EXAMPLE 1 List all the floating-point numbers that can be expressed in the form

$$x = \pm(0.b_1 b_2 b_3)_2 \times 2^{\pm k}$$

where b_1, b_2, b_3, and m are allowed to have only the value 0 or 1. Then allow only *normalized* floating-point numbers; that is, all numbers (with the exception of zero) having the form

$$x = \pm(0.b_1 b_2 b_3)_2 \times 2^{\pm k}$$

Solution There are two choices for the $\pm$, two choices for b_1, two choices for b_2, two choices for b_3, and three choices for the exponent. Thus, at first, one would expect

$$2 \times 2 \times 2 \times 2 \times 3 = 48$$

different numbers. For example, all of the possible nonnegative numbers in this system are as follows:

$$
\begin{array}{lll}
(0.000)_2 \times 2^{-1} = 0, & (0.000)_2 \times 2^{0} = 0, & (0.000)_2 \times 2^{1} = 0 \\
(0.001)_2 \times 2^{-1} = \dfrac{1}{16}, & (0.001)_2 \times 2^{0} = \dfrac{1}{8}, & (0.001)_2 \times 2^{1} = \dfrac{1}{4} \\
(0.010)_2 \times 2^{-1} = \dfrac{2}{16}, & (0.010)_2 \times 2^{0} = \dfrac{2}{8}, & (0.010)_2 \times 2^{1} = \dfrac{2}{4} \\
(0.011)_2 \times 2^{-1} = \dfrac{3}{16}, & (0.011)_2 \times 2^{0} = \dfrac{3}{8}, & (0.011)_2 \times 2^{1} = \dfrac{3}{4} \\
(0.100)_2 \times 2^{-1} = \dfrac{4}{16}, & (0.100)_2 \times 2^{0} = \dfrac{4}{8}, & (0.100)_2 \times 2^{1} = \dfrac{4}{4} \\
(0.101)_2 \times 2^{-1} = \dfrac{5}{16}, & (0.101)_2 \times 2^{0} = \dfrac{5}{8}, & (0.101)_2 \times 2^{1} = \dfrac{5}{4} \\
(0.110)_2 \times 2^{-1} = \dfrac{6}{16}, & (0.110)_2 \times 2^{0} = \dfrac{6}{8}, & (0.110)_2 \times 2^{1} = \dfrac{6}{4} \\
(0.111)_2 \times 2^{-1} = \dfrac{7}{16}, & (0.111)_2 \times 2^{1} = \dfrac{7}{4}, & (0.111)_2 \times 2^{0} = \dfrac{7}{8}
\end{array}
$$

Here there are many duplications! So we obtain only these nonnegative numbers $\frac{1}{16}, \frac{3}{16}, \frac{5}{16}, \frac{7}{16}; \frac{1}{8}, \frac{3}{8}, \frac{5}{8}, \frac{7}{8}; \frac{1}{4}, \frac{3}{4}, \frac{5}{4}, \frac{7}{4}; \frac{1}{2}, \frac{3}{2}$; 0, 1. Altogether there are 31 distinct numbers in the system. The nonnegative numbers obtained are shown as dots on a line in Figure 1.6.

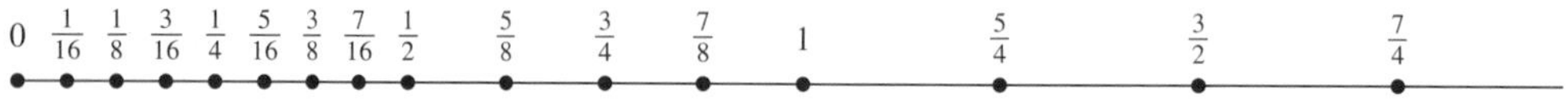

FIGURE 1.6 Example 1: Nonnegative machine numbers

Observe that the numbers are symmetrically, but unevenly distributed, about zero.

Now allowing only normalized floating-point numbers ($b_2 = 1$), we cannot represent $\frac{1}{16}, \frac{1}{8}$, and $\frac{3}{16}$. Hence, there are only 25 distinct numbers, and the nonnegative machine numbers are now distributed as in Figure 1.7.

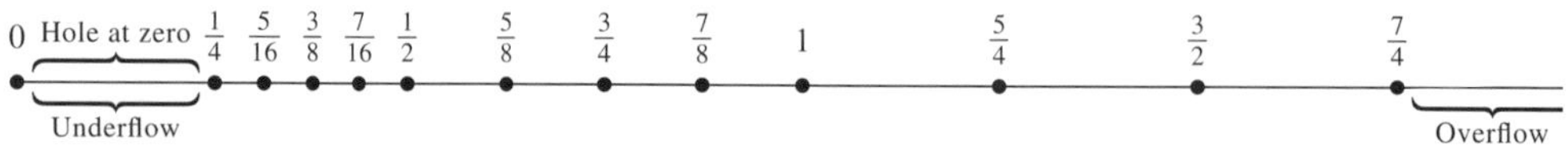

FIGURE 1.7 Example 1: Nonnegative normalized machine numbers ■

If, in the course of a computation, a number x is produced which is outside the computer's permissible range, then we say that an **overflow** has occurred or that x is **outside the range of the computer**. Generally, an overflow results in a fatal error (or **exception**), and the normal execution of the program stops! An **underflow** is usually treated automatically by setting x to zero without any interruption of the program and without a warning message in most computers.

In a computer whose floating-point numbers are restricted to the form in Example 1, any number closer to zero than $\frac{1}{4}$ would *underflow* to zero, and any number outside the range to the left of -1.75 to the right of $+1.75$ would *overflow* to machine $\pm$ infinity, respectively. Notice that there is a relatively wide gap between zero and the smallest positive machine number, which is $(0.100)_2 \times 2^{-1} = \frac{1}{4}$. This creates a phenomenon known as the

hole at zero. Figure 1.7 illustrates concepts such as the hole-at-zero, underflow, and overflow for Example 1.

We can store the normalized floating-point numbers from Example 1 in a five-bit computer with one bit for the sign of the number, two bits for the exponent, and two bits for the mantissa:

$$\boxed{\pm}\boxed{e_1}\boxed{e_2}\boxed{b_2}\boxed{b_3}$$

All possible combinations of positive normalized floating-point numbers are

$$(0.1b_2b_3)_2 \times 2^m = \begin{Bmatrix} (0.100)_2 = \frac{1}{2} \\ (0.101)_2 = \frac{5}{8} \\ (0.110)_2 = \frac{3}{4} \\ (0.111)_2 = \frac{7}{8} \end{Bmatrix} \times 2^{-1,0,1} = \begin{Bmatrix} \frac{1}{4}, \frac{1}{2}, 1 \\ \frac{5}{16}, \frac{5}{8}, \frac{5}{4} \\ \frac{3}{8}, \frac{3}{4}, \frac{3}{2} \\ \frac{7}{16}, \frac{7}{8}, \frac{7}{4} \end{Bmatrix}$$

A machine number in floating-point single-precision is of the form

$$(-1)^s q \times 2^m = (-1)^s \times 2^{c-1} \times (1.b_2b_3)_2$$

Here we set the exponent to $m = c - 1$. (In a 32-bit computer with an 8-bit exponent, $m = c - 127$. Why?) Some special cases are for ± 0, $\pm\infty$, and so on.

Floating-Point Representation

Before the Standard for Floating-Point Arithmetic (IEEE-754) was established in the early 1980s, computers used many different forms of floating-point representation, differing in the word length, the format of the representation, and the rounding used between operations! Now IEEE-754 has been accepted by almost all hardware and software manufactures worldwide. It defines the floating-point number system used by computers and it offers several rounding schemes, which affects the accuracy, among other things. In most computers, there are three common levels of precision for floating-point numbers, with the number of bits allocated for each organized as shown in the following table.

Precision	Bits	Sign	Exponent	Mantissa
Single	32	1	8	23
Double	64	1	11	52
Long Double	80	1	15	64

A computer that operates in floating-point mode represents numbers as described earlier except for the limitations imposed by the finite word length. Many binary computers have a word length of 32 or 64 bits (binary digits). We shall describe a machine of this type whose features mimic many workstations and personal computers in widespread use. The internal representation of numbers and their storage is **standard floating-point form**, which is used in almost all computers. For simplicity, we have left out a discussion of some of the details and features. Fortunately, one need not know all the details of the floating-point arithmetic system used in a computer to use it intelligently. Nevertheless, it is generally helpful in debugging a program to have a basic understanding of the representation of numbers in your computer.

By **single-precision floating-point numbers**, we mean all acceptable numbers in a computer using the standard single-precision floating-point arithmetic format. (In this discussion, we are assuming that such a computer stores these numbers in 32-bit words.) This set is a finite subset of the real numbers. It consists of ± 0, $\pm\infty$, normal and subnormal single-precision floating-point numbers, and even NotaNumber (`NaN`) values. (More detail on these subjects are in Appendix B and in the references.)

Recall that most real numbers *cannot* be represented exactly as floating-point numbers, since they have infinite decimal or binary expansions (all irrational numbers and some rational numbers); for example, π, e, $\frac{1}{3}$, 0.1, and so on.

Because of the 32-bit word-length, as much as possible of the normalized floating-point number

$$\pm q \times 2^m$$

must be contained in those 32 bits. One way of allocating the 32 bits is as follows:

sign of q	1 bit
integer $\lvert m\rvert$	8 bits
number q	23 bits

Information on the sign of m is contained in the 8 bits allocated for the integer $|m|$. In such a scheme, we can represent real numbers with $|m|$ as large as $2^7 - 1 = 127$. The exponent represents numbers from -127 through 128.

Single-Precision Floating-Point Form

We now describe a machine number of the following form in **standard single-precision floating-point** representation:

$$(-1)^s \times 2^{c-127} \times (1.f)_2$$

The leftmost bit is used for the sign of the mantissa, where $s = 0$ corresponds to $+$ and $s = 1$ corresponds to $-$. The next 8 bits are used to represent the number c in the exponent of 2^{c-127}, which is interpreted as an *excess-127 code*. Finally, the last 23 bits represent f from the fractional part of the mantissa in the 1-plus form: $(1.f)_2$. Each floating-point single-precision word is partitioned as in Figure 1.8.

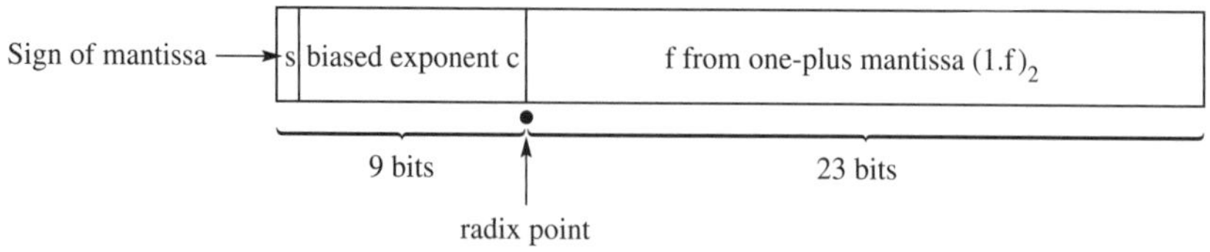

FIGURE 1.8 Partitioned floating-point single-precision computer word

In the normalized representation of a nonzero floating-point number, the first bit in the mantissa is *always* 1 so that this bit does not have to be stored. This can be accomplished by shifting the binary point to a "1-plus" form $(1.f)_2$. The mantissa is the rightmost 23 bits and contains f with an understood binary point as in Figure 1.8. So the mantissa **(significand)** *actually* corresponds to 24 binary digits since there is a **hidden bit**. (An important exception is the number ± 0.)

We now outline the procedure for determining the representation of a real number x. If x is zero, it is represented by a full word of zero bits with the possible exception of the sign bit. For a nonzero x, first assign the sign bit for x and consider $|x|$. Then convert both the integer and fractional parts of $|x|$ from decimal to binary. Next one-plus normalize $(|x|)_2$ by shifting the binary point so that the first bit to the left of the binary point is a 1 and all bits to the left of this 1 are 0. To compensate for this shift of the binary point, adjust the exponent of 2; that is, multiply by the appropriate power of 2. The 24-bit one-plus-normalized mantissa in binary is thus found. Now the current exponent of 2 should be set equal to $c - 127$ to determine c, which is then converted from decimal to binary. The sign bit of the mantissa is combined with $(c)_2$ and $(f)_2$. Finally, write the 32-bit representation of x as eight hexadecimal digits.

The value of c in the representation of a floating-point number in single precision is restricted by the inequality

$$0 < c < (11\,111\,111)_2 = 255$$

The values 0 and 255 are reserved for special cases, including ± 0 and $\pm\infty$, respectively. Hence, the actual exponent of the number is restricted by the inequality

$$-126 \leqq c - 127 \leqq 127$$

Likewise, we find that the mantissa of each nonzero number is restricted by the inequality

$$1 \leqq (1.f)_2 \leqq (1.111\,111\,111\,111\,111\,111\,111\,11)_2 = 2 - 2^{-23}$$

The largest number representable is therefore $(2 - 2^{-23})2^{127} \approx 2^{128} \approx 3.4 \times 10^{38}$. The smallest positive number is $2^{-126} \approx 1.2 \times 10^{-38}$.

The binary machine floating-point number $\varepsilon = 2^{-24}$ is called the **machine epsilon** when using single precision. It is the smallest positive machine number ε such that $1 + \varepsilon \neq 1$. Because $2^{-24} \approx 5.96 \times 10^{-8}$, we infer that in a simple computation, approximately **7 significant decimal digits** of accuracy may be obtained in single precision. Recall that 23 bits are allocated for the mantissa plus the hidden bit.

Double-Precision Floating-Point Form

When more precision is needed, **double precision** can be used, in which case each double-precision floating-point number is stored in two computer words in memory. In double precision, there are 52 bits allocated for the mantissa. The double precision **machine epsilon** is $2^{-53} \approx 1.11 \times 10^{-16}$, so approximately **15 significant decimal digits** of precision are available. There are 11 bits allowed for the exponent, which is biased by 1023. The exponent represents numbers from -1022 through 1023. A machine number in **standard double-precision floating-point form** corresponds to

$$(-1)^s \times 2^{c-1023} \times (1.f)_2$$

The leftmost bit is used for the sign of the mantissa with $s = 0$ for $+$ and $s = 1$ for $-$. The next 11 bits are used to represent the exponent c corresponding to 2^{c-1023}. Finally, 52 bits represent f from the fractional part of the mantissa in the one-plus form: $(1.f)_2$.

The value of c in the representation of a floating-point number in double precision is restricted by the inequality

$$0 < c < (1\,111\,111\,111)_2 = 2047$$

As in single precision, the values at the ends of this interval are reserved for special cases. Hence, the actual exponent of the number is restricted by the inequality

$$-1022 \leqq c - 1023 \leqq 1023$$

We find that the mantissa of each nonzero number is restricted by the inequality

$$1 \leqq (1.f)_2 \leqq (1.111\,111\,111 \cdots 111\,111\,111\,1)_2 = 2 - 2^{-52}$$

Because $2^{-52} \approx 1.2 \times 10^{-16}$, we infer that in a simple computation approximately 15 significant decimal digits of accuracy may be obtained in double precision.

Recall that 52 bits are allocated for the mantissa. The largest double-precision machine number is $(2 - 2^{-52})2^{1023} \approx 2^{1024} \approx 1.8 \times 10^{308}$. The smallest double-precision positive machine number is $2^{-1022} \approx 2.2 \times 10^{-308}$.

Single precision on a 64-bit computer is comparable to double precision on a 32-bit computer, whereas double precision on a 64-bit computer gives four times the precision available on a 32-bit computer.

In single precision, 31 bits are available for an integer because only 1 bit is needed for the sign. Consequently, the range for integers is from $-(2^{31}-1)$ to $(2^{31}-1) = 21474\,83647$. In double precision, 63 bits are used for integers, giving integers in the range $-(2^{63}-1)$ to $(2^{63}-1)$. In using integer arithmetic, accurate calculations can result in only approximately 9 digits in single precision and 18 digits in double precision! For high accuracy, most computations should be done by using double-precision floating-point arithmetic.

At this point, some students may wish to read Appendix B for a review of representing numbers in different bases.

EXAMPLE 2 Determine the machine representation of the decimal number $-52.23437\,5$ in both single precision and double precision.

Solution Converting the integer part to binary, we have $(52.)_{10} = (64.)_8 = (110\,100.)_2$. Next, converting the fractional part, we have $(.23437\,5)_{10} = (.17)_8 = (.001\,111)_2$. Now

$$(52.23437\,5)_{10} = (110\,100.001\,111)_2 = (1.101\,000\,011\,110)_2 \times 2^5$$

is the corresponding one-plus form in base 2, and $(.101\,000\,011\,110)_2$ is the stored mantissa. Next the exponent is $(5)_{10}$, and since $c - 127 = 5$, we immediately see that $(132)_{10} = (204)_8 = (10\,000\,100)_2$ is the stored exponent. Thus, the single-precision machine representation of $-52.23437\,5$ is

$$[1\;10\,000\,100\;101\,000\,011\,110\,000\,000\,000\,00]_2 =$$
$$[1100\;0010\;0101\;0000\;1111\;0000\;0000\;0000]_2 = [\text{C250F000}]_{16}$$

Here is the bit pattern for $-52, 23437\,5$ in single-precision floating-point using 32 bits:

↓ 8 bit exp. ↓

1	1	0	0	0	0	1	0	0	1	0	1	0	0	0	0	1	1	1	1	0	0	0	0	0	0	0	0	0	0	0	0

↑ sign bit ↑ 23 bit mantissa ↑

In double precision, for the exponent $(5)_{10}$, we let $c - 1023 = 5$, and we have $(1028)_{10} = (2004)_8 = (10\,000\,000\,100)_2$, which is the stored exponent. Thus, the double-precision machine representation of $-52.23437\,5$ is

$$[1\;10\,000\,000\,100\;101\,000\,011\,110\,000 \cdots 00]_2 =$$
$$[1100\;0000\;0100\;1010\;0001\;1110\;0000 \cdots 0000]_2 = [\text{C04A1E0000000000}]_{16}$$

Here $[\cdots]_k$ is the bit pattern of the machine word(s) that represents floating-point numbers, which is displayed in base-k. Here is the bit pattern for $-52, 23437\,5$ in double-precision floating-point using 64 bits:

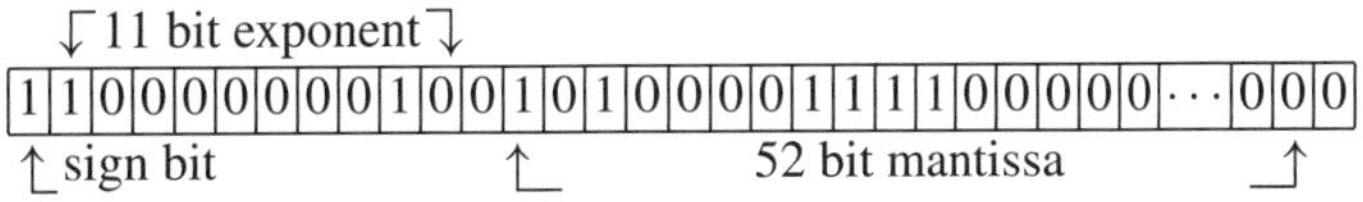

■

Mathematical software can be used to display the range of the numbers in a computer. In MATLAB, commands `realmin('single')` and `realmax('single')` return the smallest and largest finite floating-point numbers in single precision—double precision is similar. Here is a table of the range of floating-point normalized numbers with the hole at zero. Because this number range is not a continuum, there are lots of holes or gaps (*discontinuities*) throughout.

	Range
Single Precision	$-2^{128} \approx -3.4 \times 10^{38} \leqq x \leqq -1.2 \times 10^{-38} \approx -2^{-126}$
	0
	$2^{-126} \approx 1.2 \times 10^{-38} \leqq x \leqq 3.4 \times 10^{38} \approx 2^{128}$
Double Precision	$-2^{1024} \approx -1.8 \times 10^{318} \leqq x \leqq -2.2 \times 10^{-308} \approx 2^{-1022}$
	0
	$2^{-1022} \approx 2.2 \times 10^{-308} \leqq x \leqq 1.8 \times 10^{308} \approx 2^{1024}$

EXAMPLE 3 Determine the decimal numbers that correspond to these machine words:

$$[45DE4000]_{16} \qquad [BA390000]_{16}$$

Solution The first number in binary is

$$[0100\,0101\,1101\,1110\,0100\,0000\,0000\,0000]_2$$

The stored exponent is $(10\,001\,011)_2 = (213)_8 = (139)_{10}$, so $139 - 127 = 12$. The mantissa is positive and represents the number

$$\begin{aligned}(1.101\,111\,001)_2 \times 2^{12} &= (1\,101\,111\,001\,000.)_2\\ &= (15710.)_8\\ &= 0 \times 1 + 1 \times 8 + 7 \times 8^2 + 5 \times 8^3 + 1 \times 8^4\\ &= 8(1 + 8(7 + 8(5 + 8(1))))\\ &= 7112\end{aligned}$$

Similarly, the second word in binary is

$$[1011\,1010\,0011\,1001\,0000\,0000\,0000\,0000]_2$$

The exponential part of the word is $(01\,110\,100)_2 = (164)_8 = 116$, so the exponent is $116 - 127 = -11$. The mantissa is negative and corresponds to the following floating-point number:

$$\begin{aligned}-(1.011\,100\,100)_2 \times 2^{-11} &= -(0.000\,000\,000\,010\,111\,001)_2\\ &= -(0.00027\,1)_8\\ &= -2\times 8^{-4} - 7\times 8^{-5} - 1\times 8^{-6}\\ &= -8^{-6}(1+8(7+8(2)))\\ &= -\frac{185}{262144} \approx -7.05718\,99\times 10^{-4}\end{aligned}$$ ■

Computer Errors in Representing Numbers

We turn now to the errors that can occur when we attempt to represent a given real number x in the computer. We use a model computer with a 32-bit word length. Suppose first that we let $x = 2^{53\,21697}$ or $x = 2^{-32591}$. The exponents of these numbers far exceed the limitations of the machine (as described above). These numbers would overflow and underflow, respectively, and the relative error in replacing x by the closest machine number would be very large. Such numbers are *outside the range* of a 32-bit word-length computer.

Consider next a positive real number x in normalized floating-point form

$$x = q \times 2^m \qquad \left(\tfrac{1}{2} \leqq q < 1,\ -125 \leqq m \leqq 128\right)$$

The process of replacing x by its nearest machine number is called **correct rounding**, and the error involved is called **roundoff error**. We want to know how large it can be. We suppose that q is expressed in normalized binary notation, so

$$x = (0.1b_2b_3b_4\dots b_{24}b_{25}b_{26}\dots)_2 \times 2^m$$

One nearby machine number can be obtained by *rounding down* or by simply dropping the excess bits $b_{25}b_{26}\dots$, since only 23 bits have been allocated to the stored mantissa. This machine number is

$$x_- = (0.1b_2b_3b_4\dots b_{24})_2 \times 2^m$$

It lies to the left of x on the real-number axis. Another machine number, x_+, is just to the right of x on the real axis and is obtained by *rounding up*. It is found by adding one unit to b_{24} in the expression for x_-. Thus, we have

$$x_+ = \left[(0.1b_2b_3b_4\dots b_{24})_2 + 2^{-24}\right] \times 2^m$$

The closer of these machine numbers is the one chosen to represent x.

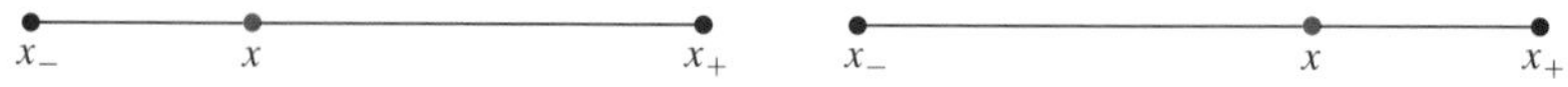

FIGURE 1.9 A possible relationship between x_-, x_+, and x.

The two situations are illustrated by the simple diagrams in Figure 1.9. If x lies closer to x_- than to x_+, then

$$|x - x_-| \leqq \tfrac{1}{2}|x_+ - x_-| = 2^{-25+m}$$

In this case, the relative error is bounded as follows:

$$\left|\frac{x - x_-}{x}\right| \leqq \frac{2^{-25+m}}{(0.1b_2b_3b_4\dots)_2 \times 2^m} \leqq \frac{2^{-25}}{1/2} = 2^{-24} = u$$

where $u = 2^{-24}$ is the **unit roundoff error** for a 32-bit binary computer with standard floating-point arithmetic.

Mathematical software can be used to display the precision of a computer. For example in MATLAB, command `eps('single')` returns the distance from 1.0 to the next largest single-precision floating-point number—double precision is similar.

	Precision
Single Precision	$2^{-23} \approx 1.2 \times 10^{-7}$
Double Precision	$2^{-52} \approx 2.2 \times 10^{-16}$

Recall that machine epsilon is $\varepsilon = 2^{-24}$, so $u = \varepsilon$. Moreover, $u = 2^{-k}$, where k is the number of binary digits used in the mantissa, including the hidden bit ($k = 24$ in single precision and $k = 53$ in double precision). On the other hand, if x lies closer to x_+ than to x_-, then

$$|x - x_+| \leqq \tfrac{1}{2}|x_+ - x_-|$$

and the same analysis shows that the relative error is no greater than $2^{-24} = u$. So in the case of rounding to the nearest machine number, the relative error is bounded by u. We note in passing that when *all* excess digits or bits are discarded, the process is called **chopping**. If a 32-bit word-length computer has been designed to chop numbers, the relative error bound would be twice as large as above.

The terms **machine epsilon** (ϵ) and **unit roundoff error** (u) are used interchangeably. Machine epsilon is used to study the effect of rounding errors because the actual errors of machine arithmetic are extremely complicated. Program libraries may provide precomputed values for these and other standard numerical quantities.

Often students are assigned the textbook exercise to compute an approximate value for machine epsilon. It is done in the sense of the spacing of the floating-point numbers at 1 rather than in the sense of the unit roundoff error. The following pseudocode produces an approximation to machine epsilon (within a factor of 2).

Machine Epsilon Pseudocode

```
epsi ← 1.0
while (1.0 + eps ≧ 1.0)
    epsi ← epsi/2.0
end for
epsi ← 2.0 * epsi
```

As with any computational results, it depends on the particular computer platform used as well as the programming language, the floating-point format (float, double, long double, etc.), and the runtime library.

Notation fl(x) and Backward Error Analysis

Next let us turn to the errors that are produced in the course of elementary arithmetic operations. To illustrate the principles, suppose that we are working with a five-place decimal machine and wish to add numbers. Two typical machine numbers in normalized floating-point form are

$$x = 0.37218 \times 10^4, \qquad y = 0.71422 \times 10^{-1}$$

Many computers perform arithmetic operations in a double-length work area, so assume that our computer has a ten-place accumulator. First, the exponent of the smaller number is adjusted so that both exponents are the same. Then the numbers are added in the accumulator, and the rounded result is placed in a computer word:

$$\begin{aligned} x &= 0.37218\,00000 \times 10^4 \\ y &= 0.00000\,71422 \times 10^4 \\ \hline x + y &= 0.37218\,71422 \times 10^4 \end{aligned}$$

The nearest machine number is $z = 0.37219 \times 10^4$, and the relative error involved in this machine addition is

$$\frac{|x + y - z|}{|x + y|} = \frac{0.00000\,28578 \times 10^4}{0.37218\,71422 \times 10^4} \approx 0.77 \times 10^{-5}$$

This relative error would be regarded as acceptable on a machine of such low precision.

To facilitate the analysis of such errors, it is convenient to introduce the notation $\text{fl}(x)$ to denote the **floating-point machine number** that corresponds to the real number x. Of course, the function fl depends on the particular computer involved. Our hypothetical five-decimal-digit machine would give

$$\text{fl}(0.37218\,71422 \times 10^4) = 0.37219 \times 10^4$$

For a 32-bit word-length computer, we established previously that if x is any real number within the range of the computer, then

$$\frac{|x - \text{fl}(x)|}{|x|} \leqq u \qquad \left(u = 2^{-24}\right) \tag{1}$$

Here and throughout, we assume that correct rounding is used. This inequality can also be expressed in the more useful form

$$\text{fl}(x) = x(1 + \delta) \qquad \left(|\delta| \leqq 2^{-24}\right)$$

To see that these two inequalities are equivalent, simply let $\delta = [\text{fl}(x) - x]/x$. Then, by Inequality (1), we have $|\delta| \leqq 2^{-24}$ and solving for $\text{fl}(x)$ yields $\text{fl}(x) = x(1 + \delta)$.

By considering the details in the addition $1+\varepsilon$, we see that if $\varepsilon \geqq 2^{-23}$, then $\text{fl}(1+\varepsilon) > 1$, whereas if $\varepsilon < 2^{-23}$, then $\text{fl}(1 + \varepsilon) = 1$. Consequently, if **machine epsilon** is the smallest positive machine number ε such that

Machine Epsilon

$$\text{fl}(1 + \varepsilon) > 1$$

then $\varepsilon = 2^{-23}$. Sometimes it is necessary to furnish the machine epsilon to a program. Because it is a machine-dependent constant, it can be found by either calling a system routine or by writing a simple program that finds the smallest positive number $x = 2^m$ such that $1 + x > 1$ in the machine.

Now let the symbol $\odot$ denote any one of the arithmetic operations $+$, $-$, $\times$, or $\div$. Suppose a 32-bit word-length computer has been designed so that whenever two *machine numbers* x and y are to be combined arithmetically, the computer produces $\text{fl}(x \odot y)$ instead of $x \odot y$. We can imagine that $x \odot y$ is first *correctly* formed, then normalized, and finally rounded to become a machine number. Under this assumption, the relative error does not exceed 2^{-24} by the previous analysis:

$$\text{fl}(x \odot y) = (x \odot y)(1 + \delta) \qquad \left(|\delta| \leqq 2^{-24}\right)$$

Special cases of this are, of course,

fl Properties

$$\begin{aligned} \text{fl}(x \pm y) &= (x \pm y)(1 + \delta) \\ \text{fl}(xy) &= xy(1 + \delta) \\ \text{fl}\left(\frac{x}{y}\right) &= \left(\frac{x}{y}\right)(1 + \delta) \end{aligned}$$

In these equations, δ is variable but satisfies $-2^{-24} \leqq \delta \leqq 2^{-24}$. The assumptions that we have made about a model 32-bit word-length computer are not quite true for a real computer. For example, it is possible for x and y to be machine numbers and for $x \odot y$ to overflow or underflow. Nevertheless, the assumptions should be realistic for most computing machines.

The equations given can be written in a variety of ways, some of which suggest alternative interpretations of roundoff. For example, we have

$$\text{fl}(x + y) = x(1 + \delta) + y(1 + \delta)$$

This says that the result of adding machine numbers x and y is not in general $x + y$ but is the true sum of $x(1 + \delta)$ and $y(1 + \delta)$. We can think of $x(1 + \delta)$ as the result of slightly perturbing x. Thus, the machine version of $x + y$, which is $\text{fl}(x + y)$, is the *exact* sum of a slightly perturbed x and a slightly perturbed y. The reader can supply similar interpretations in the examples given in the exercises.

This interpretation is an example of **backward error analysis**. It attempts to determine what perturbation of the original data would cause the *computer results* to be the exact results for a perturbed problem. In contrast, a **direct error analysis** attempts to determine how computed answers differ from exact answers based on the same data. In this aspect of scientific computing, computers have stimulated a new way of looking at computational errors.

Because the set of machine numbers is finite, some of the basic mathematical operations are not well defined and may breakdown in floating-point arithmetic. For example, we have

$$\text{fl}(x) = x(1 + \epsilon)$$

in which $|\epsilon| \leqq \epsilon_m$ where ϵ_m is **machine epsilon**, which is the smallest number such that $\text{fl}(1 + \epsilon_m) \neq 1$. Moreover, we obtain

$$\text{fl}(x \odot y) = (x \odot y)(1 + \epsilon_{\odot})$$

where $|\epsilon_{\odot}| \leqq \epsilon_m$ for the operations $\odot = +, -, *, /$ and

$$\text{fl}(x \odot y) = \text{fl}(y \odot x)$$

for operations $\odot = +, *$.

EXAMPLE 4 If x, y, and z are machine numbers in a 32-bit word-length computer, what upper bound can be given for the relative roundoff error in computing $z(x + y)$?

Solution In the computer, the calculation of $x + y$ would be done first. This arithmetic operation produces the machine number $\text{fl}(x + y)$, which differs from $x + y$ because of roundoff. By the principles established above, there is a δ_1 such that

$$\text{fl}(x + y) = (x + y)(1 + \delta_1) \qquad \left(|\delta_1| \leqq 2^{-24}\right)$$

When z multiplies the machine number $\mathrm{fl}(x+y)$, the result is the machine number $\mathrm{fl}[z\,\mathrm{fl}(x+y)]$ because z is already a machine number. This, too, differs from its exact counterpart, and we have, for some δ_2,

$$\mathrm{fl}[z\,\mathrm{fl}(x+y)] = z\,\mathrm{fl}(x+y)(1+\delta_2) \qquad \left(|\delta_2| \leqq 2^{-24}\right)$$

Putting both of our equations together, we have

$$\begin{aligned}\mathrm{fl}[z\,\mathrm{fl}(x+y)] &= z(x+y)(1+\delta_1)(1+\delta_2)\\ &= z(x+y)(1+\delta_1+\delta_2+\delta_1\delta_2)\\ &\approx z(x+y)(1+\delta_1+\delta_2)\\ &= z(x+y)(1+\delta) \qquad \left(|\delta| \leqq 2^{-23}\right)\end{aligned}$$

In this calculation, $|\delta_1\delta_2| \leqq 2^{-48}$, and so we ignore it. Also, we put $\delta = \delta_1 + \delta_2$ and then reason that $|\delta| = |\delta_1+\delta_2| \leqq |\delta_1| + |\delta_2| \leqq 2^{-24} + 2^{-24} = 2^{-23}$. ■

EXAMPLE 5 Critique the following attempt to estimate the relative roundoff error in computing the sum of two real numbers, x and y. In a 32-bit word-length computer, the calculation yields

$$\begin{aligned} z &= \mathrm{fl}[\mathrm{fl}(x) + \mathrm{fl}(y)]\\ &= [x(1+\delta) + y(1+\delta)](1+\delta)\\ &= (x+y)(1+\delta)^2\\ &\approx (x+y)(1+2\delta)\end{aligned}$$

Therefore, the relative error is bounded as follows:

$$\left|\frac{(x+y)-z}{(x+y)}\right| = \left|\frac{2\delta(x+y)}{(x+y)}\right| = |2\delta| \leqq 2^{-23}$$

Why is this calculation *not* correct?

Solution The quantities δ that occur in such calculations are not, in general, equal to each other. The correct calculation is

$$\begin{aligned} z &= \mathrm{fl}[\mathrm{fl}(x) + \mathrm{fl}(y)]\\ &= [x(1+\delta_1) + y(1+\delta_2)](1+\delta_3)\\ &= [(x+y) + \delta_1 x + \delta_2 y](1+\delta_3)\\ &= (x+y) + \delta_1 x + \delta_2 y + \delta_3 x + \delta_3 y + \delta_1\delta_3 x + \delta_2\delta_3 y\\ &\approx (x+y) + x(\delta_1+\delta_3) + y(\delta_2+\delta_3)\end{aligned}$$

Therefore, the relative roundoff error is

$$\begin{aligned}\left|\frac{(x+y)-z}{(x+y)}\right| &= \left|\frac{x(\delta_1+\delta_3)+y(\delta_2+\delta_3)}{(x+y)}\right|\\ &= \left|\frac{(x+y)\delta_3 + x\delta_1 + y\delta_2}{(x+y)}\right|\\ &= \left|\delta_3 + \frac{x\delta_1 + y\delta_2}{(x+y)}\right|\end{aligned}$$

This cannot be bounded, because the second term has a denominator that can be zero or close to zero. Notice that if x and y are machine numbers, then δ_1 and δ_2 are zero, and a useful bound results—namely, δ_3. But we do not need this calculation to know that! It has been assumed that when machine numbers are combined with any of the four arithmetic operations, the relative roundoff error would not exceed 2^{-24} in magnitude (on a 32-bit word-length computer). ■

Historical Notes

Examples of Numerical Computer Failures

In the 1991 Gulf War, a failure of the Patriot missile defense system was the result of a software conversion error. The system clock measured time in tenths of a second, but it was stored as a 24-bit floating-point number, resulting in rounding errors. The system failed to intercept an incoming Iraqi Scud missile, which resulted in the death of 28 American soldiers in a barracks in Dhahran, Saudi Arabia. (Field data had shown that the system would fail to track and intercept an incoming missile after being on for 20 consecutive hours and would need to be rebooted—it had been on for 100 hours!)

In 1994, Professor Thomas R. Nicely discovered that the Intel Pentium floating-point processor returned erroneous results for certain division operations. For example, 0001/824633702441.0 was calculated incorrectly for all digits beyond the eighth significant digit. After initially declaring that it would not impact many users, Intel eventually set aside \$420 million dollars to fix the problem and replaced the chip for anyone that requested it!

In 1996, the Ariane 5 rocket launched by the European Space Agency exploded 40 seconds after lift-off from Kourou, French Guiana. An investigation determined that the horizontal velocity required the conversion of a 64-bit floating-point number to a 16-bit signed integer. It failed because the number was larger than 32,767, which was the largest integer of this type that could be stored in memory. The rocket and its cargo were valued at \$500 million.

Further details about these disasters can be found by searching the World Wide Web on the Internet. There are other interesting accounts of calamities that could have been averted by more careful computer programming, especially in using floating-point arithmetic.

Summary 1.3

- A **single-precision floating-point number** in a 32-bit word-length computer with standard floating-point representation is stored in a single word with the bit pattern

$$\boxed{b_1 b_2 b_3 \cdots b_9 b_{10} b_{11} \cdots b_{32}}$$

which is interpreted as the real number

$$(-1)^{b_1} \times 2^{(b_2 b_3 \ldots b_9)_2} \times 2^{-127} \times (1.b_{10} b_{11} \ldots b_{32})_2$$

- A **double-precision floating-point number** in a 32-bit word-length computer with standard floating-point representation is stored in two words with the bit pattern

$$\boxed{b_1 b_2 b_3 \cdots b_9 b_{10} b_{11} b_{12} b_{13} \cdots b_{32} b_{33} b_{34} b_{35} \cdots\cdots b_{64}}$$

which is interpreted as the real number

$$(-1)^{b_1} \times 2^{(b_2 b_3 \ldots b_{12})_2} \times 2^{-1023} \times (1.b_{13} b_{14} \ldots b_{64})_2$$

- The relationship between a real number x and the **floating-point machine number** $\text{fl}(x)$ can be written as

$$\text{fl}(x) = x(1+\delta) \qquad \left(|\delta| \leqq 2^{-24}\right)$$

If $\odot$ denotes any one of the arithmetic operations, then we write

$$\text{fl}(x \odot y) = (x \odot y)(1+\delta)$$

In these equations, δ depends on x and y.

Exercises 1.3

1. Determine the machine representation in single precision on a 32-bit word-length computer for the following decimal numbers.

a. 2^{-30} **b.** 64.015625 [a]**c.** -8×2^{-24}

2. Determine the single-precision and double-precision machine representation in a 32-bit word-length computer of the following decimal numbers:

a. 0.5, −0.5 **b.** 0.125, −0.125
c. e0.0625, −0.0625 [a]**d.** 0.03125, −0.03125

3. Which of these are machine numbers?

a. 10^{403} **b.** $1 + 2^{-32}$ **c.** $1/5$
d. $1/10$ **e.** $1/256$

4. Determine the single-precision and double-precision machine representation of the following decimal numbers:

a. 1.0, −1.0 **b.** +0.0, −0.0 **c.** −9876.54321
[a]**d.** 0.234375 [a]**e.** 492.78125 **f.** 64.37109375
g. −285.75 **h.** 10^{-2}

5. Identify the floating-point numbers corresponding to the following bit strings:

a. 0 00000000 00000000000000000000000
b. 1 00000000 00000000000000000000000
c. 0 11111111 00000000000000000000000
[a]**d.** 1 11111111 00000000000000000000000
e. 0 00000001 00000000000000000000000
f. 0 10000001 01100000000000000000000
g. 0 01111111 00000000000000000000000
h. 0 01111011 10011001100110011001100

6. What are the bit-string machine representations for the following subnormal numbers?

a. $2^{-127} + 2^{-128}$ **b.** $2^{-127} + 2^{-150}$
c. $2^{-127} + 2^{-130}$ **d.** $\sum_{k=127}^{150} 2^{-k}$

7. Determine the decimal numbers that have the following machine representations:

a. $[3F27E520]_{16}$ **b.** $[3BCDCA00]_{16}$
c. $[BF4F9680]_{16}$ **d.** $[CB187ABC]_{16}$

8. Determine the decimal numbers that have the following machine representations:

[a]**a.** $[CA3F2900]_{16}$ **b.** $[C705A700]_{16}$
c. $[494F96A0]_{16}$ [a]**d.** $[4B187ABC]_{16}$
e. $[45223000]_{16}$ **f.** $[45607000]_{16}$
[a]**g.** $[C553E000]_{16}$ **h.** $[437F0000]_{16}$

9. Are these machine representations? Why or why not?

a. $[4BAB2BEB]_{16}$ **b.** $[1A1AIA1A]_{16}$
c. $[FADEDEAD]_{16}$ **d.** $[CABE6G94]_{16}$

10. The computer word associated with the variable Δ appears as $[7F7FFFFF]_{16}$, which is the largest representable floating-point single-precision number. What is the decimal value of Δ? The variable ε appears as $[00800000]_{16}$, which is the smallest positive number. What is the decimal value of ε?

11. Enumerate the set of numbers in the floating-point number system that have binary representations of the form $\pm(0.b_1b_2) \times 2^k$, where

a. $k \in \{-1, 0\}$ **b.** $k \in \{-1, 1\}$
[a]**c.** $k \in \{-1, 0, 1\}$

12. What are the machine numbers immediately to the right and left of 2^m? How far is each from 2^m?

13. Generally, when a list of floating-point numbers is added, less roundoff error will occur if the numbers are added in order of increasing magnitude. Give some examples to illustrate this principle.

14. (Continuation) The principle of the preceding exercise is *not universally* valid. Consider a decimal machine with two decimal digits allocated to the mantissa. Show that the four numbers 0.25, 0.0034, 0.00051, and 0.061 can be added with less roundoff error if *not* added in ascending order.

[a]**15.** In the case of machine underflow, what is the relative error involved in replacing a number x by zero?

16. Consider a computer that operates in base β and carries n digits in the mantissa of its floating-point number system. Show that the rounding of a real number x to the nearest machine number $\widetilde{x}$ involves a relative error of at most $\frac{1}{2}\beta^{1-n}$.
Hint: Imitate the argument in the text.

[a]**17.** Consider a decimal machine in which five decimal digits are allocated to the mantissa. Give an example, avoiding overflow or underflow, of a real number x whose closest machine number $\widetilde{x}$ involves the greatest possible relative error.

[a]**18.** In a five-decimal machine that correctly rounds numbers to the nearest machine number, what real numbers x have the property $\text{fl}(1.0 + x) = 1.0$?

[a]**19.** Consider a computer operating in base β. Suppose that it chops numbers instead of correctly rounding them. If its floating-point numbers have a mantissa of n digits, how large is the relative error in storing a real number in machine format?

20. What is the roundoff error when we represent $2^{-1}+2^{-25}$ by a machine number?
Note: This refers to absolute error, not relative error.

[a]**21.** (Continuation) What is the relative roundoff error when we round off $2^{-1} + 2^{-26}$ to get the closest machine number?

22. If x is a real number within the range of a 32-bit word-length computer that is rounded and stored, what can happen when x^2 is computed? Explain the difference between $\text{fl}[\text{fl}(x)\text{fl}(x)]$ and $\text{fl}(x\,x)$.

23. A binary machine that carries 30 bits in the fractional part of each floating-point number is designed to round a number up or down correctly to get the closest floating-point number. What simple upper bound can be given for the relative error in this rounding process?

24. A decimal machine that carries 15 decimal places in its floating-point numbers is designed to chop numbers. If x is a real number in the range of this machine and $\widehat{x}$ is its machine representation, what upper bound can be given for $|x - \widehat{x}|/|x|$?

[a]**25.** If x and y are real numbers within the range of a 32-bit word-length computer and if xy is also within the range, what relative error can there be in the machine computation of xy?
Hint: The machine produces $\text{fl}[\text{fl}(x)\text{fl}(y)]$.

[a]**26.** Let x and y be positive real numbers that are not machine numbers but are within the exponent range of a 32-bit word-length computer. What is the largest possible relative error in the machine representation of $x+y^2$? Include errors made to get the numbers in the machine as well as errors in the arithmetic.

27. Show that if x and y are positive real numbers that have the same first n digits in their decimal representations, then y approximates x with relative error less than 10^{1-n}. Is the converse true?

28. Show that a rough bound on the relative roundoff error when n machine numbers are multiplied in a 32-bit word-length computer is $(n-1)2^{-24}$.

29. Show that $\text{fl}(x + y) = y$ on a 32-bit word-length computer if x and y are positive machine numbers and $x < y \times 2^{-25}$.

[a]**30.** If 1000 nonzero machine numbers are added in a 32-bit word-length computer, what upper bound can be given for the relative roundoff error in the result? How many decimal digits in the answer can be trusted?

31. Suppose that $x = \sum_{i=1}^{n} a_i 2^{-i}$, where $a_i \in \{-1, 0, 1\}$ is a positive number. Show that x can also be written in the form $\sum_{i=1}^{n} b_i 2^{-i}$, where $b_i \in \{0, 1\}$.

32. If x and y are machine numbers in a 32-bit word-length computer and if $\text{fl}(x/y) = x/[y(1 + \delta)]$, what upper bound can be placed on $|\delta|$?

33. How big is the hole at zero in a 32-bit word-length computer?

34. How many machine numbers are there in a 32-bit length computer? (Consider only normalized floating-point numbers.)

35. How many normalized floating-point numbers are available in a binary machine if n bits are allocated to the mantissa and m bits are allocated to the exponent? Assume that two additional bits are used for signs, as in a 32-bit length computer.

36. Show by an example that in computer arithmetic $a + (b + c)$ may differ from $(a + b) + c$.

[a]**37.** Consider a decimal machine in which floating-point numbers have 13 decimal places. Suppose that numbers are correctly rounded up or down to the nearest machine number. Give the best bound for the roundoff error, assuming

no underflow or overflow. Use relative error, of course. What if the numbers are always chopped?

[a]**38.** Consider a computer that uses five-decimal-digit numbers. Let $\mathrm{fl}(x)$ denote the floating-point machine number closest to x. Show that if $x = 0.53214\,87513$ and $y = 0.53213\,04421$, then the operation $\mathrm{fl}(x) - \mathrm{fl}(y)$ involves a large relative error. Compute it.

[a]**39.** Two numbers x and y that are not machine numbers are read into a 32-bit word-length computer. The machine computes xy^2. What sort of relative error can be expected? Assume no underflow or overflow.

40. Let x, y, and z be three machine numbers in a 32-bit word-length computer. By analyzing the relative error in the worst case, determine how much roundoff error should be expected in forming $(xy)z$.

41. Let x and y be machine numbers in a 32-bit word-length computer. What relative roundoff error should be expected in the computation of $x + y$? If x is around 30 and y is around 250, what absolute error should be expected in the computation of $x + y$?

[a]**42.** Every machine number in a 32-bit word-length computer can be interpreted as the correct machine representation of an entire *interval* of real numbers. Describe this interval for the machine number $q \times 2^m$.

43. Is every machine number on a 32-bit word-length computer the average of two other machine numbers? If not, describe those that are not averages.

44. Let x and y be machine numbers in a 32-bit word-length computer. Let u and v be real numbers in the range of a 32-bit word-length computer but not machine numbers. Find a realistic upper bound on the relative roundoff error when u and v are read into the computer and then used to compute $(x + y)/(uv)$. As usual, ignore products of two or more numbers having magnitudes as small as 2^{-24}. Assume that no overflow or underflow occurs in this calculation.

45. Interpret the following:

a. $\mathrm{fl}(x) = x(1 - \delta)$

b. $\mathrm{fl}(xy) = [x(1 + \delta)]y$

c. $\mathrm{fl}(xy) = x[y(1 + \delta)]$

d. $\mathrm{fl}(xy) = \left(x\sqrt{1+\delta}\right)\left(y\sqrt{1+\delta}\right)$

e. $\mathrm{fl}\left(\dfrac{x}{y}\right) = \dfrac{x(1+\delta)}{y}$

f. $\mathrm{fl}\left(\dfrac{x}{y}\right) = \dfrac{x\sqrt{1+\delta}}{y/\sqrt{1+\delta}}$

g. $\mathrm{fl}\left(\dfrac{x}{y}\right) \approx \dfrac{x}{y(1-\delta)}$

46. Let x and y be real numbers that are not machine numbers for a 32-bit word-length computer and have to be rounded to get them into the machine. Assume that there is no overflow or underflow in getting their (rounded) values into the machine. (Thus, the numbers are within the *range* of a 32-bit word-length computer, although they are not machine numbers.) Find a rough upper bound on the relative error in computing x^2y^3.
Hint: We say *rough upper bound* because you may use $(1 + \delta_1)(1 + \delta_2) \approx 1 + \delta_1 + \delta_2$ and similar approximations. Be sure to include errors involved in getting the numbers into the machine as well as errors that arise from the arithmetic operations.

47. (**Student Research Project**) Write a research paper on the standard floating-point number system, providing additional details on

a. Types of rounding

b. Subnormal floating-point numbers

c. Long-double precision

d. Handling exceptional situations

Computer Exercises 1.3

1. Print several numbers, both integers and reals, in octal format and try to explain the machine representation used in your computer. For example, examine $(0.1)_{10}$ and compare to the results given at the beginning of this chapter.

2. Use your computer to construct a table of three functions f, g, and h defined as follows. For each integer n in the range 1 to 50, let $f(n) = 1/n$. Then $g(n)$ is computed by adding $f(n)$ to itself $n - 1$ times. Finally, set $h(n) = nf(n)$. We want to see the effects of roundoff error in these computations. Use the function real(n) to convert an integer variable n to its real (floating-point) form. Print the table with all the precision of which your computer is capable (in single-precision mode).

[a]**3.** Predict and then show what value your computer will print for $\sqrt{2}$ computed in single precision. Repeat for double or long-double precision. Explain.

4. Write a program to determine the machine epsilon ε within a factor of 2 for single, double, and long-double precision.

5. Let $\mathcal{A}$ denote the set of positive integers whose decimal representation does not contain the digit 0. The sum of the reciprocals of the elements in $\mathcal{A}$ is known to be 23.10345. Can you verify this numerically?

6. Write a computer code

```
integer function nDigit(n, x)
```

which returns the nth nonzero digit in the decimal expression for the real number x.

[a]7. The **harmonic series** $1 + \frac{1}{2} + \frac{1}{3} + \frac{1}{4} + \cdots$ is known to diverge to $+\infty$. The nth partial sum approaches $+\infty$ at the same rate as $\ln(n)$. **Euler's constant** is defined to be

$$\gamma = \lim_{n\to\infty}\left[\sum_{k=1}^{n}\frac{1}{k} - \ln(n)\right] \approx 0.57721$$

If your computer ran a program for a week based on the pseudocode

```
real s, x
x ← 1.0;
s ← 1.0
repeat
    x ← x + 1.0;
    s ← s + 1.0/x
end repeat
```

what is the largest value of s it would obtain? Write and test a program that uses a loop of 5000 steps to estimate Euler's constant. Print intermediate answers at every 100 steps.

[a]8. (Continuation) Prove that **Euler's constant**, γ, can also be represented by

$$\gamma = \lim_{m\to\infty}\left[\sum_{k=1}^{m}\frac{1}{k} - \ln\left(m + \frac{1}{2}\right)\right]$$

Write and test a program that uses $m = 1, 2, 3, \ldots, 5000$ to compute γ by this formula. The convergence should be more rapid than that in the preceding computer exercise. (See the article by de Temple [1993].)

9. Determine the binary form of $\frac{1}{3}$. What is the correctly rounded machine representation in single precision on a 32-bit word-length computer? Check your answer on an actual machine with the instructions

```
x ← 1.0/3.0; output x
```

using a long format of 16 digits for the output statement.

10. Owing to its gravitational pull, the Earth gains weight and volume slowly over time from space dust, meteorites, and comets. Suppose the Earth is a sphere. Let the radius be $r_a = 7000$ kilometers at the beginning of the year 1900, and let r_b be its radius at the end of the year 2000. Assume that $r_b = r_a + 0.000001$, an increase of 1 millimeter. Using a computer, calculate how much the Earth's volume and surface area has increased during the last century by the following three procedures (exactly as given):

   a. Difference in **spherical volume**

   $$V_a = \frac{4}{3}\pi r_a^3, \quad V_b = \frac{4}{3}\pi r_b^3, \delta_1 = V_b - V_a$$

   b. Difference in **spherical volume**

   $$\delta_2 = \frac{4}{3}\pi(r_b - r_a)(r_b^2 + r_b r_a + r_a^2)$$

   c. Difference in **spherical surface area**

   $$h = r_b - r_a, \quad \delta_3 = 4\pi r_a^2 h$$

   First use single precision and then double precision. Compare and analyze your results.

11. (**Student Research Project**) Explore recent developments in floating-point arithmetic. In particular, learn about long-double precision for both real numbers and integers as well as for complex numbers.

12. What is the largest integer your computer can handle?

13. Program each of the following pseudocodes in MATLAB or using some other mathematical software package or programming language. What is the purpose of each of them? What happens when each program is run on a computer? Explain what (if anything) is significant about the output. For example, how many iterations until the program stops, and what are the final values.

```
a. x = 0.0
   while x ≠ 0
       x = x + 0.1
   end while
   output x
```

```
b. x = 1.0
   while (x + 1.0) > 1.0
       x = x/2.0
   end while
   y = 2.0*x
   output y
```

c. $x = 1.0$
  while $x > 0.0$
    $y = x$
    $x = x/2.0$
  end while
  output y

14. MATLAB `dec2bin` and `bin2dec` for converting numbers from decimal to binary and vise versa as well as `hex2dec` and `dec2hex` for converting a hexadecimal numbers to double-precision and vise versa. Test these commands with these numbers:

a. $(11)_{10}$
b. $(197)_{10}$
c. $(0.625)_{10}$
d. $(0.2)_{10}$

1.4 Loss of Significance

In this section, we show how loss of significance in subtraction can often be reduced or eliminated by various techniques, such as the use of rationalization, Taylor series, trigonometric identities, logarithmic properties, double precision, and/or range reduction. These are some of the techniques that can be used when one wants to guard against the degradation of precision in a calculation. Of course, we cannot always know when a loss of significance has occurred in a long computation, but we should be alert to the possibility and take steps to avoid it, if possible.

Significant Digits

We first address the elusive concept of **significant digits** in a number. Suppose that x is a real number expressed in normalized scientific notation in the decimal system

$$x = \pm r \times 10^n \qquad \left(\tfrac{1}{10} \leqq r < 1\right)$$

For example, x might be

An Example of Significant Digits

$$x = 0.37214\,98 \times 10^{-5}$$

The digits 3, 7, 2, 1, 4, 9, 8 used to express r do not all have the same significance because they represent different powers of 10. Thus, we say that 3 is the *most* significant digit, and the significance of the digits diminishes from left to right. In this example, 8 is the *least* significant digit.

If x is a *mathematically exact* real number, then its approximate decimal form can be given with as many significant digits as we wish. Thus, we may write

$$\frac{\pi}{10} \approx 0.31415\,92653\,58979$$

Examples of Measured Quantities

and all the digits given are correct. If x is a *measured quantity*, however, the situation is quite different. Every measured quantity involves an error whose magnitude depends on the nature of the measuring device. Thus, if a meter stick is used, it is not reasonable to measure any length with precision better than 1 millimeter. Therefore, the result of measuring, say, a plate glass window with a meter stick should not be reported as 2.73594 meters. That would be misleading! Only digits that are believed to be correct or in error by at most a few units should be reported. It is a scientific convention that the least significant digit given in a measured quantity should be in error by at most five units; that is, the result is rounded correctly.

Similar remarks pertain to quantities computed from measured quantities. For example, if the side of a square is reported to be $s = 0.736$ meter, then one can assume that the error does not exceed a few units in the third decimal place. The diagonal of that square is then

$$s\sqrt{2} \approx 0.10408\,61182 \times 10^1$$

but should be reported as 0.1041×10^1 or (more conservatively) 0.104×10^1. The infinite precision available in $\sqrt{2}$,

$$\sqrt{2} = 1.41421\,35623\,73095\ldots$$

does *not* convey any more precision to $s\sqrt{2}$ than was already present in s.

Computer-Caused Loss of Significance

Perhaps it is surprising that a loss of significance can occur within a computer. It is essential to understand this process so that blind trust will not be placed in numerical output from a computer. One of the most common causes for a deterioration in precision is the subtraction of one quantity from another nearly equal quantity. This effect is potentially quite serious and can be catastrophic. The closer these two numbers are to each other, the more pronounced is the effect.

Loss of Significance $x - \sin(x)$

To illustrate this phenomenon, consider the assignment statement

$$y \leftarrow x - \sin(x)$$

and suppose that at some point in a computer program this statement is executed with an x value of $\frac{1}{15}$. Assume further that our computer works with floating-point numbers that have ten decimal digits. Then

$$\begin{aligned} x &\leftarrow 0.66666\,66667 \times 10^{-1} \\ \sin(x) &\leftarrow 0.66617\,29492 \times 10^{-1} \\ x - \sin(x) &\leftarrow 0.00049\,37175 \times 10^{-1} \\ x - \sin(x) &\leftarrow 0.49371\,75000 \times 10^{-4} \end{aligned}$$

In the last step, the result has been shifted to normalized floating-point form. Three zeros have then been supplied by the computer in the three *least* significant decimal places. We refer to these as **spurious zeros**; they are *not* significant digits. In fact, the ten-decimal-digit correct value is

$$\frac{1}{15} - \sin\left(\frac{1}{15}\right) \approx 0.49371\,74327 \times 10^{-4}$$

Another way of interpreting this is to note that the final digit in $x - \sin(x)$ is derived from the tenth digits in x and $\sin(x)$. When the eleventh digit in either x or $\sin(x)$ is 5, 6, 7, 8, or 9, the numerical values are rounded up to ten digits so that their tenth digits may be altered by plus one unit. Since these tenth digits may be in error, the final digit in $x - \sin(x)$ may also be in error—which it is!

EXAMPLE 1 If $x = 0.37214\,48693$ and $y = 0.37202\,14371$, what is the relative error in the computation of $x - y$ in a computer that has five decimal digits of accuracy?

Solution The numbers would first be rounded to $\widetilde{x} = 0.37214$ and $\widetilde{y} = 0.37202$. Then we have $\widetilde{x} - \widetilde{y} = 0.00012$, while the correct answer is $x - y = 0.00012\,34322$. The relative error involved is

$$\frac{|(x-y) - (\widetilde{x} - \widetilde{y})|}{|x-y|} = \frac{0.00000\,34322}{0.00012\,34322} \approx 3 \times 10^{-2}$$

This magnitude of relative error must be judged quite large when compared with the relative error of $\widetilde{x}$ and $\widetilde{y}$. (They cannot exceed $\frac{1}{2} \times 10^{-4}$ by the coarsest estimates, and in this example, they are, in fact, approximately 1.3×10^{-5}.) ■

It should be emphasized that this discussion pertains not to the operation

$$\text{fl}(x - y) \leftarrow x - y$$

but rather to the operation

fl Notation

$$\text{fl}[\text{fl}(x) - \text{fl}(y)] \leftarrow x - y$$

Roundoff error in the former case is governed by the equation

$$\text{fl}(x - y) = (x - y)(1 + \delta)$$

where $|\delta| \leqq 2^{-24}$ on a 32-bit word-length computer, and on a five-decimal-digit computer in the preceding example $|\delta| \leqq \frac{1}{2} \times 10^{-4}$.

In Example 1 above, we observe that the computed difference of 0.00012 has only two significant figures of accuracy, whereas in general, one expects the numbers and calculations in this computer to have five significant figures of accuracy.

The remedy for this difficulty is first to anticipate that it may occur and then to reprogram. The simplest technique may be to carry out part of a computation in double- or long-double-precision arithmetic (that means roughly twice as many significant digits), but often a slight change in the formulas is required. Several illustrations of this will be given, and additional examples are found among the exercises.

Again consider Example 1, but imagine that the calculations to obtain x, y, and $x - y$ are being done in double precision. Suppose that single-precision arithmetic is used thereafter. In the computer, all ten digits of x, y, and $x - y$ are retained, but at the end, $x - y$ is rounded to its five-digit form, which is 0.12343×10^{-3}. This answer has five significant digits of accuracy, as we would like. Of course, the programmer or analyst must know in advance where double-precision arithmetic may be necessary in the computation. Programming everything in double precision is very wasteful if it is not needed. This approach has another drawback: There may be such serious cancellation of significant digits that even double precision might not help!

Theorem on Loss of Precision

Before considering other techniques for avoiding this problem, we ask the following question:

> *Exactly how many significant binary digits are lost in the subtraction $x - y$ when x is close to y?*

The closeness of x and y is conveniently measured by $|1 - (y/x)|$. Here is the result:

■ Theorem 1 **Loss of Precision Theorem**

Let x and y be normalized floating-point machine numbers, where $x > y > 0$. If $2^{-p} \leqq 1 - (y/x) \leqq 2^{-q}$ for some positive integers p and q, then at most p and at least q significant binary bits are lost in the subtraction $x - y$.

Proof We prove the second part of the theorem and leave the first as an exercise. To this end, let $x = r \times 2^n$ and $y = s \times 2^m$, where $\frac{1}{2} \leqq r, s < 1$. (This is the normalized binary floating-point form.) Since $y < x$, the computer may have to *shift* y before carrying out the subtraction. In any case, y must first be expressed with the same exponent as x. Hence, $y = (s2^{m-n}) \times 2^n$ and

$$x - y = (r - s2^{m-n}) \times 2^n$$

The mantissa of this number satisfies

$$r - s2^{m-n} = r\left(1 - \frac{s2^m}{r2^n}\right) = r\left(1 - \frac{y}{x}\right) < 2^{-q}$$

Hence, to normalize the representation of $x - y$, a shift of at least q bits to the left is necessary. Then at least q (spurious) zeros are supplied on the right-hand end of the mantissa. This means that at least q bits of precision have been lost. ■

EXAMPLE 2 In the subtraction 37.59362 1 − 37.58421 6, how many bits of significants are lost?

Solution Let x denote the first number and y the second. Then

$$1 - \frac{y}{x} = 0.00025\,01754$$

This lies between 2^{-12} and 2^{-11}. These two numbers are 0.00024 4 and 0.00048 8. Hence, at least 11 but not more than 12 bits are lost. ■

Here is an example in decimal form.

EXAMPLE 3 In the subtraction of $y = .6311$ from $x = .6353$, how many significance are lost?

Solution These numbers are close, and $1 - y/x = .00661 < 10^{-2}$. In the subtraction, we have $x - y = .0042$. There are two significant figures in the answer, although there were four significant figures in x and y. ■

Avoiding Loss of Significance in Subtraction

Now we take up various techniques that can be used to avoid the loss of significance that may occur in subtraction.

EXAMPLE 4 Explore the function

$$f(x) = \sqrt{x^2+1} - 1 \tag{1}$$

whose values may be required for x near zero.

Solution Since $\sqrt{x^2+1} \approx 1$ when $x \approx 0$, we see that there is a potential loss of significance in the subtraction. However, the function can be rewritten in the form

$$f(x) = \left(\sqrt{x^2+1} - 1\right)\left(\frac{\sqrt{x^2+1}+1}{\sqrt{x^2+1}+1}\right) = \frac{x^2}{\sqrt{x^2+1}+1} \tag{2}$$

by **rationalizing** the numerator—that is, removing the radical in the numerator. This procedure allows terms to be canceled and thereby removes the subtraction. For example, if we use five-decimal-digit arithmetic and if $x = 10^{-3}$, then $f(x)$ is computed incorrectly as zero by the first formula, but as $\frac{1}{2} \times 10^{-6}$ by the second. If we use the first formula together with double precision, the difficulty is ameliorated, but *not* circumvented altogether. For example, in double precision, we have the same problem when $x = 10^{-6}$. ■

EXAMPLE 5 How can accurate values of the function

$$f(x) = x - \sin x \tag{3}$$

be computed near $x = 0$.

Solution A careless programmer might code this function just as indicated in Equation (3), not realizing that a serious loss of accuracy occurs. Recall from calculus that

$$\lim_{x\to 0} \frac{\sin x}{x} = 1$$

to see that $\sin x \approx x$ when $x \approx 0$. One cure for this problem is to use the Taylor series for $\sin x$:

$$\sin x = x - \frac{x^3}{3!} + \frac{x^5}{5!} - \frac{x^7}{7!} + \cdots$$

This series is known to represent $\sin x$ for all real values of x. For x near zero, it converges quite rapidly. Using this series, we can write the function f as

$$f(x) = x - \left(x - \frac{x^3}{3!} + \frac{x^5}{5!} - \frac{x^7}{7!} - \cdots\right) = \frac{x^3}{3!} - \frac{x^5}{5!} + \frac{x^7}{7!} - \cdots \tag{4}$$

We see in this equation where the original difficulty arose; namely, for small values of x, the term x in the sine series is much larger than $x^3/3!$ and thus more important. But when $f(x)$ is formed, this dominant x term disappears, leaving only the lesser terms. The series that starts with $x^3/3!$ is very effective for calculating $f(x)$ when x is small. ■

In Example 5, further analysis is needed to determine the range in which Series (4) should be used and the range in which Formula (3) can be used. Using the Theorem on Loss of Precision, we see that the loss of bits in the subtraction of Formula (3) can be limited to at most 1 bit by restricting x so that $\frac{1}{2} \leqq 1 - \sin x/x$. (Here we are considering only the case when $\sin x > 0$.) With a calculator, it is easy to see that x must be at least 1.9. Thus, for $|x| < 1.9$, we use the first few terms in Series (4), and for $|x| \geqq 1.9$, we use $f(x) = x - \sin x$. We can verify that for the worst case ($x = 1.9$), ten terms in the series give $f(x)$ with an error of at most 10^{-16}. (That is good enough for double precision on a 32-bit word-length computer.)

Now we can construct a function procedure for computing $f(x) = x - \sin x$ and write its pseudocode. Notice that the terms in the series can be obtained inductively by the algorithm

$$\begin{cases} t_1 = \dfrac{x^3}{6} \\ t_{n+1} = \dfrac{-t_n x^2}{(2n+2)(2n+3)} \qquad (n \geqq 1) \end{cases}$$

Then the partial sums can be obtained inductively by

$$\begin{cases} s_1 = t_1 \\ s_{n+1} = s_n + t_{n+1} \qquad (n \geqq 1) \end{cases}$$

so that

$$s_n = \sum_{k=1}^{n} t_k = \sum_{k=1}^{n} (-1)^{k+1} \left[\frac{x^{2k+1}}{(2k+1)!} \right]$$

A suitable pseudocode for the function in Example 5 is given here:

$f(x) = x - \sin x$ Function Pseudocode

```
real function f(x)
integer i, n ← 10;   real s, t, x
if |x| ≧ 1.9 then
    s ← x − sin x
    else
    t ← x³/6
    s ← t
    for i = 2 to n
        t ← −tx²/[(2i + 2)(2i + 3)]
        s ← s + t
    end for
end if
f ← s
end function f
```

EXAMPLE 6 How can accurate values of the function

$$f(x) = e^x - e^{-2x}$$

be computed in the vicinity of $x = 0$?

Solution Since e^x and e^{-2x} are both equal to 1 when $x = 0$, there is a loss of significance in the subtraction when x is close to zero. Inserting the appropriate Taylor series, we obtain

$$\begin{aligned} f(x) &= \left(1 + x + \frac{x^2}{2!} + \frac{x^3}{3!} + \cdots\right) - \left(1 - 2x + \frac{4x^2}{2!} - \frac{8x^3}{3!} + \cdots\right) \\ &= 3x - \frac{3}{2}x^2 + \frac{3}{2}x^3 - \cdots \end{aligned}$$

An alternative approach is to write

$$f(x) = e^{-2x}\left(e^{3x} - 1\right)$$
$$= e^{-2x}\left(3x + \frac{9}{2!}x^2 + \frac{27}{3!}x^3 + \cdots\right)$$

By using the Theorem on Loss of Precision, we find that at most 1 bit is lost in the subtraction $e^x - e^{-2x}$ when $x > 0$ and

$$\frac{1}{2} \leqq 1 - e^{-3x}$$

This inequality is valid when $x \geqq \frac{1}{3}\ln 2 = 0.23105$. Similar reasoning when $x < 0$ shows that for $x \leqq -0.23105$ and at most 1 bit is lost. Hence, the series should be used for $|x| < 0.23105$. ■

EXAMPLE 7 Criticize the assignment statement

$$y \leftarrow \cos^2(x) - \sin^2(x)$$

Solution When $\cos^2(x) - \sin^2(x)$ is computed, there is a loss of significance at $x = \pi/4$ (and at other points). The simple trigonometric identity

$$\cos 2\theta = \cos^2\theta - \sin^2\theta$$

can be used. Thus, the assignment statement can be replaced by

$$y \leftarrow \cos(2x)$$ ■

EXAMPLE 8 Criticize the assignment statement

$$y \leftarrow \ln(x) - 1$$

Solution If the expression $\ln x - 1$ is used for x near e, there is a cancellation of digits and a loss of accuracy. Use elementary facts about logarithms to overcome the difficulty. Thus, we have $y = \ln x - 1 = \ln x - \ln e = \ln(x/e)$. Here is a suitable assignment statement

$$y \leftarrow \ln\left(\frac{x}{e}\right)$$ ■

Range Reduction

Another cause of loss of significant figures is the evaluation of various library functions with large arguments. This problem is more subtle than those previously discussed. We illustrate with the sine function.

A basic property of the function $\sin x$ is its **periodicity**:

$$\sin x = \sin(x + 2n\pi)$$

for all real values of x and for all integer values of n. Because of this relationship, we need to know only the values of $\sin x$ in some fixed interval of length 2π to compute $\sin x$ for arbitrary x. This property can be used in the computer evaluation of $\sin x$ and is called **range reduction**.

EXAMPLE 9 Discuss how to evaluate $\sin(12532.14)$, by subtracting integer multiples of 2π. Show that it equals $\sin(3.47)$, if we retain only two decimal digits of accuracy.

Solution From $\sin(12532.14) = \sin(12532.14 - 2k\pi)$, we want $12532 = 2k\pi$ and $k = 3989/2\pi \approx 1994$. Consequently, we obtain $12532.14 - 2(1994)\pi = 3.49$ and $\sin(12532.14) \approx \sin(3.49)$. Thus, although our original argument 12532.14 had seven significant figures, the reduced argument has only three. The remaining digits disappeared in the subtraction of 3988π. Since 3.47 has only three significant figures, our computed value of $\sin(12532.14)$ has *no more than* three significant figures. This decrease in precision is unavoidable, if there is no way of increasing the precision of the original argument. If the original argument (12532.14) can be obtained with more significant figures, these additional figures are present in the *reduced* argument (3.47). In some cases, double- or long-double-precision programming may be helpful. ■

EXAMPLE 10 For $\sin x$, how many binary bits of significance are lost in range reduction to the interval $[0, 2\pi)$?

Solution Given an argument $x > 2\pi$, we determine an integer n that satisfies the inequality $0 \leqq x - 2n\pi < 2\pi$. Then in evaluating elementary trigonometric functions, we use $f(x) = f(x - 2n\pi)$. In the subtraction $x - 2n\pi$, there is a loss of significance. By the Theorem on Loss of Precision, at least q bits are lost if

$$1 - \frac{2n\pi}{x} \leqq 2^{-q}$$

Since

$$1 - \frac{2n\pi}{x} = \frac{x - 2n\pi}{x} < \frac{2\pi}{x}$$

we conclude that at least q bits are lost if $2\pi/x \leqq 2^{-q}$. Stated otherwise, at least q bits are lost if $2^q \leqq x/2\pi$. ■

Summary 1.4

- To avoid loss of significance in subtraction, one may be able to reformulate the expression using rationalizing, series expansions, or mathematical identities.
- If x and y are positive normalized floating-point machine numbers with

$$2^{-p} \leqq 1 - \frac{y}{x} \leqq 2^{-q}$$

then at most p and at least q significant binary bits are lost in computing $x - y$. (*Note*: It is permissible to leave out the hypothesis $x > y$ here.)

Exercises 1.4

1. How can values of the function $f(x) = \sqrt{x+4} - 2$ be computed accurately when x is small?

2. Calculate $f(10^{-2})$ for the function

$$f(x) = e^x - x - 1$$

The answer should have five significant figures and can easily be obtained with pencil and paper. Contrast it with the straightforward evaluation of $f(10^{-2})$ using $e^{0.01} \approx 1.0101$.

3. What is a good way to compute values of the function $f(x) = e^x - e$ if full machine precision is needed? *Note*: There is some difficulty when $x = 1$.

[a]**4.** What difficulty could the following assignment cause?

$$y \leftarrow 1 - \sin x$$

Circumvent it without resorting to a Taylor series if possible.

5. The hyperbolic sine function is defined by $\sinh x = \frac{1}{2}(e^x - e^{-x})$. What drawback could there be in using this formula to obtain values of the function? How can values of $\sinh x$ be computed to full machine precision when $|x| \leqq \frac{1}{2}$?

[a]**6.** Determine the first two nonzero terms in the expansion about zero for the function

$$f(x) = \frac{\tan x - \sin x}{x - \sqrt{1 + x^2}}$$

Give an approximate value for $f(0.0125)$.

7. Find a method for computing

$$y \leftarrow \frac{1}{x}(\sinh x - \tanh x)$$

that avoids loss of significance when x is small. Find appropriate identities to solve this problem without using Taylor series.

[a]**8.** Find a way to calculate accurate values for

$$f(x) = \frac{\sqrt{1 + x^2} - 1}{x^2} - \frac{x^2 \sin x}{x - \tan x}$$

Determine $\lim_{x \to 0} f(x)$.

9. For some values of x, the assignment statement $y \leftarrow 1 - \cos x$ involves a difficulty. What is it, what values of x are involved, and what remedy do you propose?

[a]**10.** For some values of x, the function $f(x) = \sqrt{x^2 + 1} - x$ cannot be accurately computed by using this formula. Explain and find a way around the difficulty.

[a]**11.** The inverse hyperbolic sine is given by $f(x) = \ln\left(x + \sqrt{x^2 + 1}\right)$. Show how to avoid loss of significance in computing $f(x)$ when x is negative.
Hint: Find and exploit the relationship between $f(x)$ and $f(-x)$.

12. On most computers, a highly accurate routine for $\cos x$ is provided. It is proposed to base a routine for $\sin x$ on the formula $\sin x = \pm\sqrt{1 - \cos^2 x}$. From the standpoint of precision (not efficiency), what problems do you foresee and how can they be avoided if we insist on using the routine for $\cos x$?

[a]**13.** Criticize and recode the assignment statement

$$z \leftarrow \sqrt{x^4 + 4} - 2$$

assuming that z is sometimes needed for an x close to zero.

14. How can values of the function $f(x) = \sqrt{x + 2} - \sqrt{x}$ be computed accurately when x is large?

15. Write a function that computes accurate values of $f(x) = \sqrt[4]{x + 4} - \sqrt[4]{x}$ for positive x.

[a]**16.** Find a way to calculate

$$f(x) = (\cos x - e^{-x}) / \sin x$$

correctly. Determine $f(0.008)$ correctly to ten decimal places (rounded).

17. Without using series, how could the function

$$f(x) = \frac{\sin x}{x - \sqrt{x^2 - 1}}$$

be computed to avoid loss of significance?

18. Write a function procedure that returns accurate values of the hyperbolic tangent function

$$\tanh x = \frac{e^x - e^{-x}}{e^x + e^{-x}}$$

for all values of x. Notice the difficulty when $|x| < \frac{1}{2}$.

19. Find a good way to compute $\sin x + \cos x - 1$ for x near zero.

[a]**20.** Find a good way to compute $\arctan x - x$ for x near zero.

21. Find a good bound for $|\sin x - x|$ using Taylor series and assuming that $|x| < \frac{1}{10}$.

[a]**22.** How would you compute $(e^{2x} - 1)/(2x)$ to avoid loss of significance near zero?

23. For any $x_0 > -1$, the sequence defined recursively by

$$x_{n+1} = 2^{n+1}\left(\sqrt{1 + 2^{-n}x_n} - 1\right) \qquad (n \geqq 0)$$

converges to $\ln(x_0 + 1)$. Arrange this formula in a way that avoids loss of significance.

24. Indicate how the following formulas may be useful for arranging computations to avoid loss of significant digits.

[a]**a.** $\sin x - \sin y = 2 \sin \frac{1}{2}(x - y) \cos \frac{1}{2}(x + y)$

b. $\log x - \log y = \log(x/y)$

c. $e^{x-y} = e^x / e^y$

d. $1 - \cos x = 2 \sin^2(x/2)$

e. $\arctan x - \arctan y = \arctan\left(\dfrac{x - y}{1 + xy}\right)$

25. What is a good way to compute $\tan x - x$ when x is near zero?

26. Find ways to compute these functions without serious loss of significant figures:

a. $e^x - \sin x - \cos x$ [a]**b.** $\ln(x) - 1$

c. $\log x - \log(1/x)$

[a]**d.** $x^{-2}(\sin x - e^x + 1)$ **e.** $x - \operatorname{arctanh} x$

27. Let

$$a(x) = \frac{1 - \cos x}{\sin x}$$
$$b(x) = \frac{\sin x}{1 + \cos x}$$
$$c(x) = \frac{x}{2} + \frac{x^3}{24}$$

Show that $b(x)$ is identical to $a(x)$ and that $c(x)$ approximates $a(x)$ in a neighborhood of zero.

[a]**28.** On your computer determine the range of x for which $(\sin x)/x \approx 1$ with full machine precision.
Hint: Use Taylor series.

[a]**29.** The familiar quadratic formula

$$x = \frac{1}{2a}\left(-b \pm \sqrt{b^2 - 4ac}\right)$$

will cause a problem when the quadratic equation $x^2 - 10^5 x + 1 = 0$ is solved with a machine that carries only eight decimal digits. Investigate the example, observe the difficulty, and propose a remedy.
Hint: An example in the text is similar.

[a]**30.** When accurate values for the roots of a quadratic equation are desired, some loss of significance may occur if $b^2 \approx 4ac$. What (if anything) can be done to overcome this when writing a computer routine?

31. Refer to the discussion of the function $f(x) = x - \sin x$ given in the text. Show that when $0 < x < 1.9$, there will be no undue loss of significance from subtraction in Equation (3).

32. Discuss the exercise of computing $\tan(10^{100})$. (See Gleick [1992], p. 178.)

33. Let x and y be two normalized binary floating-point machine numbers. Assume that $x = q \times 2^n$, $y = r \times 2^{n-1}$, $\frac{1}{2} \leqq r$, $q < 1$, and $2q - 1 \geqq r$. How much loss of significance occurs in subtracting $x - y$? Answer the same question when $2q - 1 < r$. Observe that the Theorem on Loss of Precision is not strong enough to solve this exercise precisely.

34. Prove the first part of the Theorem on Loss of Precision.

35. Show that if x is a machine number on a 32-bit computer that satisfies the inequality $x > \pi 2^{25}$, then $\sin x$ will be computed with *no* significant digits.

36. Let x and y be two positive normalized floating-point machine numbers in a 32-bit computer. Let $x = q \times 2^m$ and $y = r \times 2^n$ with $\frac{1}{2} \leqq r, q < 1$. Show that if $n = m$, then at least 1 bit of significance is lost in the subtraction $x - y$.

37. (**Student Research Project**) Read about and discuss the difference between **cancellation error**, a **bad algorithm**, and an **ill-conditioned problem**.
Suggestion: One example involves the quadratic equation. Read Stewart [1996].

38. On a three-significant-digit computer, calculate $\sqrt{9.01} - 3.00$, with as much accuracy as possible.

Computer Exercises 1.4

[a]**1.** Write a routine for computing the two roots x_1 and x_2 of the quadratic equation $f(x) = ax^2 + bx + c = 0$ with real constants a, b, and c and for evaluating $f(x_1)$ and $f(x_2)$. Use formulas that reduce roundoff errors and write efficient code. Test your routine on the following (a, b, c) values: (0, 0, 1); (0, 1, 0); (1, 0, 0); (0, 0, 0); (1, 1, 0); (2, 10, 1); (1, −4, 3.99999); (1, −8.01, 16.004); $(2 \times 10^{17}, 10^{18}, 10^{17})$; and $(10^{-17}, -10^{17}, 10^{17})$.

2. (Continuation) Write and test a routine for solving a quadratic equation that may have complex roots.

3. Alter and test the pseudocode in the text for computing $x - \sin x$ by using nested multiplication to evaluate the series.

4. Write a routine for the function $f(x) = e^x - e^{-2x}$ using the examples in the text for guidance.

5. Write code using double or extended precision to evaluate $f(x) = \cos(10^4 x)$ on the interval [0, 1]. Determine how many significant figures the values of $f(x)$ will have.

6. Write a procedure to compute $f(x) = \sin x - 1 + \cos x$. The routine should produce nearly full machine precision for all x in the interval $[0, \pi/4]$.
Hint: The trigonometric identity $\sin^2\theta = \frac{1}{2}(1 - \cos 2\theta)$ may be useful.

7. Write a procedure to compute $f(x, y) = \int_1^x t^y \, dt$ for arbitrary x and y.

Note: Notice the exceptional case $y = -1$ and the numerical problem *near* the exceptional case.

8. Suppose that we wish to evaluate the function $f(x) = (x - \sin x)/x^3$ for values of x close to zero.

a. Write a routine for this function. Evaluate $f(x)$ 16 times. Initially, let $x \leftarrow 1$, and then let $x \leftarrow \frac{1}{10}x$ 15 times. Explain the results.
Note: L'Hôpital's Rule indicates that $f(x)$ should tend to $\frac{1}{6}$. Test this code.

b. Write a function procedure that produces more accurate values of $f(x)$ for all values of x. Test this code.

9. Write a program to print a table of the function $f(x) = 5 - \sqrt{25 + x^2}$ for $x = 0$ to 1 with steps of 0.01. Be sure that your program yields full machine precision, but do not program the exercise in double precision. Explain the results.

[a]**10.** Write a routine that computes e^x by summing n terms of the Taylor series until the $n + 1$st term t is such that $|t| < \varepsilon = 10^{-6}$. Use the reciprocal of e^x for negative values of x. Test on the following data: 0, +1, −1, 0.5, −0.123, −25.5, −1776, 3.14159. Compute the relative error, the absolute error, and n for each case, using the exponential function on your computer system for the exact value. Sum no more than 25 terms.

11. (Continuation) The computation of e^x can be reduced to computing e^u for $|u| < (\ln 2)/2$ only. This algorithm removes powers of 2 and computes e^u in a range where the series converges very rapidly. It is given by

$$e^x = 2^m e^u$$

where m and u are computed by the steps

$$z \leftarrow x/\ln 2;\ m \leftarrow \text{integer}\,(z \pm \tfrac{1}{2})$$
$$w \leftarrow z - m;\ u \leftarrow w \ln 2$$

Here the minus sign is used if $x < 0$ because $z < 0$. Incorporate this range reduction technique into the code.

12. (Continuation) Write a routine that uses range reduction $e^x = 2^m e^u$ and computes e^u from the even part of the **Gaussian continued fraction**; that is,

$$e^u = \frac{s+u}{s-u}, \qquad s = 2 + u^2\left(\frac{2520 + 28u^2}{15120 + 420u^2 + u^4}\right)$$

Test on the data given in Computer Exercise 1.4.10.
Note: Some of the computer exercises in this section contain rather complicated algorithms for computing various intrinsic functions that correspond to those actually used on a large mainframe computer system. Descriptions of these and other similar library functions are frequently found in the supporting documentation of your computer system.

13. Quite important in many numerical calculations is the accurate computation of the absolute value $|z|$ of a complex number $z = a + bi$. Design and carry out a computer experiment to compare the following three schemes:

a. $|z| = (a^2 + b^2)^{1/2}$ **b.** $|z| = v\left[1 + \left(\frac{w}{v}\right)^2\right]^{1/2}$

c. $|z| = 2v\left[\frac{1}{4} + \left(\frac{w}{2v}\right)^2\right]^{1/2}$

where $v = \max\{|a|, |b|\}$ and $w = \min\{|a|, |b|\}$. Use very small and large numbers for the experiment.

[a]**14.** For what range of x is the approximation $(e^x - 1)/2x \approx 0.5$ correct to 15 decimal digits of accuracy? Using this information, write a function procedure for $(e^x - 1)/2x$, producing 15 decimals of accuracy throughout the interval $[-10, 10]$.

[a]**15.** In the theory of Fourier series, some numbers known as **Lebesgue constants** play a role. A formula for them is

$$\rho_n = \frac{1}{2n+1} + \frac{2}{\pi}\sum_{k=1}^{n}\frac{1}{k}\tan\frac{\pi k}{2n+1}$$

Write and run a program to compute $\rho_1, \rho_2, \dots, \rho_{100}$ with eight decimal digits of accuracy. Then test the validity of the inequality

$$0 \leqq \frac{4}{\pi^2}\ln(2n+1) + 1 - \rho_n \leqq 0.0106$$

16. Compute in double or extended precision the following number:

$$x = \left[\frac{1}{\pi}\ln(640320^3 + 744)\right]^2$$

What is the point of this exercise?

17. Write a routine to compute $\sin x$ for x in radians as follows. First, using properties of the sine function, reduce the range so that $-\pi/2 \leqq x \leqq \pi/2$. Then if $|x| < 10^{-8}$, set $\sin x \approx x$; if $|x| > \pi/6$, set $u = x/3$, compute $\sin u$ by the formula below, and then set $\sin x \approx [3 - 4\sin^2 u]\sin u$; if $|x| \leqq \pi/6$, set $u = x$ and compute $\sin u$ as follows:

$$\sin u \approx u\left[\frac{1 - \left(\frac{29593}{207636}\right)u^2 + \left(\frac{34911}{7613320}\right)u^4 - \left(\frac{479249}{11511339840}\right)u^6}{1 + \left(\frac{1671}{69212}\right)u^2 + \left(\frac{97}{351384}\right)u^4 + \left(\frac{2623}{1644477120}\right)u^6}\right]$$

Try to determine whether the sine function on your computer system uses this algorithm.
Note: This is the **Padé rational approximation** for sine.

18. Write a routine to compute the natural logarithm by the algorithm outlined here based on **telescoped rational** and **Gaussian continued fractions** for $\ln x$ and test for several values of x. First check whether $x = 1$ and return zero if so. Reduce the range of x by determining n and r such that $x = r \times 2^n$ with $\frac{1}{2} \leqq r < 1$. Next, set $u = (r - \sqrt{2}/2)/(r + \sqrt{2}/2)$, and compute $\ln[(1+u)/(1-u)]$ by the approximation

$$\ln\left(\frac{1+u}{1-u}\right) \approx u\left(\frac{20790 - 21545.27u^2 + 4223.9187u^4}{10395 - 14237.635u^2 + 4778.8377u^4 - 230.41913u^6}\right)$$

which is valid for $|u| < 3 - 2\sqrt{2}$. Finally, set

$$\ln x \approx \left(n - \frac{1}{2}\right)\ln 2 + \ln\left[\frac{1+u}{1-u}\right]$$

19. Write a routine to compute the tangent of x in radians, using the algorithm following. Test the resulting routine over a range of values of x. First, the argument x is reduced to $|x| \leqq \pi/2$ by adding or subtracting multiples of π. If we have $0 \leqq |x| \leqq 1.7 \times 10^{-9}$, set $\tan x \approx x$. If $|x| > \pi/4$, set $u = \pi/2 - x$; otherwise, set $u = x$. Now compute the approximation

$$\tan u \approx u\left(\frac{135135 - 17336.106u^2 + 379.23564u^4 - 1.0118625u^6}{135135 - 62381.106u^2 + 3154.9377u^4 - 28.17694u^6}\right)$$

Finally, if $|x| > \pi/4$, set $\tan x \approx 1/\tan u$; if $|x| \leqq \pi/4$, set $\tan x \approx \tan u$.
Note: This algorithm is obtained from the **telescoped rational** and **Gaussian continued fraction** for the tangent function.

20. Write a routine to compute $\arcsin x$ based on the following algorithm, using telescoped polynomials for the arcsine. If $|x| < 10^{-8}$, set $\arcsin x \approx x$. Otherwise, if $0 \leqq x \leqq \frac{1}{2}$, set $u = x$, $a = 0$, and $b = 1$; if $\frac{1}{2} < x \leqq \frac{1}{2}\sqrt{3}$, set $u = 2x^2 - 1$, $a = \pi/4$, and $b = \frac{1}{2}$; if $\frac{1}{2}\sqrt{3} < x \leqq \frac{1}{2}\sqrt{2+\sqrt{3}}$, set $u = 8x^4 - 8x^2 + 1$, $a = 3\pi/8$, and $b = \frac{1}{4}$; if $\frac{1}{2}\sqrt{2+\sqrt{3}} < x \leqq 1$, set $u = \sqrt{\frac{1}{2}(1-x)}$, $a = \pi/2$, and $b = -2$. Now compute the approximation

$$\begin{aligned}\arcsin u \approx u\Big(1.0 + \frac{1}{6}u^2 &+ 0.075u^4 + 0.04464286u^6 \\ &+ 0.03038182u^8 + 0.022375u^{10} \\ &+ 0.01731276u^{12} + 0.01433124u^{14} \\ &+ 0.009342806u^{16} + 0.01835667u^{18} \\ &- 0.01186224u^{20} + 0.03162712u^{22}\Big)\end{aligned}$$

Finally, set $\arcsin x \approx a + b \arcsin u$. Test this routine for various values of x.

21. Write and test a routine to compute $\arctan x$ for x in radians as follows. If $0 \leqq x \leqq 1.7 \times 10^{-9}$, set $\arctan x \approx x$. If $1.7 \times 10^{-9} < x \leqq 2 \times 10^{-2}$, use the series approximation

$$\arctan x \approx x - \frac{x^3}{3} + \frac{x^5}{5} - \frac{x^7}{7}$$

Otherwise, set $y = x$, $a = 0$, and $b = 1$ if $0 \leqq x \leqq 1$; set $y = 1/x$, $a = \pi/2$, and $b = -1$ if $1 < x$. Then set $c = \pi/16$ and $d = \tan c$ if $0 \leqq y \leqq \sqrt{2} - 1$ and $c = 3\pi/16$ and $d = \tan c$ if $\sqrt{2} - 1 < y \leqq 1$. Compute $u = (y - d)/(1 + dy)$ and the approximation

$$\arctan u \approx u\left(\frac{135135 + 171962.46u^2 + 52490.4832u^4 + 2218.1u^6}{135135 + 217007.46u^2 + 97799.3033u^4 + 10721.3745u^6}\right)$$

Finally, set $\arctan x \approx a + b(c + \arctan u)$.
Note: This algorithm uses **telescoped rational** and **Gaussian continued fractions**.

22. A fast algorithm for computing $\arctan x$ to n-bit precision for x in the interval $(0, 1]$ is as follows: Set $a = 2^{-n/2}$, $b = x/(1 + \sqrt{1 + x^2})$, $c = 1$, and $d = 1$. Then repeatedly update these variables by these formulas (in order from left to right and top to bottom):

real a, b, c, d

$$c \leftarrow \frac{2c}{1+a}; \quad d \leftarrow \frac{2ab}{1+b^2}; \quad d \leftarrow \frac{d}{1+\sqrt{1-d^2}}$$

$$d \leftarrow \frac{b+d}{1-bd}; \quad b \leftarrow \frac{d}{1+\sqrt{1+d^2}}; \quad a \leftarrow \frac{2\sqrt{a}}{1+a}$$

After each sweep, print $f = c\ln[(1+b)/(1-b)]$. Stop when $1 - a \leqq 2^{-n}$. Write a double-precision routine to implement this algorithm and test it for various values of x. Compare the results to those obtained from the arctangent function on your computer system.
Note: This fast multiple-precision algorithm depends on the theory of **elliptic integrals**, using the arithmetic-geometric mean iteration and **ascending Landen transformations**. Other fast algorithms for trigonometric functions are discussed in Brent [1976].

23. On your computer, show that in single precision, you have only six decimal digits of accuracy if you enter 20 digits. Show that going to double precision is effective only if all work is done in double precision. For example, if you use `pi = 3.14` or `pi = 22/7`, you will lose all the precision that you have gained by using double precision. Remember that the number of significant digits in the final results is what counts!

24. In some programming languages such as Java and C++, show that mixed-mode arithmetic can lead to results such as `(4/3)*pi=pi` when `pi` is a floating-point number because the fraction inside the parentheses is computed in integer mode.

25. (**Student Research Project**) Investigate interval arithmetic, which has the goal of obtaining results with a guaranteed precision.

26. The smallest root in absolute value of the equation $f(x) = x^2 + 2px - q = 0$ can be computed using either $y_1 = -p + \sqrt{p^2 + q}$ or $y_2 = q/[p + \sqrt{p^2 + q}]$. Show that severe cancellation may occur when p is positive and much greater than q. Program and test the following pseudocode for carrying out these computations step-by-step. Consider the case when $p = 10^4$ and $q = 1$ as well as when $p = -(1 + \frac{1}{2}10^{-8})$ and $q = -(1 + 10^8)$. Check the results by substitution back into the original equation. Explain your results.

$$
\begin{aligned}
&s = p^2\\
&t = s + q\\
&u = \sqrt{t}\\
&y_1 = -p + u\\
&v = p + u\\
&y_2 = q/v
\end{aligned}
$$

2 Linear Systems

A simple electrical network contains a number of resistances and a single source of electromotive force (a battery), as shown in Figure 2.1. Using Kirchhoff's laws and Ohm's law, we can write a system of linear equations that govern this circuit. If x_1, x_2, x_3, and x_4 are the loop currents as shown, then the equations are

$$\begin{cases} 15x_1 - 2x_2 - 6x_3 = 300 \\ -2x_1 + 12x_2 - 4x_3 - x_4 = 0 \\ -6x_1 - 4x_2 + 19x_3 - 9x_4 = 0 \\ - x_2 - 9x_3 + 21x_4 = 0 \end{cases}$$

Systems of equations like this, even those that contain hundreds of unknowns, can be solved by using the methods developed in this chapter. The solution to the preceding system is

$$x_1 = 26.5, \qquad x_2 = 9.35, \qquad x_3 = 13.3, \qquad x_4 = 6.13$$

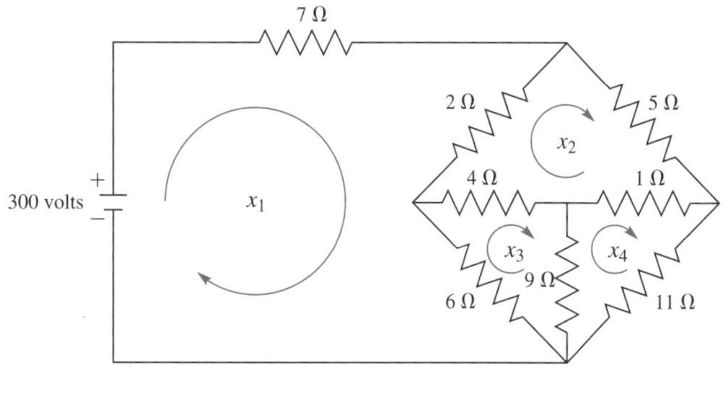

FIGURE 2.1
Electrical network

2.1 Naive Gaussian Elimination

One of the fundamental problems in many scientific and engineering applications is to solve an algebraic linear system

$$\boldsymbol{A}\boldsymbol{x} = \boldsymbol{b}$$

for the unknown vector $\boldsymbol{x}$ when the coefficient matrix $\boldsymbol{A}$ and right-hand side vector $\boldsymbol{b}$ are known. Such systems arise naturally in various applications, such as approximating nonlinear equations by linear equations or differential equations by algebraic equations. The cornerstone of many numerical methods for solving a variety of practical computational problems is the efficient and accurate solution of linear systems. The system of linear algebraic equations $\boldsymbol{Ax} = \boldsymbol{b}$ may or may not have a solution, and if it has a solution, it may or may not be unique. Gaussian elimination is the standard method for solving the linear system by using a calculator or a computer. This method is undoubtedly familiar to most readers, since it is the simplest way to solve a linear system by hand. When the system has no solution, other approaches are used, such as linear least squares, which is discussed in Chapter 9. In this chapter, we assume that the coefficient matrix $\boldsymbol{A}$ is $n \times n$ and invertible (nonsingular).

In a pure mathematical approach, the solution to the problem $\boldsymbol{Ax} = \boldsymbol{b}$ is simply

$$\boldsymbol{x} = \boldsymbol{A}^{-1}\boldsymbol{b}$$

where $\boldsymbol{A}^{-1}$ is the **inverse matrix**. But in most applications, it is advisable to solve the system directly for the unknown vector $\boldsymbol{x}$ rather than explicitly computing the inverse matrix.

In applied mathematics and in many applications, it can be a daunting task for even the largest and fastest computers to solve accurately extremely large systems involving thousands or millions of unknowns. Some important questions are the following:

Important Question

- How do we store such large systems in the computer?
- How do we know that the computed answers are correct?
- What is the precision of the computed results?
- Can the algorithm fail?
- How long will it take to compute the answers?
- What is the asymptotic operation count of the algorithm?
- Will the algorithm be unstable for certain systems?
- Can instability be controlled by pivoting? (Permuting the order of the rows of the matrix is called **pivoting**.)
- Which strategy of pivoting should be used?
- How do we know whether the matrix is ill-conditioned and whether the answers are accurate?

Factorization

Gaussian elimination transforms a linear system into an upper triangular form, which is easier to solve. This process, in turn, is equivalent to finding the **factorization**

$$\boldsymbol{A} = \boldsymbol{LU}$$

where $\boldsymbol{L}$ is a unit lower triangular matrix and $\boldsymbol{U}$ is an upper triangular matrix. This factorization is especially useful when solving many linear systems involving the same coefficient matrix but different right-hand sides, which occurs in various applications.

When the coefficient matrix $\boldsymbol{A}$ has a special structure, such as being symmetric, positive definite, triangular, banded, block, or sparse, the general approach of Gaussian elimination with partial pivoting needs to be modified or rewritten specifically for the system. When the coefficient matrix has predominantly zero entries, the system is **sparse**, and iterative methods can involve much less computer memory than Gaussian elimination. We address many of these issues in this chapter and in Chapter 8.

Our objective in this chapter is to develop a good computer program for solving a system of n linear equations in n unknowns:

General Linear System

$$\begin{cases} a_{11}x_1 + a_{12}x_2 + a_{13}x_3 + \cdots + a_{1n}x_n = b_1 \\ a_{21}x_1 + a_{22}x_2 + a_{23}x_3 + \cdots + a_{2n}x_n = b_2 \\ a_{31}x_1 + a_{32}x_2 + a_{33}x_3 + \cdots + a_{3n}x_n = b_3 \\ \vdots \qquad \vdots \qquad \vdots \qquad \qquad \vdots \quad \vdots \\ a_{i1}x_1 + a_{i2}x_2 + a_{i3}x_3 + \cdots + a_{in}x_n = b_i \\ \vdots \qquad \vdots \qquad \vdots \qquad \qquad \vdots \quad \vdots \\ a_{n1}x_1 + a_{n2}x_2 + a_{n3}x_3 + \cdots + a_{nn}x_n = b_n \end{cases} \tag{1}$$

In compact form, System (1) can be written simply as

$$\sum_{j=1}^{n} a_{ij}x_j = b_i \qquad (1 \leqq i \leqq n)$$

In these equations, a_{ij} and b_i are prescribed real numbers (data), and the unknowns x_j are to be determined. Subscripts on the letter a are separated by a comma only if necessary for clarity—for example, in $a_{32,75}$ but not in a_{ij}.

A Larger Numerical Example

In this section, the simplest form of Gaussian elimination is explained. The adjective **naive** applies because this form is not usually suitable for automatic computation unless essential modifications are made, as in Section 2.2. We illustrate naive Gaussian elimination with a specific example that has four equations and four unknowns:

Sample Linear System

$$\begin{cases} 6x_1 - 2x_2 + 2x_3 + 4x_4 = 16 \\ 12x_1 - 8x_2 + 6x_3 + 10x_4 = 26 \\ 3x_1 - 13x_2 + 9x_3 + 3x_4 = -19 \\ -6x_1 + 4x_2 + x_3 - 18x_4 = -34 \end{cases} \tag{2}$$

Forward Elimination

In the first step of the elimination procedure, certain multiples of the first equation are subtracted from the second, third, and fourth equations so as to eliminate x_1 from these equations. Thus, we want to create 0's as coefficients for each x_1 below the first (where 12, 3, and -6 now stand). It is clear that we should subtract 2 times the first equation from the second. (This **multiplier** is simply the quotient $\frac{12}{6}$.) Likewise, we should subtract $\frac{1}{2}$ times the first equation from the third. (Again, this **multiplier** is just $\frac{3}{6}$.) Finally, we should subtract -1 times the first equation from the fourth. When all of this has been done, the result is

$$\begin{cases} 6x_1 - 2x_2 + 2x_3 + 4x_4 = 16 \\ - 4x_2 + 2x_3 + 2x_4 = -6 \\ - 12x_2 + 8x_3 + x_4 = -27 \\ 2x_2 + 3x_3 - 14x_4 = -18 \end{cases} \tag{3}$$

Note that the first equation was not altered in this process, although it was used to produce the 0 coefficients in the other equations. In this context, it is called the **pivot equation**.

Notice also that Systems (2) and (3) are *equivalent* in the following technical sense: Any solution of (2) is also a solution of (3), and vice versa. This follows at once from the fact that if equal quantities are added to equal quantities, the resulting quantities are equal. One can get System (2) from System (3) by subtracting 2 times the first equation to the second, and so on.

Forward Elimination

In the second step of the process, we mentally ignore the first equation and the first column of coefficients. This leaves a system of three equations with three unknowns. The same process is now repeated using the top equation in the smaller system as the current pivot equation. Thus, we begin by subtracting 3 times the second equation from the third. (The **multiplier** is just the quotient $\frac{-12}{-4}$.) Then we subtract $-\frac{1}{2}$ times the second equation from the fourth. After doing the arithmetic, we arrive at

$$\begin{cases} 6x_1 - 2x_2 + 2x_3 + 4x_4 = 16 \\ - 4x_2 + 2x_3 + 2x_4 = -6 \\ 2x_3 - 5x_4 = -9 \\ 4x_3 - 13x_4 = -21 \end{cases} \tag{4}$$

The final step consists of subtracting 2 times the third equation from the fourth. The result is

Sample Upper Triangular System

$$\begin{cases} 6x_1 - 2x_2 + 2x_3 + 4x_4 = 16 \\ - 4x_2 + 2x_3 + 2x_4 = -6 \\ 2x_3 - 5x_4 = -9 \\ - 3x_4 = -3 \end{cases} \tag{5}$$

This system is said to be in **upper triangular** form. It is equivalent to System (2).

This completes the first phase (**forward elimination**) in the Gaussian algorithm.

Back Substitution

The second phase (**back substitution**) solves System (5) for the unknowns *starting at the bottom*. Thus, from the fourth equation, we obtain the last unknown

$$x_4 = \frac{-3}{-3} = 1$$

Putting $x_4 = 1$ in the third equation gives us

$$2x_3 - 5 = -9$$

and we find the next to last unknown

$$x_3 = \frac{-4}{2} = -2$$

and so on. The solution is

True Solution

$$x_1 = 3, \qquad x_2 = 1, \qquad x_3 = -2, \qquad x_4 = 1$$

Algorithm

To simplify the discussion, we write System (1) in matrix-vector form. The coefficient elements a_{ij} form an $n \times n$ square array, or matrix. The unknowns x_i and the right-hand side elements b_i form $n \times 1$ arrays, or vectors.* (See Appendix D for linear algebra notation and concepts.) Hence, we have

General $n \times n$ Linear System in Matrix-Vector Form

$$\begin{bmatrix} a_{11} & a_{12} & a_{13} & \cdots & a_{1n} \\ a_{21} & a_{22} & a_{23} & \cdots & a_{2n} \\ a_{31} & a_{32} & a_{33} & \cdots & a_{3n} \\ \vdots & \vdots & \vdots & & \vdots \\ a_{i1} & a_{i2} & a_{i3} & \cdots & a_{in} \\ \vdots & \vdots & \vdots & & \vdots \\ a_{n1} & a_{n2} & a_{n3} & \cdots & a_{nn} \end{bmatrix} \begin{bmatrix} x_1 \\ x_2 \\ x_3 \\ \vdots \\ x_i \\ \vdots \\ x_n \end{bmatrix} = \begin{bmatrix} b_1 \\ b_2 \\ b_3 \\ \vdots \\ b_i \\ \vdots \\ b_n \end{bmatrix} \tag{6}$$

or

$$\boldsymbol{Ax} = \boldsymbol{b}$$

Operations between *equations* correspond to operations between *rows* in this notation. We shall use these two words interchangeably.

Now let us organize the naive Gaussian elimination algorithm for the general system, which contains n equations and n unknowns. In this algorithm, the original data are overwritten with new computed values. In the forward elimination phase of the process, there are $n - 1$ principal steps. The first of these steps uses the first equation to produce $n - 1$ zeros as coefficients for each x_1 in all but the first equation. This is done by subtracting appropriate multiples of the first equation from the others. In this process, we refer to the first equation as the first **pivot equation** and to a_{11} as the first **pivot element**. For each of the remaining equations $(2 \leqq i \leqq n)$, we compute

1st Step

$$\begin{cases} a_{ij} \leftarrow a_{ij} - \left(\dfrac{a_{i1}}{a_{11}}\right) a_{1j} & (1 \leqq j \leqq n) \\[2ex] b_i \leftarrow b_i - \left(\dfrac{a_{i1}}{a_{11}}\right) b_1 \end{cases}$$

The symbol $\leftarrow$ indicates a **replacement**. Thus, the content of the memory location allocated to a_{ij} is replaced by $a_{ij} - (a_{i1}/a_{11})a_{1j}$, and so on. This is accomplished by the following line of pseudocode:

$$a_{ij} \leftarrow a_{ij} - (a_{i1}/a_{11})a_{1j}$$

Note that the quantities (a_{i1}/a_{11}) are the **multipliers**. The new coefficient of x_1 in the ith equation becomes 0 because $a_{i1} - (a_{i1}/a_{11})a_{11} = 0$.

---

*To save space, we occasionally write a column vector as $[x_1, x_2, \dots, x_n]^T$, where the T stands for **transpose**. It tells us that this is an $n \times 1$ array or vector and *not* $1 \times n$, as would be indicated without the transpose symbol.

2nd Step

After the first step, the system will be of the form

$$
\begin{bmatrix}
a_{11} & a_{12} & a_{13} & \cdots & a_{1n} \\
0 & a_{22} & a_{23} & \cdots & a_{2n} \\
0 & a_{23} & a_{33} & \cdots & a_{3n} \\
\vdots & \vdots & \vdots & & \vdots \\
0 & a_{i2} & a_{i3} & \cdots & a_{in} \\
\vdots & \vdots & \vdots & & \vdots \\
0 & a_{n2} & a_{n3} & \cdots & a_{nn}
\end{bmatrix}
\begin{bmatrix} x_1 \\ x_2 \\ x_3 \\ \vdots \\ x_i \\ \vdots \\ x_n \end{bmatrix}
=
\begin{bmatrix} b_1 \\ b_2 \\ b_3 \\ \vdots \\ b_i \\ \vdots \\ b_n \end{bmatrix}
$$

From here on, we do not alter the first equation, nor do we alter any of the coefficients for x_1 (since a multiplier times 0 subtracted from 0 is still 0). Thus, we can mentally ignore the first row and the first column and repeat the process on the smaller system. With the second equation as the **pivot equation**, we compute for each remaining equation $(3 \leqq i \leqq n)$

$$
\begin{cases}
a_{ij} \leftarrow a_{ij} - \left(\dfrac{a_{i2}}{a_{22}}\right) a_{2j} & (2 \leqq j \leqq n) \\[2ex]
b_i \leftarrow b_i - \left(\dfrac{a_{i2}}{a_{22}}\right) b_2
\end{cases}
$$

***k*th Step**

Just prior to the kth step in the forward elimination, the system appears as follows:

$$
\begin{bmatrix}
a_{11} & a_{12} & a_{13} & \cdots & & \cdots & & \cdots & a_{1n} \\
0 & a_{22} & a_{23} & \cdots & & \cdots & & \cdots & a_{2n} \\
0 & 0 & a_{33} & \cdots & & \cdots & & \cdots & a_{3n} \\
\vdots & \vdots & \vdots & \ddots & & & & & \vdots \\
0 & 0 & 0 & \cdots & a_{kk} & \cdots & a_{kj} & \cdots & a_{kn} \\
\vdots & \vdots & \vdots & \vdots & \vdots & & \vdots & & \vdots \\
0 & 0 & 0 & \cdots & a_{ik} & \cdots & a_{ij} & \cdots & a_{in} \\
\vdots & \vdots & \vdots & \vdots & \vdots & & \vdots & & \vdots \\
0 & 0 & 0 & \cdots & a_{nk} & \cdots & a_{nj} & \cdots & a_{nn}
\end{bmatrix}
\begin{bmatrix} x_1 \\ x_2 \\ x_3 \\ \vdots \\ x_k \\ \vdots \\ x_i \\ \vdots \\ x_n \end{bmatrix}
=
\begin{bmatrix} b_1 \\ b_2 \\ b_3 \\ \vdots \\ b_k \\ \vdots \\ b_i \\ \vdots \\ b_n \end{bmatrix}
$$

Here, a **wedge** of 0 coefficients has been created, and the first k equations have been processed and are now fixed. Using the kth equation as the **pivot equation**, we select multipliers to create 0's as coefficients below the a_{kk} coefficient.

$$
\begin{cases}
a_{ij} \leftarrow a_{ij} - \left(\dfrac{a_{ik}}{a_{kk}}\right) a_{kj} & (k \leqq j \leqq n) \\[2ex]
b_i \leftarrow b_i - \left(\dfrac{a_{ik}}{a_{kk}}\right) b_k
\end{cases}
$$

Obviously, we must assume that all the divisors in this algorithm are nonzero. Similarly, we compute zero entries for each remaining ith pivoting equation $(k + 1 \leqq i \leqq n - 1)$.

Pseudocode

We now consider the pseudocode for forward elimination. The coefficient array is stored as an $n \times n$ double-subscripted array (a_{ij}); the right-hand side of the system of equations is stored as an $n \times 1$ single-subscripted array (b_i); the solution is computed and stored in an $n \times 1$ single-subscripted array (x_i). It is easy to see that the following lines of pseudocode carry out the **forward elimination phase** of naive Gaussian elimination:

Forward Elimination Pseudocode

```
integer i, j, k;   real array (a_ij)_{1:n×1:n}, (b_i)_{1:n}
for k = 1 to n − 1
    for i = k + 1 to n
        for j = k to n
            a_ij ← a_ij − (a_ik/a_kk)a_kj
        end for
        b_i ← b_i − (a_ik/a_kk)b_k
    end for
end for
```

Since the multiplier a_{ik}/a_{kk} does not depend on j, it can be moved outside the j loop. Notice also that the new values in column k will be 0, at least theoretically, because when $j = k$, we have

Key Step

$$a_{ik} \leftarrow a_{ik} - (a_{ik}/a_{kk})a_{kk}$$

Since we expect this to be 0, *no* purpose is served in computing it. The location where the 0 is being created is a good place to store the multiplier. If these remarks are put into practice, the pseudocode looks like this:

Improved Forward Elimination Pseudocode

```
integer i, j, k;   real xmult;   real array (a_ij)_{1:n×1:n}, (b_i)_{1:n}
for k = 1 to n − 1
    for i = k + 1 to n
        xmult ← a_ik/a_kk
        a_ik ← xmult
        for j = k + 1 to n
            a_ij ← a_ij − (xmult)a_kj
        end for
        b_i ← b_i − (xmult)b_k
    end for
end for
```

Here, the multipliers are stored because they are part of the ***LU*** factorization that can be useful in some applications. This matter is discussed in Section 8.1.

At the beginning of the **back substitution phase**, the linear system is of the form

General Upper Triangular System

$$\begin{cases} a_{11}x_1 + a_{12}x_2 + a_{13}x_3 + \cdots & \cdots + \quad a_{1n}x_n = b_1 \\ \qquad a_{22}x_2 + a_{23}x_3 + \cdots & \cdots + \quad a_{2n}x_n = b_2 \\ \qquad\qquad a_{33}x_3 + \cdots & + \quad a_{3n}x_n = b_3 \\ \qquad\qquad\qquad \ddots & \vdots \qquad\quad \vdots \\ \qquad\qquad a_{ii}x_i + a_{i,i+1}x_{i+1} + & \cdots + \quad a_{in}x_n = b_i \\ \qquad\qquad\qquad \ddots & \vdots \qquad\quad \vdots \\ & a_{n-1,n-1}x_{n-1} + a_{n-1,n}x_n = b_{n-1} \\ & a_{nn}x_n = b_n \end{cases}$$

where the a_{ij}'s and b_i's are *not* the original ones from System (6), but instead are the ones that have been altered by the elimination process.

The back substitution starts by solving the nth equation for x_n:

$$x_n = \frac{b_n}{a_{nn}}$$

Then, using the $(n-1)$th equation, we solve for x_{n-1}:

$$x_{n-1} = \frac{1}{a_{n-1,n-1}}\left(b_{n-1} - a_{n-1,n}x_n\right)$$

We continue working upward, recovering each x_i by the formula

Back Substitution Formula

$$x_i = \frac{1}{a_{ii}}\left(b_i - \sum_{j=i+1}^{n} a_{ij}x_j\right) \qquad (i = n-1, n-2, \ldots, 1) \tag{7}$$

Here is pseudocode to do this:

Back Substitution Pseudocode

```
integer i, j, n;   real sum;   real array (a_ij)_{1:n×1:n}, (x_i)_{1:n}
x_n ← b_n/a_nn
for i = n − 1 to 1
    sum ← b_i
    for j = i + 1 to n
        sum ← sum − a_ij x_j
    end for
    x_i ← sum/a_ii
end for
```

Now we put these segments of pseudocode together to form a procedure, called *Naive_Gauss*, which is intended to solve an invertible system of n linear equations in n unknowns by the method of naive Gaussian elimination. This pseudocode serves a didactic purpose only; a more robust pseudocode is developed in the next section.

Naive_Gauss **Pseudocode**

```
procedure Naive_Gauss(n, (a_ij), (b_i), (x_i))
integer i, j, k, n;   real sum, xmult
real array (a_ij)_{1:n×1:n}, (b_i)_{1:n}, (x_i)_{1:n}
for k = 1 to n − 1
    for i = k + 1 to n
        xmult ← a_ik/a_kk
        a_ik ← xmult
        for j = k + 1 to n
            a_ij ← a_ij − (xmult)a_kj
        end for
        b_i ← b_i − (xmult)b_k
    end for
end for
x_n ← b_n/a_nn
for i = n − 1 to 1
    sum ← b_i
    for j = i + 1 to n
        sum ← sum − a_ij x_j
    end for
    x_i ← sum/a_ii
end for
end procedure Naive_Gauss
```

Before giving a test example, let us examine the crucial computation in our pseudocode, namely, a triply nested for-loop containing a replacement operation:

Key Loops

```
for k .........
    for i .........
        for j .........
            a_ij ← a_ij − (a_ik/a_kk)a_kj
        end do
    end do
end do
```

Here, we must expect all quantities to be infected with roundoff error. Such a roundoff error in a_{kj} is multiplied by the factor (a_{ik}/a_{kk}). This factor is large if the pivot element $|a_{kk}|$ is small relative to $|a_{ik}|$. Hence, we conclude, tentatively, that small pivot elements lead to large multipliers and to *worse* roundoff errors!

Testing the Pseudocode

One good way to test a procedure is to set up an artificial problem whose solution is known beforehand. Sometimes the test problem includes a parameter that can be changed to vary the difficulty. The next example illustrates this.

Fixing a value of n, we define the polynomial

Sample Polynomial

$$p(t) = 1 + t + t^2 + \cdots + t^{n-1} = \sum_{j=1}^{n} t^{j-1}$$

The coefficients in this polynomial are all equal to 1. We shall try to recover these known coefficients from n values of the polynomial. We use the values of $p(t)$ at the integers $t = 1+i$ for $i = 1, 2, \ldots, n$. If the coefficients in the polynomial are denoted by $x_1, x_2, \ldots, x_n$, we have

$$\sum_{j=1}^{n}(1+i)^{j-1}x_j = \frac{1}{i}\left[(1+i)^n - 1\right] \qquad (1 \leqq i \leqq n) \tag{8}$$

Here, we have used the formula for the **sum of a geometric series** on the right-hand side; that is,

$$p(1+i) = \sum_{j=1}^{n}(1+i)^{j-1} = \frac{(1+i)^n - 1}{(1+i) - 1} = \frac{1}{n}\left[(1+i)^n - 1\right] \tag{9}$$

Letting $a_{ij} = (1+i)^{j-1}$ and $b_i = [(1+i)^n - 1]/i$ in Equation (8), we have a linear system $\boldsymbol{Ax} = \boldsymbol{b}$.

EXAMPLE 1 We write a pseudocode for a specific test case that solves the system of Equation (8) for various values of n.

Solution Since the naive Gaussian elimination procedure *Naive_Gauss* can be used, all that is needed is a calling program. We decide to use $n = 4, 5, 6, 7, 8, 9, 10$ in this test. Here is a suitable pseudocode:

***Test_NGE* Pseudocode**

```
program Test_NGE
integer parameter m ← 10
integer i, j, n;   real array, (a_ij)_1:m×1:m, (b_i)_1:m, (x_i)_1:m
for n = 4 to 10
    for i = 1 to n
        for j = 1 to n
            a_ij ← (i + 1)^(j−1)
        end for
        b_i ← [(i + 1)^n − 1]/i
    end for
    call Naive_Gauss(n, (a_ij), (b_i), (x_i))
    output n, (x_i)_1:n
end for
end program Test_NGE
```

Failure!

When this pseudocode was run on a machine that carries approximately seven decimal digits of accuracy, the solution was obtained with near full precision until n reached 9, and then the computed solution was worthless because one component exhibited a relative error of 16,120%! (Write and run a computer program to see for yourself!) ■

Ill-Conditioned Vandermonde Matrix

The coefficient matrix for this linear system is an example of a well-known *ill-conditioned* matrix called the **Vandermonde matrix**. This accounts for the fact that the system cannot be solved accurately using naive Gaussian elimination! What is amazing is that the trouble happens so suddenly! When $n \geqq 9$, the roundoff error that is present in computing x_i is propagated and magnified throughout the back substitution phase so that most of the computed values for x_i are worthless! Insert some intermediate print statements in the code to see for yourself what is going on here. (See Gautschi [1990] for more information on the Vandermonde matrix and its ill-conditioned nature.)

Residual and Error Vectors

For a linear system $\boldsymbol{Ax} = \boldsymbol{b}$ having the *true solution* $\boldsymbol{x}$ and a *computed solution* $\widetilde{\boldsymbol{x}}$, we define

Error/Residual Vectors

$$\boldsymbol{e} = \widetilde{\boldsymbol{x}} - \boldsymbol{x} \qquad \textbf{(error vector)}$$
$$\boldsymbol{r} = \boldsymbol{A}\widetilde{\boldsymbol{x}} - \boldsymbol{b} \qquad \textbf{(residual vector)}$$

An important relationship between the error vector and the residual vector is

$$\boldsymbol{Ae} = \boldsymbol{r}$$

Suppose that two students using different computer systems solve the same linear system, $\boldsymbol{Ax} = \boldsymbol{b}$. What algorithm and what precision each student used are *not* known. Each vehemently claims to have the correct answer, but the two computer solutions $\widetilde{\boldsymbol{x}}$ and $\widehat{\boldsymbol{x}}$ are totally different! *How do we determine which, if either, computed solution is correct?*

We can *check* the solutions by substituting them into the original system, which is the same as computing the **residual vectors**

$$\widetilde{\boldsymbol{r}} = \boldsymbol{A}\widetilde{\boldsymbol{x}} - \boldsymbol{b}, \qquad \widehat{\boldsymbol{r}} = \boldsymbol{A}\widehat{\boldsymbol{x}} - \boldsymbol{b}$$

Of course, the computed solutions are not exact because each must contain some roundoff errors. So we would want to accept the solution with the *smaller* residual vector. However, if we knew the exact solution $\boldsymbol{x}$, then we would just compare the computed solutions with the exact solution, which is the same as computing the **error vectors**

$$\widetilde{\boldsymbol{e}} = \widetilde{\boldsymbol{x}} - \boldsymbol{x}, \qquad \widehat{\boldsymbol{e}} = \widehat{\boldsymbol{x}} - \boldsymbol{x}$$

Now the computed solution that produces the smaller error vector would most assuredly be the better answer!

Since the exact solution is usually not known in applications, one would tend to accept the computed solution that has the smaller residual vector. But this may not be the best computed solution if the original problem is sensitive to roundoff errors—that is, it is *ill-conditioned*! In fact, the question of whether a computed solution to a linear system is a good solution is extremely difficult and beyond the scope of this book. Some of the exercises may give some insight into the difficulty of assessing the accuracy of computed solutions of linear systems.

Summary 2.1

- The basic **forward elimination** procedure over writes these value for $1 \leqq k \leqq n - 1$

$$\begin{cases} a_{ij} \leftarrow a_{ij} - (a_{ik}/a_{kk})\, a_{kj} & (k+1 \leqq i \leqq n, k+1 \leqq j \leqq n) \\ b_i \leftarrow b_i - (a_{ik}/a_{kk})\, b_k \end{cases}$$

Here we assume $a_{kk} \neq 0$.

- The basic **back substitution** procedure uses $x_n = b_n/a_n$ and then

$$x_i = \frac{1}{a_{ii}}\left(b_i - \sum_{j=i+1}^{n} a_{ij}x_j\right) \qquad (i = n-1, n-2, \dots, 1)$$

- When solving the linear system $\boldsymbol{Ax} = \boldsymbol{b}$, if the true or exact solution is $\boldsymbol{x}$ and the approximate or computed solution is $\widetilde{\boldsymbol{x}}$, then important quantities are

$$\boldsymbol{e} = \widetilde{\boldsymbol{x}} - \boldsymbol{x} \qquad \textbf{(error vectors)}$$
$$\boldsymbol{r} = \boldsymbol{A}\widetilde{\boldsymbol{x}} - \boldsymbol{b} \qquad \textbf{(residual vectors)}$$

Exercises 2.1

[a]**1.** Show that the system of equations

$$\begin{cases} x_1 + 4x_2 + \alpha x_3 = 6 \\ 2x_1 - x_2 + 2\alpha x_3 = 3 \\ \alpha x_1 + 3x_2 + x_3 = 5 \end{cases}$$

possesses a unique solution when $\alpha = 0$, no solution when $\alpha = -1$, and infinitely many solutions when $\alpha = 1$. Also, investigate the corresponding situation when the right-hand side is replaced by 0's.

[a]**2.** For what values of α does naive Gaussian elimination produce erroneous answers for this system?

$$\begin{cases} x_1 + x_2 = 2 \\ \alpha x_1 + x_2 = 2 + \alpha \end{cases}$$

Explain what happens in the computer.

3. Apply naive Gaussian elimination to these examples and account for the failures. Solve the systems by other means if possible.

[a]**a.** $\begin{cases} 3x_1 + 2x_2 = 4 \\ -x_1 - \frac{2}{3}x_2 = 1 \end{cases}$ [a]**b.** $\begin{cases} 6x_1 - 3x_2 = 6 \\ -2x_1 + x_2 = -2 \end{cases}$

c. $\begin{cases} 0x_1 + 2x_2 = 4 \\ x_1 - x_2 = 5 \end{cases}$ **d.** $\begin{cases} x_1 + x_2 + 2x_3 = 4 \\ x_1 + x_2 + 0x_3 = 2 \\ 0x_1 + x_2 + x_3 = 0 \end{cases}$

[a]**4.** Solve the following system of equations, retaining only four significant figures in each step of the calculation, and compare your answer with the solution obtained when eight significant figures are retained. Be consistent by either always rounding to the number of significant figures that are being carried or always chopping.

$$\begin{cases} 0.1036x_1 + 0.2122x_2 = 0.7381 \\ 0.2081x_1 + 0.4247x_2 = 0.9327 \end{cases}$$

[a]**5.** Consider

$$\boldsymbol{A} = \begin{bmatrix} 0.780 & 0.563 \\ 0.913 & 0.659 \end{bmatrix}, \qquad \boldsymbol{b} = \begin{bmatrix} 0.217 \\ 0.254 \end{bmatrix}$$

$$\widetilde{\boldsymbol{x}} = \begin{bmatrix} 0.999 \\ -1.001 \end{bmatrix}, \qquad \widehat{\boldsymbol{x}} = \begin{bmatrix} 0.341 \\ -0.087 \end{bmatrix}$$

Compute residual vectors $\widetilde{\boldsymbol{r}} = \boldsymbol{A}\widetilde{\boldsymbol{x}} - \boldsymbol{b}$ and $\widehat{\boldsymbol{r}} = \boldsymbol{A}\widehat{\boldsymbol{x}} - \boldsymbol{b}$ and decide which of $\widetilde{\boldsymbol{x}}$ and $\widehat{\boldsymbol{x}}$ is the better solution vector. Now compute the error vectors $\boldsymbol{e} = \widetilde{\boldsymbol{x}} - \boldsymbol{x}$ and $\widehat{\boldsymbol{e}} = \widehat{\boldsymbol{x}} - \boldsymbol{x}$, where $\boldsymbol{x} = [1, -1]^T$ is the exact solution. Discuss the implications of this example.

6. Consider the system

$$\begin{cases} 10^{-4}x_1 + x_2 = b_1 \\ x_1 + x_2 = b_2 \end{cases}$$

where $b_1 \neq 0$ and $b_2 \neq 0$. Its exact solution is

$$x_1 = \frac{-b_1 + b_2}{1 - 10^{-4}}, \qquad x_2 = \frac{b_1 - 10^{-4}b_2}{1 - 10^{-4}}$$

[a]**a.** Let $b_1 = 1$ and $b_2 = 2$. Solve this system using naive Gaussian elimination with three-digit (rounded) arithmetic and compare with the exact solution $x_1 = 1.00010\dots$ and $x_2 = 0.99989\,9\dots$.

[a]**b.** Repeat the preceding part after interchanging the order of the two equations.

[a]**c.** Find values of b_1 and b_2 in the original system so that naive Gaussian elimination does not give poor answers.

7. Solve each of the following systems using naive Gaussian elimination—that is, forward elimination and back substitution. Carry four significant figures.

[a]**a.** $\begin{cases} 3x_1 + 4x_2 + 3x_3 = 10 \\ x_1 + 5x_2 - x_3 = 7 \\ 6x_1 + 3x_3 + 7x_3 = 15 \end{cases}$

[a]**b.** $\begin{cases} 3x_1 + 2x_2 - 5x_3 = 0 \\ 2x_1 - 3x_2 + x_3 = 0 \\ x_1 + 4x_2 - x_3 = 4 \end{cases}$

[a] **c.** $\begin{bmatrix} 1 & -1 & 2 & 1 \\ 3 & 2 & 1 & 4 \\ 5 & 8 & 6 & 3 \\ 4 & 2 & 5 & 3 \end{bmatrix} \begin{bmatrix} x_1 \\ x_2 \\ x_3 \\ x_4 \end{bmatrix} = \begin{bmatrix} 1 \\ 1 \\ 1 \\ -1 \end{bmatrix}$

d. $\begin{cases} 3x_1 + 2x_2 - x_3 = 7 \\ 5x_1 + 3x_2 + 2x_3 = 4 \\ -x_1 + x_2 - 3x_3 = -1 \end{cases}$

e. $\begin{cases} x_1 + 3x_2 + 2x_3 + x_4 = -2 \\ 4x_1 + 2x_2 + x_3 + 2x_4 = 2 \\ 2x_1 + x_2 + 2x_3 + 3x_4 = 1 \\ x_1 + 2x_2 + 4x_3 + x_4 = -1 \end{cases}$

f. $\left[\begin{array}{cccc|c} 2 & 4 & -2 & -2 & -4 \\ 1 & 2 & 4 & -3 & 5 \\ -3 & -3 & 8 & -2 & 7 \\ -1 & 1 & 6 & -3 & 7 \end{array}\right]$

Computer Exercises 2.1

1. Program and run the Example 1 and insert some print statements to see what is happening.

2. Rewrite and test procedure *Naive_Gauss* so that it is column oriented; that is, the first index of a_{ij} varies on the innermost loop.

3. Define an $n \times n$ matrix $\boldsymbol{A}$ by the equation $a_{ij} = i + j$. Define $\boldsymbol{b}$ by the equation $b_i = i + 1$. Solve $\boldsymbol{Ax} = \boldsymbol{b}$ by using procedure *Naive_Gauss*. What should $\boldsymbol{x}$ be?

4. Define an $n \times n$ array by $a_{ij} = -1 + 2\min\{i, j\}$. Then set up the array (b_i) in such a way that the solution of the system $\sum_{j=1}^{n} a_{ij}x_j = b_i$ $(1 \leqq i \leqq n)$ is $x_j = 1$ $(1 \leqq j \leqq n)$. Test procedure *Naive_Gauss* on this system for a moderate value of n, say $n = 15$.

5. Write and test a version of *Naive_Gauss* in which

   a. An attempted division by 0 is signaled by an error return.

   b. The solution $\boldsymbol{x}$ is placed in array (b_i).

[a]6. Write and run a complex arithmetic version of *Naive_Gauss* by declaring certain variables complex and making other necessary changes to the code. Consider the complex linear system $\boldsymbol{Az} = \boldsymbol{b}$ where

$$\boldsymbol{A} = \begin{bmatrix} 5+9i & 5+5i & -6-6i & -7-7i \\ 3+3i & 6+10i & -5-5i & -6-6i \\ 2+2i & 3+3i & -1+3i & -5-5i \\ 1+i & 2+2i & -3-3i & 4i \end{bmatrix}$$

Solve this system four times with the following vectors $\boldsymbol{b}$:

$$\begin{bmatrix} -10+2i \\ -5+i \\ -5+i \\ -5+i \end{bmatrix}, \quad \begin{bmatrix} 2+6i \\ 4+12i \\ 2+6i \\ 2+6i \end{bmatrix},$$

$$\begin{bmatrix} 7-3i \\ 7-3i \\ 0 \\ 7-3i \end{bmatrix}, \quad \begin{bmatrix} -4-8i \\ -4-8i \\ -4-8i \\ 0 \end{bmatrix}$$

Verify that the solutions are $\boldsymbol{z} = \lambda^{-1}\boldsymbol{b}$ for scalars λ. The numbers λ are called **eigenvalues**, and the solutions $\boldsymbol{z}$ are **eigenvectors** of $\boldsymbol{A}$. Usually, the $\boldsymbol{b}$ vector is not known, and the solution of the problem $\boldsymbol{Az} = \lambda\boldsymbol{z}$ cannot be obtained by using a linear equation solver.

7. (Continuation) A common electrical engineering problem is to calculate currents in an electric circuit. For example, the circuit shown in the figure with R_i (ohms), C_i (farads), L (henries), and ω (hertz)

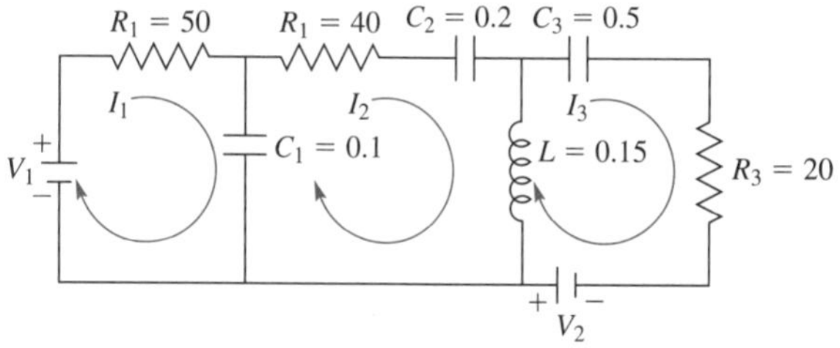

leads to the complex linear system

$$\begin{cases} (50-10i)I_1 + (10i)I_2 = V_1 \\ (10i)I_1 + (40+3i)I_2 - (15i)I_3 = 0 \\ -(15i)I_2 + (20+10i)I_3 = -V_2 \end{cases}$$

Select $V_1 = 100$ volts, $\omega = 1$ hertz, and solve these two cases using the complex arithmetic version of *Naive_Gauss*:

[a]**a.** The two voltages are in phase; that is, $V_2 = V_1$.

[a]**b.** The second voltage is a quarter of a cycle ahead of the first; that is, $V_2 = iV_1$.

8. Select a reasonable value of n, and generate a random $n \times n$ array a using a random-number generator. Define the array b such that the solution of the system

$$\sum_{j=1}^{n} a_{ij}x_j = b_i \qquad (1 \leqq i \leqq n)$$

is $x_j = j$, where $1 \leqq j \leqq n$. Test the naive Gaussian algorithm on this system.

Hint: You may use the function *Random*, which is discussed in Chapter 10, to generate the random elements of the (a_{ij}) array.

9. Carry out the test described in the text for procedure *Naive_Gauss* but *reverse* the order of the equations.
Hint: It suffices, in the code, to replace i by $n - i + 1$ in appropriate places.

10. Solve the linear system given in the leadoff example to this chapter using *Naive_Gauss*.

11. Use mathematical software such as the built-in routines in MATLAB, Maple, or Mathematica to directly solve linear system (2).

2.2 Gaussian Elimination with Scaled Partial Pivoting

Naive Gaussian Elimination Can Fail

To see why the naive Gaussian elimination algorithm is unsatisfactory, consider the following system:

Sample 2 × 2 Systems

$$\begin{cases} 0x_1 + x_2 = 1 \\ x_1 + x_2 = 2 \end{cases} \tag{1}$$

The pseudocode that we constructed in Section 2.1 would attempt to subtract some multiple of the first equation from the second to produce 0 as the coefficient for x_1 in the second equation. This, of course, is impossible, so the algorithm fails if $a_{11} = 0$.

If a numerical procedure actually fails for some values of the data, then the procedure is probably untrustworthy for values of the data *near* the failing values. To test this dictum, consider the system

$$\begin{cases} \varepsilon x_1 + x_2 = 1 \\ x_1 + x_2 = 2 \end{cases} \tag{2}$$

in which ε is a small number different from 0. Now the naive Gaussian elimination algorithm of Section 2.1 works, and after forward elimination it produces the system

$$\begin{cases} \varepsilon x_1 + \phantom{(1 - \varepsilon^{-1})}x_2 = 1 \\ (1 - \varepsilon^{-1})x_2 = 2 - \varepsilon^{-1} \end{cases} \tag{3}$$

In the back substitution, the arithmetic is as follows:

$$x_2 = \frac{2 - \varepsilon^{-1}}{1 - \varepsilon^{-1}} \approx 1, \qquad x_1 = \varepsilon^{-1}(1 - x_2) \approx 0$$

But ε^{-1} may be large! So if this calculation is performed by a computer that has a fixed word length, then for small values of ε, both $(2 - \varepsilon^{-1})$ and $(1 - \varepsilon^{-1})$ would be computed as $-\varepsilon^{-1}$.

For example, in an 8-digit decimal machine with a 16-digit accumulator, when $\varepsilon = 10^{-9}$, it follows that $\varepsilon^{-1} = 10^9$. To subtract, the computer must interpret the numbers as

On 8-Digit Decimal Computer

$$\begin{aligned} \varepsilon^{-1} = 10^9 = 0.10000\,000 \times 10^{10} &= 0.10000\,00000\,00000\,0 \times 10^{10} \\ 2 = 0.20000\,000 \times 10^1 &= 0.00000\,00002\,00000\,0 \times 10^{10} \end{aligned}$$

Thus, $(\varepsilon^{-1} - 2)$ is initially computed as $0.09999\,99998\,00000\,0 \times 10^{10}$ and then rounded to $0.10000\,000 \times 10^{10} = \varepsilon^{-1}$.

We conclude that for values of ε sufficiently close to 0, the computer calculates x_2 as 1 and then x_1 as 0. Since the *correct* solution is

$$x_1 = \frac{1}{1-\varepsilon} \approx 1, \qquad x_2 = \frac{1-2\varepsilon}{1-\varepsilon} \approx 1$$

the relative error in the computed solution for x_1 is extremely large: 100%.

Actually, the naive Gaussian elimination algorithm works well on Systems (1) and (2) if the equations are first permuted:

$$\begin{cases} x_1 + x_2 = 2 \\ 0x_1 + x_2 = 1 \end{cases}$$

Reordered Sample 2 × 2 Systems

and

$$\begin{cases} x_1 + x_2 = 2 \\ \varepsilon x_1 + x_2 = 1 \end{cases}$$

System (1') is easily solved obtaining $x_2 = 1$ and $x_1 = 2 - x_2 = 1$. Moreover, System (2') becomes

$$\begin{cases} x_1 + \quad x_2 = 2 \\ \qquad (1-\varepsilon)x_2 = 1 - 2\varepsilon \end{cases}$$

after the forward elimination. Then from the back substitution, the solution is computed as

$$x_2 = \frac{1-2\varepsilon}{1-\varepsilon} \approx 1, \qquad x_1 = 2 - x_2 \approx 1$$

Notice that we do *not* have to rearrange the equations in the system: it is necessary only to select a different pivot row. The difficulty in System (2) is *not* due simply to ε being small, but rather to its being small relative to other coefficients in the same row. To verify this, consider

Modified Sample 2 × 2 System

$$\begin{cases} x_1 + \varepsilon^{-1}x_2 = \varepsilon^{-1} \\ x_1 + \qquad x_2 = 2 \end{cases} \tag{4}$$

System (4) is mathematically equivalent to (2). The naive Gaussian elimination algorithm fails here! It produces the triangular system

$$\begin{cases} x_1 + \qquad \varepsilon^{-1}x_2 = \varepsilon^{-1} \\ \qquad (1-\varepsilon^{-1})x_2 = 2 - \varepsilon^{-1} \end{cases}$$

and then, in the back substitution, it produces the erroneous result

$$x_2 = \frac{2-\varepsilon^{-1}}{1-\varepsilon^{-1}} \approx 1, \qquad x_1 = \varepsilon^{-1} - \varepsilon^{-1}x_2 \approx 0$$

This situation can be resolved by interchanging the two equations in System (4):

Another Sample 2 × 2 System

$$\begin{cases} x_1 + \qquad x_2 = 2 \\ x_1 + \varepsilon^{-1}x_2 = \varepsilon^{-1} \end{cases}$$

Now the naive Gaussian elimination algorithm can be applied, resulting in the system

$$\begin{cases} \qquad x_1 + x_2 = 2 \\ (\varepsilon^{-1} - 1)x_2 = \varepsilon^{-1} - 2 \end{cases}$$

The solution is

$$x_2 = \frac{\varepsilon^{-1} - 2}{\varepsilon^{-1} - 1} \approx 1, \qquad x_1 = 2 - x_2 \approx 1$$

which is the correct solution!

Partial Pivoting and Full (Complete) Pivoting

Partial/Full Pivoting

Gaussian elimination with **partial pivoting** selects the pivot row to be the one with the maximum pivot entry in absolute value from those in the leading column of the reduced submatrix. Two rows are interchanged to move the designated row into the pivot row position. Gaussian elimination with **full (complete) pivoting** selects the pivot entry as the maximum pivot entry from all entries in the submatrix. (This complicates things because some of the unknowns are rearranged.) Two rows and two columns are interchanged to accomplish this. In practice, partial pivoting is almost as good as full pivoting and involves significantly less work. See Wilkinson [1963] for more details on this matter. He presents an analysis of full pivoting with regard to **growth factors**, even if it was *not* proven there. In fact, there is little if any *theory* to mathematically justify partial pivoting—it's primarily heuristic! (See, for example, Higham [2002].)

Simply picking the largest number in magnitude as is done in partial pivoting may work well. In this case, row scaling does *not* play a role—the relative sizes of entries in a row are *not* considered. Systems with equations having coefficients of disparate sizes may cause difficulties and should be viewed with suspicion. Sometimes a scaling strategy may ameliorate these problems. We present Gaussian elimination with scaled partial pivoting, and the pseudocode contains an implicit pivoting scheme.

Simple Partial Pivoting Strategy

In certain situations, Gaussian elimination with the simple partial pivoting strategy may lead to an incorrect solution. Consider the augmented matrix

$$\left[\begin{array}{cc|c} 2 & 2c & 2c \\ 1 & 1 & 2 \end{array}\right]$$

where c is a parameter that can take on very large numerical values and the variables are x and y. The first row is selected as the pivot row by choosing the larger number in the first column. Since the multiplier is $1/2$, one step in the row reduction process brings us to

$$\left[\begin{array}{cc|c} 2 & 2c & 2c \\ 0 & 1-c & 2-c \end{array}\right]$$

Now suppose that we are working with a computer of limited word length. So in this computer, we obtain $1-c \approx -c$ and $2-c \approx -c$. Consequently, the computer returns these numbers:

$$\left[\begin{array}{cc|c} 2 & 2c & 2c \\ 0 & -c & -c \end{array}\right]$$

Thus, we obtain $y = 1$ and $x = 0$ as the solution; whereas the correct solution is $x = y = 1$.

On the other hand, Gaussian elimination with scaled partial pivoting selects the second row as the pivot row. The scaling constants are $(2c, 1)$. The larger of the two ratios for selecting the pivot row from $\{2/(2c), 1\}$ is the second one. Now the multiplier is 2, and one step in the row reduction process brings us to

$$\left[\begin{array}{cc|c} 0 & 2c-2 & 2c-4 \\ 1 & 1 & 2 \end{array}\right]$$

On our computer of limited word length, we find $2c - 2 \approx 2c$ and $2c - 4 \approx 2c$. Consequently, the computer returns these numbers:

$$\left[\begin{array}{cc|c} 0 & 2c & 2c \\ 1 & 1 & 2 \end{array}\right]$$

Now we obtain the correct solution, $y = 1$ and $x = 1$.

Gaussian Elimination with Scaled Partial Pivoting

These simple examples should make it clear that the *order* in which we treat the equations significantly affects the accuracy of the elimination algorithm in the computer. In the naive Gaussian elimination algorithm, we use the first equation to eliminate x_1 from the equations that follow it. Then we use the second equation to eliminate x_2 from the equations that follow it, and so on. The order in which the equations are used as pivot equations is the **natural** order $\{1, 2, \ldots, n\}$. Note that the last equation (equation number n) is *not* used as an operating equation in the natural ordering: at no time are multiples of it subtracted from other equations in the naive algorithm.

From the previous examples, it is clear that a strategy is needed for selecting new pivots at each stage in Gaussian elimination. Perhaps the best approach is **complete pivoting**, which involves searches over all entries in the submatrices for the largest entry in absolute value and then interchanges rows and columns to move it into the pivot position. This would be quite expensive, since it involves a great amount of searching and data movement. However, searching just the first column in the submatrix at each stage accomplishes most of what is needed (avoiding small or zero pivots). This is **partial pivoting**, and it is the most common approach. It does not involve an examination of the elements in the rows, since it looks only at column entries. We advocate a strategy that simulates a scaling of the row vectors and then selects as a pivot element the relatively largest entry in a column. Also, rather than interchanging rows to move the desired element into the pivot position, we use an indexing array to avoid the data movement. This procedure is not as expensive as complete pivoting, and it goes beyond partial pivoting to include an examination of all elements in the original matrix. Of course, other strategies for selecting pivot elements could be used.

Types of Pivoting

The Gaussian elimination algorithm we develop uses the equations in an order that is determined by the actual system being solved. For instance, if the algorithm were asked to solve System (1) or (2), the order in which the equations would be used as pivot equations would *not* be the natural order $\{1, 2\}$, but rather $\{2, 1\}$. This order is automatically determined by the computer program. The order in which the equations are employed is denoted by the row vector $[\ell_1, \ell_2, \ldots, \ell_n]$, where ℓ_n is *not* actually being used in the forward elimination phase. Here, the ℓ_i are integers from 1 to n in a possibly different order. We call $\boldsymbol{\ell} = [\ell_1, \ell_2, \ldots, \ell_n]$ the **index vector**. The strategy to be described now for determining the index vector is termed **scaled partial pivoting** (SPP).

Scaled Partial Pivoting

At the beginning, a **scale factor** must be computed for each equation in the system. Referring to the notation in Section 2.1, we define

Scale Vector

$$s_i = \max_{1 \leqq j \leqq n} |a_{ij}| \qquad (1 \leqq i \leqq n)$$

These n numbers are recorded in the **scale vector** $\boldsymbol{s} = [s_1, s_2, \ldots, s_n]$.

In starting the forward elimination process, we do *not* arbitrarily use the first equation as the pivot equation. Instead, we use the equation for which the ratio $|a_{i,1}|/s_i$ is greatest. Let ℓ_1 be the first index for which this ratio is greatest. Now appropriate multiples of equation ℓ_1 are subtracted from the other equations to create 0's as coefficients for each x_1 except in the pivot equation.

GE (SPP)

The best way of keeping track of the indices is as follows: At the beginning, define the index vector $\boldsymbol{\ell}$ to be $[\ell_1, \ell_2, \ldots, \ell_n] = [1, 2, \ldots, n]$. In Step 1, select j to be the first index associated with the largest ratio in the set:

Step 1

$$\left\{ \frac{|a_{\ell_i 1}|}{s_{\ell_i}} : 1 \leqq i \leqq n \right\}$$

Now interchange ℓ_j with ℓ_1 in the index vector $\boldsymbol{\ell}$. Next, use multipliers

$$\frac{a_{\ell_i 1}}{a_{\ell_1 1}}$$

times row ℓ_1, and subtract from equations ℓ_i for $2 \leqq i \leqq n$. It is important to note that only entries in $\boldsymbol{\ell}$ are being interchanged and *not* the equations. This eliminates the time-consuming and unnecessary process of moving the coefficients of equations around in the computer memory!

In Step 2, the ratios

Step 2

$$\left\{ \frac{|a_{\ell_i,2}|}{s_{\ell_i}} : 2 \leqq i \leqq n \right\}$$

are scanned. If j is the first index for the largest ratio, interchange ℓ_j with ℓ_2 in $\boldsymbol{\ell}$. Then multipliers

$$\frac{a_{\ell_i 2}}{a_{\ell_2 2}}$$

times equation ℓ_2 are subtracted from equations ℓ_i for $3 \leqq i \leqq n$.

At Step k, select j to be the first index corresponding to the largest of the ratios,

Step *k*

$$\left\{ \frac{|a_{\ell_i k}|}{s_{\ell_i}} : k \leqq i \leqq n \right\}$$

and interchange ℓ_j and ℓ_k in index vector $\boldsymbol{\ell}$. Then multipliers

$$\frac{a_{\ell_i k}}{a_{\ell_k k}}$$

times pivot equation ℓ_k are subtracted from equations ℓ_i for $k + 1 \leqq i \leqq n$.

Notice that the scale factors are *not* changed after each pivot step. Intuitively, one might think that after each step in the Gaussian algorithm, the remaining (modified) coefficients should be used to recompute the scale factors instead of using the original scale vector. Of course, this could be done, but it is generally believed that the extra computations involved in this procedure are *not* worthwhile for the majority of linear systems. The reader is encouraged to explore this question. (See Computer Exercise 2.2.16.)

EXAMPLE 1 Solve this system of linear equations:

$$\begin{cases} 0.0001x + y = 1 \\ x + y = 2 \end{cases}$$

using in order no pivoting, partial pivoting, and scaled partial pivoting. Carry at most five significant digits of precision (rounding) to see how finite precision computations and roundoff errors can affect the calculations.

Solution By direct substitution, it is easy to verify that the true solution is $x = 1.0001$ and $y = 0.99990$ to five significant digits.

No Pivoting

For **no pivoting**, the first equation in the original system is the pivot equation, and the multiplier is xmult $= 1/0.0001 = 10000$. Multiplying the first equation by this multiplier and subtracting the result from the second equation, the necessary calculations are $(10000)(0.0001) - 1 = 0$, $(10000)(1) - 1 = 9999$, and $(10000)(1) - 2 = 9998$. The new system of equations is

$$\begin{cases} 0.0001x + y = 1 \\ 9999y = 9998 \end{cases}$$

From the second equation, we obtain $y = 9998/9999 \approx 0.99990$. Using this result and the first equation, we find $0.0001x = 1 - y = 1 - 0.999900 = 0.0001$ and $x = 0.0001/0.0001 = 1$. Notice that we have lost the right-most significant digit in the correct value of x!

Partial Pivoting

We repeat the solution using **partial pivoting** in the original system. Examining the first column of x coefficients (0.0001, 1), we see that the second is larger, so the second equation is used as the pivot equation. We can interchange the two equations, obtaining

$$\begin{cases} x + y = 2 \\ 0.0001x + y = 1 \end{cases}$$

The multiplier is xmult $= 0.0001/1 = 0.0001$. This multiple of the first equation is subtracted from the second equation. The calculations are $(-0.0001)(1) + 0.0001 = 0$, $(0.0001)(1) - 1 = 0.99990$, and $(0.0001)(2) - 1 = 0.99980$. The new system of equations is

$$\begin{cases} x + y = 2 \\ 0.99990y = 0.99980 \end{cases}$$

We obtain $y = 0.99980/0.99990 \approx 0.99990$. Now, using the second equation and this value, we find $x = 2 - y = 2 - 0.99990 = 1.0001$. Both computed values of x and y are correct to five significant digits!

Scale Partial Pivoting

We repeat the solution using **scaled partial pivoting** on the original system. Since the scaling constants are $s = [1, 1]$ and the ratios for determining the pivot equation are $(0.0001/1, 1/1)$, the second equation is now the pivot equation. We do not actually interchange the equations, but work with an index array $\ell = [2, 1]$ that tells us to use the second equation as the first pivot equation. The rest of the calculations are the same as those for partial pivoting with the results correct to five significant digits.

We cannot promise that scaled partial pivoting is better than partial pivoting in all cases, but it clearly has some advantages. For example, suppose that we want to force the first equation in the original system to be the pivot equation and multiply it by a large number such as 20,000, obtaining

Sample 2 × 2 System

$$\begin{cases} 2x + 20000y = 20000 \\ x + y = 2 \end{cases}$$

Partial pivoting ignores the fact that the coefficients in the first equation differ by orders of magnitude and selects the first equation as the pivot equation. However, scaled partial pivoting uses the scaling constants (20000, 1), and the ratios for determining the pivot equations are $(2/20000, 1/1)$. Scaled partial pivoting continues to select the second equation as the pivot equation! ■

A Larger Numerical Example

We are *not* quite ready to write pseudocode. First, let us consider a concrete example:

Sample 5 × 5 System

$$\begin{bmatrix} 3 & -13 & 9 & 3 \\ -6 & 4 & 1 & -18 \\ 6 & -2 & 2 & 4 \\ 12 & -8 & 6 & 10 \end{bmatrix} \begin{bmatrix} x_1 \\ x_2 \\ x_3 \\ x_4 \end{bmatrix} = \begin{bmatrix} -19 \\ -34 \\ 16 \\ 26 \end{bmatrix} \tag{5}$$

The index vector is $\boldsymbol{\ell} = [1, 2, 3, 4]$ at the beginning. The scale vector does *not* change throughout the procedure and is $\boldsymbol{s} = [13, 18, 6, 12]$.

Step 1

To determine the first pivot row, we look at four ratios:

$$\left\{ \frac{|a_{\ell_i,1}|}{s_{\ell_i}} : i = 1, 2, 3, 4 \right\} = \left\{ \frac{3}{13}, \frac{6}{18}, \frac{6}{6}, \frac{12}{12} \right\} \approx \{0.23, 0.33, 1.0, 1.0\}$$

We select the index j as the *first* occurrence of the largest value of these ratios. In this example, the largest of these occurs for the index $j = 3$. So row three is to be the pivot equation in Step $k = 1$ of the elimination process. In the index vector ℓ, entries ℓ_k and ℓ_j are interchanged so that the new index vector is $\boldsymbol{\ell} = [3, 2, 1, 4]$. Thus, the pivot equation is ℓ_k, which is $\ell_1 = 3$. Now appropriate multiples of the third equation are subtracted from the other equations so as to create 0's as coefficients for x_1 in each of those equations. Explicitly, $\frac{1}{2}$ times row three is subtracted from row one, -1 times row three is subtracted from row two, and 2 times row three is subtracted from row four. The result is

$$\begin{bmatrix} 0 & -12 & 8 & 1 \\ 0 & 2 & 3 & -14 \\ 6 & -2 & 2 & 4 \\ 0 & -4 & 2 & 2 \end{bmatrix} \begin{bmatrix} x_1 \\ x_2 \\ x_3 \\ x_4 \end{bmatrix} = \begin{bmatrix} -27 \\ -18 \\ 16 \\ -6 \end{bmatrix}$$

Step 2

In Step $k = 2$, we use the index vector $\boldsymbol{\ell} = [3, 2, 1, 4]$ and scan the ratios corresponding to rows two, one, and four:

$$\left\{ \frac{|a_{\ell_i,2}|}{s_{\ell_i}} : i = 2, 3, 4 \right\} = \left\{ \frac{2}{18}, \frac{12}{13}, \frac{4}{12} \right\} \approx \{0.11, 0.92, 0.33\}$$

looking for the largest value. We find that the largest is the second ratio, and we therefore set $j = 3$ and interchange ℓ_k with ℓ_j in the index vector. Thus, the index vector becomes $\boldsymbol{\ell} = [3, 1, 2, 4]$. The pivot equation for Step 2 in the elimination is now row one, and $\ell_2 = 1$. Next, multiples of the first equation are subtracted from the second equation and the fourth equation. The appropriate multiples are $-\frac{1}{6}$ and $\frac{1}{3}$, respectively. The result is

$$\begin{bmatrix} 0 & -12 & 8 & 1 \\ 0 & 0 & 13/3 & -83/6 \\ 6 & -2 & 2 & 4 \\ 0 & 0 & -2/3 & 5/3 \end{bmatrix} \begin{bmatrix} x_1 \\ x_2 \\ x_3 \\ x_4 \end{bmatrix} = \begin{bmatrix} -27 \\ -45/2 \\ 16 \\ 3 \end{bmatrix}$$

Step 3

The final Step $k = 3$ is to examine the ratios corresponding to rows two and four:

$$\left\{ \frac{|a_{\ell_i,3}|}{s_{\ell_i}} : i = 3, 4 \right\} = \left\{ \frac{13/3}{18}, \frac{2/3}{12} \right\} \approx \{0.24, 0.06\}$$

with the index vector $\boldsymbol{\ell} = [3, 1, 2, 4]$. The larger value is the first, so we set $j = 3$. Since this is Step $k = 3$, interchanging ℓ_k with ℓ_j leaves the index vector unchanged, $\boldsymbol{\ell} = [3, 1, 2, 4]$. The pivot equation is row two and $\ell_3 = 2$, and we subtract $-\frac{2}{13}$ times the second equation from the fourth equation.

So the forward elimination phase ends with the final system

Forward Elimination Results

$$\begin{bmatrix} 0 & -12 & 8 & 1 \\ 0 & 0 & 13/3 & -83/6 \\ 6 & -2 & 2 & 4 \\ 0 & 0 & 0 & -6/13 \end{bmatrix} \begin{bmatrix} x_1 \\ x_2 \\ x_3 \\ x_4 \end{bmatrix} = \begin{bmatrix} -27 \\ -45/2 \\ 16 \\ -6/13 \end{bmatrix}$$

The order in which the pivot equations were selected is displayed in the final index vector $\boldsymbol{\ell} = [3, 1, 2, 4]$.

Now, reading the entries in the index vector from the last to the first, we have the order in which the back substitution is to be performed. The solution is obtained by using equation $\ell_4 = 4$ to determine x_4, and then equation $\ell_3 = 2$ to find x_3, and so on. Carrying out the calculations, we have

Back Substitution Results

$$
\begin{aligned}
x_4 &= \frac{1}{-6/13}[-6/13] = 1 \\
x_3 &= \frac{1}{13/3}[(-45/2) + (83/6)(1)] = -2 \\
x_2 &= \frac{1}{-12}[-27 - 8(-2) - 1(1)] = 1 \\
x_1 &= \frac{1}{6}[16 + 2(1) - 2(-2) - 4(1)] = 3
\end{aligned}
$$

Hence, the solution is

Solution

$$
\boldsymbol{x} = \begin{bmatrix} 3 & 1 & -2 & 1 \end{bmatrix}^T
$$

Pseudocode

The algorithm as programmed carries out the forward elimination phase on the coefficient array (a_{ij}) only. The right-hand side array (b_i) is treated in the back substitution phase. This method is adopted because it is more efficient if several systems must be solved with the same array (a_{ij}), but differing arrays (b_i). Because we wish to treat (b_i) later, it is necessary to store not only the index array but also the various multipliers that are used. These multipliers are conveniently stored in array (a_{ij}) in the positions where the 0 entries would have been created. These multipliers are useful in constructing the ***LU*** factorization of the matrix $\boldsymbol{A}$, as we explain in Section 8.1.

We are now ready to write a procedure for forward elimination with scaled partial pivoting. Our approach is to modify procedure *Naive_Gauss* of Section 2.1 by introducing scaling and indexing arrays. The procedure that carries out Gaussian elimination with scaled partial pivoting on the square array (a_{ij}) is called *Gauss*. Its calling sequence is $(n, (a_{ij}), (\ell_i))$, where (a_{ij}) is the $n \times n$ coefficient array and (ℓ_i) is the index array $\boldsymbol{\ell}$. In the pseudocode, (s_i) is the scale array, $\boldsymbol{s}$.

```
procedure Gauss(n, (a_ij), (ℓ_i))
integer i, j, k, n;   real r, rmax, smax, xmult
real array (a_ij)_{1:n×1:n}, (ℓ_i)_{1:n};   real array allocate (s_i)_{1:n}
for i = 1 to n
    ℓ_i ← i
    smax ← 0
    for j = 1 to n
        smax ← max(smax, |a_ij|)
    end for
    s_i ← smax
end for
```

(Continued)

Gauss Procedure Pseudocode

```
for k = 1 to n − 1
    rmax ← 0
    for i = k to n
        r ← |a_{ℓ_i,k}/s_{ℓ_i}|
        if (r > rmax) then
            rmax ← r
            j ← i
        end if
    end for
    ℓ_j ↔ ℓ_k
    for i = k + 1 to n
        xmult ← a_{ℓ_i,k}/a_{ℓ_k,k}
        a_{ℓ_i,k} ← xmult
        for j = k + 1 to n
            a_{ℓ_i,j} ← a_{ℓ_i,j} − (xmult)a_{ℓ_k,j}
        end for
    end for
end for
deallocate array (s_i)
end procedure Gauss
```

Discussion of Pseudocode

A detailed explanation of the above procedure is now presented. In the first loop, the initial form of the index array is being established, namely, $\ell_i = i$. Then the scale array (s_i) is computed.

The statement **for** $k = 1$ **to** $n - 1$ initiates the principal outer loop. The index k is the subscript of the variable whose coefficients will be made 0 in the array (a_{ij}); that is, k is the index of the column in which new 0's are to be created. Remember that the 0's in the array (a_{ij}) do *not* actually appear because those storage locations are used for the multipliers. This fact can be seen in the line of the procedure where *xmult* is stored in the array (a_{ij}). (See Section 8.1 on the ***LU*** factorization of ***A*** for why this is done.)

Once k has been set, the first task is to select the correct pivot row, which is done by computing $|a_{\ell_i k}|/s_{\ell_i}$ for $i = k, k+1, \ldots, n$. The next set of lines in the pseudocode is calculating this greatest ratio, called *rmax* in the routine, and the index j where it occurs. Next, ℓ_k and ℓ_j are interchanged in the array (ℓ_i).

The arithmetic modifications in the array (a_{ij}) due to subtracting multiples of row ℓ_k from rows $\ell_{k+1}, \ell_{k+2}, \ldots, \ell_n$ all occur in the final lines. First the multiplier is computed and stored; then the subtraction occurs in a loop.

Warning

Caution: Values in array (a_{ij}) that result as *output* from procedure *Gauss* are *not* the same as those in array (a_{ij}) at *input*. If the original array must be retained, store a duplicate of it in another array.

In the procedure *Naive_Gauss* for naive Gaussian elimination from Section 2.1, the right-hand side ***b*** was modified during the forward elimination phase; however, this was *not* done in the procedure *Gauss*. Therefore, we need to update ***b*** before considering the back substitution phase. For simplicity, we discuss updating ***b*** for the naive forward elimination first. Stripping out the pseudocode from *Naive_Gauss* that involves the (b_i) array in the

forward elimination phase, we obtain

Naive Forward Elimination on rhs

for $k = 1$ **to** $n - 1$
  for $i = k + 1$ **to** n
    $b_i = b_i - a_{ik} b_k$
  end for
end for

This updates the (b_i) array based on the stored multipliers from the (a_{ij}) array. When scaled partial pivoting is done in the forward elimination phase, such as in procedure *Gauss*, the multipliers for each step are *not* one below another in the (a_{ij}) array, but are jumbled around. To unravel this situation, all we have to do is introduce the index array (ℓ_i) into the above pseudocode:

Modified Forward Elimination on rhs

for $k = 1$ **to** $n - 1$
  for $i = k + 1$ **to** n
    $b_{\ell_i} = b_{\ell_i} - a_{\ell_i k} b_{\ell_k}$
  end for
end for

After the array b has been processed in the forward elimination, the **back substitution** process is carried out. It begins by solving the equation

$$a_{\ell_n,n} x_n = b_{\ell_n} \tag{6}$$

whence

$$x_n = \frac{b_{\ell_n}}{a_{\ell_n,n}}$$

Then the equation

$$a_{\ell_{n-1},n-1} x_{n-1} + a_{\ell_{n-1},n} x_n = b_{\ell_{n-1}}$$

is solved for x_{n-1}:

Back Substitution Process

$$x_{n-1} = \frac{1}{a_{\ell_{n-1},n-1}} \left(b_{\ell_{n-1}} - a_{\ell_{n-1},n} x_n \right)$$

After $x_n, x_{n-1}, \ldots, x_{i+1}$ have been determined, x_i is found from the equation

$$a_{\ell_i,i} x_i + a_{\ell_i,i+1} x_{i+1} + \cdots + a_{\ell_i,n} x_n = b_{\ell_i}$$

whose solution is

$$x_i = \frac{1}{a_{\ell_i,i}} \left(b_{\ell_i} - \sum_{j=i+1}^{n} a_{\ell_i,j} x_j \right) \tag{7}$$

Except for the presence of the index array ℓ_i, this is similar to the back substitution formula (7) in Section 2.1 (p. 76) obtained for naive Gaussian elimination.

The procedure for processing the array b and performing the **back substitution** phase is given next:

***Solve* Procedure Pseudocode**

```
procedure Solve (n, (a_ij), (ℓ_i), (b_i), (x_i))
integer i, k, n;      real sum
real array (a_ij)_{1:n×1:n}, (ℓ_i)_{1:n}, (b_i)_{1:n}, (x_i)_{1:n}
for k = 1 to n − 1
    for i = k + 1 to n
        b_{ℓ_i} ← b_{ℓ_i} − a_{ℓ_i,k} b_{ℓ_k}
    end for
end for
x_n ← b_{ℓ_n} / a_{ℓ_n,n}
for i = n − 1 to 1
    sum ← b_{ℓ_i}
    for j = i + 1 to n do
        sum ← sum − a_{ℓ_i,j} x_j
    end for
    x_i ← sum / a_{ℓ_i,i}
end for
end procedure Solve
```

Here, the first loop carries out the forward elimination process on array (b_i), using arrays (a_{ij}) and (ℓ_i) that result from procedure *Gauss*. The next line carries out the solution of Equation (6). The final part carries out Equation (7). The variable *sum* is a temporary variable for accumulating the terms in parentheses.

As with most pseudocode in this book, those in this chapter contain only the basic ingredients for good mathematical software. They are not suitable as *production* code for various reasons. For example, procedures for optimizing code are ignored. Furthermore, the procedures do *not* give warnings for difficulties that may be encountered, such as division by zero! General-purpose software should be **robust**; that is, it should anticipate every possible situation and deal with each in a prescribed way. (See Computer Exercise 2.2.11.)

Long Operation Count

Solving large systems of linear equations can be expensive in terms of computer time. To understand why, let us perform an operation count on the two algorithms whose codes have been given. We count only multiplications and divisions (long operations) because they are more time consuming than addition/subtraction. Furthermore, we lump multiplications and divisions together even though division is slower than multiplication. In modern computers, all floating-point operations are done in hardware, so long operations may *not* be as significant, but this still gives an indication of the operational cost of Gaussian elimination.

Consider first procedure *Gauss*. In Step 1, the choice of a pivot element requires the calculation of n ratios—that is, n divisions. Then for rows $\ell_2, \ell_3, \dots, \ell_n$, we first compute a multiplier and then subtract from row ℓ_i that multiplier times row ℓ_1. The zero that is being created in this process is *not* computed. So the elimination requires $n-1$ multiplications per row. If we include the calculation of the multiplier, there are n long operations (divisions or multiplications) per row. There are $n-1$ rows to be processed for a total of $n(n-1)$ operations. If we add the cost of computing the ratios, a total of n^2 operations is needed for Step 1.

Step 2 is like Step 1 except that row ℓ_1 is not affected, nor is the column of multipliers created and stored in Step 1. So Step 2 requires $(n-1)^2$ multiplications or divisions because it operates on a system without row ℓ_1 and without column 1. Continuing this reasoning, we conclude that the total number of long operations for procedure *Gauss* is

***Gauss* ops Count**

$$n^2 + (n-1)^2 + (n-2)^2 + \cdots + 4^2 + 3^2 + 2^2 = \frac{n}{6}(n+1)(2n+1) - 1 \approx \frac{n^3}{3} = \mathcal{O}(n^3)$$

(The derivation of this formula is outlined in Exercise 2.2.16.) Note that the number of long operations in this procedure grows like $n^3/3$, the dominant term.

Now consider procedure *Solve*. The **forward processing** of the array (b_i) involves $n-1$ steps. Step 1 contains $n-1$ multiplications, Step 2 contains $n-2$ multiplications, and so on. The total of the forward processing of array (b_i) is thus

***Solve* ops Count**

$$(n-1) + (n-2) + \cdots + 3 + 2 + 1 = \frac{n}{2}(n-1)$$

(See Exercise 2.2.15.) In the **back substitution** procedure, one long operation is involved in Step 1, two in Step 2, and so on. The total is

$$1 + 2 + 3 + \cdots + n = \frac{n}{2}(n+1) = \mathcal{O}(n^2)$$

Thus, procedure *Solve* involves altogether n^2 long operations. To summarize:

■ Theorem 1 **Theorem on Long Operations**

The forward elimination phase of the Gaussian elimination algorithm with scaled partial pivoting, if applied only to the $n \times n$ coefficient array, involves approximately $n^3/3$ long operations (multiplications or divisions). Solving for x requires an additional n^2 long operations.

An intuitive way to think of this result is that the Gaussian elimination algorithm involves a triply nested for-loop. So an $\mathcal{O}(n^3)$ algorithmic structure is driving the elimination process, and the work is heavily influenced by the cube of n (the number of equation unknowns).

Numerical Stability

Remarks on Numerical Stability

The **numerical stability** of a numerical algorithm is related to the accuracy of the procedure. An algorithm can have different levels of numerical stability because many computations can be achieved in various ways that are algebraically equivalent, but may produce different results. A robust numerical algorithm with a high level of numerical stability is desirable. Gaussian elimination is numerically stable for strictly diagonally dominant matrices or symmetric positive definite matrices. (These properties are discussed in Section 2.3 and Chapter 8, respectively.) For matrices with a general dense structure, Gaussian elimination with partial pivoting is usually numerically stable in practice. Nevertheless, there exist unstable pathological examples in which it may fail. For additional details, see Golub and Van Loan [1996] and Highman [2002].

Brief History of Gaussian Elimination

An early version of Gaussian elimination was found in the Chinese mathematics text (*jiuzhang suanshu* or *The Nine Chapters on the Mathematical Art*, Chapter 8, *Rectangular Arrays*). The method was illustrated in 18 problems with two to five equations. The first reference to this book is dated 179 C.E., but parts of it were written as early as approximately 150 B.C.E.. In Europe, Isaac Newton's notes on solving simultaneous equations were published in 1707. In 1816, Carl Friedrich Gauss devised a notation for symmetric elimination. Because of some confusion over its history, the Gaussian elimination method was named for Gauss in the 1950s.

Scaling

Readers should *not* confuse *scaling* in Gaussian elimination (which is *not* recommended) with our discussion of *scaled* partial pivoting in Gaussian elimination.

Discussion on Scaling

The word **scaling** has more than one meaning. It could mean actually dividing each row by its maximum element in absolute value. We certainly do *not* advocate that. In other words, we do *not* recommend scaling of the matrix at all! However, we do compute a scale array and use it in selecting the pivot element in Gaussian elimination with scaled partial pivoting. We do not actually scale the rows; we just keep a vector of the "row infinity norms," that is, the maximum element in absolute value for each row. This and the need for a vector of indices to keep track of the pivot rows makes the algorithm somewhat complicated, but that is the price to be paid for some degree of robustness in the procedure.

The simple 2×2 system in Equations (2) and (4) show that scaling does *not* help in choosing a good pivot row. In this example, scaling is of no use. Scaling of the rows is contemplated in Exercises 2.2.23 and Computer Exercise 2.2.17. Notice that this procedure requires at least n^2 arithmetic operations. Again, we are *not* recommending it for a general-purpose code.

Some codes actually move the rows around in storage. Because that should *not* be done in practice, we do *not* do it in the code, since it might be misleading. Also, to avoid misleading the casual reader, we called our initial algorithm (in Section 2.1) *naive*, hoping that nobody would mistake it for a reliable code.

Variants of Gaussian Eliminations

There are four ways to view Gaussian elimination:

Variant of Gaussian Elimination

- as the elimination of variables in a linear system,
- as row operations on a linear system,
- as a transformation of a matrix into triangular form by using elementary lower triangular matrices,
- as the factorization (or decomposition) of a matrix into the product of lower and upper triangular factors.

Since each of these approaches are related, they yield variants of the Gaussian elimination algorithm, with each having its own advantages and disadvantages in specific applications.

As we have seen, it is easy to incorporate pivoting into Gaussian elimination. Some of the standard terminology used is as follows. At the k-th stage of the algorithm, the element a_{kk} is the **pivot element** or simply the **pivot**. The process of performing interchanges of rows or columns is **pivoting**, and it alters the selection of the **pivot**.

Pivot/Pivoting

The process of selecting pivots has two aspects: where the pivots come from and how the pivots are chosen. The details of pivoting depend on the algorithm used and its application. It is useful to note that Gaussian elimination with pivoting is equivalent to making all the interchanging in the original matrix first and then performing Gaussian elimination without pivoting of the resulting matrix. This is nice in theory, but in practice, this is easier said than done!

Since the basic Gaussian elimination algorithm can fail when a division by zero is encountered, row and/or column interchanges may be used to avoid this difficulty; in other words, *pivoting*. Although pivoting is a simple idea, it is a nontrivial matter to decide which element to use as a pivot! Of all the pivoting strategies, by far the most common is **partial pivoting for size**, which means selecting a pivot from a set of candidates by choosing the one that is largest in magnitude. (This usage is natural for dense matrices where pivoting for size utilizes a norm.) In fact, Gaussian elimination and variations of it are some of the most frequently used algorithms in computational mathematics!

Partial Pivoting for Size

A major virtue of Gaussian elimination is its ability to be adapted to special structured matrices such as sparse or banded matrices. The introduction of nonzero elements in the place of a zero element is called **fill-in**. In many applications involving large sparse linear systems, the coefficient matrix has predominantly zero elements. Clearly, the choice of the pivot strategy influences the amount of fill-in. Most algorithms for sparse matrices use a pivoting strategy that reduces fill-in, called **pivoting for sparsity**. Unfortunately, pivoting for size and pivoting for sparsity can be at odds with one another!

Pivoting for Sparsity

Classical Gaussian Elimination

Classical Gaussian elimination can be thought of as expanding the LU decomposition or factorization of the coefficient matrix **A** in a linear system. In Figure 2.2, the shaded area represents the part of the LU decomposition that has already been computed with **L** and **U** separated by a diagonal line. The thin horizontal rectangular area represents the next partial row of **U** to be computed; whereas the thin vertical rectangular area represents the corresponding partial column elements of **L**. These two rectangles overlap, reflecting the fact that the diagonal elements of **L** are 1's, which are *not* stored. (We discuss LU factorization more in Section 8.1.) For a given matrix, the operations for classical Gaussian elimination are well known. However, there is considerable freedom in how these operations can be interleaved one with another. Moreover, each style of interleaving gives rise to a variant of the basic algorithm.

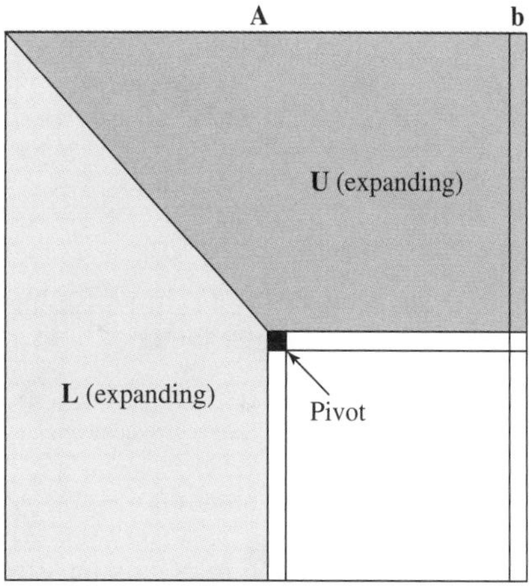

FIGURE 2.2 Classical Gaussian elimination $A = LU$ (no pivoting)

Partial Pivoting

Partial pivoting involves row interchanges. To minimize the roundoff errors, the row that moves the largest pivot to the diagonal position is chosen. A pivot element is selected from the left thin vertical rectangular area in Figure 2.3 (p. 96), and an interchange of that entire row in the **A** array is done with the pivot row (as indicated by the arrows).

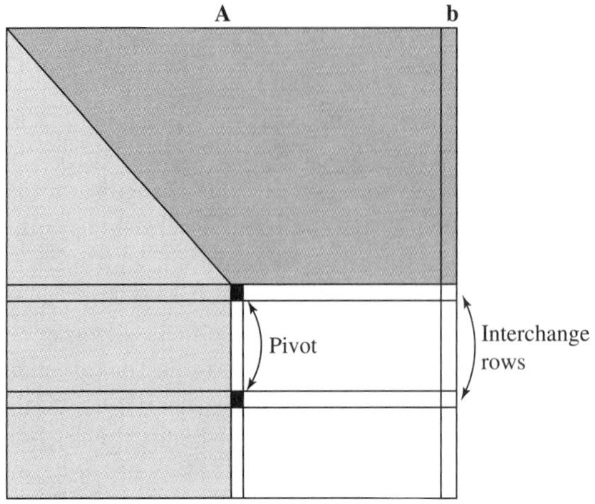

FIGURE 2.3
Partial pivoting Gaussian elimination

Scaled Partial Pivoting

Scaled partial pivoting is a rather clever modification of partial pivoting that simulates full pivoting by using an index vector and a scale vector containing information about the relative sizes of elements in each row.

Full Pivoting

Complete (full) pivoting is more complicated, involving exchanges of both rows and columns, which can change the order of the unknowns. For increased stability, the largest possible pivot is sought, requiring a search in the entire submatrix as shown in Figure 2.4. Full pivoting is less susceptible to roundoff errors, but this increase in numerical stability comes at the cost of an increase in the work associated with searching and in the amount of data movement involved. The general feeling is that the benefits of full pivoting are not worth the extra effort!

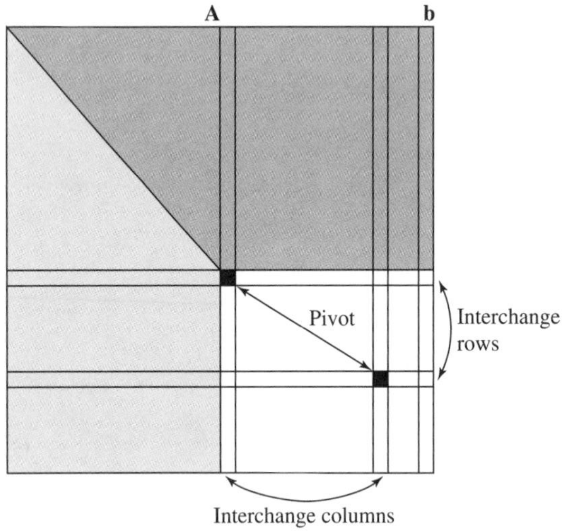

FIGURE 2.4
Complete pivoting Gaussian elimination

In practice, applying Gaussian elimination with partial pivoting and back substitution gives the exact solution to a *nearby* problem, which is *exactly the right answer to nearly the right question!* (See Trefethen and Bau [1997].) Such an algorithm is called **backward stable**. Gaussian elimination without partial pivoting is *not* backward stable for a linear system with a general coefficient matrix **A**, but it is if **A** is symmetric and positive definite.

Backward Stable

In 1810, Gauss gave the first layout, up to diagonal scaling, of the algorithm now known as "classical Gaussian elimination." The terms "partial pivoting" and "complete pivoting" are attributable to Wilkinson [1965]. There are four or more algorithmic variants for classical Gaussian elimination. For additional details, see Stewart [1998b].

Condition Number

Often mathematical software for solving linear systems return not only the approximate solution but the **condition number** of the linear system. Use the following rules to interpret these results.

■ **Rule**

Rules of Thumb

1. The **condition number** $\kappa(A)$ (p. 406) indicates how close $\boldsymbol{A}$ is to being numerically singular (non-invertible).
2. In practice, applying Gaussian elimination with a variant of partial pivoting and back substitution to solve $\boldsymbol{Ax} = \boldsymbol{b}$ yields a numerical solution such that the **residual vector** $\boldsymbol{r} = \boldsymbol{b} - \boldsymbol{A}$ is small even if the condition number $\kappa(\boldsymbol{A})$ is large.
3. If $\kappa(\boldsymbol{A})$ is *large*, $\boldsymbol{A}$ is **ill-conditioned**, and even the *best* numerical algorithm produces a solution that *cannot* be guaranteed to be *close* to the true solution.
4. If $\boldsymbol{A}$ and $\boldsymbol{b}$ are stored to machine precision ε_m, the numerical solution to $\boldsymbol{Ax} = \boldsymbol{b}$ by *any* variant of Gaussian elimination is correct to $d = |\log_{10} \varepsilon_m| - \log \kappa(\boldsymbol{A})$ digits.

See Section 8.4 for more details on these topics.

Backslash Operator in MATLAB

MATLAB Backslash

The system of equations $\boldsymbol{Ax} = \boldsymbol{b}$ has the *formal* solution $\boldsymbol{x} = \boldsymbol{A}^{-1}\boldsymbol{b}$. In MATLAB notation, the system is solved with the backslash command: `x = A\b`. The software attempts to solve the system with the method that gives the least roundoff error and fewest operations. When $\boldsymbol{A}$ is an $n \times n$ matrix, MATLAB examines $\boldsymbol{A}$ to see:

1. If it is a permutation of a triangular system—if so, the appropriate triangular solve is used.
2. If it *appears* to be symmetric and positive definite—if so, a Cholesky factorization and two triangular solves are attempted.
3. If Cholesky factorization fails or if $\boldsymbol{A}$ does *not* appear to be symmetric, an LU factorization and two triangular solves are attempted.

If $\boldsymbol{A}$ is $n \times m$ with $n \neq m$, MATLAB attempts to solve the system using the appropriate algorithms—some of which are discussed later.

Other mathematical software systems such as Maple and Mathematica have similar multi-algorithmic procedures or hybrid schemes.

Summary 2.2

- In performing Gaussian elimination, partial pivoting is highly recommended to avoid zero pivots and small pivots. In Gaussian elimination with scaled partial pivoting, we use a **scale vector** $\boldsymbol{s} = [s_1, s_2, \dots, s_n]^T$ in which

$$s_i = \max_{1 \leqq j \leqq n} |a_{ij}| \qquad (1 \leqq i \leqq n)$$

and an **index vector** $\boldsymbol{\ell} = [\ell_1, \ell_2, \dots, \ell_n]$, initially set as $\boldsymbol{\ell} = [1, 2, \dots, n]$. The scale array is set once at the beginning of the algorithm. The elements in the index array are interchanged rather than the rows of the matrix $\boldsymbol{A}$, which reduces the amount of data movement considerably. The key step in the pivoting procedure is to select j to be the first index associated with the largest ratio in the set

$$\left\{ \frac{|a_{\ell_i,k}|}{s_{\ell_i}} : k \leqq i \leqq n \right\}$$

and interchange ℓ_j with ℓ_k in the index array $\boldsymbol{\ell}$. Then use multipliers

$$\frac{a_{\ell_i,k}}{a_{\ell_k,k}}$$

times row ℓ_k and subtract from equations ℓ_i for $k + 1 \leqq i \leqq n$.

- The **forward elimination** from equation ℓ_i for $\ell_{k+1} \leqq \ell_i \leqq \ell_n$ is

$$\begin{cases} a_{\ell_i,j} \leftarrow a_{\ell_i,j} - (a_{\ell_i,k}/a_{\ell_k k})a_{kj} & (\ell_k \leqq \ell_j \leqq \ell_n) \\ b_{\ell_i} \leftarrow b_{\ell_i} - (a_{\ell_i,k}/a_{\ell_k k})b_{\ell_k} \end{cases}$$

The steps involving the vector $\boldsymbol{b}$ are usually done separately just before the back substitution phase, which we call **updating** the right-hand side.

- The **back substitution** is

$$x_i = \frac{1}{a_{\ell_i,i}}\left(b_{\ell_i} - \sum_{j=i+1}^{n} a_{\ell_i,j} x_j \right) \qquad (i = n, n-1, n-2, \dots, 1)$$

- For an $n \times n$ system of linear equations $\boldsymbol{Ax} = \boldsymbol{b}$, the forward elimination phase of the Gaussian elimination with scaled partial pivoting involves approximately $n^3/3$ long operations (multiplications or divisions), whereas the back substitution requires only n^2 long operations.

Exercises 2.2

[a]**1.** Show how Gaussian elimination with scaled partial pivoting works on the following matrix $\boldsymbol{A}$:

$$\begin{bmatrix} 2 & 3 & -4 & 1 \\ 1 & -1 & 0 & -2 \\ 3 & 3 & 4 & 3 \\ 4 & 1 & 0 & 4 \end{bmatrix}$$

[a]**2.** Solve the following system using Gaussian elimination with scaled partial pivoting:

$$\begin{bmatrix} 1 & -1 & 2 \\ -2 & 1 & -1 \\ 4 & -1 & 2 \end{bmatrix} \begin{bmatrix} x_1 \\ x_2 \\ x_3 \end{bmatrix} = \begin{bmatrix} -2 \\ 2 \\ -1 \end{bmatrix}$$

Show intermediate matrices at each step.

[a]**3.** Carry out Gaussian elimination with scaled partial pivoting on the matrix

$$\begin{bmatrix} 1 & 0 & 3 & 0 \\ 0 & 1 & 3 & -1 \\ 3 & -3 & 0 & 6 \\ 0 & 2 & 4 & -6 \end{bmatrix}$$

Show intermediate matrices.

4. Consider the matrix

$$\begin{bmatrix} -0.0013 & 56.4972 & 123.4567 & 987.6543 \\ 0.0000 & -0.0145 & 8.8990 & 833.3333 \\ 0.0000 & 102.7513 & -7.6543 & 69.6869 \\ 0.0000 & -1.3131 & -9876.5432 & 100.0001 \end{bmatrix}$$

Identify the entry that is used as the next pivot element for naive Gaussian elimination, for Gaussian elimination with partial pivoting (the scale vector is [1, 1, 1, 1]), and for Gaussian elimination with scaled partial pivoting (the scale vector is [987.6543, 46.79, 256.29, 1.096]).

[a]**5.** Without using a computer, determine the final contents of the array (a_{ij}) after procedure *Gauss* has processed the following array. Indicate the multipliers by underlining them.

$$\begin{bmatrix} 1 & 3 & 2 & 1 \\ 4 & 2 & 1 & 2 \\ 2 & 1 & 2 & 3 \\ 1 & 2 & 4 & 1 \end{bmatrix}$$

[a]**6.** If the Gaussian elimination algorithm with scaled partial pivoting is used on the matrix shown, what is the scale vector? What is the second pivot row?

$$\begin{bmatrix} 4 & 7 & 3 \\ 1 & 3 & 2 \\ 2 & -4 & -1 \end{bmatrix}$$

7. If the Gaussian elimination algorithm with scaled partial pivoting is used on the example shown, which row is selected as the third pivot row?

$$\begin{bmatrix} 8 & -1 & 4 & 9 & 2 \\ 1 & 0 & 3 & 9 & 7 \\ -5 & 0 & 1 & 3 & 5 \\ 4 & 3 & 2 & 2 & 7 \\ 3 & 0 & 0 & 0 & 9 \end{bmatrix}$$

[a]**8.** Solve the system

$$\begin{cases} 2x_1 + 4x_2 - 2x_3 = 6 \\ x_1 + 3x_2 + 4x_3 = -1 \\ 5x_1 + 2x_2 = 2 \end{cases}$$

using Gaussian elimination with scaled partial pivoting. Show intermediate results at each step; in particular, display the scale and index vectors.

9. Consider the linear system

$$\begin{cases} 2x_1 + 3x_2 = 8 \\ -x_1 + 2x_2 - x_3 = 0 \\ 3x_1 + 2x_3 = 9 \end{cases}$$

Solve for x_1, x_2, and x_3 using Gaussian elimination with scaled partial pivoting. Show intermediate matrices and vectors.

[a]**10.** Consider the linear system of equations

$$\begin{cases} -x_1 + x_2 - 3x_4 = 4 \\ x_1 + 3x_3 + x_4 = 0 \\ x_2 - x_3 - x_4 = 3 \\ 3x_1 + x_3 + 2x_4 = 1 \end{cases}$$

Solve this system using Gaussian elimination with scaled partial pivoting. Show all intermediate steps, and write down the index vector at each step.

11. Consider Gaussian elimination with scaled partial pivoting applied to the coefficient matrix

$$\begin{bmatrix} \# & \# & \# & \# & 0 \\ \# & \# & \# & 0 & \# \\ 0 & \# & \# & \# & 0 \\ 0 & \# & 0 & \# & 0 \\ \# & 0 & 0 & \# & \# \end{bmatrix}$$

where each # denotes a different nonzero element. Circle the locations of elements in which multipliers are stored and mark with an f those where fill-in occurs. The final index vector is $\boldsymbol{\ell} = [2, 3, 1, 5, 4]$.

12. Repeat Exercise 2.1.6a using Gaussian elimination with scaled partial pivoting.

13. Solve each of the following systems using Gaussian elimination with scaled partial pivoting. Carry four significant figures. What are the contents of the index array at each step?

a. $\begin{cases} 3x_1 + 4x_2 + 3x_3 = 10 \\ x_1 + 5x_2 - x_3 = 7 \\ 6x_1 + 3x_3 + 7x_3 = 15 \end{cases}$

[a]**b.** $\begin{cases} 3x_1 + 2x_2 - 5x_3 = 0 \\ 2x_1 - 3x_2 + x_3 = 0 \\ x_1 + 4x_2 - x_3 = 4 \end{cases}$

c. $\begin{bmatrix} 1 & -1 & 2 & 1 \\ 3 & 2 & 1 & 4 \\ 5 & 8 & 6 & 3 \\ 4 & 2 & 5 & 3 \end{bmatrix} \begin{bmatrix} x_1 \\ x_2 \\ x_3 \\ x_4 \end{bmatrix} = \begin{bmatrix} 1 \\ 1 \\ 1 \\ -1 \end{bmatrix}$

[a]**d.** $\begin{cases} 3x_1 + 2x_2 - x_3 = 7 \\ 5x_1 + 3x_2 + 2x_3 = 4 \\ -x_1 + x_2 - 3x_3 = -1 \end{cases}$

e. $\begin{cases} x_1 + 3x_2 + 2x_3 + x_4 = -2 \\ 4x_1 + 2x_2 + x_3 + 2x_4 = 2 \\ 2x_1 + x_2 + 2x_3 + 3x_4 = 1 \\ x_1 + 2x_2 + 4x_3 + x_4 = -1 \end{cases}$

14. Using scaled partial pivoting, show how a computer would solve the following system of equations. Show the scale array, tell how the *pivot* rows are selected, and carry out the computations. Include the index array for each step. There are no fractions in the correct solution, except for certain ratios that must be looked at to select pivots. You should follow exactly the scaled-partial-pivoting code, except that you can include the right-hand side of the system in your calculations as you go along.

$$\begin{cases} 2x_1 - x_2 + 3x_3 + 7x_4 = 15 \\ 4x_1 + 4x_2 + 7x_4 = 11 \\ 2x_1 + x_2 + x_3 + 3x_4 = 7 \\ 6x_1 + 5x_2 + 4x_3 + 17x_4 = 31 \end{cases}$$

15. Derive the formula

$$\sum_{k=1}^{n} k = \frac{n}{2}(n+1)$$

Hint: Set $S = \sum_{k=1}^{n} k$; also observe that

$$\begin{aligned} 2S &= (1+2+\cdots+n) + [n+(n-1)+\cdots+2+1] \\ &= (n+1)+(n+1)+\cdots \end{aligned}$$

or use induction.

16. Derive the formula

$$\sum_{k=1}^{n} k^2 = \frac{n}{6}(n+1)(2n+1)$$

Hint: Induction is probably easiest.

[a]**17.** Count the number of operations in the following pseudocode:

```
real array (a_ij)1:n×1:n, (x_ij)1:n×1:n
real z;  integer i, j, n
for i = 1 to n
    for j = 1 to i
        z = z + a_ij x_ij
    end for
end for
```

[a]**18.** Count the number of divisions in procedure *Gauss*. Count the number of multiplications. Count the number of additions or subtractions. Using execution times in microseconds (multiplication 1, division 2.9, addition 0.4, subtraction 0.4), write a function of n that represents the time used in these arithmetic operations.

[a]**19.** Considering long operations only and assuming 1-microsecond execution time for all long operations, give the approximate execution times and costs for procedure *Gauss* when $n = 10, 10^2, 10^3, 10^4$. Use only the dominant term in the operation count. Estimate costs at \$500 per hour.

20. (Continuation) How much time would be used on the computer to solve 2000 equations using Gaussian elimination with scaled partial pivoting? How much would it cost? Give a rough estimate based on operation times.

[a]**21.** After processing a matrix $\boldsymbol{A}$ by procedure *Gauss*, how can the results be used to solve a system of equations of form $\boldsymbol{A}^T\boldsymbol{x} = \boldsymbol{b}$?

22. What modifications would make procedure *Gauss* more efficient if division were *much* slower than multiplication?

23. The matrix $\boldsymbol{A} = (a_{ij})_{n\times n}$ is **row-equilibrated** if it is scaled so that

$$\max_{1\leqq j\leqq n} |a_{ij}| = 1 \qquad (1 \leqq i \leqq n)$$

In solving a system of equations $\boldsymbol{Ax} = \boldsymbol{b}$, we can produce an equivalent system in which the matrix is row-equilibrated by dividing the ith equation by $\max_{1\leqq j\leqq n} |a_{ij}|$.

[a]**a.** Solve the system of equations

$$\begin{bmatrix} 1 & 1 & 2\times 10^9 \\ 2 & -1 & 10^9 \\ 1 & 2 & 0 \end{bmatrix} \begin{bmatrix} x_1 \\ x_2 \\ x_3 \end{bmatrix} = \begin{bmatrix} 1 \\ 1 \\ 1 \end{bmatrix}$$

by Gaussian elimination with scaled partial pivoting.

b. Solve by using row-equilibrated naive Gaussian elimination. Are the answers the same? Why or why not?

24. Solve each system using partial pivoting and scaled partial pivoting carrying four significant digits. Also, find the true solutions.

a. $\begin{cases} 0.004000x + 69.13y = 69.17 \\ 4.281x - 5.230y = 41.91 \end{cases}$

b. $\begin{cases} 40.00x + 691300y = 691700 \\ 4.281x - 5.230y = 41.91 \end{cases}$

c. $\begin{cases} 0.003000x + 59.14y = 59.17 \\ 5.291x - 6.130y = 46.78 \end{cases}$

d. $\begin{cases} 30.00x + 591400y = 591700 \\ 5.291x - 6.130y = 46.78 \end{cases}$

e. $\begin{cases} 0.7000x + 1725y = 1739 \\ 0.4352x - 5.433y = 5.278 \end{cases}$

f. $\begin{cases} 0.8000x + 1825y = 2040 \\ 0.4321x - 5.432y = 7.531 \end{cases}$

Computer Exercises 2.2

1. Test numerical example (5) in the text using naive Gaussian algorithm and Gaussian algorithm with scaled partial pivoting.

[a]**2.** Consider the augmented matrix

$$\left[\begin{array}{cccc|c} 0.4096 & 0.1234 & 0.3678 & 0.2943 & 0.4043 \\ 0.2246 & 0.3872 & 0.4015 & 0.1129 & 0.1550 \\ 0.3645 & 0.1920 & 0.3781 & 0.0643 & 0.4240 \\ 0.1784 & 0.4002 & 0.2786 & 0.3927 & 0.2557 \end{array}\right]$$

Solve it by Gaussian elimination with scaled partial pivoting using procedures *Gauss* and *Solve*.

[a]**3.** (Continuation) Assume that an error was made when the coefficient matrix in Computer Exercise 2.2.2 was typed and that a single digit was mistyped—namely, 0.3645 became 0.3345. Solve this system, and notice the effect of this small change. Explain.

[a]**4.** The **Hilbert matrix** of order n is defined by $a_{ij} = (i + j - 1)^{-1}$ for $1 \leqq i, j \leqq n$. It is often used for test purposes because of its **ill-conditioned** nature. Define $b_i = \sum_{j=1}^{n} a_{ij}$. Then the solution of the system of equations $\sum_{j=1}^{n} a_{ij}x_j = b_i$ for $1 \leqq i \leqq n$ is $\boldsymbol{x} = [1, 1, \dots, 1]^T$. Verify this. Select some values of n in the range $2 \leqq n \leqq 15$, solve the system of equations for $\boldsymbol{x}$ using procedures *Gauss* and *Solve*, and see whether the result is as predicted. Do the case $n = 2$ by hand to see what difficulties occur in the computer.

[a]**5.** Define the $n \times n$ array (a_{ij}) by $a_{ij} = -1 + 2\max\{i, j\}$. Set up array (b_i) in such a way that the solution of the system $\boldsymbol{Ax} = \boldsymbol{b}$ is $x_i = 1$ for $1 \leqq i \leqq n$. Test procedures *Gauss* and *Solve* on this system for a moderate value of n, say, $n = 30$.

[a]**6.** Select a modest value of n, say, $5 \leqq n \leqq 20$, and let $a_{ij} = (i - 1)^{j-1}$ and $b_i = i - 1$. Solve the system $\boldsymbol{Ax} = \boldsymbol{b}$ on the computer. By looking at the output, guess what the correct solution is. Establish algebraically that your guess is correct. Account for the errors in the computed solution.

7. For a fixed value of n from 2 to 4, let

$$a_{ij} = (i+j)^2, \qquad b_i = ni(i+n+1)+\tfrac{1}{6}n(1+n(2n+3))$$

Show that the vector $\boldsymbol{x} = [1, 1, \dots, 1]^T$ solves the system $\boldsymbol{Ax} = \boldsymbol{b}$. Test whether procedures *Gauss* and *Solve* can compute $\boldsymbol{x}$ correctly for $n = 2, 3, 4$. Explain what happens.

8. Using each value of n from 2 to 9, solve the $n \times n$ system $\boldsymbol{Ax} = \boldsymbol{b}$, where $\boldsymbol{A}$ and $\boldsymbol{b}$ are defined by

$$a_{ij} = (i + j - 1)^7, \qquad b_i = p(n + i - 1) - p(i - 1)$$

where

$$p(x) = \frac{x^2}{24}(2 + x^2(-7 + n^2(14 + n(12 + 3n))))$$

Explain what happens.

9. Solve the following augmented matrix using procedures *Gauss* and *Solve* and then using procedure *Naive_Gauss*. Compare the results and explain.

$$\left[\begin{array}{cccc|c} 0.0001 & -5.0300 & 5.8090 & 7.8320 & 9.5740 \\ 2.2660 & 1.9950 & 1.2120 & 8.0080 & 7.2190 \\ 8.8500 & 5.6810 & 4.5520 & 1.3020 & 5.7300 \\ 6.7750 & -2.2530 & 2.9080 & 3.9700 & 6.2910 \end{array}\right]$$

10. Without changing the parameter list, rewrite and test procedure *Gauss* so that it does both forward elimination and back substitution. Increase the size of array (a_{ij}), and store the right-hand side array (b_i) in the $n + 1$st column of (a_{ij}). Also, return the solution in this column.

11. Modify procedures *Gauss* and *Solve* so that they are more robust. Two suggested changes are as follows: (i) skip elimination if $a_{\ell_i,k} = 0$ and (ii) add an error parameter *ierr* to the parameter list and perform error checking (e.g., on division by zero or a row of zeros). Test the modified code on linear systems of varying sizes.

12. Rewrite procedures *Gauss* and *Solve* so that they are column oriented—that is, so that all inner loops vary the first index of (a_{ij}). On some computer systems, this implementation may avoid paging or swapping between high-speed and secondary memory and be more efficient for large matrices.

13. Computer memory can be minimized by using a different storage mode when the coefficient matrix is symmetric. An $n \times n$ symmetric matrix $\boldsymbol{A} = (a_{ij})$ has the property that $a_{ij} = a_{ji}$, so only the elements on and below the main diagonal need to be stored in a vector of length $n(n + 1)/2$. The elements of the matrix $\boldsymbol{A}$ are placed in a vector $\boldsymbol{v} = (v_k)$ in this order: $a_{11}, a_{21}, a_{22}, a_{31}, a_{32}, a_{33}, \dots, a_{n,n}$. Storing a matrix in this way is known as **symmetric storage mode** and affects a savings of $n(n - 1)/2$ memory locations. Here, $a_{ij} = v_k$, where $k = \frac{1}{2}i(i - 1) + j$ for $i \geqq j$. Verify these statements.

Write and test procedures

$$Gauss\_Sym(n, (v_i), (\ell_i))$$
$$Solve\_Sym(n, (v_i), (\ell_i), (b_i))$$

which are analogous to procedures *Gauss* and *Solve*, except that the coefficient matrix is stored in symmetric storage mode in a one-dimensional array (v_i) and the solution is returned in array (b_i).

14. The **determinant** of a square matrix can be easily computed with the help of procedure *Gauss*. We require three facts about determinants. First, the determinant of a triangular matrix is the product of the elements on its diagonal. Second, if a multiple of one row is added to another row, the determinant of the matrix does not change. Third, if two rows in a matrix are interchanged, the determinant changes sign. Procedure *Gauss* can be *interpreted* as a procedure for reducing a matrix to upper triangular form by interchanging rows and adding multiples of one row to another. Write a function $\det(n, (a_{ij}))$ that computes the determinant of an $n \times n$ matrix. It will call procedure *Gauss* and utilize the arrays (a_{ij}) and (ℓ_i) that result from that call. Numerically verify function det by using the following test matrices with several values of n:

a. $a_{ij} = |i - j| \qquad \text{Det}(\boldsymbol{A}) = (-1)^{n-1}(n-1)2^{n-2}$

b. $a_{ij} = \begin{cases} 1 & j \geqq i \\ -j & j < i \end{cases} \qquad \text{Det}(\boldsymbol{A}) = n!$

c. $\begin{cases} a_{ij} = a_{j1} = n^{-1} & j \geqq 1 \\ a_{ij} = a_{i-1,\,j} + a_{i,\,j-1} & i, j \geqq 2 \end{cases} \qquad \text{Det}(\boldsymbol{A}) = n^{-n}$

15. (Continuation) Overflow and underflow may occur in evaluating determinants by this procedure. To avoid this, one can compute $\log|\text{Det}(\boldsymbol{A})|$ as the sum of terms $\log|a_{\ell_i,i}|$ and use the exponential function at the end. Repeat the numerical experiments in Computer Exercise 2.2.14 using this idea.

16. Test a modification of procedure *Gauss* in which the scale array is recomputed at each step (each new value of k) of the forward elimination phase. Try to construct an example for which this procedure would produce less roundoff error than the scaled partial pivoting method given in the text with fixed-scale array. It is generally believed that the extra computations that are involved in this procedure are not worthwhile for most linear systems.

17. (Continuation) Modify and test procedure *Gauss* so that the original system is initially row-equilibrated; that is, it is scaled so that the maximum element in every row is 1.

18. Modify and test procedures *Gauss* and *Solve* so that they carry out scaled *complete* pivoting; that is, the pivot element is selected from all elements in the submatrix, not just those in the kth column. Keep track of the order of the unknowns in the solution array in another index array because they will not be determined in the order $x_n, x_{n-1}, \ldots, x_1$.

19. Compare the computed numerical solutions of two 5×5 linear systems with coefficent matrices:

$$\begin{bmatrix} 1 & 1/2 & 1/3 & 1/4 & 1/5 \\ 1/2 & 1/3 & 1/4 & 1/5 & 1/6 \\ 1/3 & 1/4 & 1/5 & 1/6 & 1/7 \\ 1/4 & 1/5 & 1/6 & 1/7 & 1/8 \\ 1/5 & 1/6 & 1/7 & 1/8 & 1/9 \end{bmatrix}$$

$$\begin{bmatrix} 1.0 & 0.5 & 0.333333 & 0.25 & 0.2 \\ 0.5 & 0.333333 & 0.25 & 0.2 & 0.166667 \\ 0.333333 & 0.25 & 0.2 & 0.166667 & 0.142857 \\ 0.25 & 0.2 & 0.166667 & 0.142857 & 0.125 \\ 0.2 & 0.166667 & 0.142857 & 0.125 & 0.111111 \end{bmatrix}$$

and both with the right-hand side vector $\mathbf{b} = [1, 0, 0, 0, 0]^T$. Solve both systems using single-precision Gaussian elimination with scaled partial pivoting. For each system, compute the ℓ_2-norms $||\boldsymbol{u}||_2 = \sqrt{\sum_{i=1}^{n} u_i^2}$ of the **residual vector** $\widetilde{\boldsymbol{r}} = \boldsymbol{A}\widetilde{\boldsymbol{x}} - \boldsymbol{b}$ and of the **error vector** $\widetilde{\boldsymbol{e}} = \widetilde{\boldsymbol{x}} - \boldsymbol{x}$, where $\widetilde{\boldsymbol{x}}$ is the computed solution and $\boldsymbol{x}$ is the true, or exact, solution. For the first system, the exact solution is $\boldsymbol{x} = [25, -300, 1050, -1400, 630]^T$, and for the second system, the exact solution, to six decimal digits of accuracy, is $\boldsymbol{x} = [26.9314, -336.018, 1205.11, -1634.03, 744.411]^T$. Do not change the input data of the second system to include more than the number of digits shown. Analyze the results. What have you learned?

20. (Continuation) Repeat the preceding computer problem, but set

$$a_{ij} \leftarrow 7560a_{ij}; \qquad b_i \leftarrow 7560b_i$$

for each system before solving.

21. Write complex arithmetic versions of procedures *Gauss* and *Solve* by declaring certain variables complex and making other necessary changes in the code. Test them on the complex linear systems given in Computer Exercise 2.1.6.

22. (Continuation) Solve the complex linear systems given in Computer Exercises 2.1.7.

23. The fact that in the previous two problem solutions of complex linear systems were asked for may lead you to think that you *must* have complex versions of procedures *Gauss* and *Solve*. This is not the case. A complex system $\boldsymbol{A}\boldsymbol{x} = \boldsymbol{b}$ can also be written as a $2n \times 2n$ real system $(1 \leqq i \leqq n)$:

$$\sum_{j=1}^{n} \left[\text{Re}(a_{ij})\text{Re}(x_j) - \text{Im}(a_{ij})\text{Im}(x_j)\right] = \text{Re}(b_i)$$

$$\sum_{j=1}^{n} \left[\text{Re}(a_{ij})\text{Im}(x_j) + \text{Im}(a_{ij})\text{Re}(x_j)\right] = \text{Im}(b_i)$$

Repeat these two problems using this idea and the two procedures of this section. (Here, Re denotes the real part and Im the imaginary part.)

24. (**Student Research Project**) The **Gauss-Huard algorithm** is a variant of the **Gauss-Jordan algorithm** for solving dense linear systems. Both algorithms reduce the system to an equivalent diagonal system. However, the Gauss-Jordan method does more floating-point operations than Gaussian elimination, whereas the Gauss-Huard method does not. To preserve stability, the Gauss-Huard method incorporates a pivoting strategy using column interchanges. An error analysis shows that the Gauss-Huard method is as stable as Gauss-Jordan elimination with an appropriate pivoting strategy. Read about these algorithms in papers by Dekker and Hoffmann [1989], Dekker, Hoffmann, and Potma [1997], Hoffmann [1989], and Huard [1979]. Carry out some numerical experiments by programming and testing the Gauss-Jordan and Gauss-Huard algorithms on some dense linear systems.

25. Solve System (5) using mathematical software routines based on Gaussian elimination such as found in MATLAB, Maple, or Mathematica. There are many computer programs and software packages for solving linear systems, each of which may use a slightly different pivoting strategy.

2.3 Tridiagonal and Banded Systems

In many applications, including several that are considered later on, we will encounter extremely large linear systems that have a **banded** structure. Banded matrices often occur in solving ordinary and partial differential equations. It is advantageous to develop computer codes specifically designed for such linear systems, because they reduce the amount of storage used.

Of practical importance is the **tridiagonal** system. Here, all the nonzero elements in the coefficient matrix must be on the main diagonal or on the two diagonals just above and below the main diagonal (usually called **superdiagonal** and **subdiagonal**, respectively):

Tridiagonal $n \times n$ System

$$\begin{bmatrix} d_1 & c_1 & & & & & & \\ a_1 & d_2 & c_2 & & & & & \\ & a_2 & d_3 & c_3 & & & & \\ & & \ddots & \ddots & \ddots & & & \\ & & & a_{i-1} & d_i & c_i & & \\ & & & & \ddots & \ddots & \ddots & \\ & & & & & a_{n-2} & d_{n-1} & c_{n-1} \\ & & & & & & a_{n-1} & d_n \end{bmatrix} \begin{bmatrix} x_1 \\ x_2 \\ x_3 \\ \vdots \\ x_i \\ \vdots \\ x_{n-1} \\ x_n \end{bmatrix} = \begin{bmatrix} b_1 \\ b_2 \\ b_3 \\ \vdots \\ b_i \\ \vdots \\ b_{n-1} \\ b_n \end{bmatrix} \tag{1}$$

(In the coefficient matrix, all elements *not* in the displayed diagonals are 0's.) A **tridiagonal** matrix is characterized by the condition $a_{ij} = 0$ if $|i - j| \geqq 2$. In general, a matrix is said to have a **banded structure** if there is an integer k (less than n) such that $a_{ij} = 0$ whenever $|i - j| \geqq k$ with **bandwidth** $2k - 1$.

The storage requirements for a banded matrix are less than those for a general matrix of the same size. Thus, an $n \times n$ **diagonal** matrix requires only n memory locations in the computer, and a **tridiagonal** matrix requires only $3n - 2$. This fact is important if banded matrices of very large order are being used.

For banded matrices, the Gaussian elimination algorithm can be made very efficient, if it is known beforehand that pivoting is unnecessary. This situation occurs often enough to justify special procedures. Here, we develop a code for the tridiagonal system and give a listing for the **pentadiagonal** system (in which $a_{ij} = 0$ if $|i - j| \geqq 3$).

Tridiagonal Systems

Now we describe a procedure called *Tri*. It is designed to solve a system of n linear equations in n unknowns, as shown in Equation (1). Both the forward elimination phase and the back substitution phase are incorporated into the same procedure, and *no* pivoting is used; that is, the pivot equations are those given by the natural ordering $\{1, 2, \dots, n\}$. Thus, naive Gaussian elimination is used.

In **Step 1**, we subtract a_1/d_1 times row 1 from row 2, thus creating a 0 in the a_1 position. Only the entries d_2 and b_2 are altered. Observe that c_2 is *not* altered.

In **Step 2**, the process is repeated, using the new row 2 as the pivot row. Here is how the d_i's and b_i's are altered in each step:

$$\begin{cases} d_2 \leftarrow d_2 - \left(\dfrac{a_1}{d_1}\right) c_1 \\[2ex] b_2 \leftarrow b_2 - \left(\dfrac{a_1}{d_1}\right) b_1 \end{cases}$$

In general, we obtain

Forward Eliminations

$$\begin{cases} d_i \leftarrow d_i - \left(\dfrac{a_{i-1}}{d_{i-1}}\right) c_{i-1} \\[2ex] b_i \leftarrow b_i - \left(\dfrac{a_{i-1}}{d_{i-1}}\right) b_{i-1} \qquad (2 \leqq i \leqq n) \end{cases}$$

At the end of the forward elimination phase, the form of the system is as follows:

Upper Bi-Diagonal $n \times n$ System

$$\begin{bmatrix} d_1 & c_1 & & & & & \\ & d_2 & c_2 & & & & \\ & & d_3 & c_3 & & & \\ & & & \ddots & \ddots & & \\ & & & & d_i & c_i & \\ & & & & & \ddots & \ddots \\ & & & & & d_{n-1} & c_{n-1} \\ & & & & & & d_n \end{bmatrix} \begin{bmatrix} x_1 \\ x_2 \\ x_3 \\ \vdots \\ x_i \\ \vdots \\ x_{n-1} \\ x_n \end{bmatrix} = \begin{bmatrix} b_1 \\ b_2 \\ b_3 \\ \vdots \\ b_i \\ \vdots \\ b_{n-1} \\ b_n \end{bmatrix}$$

Of course, the b_i's and d_i's are *not* as they were at the beginning of this process, but the c_i's are.

The **back substitution** phase solves for $x_n, x_{n-1}, \dots, x_1$ as follows:

Back Substitution

$$\begin{aligned} x_n &\leftarrow \frac{b_n}{d_n} \\ x_{n-1} &\leftarrow \frac{1}{d_{n-1}}(b_{n-1} - c_{n-1}x_n) \\ &\vdots \\ x_1 &\leftarrow \frac{1}{d_1}(b_1 - c_1 x_2) \end{aligned}$$

In general we obtain

Solution

$$
\begin{aligned}
x_n &\leftarrow b_n/d_n \\
x_i &\leftarrow \frac{1}{d_i}(b_i - c_i x_{i+1}) \qquad (i = n-1, n-2, \dots, 1)
\end{aligned}
$$

In procedure *Tri* for a tridiagonal system, we use $n \times 1$ single-dimensioned arrays (a_i), (d_i), and (c_i) for the diagonals in the coefficient matrix and $n \times 1$ array (b_i) for the right-hand side, and store the solution in $n \times 1$ array (x_i).

***Tri* Procedure Pseudocode**

```
procedure Tri(n, (a_i), (d_i), (c_i), (b_i), (x_i))
integer i, n;   real xmult
real array (a_i)_{1:n}, (d_i)_{1:n}, (c_i)_{1:n}, (b_i)_{1:n}, (x_i)_{1:n}
for i = 2 to n
    xmult ← a_{i-1}/d_{i-1}
    d_i ← d_i − (xmult)c_{i-1}
    b_i ← b_i − (xmult)b_{i-1}
end for
x_n ← b_n/d_n
for i = n − 1 to 1
    x_i ← (b_i − c_i x_{i+1})/d_i
end for
end procedure Tri
```

Notice that the original data in arrays (d_i) and (b_i) have been changed.

A symmetric tridiagonal system arises in the cubic spline development of Chapter 6 and elsewhere. A general **symmetric tridiagonal** system has the form

Symmetric $n \times n$ Tridiagonal System

$$
\begin{bmatrix}
d_1 & c_1 & & & & & & \\
c_1 & d_2 & c_2 & & & & & \\
 & c_2 & d_3 & c_3 & & & & \\
 & & \ddots & \ddots & \ddots & & & \\
 & & & c_{i-1} & d_i & c_i & & \\
 & & & & \ddots & \ddots & \ddots & \\
 & & & & & c_{n-2} & d_{n-1} & c_{n-1} \\
 & & & & & & c_{n-1} & d_n
\end{bmatrix}
\begin{bmatrix} x_1 \\ x_2 \\ x_3 \\ \vdots \\ x_i \\ \vdots \\ x_{n-1} \\ x_n \end{bmatrix}
=
\begin{bmatrix} b_1 \\ b_2 \\ b_3 \\ \vdots \\ b_i \\ \vdots \\ b_{n-1} \\ b_n \end{bmatrix}
\qquad (2)
$$

One could overwrite the right-hand side vector $\boldsymbol{b}$ with the solution vector $\boldsymbol{x}$ as well. Thus, a symmetric linear system can be solved with a procedure call of the form

Key Procedure Call

```
call Tri(n, (c_i), (d_i), (c_i), (b_i), (b_i))
```

which reduces the number of linear arrays from five to three!

Strictly Diagonal Dominance

Because procedure *Tri* does *not* involve pivoting, it is natural to ask whether it is likely to fail. Simple examples can be given to illustrate failure because of attempted division by zero even though the coefficient matrix in Equation (1) is invertible (nonsingular). On the other hand, it is *not* easy to give the weakest possible conditions on this matrix to guarantee the success of the algorithm. We content ourselves with one property that is easily checked and commonly encountered. If the tridiagonal coefficient matrix is diagonally dominant, then procedure *Tri* will *not* encounter zero divisors.

■ Definition 1 **Strictly Diagonal Dominance**

Strictly Diagonally Dominant

A general matrix $\boldsymbol{A} = (a_{ij})_{n\times n}$ is **strictly diagonally dominant** if

$$|a_{ii}| > \sum_{\substack{j=1\\ j\neq i}}^{n} |a_{ij}| \qquad (1 \leqq i \leqq n)$$

Tridiagonal System Case In the case of the tridiagonal system of Equation (1), strict diagonal dominance means simply that (with $a_0 = a_n = 0$)

$$|d_i| > |a_{i-1}| + |c_i| \qquad (1 \leqq i \leqq n)$$

Let us verify that the forward elimination phase in procedure *Tri* preserves strictly diagonal dominance. The new coefficient matrix produced by Gaussian elimination has 0 elements where the a_i's originally stood, and new diagonal elements are determined recursively by

$$\begin{cases} \widehat{d}_1 = d_1 \\ \widehat{d}_i = d_i - \left(\dfrac{a_{i-1}}{\widehat{d}_{i-1}}\right) c_{i-1} \qquad (2 \leqq i \leqq n) \end{cases}$$

where $\widehat{d}_i$ denotes new diagonal elements. The c_i elements are unaltered. Now we assume that $|d_i| > |a_{i-1}| + |c_i|$, and we want to be sure that $|\widehat{d}_i| > |c_i|$. Obviously, this is true for $i = 1$ because $\widehat{d}_1 = d_1$. If it is true for index $i - 1$ (that is, $|\widehat{d}_{i-1}| > |c_{i-1}|$), then it is true for index i because

$$\begin{aligned} \left|\widehat{d}_i\right| &= \left|d_i - \left(\frac{a_{i-1}}{\widehat{d}_{i-1}}\right) c_{i-1}\right| \\ &\geqq |d_i| - |a_{i-1}| \frac{|c_{i-1}|}{\left|\widehat{d}_{i-1}\right|} \\ &> |a_{i-1}| + |c_i| - |a_{i-1}| = |c_i| \end{aligned}$$

Although the number of long operations in Gaussian elimination on full matrices is $\mathcal{O}(n^3)$, it is only $\mathcal{O}(n)$ for tridiagonal matrices. Also, the scaled pivoting strategy is *not* needed on strictly diagonally dominant tridiagonal systems.

Pentadiagonal Systems

The principles illustrated by procedure *Tri* can be applied to matrices that have wider bands of nonzero elements. A procedure called *Penta* is given here to solve the five-diagonal system:

Pentadiagonal $n \times n$ System

$$\begin{bmatrix} d_1 & c_1 & f_1 & & & & & & \\ a_1 & d_2 & c_2 & f_2 & & & & & \\ e_1 & a_2 & d_3 & c_3 & f_3 & & & & \\ & e_2 & a_3 & d_4 & c_4 & f_4 & & & \\ & & \ddots & \ddots & \ddots & \ddots & \ddots & & \\ & & & e_{i-2} & a_{i-1} & d_i & c_i & f_i & \\ & & & & \ddots & \ddots & \ddots & \ddots & \ddots \\ & & & & & e_{n-4} & a_{n-3} & d_{n-2} & c_{n-2} & f_{n-2} \\ & & & & & & e_{n-3} & a_{n-2} & d_{n-1} & c_{n-1} \\ & & & & & & & e_{n-2} & a_{n-1} & d_n \end{bmatrix} \begin{bmatrix} x_1 \\ x_2 \\ x_3 \\ x_4 \\ \vdots \\ x_i \\ \vdots \\ x_{n-2} \\ x_{n-1} \\ x_n \end{bmatrix} = \begin{bmatrix} b_1 \\ b_2 \\ b_3 \\ b_4 \\ \vdots \\ b_i \\ \vdots \\ b_{n-2} \\ b_{n-1} \\ b_n \end{bmatrix}$$

In the pseudocode, the solution vector is placed in an $n \times 1$ array (x_i). Also, one should *not* use this routine if $n \leqq 4$. (Why?)

***Penta* Procedure Pseudocode**

procedure *Penta*$(n, (e_i), (a_i), (d_i), (c_i), (f_i), (b_i), (x_i))$
integer i, n; **real** $r, s, xmult$
real array $(e_i)_{1:n}, (a_i)_{1:n}, (d_i)_{1:n}, (c_i)_{1:n}, (f_i)_{1:n}, (b_i)_{1:n}, (x_i)_{1:n}$
$r \leftarrow a_1$
$s \leftarrow a_2$
$t \leftarrow e_1$
for $i = 2$ **to** $n - 1$
  $xmult \leftarrow r/d_{i-1}$
  $d_i \leftarrow d_i - (xmult)c_{i-1}$
  $c_i \leftarrow c_i - (xmult)f_{i-1}$
  $b_i \leftarrow b_i - (xmult)b_{i-1}$
  $xmult \leftarrow t/d_{i-1}$
  $r \leftarrow s - (xmult)c_{i-1}$
  $d_{i+1} \leftarrow d_{i+1} - (xmult)f_{i-1}$
  $b_{i+1} \leftarrow b_{i+1} - (xmult)b_{i-1}$
  $s \leftarrow a_{i+1}$
  $t \leftarrow e_i$
end for
$xmult \leftarrow r/d_{n-1}$
$d_n \leftarrow d_n - (xmult)c_{n-1}$
$x_n \leftarrow (b_n - (xmult)b_{n-1})/d_n$
$x_{n-1} \leftarrow (b_{n-1} - c_{n-1}x_n)/d_{n-1}$
for $i = n - 2$ **to** 1
  $x_i \leftarrow (b_i - f_i x_{i+2} - c_i x_{i+1})/d_i$
end for
end procedure *Penta*

To be able to solve symmetric pentadiagonal systems with the same code and with a minimum of storage, we have used variables r, s, and t to store temporarily some information rather than overwriting into arrays. This allows us to solve a symmetric pentadiagonal system with a procedure call of the form

Symmetric *Penta* Key Call

call $Penta(n, (f_i), (c_i), (d_i), (c_i), (f_i), (b_i), (b_i))$

This reduces the number of linear arrays from seven to four! Of course, the original data in some of these arrays may be corrupted. The computed solution may be stored in the (b_i) array. Here, we assume that all linear arrays are padded with zeros to length n in order *not* to exceed the array dimensions in the pseudocode.

Block Pentadiagonal Systems

Many mathematical problems involve matrices with block structures. In many cases, there are advantages in exploiting the block structure in the numerical solution. This is particularly true in solving partial differential equations numerically as in Section 12.3.

We can consider a pentadiagonal system as a block tridiagonal system

Block Pentadiagonal $n \times n$ System

$$\begin{bmatrix} \boldsymbol{D}_1 & \boldsymbol{C}_1 & & & & & & \\ \boldsymbol{A}_1 & \boldsymbol{D}_2 & \boldsymbol{C}_2 & & & & & \\ & \boldsymbol{A}_2 & \boldsymbol{D}_3 & \boldsymbol{C}_3 & & & & \\ & & \ddots & \ddots & \ddots & & & \\ & & & \boldsymbol{A}_{i-1} & \boldsymbol{D}_i & \boldsymbol{C}_i & & \\ & & & & \ddots & \ddots & \ddots & \\ & & & & & \boldsymbol{A}_{n-2} & \boldsymbol{D}_{n-1} & \boldsymbol{C}_{n-1} \\ & & & & & & \boldsymbol{A}_{n-1} & \boldsymbol{D}_n \end{bmatrix} \begin{bmatrix} \boldsymbol{X}_1 \\ \boldsymbol{X}_2 \\ \boldsymbol{X}_3 \\ \vdots \\ \boldsymbol{X}_i \\ \vdots \\ \boldsymbol{X}_{n-1} \\ \boldsymbol{X}_n \end{bmatrix} = \begin{bmatrix} \boldsymbol{B}_1 \\ \boldsymbol{B}_2 \\ \boldsymbol{B}_3 \\ \vdots \\ \boldsymbol{B}_i \\ \vdots \\ \boldsymbol{B}_{n-1} \\ \boldsymbol{B}_n \end{bmatrix}$$

where

$$\boldsymbol{D}_i = \begin{bmatrix} d_{2i-1} & c_{2i-1} \\ a_{2i-1} & d_{2i} \end{bmatrix}, \qquad \boldsymbol{A}_i = \begin{bmatrix} e_{2i-1} & c_{2i-1} \\ 0 & e_{2i} \end{bmatrix}, \qquad \boldsymbol{C}_i = \begin{bmatrix} f_{2i-1} & 0 \\ c_{2i-1} & f_{2i} \end{bmatrix}$$

Here, we assume that n is even, say $n = 2m$. If n is *not* even, then the system can be padded with an extra equation $x_{n+1} = 1$ so that the number of rows is even.

The algorithm for this block tridiagonal system is similar to the one for tridiagonal systems. Hence, we have the **forward elimination** phase

Forward Elimination

$$\begin{cases} \boldsymbol{D}_i \leftarrow \boldsymbol{D}_i - \boldsymbol{A}_{i-1}\boldsymbol{D}_{i-1}^{-1}\boldsymbol{C}_{i-1} \\ \boldsymbol{B}_i \leftarrow \boldsymbol{B}_i - \boldsymbol{A}_{i-1}\boldsymbol{D}_{i-1}^{-1}\boldsymbol{B}_{i-1} \qquad (2 \leqq i \leqq m) \end{cases}$$

and the **back substitution** phase

Back Substitution

$$\begin{cases} \boldsymbol{X}_n \leftarrow \boldsymbol{D}_n^{-1}\boldsymbol{B}_n \\ \boldsymbol{X}_i \leftarrow \boldsymbol{D}_i^{-1}(\boldsymbol{B}_i - \boldsymbol{C}_i\boldsymbol{X}_{i+1}) \qquad (m-1 \leqq i \leqq 1) \end{cases}$$

Here, we let

$$D_i^{-1} = \frac{1}{\Delta}\begin{bmatrix} d_{2i} & -c_{2i-1} \\ -a_{2i-1} & d_{2i-1} \end{bmatrix}$$

where $\Delta = d_{2i}d_{2i-1} - a_{2i-1}c_{2i-1}$.

Code for solving a pentadiagonal system using this block procedure is left as Computer Exercise 2.3.21. The results from the block pentadiagonal code are the same as those from the procedure *Penta*, except for roundoff error. Also, this procedure can be used for symmetric pentadiagonal systems (in which the subdiagonals are the same as the superdiagonals).

In Section 12.3, we discuss two-dimensional elliptic partial differential equations. For example, the **Laplace equation** is defined on the unit square with a 3×3 mesh of grid points placed over the unit square region which are ordered in the natural ordering (left-to-right and up) as shown in Figure 2.5.

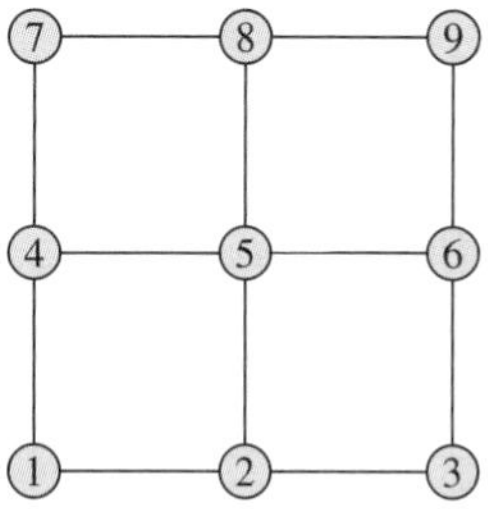

FIGURE 2.5
Mesh points in natural order

In the Laplace equation, the second-order partial derivatives are approximated by second-order centered finite difference formulas. This results in a 9×9 system of linear equations having a sparse coefficient matrix with this nonzero pattern:

Sample Sparse 9×9 System

$$A = \left[\begin{array}{ccc|ccc|ccc}
\times & \times & & \times & & & & & \\
\times & \times & \times & & \times & & & & \\
& \times & \times & & & \times & & & \\
\hline
\times & & & \times & \times & & \times & & \\
& \times & & \times & \times & \times & & \times & \\
& & \times & & \times & \times & & & \times \\
\hline
& & & \times & & & \times & \times & \\
& & & & \times & & \times & \times & \times \\
& & & & & \times & & \times & \times
\end{array}\right]$$

Here, the nonzero entries in the matrix are indicated by the $\times$ symbol, and the zero entries are a blank. This matrix is block tridiagonal, and each nonzero block is either tridiagonal or diagonal. Other orderings of the mesh points result in sparse matrices with different patterns.

Summary 2.3

- In many applications, tridiagonal, pentadiagonal, and other banded systems are solved using special algorithms use Gaussian elimination without pivoting. The **forward elimination** procedure for a tridiagonal linear system $A = \text{Tridiagonal}[(a_i), (d_i), (c_i)]$ is

$$\begin{cases} d_i \leftarrow d_i - \left(\dfrac{a_{i-1}}{d_{i-1}}\right) c_{i-1} \\ b_i \leftarrow b_i - \left(\dfrac{a_{i-1}}{d_{i-1}}\right) b_{i-1} \qquad (2 \leqq i \leqq n) \end{cases}$$

The **back substitution** procedure is

$$x_i \leftarrow \frac{1}{d_i}(b_i - c_i x_{i+1}) \qquad (i = n-1, n-2, \dots, 1)$$

- A **strictly diagonally dominant** matrix $A = (a_{ij})_{n \times n}$ is one in which the magnitude of the diagonal entry is larger than the sum of the magnitudes of the off-diagonal entries in the same row, and this is true for all rows, namely,

$$|a_{ii}| > \sum_{\substack{j=1 \\ j \neq i}}^{n} |a_{ij}| \qquad (1 \leqq i \leqq n)$$

For strictly diagonally dominant tridiagonal coefficient matrices, partial pivoting is *not* necessary because zero divisors will *not* be encountered.

- The forward elimination and back substitution procedures for a pentadiagonal linear system $A =$ Pentadiagonal $[(e_i), (a_i), (d_i), (c_i), (f_i)]$ is similar to that for a tridiagonal system.

Exercises 2.3

1. What happens to the tridiagonal System (1) if Gaussian elimination with partial pivoting is used to solve it? In general, what happens to a banded system?

2. Count the long arithmetic operations involved in procedures:

   [a]**a.** *Tri* **b.** *Penta*

[a]**3.** How many storage locations are needed for a system of n linear equations if the coefficient matrix has banded structure in which $a_{ij} = 0$ for $|i - j| \geqq k + 1$?

4. Give an example of a system of linear equations in tridiagonal form that cannot be solved without pivoting.

5. What is the appearance of a matrix A if its elements satisfy $a_{ij} = 0$ when:

   a. $j < i - 2$ **b.** $j > i + 1$

[a]**6.** Consider a strictly diagonally dominant matrix A whose elements satisfy $a_{ij} = 0$ when $i > j + 1$. Does Gaussian elimination without pivoting preserve the strictly diagonal dominance? Why or why not?

[a]**7.** Let A be a matrix of form (1) such that $a_i c_i > 0$ for $1 \leqq i \leqq n - 1$. Find the general form of the diagonal matrix $D = \text{Diag}(\alpha_i)$ with $\alpha_i \neq 0$ such that $D^{-1}AD$ is symmetric. What is the general form of $D^{-1}AD$?

Computer Exercises 2.3

1. Rewrite procedure *Tri* using only four arrays, (a_i), (d_i), (c_i), and (b_i), and storing the solution in the (b_i) array. Test the code with both a nonsymmetric and a symmetric tridiagonal system.

2. Repeat the previous computer problem for procedure *Penta* with six arrays (e_i), (a_i), (d_i), (c_i), (f_i), and (b_i). Use the example that begins this chapter as one of the test cases.

[a]**3.** Write and test a special procedure to solve the tridiagonal system in which $a_i = c_i = 1$ for all i.

[a]**4.** Use procedure *Tri* to solve the following system of 100 equations. Compare the numerical solution to the obvious exact solution.

$$\begin{cases} x_1 \quad + 0.5x_2 \quad = 1.5 \\ 0.5x_{i-1} + \quad x_i \quad + 0.5x_{i+1} = 2.0 \quad (2 \leqq i \leqq 99) \\ \quad 0.5x_{99} + \quad x_{100} = 1.5 \end{cases}$$

5. Solve the system

$$\begin{cases} 4x_1 \quad - \quad x_2 \quad = -20 \\ x_{j-1} - 4x_j \quad + \quad x_{j+1} = \quad 40 \quad (2 \leqq j \leqq n-1) \\ \quad - \quad x_{n-1} + 4x_n \quad = -20 \end{cases}$$

using procedure *Tri* with $n = 100$.

6. Let $\boldsymbol{A}$ be the 50×50 tridiagonal matrix

$$\begin{bmatrix} 5 & -1 & & & & \\ -1 & 5 & -1 & & & \\ & -1 & 5 & -1 & & \\ & & \ddots & \ddots & \ddots & \\ & & & -1 & 5 & -1 \\ & & & & -1 & 5 \end{bmatrix}$$

Consider the problem $\boldsymbol{Ax} = \boldsymbol{b}$ for 50 different vectors $\boldsymbol{b}$ of the form

$$[1, 2, \ldots, 49, 50]^T, \quad [2, 3, \ldots, 50, 1]^T,$$

$$[3, 4, \ldots, 50, 1, 2]^T, \quad \ldots$$

Write and test an efficient code for solving this problem. *Hint*: Rewrite procedure *Tri*.

7. Rewrite and test procedure *Tri* so that it performs Gaussian elimination with scaled partial pivoting.
Hint: Additional temporary storage arrays may be needed.

8. Rewrite and test *Penta* so that it does Gaussian elimination with scaled partial pivoting. Is this worthwhile?

9. Using the ideas illustrated in *Penta*, write a procedure for solving seven-diagonal systems. Test it on several such systems.

10. Consider the system of equations ($n = 7$)

$$\begin{bmatrix} d_1 & & & & & & a_7 \\ & d_2 & & & & a_6 & \\ & & d_3 & & a_5 & & \\ & & & d_4 & & & \\ & & a_3 & & d_5 & & \\ & a_2 & & & & d_6 & \\ a_1 & & & & & & d_7 \end{bmatrix} \begin{bmatrix} x_1 \\ x_2 \\ x_3 \\ x_4 \\ x_5 \\ x_6 \\ x_7 \end{bmatrix} = \begin{bmatrix} b_1 \\ b_2 \\ b_3 \\ b_4 \\ b_5 \\ b_6 \\ b_7 \end{bmatrix}$$

For n odd, write and test

procedure *X_Gauss*$(n, (a_i), (d_i), (b_i))$

that does the forward elimination phase of Gaussian elimination (without scaled partial pivoting) and

procedure *X_Solve*$(n, (a_i), (d_i), (b_i), (x_i))$

that does the back substitution for cross-systems of this form.

11. Consider the $n \times n$ lower-triangular system $\boldsymbol{Ax} = \boldsymbol{b}$, where $\boldsymbol{A} = (a_{ij})$ and $a_{ij} = 0$ for $i < j$.

[a]**a.** Write an algorithm (in mathematical terms) for solving for $\boldsymbol{x}$ by forward substitution.

b. Write

procedure *Forward_Sub*$(n, (a_i), (b_i), (x_i))$

which uses this algorithm.

c. Determine the number of divisions, multiplications, and additions (or subtractions) in using this algorithm to solve for $\boldsymbol{x}$.

d. Should Gaussian elimination with partial pivoting be used to solve such a system?

[a]**12.** (**Normalized Tridiagonal Algorithm**) Construct an algorithm for handling tridiagonal systems in which the normalized Gaussian elimination procedure without pivoting is used. In this process, each pivot row is divided by the diagonal element before a multiple of the row is subtracted from the successive rows. Write the equations involved in the forward elimination phase and store the upper diagonal entries back in array (c_i) and the right-hand side entries back in array (b_i). Write the equations for the back substitution phase, storing the solution in array (b_i). Code and test this procedure. What are its advantages and disadvantages?

13. For a $(2n) \times (2n)$ tridiagonal system, write and test a procedure that proceeds as follows: In the forward elimination phase, the routine simultaneously eliminates the elements in the subdiagonal from the top to the middle and in the superdiagonal from the bottom to the middle. In the back substitution phase, the unknowns are determined two at a time from the middle outward.

14. (Continuation) Rewrite and test the procedure in the preceding computer problem for a general $n \times n$ tridiagonal matrix.

15. Suppose

> **procedure** *Tri_Normal*$(n, (a_i), (d_i), (c_i), (b_i), (x_i))$

performs the normalized Gaussian elimination algorithm of Computer Exercise 2.3.12 and

> **procedure** *Tri_2n*$(n, (a_i), (d_i), (c_i), (b_i), (x_i))$

performs the algorithm outlined in Computer Exercise 2.3.13. Using a timing routine on your computer, compare *Tri*, *Tri_Normal*, and *Tri_2n* to determine which of them is fastest for the tridiagonal system

$$\begin{cases} a_i = i(n-i+1), & c_i = (i+1)(n-i-1), \\ d_i = (2i+1)n - i - 2i, & b_i = i \end{cases}$$

with a large even value of n.
Note: Mathematical algorithms may behave differently on parallel and vector computers. Generally speaking, parallel computations completely alter our conventional notions about what's best or most efficient.

16. Consider a special bidiagonal linear system of the following form (illustrated with $n = 7$) with nonzero diagonal elements:

$$\begin{bmatrix} d_1 & & & & & & \\ a_1 & d_2 & & & & & \\ & a_2 & d_3 & & & & \\ & & a_3 & d_4 & a_4 & & \\ & & & & d_5 & a_5 & \\ & & & & & d_6 & a_6 \\ & & & & & & d_7 \end{bmatrix} \begin{bmatrix} x_1 \\ x_2 \\ x_3 \\ x_4 \\ x_5 \\ x_6 \\ x_7 \end{bmatrix} = \begin{bmatrix} b_1 \\ b_2 \\ b_3 \\ b_4 \\ b_5 \\ b_6 \\ b_7 \end{bmatrix}$$

Write and test

> **procedure** *Bi_Diagonal*$(n, (a_i), (d_i), (b_i))$

to solve the general system of order n (odd). Store the solution in array b, and assume that all arrays are of length n. *Do not* use forward elimination because the system can be solved quite easily without it.

17. Write and test

> **procedure** *Backward_Tri*$(n, (a_i), (d_i), (c_i), (b_i), (x_i))$

for solving a backward tridiagonal system of linear equations of the form

$$\begin{bmatrix} & & & & a_1 & d_1 \\ & & & a_2 & d_2 & c_1 \\ & & a_3 & d_3 & c_2 & \\ & \iddots & \iddots & \iddots & & \\ a_{n-1} & d_{n-1} & c_{n-1} & & & \\ d_n & c_{n-1} & & & & \end{bmatrix} \begin{bmatrix} x_1 \\ x_2 \\ x_3 \\ \vdots \\ x_{n-1} \\ x_n \end{bmatrix} = \begin{bmatrix} b_1 \\ b_2 \\ b_3 \\ \vdots \\ b_{n-1} \\ b_n \end{bmatrix}$$

using Gaussian elimination without pivoting.

18. An upper **Hessenberg** matrix is of the form

$$\begin{bmatrix} a_{11} & a_{12} & a_{13} & \cdots & a_{1n} \\ a_{21} & a_{22} & a_{23} & \cdots & a_{2n} \\ & a_{32} & a_{33} & \cdots & a_{3n} \\ & & \ddots & \ddots & \vdots \\ & & & a_{n,n-1} & a_{nn} \end{bmatrix} \begin{bmatrix} x_1 \\ x_2 \\ x_3 \\ \vdots \\ x_n \end{bmatrix} = \begin{bmatrix} b_1 \\ b_2 \\ b_3 \\ \vdots \\ b_n \end{bmatrix}$$

Write a procedure for solving such a system, and test it on a system having 10 or more equations.

19. An $n \times n$ banded coefficient matrix with ℓ subdiagonals and m superdiagonals can be stored in **banded storage mode** in an $n \times (\ell + m + 1)$ array. The matrix is stored with the row and diagonal structure preserved with almost all 0 elements unstored. If the original $n \times n$ banded matrix had the form shown in the figure, then the $n \times (\ell + m + 1)$ array in banded storage mode would be as shown. The main diagonal would be the $\ell + 1$st column of the new array. Write and test a procedure for solving a linear system with the coefficient matrix stored in banded storage mode.

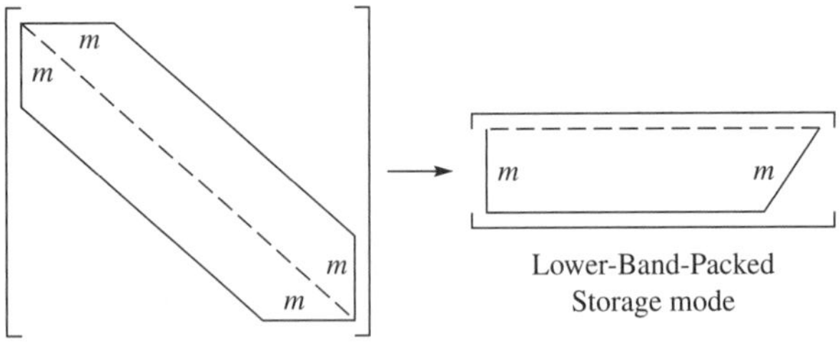

Symmetric banded array

Lower-Band-Packed Storage mode

20. An $n \times n$ symmetric banded coefficient matrix with m subdiagonals and m superdiagonals can be stored in **symmetric banded storage mode** in an $n \times (m + 1)$ array. Only the main diagonal and subdiagonals are stored so that the main diagonal is the last column in the new array. Write and test a procedure for solving a linear system with the coefficient matrix stored in symmetric banded storage mode.

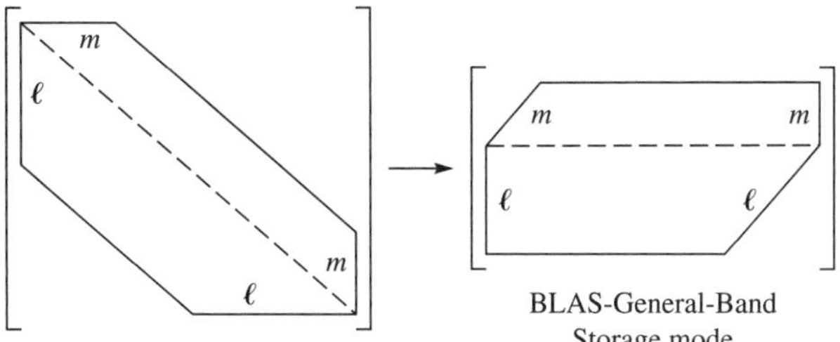

Banded array

BLAS-General-Band Storage mode

21. Write code for solving block pentadiagonal systems and test it on the systems with block submatrices. Compare the code to *Penta* using symmetric and nonsymmetric systems.

22. (**Nonperiodic Spline Filter**) The filter equation for the nonperiodic spline filter is given by the $n \times n$ system

$$\left(I + \alpha^4 \boldsymbol{Q}\right) \boldsymbol{w} = \boldsymbol{z}$$

where the matrix is

$$\boldsymbol{Q} = \begin{bmatrix} 1 & -2 & 1 & & & & & \\ -2 & 5 & -4 & 1 & & & & \\ 1 & -4 & 6 & -4 & 1 & & & \\ & \ddots & \ddots & \ddots & \ddots & \ddots & & \\ & & & 1 & -4 & 6 & -4 & 1 \\ & & & & 1 & -4 & 5 & -2 \\ & & & & & 1 & -2 & 1 \end{bmatrix}$$

Here the parameter $\alpha = 1/[2\sin(\pi\,\Delta x/\lambda_c)]$ involves measurement values of the profile, dimensions, and wavelength over a sampling interval. The solution $\boldsymbol{w}$ gives the profile values for the long wave components, and $\boldsymbol{z} - \boldsymbol{w}$ are those for the short wave components. Use this system to test the *Penta* code using various values of α.
Hint: For test systems, select a simple solution vector $\boldsymbol{w} = [1, -1, 1, -1, \ldots, 1]^T$ with a modest value for n, and then compute the right-hand side using matrix-vector multiplication $\boldsymbol{z} = (I + \alpha^4 \boldsymbol{Q})\boldsymbol{w}$.

23. (Continuation, **Periodic Spline Filter**) The filter equation for the periodic spline filter is given by the $n \times n$ system

$$\left(I + \alpha^4 \widehat{\boldsymbol{Q}}\right) \widehat{\boldsymbol{w}} = \widehat{\boldsymbol{z}}$$

where the matrix is

$$\widehat{\boldsymbol{Q}} = \begin{bmatrix} 6 & -4 & 1 & & & & 1 & -4 \\ -4 & 6 & -4 & 1 & & & & 1 \\ 1 & -4 & 6 & -4 & 1 & & & \\ & \ddots & \ddots & \ddots & \ddots & \ddots & & \\ & & 1 & -4 & 6 & -4 & 1 & \\ 1 & & & 1 & -4 & 6 & -4 & \\ -4 & 1 & & & 1 & -4 & 6 & \end{bmatrix}$$

Periodic spline filters are used in cases of filtering closed profiles. Making use of the symmetry, modify the *Penta* pseudocode to handle this system and then code and test it.

24. Use mathematical software such as MATLAB, Maple, or Mathematica to generate a tridiagonal system and solve it. For example, use the 5×5 tridiagonal system $\boldsymbol{A} = \text{Band\_Matrix}(-1, 2, 1)$ with right-hand side $\boldsymbol{b} = [1, 4, 9, 16, 25]^T$.

3 Nonlinear Equations

An electric power cable is suspended (at points of equal height) from two towers that are 100 meters apart. The cable is allowed to dip 10 meters in the middle. How long is the cable?

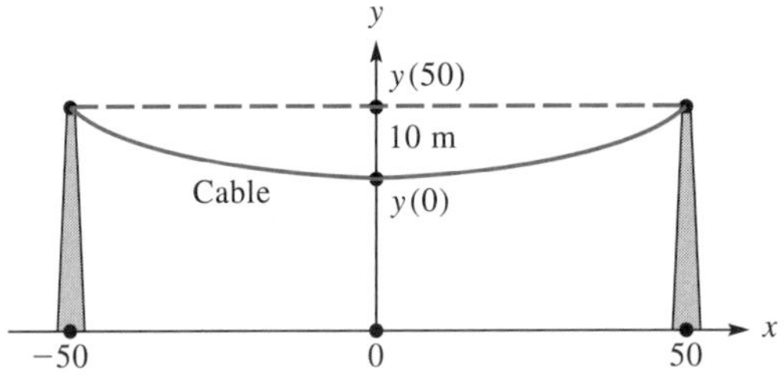

It is known that the curve assumed by a suspended cable is a **catenary**. When the y-axis passes through the lowest point, we can assume an equation of the form $y = \lambda \cosh(x/\lambda)$. Here λ is a parameter to be determined. The conditions of the problem are that $y(50) = y(0) + 10$. Hence, we obtain

$$\lambda \cosh\left(\frac{50}{\lambda}\right) = \lambda + 10$$

By the methods of this chapter, the parameter is found to be $\lambda = 126.632$. After this value is substituted into the arc length formula of the catenary, the length is determined to be 102.619 meters. (See Computer Exercise 5.1.4.)

3.1 Bisection Method

Introduction

Let f be a real- or complex-valued function of a real or complex variable. A number r, real or complex, for which $f(r) = 0$ is called a **root** of that equation or a **zero** of f. For example, the function

Sample Functions

$$f(x) = 6x^2 - 7x + 2$$

has $\frac{1}{2}$ and $\frac{2}{3}$ as zeros, as can be verified by direct substitution or by writing f in its factored form:

$$f(x) = (2x - 1)(3x - 2)$$

For another example, the function

$$g(x) = \cos 3x - \cos 7x$$

has not only the obvious zero $x = 0$, but every integer multiple of $\pi/5$ and of $\pi/2$ as well, which we discover by applying the trigonometric identity

$$\cos A - \cos B = 2\sin\left[\frac{1}{2}(a+b)\right]\sin\left[\frac{1}{2}(b-a)\right]$$

Consequently, we find

$$g(x) = 2\sin(5x)\sin(2x)$$

Why is locating roots important?

Frequently, the solution to a scientific problem is a number about which we have little information other than that it satisfies some equation. Since every equation can be written so that a function stands on one side and zero on the other, the desired number must be a zero of the function. Thus, if we possess an arsenal of methods for locating zeros of functions, we shall be able to solve such problems.

We illustrate this claim by use of a specific engineering problem whose solution is the root of an equation. In a certain electrical circuit, the voltage V and current I are related by two equations of the form

Engineering Problem

$$\begin{cases} I = a(e^{bV} - 1) \\ c = dI + V \end{cases}$$

in which a, b, c, and d are constants. For our purpose, these four numbers are assumed to be known. When these equations are combined by eliminating I between them, the result is a single equation:

$$c = ad(e^{bV} - 1) + V$$

In a concrete case, this might reduce to

$$12 = 14.3(e^{2V} - 1) + V$$

and its solution is required. (It turns out that $V \approx 0.299$ in this case.)

In some problems in which a root of an equation is sought, we can perform the required calculation with a hand calculator. But how can we locate zeros of complicated functions such as these?

More Samples Functions

$$\begin{aligned} f(x) &= 3.24x^8 - 2.42x^7 + 10.34x^6 + 11.01x^2 + 47.98 \\ g(x) &= 2^{x^2} - 10x + 1 \\ h(x) &= \cosh\left(\sqrt{x^2+1} - e^x\right) + \log|\sin x| \end{aligned}$$

What is needed is a general numerical method that does not depend on special properties of our functions. Of course, continuity and differentiability are special properties, but they are common attributes of functions that are usually encountered. The sort of special property that we probably *cannot* easily exploit in general-purpose codes is typified by the trigonometric identity mentioned previously.

Hundreds of methods are available for locating zeros of functions, and three of the most useful have been selected for study here: the bisection method, Newton's method, and the secant method.

Let f be a function that has values of opposite sign at the two ends of an interval. Suppose also that f is continuous on that interval. To fix the notation, let $a < b$ and $f(a)f(b) < 0$. It then follows that f has a root in the interval (a, b). In other words, there must exist a number r that satisfies the two conditions $a < r < b$ and $f(r) = 0$. *How is this conclusion reached?* One must recall the **Intermediate-Value Theorem**.* If x traverses an interval $[a, b]$, then the values of $f(x)$ completely fill out the interval between $f(a)$ and $f(b)$. No intermediate values can be skipped. Hence, a specific function f must take on the value zero somewhere in the interval (a, b) because $f(a)$ and $f(b)$ are of opposite signs.

Bisection Algorithm

Method Description

The **bisection method** exploits this property of continuous functions. At each step in this algorithm, we have an interval $[a, b]$ and the values $u = f(a)$ and $v = f(b)$. The numbers u and v satisfy $uv < 0$. Next, we construct the midpoint of the interval, $c = \frac{1}{2}(a + b)$, and compute $w = f(c)$. It can happen fortuitously that $f(c) = 0$. If so, the objective of the algorithm has been fulfilled. In the usual case, $w \neq 0$, and either $wu < 0$ or $wv < 0$. (Why?) If $wu < 0$, we can be sure that a root of f exists in the interval $[a, c]$. Consequently, we store the value of c in b and w in v. If $wu > 0$, then we cannot be sure that f has a root in $[a, c]$, but since $wv < 0$, f must have a root in $[c, b]$. In this case, we store the value of c in a and w in u. In either case, the situation at the end of this step is just like that at the beginning except that the final interval is *half* as large as the initial interval. This step can now be repeated until the interval is satisfactorily small, say $|b - a| < \frac{1}{2} \times 10^{-6}$. At the end, the best estimate of the root would be $(a+b)/2$, where $[a, b]$ is the last interval in the procedure.

Pseudocode

Now let's construct pseudocode to carry out this procedure. We shall not try to create a piece of high-quality software with many "bells and whistles," but we write the pseudocode in the form of a procedure for general use. This allows the reader an opportunity to review how a main program and one or more procedures can be connected.

As a general rule, in programming routines to locate the roots of arbitrary functions, unnecessary evaluations of the function should be avoided because a given function may be costly to evaluate in terms of computer time. Thus, any value of the function that may be needed later should be stored rather than recomputed. A careless programming of the bisection method might violate this principle.

The procedure to be constructed operates on an arbitrary function f. An interval $[a, b]$ is also specified, and the number of steps to be taken, *nmax*, is given. Pseudocode to perform *nmax* steps of the bisection algorithm follows:

```
procedure Bisection(f, a, b, nmax, ε)
integer n, nmax;   real a, b, c, fa, fb, fc, error
fa ← f(a)
fb ← f(b)
if sign(fa) = sign(fb) then
    output a, b, fa, fb
    output "function has same signs at a and b"
    return
```

(*Continued*)

*__Intermediate-Value Theorem__: If the function f is continuous on the closed interval $[a, b]$, and if $f(a) \leqq y \leqq f(b)$ or $f(b) \leqq y \leqq f(a)$, then there exists a point c such that $a \leqq c \leqq b$ and $f(c) = y$.

Bisection Pseudocode

```
end if
error ← b − a
for n = 0 to nmax
    error ← error/2
    c ← a + error
    fc ← f(c)
    output n, c, fc, error
    if |error| < ε then
        output "convergence"
        return
    end if
    if sign(fa) ≠ sign(fc) then
        b ← c
        fb ← fc
    else
        a ← c
        fa ← fc
    end if
end for
end procedure Bisection
```

Here many modifications are incorporated to enhance the pseudocode. For example, we use *fa*, *fb*, *fc* as mnemonics for u, v, w, respectively. Also, we illustrate some techniques of structured programming and some other alternatives, such as a test for convergence. For example, if u, v, or w is close to zero, then uv or wu may underflow. Similarly, an overflow situation may arise. A test involving the intrinsic function *sign* could be used to avoid these difficulties, such as a test that determines whether $\text{sign}(u) \neq \text{sign}(v)$. Here, the iterations terminate if they exceed *nmax* or if the error bound (discussed later in this section) is less than ε. The reader should trace the steps in the routine to see that it does what is claimed!

Numerical Examples

Now we want to illustrate how the bisection pseudocode can be used. Suppose that we have two functions, and for each, we seek a zero in a specified interval:

Find a Root for Each Function f and g

$$f(x) = x^3 - 3x + 1, \quad \text{on } [0, 1]$$
$$g(x) = x^3 - 2\sin x, \quad \text{on } [0.5, 2]$$

First, we write two procedure functions to compute $f(x)$ and $g(x)$. Then we input the initial intervals and the number of steps to be performed in a main program. Since this is a rather simple example, this information can be assigned directly in the main program or by way of statements in the subprograms rather than being read into the program. Also, depending on the computer language being used, an external or interface statement is needed to tell the compiler that the parameter f in the bisection procedure is *not* an ordinary variable with numerical values, but the name of a function procedure defined externally to the main program. In this example, there are two function procedures and two calls to the bisection procedure.

A main program follows which calls the bisection routine for each of these functions:

***Test_Bisection* Pseudocode**

```
program Test_Bisection
integer n, nmax ← 20
real a, b, ε ← ½10^−6
external function f, g
a ← 0.0
b ← 1.0
call Bisection (f, a, b, nmax, ε)
a ← 0.5
b ← 2.0
call Bisection (g, a, b, nmax, ε)
end program Test_Bisection

real function f(x)
real x
f ← x^3 − 3x + 1
end function f

real function g(x)
real x
g ← x^3 − 2 sin x
end function g
```

Here are the computer results with the iterative steps of the bisection method for $f(x)$:

$f(x)$ Output

n	c_n	$f(c_n)$	error
0	0.5	−0.375	0.5
1	0.25	0.266	0.25
2	0.375	-7.23×10^{-2}	0.125
3	0.3125	9.30×10^{-2}	6.25×10^{-2}
4	0.34375	9.37×10^{-3}	3.125×10^{-2}
⋮			
19	0.34729 67	-9.54×10^{-7}	9.54×10^{-7}
20	0.34729 62	3.58×10^{-7}	4.77×10^{-7}

Also, the results for $g(x)$ are as follows:

$g(x)$ Output

n	c_n	$g(c_n)$	error
0	1.25	5.52×10^{-2}	0.75
1	0.875	−0.865	0.375
2	1.0625	−0.548	0.188
3	1.15625	−0.285	9.38×10^{-2}
4	1.20312 5	−0.125	4.69×10^{-2}
⋮			
19	1.23618 27	-4.88×10^{-6}	1.43×10^{-6}
20	1.23618 34	-2.15×10^{-6}	7.15×10^{-7}

Mathematical Software

To verify these results, we use sophisticated procedures in mathematical software such as MATLAB, Mathematica, or Maple to find the desired roots of f and g to be 0.34729 63553 and 1.23618 3928, respectively. Since f is a polynomial, we can use a routine for finding numerical approximations to all the zeros of a polynomial function. However, when more complicated nonpolynomial functions are involved, there is generally no systematic procedure for finding all zeros. In this case, a routine can be used to search for zeros (one at a time), but we have to specify a point at which to start the search, and different starting points may result in the same or different zeros. It may be particularly troublesome to find all the zeros of a function whose behavior is unknown.

Convergence Analysis

Now let us investigate the *accuracy* with which the bisection method determines a root of a function. Suppose that f is a continuous function that takes values of opposite sign at the ends of an interval $[a_0, b_0]$. Then there is a root r in $[a_0, b_0]$, and if we use the midpoint $c_0 = (a_0 + b_0)/2$ as our estimate of r, we have

$$|r - c_0| \leqq \frac{b_0 - a_0}{2}$$

as illustrated in Figure 3.1.

FIGURE 3.1
Bisection method: Illustrating error upper bound

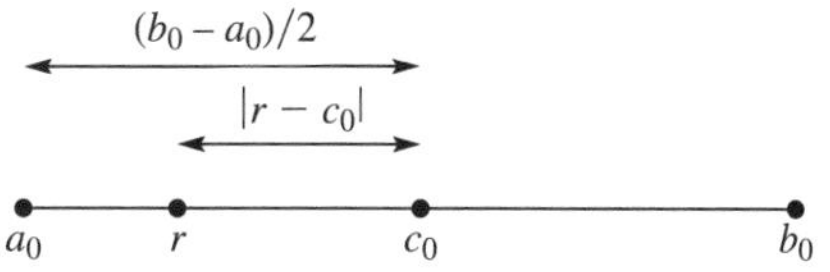

If the bisection algorithm is now applied and if the computed quantities are denoted by $a_0, b_0, c_0, a_1, b_1, c_1$, and so on, then by the same reasoning,

$$|r - c_n| \leqq \frac{b_n - a_n}{2} \qquad (n \geqq 0)$$

Since the widths of the intervals are divided by 2 in each step, we conclude that

Error Bound

$$|r - c_n| \leqq \frac{b_0 - a_0}{2^{n+1}} \tag{1}$$

To summarize, a theorem can be written as follows:

■ **Theorem 1**

Bisection Method Theorem

If the bisection algorithm is applied to a continuous function f on an interval $[a, b]$, where $f(a)f(b) < 0$, then, after n steps, an approximate root will have been computed with error at most $(b - a)/2^{n+1}$.

If an error tolerance has been prescribed in advance, it is possible to determine the number of steps required in the bisection method. Suppose that we want

$$|r - c_n| < \varepsilon$$

Then it is necessary to solve the following inequality for n:

$$\frac{b-a}{2^{n+1}} < \varepsilon$$

By taking logarithms (with any convenient base), we obtain

$$n > \frac{\log(b-a) - \log(2\varepsilon)}{\log 2} \tag{2}$$

EXAMPLE 1 How many steps of the bisection algorithm are needed to compute a root of f to full machine single precision on a 32-bit word-length computer if $a = 16$ and $b = 17$?

Solution The root is between the two binary numbers $a = (10\,000.0)_2$ and $b = (10\,001.0)_2$. Thus, we already know five of the binary digits in the answer. Since we can use only 24 bits altogether, that leaves 19 bits to determine. We want the *last* one to be correct, so we want the error to be less than 2^{-19} or 2^{-20} (being conservative). Since a 32-bit word-length computer has a 24-bit mantissa, we can expect the answer to have an accuracy of only 2^{-20}. From the equation above, we want

$$(b-a)/2^{n+1} < \varepsilon$$

Since $b - a = 1$ and $\varepsilon = 2^{-20}$, we have $1/2^{n+1} < 2^{-20}$. Taking reciprocals gives $2^{n+1} > 2^{20}$, or $n \geqq 20$.

Alternatively, we can use Inequality (2), which in this case is

$$n > \frac{\log 1 - \log 2^{-19}}{\log 2}$$

Using a basic property of logarithms ($\log x^y = y \log x$), we find that $n \geqq 20$. In this example, each step of the algorithm determines the root with one additional binary digit of precision. ■

A sequence $\{x_n\}$ exhibits **linear convergence** to a limit x if there is a constant C in the interval $[0, 1)$ such that

Linear Convergence

$$|x_{n+1} - x| \leqq C|x_n - x| \qquad (n \geqq 1) \tag{3}$$

If this inequality is true for all n, then

$$|x_{n+1} - x| \leqq C|x_n - x| \leqq C^2|x_{n-1} - x| \leqq \cdots \leqq C^n|x_1 - x|$$

Thus, it is a consequence of linear convergence that

Error Bound (Linear Convergence)

$$|x_{n+1} - x| \leqq AC^n \qquad (0 \leqq C < 1) \tag{4}$$

The sequence produced by the bisection method obeys Inequality (4), as we see from Inequality (1). However, the sequence need *not* obey Inequality (3).

The bisection method is the simplest way to solve a nonlinear equation $f(x) = 0$. It arrives at the root by constraining the interval in which a root lies, and it eventually makes the interval quite small. Because the bisection method halves the width of the interval at each step, one can predict exactly how long it takes to find the root within any desired degree of accuracy. In the bisection method, not every guess is closer to the root than the previous guess because the bisection method does not use the nature of the function itself. Often the bisection method is used to get close to the root before switching to a faster method. (Root finding by the bisection method uses the same idea as in the **binary search** method taught in data structures.)

False Position (Regula Falsi) Method

The **false position method** retains the main feature of the bisection method: that a root is trapped in a sequence of intervals of decreasing size. Rather than selecting the midpoint of each interval, this method uses the point where the **secant lines** intersect the x-axis.

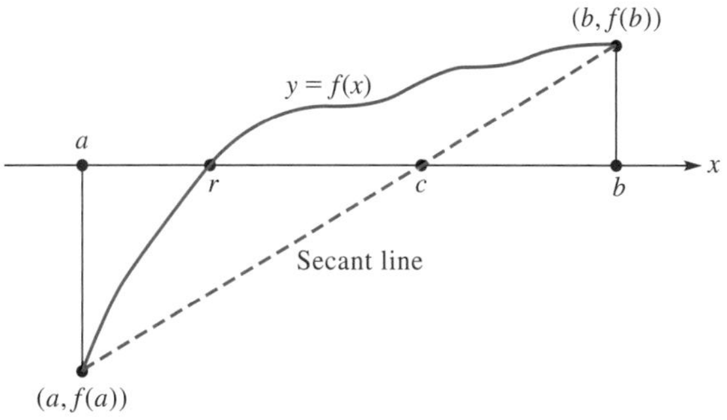

FIGURE 3.2
False position method

In Figure 3.2, the secant line over the interval $[a, b]$ is the chord between $(a, f(a))$ and $(b, f(b))$. The two right triangles in the figure are *similar*, which means that

$$\frac{b-c}{f(b)} = \frac{c-a}{-f(a)}$$

It is easy to show that

$$c = b - f(b)\left[\frac{a-b}{f(a)-f(b)}\right] = a - f(a)\left[\frac{b-a}{f(b)-f(a)}\right] = \frac{af(b)-bf(a)}{f(b)-f(a)}$$

We then compute $f(c)$ and proceed to the next step with the interval $[a, c]$ if $f(a)f(c) < 0$ or to the interval $[c, b]$ if $f(c)f(b) < 0$.

FP Method Description

In the general case, the **false position method** starts with the interval $[a_0, b_0]$ containing a root: $f(a_0)$ and $f(b_0)$ are of opposite signs. The false position method uses intervals $[a_k, b_k]$ that contain roots in almost the same way that the bisection method does. However, instead of finding the midpoint of the interval, it finds where the secant line joining $(a_k, f(a_k))$ and $(b_k, f(b_k))$ crosses the x-axis and then selects it to be the new endpoint. At the kth step, it computes

$$c_k = \frac{a_k f(b_k) - b_k f(a_k)}{f(b_k) - f(a_k)}$$

If $f(a_k)$ and $f(c_k)$ have the same sign, then set $a_{k+1} = c_k$ and $b_{k+1} = b_k$; otherwise, set $a_{k+1} = a_k$ and $b_{k+1} = c_k$. The process is repeated until the root is approximated sufficiently well.

Modified False Position Method

For some functions, the false position method may repeatedly select the same endpoint, and the process may degrade to linear convergence. There are various approaches to rectify this.

For example, when the same endpoint is to be retained twice, the **modified false position method** uses

MFP Method Description

$$c_k^{(m)} = \begin{cases} \dfrac{a_k f(b_k) - 2b_k f(a_k)}{f(b_k) - 2f(a_k)}, & \text{if } f(a_k)f(b_k) < 0 \\ \dfrac{2a_k f(b_k) - b_k f(a_k)}{2f(b_k) - f(a_k)}, & \text{if } f(a_k)f(b_k) > 0 \end{cases}$$

So rather than selecting points on the same side of the root as the regular false position method does, the modified false position method changes the slope of the straight line so that it is closer to the root. See Figure 3.3.

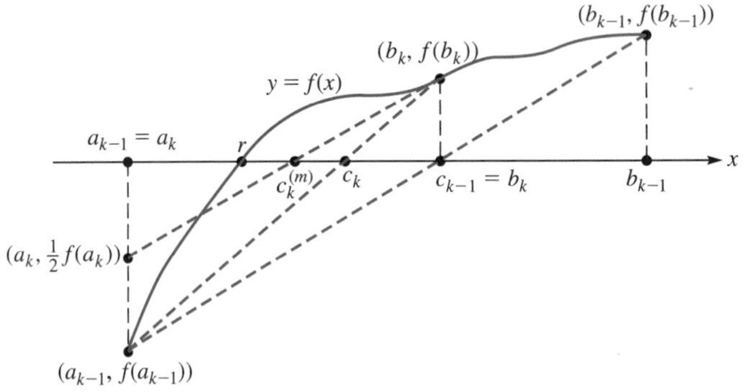

FIGURE 3.3 Modified false position method

The bisection method uses only the fact that $f(a)f(b) < 0$ for each new interval $[a, b]$, but the false position method uses the *values* of $f(a)$ and $f(b)$. This is an example showing how to include additional information in an algorithm to build a better one. In the next section, Newton's method uses not only the function, but also its first derivative.

Some variants of the modified false position procedure have superlinear convergence, which we discuss in Section 3.3. (See, for example, Ford [1995].) Another modified false position method replaces the secant lines by straight lines with ever smaller slope until the iterate falls to the opposite side of the root. (See Conte and de Boor [1980].) Early versions of the false position method date back to a Chinese mathematical text (200 B.C.E. to 100 C.E.) and an Indian mathematical text (3 B.C.E.).

Summary 3.1

- For finding a zero r of a given continuous function f in an interval $[a, b]$, n steps of the bisection method produces a sequence of intervals $[a, b] = [a_0, b_0], [a_1, b_1], [a_2, b_2], \ldots, [a_n, b_n]$ with each containing the desired root of the function. The midpoints of these intervals $c_0, c_1, c_2, \ldots, c_n$ form a sequence of approximations to the root, namely, $c_i = \frac{1}{2}(a_i + b_i)$. On each interval $[a_i, b_i]$, the error $e_i = r - c_i$ obeys the inequality

$$|e_i| \leqq \frac{1}{2}(b_i - a_i)$$

and after n steps we have

$$|e_n| \leqq \frac{1}{2^{n+1}}(b_0 - a_0)$$

- For an error tolerance ε such that $|e_n| < \varepsilon$, n steps are needed, where n satisfies the inequality

$$n > \frac{\log(b-a) - \log 2\varepsilon}{\log 2}$$

- For the kth step of the **false position method** over the interval $[a_k, b_k]$, let

$$c_k = \frac{a_k f(b_k) - b_k f(a_k)}{f(b_k) - f(a_k)}$$

If $f(a_k)f(c_k) > 0$, set $a_{k+1} = c_k$ and $b_{k+1} = b_k$; otherwise, set $a_{k+1} = a_k$ and $b_{k+1} = c_k$.

Exercises 3.1

[a]**1.** Find where the graphs of $y = 3x$ and $y = e^x$ intersect by finding roots of $e^x - 3x = 0$ correct to four decimal digits.

2. Give a graphical demonstration that the equation $\tan x = x$ has infinitely many roots. Determine one root precisely and another approximately by using a graph. *Hint*: Use the approach of the preceding exercise.

3. Demonstrate graphically that the equation $50\pi + \sin x = 100 \arctan x$ has infinitely many solutions.

[a]**4.** By graphical methods, locate approximations to all roots of the nonlinear equation $\ln(x+1) + \tan(2x) = 0$.

5. Give an example of a function for which the bisection method does *not* converge linearly.

6. Draw a graph of a function that is discontinuous yet the bisection method converges. Repeat, getting a function for which it diverges.

7. Prove Inequality (1).

8. If $a = 0.1$ and $b = 1.0$, how many steps of the bisection method are needed to determine the root with an error of at most $\frac{1}{2} \times 10^{-8}$?

[a]**9.** Find all the roots of $f(x) = \cos x - \cos 3x$. Use two different methods.

[a]**10.** (Continuation) Find the root or roots of $\ln[(1+x)/(1-x^2)] = 0$.

11. If f has an inverse, then the equation $f(x) = 0$ can be solved by simply writing $x = f^{-1}(0)$. Does this remark eliminate the problem of finding roots of equations? Illustrate with $\sin x = 1/\pi$.

[a]**12.** How many binary digits of precision are gained in each step of the bisection method? How many steps are required for each decimal digit of precision?

13. Try to devise a stopping criterion for the bisection method to guarantee that the root is determined with *relative* error at most ε.

14. Denote the successive intervals that arise in the bisection method by $[a_0, b_0]$, $[a_1, b_1]$, $[a_2, b_2]$, and so on. Show that

a. $a_0 \leqq a_1 \leqq a_2 \leqq \cdots$ and $b_0 \geqq b_1 \geqq b_2 \geqq \cdots$.

b. $b_n - a_n = 2^{-n}(b_0 - a_0)$.

c. $a_n b_n + a_{n-1} b_{n-1} = a_{n-1} b_n + a_n b_{n-1}$, for all n.

15. (Continuation) Can $a_0 = a_1 = a_2 = \cdots$ happen?

16. (Continuation) Let $c_n = (a_n + b_n)/2$. Show that

$$\lim_{n\to\infty} c_n = \lim_{n\to\infty} a_n = \lim_{n\to\infty} b_n$$

[a]**17.** (Continuation) Consider the bisection method with the initial interval $[a_0, b_0]$. Show that after 10 steps with this method,

$$\left|\frac{1}{2}(a_{10} + b_{10}) - \frac{1}{2}(a_9 + b_9)\right| = 2^{-11}(b_0 - a_0)$$

Also, determine how many steps are required to guarantee an approximation of a root to six decimal places (rounded).

18. (**True-False**) If the bisection method generates intervals $[a_0, b_0]$, $[a_1, b_1]$, and so on, which of these inequalities are true for the root r that is being calculated? Give proofs or counterexamples in each case.

a. $|r - a_n| \leqq 2|r - b_n|$

[a]**b.** $|r - a_n| \leqq 2^{-n-1}(b_0 - a_0)$

c. $|r - \frac{1}{2}(a_n + b_n)| \leqq 2^{-n-2}(b_0 - a_0)$

[a]**d.** $0 \leqq r - a_n \leqq 2^{-n}(b_0 - a_0)$

e. $|r - b_n| \leqq 2^{-n-1}(b_0 - a_0)$

19. (**True-False**) Using the notation of the text, determine which of these assertions are true and which are generally false:

[a]**a.** $|r - c_n| < |r - c_{n-1}|$ **b.** $a_n \leqq r \leqq c_n$

c. $c_n \leqq r \leqq b_n$ **d.** $|r - a_n| \leqq 2^{-n}$

[a]**e.** $|r - b_n| \leqq 2^{-n}(b_0 - a_0)$

20. Prove that $|c_n - c_{n+1}| = 2^{-n-2}(b_0 - a_0)$.

[a]21. If the bisection method is applied with starting interval $[a, a + 1]$ and $a = 2^m$, where $n \geqq 24 - m0$, what is the correct number of steps to compute the root with full machine precision on a 32-bit word-length computer?

22. If the bisection method is applied with starting interval $[2^m, 2^{m+1}]$, where m is a positive or negative integer, how many steps should be taken to compute the root to full machine precision on a 32-bit word-length computer?

[a]23. Every polynomial of degree n has n zeros (counting multiplicities) in the complex plane. Does every real polynomial have n real zeros? Does every *polynomial of infinite degree* $f(x) = \sum_{n=0}^{\infty} a_n x^n$ have infinitely many zeros?

Computer Exercises 3.1

1. Using the bisection method, determine the point of intersection of the curves given by $y = x^3 - 2x + 1$ and $y = x^2$.

2. Find a root of the following equation in the interval $[0, 1]$ by using the bisection method: $9x^4 + 18x^3 + 38x^2 - 57x + 14 = 0$.

3. Find a root of the equation $\tan x = x$ on the interval $[4, 5]$ by using the bisection method. What happens on the interval $[1, 2]$?

4. Find a root of the equation $6(e^x - x) = 6 + 3x^2 + 2x^3$ between -1 and $+1$ using the bisection method.

5. Use the bisection method to find a zero of the equation $\lambda \cosh(50/\lambda) = \lambda + 10$ that begins this chapter.

6. Program the bisection method as a recursive procedure and test it on one or two of the examples in the text.

7. Use the bisection method to determine roots of these functions on the intervals indicated. Process all three functions in *one* computer run.

$$\begin{cases} f(x) = x^3 + 3x - 1 & \text{on } [0, 1] \\ g(x) = x^3 - 2\sin x & \text{on } [0.5, 2] \\ h(x) = x + 10 - x\cosh(50/x) & \text{on } [120, 130] \end{cases}$$

Find each root to full machine precision. Use the correct number of steps, at least approximately. Repeat using the false position method.

8. Test the three bisection routines on $f(x) = x^3 + 2x^2 + 10x - 20$, with $a = 1$ and $b = 2$. The zero is 1.36880 8108. In programming this polynomial function, use nested multiplication. Repeat using the modified false position method.

9. Write a program to find a zero of a function f in the following way: In each step, an interval $[a, b]$ is given and $f(a)f(b) < 0$. Then c is computed as the root of the linear function that agrees with f at a and b. We retain either $[a, c]$ or $[c, b]$, depending on whether $f(a)f(c) < 0$ or $f(c)f(b) < 0$. Test your program on several functions.

[a]10. Select a routine from your program library to solve polynomial equations and use it to find the roots of the equation

$$x^8 - 36x^7 + 546x^6 - 4536x^5 + 22449x^4 - 67284x^3 + 118124x^2 - 109584x + 40320 = 0$$

The correct roots are the integers $1, 2, \ldots, 8$. Next, solve the same equation when the coefficient of x^7 is changed to -37. Observe how a minor perturbation in the coefficients can cause massive changes in the roots. Thus, the roots are **unstable** functions of the coefficients. (Be sure to program the exercise to allow for complex roots.) *Cultural Note*: This is a simplified version of **Wilkinson's polynomial**, which is found in Computer Exercise 3.3.9.

[a]11. A **circular metal shaft** is being used to transmit power. It is known that at a certain critical angular velocity ω, any jarring of the shaft during rotation will cause the shaft to deform or buckle. This is a dangerous situation because the shaft might shatter under the increased centrifugal force. To find this critical velocity ω, we must first compute a number x that satisfies the equation

$$\tan x + \tanh x = 0$$

This number is then used in a formula to obtain ω. Solve for x $(x > 0)$.

12. Using built-in routines in mathematical software systems such as MATLAB, Mathematica, or Maple, find the roots for $f(x) = x^3 - 3x + 1$ on $[0, 1]$ and $g(x) = x^3 - \sin x$ on $[0.5, 2]$ to more digits of accuracy than shown in the text.

13. (**Engineering Problem**) Nonlinear equations occur in almost all fields of engineering. For example, suppose a given task is expressed in the form $f(x) = 0$ and the objective is to find values of x that satisfy this condition. It is often difficult to find an explicit solution, and an approximate solution is sought with the aid of mathematical software. Find a solution of

$$f(x) = \frac{1}{\sqrt{2\pi}} e^{-(1/2)x^2} + \frac{1}{10}\sin(\pi x)$$

Plot the curve in the range $[-3.5, 3.5]$ for x values and $[-0.5, 0.5]$ for $y = f(x)$ values.

14. (**Circuit Problem**) A simple circuit with resistance R, capacitance C in series with a battery of voltage V is given by $Q = CV[1 - e^{-T/(RC)}]$, where Q is the charge of the capacitor and T is the time needed to obtain the charge. We wish to solve for the unknown C. For example, solve this exercise

$$f(x) = \left[10x\left(1 - e^{-0.004/(2000x)}\right) - 0.00001\right]$$

Plot the curve.
Hint: You may wish to magnify the vertical scale by using $y = 10^5 f(x)$.

15. (**Engineering Polynomials**) Equations such as

$$\begin{aligned} A + Bx^2 e^{Cx} &= 0 \\ A + Bx + Cx^2 + Dx^3 + Ex^4 &= 0 \end{aligned}$$

occur in engineering problems. Using mathematical software, find one or more solutions to the following equations and plot their curves:

a. $2 - x^2 e^{-0.385x} = 0$

b. $1 - 32x + 160x^2 - 256x^3 + 128x^4 = 0$

16. (**Reinforced Concrete**) In the design of reinforced concrete with regard to stress, one needs to solve numerically a quadratic equation such as

$$24147\,07.2x[450 - 0.822x(225)] - 265{,}000{,}000 = 0$$

Find approximate values of the roots.

17. (**Board Across Hallways**) In a building, two intersecting halls with widths $w_1 = 9$ feet and $w_2 = 7$ feet meet at an angle $\alpha = 125°$, as shown:

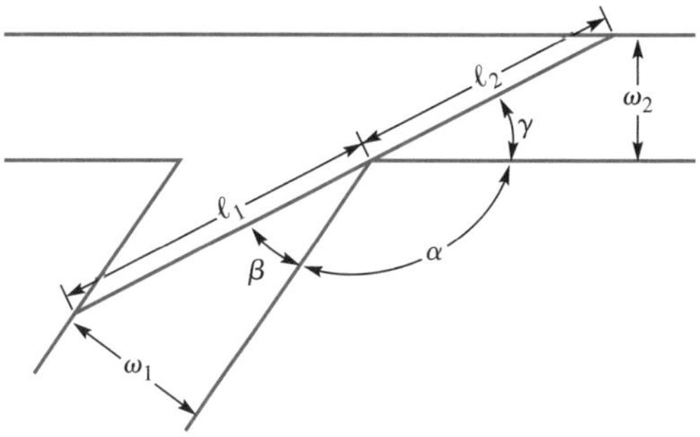

Assuming a two-dimensional situation, what is the longest board that can negotiate the turn? Ignore the thickness of the board. The relationship between the angles θ and the length of the board $\ell = \ell_1 + \ell_2$ is $\ell_1 = w_1 \csc(\beta)$, $\ell_2 = w_2 \csc(\gamma)$, $\beta = \pi - \alpha - \gamma$ and $\ell = w_1 \csc(\pi - \alpha - \gamma) + w_2 \csc(\gamma)$. The maximum length of the board that can make the turn is found by minimizing ℓ as a function of γ. Taking the derivative and setting $d\ell/d\gamma = 0$, we obtain

$$w_1 \cot(\pi - \alpha - \gamma)\csc(\pi - \alpha - \gamma) - w_2 \cot(\gamma)\csc(\gamma) = 0$$

Substitute in the known values and numerically solve the nonlinear equation.

18. Find the rectangle of maximum area if its vertices are at $(0, 0)$, $(x, 0)$, $(x, \cos x)$, $(0, \cos x)$. Assume that $0 \leqq x \leqq \pi/2$.

19. Program the false position algorithm and test it on some examples such as some of the nonlinear exercises in the text or in the computer exercises. Compare your results with those given for the bisection method.

20. Program the modified false position method, test it, and compare it to the false position method when using some sample functions.

3.2 Newton's Method

The procedure known as Newton's method is also called the **Newton-Raphson iteration**. It has a more general form than the one seen here, and the more general form can be used to find roots of systems of equations. Indeed, it is one of the more important procedures in numerical analysis, and its applicability extends to differential equations and integral equations. Here it is being applied to a single equation of the form $f(x) = 0$. As before, we seek one or more points at which the value of the function f is zero.

Interpretations of Newton's Method

In Newton's method, it is assumed at once that the function f is differentiable. This implies that the graph of f has a definite *slope* at each point and hence a unique **tangent line**. Now let's pursue the following simple idea. At a certain point $(x_0, f(x_0))$ on the graph of f, there is a tangent, which is a rather good approximation to the curve in the vicinity of that point. Analytically, it means that the linear function

Tangent Line Approach

$$\ell(x) = f'(x_0)(x - x_0) + f(x_0)$$

is close to the given function f near x_0. At x_0, the two functions ℓ and f agree. We take the zero of ℓ as an approximation to the zero of f. The zero of ℓ is easily found:

$$x_1 = x_0 - \frac{f(x_0)}{f'(x_0)}$$

Thus, starting with point x_0 (which we may interpret as an approximation to the root sought), we pass to a new point x_1 obtained from the preceding formula. Naturally, this process can be repeated (iterated) to produce a sequence of points:

$$x_2 = x_1 - \frac{f(x_1)}{f'(x_1)}, \qquad x_3 = x_2 - \frac{f(x_2)}{f'(x_2)}, \qquad \text{and so on.}$$

Under favorable conditions, the sequence of points approaches a zero of f.

Geometric Approach

The geometry of Newton's method is shown in Figure 3.4. The line $y = l(x)$ is tangent to the curve $y = f(x)$. It intersects the x-axis at a point x_1. The slope of $\ell(x)$ is $f'(x_0)$.

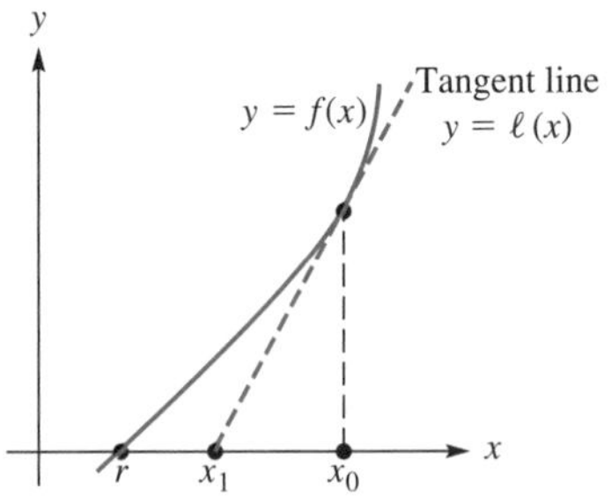

FIGURE 3.4 Newton's method

Taylor Series Approach

There are other ways of interpreting Newton's method. Suppose again that x_0 is an initial approximation to a root of f. We ask:

> *What* correction *h should be added to x_0 to obtain the root precisely?*

Obviously, we want

$$f(x_0 + h) = 0$$

If f is a sufficiently well-behaved function, it has a Taylor series at x_0. (See Equation (11) in Section 1.2.) Thus, we could write

$$f(x_0) + hf'(x_0) + \frac{h^2}{2} f''(x_0) + \cdots = 0$$

Determining h from this equation is, of course, not easy. Therefore, we give up the expectation of arriving at the true root in one step and seek only an approximation to h. This can be obtained by ignoring all but the first two terms in the series:

$$f(x_0) + hf'(x_0) = 0$$

The h that solves this is *not* the h that solves $f(x_0 + h) = 0$, but it is the easily computed number

$$h = -\frac{f(x_0)}{f'(x_0)}$$

Our new approximation is then

$$x_1 = x_0 + h = x_0 - \frac{f(x_0)}{f'(x_0)}$$

and the process can be repeated. In retrospect, we see that the Taylor series was not needed after all because we used only the first two terms. In the analysis to be given later, it is assumed that f'' is continuous in a neighborhood of the root. This assumption enables us to estimate the errors in the process.

If Newton's method is described in terms of a sequence $x_0, x_1, \ldots$, then the following **recursive** or **inductive** definition applies:

Newton's Method Formula

$$x_{n+1} = x_n - \frac{f(x_n)}{f'(x_n)}$$

Naturally, the interesting question is whether

$$\lim_{n\to\infty} x_n = r$$

where r is the desired root.

EXAMPLE 1 If $f(x) = x^3 - x + 1$ and $x_0 = 1$, what are x_1 and x_2 in the Newton iteration?

Solution From the basic formula, $x_1 = x_0 - f(x_0)/f'(x_0)$. Now $f'(x) = 3x^2 - 1$, and so $f'(1) = 2$. Also, we find $f(1) = 1$. Hence, we have $x_1 = 1 - \frac{1}{2} = \frac{1}{2}$. Similarly, we obtain $f\left(\frac{1}{2}\right) = \frac{5}{8}$, $f'\left(\frac{1}{2}\right) = -\frac{1}{4}$, and $x_2 = 3$. ■

Pseudocode

A pseudocode procedure for Newton's method can be written as follows:

***Newton* Pseudocode**

```
procedure Newton(f, f', x, nmax, ε, δ)
integer n, nmax;    real x, fx, fp, ε, δ
external function f, f'
fx ← f(x)
output 0, x, fx
for n = 1 to nmax
    fp ← f'(x)
    if |fp| < δ then
        output "small derivative"
        return
    end if
    d ← fx/fp
    x ← x − d
    fx ← f(x)
```

(Continued)

```
        output n, x, fx
        if |d| < ε then
            output "convergence"
            return
        end if
    end for
end procedure Newton
```

Using the initial value of x as the starting point, we carry out a maximum of $nmax$ iterations of Newton's method. Procedures must be supplied for the two external functions $f(x)$ and $f'(x)$. The parameters ε and δ are used to control the convergence and are related to the accuracy desired or to the machine precision available.

Illustration

Now we illustrate Newton's method by locating a root of $x^3 + x = 2x^2 + 3$. We apply the method to the function $f(x) = x^3 - 2x^2 + x - 3$, starting with $x_0 = 3$. Of course, $f'(x) = 3x^2 - 4x + 1$, and these two functions should be arranged in nested form for efficiency:

Sample Problem

$$f(x) = ((x - 2)x + 1)x - 3$$
$$f'(x) = (3x - 4)x + 1$$

To see in greater detail the rapid convergence of Newton's method, we use arithmetic with double the normal precision in the program and obtain the following results:

n	x_n	$f(x_n)$
0	3.0	9.0
1	2.4375	2.04
2	2.21303 27224 73144 5	2.56×10^{-1}
3	2.17555 49386 14368 4	6.46×10^{-3}
4	2.17456 01006 55071 4	4.48×10^{-6}
5	2.17455 94102 93284 1	1.97×10^{-12}

Notice the doubling of the accuracy in $f(x)$ (and also in x) until the maximum precision of the computer is encountered. Figure 3.5 shows a computer plot of three iterations of Newton's method for this sample problem.

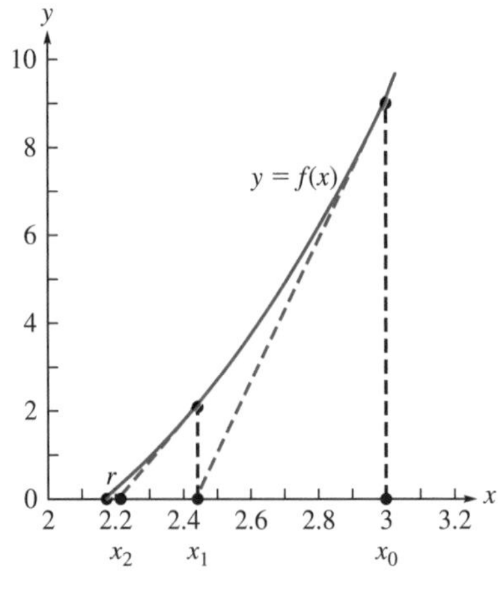

FIGURE 3.5 Three steps of Newton's method $f(x) = x^3 - 2x^2 + x - 3$

Using mathematical software that allows for complex roots such as in MATLAB,

Mathematical Software

Maple, or Mathematica, we find that this polynomial has a single real root, 2.17456, and a pair of complex conjugate roots, $-0.0872797 \pm 1.17131i$.

Convergence Analysis

Anyone who has experimented with Newton's method—for instance, by working some of the exercises in this section—has observed the remarkable rapidity in the convergence of the sequence to the root. This phenomenon is also noticeable in the example just given. Usually, the number of correct figures in the answer nearly *doubles* at each successive step! Indeed, in the preceding example, we have first 0 and then $1, 2, 3, 6, 12, 24, \ldots$ accurate digits from each Newton iteration. Five or six steps of Newton's method often suffice to yield full machine precision in the determination of a root. There is a theoretical basis for this dramatic performance, as we shall now see.

Let the function f, whose zero we seek, possess two continuous derivatives f' and f'', and let r be a zero of f. Assume further that r is a **simple zero**; that is, $f'(r) \neq 0$. Then Newton's method, if started sufficiently close to r, **converges quadratically** to r. This means that the errors in successive steps obey an inequality of the form

Error Bound Quadratic Convergence

$$|r - x_{n+1}| \leqq c|r - x_n|^2$$

We shall establish this fact presently, but first, an informal interpretation of the inequality may be helpful.

Suppose, for simplicity, that $c = 1$. Suppose also that x_n is an estimate of the root r that differs from it by at most one unit in the kth decimal place. This means that

Informal Interpretation

$$|r - x_n| \leqq 10^{-k}$$

The two inequalities above imply that

$$|r - x_{n+1}| \leqq 10^{-2k}$$

In other words, x_{n+1} differs from r by at most one unit in the $(2k)$th decimal place. So x_{n+1} has approximately twice as many correct digits as x_n! This is the doubling of significant digits alluded to previously.

■ **Theorem 1**

Newton's Method Theorem

If f, f', and f'' are continuous in a neighborhood of a root r of f and if $f'(r) \neq 0$, then there is a positive δ with the following property: If the initial point in Newton's method satisfies $|r - x_0| \leqq \delta$, then all subsequent points x_n satisfy the same inequality, converge to r, and do so quadratically; that is,

$$|r - x_{n+1}| \leqq c(\delta)|r - x_n|^2$$

where $c(\delta)$ is given by Equation (2).

Proof To establish the quadratic convergence of Newton's method, let $e_n = r - x_n$. The formula that defines the sequence $\{x_n\}$ then gives

$$e_{n+1} = r - x_{n+1} = r - x_n + \frac{f(x_n)}{f'(x_n)} = e_n + \frac{f(x_n)}{f'(x_n)} = \frac{e_n f'(x_n) + f(x_n)}{f'(x_n)}$$

By Taylor's Theorem (see Section 1.2), there exists a point ξ_n situated between x_n and r for which

$$0 = f(r) = f(x_n + e_n) = f(x_n) + e_n f'(x_n) + \frac{1}{2} e_n^2 f''(\xi_n)$$

(The subscript on ξ_n emphasizes the dependence on x_n.) This last equation can be rearranged to read

$$e_n f'(x_n) + f(x_n) = -\frac{1}{2} e_n^2 f''(\xi_n)$$

and if this is used in the previous equation for e_{n+1}, the result is

Error Vector Recursive Relation

$$e_{n+1} = -\frac{1}{2}\left(\frac{f''(\xi_n)}{f'(x_n)}\right) e_n^2 \tag{1}$$

This is, at least qualitatively, the sort of equation we want. Continuing the analysis, we define a function

$$c(\delta) = \frac{1}{2} \frac{\displaystyle\max_{|x-r| \leqq \delta} |f''(x)|}{\displaystyle\min_{|x-r| \leqq \delta} |f'(x)|} \qquad (\delta > 0) \tag{2}$$

By virtue of this definition, we can assert that, for any two points x and ξ within distance δ of the root r, the inequality $\frac{1}{2}|f''(\xi)/f'(x)| \leqq c(\delta)$ is true. Now select δ so small that $\delta c(\delta) < 1$. This is possible because as δ approaches 0, $c(\delta)$ converges to $\frac{1}{2}|f''(r)/f'(r)|$, and so $\delta c(\delta)$ converges to 0. Recall that we assumed that $f'(r) \neq 0$. Let $\rho = \delta c(\delta)$. In the remainder of this argument, we hold δ, $c(\delta)$, and ρ fixed with $\rho < 1$.

Suppose now that some iterate x_n lies within distance δ from the root r. We have

$$|e_n| = |r - x_n| \leqq \delta \qquad \text{and} \qquad |\xi_n - r| \leqq \delta$$

By the definition of $c(\delta)$, it follows that $\frac{1}{2}|f''(\xi_n)|/|f'(x_n)| \leqq c(\delta)$. From Equation (1), we now have

$$|e_{n+1}| = \frac{1}{2}\left|\frac{f''(\xi_n)}{f'(x_n)}\right| e_n^2 \leqq c(\delta) e_n^2 \leqq \delta c(\delta)|e_n| = \rho |e_n|$$

Consequently, x_{n+1} is also within distance δ of r because

$$|r - x_{n+1}| = |e_{n+1}| \leqq \rho |e_n| \leqq |e_n| \leqq \delta$$

If the initial point x_0 is chosen within distance δ of r, then

$$|e_n| \leqq \rho |e_{n-1}| \leqq \rho^2 |e_{n-1}| \leqq \cdots \leqq \rho^n |e_0|$$

Since $0 < \rho < 1$, $\lim_{n\to\infty} \rho^n = 0$ and $\lim_{n\to\infty} e_n = 0$. In other words, we obtain

$$\lim_{n\to\infty} x_n = r$$

In this process, we have

Error Bound (Quadratic Convergence)

$$|e_{n+1}| \leqq c(\delta) e_n^2$$ ■

Discussion of Newton's Method

In the use of Newton's method, consideration must be given to the proper choice of a starting point. Usually, one must have some insight into the shape of the graph of the function. Sometimes a coarse graph is adequate, but in other cases, a step-by-step evaluation of the

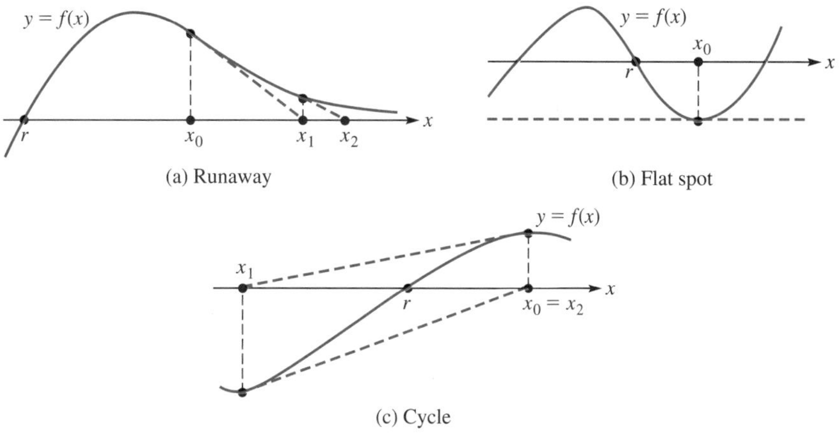

FIGURE 3.6 Failure of Newton's method due to bad starting points

function at various points may be necessary to find a point near the root. Often several steps of the bisection method are used to obtain a suitable starting point, so that Newton's method converges more rapidly.

Poor Choices of x_0

Although Newton's method is truly a marvelous invention, its convergence depends upon hypotheses that are difficult to verify a priori. Some graphical examples show what can happen. In Figure 3.6a, the tangent to the graph of the function f at x_0 intersects the x-axis at a point remote from the root r, and successive points in Newton's iteration *recede* from r instead of converging to r. The difficulty can be ascribed to a poor choice of the initial point x_0; it is *not* sufficiently close to r. In Figure 3.6b, the tangent to the curve is parallel to the x-axis and $x_1 = \pm\infty$, or it is assigned the value of machine infinity in a computer. In Figure 3.6c, the iteration values *cycle* because $x_2 = x_0$. In a computer, roundoff errors or limited precision may eventually cause this situation to become unbalanced such that the iterates either spiral inward and converge or spiral outward and diverge.

Multiple Zero

The analysis that establishes the quadratic convergence discloses another troublesome hypothesis; namely, $f'(r) \neq 0$. If $f'(r) = 0$, then r is a zero of f and f'. Such a zero is termed a **multiple zero** of f—in this case, at least a double zero. Newton's iteration for a multiple zero converges only linearly! Ordinarily, one would not know in advance that the zero sought was a multiple zero. If one knew that the multiplicity was m, however, Newton's method could be accelerated by modifying the equation to read

Modified Newton's Method

$$x_{n+1} = x_n - m\frac{f(x_n)}{f'(x_n)}$$

in which m is the *multiplicity* of the zero in question. The **multiplicity** of the zero r is the least m such that $f^{(k)}(r) = 0$ for $0 \leq k < m$, but $f^{(m)}(r) \neq 0$. (See Exercise 3.2.35.)

Special Cases

As is shown in Figure 3.7 (p. 132), the equation $p_2(x) = x^2 - 2x + 1 = 0$ has a root at 1 of multiplicity 2, and the equation $p_3(x) = x^3 - 3x^2 + 3x - 1 = 0$ has a root at 1 of multiplicity 3. It is instructive to plot these curves. Both curves are rather flat at the roots, which slows down the convergence of the regular Newton's method. Also, the figures illustrate the curves of two nonlinear functions with multiplicities as well as their regions of uncertainty about the curves. So the computed solutions could be anywhere within the indicated intervals on the x-axis. This is an indication of the difficulty in obtaining precise solutions of nonlinear functions with multiplicities.

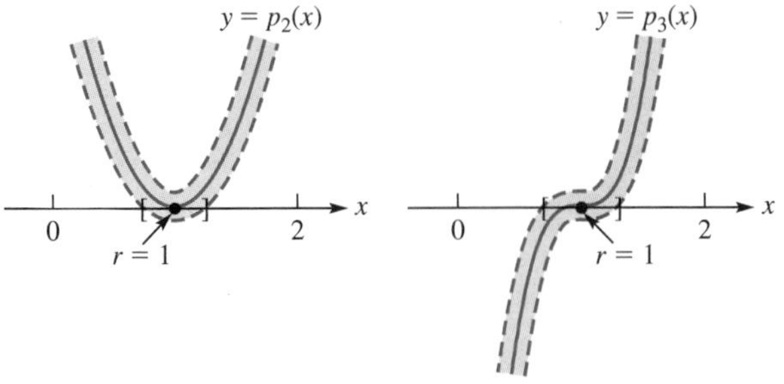

FIGURE 3.7 Curves p_2 and p_3 with multiplicity 2 and 3

Systems of Nonlinear Equations

Some physical problems involve the solution of systems of N nonlinear equations in N unknowns. One approach is to **linearize** and **solve**, repeatedly. This is the same strategy used by Newton's method in solving a single nonlinear equation. Not surprisingly, a natural extension of Newton's method for nonlinear systems can be found. The topic of systems of nonlinear equations requires some familiarity with matrices and their inverses.

In the general case, a system of N nonlinear equations in N unknowns x_i can be displayed in the form

Nonlinear Systems $N \times N$

$$\begin{cases} f_1(x_1, x_2, \ldots, x_N) = 0 \\ f_2(x_1, x_2, \ldots, x_N) = 0 \\ \qquad\qquad \vdots \\ f_N(x_1, x_2, \ldots, x_N) = 0 \end{cases}$$

Using vector notation, we can write this system in a more elegant form:

$$\mathbf{F}(\mathbf{X}) = \mathbf{0}$$

by defining column vectors as

$$\mathbf{F} = [f_1, f_2, \ldots, f_N]^T$$
$$\mathbf{X} = [x_1, x_2, \ldots, x_N]^T$$

The extension of Newton's method for nonlinear systems is

$$\mathbf{X}^{(k+1)} = \mathbf{X}^{(k)} - \left[\mathbf{F}'\left(\mathbf{X}^{(k)}\right)\right]^{-1}\mathbf{F}\left(\mathbf{X}^{(k)}\right)$$

where $\mathbf{F}'\left(\mathbf{X}^{(k)}\right)$ is the **Jacobian matrix**, which will be defined presently. It comprises partial derivatives of $\mathbf{F}$ evaluated at $\mathbf{X}^{(k)} = \left[x_1^{(k)}, x_2^{(k)}, \ldots, x_N^{(k)}\right]^T$. This formula is similar to the previously seen version of Newton's method, except that the derivative expression is not in the denominator but in the numerator as the inverse of a matrix. In the computational form of the formula, $\mathbf{X}^{(0)} = \left[x_1^{(0)}, x_2^{(0)}, \ldots, x_N^{(0)}\right]^T$ is an initial approximation vector, taken to be close to the solution of the nonlinear system, and the inverse of the Jacobian matrix is not computed, but rather a related system of equations is solved.

We illustrate the development of this procedure using three nonlinear equations

Nonlinear System 3 × 3

$$\begin{cases} f_1(x_1, x_2, x_3) = 0 \\ f_2(x_1, x_2, x_3) = 0 \\ f_3(x_1, x_2, x_3) = 0 \end{cases} \tag{3}$$

Recall the Taylor expansion in three variables for $i = 1, 2, 3$:

$$f_i(x_1 + h_1, x_2 + h_2, x_3 + h_3) = f_i(x_1, x_2, x_3) + h_1 \frac{\partial f_i}{\partial x_1} + h_2 \frac{\partial f_i}{\partial x_2} + h_3 \frac{\partial f_i}{\partial x_3} + \cdots \tag{4}$$

where the partial derivatives are evaluated at the point (x_1, x_2, x_3). Here only the linear terms in step sizes h_i are shown. Suppose that the vector $\mathbf{X}^{(0)} = \left(x_1^{(0)}, x_2^{(0)}, x_3^{(0)}\right)^T$ is an approximate solution to (3). Let $\mathbf{H} = \left[h_1, h_2, h_3\right]^T$ be a computed **correction** to the initial guess so that $\mathbf{X}^{(0)} + \mathbf{H} = \left[x_1^{(0)} + h_1, x_2^{(0)} + h_2, x_3^{(0)} + h_3\right]^T$ is a better approximate solution. Discarding the higher-order terms in the Taylor expansion (4), we have in vector notation

$$\mathbf{0} \approx \mathbf{F}\left(\mathbf{X}^{(0)} + \mathbf{H}\right) \approx \mathbf{F}\left(\mathbf{X}^{(0)}\right) + \mathbf{F}'\left(\mathbf{X}^{(0)}\right)\mathbf{H} \tag{5}$$

where the **Jacobian matrix** is defined by

Jacobian Matrix

$$\mathbf{F}'\left(\mathbf{X}^{(0)}\right) = \begin{bmatrix} \dfrac{\partial f_1}{\partial x_1} & \dfrac{\partial f_1}{\partial x_2} & \dfrac{\partial f_1}{\partial x_3} \\[2ex] \dfrac{\partial f_2}{\partial x_1} & \dfrac{\partial f_2}{\partial x_2} & \dfrac{\partial f_2}{\partial x_3} \\[2ex] \dfrac{\partial f_3}{\partial x_1} & \dfrac{\partial f_3}{\partial x_2} & \dfrac{\partial f_3}{\partial x_3} \end{bmatrix}$$

Here all of the partial derivatives are evaluated at $\mathbf{X}^{(0)}$; namely,

$$\frac{\partial f_i}{\partial x_j} = \frac{\partial f_i\left(\mathbf{X}^{(0)}\right)}{\partial x_j}$$

Also, we assume that the Jacobian matrix $\mathbf{F}'\left(\mathbf{X}^{(0)}\right)$ is nonsingular, so its inverse exists. Solving for $\mathbf{H}$ in (5), we have

$$\mathbf{H} \approx -\left[\mathbf{F}'\left(\mathbf{X}^{(0)}\right)\right]^{-1}\mathbf{F}\left(\mathbf{X}^{(0)}\right)$$

Let $\mathbf{X}^{(1)} = \mathbf{X}^{(0)} + \mathbf{H}$ be the better approximation after the correction; we then arrive at the first iteration of Newton's method for nonlinear systems

$$\mathbf{X}^{(1)} = \mathbf{X}^{(0)} - \left[\mathbf{F}'\left(\mathbf{X}^{(0)}\right)\right]^{-1}\mathbf{F}\left(\mathbf{X}^{(0)}\right)$$

In general, Newton's method uses this iteration:

$$\mathbf{X}^{(k+1)} = \mathbf{X}^{(k)} - \left[\mathbf{F}'\left(\mathbf{X}^{(k)}\right)\right]^{-1}\mathbf{F}\left(\mathbf{X}^{(k)}\right)$$

In practice, the computational form of Newton's method does not involve inverting the Jacobian matrix but rather solves the **Jacobian linear systems**

Jacobian Linear Systems

$$\left[\mathbf{F}'\left(\mathbf{X}^{(k)}\right)\right]\mathbf{H}^{(k)} = -\mathbf{F}\left(\mathbf{X}^{(k)}\right) \tag{6}$$

The next iteration of Newton's method is then

Newton's Method for Nonlinear Systems

$$\mathbf{X}^{(k+1)} = \mathbf{X}^{(k)} + \mathbf{H}^{(k)} \tag{7}$$

This is **Newton's method for nonlinear systems**. The linear system (6) can be solved by procedures *Gauss* and *Solve* as discussed in Chapter 2. Small systems of order 2 can be solved easily. (See Exercise 2.2.39.)

EXAMPLE 2 As an illustration, we can write a pseudocode to solve the following nonlinear system of equations using a variant of Newton's method given by (6) and (7):

$$\begin{cases} x+y+z &= 3 \\ x^2+y^2+z^2 &= 5 \\ e^x+xy-xz &= 1 \end{cases} \tag{8}$$

Solution With a sharp eye, the reader immediately sees that the solution of this system is $x = 0$, $y = 1, z = 2$. But in most realistic problems, the solution is not so obvious. We wish to develop a numerical procedure for finding such a solution. Here is the pseudocode:

$\mathbf{X} = \left[0.1,\ 1.2,\ 2.5\right]^T$
for $k = 1$ **to** 10

$$\mathbf{F} = \begin{bmatrix} x_1+x_2+x_3-3 \\ x_1^2+x_2^2+x_3^2-5 \\ e^{x_1}+x_1x_2-x_1x_3-1 \end{bmatrix}$$

$$\mathbf{J} = \begin{bmatrix} 1 & 1 & 1 \\ 2x_1 & 2x_2 & 2x_3 \\ e^{x_1}+x_2-x_3 & x_1 & -x_1 \end{bmatrix}$$

solve JH = F
X = X − H
end for

When programmed and executed on a computer, we found that it converges to $\boldsymbol{x} = (0, 1, 2)$, but when we change to a different starting vector, $(1, 0, 1)$, it converges to another root, $(1.2244, -0.0931, 1.8687)$. (Why?) ■

We can use mathematical software such as in MATLAB, Maple, or Mathematica and their procedures for solving the system of nonlinear equations (8). An important application area of solving systems of nonlinear equations is used in Chapter 13 on minimization of functions.

Fractal Basins of Attraction

The applicability of Newton's method for finding complex roots is one of its outstanding strengths. One need only program Newton's method using complex arithmetic.

The frontiers of numerical analysis and nonlinear dynamics overlap in some intriguing ways. Computer-generated displays with fractal patterns, such as in Figure 3.8, can easily be created with the help of the Newton iteration. The resulting pictures show intricately interwoven sets in the plane that are quite beautiful and colorful when displayed on a computer. One begins with a polynomial in the complex variable z. For example, $p(z) = z^4 - 1$ is suitable. This polynomial has four zeros, which are the fourth roots of unity. Each of these zeros has a **basin of attraction**, that is, the set of all points z_0 such that Newton's iteration, started at z_0, will converge to that zero. These four basins of attraction are disjoint from each other, because if the Newton iteration starting at z_0 converges to one zero, then it cannot also converge to another zero. One would naturally expect each basin to be a simple set surrounding the zero in the complex plane. But they turn out to be far from simple. To

Basin of Attraction of Fractals

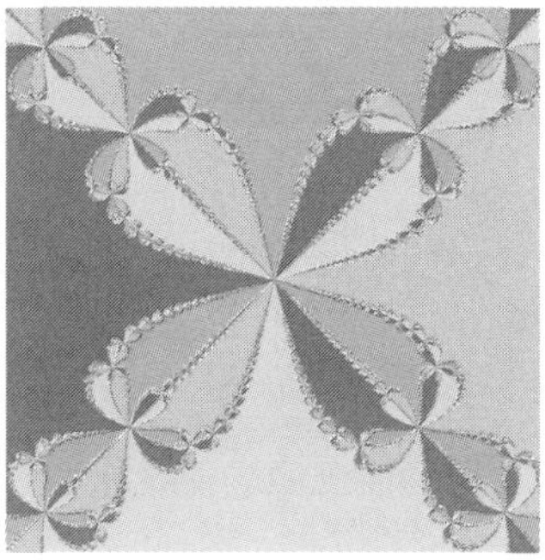

FIGURE 3.8
Basins of attraction

see what they are, we can systematically determine, for a large number of points, which zero of p the Newton iteration converges to if started at z_0. Points in each basin can be assigned different colors. The (rare) points for which the Newton iteration does not converge can be left uncolored. Computer Exercise 3.2.27 suggests how to do this.

Summary 3.2

- For finding a zero of a continuous and differentiable function f, **Newton's method** is given by

$$x_{n+1} = x_n - \frac{f(x_n)}{f'(x_n)} \qquad (n \geqq 0)$$

  It requires a given initial value x_0 and two function evaluations (for f and f') per step.

- The errors are related by

$$e_{n+1} = -\frac{1}{2}\left(\frac{f''(\xi_n)}{f'(x_n)}\right)e_n^2$$

  which leads to the inequality

$$|e_{n+1}| \leqq c|e_n|^2$$

  This means that Newton's method has **quadratic convergence** behavior for x_0 sufficiently close to the root r.

- For an $N \times N$ system of nonlinear equations $\mathbf{F}(\mathbf{X}) = \mathbf{0}$, **Newton's method** is written as

$$\mathbf{X}^{(k+1)} = \mathbf{X}^{(k)} - \left[\mathbf{F}'\left(\mathbf{X}^{(k)}\right)\right]^{-1}\mathbf{F}\left(\mathbf{X}^{(k)}\right) \qquad (k \geqq 0)$$

  which involves the **Jacobian matrix** $\mathbf{F}'\left(\mathbf{X}^{(k)}\right) = \mathbf{J} = \left[\left(\partial f_i\left(\mathbf{X}^{(k)}\right)/\partial x_j\right)\right]_{N\times N}$. In practice, one solves the **Jacobian linear system**

$$\left[\mathbf{F}'(\mathbf{X}^{(k)})\right]\mathbf{H}^{(k)} = -\mathbf{F}\left(\mathbf{X}^{(k)}\right)$$

  using Gaussian elimination and then finds the next iterate from the equation

$$\mathbf{X}^{(k+1)} = \mathbf{X}^{(k)} + \mathbf{H}^{(k)}$$

Exercises 3.2

1. Verify that when Newton's method is used to compute $\sqrt{R}$ (by solving the equation $x^2 = R$), the sequence of iterates is defined by

$$x_{n+1} = \frac{1}{2}\left(x_n + \frac{R}{x_n}\right)$$

2. (Continuation) Show that if the sequence $\{x_n\}$ is defined as in the preceding exercise, then

$$x_{n+1}^2 - R = \left[\frac{x_n^2 - R}{2x_n}\right]^2$$

Interpret this equation in terms of quadratic convergence.

[a]3. Write Newton's method in simplified form for determining the reciprocal of the square root of a positive number. Perform two iterations to approximate $1/\pm\sqrt{5}$, starting with $x_0 = 1$ and $x_0 = -1$.

[a]4. Two of the four zeros of $x^4 + 2x^3 - 7x^2 + 3$ are positive. Find them by Newton's method, correct to two significant figures.

5. The equation $x - Rx^{-1} = 0$ has $x = \pm R^{1/2}$ for its solution. Establish Newton's iterative scheme, in simplified form, for this situation. Carry out five steps for $R = 25$ and $x_0 = 1$.

6. Using a calculator, observe the sluggishness with which Newton's method converges in the case of $f(x) = (x-1)^m$ with $m = 8$ or 12. Reconcile this with the theory. Use $x_0 = 1.1$.

[a]7. What linear function $y = ax + b$ approximates $f(x) = \sin x$ best in the vicinity of $x = \pi/4$? How does this exercise relate to Newton's method?

8. In Exercises 1.2.10–1.2.12, several methods are suggested for computing $\ln 2$. Compare them with the use of Newton's method applied to the equation $e^x = 2$.

[a]9. Define a sequence $x_{n+1} = x_n - \tan x_n$ with $x_0 = 3$. What is $\lim_{n\to\infty} x_n$?

10. The iteration formula

$$x_{n+1} = x_n - (\cos x_n)(\sin x_n) + R\cos^2 x_n$$

where R is a positive constant, was obtained by applying Newton's method to some function $f(x)$. What was $f(x)$? What can this formula be used for?

[a]11. Establish Newton's iterative scheme in simplified form, not involving the reciprocal of x, for the function $f(x) = xR - x^{-1}$. Carry out three steps of this procedure using $R = 4$ and $x_0 = -1$.

12. Consider the following procedures:

[a]**a.** $x_{n+1} = \dfrac{1}{3}\left(2x_n - \dfrac{r}{x_n^2}\right)$ **b.** $x_{n+1} = \dfrac{1}{2}x_n + \dfrac{1}{x_n}$

Do they converge for any nonzero initial point? If so, to what values?

13. Each of the following functions has $\sqrt[3]{R}$ as a zero for any positive real number R. Determine the formulas for Newton's method for each and any necessary restrictions on the choice for x_0.

[a]**a.** $a(x) = x^3 - R$ **b.** $b(x) = 1/x^3 - 1/R$
[a]**c.** $c(x) = x^2 - R/x$ **d.** $d(x) = x - R/x^2$
[a]**e.** $e(x) = 1 - R/x^3$ **f.** $f(x) = 1/x - x^2/R$
[a]**g.** $g(x) = 1/x^2 - x/R$ **h.** $h(x) = 1 - x^3/R$

14. Determine the formulas for Newton's method for finding a root of the function $f(x) = x - e/x$. What is the behavior of the iterates?

[a]15. If Newton's method is used on $f(x) = x^3 - x + 1$ starting with $x_0 = 1$, what will x_1 be?

16. Locate the root of $f(x) = e^{-x} - \cos x$ that is nearest $\pi/2$.

[a]17. If Newton's method is used on $f(x) = x^5 - x^3 + 3$ and if $x_n = 1$, what is x_{n+1}?

18. Determine Newton's iteration formula for computing the cube root of N/M for nonzero integers N and M.

[a]19. For what starting values will Newton's method converge if the function f is $f(x) = x^2/(1+x^2)$?

20. Starting at $x = 3$, $x < 3$, or $x > 3$, analyze what happens when Newton's method is applied to the function $f(x) = 2x^3 - 9x^2 + 12x + 15$.

[a]21. (Continuation) Repeat for $f(x) = \sqrt{|x|}$, starting with $x < 0$ or $x > 0$.

[a]22. To determine $x = \sqrt[3]{R}$, we can solve the equation $x^3 = R$ by Newton's method. Write the loop that carries out this process, starting from the initial approximation $x_0 = R$.

23. The **reciprocal** of a number R can be computed without division by the iterative formula

$$x_{n+1} = x_n(2 - x_n R)$$

Establish this relation by applying Newton's method to some $f(x)$. Beginning with $x_0 = 0.2$, compute the reciprocal of 4 correct to six decimal digits or more by this rule.

Tabulate the error at each step and observe the quadratic convergence.

24. On a certain modern computer, floating-point numbers have a 48-bit mantissa. Moreover, floating-point hardware can perform addition, subtraction, multiplication, and reciprocation, but not division. Unfortunately, the reciprocation hardware produces a result accurate to less than full precision, whereas the other operations produce results accurate to full floating-point precision.

a. Show that Newton's method can be used to find a zero of the function $f(x) = 1 - 1/(ax)$. This will provide an approximation to $1/a$ that is accurate to full floating-point precision. How many iterations are required?

b. Show how to obtain an approximation to b/a that is accurate to full floating-point precision.

25. Newton's method for finding $\sqrt{R}$ is

$$x_{n+1} = \frac{1}{2}\left(x_n + \frac{R}{x_n}\right)$$

Perform three iterations of this scheme for computing $\sqrt{2}$, starting with $x_0 = 1$, and of the bisection method for $\sqrt{2}$, starting with interval $[1, 2]$. How many iterations are needed for each method in order to obtain 10^{-6} accuracy?

26. (Continuation) Newton's method for finding $\sqrt{R}$, where $R = AB$, gives this approximation:

$$\sqrt{AB} \approx \frac{A+B}{4} + \frac{AB}{A+B}$$

Show that if $x_0 = A$ or B, then two iterations of Newton's method are needed to obtain this approximation, whereas if $x_0 = \frac{1}{2}(A+B)$, then only one iteration is needed.

[a]**27.** Show that Newton's method applied to $x^m - R$ and to $1-(R/x^m)$ for determining $\sqrt[m]{R}$ results in two similar yet different iterative formulas. Here $R > 0$, $m \geqq 2$. Which formula is better and why?

28. Using a handheld calculator, carry out three iterations of Newton's method using $x_0 = 1$ and $f(x) = 3x^3 + x^2 - 15x + 3$.

[a]**29.** What happens if the Newton iteration is applied to $f(x) = \arctan x$ with $x_0 = 2$? For what starting values will Newton's method converge? (See Computer Exercise 3.2.7.)

30. Newton's method can be interpreted as follows: Suppose that $f(x+h) = 0$. Then $f'(x) \approx [f(x+h) - f(x)]/h = -f(x)/h$. Continue this argument.

[a]**31.** Derive a formula for Newton's method for the function $F(x) = f(x)/f'(x)$, where $f(x)$ is a function with simple zeros that is three times continuously differentiable. Show that the convergence of the resulting method to any zero r of $f(x)$ is at least quadratic.
Hint: Apply the result in the text to F, making sure that F has the required properties.

[a]**32.** The Taylor series for a function f looks like this:

$$f(x+h) = f(x) + hf'(x) + \frac{h^2}{2} f''(x) + \frac{h^3}{6} f'''(x) + \cdots$$

Suppose that $f(x)$, $f'(x)$, and $f''(x)$ are easily computed. Derive an algorithm like Newton's method that uses three terms in the Taylor series. The algorithm should take as input an approximation to the root and produce as output a better approximation to the root. Show that the method is cubically convergent.
Hint: Use $e_n = e_{n+1} - h$ and ignore e_{n+1}^2 terms as being negligible.

33. To avoid computing the derivative at each step in Newton's method, it has been proposed to replace $f'(x_n)$ by $f'(x_0)$. Derive the rate of convergence for this method.

34. Refer to the discussion of Newton's method and establish that

$$\lim_{n\to\infty} \left(e_{n+1} e_n^{-2}\right) = -\frac{1}{2}\left[\frac{f''(r)}{f'(r)}\right]$$

How can this be used in a practical case to test whether the convergence is quadratic? Devise an example in which r, $f'(r)$, and $f''(r)$ are all known, and test numerically the convergence of $e_{n+1}e_n^{-2}$.

[a]**35.** Show that in the case of a zero of multiplicity m, the **modified Newton's method**

$$x_{n+1} = x_n - m\frac{f(x_n)}{f'(x_n)}$$

is quadratically convergent.
Hint: Use Taylor series for each of $f(r + e_n)$ and $f'(r + e_n)$.

[a]**36.** The **Steffensen method** for solving the equation $f(x) = 0$ uses the formula

$$x_{n+1} = x_n - \frac{f(x_n)}{g(x_n)}$$

in which $g(x) = \{f[x + f(x)] - f(x)\}/f(x)$. It is quadratically convergent, like Newton's method. How many function evaluations are necessary per step? Using Taylor series, show that $g(x) \approx f'(x)$ if $f(x)$ is small and thus relate Steffensen's iteration to Newton's. What advantage does Steffensen's have? Establish the quadratic convergence.

[a]**37.** A proposed **Generalization of Newton's method** is

$$x_{n+1} = x_n - \omega\frac{f(x_n)}{f'(x_n)}$$

where the constant ω is an acceleration factor chosen to increase the rate of convergence. For what range of values of ω is a simple root r of $f(x)$ a **point of attraction**; that is, $|g'(r)| < 1$, where $g(x) = x - \omega f(x)/f'(x)$? This method is quadratically convergent *only* if $\omega = 1$ because $g'(r) \neq 0$ when $\omega \neq 1$.

38. Suppose that r is a double root of $f(x) = 0$; that is, $f(r) = f'(r) = 0$ but $f''(r) \neq 0$, and suppose that f and all derivatives up to and including the second are continuous in some neighborhood of r. Show that $e_{n+1} \approx \frac{1}{2}e_n$ for Newton's method and thereby conclude that the rate of convergence is *linear* near a double root. (If the root has multiplicity m, then $e_{n+1} \approx [(m-1)/m]e_n$.)

39. (**Simultaneous Nonlinear Equations**) Using the Taylor series in two variables (x, y) of the form

$$f(x+h, y+k) = f(x, y) + hf_x(x, y) + kf_y(x, y) + \cdots$$

where $f_x = \partial f/\partial x$ and $f_y = \partial f/\partial y$, establish that Newton's method for solving the two simultaneous nonlinear equations

$$\begin{cases} f(x, y) = 0 \\ g(x, y) = 0 \end{cases}$$

can be described with the formulas

$$\begin{cases} x_{n+1} = x_n - \dfrac{fg_y - gf_y}{f_x g_y - g_x f_y} \\[2ex] y_{n+1} = y_n - \dfrac{f_x g - g_x f}{f_x g_y - g_x f_y} \end{cases}$$

Here the functions f, f_x, and so on are evaluated at (x_n, y_n).

40. Newton's method can be defined for the equation $f(z) = g(x, y) + ih(x, y)$, where $f(z)$ is an analytic function of the complex variable $z = x + iy$ (x and y real) and $g(x, y)$ and $h(x, y)$ are real functions for all x and y. The derivative $f'(z)$ is given by $f'(z) = g_x + ih_x = h_y - ig_y$ because the **Cauchy-Riemann equations** $g_x = h_y$ and $h_x = -g_y$ hold. Here the partial derivatives are defined as $g_x = \partial g/\partial x$, $g_y = \partial g/\partial y$, and so on. Show that Newton's method

$$z_{n+1} = z_n - \frac{f(z_n)}{f'(z_n)}$$

can be written in the form

$$\begin{cases} x_{n+1} = x_n - \dfrac{gh_y - hg_y}{g_x h_y - g_y h_x} \\[2ex] y_{n+1} = y_n - \dfrac{hg_x - gh_x}{g_x h_y - g_y h_x} \end{cases}$$

Here all functions are evaluated at $z_n = x_n + iy_n$.

[a]**41.** Consider the algorithm of which *one* step consists of two steps of Newton's method. What is its order of convergence?

42. (Continuation) Using the idea of the preceding exercise, show how we can easily create methods of arbitrarily high order for solving $f(x) = 0$. Why is the order of a method not the only criterion that should be considered in assessing its merits?

43. If we want to solve the equation $2 - x = e^x$ using Newton's iteration, what are the equations and functions that must be coded? Give a pseudocode for doing this exercise. Include a suitable starting point and a suitable stopping criterion.

44. Suppose that we want to compute $\sqrt{2}$ by using Newton's Method on the equation $x^2 = 2$ (in the obvious, straightforward way). If the starting point is $x_0 = \frac{7}{5}$, what is the numerical value of the correction that must be added to x_0 to get x_1?
Hint: The arithmetic is quite easy if you do it using ratios of integers.

45. Apply Newton's method to the equation $f(x) = 0$ with $f(x)$ as given below. Find out what happens and why.

a. $f(x) = e^x$ **b.** $f(x) = e^x + x^2$

46. Consider Newton's method $x_{n+1} = x_n - f(x_n)/f'(x_n)$. If the sequence converges then the limit point is a solution. Explain why or why not.

Computer Exercises 3.2

1. Using the procedure *Newton* and a single computer run, test your code on these examples: $f(t) = \tan t - t$ with $x_0 = 7$ and $g(t) = e^t - \sqrt{t+9}$ with $x_0 = 2$. Print each iterate and its accompanying function value.

2. Write a simple, self-contained program to apply Newton's method to the equation $x^3 + 2x^2 + 10x = 20$, starting with $x_0 = 2$. Evaluate the appropriate $f(x)$ and $f'(x)$, using nested multiplication. Stop the computation when two successive points differ by $\frac{1}{2} \times 10^{-5}$ or some other convenient tolerance close to your machine's capability. Print all intermediate points and function values. Put an upper limit of ten on the number of steps.

3. (Continuation) Repeat using double precision and more steps.

[a]**4.** Find the root of the equation

$$2x(1 - x^2 + x)\ln x = x^2 - 1$$

in the interval [0, 1] by Newton's method using double precision. Make a table that shows the number of correct digits in each step.

[a]**5.** In 1685, John Wallis published a book called *Algebra*, in which he described a method devised by Newton for solving equations. In slightly modified form, this method was also published by Joseph Raphson in 1690. This form is the one now commonly called Newton's method or the Newton-Raphson method. Newton himself discussed the method in 1669 and illustrated it with the equation $x^3 - 2x - 5 = 0$. Wallis used the same example. Find a root of this equation in double precision, thus continuing the tradition that every numerical analysis student should solve this venerable equation.

6. In celestial mechanics, **Kepler's equation** is important. It reads $x = y - \varepsilon \sin y$, in which x is a planet's mean anomaly, y its eccentric anomaly, and ε the eccentricity of its orbit. Taking $\varepsilon = 0.9$, construct a table of y for 30 equally spaced values of x in the interval $0 \leqq x \leqq \pi$. Use Newton's method to obtain each value of y. The y corresponding to an x can be used as the starting point for the iteration when x is changed slightly.

7. In Newton's method, we progress in each step from a given point x to a new point $x - h$, where $h = f(x)/f'(x)$. A refinement that is easily programmed is this: If $|f(x - h)|$ is not smaller than $|f(x)|$, then reject this value of h and use $h/2$ instead. Test this refinement.

[a]**8.** Write a brief program to compute a root of the equation $x^3 = x^2 + x + 1$, using Newton's method. Be careful to select a suitable starting value.

[a]**9.** Find the root of the equation $5(3x^4 - 6x^2 + 1) = 2(3x^5 - 5x^3)$ that lies in the interval [0, 1] by using Newton's method and a short program.

10. For each equation, write a brief program to compute and print eight steps of Newton's method for finding a positive root.

[a]**a.** $x = 2\sin x$

[a]**b.** $x^3 = \sin x + 7$

[a]**c.** $\sin x = 1 - x$

[a]**d.** $x^5 + x^2 = 1 + 7x^3$ for $x \geqq 2$

11. Write and test a recursive procedure for Newton's method.

12. Rewrite and test the *Newton* procedure so that it is a character function and returns key words such as `iterating`, `success`, `near-zero`, `max-iteration`. Then a case statement can be used to print the results.

13. Would you like to see the number 0.55887 766 come out of a calculation? Take three steps in Newton's method on $10 + x^3 - 12\cos x = 0$ starting with $x_0 = 1$.

[a]**14.** Write a short program to solve for a root of the equation $e^{-x^2} = \cos x + 1$ on [0, 4]. What happens in Newton's method if we start with $x_0 = 0$ or with $x_0 = 1$?

15. Find the root of the equation $\frac{1}{2}x^2 + x + 1 - e^x = 0$ by Newton's method, starting with $x_0 = 1$, and account for the slow convergence.

16. Using $f(x) = x^5 - 9x^4 - x^3 + 17x^2 - 8x - 8$ and $x_0 = 0$, study and explain the behavior of Newton's method.
Hint: The iterates are initially cyclic.

17. Find the zero of the function $f(x) = x - \tan x$ that is closest to 99 (radians) by both the bisection method and Newton's method.
Hint: Extremely accurate starting values are needed for this function. Use the computer to construct a table of values of $f(x)$ around 99 to determine the nature of this function.

18. Using the bisection method, find the positive root of $2x(1 + x^2)^{-1} = \arctan x$. Using the root as x_0, apply Newton's method to the function $\arctan x$. Interpret the results.

19. If the root of $f(x) = 0$ is a double root, then Newton's method can be accelerated by using

$$x_{n+1} = x_n - 2\frac{f(x_n)}{f'(x_n)}$$

Numerically compare the convergence of this scheme with Newton's method on a function with a known double root.

20. Program and test Steffensen's method, as described in Exercise 3.2.36.

21. Consider the nonlinear system

$$\begin{cases} f(x, y) = x^2 + y^2 - 25 = 0 \\ g(x, y) = x^2 - y - 2 = 0 \end{cases}$$

Using a software package that has 2D plotting capabilities, illustrate what is going on in solving such a system by plotting $f(x, y)$, $g(x, y)$, and show their intersection with the (x, y)-plane. Determine approximate roots of these equations from the graphical results.

22. Solve this pair of simultaneous nonlinear equations by first eliminating y and then solving the resulting equation in x by Newton's method. Start with the initial value $x_0 = 1.0$.

$$\begin{cases} x^3 - 2xy + y^7 - 4x^3y = 5 \\ y \sin x + 3x^2y + \tan x = 4 \end{cases}$$

23. Using Equations (6) and (7), code Newton's methods for nonlinear systems. Test your program by solving one or more of the following systems:

a. System in Computer Exercise 3.2.21.

b. System in Computer Exercise 3.2.22.

c. System (3) using starting values (0, 0, 0).

d. Using starting values $\left(\frac{3}{4}, \frac{1}{2}, -\frac{1}{2}\right)$, solve

$$\begin{cases} x + y + z = 0 \\ x^2 + y^2 + z^2 = 2 \\ x(y + z) = -1 \end{cases}$$

e. Using starting values $(-0.01, -0.01)$, solve

$$\begin{cases} 4y^2 + 4y + 52x - 19 = 0 \\ 169x^2 + 3y^2 + 111x - 10y - 10 = 0 \end{cases}$$

f. Select starting values, and solve

$$\begin{cases} \sin(x + y) = e^{x-y} \\ \cos(x + 6) = x^2y^2 \end{cases}$$

24. Investigate the behavior of Newton's method for finding complex roots of polynomials with real coefficients. For example, the polynomial $p(x) = x^2 + 1$ has the complex conjugate pair of roots $\pm i$ and Newton's method is $x_{n+1} = \frac{1}{2}(x_n - 1/x_n)$. First, program this method using real arithmetic and real numbers as starting values. Second, modify the program using complex arithmetic but still using only real starting values. Finally, use complex numbers as starting values. Observe the behavior of the iterates in each case.

25. Using Exercise 3.2.40, find a complex root of each of the following:

a. $z^3 - z - 1 = 0$ **b.** $z^4 - 2z^3 - 2iz^2 + 4iz = 0$

c. $2z^3 - 6(1 + i)z^2 - 6(1 - i) = 0$ **d.** $z = e^z$

Hint: For the last part, use Euler's relation $e^{iy} = \cos y + i \sin y$.

26. In the Newton method for finding a root r of $f(x) = 0$, we start with x_0 and compute the sequence $x_1, x_2, \dots$ using the formula $x_{n+1} = x_n - f(x_n)/f'(x_n)$. To avoid computing the derivative at each step, it has been proposed to replace $f'(x_n)$ with $f'(x_0)$ in all steps. It has also been suggested that the derivative in Newton's formula be computed only every other step. This method is given by

$$\begin{cases} x_{2n+1} = x_{2n} - \dfrac{f(x_{2n})}{f'(x_{2n})} \\[2ex] x_{2n+2} = x_{2n+1} - \dfrac{f(x_{2n+1})}{f'(x_{2n})} \end{cases}$$

Numerically compare both proposed methods to Newton's method for several simple functions that have known roots. Print the error of each method on every iteration to monitor the convergence. How well do the proposed methods work?

27. **(Basin of Attraction)** Consider the complex polynomial $z^3 - 1$, whose zeros are the three cube roots of unity. Generate a picture showing three basins of attraction in the complex plane in the square region defined by $-1 \leqq \text{Real}(z) \leqq 1$ and $-1 \leqq \text{Imaginary}(z) \leqq 1$. To do this, use a mesh of 1000×1000 pixels inside the square. The center point of each pixel is used to start the iteration of Newton's method. Assign a particular basin color to each pixel if convergence to a root is obtained with $nmax = 10$ iterations. The large number of iterations suggested can be avoided by doing some analysis with the aid of Theorem 1, since the iterates get within a certain neighborhood of the root and the iteration can be stopped. The criterion for convergence is to check both $|z_{n+1} - z_n| < \varepsilon$ and $|z_{n+1}^3 - 1| < \varepsilon$ with a small value such as $\varepsilon = 10^{-4}$ as well as a maximum number of iterations.
Hint: It is best to debug your program and get a crude picture with only a small number of pixels such as 10×10.

28. (Continuation) Repeat for the polynomial $z^4 - 1 = 0$.

29. Write **real function** *Sqrt*(x) to compute the square root of a real argument x by the following algorithm: First, reduce the range of x by finding a real number r and an integer m such that $x = 2^{2m}r$ with $\frac{1}{4} \leqq r < 1$. Next, compute x_2 by using three iterations of Newton's method given by

$$x_{n+1} = \frac{1}{2}\left(x_n + \frac{r}{x_n}\right)$$

with the special initial approximation

$$x_0 = 1.27235\,367 + 0.24269\,3281r - \frac{1.02966\,039}{1 + r}$$

Then set $\sqrt{x} \approx 2^m x_2$. Test this algorithm on various values of x. Obtain a listing of the code for the square-root function on your computer system. By reading the comments, try to determine what algorithm it uses.

30. The following method has third-order convergence for computing $\sqrt{R}$:

$$x_{n+1} = \frac{x_n\left(x_n^2 + 3R\right)}{3x_n^2 + R}$$

Carry out some numerical experiments using this method and the method of the preceding exercise to see whether you observe a difference in the rate of convergence. Use the same starting procedures of range reduction and initial approximation.

31. Write **real function** *CubeRoot*(x) to compute the cube root of a real argument x by the following procedure: First, determine a real number r and an integer m such that $x = r2^{3m}$ with $\frac{1}{8} \leq r < 1$. Compute x_4 using four iterations of Newton's method:

$$x_{n+1} = \frac{2}{3}\left(x_n + \frac{r}{2x_n^2}\right)$$

with the special starting value

$$x_0 = 2.502926 - \frac{8.0451 25(r + 0.3877552)}{(r + 4.6122 44)(r + 0.3877552) - 0.3598496}$$

Then set $\sqrt[3]{x} \approx 2^m x_4$. Test this algorithm on a variety of x values.

32. Use mathematical software in MATLAB, Maple, or Mathematica to compute ten iterates of Newton's method starting with $x_0 = 0$ for $f(x) = x^3 - 2x^2 + x - 3$. With 100 decimal places of accuracy and after nine iterations, show that the value of x is

2.17455 94102 92980 07420 23189 88695 65392 56759 48725 33708 24983 36733 92030 23647 64792 75760 66115 28969 38832 0640

Show that the values of the function at each iteration are 9.0, 2.0, 0.26, 0.0065, 0.45×10^{-5}, 0.22×10^{-11}, 0.50×10^{-24}, 0.27×10^{-49}, 0.1×10^{-98}, and 0.1×10^{-98}. Again notice that the number of digits of accuracy in Newton's method doubles (approximately) with each iteration once they are sufficiently close to the root. (Also, see Bornemann, Wagon, and Waldvogel [2004] for a 100-Digit Challenge, which is a study in high-accuracy numerical computing.)

33. (Continuation) Use MATLAB, Maple or Mathematica to discover that this root is exactly

$$\sqrt[3]{\frac{79}{54} + \frac{1}{6}\sqrt{77}} + \frac{1}{9\sqrt[3]{\frac{79}{54} + \frac{1}{6}\sqrt{77}}} + \frac{2}{3}$$

Clearly, the decimal results are of more interest to us in our study of numerical methods.

34. (Continuation) Find all the roots including complex roots.

35. Numerically, find all the roots of the following systems of nonlinear equations. Then plot the curves to verify your results:

a. $y = 2x^2 + 3x - 4$, $y = x^2 + 2x + 3$
b. $y + x + 3 = 0$, $x^2 + y^2 = 17$
c. $y = \frac{1}{2}x - 5$, $y = x^2 + 2x - 15$
d. $xy = 1$, $x + y = 2$
e. $y = x^2$, $x^2 + (y - 2)^2 = 4$
f. $3x^2 + 2y^2 = 35$, $4x^2 - 3y^2 = 24$
g. $x^2 - xy + y^2 = 21$, $x^2 + 2xy - 8y^2 = 0$

36. Apply Newton's method on these test problems:

a. $f(x) = x^2$.
Hint: The first derivative is zero at the root, and convergence may not be quadratic.

b. $f(x) = x + x^{4/3}$.
Hint: There is no second derivative at the root, and convergence may fail to be quadratic.

c. $f(x) = x + x^2 \sin(2/x)$ for $x \neq 0$ and $f(0) = 0$ and $f'(x) = 1 + 2x \sin(2/x) - 2\cos(2/x)$ for $x \neq 0$ and $f'(0) = 1$.
Hint: The derivative of this function is not continuous at the root, and convergence may fail.

37. Let $\mathbf{F}(\mathbf{X}) = \begin{bmatrix} x_1^2 - x_2 + c \\ x_2^2 - x_1 + c \end{bmatrix} = \begin{bmatrix} 0 \\ 0 \end{bmatrix}$.

Each component equation $f_1(x) = 0$ and $f_2(x) = 0$ describes a parabola. Any point (x^*, y^*) where these two parabolas intersect is a solution to the nonlinear system of equations. Using Newton's method for systems of nonlinear equations, find the solutions for each of these values of the parameter $c = \frac{1}{2}, \frac{1}{4}, -\frac{1}{2}, -1$. Give the Jacobian matrix for each. Also for each of these values, plot the resulting curves showing the points of intersection (Heath [2000], p. 218).

38. Let $\mathbf{F}(\mathbf{X}) = \begin{bmatrix} x_1^2 + 2x_2 - 2 \\ x_1 + 4x_2^2 - 4 \end{bmatrix} = \begin{bmatrix} 0 \\ 0 \end{bmatrix}$.

Solve this nonlinear system starting with $\mathbf{X}^{(0)} = (1, 2)$. Give the Jacobian matrix. Also plot the resulting curves showing the point(s) of intersection.

39. Using Newton's method, find the zeros of $f(z) = z^3 - z$ with these starting values $z^{(0)} = 1 + 1.5i$, $1 + 1.1i$, $1 + 1.2i$, $1 + 1.3i$.

40. Use Halley's method to produce a plot of the basins of attraction for $p(z) = z^6 - 1$. Compare to Figure 3.8.

41. (**Global Positioning System Project**) Each time a GPS is used, a system of nonlinear equations of the form

$$\begin{cases} (x-a_1)^2+(y-b_1)^2+(z-c_i)^2 = [(C(t_1-D)]^2 \\ (x-a_2)^2+(y-b_2)^2+(z-c_i)^2 = [(C(t_2-D)]^2 \\ (x-a_3)^2+(y-b_3)^2+(z-c_i)^2 = [(C(t_3-D)]^2 \\ (x-a_4)^2+(y-b_4)^2+(z-c_i)^2 = [(C(t_4-D)]^2 \end{cases}$$

is solved for the (x, y, z) coordinates of the receiver. For each satellite i, the locations are (a_i, b_i, c_i), and t_i is the synchronized transmission time from the satellite. Further, C is the speed of light, and D is the difference between the synchronized time of the satellite clocks and the earth-bound receiver clock. Although there are only two points on the intersection of three spheres (one of which can be determined to be the desired location), a fourth sphere (satellite) must be used to resolve the inaccuracy in the clock contained in the low-cost receiver on earth. Explore various ways for solving such a nonlinear system. (See Hofmann-Wellenhof, Lichtenegger, and Collins [2001], Sauer [2012], and Strang and Borre [1997].)

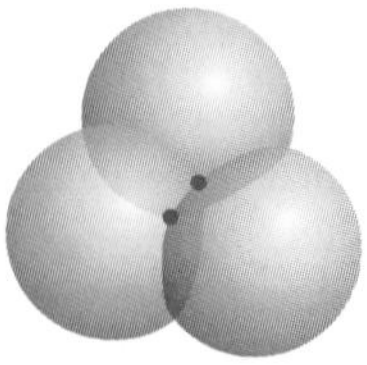

42. Use mathematical software such as in MATLAB, Maple, or Mathematica and their built-in procedures to solve the system of nonlinear equations (8) in Example 2. Also, plot the given surfaces and the solution obtained.
Hint: You may need to use a slightly perturbed starting point (0.5, 1.5, 0.5) to avoid a singularity in the Jacobian matrix.

3.3 Secant Method

We now consider a general-purpose procedure that converges almost as fast as Newton's method. This method mimics Newton's method, but avoids the calculation of derivatives.

Method Description

Recall that Newton's iteration defines x_{n+1} in terms of x_n via the formula

$$x_{n+1} = x_n - \frac{f(x_n)}{f'(x_n)} \tag{1}$$

In the secant method, we replace $f'(x_n)$ in Formula (1) by an approximation that is easily computed. Since the derivative is *defined* by

$$f'(x) = \lim_{h\to 0} \frac{f(x+h)-f(x)}{h}$$

we can say that for small h,

$$f'(x) \approx \frac{f(x+h)-f(x)}{h}$$

(In Section 4.3, we revisit this subject and learn that this is a finite difference approximation to the first derivative.) In particular, if $x = x_n$ and $h = x_{n-1} - x_n$, we have

$$f'(x_n) \approx \frac{f(x_{n-1})-f(x_n)}{x_{n-1}-x_n} \tag{2}$$

When this is used in Equation (1), the result defines the **secant method**:

Secant Method Formula

$$x_{n+1} = x_n - \left(\frac{x_n - x_{n-1}}{f(x_n)-f(x_{n-1})}\right) f(x_n) \tag{3}$$

The secant method (like Newton's) can be used to solve systems of equations as well.

The name of the method is taken from the fact that the right member of Equation (2) is the slope of a secant line to the graph of f (see Figure 3.9). Of course, the left member is the slope of a *tangent* line to the graph of f. (Similarly, Newton's method could be called the "tangent method.")

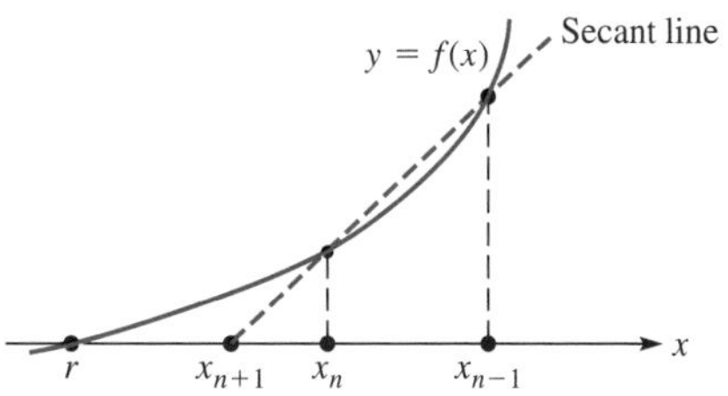

FIGURE 3.9
Secant method

A few remarks about Equation (3) are in order. Clearly, x_{n+1} depends on *two* previous elements of the sequence. So to start, two points (x_0 and x_1) must be provided. Equation (3) can then generate $x_2, x_3, \ldots$. In programming the secant method, we could calculate and test the quantity $f(x_n) - f(x_{n-1})$. If it is nearly zero, an overflow can occur in Equation (3). Of course, if the method is succeeding, the points x_n will be approaching a zero of f, so $f(x_n)$ will be converging to zero. (We are assuming that f is continuous.) Also, $f(x_{n-1})$ will be converging to zero, and, a fortiori, $f(x_n) - f(x_{n-1})$ will approach zero. If the terms $f(x_n)$ and $f(x_{n-1})$ have the same sign, additional significant digits are canceled in the subtraction. So we could perhaps halt the iteration when $|f(x_n) - f(x_{n-1})| \leqq \delta|f(x_n)|$ for some specified tolerance δ, such as $\frac{1}{2} \times 10^{-6}$. (See Computer Exercise 3.3.18.)

Secant Algorithm

A procedure pseudocode for *nmax* steps of the secant method applied to the function f starting with the interval $[a, b] = [x_0, x_1]$ can be written as follows:

Secant **Pseudocode**

```
procedure Secant(f, a, b, nmax, ε)
integer n, nmax;   real a, b, fa, fb, ε, d
external function f
fa ← f(a)
fb ← f(b)
if |fa| > |fb| then
    a ⟷ b
    fa ⟷ fb
end if
output 0, a, fa
output 1, b, fb
for n = 2 to nmax
    if |fa| > |fb| then
        a ⟷ b
        fa ⟷ fb
    end if
    d ← (b − a)/(fb − fa)
    b ← a
    fb ← fa
    d ← d · fa
    if |d| < ε then
        output "convergence"
        return
    end if
```

(Continued)

```
        a ← a − d
        fa ← f(a)
        output n, a, fa
    end for
    end procedure Secant
```

Here $\longleftrightarrow$ means interchange values. The endpoints $[a, b]$ are interchanged, if necessary, to keep $|f(a)| \leqq |f(b)|$. Consequently, the absolute values of the function are nonincreasing; thus, we have $|f(x_n)| \geqq |f(x_{n+1})|$ for $n \geqq 1$.

EXAMPLE 1 If the secant method is used on $p(x) = x^5 + x^3 + 3$ with $x_0 = -1$ and $x_1 = 1$, what is x_8?

Solution The output from the computer program corresponding to the pseudocode for the secant method is as follows. (We used a 32-bit word-length computer.)

n	x_n	$p(x_n)$
0	−1.0	1.0
1	1.0	5.0
2	−1.5	−7.97
3	−1.05575	0.512
4	−1.11416	-9.991×10^{-2}
5	−1.10462	7.593×10^{-3}
6	−1.10529	1.011×10^{-4}
7	−1.10530	2.990×10^{-7}
8	−1.10530	2.990×10^{-7}

We can use mathematical software to find the single real root, -1.1053, and the two pairs of complex roots, $-0.319201 \pm 1.35008i$ and $0.871851 \pm 0.806311i$. ■

Convergence Analysis

The advantages of the secant method are that (after the first step) only one function evaluation is required per step (in contrast to Newton's iteration, which requires two) and that it is almost as rapidly convergent. It can be shown that the basic secant method defined by Equation (3) obeys an equation of the form

Error Vector Approximation

$$e_{n+1} = -\frac{1}{2}\left(\frac{f''(\xi_n)}{f'(\zeta_n)}\right)e_n e_{n-1} \approx -\frac{1}{2}\left(\frac{f''(r)}{f'(r)}\right)e_n e_{n-1} \tag{4}$$

where ξ_n and ζ_n are in the smallest interval that contains r, x_n, and x_{n-1}. Thus, the ratio $e_{n+1}(e_n e_{n-1})^{-1}$ converges to $-\frac{1}{2}f''(r)/f'(r)$. The rapidity of convergence of this method is, in general, between those for bisection and for Newton's method.

To prove the second part of Equation (4), we begin with the definition of the secant method in Equation (3) and the error

$$\begin{aligned} e_{n+1} &= r - x_{n+1} \\ &= r - \frac{f(x_n)x_{n-1} - f(x_{n-1})x_n}{f(x_n) - f(x_{n-1})} \end{aligned}$$

$$= \frac{f(x_n)e_{n-1} - f(x_{n-1})e_n}{f(x_n) - f(x_{n-1})}$$

$$= \left[\frac{x_n - x_{n-1}}{f(x_n) - f(x_{n-1})}\right] \left[\frac{\dfrac{f(x_n)}{e_n} - \dfrac{f(x_{n-1})}{e_{n-1}}}{x_n - x_{n-1}}\right] e_n e_{n-1} \tag{5}$$

By Taylor's Theorem, we establish

$$f(x_n) = f(r - e_n) = f(r) - e_n f'(r) + \frac{1}{2}e_n^2 f''(r) + \mathcal{O}\big(e_n^3\big)$$

Since $f(r) = 0$, this gives us

$$\frac{f(x_n)}{e_n} = -f'(r) + \frac{1}{2}e_n f''(r) + \mathcal{O}\big(e_n^2\big)$$

Changing the index to $n - 1$ yields

$$\frac{f(x_{n-1})}{e_{n-1}} = -f'(r) + \frac{1}{2}e_{n-1} f''(r) + \mathcal{O}\big(e_{n-1}^2\big)$$

By subtraction between these equations, we arrive at

$$\frac{f(x_n)}{e_n} - \frac{f(x_{n-1})}{e_{n-1}} = \frac{1}{2}(e_n - e_{n-1})f''(r) + \mathcal{O}\big(e_{n-1}^2\big)$$

Since $x_n - x_{n-1} = e_{n-1} - e_n$, we reach the equation

$$\frac{\dfrac{f(x_n)}{e_n} - \dfrac{f(x_{n-1})}{e_{n-1}}}{x_n - x_{n-1}} \approx -\frac{1}{2}f''(r)$$

The first bracketed expression in Equation (5) can be written as

$$\frac{x_n - x_{n-1}}{f(x_n) - f(x_{n-1})} \approx \frac{1}{f'(r)}$$

Hence, we have shown the second part of Equation (4). We leave the establishment of the first part of Equation (4) as an exercise because it depends on some material to be covered in Chapter 4. (See Exercise 3.3.18.)

From Equation (4), the order of convergence for the secant method can be expressed in terms of the inequality

Error Bound (Superlinear Convergence)

$$|e_{n+1}| \leqq C|e_n|^{\alpha} \tag{6}$$

where $\alpha = \frac{1}{2}\big(1+\sqrt{5}\,\big) \approx 1.62$ is the **golden ratio**. Since $\alpha > 1$, we say that the convergence is **superlinear**. Assuming that Inequality (6) is true, we can show that the secant method converges under certain conditions.

Let $c = c(\delta)$ be defined as in Equation (2) of Section 3.2. If $|r - x_n| \leqq \delta$ and $|r - x_{n-1}| \leqq \delta$, for some root r, then Equation (4) yields

$$|e_{n+1}| \leqq c|e_n||e_{n-1}| \tag{7}$$

Suppose that the initial points x_0 and x_1 are sufficiently close to r that $c|e_0| \leqq D$ and $c|e_1| \leqq D$ for some $D < 1$. Then

$$c|e_1| \leqq D, \quad c|e_0| \leqq D$$
$$c|e_2| \leqq c|e_1|\,c|e_0| \leqq D^2$$

$$
\begin{aligned}
c|e_3| &\leqq c|e_2|\,c|e_1| \leqq D^3\\
c|e_4| &\leqq c|e_3|\,c|e_2| \leqq D^5\\
c|e_5| &\leqq c|e_4|\,c|e_3| \leqq D^8\\
&\text{etc.}
\end{aligned}
$$

In general, we have

$$|e_n| \leqq c^{-1} D^{\lambda_{n+1}} \tag{8}$$

where inductively,

Fibonacci Sequence

$$\begin{cases} \lambda_1 = 1, \quad \lambda_2 = 1 \\ \lambda_n = \lambda_{n-1} + \lambda_{n-2} \qquad (n \geqq 3) \end{cases} \tag{9}$$

This is the recurrence relation for generating the famous **Fibonacci sequence**, 1, 1, 2, 3, 5, 8, ... It can be shown to have the surprising explicit form

$$\lambda_n = \frac{1}{\sqrt{5}}\left(\alpha^n - \beta^n\right) \tag{10}$$

where $\alpha = \frac{1}{2}(1+\sqrt{5})$ and $\beta = \frac{1}{2}(1-\sqrt{5})$. Since $D < 1$ and $\lambda_n \to \infty$, we conclude from Inequality (8) that $e_n \to 0$. Hence, $x_n \to r$ as $n \to \infty$, and the secant method converges to the root r if x_0 and x_1 are sufficiently close to it.

Next, we show that Inequality (6) is in fact *reasonable*—not a proof. From Inequality (7), we now have

$$
\begin{aligned}
|e_{n+1}| &\leqq c|e_n||e_{n-1}|\\
&= c|e_n|^{\alpha}|e_n|^{1-\alpha}|e_{n-1}|\\
&\approx c|e_n|^{\alpha}\left(c^{-1}D^{\lambda_{n+1}}\right)^{1-\alpha}\left(c^{-1}D^{\lambda_n}\right)\\
&= |e_n|^{\alpha}c^{\alpha-1}D^{\lambda_{n+1}(1-\alpha)+\lambda_n}\\
&= |e_n|^{\alpha}c^{\alpha-1}D^{\lambda_{n+2}-\alpha\lambda_{n+1}}
\end{aligned}
$$

by using an approximation to Inequality (8). In the last line, we used the recurrence relation (9). Now $\lambda_{n+2} - \alpha\lambda_{n+1}$ converges to zero. (See Exercise 3.3.6.). Hence, $c^{\alpha-1}D^{\lambda_{n+2}-\alpha\lambda_{n+1}}$ is bounded, say by C, as a function of n. Thus, we have

Error Vector Approximation

$$|e_{n+1}| \approx C|e_n|^{\alpha}$$

which is a reasonable approximation to Inequality (6).

Alternative Convergence Analysis

Another derivation (with a bit of *hand waving*) for the order of convergence of the secant method can be given by using a general recurrence relation. Equation (4) gives us

$$e_{n+1} \approx K e_n e_{n-1}$$

where $K = -\frac{1}{2}f''(r)/f'(r)$. We can write this as

$$|K e_{n+1}| \approx |K e_n|\,|K e_{n-1}|$$

Let $z_i = \log|K e_i|$. Then we want to solve the recurrence equation

$$z_{n+1} = z_n + z_{n-1}$$

where z_0 and z_1 are arbitrary. This is a linear recurrence relation with constant coefficients similar to the one for the Fibonacci numbers (9) except that the first two values z_0 and z_1

are unknown. The solution is of the form

$$z_n = A\alpha^n + B\beta^n \tag{11}$$

where $\alpha = \frac{1}{2}\left(1+\sqrt{5}\right)$ and $\beta = \frac{1}{2}\left(1-\sqrt{5}\right)$. These are the roots of the quadratic equation $\lambda^2 - \lambda - 1 = 0$. Since $|\alpha| > |\beta|$, the term $A\alpha^n$ dominates, and we can say that

$$z_n \approx A\alpha^n$$

for large n and for some constant A. Consequently, we have

$$|Ke_n| \approx 10^{A\alpha^n}$$

Then it follows that

$$|Ke_{n+1}| \approx 10^{A\alpha^{n+1}} = \left(10^{A\alpha^n}\right)^\alpha = |Ke_n|^\alpha$$

Hence, we have

Error Vectors Approximation

$$|e_{n+1}| \approx C|e_n|^\alpha \tag{12}$$

for large n and for some constant C. Again, Inequality (6) is *essentially* established! A rigorous proof of Inequality (6) is tedious and quite long.

Comparison of Methods

Pros and Cons of Bisection, Newton's, and Secant Methods

In this chapter, three primary methods for solving an equation $f(x) = 0$ have been presented. The bisection method is reliable but slow. Newton's method is fast but often only near the root and requires f'. The secant method is nearly as fast as Newton's method and does not require knowledge of the derivative f', which may not be available or may be too expensive to compute. The user of the bisection method must provide two points at which the signs of $f(x)$ differ, and the function f need only be continuous. In using Newton's method, one must specify a starting point near the root, and f must be differentiable. The secant method requires two good starting points. Newton's procedure can be interpreted as the repetition of a two-step procedure summarized by the prescription *linearize and solve*. This strategy is applicable in many other numerical problems, and its importance cannot be overemphasized. Both Newton's method and the secant method fail to bracket a root. The modified false position method can retain the advantages of both methods.

The secant method is often faster at approximating roots of nonlinear functions in comparison to bisection and false position. Unlike these two methods, the intervals $[a_k, b_k]$ do not have to be on opposite sides of the root and have a change of sign. Moreover, the slope of the secant line can become quite small, and a step can move far from the current point. The secant method can fail to find a root of a nonlinear function that has a small slope near the root because the secant line can jump a large amount.

For nice functions and guesses relatively close to the root, most of these methods require relatively few iterations before coming close to the root. However, there are pathological functions that can cause troubles for any of those methods. When selecting a method for solving a given nonlinear problem, one must consider many issues such as what you know about the behavior of the function, an interval $[a, b]$ satisfying $f(a)f(b) < 0$, the first derivative of the function, a good initial guess to the desired root, and so on.

Hybrid Schemes

Combining Schemes

In an effort to find the *best* algorithm for finding a zero of a given function, various hybrid methods have been developed. Some of these procedures combine the bisection method (used during the early iterations) with either the secant method or the Newton method. Also, adaptive schemes are used for monitoring the iterations and for carrying out stopping rules. More information on some hybrid secant-bisection methods and hybrid Newton-bisection methods with adaptive stopping rules can be found in Bus and Dekker [1975], Dekker [1969], Kahaner, Moler, and Nash [1989], and Novak, Ritter, and Woźniakowski [1995].

Fixed-Point Iteration

Yet Another Approach

For a nonlinear equation $f(x) = 0$, we seek a point where the curve f intersects the x-axis ($y = 0$). An alternative approach is to recast the problem as a fixed-point problem $x = g(x)$ for a related nonlinear function g. For the fixed point problem, we seek a point where the curve g intersects the diagonal line $y = x$. A value of x such that $x = g(x)$ is a **fixed point** of g because x is unchanged when g is applied to it. Many iterative algorithms for solving a nonlinear equation $f(x) = 0$ are based on a fixed-point iterative method

$$x^{(n+1)} = g\left(x^{(n)}\right)$$

where g has fixed points that are solutions of $f(x) = 0$. An initial starting value $x^{(0)}$ is selected, and the iterative method is applied repeatedly until it converges sufficiently well.

EXAMPLE 2 Apply the fixed-point procedure, where $g(x) = 1 + 2/x$, starting with $x^{(0)} = 1$, to compute a zero of the nonlinear function $f(x) = x^2 - x - 2$. Graphically, trace the convergence process.

Solution The fixed-point method is

$$x^{(n+1)} = 1 + \frac{2}{x^{(n)}}$$

Eight steps of the iterative algorithm are $x^{(0)} = 1$, $x^{(1)} = 3$, $x^{(2)} = 5/3$, $x^{(3)} = 11/5$, $x^{(4)} = 21/11$, $x^{(5)} = 43/21$, $x^{(6)} = 85/43$, $x^{(7)} = 171/85$, and $x^{(8)} = 341/171 \approx 1.99415$. In Figure 3.10, we see that these steps spiral into the fixed point 2.

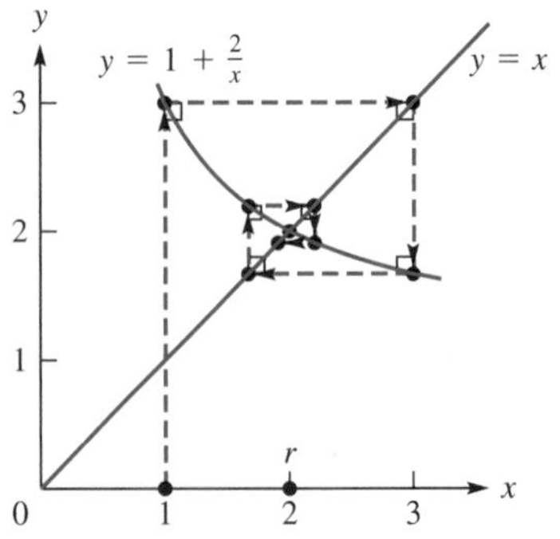

FIGURE 3.10 Fixed-point iterations for $f(x) = x^2 - x - 2$

■

For a given nonlinear equation $f(x) = 0$, there may be many equivalent fixed-point problems $x = g(x)$ with different functions g, some better than others. A simple way to

Locally Convergence

characterize the behavior of an iterative method $x^{(n+1)} = g\left(x^{(n)}\right)$ is *locally convergent* for x^* if $x^* = g(x^*)$ and $|g'(x^*)| < 1$. By *locally convergent*, we mean that there is an interval containing $x^{(0)}$ such that the fixed-point method converges for any starting value $x^{(0)}$ within that interval. If $|g'(x^*)| > 1$, then the fixed-point method diverges for any starting point $x^{(0)}$ other than x^*. Fixed-point iterative methods are used in standard practice for solving many science and engineering problems. In fact, the fixed-point theory can simplify the proof of the convergence of Newton's method.

For additional details and sample plots, see Kincaid and Cheney [2002] or Epureanu and Greenside [1998]. Also, look at other items in the Bibliography. For example, an expository paper by Ypma [1995] traces the historical development of Newton's method through notes, letters, and publications by Isaac Newton, Joseph Raphson, and Thomas Simpson.

Summary 3.3

- The **secant method** for finding a zero r of a function $f(x)$ is written as

$$x_{n+1} = x_n - \left(\frac{x_n - x_{n-1}}{f(x_n) - f(x_{n-1})}\right) f(x_n)$$

  for $n \geqq 1$, which requires two initial values x_0 and x_1. After the first step, only one new function evaluation per step is needed.

- After $n+1$ steps of the secant method, the error iterates $e_i = r - x_i$ obey the equation

$$e_{n+1} = -\frac{1}{2}\left(\frac{f''(\xi_n)}{f'(\zeta_n)}\right) e_n e_{n-1}$$

  which leads to the approximation

$$|e_{n+1}| \approx C|e_n|^{(1+\sqrt{5})/2} \approx C|e_n|^{1.62}$$

  Therefore, the secant method has **superlinear convergence** behavior.

Exercises 3.3

[a]**1.** Calculate an approximate value for $4^{3/4}$ using one step of the secant method with $x_0 = 3$ and $x_1 = 2$.

2. If we use the secant method on $f(x) = x^3 - 2x + 2$ starting with $x_0 = 0$ and $x_1 = 1$, what is x_2?

[a]**3.** If the secant method is used on $f(x) = x^5 + x^3 + 3$ and if $x_{n-2} = 0$ and $x_{n-1} = 1$, what is x_n?

[a]**4.** If $x_{n+1} = x_n + (2 - e^{x_n})(x_n - x_{n-1})/(e^{x_n} - e^{x_{n-1}})$ with $x_0 = 0$ and $x_1 = 1$, what is $\lim_{n\to\infty} x_n$?

5. Using the bisection method, Newton's method, and the secant method, find the largest positive root correct to three decimal places of $x^3 - 5x + 3 = 0$. (All roots are in $[-3, +3]$.)

6. Prove that in the first analysis of the secant method, $\lambda_{n+1} - \alpha\lambda_n$ converges to zero as $n \to \infty$.

7. Establish Equation (10).

8. Write out the derivation of the order of convergence of the secant method that uses recurrence relations; that is, find the constants A and B in Equation (11), and fill in the details in arriving at Equation (12).

[a]**9.** What is the appropriate formula for finding square roots using the secant method? (Refer to Exercise 3.2.1.)

10. Establish that the formula for the secant method can be written as

$$x_{n+1} = \frac{x_{n-1} f(x_n) - x_n f(x_{n-1})}{f(x_n) - f(x_{n-1})}$$

Explore whether it or Equation (3) is more numerically stable with better achievable accuracy in a computer program.

11. Show that if the iterates in Newton's method converge to a point r for which $f'(r) \neq 0$, then $f(r) = 0$. Establish the same assertion for the secant method.
Hint: In the latter, the Mean-Value Theorem of Differential Calculus is useful. This is the case $n = 0$ in Taylor's Theorem.

[a]**12.** A method of finding a zero of a given function f proceeds as follows. Two initial approximations x_0 and x_1 to the zero are chosen, the value of x_0 is fixed, and successive iterations are given by

$$x_{n+1} = x_n - \left(\frac{x_n - x_0}{f(x_n) - f(x_0)}\right) f(x_n)$$

This process will converge to a zero of f under certain conditions. Show that the rate of convergence to a simple zero is *linear* under some conditions.

13. Test the following sequences for different types of convergence (i.e., linear, superlinear, or quadratic), where $n = 1, 2, 3 \ldots.$

[a]**a.** $x_n = n^{-2}$ **b.** $x_n = 2^{-n}$ [a]**c.** $x_n = 2^{-2^n}$

d. $x_n = 2^{-a_n}$ with $a_0 = a_1 = 1$ and $a_{n+1} = a_n + a_{n-1}$ for $n \geqq 2$

14. This exercise and the next three deal with the method of **functional iteration**. The method of functional iteration is as follows: Starting with any x_0, we define $x_{n+1} = f(x_n)$, where $n = 0, 1, 2, \ldots.$ Show that if f is continuous and if the sequence $\{x_n\}$ converges, then its limit is a fixed point of f.

[a]**15.** (Continuation) Show that if f is a function defined on the whole real line whose derivative satisfies $|f'(x)| \leqq c$ with a constant c less than 1, then the method of functional iteration produces a fixed point of f.
Hint: In establishing this, the Mean-Value Theorem from Section 1.2 is helpful.

[a]**16.** (Continuation) With a calculator, try the method of functional iteration with $f(x) = x/2 + 1/x$, taking $x_0 = 1$. What is the limit of the resulting sequence?

[a]**17.** (Continuation) Using functional iteration, show that the equation $10 - 2x + \sin x = 0$ has a root. Locate the root approximately by drawing a graph. Starting with your approximate root, use functional iteration to obtain the root accurately by using a calculator.
Hint: Write the equation in the form $x = 5 + \frac{1}{2}\sin x$.

18. Establish the first part of Equation (4) using Equation (5).
Hint: Use the relationship between divided differences and derivatives from Section 4.3.

Computer Exercises 3.3

[a]**1.** Use the secant method to find the zero near -0.5 of $f(x) = e^x - 3x^2$. This function also has a zero near 4. Find this positive zero by Newton's method.

2. Write

procedure *Secant(f, x1, x2, epsi, delta, maxf, x, ierr)*

which uses the secant method to solve $f(x) = 0$. The input parameters are as follows: f is the name of the given function; $x1$ and $x2$ are the initial estimates of the solution; *epsi* is a positive tolerance such that the iteration stops if the difference between two consecutive iterates is smaller than this value; *delta* is a positive tolerance such that the iteration stops if a function value is smaller in magnitude than this value; and *maxf* is a positive integer bounding the number of evaluations of the function allowed. The output parameters are as follows: x is the final estimate of the solution, and *ierr* is an integer error flag that indicates whether a tolerance test was violated. Test this routine using the function of Computer Exercise 3.3.1. Print the final estimate of the solution and the value of the function at this point.

3. Find a zero of one of the functions given in the introduction of this chapter using one of the methods introduced in this chapter.

4. Write and test a recursive procedure for the secant method.

5. Rerun the example in this section with $x_0 = 0$ and $x_1 = 1$. Explain any unusual results.

6. Write a simple program to compare the secant method with Newton's method for finding a root of each function.

[a]**a.** $x^3 - 3x + 1$ with $x_0 = 2$

[a]**b.** $x^3 - 2\sin x$ with $x_0 = \frac{1}{2}$

Use the x_1 value from Newton's method as the second starting point for the secant method. Print out each iteration for both methods.

[a]**7.** Write a simple program to find the root of $f(x) = x^3+2x^2+10x-20$ using the secant method with starting values $x_0 = 2$ and $x_1 = 1$. Let it run at most 20 steps, and include a stopping test as well. Compare the number of steps needed here to the number needed in Newton's method. Is the convergence quadratic?

8. Test the secant method on the set of functions $f_k(x) = 2e^{-k}x+1-3e^{-kx}$ for $k = 1, 2, 3, \dots, 10$. Use the starting points 0 and 1 in each case.

[a]**9.** An example by Wilkinson [1963] shows that minute alterations in the coefficients of a polynomial may have massive effects on the roots. Let

$$f(x) = (x-1)(x-2)\cdots(x-20)$$

which has become known as the **Wilkinson polynomial**. The zeros of f are, of course, the integers $1, 2, \dots, 20$. Try to determine what happens to the zero $r = 20$ when the function is altered to $f(x) - 10^{-8}x^{19}$.
Hint: The secant method in double precision will locate a zero in the interval [20, 21].

10. Test the secant method on an example in which r, $f'(r)$, and $f''(r)$ are known in advance. Monitor the ratios $e_{n+1}/(e_n e_{n-1})$ to see whether they converge to $-\frac{1}{2}f''(r)/f'(r)$. The function $f(x) = \arctan x$ is suitable for this experiment.

11. Using a function of your choice, verify numerically that the iterative method

$$x_{n+1} = x_n - \frac{f(x_n)}{\sqrt{[f'(x_n)]^2 - f(x_n)f''(x_n)}}$$

is cubically convergent at a simple root but only linearly convergent at a multiple root.

12. Test numerically whether **Olver's method**, given by

$$x_{n+1} = x_n - \frac{f(x_n)}{f'(x_n)} - \frac{1}{2}\frac{f''(x_n)}{f'(x_n)}\left[\frac{f(x_n)}{f'(x_n)}\right]^2$$

is cubically convergent to a root of f. Try to establish that it is.

13. (Continuation) Repeat for **Halley's method**

$$x_{n+1} = x_n - \frac{1}{a_n} \quad \text{with} \quad a_n = \frac{f'(x_n)}{f(x_n)} - \frac{1}{2}\left[\frac{f''(x_n)}{f'(x_n)}\right]$$

14. (**Moler-Morrison Algorithm**) Computing an approximation for $\sqrt{x^2+y^2}$ does not require square roots. It can be done as follows:

```
real function f(x, y)
integer n;   real a, b, c, x, y
f ← max {|x|, |y|}
a ← min {|x|, |y|}
for n = 1 to 3
    b ← (a/f)^2
    c ← b/(4 + b)
    f ← f + 2cf
    a ← ca
end for
end function f
```

Test the algorithm on some simple cases such as $(x, y) = (3, 4)$, $(-5, 12)$, and $(7, -24)$. Then write a routine that uses the function $f(x, y)$ for approximating the **Euclidean norm** of a vector $x = (x_1, x_2, \dots, x_n)$; that is, the nonnegative number $\|x\| = \left(x_1^2 + x_2^2 + \cdots + x_n^2\right)^{1/2}$.

15. Study the following functions by starting with any initial value of x_0 in the domain [0, 2] and iterating $x_{n+1} = F(x_n)$. First use a calculator and then a computer. Explain the results.

a. Use the **tent function**

$$F(x) = \begin{cases} 2x & \text{if } 2x < 1 \\ 2x - 1 & \text{if } 2x \geqq 1 \end{cases}$$

b. Repeat using the function

$$F(x) = 10x \text{ (modulo 1)}$$

Hint: Don't be surprised by chaotic behavior. The interested reader can learn more about the dynamics of one-dimensional maps by reading papers such as the one by Bassien [1998].

16. Show how the secant method can be used to solve systems of equations such as those in Computer Exercises 3.2.21–3.2.23.

17. (**Student Research Project**) **Muller's method** is an algorithm for computing solutions of an equation $f(x) = 0$. It is similar to the secant method in that it replaces f locally by a simple function and finds a root of it. Naturally, this step is repeated. The simple function chosen in Muller's method is a quadratic polynomial, p, that interpolates f at the three most recent points. After p has been determined, its roots are computed, and one of them is chosen as the next point in the sequence. Since this quadratic function may have complex roots, the algorithm should be programmed with this in mind. Suppose that points x_{n-2}, x_{n-1}, and x_n have been computed. Set

$$p(x) = a(x - x_n)(x - x_{n-1}) + b(x - x_n) + c$$

where a, b, and c are determined so that p interpolates f at the three points mentioned previously. Then find the roots of p and take x_{n+1} to be the root of p closest to x_n. At the beginning, three points must be furnished by the user. Program the method, allowing for complex numbers throughout. Test your program on the example

$$p(x) = x^3 + x^2 - 10x - 10$$

If the first three points are $1, 2, 3$, then you should find that the polynomial is $p(x) = 7(x-3)(x-2) + 14(x-3) - 4$ and $x_4 = 3.17971\,086$. Next, test your code on a polynomial having real coefficients but some complex roots.

18. Program and test the code for the secant algorithm after incorporating the stopping criterion described in the text.

19. Using mathematical software such as MATLAB, Mathematica, and Maple, find the real zero of the polynomial $p(x) = x^5 + x^3 + 3$. Attain more digits of accuracy than shown in the solution to Example 1 in the text.

20. (Continuation) Using mathematical software that allows for complex roots, find all zeros of the polynomial.

21. Program a hybrid method for solving several of the nonlinear problems given as examples in the text, and compare your results with those given.

22. Find the fixed points for each of the following functions:

a. $e^x + 1$ **b.** $e^{-x} - x$ **c.** $x^2 - 4\sin x$

d. $x^3 + 6x^2 + 11x - 6$ **e.** $\sin x$

23. For the nonlinear equation $f(x) = x^2 - x - 2 = 0$ with roots 1 and 2, write four fixed-point problems $x = g(x)$ that are equivalent. Plot all of these, and show that they all intersect the line $x = y$. Also, plot the convergence steps of each of these fixed-point iterations for different starting values $x^{(0)}$. Show that the behavior of these fixed-point schemes can vary wildly: slow convergence, fast convergence, and divergence.

4 Interpolation and Numerical Differentiation

The viscosity of water has been experimentally determined at different temperatures, as indicated in the following table:

Temperature	0°	5°	10°	15°
Viscosity	1.792	1.519	1.308	1.140

From this table, how can we estimate a reasonable value for the viscosity at temperature 8°?

The method of polynomial interpolation, described in Section 4.1, can be used to create a polynomial of degree 3 that assumes the values in the table. This polynomial should provide acceptable intermediate values for temperatures not tabulated. The value of that polynomial at the point 8° turns out to be 1.386.

4.1 Polynomial Interpolation

Preliminary Remarks

We pose three problems concerning the representation of functions to give an indication of the subject matter in this chapter, in Chapter 6 (on splines), and in Chapter 9 (on least squares).

First, suppose that we have a table of numerical values of a function:

Problem 1

x	x_0	x_1	$\cdots$	x_n
y	y_0	y_1	$\cdots$	y_n

Is it possible to find a simple and convenient formula that reproduces the given points exactly?

Problem 2

The second problem is similar, but it is assumed that the given table of numerical values is contaminated by errors, as might occur if the values came from a physical experiment. Now we ask for a formula that represents the data (approximately) and, if possible, filters out the errors.

Problem 3

As a third problem, a function f is given, perhaps in the form of a computer procedure, but it is an expensive function to evaluate. In this case, we ask for another function g that is simpler to evaluate and produces a reasonable approximation to f. Sometimes in this problem, we want g to approximate f with full machine precision.

A Simple Function

In all of these problems, a simple function p can be obtained that represents or approximates the given table or function f. The representation p can always be taken to be a polynomial, although many other types of simple functions can also be used. Once a simple function p has been obtained, it can be used in place of f in many situations. For example, the integral of f could be estimated by the integral of p, and the latter should generally be easier to evaluate.

In many situations, a polynomial solution to the problems outlined above are unsatisfactory from a practical point of view, and other classes of functions must be considered. In this book, one other class of versatile functions is discussed: the spline functions (see Chapter 6). The present chapter concerns polynomials exclusively, and Chapter 9 discusses general linear families of functions, of which splines and polynomials are important examples.

High-Degree Polynomial

The obvious way in which a polynomial can *fail* as a practical solution to one of the preceding problems is that its degree may be unreasonably high. For instance, if the table considered contains 1,000 entries, a polynomial of degree 999 may be required to represent it. Polynomials also may have the surprising defect of being highly oscillatory. If the table is precisely represented by a polynomial p, then $p(x_i) = y_i$ for $0 \leqq i \leqq n$. For points other than the given x_i, however, $p(x)$ may be a very poor representation of the function from which the table arose. The example in Section 4.2 involving the Runge function illustrates this phenomenon.

Polynomial Interpolation

We begin again with a table of values:

Polynomial Interpolations Table

x	x_0	x_1	$\cdots$	x_n
y	y_0	y_1	$\cdots$	y_n

and assume that the x_i's form a set of $n + 1$ distinct points. The table represents $n + 1$ points in the Cartesian plane, and we want to find a polynomial curve that passes through all points. Thus, we seek to determine a polynomial that is defined for *all* x and takes on the corresponding values of y_i for each of the $n + 1$ distinct x_i's in this table. A polynomial p for which $p(x_i) = y_i$ when $0 \leqq i \leqq n$ is said to **interpolate** the table. The points x_i are called **nodes**.

Constant Case

Consider the first and simplest case, $n = 0$. Here, a constant function solves the problem. In other words, the polynomial p of degree 0 defined by the equation $p(x) = y_0$ reproduces the one-node table.

The next simplest case occurs when $n = 1$. Since a straight line can be passed through two points, a linear function is capable of solving the problem. Explicitly, the polynomial p defined by

Linear Case

$$\begin{aligned} p(x) &= \left(\frac{x - x_1}{x_0 - x_1}\right)y_0 + \left(\frac{x - x_0}{x_1 - x_0}\right)y_1 \\ &= y_0 + \left(\frac{y_1 - y_0}{x_1 - x_0}\right)(x - x_0) \end{aligned}$$

is of first degree (at most) and reproduces the table. That means (in this case) that $p(x_0) = y_0$ and $p(x_1) = y_1$, as is easily verified. This p is used for **linear interpolation**.

EXAMPLE 1 Find the polynomial of least degree that interpolates this table:

x	1.4	1.25
y	3.7	3.9

Solution By the linear case equation, the polynomial that is sought is

$$p(x) = \left(\frac{x - 1.25}{1.4 - 1.25}\right)3.7 + \left(\frac{x - 1.4}{1.25 - 1.4}\right)3.9$$
$$= 3.7 + \left(\frac{3.9 - 3.7}{1.25 - 1.4}\right)(x - 1.4)$$
$$= 3.7 - \frac{4}{3}(x - 1.4)$$

■

As we can see, an interpolating polynomial can be written in a variety of forms, including the Newton form and the Lagrange form. The Newton form is probably the most convenient and efficient; however, conceptually, the Lagrange form has several advantages. We begin with the Lagrange form, since it may be easier to understand.

Interpolating Polynomial: Lagrange Form

Suppose that we wish to interpolate arbitrary functions at a set of fixed nodes $x_0, x_1, \ldots, x_n$. We first define a system of $n + 1$ special polynomials of degree n known as **cardinal polynomials** in interpolation theory. These are denoted by $\ell_0, \ell_1, \ldots, \ell_n$ and have the property

Kronecker Delta Property

$$\ell_i(x_j) = \delta_{ij} = \begin{cases} 0 & \text{if } i \neq j \\ 1 & \text{if } i = j \end{cases}$$

Once these are available, we can interpolate *any* function f by the **Lagrange form of the interpolation polynomial**:

Lagrange Form

$$p_n(x) = \sum_{i=0}^{n} \ell_i(x) f(x_i) \tag{1}$$

This function p_n, being a linear combination of the polynomials ℓ_i, is itself a polynomial of degree at most n. Furthermore, when we evaluate p_n at x_j, we get $f(x_j)$:

$$p_n(x_j) = \sum_{i=0}^{n} \ell_i(x_j) f(x_i) = \ell_j(x_j) f(x_j) = f(x_j)$$

Thus, p_n is the interpolating polynomial for the function f at nodes $x_0, x_1, \ldots, x_n$. It remains now only to write the formula for the **cardinal polynomial** ℓ_i, which is

Cardinal Polynomial

$$\ell_i(x) = \prod_{\substack{j \neq i \\ j=0}}^{n} \left(\frac{x - x_j}{x_i - x_j}\right) \qquad (0 \leqq i \leqq n) \tag{2}$$

This formula indicates that $\ell_i(x)$ is the product of n linear factors:

Expanded Form

$$\ell_i(x) = \left(\frac{x - x_0}{x_i - x_0}\right)\left(\frac{x - x_1}{x_i - x_1}\right) \cdots \left(\frac{x - x_{i-1}}{x_i - x_{i-1}}\right)\left(\frac{x - x_{i+1}}{x_i - x_{i+1}}\right) \cdots \left(\frac{x - x_n}{x_i - x_n}\right)$$

(The denominators are just numbers; the variable x occurs only in the numerators.) Thus, ℓ_i is a polynomial of degree n. Notice that when $\ell_i(x)$ is evaluated at $x = x_i$, each factor in the preceding equation becomes 1. Hence, we obtain $\ell_i(x_i) = 1$. But when $\ell_i(x)$ is evaluated at any *other* node, say, x_j, one of the factors in the preceding equation is 0, and $\ell_i(x_j) = 0$, for $i \neq j$.

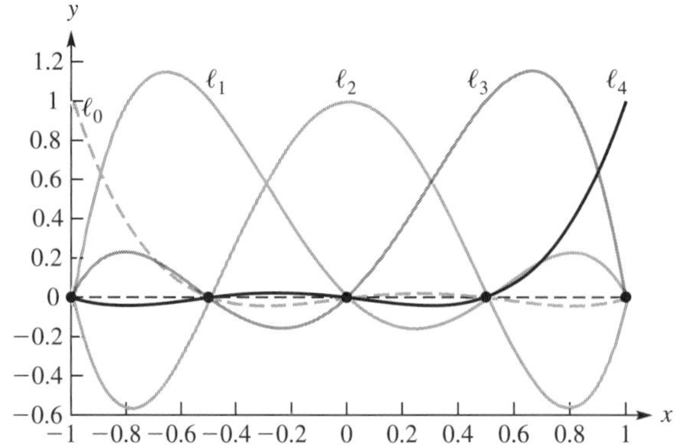

FIGURE 4.1 First few Lagrange cardinal polynomials

Figure 4.1 shows the first few Lagrange cardinal polynomials: $\ell_0(x)$, $\ell_1(x)$, $\ell_2(x)$, $\ell_3(x)$, and $\ell_4(x)$.

EXAMPLE 2 Write out the cardinal polynomials appropriate to the problem of interpolating the following table, and give the Lagrange form of the interpolating polynomial:

x	$\frac{1}{3}$	$\frac{1}{4}$	1
$f(x)$	2	−1	7

Solution Using Equation (2), we have

$$\ell_0(x) = \frac{\left(x-\frac{1}{4}\right)(x-1)}{\left(\frac{1}{3}-\frac{1}{4}\right)\left(\frac{1}{3}-1\right)} = -18\left(x-\frac{1}{4}\right)(x-1)$$

$$\ell_1(x) = \frac{\left(x-\frac{1}{3}\right)(x-1)}{\left(\frac{1}{4}-\frac{1}{3}\right)\left(\frac{1}{4}-1\right)} = 16\left(x-\frac{1}{3}\right)(x-1)$$

$$\ell_2(x) = \frac{\left(x-\frac{1}{3}\right)\left(x-\frac{1}{4}\right)}{\left(1-\frac{1}{3}\right)\left(1-\frac{1}{4}\right)} = 2\left(x-\frac{1}{3}\right)\left(x-\frac{1}{4}\right)$$

Therefore, the interpolating polynomial in Lagrange's form is

$$p_2(x) = -36\left(x-\frac{1}{4}\right)(x-1) - 16\left(x-\frac{1}{3}\right)(x-1) + 14\left(x-\frac{1}{3}\right)\left(x-\frac{1}{4}\right)$$ ■

Existence of Interpolating Polynomial

The Lagrange interpolation formula proves the existence of an interpolating polynomial for any table of values. There is another constructive way of proving this fact, and it leads to a different formula.

Existence of $p(x)$

Suppose that we have succeeded in finding a polynomial p that reproduces *part* of the table. Assume, say, that $p(x_i) = y_i$ for $0 \leqq i \leqq k$. We shall attempt to add to p another term that enables the new polynomial to reproduce one more entry in the table. We consider

$$p(x) + c(x - x_0)(x - x_1)\cdots(x - x_k)$$

where c is a constant to be determined. This is surely a polynomial. It also reproduces the first k points in the table because p itself does so, and the added portion takes the value 0 at each of the points $x_0, x_1, \ldots, x_k$. (Its form is chosen for precisely this reason.) Now we adjust the parameter c so that the new polynomial takes the value y_{k+1} at x_{k+1}. Imposing this condition, we obtain

$$p(x_{k+1}) + c(x_{k+1} - x_0)(x_{k+1} - x_1) \cdots (x_{k+1} - x_k) = y_{k+1}$$

The proper value of c can be obtained from this equation because none of the factors $x_{k+1} - x_i$, for $0 \leqq i \leqq k$, can be zero. Remember our original assumption that the x_i's are all distinct.

Inductive Reasoning

This analysis is an example of inductive reasoning. We have shown that the process can be started and that it can be continued. Hence, the following formal statement has been partially justified:

■ **Theorem 1**

Theorem on Existence of Polynomial Interpolation

If points $x_0, x_1, \ldots, x_n$ are distinct, then for arbitrary real values $y_0, y_1, \ldots, y_n$, there is a unique polynomial p of degree at most n such that $p(x_i) = y_i$ for $0 \leqq i \leqq n$.

Degree of $p(x)$

Two parts of this formal statement must still be established. First, the degree of the polynomial increases by at most 1 in each step of the inductive argument. At the beginning, the degree was at most 0, so at the end, the degree is at most n.

Uniqueness of $p(x)$

Second, we establish the uniqueness of the polynomial p. Suppose that another polynomial q claims to accomplish what p does; that is, q is also of degree at most n and satisfies $q(x_i) = y_i$ for $0 \leqq i \leqq n$. Then the polynomial $p - q$ is of degree at most n and takes the value 0 at $x_0, x_1, \ldots, x_n$. Recall, however, that a *nonzero* polynomial of degree n can have at most n roots. We conclude that $p = q$, which establishes the uniqueness of p.

Interpolating Polynomial: Newton Form

In Example 2, we found the Lagrange form of the interpolating polynomial:

$$p_2(x) = -36\left(x - \frac{1}{4}\right)(x - 1) - 16\left(x - \frac{1}{3}\right)(x - 1) + 14\left(x - \frac{1}{3}\right)\left(x - \frac{1}{4}\right)$$

It can be simplified to

$$p_2(x) = -\frac{79}{6} + \frac{349}{6}x - 38x^2$$

Now, we learn that this polynomial can be written in another form called the nested Newton form:

Nested Newton's Form for $p_2(x)$

$$p_2(x) = 2 + \left(x - \frac{1}{3}\right)\left[36 + \left(x - \frac{1}{4}\right)(-38)\right]$$

It involves the fewest arithmetic operations and is recommended for evaluating $p_2(x)$. It cannot be overemphasized that the Newton and Lagrange forms are just two different derivations for precisely the same polynomial. The Newton form has the advantage of easy extensibility to accommodate additional data points.

The preceding discussion provides a method for constructing an interpolating polynomial. The method is known as the **Newton algorithm**, and the resulting polynomial is the **Newton form of the interpolating polynomial**.

EXAMPLE 3 Using the Newton algorithm, find the interpolating polynomial of least degree for this table:

x	0	1	-1	2	-2
y	-5	-3	-15	39	-9

Solution In the construction, five successive polynomials appear; these are labeled p_0, p_1, p_2, p_3, and p_4. The polynomial p_0 is defined to be

Polynomials p_0, p_1, p_2, p_3, p_4

$$p_0(x) = -5$$

The polynomial p_1 has the form

$$p_1(x) = p_0(x) + c(x - x_0) = -5 + c(x - 0)$$

The interpolation condition placed on p_1 is that $p_1(1) = -3$. Therefore, we have $-5 + c(1 - 0) = -3$. Hence, $c = 2$, and p_1 is

$$p_1(x) = -5 + 2x$$

The polynomial p_2 has the form

$$p_2(x) = p_1(x) + c(x - x_0)(x - x_1) = -5 + 2x + cx(x - 1)$$

The interpolation condition placed on p_2 is that $p_2(-1) = -15$. Hence, we have $-5 + 2(-1) + c(-1)(-1 - 1) = -15$. This yields $c = -4$, so

$$p_2(x) = -5 + 2x - 4x(x - 1)$$

The remaining steps for $p_3(x)$ are similar. The final result is the Newton form of the interpolating polynomial:

$$p_4(x) = -5 + 2x - 4x(x - 1) + 8x(x - 1)(x + 1) + 3x(x - 1)(x + 1)(x - 2) \quad \blacksquare$$

Adding a New Term

Later, we develop a better algorithm for constructing the Newton interpolating polynomial. Nevertheless, the method just explained is a systematic one and involves very little computation. An important feature to notice is that each new polynomial in the algorithm is obtained from its predecessor by adding a new term. Thus, at the end, the final polynomial exhibits all the previous polynomials as constituents.

Nested Form

Before continuing, let's rewrite the Newton form of the interpolating polynomial for efficient evaluation.

EXAMPLE 4 Write the polynomial p_4 of Example 3 in *nested* form and use it to evaluate $p_4(3)$.

Solution We write p_4 as

Nested Form $p_4(x)$

$$p_4(x) = -5 + x\big(2 + (x - 1)\big(-4 + (x + 1)\big(8 + (x - 2)3\big)\big)\big)$$

Therefore, we obtain

$$\begin{aligned} p_4(3) &= -5 + 3\big(2 + 2\big(-4 + 4(8 + 3)\big)\big) \\ &= 241 \end{aligned}$$

Another solution, also in nested form, is

$$p_4(x) = -5 + x\big(4 + x\big(-7 + x(2 + 3x)\big)\big)$$

from which we obtain

Nested Multiplication $p_4(x)$

$$p_4(3) = -5 + 3\big(4 + 3\big(-7 + 3(2 + 3 \cdot 3)\big)\big) = 241$$

This form is obtained by expanding and systematic factoring of the original polynomial. It is also known as a **nested form**, and its evaluation is by **nested multiplication**. ■

To describe nested multiplication in a formal way (so that it can be translated into a pseudocode), consider a general polynomial in the Newton form. It might be

$$p_n(x) = a_0 + a_1(x - x_0) + a_2(x - x_0)(x - x_1) + \cdots + a_n(x - x_0)(x - x_1)\cdots(x - x_{n-1})$$

The nested form of $p_n(x)$ is

$$p_n(x) = a_0 + (x - x_0)\big(a_1 + (x - x_1)\big(a_2 + \cdots + (x - x_{n-2})\big(a_{n-1} + (x - x_{n-1})(a_n)\big)\cdots\big)\big)$$

The **Newton interpolation polynomial** can be written succinctly as

Newton Interpolation Polynomial

$$p_n(x) = \sum_{i=0}^{n} a_i \prod_{j=0}^{i-1}(x - x_j) \tag{3}$$

Here $\prod_{j=0}^{-1}(x - x_j)$ is interpreted to be 1. Also, we can write it as

$$p_n(x) = \sum_{i=0}^{n} a_i\, \pi_i(x)$$

where

Newton Polynomials

$$\pi_i(x) = \prod_{j=0}^{i-1}(x - x_j) \tag{4}$$

Figure 4.2 shows the first few Newton polynomials: $\pi_0(x)$, $\pi_1(x)$, $\pi_2(x)$, $\pi_3(x)$, $\pi_4(x)$, and $\pi_5(x)$.

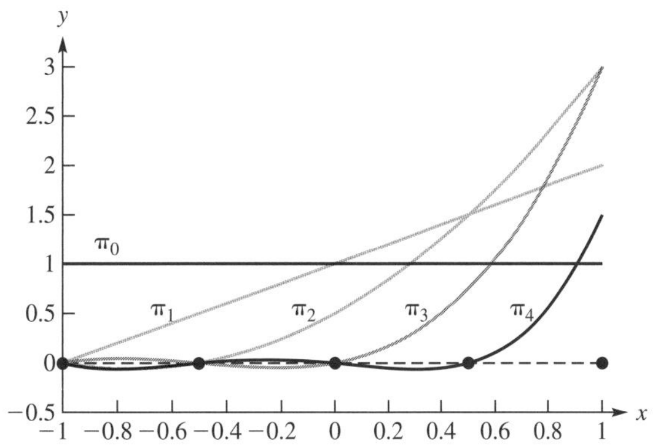

FIGURE 4.2 First few Newton polynomials

In evaluating $p(t)$ for a given numerical value of t, we naturally start with the innermost parentheses, forming successively the following quantities:

$$\begin{aligned}
v_0 &= a_n \\
v_1 &= v_0(t - x_{n-1}) + a_{n-1} \\
v_2 &= v_1(t - x_{n-2}) + a_{n-2} \\
&\vdots \\
v_n &= v_{n-1}(t - x_0) + a_0
\end{aligned}$$

The quantity v_n is now $p(t)$. In the following pseudocode, a subscripted variable is *not* needed for v_i. Instead, we can write

Evaluation of Interpolation Polynomial Pseudocode

```
integer i, n;   real t, v;   real array (a_i)_{0:n}, (x_i)_{0:n}
v ← a_n
for i = n − 1 to 0 step −1
    v ← v(t − x_i) + a_i
end for
```

Here, the array $(a_i)_{0:n}$ contains the $n+1$ coefficients of the Newton form of the interpolating polynomial (3) of degree at most n, and the array $(x_i)_{0:n}$ contains the $n+1$ nodes x_i.

Calculating Coefficients a_i Using Divided Differences

We turn now to the problem of determining the coefficients $a_0, a_1, \ldots, a_n$ efficiently. Again we start with a table of values of a function f:

x	x_0	x_1	x_2	$\cdots$	x_n
$f(x)$	$f(x_0)$	$f(x_1)$	$f(x_2)$	$\cdots$	$f(x_n)$

The points $x_0, x_1, \ldots, x_n$ are assumed to be distinct, but *no* assumption is made about their positions on the real line.

Previously, we established that for each $n = 0, 1, \ldots,$ there exists a unique polynomial p_n such that

Unique Polynomial p_n

- The degree of p_n is at most n.
- $p_n(x_i) = f(x_i)$ for $i = 0, 1, \ldots, n$.

It was shown that p_n can be expressed in the Newton form

$$p_n(x) = a_0 + a_1(x - x_0) + a_2(x - x_0)(x - x_1) + \cdots + a_n(x - x_0)\cdots(x - x_{n-1})$$

A crucial observation about p_n is that the coefficients $a_0, a_1, \ldots$ do not depend on n. In other words, p_n is obtained from p_{n-1} by adding one more term, without altering the coefficients already present in p_{n-1} itself. This is because we began with the hope that p_n could be expressed in the form

Adding One Term at a Time

$$p_n(x) = p_{n-1}(x) + a_n(x - x_0)\cdots(x - x_{n-1})$$

and discovered that it was indeed possible.

A way of systematically determining the unknown coefficients $a_0, a_1, \dots, a_n$ is to set x equal in turn to $x_0, x_1, \dots, x_n$ in the Newton form (3) and to write down the resulting equations:

$$\begin{cases} f(x_0) = a_0 \\ f(x_1) = a_0 + a_1(x_1 - x_0) \\ f(x_2) = a_0 + a_1(x_2 - x_0) + a_2(x_2 - x_0)(x_2 - x_1) \\ \quad \text{etc.} \end{cases} \tag{5}$$

The compact form of Equations (5) is

$$f(x_k) = \sum_{i=0}^{k} a_i \prod_{j=0}^{i-1}(x_k - x_j) \qquad (0 \leqq k \leqq n) \tag{6}$$

Equations (5) can be solved for the a_i's in turn, starting with a_0. Then we see that a_0 depends on $f(x_0)$, that a_1 depends on $f(x_0)$ and $f(x_1)$, and so on. In general, a_k depends on $f(x_0)$, $f(x_1), \dots, f(x_k)$. In other words, a_k depends on the values of f at the nodes $x_0, x_1, \dots, x_k$. The traditional notation is

a_k Divided Difference of Order k

$$a_k = f[x_0, x_1, \dots, x_k] \tag{7}$$

This equation defines $f[x_0, x_1, \dots, x_k]$. The quantity $f[x_0, x_1, \dots, x_k]$ is called the **divided difference of order k** for f. Notice also that the coefficients $a_0, a_1, \dots, a_k$ are *uniquely* determined by System (6). Indeed, there is no possible choice for a_0 other than $a_0 = f(x_0)$. Similarly, there is now no choice for a_1 other than $[f(x_1) - a_0]/(x_1 - x_0)$ and so on. Using Equations (5), we see that the first few divided differences can be written as

a_0, a_1, a_2 Divided Differences

$$a_0 = f(x_0)$$

$$a_1 = \frac{f(x_1) - a_0}{x_1 - x_0} = \frac{f(x_1) - f(x_0)}{x_1 - x_0}$$

$$a_2 = \frac{f(x_2) - a_0 - a_1(x_2 - x_0)}{(x_2 - x_0)(x_2 - x_1)} = \frac{\dfrac{f(x_2) - f(x_1)}{x_2 - x_1} - \dfrac{f(x_1) - f(x_0)}{x_1 - x_0}}{x_2 - x_0}$$

EXAMPLE 5 For the table:

x	1	−4	0
$f(x)$	3	13	−23

determine the quantities $f[x_0]$, $f[x_0, x_1]$, and $f[x_0, x_1, x_2]$.

Solution We write out the system of Equations (5) for this concrete case:

$$\begin{cases} 3 = a_0 \\ 13 = a_0 + a_1(-5) \\ -23 = a_0 + a_1(-1) + a_2(-1)(4) \end{cases}$$

The solution is $a_0 = 3$, $a_1 = -2$, and $a_2 = 7$. Hence, for this function, $f[1] = 3$, $f[1, -4] = -2$, and $f[1, -4, 0] = 7$. ■

With this new notation, the **Newton form of the interpolating polynomial** takes the form

Newton Form of Interpolating Polynomial

$$p_n(x) = \sum_{i=0}^{n} \left\{ f[x_0, x_1, \dots, x_i] \prod_{j=0}^{i-1}(x - x_j) \right\} \tag{8}$$

with the usual convention that $\prod_{j=0}^{-1}(x - x_j) = 1$. Notice that the coefficient of x^n in p_n is $f[x_0, x_1, \ldots, x_n]$ because the term x^n occurs only in $\prod_{j=0}^{n-1}(x - x_j)$. It follows that if f is a polynomial of degree $\leq n - 1$, then $f[x_0, x_1, \ldots, x_n] = 0$.

We return to the question of how to compute the required divided differences $f[x_0, x_1, \ldots, x_k]$. From System (5) or System (6), it is evident that this computation can be performed *recursively*. We simply solve Equation (6) for a_k as follows:

$$f(x_k) = a_k \prod_{j=0}^{k-1}(x_k - x_j) + \sum_{i=0}^{k-1} a_i \prod_{j=0}^{i-1}(x_k - x_j)$$

and

$$a_k = \frac{f(x_k) - \sum_{i=0}^{k-1} a_i \prod_{j=0}^{i-1}(x_k - x_j)}{\prod_{j=0}^{k-1}(x_k - x_j)}$$

Using Equation (7), we have

$$f[x_0, x_1, \ldots, x_k] = \frac{f(x_k) - \sum_{i=0}^{k-1} f[x_0, x_1, \ldots, x_i] \prod_{j=0}^{i-1}(x_k - x_j)}{\prod_{j=0}^{k-1}(x_k - x_j)} \tag{9}$$

■ Algorithm **Computing the Divided Differences of f**

- Set $f[x_0] = f(x_0)$.
- For $k = 1, 2, \ldots, n$, compute $f[x_0, x_1, \ldots, x_k]$ by Equation (9). (10)

EXAMPLE 6 Using Algorithm (10), write out the divided differences formulas for $f[x_0]$, $f[x_0, x_1]$, $f[x_0, x_1, x_2]$, and $f[x_0, x_1, x_2, x_3]$.

Solution

First Four Divided Differences

$$f[x_0] = f(x_0)$$

$$f[x_0, x_1] = \frac{f(x_1) - f[x_0]}{x_1 - x_0}$$

$$f[x_0, x_1, x_2] = \frac{f(x_2) - f[x_0] - f[x_0, x_1](x_2 - x_0)}{(x_2 - x_0)(x_2 - x_1)}$$

$$f[x_0, x_1, x_2, x_3] = \frac{f(x_3) - f[x_0] - f[x_0, x_1](x_3 - x_0) - f[x_0, x_1, x_2](x_3 - x_0)(x_3 - x_1)}{(x_3 - x_0)(x_3 - x_1)(x_3 - x_2)}$$

■

Operation Count

Algorithm (10) is easily programmed and is capable of computing the divided differences $f[x_0]$, $f[x_0, x_1], \ldots, f[x_0, x_1, \ldots, x_n]$ at the cost of $\frac{1}{2}n(3n + 1)$ additions, $(n - 1)(n - 2)$ multiplications, and n divisions excluding arithmetic operations on the indices. Now a more refined method is presented for which the pseudocode requires only three statements (!) and costs only $\frac{1}{2}n(n + 1)$ divisions and $n(n + 1)$ additions.

At the heart of the new method is the following remarkable theorem:

■ **Theorem 2**

Recursive Property of Divided Differences

Recursive Property

The divided differences obey the formula

$$f[x_0, x_1, \dots, x_k] = \frac{f[x_1, x_2, \dots, x_k] - f[x_0, x_1, \dots, x_{k-1}]}{x_k - x_0} \tag{11}$$

Proof Since $f[x_0, x_1, \dots, x_k]$ was defined to be equal to the coefficient a_k in the Newton form of the interpolating polynomial p_k of Equation (3), we can say that $f[x_0, x_1, \dots, x_k]$ is the coefficient of x^k in the polynomial p_k of degree $\leqq k$, which interpolates f at $x_0, x_1, \dots, x_k$. Similarly, $f[x_1, x_2, \dots, x_k]$ is the coefficient of x^{k-1} in the polynomial q of degree $\leqq k-1$, which interpolates f at $x_1, x_2, \dots, x_k$. Likewise, $f[x_0, x_1, \dots, x_{k-1}]$ is the coefficient of x^{k-1} in the polynomial p_{k-1} of degree $\leqq k-1$, which interpolates f at $x_0, x_1, \dots, x_{k-1}$. The three polynomials p_k, q, and p_{k-1} are intimately related. In fact,

$$p_k(x) = q(x) + \frac{x - x_k}{x_k - x_0}[q(x) - p_{k-1}(x)] \tag{12}$$

To establish Equation (12), observe that the right side is a polynomial of degree at most k. Evaluating it at x_i, for $1 \leqq i \leqq k-1$, results in $f(x_i)$:

$$\begin{aligned} q(x_i) + \frac{x_i - x_k}{x_k - x_0}[q(x_i) - p_{k-1}(x_i)] &= f(x_i) + \frac{x_i - x_k}{x_k - x_0}[f(x_i) - f(x_i)] \\ &= f(x_i) \end{aligned}$$

Similarly, evaluating it at x_0 and x_k gives $f(x_0)$ and $f(x_k)$, respectively. By the uniqueness of interpolating polynomials, the right side of Equation (12) must be $p_k(x)$, and Equation (12) is established.

Completing the argument to justify Equation (11), we take the coefficient of x^k on both sides of Equation (12). The result is Equation (11). Indeed, we see that $f[x_1, x_2, \dots, x_k]$ is the coefficient of x^{k-1} in q, and $f[x_0, x_1, \dots, x_{k-1}]$ is the coefficient of x^{k-1} in p_{k-1}. ■

Notice that $f[x_0, x_1, \dots, x_k]$ is *not* changed if the nodes $x_0, x_1, \dots, x_k$ are permuted. Thus, for example, we have

$$f[x_0, x_1, x_2] = f[x_1, x_2, x_0]$$

The reason is that $f[x_0, x_1, x_2]$ is the coefficient of x^2 in the quadratic polynomial interpolating f at x_0, x_1, x_2, whereas $f[x_1, x_2, x_0]$ is the coefficient of x^2 in the quadratic polynomial interpolating f at x_1, x_2, x_0. These two polynomials are, of course, the same! A formal statement in mathematical language is as follows:

■ **Theorem 3**

Invariance Theorem

Invariance Property

The divided difference $f[x_0, x_1, \dots, x_k]$ is invariant under all permutations of the arguments $x_0, x_1, \dots, x_k$.

Since the variables $x_0, x_1, \ldots, x_k$ and k are arbitrary, the recursive Formula (11) can also be written as

$$f[x_i, x_{i+1}, \ldots, x_{j-1}, x_j] = \frac{f[x_{i+1}, x_{i+2}, \ldots, x_j] - f[x_i, x_{i+1}, \ldots, x_{j-1}]}{x_j - x_i} \tag{13}$$

The first three divided differences are thus

First Three Divided Differences

$$f[x_i] = f(x_i)$$

$$f[x_i, x_{i+1}] = \frac{f[x_{i+1}] - f[x_i]}{x_{i+1} - x_i}$$

$$f[x_i, x_{i+1}, x_{i+2}] = \frac{f[x_{i+1}, x_{i+2}] - f[x_i, x_{i+1}]}{x_{i+2} - x_i}$$

Using Formula (13), we can construct a divided-difference table for a function f. It is customary to arrange it as follows (here $n = 3$):

Divided-Difference Table

x	$f[\ ]$	$f[\ ,\ ]$	$f[\ ,\ ,\ ]$	$f[\ ,\ ,\ ,\ ]$
x_0	$\boxed{f[x_0]}$			
		$\boxed{f[x_0, x_1]}$		
x_1	$f[x_1]$		$\boxed{f[x_0, x_1, x_2]}$	
		$f[x_1, x_2]$		$\boxed{f[x_0, x_1, x_2, x_3]}$
x_2	$f[x_2]$		$f[x_1, x_2, x_3]$	
		$f[x_2, x_3]$		
x_3	$f[x_3]$			

In the table, the coefficients along the top diagonal are the ones needed to form the Newton form of the interpolating polynomial (3).

EXAMPLE 7 Construct a divided-difference diagram for the function f given in the following table, and write out the Newton form of the interpolating polynomial.

x	1	$\frac{3}{2}$	0	2
$f(x)$	3	$\frac{13}{4}$	3	$\frac{5}{3}$

Solution The first entry is

$$f[x_0, x_1] = \frac{\left(\frac{13}{4} - 3\right)}{\left(\frac{3}{2} - 1\right)} = \frac{1}{2}$$

After completion of column 3, the first entry in column 4 is

$$f[x_0, x_1, x_2] = \frac{f[x_1, x_2] - f[x_0, x_1]}{x_2 - x_0} = \frac{\frac{1}{6} - \frac{1}{2}}{0 - 1} = \frac{1}{3}$$

The complete diagram is

Sample Divided-Difference Table

x	$f[\ ]$	$f[\ ,\ ]$	$f[\ ,\ ,\ ]$	$f[\ ,\ ,\ ,\ ]$
1	$\boxed{3}$			
		$\boxed{\frac{1}{2}}$		
$\frac{3}{2}$	$\frac{13}{4}$		$\boxed{\frac{1}{3}}$	
		$\frac{1}{6}$		$\boxed{-2}$
0	3		$-\frac{5}{3}$	
		$-\frac{2}{3}$		
2	$\frac{5}{3}$			

Thus, we obtain

$$p_3(x) = 3 + \tfrac{1}{2}(x-1) + \tfrac{1}{3}(x-1)\left(x-\tfrac{3}{2}\right) - 2(x-1)\left(x-\tfrac{3}{2}\right)x \qquad \blacksquare$$

Algorithms and Pseudocode

Turning next to algorithms, we suppose that a table for f is given at points $x_0, x_1, \ldots, x_n$ and that all the divided differences $a_{ij} \equiv f[x_i, x_{i+1}, \ldots, x_j]$ are to be computed. The following pseudocode accomplishes this:

Divided Differences Pseudocode

```
integer i, j, n;   real array (a_ij)_{0:n×0:n}, (x_i)_{0:n}
for i = 0 to n
    a_i0 ← f(x_i)
end for
for j = 1 to n
    for i = 0 to n − j
        a_ij ← (a_{i+1, j−1} − a_{i, j−1})/(x_{i+j} − x_i)
    end for
end for
```

Observe that the coefficients of the interpolating polynomial (3) are stored in the first row of the array $(a_{ij})_{0:n\times 0:n}$.

If the divided differences are being computed for use only in constructing the Newton form of the interpolation polynomial

$$p_n(x) = \sum_{i=0}^{n} a_i \prod_{j=0}^{i-1}(x - x_j)$$

where $a_i = f[x_0, x_1, \ldots, x_i]$, there is *no* need to store all of them. Only $f[x_0], f[x_0, x_1], \ldots, f[x_0, x_1, \ldots, x_n]$ need to be stored.

When a one-dimensional array $(a_i)_{0:n}$ is used, the divided differences can be overwritten each time from the last storage location backward so that, finally, only the desired coefficients remain. In this case, the amount of computing is the same as in the preceding case, but the storage requirements are less. (Why?) Here is a pseudocode to do this:

Improved Pseudocode

```
integer i, j, n;   real array (a_i)_{0:n}, (x_i)_{0:n}
for i = 0 to n
    a_i ← f(x_i)
end for
for j = 1 to n
    for i = n to j step −1
        a_i ← (a_i − a_{i−1})/(x_i − x_{i−j})
    end for
end for
```

This algorithm is more intricate, and the reader is invited to verify it—say, in the case $n = 3$.

For the numerical experiments suggested in the computer problems, the following two procedures should be satisfactory. The first is called *Coef.* It requires as input the number n and tabular values in the arrays (x_i) and (y_i). Remember that the number of points in the table is $n+1$. The procedure then computes the coefficients required in the Newton interpolating polynomial, storing them in the array (a_i).

***Coef* Pseudocode**

```
procedure Coef(n, (x_i), (y_i), (a_i))
integer i, j, n;   real array (x_i)_{0:n}, (y_i)_{0:n}, (a_i)_{0:n}
for i = 0 to n
    a_i ← y_i
end for
for j = 1 to n
    for i = n to j step −1
        a_i ← (a_i − a_{i−1})/(x_i − x_{i−j})
    end for
end for
end procedure Coef
```

The second is function *Eval*. It requires as input the array (x_i) from the original table and the array (a_i), which is *output* from *Coef*. The array (a_i) contains the coefficients for the Newton form of the interpolation polynomial. Finally, as input, a single real value for t is given. The function then returns the value of the interpolating polynomial at t.

***Eval* Pseudocode**

```
real function Eval(n, (x_i), (a_i), t)
integer i, n;   real t, temp;   real array (x_i)_{0:n}, (a_i)_{0:n}
temp ← a_n
for i = n − 1 to 0 step −1
    temp ← (temp)(t − x_i) + a_i
end for
Eval ← temp
end function Eval
```

Since the coefficients of the interpolating polynomial need be computed only once, we call *Coef* first, and then all subsequent calls for evaluating this polynomial are accomplished with *Eval*. Notice that only the t argument should be changed between successive calls to function *Eval*.

EXAMPLE 8 Write pseudocode for the Newton form of the interpolating polynomial p for $\sin x$ at ten equidistant points in the interval $[0, 1.6875]$. The code finds the maximum value of $|\sin x - p(x)|$ over a finer set of equally spaced points in the same interval.

Solution If we take 10 points, including the ends of the interval, then we create 9 subintervals, each of length $h = 0.1875$. The points are then $x_i = ih$ for $i = 0, 1, \ldots, 9$. After obtaining the polynomial, we divide each subinterval into four panels, and we evaluate $|\sin x - p(x)|$ at 37 points (called t in the pseudocode). These are $t_j = jh/4$ for $j = 0, 1, \ldots, 36$. Here is a suitable main program in pseudocode that calls routines *Coef* and *Eval* previously given:

```
program Test_Coef_Eval
integer j, k, n, j_max;   real e, h, p, e_max, p_max, t_max,
real array (x_i)_{0:n}, (y_i)_{0:n}, (a_i)_{0:n}
```

(Continued)

***Test_Coef_Eval* Pseudocode**

```
n ← 9
h ← 1.6875/n
for k = 0 to n
    x_k ← kh
    y_k ← sin(x_k)
end for
call Coef(n, (x_i), (y_i), (a_i))
output (a_i); e_max ← 0
for j = 0 to 4n
    t ← jh/4
    p ← Eval(n, (x_i)_n, (a_i)_n, t)
    e ← |sin(t) − p|
    output j, t, p, e
    if e > e_max then
        j_max ← j; t_max ← t; p_max ← p; e_max ← e
    end if
end for
output j_max, t_max, p_max, e_max
end program Test_Coef_Eval
```

The first coefficient in the Newton form of the interpolating polynomial is 0 (why?), and the others range in magnitude from approximately 0.99 to 0.18×10^{-5}. The deviation between $\sin x$ and $p(x)$ is practically zero at each interpolation node. (Because of roundoff errors, they are *not* precisely zero.) From the computer output, the largest error is at $j_{\max} = 35$, where $\sin(1.64062\,5) \approx 0.99756\,31$ with an error of 1.19×10^{-7}. ■

Vandermonde Matrix

Another view of interpolation is that for a given set of $n + 1$ data points (x_0, y_0), (x_1, y_1), $\dots$, (x_n, y_n), we want to express an interpolating function $f(x)$ as a linear combination of a set of **basis functions** $\varphi_0, \varphi_1, \varphi_2, \dots, \varphi_n$ so that

$f(x)$ Approximation with Basis Functions

$$f(x) \approx c_0\varphi_0(x) + c_1\varphi_1(x) + c_2\varphi_2(x) + \cdots + c_n\varphi_n(x)$$

Here the coefficients $c_0, c_1, c_2, \dots, c_n$ are to be determined. We want the function f to interpolate the data (x_i, y_i). This means that we have linear equations of the form

$$f(x_i) = c_0\varphi_0(x_i) + c_1\varphi_1(x_i) + c_2\varphi_2(x_i) + \cdots + c_n\varphi_n(x_i) = y_i$$

for each $i = 0, 1, 2, \dots, n$. This is a system of linear equations

Linear System

$$\mathbf{A}\mathbf{c} = \mathbf{y}$$

Here, the entries in the coefficient matrix $\mathbf{A}$ are given by $a_{ij} = \varphi_j(x_i)$, which is the value of the jth basis function evaluated at the ith data point. The right-hand side vector $\mathbf{y}$ contains the known data values y_i, and the components of the vector $\mathbf{c}$ are the unknown coefficients c_i. Systems of linear equations are discussed in Chapters 2 and 8.

Polynomials are the simplest and most common basis functions. The natural basis for $\mathbb{P}_n$ consists of the **monomials**

Monomials

$$\varphi_0(x) = 1, \quad \varphi_1(x) = x, \quad \varphi_2(x) = x^2, \quad \dots, \quad \varphi_n(x) = x^n$$

Figure 4.3 (p. 168) shows the first few monomials: $1, x, x^2, x^3, x^4$, and x^5.

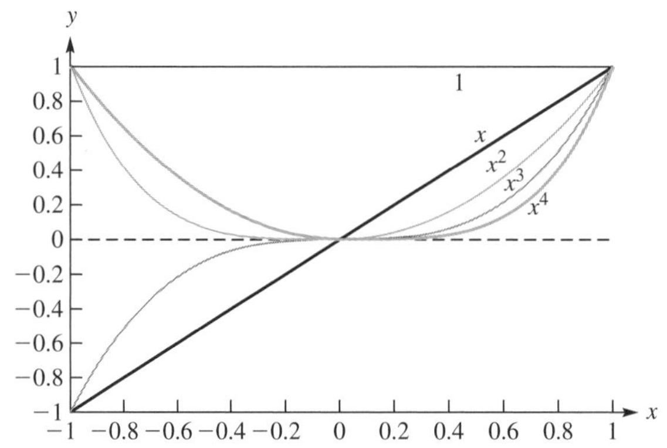

FIGURE 4.3
First few monomials

Consequently, a given polynomial p_n has the form

$$p_n(x) = c_0 + c_1 x + c_2 x^2 + \cdots + c_n x^n$$

The corresponding linear system $\mathbf{Ac} = \mathbf{y}$ has the form

Vandermonde Matrix Linear System

$$\begin{bmatrix} 1 & x_0 & x_0^2 & \cdots & x_0^n \\ 1 & x_1 & x_1^2 & \cdots & x_1^n \\ 1 & x_2 & x_2^2 & \cdots & x_2^n \\ \vdots & \vdots & \vdots & \ddots & \vdots \\ 1 & x_n & x_n^2 & \cdots & x_n^n \end{bmatrix} \begin{bmatrix} c_0 \\ c_1 \\ c_2 \\ \vdots \\ c_n \end{bmatrix} = \begin{bmatrix} y_0 \\ y_1 \\ y_2 \\ \vdots \\ y_n \end{bmatrix}$$

The coefficient matrix is called a **Vandermonde matrix**. It can be shown that this matrix is invertible provided that the points $x_0, x_1, x_2, \ldots, x_n$ are distinct. So we can, in theory, solve the system for the polynomial interpolant. Although the Vandermonde matrix is invertible, it is ill-conditioned as n increases. For large n, the monomials are less distinguishable from one another, as shown in Figure 4.3. Moreover, the columns of the Vandermonde become nearly linearly dependent in this case. High-degree polynomials often oscillate wildly and are highly sensitive to small changes in the data.

As Figures 4.1–4.3 show, we have discussed three choices for the basis functions: the Lagrange cardinal polynomials $\ell_i(x)$, the Newton polynomials $\pi_i(x)$, and the monomials. It turns out that there are better choices for the basis functions; namely, the Chebyshev polynomials have more desirable features.

The **Chebyshev polynomials** play an important role in mathematics because they have several special properties such as the recursive relation

Chebyshev Recursive Relation

$$\begin{cases} T_0(x) = 1, \quad T_1(x) = x \\ T_i(x) = 2xT_{i-1}(x) - T_{i-2}(x) \end{cases}$$

for $i = 2, 3, 4$, and so on. Thus, the first six Chebyshev polynomials are

First Six Chebyshev Polynomials

$$T_0(x) = 1, \quad T_1(x) = x, \quad T_2(x) = 2x^2 - 1, \quad T_3(x) = 4x^3 - 3x$$
$$T_4(x) = 8x^4 - 8x^2 + 1, \quad T_5(x) = 16x^5 - 20x^3 + 5x$$

These curves for these polynomials, as is shown in Figure 4.4, are quite different from one another. The Chebyshev polynomials are usually employed on the interval $[-1, 1]$. With changes of variable, they can be used on any interval, but the results will be more complicated.

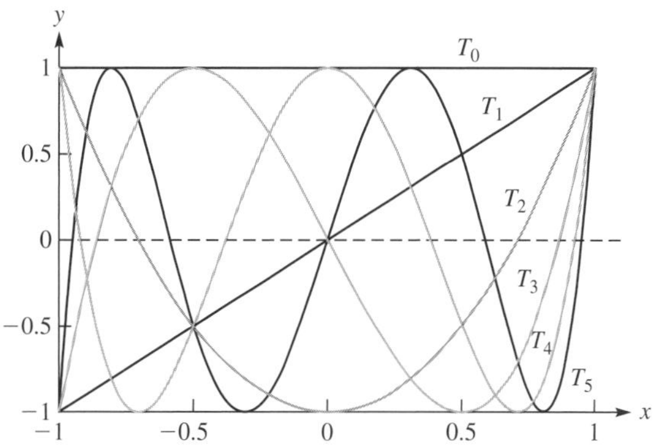

FIGURE 4.4 First few Chebyshev polynomials

One of the important properties of the Chebyshev polynomials is the **equal oscillation property**. Notice in Figure 4.4 that successive extreme points of the Chebyshev polynomials are equal in magnitude and alternate in sign. This property tends to distribute the error uniformly when the Chebyshev polynomials are used as the basis functions. In polynomial interpolation for continuous functions, it is particularly advantageous to select as the interpolation points the roots or the extreme points of a Chebyshev polynomial. This causes the maximum error over the interval of interpolation to be minimized. An example of this is given in Section 4.2. In Section 9.2, we discuss Chebyshev polynomials in more detail.

Inverse Interpolation

Inverse Interpolation

A process called **inverse interpolation** is often used to approximate an inverse function. Suppose that values $y_i = f(x_i)$ have been computed at $x_0, x_1, \ldots, x_n$. Using the table

y	y_0	y_1	$\cdots$	y_n
x	x_0	x_1	$\cdots$	x_n

we form the interpolation polynomial

$$p(y) = \sum_{i=0}^{n} c_i \prod_{j=0}^{i-1} (y - y_j)$$

The original relationship, $y = f(x)$, has an inverse, under certain conditions. This inverse is being approximated by $x = p(y)$. *Coef* and *Eval* can be used to carry out the inverse interpolation by reversing the arguments x and y in the calling sequence for *Coef*.

Finding a Root

Inverse interpolation can be used to find where a given function f has a root or *zero*. This means inverting the equation $f(x) = 0$. We propose to do this by creating a table of values $(f(x_i), x_i)$ and interpolating with a polynomial, p. Thus, we obtain $p(y_i) = x_i$. The points x_i should be chosen near the unknown root, r. The approximate root is then given by $r \approx p(0)$. See Figure 4.5 for an example of function $y = f(x)$ and its inverse function $x = g(y)$ with the root $r = g(0)$.

EXAMPLE 9 For a concrete case, let the table of known values be

y	−0.57892 00	−0.36263 70	−0.18491 60	−0.03406 42	0.09698 58
x	1.0	2.0	3.0	4.0	5.0

Find the inverse interpolation polynomial.

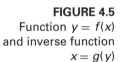

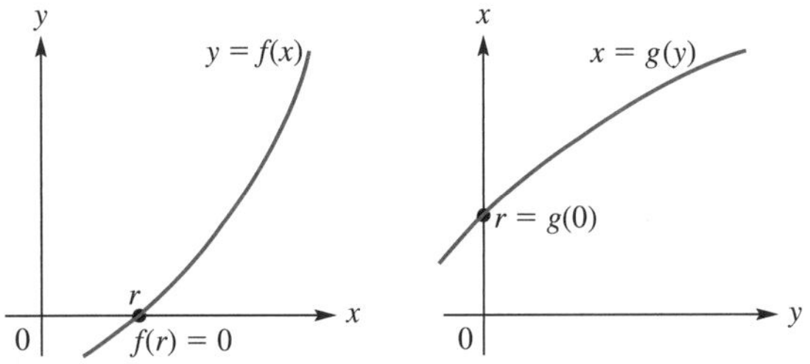

FIGURE 4.5 Function $y = f(x)$ and inverse function $x = g(y)$

Solution The nodes in this problem are the points in the row of the table headed y, and the function values being interpolated are in the x row. The resulting polynomial is

Inverse Interpolation Polynomial

$$p(y) = 0.25y^4 + 1.2y^3 + 3.69y^2 + 7.39y + 4.24747\,0086$$

and $p(0) = 4.24747\,0086$. Only the last coefficient is shown with all the digits carried in the calculation, as it is the only one needed for the problem at hand. ■

Polynomial Interpolation by Neville's Algorithm

Another method of obtaining a polynomial interpolant from a given table of values

x	x_0	x_1	$\cdots$	x_n
y	y_0	y_1	$\cdots$	y_n

was given by Neville. It builds the polynomial in steps, just as the Newton algorithm does. The constituent polynomials have interpolating properties of their own.

Let $P_{a,b,\ldots,s}(x)$ be the polynomial interpolating the given data at a sequence of nodes $x_a, x_b, \ldots, x_s$. We start with constant polynomials $P_i(x) = f(x_i)$. Selecting two nodes x_i and x_j with $i > j$, we define recursively

Neville: 1st Recursive Relation

$$P_{u,\ldots,v}(x) = \left(\frac{x - x_j}{x_i - x_j}\right) P_{u,\ldots,j-1,j+1,\ldots,v}(x) + \left(\frac{x_i - x}{x_i - x_j}\right) P_{u,\ldots,i-1,i+1,\ldots,v}(x)$$

Using this formula repeatedly, we can create an array of polynomials:

x_0	$P_0(x)$				
x_1	$P_1(x)$	$P_{0,1}(x)$			
x_2	$P_2(x)$	$P_{1,2}(x)$	$P_{0,1,2}(x)$		
x_3	$P_3(x)$	$P_{2,3}(x)$	$P_{1,2,3}(x)$	$P_{0,1,2,3}(x)$	
x_4	$P_4(x)$	$P_{3,4}(x)$	$P_{2,3,4}(x)$	$P_{1,2,3,4}(x)$	$P_{0,1,2,3,4}(x)$

Here, each successive polynomial can be determined from two adjacent polynomials in the previous column.

We can simplify the notation by letting

$$S_{ij}(x) = P_{i-j,i-j+1,\ldots,i-1,i}(x)$$

where $S_{ij}(x)$ for $i \geqq j$ denotes the interpolating polynomial of degree j on the $j+1$ nodes $x_{i-j}, x_{i-j+1}, \ldots, x_{i-1}, x_i$. Next we can rewrite the recurrence relation above as

Neville: 2nd Recursive Relation

$$S_{ij}(x) = \left(\frac{x - x_{i-j}}{x_i - x_{i-j}}\right) S_{i,j-1}(x) + \left(\frac{x_i - x}{x_i - x_{i-j}}\right) S_{i-1,j-1}(x)$$

So the displayed array becomes

$$
\begin{array}{c|ccccc}
x_0 & S_{00}(x) \\
x_1 & S_{10}(x) & S_{11}(x) \\
x_2 & S_{20}(x) & S_{21}(x) & S_{22}(x) \\
x_3 & S_{30}(x) & S_{31}(x) & S_{32}(x) & S_{33}(x) \\
x_4 & S_{40}(x) & S_{41}(x) & S_{42}(x) & S_{43}(x) & S_{44}(x)
\end{array}
$$

To prove some theoretical results, we change the notation by making the superscript the degree of the polynomial. At the beginning, we define constant polynomials (i.e., polynomials of degree 0) as $P_i^0(x) = y_i$ for $0 \leqq i \leqq n$. Then we define

Neville: 3rd Recursive Relation

$$P_i^j(x) = \left(\frac{x - x_{i-j}}{x_i - x_{i-j}}\right) P_i^{j-1}(x) + \left(\frac{x_i - x}{x_i - x_{i-j}}\right) P_{i-1}^{j-1}(x) \tag{14}$$

In this equation, the superscripts are simply indices, not exponents. The range of j is $1 \leqq j \leqq n$, while that of i is $j \leqq i \leqq n$. Formula (14) is seen again, in slightly different form, in the theory of B splines in Section 6.3.

The interpolation properties of these polynomials are given in the next result.

■ **Theorem 4** **Interpolation Properties**

The polynomials P_i^j defined above interpolate as follows:

$$P_i^j(x_k) = y_k \qquad (0 \leqq i - j \leqq k \leqq i \leqq n) \tag{15}$$

Proof We use induction on j. When $j = 0$, the assertion in Equation (15) reads

$$P_i^0(x_k) = y_k \qquad (0 \leqq i \leqq k \leqq i \leqq n)$$

In other words, $P_i^0(x_i) = y_i$, which is true by the definition of P_i^0.

Now assume, as an induction hypothesis, that for some $j \geqq 1$,

$$P_i^{j-1}(x_k) = y_k \qquad (0 \leqq i - j + 1 \leqq k \leqq i \leqq n)$$

To prove the next case in Equation (15), we begin by verifying the two extreme cases for k, namely, $k = i - j$ and $k = i$. We have, by Equation (14),

$$
\begin{aligned}
P_i^j(x_{i-j}) &= \left(\frac{x_i - x_{i-j}}{x_i - x_{i-j}}\right) P_{i-1}^{j-1}(x_{i-j}) \\
&= P_{i-1}^{j-1}(x_{i-j}) = y_{i-j}
\end{aligned}
$$

The last equality is justified by the induction hypothesis. It is necessary to observe that $0 \leqq i - 1 - j + 1 \leqq i - j \leqq i - 1 \leqq n$. In the same way, we compute

$$
\begin{aligned}
P_i^j(x_i) &= \left(\frac{x_i - x_{i-j}}{x_i - x_{i-j}}\right) P_i^{j-1}(x_i) \\
&= P_i^{j-1}(x_i) = y_i
\end{aligned}
$$

Here, in using the induction hypothesis, observe that $0 \leqq i - j + 1 \leqq i \leqq i \leqq n$.

Now let $i - j < k < i$. Then

$$P_i^j(x_k) = \left(\frac{x_k - x_{i-j}}{x_i - x_{i-j}}\right) P_i^{j-1}(x_k) + \left(\frac{x_i - x_k}{x_i - x_{i-j}}\right) P_{i-1}^{j-1}(x_k)$$

In this equation, $P_i^{j-1}(x_k) = y_k$ by the induction hypothesis, because $0 \leqq i-j+1 \leqq k \leqq i \leqq n$. Likewise, $P_{i-1}^{j-1}(x_k) = y_k$ because $0 \leqq i-1-j+1 \leqq k \leqq i-1 \leqq n$. Thus, we have

$$P_i^j(x_k) = \left(\frac{x_k - x_{i-j}}{x_i - x_{i-j}}\right)y_k + \left(\frac{x_i - x_k}{x_i - x_{i-j}}\right)y_k = y_k \qquad \blacksquare$$

An algorithm follows in pseudocode to evaluate $P_n^n(t)$ when a table of values is given:

Evaluate $P_n^n(t)$ Pseudocode

integer i, j, n; **real array** $(x_i)_{0:n}, (y_i)_{0:n}, (S_{ij})_{0:n\times 0:n}$
for $i = 0$ **to** n
    $S_{i0} \leftarrow y_i$
end for
for $j = 1$ **to** n
    for $i = j$ **to** n
        $S_{ij} \leftarrow \left[(t - x_{i-j})S_{i,j-1} + (x_i - t)S_{i-1,j-1}\right] / (x_i - x_{i-j})$
    end for
end for
return S_{nn}

We begin the algorithm by finding the node nearest the point t at which the evaluation is to be made. In general, interpolation is more accurate when this is done.

Interpolation of Bivariate Functions

The methods we have discussed for interpolating functions of one variable by polynomials extend to *some* cases of functions of two or more variables. An important case occurs when a function $(x, y) \mapsto f(x, y)$ is to be approximated on a rectangle. This leads to what is known as **tensor-product interpolation**.

Suppose the rectangle is the Cartesian product of two intervals: $[a, b] \times [\alpha, \beta]$. That is, the variables x and y run over the intervals $[a, b]$, and $[\alpha, \beta]$, respectively. Select n nodes x_i in $[a, b]$, and define the **Lagrangian polynomials**

$$\ell_i(x) = \prod_{\substack{j\neq i\\ j=1}}^{n} \frac{x - x_j}{x_i - x_j} \qquad (1 \leqq i \leqq n)$$

Similarly, we select m nodes y_i in $[\alpha, \beta]$ and define

$$\overline{\ell}_i(y) = \prod_{\substack{j\neq i\\ j=1}}^{m} \frac{y - y_j}{y_i - y_j} \qquad (1 \leqq i \leqq m)$$

Then the function

Tensor-Product Interpolation

$$P(x, y) = \sum_{i=1}^{n}\sum_{j=1}^{m} f(x_i, y_j)\ell_i(x)\overline{\ell}_j(y)$$

is a polynomial in two variables that interpolates f at the *grid points* (x_i, y_j). There are nm such points of interpolation.

The proof of the interpolation property is quite simple because $\ell_i(x_q) = \delta_{iq}$ and $\overline{\ell}_j(y_p) = \delta_{jp}$. Consequently, we have

$$\begin{aligned} P(x_q, y_p) &= \sum_{i=1}^{n}\sum_{j=1}^{m} f(x_i, y_j)\ell_i(x_q)\overline{\ell}_j(y_p) \\ &= \sum_{i=1}^{n}\sum_{j=1}^{m} f(x_i, y_j)\delta_{iq}\delta_{jp} = f(x_q, y_p) \end{aligned}$$

The same procedure can be used with spline interpolants (or indeed any other type of function).

Summary 4.1

- The **Lagrange form** of the interpolation polynomial is

$$p_n(x) = \sum_{i=0}^{n} \ell_i(x) f(x_i)$$

with **cardinal polynomials**

$$\ell_i(x) = \prod_{\substack{j \neq i \\ j=0}}^{n} \left(\frac{x - x_j}{x_i - x_j}\right) \qquad (0 \leqq i \leqq n)$$

that obey the **Kronecker delta equation**

$$\ell_i(x_j) = \delta_{ij} = \begin{cases} 0 & \text{if } i \neq j \\ 1 & \text{if } i = j \end{cases}$$

- The **Newton form** of the interpolation polynomial is

$$p_n(x) = \sum_{i=0}^{n} a_i \prod_{j=0}^{i-1} (x - x_j)$$

with **divided differences**

$$a_i = f[x_0, x_1, \ldots, x_i] = \frac{f[x_1, x_2, \ldots, x_i] - f[x_0, x_1, \ldots, x_{i-1}]}{x_i - x_0}$$

These are two different forms of the unique polynomial p of degree n that interpolates a table of $n + 1$ pairs of points $(x_i, f(x_i))$ for $0 \leqq i \leqq n$.

- We can illustrate this with a small table for $n = 2$:

x	x_0	x_1	x_2
$f(x)$	$f(x_0)$	$f(x_1)$	$f(x_2)$

The Lagrange interpolating polynomial is

$$\begin{aligned} p_2(x) = \frac{(x - x_1)(x - x_2)}{(x_0 - x_1)(x_0 - x_2)} f(x_0) &+ \frac{(x - x_0)(x - x_2)}{(x_1 - x_0)(x_1 - x_2)} f(x_1) \\ &+ \frac{(x - x_0)(x - x_1)}{(x_2 - x_0)(x_2 - x_1)} f(x_2) \end{aligned}$$

Clearly, $p_2(x_0) = f(x_0)$, $p_2(x_1) = f(x_1)$, and $p_2(x_2) = f(x_2)$. Next, we form the divided-difference table:

x_0	$\boxed{f(x_0)}$		
		$\boxed{f[x_0, x_1]}$	
x_1	$f(x_1)$		$\boxed{f[x_0, x_1, x_2]}$
		$f[x_1, x_2]$	
x_2	$f(x_2)$		

Using the divided-difference entries from the top diagonal, we have

$$p_n(x) = f(x_0) + f[x_0, x_1](x - x_0) + f[x_0, x_1, x_2](x - x_0)(x - x_1)$$

Again, it can be easily shown that $p_2(x_0) = f(x_0)$, $p_2(x_1) = f(x_1)$, and $p_2(x) = f(x_2)$.

- We can use **inverse polynomial interpolation** to find an approximate value of a root r of the equation $f(x) = 0$ from a table of values (x_i, y_i) for $1 \leq i \leq n$. Here we are assuming that the table values are in the vicinity of this zero of the function f. Flipping the table values, we use the reversed table values (y_i, x_i) to determine the interpolating polynomial called $p_n(y)$. Now evaluating it at 0, we find a value that approximates the desired zero, namely, $r \approx p_n(0)$ and $f(p_n(0)) \approx f(r) = 0$.
- Other advanced polynomial interpolation methods discussed are **Neville's algorithm** and **bivariate function interpolation**.

Exercises 4.1

[a]**1.** Use the Lagrange interpolation process to obtain a polynomial of least degree that assumes these values:

x	0	2	3	4
y	7	11	28	63

2. (Continuation) Rearrange the points in the table of the preceding problem and find the Newton form of the interpolating polynomial. Show that the polynomials obtained are identical, although their forms may differ.

[a]**3.** For the four interpolation nodes $-1, 1, 3, 4$, what are the ℓ_i Functions (2) required in the Lagrange interpolation procedure? Draw the graphs of these four functions to show their essential properties.

4. Verify that the polynomials

$$p(x) = 5x^3 - 27x^2 + 45x - 21$$
$$q(x) = x^4 - 5x^3 + 8x^2 - 5x + 3$$

interpolate the data

x	1	2	3	4
y	2	1	6	47

and explain why this does not violate the uniqueness part of the theorem on existence of polynomial interpolation.

5. Verify that the polynomials

$$p(x) = 3 + 2(x - 1) + 4(x - 1)(x + 2)$$
$$q(x) = 4x^2 + 6x - 7$$

are both interpolating polynomials for the following table, and explain why this does not violate the uniqueness part of the existence theorem for polynomial interpolation.

x	1	-2	0
y	3	-3	-7

6. Find the polynomial p of least degree that takes these values: $p(0) = 2$, $p(2) = 4$, $p(3) = -4$, $p(5) = 82$. Use divided differences to get the correct polynomial. It is *not* necessary to write the polynomial in the standard form $a_0 + a_1x + a_2x^2 + \cdots$.

7. Complete the following divided-difference tables, and use them to obtain polynomials of degree 3 that interpolate the function values indicated:

[a]**a.**

x	$f[\ ]$	$f[\ ,\ ]$	$f[\ ,\ ,\ ]$	$f[\ ,\ ,\ ,\ ]$
−1	2			
1	−4		2	
3	6			
		2		
5	10			

b.

x	$f[\ ]$	$f[\ ,\ ]$	$f[\ ,\ ,\ ]$	$f[\ ,\ ,\ ,\ ]$
−1	2			
1	−4			
3	46			
		53.5		
4	99.5			

Write the final polynomials in a form most efficient for computing.

[a]**8.** Find an interpolating polynomial for this table:

x	1	2	2.5	3	4
y	−1	$-\frac{1}{3}$	$\frac{3}{32}$	$\frac{4}{3}$	25

9. Given the data

x	0	1	2	4	6
$f(x)$	1	9	23	93	259

do the following.

[a]**a.** Construct the divided-difference table.

[a]**b.** Using Newton's interpolation polynomial, find an approximation to $f(4.2)$.
Hint: Use polynomials starting with 93 and involving factors $(x - 4)$.

10. a. Construct Newton's interpolation polynomial for the data shown.

x	0	2	3	4
y	7	11	28	63

b. Without simplifying it, write the polynomial obtained in nested form for easy evaluation.

11. From census data, the approximate population of the United States was 150.7 million in 1950, 179.3 million in 1960, 203.3 million in 1970, 226.5 million in 1980, and 249.6 million in 1990. Using Newton's interpolation polynomial for these data, find an approximate value for the population in 2000. Then use the polynomial to estimate the population in 1920 based on these data. What conclusion should be drawn?

[a]**12.** The polynomial $p(x) = x^4 - x^3 + x^2 - x + 1$ has the following values:

x	−2	−1	0	1	2	3
$p(x)$	31	5	1	1	11	61

Find a polynomial q that takes these values:

x	−2	−1	0	1	2	3
$q(x)$	31	5	1	1	11	30

Hint: This can be done with little work.

13. Use the divided-difference method to obtain a polynomial of least degree that fits the values shown.

[a]**a.**

x	0	1	2	−1	3
y	−1	−1	−1	−7	5

b.

x	1	3	−2	4	5
y	2	6	−1	−4	2

[a]**14.** Find the interpolating polynomial for these data:

x	1.0	2.0	2.5	3.0	4.0
$f(x)$	−1.5	−0.5	0.0	0.5	1.5

15. It is suspected that the table

x	−2	−1	0	1	2	3
y	1	4	11	16	13	−4

comes from a cubic polynomial. How can this be tested? Explain.

[a]**16.** There exists a unique polynomial $p(x)$ of degree 2 or less such that $p(0) = 0$, $p(1) = 1$, and $p'(\alpha) = 2$ for any value of α between 0 and 1 (inclusive) except one value of α, say, α_0. Determine α_0, and give this polynomial for $\alpha \neq \alpha_0$.

17. Determine by two methods the polynomial of degree 2 or less whose graph passes through the points (0, 1.1), (1, 2), and (2, 4.2). Verify that they are the same.

[a]**18.** Develop the divided-difference table from the given data. Write down the interpolating polynomial, and rearrange it for fast computation without simplifying.

x	0	1	3	2	5
$f(x)$	2	1	5	6	−183

Checkpoint: $f[1, 3, 2, 5] = -7$.

[a]**19.** Let $f(x) = x^3 + 2x^2 + x + 1$. Find the polynomial of degree 4 that interpolates the values of f at

$x = -2, -1, 0, 1, 2$. Find the polynomial of degree 2 that interpolates the values of f at $x = -1, 0, 1$.

20. Without using a divided-difference table, derive and simplify the polynomial of least degree that assumes these values:

x	-2	-1	0	1	2
y	2	14	4	2	2

21. (Continuation) Find a polynomial that takes the values shown in the preceding problem and has at $x = 3$ the value 10.
Hint: Add a suitable polynomial to the $p(x)$ of the previous problem.

[a]**22.** Find a polynomial of least degree that takes these values:

x	1.73	1.82	2.61	5.22	8.26
y	0	0	7.8	0	0

Hint: Rearrange the table so that the nonzero value of y is the *last* entry, or think of some better way.

23. Form a divided-difference table for the following and explain what happened.

x	1	2	3	1
y	3	5	5	7

24. Simple polynomial interpolation in two dimensions is not always possible. For example, suppose that the following data are to be represented by a polynomial of first degree in x and y, $p(t) = a + bx + cy$, where $t = (x, y)$:

t	$(1, 1)$	$(3, 2)$	$(5, 3)$
$f(t)$	3	2	6

Show that it is not possible.

[a]**25.** Consider a function $f(x)$ such that $f(2) = 1.5713$, $f(3) = 1.5719$, $f(5) = 1.5738$, and $f(6) = 1.5751$. Estimate $f(4)$ using a second-degree interpolating polynomial and a third-degree polynomial. Round the final results off to four decimal places. Is there any advantage here in using a third-degree polynomial?

26. Use inverse interpolation to find an approximate value of x such that $f(x) = 0$ given the following table of values for f. Look into what happens and draw a conclusion.

x	-2	-1	1	2	3
$f(x)$	-31	5	1	11	61

[a]**27.** Find a polynomial $p(x)$ of degree at most 3 such that $p(0) = 1$, $p(1) = 0$, $p'(0) = 0$, and $p'(-1) = -1$.

[a]**28.** From a table of logarithms, we obtain the following values of $\log x$ at the indicated tabular points:

x	1	1.5	2	3	3.5	4
$\log x$	0	0.17609	0.30103	0.47712	0.54407	0.60206

Form a divided-difference table based on these values. Interpolate for log 2.4 and log 1.2 using third-degree interpolation polynomials in Newton form.

29. Show that the divided differences are linear maps; that is,

$$(\alpha f + \beta g)[x_0, x_1, \ldots, x_n] = \alpha f[x_0, x_1, \ldots, x_n] + \beta g[x_0, x_1, \ldots, x_n]$$

Hint: Use induction.

30. Show that another form for the polynomial p_n of degree at most n that takes values $y_0, y_1, \ldots, y_n$ at abscissas $x_0, x_1, \ldots, x_n$ is

$$\sum_{i=0}^{n} f[x_n, x_{n-1}, \ldots, x_{n-i}] \prod_{j=0}^{i-1} (x - x_{n-j})$$

31. Use the uniqueness of the interpolating polynomial to verify that

$$\sum_{i=0}^{n} f(x_i)\ell_i(x) = \sum_{i=0}^{n} f[x_0, x_1, \ldots, x_i] \prod_{j=0}^{i-1} (x - x_j)$$

32. (Continuation) Show that the following explicit formula is valid for divided differences:

$$f[x_0, x_1, \ldots, x_n] = \sum_{i=0}^{n} f(x_i) \prod_{\substack{j \neq i \\ j=0}}^{n} (x_i - x_j)^{-1}$$

Hint: If two polynomials are equal, the coefficients of x^n in each are equal.

33. Verify directly that

$$\sum_{i=0}^{n} \ell_i(x) = 1$$

for the case $n = 1$. Then establish the result for arbitrary values of n.

34. Write the Lagrange form (1) of the interpolating polynomial of degree at most 2 that interpolates $f(x)$ at x_0, x_1, and x_2, where $x_0 < x_1 < x_2$.

35. (Continuation) Write the Newton form of the interpolating polynomial $p_2(x)$, and show that it is equivalent to the Lagrange form.

36. (Continuation) Show directly that

$$p_2''(x) = 2f[x_0, x_1, x_2]$$

37. (Continuation) Show directly for uniform spacing $h = x_1 - x_0 = x_2 - x_1$ that

$$f[x_0, x_1] = \frac{\Delta f_0}{h}, \qquad f[x_0, x_1, x_2] = \frac{\Delta^2 f_0}{2h^2}$$

where $\Delta f_i = f_{i+1} - f_i$, $\Delta^2 f_i = \Delta f_{i+1} - \Delta f_i$, and $f_i = f(x_i)$.

38. (Continuation) Establish **Newton's forward-difference** form of the interpolating polynomial with uniform spacing

$$p_2(x) = f_0 + \binom{s}{1} \Delta f_0 + \binom{s}{2} \Delta^2 f_0$$

where $x = x_0 + sh$. Here, $\binom{s}{m}$ is the binomial coefficient $[s!]/[(s-m)!\,m!]$, and $s!/(s-m)! = s(s-1)(s-2)\cdots(s-m+1)$ because s can be any real number and $m!$ has the usual definition because m is an integer.

[a]**39.** (Continuation) From the following table of values of $\ln x$, interpolate to obtain $\ln 2.352$ and $\ln 2.387$ using the Newton forward-difference form of the interpolating polynomial:

x	$f(x)$	Δf	$\Delta^2 f$
2.35	0.85442		
		0.00424	
2.36	0.85866		−0.00001
		0.00423	
2.37	0.86289		−0.00002
		0.00421	
2.38	0.86710		−0.00002
		0.00419	
2.39	0.87129		

Using the correctly rounded values $\ln 2.352 \approx 0.85527$ and $\ln 2.387 \approx 0.87004$, show that the forward-difference formula is more accurate near the top of the table than it is near the bottom.

[a]**40.** Count the number of multiplications, divisions, and additions/subtractions in the generation of the divided-difference table that has $n + 1$ points.

41. Verify directly that for any three distinct points x_0, x_1, and x_2,

$$f[x_0, x_1, x_2] = f[x_2, x_0, x_1] = f[x_1, x_2, x_0]$$

Compare this argument to the one in the text.

[a]**42.** Let p be a polynomial of degree n. What is $p[x_0, x_1, \ldots, x_{n+1}]$?

43. Show that if f is continuously differentiable on the interval $[x_0, x_1]$, then $f[x_0, x_1] = f'(c)$ for some c in (x_0, x_1).

44. If f is a polynomial of degree n, show that in a divided-difference table for f, the nth column has a single constant value—a column containing entries $f[x_i, x_{i+1}, \ldots, x_{i+n}]$.

[a]**45.** Determine whether the following assertion is true or false. If $x_0, x_1, \ldots, x_n$ are distinct, then for arbitrary real values $y_0, y_1, \ldots, y_n$, there is a unique polynomial p_{n+1} of degree $\leq n + 1$ such that $p_{n+1}(x_i) = y_i$ for all $i = 0, 1, \ldots, n$.

46. Show that if a function g interpolates the function f at $x_0, x_1, \ldots, x_{n-1}$ and h interpolates f at $x_1, x_2, \ldots, x_n$, then

$$g(x) + \left(\frac{x_0 - x}{x_n - x_0}\right)[g(x) - h(x)]$$

interpolates f at $x_0, x_1, \ldots, x_n$.

47. (**Vandermonde Determinant**) Using $f_i = f(x_i)$, show the following:

a. $$f[x_0, x_1] = \frac{\begin{vmatrix} 1 & f_0 \\ 1 & f_1 \end{vmatrix}}{\begin{vmatrix} 1 & x_0 \\ 1 & x_1 \end{vmatrix}}$$

b. $$f[x_0, x_1, x_2] = \frac{\begin{vmatrix} 1 & x_0 & f_0 \\ 1 & x_1 & f_1 \\ 1 & x_2 & f_2 \end{vmatrix}}{\begin{vmatrix} 1 & x_0 & x_0^2 \\ 1 & x_1 & x_1^2 \\ 1 & x_2 & x_2^2 \end{vmatrix}}$$

Computer Exercises 4.1

[a]**1.** Test the procedure given in the text for determining the Newton form of the interpolating polynomial. For example, consider this table:

x	1	2	3	−4	5
y	2	48	272	1182	2262

Find the interpolating polynomial and verify that $p(-1) = 12$.

2. Find the polynomial of degree 10 that interpolates the function $\arctan x$ at 11 equally spaced points in the interval $[1, 6]$. Print the coefficients in the Newton form of the polynomial. Compute and print the difference between the polynomial and the function at 33 equally spaced points in the interval $[0, 8]$. What conclusion can be drawn?

3. Write a simple program using procedure *Coef* that interpolates e^x by a polynomial of degree 10 on $[0, 2]$ and then compares the polynomial to exp at 100 points.

4. Use as input data to procedure *Coef* the annual rainfall in your town for each of the last five years. Using function *Eval*, predict the rainfall for this year. Is the answer reasonable?

5. A table of values of a function f is given at the points $x_i = i/10$ for $0 \leqq i \leqq 100$. In order to obtain a graph of f with the aid of an automatic plotter, the values of f are required at the points $z_i = i/20$ for $0 \leqq i \leqq 200$. Write a procedure to do this, using a cubic interpolating polynomial with nodes x_i, x_{i+1}, x_{i+2}, and x_{i+3} to compute f at $\frac{1}{2}(x_{i+1} + x_{i+2})$. For z_1 and z_{199}, use the cubic polynomial associated with z_3 and z_{197}, respectively. Compare this routine to *Coef* for a given function.

6. Write routines analogous to *Coef* and *Eval* using the Lagrange form of the interpolation polynomial. Test on the example given in this section at 20 points with $h/2$. Does the Lagrange form have any advantage over the Newton form?

7. (Continuation) Design and carry out a numerical experiment to compare the accuracy of the Newton and Lagrange forms of the interpolation polynomials at values throughout the interval $[x_0, x_n]$.

8. Rewrite and test routines *Coef* and *Eval* so that the array (a_i) is not used.
Hint: When the elements in the array (y_i) are no longer needed, store the divided differences in their places.

9. Write a procedure for carrying out inverse interpolation to solve equations of the form $f(x) = 0$. Test it on the introductory example at the beginning of this chapter.

10. For Example 8, compare the results from your code with that in the text. Redo using linear interpolation based on the 10 equidistant points. How do the errors compare at intermediate points? Plot curves to visualize the difference between linear interpolation and a higher-degree polynomial interpolation.

11. Use mathematical software such as MATLAB, Maple, or Mathematica to find an interpolation polynomial for the points $(0, 0)$, $(1, 1)$, $(2, 2.001)$, $(3, 3)$, $(4, 4)$, $(5, 5)$. Evaluate the polynomial at the point $x = 14$ or $x = 20$ to show that slight roundoff errors in the data can lead to suspicious results in extrapolation.

12. Use symbolic mathematical software such as MATLAB, Maple, or Mathematica to generate the interpolation polynomial for the data points in Example 3. Plot the polynomial and the data points.

13. (Continuation.) Repeat these instructions using Example 7.

14. Carry out the details in Example 8 by writing a computer program that plots the data points and the curve for the interpolation polynomial.

15. (Continuation.) Repeat the instructions using Example 9.

16. Using mathematical software, carry out the details and verify the results in the introductory example to this chapter.

17. (**Padé Interpolation**) Find a rational function of the form

$$g(x) = \frac{a + bx}{1 + cx}$$

that interpolates the function $f(x) = \arctan(x)$ at the points $x_0 = 1$, $x_1 = 2$, and $x_2 = 3$. On the same axes, plot the graphs of f and g, using dashed and dotted lines, respectively.

4.2 Errors in Polynomial Interpolation

Introduction

When a function f is approximated on an interval $[a, b]$ by means of an interpolating polynomial p, the discrepancy between f and p will (theoretically) be 0 at each node of interpolation. A natural expectation is that the function f is well approximated at all intermediate points and that as the number of nodes increases, this agreement will become better and better.

In the history of numerical mathematics, it was a severe shock to realize that this expectation was ill-founded! Of course, if the function being approximated is *not* required to be continuous, then there may be no agreement at all between $p(x)$ and $f(x)$ except at the nodes.

EXAMPLE 1 Consider these five data points: (0, 8), (1, 12), (3, 2), (4, 6), (8, 0). Construct and plot the interpolation polynomial using the two outermost points. Repeat this process by adding one additional point at a time until all the points are included. What conclusions can you draw?

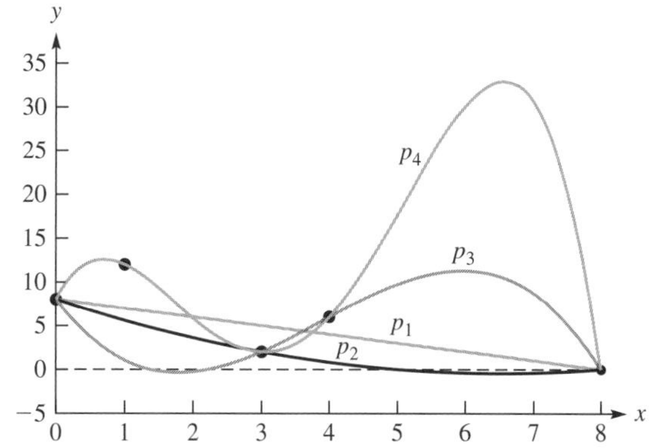

FIGURE 4.6 Interpolant polynomials over data points

Solution The first interpolation polynomial is the line between the outermost points (0, 8) and (8, 0). Then we added the points (3, 2), (4, 6), and (1, 12) in that order and plotted a curve for each additional point. All of these polynomials are shown in Figure 4.6. We were hoping for a smooth curve going through these points without wide fluctuations, but this did not happen! (Why?) It may seem counterintuitive, but as we added more points, the situation became worse instead of better! The reason for this comes from the nature of high-degree polynomials. A polynomial of degree n has n zeros. If all of these zero points are real, then the curve crosses the x-axis n times. The resulting curve must make many turns for this to happen, resulting in wild oscillations. In Chapter 6, we discuss fitting the data points with spline curves. ■

Poorly Fitting Polynomials

Dirichlet Function

Dirichlet Function Numerical Experiment

As a pathological example, consider the so-called **Dirichlet function** f, defined to be 1 at each irrational point and 0 at each rational point. If we choose nodes that are rational numbers, then $p(x) \equiv 0$ and $f(x) - p(x) = 0$ for all rational values of x, but $f(x) - p(x) = 1$ for all irrational values of x.

However, if the function f is well behaved, can we not assume that the differences $|f(x) - p(x)|$ are small when the number of interpolating nodes is large? The answer is still *no*, even for functions that possess continuous derivatives of all orders on the interval!

Runge Function

A specific example of this remarkable phenomenon is provided by the **Runge function**:

Runge Function

$$f(x) = (1 + x^2)^{-1} \tag{1}$$

on the interval $[-5, 5]$. Let p_n be the polynomial that interpolates this function at $n + 1$ equally spaced points on the interval $[-5, 5]$, including the endpoints. Then

Runge Function Numerical Experiment

$$\lim_{n\to\infty} \max_{-5\leqq x\leqq 5} |f(x) - p_n(x)| = \infty$$

Thus, the effect of requiring the agreement of f and p_n at more and more points is to *increase* the error at nonnodal points, and the error actually increases beyond all bounds!

The moral of this example, then, is that polynomial interpolation of high degree with many nodes is a risky operation; the resulting polynomials may be very unsatisfactory as representations of functions unless the set of nodes is chosen with great care.

The reader can easily observe the phenomenon just described by using the pseudocodes already developed in this chapter. See Computer Exercise 4.2.1 for a suggested numerical experiment. In a more advanced study of this topic, it would be shown that the divergence of the polynomials can often be ascribed to the fact that the nodes are equally spaced. Again, contrary to intuition, equally distributed nodes are usually a very poor choice in interpolation. A much better choice for $n + 1$ nodes in $[-1, 1]$ is the set of **Chebyshev nodes**:

Chebyshev Nodes

$$x_i = \cos\left[\left(\frac{2i+1}{2n+2}\right)\pi\right] \qquad (0 \leqq i \leqq n)$$

The corresponding set of nodes on an arbitrary interval $[a, b]$ would be derived from a linear mapping to obtain

Better Fitting with Chebyshev Polynomials

$$x_i = \frac{1}{2}(a+b) + \frac{1}{2}(b-a)\cos\left[\left(\frac{2i+1}{2n+2}\right)\pi\right] \qquad (0 \leqq i \leqq n)$$

Notice that these nodes are numbered from right to left. Since the theory does not depend on any particular ordering of the nodes, this is not troublesome.

A simple graph illustrates this phenomenon best. Again, consider Equation (1) on the interval $[-5, 5]$. First, we select nine equally spaced nodes and use routines *Coef* and *Eval* with an automatic plotter to graph p_8. As shown in Figure 4.7, the resulting curve assumes negative values, which, of course, $f(x)$ does *not* have! Adding more equally spaced nodes—and thereby obtaining a higher-degree polynomial—only makes matters worse with wilder oscillations. In Figure 4.8, nine Chebyshev nodes are used, and the resulting polynomial

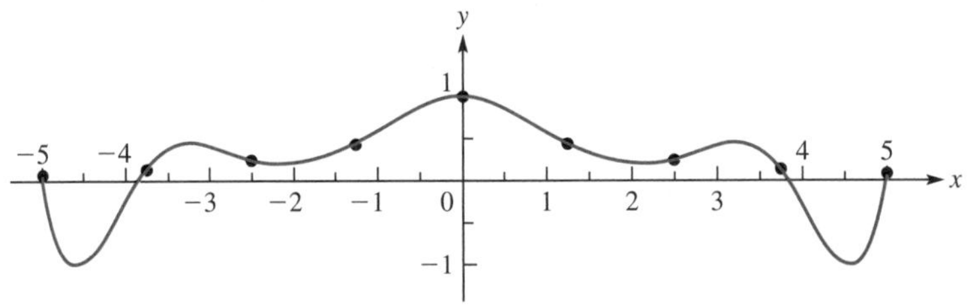

FIGURE 4.7 Polynomial interpolant with nine equally spaced nodes

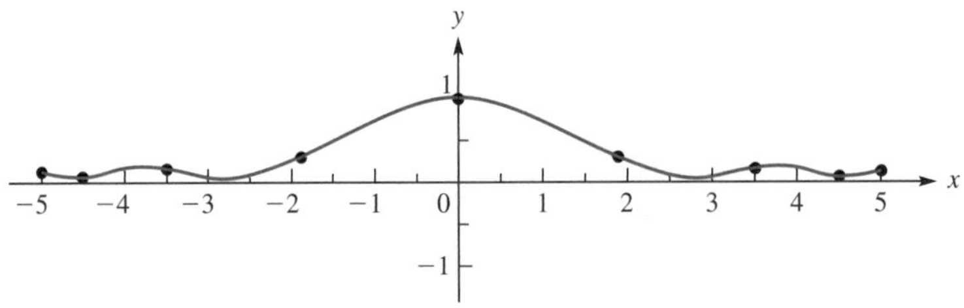

FIGURE 4.8 Polynomial interpolant with nine Chebyshev nodes

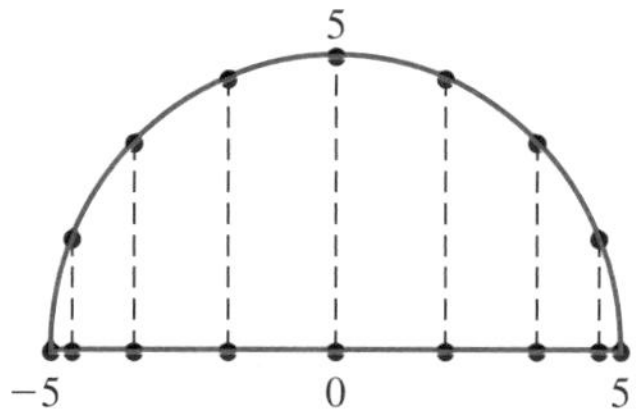

FIGURE 4.9 Chebyshev nodes of T_9

curve is smoother. However, cubic splines (discussed in Section 6.2) produce an even better curve fit.

The Chebyshev nodes of T_9 are obtained by taking equally spaced points on the unit circle and projecting them onto the horizontal axis, as in Figure 4.9.

Theorems on Interpolation Errors

It is possible to assess the errors of interpolation by means of a formula that involves the $(n+1)$st derivative of the function being interpolated. Here is the formal statement:

■ **Theorem 1** **First Interpolation Error Theorem**

If p is the polynomial of degree at most n that interpolates f at the $n+1$ distinct nodes $x_0, x_1, \dots, x_n$ belonging to an interval $[a, b]$ and if $f^{(n+1)}$ is continuous, then for each x in $[a, b]$, there is a ξ in (a, b) for which

$$f(x) - p(x) = \frac{1}{(n+1)!} f^{(n+1)}(\xi) \prod_{i=0}^{n} (x - x_i) \tag{2}$$

Proof Observe first that Equation (2) is obviously valid if x is one of the nodes x_i because then both sides of the equation reduce to zero. If x is not a node, let it be fixed in the remainder of the discussion, and define

$$\begin{aligned} w(t) &= \prod_{i=0}^{n}(t - x_i) && \text{(polynomial in the variable } t) \\ c &= \frac{f(x) - p(x)}{w(x)} && \text{(constant)} \\ \varphi(t) &= f(t) - p(t) - cw(t) && \text{(function in the variable } t) \end{aligned} \tag{3}$$

Observe that c is well defined because $w(x) \neq 0$ (x is not a node). Note also that φ takes the value 0 at the $n+2$ points $x_0, x_1, \dots, x_n$, and x. Now invoke **Rolle's Theorem**,* which states that between any two roots of φ, there must occur a root of φ'. Thus, φ' has at least $n+1$ roots. By similar reasoning, φ'' has at least n roots, φ''' has at least $n-1$ roots, and so on. Finally, it can be inferred that $\varphi^{(n+1)}$ must have at least one root. Let ξ be a root of $\varphi^{(n+1)}$. All the roots being counted in this argument are in (a, b). Thus, we obtain

$$0 = \varphi^{(n+1)}(\xi) = f^{(n+1)}(\xi) - p^{(n+1)}(\xi) - cw^{(n+1)}(\xi)$$

---

*__Rolle's Theorem__: Let f be a function that is continuous on $[a, b]$ and differentiable on (a, b). If $f(a) = f(b) = 0$, then $f'(c) = 0$ for some point c in (a, b).

In this equation, $p^{(n+1)}(\xi) = 0$ because p is a polynomial of degree $\leqq n$. Also, $w^{(n+1)}(\xi) = (n+1)!$ because $w(t) = t^{n+1} +$ (lower-order terms in t). Thus, we have

$$0 = f^{(n+1)}(\xi) - c(n+1)! = f^{(n+1)}(\xi) - \frac{(n+1)!}{w(x)}[f(x) - p(x)]$$

This equation is a rearrangement of Equation (2). ■

A special case that often arises is the one in which the interpolation nodes are equally spaced.

■ **Lemma 1**

Upper Bound Lemma

Suppose that $x_i = a + ih$ for $i = 0, 1, \ldots, n$ and that $h = (b-a)/n$. Then for any $x \in [a, b]$

$$\prod_{i=0}^{n} |x - x_i| \leqq \frac{1}{4}h^{n+1}n! \tag{4}$$

Proof To establish this inequality, fix x and select j so that $x_j \leqq x \leqq x_{j+1}$. It is an exercise in calculus (Exercise 5.2.2) to show that

$$|x - x_j||x - x_{j+1}| \leqq \frac{h^2}{4} \tag{5}$$

Using Equation (5), we have

$$\prod_{i=0}^{n} |x - x_i| \leqq \frac{h^2}{4} \prod_{i=0}^{j-1}(x - x_i) \prod_{i=j+2}^{n} (x_i - x)$$

The sketch in Figure 4.10, showing a typical case of equally spaced nodes, may be helpful.

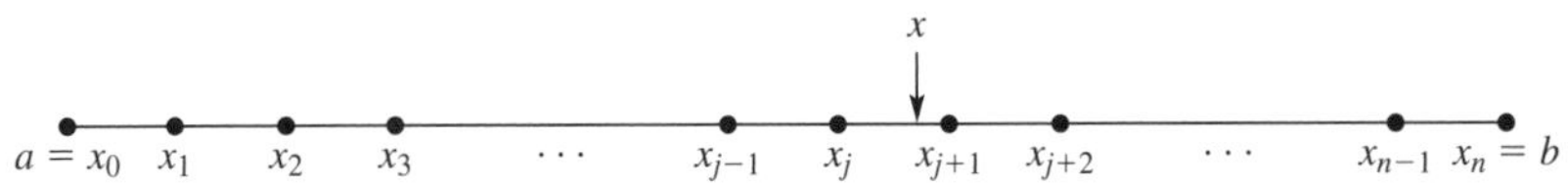

FIGURE 4.10 Typical location of x in equally spaced nodes

Since $x_j \leqq x \leqq x_{j+1}$, we have further

$$\prod_{i=0}^{n} |x - x_i| \leqq \frac{h^2}{4} \prod_{i=0}^{j-1}(x_{j+1} - x_i) \prod_{i=j+2}^{n} (x_i - x_j)$$

Now use the fact that $x_i = a + ih$. Then we have $x_{j+1} - x_i = (j - i + 1)h$ and $x_i - x_j = (i - j)h$. Therefore, we obtain

$$\begin{aligned}\prod_{i=0}^{n} |x - x_i| &\leqq \frac{h^2}{4} h^j h^{n-(j+2)+1} \prod_{i=0}^{j-1}(j - i + 1) \prod_{i=j+2}^{n} (i - j) \\ &\leqq \frac{1}{4}h^{n+1}(j+1)!(n-j)! \leqq \frac{1}{4}h^{n+1}n!\end{aligned}$$

In the last step, we use the fact that if $0 \leqq j \leqq n-1$, then $(j+1)!(n-j)! \leqq n!$. This, too, is left as an exercise (Exercise 5.2.3). Hence, Inequality (4) is established. ■

We can now find a bound on the interpolation error.

Theorem 2 **Second Interpolation Error Theorem**

Let f be a function such that $f^{(n+1)}$ is continuous on $[a, b]$ and satisfies $|f^{(n+1)}(x)| \leqq M$. Let p be the polynomial of degree $\leqq n$ that interpolates f at $n+1$ equally spaced nodes in $[a, b]$, including the endpoints. Then on $[a, b]$,

$$|f(x) - p(x)| \leqq \frac{1}{4(n+1)} M h^{n+1} \tag{6}$$

where $h = (b-a)/n$ is the spacing between nodes.

Proof Use Theorem 1 on interpolation errors and Inequality (4) in Lemma 1. ■

This theorem gives loose upper bounds on the interpolation error for different values of n. By other means, one can find tighter upper bounds for small values of n (Cf. Exercise 4.2.5). If the nodes are not uniformly spaced, then a better bound can be found by use of the Chebyshev nodes.

EXAMPLE 2 Assess the error if $\sin x$ is replaced by an interpolation polynomial that has 10 equally spaced nodes in [0, 1.6875]. (See the related Example 8 in Section 4.1.)

Solution We use Theorem 2 on interpolation errors, taking $f(x) = \sin x$, $n = 9$, $a = 0$, and $b = 1.6875$. Since $f^{(10)}(x) = -\sin x$, $|f^{(10)}(x)| \leqq 1$. Hence, in Equation (6), we can let $M = 1$. The result is

$$|\sin x - p(x)| \leqq 1.34 \times 10^{-9}$$

Thus, $p(x)$ represents $\sin x$ on this interval with an error of at most two units in the ninth decimal place. Therefore, the interpolation polynomial that has 10 equally spaced nodes on the interval [0, 1.6875] approximates $\sin x$ to at least eight decimal digits of accuracy. In fact, a careful check on a computer would reveal that the polynomial is accurate to even more decimal places. (Why?) ■

The error expression in polynomial interpolation can also be given in terms of divided differences:

Theorem 3 **Third Interpolation Error Theorem**

If p is the polynomial of degree n that interpolates the function f at nodes $x_0, x_1, \ldots, x_n$, then for any x that is not a node,

$$f(x) - p(x) = f[x_0, x_1, \ldots, x_n, x] \prod_{i=0}^{n} (x - x_i)$$

Proof Let t be any point, other than a node, where $f(t)$ is defined. Let q be the polynomial of degree $\leqq n+1$ that interpolates f at $x_0, x_1, \ldots, x_n, t$. By the Newton form of the interpolation formula (Equation (8) in Section 4.1), we have

$$q(x) = p(x) + f[x_0, x_1, \ldots, x_n, t] \prod_{i=0}^{n} (x - x_i)$$

Since $q(t) = f(t)$, this yields at once

$$f(t) = p(t) + f[x_0, x_1, \ldots, x_n, t] \prod_{i=0}^{n} (t - x_i)$$ ■

The following theorem shows that there is a relationship between divided differences and derivatives.

■ **Theorem 4** **Divided Differences and Derivatives**

If $f^{(n)}$ is continuous on $[a, b]$ and if $x_0, x_1, \ldots, x_n$ are any $n + 1$ distinct points in $[a, b]$, then for some ξ in (a, b),

$$f[x_0, x_1, \ldots, x_n] = \frac{1}{n!} f^{(n)}(\xi)$$

Proof Let p be the polynomial of degree $\leqq n - 1$ that interpolates f at $x_0, x_1, \ldots, x_{n-1}$. By Theorem 1 on interpolation errors, there is a point ξ such that

$$f(x_n) - p(x_n) = \frac{1}{n!} f^{(n)}(\xi) \prod_{i=0}^{n-1} (x_n - x_i)$$

By Theorem 3 on interpolation errors, we obtain

$$f(x_n) - p(x_n) = f[x_0, x_1, \ldots, x_{n-1}, x_n] \prod_{i=0}^{n-1} (x_n - x_i)$$ ■

As an immediate consequence of this theorem, we observe that all high-order divided differences are zero for a polynomial.

■ **Corollary 1** **Divided Differences Corollary**

If f is a polynomial of degree n, then all of the divided differences $f[x_0, x_1, \ldots, x_i]$ are zero for $i \geqq n + 1$.

EXAMPLE 3 Is there a cubic polynomial that takes these values?

x	1	−2	0	3	−1	7
y	−2	−56	−2	4	−16	376

Solution If such a polynomial exists, its fourth-order divided differences $f[\ ,\ ,\ ,\ ,\ ]$ would all be zero. We form a divided-difference table to check this possibility:

Sample Divided-Difference Table

x	$f[\ ]$	$f[\ ,\ ]$	$f[\ ,\ ,\ ]$	$f[\ ,\ ,\ ,\ ]$	$f[\ ,\ ,\ ,\ ,\ ]$
1	$\boxed{-2}$				
		$\boxed{18}$			
−2	−56		$\boxed{-9}$		
		27		$\boxed{2}$	
0	−2		−5		0
		2		2	
3	4		−3		0
		5		2	
−1	−16		11		
		49			
7	376				

The data can be represented by a cubic polynomial because the fourth-order divided differences $f[\ ,\ ,\ ,\ ,\ ]$ are zero. From the Newton form of the interpolation formula, this polynomial is

Newton Interpolation Polynomials

$$p_3(x) = -2 + 18(x-1) - 9(x-1)(x+2) + 2(x-1)(x+2)x \qquad \blacksquare$$

Summary 4.2

- The **Runge function** $f(x) = 1/(1+x^2)$ on the interval $[-5, 5]$ shows that high-degree polynomial interpolation and uniform spacing of nodes may not be satisfactory. The **Chebyshev nodes** for the interval $[a, b]$ are given by

$$x_i = \frac{1}{2}(a+b) + \frac{1}{2}(b-a)\cos\left[\left(\frac{2i+1}{2n+2}\right)\pi\right]$$

- There is a relationship between differences and derivatives:

$$f[x_0, x_1, \ldots, x_n] = \frac{1}{n!} f^{(n)}(\xi)$$

- Expressions for errors in polynomial interpolation are

$$f(x) - p(x) = \frac{1}{(n+1)!} f^{(n+1)}(\xi) \prod_{i=0}^{n} (x - x_i)$$

$$f(x) - p(x) = f[x_0, x_1, \ldots, x_n, x] \prod_{i=0}^{n} (x - x_i)$$

- For $n+1$ equally spaced nodes, an upper bound on the error is given by

$$|f(x) - p(x)| \leqq \frac{M}{4(n+1)} \left(\frac{b-a}{n}\right)^{n+1}$$

  Here M is an upper bound on $\left|f^{(n+1)}(x)\right|$ when $a \leqq x \leqq b$.

- If f is a polynomial of degree n, then all of the divided differences $f[x_0, x_1, \ldots, x_i]$ are zero for $i \geqq n+1$.

Exercises 4.2

[a]1. Use a divided-difference table to show that the following data can be represented by a polynomial of degree 3:

x	-2	-1	0	1	2	3
y	1	4	11	16	13	-4

2. Fill in a detail in the proof of Inequality (4) by proving Inequality (5).

3. (Continuation) Fill in another detail in the proof of Inequality (4) by showing that $(j+1)!(n-j)! \leqq n!$ if $0 \leqq j \leqq n-1$. Induction and a symmetry argument can be used.

4. For nonuniformly distributed nodes $a = x_0 < x_1 < \cdots < x_n = b$, where $h = \max_{1 \leqq i \leqq n}\{(x_i - x_{i-1})\}$, show that Inequality (4) is true.

5. Using Theorem 1, show directly that the maximum interpolation error is bounded by the following expressions and compare them to the bounds given by Theorem 2:

   a. $\frac{1}{8}h^2 M$ for linear interpolation, where $h = x_1 - x_0$ and $M = \max_{x_0 \leqq x \leqq x_1} |f''(x)|$.

   b. $\frac{1}{9\sqrt{3}}h^3 M$ for quadratic interpolation, where $h = x_1 - x_0 = x_2 - x_1$ and $M = \max_{x_0 \leqq x \leqq x_2} |f'''(x)|$.

   c. $\frac{3}{128}h^4 M$ for cubic interpolation, where $h = x_1 - x_0 = x_2 - x_1 = x_3 = x_2$ and $M = \max_{x_0 \leqq x \leqq x_3} |f''(x)|$.

[a]6. How accurately can we determine $\sin x$ by linear interpolation, given a table of $\sin x$ to 10 decimal places, for x in $[0, 2]$ with $h = 0.01$?

[a]7. (Continuation) Given the data

x	$\sin x$	$\cos x$
0.70	0.64421 76872	0.76484 21873
0.71	0.65183 37710	0.75836 18760

find approximate values of $\sin 0.705$ and $\cos 0.702$ by linear interpolation. What is the error?

[a]8. **Linear interpolation** in a table of function values means the following: If $y_0 = f(x_0)$ and $y_1 = f(x_1)$ are tabulated values, and if $x_0 < x < x_1$, then an interpolated value of $f(x)$ is $y_0 + [(y_1 - y_0)/(x_1 - x_0)](x - x_0)$, as explained at the beginning of Section 4.1. A table of values of $\cos x$ is required so that the linear interpolation will yield five-decimal-place accuracy for any value of x in $[0, \pi]$. Assume that the tabular values are equally spaced, and determine the minimum number of entries needed in this table.

[a]9. An interpolating polynomial of degree 20 is to be used to approximate e^{-x} on the interval $[0, 2]$. How accurate will it be? (Use 21 uniform nodes, including the endpoints of the interval. Compare results, using Theorems 1 and 2.)

[a]10. Let the function $f(x) = \ln x$ be approximated by an interpolation polynomial of degree 9 with 10 nodes uniformly distributed in the interval $[1, 2]$. What bound can be placed on the error?

11. In the first theorem on interpolation errors, show that if $x_0 < x_1 < \cdots < x_n$ and $x_0 < x < x_n$, then $x_0 < \xi < x_n$.

12. (Continuation) In the same theorem, considering ξ as a function of x, show that $f^{(n)}[\xi(x)]$ is a continuous function of x.
Note: $\xi(x)$ need not be a continuous function of x.

[a]13. Suppose $\cos x$ is to be approximated by an interpolating polynomial of degree n, using $n+1$ equally spaced nodes in the interval $[0, 1]$. How accurate is the approximation? (Express your answer in terms of n.) How accurate is the approximation when $n = 9$? For what values of n is the error less than 10^{-7}?

[a]14. In interpolating with $n+1$ equally spaced nodes on an interval, we could use $x_i = a + (2i+1)h/2$, where $0 \leqq i \leqq n-1$ and $h = (b-a)/n$. What bound can be given now for $\prod_{i=0}^{n} |x - x_i|$ when $a \leqq x \leqq b$?
Note: We are not requiring the endpoints to be nodes.

15. Using Equation (3), show that

$$w'(t) = \sum_{i=0}^{n} \prod_{\substack{j \neq i \\ j=0}}^{n} (t - x_j), \qquad w'(x_i) = \prod_{\substack{j \neq i \\ j=0}}^{n} (x_i - x_j)$$

[a]16. Does every polynomial p of degree at most n obey the following equation? Explain why or why not.

$$p(x) = \sum_{i=0}^{n} p[x_0, x_1, \ldots, x_i] \prod_{j=0}^{i-1} (x - x_j)$$

Hint: Use the uniqueness of the interpolating polynomial.

17. Find a polynomial p that takes these values: $p(1) = 3$, $p(2) = 1$, $p(0) = -5$. You may use any method you wish. You may leave the polynomial in any convenient form, not necessarily in the *standard* form, $\sum_{k=1}^{n} c_k x^k$. Next, find a new polynomial q that takes those same three values *and* $q(3) = 7$.

18. For the case $n = 2$, establish Theorem 4 and Corollory 1 directly.

Computer Exercises 4.2

1. Using 21 equally spaced nodes on the interval $[-5, 5]$, find the interpolating polynomial p of degree 20 for the function $f(x) = (x^2+1)^{-1}$. Print the values of $f(x)$ and $p(x)$ at 41 equally spaced points, including the nodes. Observe the large discrepancy between $f(x)$ and $p(x)$.

2. (Continuation) Perform the experiment in the preceding computer problem, using Chebyshev nodes $x_i = 5\cos(i\pi/20)$, where $0 \leqq i \leqq 20$, and nodes $x_i = 5\cos[(2i+1)\pi/42]$, where $0 \leqq i \leqq 20$. Record your conclusions.

3. Using procedures corresponding to the pseudocode in the text, find a polynomial of degree 13 that interpolates $f(x) = \arctan x$ on the interval $[-1, 1]$. Test numerically by taking 100 points to determine how accurate the polynomial approximation is.

4. (Continuation) Write a function for $\arctan x$ that uses the polynomial of the preceding computer problem. If x is not in the interval $[-1, 1]$, use the formula $1/\tan\theta = \cot\theta = \tan(\pi/2 - \theta)$.

5. Approximate $\arcsin x$ on the interval $\left[-1/\sqrt{2}, 1/\sqrt{2}\right]$ by an interpolating polynomial of degree 15. Determine how accurate the approximation is by numerical tests. Use equally spaced nodes.

6. (Continuation) Write a function for $\arcsin x$, using the polynomial of the previous computer problem. Use $\sin(\pi/2 - \theta) = \cos\theta = \sqrt{1 - \sin^2\theta}$ if x is in the interval $|x| > 1/\sqrt{2}$.

7. Let $f(x) = \max\{0, 1 - x\}$. Sketch the function f. Then find interpolating polynomials p of degrees 2, 4, 8, 16, and 32 to f on the interval $[-4, 4]$, using equally spaced nodes. Print out the discrepancy $f(x) - p(x)$ at 128 equally spaced points. Then redo the problem using Chebyshev nodes.

8. Using *Coef* and *Eval* and an automatic plotter, fit a polynomial through the following data:

x	0.0	0.60	1.50	1.70	1.90	2.1	2.30	2.60	2.8	3.00
y	−0.8	−0.34	0.59	0.59	0.23	0.1	0.28	1.03	1.5	1.44

Does the resulting curve look like a good fit? Explain.

9. Find the polynomial p of degree $\leqq 10$ that interpolates $|x|$ on $[-1, 1]$ at 11 equally spaced points. Print the difference $|x| - p(x)$ at 41 equally spaced points. Then do the same with Chebyshev nodes. Compare.

[a]10. Why are the Chebyshev nodes generally better than equally spaced nodes in polynomial interpolation? The answer lies in the term $\prod_{i=0}^{n}(x - x_i)$ that occurs in the error formula. If $x_i = \cos[(2i+1)\pi/(2n+2)]$, then

$$\left|\prod_{i=0}^{n}(x - x_i)\right| \leqq 2^{-n}$$

for all x in $[-1, 1]$. Carry out a numerical experiment to test the given inequality for $n = 3, 7, 15$.

11. (**Student Research Project**) Explore the topic of interpolation of multivariate scattered data, such as those in geophysics and other areas.

12. Use mathematical software such as found in MATLAB, Maple, or Mathematica to reproduce Figures 4.7–4.8.

13. Use symbolic mathematical software such as MATLAB, Maple or Mathematica to generate the interpolation polynomial for the data points in Example 2. Plot the polynomial and the data points.

14. Use graphical software to plot four or five points that happen to generate an interpolating polynomial that exhibits a great deal of oscillations. This piece of software should let you use your computer mouse to *click* on three or four points that visually appear to be part of a smooth curve. Next it uses Newton's interpolating polynomial to sketch the curve through these points. Then add another point that is somewhat remote from the curve and refit all the points. Repeat, adding other points. After a few points have been added in this way, you should have evidence that polynomials can oscillate wildly.

4.3 Estimating Derivatives and Richardson Extrapolation

A numerical experiment outlined in Section 1.1 (p. 12) showed that determining the derivative of a function f at a point x is *not* a trivial numerical problem. Specifically, if $f(x)$ can be computed with only n digits of precision, it is difficult to calculate $f'(x)$ numerically with n digits of precision. This difficulty can be traced to the subtraction between quantities that are nearly equal. In this section, several alternatives are offered for the numerical computation of $f'(x)$ and $f''(x)$.

First-Derivative Formulas via Taylor Series

First, consider again the obvious method based on the definition of $f'(x)$. It consists of selecting one or more small values of h and writing

$f'(x)$ Approximation

$$f'(x) \approx \frac{1}{h}[f(x+h) - f(x)] \tag{1}$$

What truncation error is involved in this formula?

To find out, use Taylor's Theorem:

$$f(x+h) = f(x) + hf'(x) + \frac{1}{2}h^2 f''(\xi)$$

Rearranging this equation gives

$f'(x)$ Forward Difference Rule with Error Term

$$f'(x) = \frac{1}{h}[f(x+h) - f(x)] - \frac{1}{2}hf''(\xi) \tag{2}$$

Hence, we see that approximation (1) has error term $-\frac{1}{2}hf''(\xi) = \mathcal{O}(h)$, where ξ is in the interval having endpoints x and $x+h$.

Equation (2) shows that in general, as $h \to 0$, the difference between $f'(x)$ and the estimate $h^{-1}[f(x+h) - f(x)]$ approaches zero at the same rate that h does—that is, $\mathcal{O}(h)$. Of course, if $f''(x) = 0$, then the error term is $\frac{1}{6}h^2 f'''(\gamma)$, which converges to zero somewhat faster at $\mathcal{O}(h^2)$. But usually, $f''(x)$ is *not* zero.

Error Analysis

Equation (2) gives the **truncation error** for this numerical procedure, namely, $-\frac{1}{2}hf''(\xi)$. This error is present even if the calculations are performed with *infinite* precision; it is due to our imitating the mathematical limit process by means of an approximation formula. Additional (and worse) errors must be expected when calculations are performed on a computer with finite word length.

EXAMPLE 1 In Section 1.1 (p. 12), the program named *First* used the forward difference rule (1) to approximate the first derivative of the function $f(x) = \sin x$ at $x = 0.5$. Explain what happens when a large number of iterations are performed, say $n = 50$.

Solution There is a total loss of all significant digits! When we examine the computer output closely, we find that, in fact, a good approximation $f'(0.5) \approx 0.87758$ was found, but it deteriorated as the process continued. This was caused by the subtraction of two nearly equal quantities $f(x+h)$ and $f(x)$, resulting in a loss of significant digits as well as a magnification of this effect from dividing by a small value of h. We need to stop the iterations sooner! When to stop an iterative process is a common question in numerical algorithms. In this case, monitor the iterations to determine when they settle down, namely, when two successive ones are within a prescribed tolerance. Alternatively, we can use the truncation error term. If we want six significant digits of accuracy in the results, we set

$$\left| -\frac{1}{2}hf''(\xi) \right| \leqq \frac{1}{2}4^{-n} < \frac{1}{2}10^{-6}$$

since $|f''(x)| < 1$ and $h = 1/4^n$. We find $n > 6/\log 4 \approx 9.97$. So we should stop after about 10 steps in the process. (The least error of 3.1×10^{-9} was found at iteration 14.) ■

As we saw in Newton's method (Section 3.2) and will see in the Romberg method (Section 5.2), it is advantageous to have the convergence of numerical processes occur with higher powers of some quantity approaching zero. In the present situation, we want an approximation to $f'(x)$ in which the error behaves like $\mathcal{O}(h^2)$. One such method is easily obtained with the aid of the following two Taylor series:

$$\begin{cases} f(x+h) = f(x) + hf'(x) + \dfrac{1}{2!}h^2 f''(x) + \dfrac{1}{3!}h^3 f'''(x) + \dfrac{1}{4!}h^4 f^{(4)}(x) + \cdots \\ f(x-h) = f(x) - hf'(x) + \dfrac{1}{2!}h^2 f''(x) - \dfrac{1}{3!}h^3 f'''(x) + \dfrac{1}{4!}h^4 f^{(4)}(x) - \cdots \end{cases} \tag{3}$$

By subtraction, we obtain

$$f(x+h) - f(x-h) = 2hf'(x) + \frac{2}{3!}h^3 f'''(x) + \frac{2}{5!}h^5 f^{(5)}(x) + \cdots$$

This leads to a very important formula for approximating $f'(x)$:

$$f'(x) = \frac{1}{2h}[f(x+h) - f(x-h)] - \frac{h^2}{3!}f'''(x) - \frac{h^4}{5!}f^{(5)}(x) - \cdots \tag{4}$$

Expressed otherwise,

$f'(x)$ Central Difference Rule

$$f'(x) \approx \frac{1}{2h}[f(x+h) - f(x-h)] \tag{5}$$

with an error whose leading term is $-\frac{1}{6}h^2 f'''(x)$, which makes it $\mathcal{O}(h^2)$.

By using Taylor's Theorem with its error term, we could have obtained the following two expressions:

$$\begin{aligned} f(x+h) &= f(x) + hf'(x) + \frac{1}{2}h^2 f''(x) + \frac{1}{6}h^3 f'''(\xi_1) \\ f(x-h) &= f(x) - hf'(x) + \frac{1}{2}h^2 f''(x) - \frac{1}{6}h^3 f'''(\xi_2) \end{aligned}$$

Then the subtraction would lead to

$$f'(x) = \frac{1}{2h}[f(x+h) - f(x-h)] - \frac{1}{6}h^2\left[\frac{f'''(\xi_1) + f'''(\xi_2)}{2}\right]$$

The error term here can be simplified by the following reasoning: The expression $\frac{1}{2}[f'''(\xi_1) + f'''(\xi_2)]$ is the average of two values of f''' on the interval $[x-h, x+h]$. It therefore lies between the least and greatest values of f''' on this interval. If f''' is continuous on this interval, then this average value is assumed at some point ξ. Hence, the formula with its error term can be written as

$f'(x)$ Central Difference Rule with Error Term

$$f'(x) = \frac{1}{2h}[f(x+h) - f(x-h)] - \frac{1}{6}h^2 f'''(\xi)$$

This is based on the sole assumption that f''' is continuous on $[x-h, x+h]$. This formula for numerical differentiation turns out to be very useful in the numerical solution of certain differential equations, as we shall see in Chapter 11 (on boundary value problems) and Chapter 12 (on partial differential equations).

EXAMPLE 2 Modify program *First* in Section 1.1 so that it uses the central difference formula (5) to approximate the first derivative of the function $f(x) = \sin x$ at $x = 0.5$. Determine how many iterations are needed for error less than $\frac{1}{2}10^{-6}$.

Solution Using the truncation error term for the central difference formula (5), we set

$$\left| -\frac{1}{6}h^2 f'''(\xi) \right| \leqq \frac{1}{6}4^{-2n} < \frac{1}{2}10^{-6}$$

or $n > (6 - \log 3)/\log 16 \approx 4.59$. We obtain a good approximation after about five iterations with this higher-order formula. (The least error of 3.6×10^{-12} was at step 9.) ■

Richardson Extrapolation

Returning now to Equation (4), we write it in a simpler form:

$f'(x)$ Central Difference Rule with Error Series

$$f'(x) = \frac{1}{2h}[f(x+h) - f(x-h)] + a_2h^2 + a_4h^4 + a_6h^6 + \cdots \tag{6}$$

in which the constants $a_2, a_4, \ldots$ depend on f and x. When such information is available about a numerical process, it is possible to use a powerful technique known as *Richardson extrapolation* to wring more accuracy out of the method. This procedure is explained here, using Equation (6) as our model.

Holding f and x fixed, we define a function of h by the formula

$\varphi(h)$ Formula

$$\varphi(h) = \frac{1}{2h}[f(x+h) - f(x-h)] \tag{7}$$

From Equation (6), we see that $\varphi(h)$ is an approximation to $f'(x)$ with error of order $\mathcal{O}(h^2)$. Our objective is to compute $\lim_{h\to 0}\varphi(h)$ because this is the quantity $f'(x)$ that we wanted in the first place. If we select a function f and plot $\varphi(h)$ for $h = 1, \frac{1}{2}, \frac{1}{4}, \frac{1}{8}, \ldots$, then we get a graph (Computer Exercise 4.3.5). Near zero, where we cannot actually calculate the value of φ from Equation (7), φ is approximately a quadratic function of h, since the higher-order terms from Equation (6) are negligible. Richardson extrapolation seeks to estimate the limiting value at 0 from some computed values of $\varphi(h)$ near 0. Obviously, we can take any convenient sequence h_n that converges to zero, calculate $\varphi(h_n)$ from Equation (7), and use these as approximations to $f'(x)$.

But something much more clever can be done. Suppose we compute $\varphi(h)$ for some h and then compute $\varphi(h/2)$. By Equation (6), we have

$\varphi(h)$ with Error Series

$$\varphi(h) = f'(x) - a_2h^2 - a_4h^4 - a_6h^6 - \cdots$$

$$\varphi\left(\frac{h}{2}\right) = f'(x) - a_2\left(\frac{h}{2}\right)^2 - a_4\left(\frac{h}{2}\right)^4 - a_6\left(\frac{h}{2}\right)^6 - \cdots$$

We can eliminate the dominant term in the error series by simple algebra. To do so, multiply the second equation by 4 and subtract it from the first equation. The result is

$$\varphi(h) - 4\varphi\left(\frac{h}{2}\right) = -3f'(x) - \frac{3}{4}a_4h^4 - \frac{15}{16}a_6h^6 - \cdots$$

We divide by -3 and rearrange this to get

Linear Combination of φ

$$\varphi\left(\frac{h}{2}\right) + \frac{1}{3}\left[\varphi\left(\frac{h}{2}\right) - \varphi(h)\right] = f'(x) + \frac{1}{4}a_4h^4 + \frac{5}{16}a_6h^6 + \cdots$$

This is a marvelous discovery. Simply by adding $\frac{1}{3}[\varphi(h/2) - \varphi(h)]$ to $\varphi(h/2)$, we have apparently improved the precision to $\mathcal{O}(h^4)$ because the error series that accompanies this new combination begins with $\frac{1}{4}a_4h^4$. When h is small, this is a dramatic improvement!

We can repeat this process by letting

$\Phi(h)$ Formula

$$\Phi(h) = \frac{4}{3}\varphi\left(\frac{h}{2}\right) - \frac{1}{3}\varphi(h)$$

Then we have from the previous derivation that

$$\Phi(h) = f'(x) + b_4 h^4 + b_6 h^6 + \cdots$$

$$\Phi\left(\frac{h}{2}\right) = f'(x) + b_4\left(\frac{h}{2}\right)^4 + b_6\left(\frac{h}{2}\right)^6 + \cdots$$

We can combine these equations to eliminate the first term in the error series

$$\Phi(h) - 16\Phi\left(\frac{h}{2}\right) = -15f'(x) + \frac{3}{4}b_6 h^6 + \cdots$$

Hence, we have

Linear Combination of Φ

$$\Phi\left(\frac{h}{2}\right) + \frac{1}{15}\left[\Phi\left(\frac{h}{2}\right) - \Phi(h)\right] = f'(x) - \frac{1}{20}b_6 h^6 + \cdots$$

This is yet another apparent improvement in the precision to $\mathcal{O}(h^6)$. And now, to top it off, note that the same procedure can be repeated over and over again to *kill* higher and higher terms in the error. This is **Richardson extrapolation**.

Essentially the same situation arises in the derivation of Romberg's algorithm in Section 5.2. We begin a general discussion of the procedure here. We start with an equation that includes both situations. Let φ be a function such that

$\varphi(h)$ Property

$$\varphi(h) = L - \sum_{k=1}^{\infty} a_{2k} h^{2k} \tag{8}$$

where the coefficients a_{2k} are not known. Equation (8) is *not* interpreted as the *definition* of φ, but rather as a *property* that φ possesses. It is assumed that $\varphi(h)$ can be computed for any $h > 0$ and that our objective is to approximate L accurately using φ.

Select a convenient h, and compute the numbers

$$D(n, 0) = \varphi\left(\frac{h}{2^n}\right) \qquad (n \geqq 0) \tag{9}$$

Because of Equation (8), we have

$D(n, 0)$ Formula

$$D(n, 0) = L + \sum_{k=1}^{\infty} A(k, 0)\left(\frac{h}{2^n}\right)^{2k}$$

where $A(k, 0) = -a_{2k}$. These quantities $D(n, 0)$ give a crude estimate of the unknown number $L = \lim_{x\to 0}\varphi(x)$. More accurate estimates are obtained via Richardson extrapolation. The extrapolation formula is

Extrapolation Formula $D(n, m)$

$$D(n, m) = \frac{4^m}{4^m - 1}D(n, m-1) - \frac{1}{4^m - 1}D(n-1, m-1) \qquad (1 \leqq m \leqq n) \tag{10}$$

■ **Theorem 1**

Richardson Extrapolation Theorem

The quantities $D(n, m)$ defined in the Richardson extrapolation process (10) obey the equation

$$D(n, m) = L + \sum_{k=m+1}^{\infty} A(k, m)\left(\frac{h}{2^n}\right)^{2k} \qquad (0 \leqq m \leqq n) \tag{11}$$

Proof Equation (11) is true by hypothesis if $m = 0$. For the purpose of an inductive proof, we *assume* that Equation (11) is valid for an arbitrary value of $m - 1$, and we prove that Equation (11) is then valid for m. Now from Equations (10) and (11) for a fixed value m, we have

$$D(n,m) = \frac{4^m}{4^m - 1}\left[L + \sum_{k=m}^{\infty} A(k, m-1)\left(\frac{h}{2^n}\right)^{2k}\right] - \frac{1}{4^m - 1}\left[L + \sum_{k=m}^{\infty} A(k, m-1)\left(\frac{h}{2^{n-1}}\right)^{2k}\right]$$

After simplification, this becomes

$$D(n,m) = L + \sum_{k=m}^{\infty} A(k, m-1)\left(\frac{4^m - 4^k}{4^m - 1}\right)\left(\frac{h}{2^n}\right)^{2k} \tag{12}$$

Thus, we are led to define

$$A(k,m) = A(k, m-1)\left(\frac{4^m - 4^k}{4^m - 1}\right)$$

At the same time, we notice that $A(m, m) = 0$. Hence, Equation (12) can be written as

$D(n, m)$ Formula

$$D(n,m) = L + \sum_{k=m+1}^{\infty} A(k,m)\left(\frac{h}{2^n}\right)^{2k}$$

Equation (11) is true for m, and the induction is complete. ■

The significance of Equation (11) is that the summation *begins* with the term $(h/2^n)^{2m+2}$. Since $h/2^n$ is small, this indicates that the numbers $D(n, m)$ are approaching L very rapidly, namely,

Rule of Converging of $D(n, m)$

$$D(n,m) = L + \mathcal{O}\left(\frac{h^{2(m+1)}}{2^{2n(m+1)}}\right)$$

In practice, one can arrange the quantities in a two-dimensional triangular array as follows:

2D Triangular Array

$$\begin{array}{lllll}
D(0,0) & & & & \\
D(1,0) & D(1,1) & & & \\
D(2,0) & D(2,1) & D(2,2) & & \\
\vdots & \vdots & \vdots & \ddots & \\
D(N,0) & D(N,1) & D(N,2) & \cdots & D(N,N)
\end{array} \tag{13}$$

The main tasks to generate such an array are as follows:

■ **Algorithm**

Richardson Extrapolation

1. Write a function for φ.
2. Decide on suitable values for N and h.
3. For $i = 0, 1, \dots, N$, compute $D(i, 0) = \varphi(h/2^i)$.
4. For $1 \leqq i \leqq j \leqq N$, compute

$$D(i, j) = D(i, j-1) + (4^j - 1)^{-1}[D(i, j-1) - D(i-1, j-1)]$$

Notice that in this algorithm, the computation of $D(i, j)$ follows Equation (10) but has been rearranged slightly to improve its numerical properties.

EXAMPLE 3 Write a procedure to compute the derivative of a function at a point by using Equation (5) and Richardson extrapolation.

Solution The input to the procedure are a function f, a specific point x, a value of h, and a number n signifying how many rows in the array (13) are to be computed. The output is the array (13). Here is a suitable pseudocode:

Derivative **Pseudocode**

```
procedure Derivative(f, x, n, h, (d_ij))
integer i, j, n;   real h, x;   real array (d_ij)_{0:n×0:n}
external function f
for i = 0 to n
    d_i0 ← [f(x + h) − f(x − h)]/(2h)
    for j = 1 to i
        d_{i,j} ← d_{i,j−1} + (d_{i,j−1} − d_{i−1,j−1})/(4^j − 1)
    end for
    h ← h/2
end for
end procedure Derivative
```

To test the procedure, choose $f(x) = \sin x$, where $x_0 = 1.23095\,94154$ and $h = 1$. Then $f'(x) = \cos x$ and $f'(x_0) = \frac{1}{3}$. A pseudocode is written as follows:

Test_Derivative **Pseudocode**

```
program Test_Derivative
real array (d_ij)_{0:n×0:n};   external function f
integer n ← 10;   real h ← 1;   x ← 1.23095 94154
call Derivative(f, x, n, h, (d_ij))
output (d_ij)
end program Test_Derivative

real function f(x)
real x
f ← sin(x)
end function f
```

We invite the reader to program the pseudocode and execute it on a computer. The computer output is the triangular array (d_{ij}) with indices $0 \leqq j \leqq i \leqq 10$. The most accurate value is $(d_{4,1}) = 0.33333\,33433$. The values d_{i0}, which are obtained solely by Equations (7) and (9) without any extrapolation, are not as accurate, having no more than four correct digits. ■

Derivatives Symbolically

Mathematical software is now available with algebraic manipulation capabilities. Using them, we could write a computer program to find derivatives symbolically for a rather large class of functions—probably all those you would encounter in a calculus course. For example, we could verify the numerical results above by first finding the derivative exactly and then evaluating the numerical answer $\cos(1.23095\,94154) \approx 0.33333\,33355$ since $\arccos\left(\frac{1}{3}\right) \approx 1.23095\,941543$. Of course, the procedures discussed in this section are for approximating derivatives that cannot be determined exactly.

First-Derivative Formulas via Interpolation Polynomials

An important general stratagem can be used to approximate derivatives (as well as integrals and other quantities). The function f is first approximated by a polynomial p so that $f \approx p$. Then we simply proceed to the approximation $f'(x) \approx p'(x)$ as a consequence. Of course, this strategy should be used *very cautiously* because the behavior of the interpolating polynomial can be oscillatory.

In practice, the approximating polynomial p is often determined by interpolation at a few points. For example, suppose that p is the polynomial of degree at most 1 that interpolates f at two nodes, x_0 and x_1. Then from Equation (8) in Section 4.1 with $n = 1$, we have

$$p_1(x) = f(x_0) + f[x_0, x_1](x - x_0)$$

Consequently, we have

$f'(x)$ Finite Difference Approximation

$$f'(x) \approx p_1'(x) = f[x_0, x_1] = \frac{f(x_1) - f(x_0)}{x_1 - x_0} \tag{14}$$

If $x_0 = x$ and $x_1 = x + h$ (see Figure 4.11), this formula is one previously considered, namely, Equation (1):

$$f'(x) \approx \frac{1}{h}[f(x + h) - f(x)] \tag{15}$$

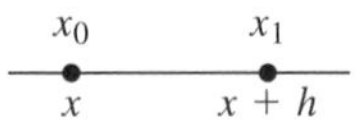

FIGURE 4.11 Forward difference: two nodes

If $x_0 = x - h$ and $x_1 = x + h$ (see Figure 4.12), the resulting formula is Equation (5):

$$f'(x) \approx \frac{1}{2h}[f(x + h) - f(x - h)] \tag{16}$$

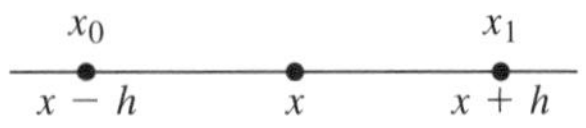

FIGURE 4.12 Central difference: two nodes

Now consider interpolation with three nodes, x_0, x_1, and x_2. The interpolating polynomial is obtained from Equation (8) in Section 4.1:

$$p_2(x) = f(x_0) + f[x_0, x_1](x - x_0) + f[x_0, x_1, x_2](x - x_0)(x - x_1)$$

and its derivative is

$$p_2'(x) = f[x_0, x_1] + f[x_0, x_1, x_2](2x - x_0 - x_1) \tag{17}$$

Here the right-hand side consists of two terms. The first is the previous estimate in Equation (14), and the second is a refinement or correction term.

If Equation (17) is used to evaluate $f'(x)$ when $x = \frac{1}{2}(x_0 + x_1)$, as in Equation (16), then the correction term in Equation (17) is zero. Thus, the first term in this case must be more accurate than those in other cases because the correction term adds nothing. This is why Equation (16) is more accurate than (15).

An analysis of the errors in this general procedure goes as follows: Suppose that p_n is the polynomial of least degree that interpolates f at the nodes $x_0, x_1, \ldots, x_n$. Then according to Theorem 1 on interpolating errors in Section 4.2 (p. 181),

$$f(x) - p_n(x) = \frac{1}{(n+1)!} f^{(n+1)}(\xi) w(x)$$

where ξ is dependent on x, and $w(x) = (x - x_0)(x - x_1) \cdots (x - x_n)$. Differentiating gives

$$f'(x) - p_n'(x) = \frac{1}{(n+1)!} w(x) \frac{d}{dx} f^{(n+1)}(\xi) + \frac{1}{(n+1)!} f^{(n+1)}(\xi) w'(x) \tag{18}$$

Here, we had to assume that $f^{(n+1)}(\xi)$ is differentiable as a function of x, a fact that is known if $f^{(n+2)}$ exists and is continuous.

The first observation to make about the error formula in Equation (18) is that $w(x)$ vanishes at each node, so if the evaluation is at a node x_i, the resulting equation is simpler:

$$f'(x_i) = p_n'(x_i) + \frac{1}{(n+1)!} f^{(n+1)}(\xi) w'(x_i)$$

For example, taking just two points x_0 and x_1, we obtain with $n = 1$ and $i = 0$,

$$\begin{aligned} f'(x_0) &= f[x_0, x_1] + \frac{1}{2} f''(\xi) \frac{d}{dx}[(x - x_0)(x - x_1)]\Big|_{x=x_0} \\ &= f[x_0, x_1] + \frac{1}{2} f''(\xi)(x_0 - x_1) \end{aligned}$$

This is Equation (2) in disguise when $x_0 = x$ and $x_1 = x + h$. Similar results follow with $n = 1$ and $i = 1$.

The second observation to make about Equation (18) is that it becomes simpler if x is chosen as a point where $w'(x) = 0$. For instance, if $n = 1$, then w is a quadratic function that vanishes at the two nodes x_0 and x_1. Because a parabola is symmetric about its axis, $w'[(x_0 + x_1)/2] = 0$. The resulting formula is

$f'\left(\frac{x_0+x_1}{2}\right)$ with Finite Difference Formula

$$f'\left(\frac{x_0 + x_1}{2}\right) = f[x_0, x_1] - \frac{1}{8}(x_1 - x_0)^2 \frac{d}{dx} f''(\xi)$$

As a final example, consider four interpolation points: x_0, x_1, x_2, and x_3. The interpolating polynomial from Equation (8) in Section 4.1 with $n = 3$ is

$$\begin{aligned} p_3(x) = f(x_0) + f[x_0, x_1](x - x_0) &+ f[x_0, x_1, x_2](x - x_0)(x - x_1) \\ &+ f[x_0, x_1, x_2, x_3](x - x_0)(x - x_1)(x - x_2) \end{aligned}$$

Its derivative is

$$\begin{aligned} p_3'(x) = {} & f[x_0, x_1] + f[x_0, x_1, x_2](2x - x_0 - x_1) \\ & + f[x_0, x_1, x_2, x_3]((x - x_1)(x - x_2) \\ & + (x - x_0)(x - x_2) + (x - x_0)(x - x_1)) \end{aligned}$$

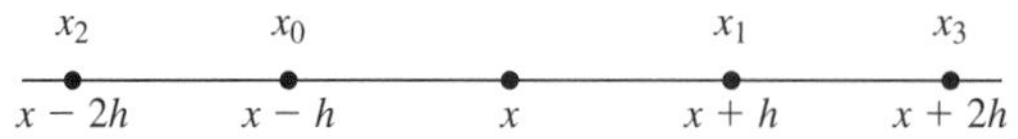

FIGURE 4.13 Central difference: four nodes

A useful special case occurs if $x_0 = x - h$, $x_1 = x + h$, $x_2 = x - 2h$, and $x_3 = x + 2h$ (see Figure 4.13). The resulting formula is

$$f'(x) \approx -\frac{2}{3h}[f(x+h) - f(x-h)] - \frac{1}{12h}[f(x+2h) - f(x-2h)]$$

This can be arranged in a form in which it is computed with a principal term plus a correction or refining term:

$f'(x)$ Finite Difference Formula at Four Nodes

$$f'(x) \approx \frac{1}{2h}[f(x+h) - f(x-h)] - \frac{1}{12h}\{f(x+2h) - 2[f(x+h) - f(x-h)] - f(x-2h)\} \tag{19}$$

The error term is $-\frac{1}{30}h^4 f^{(5)}(\xi) = \mathcal{O}(h^5)$.

Second-Derivative Formulas via Taylor Series

In the numerical solution of differential equations, it is often necessary to approximate second derivatives. We shall derive the most important formula for accomplishing this. Simply *add* the two Taylor series (3) for $f(x+h)$ and $f(x-h)$. The result is

$$f(x+h) + f(x-h) = 2f(x) + h^2 f''(x) + 2\left[\frac{1}{4!}h^4 f^{(4)}(x) + \cdots\right]$$

When this is rearranged, we get

$$f''(x) = \frac{1}{h^2}[f(x+h) - 2f(x) + f(x-h)] + E$$

where the error series is

$$E = -2\left[\frac{1}{4!}h^2 f^{(4)}(x) + \frac{1}{6!}h^4 f^{(6)}(x) + \cdots\right]$$

By carrying out the same process using Taylor's formula with a remainder, one can show that E is also given by

$$E = -\frac{1}{12}h^2 f^{(4)}(\xi) = \mathcal{O}(h^2)$$

for some ξ in the interval $(x - h, x + h)$. Hence, we have the approximation

$f''(x)$ Central Difference Formula and Error Term

$$f''(x) \approx \frac{1}{h^2}[f(x+h) - 2f(x) + f(x-h)] \tag{20}$$

with error $\mathcal{O}(h^2)$.

EXAMPLE 4 Repeat Example 2, using the central difference formula (20) to approximate the second derivative of the function $f(x) = \sin x$ at the given point $x = 0.5$.

Solution Using the truncation error term, we set

$$\left| -\frac{1}{12}h^2 f^{(4)}(\xi)\right| \leqq \frac{1}{12}4^{-2n} < \frac{1}{2}10^{-6}$$

and we obtain $n > (6 - \log 6)/\log 16 \approx 4.34$. Hence, the modified program *First* finds a good approximation of $f''(0.5) \approx -0.47942$ after about four iterations. (The least error of 3.1×10^{-9} was obtained at iteration 6.) ■

Approximate derivative formulas of high order can be obtained by using unequally spaced points such as at Chebyshev nodes. Recently, software packages have been developed for automatic differentiation of functions that are expressible by a computer program. They produce true derivatives with only rounding errors and no discretization errors.

Noise in Computation

An interesting question is how noise in the evaluation of $f(x)$ affects the computation of derivatives when using the standard formulas.

The formulas for derivatives are derived with the expectation that evaluation of the function at any point is possible, with *complete precision*. Then the approximate derivative produced by the formula differs from the actual derivative by a quantity called the **error term**, which involves the spacing of the sample points and some higher derivative of the function.

Noise and Truncation Error

If there are errors in the values of the function (**noise**), they can vitiate the whole process! Those errors could overwhelm the error inherent in the formulas. The inherent error arises from the fact that in deriving the formulas, a Taylor series was truncated after only a few terms. It is called the **truncation error**. It is present even if the evaluation of the function at the required sample points is absolutely correct.

$f'(x)$ Central Difference Formula with Error Term

For example, consider the formula

$$f'(x) = \frac{f(x+h) - f(x-h)}{2h} - \frac{h^2}{6} f'''(\xi)$$

The term with h^2 is the error term. The point ξ is a nearby point (unknown). If $f(x+h)$ and $f(x-h)$ are in error by at most d, then one can see that the formula produces a value for $f'(x)$ that is in error by d/h, which is large when h is small. Noise completely spoils the process if d is large.

For a specific numerical case, suppose that $h = 10^{-2}$ and $|f'''(s)| \leqq 6$. Then the truncation error, E, satisfies $|E| \leqq 10^{-4}$. The derivative computed from the formula with complete precision is within 10^{-4} of the actual derivative. Suppose, however, that there is noise in the evaluation of $f(x \pm h)$ of magnitude $d = h$. The correct value of $[f(x+h) - f(x-h)]/(2h)$ may differ from the noisy value by $(2d)/(2h) = 1$.

Summary 4.3

- A forward difference formula is

$$f'(x) \approx \frac{1}{h}[f(x+h) - f(x)]$$

with error term $-\frac{1}{2}hf''(\xi)$.

- A central difference formula is

$$f'(x) \approx \frac{1}{2h}[f(x+h) - f(x-h)]$$

with error $-\frac{1}{6}h^2 f'''(\xi) = \mathcal{O}(h^2)$.

- A central difference formula with a correction term is
$$f'(x) \approx \frac{1}{2h}[f(x+h) - f(x-h)] - \frac{1}{12h}[f(x+2h) - 2f(x+h) + 2f(x-h) - f(x-2h)]$$
with error term $-\frac{1}{30}h^4 f^{(5)}(\xi) = \mathcal{O}(h^4)$.
- For $f''(x)$, a central difference formula is
$$f''(x) \approx \frac{1}{h^2}[f(x+h) - 2f(x) + f(x-h)]$$
with error term $-\frac{1}{12}h^2 f^{(4)}(\xi)$.
- If $\varphi(h)$ is one of these formulas with error series $a_2h^2 + a_4h^4 + a_6h^6 + \cdots$, then we can apply **Richardson extrapolation** as follows
$$\begin{cases} D(n,0) &= \varphi(h/2^n) \\ D(n,m) &= D(n,m-1) + [D(n,m-1) - D(n-1,m-1)]/(4^m - 1) \end{cases}$$
with error terms
$$D(n,m) = L + \mathcal{O}\left(\frac{h^{2(m+1)}}{2^{2n(m+1)}}\right)$$

Exercises 4.3

[a]1. Determine the error term for the formula
$$f'(x) \approx \frac{1}{4h}[f(x+3h) - f(x-h)]$$

[a]2. Using Taylor series, establish the error term for the formula
$$f'(0) \approx \frac{1}{2h}[f(2h) - f(0)]$$

3. Derive the approximation formula
$$f'(x) \approx \frac{1}{2h}[4f(x+h) - 3f(x) - f(x+2h)]$$
Show that its error term is of the form $\frac{1}{3}h^2 f'''(\xi)$.

[a]4. Can you find an approximation formula for $f'(x)$ that has error term $\mathcal{O}(h^3)$ and involves only two evaluations of the function f? Prove or disprove.

5. Averaging the forward-difference formula $f'(x) \approx [f(x+h) - f(x)]/h$ and the backward-difference formula $f'(x) \approx [f(x) - f(x-h)]/h$, each with error term $\mathcal{O}(h)$, results in the central-difference formula $f'(x) \approx [f(x+h) - f(x-h)]/(2h)$ with error $\mathcal{O}(h^2)$. Show why.
Hint: Determine at least the first term in the error series for each formula.

[a]6. Criticize the following analysis. By Taylor's formula, we have
$$f(x+h) - f(x) = hf'(x) + \frac{h^2}{2}f''(x) + \frac{h^3}{6}f'''(\xi)$$
$$f(x-h) - f(x) = -hf'(x) + \frac{h^2}{2}f''(x) - \frac{h^3}{6}f'''(\xi)$$
So by adding, we obtain an *exact* expression for $f''(x)$:
$$f(x+h) + f(x-h) - 2f(x) = h^2 f''(x)$$

7. Criticize the following analysis. By Taylor's formula, we have
$$f(x+h) - f(x) = hf'(x) + \frac{h^2}{2}f''(x) + \frac{h^3}{6}f'''(\xi_1)$$
$$f(x-h) - f(x) = -hf'(x) + \frac{h^2}{2}f''(x) - \frac{h^3}{6}f'''(\xi_2)$$
Therefore, we have
$$\frac{1}{h^2}[f(x+h) - 2f(x) + f(x-h)] = f''(x) + \frac{h}{6}[f'''(\xi_1) - f'''(\xi_2)]$$
The error in the approximation formula for f'' is thus $\mathcal{O}(h)$.

8. Derive the two formulas

[a]**a.** $f'(x) \approx \dfrac{1}{4h}[f(x+2h) - f(x-2h)]$

b. $f''(x) \approx \dfrac{1}{4h^2}[f(x+2h) - 2f(x) + f(x-2h)]$

Establish formulas for the errors in using them.

9. Derive the following rules for these estimating derivatives and establish their error terms. Which is more accurate?

[a]**a.** $f'''(x) \approx \dfrac{1}{2h^3}[f(x+2h) - 2f(x+h) + 2f(x-h) - f(x-2h)]$

[a]**b.** $f^{(4)}(x) \approx \dfrac{1}{h^4}[f(x+2h) - 4f(x+h) + 6f(x) - 4f(x-h) + f(x-2h)]$

Hint: Consider the Taylor series for $D(h) \equiv f(x+h) - f(x-h)$ and $S(h) \equiv f(x+h) + f(x-h)$.

10. Establish the formula

$$f''(x) \approx \frac{2}{h^2}\left[\frac{f(x_0)}{(1+\alpha)} - \frac{f(x_1)}{\alpha} + \frac{f(x_2)}{\alpha(\alpha+1)}\right]$$

in the following two ways, using the unevenly spaced points $x_0 < x_1 < x_2$, where $x_1 - x_0 = h$ and $x_2 - x_1 = \alpha h$. Notice that this formula reduces to the standard central-difference formula (20) when $\alpha = 1$.

a. Approximate $f(x)$ by the Newton form of the interpolating polynomial of degree 2.

b. Calculate the undetermined coefficients A, B, and C in the expression $f''(x) \approx Af(x_0)+Bf(x_1)+Cf(x_2)$ by making it exact for the three polynomials $1, x - x_1$, and $(x - x_1)^2$ and thus exact for all polynomials of degree ≤ 2.

[a]**11.** (Continuation) Using Taylor series, show that

$$f'(x_1) = \frac{f(x_2) - f(x_0)}{x_2 - x_0} + (\alpha - 1)\frac{h}{2}f''(x_1) + \mathcal{O}(h^2)$$

Establish that the error for approximating $f'(x_1)$ by $[f(x_2) - f(x_0)]/(x_2 - x_0)$ is $\mathcal{O}(h^2)$, when x_1 is midway between x_0 and x_2, but only $\mathcal{O}(h)$ otherwise.

[a]**12.** A certain calculation requires an approximation formula for $f'(x) + f''(x)$. How well does the expression

$$\left(\frac{2+h}{2h^2}\right)f(x+h) - \left(\frac{2}{h^2}\right)f(x) + \left(\frac{2-h}{2h^2}\right)f(x-h)$$

serve? Derive this approximation and its error term.

[a]**13.** The values of a function f are given at three points x_0, x_1, and x_2. If a quadratic interpolating polynomial is used to estimate $f'(x)$ at $x = \frac{1}{2}(x_0 + x_1)$, what formula will result?

14. Consider Equation (19).

a. Fill in the details in its derivation.

b. Using Taylor series, derive its error term.

15. Show how Richardson extrapolation would work on Formula (20).

[a]**16.** If $\varphi(h) = L - c_1h - c_2h^2 - c_3h^3 - \cdots$, then what combination of $\varphi(h)$ and $\varphi(h/2)$ should give an accurate estimate of L?

17. (Continuation) State and prove a theorem analogous to the theorem on Richardson extrapolation for the situation of the preceding problem.

18. If $\varphi(h) = L - c_1h^{1/2} - c_2h^{2/2} - c_3h^{3/2} - \cdots$, then what combination of $\varphi(h)$ and $\varphi(h/2)$ should give an accurate estimate of L?

19. Show that Richardson extrapolation can be carried out for any two values of h. Thus, if $\varphi(h) = L - \mathcal{O}(h^p)$, then from $\varphi(h_1)$ and $\varphi(h_2)$, a more accurate estimate of L is given by

$$\varphi(h_2) + \frac{h_2^p}{h_1^p - h_2^p}[\varphi(h_2) - \varphi(h_1)]$$

[a]**20.** Consider a function φ such that $\lim_{h\to 0}\varphi(h) = L$ and $L - \varphi(h) \approx ce^{-1/h}$ for some constant c. By combining $\varphi(h)$, $\varphi(h/2)$, and $\varphi(h/3)$, find an accurate estimate of L.

21. Consider the approximate formula

$$f'(x) \approx \frac{3}{2h^3}\int_{-h}^{h} tf(x+t)\,dt$$

Determine its error term. Does the function f have to be differentiable for the formula to be meaningful?
Hint: This is a novel method of doing numerical differentiation. Read more about **Lanczos' generalized derivative** in Groetsch [1998].

22. Derive the error terms for $D(3, 0)$, $D(3, 1)$, $D(3, 2)$, and $D(3, 3)$.

23. Differentiation and integration are mutual inverse processes. Differentiation is an inherently sensitive problem in which small changes in the data can cause large changes in the results. Integration is a smoothing process and is inherently stable. Display two functions that have very different derivatives but equal definite integrals and vice versa.

24. Establish the error terms for these rules:

a. $f'''(x) \approx \dfrac{1}{2h^3}[3f(x+h) - 10f(x) + 12f(x-h) - 6f(x-2h) + f(x-3h)]$

b. $f'(x) + \dfrac{h}{2}f'' \approx \dfrac{1}{h}[f(x+h) - f(x)]$

c. $f^{(iv)}(x) \approx \dfrac{1}{h^4}\left[\dfrac{4}{3}f(x+3h) - 6f(x+2h) + 12f(x+h)\right]$

if $f(x) = f'(x) = 0$.

Computer Exercises 4.3

1. Test procedure *Derivative* on the following functions at the points indicated in a single computer run. Interpret the results.

a. $f(x) = \cos x$ at $x = 0$

b. $f(x) = \arctan x$ at $x = 1$

c. $f(x) = |x|$ at $x = 0$

2. (Continuation) Write and test a procedure similar to *Derivative* that computes $f''(x)$ with repeated Richardson extrapolation.

[a]**3.** Find $f'(0.25)$ as accurately as possible, using only the function corresponding to the following pseudocode and a method for numerical differentiation:

```
real function f(x)
integer i;   real a, b, c, x
a ← 1; b ← cos(x)
for i = 1 to 5
    c ← b
    b ← √(ab)
    a ← (a + c)/2
end for
f ← 2 arctan(1)/a
end function f
```

4. Carry out a numerical experiment to compare the accuracy of Formulas (5) and (19) on a function f whose derivative can be computed precisely. Take a sequence of values for h, such as 4^{-n} with $0 \leqq n \leqq 12$.

5. Using the discussion of the geometric interpretation of Richardson extrapolation, produce a graph to show that $\varphi(h)$ looks like a quadratic curve in h.

6. Use symbolic mathematical software such as MATLAB, Maple or Mathematica to establish the first term in the error series for Equation (19).

7. Use mathematical software such as found in MATLAB, Maple, or Mathematica to redo Example 1.

5 Numerical Integration

In electrical field theory, it is proved that the magnetic field induced by a current flowing in a circular loop of wire has intensity

$$H(x) = \frac{4Ir}{r^2 - x^2} \int_0^{\pi/2} \left[1 - \left(\frac{x}{r}\right)^2 \sin^2\theta\right]^{1/2} d\theta$$

where I is the current, r is the radius of the loop, and x is the distance from the center to the point where the magnetic intensity is being computed ($0 \leqq x \leqq r$). If I, r, and x are given, we have a formidable integral to evaluate. It is an **elliptic integral** and not expressible in terms of familiar functions. But H *can* be computed precisely by the methods of this chapter. For example, if $I = 15.3$, $r = 120$, and $x = 84$, we find $H = 1.35566\,1135$ accurate to nine decimals.

5.1 Trapezoid Method

One of the main mathematical procedures that is the focus of elementary calculus is *integration*, which is examined from the standpoint of numerical mathematics in this chapter.

Definite and Indefinite Integrals

It is customary to distinguish two types of integrals: the definite and the indefinite integral. The **indefinite integral** of a function is another *function* or a class of functions, whereas the **definite integral** of a function over a fixed interval is a *number*. For example, we have

Indefinite/Definite Integral

$$\text{Indefinite integral:} \quad \int x^2\,dx = \frac{1}{3}x^3 + C$$

$$\text{Definite integral:} \quad \int_0^2 x^2\,dx = \frac{8}{3}$$

Actually, a function has *not* just one, but many indefinite integrals. These differ from each other by constants. Thus, in the preceding example, any constant value may be assigned to C, and the result is still an indefinite integral. In elementary calculus, the concept of an indefinite integral is identical with the concept of an antiderivative. An **antiderivative** of a function f is any function F having the property that $F' = f$.

The definite and indefinite integrals are related by the **Fundamental Theorem of Calculus**,* which states that $\int_a^b f(x)\,dx$ can be computed by first finding an antiderivative F of f and then evaluating $F(b) - F(a)$. Thus, using traditional notation, we have

$$\int_1^3 (x^2 - 2)\,dx = \left(\frac{x^3}{3} - 2x\right)\Bigg|_1^3 = \left(\frac{27}{3} - 6\right) - \left(\frac{1}{3} - 2\right) = \frac{14}{3}$$

As another example of the Fundamental Theorem of Calculus, we can write

Examples of Fundamental Theorem of Calculus

$$\int_a^b F'(x)\,dx = F(b) - F(a)$$

$$\int_a^x F'(t)\,dt = F(x) - F(a)$$

If this second equation is differentiated with respect to x, the result is (and here we have put $f = F'$)

Derivative of an Integral

$$\frac{d}{dx}\int_a^x f(t)\,dt = f(x)$$

This last equation shows that $\int_a^x f(t)\,dt$ must be an antiderivative (indefinite integral) of f.

The foregoing technique for computing definite integrals is virtually the only one emphasized in elementary calculus. The definite integral of a function, however, has an interpretation as the area under a curve, and so the existence of a numerical value for $\int_a^b f(x)\,dx$ should *not* depend logically on our limited ability to find antiderivatives. Thus, for instance,

Antiderivative versus Area Under Curves

$$\int_0^1 e^{x^2}\,dx$$

has a precise numerical value despite the fact that there is *no* elementary function F such that $F'(x) = e^{x^2}$. By the preceding remarks, e^{x^2} does have antiderivatives, one of which is

$$F(x) = \int_0^x e^{t^2}\,dt$$

However, this form of the function F is of *no* help in determining the numerical value sought.

Trapezoid Rule: Nonuniform Spacing

The **trapezoid rule** is based on an estimation of the area beneath a curve using trapezoids. Moreover, it is an important ingredient of the Romberg algorithm of the next section. The estimation of $\int_a^b f(x)\,dx$ is approached by first dividing the interval $[a, b]$ into subintervals according to the *partition*

Nonuniform Partition

$$P = \{a = x_0 < x_1 < x_2 < \cdots < x_n = b\}$$

For each such partition of the interval (the partition points x_i need *not* be uniformly spaced), an estimation of the integral by the trapezoid rule is obtained. We denote it by $T(f; P)$.

---

*__Fundamental Theorem of Calculus__: If f is continuous on the interval $[a, b]$ and F is an antiderivative of f, then

$$\int_a^b f(x)\,dx = F(b) - F(a)$$

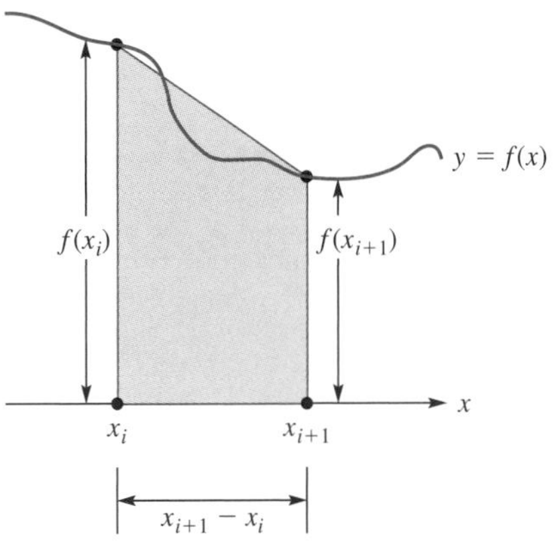

FIGURE 5.1
Typical trapezoid

Figure 5.1 shows a typical trapezoid which has the subinterval $[x_i, x_{i+1}]$ as its base, and the two vertical sides are $f(x_i)$ and $f(x_{i+1})$. The area is equal to the base times the average height, and we have the **basic trapezoid rule** for the subinterval $[x_i, x_{i+1}]$:

Basic Trapezoid Rule

$$\int_{x_i}^{x_{i+1}} f(x)\,dx \approx \frac{1}{2}(x_{i+1} - x_i)[f(x_i) + f(x_{i+1})]$$

Hence, the total area of all the trapezoids is

Composite Trapezoid Rule

$$\int_a^b f(x)\,dx \approx \frac{1}{2}\sum_{i=0}^{n-1}(x_{i+1} - x_i)[f(x_i) + f(x_{i+1})]$$

This formula is called the **composite trapezoid rule**, which is easy to understand: it is the sum of the individual applications of the basic trapezoid rule on each of the subintervals. (See Figure 5.2.)

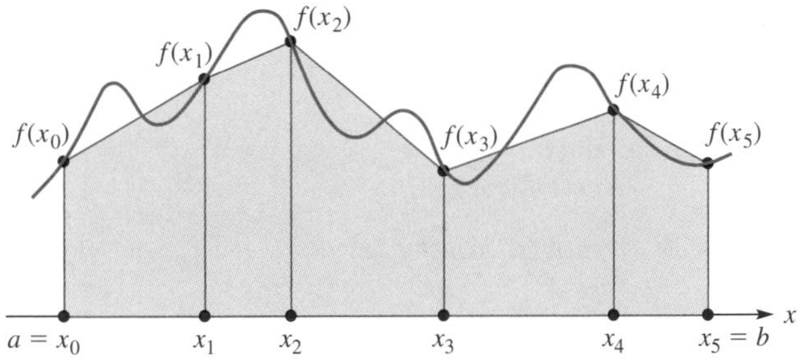

FIGURE 5.2
Composite trapezoid rule nonuniform spacing

Composite Trapezoid Rule: Uniform Spacing

In practice and in the Romberg algorithm (discussed in the next section), the trapezoid rule is used with a *uniform* partition of the interval. This means that the division points x_i are equally spaced: $x_i = a + ih$, where $h = (b - a)/n$ and $0 \leqq i \leqq n$. Think of h as the step size in the process. In this case, the formula for $T(f; P)$ can be given in simpler form because $x_{i+1} - x_i = h$. Thus, we obtain

$$T(f; P) = \frac{h}{2}\sum_{i=0}^{n-1}[f(x_i) + f(x_{i+1})]$$

To economize the amount of arithmetic, the computationally preferable formula for the **composite trapezoid rule** is

Composite Trapezoid Rule Uniform Spacing

$$\int_a^b f(x)\,dx \approx T(f; P) = \frac{h}{2}[f(x_0) + f(x_n)] + h \sum_{i=1}^{n-1} f(x_i) \tag{1}$$

Here, we have expanded the summation and gathered similar terms in the new summation.

Pseudocode

The process just described can easily be carried out on a computer. To illustrate, we select the function $f(x) = e^{-x^2}$ and the interval [0, 1]; that is, we consider

Sample Integral

$$\int_0^1 e^{-x^2}\,dx \tag{2}$$

This function is of great importance in statistics, but its indefinite integral cannot be obtained by the elementary techniques of calculus. For partitions, we take equally spaced points in [0, 1]. Thus, if there are to be n subintervals in P_n, then we define $P_n = \{x_0, x_1, \ldots, x_n\}$, where $x_i = ih$ for $0 \leqq i \leqq n$ and $h = 1/n$.

Here is a pseudocode for using the composite trapezoid rule (1) on integral (2) with $n = 60$ and $f(x) = e^{-x^2}$:

Trapezoid **Pseudocode**

```
program Trapezoid
integer i;   real h, sum, x
integer n ← 60;   real a ← 0, b ← 1
h ← (b − a)/n
sum ← ½[f(a) + f(b)]
for i = 1 to n − 1
    x ← a + ih
    sum ← sum + f(x)
end for
sum ← (sum)h
output sum
end Trapezoid

real function f(x)
real x
f ← 1/e^(x^2)
end function f
```

The computer output for the approximate value of the integral is 0.74681.

A few comments about this pseudocode may be helpful. First, a subscripted variable is *not* needed for the points x_i. Each point is labeled x. After it has been defined and used, it need not be saved. Next, observe that the pseudocode has been written so that only one line of code must be changed if another value of n is required.

At this juncture, the reader is urged to program this experiment or one like it. The experiment shows how the computer can mimic the abstract definition of the Riemann integral. Another conclusion that can be drawn from the experiment is that the direct translation of a definition into a computer algorithm may leave much to be desired in *precision*. We shall soon see that more sophisticated algorithms (such as Romberg's) improve this situation dramatically.

A good approximate value for the integral $\int_0^1 e^{-x^2}\,dx$ can be computed from knowing that this integral is related to the **error function**

Error Function erf(x)

$$\text{erf}(x) = \frac{2}{\sqrt{\pi}} \int_0^x e^{-t^2}\,dt$$

Using appropriate mathematical software, we obtain

$$\int_0^1 e^{-x^2}\,dx = \frac{1}{2}\sqrt{\pi}\,\text{erf}(1) \approx 0.74682\,41330$$

Mathematical software systems such as Maple, Mathematica, and MATLAB contain the error function. However, we are interested in learning about algorithms for approximating integrals that can only be evaluated numerically.

EXAMPLE 1 Compute

$$\int_0^1 \left(\frac{\sin x}{x}\right) dx$$

by using the composite trapezoid rule with six uniform points.

Solution The function values are arranged in a table as follows:

x_i	$f(x_i)$
0.0	1.00000
0.2	0.99335
0.4	0.97355
0.6	0.94107
0.8	0.89670
1.0	0.84147

Notice that we have assigned the value $(\sin x)/x = 1$ at $x = 0$. Then

$$\begin{aligned} T(f; P) &= 0.2 \sum_{i=1}^{4} f(x_i) + (0.1)[f(x_0) + f(x_5)] \\ &= (0.2)(3.80467) + (0.1)(1.84147) \\ &= 0.94508 \end{aligned}$$

This result is *not* accurate to all the digits shown, as might be expected because only five subintervals were used. The given integral is the **sine integral** value S(1).

Sine Integral

Using mathematical software, we obtain Si(1) $\approx$ 0.94608 30704. (Refer to Computer Exercise 5.1.9.) We shall see later how to determine a suitable value for n to obtain a desired accuracy using the trapezoid rule. ■

Error Analysis

The next task is to analyze the error incurred in using the trapezoid rule to estimate an integral. We shall establish the following result.

■ **Theorem 1**

Theorem on Precision of Composite Trapezoid Rule

If f'' exists and is continuous on the interval $[a, b]$ and if the composite trapezoid rule T with uniform spacing h is used to estimate the integral $I = \int_a^b f(x)\,dx$, then for some ζ in (a, b),

$$I - T = -\frac{1}{12}(b-a)h^2 f''(\zeta) = \mathcal{O}(h^2)$$

Proof The first step in the analysis is to prove the preceding result when $a = 0$, $b = 1$, and $h = 1$. In this case, we have to show that

$$\int_0^1 f(x)\,dx - \frac{1}{2}[f(0) + f(1)] = -\frac{1}{12}f''(\zeta) \tag{3}$$

This is easily established with the aid of the error formula for polynomial interpolation. (See the First Interpolation Error Theorem in Section 4.2.) To use this formula, let p be the polynomial of degree 1 that interpolates f at 0 and 1. Then p is given by

$$p(x) = f(0) + [f(1) - f(0)]x$$

Hence, we have

$$\begin{aligned}\int_0^1 p(x)\,dx &= f(0) + \frac{1}{2}[f(1) - f(0)]\\ &= \frac{1}{2}[f(0) + f(1)]\end{aligned}$$

By the error formula that governs polynomial interpolation [Equation (2) in Section 4.2], we have (here, of course, $n = 1$, $x_0 = 0$, and $x_1 = 1$)

$$f(x) - p(x) = \tfrac{1}{2}f''[\xi(x)]x(x-1) \tag{4}$$

where $\xi(x)$ depends on x in $(0, 1)$. From Equation (4), it follows that

$$\int_0^1 f(x)\,dx - \int_0^1 p(x)\,dx = \frac{1}{2}\int_0^1 f''[\xi(x)]x(x-1)\,dx$$

That $f''[\xi(x)]$ is continuous can be proved by solving Equation (4) for $f''[\xi(x)]$ and verifying the continuity. (See Exercise 5.2.12.) Notice that $x(x-1)$ does not change sign in the interval $[0, 1]$. Hence, by the **Mean-Value Theorem for Integrals**,* there is a point $x = s$ for which $\xi = \xi(s)$ and

$$\begin{aligned}\int_0^1 f''[\xi(x)]x(x-1)\,dx &= f''[\xi(s)]\int_0^1 x(x-1)\,dx\\ &= -\frac{1}{6}f''(\zeta)\end{aligned}$$

---

*__Mean-Value Theorem for Integrals__: Let f be continuous on $[a, b]$ and assume that g is Riemann-integrable on $[a, b]$. If $g(x) \geqq 0$ on $[a, b]$, then there exists a point ξ such that $a \leqq \xi \leqq b$ and $\int_a^b f(x)g(x)\,dx = f(\xi)\int_a^b g(x)\,dx$.

By putting all these equations together, we obtain Equation (3). Then, by making a change of variable, we obtain the **basic trapezoid rule** with its error term:

Basic Trapezoid Rule with Error Term

$$\int_a^b f(x)\,dx = \frac{b-a}{2}[f(a)+f(b)] - \frac{1}{12}(b-a)^3 f''(\xi) \tag{5}$$

The details of this are as follows: Let $g(t) = f(a + t(b-a))$ and $x = a + (b-a)t$. Thus, as t traverses the interval $[0, 1]$, x traverses the interval $[a, b]$. Also, $dx = (b-a)\,dt$, $g'(t) = f'[a + t(b-a)](b-a)$ and $g''(t) = f''[a + t(b-a)](b-a)^2$. Hence, by Equation (3), we have

$$\begin{aligned}
\int_a^b f(x)\,dx &= (b-a)\int_0^1 f[a + t(b-a)]\,dt \\
&= (b-a)\int_0^1 g(t)\,dt \\
&= (b-a)\left\{\frac{1}{2}[g(0)+g(1)] - \frac{1}{12}g''(\zeta)\right\} \\
&= \frac{b-a}{2}[f(a)+f(b)] - \frac{(b-a)^3}{12}f''(\xi)
\end{aligned}$$

This is the trapezoid rule and error term for the interval $[a, b]$ with only one subinterval, which is the entire interval. Thus, the error term is $\mathcal{O}(h^3)$, where $h = b - a$. Here, ξ is in (a, b).

Now let the interval $[a, b]$ be divided into n equal subintervals by points $x_0, x_1, \dots, x_n$ with spacing h. Applying Formula (5) to subinterval $[x_i, x_{i+1}]$, we have

$$\int_{x_i}^{x_{i+1}} f(x)\,dx = \frac{h}{2}[f(x_i) + f(x_{i+1})] - \frac{1}{12}h^3 f''(\xi_i) \tag{6}$$

where $x_i < \xi_i < x_{i+1}$. We use this result over the interval $[a, b]$, obtaining the **composite trapezoid rule**:

Composite Trapezoid Rule with Error Term

$$\begin{aligned}
\int_a^b f(x)\,dx &= \sum_{i=0}^{n-1}\int_{x_i}^{x_{i+1}} f(x)\,dx \\
&= \frac{h}{2}\sum_{i=0}^{n-1}[f(x_i)+f(x_{i+1})] - \frac{h^3}{12}\sum_{i=0}^{n-1} f''(\xi_i)
\end{aligned} \tag{7}$$

The final term in Equation (7) is the error term, and it can be simplified in the following way: Since $h = (b-a)/n$, the error term for the composite trapezoid rule is

Error Term Composite Trapezoid Rule

$$-\frac{h^3}{12}\sum_{i=0}^{n-1} f''(\xi_i) = -\frac{b-a}{12}h^2\left[\frac{1}{n}\sum_{i=0}^{n-1} f''(\xi_i)\right] = -\frac{b-a}{12}h^2 f''(\zeta) = \mathcal{O}(h^2)$$

Here, we have reasoned that the average $[1/n]\sum_{i=0}^{n-1} f''(\xi_i)$ lies between the least and greatest values of f'' on the interval (a, b). Hence, by the **Intermediate-Value Theorem**,* it is $f''(\zeta)$ for some point ζ in (a, b). This completes our proof of the error formula. ■

EXAMPLE 2 Use Taylor series to represent the error in the basic trapezoid rule (5) by an infinite series.

Solution Equation (5) is equivalent to

$$\int_a^{a+h} f(x)\,dx = \frac{h}{2}[f(a) + f(a+h)] - \frac{1}{12}h^3 f''(\xi)$$

Let

$$F(t) = \int_a^t f(x)\,dx$$

The Taylor series for F is

$$F(a+h) = F(a) + hF'(a) + \frac{h^2}{2}F''(a) + \frac{h^3}{3!}F'''(a) + \cdots$$

By the Fundamental Theorem of Calculus, $F' = f$, and we observe that $F(a) = 0$, $F'' = f'$, $F''' = f''$, and so on. Hence, we have

$$\int_a^{a+h} f(x)\,dx = hf(a) + \frac{h^2}{2}f'(a) + \frac{h^3}{3!}f''(a) + \cdots$$

The Taylor series for f is

$$f(a+h) = f(a) + hf'(a) + \frac{h^2}{2}f''(a) + \frac{h^3}{3!}f'''(a) + \cdots$$

Adding $f(a)$ to both sides of this equation and then multiplying by $h/2$, we get

$$\frac{h}{2}[f(a) + f(a+h)] = hf(a) + \frac{h^2}{2}f'(a) + \frac{h^3}{4}f''(a) + \cdots$$

Subtracting, we have

$$\int_a^{a+h} f(x)\,dx - \frac{h}{2}[f(a) + f(a+h)] = -\frac{1}{12}h^3 f''(a) + \cdots$$ ■

Applying the Error Term

How can an error term like the one just derived be used? Our first application is in predicting how small h must be to attain a specified precision in the trapezoid rule.

EXAMPLE 3 If the composite trapezoid rule is to be used to compute

$$\int_0^1 e^{-x^2}\,dx$$

with an error of at most $\frac{1}{2} \times 10^{-4}$, how many points should be used?

***Intermediate-Value Theorem**: If the function g is continuous on an interval $[a, b]$, then for each c between $g(a)$ and $g(b)$, there is a point ξ in $[a, b]$ for which $g(\xi) = c$.

Solution The error formula is

$$-\frac{b-a}{12}h^2 f''(\zeta)$$

In this example, $f(x) = e^{-x^2}$, $f'(x) = -2xe^{-x^2}$, and $f''(x) = (4x^2 - 2)e^{-x^2}$. Thus, $|f''(x)| \leqq 2$ on the interval [0, 1], and the error in absolute value is no greater than $\frac{1}{6}h^2$. To have an error of at most $\frac{1}{2} \times 10^{-4}$, we require

$$\frac{1}{6}h^2 \leqq \frac{1}{2} \times 10^{-4} \quad \text{or} \quad h \leqq 0.01732$$

Since $h = 1/n$, we require $n \geqq 58$. Hence, 59 or more points are certainly needed to produce the desired accuracy. ■

EXAMPLE 4 If the integral

$$\int_0^{\pi} e^{\cos x}\, dx$$

is to be computed with absolute error less than $\frac{1}{2} \times 10^{-3}$, and if we are going to use the composite trapezoid rule with a uniform partition, how many subintervals are needed?

Solution The integrand $f(x) = e^{\cos x}$ is a decreasing function on $[0, \pi]$. Let P denote the partition of $[0, \pi]$ by equally spaced points, $0 = x_0 < x_1 < \cdots < x_{n-1} < x_n = \pi$. Then there are n subintervals, all of width $h = \pi/n$. Here $f(x) = e^{\cos x}$, $f'(x) = (\sin x)e^{\cos x}$, and $f''(x) = \cos x(1 - \sin x)e^{\cos x}$. So on the interval $[0, \pi]$, we want the error term in absolute value to satisfy

$$\left|\frac{b-a}{12}h^2 f''(\zeta)\right| \leqq \frac{\pi}{12}h^2 |\cos x||1 - \sin x|e^{\cos x} \leqq \frac{\pi e}{12}h^2 \leqq \frac{1}{2}10^{-3}$$

Since $h = \pi/n$, we have $\sqrt{e(10\pi)^3/6} \leqq n$. With the aid of a calculator, we determine that n must be at least 119. ■

EXAMPLE 5 How many subintervals are needed to approximate

$$\int_0^1 \frac{\sin x}{x}\, dx$$

with error not to exceed $\frac{1}{2} \times 10^{-5}$ using the composite trapezoid rule? Here, the integrand, $f(x) = x^{-1} \sin x$, is defined to be 1 when x is 0.

Solution We wish to establish a bound on $f''(x)$ for x in the range [0, 1]. Taking derivatives in the usual way is not satisfactory because each term contains x with a negative power, and it is difficult to find an upper bound on $|f''(x)|$. However, using Taylor series, we have

$$f(x) = 1 - \frac{x^2}{3!} + \frac{x^4}{5!} - \frac{x^6}{7!} + \frac{x^8}{9!} - \cdots$$

$$f'(x) = -\frac{2x}{3!} + \frac{4x^3}{5!} - \frac{6x^5}{7!} + \frac{8x^7}{9!} - \cdots$$

$$f''(x) = -\frac{2}{3!} + \frac{3 \times 4x^2}{5!} - \frac{5 \times 6x^4}{7!} + \frac{7 \times 8x^6}{9!} - \cdots$$

Thus, on the interval [0, 1], $|f''(x)|$ cannot exceed $\frac{1}{2}$ because

$$\frac{2}{3!} + \frac{3 \times 4}{5!} + \frac{5 \times 6}{7!} + \frac{7 \times 8}{9!} + \cdots < \frac{1}{3} + \frac{1}{10} + \frac{1}{24}\left(\frac{1}{2} + \frac{1}{4} + \frac{1}{8} + \cdots\right) < \frac{1}{2}$$

Therefore, the error term $|(b-a)h^2 f''(\zeta)/12|$ cannot exceed $h^2/24$. For this to be less than $\frac{1}{2} \times 10^{-5}$, it suffices to take $h < \sqrt{1.2} \times 10^{-2}$ or $n > (1/\sqrt{1.2})10^2 = 91.3$. This analysis induces us to take 92 subintervals. ■

Recursive Trapezoid Formula

In the next section, we require a formula for the composite trapezoid rule when the interval $[a, b]$ is subdivided into 2^n equal parts. By the composite trapezoid rule (1), we have

$$\begin{aligned} T(f; P) &= h \sum_{i=1}^{n-1} f(x_i) + \frac{h}{2}[f(x_0) + f(x_n)] \\ &= h \sum_{i=1}^{n-1} f(a + ih) + \frac{h}{2}[f(a) + f(b)] \end{aligned}$$

If we now replace n by 2^n and use $h = (b-a)/2^n$, the preceding formula becomes

$$R(n, 0) = h \sum_{i=1}^{2^n - 1} f(a + ih) + \frac{h}{2}[f(a) + f(b)] \tag{8}$$

Here, we have introduced the notation that is used in Section 5.2 on the Romberg algorithm, namely, $R(n, 0)$. It denotes the result of applying the composite trapezoid rule with 2^n equal subintervals.

In the Romberg algorithm, it is necessary to have a means of computing $R(n, 0)$ from $R(n-1, 0)$ without involving unneeded evaluations of f. For example, the computation of $R(2, 0)$ utilizes the values of f at the five points a, $a + (b-a)/4$, $a + 2(b-a)/4$, $a + 3(b-a)/4$, and b. In computing $R(3, 0)$, we need values of f at these five points, as well as at four new points: $a + (b-a)/8$, $a + 3(b-a)/8$, $a + 5(b-a)/8$, and $a + 7(b-a)/8$ (see Figure 5.3). The computation should take advantage of the previously computed result. The manner of doing so is explained below.

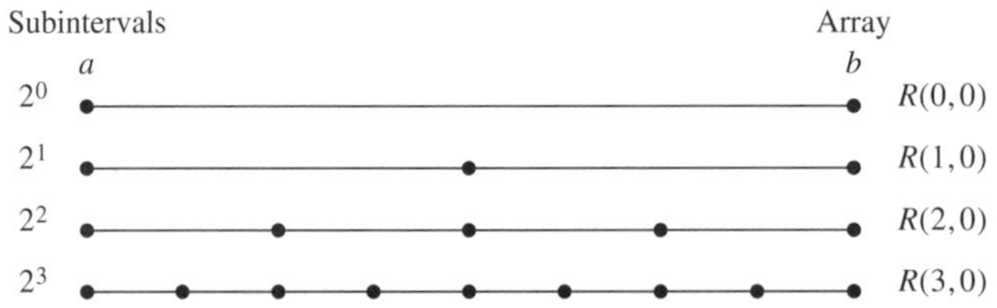

FIGURE 5.3
2^n equal subintervals

If $R(n-1, 0)$ has been computed and $R(n, 0)$ is to be computed, we use the identity

$$R(n, 0) = \tfrac{1}{2}R(n-1, 0) + \left[R(n, 0) - \tfrac{1}{2}R(n-1, 0)\right]$$

It is desirable to compute the bracketed expression with as little additional work as possible. Fixing $h = (b-a)/2^n$ for the analysis and putting

$$C = \frac{h}{2}[f(a) + f(b)]$$

we have, from Equation (8),

$$R(n,0) = h\sum_{i=1}^{2^n-1} f(a+ih) + C \tag{9}$$

$$R(n-1,0) = 2h\sum_{j=1}^{2^{n-1}-1} f(a+2jh) + 2C \tag{10}$$

Notice that the subintervals for $R(n-1, 0)$ are *twice* the size of those for $R(n, 0)$. Now from Equations (9) and (10), we have

$$\begin{aligned} R(n,0) - \frac{1}{2}R(n-1,0) &= h\sum_{i=1}^{2^n-1} f(a+ih) - h\sum_{j=1}^{2^{n-1}-1} f(a+2jh) \\ &= h\sum_{k=1}^{2^{n-1}} f[a+(2k-1)h] \end{aligned}$$

Here, we have taken account of the fact that each term in the first sum that corresponds to an *even* value of i is *canceled* by a term in the second sum. This leaves only terms that correspond to *odd* values of i.

To summarize, we have the Recursive Trapezoid Formula.

Theorem 2

Recursive Trapezoid Formula

If $R(n-1, 0)$ is available, then $R(n, 0)$ can be computed by the formula

$$R(n,0) = \frac{1}{2}R(n-1,0) + h\sum_{k=1}^{2^{n-1}} f[a+(2k-1)h] \qquad (n \geqq 1) \tag{11}$$

using $h = (b-a)/2^n$. Here, $R(0, 0) = \frac{1}{2}(b-a)[f(a) + f(b)]$.

This formula allows us to compute a sequence of approximations to a definite integral using the trapezoid rule without reevaluating the integrand at points where it has already been evaluated.

Multidimensional Integration

Here, we give a brief account of **multidimensional numerical integration**. For simplicity, we illustrate with the trapezoid rule for the interval [0, 1], using $n+1$ equally spaced points. The step size is therefore $h = 1/n$. The composite trapezoid rule is then

$$\int_0^1 f(x)\,dx \approx \frac{1}{2n}\left[f(0) + 2\sum_{i=1}^{n-1} f\left(\frac{i}{n}\right) + f(1)\right]$$

We write this in the form

$$\int_0^1 f(x)\,dx \approx \sum_{i=0}^{n} C_i f\left(\frac{i}{n}\right)$$

where

$$C_i = \begin{cases} 1/(2n), & i = 0 \\ 1/n, & 0 < i < n \\ 1/(2n), & i = n \end{cases}$$

The error is $\mathcal{O}(h^2) = \mathcal{O}(n^{-2})$ for functions having a continuous second derivative.

If one is faced with a **two-dimensional integration over the unit square**, then the trapezoid rule can be applied twice:

$$\begin{aligned}
\int_0^1 \int_0^1 f(x, y)\,dx\,dy &\approx \int_0^1 \sum_{\alpha_1=0}^{n} C_{\alpha_1} f\left(\frac{\alpha_1}{n}, y\right) dy \\
&= \sum_{\alpha_1=0}^{n} C_{\alpha_1} \int_0^1 f\left(\frac{\alpha_1}{n}, y\right) dy \\
&\approx \sum_{\alpha_1=0}^{n} C_{\alpha_1} \sum_{\alpha_2=0}^{n} C_{\alpha_2} f\left(\frac{\alpha_1}{n}, \frac{\alpha_2}{n}\right) \\
&= \sum_{\alpha_1=0}^{n} \sum_{\alpha_2=0}^{n} C_{\alpha_1} C_{\alpha_2} f\left(\frac{\alpha_1}{n}, \frac{\alpha_2}{n}\right)
\end{aligned}$$

The error here is again $\mathcal{O}(h^2)$, because each of the two applications of the trapezoid rule entails an error of $\mathcal{O}(h^2)$.

In the same way, we can integrate a function of k variables. Suitable notation is the vector $x = (x_1, x_2, \dots, x_k)^T$ for the independent variable. The region now is taken to be the k-dimensional cube $[0, 1]^k \equiv [0, 1] \times [0, 1] \times \cdots \times [0, 1]$. Then we obtain a **multidimensional numerical integration rule**

$$\int_{[0,1]^k} f(\mathbf{x})\,d\mathbf{x} \approx \sum_{\alpha_1=0}^{n} \sum_{\alpha_2=0}^{n} \cdots \sum_{\alpha_k=0}^{n} C_{\alpha_1} C_{\alpha_2} \cdots C_{\alpha_k} f\left(\frac{\alpha_1}{n}, \frac{\alpha_2}{n}, \dots, \frac{\alpha_k}{n}\right)$$

The error is still $\mathcal{O}(h^2) = \mathcal{O}(n^{-2})$, provided that f has continuous partial derivatives $\partial^2 f / \partial x_i^2$.

Besides the error involved, one must consider the effort, or work, required to attain a desired level of accuracy. The work in the one-variable case is $\mathcal{O}(n)$. In the two-variable case, it is $\mathcal{O}(n^2)$, and it is $\mathcal{O}(n^k)$ for k variables. The error, now expressed as a function of the number of nodes $N = n^k$, is

$$\mathcal{O}(h^2) = \mathcal{O}(n^{-2}) = \mathcal{O}\left((n^k)^{-2/k}\right) = \mathcal{O}(N^{-2/k})$$

Thus, the quality of the numerical approximation of the integral declines very quickly as the number of variables, k, increases. Expressed in other terms, if a constant order of accuracy is to be retained while the number of variables, k, goes up, the number of nodes must go up like n^k. These remarks indicate why the Monte Carlo method for numerical integration becomes more attractive for high-dimensional integration. (This subject is discussed in Chapter 10.)

Remarks

For historical reasons, formulas for approximating definite integrals are called **rules** such as the trapezoid rule and many other rules, some of which are found in the exercises and subsequent chapters of this book. A large collection of these **quadrature rules** can be found in Abramowitz and Stegun [1964], *Standard Mathematical Tables*, which had its origins in a U.S. Federal Government public works project during the Great Depression in the 1930s.

The word **quadrature** has several meanings both in mathematics and in astronomy. In the dictionary, the first mathematical meaning is the process of finding a square whose area is equal to the area enclosed by a given curve. The general mathematical meaning is the process of determining the area of a surface, especially one bounded by a curve. We use it primarily to mean the approximation of the area under a curve in numerical integration.

In the exercises of this chapter, we have used various well-known integrals to illustrate numerical integration. Many of these integrals have been thoroughly investigated and tabulated. Examples are elliptic integrals, the sine integral, the Fresnel integral, the logarithmic integral, the error function, and Bessel functions. When one is faced with a daunting integral, the first question to raise is whether the integral has already been studied and perhaps tabulated. The first places to search are websites on the Internet and books such as M. Abramowitz and I. Stegun [1964]. In modern scientific computing, such tables are of limited use because of the ready availability of software packages such as MATLAB, Maple, and Mathematica. Nevertheless, one needs a basic understanding of numerical methods to intelligently use mathematical software.

Summary 5.1

- To estimate $\int_a^b f(x)\,dx$, divide the interval $[a, b]$ into subintervals according to the partition $P = \{a = x_0 < x_1 < x_2 < \cdots < x_n = b\}$.
- The **basic trapezoid rule** for the subinterval $[x_i, x_{i+1}]$ is

$$\int_{x_i}^{x_{i+1}} f(x)\,dx \approx A_i = \frac{1}{2}(x_{i+1} - x_i)[f(x_i) + f(x_{i+1})]$$

  where the error is $-\frac{1}{12}(x_{i+1} - x_i)^3 f''(\xi_i)$.
- The **composite trapezoid rule with nonuniform spacing** is

$$\int_a^b f(x)\,dx \approx T(f; P) = \sum_{i=0}^{n-1} A_i = \frac{1}{2}\sum_{i=0}^{n-1}(x_{i+1} - x_i)[f(x_i) + f(x_{i+1})]$$

  where the error is $-\frac{1}{12}\sum_{i=1}^{n}(x_{i+1} - x_i)^2 f''(\xi_i)$.
- For uniform spacing of nodes in the interval $[a, b]$, we let $x_i = a + ih$, where $h = (b - a)/n$ and $0 \leqq i \leqq n$. The **composite trapezoid rule with uniform spacing** is

$$\int_a^b f(x)\,dx \approx T(f; P) = \frac{h}{2}[f(x_0) + f(x_n)] + h\sum_{i=1}^{n-1} f(x_i)$$

  where the error is $-\frac{1}{12}(b - a)h^2 f''(\zeta)$.
- For uniform spacing of nodes in the interval $[a, b]$ with 2^n subintervals, we let $h = (b - a)/2^n$, and we have

$$\begin{cases} R(0, 0) = \frac{1}{2}(b - a)[f(a) + f(b)] \\ R(n, 0) = h\displaystyle\sum_{i=1}^{2^n - 1} f(a + ih) + \frac{h}{2}[f(a) + f(b)] \end{cases}$$

- We can compute the first column of the array $R(n, 0)$ by the **recursive trapezoid formula**:

$$R(n, 0) = \frac{1}{2}R(n-1, 0) + h\sum_{k=1}^{2^{n-1}} f[a + (2k-1)h]$$

- For **two-dimensional integration over the unit square**, the trapezoid rule can be applied twice:

$$\int_0^1\int_0^1 f(x, y)\,dx\,dy \approx \sum_{\alpha_1=0}^{n}\sum_{\alpha_2=0}^{n} C_{\alpha_1}C_{\alpha_2} f\left(\frac{\alpha_1}{n}, \frac{\alpha_2}{n}\right)$$

with error $\mathcal{O}(h^2)$. For a k-dimensional cube $[0, 1]^k \equiv [0, 1] \times [0, 1] \times \cdots \times [0, 1]$, a **multidimensional numerical integration rule** is

$$\int_{[0,1]^k} f(\boldsymbol{x})\,d\boldsymbol{x} \approx \sum_{\alpha_1=0}^{n}\sum_{\alpha_2=0}^{n}\cdots\sum_{\alpha_k=0}^{n} C_{\alpha_1}C_{\alpha_2}\cdots C_{\alpha_k} f\left(\frac{\alpha_1}{n}, \frac{\alpha_2}{n}, \ldots, \frac{\alpha_k}{n}\right)$$

with error $\mathcal{O}(h^2) = \mathcal{O}(n^{-2})$.

Exercises 5.1

[a]**1.** If we estimate $\int_0^1 (x^2+2)^{-1}\,dx$ by means of the trapezoid rule using the partition $P = \left\{0, \frac{1}{2}, 1\right\}$, what is the result?

2. What is the result if we estimate $\int_1^2 x^{-1}\,dx$ by means of the trapezoid rule using the partition $P = \left\{1, \frac{3}{2}, 2\right\}$?

[a]**3.** Calculate an approximate value of $\int_0^\alpha \left[(e^x - 1)/x\right]\,dx$ for $\alpha = 10^{-4}$ correct to 14 decimal places (rounded). *Hint*: Use Taylor series.

4. If f is a (strictly) increasing function on $[a, b]$, and if $\alpha = f(a)$ and $\beta = f(b)$, then $f^{-1}(x)$ is well defined for $\alpha \leqq x \leqq \beta$. Discover the relationship between $\int_a^b f(x)\,dx$ and $\int_\alpha^\beta f^{-1}(x)\,dx$.

5. Show that if $\theta_i \geqq 0$ and $\sum_{i=0}^n \theta_i = 1$, then $\sum_{i=0}^n \theta_i a_i$ lies between the least and the greatest of the numbers a_i.

6. Establish that the **composite midpoint rule** for estimating an integral is

$$\int_a^b f(x)\,dx \approx \sum_{i=0}^{n-1}(x_{i+1} - x_i)\, f\left[\frac{1}{2}(x_{i+1} + x_i)\right]$$

7. (Continuation) Establish that the **composite midpoint rule with equal subintervals** is given by

$$\int_a^b f(x)\,dx \approx h\sum_{i=0}^{n-1} f\left(x_i + \frac{1}{2}h\right)$$

where $h = (b-a)/n$, $x_i = a + ih$, and $0 \leqq i \leqq n$.

[a]**8.** What is the numerical value of the composite trapezoid rule applied to the reciprocal function $f(x) = x^{-1}$ using the points 1, $\frac{4}{3}$, and 2?

[a]**9.** Compute an approximate value of $\int_0^1 (x^2+1)^{-1}\,dx$ by using the composite trapezoid rule with three points. Then compare with the actual value of the integral. Next, determine the error formula and numerically verify an upper bound on it.

10. (Continuation) Having computed $R(1, 0)$ in the preceding problem, compute $R(2, 0)$ by using recursive trapezoid formula (11).

11. Obtain an upper bound on the absolute error when we compute $\int_0^6 \sin x^2\,dx$ by means of the composite trapezoid rule using 101 equally spaced points.

12. If the composite trapezoid rule is used to compute $\int_{-1}^2 \sin x\,dx$ with $h = 0.01$, give a realistic bound on the error.

[a]**13.** Consider the function $f(x) = |x|$ on the interval $[-1, 1]$. Calculate the results of applying the composite trapezoid rule to approximate $\int_{-1}^1 f(x)dx$. Account for the differences in the results and compare with the true solution.

[a]**14.** How large must n be if the composite trapezoid rule in Equation (1) is being used to estimate $\int_0^\pi \sin x\,dx$ with error $\leqq 10^{-12}$? Will the estimate be too big or too small?

[a]**15.** What formula results from using the composite trapezoid rule on $f(x) = x^2$, with interval [0, 1] and $n + 1$ equally spaced points? Simplify your result by using the fact that $1^2 + 2^2 + 3^2 + \cdots + n^2 = \frac{1}{6}n(2n+1)(n+1)$. Show that as $n \to \infty$, the trapezoidal estimate converges to the correct value, $\frac{1}{3}$.

16. Prove that if a function is concave downward, then the trapezoid rule underestimates the integral.

17. Compute two approximate values for $\int_1^2 dx/x^2$ using $h = \frac{1}{2}$ with the composite trapezoid rule.

18. Consider $\int_1^2 dx/x^3$. What is the result of using the composite trapezoid rule with the partition points 1, $\frac{3}{2}$, and 2?

[a]**19.** If the composite trapezoid rule is used with $h = 0.01$ to compute $\int_2^5 \sin x\, dx$, what numerical value will the error not exceed? (Use the absolute value of error.) Give the best answer based on the error formula.

[a]**20.** Approximate $\int_0^2 2^x\, dx$ using the composite trapezoid rule with $h = \frac{1}{2}$.

[a]**21.** Consider $\int_0^1 dx/(x^2+2)$. What is the result of using the composite trapezoid rule with 0, $\frac{1}{2}$, and 1 as partition points?

[a]**22.** What is a reasonable bound on the error when we use the composite trapezoid rule on $\int_0^4 \cos x^3\, dx$ taking 201 equally spaced points (including endpoints)?

[a]**23.** We want to approximate $\int_1^2 f(x)\, dx$ given the table of values

x	1	$\frac{5}{4}$	$\frac{3}{2}$	$\frac{7}{4}$	2
$f(x)$	10	8	7	6	5

Compute an estimate by the composite trapezoid rule.

24. Consider the integral $I(h) \equiv \int_a^{a+h} f(x)\, dx$. Establish an expression for the error term for each of the following rules:

[a]**a.** $I(h) \approx hf(a+h)$

[a]**b.** $I(h) \approx hf(a+h) - \frac{1}{2}h^2 f'(a)$

c. $I(h) \approx hf(a)$ **d.** $I(h) \approx hf(a) + \frac{1}{2}h^2 f'(a)$

For each, determine the corresponding general rule and error terms for the integral $\int_a^b f(x)\, dx$, where the partition is uniform; that is, $x_i = a + ih$ and $h = (b-a)/n$ for $0 \leqq i \leqq n$.

25. Obtain the following expressions for the **midpoint rule** error terms

[a]**a.** $$\int_a^{a+h} f(x)\, dx \approx hf\left(a + \frac{1}{2}h\right)$$
(one subinterval)

[a]**b.** $$\int_a^b f(x)\, dx \approx \sum_{i=0}^{n-1} h_i f\left(x_i + \frac{1}{2}h_i\right)$$
(n unequal subintervals)

[a]**c.** $$\int_a^b f(x)\, dx \approx h \sum_{i=0}^{n-1} f\left[a + \left(i + \frac{1}{2}\right)h\right]$$
(n uniform subintervals)

where $h_i = x_{i+1} - x_i$ and $h = (b-a)/n$.

The midpoint rule was introduced in Exercises 5.1.6–7.

26. Show that there exist coefficients $w_0, w_1, \ldots, w_n$ depending on $x_0, x_1, \ldots, x_n$ and on a, b such that

$$\int_a^b p(x)\, dx = \sum_{i=0}^{n} w_i p(x_i)$$

for all polynomials p of degree $\leqq n$.
Hint: Use the Lagrange form of the interpolating polynomials from Section 4.1.

27. Show that when the composite trapezoid rule is applied to $\int_a^b e^x\, dx$ using equally spaced points, the relative error is exactly $1 - (h/2) - [h/(e^h - 1)]$.

28. Let f be a continuous function and let P_n, for $n = 0, 1, \ldots$, be partitions of $[a, b]$ such that the width of the largest subinterval in P_n converges to zero as $n \to \infty$. Show that $T(f; P_n)$ converges to $\int_a^b f(x)\, dx$ as $n \to \infty$.
Hint: Use the preceding exercise and known facts about upper and lower sums.

[a]**29.** Can you find an example of a function f and a partition P for which $T(f; P)$ is a better estimate of $\int_a^b f(x)\, dx$ than is $R(n, 0)$?

[a]**30.** A function is said to be **convex** if its graph between any two points lies beneath the chord drawn between those two points. Can you find a relationship between $T(f; P)$ and $\int_a^b f(x)\, dx$ for such a function?

[a]**31.** How large must n be if the composite trapezoid rule with equal subintervals is to estimate $\int_0^2 e^{-x^2}\, dx$ with an error not exceeding 10^{-6}? First find a crude estimate of n by using the error formula. Then determine the least possible value for n.

32. Show that

$$\int_a^b f(x)\,dx - \frac{b-a}{2}[f(a)+f(b)] = -\sum_{k=3}^{\infty} \frac{k-2}{2\times k!}(b-a)^k f^{(k-1)}(a)$$

33. The **composite (left) rectangle rule with nonuniform spacing** for numerical integration is

$$\int_a^b f(x)\,dx \approx \sum_{i=0}^{n-1}(x_{i+1}-x_i)f(x_i)$$

The partition is $P=\{a=x_0<x_1<x_2<\cdots<x_n=b\}$. Show that the rectangle rule converges to the integral as $n\to\infty$.

[a]**34.** (Continuation) The **composite rectangle rule with uniform spacing** reads as follows:

$$\int_a^b f(x)\,dx \approx h\sum_{i=0}^{n-1} f(x_i)$$

where $h=(b-a)/n$ and $x_i=a+ih$ for $0\leqq i\leqq n$. Find an expression for the error involved in this latter formula.

[a]**35.** From the previous two problems, the basic rectangle rule for a single interval is given by

$$\int_a^b f(x)\,dx = (b-a)f(a)+\frac{1}{2}(b-a)^2 f'(\zeta)$$

Establish the rectangle rule and its error term when the interval $[a,b]$ is partitioned into 2^n uniform subintervals, each of width h. Simplify the results.

36. In the composite trapezoid rule, the spacing need not be uniform. Establish the formula

$$\int_a^b f(x)\,dx \approx \frac{1}{2}\sum_{i=1}^{n-1}(h_{i-1}+h_i)f(x_i) + \frac{1}{2}[h_0 f(x_0)+h_{n-1}f(x_n)]$$

where $h_i = x_{i+1}-x_i$ and $a=x_0<x_1<x_2<\cdots<x_n=b$.

37. (Continuation) Establish the following error formula for the **composite trapezoid rule with unequal spacing** of points:

$$\int_a^b f(x)\,dx = \sum_{i=0}^{n-1}\frac{h_i}{2}[f(x_i)+f(x_{i+1})] - \frac{1}{12}(b-a)h^2 f''(\xi)$$

where $\xi\in(a,b)$, $h_i=x_{i+1}-x_i$, and $\min_i h_i \leqq h \leqq \max_i h_i$. (The composite trapezoid rule with nonuniform spacing was introduced in the preceding problem.)

38. How many points should we use in the trapezoid rule in computing an approximate value of $\int_0^1 e^{x^2}\,dx$ if the answer is to be within 10^{-6} of the correct value? *Hint*: Recall the error formula for the trapezoid rule: $-\frac{1}{12}h^2(b-a)f''(\xi)$. You may use coarse estimates, such as $2<e<3$. Explain what you are doing. In the end, we want a suitable value of n, the number of points.

Computer Exercises 5.1

1. Write

real function *Trapezoid_Uniform*(f,a,b,n)

to calculate $\int_a^b f(x)\,dx$ using the composite trapezoid rule with n equal subintervals.

2. (Continuation) Test the code written in the preceding computer problem on the following functions. In each case, compare with the correct answer.

[a]**a.** $\int_0^{\pi}\sin x\,dx$ [a]**b.** $\int_0^1 e^x\,dx$ [a]**c.** $\int_0^1 \arctan x\,dx$

3. Compute π from an integral of the form $c\int_a^b dx/(1+x^2)$.

4. Compute an approximate value for the integral $\int_0^{0.8}(\sin x/x)\,dx$.

5. Compute these integrals by using small and large values for the lower and upper limits and applying a numerical method. Then compute them by first making the indicated change of variable.

a. $\displaystyle\int_0^{\infty} e^{-x^2}\,dx = \sqrt{\frac{\pi}{2}},$ using $x=-\ln t$

(Gaussian Probability Integral)

b. $\displaystyle\int_0^{\infty} x^{-1}\sin x\,dx = \frac{\pi}{2},$ using $x=t^{-1}$

(Sine Integral)

c. $\displaystyle\int_0^{\infty}\sin x^2\,dx = \frac{1}{2}\sqrt{\frac{\pi}{2}},$ using $x=\tan t$

(Fresnel Sine Integral)

Here and elsewhere, we have used various well-known integrals as examples in testing numerical integration schemes. Some of these integrals are tabulated and can be found in tables in Abramowitz and Stegun [1964].

6. Using a numerical integration routine in a mathematical software system, find an approximate value for the sine integral $\int_0^1 (\sin x/x)\,dx$. Compare the approximate value obtained to the value of Si(1) if the system contains this function. Make a plot of the integrand.

7. Use the composite trapezoid rule with 59 subintervals to verify numerically that the approximation obtained agrees with results from Example 3.

8. Using a mathematical software system, verify the numerical approximation to the elliptic integral (p. 201).

^{a}**9.** Estimate the definite integral $\int_0^1 x^{-1}\sin x\,dx$ by computing the trapezoid rule, using 800 points in the interval. The integrand is defined to be 1 at $x = 0$. The function is decreasing, and this fact should be shown by calculus. (For a decreasing function f, $f' < 0$.)
Note: The function

$$\mathrm{Si}(x) = \int_0^x t^{-1}\sin t\,dt$$

is an important special function known as the **sine integral**. It is represented by a Taylor series that converges for all real or complex values of x. The easiest way to obtain this series is to start with the series for $\sin t$, divide by t, and integrate term by term:

$$\begin{aligned}\mathrm{Si}(x) &= \int_0^x t^{-1}\sin t\,dt = \int_0^x \sum_{n=0}^{\infty}(-1)^n \frac{t^{2n}}{(2n+1)!}\,dt\\ &= \sum_{n=0}^{\infty}(-1)^n \frac{x^{2n+1}}{(2n+1)!(2n+1)}\\ &= x - \frac{x^3}{18} + \frac{x^5}{600} - \frac{x^7}{35280} + \cdots\end{aligned}$$

This series is rapidly convergent. For example, from only the terms shown, Si(1) is computed to be 0.94608 27 with an error of at most four units in the last digit shown.

10. The **logarithmic integral** is a special mathematical function defined by the equation

$$\mathrm{li}(x) = \int_2^x \frac{dt}{\ln t}$$

For large x, the number of prime integers less than or equal to x is closely approximated by li(x). For example, there are 46 primes less than 200, and li(200) is around 50. Find li(200) with three significant figures by means of the composite trapezoid rule. Determine the number of partition points needed *prior* to executing the program.

^{a}**11.** From calculus, the length of a curve is

$$\int_a^b \sqrt{1 + [f'(x)]^2}\,dx$$

where f is a function whose graph is the curve on the interval $a \le x \le b$.

a. Find the length of the ellipse $y^2 + 4x^2 = 1$. Use the symmetry of the ellipse.

b. Verify the numerical approximation given for the arc length in the introductory example at the beginning of Chapter 3.

^{a}**12.** Using a mathematical software system that contains the error function erf, find a numerical approximation of $\int_0^1 e^{-x^2}\,dx$ to the full precision available. Also, plot the error function.

13. (Continuation) Evaluate the integral $\int_0^1 e^{-x^2}\,dx$ using a numerical integration routine in a mathematical software system such as MATLAB, Maple, or Mathematica. Compare the results to those obtained previously.

5.2 Romberg Algorithm

Description

The *Romberg algorithm* produces a triangular array of numbers, all of which are numerical estimates of the definite integral $\int_a^b f(x)\,dx$. The array is denoted here by the notation

Romberg Array

$$\begin{array}{lll} R(0,0) & & \\ R(1,0) & R(1,1) & \\ R(2,0) & R(2,1) & R(2,2) \end{array}$$

$$\begin{array}{cccccc} R(3,0) & R(3,1) & R(3,2) & R(3,3) & & \\ \vdots & \vdots & \vdots & \vdots & \ddots & \\ R(n,0) & R(n,1) & R(n,2) & R(n,3) & \cdots & R(n,n) \end{array}$$

The first column of this table contains estimates of the integral obtained by the recursive trapezoid formula with decreasing values of the step size. Explicitly, $R(n, 0)$ is the result of applying the trapezoid rule with 2^n equal subintervals. The first of them, $R(0, 0)$, is obtained with just one trapezoid:

$R(0, 0)$

$$R(0,0) = \frac{1}{2}(b-a)[f(a) + f(b)]$$

Similarly, $R(1, 0)$ is obtained with two trapezoids:

$R(1, 0)$

$$\begin{aligned} R(1,0) &= \frac{1}{4}(b-a)\left[f(a) + f\left(\frac{a+b}{2}\right)\right] + \frac{1}{4}(b-a)\left[f\left(\frac{a+b}{2}\right) + f(b)\right] \\ &= \frac{1}{4}(b-a)[f(a) + f(b)] + \frac{1}{2}(b-a)\, f\left(\frac{a+b}{2}\right) \\ &= \frac{1}{2}R(0,0) + \frac{1}{2}(b-a)\, f\left(\frac{a+b}{2}\right) \end{aligned}$$

These formulas agree with those developed in the preceding section. In particular, note that $R(n, 0)$ is obtained easily from $R(n-1, 0)$ if Equation (11) in Section 5.1 is used; that is,

$R(n, 0)$

$$R(n,0) = \frac{1}{2}R(n-1,0) + h\sum_{k=1}^{2^{n-1}} f[a + (2k-1)h] \tag{1}$$

where $h = (b-a)/2^n$ and $n \geqq 1$.

The second and successive columns in the Romberg array are generated by the extrapolation formula

$R(n, m)$

$$R(n,m) = R(n,m-1) + \frac{1}{4^m - 1}[R(n,m-1) - R(n-1,m-1)] \tag{2}$$

with $n \geqq 1$ and $m \geqq 1$. This formula is derived later using the theory of Richardson extrapolation from Section 4.3.

EXAMPLE 1 If $R(4, 2) = 8$ and $R(3, 2) = 1$, what is $R(4, 3)$?

Solution From Equation (2), we have

$$\begin{aligned} R(4,3) &= R(4,2) + \frac{1}{63}[R(4,2) - R(3,2)] \\ &= 8 + \frac{1}{63}(8-1) = \frac{73}{9} \end{aligned}$$

■

Pseudocode

The objective now is to develop computational formulas for the **Romberg algorithm**. By replacing n with i and m with j in Equation (2), we obtain, for $i \geqq 1$ and $j \geqq 1$,

$$R(i,j) = R(i,j-1) + \frac{1}{4^j - 1}[R(i,j-1) - R(i-1,j-1)]$$

and

$$R(i,0) = \frac{1}{2}R(i-1,0) + h\sum_{k=1}^{2^{i-1}} f[a + (2k-1)h]$$

The range of the summation is $1 \leqq k \leqq 2^{i-1}$, so that $1 \leqq 2k - 1 \leqq 2^i - 1$.

One way to generate the Romberg array is to compute a reasonable number of terms in the first column, $R(0,0)$ up to $R(n,0)$, and then use the extrapolation Formula (2) to construct columns $1, 2, \dots, n$ in order. Another way is to compute the array row by row. Observe, for example, that $R(1,1)$ can be computed by the extrapolation formula as soon as $R(1,0)$ and $R(0,0)$ are available. The procedure *Romberg* computes, row by row, n rows and columns of the Romberg array for a function f and a specified interval $[a, b]$:

***Romberg* Pseudocode**

```
procedure Romberg(f, a, b, n, (r_ij))
integer i, j, k, n;   real a, b, h, sum;   real array (r_ij)_{0:n×0:n}
external function f
h ← b − a
r_00 ← (h/2)[f(a) + f(b)]
for i = 1 to n
    h ← h/2
    sum ← 0
    for k = 1 to 2^i − 1 step 2
        sum ← sum + f(a + kh)
    end for
    r_i0 ← (1/2) r_{i−1,0} + (sum)h
    for j = 1 to i
        r_ij ← r_{i,j−1} + (r_{i,j−1} − r_{i−1,j−1})/(4^j − 1)
    end for
end for
end procedure Romberg
```

This procedure is used with a main program and a function procedure (for computing values of the function f). In the main program and perhaps in the procedure *Romberg*, some language-specific interface must be included to indicate that the first argument is an external function. Remember that in the Romberg algorithm as described, the number of subintervals is 2^n. Thus, a modest value of n should be chosen—for example, $n = 5$. A more sophisticated program would include an automatic stopping test to terminate the calculation as soon as the error reaches a preassigned tolerance.

As an example, one can approximate π by using the procedure *Romberg* with $n = 5$ to obtain a numerical approximation for the integral

Sample Integral

$$\int_0^1 \frac{4}{1+x^2}\,dx$$

We obtain the following results:

3.00000 00000 000				
3.09999 99046 326	3.13333 32061 768			
3.13117 64717 102	3.14156 86607 361	3.14211 77387 238		
3.13898 84948 730	3.14159 25025 940	3.14159 41715 240	3.14158 58268 738	
3.14094 16198 730	3.14159 27410 126	3.14159 27410 126	3.14159 27410 126	3.14159 27410 126

Euler-Maclaurin Formula

Here we explain the source of Equation (2), which is used for constructing the successive columns of the Romberg array. We begin with a formula that expresses the error in the trapezoid rule over 2^{n-1} subintervals:

$$\int_a^b f(x)\,dx = R(n-1,0) + a_2h^2 + a_4h^4 + a_6h^6 + \cdots \tag{3}$$

Here, $h = (b-a)/2^{n-1}$ and the coefficients a_i depend on f, but not on h. This equation is one form of the **Euler-Maclaurin formula** and is given here without proof. (See Young and Gregory [1972].) In this equation, $R(n-1,0)$ denotes a typical element of the first column in the Romberg array; hence, it is one of the trapezoidal estimates of the integral. Notice particularly that the error is expressed in powers of h^2, and the error series is $\mathcal{O}(h^2)$. For our purposes, it is not necessary to know the coefficients, but, in fact, they have definite expressions in terms of f and its derivatives. For the theory to work smoothly, it is assumed that f possesses derivatives of all orders on the interval $[a,b]$.

The reader should now recall the theory of Richardson extrapolation as outlined in Section 4.3. That theory is applicable because of Equation (3). In Equation (8) of Section 4.3, $L = \phi(h) + \sum_{k=1}^{\infty} a_{2k}h^{2k}$. Here, L is the value of the integral and $\phi(h)$ is $R(n-1,0)$, the trapezoidal estimate of L using subintervals of size h. Equation (10) of Section 4.3 gives the approximate extrapolation formula, which in this situation is Equation (2).

We briefly review this procedure. Replacing n with $n+1$ and h with $h/2$ in Equation (3), we have

$$\int_a^b f(x)\,dx = R(n,0) + \frac{1}{4}a_2h^2 + \frac{1}{16}a_4h^4 + \frac{1}{64}a_6h^6 + \cdots \tag{4}$$

Subtract Equation (3) from 4 times Equation (4) to obtain

$$\int_a^b f(x)\,dx = R(n,1) - \frac{1}{4}a_4h^4 - \frac{5}{16}a_6h^6 - \cdots \tag{5}$$

where

$R(n,1)$

$$R(n,1) = R(n,0) + \frac{1}{3}[R(n,0) - R(n-1,0)] \qquad (n \geqq 1)$$

Note that this is the first case ($m = 1$) of the extrapolation Formula (2). Now $R(n,1)$ should be considerably more accurate than $R(n,0)$ or $R(n-1,0)$ because its error formula begins with an h^4 term. Hence, the error series is now $\mathcal{O}(h^4)$. This process can be repeated using Equation (5) slightly modified as the starting point—that is, with n replaced by $n-1$ and with h replaced by $2h$. Then combine the two equations appropriately to eliminate the h^4 term. The result is a new combination of elements from column 2 in the Romberg array:

$$\int_a^b f(x)\,dx = R(n,2) + \frac{1}{4^3}a_6h^6 + \frac{21}{4^5}a_8h^8 + \cdots \tag{6}$$

where

$R(n,2)$

$$R(n,2) = R(n,1) + \frac{1}{15}[R(n,1) - R(n-1,1)] \qquad (n \geqq 2)$$

which agrees with Equation (2) when $m = 2$. Thus, $R(n,2)$ is an even more accurate approximation to the integral because its error series is $\mathcal{O}(h^6)$.

The basic assumption on which of all this analysis depends is that Equation (3) is valid for the function f being integrated. Of course, in practice, we use a modest number of rows in the Romberg algorithm, and only this number of terms in Equation (3) is needed. Here is the theorem that governs the situation:

Theorem 1 **Euler-Maclaurin Formula and Error Term**

If $f^{(2m)}$ exists and is continuous on the interval $[a, b]$, then

$$\int_a^b f(x)\,dx = \frac{h}{2}\sum_{i=0}^{n-1}[f(x_i) + f(x_{i+1})] + E$$

where $h = (b - a)/n$, $x_i = a + ih$ for $0 \leqq i \leqq n$, and

$$E = \sum_{k=1}^{m-1} A_{2k}h^{2k}[f^{(2k-1)}(a) - f^{(2k-1)}(b)] - A_{2m}(b-a)h^{2m}f^{(2m)}(\xi)$$

for some ξ in the interval (a, b).

In this theorem, the A_k's are constants (related to the **Bernoulli numbers**) and ξ is some point in the interval (a, b). Refer to Young and Gregory [1972, vol. 1, p. 374]. It turns out that the A_k's can be defined by the equation

$$\frac{x}{e^x - 1} = \sum_{k=0}^{\infty} A_k x^k \tag{7}$$

Observe that in the Euler-Maclaurin formula, the right-hand side contains the trapezoid rule and an error term, E. Furthermore, E can be expressed as a finite sum in ascending powers of h^2. This theorem gives the formal justification (and the details) of Equation (3).

If the integrand f does not possess a large number of derivatives, but is at least Riemann-integrable, then the Romberg algorithm still converges in the following sense:
The limit of each *column* in the array equals the integral:

$$\lim_{n\to\infty} R(n, m) = \int_a^b f(x)\,dx \qquad (m \geqq 0)$$

The convergence of the first column is easily justified by referring to the upper and lower sums. (See Exercise 5.2.28.) After the convergence of the first column has been established, the convergence of the remaining columns can be proved by using Equation (2). (See Exercises 5.3.24–5.3.25.)

In practice, we may not know whether the function f whose integral we seek satisfies the smoothness criterion upon which the theory depends. Then it would not be known whether Equation (3) is valid for f. One way of testing this in the course of the Romberg algorithm is to compute the ratios

Ratios

$$\frac{R(n, m) - R(n-1, m)}{R(n+1, m) - R(n, m)} \approx 4^{m+1} \qquad \text{as } h \to 0$$

and establish that they are close to 4^{m+1}. Let us verify, at least for the case $m = 0$, that this ratio is near 4 for a function that obeys Equation (3).

If we subtract Equation (3) from (4), the result is

$$R(n, 0) - R(n-1, 0) = \frac{3}{4}a_2h^2 + \frac{15}{16}a_4h^4 + \frac{63}{64}a_6h^6 + \cdots \tag{8}$$

If we write down the same equation for the *next* value of n, then the h of that equation is half the value of h used in Equation (8). Hence, we obtain

$$R(n+1,0) - R(n,0) = \frac{3}{4^2}a_2h^2 + \frac{15}{16^2}a_4h^4 + \frac{63}{64^2}a_6h^6 + \cdots \tag{9}$$

Equations (8) and (9) are now used to express the ratio mentioned previously:

Ratios ($m = 0$)

$$\frac{R(n,0) - R(n-1,0)}{R(n+1,0) - R(n,0)} = 4\left[\frac{1 + \frac{5}{4}\left(\frac{a_4}{a_2}\right)h^2 + \frac{21}{16}\left(\frac{a_6}{a_2}\right)h^4 + \cdots}{1 + \frac{5}{4^2}\left(\frac{a_4}{a_2}\right)h^2 + \frac{21}{16^2}\left(\frac{a_6}{a_2}\right)h^4 + \cdots}\right]$$

$$= 4\left[1 + \frac{15}{4^2}\left(\frac{a_4}{a_2}\right)h^2 + \cdots\right] \approx 4 \qquad \text{as } h \to 0$$

For small values of h, this expression is close to 4.

General Extrapolation

In closing, we return to the extrapolation process that is the heart of the Romberg algorithm. The process is Richardson extrapolation, which was discussed in Section 4.3. It is an example of a general dictum in numerical mathematics that if anything is known about the errors in a process, then that knowledge can be exploited to improve the process.

The only type of extrapolation illustrated so far (in this section and Section 4.3) has been *Richardson* extrapolation. It applies to a numerical process in which the error series is of the form

$$E = a_2h^2 + a_4h^4 + a_6h^6 + \cdots$$

In this case, the errors behave like $\mathcal{O}(h^2)$ as $h \to 0$, but the basic idea of Richardson extrapolation has much wider applicability. We could apply extrapolation if we knew, for example, that

$$E = ah^\alpha + bh^\beta + ch^\gamma + \cdots$$

provided that $0 < \alpha < \beta < \gamma < \cdots$. It is sufficient to see how to annihilate the first term of the error expansion because the succeeding steps would be similar.

Suppose therefore that

$$L = \varphi(h) + ah^\alpha + bh^\beta + ch^\gamma + \cdots \tag{10}$$

Here, L is a mathematical entity that is approximated by a formula $\varphi(h)$ depending on h with the error series $ah^\alpha + bh^\beta + \cdots$. It follows that

$$L = \varphi\left(\frac{h}{2}\right) + a\left(\frac{h}{2}\right)^\alpha + b\left(\frac{h}{2}\right)^\beta + c\left(\frac{h}{2}\right)^\gamma + \cdots$$

Hence, if we multiply this by 2^α, we get

$$2^\alpha L = 2^\alpha\varphi\left(\frac{h}{2}\right) + ah^\alpha + 2^\alpha b\left(\frac{h}{2}\right)^\beta + 2^\alpha c\left(\frac{h}{2}\right)^\gamma + \cdots \tag{11}$$

By subtracting Equation (10) from (11), we rid ourselves of the h^α term:

$$(2^\alpha - 1)L = 2^\alpha\varphi\left(\frac{h}{2}\right) - \varphi(h) + (2^{\alpha-\beta} - 1)bh^\beta + (2^{\alpha-\gamma} - 1)ch^\gamma + \cdots$$

We rewrite this as

$$L = \frac{2^\alpha}{2^\alpha - 1}\varphi\left(\frac{h}{2}\right) - \frac{1}{2^\alpha - 1}\varphi(h) + \widetilde{b}h^\beta + \widetilde{c}h^\gamma + \cdots \tag{12}$$

Thus, the special linear combination

$$\frac{2^\alpha}{2^\alpha - 1}\varphi\left(\frac{h}{2}\right) - \frac{1}{2^\alpha - 1}\varphi(h) = \varphi\left(\frac{h}{2}\right) + \frac{1}{2^\alpha - 1}\left[\varphi\left(\frac{h}{2}\right) - \varphi(h)\right] \tag{13}$$

should be a more accurate approximation to L than either $\varphi(h)$ or $\varphi(h/2)$ because their error series, in Equations (10) and (12), improve from $\mathcal{O}(h^\alpha)$ to $\mathcal{O}(h^\beta)$ as $h \to 0$ and $\beta > \alpha > 0$. Notice that when $\alpha = 2$, the combination in Equation (13) is the one we have already used for the second column in the Romberg array, namely, $R(n, 1)$.

Extrapolation of the same type can be used in still more general situations, as is illustrated next (and in the exercises).

EXAMPLE 2 If φ is a function with the property

$$\varphi(x) = L + a_1x^{-1} + a_2x^{-2} + a_3x^{-3} + \cdots$$

how can L be estimated using Richardson extrapolation?

Solution Obviously, $L = \lim_{x\to\infty} \varphi(x)$. Thus, L can be estimated by evaluating $\varphi(x)$ for a succession of ever larger values of x. To use extrapolation, we write

$$\begin{aligned}
\varphi(x) &= L + a_1x^{-1} + a_2x^{-2} + a_3x^{-3} + \cdots \\
\varphi(2x) &= L + 2^{-1}a_1x^{-1} + 2^{-2}a_2x^{-2} + 2^{-3}a_3x^{-3} + \cdots \\
2\varphi(2x) &= 2L + a_1x^{-1} + 2^{-1}a_2x^{-2} + 2^{-2}a_3x^{-3} + \cdots \\
\psi(x) = 2\varphi(2x) - \varphi(x) &= L - 2^{-1}a_2x^{-2} - 3\cdot 2^{-2}a_3x^{-3} - \cdots
\end{aligned}$$

Having computed $\varphi(x)$ and $\varphi(2x)$, we can compute a new function $\psi(x)$, which should be a better approximation to L because its error series begins with x^{-2} and is $\mathcal{O}(x^{-2})$ as $x \to \infty$. This process can be repeated, as in the Romberg algorithm. ■

Here is a concrete illustration of the preceding example. We want to estimate $\lim_{x\to\infty} \varphi(x)$ from the following table of numerical values:

x	1	2	4	8	16	32	64	128
$\varphi(x)$	21.1100	16.4425	14.3394	13.3455	12.8629	12.6253	12.5073	12.4486

A tentative hypothesis is that φ has the form in the preceding example. When we compute the values of the function $\psi(x) = 2\varphi(2x) - \varphi(x)$, we get a new table of values:

x	1	2	4	8	16	32	64
$\psi(x)$	11.7750	12.2363	12.3516	12.3803	12.3877	12.3893	12.3899

It seems reasonable to believe that the value of $\lim_{x\to\infty} \varphi(x)$ is approximately 12.3899. If we do another extrapolation, we should compute $\theta(x) = [4\psi(2x) - \psi(x)]/3$; values for this table are

x	1	2	4	8	16	32
$\theta(x)$	12.3901	12.3900	12.3899	12.3902	12.3898	12.3901

For the precision of the given data, we conclude that $\lim_{x\to\infty} \varphi(x) = 12.3900$ to within roundoff error.

Summary 5.2

- By using the recursive trapezoid rule, we find that the first column of the **Romberg algorithm** is

$$R(n,0) = \frac{1}{2}R(n-1,0) + h\sum_{k=1}^{2^{n-1}} f[a + (2k-1)h]$$

where $h = (b-a)/2^n$ and $n \geqq 1$. The second and successive columns in the Romberg array are generated by the Richardson extrapolation formula and are

$$R(n,m) = R(n,m-1) + \frac{1}{4^m - 1}[R(n,m-1) - R(n-1,m-1)]$$

with $n \geqq 1$ and $m \geqq 1$. The error is $\mathcal{O}(h^2)$ for the first column, $\mathcal{O}(h^4)$ for the second column, $\mathcal{O}(h^6)$ for the third column, and so on. Check the ratios

$$\frac{R(n,m) - R(n-1,m)}{R(n+1,m) - R(n,m)} \approx 4^{m+1} \qquad \text{as } h \to 0$$

to test whether the algorithm is working.

- If the expression L is approximated by $\varphi(h)$ and if these entities are related by the error series

$$L = \varphi(h) + ah^\alpha + bh^\beta + ch^\gamma + \cdots$$

then a more accurate approximation is

$$L \approx \varphi\left(\frac{h}{2}\right) + \frac{1}{2^\alpha - 1}\left[\varphi\left(\frac{h}{2}\right) - \varphi(h)\right]$$

with error $\mathcal{O}(h^\beta)$.

Exercises 5.2

[a]**1.** What is $R(5,3)$ if $R(5,2) = 12$ and $R(4,2) = -51$, in the Romberg algorithm?

2. If $R(3,2) = -54$ and $R(4,2) = 72$, what is $R(4,3)$?

[a]**3.** Compute $R(5,2)$ from $R(3,0) = R(4,0) = 8$ and $R(5,0) = -4$.

4. Let $f(x) = 2^x$. Approximate $\int_0^4 f(x)\,dx$ by the trapezoid rule using partition points 0, 2, and 4. Repeat by using partition points 0, 1, 2, 3, and 4. Now apply Romberg extrapolation to obtain a better approximation.

[a]**5.** By the Romberg algorithm, approximate $\int_0^2 4\,dx/(1+x^2)$ by evaluating $R(1,1)$.

6. Using the Romberg scheme, establish a numerical value for the approximation

$$\int_0^1 e^{-(10x)^2}\,dx \approx R(1,1)$$

Compute the approximation to only three decimal places of accuracy.

[a]**7.** We are going to use the Romberg method to estimate $\int_0^1 \sqrt{x}\cos x\,dx$. Will the method work? Will it work well? Explain.

[a]**8.** By combining $R(0,0)$ and $R(1,0)$ for the partition $P = \{-h < 0 < h\}$, determine $R(1,1)$.

9. In calculus, a technique of integration by substitution is developed. For example, if the substitution $x = z^2$ is made in the integral $\int_0^1 (e^x/\sqrt{x})\,dx$, the result is $2\int_0^1 e^{z^2}\,dz$. Verify this and discuss the numerical aspects of this example. Which form is likely to produce a more accurate answer by the Romberg method?

[a]**10.** How many evaluations of the function (integrand) are needed if the Romberg array with n rows and n columns is to be constructed?

11. Using Equation (2), fill in the circles in the following diagram with coefficients used in the Romberg algorithm:

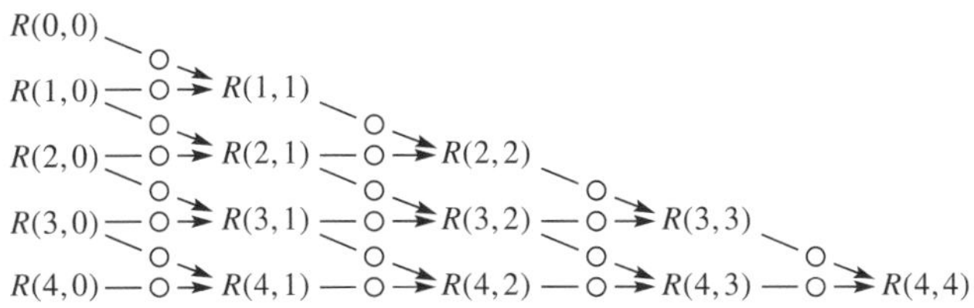

12. Derive the quadrature rule for $R(1, 1)$ in terms of the function f evaluated at partition points a, $a + h$, and $a + 2h$, where $h = (b - a)/2$. Do the same for $R(n, 1)$ with $h = (b - a)/2^n$.

[a]**13.** (Continuation) Derive the quadrature rule $R(2, 2)$ in terms of the function f evaluated at a, $a + h$, $a + 2h$, $a + 3h$, and b, where $h = (b - a)/4$.

[a]**14.** We want to compute $X = \lim_{n\to\infty} S_n$, and we have already computed the two numbers $u = S_{10}$ and $v = S_{30}$. It is known that $X = S_n + Cn^{-3}$. What is X in terms of u and v?

[a]**15.** Suppose that we want to estimate $Z = \lim_{h\to 0} f(h)$ and that we calculate $f(1)$, $f(2^{-1})$, $f(2^{-2})$, $f(2^{-3}), \dots, f(2^{-10})$. Then suppose also that it is known that $Z = f(h) + ah^2 + bh^4 + ch^6$. Show how to obtain an improved estimate of Z from the 11 numbers already computed. Show how Z can be determined exactly from any 4 of the 11 computed numbers.

16. Show how Richardson extrapolation works on a sequence $x_1, x_2, x_3, \dots$ that converges to L as $n \to \infty$ in such a way that $L - x_n = a_2 n^{-2} + a_3 n^{-3} + a_4 n^{-4} + \cdots$.

[a]**17.** Let x_n be a sequence that converges to L as $n \to \infty$. If $L - x_n$ is known to be of the form $a_3 n^{-3} + a_4 n^{-4} + \cdots$ (in which the coefficients are unknown), how can the convergence of the sequence be accelerated by taking combinations of x_n and x_{n+1}?

[a]**18.** If the Romberg algorithm is operating on a function that possesses continuous derivatives of all orders on the interval of integration, then what is a bound on the quantity $|\int_a^b f(x)\,dx - R(n, m)|$ in terms of h?

19. Show that the precise form of Equation (5) is

$$\int_a^b f(x)\,dx = R(n,1) - \sum_{j=1}^{\infty}\left(\frac{4^j - 1}{3\times 4^j}\right) a_{2j+2}h^{2j+2}$$

20. Derive Equation (6), and show that its precise form is

$$\int_a^b f(x)\,dx = R(n,2) + \sum_{j=2}^{\infty}\left(\frac{4^j - 1}{3\times 4^j}\right)\left(\frac{4^{j-1} - 1}{15\times 4^{j-1}}\right) a_{2j+2}h^{2j+2}$$

21. Use the fact that the coefficients in Equation (3) have the form

$$a_k = c_k[f^{(k-1)}(b) - f^{(k-1)}(a)]$$

to prove that $\int_a^b f(x)\,dx = R(n, m)$ if f is a polynomial of degree $\leq 2m - 2$.

[a]**22.** In the Romberg algorithm, $R(n, 0)$ denotes an estimate of $\int_a^b f(x)\,dx$ with subintervals of size $h = (b - a)/2^n$. If it were known that

$$\int_a^b f(x)\,dx = R(n,0) + a_3h^3 + a_6h^6 + \cdots$$

how would we have to modify the Romberg algorithm?

[a]**23.** Show that if f'' is continuous, then the first column in the Romberg array converges to the integral in such a way that the error at the nth step is bounded in magnitude by a constant times 4^{-n}.

[a]**24.** Assuming that the first column of the Romberg array converges to $\int_a^b f(x)\,dx$, show that the second column does also.

25. (Continuation) In the preceding problem, we established the elementary property that if $\lim_{n\to\infty} R(n, 0) = \int_a^b f(x)\,dx$, then $\lim_{n\to\infty} R(n, 1) = \int_a^b f(x)\,dx$. Show that

$$\begin{aligned}\lim_{n\to\infty} R(n,2) &= \lim_{n\to\infty} R(n,3) = \cdots\\ &= \lim_{n\to\infty} R(n,n) = \int_a^b f(x)\,dx\end{aligned}$$

26. a. Using Formula (7), prove that Euler-Maclaurin coefficients can be generated recursively:

$$A_0 = 1, \qquad A_k = -\sum_{j=1}^{k} \frac{A_{k-j}}{(j+1)!}$$

b. Determine A_k for $1 \leq k \leq 6$.

[a]**27.** Evaluate E in the theorem on the Euler-Maclaurin formula for this special case: $a = 0$, $b = 2\pi$, $f(x) = 1 + \cos 4x$, $n = 4$, and m arbitrary.

Computer Exercises 5.2

[a]**1.** Compute eight rows and columns in the Romberg array for $\int_{1.3}^{2.19} x^{-1} \sin x \, dx$.

2. Design and carry out an experiment using the Romberg algorithm. *Suggestions*: For a function that possesses many continuous derivatives on the interval, the method should work well. Try such a function first. If you choose one whose integral you can compute by other means, you will acquire a better understanding of the accuracy in the Romberg algorithm. For example, try definite integrals for each of these:

$$\int (1+x)^{-1}\,dx = \ln(1+x),$$

$$\int e^x\,dx = e^x, \qquad \int (1+x^2)^{-1}\,dx = \arctan x$$

3. Test the Romberg algorithm on a *bad* function, such as $\sqrt{x}$ on $[0, 1]$. Why is it bad?

4. The transcendental number π is the area of a circle whose radius is 1. Show that

$$8\int_0^{1/\sqrt{2}} \left(\sqrt{1-x^2} - x\right) dx = \pi$$

with the help of a diagram, and use this integral to approximate π by the Romberg method.

[a]**5.** Apply the Romberg method to estimate $\int_0^{\pi} (2 + \sin 2x)^{-1}\,dx$. Observe the high precision obtained in the first column of the array, that is, by the simple trapezoidal estimates.

[a]**6.** Compute $\int_0^{\pi} x \cos 3x \, dx$ by the Romberg algorithm using $n = 6$. What is the correct answer?

[a]**7.** An integral of the form $\int_0^{\infty} f(x)\,dx$ can be transformed into an integral on a finite interval by making a change of variable. Verify, for instance, that the substitution $x = -\ln y$ changes the integral $\int_0^{\infty} f(x)\,dx$ into $\int_0^1 y^{-1} f(-\ln y)\,dy$. Use this idea to compute $\int_0^{\infty} [e^{-x}/(1+x^2)]\,dx$ by means of the Romberg algorithm, using 128 evaluations of the transformed function.

8. By the Romberg algorithm, calculate

$$\int_0^{\infty} e^{-x}\sqrt{1-\sin x}\,dx$$

9. Calculate

$$\int_0^1 \frac{\sin x}{\sqrt{x}}\,dx$$

by the Romberg algorithm.
Hint: Consider making a change of variable.

10. Compute log 2 by using the Romberg algorithm on a suitable integral.

[a]**11.** The **Bessel function** of order 0 is defined by the equation

$$J_0(x) = \frac{1}{\pi}\int_0^{\pi} \cos(x \sin\theta)\,d\theta$$

Calculate $J_0(1)$ by applying the Romberg algorithm to the integral.

12. Recode the Romberg procedure so that *all* the trapezoid rule results are computed *first* and stored in the first column. Then in a separate procedure,

procedures *Extrapolate*$(n, (r_{ij}))$

carry out Richardson extrapolation, and store the results in the lower triangular part of the (r_{ij}) array. What are the advantages and disadvantages of this procedure over the routine given in the text? Test on the two integrals $\int_0^4 dx/(1+x)$ and $\int_{-1}^1 e^x\,dx$ using only one computer run.

13. (**Student Research Project**) Study the Clenshaw-Curtis method for numerical quadrature. If possible, read the original paper by Clenshaw and Curtis [1960] and then program the method. If programmed well, it should be superior to the Romberg method in many cases. For further information on it, consult papers by Dixon [1974], Fraser and Wilson [1966], Gentleman [1972], Havie [1969], Kahaner [1971], and O'Hara and Smith [1968].

14. (**Student Research Project**) Numerical integration is an ideal problem for use on a *parallel computer*, since the interval of integration can be subdivided into subintervals on each of which the integral can be approximated simultaneously and independently of each other. Investigate how numerical integration can be done in parallel. If you have access to a parallel computer or can simulate a parallel computer on a collection of PCs, write a parallel program to approximate π by using the standard example

$$\int_0^1 (1+x^2)^{-1}\,dx$$

with a basic rule such as the midpoint rule. Vary the number of processors used and the number of subintervals. You can read about parallel computing in books such as Pacheco [1997], Quinn [1994], and others or at any of the numerous sites on the Internet.

15. Use a mathematical software system with symbolic capabilities to verify the relationship between A_k and the Bernoulli numbers for $k = 6$.

5.3 Simpson's Rules and Newton-Cotes Rules

Basic Trapezoid Rule

The basic trapezoid rule for approximating $\int_a^b f(x)\,dx$ is based on an estimation of the area beneath the curve over the interval $[a, b]$ using a trapezoid. The function of integration $f(x)$ is taken to be a straight line between $f(a)$ and $f(b)$. The numerical integration formula is of the form

Interval $[a, b]$

$$\int_a^b f(x)\,dx \approx Af(a) + Bf(b)$$

where the values of A and B are selected so that the resulting integration rule correctly integrates any linear function. It suffices to integrate exactly the two functions 1 and x because a polynomial of degree at most 1 is a linear combination of these two monomials. (This process is also called the **method of undetermined coefficients**.) To simplify the calculations, let $a = 0$ and $b = 1$ and find a formula of the following type:

$$\int_0^1 f(x)\,dx \approx Af(0) + Bf(1)$$

Thus, these equations should be fulfilled:

$$f(x) = 1 : \quad \int_0^1 dx = 1 = A + B$$

$$f(x) = x : \quad \int_0^1 x\,dx = \frac{1}{2} = B$$

The solution is $A = B = \frac{1}{2}$, and the integration formula is

$$\int_0^1 f(x)\,dx \approx \frac{1}{2}[f(0) + f(1)]$$

By a linear mapping $y = (b - a)x + a$ from $[0, 1]$ to $[a, b]$, the **basic trapezoid rule** for the interval $[a, b]$ is obtained:

Basic Trapezoid Rule

$$\int_a^b f(x)\,dx \approx \frac{1}{2}(b - a)[f(a) + f(b)] \tag{1}$$

See Figure 5.4 for a graphical illustration.

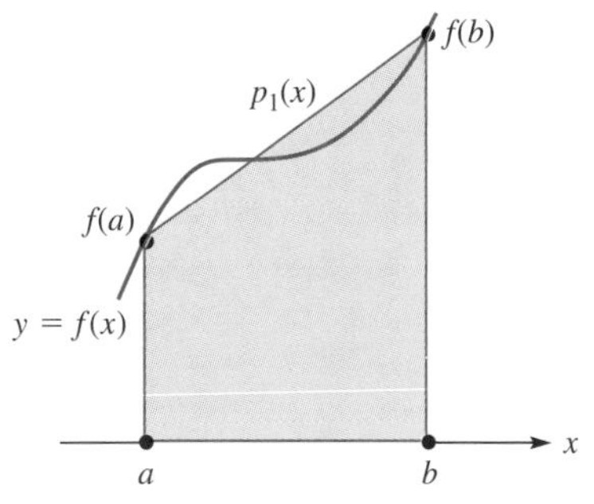

FIGURE 5.4
Basic trapezoid rule

Basic Simpson's Rule

The next obvious generalization is to take two subintervals $\left[a, \frac{a+b}{2}\right]$ and $\left[\frac{a+b}{2}, b\right]$ and to approximate $\int_a^b f(x)\,dx$ by taking the function of integration $f(x)$ to be a quadratic polynomial passing through the three points $f(a)$, $f\left(\frac{a+b}{2}\right)$, and $f(b)$. Let us seek a numerical integration rule of the following type:

Intervals $\left[a, \frac{a+b}{2}\right], \left[\frac{a+b}{2}, b\right]$

$$\int_a^b f(x)\,dx \approx Af(a) + Bf\left(\frac{a+b}{2}\right) + Cf(b)$$

The function f is assumed to be continuous on the interval $[a, b]$. The coefficients A, B, and C are chosen such that the formula above gives correct values for the integral whenever f is a quadratic polynomial. It suffices to integrate correctly the three functions 1, x, and x^2 because a polynomial of degree at most 2 is a linear combination of those 3 monomials. To simplify the calculations, let $a = -1$ and $b = 1$ and consider the formula

$$\int_{-1}^{1} f(x)\,dx \approx Af(-1) + Bf(0) + Cf(1)$$

Thus, these equations should be fulfilled:

$$f(x) = 1: \quad \int_{-1}^{1} dx = 2 = A + B + C$$

$$f(x) = x: \quad \int_{-1}^{1} x\,dx = 0 = -A + C$$

$$f(x) = x^2: \quad \int_{-1}^{1} x^2\,dx = \frac{2}{3} = A + C$$

The solution is $A = \frac{1}{3}$, $C = \frac{1}{3}$, and $B = \frac{4}{3}$. The resulting rule is

$$\int_{-1}^{1} f(x)\,dx \approx \frac{1}{3}[f(-1) + 4f(0) + f(1)]$$

Using a linear mapping $y = \frac{1}{2}(b-a)x + \frac{1}{2}(a+b)$ from $[-1, 1]$ to $[a, b]$, we obtain the **basic Simpson's rule** over the interval $[a, b]$:

Basic Simpson's Rule

$$\int_a^b f(x)\,dx \approx \frac{1}{6}(b-a)\left[f(a) + 4f\left(\frac{a+b}{2}\right) + f(b)\right] \tag{2}$$

See Figure 5.5 for an illustration.

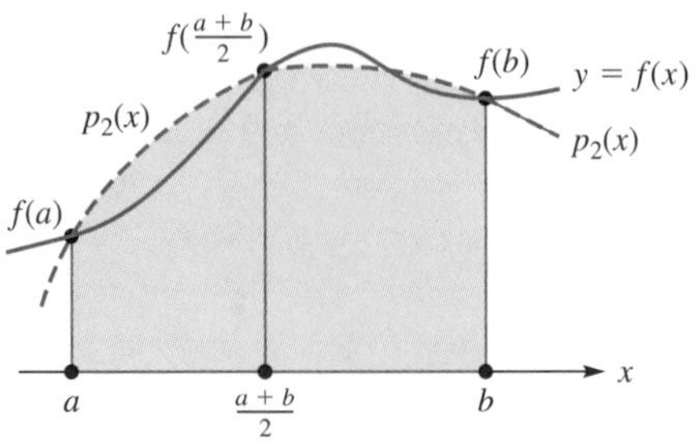

FIGURE 5.5
Basic Simpson's rule

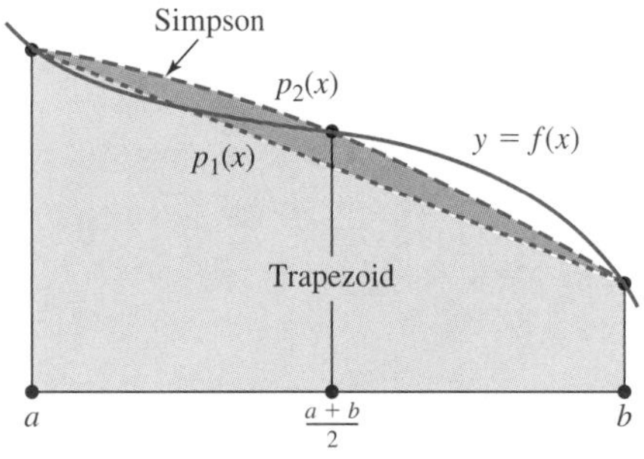

FIGURE 5.6 Example of trapezoid rule vs. Simpson's rule

Figure 5.6 shows graphically the difference between the basic trapezoid rule (1) and the basic Simpson's rule (2).

EXAMPLE 1 Find approximate values for the integral

$$\int_0^1 e^{-x^2}\,dx$$

using the basic trapezoid rule and the basic Simpson's rule. Carry five significant digits.

Solution Let $a = 0$ and $b = 1$. For the basic trapezoid rule (1), we obtain

$$\int_0^1 e^{-x^2}\,dx \approx \frac{1}{2}\left[e^0 + e^{-1}\right] \approx 0.5[1 + 0.36788] = 0.68394$$

which is correct to only one significant decimal place (rounded). For the basic Simpson's rule (2), we find

$$\int_0^1 e^{-x^2}\,dx \approx \frac{1}{6}\left[e^0 + 4e^{-0.25} + e^{-1}\right]$$
$$\approx 0.16667[1 + 4(0.77880) + 0.36788] = 0.7472$$

which is correct to three significant decimal places (rounded). Recall that $\int_0^1 e^{-x^2}\,dx = \frac{1}{2}\sqrt{\pi}\,\text{erf}(1) \approx 0.74682$. ■

Basic Simpson's Rule: Uniform Spacing

A numerical integration rule over two equal subintervals with partition points a, $a + h$, and $a + 2h = b$ is the widely used **basic Simpson's rule**:

Basic Simpson's Rule

$$\int_a^{a+h} f(x)\,dx \approx \frac{h}{3}[f(a) + 4f(a + h) + f(a + 2h)] \tag{3}$$

Simpson's rule computes exactly the integral of an interpolating quadratic polynomial over an interval of length $2h$ using three points; namely, the two endpoints and the middle point. It can be derived by integrating over the interval $[0, 2h]$ the **Lagrange quadratic polynomial** p_2 through the points $(0, f(0))$, $(h, f(h))$, and $(2h, f(2h))$:

$$\int_0^{2h} f(x)\,dx \approx \int_0^{2h} p_2(x)\,dx = \frac{h}{3}[f(0) + 4f(h) + f(2h)]$$

where

$$p_2(x) = \frac{1}{2h^2}(x - h)(x - 2h)f(0) - \frac{1}{h^2}x(x - 2h)f(h) + \frac{1}{2h^2}x(x - h)f(2h)$$

The error term in Simpson's rule can be established by using the Taylor series from Section 1.2:

$$f(a+h) = f + hf' + \frac{1}{2!}h^2 f'' + \frac{1}{3!}h^3 f''' + \frac{1}{4!}h^4 f^{(4)} + \cdots$$

where the functions $f, f', f'', \ldots$ on the right-hand side are evaluated at a. Now replacing h by $2h$, we have

$$f(a+2h) = f + 2hf' + 2h^2 f'' + \frac{4}{3}h^3 f''' + \frac{2^4}{4!}h^4 f^{(4)} + \cdots$$

Combining these two series, we obtain

$$f(a) + 4f(a+h) + f(a+2h) = 6f + 6hf' + 4h^2 f'' + 2h^3 f''' + \frac{20}{4!}h^4 f^{(4)} + \cdots$$

and, thereby, we have

$$\frac{h}{3}[f(a) + 4f(a+h) + f(a+2h)] = 2hf + 2h^2 f' + \frac{4}{3}h^3 f'' + \frac{2}{3}h^4 f''' + \frac{20}{3 \cdot 4!}h^5 f^{(4)} + \cdots \tag{4}$$

Hence, we have a series for the right-hand side of Equation (3). Now let's find one for the left-hand side. The Taylor series for $F(a+2h)$ is

$$F(a+2h) = F(a) + 2hF'(a) + 2h^2 F''(a) + \frac{4}{3}h^3 F'''(a) + \frac{2}{3}h^4 F^{(4)}(a) + \frac{2^5}{5!}h^5 F^{(5)}(a) + \cdots$$

Let

$$F(x) = \int_a^x f(t)\,dt$$

By the **Fundamental Theorem of Calculus**, $F' = f$. We observe that $F(a) = 0$ and $F(a+2h)$ is the integral on the left-hand side of Equation (3). Since $F'' = f'$, $F''' = f''$, and so on, we have

$$\int_a^{a+2h} f(x)\,dx = 2hf + 2h^2 f' + \frac{4}{3}h^3 f'' + \frac{2}{3}h^4 f''' + \frac{2^5}{5 \cdot 4!}h^5 f^{(4)} + \cdots \tag{5}$$

Subtracting Equation (4) from Equation (5), we obtain

$$\int_a^{a+2h} f(x)\,dx = \frac{h}{3}[f(a) + 4f(a+h) + f(a+2h)] - \frac{h^5}{90}f^{(4)} - \cdots$$

A more detailed analysis shows that the error term for the basic Simpson's rule (2) is $-(h^5/90)f^{(4)}(\xi) = \mathcal{O}(h^5)$ as $h \to 0$, for some ξ between a and $a + 2h$. By letting $b = a + 2h$, the **basic Simpson's rule** over the interval $[a, b]$ is

Basic Simpson's Rule

$$\int_a^b f(x)\,dx \approx \frac{(b-a)}{6}\left[f(a) + 4f\left(\frac{a+b}{2}\right) + f(b)\right]$$

with error term

Error Term

$$-\frac{1}{90}\left(\frac{b-a}{2}\right)^5 f^{(4)}(\xi) \tag{6}$$

for some ξ in (a, b).

Composite Simpson's Rule

Suppose that the interval $[a, b]$ is subdivided into an even number of subintervals, say n, each of width $h = (b - a)/n$. Then the partition points are $x_i = a + ih$ for $0 \leqq i \leqq n$, where n is divisible by 2. Now from basic calculus, we have

Interval $[a, b]$ with n (even) Subintervals of Width h

$$\int_a^b f(x)\,dx = \sum_{i=1}^{n/2} \int_{a+2(i-1)h}^{a+2ih} f(x)\,dx \tag{7}$$

Using the basic Simpson's rule, we have, for the right-hand side,

$$\approx \sum_{i=1}^{n/2} \frac{h}{3}\{f(a+2(i-1)h) + 4f(a+(2i-1)h) + f(a+2ih)\}$$

$$= \frac{h}{3}\Big\{ f(a) + \sum_{i=1}^{(n/2)-1} f(a+2ih) + 4\sum_{i=1}^{n/2} f(a+(2i-1)h) + \sum_{i=1}^{(n/2)-1} f(a+2ih) + f(b)\Big\}$$

Thus, we obtain

Composite Simpson's Rule

$$\int_a^b f(x)\,dx \approx \frac{h}{3}\Big\{[f(a)+f(b)] + 4\sum_{i=1}^{n/2} f[a+(2i-1)h] + 2\sum_{i=1}^{(n-2)/2} f(a+2ih)\Big\}$$

where $h = (b - a)/n$. The error term is

Error Term

$$-\frac{1}{180}(b-a)h^4 f^{(4)}(\xi) \tag{8}$$

Caution

Many formulas for numerical integration have error estimates that involve derivatives of the function being integrated. An important point that is frequently overlooked is that such error estimates depend on the function having derivatives. So if a piecewise function is being integrated, the numerical integration should be broken up over the region to coincide with the regions of smoothness of the function. Another important point is that *no* polynomial ever becomes infinite in the finite plane, so any integration technique that uses polynomials to approximate the integrand may fail to give good results without extra work at integrable singularities.

An Adaptive Simpson's Scheme

Now we develop an adaptive scheme based on Simpson's rule for obtaining a numerical approximation to the integral

$$\int_a^b f(x)\,dx$$

In this adaptive algorithm, the partitioning of the interval $[a, b]$ is not selected beforehand but is automatically determined. The partition is generated adaptively so that more and smaller subintervals are used in some parts of the interval and fewer and larger subintervals are used in other parts.

Adaptive Process for Subintervals

In the adaptive process, we divide the interval $[a, b]$ into two subintervals and then decide whether each of them is to be divided into more subintervals. This procedure is continued until some specified accuracy is obtained throughout the entire interval $[a, b]$.

Since the integrand f may vary in its behavior on the interval $[a, b]$, we do not expect the final partitioning to be uniform, but to vary in the density of the partition points.

It is necessary to develop the test for deciding whether subintervals should continue to be divided. One application of Simpson's rule over the interval $[a, b]$ can be written as

$$I \equiv \int_a^b f(x)\,dx = S(a,b) + E(a,b)$$

where

$$S(a,b) = \frac{(b-a)}{6}\left[f(a) + 4f\left(\frac{a+b}{2}\right) + f(b)\right]$$

and

$$E(a,b) = -\frac{1}{90}\left(\frac{b-a}{2}\right)^5 f^{(4)}(a) + \cdots$$

Letting $h = b - a$, we have

$$I = S^{(1)} + E^{(1)} \tag{9}$$

where

$$S^{(1)} = S(a,b)$$

and

$$\begin{aligned} E^{(1)} &= -\frac{1}{90}\left(\frac{h}{2}\right)^5 f^{(4)}(a) + \cdots \\ &= -\frac{1}{90}\left(\frac{h}{2}\right)^5 C \end{aligned}$$

Here we assume that $f^{(4)}$ remains a constant value C throughout the interval $[a, b]$. Now two applications of Simpson's rule over the interval $[a, b]$ give

$$I = S^{(2)} + E^{(2)} \tag{10}$$

where

$$S^{(2)} = S(a,c) + S(c,b)$$

where $c = (a + b)/2$, as in Figure 5.7, and

$$\begin{aligned} E^{(2)} &= -\frac{1}{90}\left(\frac{h/2}{2}\right)^5 f^{(4)}(a) + \cdots - \frac{1}{90}\left(\frac{h/2}{2}\right)^5 f^{(4)}(c) + \cdots \\ &= -\frac{1}{90}\left(\frac{h/2}{2}\right)^5 \left[f^{(4)}(a) + f^{(4)}(c)\right] + \cdots \\ &= -\frac{1}{90}\left(\frac{1}{2^5}\right)\left(\frac{h}{2}\right)^5 (2C) = \frac{1}{16}\left[-\frac{1}{90}\left(\frac{h}{2}\right)^5 C\right] \end{aligned}$$

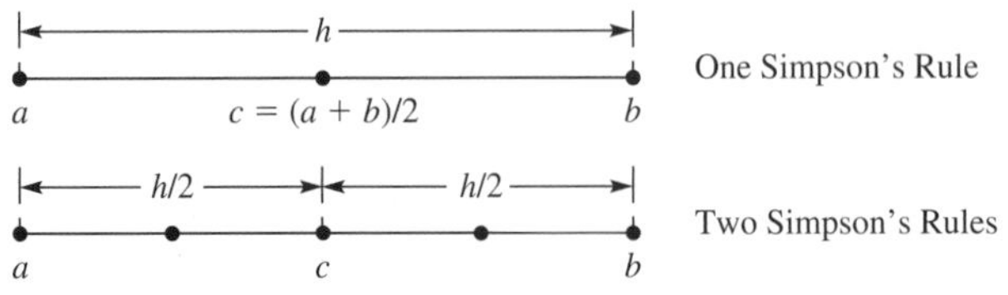

FIGURE 5.7
Simpson's rule

Again, we use the assumption that $f^{(4)}$ remains a constant value C throughout the interval $[a, b]$. We find that

$$16E^{(2)} = E^{(1)}$$

Subtracting Equation (10) from (9), we have

$$S^{(2)} - S^{(1)} = E^{(1)} - E^{(2)} = 15E^{(2)}$$

From this equation and Equation (10), we have

$$I = S^{(2)} + E^{(2)} = S^{(2)} + \frac{1}{15}\left(S^{(2)} - S^{(1)}\right)$$

We use the inequality

Stopping Test

$$\frac{1}{15}\left|S^{(2)} - S^{(1)}\right| < \varepsilon \tag{11}$$

to guide the adaptive process so that $|I - S^{(2)}| < \varepsilon$.

Recursive Process

If Test (11) is not satisfied, the interval $[a, b]$ is split into two subintervals, $[a, c]$ and $[c, b]$, where c is the midpoint $c = (a + b)/2$. On each of these subintervals, we again use Test (11) with ε replaced by $\varepsilon/2$ so that the resulting tolerance is ε over the entire interval $[a, b]$. A recursive procedure handles this quite nicely.

To see why, we take $\varepsilon/2$ on each left and right subinterval

Split Interval $[a, b]$ in Half

$$I = \int_a^b f(x)\,dx = \int_a^c f(x)\,dx + \int_c^b f(x)\,dx = I_{\text{left}} + I_{\text{right}}$$

If S is the sum of approximations $S^{(2)}_{\text{left}}$ over $[a, c]$ and $S^{(2)}_{\text{right}}$ over $[c, b]$, we have

$$\begin{aligned}
|I - S| &= \left|I_{\text{left}} + I_{\text{right}} - S^{(2)}_{\text{left}} - S^{(2)}_{\text{right}}\right| \\
&\leqq \left|I_{\text{left}} - S^{(2)}_{\text{left}}\right| + \left|I_{\text{right}} - S^{(2)}_{\text{right}}\right| \\
&= \frac{1}{15}\left|S^{(2)}_{\text{left}} - S^{(1)}_{\text{left}}\right| + \frac{1}{15}\left|S^{(2)}_{\text{right}} - S^{(1)}_{\text{right}}\right|
\end{aligned}$$

using Inequality (11). Hence, if we require

$$\frac{1}{15}\left|S^{(2)}_{\text{left}} - S^{(1)}_{\text{left}}\right| < \frac{\varepsilon}{2} \qquad \text{and} \qquad \frac{1}{15}\left|S^{(2)}_{\text{right}} - S^{(1)}_{\text{right}}\right| < \frac{\varepsilon}{2}$$

then $|I - S| < \varepsilon$ over the entire interval $[a, b]$.

We now describe an adaptive Simpson recursive procedure. The interval $[a, b]$ is partitioned into four subintervals of width $(b - a)/4$. Two Simpson approximations are computed by using two double-width subintervals and four single-width subintervals; that is,

Four Subintervals of Width $(b - a)/4$

$$\begin{aligned}
\textit{one_simpson} &\leftarrow \frac{h}{6}\left[f(a) + 4f\left(\frac{a+b}{2}\right) + f(b)\right] \\
\textit{two_simpson} &\leftarrow \frac{h}{12}\left[f(a) + 4f\left(\frac{a+c}{2}\right) + 2f(c) + 4f\left(\frac{c+b}{2}\right) + f(b)\right]
\end{aligned}$$

where $h = b - a$ and $c = (a + b)/2$.

According to Inequality (11), if *one_simpson* and *two_simpson* agree to within 15ε, then the interval $[a, b]$ does *not* need to be subdivided further to obtain an accurate approximation

Adaptive Process

to the integral $\int_a^b f(x)\,dx$. In this case, the value of $[16\,(two\_simpson) - (one\_simpson)]/15$ is used as the approximate value of the integral over the interval $[a, b]$. If the desired accuracy for the integral has not been obtained, then the interval $[a, b]$ is divided in half. The subintervals $[a, c]$ and $[c, b]$, where $c = (a + b)/2$, are used in a recursive call to the adaptive Simpson procedure with tolerance $\varepsilon/2$ on each. This procedure terminates when all subintervals satisfy Inequality (11). Alternatively, a maximum number of allowable levels of subdividing intervals is used to terminate the procedure prematurely. The recursive procedure provides an elegant and simple way to keep track of which subintervals satisfy the tolerance test and which need to be divided further.

Example Using Adaptive Simpson Procedure

The main program for calling the adaptive Simpson procedure can best be presented in terms of a concrete example. An approximate value for the integral

Sample Integral

$$\int_0^{\frac{5}{4}\pi} \left[\frac{\cos(2x)}{e^x}\right] dx \tag{12}$$

is desired with accuracy $\frac{1}{2} \times 10^{-3}$.

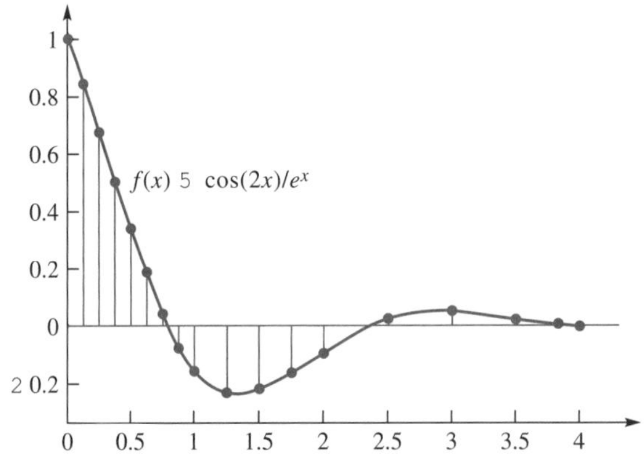

FIGURE 5.8 Adaptive Integration of $\int_0^{\frac{5}{4}\pi} \cos(2x)/e^x\,dx$

The graph of the integrand function is shown in Figure 5.8. This function has many turns and twists, so accurately determining the area under the curve may be difficult. A function procedure f is written for the integrand. Its name is the first argument in the procedure, and necessary interface statements may be needed. In the pseudocode below, other arguments are the values of the upper and lower limits a and b of the integral, the desired accuracy ε, the level of the current subinterval, and the maximum level depth.

Adaptive Procedure

```
recursive real function Simpson(f, a, b, ε, level, level_max)
        result(simpson_result)
integer level, level_max;      real a, b, c, d, e, h
external function f
level ← level + 1
h ← b − a
c ← (a + b)/2
one_simpson ← h[f(a) + 4f(c) + f(b)]/6
```

(Continued)

Simpson **Pseudocode**

```
    d ← (a + c)/2
    e ← (c + b)/2
    two_simpson ← h[f(a) + 4f(d) + 2f(c) + 4f(e) + f(b)]/12
    if level ≧ level_max then
        simpson_result ← two_simpson
        output "maximum level reached"
    else
        if |two_simpson − one_simpson| < 15ε then
            simpson_result ← two_simpson + (two_simpson − one_simpson)/15
        else
            left_simpson ← Simpson(f, a, c, ε/2, level, level_max)
            right_simpson ← Simpson(f, c, b, ε/2, level, level_max)
            simpson_result ← left_simpson + right_simpson
        end if
    end if
end function Simpson
```

By writing a driver computer program for this pseudocode and executing it on a computer, we obtain an approximate value of 0.208 for the integral. The adaptive Simpson procedure uses a different number of panels for different parts of the curve, as shown in Figure 5.8.

Newton-Cotes Rules

Newton-Cotes quadrature formulas for approximating $\int_a^b f(x)\,dx$ are obtained by approximating the function of integration $f(x)$ with interpolating polynomials. The rules are **closed** when they involve function values at the ends of the interval of integration. Otherwise, they are said to be **open**.

In the following, $a = x_0, b = x_n, x_i = x_0 + ih$, for $i = 0, 1, \dots, n$, where $h = (b-a)/n$, $f_i = f(x_i)$, and $a = x_0 < \xi < x_n = b$ in the error terms.

Some Closed Newton-Cotes Rules with Error Terms

Closed Rules

Trapezoid Rule:

$$\int_{x_0}^{x_1} f(x)\,dx = \frac{1}{2}h[f_0 + f_1] - \frac{1}{12}h^3 f''(\xi)$$

Simpson's $\frac{1}{3}$ Rule:

$$\int_{x_0}^{x_2} f(x)\,dx = \frac{1}{3}h[f_0 + 4f_1 + f_2] - \frac{1}{90}h^5 f^{(4)}(\xi)$$

Simpson's $\frac{3}{8}$ Rule:

$$\int_{x_0}^{x_3} f(x)\,dx = \frac{3}{8}h[f_0 + 3f_1 + 3f_2 + f_3] - \frac{3}{80}h^5 f^{(4)}(\xi)$$

Boole's Rule:

$$\int_{x_0}^{x_4} f(x)\,dx = \frac{2}{45}h[7f_0 + 32f_1 + 12f_2 + 32f_3 + 7f_4] - \frac{8}{945}h^7 f^{(6)}(\xi)$$

Six-Point Newton-Cotes Closed Rule:

$$\int_{x_0}^{x_5} f(x)\,dx = \frac{5}{288}h[19f_0 + 75f_1 + 50f_2 + 50f_3 + 75f_4 + 19f_5] - \frac{275}{12096}h^7 f^{(6)}(\xi)$$

Some Open Newton-Cotes Rules with Error Terms

Midpoint Rule:

$$\int_{x_0}^{x_2} f(x)\,dx = 2hf_1 + \frac{1}{24}h^3 f''(\xi)$$

Open Rules

Two-Point Newton-Cotes Open Rule:

$$\int_{x_0}^{x_3} f(x)\,dx = \frac{3}{2}h[f_1 + f_2] + \frac{1}{4}h^3 f''(\xi)$$

Three-Point Newton-Cotes Open Rule:

$$\int_{x_0}^{x_4} f(x)\,dx = \frac{4}{3}h[2f_1 - f_2 + 2f_3] + \frac{28}{90}h^5 f^{(4)}(\xi)$$

Four-Point Newton-Cotes Open Rule:

$$\int_{x_0}^{x_5} f(x)\,dx = \frac{5}{24}h[11f_1 + f_2 + f_3 + 11f_4] + \frac{95}{144}h^5 f^{(4)}(\xi)$$

Five-Point Newton-Cotes Open Rule:

$$\int_{x_0}^{x_6} f(x)\,dx = \frac{6}{20}h[11f_1 - 14f_2 + 26f_3 - 14f_4 + 11f_5] - \frac{41}{140}h^7 f^{(6)}(\xi)$$

Remarks

Over the years, many Newton-Cotes formulas have been derived and are compiled in the handbook by Abramowitz and Stegun [1964], which is available online. Rather than using a high-order Newton-Cotes rules over an entire interval, it is preferable to use a composite rule based on a low-order basic Newton-Cotes rule. There is seldom any advantage to using an open rule instead of a closed rule involving the same number of nodes. Nevertheless, open rules do have applications in integrating a function with singularities at the endpoints and in the numerical solution of ordinary differential equations, as discussed in Chapter 7.

Before the widespread use of computers, the Newton-Cotes rules were the most commonly used quadrature rules, since they involved fractions that were easy to use in hand calculations. The Gaussian quadrature rules of Section 5.4 use fewer function evaluations with higher-order error terms. The fact that they involve nodes involving irrational numbers is no longer a drawback with modern computers.

Summary 5.3

- Over the interval $[a, b]$, the **basic Simpson's rule** is

$$\int_a^b f(x)\,dx \approx S(a,b) = \frac{(b-a)}{6}\left[f(a) + 4f\left(\frac{a+b}{2}\right) + f(b)\right]$$

with error term $-\frac{1}{90}[\frac{1}{2}(b-a)]^5 f^{(4)}(\xi)$ for some ξ in (a, b). Letting $h = (b-a)/2$, another form for the **basic Simpson's rule** is

$$\int_a^{a+2h} f(x)\,dx \approx \frac{h}{3}[f(a) + 4f(a+h) + f(a+2h)]$$

with error term $-\frac{1}{90}h^5 f^{(4)}(\xi)$.

- The **composite Simpson's $\frac{1}{3}$ rule** over n (even) subintervals is

$$\int_a^b f(x)\,dx \approx \frac{h}{3}[f(a) + f(b)] + \frac{4h}{3}\sum_{i=1}^{n/2} f[a + (2i-1)h] + \frac{2h}{3}\sum_{i=1}^{(n-2)/2} f(a + 2ih)$$

where $h = (b-a)/n$ and the general error term is $-\frac{1}{180}(b-a)h^4 f^{(4)}(\xi)$.

- On the interval $[a, b]$ with $c = \frac{1}{2}(a+b)$, the test

$$\frac{1}{15}|S(a, c) + S(c, b) - S(a, b)| < \varepsilon$$

can be used in an **adaptive Simpson's algorithm**.

- Newton-Cotes quadrature rules encompass many common quadrature rules, such as the trapezoid rule, Simpson's rule, and the midpoint rule.

Exercises 5.3

[a]**1.** Compute $\int_0^1 (1+x^2)^{-1}\,dx$ by the basic Simpson's rule, using the three partition points $x = 0, 0.5$, and 1. Compare with the true solution.

2. Consider the integral $\int_0^1 \sin(\pi x^2/2)\,dx$. Suppose that we wish to integrate numerically, with an error of magnitude less than 10^{-3}.

[a]**a.** What width h is needed if we wish to use the composite trapezoid rule?

[a]**b.** Composite Simpson's rule?

c. Composite Simpson's $\frac{3}{8}$ rule?

3. A function f has the values shown.

x	1	1.25	1.5	1.75	2
$f(x)$	10	8	7	6	5

[a]**a.** Use Simpson's rule and the function values at $x = 1, 1.5$, and 2 to approximate $\int_1^2 f(x)\,dx$.

[a]**b.** Repeat the preceding part, using $x = 1, 1.25, 1.5, 1.75$, and 2.

[a]**c.** Use the results from parts **a** and **b** along with the error terms to establish an improved approximation. *Hint*: Assume constant error term Ch^4.

d. Repeat the previous parts using the trapezoid rule. Compare these results to that from Simpson's rule.

[a]**4.** Find an approximate value of $\int_1^2 x^{-1}\,dx$ using composite Simpson's rule with $h = 0.25$. Give a bound on the error.

5. Use Simpson's rule and its error formula to prove that if a cubic polynomial and a quadratic polynomial cross at three equally spaced points, then the two areas enclosed are equal.

6. For the **composite Simpson's $\frac{1}{3}$ rule** over n (even) subintervals, derive the general error term

$$-\frac{1}{180}(b-a)h^4 f^{(4)}(\xi)$$

for some $\xi \in (a, b)$.

[a]**7.** (Continuation) The composite Simpson's rule for calculating $\int_a^b f(x)\,dx$ can be written as

$$S_{n-1} = \frac{h}{3}[f(x_0) + 4f(x_1) + 2f(x_2) + \cdots + 4f(x_{n-1}) + f(x_n)]$$

where $x_i = a + ih$ for $0 \leq i \leq n$ and $h = (b - a)/n$ with n even. Its error is of the form Ch^4. Show how two values of S_k can be combined to obtain a more accurate estimate of the integral.

[a]**8.** A numerical integration scheme that is not as well known is the **basic Simpson's $\frac{3}{8}$ rule** over three subintervals:

$$\int_a^{a+3h} f(x)\,dx \approx \frac{3h}{8}[f(a) + 3f(a+h) + 3f(a+2h) + f(a+3h)]$$

Establish the error term for this rule, and explain why this rule is overshadowed by Simpson's rule.

9. (Continuation) Using the preceding exercise, establish the composite Simpson's $\frac{3}{8}$ rule over n (divisible by 3) subintervals. Derive the general error term.

10. Write out the details in the derivation of Simpson's rule.

11. Find a formula of the type

$$\int_0^1 f(x)\,dx \approx \alpha f(0) + \beta f(1)$$

that gives correct values for $f(x) = 1$ and $f(x) = x^2$. Does your formula give the correct value when $f(x) = x$?

12. If possible, find a formula

$$\int_{-1}^1 f(x)\,dx \approx \alpha f(-1) + \beta f(0) + \gamma f(1)$$

that gives the correct value for $f(x) = x, x^2$, and x^3. Does it correctly integrate the functions $x \mapsto 1, x^4$, and x^5?

13. Use linear mappings from $[0, 1]$ to $[a, b]$ and from $[-1, 1]$ to $[a, b]$ to justify the basic trapezoid rule and the basic Simpson's rule in general terms, respectively.

Computer Exercises 5.3

[a]**1.** Find approximate values for the two integrals

$$4\int_0^1 \frac{dx}{1+x^2}, \qquad 8\int_0^{1/\sqrt{2}} (\sqrt{1-x^2} - x)\,dx$$

Use recursive function *Simpson* with $\varepsilon = \frac{1}{2} \times 10^{-5}$ and $level\_max = 4$. Sketch the curves of the integrand $f(x)$ in each case, and show how *Simpson* partitions the intervals. You may want to print the intervals at which new values are added to *simpson_result* in function *Simpson* and also to print values of $f(x)$ over the entire interval $[a, b]$ in order to sketch the curves.

[a]**2.** Discover how to save function evaluations in function *Simpson* so that the integrand $f(x)$ is evaluated only once at each partition point. Test the modified code using the example in the text; that is,

$$\int_0^{2\pi} \cos(2x)e^{-x}\,dx$$

with $\varepsilon = 5.0 \times 10^{-5}$ and $level\_max = 4$.

3. Modify and test the pseudocode in this section so that it stores the partition points and function values. Using the modified code, repeat the preceding computer exercise, and plot the resulting partition points and function values.

4. Write and test code similar to that in this section but based on a different Newton-Cotes rule.

5. Using mathematical software such as MATLAB, Maple, or Mathematica, write and execute a computer program for finding an approximate value for the integral in Equation (12). Interpret warning messages. Try to obtain a more accurate approximation with more digits of precision by using additional (optional) parameters in the procedure.

6. Code and execute the recursive Simpson algorithm. Use integral (12) for one test.

7. Consider the integral

$$\int_{-1}^1 \frac{1}{\sqrt{1-x^2}}\,dx$$

Because it has singularities at the endpoints of the interval $[-1, 1]$, closed rules cannot be used. Apply all of the Newton-Cote open rules. Compare and explain these numerical results to the true solution, which is $\int_{-1}^1 (1-x^2)^{-1/2}\,dx = \arcsin x\big|_{-1}^1 = \pi$.

5.4 Gaussian Quadrature Formulas

Description

Most numerical integration formulas conform to the following pattern:

Nodes and Weights

$$\int_a^b f(x)\,dx \approx A_0 f(x_0) + A_1 f(x_1) + \cdots + A_n f(x_n) \tag{1}$$

To use such a formula, it is necessary only to know the **nodes** $x_0, x_1, \dots, x_n$ and the **weights** $A_0, A_1, \dots, A_n$. There are tables that list the numerical values of the nodes and weights for important special cases.

Where do formulas such as Formula (1) come from? One major source is the theory of polynomial interpolation as presented in Chapter 4. If the nodes have been fixed, then there is a corresponding Lagrange interpolation formula:

$$p(x) = \sum_{i=0}^{n} f(x_i)\,\ell_i(x) \quad \text{where} \quad \ell_i(x) = \prod_{\substack{j=0\\ j\neq i}}^{n} \left(\frac{x - x_j}{x_i - x_j}\right)$$

This formula (Equations (1) and (2) from Section 4.1) provides a polynomial p of degree at most n that interpolates f at the nodes; that is, $p(x_i) = f(x_i)$ for $0 \leqq i \leqq n$. When the circumstances are favorable, p is a good approximation to f, and $\int_a^b p(x)\,dx$ is a good approximation to $\int_a^b f(x)\,dx$. Therefore, we obtain

$$\int_a^b f(x)\,dx \approx \int_a^b p(x)\,dx = \sum_{i=0}^{n} f(x_i) \int_a^b \ell_i(x)\,dx = \sum_{i=0}^{n} A_i f(x_i) \tag{2}$$

where we have put

Weight Integral

$$A_i = \int_a^b \ell_i(x)\,dx$$

From the way in which Formula (2) has been derived, we know that it gives correct values for the integral of every polynomial of degree at most n.

EXAMPLE 1 Determine the weights in the quadrature formula of the form (1) when the interval is $[-2, 2]$ and the nodes are -1, 0, and 1.

Solution The functions ℓ_i are given above. Thus, we have

$$\ell_0(x) = \prod_{j=1}^{2} \left(\frac{x - x_j}{x_0 - x_j}\right) = \frac{1}{2}x(x-1)$$

Similarly, $\ell_1(x) = -(x+1)(x-1)$ and $\ell_2(x) = \frac{1}{2}x(x+1)$. The weights are obtained by integrating these functions. For example, we obtain

$$A_0 = \int_{-2}^{2} \ell_0(x)\,dx = \frac{1}{2}\int_{-2}^{2} (x^2 - x)\,dx = \frac{8}{3}$$

Similarly, $A_1 = -\frac{4}{3}$ and $A_2 = \frac{8}{3}$. Therefore, the quadrature formula is

$$\int_{-2}^{2} f(x)\,dx \approx \frac{8}{3}f(-1) - \frac{4}{3}f(0) + \frac{8}{3}f(1)$$

As a check on the work, one can verify that the formula gives exact values for the three monomials $f(x) = 1$, x, and x^2. By linear algebra, the formula provides correct values for any quadratic polynomial. ■

Change of Intervals

Gaussian rules for numerical integration are usually given on an interval such as [0, 1] or [−1, 1]. Often, we want to use these rules over a different interval! We can derive a formula for any other interval by making a linear change of variables. If the first formula is exact for polynomials of a certain degree over the first interval, the same is true of the second interval. Let's see how this is accomplished.

Suppose that a numerical integration formula of this form is given:

Quadrature Rule Over [c, d]

$$\int_c^d f(t)\,dt \approx \sum_{i=0}^{n} A_i f(t_i) \tag{3}$$

Let's assume that it is exact for all polynomials of degree at most m. If a formula is needed for some other interval, say, $[a, b]$, we first define a linear function λ of t such that if t traverses $[c, d]$, then $\lambda(t)$ traverses $[a, b]$. The function λ is given explicitly by

Linear Transformation

$$x = \lambda(t) = \left(\frac{b-a}{d-c}\right)t + \left(\frac{ad-bc}{d-c}\right) \tag{4}$$

Now in the integral

$$\int_a^b f(x)\,dx$$

we change the variable, $x = \lambda(t)$. Then $dx = \lambda'(t)\,dt = (b-a)(d-c)^{-1}\,dt$, and so we have

$$\begin{aligned}\int_a^b f(x)\,dx &= \left(\frac{b-a}{d-c}\right)\int_c^d f(\lambda(t))\,dt\\ &\approx \left(\frac{b-a}{d-c}\right)\sum_{i=0}^{n} A_i f(\lambda(t_i))\end{aligned}$$

Hence, we have

Quadrature Rule Over [a, b]

$$\int_a^b f(x)\,dx \approx \left(\frac{b-a}{d-c}\right)\sum_{i=0}^{n} A_i f\left(\left(\frac{b-a}{d-c}\right)t_i + \left(\frac{ad-bc}{d-c}\right)\right) \tag{5}$$

Observe that because λ is linear, $f(\lambda(t))$ is a polynomial in t if f is a polynomial, and the degrees are the same. Hence, the new formula is exact for polynomials of degree at most m.

As an example, we can use Formulas (3), (4), and (5) to change intervals from [−1, 1] to $[a, b]$. With the transformation $x = \frac{1}{2}(b-a)t + (a+b)$, a Gaussian quadrature rule of the form

$$\int_{-1}^{1} f(t)\,dt \approx \sum_{i=0}^{n} A_i f(t_i)$$

can be used over the interval $[a, b]$;

$$\int_a^b f(x)\,dx = \frac{1}{2}(b-a)\int_{-1}^{1} f\left[\frac{1}{2}(b-a)t + \frac{1}{2}(b+a)\right]dt$$

Gaussian Quadrature Rules

In the preceding discussion, the nodes were arbitrary, although for practical reasons, they should belong to the interval in which the integration is to be carried out. The great mathematician Karl Friedrich Gauss (1777–1855) discovered that by a special placement of the nodes, the accuracy of the numerical integration process could be greatly increased. Here is Gauss's remarkable result!

Theorem 1 **Gaussian Quadrature Theorem**

Let q be a nontrivial polynomial of degree $n+1$ such that

$$\int_a^b x^k q(x)\,dx = 0 \qquad (0 \leqq k \leqq n)$$

Let $x_0, x_1, \ldots, x_n$ be the zeros of q. Then the formula

$$\int_a^b f(x)\,dx \approx \sum_{i=0}^{n} A_i f(x_i) \quad \text{where} \quad A_i = \int_a^b \ell_i(x)\,dx \tag{6}$$

with these x_i's as nodes will be exact for all polynomials of degree at most $2n+1$. Furthermore, the nodes lie in the open interval (a, b).

Proof (We prove only the first assertion.) Let f be any polynomial of degree $\leqq 2n+1$. Dividing f by q, we obtain a quotient p and a remainder r, both of which have degree at most n:

$$f = pq + r$$

By our hypothesis, $\int_a^b q(x)p(x)\,dx = 0$. Furthermore, because each x_i is a root of q, we have $f(x_i) = p(x_i)q(x_i) + r(x_i) = r(x_i)$. Finally, since r has degree at most n, Formula (3) gives $\int_a^b r(x)\,dx$ precisely. Hence, we obtain

$$\int_a^b f(x)\,dx = \int_a^b p(x)q(x)\,dx + \int_a^b r(x)\,dx = \int_a^b r(x)\,dx$$
$$= \sum_{i=0}^{n} A_i r(x_i) = \sum_{i=0}^{n} A_i f(x_i) \qquad \blacksquare$$

Summary

With arbitrary nodes, Formula (6) is exact for all polynomials of degree $\leqq n$. With the Gaussian nodes, Formula (6) is exact for all polynomials of degree $\leqq 2n+1$.

The quadrature formulas that arise as applications of this theorem are called **Gaussian rules** or **Gauss-Legendre quadrature rules**. There is a different formula for each interval $[a, b]$ and each value of n. There are also more general Gaussian formulas to give approximate values of integrals, such as these

$$\int_0^\infty f(x)e^{-x}\,dx, \qquad \int_{-1}^{1} f(x)(1-x^2)^{1/2}\,dx, \qquad \int_{-\infty}^{\infty} f(x)e^{-x^2}\,dx, \qquad \text{etc.}$$

First, we derive a Gaussian formula that is not so complicated.

EXAMPLE 2 Determine the Gaussian quadrature formula with three Gaussian nodes and three weights for the integral

$$\int_{-1}^{1} f(x)\,dx$$

Solution We must find the polynomial q referred to in the Gaussian Quadrature Theorem and then compute its roots. The degree of q is 3, so q has the form

Finding Roots of a Degree 3 Polynomial

$$q(x) = c_0 + c_1 x + c_2 x^2 + c_3 x^3$$

The conditions that q must satisfy are

$$\int_{-1}^{1} q(x)\,dx = \int_{-1}^{1} x q(x)\,dx = \int_{-1}^{1} x^2 q(x)\,dx = 0$$

If we let $c_0 = c_2 = 0$, then $q(x) = c_1 x + c_3 x^3$, and so

$$\int_{-1}^{1} q(x)\,dx = \int_{-1}^{1} x^2 q(x)\,dx = 0$$

because the integral of an odd function over a symmetric interval is 0. To obtain c_1 and c_3, we impose the condition

$$\int_{-1}^{1} x(c_1 x + c_3 x^3)\,dx = 0$$

A convenient solution of this is $c_1 = -3$ and $c_3 = 5$. (Because it is a homogeneous equation, any multiple of a solution is another solution. We take the smallest integers that work.) Hence, we obtain

$$q(x) = 5x^3 - 3x$$

The roots of q are $-\sqrt{3/5}$, 0, and $\sqrt{3/5}$. These, then, are the Gaussian nodes for the desired quadrature formula.

To obtain the weights A_0, A_1, and A_2, we use a procedure known as the **method of undetermined coefficients**. We want to select values of A_0, A_1, and A_2 in the formula

Method of Undetermined Coefficients

$$\int_{-1}^{1} f(x)\,dx \approx A_0 f\left(-\sqrt{\frac{3}{5}}\right) + a_1 f(0) + a_2 f\left(\sqrt{\frac{3}{5}}\right) \tag{7}$$

so that the approximate equality ($\approx$) is an exact equality ($=$) whenever f is of the form $ax^2 + bx + c$. Since integration is a linear process, Formula (7) is exact for all polynomials of degree ≤ 2 if it is exact for these three monomials: 1, x, and x^2. We arrange the calculations in a tabular form.

f	Left-hand side	Right-hand side
1	$\int_{-1}^{1} dx = 2$	$A_0 + A_1 + A_2$
x	$\int_{-1}^{1} x\,dx = 0$	$-\sqrt{\frac{3}{5}}A_0 + \sqrt{\frac{3}{5}}A_2$
x^2	$\int_{-1}^{1} x^2\,dx = \frac{2}{3}$	$\frac{3}{5}A_0 + \frac{3}{5}A_2$

The left-hand side of Formula (7) equals the right-hand side for all quadratic polynomials when A_0, A_1, and A_2 satisfy these equations

$$\begin{cases} A_0 + A_1 + A_2 = 2 \\ A_0 \qquad - A_2 = 0 \\ A_0 \qquad + A_2 = 10/9 \end{cases}$$

The weights are $A_0 = A_2 = \frac{5}{9}$ and $A_1 = \frac{8}{9}$. Therefore, the final formula is

$$\int_{-1}^{1} f(x)\,dx \approx \frac{5}{9} f\left(-\sqrt{\frac{3}{5}}\right) + \frac{8}{9} f(0) + \frac{5}{9} f\left(\sqrt{\frac{3}{5}}\right) \tag{8}$$

It integrates correctly all polynomials up to and including quintic ones. For example, $\int_{-1}^{1} x^4\,dx = \frac{2}{5}$, and the formula also yields the value $\frac{2}{5}$ for this polynomial. ■

EXAMPLE 3 Use Formulas (3), (5), and (8) to approximate the integral

$$\int_0^1 e^{-x^2}\,dx$$

Solution Since $a = 0$ and $b = 1$, we have

$$\begin{aligned} \int_0^1 f(x)\,dx &= \frac{1}{2}\int_{-1}^{1} f\left(\frac{1}{2}t + \frac{1}{2}\right) dt \\ &\approx \frac{1}{2}\left[\frac{5}{9} f\left(\frac{1}{2} - \frac{1}{2}\sqrt{\frac{3}{5}}\right) + \frac{8}{9} f\left(\frac{1}{2}\right) + \frac{5}{9} f\left(\frac{1}{2} + \frac{1}{2}\sqrt{\frac{3}{5}}\right)\right] \end{aligned}$$

Letting $f(x) = e^{-x^2}$, we have

$$\begin{aligned} \int_0^1 e^{-x^2}\,dx &\approx \frac{5}{18} e^{-0.11270\,1665^2} + \frac{4}{9} e^{-0.5^2} + \frac{5}{18} e^{-0.88729\,8335^2} \\ &\approx 0.74681\,4584 \end{aligned}$$

Comparing against the true solution $\frac{1}{2}\sqrt{\pi}\,\mathrm{erf}(1) \approx 0.74682\,41330$, we find that the error in the computed solution is approximately 10^{-5}, which is excellent, considering that there were only three function evaluations. ■

Legendre Polynomials

Much more could be said about Gaussian quadrature formulas. In particular, there are efficient methods for generating the special polynomials whose roots are used as nodes in the quadrature formula. If we specialize to the integral

$$\int_{-1}^{1} f(x)\,dx$$

and standardize q_n so that $q_n(1) = 1$, then these polynomials are called **Legendre polynomials**. Thus, the roots of the Legendre polynomials are the nodes for Gaussian quadrature on the interval $[-1, 1]$. The first few Legendre polynomials are

First Three Legendre Polynomials

$$\begin{aligned} q_0(x) &= 1 \\ q_1(x) &= x \\ q_2(x) &= \frac{3}{2}x^2 - \frac{1}{2} \end{aligned}$$

$$q_3(x) = \frac{5}{2}x^3 - \frac{3}{2}$$

$$\vdots$$

They can be generated by a three-term recurrence relation:

Recurrence Relation

$$q_n(x) = \left(\frac{2n-1}{n}\right)xq_{n-1}(x) - \left(\frac{n-1}{n}\right)q_{n-2}(x) \qquad (n \geqq 2) \tag{9}$$

With no new ideas, we can treat integrals of the form

$$\int_a^b f(x)w(x)\,dx$$

Here, $w(x)$ should be a fixed positive function on (a, b) for which the integrals $\int_a^b x^n w(x)\,dx$ all exist, for $n = 0, 1, 2, \ldots$. Important examples for the interval $[-1, 1]$ are given by

$$w(x) = (1-x^2)^{-1/2}, \qquad w(x) = (1-x^2)^{1/2}$$

The corresponding theorem is as follows:

■ **Theorem 2** **Weighted Gaussian Quadrature Theorem**

Let q be a nonzero polynomial of degree $n+1$ such that

$$\int_a^b x^k q(x)w(x)\,dx = 0 \qquad (0 \leqq k \leqq n)$$

Let $x_0, x_1, \ldots, x_n$ be the roots of q. Then the formula

$$\int_a^b f(x)w(x)\,dx \approx \sum_{i=0}^{n} A_i f(x_i)$$

where

$$\ell_i(x) = \prod_{\substack{j=0\\ j\neq i}}^{n} \frac{x-x_j}{x_i-x_j} \qquad \text{and} \qquad A_i = \int_a^b \ell_i(x)w(x)\,dx$$

will be exact whenever f is a polynomial of degree at most $2n+1$.

Gaussian Quadrature Nodes and Weights

The nodes and weights for several values of n in the Gaussian quadrature formula

$$\int_{-1}^{1} f(x)\,dx \approx \sum_{i=0}^{n} A_i f(x_i)$$

are given in Table 5.1. The numerical values of nodes and weights for various values of n up to 95 can be found in Abramowitz and Stegun [1964]. (See also Stroud and Secrest [1966].) Since these nodes and weights are mostly irrational numbers, they are *not* used in computations by hand as much as are simpler rules that involve integer and rational values. However, in a computer programs, it does not matter whether an integration formula is elegant, and the Gaussian quadrature formulas usually give greater accuracy with

TABLE 5.1 Gaussian Quadrature Nodes and Weights

n	Nodes x_i	Weights A_i
1	$-\sqrt{\frac{1}{3}}$	1
	$+\sqrt{\frac{1}{3}}$	1
2	$-\sqrt{\frac{3}{5}}$	$\frac{5}{9}$
	0	$\frac{8}{9}$
	$+\sqrt{\frac{3}{5}}$	$\frac{5}{9}$
3	$-\sqrt{\frac{1}{7}\left(3-4\sqrt{0.3}\right)}$	$\frac{1}{2}+\frac{1}{12}\sqrt{\frac{10}{3}}$
	$-\sqrt{\frac{1}{7}\left(3+4\sqrt{0.3}\right)}$	$\frac{1}{2}-\frac{1}{12}\sqrt{\frac{10}{3}}$
	$+\sqrt{\frac{1}{7}\left(3-4\sqrt{0.3}\right)}$	$\frac{1}{2}+\frac{1}{12}\sqrt{\frac{10}{3}}$
	$+\sqrt{\frac{1}{7}\left(3+4\sqrt{0.3}\right)}$	$\frac{1}{2}-\frac{1}{12}\sqrt{\frac{10}{3}}$
4	$-\sqrt{\frac{1}{9}\left(5-2\sqrt{\frac{10}{7}}\right)}$	$0.3\left(\frac{-0.7+5\sqrt{0.7}}{-2+5\sqrt{0.7}}\right)$
	$-\sqrt{\frac{1}{9}\left(5+2\sqrt{\frac{10}{7}}\right)}$	$0.3\left(\frac{0.7+5\sqrt{0.7})}{2+5\sqrt{0.7}}\right)$
	0	$\frac{128}{225}$
	$+\sqrt{\frac{1}{9}\left(5-2\sqrt{\frac{10}{7}}\right)}$	$0.3\left(\frac{-0.7+5\sqrt{0.7}}{-2+5\sqrt{0.7}}\right)$
	$+\sqrt{\frac{1}{9}\left(5+2\sqrt{\frac{10}{7}}\right)}$	$0.3\left(\frac{0.7+5\sqrt{0.7}}{2+5\sqrt{0.7}}\right)$

fewer function evaluations. The choice of quadrature formulas depends on the specific application being considered, and the reader should consult more advanced references for guidelines. See, for example, Davis and Rabinowitz [1984], Ghizetti and Ossiccini [1970], or Krylov [1962].

Integrals with Singularities

If either the interval of integration is unbounded or the function of integration is unbounded, then special procedures must be used to obtain accurate approximations to the integrals.

One approach for handling a singularity in the function of integration is to change variables to remove the singularity and then use a standard approximation technique. For example, we obtain

Remove Singularity

$$\int_0^1 \frac{dx}{e^x\sqrt{x}} = 2\int_0^1 \frac{dt}{e^{t^2}}$$

and

$$\int_0^{\pi/2} \frac{\cos x}{\sqrt{x}}\,dx = 2\int_0^{\sqrt{\pi/2}} \cos t^2\,dt$$

using $x = t^2$. Some other useful transformations are

$$x = -\log t, \quad x = t/(1-t), \quad x = \tan t, \quad x = \sqrt{(1+t)/(1-t)}$$

An important case where Gaussian formulas have an advantage occurs in integrating a function that is infinite at one end of the interval. The reason for this advantage is that the nodes in Gaussian quadrature are always *interior* points of the interval. Thus, for example, in computing the integral

Integrals with Singularities

$$\int_0^1 \frac{\sin x}{x}\,dx$$

we can safely use the pseudocode statement $y \leftarrow (\sin x)/x$ with a Gaussian formula because the value at $x = 0$ is not required. More difficult integrals such as

$$\int_0^1 \frac{\sqrt[3]{x^2-1}}{\sqrt{\sin(e^x-1)}}\,dx$$

can be computed directly with a Gaussian formula in spite of the singularity at 0. Of course, we are referring to integrals that are well defined and finite in spite of a singularity. A typical case is

$$\int_0^1 \frac{dx}{\sqrt{x}}$$

Summary 5.4

- **Gaussian quadrature rules** with **nodes** x_i and **weights** A_i are of the form

$$\int_a^b f(x)\,dx \approx \sum_{i=0}^{n} A_i f(x_i)$$

where the weights are

$$A_i = \int_a^b \ell_i(x)\,dx, \qquad \ell_i(x) = \prod_{\substack{j=0\\ j\neq i}}^{n} \left(\frac{x - x_j}{x_i - x_j}\right)$$

If q is a nontrivial polynomial of degree $n + 1$ such that

$$\int_a^b x^k q(x)\,dx = 0 \qquad (0 \leqq k \leqq n)$$

then the nodes $x_0, x_1, \ldots, x_n$ are the zeros of q. Furthermore, the nodes lie in the open interval (a, b). The rule is exact for all polynomials of degree at most $2n + 1$.

- To change intervals from $[c, d]$ to $[a, b]$, use the transformation

$$x = \lambda(t) = \left(\frac{b-a}{d-c}\right)t + \left(\frac{ad-bc}{d-c}\right)$$

The numerical integration formula

$$\int_c^d f(t)dt \approx \sum_{i=0}^{n} A_i f(t_i)$$

becomes

$$\int_a^b f(x)dx \approx \left(\frac{b-a}{d-c}\right)\sum_{i=0}^{n} A_i f(\lambda(t_i))$$

So we use the following formula to change an integration rule from the interval $[c, d]$ to $[a, b]$:

$$\int_a^b f(x)\,dx \approx \left(\frac{b-a}{d-c}\right)\sum_{i=0}^{n} A_i f\left(\left(\frac{b-a}{d-c}\right)x_i + \left(\frac{ad-bc}{d-c}\right)\right)$$

- Some typical Gaussian integration rules are

$$\int_{-1}^{1} f(x)\,dx \approx f\left(-\frac{1}{\sqrt{3}}\right) + f\left(\frac{1}{\sqrt{3}}\right)$$

$$\int_{-1}^{1} f(x)\,dx \approx \frac{5}{9} f\left(-\sqrt{\frac{3}{5}}\right) + \frac{8}{9} f(0) + \frac{5}{9} f\left(\sqrt{\frac{3}{5}}\right)$$

- The **weighted Gaussian quadrature rules** are of the form

$$\int_a^b f(x)w(x)\,dx \approx \sum_{i=0}^{n} A_i f(x_i)$$

where the weights are

$$A_i = \int_a^b \ell_i(x)w(x)\,dx$$

If q is a nonzero polynomial of degree $n + 1$ such that

$$\int_a^b x^k q(x)w(x)\,dx = 0 \qquad (0 \leqq k \leqq n)$$

then nodes $x_0, x_1, \ldots, x_n$ are the roots of q. The rule is exact whenever f is a polynomial of degree at most $2n + 1$.

- If we have a basic numerical integration formula for the interval $[-1, 1]$ such as

$$\int_{-1}^{1} f(t)\,dt \approx \sum_{i=0}^{m} A_i f(t_i)$$

it can be employed on an arbitrary interval $[c, d]$ by using a change of variables. To convert to the interval $[c, d]$, change variables by writing $x = \beta t + \alpha$, where $\alpha = \frac{1}{2}(c+d)$ and $\beta = \frac{1}{2}(d-c)$. Notice that when $t = -1$ then $x = c$ and when $t = +1$ then $x = d$. Also, we must use $dx = \beta\,dt$. Putting this together, we have the following formulas:

$$\int_{c}^{d} f(x)\,dx = \beta \int_{-1}^{1} f(\beta t + \alpha)\,dt \approx \beta \sum_{i=0}^{m} A_i f(\beta t_i + \alpha)$$

If we want to find a composite rule for the interval $[a, b]$ with $m/2$ applications of the basic rule, we use

$$\int_{a}^{b} f(x)\,dx = \sum_{j=1}^{n/2} \int_{x_{2(j-1)}}^{x_{2j}} f(x)\,dx$$

and determine

$$\int_{a}^{b} f(x)\,dx \approx h \sum_{j=1}^{n/2} \sum_{i=0}^{m} A_i f\left[ht_i + t_{2i-1}\right]$$

where $h = t_{2i} - t_{2i-1} = t_{2i-1} - t_{2i-2}$.

Exercises 5.4

[a]1. A Gaussian quadrature rule for the interval $[-1, 1]$ can be used on the interval $[a, b]$ by applying a suitable linear transformation. Approximate

$$\int_{0}^{2} e^{-x^2}\,dx$$

using the transformed rule from Table 5.1 with $n = 1$.

2. Using Table 5.1, show directly that the Gaussian quadrature rule is exact for the polynomials $1, x, x^2, \ldots, x^{2n+1}$ when

a. $n = 1$ **b.** $n = 3$ **c.** $n = 4$

3. For how high a degree of polynomial is Formula (8) exact? Verify your answer by continuing the method of undetermined coefficients until an equation is not satisfied.

4. Verify parts of Table 5.1 by finding the roots of q_n and using the method of undetermined coefficients to establish the Gaussian quadrature formula on the interval $[-1, 1]$ for the following:

[a]**a.** $n = 1$ [a]**b.** $n = 3$ **c.** $n = 4$

[a]5. Construct a rule of the form

$$\int_{-1}^{1} f(x)\,dx \approx \alpha f\left(-\frac{1}{2}\right) + \beta f(0) + \gamma f\left(\frac{1}{2}\right)$$

that is exact for all polynomials of degree $\leqq 2$; that is, determine values for α, β, and γ.
Hint: Make the relation exact for 1, x, and x^2 and find a solution of the resulting equations. If it is exact for these polynomials, it is exact for all polynomials of degree $\leqq 2$.

[a]6. Establish a numerical integration formula of the form

$$\int_{a}^{b} f(x)\,dx \approx Af(a) + Bf'(b)$$

that is accurate for polynomials of as high a degree as possible.

[a]7. Derive a formula for $\int_a^{a+h} f(x)\,dx$ in terms of function evaluations $f(a)$, $f(a+h)$, and $f(a+2h)$ that is correct for polynomials of as high a degree as possible.
Hint: Use polynomials 1, $x-a$, $(x-a)^2$, and so on.

8. Derive a formula of the form

$$\int_{a}^{b} f(x)\,dx \approx w_0 f(a) + w_1 f(b) + w_2 f'(a) + w_3 f'(b)$$

that is exact for polynomials of the highest degree possible.

[a]**9.** Derive the Gaussian quadrature rule of the form

$$\int_{-1}^{1} f(x)x^2\,dx \approx af(-\alpha) + bf(0) + cf(\alpha)$$

that is exact for all polynomials of as high a degree as possible; that is, determine α, a, b, and c.

[a]**10.** Determine a formula of the form

$$\int_{0}^{h} f(x)\,dx \approx w_0 f(0)+w_1 f(h)+w_2 f''(0)+w_3 f''(h)$$

that is exact for polynomials of as high a degree as possible.

[a]**11.** Derive a numerical integration formula of the form

$$\int_{x_{n-1}}^{x_{n+1}} f(x)\,dx \approx Af(x_n) + Bf'(x_{n-1}) + Cf''(x_{n+1})$$

for uniformly spaced points x_{n-1}, x_n, and x_{n+1} with spacing h. The formula should be exact for polynomials of as high a degree as possible.
Hint: Consider

$$\int_{-h}^{h} f(x)\,dx \approx Af(0) + Bf'(-h) + Cf''(h)$$

[a]**12.** By the method of undetermined coefficients, derive a numerical integration formula of the form

$$\int_{-2}^{+2} |x| f(x)\,dx \approx Af(-1) + Bf(0) + Cf(+1)$$

that is exact for polynomials of degree $\leqq 2$. Is it exact for polynomials of degree greater than 2?

[a]**13.** Determine A, B, C, and D for a formula of the form

$$Af(-h) + Bf(0) + Cf(h) = hDf'(h) + \int_{-h}^{h} f(x)\,dt$$

that is accurate for polynomials of as high a degree as possible.

[a]**14.** The numerical integration rule

$$\int_{0}^{3h} f(x)\,dx \approx \frac{3h}{8}\,[f(0) + 3f(h) + 3f(2h) + f(3h)]$$

is exact for polynomials of degree $\leqq n$. Determine the largest value of n for which this assertion is true.

15. (**Adams-Bashforth-Moulton Formulas**) Verify that the numerical integration formulas

a. $$\int_{t}^{t+h} g(s)\,ds \approx \frac{h}{24}[55g(t) - 59g(t-h) + 37g(t-2h) - 9g(t-3h)]$$

b. $$\int_{t}^{t+h} g(s)\,ds \approx \frac{h}{24}[9g(t+h) + 19g(t) - 5g(t-h) + g(t-2h)]$$

are exact for polynomials of third degree.
Note: These two formulas can also be derived by replacing the two integrands g with two interpolating polynomials from Section 4.1 using nodes $(t,\ t-h,\ t-2h,\ t-3h)$ or nodes $(t+h,\ t,\ t-h,\ t-2h)$, respectively.

16. Let a quadrature formula be given in the form

$$\int_{-1}^{1} f(x)\,dx \approx \sum_{i=1}^{n} w_i f(x_i)$$

What is the corresponding formula for $\int_0^1 f(x)\,dx$?

17. Using the rules in Table 5.1, determine the general rules for approximating integrals of the form $\int_a^b f(x)\,dx$.

Computer Exercises 5.4

1. Write a program to evaluate an integral $\int_a^b f(x)\,dx$ using Formula (8).

2. (Continuation) By use of the same program, compute approximate values of the integrals

[a]**a.** $\int_0^1 dx/\sqrt{x}$ **b.** $\int_0^2 e^{-\cos^2 x}\,dx$

3. (Continuation) Compute $\int_0^1 x^{-1}\sin x\,dx$ by the Gaussian Formula (8) suitably modified.

4. Write a procedure for evaluating $\int_a^b f(x)\,dx$ by first subdividing the interval into n equal subintervals and then using the three-point Gaussian formula modified to apply to the n different subintervals. The function f and the integer n will be furnished to the procedure.

5. (Continuation) Test the procedure written in the preceding computer problem on these examples:

a. $\int_0^1 x^5\,dx \quad (n = 1, 2, 10)$

b. $\displaystyle\int_0^1 x^{-1} \sin x \, dx \qquad (n = 1, 2, 3, 4)$

6. Apply and compare the composite rules for trapezoid, midpoint, two-point Gaussian, and Simpson's $\frac{1}{3}$ rule for approximating the integral

$$\int_0^{2\pi} e^{-x} \cos x \, dx \approx 0.49906\,62786\,34$$

using 32 applications of each basic rule.

7. Code and test an **adaptive two-point Gaussian integration** procedure to approximate the integral

$$\int_1^3 100x^{-1} \sin(10x^{-1}) \, dx \approx -18.79829\,68367\,8703$$

Write three procedures using double precision:

a. Two-point Gauss

> **procedure** *Gauss*(*f*, *a*, *b*)

b. Nonrecursive two-point Gauss that initializes variables *sum* and *depth* to zero and calls the procedure in **c**

> **procedure** *Adaptive_Initial*(*f*, *a*, *b*)

c. Recursive two-point Gauss

> **recursive procedure** *Adaptive*(*f*, *sum*, *a*, *b*, *depth*)

that checks to see whether the maximum depth is exceeded; if so, it prints an error message and stops; if not, it continues by dividing the interval $[a, b]$ in half and calling procedure *Gauss* on the left subinterval, the right subinterval, and the whole interval, then checking to see whether the tolerance test is accepted; if it is, it adds the approximate value over the whole interval to the variable *sum*; otherwise it calls recursive procedure *Adaptive* on the left and right subintervals in addition to increasing the value of the *depth* variable. The tolerance test checks to see if the difference in absolute value between the approximate value over the whole interval and the sum of the approximate values over the left subinterval and right subinterval is less than the variable *tolerance*.

Print out the contribution of each subinterval and the depth at which the approximate value over the subinterval is accepted. Use a maximum depth of 100 subintervals, and stop subdividing subintervals when the tolerance is less than 10^{-7}.

8. Compute values of the integrals mentioned in this chapter (e.g., p. 246) and/or these:

^{a}**a.** $\displaystyle\int_0^1 \frac{dx}{\sqrt{\sin x}}$ $\qquad$ ^{a}**b.** $\displaystyle\int_0^\infty e^{-x^3} \, dx$

^{a}**c.** $\displaystyle\int_0^1 x|\sin(1/x)| \, dx$

To determine whether the computed results are accurate, use two different programs from MATLAB, Maple, or Mathematica to do these calculations.

9. (Continuation) Another approach to computing the integral $\int_0^1 x|\sin(1/x)|\,dx$ is by a change of variables. Turn it into the integral $\int_1^\infty \left[|\sin(t)|/t^3\right] dt$ and then write it as the sum of the integrals from 1 to π, π to 2π, and $2k\pi$ to $2(k+1)\pi$, for $k = 1, 2, 3, \ldots$. To get 12-decimal places of accuracy, let k run to 112,536. Adding up the subintegrals in order of smallest to largest should give better roundoff errors. Taking 10,000 steps may require several minutes of machine time, but the error should be no more than about two digits in the tenth decimal place. The first two partial integrals should be computed outside the loop and then added into the sum at the end. Using MATLAB program `quad`, integrate the original integral, and then program this alternative approach.

10. Use Gaussian quadrature formulas on these test cases:

a. $\displaystyle\int_0^1 \frac{\log(1-x)}{x} \, dx = -\frac{\pi^2}{6}$

b. $\displaystyle\int_0^1 \frac{\log(1+x)}{x} \, dx = \frac{\pi^2}{12}$

c. $\displaystyle\int_0^1 \frac{\log(1+x^2)}{x} \, dx = \frac{\pi^2}{24}$

This problem illustrates integrals with singularities at the endpoint. The integrals can be computed numerically by using Gaussian quadrature. The known values enable one to test the process. (See Haruki and Haruki [1983] and Jeffrey [2000].)

11. Suppose we want to compute $\int_a^b f(x)\,dx$. We divide the interval $[a, b]$ into n subintervals of uniform size $h = (b-a)/n$, where n is divisible by 2. Let the nodes be $x_i = a + ih$ for $0 \le i \le n$. Consider the following numerical integration rules.

a. Composite Trapezoid Rule

$$\int_a^b f(x)\,dx \approx \frac{1}{2}h\,[f(a) + f(b)] + h \sum_{i=1}^{n-1} f(x_i)$$

b. Composite Simpson's $\frac{1}{3}$-Rule (n even)

$$\int_a^b f(x)\,dx \approx \frac{1}{3}h\Big[f(a)+f(b)\Big]+\frac{4}{3}hf(b-h) + \frac{2}{3}h\sum_{i=1}^{\frac{1}{2}n-1}\Big[2f(x_{2i-1})+f(x_{2i})\Big]$$

c. Composite Gaussian Three-Point Rule (n even)

$$\int_a^b f(x)\,dx \approx h\sum_{i=1}^{n/2}\left[\frac{5}{9}f\left(x_{2i-1}-h\sqrt{\frac{3}{5}}\right)+\frac{8}{9}f(x_{2i-1}) + \frac{5}{9}f\left(x_{2i-1}+h\sqrt{\frac{3}{5}}\right)\right]$$

For each of these rules, write and run computer programs for obtaining the numerical approximation to the integral $\int_0^{2\pi}\left[\cos(2x)/e^x\right]dx$ using these rules with $n = 120$. Use the true solution $\frac{1}{5}(1-e^{-2\pi})$ computed in double precision to compute the absolute errors in these results.

12. (Continuation) Repeat the previous problem using some of the rules in Table 5.1 and compare the results.

13. (**Student Research Project**) From a practical point of view, investigate some new algorithms for numerical integration that are associated with the names Clenshaw and Curtis [1960], Kronrod [1964], and Patterson [1968]. The later two are adaptive Gaussian quadrature methods that provide error estimates based on the evaluation and reuse of the results at *Kronrod points*. See `QUADPACK` by Pessens, de Doncker, Uberhuber, and Kahaner [1983] and also Laurie [1997], Ammar, Calvetti, and Reichel [1999], and Calvetti, Golub, Gragg, and Reichel [2000] for examples.

14. Consider the integral

$$\int_{-1}^{1}\frac{1}{\sqrt{1-x^2}}\,dx$$

Because it has singularities at the endpoints of the interval $[-1, 1]$, closed rules cannot be used. Apply some of the Gaussian open rules in Table 5.1. Compare and explain these numerical results to the true solution, which is $\int_{-1}^{1}(1-x^2)^{-1/2}\,dx = \arcsin x\big|_{-1}^{1} = \pi$.

15. Use numerical integration to verify or refute each of the following conjectures:

a. $\displaystyle\int_0^1 \frac{4}{1+x^2}\,dx = \pi$

b. $\displaystyle\int_0^1 \sqrt{x}\,\log(x)\,dx = -\frac{4}{9}$

c. $\displaystyle\int_0^1 \sqrt{x^3}\,dx = \frac{2}{5}$

d. $\displaystyle\int_0^1 \frac{1}{1+10x^2}\,dx = \frac{4}{5}$

e. $\displaystyle\int_{-9}^{100} \frac{1}{\sqrt{|x|}}\,dx = 26$

f. $\displaystyle\int_0^{10} 25e^{-25x}\,dx = 1$

g. $\displaystyle\int_0^1 \log(x)\,dx = -1$

6 Spline Functions

By experimentation in a wind tunnel, an airfoil is constructed by trial and error so that it has certain desired characteristics. The cross section of the airfoil is then drawn as a curve on coordinate paper (see Figure 6.1). To study this airfoil by analytical methods or to manufacture it, it is essential to have a formula for this curve. To arrive at such a formula, one first obtains the coordinates of a finite set of points on the curve. Then a smooth curve called a **cubic interpolating spline** can be constructed to match these data points. This chapter discusses general polynomial spline functions and how they can be used in various numerical problems such as the data-fitting problem just described.

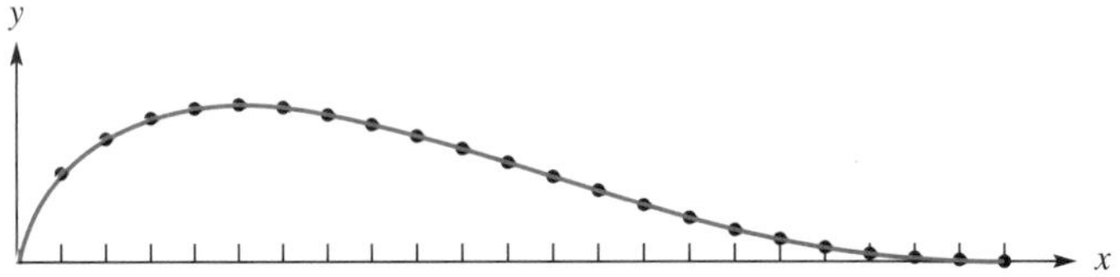

FIGURE 6.1
Airfoil cross section

6.1 First-Degree and Second-Degree Splines

Brief History of Splines

The history of spline functions is rooted in the work of draftsmen, who often needed to draw a gently turning curve between points on a drawing. This process is called **fairing** and can be accomplished with a number of ad hoc devices, such as the **French curve**, made of plastic and presenting a number of curves of different curvature for the draftsman to select. Long strips of wood were also used, being made to pass through the **control points** by weights laid on the draftsman's table and attached to the strips. The weights were called **ducks** and the strips of wood were called **splines**, even as early as 1891. The elastic nature of the wooden strips allowed them to bend only a little, while still passing through the prescribed points. The wood was, in effect, solving a differential equation and minimizing the strain energy. The latter is known to be a simple function of the curvature. The mathematical theory of these curves owes much to the early investigators, particularly Isaac Schoenberg in the 1940s and 1950s. Other important names associated with the early development of the subject (prior to 1964) are Garrett Birkhoff, C. de Boor, J. H. Ahlberg, E. N. Nilson, H. Garabedian, R. S. Johnson, F. Landis, A. Whitney, J. L. Walsh, and J. C. Holladay. The first book giving a systematic exposition of spline theory was by Ahlberg, Nilson, and Walsh [1967].

First-Degree Splines

A **spline function** is a function that consists of polynomial pieces joined together with certain smoothness conditions. A simple example is the **polygonal** function (or spline of

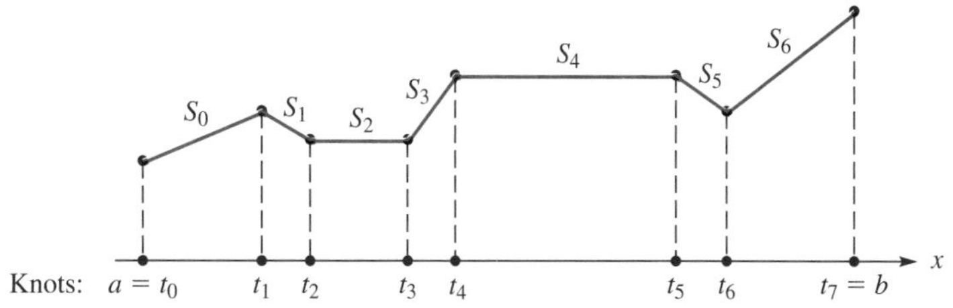

FIGURE 6.2 First-degree spline function

degree 1), whose pieces are linear polynomials joined together to achieve continuity, as in Figure 6.2. The points $t_0, t_1, \dots, t_n$ at which the function changes its character are termed **knots** in the theory of splines. Thus, the spline function shown in Figure 6.2 has eight knots.

Such a function appears somewhat complicated when defined in explicit terms. We are forced to write

$S(x)$ Piecewise Linear

$$S(x) = \begin{cases} S_0(x), & x \in [t_0, t_1] \\ S_1(x), & x \in [t_1, t_2] \\ \vdots & \vdots \\ S_{n-1}(x), & x \in [t_{n-1}, t_n] \end{cases} \tag{1}$$

where

$$S_i(x) = a_i x + b_i \tag{2}$$

because each piece of $S(x)$ is a linear polynomial. Such a function $S(x)$ is **piecewise linear**. If the knots $t_0, t_1, \dots, t_n$ were given and if the coefficients $a_0, b_0, a_1, b_1, \dots, a_{n-1}, b_{n-1}$ were all known, then the evaluation of $S(x)$ at a specific x would proceed by first determining the interval that contains x and then using the appropriate linear function for that interval.

If the function S defined by Equation (1) is continuous, we call it a **first-degree spline**. It is characterized by the following three properties.

■ **Definition 1**

Spline of Degree 1

Spline Degree 1

A function S is called a **spline of degree 1** if:

1. The domain of S is an interval $[a, b]$.
2. S is continuous on $[a, b]$.
3. There is a partitioning of the interval $a = t_0 < t_1 < \cdots < t_n = b$ such that S is a linear polynomial on each subinterval $[t_i, t_{i+1}]$.

Outside the interval $[a, b]$, $S(x)$ is usually defined to be the same function on the left of a as it is on the leftmost subinterval $[t_0, t_1]$ and the same on the right of b as it is on the rightmost subinterval $[t_{n-1}, t_n]$, namely, $S(x) = S_0(x)$ when $x < a$ and $S(x) = S_{n-1}(x)$ when $x > b$.

Continuity of a function f at a point s can be defined by the condition

Continuity Conditions

$$\lim_{x \to s^+} f(x) = f(s) = \lim_{x \to s^-} f(x)$$

Here, $\lim_{x \to s^+}$ means that the limit is taken over x values that converge to s from above s; that is, $(x - s)$ is positive for all x values. Similarly, $\lim_{x \to s^-}$ means that the x values converge to s from below.

EXAMPLE 1 Determine whether this function is a first-degree spline function:

$$S(x) = \begin{cases} x, & x \in [-1, 0] \\ 1 - x, & x \in (0, 1) \\ 2x - 2, & x \in [1, 2] \end{cases}$$

Solution The function is obviously piecewise linear, but is not a spline of degree 1 because it is discontinuous at $x = 0$. Notice that $\lim_{x \to 0^+} S(x) = \lim_{x \to 0}(1 - x) = 1$, whereas $\lim_{x \to 0^-} S(x) = \lim_{x \to 0} x = 0$. ■

The spline functions of degree 1 can be used for interpolation. Suppose the following table of function values is given:

Interpolating Spline

x	t_0	t_1	$\cdots$	t_n
y	y_0	y_1	$\cdots$	y_n

There is no loss of generality in supposing that $t_0 < t_1 < \cdots < t_n$ because this is only a matter of labeling the knots.

The table can be represented by a set of $n + 1$ points in the plane, (t_0, y_0), (t_1, y_1), ..., (t_n, y_n), and these points have distinct abscissas. Therefore, we can draw a polygonal line through the points without ever drawing a *vertical* segment. This polygonal line is the graph of a function, and this function is obviously a spline of degree 1. *What are the equations of the individual line segments that make up this graph?*

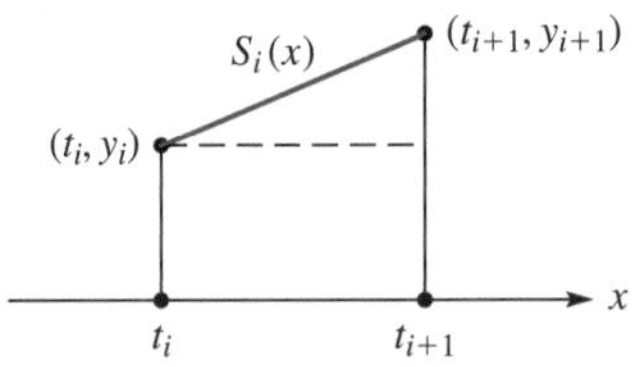

FIGURE 6.3 First-degree spline segment: linear $S_i(x)$

By referring to Figure 6.3 and using the point-slope form of a line, we obtain

Spline Segment

$$S_i(x) = y_i + m_i(x - t_i) \tag{3}$$

on the interval $[t_i, t_{i+1}]$, where m_i is the **slope** of the line and is therefore given by the formula

$$m_i = \frac{y_{i+1} - y_i}{t_{i+1} - t_i}$$

Number of Conditions

Notice that the function S that we are creating has $2n$ parameters in it: the n coefficients a_i and the n constants b_i in Equation (2). On the other hand, exactly $2n$ conditions are being imposed, since each constituent function S_i must interpolate the data at the ends of its subinterval. Thus, the number of parameters equals the number of conditions. For the higher-degree splines, we shall encounter a mismatch in these two numbers; the spline of degree k have $k - 1$ free parameters for us to use as we wish in the problem of interpolating at the knots.

The form of Equation (3) is better than that of Equation (2) for the practical evaluation of $S(x)$ because some of the quantities $x - t_i$ must be computed in any case simply to determine which subinterval contains x. If $t_0 \leq x \leq t_n$ then the interval $[t_i, t_{i+1}]$ containing x

is characterized by the fact that $x - t_i$ is the first of the quantities $x - t_{n-1}, x - t_{n-2}, \dots, x - t_0$ that is *nonnegative*.

The following is a function procedure that utilizes $n + 1$ table values (t_i, y_i) in linear arrays (t_i) and (y_i), assuming that $a = t_0 < t_1 < \cdots < t_n = b$. Given an x value, the routine returns $S(x)$ using Equations (1) and (3). If $x < t_0$, then $S(x) = y_0 + m_0(x - t_0)$; if $x > t_n$, then $S(x) = y_{n-1} + m_{n-1}(x - t_{n-1})$.

***Spline1* Pseudocode**

```
real function Spline1(n, (t_i), (y_i), x)
integer i, n;   real x;   real array (t_i)_0:n, (y_i)_0:n
for i = n − 1 to 0 step −1
    if x − t_i ≧ 0 then exit loop
end for
Spline1 ← y_i + (x − t_i)[(y_{i+1} − y_i)/(t_{i+1} − t_i)]
end function Spline1
```

Modulus of Continuity

To assess the **goodness of fit** when we interpolate a function with a first-degree spline, it is useful to have something called the **modulus of continuity** of a function f. Suppose f is defined on an interval $[a, b]$. The **modulus of continuity** of f is

Goodness of Fit

$$\omega(f; h) = \sup\{|f(u) - f(v)|: a \leqq u \leqq v \leqq b, |u - v| \leqq h\}$$

Here, sup is the **supremum**, which is the least upper bound of the given set of real numbers. The quantity $\omega(f; h)$ measures how much f can change over a small interval of width h. If f is continuous on $[a, b]$, then it is uniformly continuous, and $\omega(f; h)$ tends to zero as h tends to zero. If f is not continuous, $\omega(f; h)$ does not tend to zero. If f is differentiable on (a, b) (in addition to being continuous on $[a, b]$) and if $f'(x)$ is bounded on (a, b), then the Mean Value Theorem can be used to get an estimate of the modulus of continuity: If u and v are as described in the definition of $\omega(f; h)$, then

$$|f(u) - f(v)| = |f'(c)(u - v)| \leqq M_1|u - v| \leqq M_1 h$$

Here, M_1 denotes the maximum of $|f'(x)|$ as x runs over (a, b). For example, if $f(x) = x^3$ and $[a, b] = [1, 4]$, then we find that $\omega(f; h) \leqq 48h$.

■ Theorem 1 **First-Degree Polynomial Accuracy Theorem**

If p is the first-degree polynomial that interpolates a function f at the endpoints of an interval $[a, b]$, then with $h = b - a$, we have

$$|f(x) - p(x)| \leqq \omega(f; h) \quad (a \leqq x \leqq b)$$

Proof The linear function p is given explicitly by the formula

$$p(x) = \left(\frac{x - a}{b - a}\right) f(b) + \left(\frac{b - x}{b - a}\right) f(a)$$

Hence, we have

$$f(x) - p(x) = \left(\frac{x - a}{b - a}\right)[f(x) - f(b)] + \left(\frac{b - x}{b - a}\right)[f(x) - f(a)]$$

Then we have

$$\begin{aligned}|f(x) - p(x)| &\leqq \left(\frac{x-a}{b-a}\right)|f(x) - f(b)| + \left(\frac{b-x}{b-a}\right)|f(x) - f(a)| \\ &\leqq \left(\frac{x-a}{b-a}\right)\omega(f;h) + \left(\frac{b-x}{b-a}\right)\omega(f;h) \\ &= \left[\left(\frac{x-a}{b-a}\right) + \left(\frac{b-x}{b-a}\right)\right]\omega(f;h) = \omega(f;h)\end{aligned}$$

■

From this basic result, we can easily prove the following one, simply by applying the basic inequality to each subinterval.

■ **Theorem 2**

First-Degree Spline Accuracy Theorem

Let p be a first-degree spline having knots $a = x_0 < x_1 < \cdots < x_n = b$. If p interpolates a function f at these knots, then with $h = \max_i(x_i - x_{i-1})$, we have

$$|f(x) - p(x)| \leqq \omega(f;h) \quad (a \leqq x \leqq b)$$

If f' or f'' exist and are continuous, then more can be said, namely,

Upper Bounds

$$|f(x) - p(x)| \leqq M_1\frac{h}{2} \quad (a \leqq x \leqq b)$$

$$|f(x) - p(x)| \leqq M_2\frac{h^2}{8} \quad (a \leqq x \leqq b)$$

In these estimates, M_1 is the maximum value of $|f'(x)|$ on the interval, and M_2 is the maximum of $|f''(x)|$.

The first theorem tells us that if more knots are inserted in such a way that the maximum spacing h goes to zero, then the corresponding first-degree spline converges uniformly to f. Recall that this type of result is conspicuously lacking in the polynomial interpolation theory. In that situation, raising the degree and making the nodes fill up the interval does not necessarily ensure that convergence takes place for an arbitrary continuous function. (See Section 6.2.)

Second-Degree Splines

Splines of degree higher than 1 are more complicated. We now take up the quadratic splines. Let's use the letter Q to remind ourselves that we are considering piecewise quadratic functions. A function Q is a **second-degree spline** if it has the following properties.

■ **Definition 2**

Spline of Degree 2

A function Q is called a **spline of degree 2** if:

1. The domain of Q is an interval $[a, b]$.
2. Q and Q' are continuous on $[a, b]$.
3. There are points t_i (called **knots**) such that $a = t_0 < t_1 < \cdots < t_n = b$ and Q is a polynomial of degree at most 2 on each subinterval $[t_i, t_{i+1}]$.

Quadratic Spline

In brief, a quadratic spline is a continuously differentiable piecewise quadratic function, where *quadratic* includes all linear combinations of the basic monomials $x \mapsto 1, x, x^2$.

EXAMPLE 2 Determine whether the following function is a quadratic spline:

$$Q(x) = \begin{cases} x^2 & (-10 \leqq x \leqq 0) \\ -x^2 & (0 \leqq x \leqq 1) \\ 1 - 2x & (1 \leqq x \leqq 20) \end{cases}$$

Solution The function is obviously piecewise quadratic. Whether Q and Q' are continuous at the interior knots can be determined as follows:

$$\begin{aligned} \lim_{x \to 0^-} Q(x) &= \lim_{x \to 0^-} x^2 = 0, & \lim_{x \to 0^+} Q(x) &= \lim_{x \to 0^+} (-x^2) = 0 \\ \lim_{x \to 1^-} Q(x) &= \lim_{x \to 1^-} (-x^2) = -1, & \lim_{x \to 1^+} Q(x) &= \lim_{x \to 1^+} (1 - 2x) = -1 \\ \lim_{x \to 0^-} Q'(x) &= \lim_{x \to 0^-} 2x = 0, & \lim_{x \to 0^+} Q'(x) &= \lim_{x \to 0^+} (-2x) = 0 \\ \lim_{x \to 1^-} Q'(x) &= \lim_{x \to 1^-} (-2x) = -2, & \lim_{x \to 1^+} Q'(x) &= \lim_{x \to 1^+} (-2) = -2 \end{aligned}$$

Consequently, $Q(x)$ is a quadratic spline. ■

Interpolating Quadratic Splines

Quadratic splines are not used in applications as often as are natural cubic splines, which are developed in Section 6.2. However, the derivations of interpolating quadratic and cubic splines are similar enough that an understanding of the simpler second-degree spline theory allows us to grasp easily the more complicated third-degree spline theory. We want to emphasize that quadratic splines are rarely used for interpolation, and the discussion here is provided only as preparation for the study of higher-order splines, which are used in many applications.

Proceeding now to the interpolation problem, suppose that a table of values has been given:

Interpolating Quadratic Spline

x	t_0	t_1	t_2	$\cdots$	t_n
y	y_0	y_1	y_2	$\cdots$	y_n

We shall assume that the points $t_0, t_1, \dots, t_n$, which we think of as the **nodes** for the interpolation problem, are also the **knots** for the spline function to be constructed. Later, another quadratic spline interpolant is discussed in which the nodes for interpolation are different from the knots!

Counting Conditions

A quadratic spline, as just described, consists of n separate quadratic functions $x \mapsto a_i x^2 + b_i x + c_i$, one for each subinterval created by the $n + 1$ knots. Thus, we start with $3n$ coefficients. On each subinterval $[t_i, t_{i+1}]$, the quadratic spline function Q_i must satisfy the interpolation conditions $Q_i(t_i) = y_i$ and $Q_i(t_{i+1}) = y_{i+1}$. Since there are n such subintervals, this imposes $2n$ conditions. The continuity of Q does *not* add any additional conditions. (Why?) However, the continuity of Q' at each of the interior knots gives $n - 1$ more conditions. Thus, we have $2n + n - 1 = 3n - 1$ conditions, or *one* condition short of the $3n$ conditions required. There are a variety of ways to impose this additional condition; for example, $Q'(t_0) = 0$ or $Q''_0 = 0$.

We now derive the equations for the interpolating quadratic spline, $Q(x)$. The value of $Q'(t_0)$ is prescribed as the additional condition. We seek a piecewise quadratic function

Piecewise Quadratic Function

$$Q(x) = \begin{cases} Q_0(x) & (t_0 \leqq x \leqq t_1) \\ Q_1(x) & (t_1 \leqq x \leqq t_2) \\ \vdots & \vdots \\ Q_{n-1}(x), & (t_{n-1} \leqq x \leqq t_n) \end{cases} \tag{4}$$

which is continuously differentiable on the entire interval $[t_0, t_n]$ and which interpolates the table; that is, $Q(t_i) = y_i$ for $0 \leqq i \leqq n$.

Since Q' is continuous, we can put $z_i \equiv Q'(t_i)$. At present, we do not know the correct values of z_i; nevertheless, the following must be the formula for Q_i:

$Q_i(x)$ Linear Equation

$$Q_i(x) = \frac{z_{i+1} - z_i}{2(t_{i+1} - t_i)}(x - t_i)^2 + z_i(x - t_i) + y_i \tag{5}$$

To see that this is correct, verify that $Q_i(t_i) = y_i$, $Q_i'(t_i) = z_i$, and $Q_i'(t_{i+1}) = z_{i+1}$. These three conditions define the function Q_i uniquely on $[t_i, t_{i+1}]$ as given in Equation (5).

Now, for the quadratic spline function Q to be continuous and to interpolate the table of data, it is necessary and sufficient that $Q_i(t_{i+1}) = y_{i+1}$ for $i = 0, 1, \dots, n-1$ in Equation (5). When this equation is written out in detail and simplified, the result is

$$z_{i+1} = -z_i + 2\left(\frac{y_{i+1} - y_i}{t_{i+1} - t_i}\right) \qquad (0 \leqq i \leqq n-1) \tag{6}$$

This equation can be used to obtain the vector $[z_0, z_1, \dots, z_n]^T$, starting with an arbitrary value for z_0. We summarize with an algorithm:

■ **Algorithm**

Quadratic Spline Interpolation at the Knots

1. Determine $[z_0, z_1, \dots, z_n]^T$ by selecting z_0 arbitrarily and computing $z_1, z_2, \dots, z_n$ recursively by Formula (6).
2. The quadratic spline interpolating function Q is given by Formulas (4) and (5).

EXAMPLE 3 For the five data points (0, 8), (1, 12), (3, 2), (4, 6), (8, 0), construct the linear spline S and the quadratic spline Q.

Solution Figure 6.4 illustrates graphically these two low order spline curves. They fit better than the interpolating polynomials in Figure 4.6 (p. 179) with regard to reduced oscillations. ■

Subbotin Quadratic Splines

Counting Conditions

A useful approximation process, first proposed by Subbotin [1967], consists of interpolation with *quadratic* splines, where the nodes for interpolation are chosen to be the first and last knots and the midpoints between the knots. Remember that **knots** are defined as the points where the spline function is permitted to change in form from one polynomial to another. The **nodes** are the points where values of the spline are specified. In the Subbotin quadratic spline function, there are $n + 2$ interpolation conditions and $2(n - 1)$ conditions from the continuity of Q and Q'. Hence, we have the exact number of conditions needed, $3n$, to define the quadratic spline function completely.

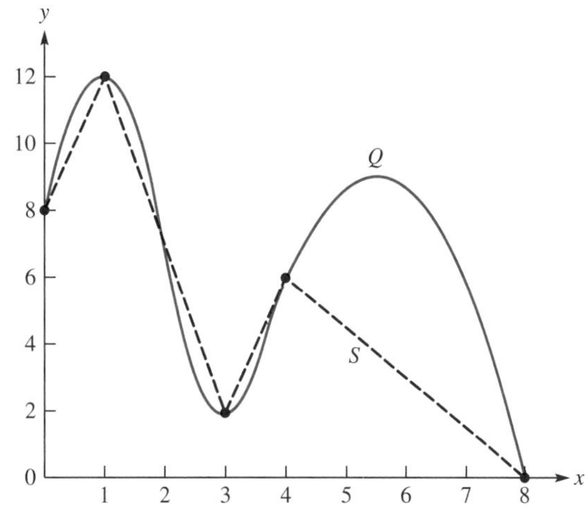

FIGURE 6.4
First-degree and second-degree spline functions

We outline the theory here, leaving details for the reader to fill in. Suppose that knots $a = t_0 < t_1 < \cdots < t_n = b$ have been specified; let the nodes be the points

Nodes

$$\begin{cases} \tau_0 = t_0, \quad \tau_{n+1} = t_n \\ \tau_i = \frac{1}{2}(t_i + t_{i-1}) \quad (1 \leqq i \leqq n) \end{cases}$$

We seek a quadratic spline function Q that has the given knots and takes prescribed values at the nodes:

Knots

$$Q(\tau_i) = y_i \quad (0 \leqq i \leqq n+1)$$

as in Figure 6.5.

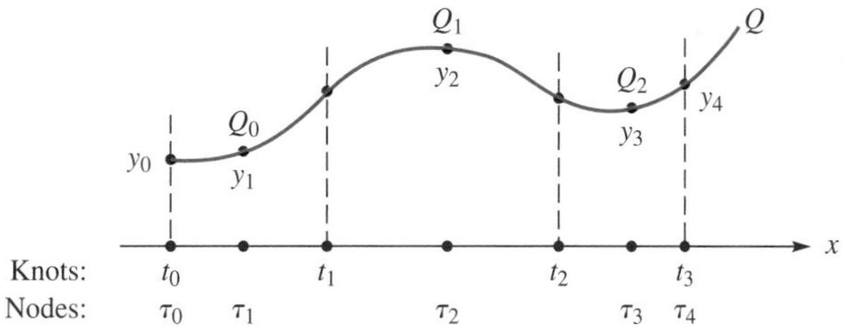

FIGURE 6.5
Subbotin quadratic splines
$(t_0 = \tau_0, t_3 = \tau_4)$

The knots create n subintervals, and in each of them, Q can be a different quadratic polynomial. Let's say that on $[t_i, t_{i+1}]$, Q is equal to the quadratic polynomial Q_i. Since Q is a quadratic spline, it and its first derivative should be continuous. Thus, $z_i \equiv Q'(t_i)$ is well defined, although as yet we do not know its values. It is easy to see that on $[t_i, t_{i+1}]$, our quadratic polynomial can be represented in the form

$Q_i(x)$ Segment

$$Q_i(x) = y_{i+1} + \frac{1}{2}(z_{i+1} + z_i)(x - \tau_{i+1}) + \frac{1}{2h_i}(z_{i+1} - z_i)(x - \tau_{i+1})^2 \tag{7}$$

in which $h_i = t_{i+1} - t_i$. To verify the correctness of Equation (7), we must check that $Q_i(\tau_{i+1}) = y_{i+1}$, $Q_i'(t_i) = z_i$, and $Q_i'(t_{i+1}) = z_{i+1}$. When the polynomial pieces Q_0, Q_1, $\ldots$, Q_{n-1} are joined together to form Q, the result may be discontinuous. Hence, we impose continuity conditions at the interior knots:

$$\lim_{x \to t_i^-} Q_{i-1}(x) = \lim_{x \to t_i^+} Q_i(x), \quad (1 \leqq i \leqq n-1)$$

The reader should carry out this analysis, which leads to

$$h_{i-1}z_{i-1} + 3(h_{i-1} + h_i)z_i + h_i z_{i+1} = 8(y_{i+1} - y_i) \quad (1 \leqq i \leqq n-1) \tag{8}$$

The first and last interpolation conditions must also be imposed:

$$Q(\tau_0) = y_0 \quad Q(\tau_{n+1}) = y_{n+1}$$

These two equations lead to

$$\begin{aligned} 3h_0 z_0 + h_0 z_1 &= 8(y_1 - y_0) \\ h_{n-1}z_{n-1} + 3h_{n-1}z_n &= 8(y_{n+1} - y_n) \end{aligned}$$

The system of equations governing the vector $\mathbf{z} = [z_0, z_1, \dots, z_n]^T$ then can be written in the matrix form

Tridiagonal Linear System

$$\begin{bmatrix} 3h_0 & h_0 & & & & & \\ h_0 & 3(h_0+h_1) & h_1 & & & & \\ & h_1 & 3(h_1+h_2) & h_2 & & & \\ & & \ddots & \ddots & \ddots & & \\ & & & h_{n-2} & 3(h_{n-2}+h_{n-1}) & h_{n-1} \\ & & & & h_{n-1} & 3h_{n-1} \end{bmatrix} \begin{bmatrix} z_0 \\ z_1 \\ z_2 \\ \vdots \\ z_{n-1} \\ z_n \end{bmatrix} = 8 \begin{bmatrix} y_1 - y_0 \\ y_2 - y_1 \\ y_3 - y_2 \\ \vdots \\ y_n - y_{n-1} \\ y_{n+1} - y_n \end{bmatrix}$$

This system of $n+1$ equations in $n+1$ unknowns can be conveniently solved by procedure *Tri* in Section 2.3. After the z vector has been obtained, values of $Q(x)$ can be computed from Equation (7). The writing of suitable code to carry out this interpolation method is left as a programming exercise.

Summary 6.1

- We are given $n+1$ pairs of points (t_i, y_i) with distinct **knots** $a = t_0 < t_1 < \cdots < t_{n-1} < t_n = b$ over the interval $[a, b]$. A **first-degree spline function** S is a piecewise linear polynomial defined on the interval $[a, b]$ so that it is continuous. It has the form

$$S(x) = \begin{cases} S_0(x), & x \in [t_0, t_1] \\ S_1(x), & x \in [t_1, t_2] \\ \vdots & \vdots \\ S_{n-1}(x), & x \in [t_{n-1}, t_n] \end{cases}$$

  where

$$S_i(x) = y_i + \left(\frac{y_{i+1} - y_i}{t_{i+1} - t_i}\right)(x - t_i)$$

  on the interval $[t_i, t_{i+1}]$. Clearly, $S(x)$ is continuous, since $S_{i-1}(t_i) = S_i(t_i) = y_i$ for $1 \leqq i \leqq n$.

- A **second-degree spline function** Q is a piecewise quadratic polynomial with Q and Q' continuous on the interval $[a, b]$. It has the form

$$Q(x) = \begin{cases} Q_0(x) & x \in [t_0, t_1] \\ Q_1(x) & x \in [t_1, t_2] \\ \vdots & \vdots \\ Q_{n-1}(x) & x \in [t_{n-1}, t_n] \end{cases}$$

where

$$Q_i(x) = \left(\frac{z_{i+1} - z_i}{2(t_{i+1} - t_i)}\right)(x - t_i)^2 + z_i(x - t_i) + y_i$$

on the interval $[t_i, t_{i+1}]$. The coefficients $z_0, z_1, \ldots, z_n$ are obtained by selecting z_0 and then using the recurrence relation

$$z_{i+1} = -z_i + 2\left(\frac{y_{i+1} - y_i}{t_{i+1} - t_i}\right) \qquad (0 \leqq i \leqq n - 1)$$

- A **Subbotin quadratic spline function** Q is a piecewise quadratic polynomial with Q and Q' continuous on the interval $[a, b]$ and with interpolation condition at the endpoints of the interval $[a, b]$ and at the midpoints of the subintervals, namely, $Q(\tau_i) = y_i$ for $0 \leqq i \leqq n + 1$, where

$$\tau_0 = t_0, \qquad \tau_i = \frac{1}{2}(t_i + t_{i-1}) \quad (1 \leqq i \leqq n), \qquad \tau_{n+1} = t_n$$

It has the form

$$Q_i(x) = y_{i+1} + \frac{1}{2}(z_{i+1} + z_i)(x - \tau_{i+1}) + \frac{1}{2h_i}(z_{i+1} - z_i)(x - \tau_{i+1})^2$$

where $h_i = t_{i+1} - t_i$. The coefficients z_i are found by solving the tridiagonal system

$$\begin{cases} 3h_0 z_0 + h_0 z_1 = 8(y_1 - y_0) \\ h_{i-1}z_{i-1} + 3(h_{i-1} + h_i)z_i + h_i z_{i+1} = 8(y_{i+1} - y_i) \quad (1 \leqq i \leqq n - 1) \\ h_{n-1}z_{n-1} + 3h_{n-1}z_n = 8(y_{n+1} - y_n) \end{cases}$$

as discussed in Section 2.3.

Exercises 6.1

[a]1. Determine whether this function is a first-degree spline:

$$S(x) = \begin{cases} x & (-1 \leqq x \leqq 0.5) \\ 0.5 + 2(x - 0.5) & (0.5 \leqq x \leqq 2) \\ x + 1.5 & (2 \leqq x \leqq 4) \end{cases}$$

2. The simplest type of spline function is the piecewise constant function, which could be defined as

$$S(x) = \begin{cases} c_0 & (t_0 \leqq x < t_1) \\ c_1 & (t_1 \leqq x < t_2) \\ \vdots & \vdots \\ c_{n-1} & (t_{n-1} \leqq x \leqq t_n) \end{cases}$$

Show that the indefinite integral of such a function is a polygonal function. What is the relationship between the piecewise constant functions and the rectangle rule of numerical integration? (See Exercise 5.1.34.)

3. Show that $f(x) - p(x) = \frac{1}{2}f''(\xi)(x-a)(x-b)$ for some ξ in the interval (a, b), where p is a linear polynomial that interpolates f at a and b.
Hint: Use a result from Section 4.2.

4. (Continuation) Show that $|f(x) - p(x)| \leqq \frac{1}{8}M\ell^2$, where $\ell = b - a$, if $|f''(x)| \leqq M$ on the interval (a, b).

5. (Continuation) Show that

$$f(x) - p(x) = \frac{(x-a)(x-b)}{b-a}\left[\frac{f(x) - f(b)}{x - b} - \frac{f(x) - f(a)}{x - a}\right]$$

[a]6. (Continuation) If $|f'(x)| \leqq C$ on (a, b), show that $|f(x) - p(x)| \leqq C\ell/2$.
Hint: Use the Mean-Value Theorem on the result of the preceding problem.

7. (Continuation) Let S be a spline function of degree 1 that interpolates f at $t_0, t_1, \ldots, t_n$. Let $t_0 < t_1 < \cdots < t_n$ and let $\delta = \max_{0 \leqq i \leqq n-1}(t_{i+1} - t_i)$. Then $|f(x) - S(x)| \leqq C\delta/2$, where C is an upper bound of $|f'(x)|$ on (t_0, t_n).

8. Let f be continuous on $[a, b]$. For a given $\varepsilon > 0$, let δ have the property that $|f(x) - f(y)| < \varepsilon$ whenever $|x - y| < \delta$ (uniform continuity principle). Let $n > 1 + (b - a)/\delta$. Show that there is a first-degree spline S having n knots such that $|f(x) - S(x)| < \varepsilon$ on $[a, b]$.
Hint: Use Exercise 6.1.5 and assume equally spaced knots.

[a]**9.** If the function $f(x) = \sin(100x)$ is to be approximated on the interval $[0, \pi]$ by an interpolating spline of degree 1, how many knots are needed to ensure that $|S(x) - f(x)| < 10^{-8}$? (Use equally spaces knots.)
Hint: Use Exercise 6.1.7.

[a]**10.** Let $t_0 < t_1 < \cdots < t_n$. Construct first-degree spline functions $G_0, G_1, \ldots, G_n$ by requiring that G_i vanish at $t_0, t_1, \ldots, t_{i-1}, t_{i+1}, \ldots, t_n$ but that $G_i(t_i) = 1$. Show that the first-degree spline function that interpolates f at $t_0, t_1, \ldots, t_n$ is $\sum_{i=0}^{n} f(t_i)G_i(x)$.

11. Show that the trapezoid rule for numerical integration (Section 5.1) results from approximating f by a first-degree spline S and then using

$$\int_a^b f(x)\,dx \approx \int_a^b S(x)\,dx$$

[a]**12.** Prove that the derivative of a quadratic spline is a first-degree spline.

13. If the knots t_i happen to be the integers $0, 1, \ldots, n$, find a good way to determine the index i for which $t_i \leqq x < t_{i+1}$. (*Note*: This problem is deceptive, for the word *good* can be given different meanings.)

14. Show that the indefinite integral of a first-degree spline is a second-degree spline.

15. Define $f(x) = 0$ if $x < 0$ and $f(x) = x^2$ if $x \geqq 0$. Show that f and f' are continuous. Show that any quadratic spline with knots $t_0, t_1, \ldots, t_n$ is of the form

$$ax^2 + bx + c + \sum_{i=1}^{n-1} d_i f(x - t_i)$$

16. Define a function g by the equation

$$g(x) = \begin{cases} 0 & (t_0 \leqq x \leqq 0) \\ x & (0 \leqq x \leqq t_n) \end{cases}$$

Prove that every first-degree spline function that has knots $t_0, t_1, \ldots, t_n$ can be written in the form

$$ax + b + \sum_{i=1}^{n-1} c_i g(x - t_i)$$

[a]**17.** Find a quadratic spline interpolant for these data:

x	-1	0	$\frac{1}{2}$	1	2	$\frac{5}{2}$
y	2	1	0	1	2	3

Assume that $z_0 = 0$.

18. (Continuation) Show that no quadratic spline Q interpolates the table of the preceding problem and satisfies $Q'(t_0) = Q'(t_5)$.

[a]**19.** What equations must be solved if a quadratic spline function Q that has knots $t_0, t_1, \ldots, t_n$ is required to take prescribed values at points $\frac{1}{2}(t_i + t_{i+1})$ for $0 \leqq i \leqq n-1$?

20. Are these functions quadratic splines? Explain why or why not.

[a]**a.** $Q(x) = \begin{cases} 0.1x^2 & (0 \leqq x \leqq 1) \\ 9.3x^2 - 18.4x + 9.2 & (1 \leqq x \leqq 1.3) \end{cases}$

[a]**b.** $Q(x) = \begin{cases} -x^2 & (-100 \leqq x \leqq 0) \\ x & (0 \leqq x \leqq 100) \end{cases}$

[a]**c.** $Q(x) = \begin{cases} x & (-50 \leqq x \leqq 1) \\ x^2 & (1 \leqq x \leqq 2) \\ 4 & (2 \leqq x \leqq 50) \end{cases}$

[a]**21.** Is $S(x) = |x|$ a first-degree spline? Why or why not?

22. Verify that Formula (5) has the three properties $Q_i(t_i) = y_i$, $Q_i'(t_i) = z_i$, and $Q_i'(t_{i+1}) = z_{i+1}$.

23. (Continuation) Impose the continuity condition on Q and derive the system of Equation (6).

24. Show by induction that the recursive Formula (6) together with Equation (5) produces an interpolating quadratic spline function.

25. Verify the correctness of the equations in the text that pertain to Subbotin's spline interpolation process.

26. Analyze the Subbotin interpolation scheme in this alternative manner. First, let $v_i = Q(t_i)$. Show that

$$Q_i(x) = A_i(x - t_i)^2 + B_i(x - t_{i+1})^2 + C_i$$

where

$$C_i = 2y_i - \frac{1}{2}v_i - \frac{1}{2}v_{i+1}, \quad B_i = \frac{v_i - C_i}{h_i^2}$$

$$A_i = \frac{v_{i+1} - C_i}{h_i^2}, \quad h_i = t_{i+1} - t_i$$

Hint: Show that $Q_i(t_i) = v_i$, $Q_i(t_{i+1}) = v_{i+1}$, and $Q_i(\tau_i) = y_i$.

27. (Continuation) When continuity conditions on Q' are imposed, show that the result is the following equation, in which $i = 1, 2, \ldots, n-1$:

$$h_i v_{i-1} + 3(h_i + h_{i+1})v_i + h_{i-1}v_{i+1} = 4h_{i-1}y_i + 4h_i y_{i-1}$$

28. (**Student Research Project**) It is commonly accepted that Schoenberg's [1946] paper is the first mathematical reference in which the word *spline* is used in connection with smooth, piecewise polynomial approximations. However, the word *spline* as a thin strip of wood used by a draftsman dates back to the 1890s at least. Many of the ideas used in spline theory have their roots in work done in various industries such as the building of aircraft, automobiles, and ships in which splines are used extensively. Research and write a paper on the history of splines. (See the mathematical history of splines in the automobile industry.)

Computer Exercises 6.1

1. Rewrite procedure *Spline1* so that ascending subintervals are considered instead of descending ones. Test the code on a table of 15 unevenly spaced data points.

2. Rewrite procedure *Spline1* so that a **binary search** is used to find the desired interval. Test the revised code. What are the advantages and disadvantages of a binary search compared to the procedure in the text? A binary search is similar to the bisection method in that we choose t_k with $k = (i + j)/2$ or $k = (i + j + 1)/2$ and determine whether x is in $[t_i, t_k]$ or $[t_k, t_j]$.

3. A **piecewise bilinear polynomial** that interpolates points (x, y) specified in a rectangular grid is given by

$$p(x, y) = \frac{(\ell_{ij} z_{i+1,\, j+1} + \ell_{i+1,\, j+1} z_{ij})}{(x_{i+1} - x_i)(y_{j+1} - y_j)} - \frac{(\ell_{i+1,\, j} z_{i,\, j+1} + \ell_{i,\, j+1} z_{i+1,\, j})}{(x_{i+1} - x_i)(y_{j+1} - y_j)}$$

where $\ell_{ij} = (x_i - x)(y_j - y)$. Here $x_i \leqq x \leqq x_{i+1}$ and $y_j \leqq y \leqq y_{j+1}$. The given grid (x_i, y_j) is specified by strictly increasing arrays (x_i) and (y_j) of length n and m, respectively. The given values z_{ij} at the grid points (x_i, y_j) are contained in the $n \times m$ array (z_{ij}), shown in the following figure. Write code for

real function *Bi_Linear*$((x_i), n, (y_j), m, (z_{ij}), x, y)$

to compute the value of $p(x, y)$. Test this routine on a set of 5×10 unequally spaced data points. Evaluate *Bi_Linear* at four grid points and five nongrid points.

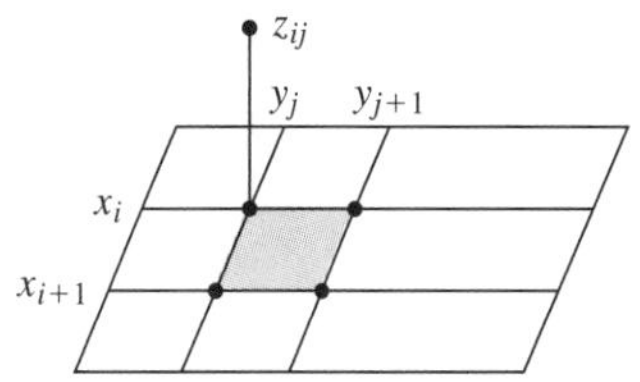

4. Write an adaptive spline interpolation procedure. The input should be a function f, an interval $[a, b]$, and a tolerance ε. The output should be a set of knots $a = t_0 < t_1 < \cdots < t_n = b$ and a set of function values $y_i = f(t_i)$ such that the first-degree spline interpolating function S satisfies $|S(x) - f(x)| \leqq \varepsilon$ whenever x is any point $x_{ij} = t_i + j(t_{i+1} - t_j)/10$ for $0 \leqq i \leqq n - 1$ and $0 \leqq j \leqq 9$.

5. Write computer code for

procedure *Spline2_Coef*$(n, t, (y_i), (z_i))$

that computes the (z_i) array in the quadratic spline interpolation process (interpolation at the knots). Then write

real function *Spline2_Eval*$(n, (t_i), (y_i), (z_i), x)$

that computes values of $Q(x)$.

6. Carry out the programming project of the preceding computer problem for the Subbotin quadratic spline.

6.2 Natural Cubic Splines

Introduction

The first- and second-degree splines discussed in the preceding section, though useful in certain applications, suffer an obvious imperfection: Their low-order derivatives are discontinuous. In the case of the first-degree spline (or polygonal line), this lack of smoothness is immediately evident because the slope of the spline may change abruptly from one value to another at each knot. For the quadratic spline, the discontinuity is in the second derivative and is therefore not so evident. But the **curvature** of the quadratic spline changes abruptly at each knot, and the curve may not be pleasing to the eye.

The general definition of spline functions of arbitrary degree is as follows.

■ **Definition 1**

Spline of Degree *k*

A function S is called a **spline of degree *k*** if:

1. The domain of S is an interval $[a, b]$.
2. $S, S', S'', \ldots, S^{(k-1)}$ are all continuous functions on $[a, b]$.
3. There are points t_i (the knots of S) such that $a = t_0 < t_1 < \cdots < t_n = b$ and such that S is a polynomial of degree at most k on each subinterval $[t_i, t_{i+1}]$.

Observe that no mention has been made of interpolation in the definition of a spline function. Indeed, splines are such versatile functions that they have many applications other than interpolation.

Finding a Continuous Piecewise Polynomial with $k - 1$ Continuous Derivatives

Higher-degree splines are used whenever more smoothness is needed in the approximating function. From the definition of a spline function of degree k, we see that such a function is continuous and has continuous derivatives $S', S'', \ldots, S^{(k-1)}$. If we want the approximating spline to have a continuous mth derivative, a spline of degree at least $m + 1$ is selected. To see why, consider a situation in which knots $t_0 < t_1 < \cdots < t_n$ have been prescribed. Suppose that a piecewise polynomial of degree m is to be defined, with its pieces joined at the knots in such a way that the resulting spline S has m continuous derivatives. At a typical interior knot t, we have the following circumstances: To the left of t, $S(x) = p(x)$; to the right of t, $S(x) = q(x)$, where p and q are mth-degree polynomials. The continuity of the mth derivative $S^{(m)}$ implies the continuity of the lower-order derivatives $S^{(m-1)}, S^{(m-2)}, \ldots, S', S$. Therefore, at the knot t, we have

$$\lim_{x \to t^-} S^{(k)}(x) = \lim_{x \to t^+} S^{(k)}(x) \quad (0 \leqq k \leqq m)$$

from which we conclude that

$$\lim_{x \to t^-} p^{(k)}(x) = \lim_{x \to t^+} q^{(k)}(x) \quad (0 \leqq k \leqq m) \tag{1}$$

Since p and q are polynomials, their derivatives of all orders are continuous, and so Equation (1) is the same as

$$p^{(k)}(t) = q^{(k)}(t) \quad (0 \leqq k \leqq m)$$

This condition forces p and q to be the *same* polynomial because by Taylor's Theorem,

$$p(x) = \sum_{k=0}^{m} \frac{1}{k!} p^{(k)}(t)(x - t)^k = \sum_{k=0}^{m} \frac{1}{k!} q^{(k)}(t)(x - t)^k = q(x)$$

This argument can be applied at each of the interior knots $t_1, t_2, \ldots, t_{n-1}$, and we see that S is simply one polynomial throughout the entire interval from t_0 to t_n. Thus, we need a piecewise polynomial of degree $m+1$ with at most m continuous derivatives to have a spline function that is not just a single polynomial throughout the entire interval. (We already know that ordinary polynomials usually do not serve well in curve fitting. See Section 4.2.)

The choice of degree most frequently made for a spline function is 3. The resulting splines are termed **cubic splines**. In this case, we join cubic polynomials together in such a

Advantages of Cubic Splines

way that the resulting spline function has two continuous derivatives everywhere. At each knot, three continuity conditions are imposed. Since S, S', and S'' are continuous, the graph of the function will appear smooth to the eye. Discontinuities, of course, may occur in the third derivative, but cannot be easily detected visually, which is one reason for choosing degree 3. Experience has shown, moreover, that using splines of degree greater than 3 seldom yields any advantage. For technical reasons, odd-degree splines behave better than even-degree splines (when interpolating at the knots). Finally, a very elegant theorem, to be proved later, shows that in a certain precise sense, the cubic interpolating spline function is the best interpolating function available. Thus, our emphasis on the cubic splines is well justified!

Natural Cubic Splines

We turn next to interpolating a given table of function values by a cubic spline whose knots coincide with the values of the independent variable in the table. As earlier, we start with the table:

Interpolating Cubic Splines

x	t_0	t_1	$\cdots$	t_n
y	y_0	y_1	$\cdots$	y_n

The t_i's are the knots and are assumed to be arranged in ascending order.

The function S that we wish to construct consists of n cubic polynomial pieces:

$$S(x) = \begin{cases} S_0(x) & (t_0 \leqq x \leqq t_1) \\ S_1(x) & (t_1 \leqq x \leqq t_2) \\ \vdots & \vdots \\ S_{n-1}(x) & (t_{n-1} \leqq x \leqq t_n) \end{cases}$$

In this formula, S_i denotes the cubic polynomial that will be used on the subinterval $[t_i, t_{i+1}]$. The interpolation conditions are

Interpolation Conditions

$$S(t_i) = y_i \quad (0 \leqq i \leqq n)$$

The continuity conditions are imposed only at the *interior* knots $t_1, t_2, \ldots, t_{n-1}$. (Why?) These conditions are written as

Continuity Conditions

$$\lim_{x \to t_i^-} S^{(k)}(t_i) = \lim_{x \to t_i^+} S^{(k)}(t_i) \quad (k = 0, 1, 2)$$

It turns out that two more conditions must be imposed to use all the degrees of freedom available. The choice that we make for these two extra conditions is

Two More Conditions

$$S''(t_0) = S''(t_n) = 0 \tag{2}$$

The resulting spline function is then termed a **natural cubic spline**.

Other Types of Splines

Additional ways to close the system of equations for the spline coefficients are **periodic cubic splines** and **clamped cubic splines**. A clamped spline is a spline curve whose slope is fixed at both end points: $S'(t_0) = d_0$ and $S'(t_n) = d_n$. A periodic cubic spline has $S(t_0) = S(t_n)$, $S'(t_0) = S'(t_n)$, and $S''(t_0) = S''(t_n)$. For all continuous differential functions, clamped and natural cubic splines yield the least oscillations about the function f that it interpolates.

We now verify that the number of conditions imposed equals the number of coefficients available. There are $n + 1$ knots and hence n subintervals. On each of these subintervals,

Counting Conditions

we shall have a different cubic polynomial. Since a cubic polynomial has four coefficients, a total of $4n$ coefficients are available. As for the conditions imposed, we have specified that within each interval the interpolating polynomial must go through two points, which gives $2n$ conditions. The continuity adds no additional conditions. The first and second derivatives must be continuous at the $n-1$ interior points, for $2(n-1)$ more conditions. The second derivatives must vanish at the two endpoints for a total of $2n+2(n-1)+2 = 4n$ conditions.

EXAMPLE 1 Derive the equations of the natural cubic interpolating spline for the following table:

x	-1	0	1
y	1	2	-1

Solution Our approach is to determine the parameters $a, b, c, d, e, f, g,$ and h so that $S(x)$ is a natural cubic spline of the form

$$S(x) = \begin{cases} S_0(s) = ax^3 + bx^2 + cx + d & x \in [-1, 0] \\ S_1(s) = ex^3 + fx^2 + gx + h & x \in [0, 1] \end{cases}$$

where the two cubic polynomials are $S_0(x)$ and $S_1(x)$. From these interpolation conditions, we have interpolation conditions $S(-1) = S_0(-1) = -a + b - c + d = 1$, $S(0) = S_0(0) = d = 2$, $S(0) = S_1(0) = h = 2$, and $S(1) = S_1(1) = e + f + g + h = -1$. Taking the first derivatives, we obtain

$$S'(x) = \begin{cases} S_0'(x) = 3ax^2 + 2bx + c \\ S_1'(x) = 3ex^2 + 2fx + g \end{cases}$$

From the continuity condition of S', we have $S_0'(0) = S_1'(0)$, and we set $c = g$. Next taking the second derivatives, we obtain

$$S''(x) = \begin{cases} S_0''(x) = 6ax + 2b \\ S_1''(s) = 6ex + 2f \end{cases}$$

From the continuity condition of S'', we have $S_0''(0) = S_1''(0)$, and we let $b = f$. For S to be a natural cubic spline, we must have $S_0''(-1) = 0$ and $S_1''(1) = 0$, and we obtain $3a = b$ and $3e = -f$. From all of these equations, we obtain $a = -1$, $b = -3$, $c = -1$, $d = 2$, $e = 1$, $f = -3$, $g = -1$, and $h = 2$. ■

Algorithm for Natural Cubic Splines

From the previous example, it is evident that we need to develop a systematic procedure for determining the formula for a natural cubic spline, given a table of interpolation values. This is our objective in the material on the next several pages.

Since S'' is continuous, the numbers

$$z_i \equiv S''(t_i) \quad (0 \leq i \leq n)$$

are unambiguously defined. We do not yet know the values $z_1, z_2, \ldots, z_{n-1}$, but, of course, $z_0 = z_n = 0$ by Equation (2).

If the z_i's were known, we could construct S as now described. On the interval $[t_i, t_{i+1}]$, S'' is a linear polynomial that takes the values z_i and z_{i+1} at the endpoints. Thus, we have

$S_i''(x)$ Linear Equation

$$S_i''(x) = \frac{z_{i+1}}{h_i}(x - t_i) + \frac{z_i}{h_i}(t_{i+1} - x) \tag{3}$$

with $h_i = t_{i+1} - t_i$ for $0 \leqq i \leqq n-1$. To verify that Equation (3) is correct, notice that $S_i''(t_i) = z_i$, $S_i''(t_{i+1}) = z_{i+1}$, and S_i'' is linear in x. If this is integrated twice, we obtain S_i itself:

$$S_i(x) = \frac{z_{i+1}}{6h_i}(x - t_i)^3 + \frac{z_i}{6h_i}(t_{i+1} - x)^3 + cx + d$$

where c and d are constants of integration. By adjusting the integration constants, we obtain a form for S_i that is easier to work with; namely,

$$S_i(x) = \frac{z_{i+1}}{6h_i}(x - t_i)^3 + \frac{z_i}{6h_i}(t_{i+1} - x)^3 + C_i(x - t_i) + D_i(t_{i+1} - x) \tag{4}$$

where C_i and D_i are constants. If we differentiate Equation (4) twice, we obtain Equation (3).

The interpolation conditions $S_i(t_i) = y_i$ and $S_i(t_{i+1}) = y_{i+1}$ can be imposed now to determine the appropriate values of C_i and D_i. The reader should do so (Exercise 6.2.27) and verify that the result is

$S_i(x)$ Cubic Equation

$$\begin{aligned} S_i(x) = {} & \frac{z_{i+1}}{6h_i}(x - t_i)^3 + \frac{z_i}{6h_i}(t_{i+1} - x)^3 \\ & + \left(\frac{y_{i+1}}{h_i} - \frac{h_i}{6}z_{i+1}\right)(x - t_i) + \left(\frac{y_i}{h_i} - \frac{h_i}{6}z_i\right)(t_{i+1} - x) \end{aligned} \tag{5}$$

When the values $z_0, z_1, \ldots, z_n$ have been determined, the spline function $S(x)$ is obtained from equations of this form for $S_0(x), S_1(x), \ldots, S_{n-1}(x)$.

At the interior knots t_i for $1 \leqq i \leqq n-1$, we must have $S_{i-1}(t_i) = S_i(t_i)$, as can be seen in Figure 6.6.

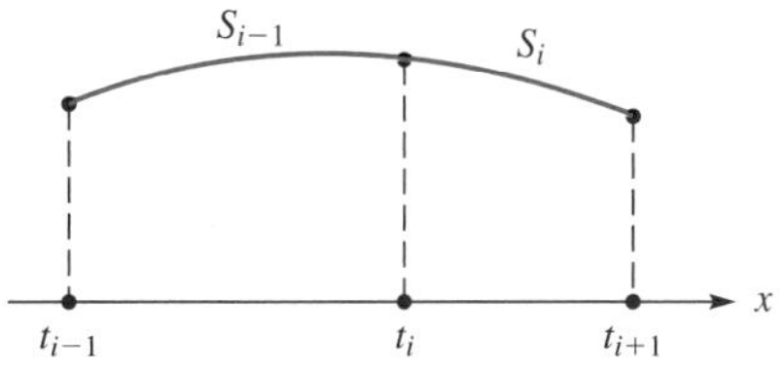

FIGURE 6.6 Cubic spline: adjacent pieces S_{i-1} and S_i

We now show how to determine the z_i's. One condition remains to be imposed—namely, the continuity of S'. From differentiating Equation (5), we obtain

$S_i'(x)$ Quadratic Equation

$$S_i'(x) = \frac{z_{i+1}}{2h_i}(x - t_i)^2 - \frac{z_i}{2h_i}(t_{i+1} - x)^2 + \frac{y_{i+1}}{h_i} - \frac{h_i}{6}z_{i+1} - \frac{y_i}{h_i} + \frac{h_i}{6}z_i$$

This gives

$$S_i'(t_i) = -\frac{h_i}{6}z_{i+1} - \frac{h_i}{3}z_i + b_i \tag{6}$$

where

$$b_i = \frac{1}{h_i}(y_{i+1} - y_i) \tag{7}$$

Analogously, we have

$$S_{i-1}'(t_i) = \frac{h_{i-1}}{6}z_{i-1} + \frac{h_{i-1}}{3}z_i + b_{i-1}$$

When these are set equal to each other, the resulting equation can be rearranged as

$$h_{i-1}z_{i-1} + 2(h_{i-1} + h_i)z_i + h_i z_{i+1} = 6(b_i - b_{i-1})$$

for $1 \leqq i \leqq n-1$. By letting

$$\begin{aligned} u_i &= 2(h_{i-1} + h_i) \\ v_i &= 6(b_i - b_{i-1}) \end{aligned} \tag{8}$$

we obtain a tridiagonal system of linear equations:

Tridiagonal System

$$\begin{cases} z_0 = 0 \\ h_{i-1} z_{i-1} + u_i z_i + h_i z_{i+1} = v_i \quad (1 \leqq i \leqq n-1) \\ z_n = 0 \end{cases} \tag{9}$$

which is to be solved for the z_i's. The simplicity of the first and last equations is a result of the natural cubic spline conditions $S''(t_0) = S''(t_n) = 0$.

EXAMPLE 2 Repeat Example 1 using the natural cubic spline algorithm with points $(-1, 1)$, $(0, 2)$, and $(1, -1)$. Also, plot the results in order to visualize the spline curve.

Solution From the given values, we have $t_0 = -1$, $t_1 = 0$, $t_2 = 1$, $y_0 = 1$, $y_1 = 2$, and $y_2 = -1$. Consequently, we obtain $h_0 = t_1 - t_0 = 1$, $h_1 = t_2 - t_1 = 1$, $b_0 = (y_1 - y_0)/h_0 = 1$, $b_1 = (y_2 - y_1)/h_1 = -3$, $u_1 = 2(h_0 + h_1) = 4$, and $v_1 = 6(b_1 - b_0) = -24$. Then the tridiagonal system of Equation (9) is

$$\begin{cases} z_0 = 0 \\ z_0 + 4z_1 + z_2 = -24 \\ z_2 = 0 \end{cases}$$

Evidently, we obtain the solution $z_0 = 0$, $z_1 = -6$, and $z_2 = 0$. From Equation (5), we have

$$S(x) = \begin{cases} S_0(x) = -(x+1)^3 + 3(x+1) - x & x \in [-1, 0] \\ S_1(x) = -(1-x)^3 - x + 3(1-x) & x \in [0, 1] \end{cases}$$

or

$$S(x) = \begin{cases} S_0(x) = -x^3 - 3x^2 - x + 2 & x \in [-1, 0] \\ S_1(x) = x^3 - 3x^2 - x + 2 & x \in [0, 1] \end{cases}$$

This agrees with the results from Example 1. The resulting natural spline curve through the given points is shown in Figure 6.7.

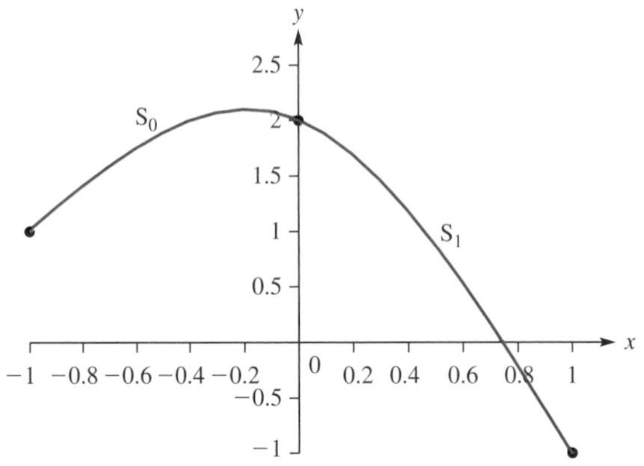

FIGURE 6.7 Natural cubic spline for Examples 1 and 2

■

Algorithm for Solving Tridiagonal System

Now consider System (9) in matrix form:

$$\begin{bmatrix} 1 & 0 & & & & \\ h_0 & u_1 & h_1 & & & \\ & h_1 & u_2 & h_2 & & \\ & & \ddots & \ddots & \ddots & \\ & & & h_{n-2} & u_{n-1} & h_{n-1} \\ & & & & 0 & 1 \end{bmatrix} \begin{bmatrix} z_0 \\ z_1 \\ z_2 \\ \vdots \\ z_{n-1} \\ z_n \end{bmatrix} = \begin{bmatrix} 0 \\ v_1 \\ v_2 \\ \vdots \\ v_{n-1} \\ 0 \end{bmatrix}$$

On eliminating the first and last equations, we have

Symmetric Tridiagonal System

$$\begin{bmatrix} u_1 & h_1 & & & \\ h_1 & u_2 & h_2 & & \\ & \ddots & \ddots & \ddots & \\ & & h_{n-3} & u_{n-2} & h_{n-2} \\ & & & h_{n-2} & u_{n-1} \end{bmatrix} \begin{bmatrix} z_1 \\ z_2 \\ \vdots \\ z_{n-2} \\ z_{n-1} \end{bmatrix} = \begin{bmatrix} v_1 \\ v_2 \\ \vdots \\ v_{n-2} \\ v_{n-1} \end{bmatrix} \tag{10}$$

which is a symmetric tridiagonal system of order $n - 1$. We could use procedure *Tri* developed in Section 2.3 to solve this system. Better yet, we can design an algorithm specifically for it (based on the ideas in Section 2.3). In Gaussian elimination *without pivoting*, the forward elimination phase modifies the u_i's and v_i's as follows:

Forward Elimination

$$\begin{cases} u_i \leftarrow u_i - \dfrac{h_{i-1}^2}{u_{i-1}} \\ v_i \leftarrow v_i - \dfrac{h_{i-1}v_{i-1}}{u_{i-1}} \qquad (i = 2, 3, \dots, n-1) \end{cases}$$

The back substitution phase yields the z's:

Back Substitution

$$\begin{cases} z_{n-1} \leftarrow \dfrac{v_{n-1}}{u_{n-1}} \\ z_i \leftarrow \dfrac{v_i - h_i z_{i+1}}{u_i} \qquad (i = n-2, n-3, \dots, 1) \end{cases}$$

Putting all this together leads to the following algorithm, designed especially for the tridiagonal System (10).

■ **Algorithm**

Solving the Natural Cubic Spline Tridiagonal System Directly

Given the interpolation points (t_i, y_i) for $i = 0, 1, \dots, n$:

1. Compute for $i = 0, 1, \dots, n - 1$:

$$\begin{cases} h_i = t_{i+1} - t_i \\ b_i = \dfrac{1}{h_i}(y_{i+1} - y_i) \end{cases}$$

2. Set

$$\begin{cases} u_1 = 2(h_0 + h_1) \\ v_1 = 6(b_1 - b_0) \end{cases}$$

and compute inductively for $i = 2, 3, \dots, n-1$:

$$\begin{cases} u_i = 2(h_i + h_{i-1}) - \dfrac{h_{i-1}^2}{u_{i-1}} \\ v_i = 6(b_i - b_{i-1}) - \dfrac{h_{i-1}v_{i-1}}{u_{i-1}} \end{cases}$$

3. Set

$$\begin{cases} z_n = 0 \\ z_0 = 0 \end{cases}$$

and compute inductively for $i = n-1, n-2, \dots, 1$:

$$z_i = \frac{v_i - h_i z_{i+1}}{u_i}$$

This algorithm conceivably could fail because of divisions by 0 in Steps 2 and 3. Therefore, let us prove that $u_i \neq 0$ for all i. It is clear that $u_1 > h_1 > 0$. If $u_{i-1} > h_{i-1}$, then $u_i > h_i$ because

$$u_i = 2(h_i + h_{i-1}) - \frac{h_{i-1}^2}{u_{i-1}} > 2(h_i + h_{i-1}) - h_{i-1} > h_i$$

Then by induction, $u_i > 0$ for $i = 1, 2, \dots, n-1$.

Equation (5) is *not* the best computational form for evaluating the cubic polynomial $S_i(x)$. We would prefer to have it in the form

$$S_i(x) = A_i + B_i(x - t_i) + C_i(x - t_i)^2 + D_i(x - t_i)^3 \tag{11}$$

because nested multiplication can then be utilized.

Notice that Equation (11) is the Taylor expansion of S_i about the point t_i. Hence, we obtain

$$A_i = S_i(t_i), \quad B_i = S_i'(t_i), \quad C_i = \tfrac{1}{2}S_i''(t_i), \quad D_i = \tfrac{1}{6}S_i'''(t_i)$$

Therefore, we have $A_i = y_i$ and $C_i = z_i/2$. The coefficient of x^3 in Equation (11) is D_i, whereas the coefficient of x^3 in Equation (5) is $(z_{i+1} - z_i)/6h_i$. Therefore, we obtain

$$D_i = \frac{1}{6h_i}(z_{i+1} - z_i)$$

Finally, Equation (6) provides the value of $S_i'(t_i)$, which is

$$B_i = -\frac{h_i}{6}z_{i+1} - \frac{h_i}{3}z_i + \frac{1}{h_i}(y_{i+1} - y_i)$$

Thus, the nested form of $S_i(x)$ is

$S_i(x)$ Cubic Equation (Nested Multiplication)

$$S_i(x) = y_i + (x - t_i)\left(B_i + (x - t_i)\left(\frac{z_i}{2} + \frac{1}{6h_i}(x - t_i)(z_{i+1} - z_i)\right)\right) \tag{12}$$

Pseudocode for Natural Cubic Splines

We now write routines for determining a natural cubic spline based on a table of values and for evaluating this function at a given value. First, we use Algorithm 1 for directly solving the tridiagonal System (10). This procedure, called *Spline3_Coef*, takes $n + 1$ table values (t_i, y_i) in arrays (t_i) and (y_i) and computes the z_i's, storing them in array (z_i). Intermediate (working) arrays (h_i), (b_i), (u_i), and (v_i) are needed.

***Spline3_Coef* Pseudocode**

```
procedure Spline3_Coef(n, (t_i), (y_i), (z_i))
integer i, n;   real array (t_i)_{0:n}, (y_i)_{0:n}, (z_i)_{0:n}
allocate real array (h_i)_{0:n−1}, (b_i)_{0:n−1}, (u_i)_{1:n−1}, (v_i)_{1:n−1}
for i = 0 to n − 1
    h_i ← t_{i+1} − t_i
    b_i ← (y_{i+1} − y_i)/h_i
end for
u_1 ← 2(h_0 + h_1)
v_1 ← 6(b_1 − b_0)
for i = 2 to n − 1
    u_i ← 2(h_i + h_{i−1}) − h_{i−1}^2/u_{i−1}
    v_i ← 6(b_i − b_{i−1}) − h_{i−1}v_{i−1}/u_{i−1}
end for
z_n ← 0
for i = n − 1 to 1 step −1
    z_i ← (v_i − h_i z_{i+1})/u_i
end for
z_0 ← 0
deallocate array (h_i), (b_i), (u_i), (v_i)
end procedure Spline3_Coef
```

Now a procedure called *Spline3_Eval* is written for evaluating Equation (12), the natural cubic spline function $S(x)$, for x a given value. The procedure *Spline3_Eval* first determines the interval $[t_i, t_{i+1}]$ that contains x and then evaluates $S_i(x)$ using the nested form of this cubic polynomial:

***Spline3_Eval* Pseudocode**

```
real function Spline3_Eval(n, (t_i), (y_i), (z_i), x)
integer i;   real h, tmp
real array (t_i)_{0:n}, (y_i)_{0:n}, (z_i)_{0:n}
for i = n − 1 to 0 step −1
    if x − t_i ≧ 0 then exit loop
end for
h ← t_{i+1} − t_i
tmp ← (z_i/2) + (x − t_i)(z_{i+1} − z_i)/(6h)
tmp ← −(h/6)(z_{i+1} + 2z_i) + (y_{i+1} − y_i)/h + (x − t_i)(tmp)
Spline3_Eval ← y_i + (x − t_i)(tmp)
end function Spline3_Eval
```

The function *Spline3_Eval* can be used repeatedly with different values of x after one call to procedure *Spline3_Coef*. For example, this would be the procedure when plotting a natural cubic spline curve. Since procedure *Spline3_Coef* stores the solution of the tridiagonal system corresponding to a particular spline function in the array (z_i), the arguments n, (t_i), (y_i), and (z_i) must *not* be altered between repeated uses of *Spline3_Eval*.

Using Pseudocode for Interpolating and Curve Fitting

To illustrate the use of the natural cubic spline routines *Spline3_Coef* and *Spline3_Eval*, we rework Example 3 from Section 4.1 (p. 166).

EXAMPLE 3 Write pseudocode for a program that determines the natural cubic spline interpolant for $\sin x$ at 10 equidistant knots in the interval [0, 1.6875]. Over the same interval, subdivide each subinterval into four equally spaced parts, and find the point where the value of $|\sin x - S(x)|$ is largest.

Solution Here is a suitable pseudocode main program, which calls procedures *Spline3_Coef* and *Spline3_Eval*:

***Test_Spline3* Pseudocode**

```
program Test_Spline3
integer i;   real e, h, x
real array (t_i)_{0:n}, (y_i)_{0:n}, (z_i)_{0:n}
integer n ← 9
real a ← 0, b ← 1.6875
h ← (b − a)/n
for i = 0 to n
    t_i ← a + ih
    y_i ← sin(t_i)
end for
call Spline3_Coef(n, (t_i), (y_i), (z_i))
temp ← 0
for j = 0 to 4n
    x ← a + jh/4
    e ← | sin(x) − Spline3_Eval(n, (t_i), (y_i), (z_i), x)|
    if e > temp then temp ← e
    output j, x, e
end for
end Test_Spline3
```

The output is $j = 19$, $x = 0.890625$, and $d = 0.930 \times 10^{-5}$. ■

We can use mathematical software to plot the cubic spline curve for this data.

Mathematical Software

Caution: MATLAB routine `spline` uses the **not-a-knot** end condition!

It dictates that S''' be a single constant in the first two subintervals and another single constant in the last two subintervals. First, the original data are generated. Next, a finer subdivision of the interval $[a, b]$ on the x-axis is made, and the corresponding y-values are obtained from the procedure `spline`. Finally, the original data points and the spline curve are plotted.

Mathematica and Maple can be used to plot cubic spline curves as well as many other computer software packages and programs.

We now illustrate the use of spline functions in fitting a curve to a set of data. Consider the following table:

Sample Data Set

x	0.0	0.6	1.5	1.7	1.9	2.1	2.3	2.6	2.8	3.0
y	−0.8	−0.34	0.59	0.59	0.23	0.1	0.28	1.03	1.5	1.44

3.6	4.7	5.2	5.7	5.8	6.0	6.4	6.9	7.6	8.0
0.74	−0.82	−1.27	−0.92	−0.92	−1.04	−0.79	−0.06	1.0	0.0

These 20 points were selected from a wiggly freehand curve drawn on graph paper. We intentionally selected more points where the curve bent sharply. A visually pleasing curve is obtained by using the cubic spline routines *Spline3_Coef* and *Spline3_Eval* and plotting the resulting natural cubic spline curve. (See Figure 6.8.)

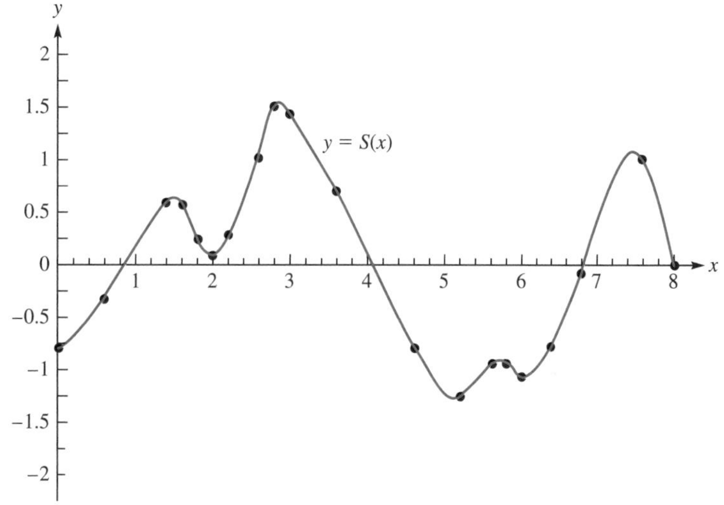

FIGURE 6.8
Natural cubic spline curve

Alternatively, we can use mathematical software such as MATLAB, Maple, or Mathematica to plot the cubic spline function for this table.

Space Curves

2D Parametric Cubic Splines

In two dimensions, two cubic spline functions can be used together to form a **parametric representation** of a complicated curve that turns and twists. Select points on the curve and label them $t = 0, 1, \ldots, n$. For each value of t, read off the x- and y-coordinates of the point, thus producing a table:

t	0	1	$\cdots$	n
x	x_0	x_1	$\cdots$	x_n
y	y_0	y_1	$\cdots$	y_n

Then fit $x = S(t)$ and $y = \overline{S}(t)$, where S and $\overline{S}$ are natural cubic spline interpolants. The two functions S and $\overline{S}$ give a parametric representation of the curve. (See Computer Exercises 6.2.6.)

EXAMPLE 4 Select 13 points on the well-known **serpentine curve** given by

$$y = \frac{x}{1/4 + x^2}$$

So that the knots will not be equally spaced, write the curve in parametric form:

$$\begin{cases} x = \frac{1}{2}\tan\theta \\ y = \sin 2\theta \end{cases}$$

and take $\theta = i(\pi/14)$, where $i = -6, -5, \dots, 5, 6$. Plot the natural cubic spline curve and the interpolation polynomial in order to compare them.

Solution This is an example of curve fitting using the polynomial interpolation routines *Coef* and *Eval* from Chapter 4 (p. 166) and the cubic spline routines *Spline3_Coef* and *Spline3_Eval*.

Figure 6.9 shows the resulting cubic spline curve and the high-degree polynomial curve (dashed line). The polynomial becomes extremely erratic after the fourth knot from the origin and oscillates wildly, whereas the spline is a near perfect fit.

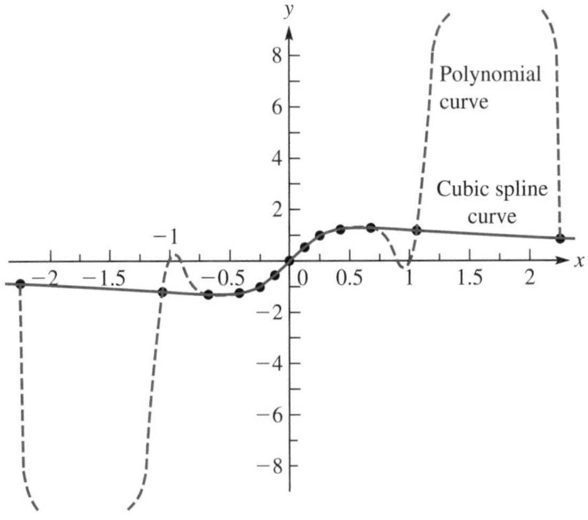

FIGURE 6.9
Serpentine curve

■

EXAMPLE 5 Use cubic spline functions to produce the curve for the following data:

t	0	1	2	3	4	5	6	7
y	1.0	1.5	1.6	1.5	0.9	2.2	2.8	3.1

It is known that the curve is continuous but its slope is not.

Solution A single cubic spline is not suitable. Instead, we can use two cubic spline interpolants, the first having knots 0, 1, 2, 3, 4 and the second having knots 4, 5, 6, 7. By carrying out two separate spline interpolation procedures, we obtain two cubic spline curves that meet at the point (4, 0.9). At this point, the two curves have different slopes. The resulting curve is shown in Figure 6.10. ■

Smoothness Property

Why do spline functions serve the needs of data fitting better than ordinary polynomials? To answer this, one should understand that interpolation by polynomials of high degree is

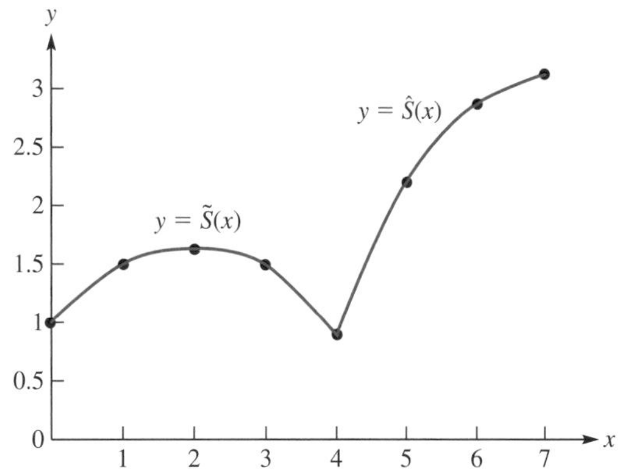

FIGURE 6.10
Two cubic splines

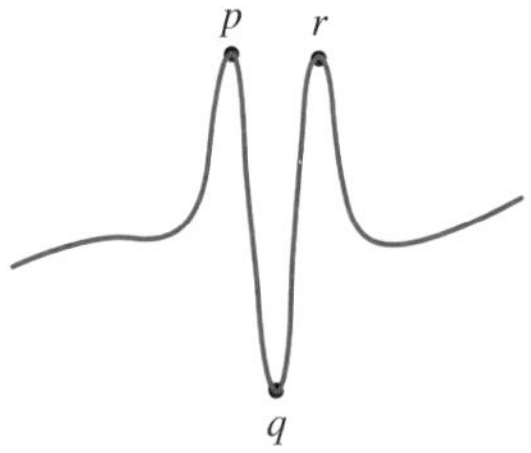

FIGURE 6.11
Wildly oscillating function

often unsatisfactory because polynomials may exhibit wild **oscillations**. Polynomials are smooth in the technical sense of possessing continuous derivatives of all orders, whereas in this sense, spline functions are *not* smooth.

Natural Cubic Splines and Avoiding Wild Oscillations

Wild oscillations in a function can be attributed to its derivatives being very large. Consider the function whose graph is shown in Figure 6.11. The slope of the chord that joins the points p and q is very large in magnitude. By the Mean-Value Theorem, the slope of that chord is the value of the derivative at some point between p and q. Thus, the derivative must attain large values. Indeed, somewhere on the curve between p and q, there is a point where $f'(x)$ is large and negative. Similarly, between q and r, there is a point where $f'(x)$ is large and positive. Hence, there is a point on the curve between p and r where $f''(x)$ is large. This reasoning can be continued to higher derivatives if there are more oscillations. This is the behavior that spline functions do *not* exhibit. In fact, the following result shows that from a certain point of view, natural cubic splines are the *best* functions to use for curve fitting!

■ Theorem 1 **Cubic Spline Smoothness Theorem**

If S is the natural cubic spline function that interpolates a twice-continuously differentiable function f at knots $a = t_0 < t_1 < \cdots < t_n = b$, then

$$\int_a^b [S''(x)]^2\,dx \leqq \int_a^b [f''(x)]^2\,dx$$

Proof To verify the assertion about $[S''(x)]^2$, we let

$$g(x) = f(x) - S(x)$$

so that $g(t_i) = 0$ for $0 \leqq i \leqq n$, and

$$f'' = S'' + g''$$

Now we obtain

$$\int_a^b (f'')^2\,dx = \int_a^b (S'')^2\,dx + \int_a^b (g'')^2\,dx + 2\int_a^b S''g''\,dx$$

If the last integral were 0, we would be finished because then

$$\int_a^b (f'')^2\,dx = \int_a^b (S'')^2\,dx + \int_a^b (g'')^2\,dx \geqq \int_a^b (S'')^2\,dx$$

We apply the technique of integration by parts to the integral in question to show that it is 0.* We have

$$\int_a^b S''g''\,dx = S''g'\Big|_a^b - \int_a^b S'''g'\,dx = -\int_a^b S'''g'\,dx$$

Here, use has been made of the fact that S is a *natural* cubic spline; that is, $S''(a) = 0$ and $S''(b) = 0$. Continuing, we have

$$\int_a^b S'''g'\,dx = \sum_{i=0}^{n-1} \int_{t_i}^{t_{i+1}} S'''g'\,dx$$

Since S is a cubic polynomial in each interval $[t_i, t_{i+1}]$, its third derivative there is a constant, say c_i. So we obtain

$$\int_a^b S'''g'\,dx = \sum_{i=0}^{n-1} c_i \int_{t_i}^{t_{i+1}} g'\,dx = \sum_{i=0}^{n-1} c_i\left[g(t_{i+1}) - g(t_i)\right] = 0$$

because g vanishes at every knot. ■

The interpretation of the integral inequality in the theorem is that the average value of $[S''(x)]^2$ on the interval $[a, b]$ is never larger than the average value of this expression with any twice-continuous function f that agrees with S at the knots. The quantity $[f''(x)]^2$ is closely related to the **curvature** of the function f.

Summary 6.2

- We are given $n + 1$ pairs of points (t_i, y_i) with distinct **knots** $a = t_0 < t_1 < \cdots < t_{n-1} < t_n = b$ over the interval $[a, b]$. A **spline function of degree *k*** is a piecewise polynomial function so that $S, S', S'', \dots, S^{(k-1)}$ are all continuous functions on $[a, b]$ and S is a polynomial of degree at most k on each subinterval $[t_i, t_{i+1}]$.
- A **natural cubic spline function** S is a piecewise cubic polynomial defined on the interval $[a, b]$ so that S, S', S'' are continuous and $S''(t_0) = S''(t_n) = 0$. It can be written

*The formula for **integration by parts** is

$$\int u\,dv = uv - \int v\,du$$

in the form

$$S(x) = \begin{cases} S_0(x), & x \in [t_0, t_1] \\ S_1(x), & x \in [t_1, t_2] \\ \vdots & \vdots \\ S_{n-1}(x), & x \in [t_{n-1}, t_n] \end{cases}$$

where on the interval $[t_i, t_{i+1}]$,

$$\begin{aligned} S_i(x) &= \frac{z_{i+1}}{6h_i}(x - t_i)^3 + \frac{z_i}{6h_i}(t_{i+1} - x)^3 \\ &\quad + \left(\frac{y_{i+1}}{h_i} - \frac{h_i}{6} z_{i+1}\right)(x - t_i) + \left(\frac{y_i}{h_i} - \frac{h_i}{6} z_i\right)(t_{i+1} - x) \end{aligned}$$

and where $h_i = t_{i+1} - t_i$. Clearly, $S(x)$ is continuous, since $S_{i-1}(t_i) = S_i(t_i) = y_i$ for $1 \leqq i \leqq n$. It can be shown that $S'_{i-1}(t_i) = S'_i(t_i)$ and $S''_{i-1}(t_i) = S''_i(t_i) = z_i$ for $1 \leqq i \leqq n$. For efficient evaluation, use the nested form of $S_i(x)$, which is

$$S_i(x) = y_i + (x - t_i)\left(B_i + (x - t_i)\left(\frac{z_i}{2} + \frac{1}{6h_i}(x - t_i)(z_{i+1} - z_i)\right)\right)$$

where $B_i = -(h_i/6)z_{i+1} - (h_i/3)z_i + (y_{i+1} - y_i)/h_i$. The coefficients $z_0, z_1, \dots, z_n$ are found by letting $b_i = (y_{i+1} - y_i)/h_i$, $u_i = 2(h_{i-1} + h_i)$, $v_i = 6(b_i - b_{i-1})$, and then solving the tridiagonal system of linear equations

$$\begin{cases} z_0 = 0 \\ h_{i-1}z_{i-1} + u_i z_i + h_i z_{i+1} = v_i & (1 \leqq i \leqq n-1) \\ z_n = 0 \end{cases}$$

This can be done efficiently by using forward substitution:

$$\begin{cases} u_i \leftarrow u_i - \dfrac{h_{i-1}^2}{u_{i-1}} \\ v_i \leftarrow v_i - \dfrac{h_{i-1}v_{i-1}}{u_{i-1}} & (i = 2, 3, \dots, n-1) \end{cases}$$

and back substitution:

$$\begin{cases} z_{n-1} \leftarrow \dfrac{v_{n-1}}{u_{n-1}} \\ z_i \leftarrow \dfrac{v_i - h_i z_{i+1}}{u_i} & (i = n-2, n-3, \dots, 1) \end{cases}$$

Exercises 6.2

[a]**1.** Do there exist a, b, c, and d such that the function

$$S(x) = \begin{cases} ax^3 + x^2 + cx & (-1 \leqq x \leqq 0) \\ bx^3 + x^2 + dx & (0 \leqq x \leqq 1) \end{cases}$$

is a natural cubic spline function that agrees with the absolute value function $|x|$ at the knots $-1, 0, 1$?

[a]**2.** Do there exist a, b, c, and d such that the function

$$S(x) = \begin{cases} -x & (-10 \leqq x \leqq -1) \\ ax^3 + bx^2 + cx + d & (-1 \leqq x \leqq 1) \\ x & (1 \leqq x \leqq 10) \end{cases}$$

is a natural cubic spline function?

3. Determine the natural cubic spline that interpolates the function $f(x) = x^6$ over the interval $[0, 2]$ using knots 0, 1, and 2.

[a]**4.** Determine the parameters a, b, c, d, and e such that S is a natural cubic spline:

$$S(x) = \begin{cases} a + b(x-1) + c(x-1)^2 + d(x-1)^3 & (x \in [0, 1]) \\ (x-1)^3 + ex^2 - 1 & (x \in [1, 2]) \end{cases}$$

[a]**5.** Determine the values of a, b, c, and d such that f is a cubic spline and such that $\int_0^2 [f''(x)]^2\,dx$ is a minimum:

$$f(x) = \begin{cases} 3 + x - 9x^3 & (0 \leqq x \leqq 1) \\ a + b(x-1) + c(x-1)^2 + d(x-1)^3 & (1 \leqq x \leqq 2) \end{cases}$$

[a]**6.** Determine whether f is a cubic spline with knots -1, 0, 1, and 2:

$$f(x) = \begin{cases} 1 + 2(x+1) + (x+1)^3 & (-1 \leqq x \leqq 0) \\ 3 + 5x + 3x^2 & (0 \leqq x \leqq 1) \\ 11 + (x-1) + 3(x-1)^2 + (x-1)^3 & (1 \leqq x \leqq 2) \end{cases}$$

7. List all the ways in which the following functions fail to be natural cubic splines:

[a]**a.** $S(x) = \begin{cases} x + 1 & (-2 \leqq x \leqq -1) \\ x^3 - 2x + 1 & (-1 \leqq x \leqq 1) \\ x - 1 & (1 \leqq x \leqq 2) \end{cases}$

b. $f(x) = \begin{cases} x^3 + x - 1 & (-1 \leqq x \leqq 0) \\ x^3 - x - 1 & (0 \leqq x \leqq 1) \end{cases}$

8. Suppose $S(x)$ is an mth-degree interpolating spline function over the interval $[a, b]$ with $n + 1$ knots $a = t_0 < t_1 < \cdots < t_n = b$.

[a]**a.** How many conditions are needed to define $S(x)$ uniquely over $[a, b]$?

[a]**b.** How many conditions are defined by the interpolation conditions at the knots?

[a]**c.** How many conditions are defined by the continuity of the derivatives?

[a]**d.** How many additional conditions are needed so that the total equals the number in Part **a**?

9. Show that

$$S(x) = \begin{cases} 28 + 25x + 9x^2 + x^3 & (-3 \leqq x \leqq -1) \\ 26 + 19x + 3x^2 - x^3 & (-1 \leqq x \leqq 0) \\ 26 + 19x + 3x^2 - 2x^3 & (0 \leqq x \leqq 3) \\ -163 + 208x - 60x^2 + 5x^3 & (3 \leqq x \leqq 4) \end{cases}$$

is a natural cubic spline function.

[a]**10.** Give an example of a cubic spline with knots 0, 1, 2, and 3 that is quadratic in $[0, 1]$, cubic in $[1, 2]$, and quadratic in $[2, 3]$.

11. Give an example of a cubic spline function S with knots 0, 1, 2, and 3 such that S is linear in $[0, 1]$ but of degree 3 in the other two intervals.

[a]**12.** Determine a, b, and c such that S is a cubic spline function:

$$S(x) = \begin{cases} x^3 & (0 \leqq x \leqq 1) \\ \frac{1}{2}(x-1)^3 + a(x-1)^2 + b(x-1) + c & (1 \leqq x \leqq 3) \end{cases}$$

[a]**13.** Is there a choice of coefficients for which the following function is a natural cubic spline? Why or why not?

$$f(x) = \begin{cases} x + 1 & (-2 \leqq x \leqq -1) \\ ax^3 + bx^2 + cx + d & (-1 \leqq x \leqq 1) \\ x - 1 & (1 \leqq x \leqq 2) \end{cases}$$

14. Determine the coefficients in the function

$$S(x) = \begin{cases} x^3 - 1 & (-9 \leqq x \leqq 0) \\ ax^3 + bx^2 + cx + d & (0 \leqq x \leqq 5) \end{cases}$$

such that it is a cubic spline that takes the value 2 when $x = 1$.

[a]**15.** Determine the coefficients such that the function

$$S(x) = \begin{cases} x^2 + x^3 & (0 \leqq x \leqq 1) \\ a + bx + cx^2 + dx^3 & (1 \leqq x \leqq 2) \end{cases}$$

is a cubic spline and has the property $S_1'''(x) = 12$.

16. Assume that $a = x_0 < x_1 < \cdots < x_m = b$. Describe the function f that interpolates a table of values (x_i, y_i), where $0 \leqq i \leqq m$, and that minimizes the expression $\int_a^b |f'(x)|dx$.

[a]**17.** How many additional conditions are needed to specify uniquely a spline of degree 4 over n knots?

18. Let knots $t_0 < t_1 < \cdots < t_n$, and let numbers y_i and z_i be given. Determine formulas for a piecewise cubic function f that has the given knots such that $f(t_i) = y_i$ $(0 \leqq i \leqq n)$, $\lim_{x \to t_i^+} f''(x) = z_i$ $(0 \leqq i \leqq n-1)$, and $\lim_{x \to t_i^-} f''(x) = z_i$ $(1 \leqq i \leqq n)$. Why is f not generally a cubic spline?

[a]**19.** Define a function f by

$$f(x) = \begin{cases} x^3 + x - 1 & (-1 \leqq x \leqq 0) \\ x^3 - x - 1 & (0 \leqq x \leqq 1) \end{cases}$$

Show that $\lim_{x\to 0^+} f(x) = \lim_{x \to 0^-} f(x)$ and that $\lim_{x\to 0^+} f''(x) = \lim_{x\to 0^-} f''(x)$. Are f and f'' continuous? Does it follow that f is a cubic spline? Explain.

20. Show that there is a unique cubic spline S with knots $t_0 < t_1 < \cdots < t_n$, interpolating data $S(t_i) = y_i$ $(0 \leqq i \leqq n)$ and satisfying the two end conditions $S'(t_0) = S'(t_n) = 0$.

21. Describe explicitly the *natural cubic spline* that interpolates a table with only two entries:

x	t_0	t_1
y	y_0	y_1

Give a formula for it. Here, t_0 and t_1 are the knots.

[a]**22.** Suppose that $f(0) = 0$, $f(1) = 1.1752$, $f'(0) = 1$, and $f'(1) = 1.5431$. Determine the cubic interpolating polynomial $p_3(x)$ for these data. Is it a natural cubic spline?

23. A **periodic cubic spline** having knots $t_0, t_1, \ldots, t_n$ is defined as a cubic spline function $S(x)$ such that $S(t_0) = S(t_n)$, $S'(t_0) = S'(t_n)$, and $S''(t_0) = S''(t_n)$. It would be used to fit data that are known to be periodic. Carry out the analysis necessary to obtain a periodic cubic spline interpolant for the table

x	t_0	t_1	$\cdots$	t_n
y	y_0	y_1	$\cdots$	y_n

assuming that $y_n = y_0$.

24. The derivatives and integrals of polynomials are polynomials. State and prove a similar result about spline functions.

25. Given a differentiable function f and knots $t_0 < t_1 < \cdots < t_n$, show how to obtain a cubic spline S that interpolates f at the knots and satisfies the end conditions $S'(t_0) = f'(t_0)$ and $S'(t_n) = f'(t_n)$.
Note: This procedure produces a better fit to f when applicable. If f' is not known, finite-difference approximations to $f'(t_0)$ and $f'(t_n)$ can be used.

[a]**26.** Let S be a cubic spline that has knots $t_0 < t_1 < \cdots < t_n$. Suppose that on the two intervals $[t_0, t_1]$ and $[t_2, t_3]$, S reduces to linear polynomials. What can be said of S on $[t_1, t_2]$?

27. In the construction of the cubic interpolating spline, carry out the evaluation of constants C_i and D_i, and thus justify Equation (5).

28. Show that S_i can also be written in the form

$$S_i(x) = y_i + A_i(x - t_i) + \frac{1}{2}z_i(x - t_i)^2 + \frac{z_{i+1} - z_i}{6h_i}(x - t_i)^3$$

with

$$A_i = -\frac{h_i}{3}z_i - \frac{h_i}{6}z_{i+1} - \frac{y_i}{h_i} + \frac{y_{i+1}}{h_i}$$

29. Carry out the details in deriving Equation (9), starting with Equation (5).

30. Verify that the algorithm for computing the (z_i) array is correct by showing that if (z_i) satisfies Equation (9), then it satisfies the equation in Step 3 of the algorithm.

31. Establish that $u_i > 2h_i + \frac{3}{2}h_{i-1}$ in the algorithm for determining the cubic spline interpolant.

[a]**32.** By hand calculation, find the natural cubic spline interpolant for this table:

x	1	2	3	4	5
y	0	1	0	1	0

[a]**33.** Find a cubic spline over knots -1, 0, and 1 such that the following conditions are satisfied: $S''(-1) = S''(1) = 0$, $S(-1) = S(1) = 0$, and $S(0) = 1$.

34. This problem and the next two lead to a more efficient algorithm for natural cubic spline interpolation in the case of equally spaced knots. Let $h_i = h$ in Equation (5), and replace the parameters z_i by $q_i = h^2 z_i/6$. Show that the new form of Equation (5) is then

$$S_i(x) = q_{i+1}\left(\frac{x - t_i}{h}\right)^3 + q_i\left(\frac{t_{i+1} - x}{h}\right)^3 + (y_{i+1} - q_{i+1})\left(\frac{x - t_i}{h}\right) + (y_i - q_i)\left(\frac{t_{i+1} - x}{h}\right)$$

35. (Continuation) Establish the new continuity conditions:

$$\begin{cases} q_0 = q_n = 0 \\ q_{i-1} + 4q_i + q_{i+1} = y_{i+1} - 2y_i + y_{i-1} \quad (1 \leqq i \leqq n - 1) \end{cases}$$

36. (Continuation) Show that the parameters q_i can be determined by backward recursion as follows:

$$\begin{cases} q_n = 0, \quad q_{n-1} = \beta_{n-1} \\ q_i = \alpha_i q_{i+1} + \beta_i \quad (i = n-2, n-3, \ldots, 0) \end{cases}$$

where the coefficients α_i and β_i are generated by ascending recursion from the formulas

$$\begin{cases} \alpha_0 = 0, \quad \alpha_i = -(\alpha_{i-1} + 4)^{-1} \quad (1 \leqq i \leqq n) \\ \beta_0 = 0, \quad \beta_i = -\alpha_i(y_{i+1} - 2y_i + y_{i-1} - \beta_{i-1}) \quad (1 \leqq i \leqq n) \end{cases}$$

(This stable and efficient algorithm is due to MacLeod [1973].)

37. Prove that if $S(x)$ is a spline of degree k on $[a, b]$, then $S'(x)$ is a spline of degree $k - 1$.

[a]**38.** How many coefficients are needed to define a piecewise quartic (fourth-degree) function with $n + 1$ knots? How many conditions will be imposed if the piecewise quartic function is to be a quartic spline? Justify your answers.

[a]**39.** Determine whether this function is a natural cubic spline:

$$S(x) = \begin{cases} x^3 + 3x^2 + 7x - 5 & (-1 \leqq x \leqq 0) \\ -x^3 + 3x^2 + 7x - 5 & (0 \leqq x \leqq 1) \end{cases}$$

40. Determine whether this function is or is *not* a natural cubic spline having knots 0, 1, and 2:

$$f(x) = \begin{cases} x^3 + x - 1 & (0 \leqq x \leqq 1) \\ -(x-1)^3 + 3(x-1)^2 + 4(x-1) + 1 & (1 \leqq x \leqq 2) \end{cases}$$

41. Show that the natural cubic spline going through the points $(0, 1)$, $(1, 2)$, $(2, 3)$, $(3, 4)$, and $(4, 5)$ must be $y = x + 1$. (The natural cubic spline interpolant to a given data set is unique, because the matrix in Equation (10) is diagonally dominant and invertible, as proven in Section 2.3.)

Computer Exercises 6.2

1. Rewrite and test procedure *Spline3_Coef* using procedure *Tri* from Section 2.3. Use the symmetry of the $(n-1) \times (n-1)$ tridiagonal system.

2. The extra storage required in Step 1 of the algorithm for solving the natural cubic spline tridiagonal system directly can be eliminated at the expense of a slight amount of extra computation—namely, by computing the h_i's and b_i's directly from the t_i's and y_i's in the forward elimination phase (Step 2) and in the back substitution phase (Step 3). Rewrite and test procedure *Spline3_Coef* using this idea.

3. Using at most 20 knots and the cubic spline routines *Spline3_Coef* and *Spline3_Eval*, plot on a computer plotter an outline of your:

a. School's mascot. **b.** Signature. **c.** Profile.

4. Let S be the cubic spline function that interpolates $f(x) = (x^2 + 1)^{-1}$ at 41 equally spaced knots in the interval $[-5, 5]$. Evaluate $S(x) - f(x)$ at 101 equally spaced points on the interval $[0, 5]$.

5. Draw a free-form curve on graph paper, making certain that the curve is the graph of a function. Then read values of your function at a reasonable number of points, say, 10–50, and compute the cubic spline function that takes those values. Compare the freely drawn curve to the graph of the cubic spline.

6. Draw a spiral (or other curve that is not a function) and reproduce it by way of parametric spline functions. (See the following figure.)

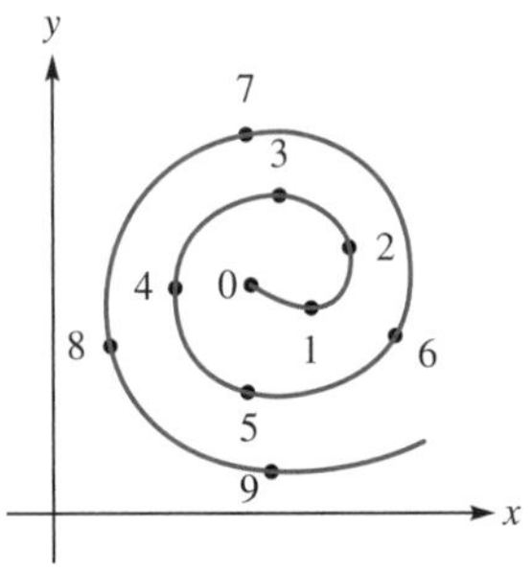

7. Write and test procedures that are *as simple as possible* to perform natural cubic spline interpolation with equally spaced knots.
Hint: See Exercises 6.2.34–6.2.36.

8. Write a program to estimate $\int_a^b f(x)\,dx$, assuming that we know the values of f at only certain prescribed knots $a = t_0 < t_1 < \cdots < t_n = b$. Approximate f first by an interpolating cubic spline, and then compute the integral of it using Equation (5).

9. Write a procedure to estimate $f'(x)$ for any x in $[a, b]$, assuming that we know only the values of f at knots $a = t_0 < t_1 < \cdots < t_n = b$.

10. Using the Runge function $f(x) = 1/(1 + x^2)$ from Section 4.2 with an increasing number of equally spaced nodes, watch the natural cubic spline curve get better with regard to curve fitting while the interpolating polynomial gets worse.

11. Use mathematical software such as MATLAB, Maple, or Mathematica to generate and plot the spline function in Example 2.

12. Learn and practice using the computer graphics found in MATLAB, Maple, or Mathematica to plot the cubic spline functions corresponding to

a. Figure 6.7. **b.** Figure 6.8.
c. Figure 6.9. **d.** Figure 6.10.

6.3 B Splines: Interpolation and Approximation

Introduction

In this section, we give an introduction to the theory of *B splines*. These are special spline functions that are well adapted to numerical tasks and are being used more frequently in production-type programs for approximating data. Thus, the intelligent user of library code should have some familiarity with them. The B splines were so named because they formed a *basis* for the set of all splines. (We prefer the more romantic name **bell splines** because of their characteristic shape.)

Throughout this section, we suppose that an infinite set of knots $\{t_i\}$ has been prescribed in such a way that

$$\begin{cases} \cdots < t_{-2} < t_{-1} < t_0 < t_1 < t_2 < \cdots \\ \lim_{i\to\infty} t_i = \infty = -\lim_{i\to\infty} t_{-i} \end{cases} \tag{1}$$

The B splines to be defined now depend on this set of knots, although the notation does not show that dependence.

B_i^0 Splines

The **B splines of degree 0** are defined by

$B_i^0(x)$ Spline

$$B_i^0(x) = \begin{cases} 1, & t_i \leqq x < t_{i+1} \\ 0, & \text{otherwise} \end{cases} \tag{2}$$

The graph of B_i^0 is shown in Figure 6.12.

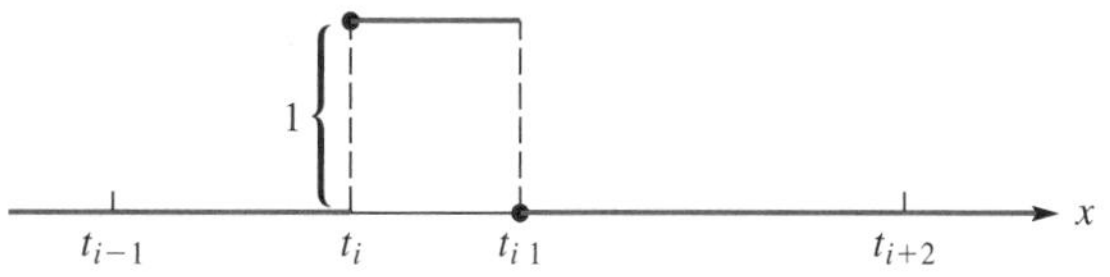

FIGURE 6.12 B_i^0 spline

Obviously, B_i^0 is discontinuous. However, it is continuous from the right at all points, even where the jumps occur. Thus, we obtain

$$\lim_{x\to t_i^+} B_i^0(x) = 1 = B_i^0(t_i), \qquad \lim_{x\to t_{i+1}^+} B_i^0(x) = 0 = B_i^0(t_{i+1})$$

If the **support** of a function f is defined as the set of points x where $f(x) \neq 0$, then we can say that the support of B_i^0 is the half-open interval $[t_i, t_{i+1})$. Since B_i^0 is a piecewise constant function, it is a spline of degree 0.

Two further observations can be made:

Properties

$$\begin{cases} B_i^0(x) \geqq 0, & \text{for all } x \text{ and for all } i \\ \sum_{i=-\infty}^{\infty} B_i^0(x) = 1, & \text{for all } x \end{cases}$$

Although the second of these assertions contains an infinite series, there is no question of convergence because for each x only one term in the series is different from 0. Indeed, for fixed x, there is a unique integer m such that $t_m \leqq x < t_{m+1}$, and then

$$\sum_{i=-\infty}^{\infty} B_i^0(x) = B_m^0(x) = 1$$

The reader should now see the reason for defining B_i^0 in the manner of Equation (2).

A final remark concerning these B splines of degree 0: Any spline of degree 0 that is continuous from the right and is based on the knots (1) can be expressed as a linear combination of the B splines B_i^0. Indeed, if S is such a function, then it can be specified by a rule such as

$$S(x) = b_i, \quad \text{if } t_i \leqq x < t_{i+1} \quad (i = 0, \pm 1, \pm 2, \dots)$$

Then S can be written as

$$S = \sum_{i=-\infty}^{\infty} b_i B_i^0$$

B_i^1 and B_i^k Splines

With the functions B_i^0 as a starting point, we now generate all the higher-degree B splines by a simple *recursive* definition:

B Spline of Degree *k* Recurrence Relation

$$B_i^k(x) = \left(\frac{x - t_i}{t_{i+k} - t_i}\right) B_i^{k-1}(x) + \left(\frac{t_{i+k+1} - x}{t_{i+k+1} - t_{i+1}}\right) B_{i+1}^{k-1}(x) \quad (k \geqq 1) \tag{3}$$

Here $k = 1, 2, \dots$, and $i = 0, \pm 1, \pm 2, \dots$.

To illustrate Equation (3), let us determine B_i^1 in an alternative form:

$B_i^1(x)$ Spline

$$B_i^1(x) = \left(\frac{x - t_i}{t_{i+1} - t_i}\right) B_i^0(x) + \left(\frac{t_{i+2} - x}{t_{i+2} - t_{i+1}}\right) B_{i+1}^0(x)$$

$$= \begin{cases} 0 & (x \geqq t_{i+2} \quad \text{or} \quad x \leqq t_i) \\ \dfrac{x - t_i}{t_{i+1} - t_i} & (t_i \leqq x < t_{i+1}) \\ \dfrac{t_{i+2} - x}{t_{i+2} - t_{i+1}} & (t_{i+1} \leqq x < t_{i+2}) \end{cases}$$

The graph of B_i^1 is shown in Figure 6.13. These are sometimes called **hat functions** or **chapeau functions** (from the French) since they resemble a triangular hat one might make from a newspaper. The support of B_i^1 is the open interval (t_i, t_{i+2}). It is true, but perhaps not so obvious, that

$$\sum_{i=-\infty}^{\infty} B_i^1(x) = 1 \quad \text{for all } x$$

and that every spline of degree 1 based on the knots (1) is a linear combination of B_i^1.

The functions B_i^k as defined by Equation (3) are called **B splines of degree *k***. Since each B_i^k is obtained by applying linear factors to B_i^{k-1} and B_{i+1}^{k-1}, we see that the degrees actually increase by 1 at each step. Therefore, B_i^1 is piecewise linear, B_i^2 is piecewise quadratic, and so on.

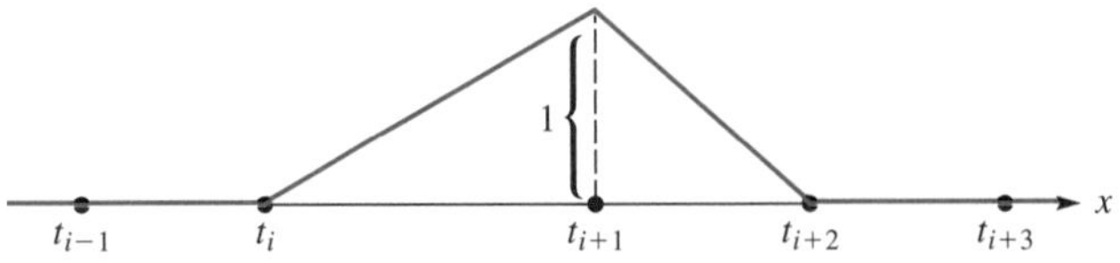

FIGURE 6.13 B_1^1 spline (Chapeau function)

It is also easily shown by induction that

$$B_i^k(x) = 0, \quad x \notin [t_i, t_{i+k+1}) \quad (k \geqq 0)$$

To establish this, we start by observing that it is true when $k = 0$ because of Definition (2). If it is true for index $k - 1$, then it is true for index k by the following reasoning. The inductive hypothesis tells us that $B_i^{k-1}(x) = 0$ if x is outside $[t_i, t_{i+k})$ and that $B_{i+1}^{k-1}(x) = 0$ if x is outside $[t_{i+1}, t_{i+k+1})$. If x is outside *both* intervals, it is outside their union, $[t_i, t_{i+k+1})$; then both terms on the right side of Equation (3) are 0. So $B_i^k(x) = 0$ outside $[t_i, t_{i+k+1})$. That $B_i^k(t_i) = 0$ follows directly from Equation (3), so we know that $B_i^k(x) = 0$ for all x outside (t_i, t_{i+k+1}) if $k \geqq 1$.

Complementary to the property just established, we can show, again by induction, that

$$B_i^k(x) > 0 \quad x \in (t_i, t_{i+k+1}) \quad (k \geqq 0)$$

By Equation (2), this assertion is true when $k = 0$. If it is true for index $k-1$, then $B_i^{k-1}(x) > 0$ on (t_i, t_{i+k}) and $B_{i+1}^{k-1}(x) > 0$ on (t_{i+1}, t_{i+k+1}). In Equation (3), the factors that multiply $B_i^{k-1}(x)$ and $B_{i+1}^{k-1}(x)$ are positive when $t_i < x < t_{i+k+1}$. Thus, $B_i^k(x) > 0$ on this interval.

Figure 6.14 shows the first four B_i^k splines plotted on the same axes.

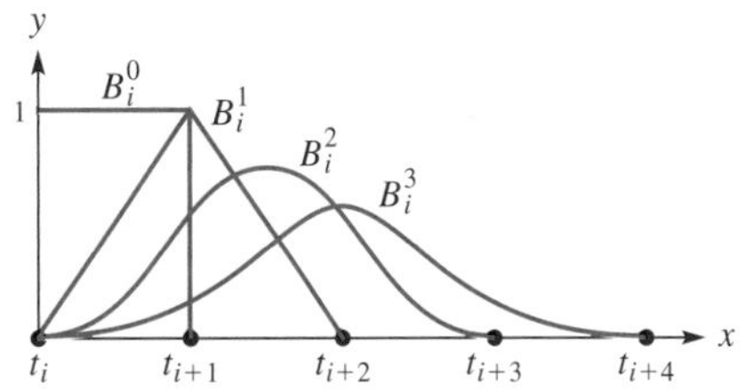

FIGURE 6.14
First four B_i^k splines

Linear Combination of B_i^k Splines

The principal use of the B splines $B_i^k (i = 0, \pm 1, \pm 2, \ldots)$ is as a basis for the set of all kth-degree splines that have the same knot sequence. Thus, linear combinations

$$\sum_{i=-\infty}^{\infty} c_i B_i^k$$

are important objects of study. (We use c_i for fixed k and C_i^k to emphasize the degree k of the corresponding B splines.)

Our first task is to develop an efficient method to evaluate a function of the form

$$f(x) = \sum_{i=-\infty}^{\infty} C_i^k B_i^k(x) \tag{4}$$

under the supposition that the coefficients C_i^k are given (as well as the knot sequence t_i). Using Definition (3) and some simple series manipulations, we have

$$\begin{aligned} f(x) &= \sum_{i=-\infty}^{\infty} C_i^k \left[\left(\frac{x - t_i}{t_{i+k} - t_i}\right) B_i^{k-1}(x) + \left(\frac{t_{i+k+1} - x}{t_{i+k+1} - t_{i+1}}\right) B_{i+1}^{k-1}(x)\right] \\ &= \sum_{i=-\infty}^{\infty} \left[C_i^k \left(\frac{x - t_i}{t_{i+k} - t_i}\right) + C_{i-1}^k \left(\frac{t_{i+k} - x}{t_{i+k} - t_i}\right)\right] B_i^{k-1}(x) \\ &= \sum_{i=-\infty}^{\infty} C_i^{k-1} B_i^{k-1}(x) \end{aligned} \tag{5}$$

where C_i^{k-1} is defined to be the appropriate coefficient from the line preceding Equation (5).

This algebraic manipulation shows how a linear combination of $B_i^k(x)$ can be expressed as a linear combination of $B_i^{k-1}(x)$. Repeating this process $k-1$ times, we eventually express $f(x)$ in the form

$$f(x) = \sum_{i=-\infty}^{\infty} C_i^0 B_i^0(x) \tag{6}$$

If $t_m \leqq x < t_{m+1}$, then $f(x) = C_m^0$. The formula by which the coefficients C_i^{j-1} are obtained is

C_i^{j-1} Coefficients Recurrance Relation

$$C_i^{j-1} = C_i^j \left(\frac{x - t_i}{t_{i+j} - t_i}\right) + C_{i-1}^j \left(\frac{t_{i+j} - x}{t_{i+j} - t_i}\right) \tag{7}$$

A nice feature of Equation (4) is that only the $k + 1$ coefficients $C_m^k, C_{m-1}^k, \ldots, C_{m-k}^k$ are needed to compute $f(x)$ if $t_m \leqq x < t_{m+1}$ (see Exercise 6.3.6). Thus, if f is defined by Equation (4) and we want to compute $f(x)$, we use Equation (7) to calculate the entries in the following triangular array:

$$\begin{array}{llll} C_m^k & C_m^{k-1} & \cdots & C_m^0 \\ C_{m-1}^k & C_{m-1}^{k-1} & \iddots & \\ \vdots & \iddots & & \\ C_{m-k}^k & & & \end{array}$$

Although our notation does not show it, the coefficients in Equation (4) are independent of x, whereas the C_i^{j-1}'s calculated subsequently by Equation (7) do depend on x.

It is now a simple matter to establish that

$$\sum_{i=-\infty}^{\infty} B_i^k(x) = 1, \qquad \text{for all } x \text{ and all } k \geqq 0$$

If $k = 0$, we already know this. If $k > 0$, we use Equation (4) with $C_i^k = 1$ for all i. By Equation (7), all subsequent coefficients $C_i^k, C_i^{k-1}, C_i^{k-2}, \ldots, C_i^0$ are also equal to 1 (induction is needed here!). Thus, at the end, Equation (6) is true with $C_i^0 = 1$, and so $f(x) = 1$. Therefore, from Equation (4), the sum of all B splines of degree k is unity.

The smoothness of the B splines B_i^k increases with the index k. In fact, we can show by induction that B_i^k has a continuous $k - 1$st derivative.

The B splines can be used as substitutes for complicated functions in many mathematical situations. Differentiation and integration are important examples.

Derivatives of B Splines

A basic result about the **derivatives of B splines** is

Derivative Property I

$$\frac{d}{dx} B_i^k(x) = \left(\frac{k}{t_{i+k} - t_i}\right) B_i^{k-1}(x) - \left(\frac{k}{t_{i+k+1} - t_{i+1}}\right) B_{i+1}^{k-1}(x) \tag{8}$$

This equation can be proved by induction using the recursive Formula (3). Once Equation (8) is established, we get the useful formula

Derivative Property II

$$\frac{d}{dx} \sum_{i=-\infty}^{\infty} c_i B_i^k(x) = \sum_{i=-\infty}^{\infty} d_i B_i^{k-1}(x) \tag{9}$$

where

$$d_i = k\left(\frac{c_i - c_{i-1}}{t_{i+k} - t_i}\right)$$

The verification is as follows. By Equation (8), we obtain

$$\begin{aligned}
&\frac{d}{dx}\sum_{i=-\infty}^{\infty} c_i B_i^k(x) \\
&= \sum_{i=-\infty}^{\infty} c_i \frac{d}{dx} B_i^k(x) \\
&= \sum_{i=-\infty}^{\infty} c_i \left[\left(\frac{k}{t_{i+k} - t_i}\right) B_i^{k-1}(x) - \left(\frac{k}{t_{i+k+1} - t_{i+1}}\right) B_{i+1}^{k-1}(x)\right] \\
&= \sum_{i=-\infty}^{\infty} \left[\left(\frac{c_i k}{t_{i+k} - t_i}\right) - \left(\frac{c_{i-1} k}{t_{i+k} - t_i}\right)\right] B_i^{k-1}(x) \\
&= \sum_{i=-\infty}^{\infty} d_i B_i^{k-1}(x)
\end{aligned}$$

Integration of B Splines

For numerical integration, the B splines are also recommended, especially for indefinite integration. Here is the basic result needed for integration:

Integration Property I

$$\int_{-\infty}^{x} B_i^k(s)\,ds = \left(\frac{t_{i+k+1} - t_i}{k+1}\right)\sum_{j=i}^{\infty} B_j^{k+1}(x) \tag{10}$$

This equation can be verified by differentiating both sides with respect to x and simplifying by the use of Equation (9). To be sure that the two sides of Equation (10) do not differ by a constant, we note that for any $x < t_i$, both sides reduce to zero.

The basic result (10) produces this useful formula:

Integration Property II

$$\int_{-\infty}^{x} \sum_{i=-\infty}^{\infty} c_i B_i^k(s)\,ds = \sum_{i=-\infty}^{\infty} e_i B_i^{k+1}(x) \tag{11}$$

where

$$e_i = \frac{1}{k+1}\sum_{j=-\infty}^{i} c_j (t_{j+k+1} - t_j)$$

It should be emphasized that this formula gives an indefinite integral (antiderivative) of any function expressed as a linear combination of B splines. Any definite integral can be obtained by selecting a specific value of x. For example, if x is a knot, say, $x = t_m$, then

$$\int_{-\infty}^{t_m} \sum_{i=-\infty}^{\infty} c_i B_i^k(s)\,ds = \sum_{i=-\infty}^{\infty} e_i B_i^{k+1}(t_m) = \sum_{i=m-k-1}^{m} e_i B_i^{k+1}(t_m)$$

Mathematical Software

MATLAB has a *Spline* Toolbox, developed by Carl de Boor, that can be used for many tasks involving splines. For example, there are routines for interpolating data by splines with diverse end conditions and routines for least-squares fits to data. There are many demonstration routines in this Toolbox that exhibit plots and provide models for programming MATLAB M-files. These demonstrations are quite instructive for visualizing and learning the concepts in spline theory, especially B splines.

Maple has a `BSpline` package for constructing B spline basis functions of degree k from a given knot list, which may include multiple knots. It is based on a divided-difference implementation found in Bartels, Beatty, and Barskey [1987].

Mathematica has a splines packages also.

Interpolation and Approximation by B Splines

We developed a number of properties of B splines and showed how B splines are used in various numerical tasks. The problem of obtaining a B spline representation of a given function was not discussed. Here, we consider the problem of interpolating a table of data; later, a noninterpolatory method of approximation is described.

A basic question is how to determine the coefficients in the expression

Basic Property

$$S(x) = \sum_{i=-\infty}^{\infty} A_i B_{i-k}^k(x) \tag{12}$$

so that the resulting spline function interpolates a prescribed table:

x	t_0	t_1	$\cdots$	t_n
y	y_0	y_1	$\cdots$	y_n

We mean by *interpolate* that

Interpolation Conditions

$$S(t_i) = y_i \quad (0 \leqq i \leqq n) \tag{13}$$

The natural starting point is with the simplest splines, corresponding to $k = 0$. Since

$$B_i^0(t_j) = \delta_{ij} = \begin{cases} 1 & (i = j) \\ 0 & (i \neq j) \end{cases}$$

the solution to the problem is immediate: Just set $A_i = y_i$ for $0 \leqq i \leqq n$. All other coefficients in Equation (12) are arbitrary. In particular, they can be zero. We arrive then at this result: The B spline of degree 0

B Splines of Degree 0

$$S(x) = \sum_{i=0}^{n} y_i B_i^0(x)$$

has the interpolation property (13).

The next case, $k = 1$, also has a simple solution. We use the fact that

$$B_{i-1}^1(t_j) = \delta_{ij}$$

Hence, the following is true: The B spline of degree 1

B Spline of Degree 1

$$S(x) = \sum_{i=0}^{n} y_i B_{i-1}^1(x)$$

has the interpolation property (13). So $A_i = y_i$ again.

If the table has four entries ($n = 3$), for instance, we use B_{-1}^1, B_0^1, B_1^1, and B_2^1. They, in turn, require for their definition knots $t_{-1}, t_0, t_1, \ldots, t_4$. Knots t_{-1} and t_4 can be arbitrary. Figure 6.15 shows the graphs of the four B^1 splines. In such a problem, if t_{-1} and t_4 are not prescribed, it is natural to define them in such a way that t_0 is the midpoint of the interval $[t_{-1}, t_1]$ and t_3 is the midpoint of $[t_2, t_4]$.

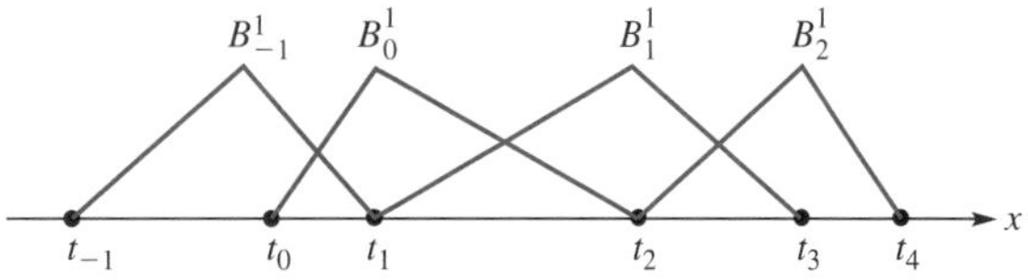

FIGURE 6.15 B^1_i splines

In both elementary cases considered, the unknown coefficients $A_0, A_1, \ldots, A_n$ in Equation (12) were uniquely determined by the interpolation conditions (13). If terms were present in Equation (12) corresponding to values of i *outside* the range $\{0, 1, \ldots, n\}$, then they would have no influence on the values of $S(x)$ at $t_0, t_1, \ldots, t_n$.

Higher-Degree B Splines

For higher-degree splines, we shall see that some arbitrariness exists in choosing coefficients. In fact, *none* of the coefficients is uniquely determined by the interpolation conditions. This fact can be advantageous if other properties are desired of the solution. In the quadratic case, we begin with the equation

Quadratic Case

$$\sum_{i=-\infty}^{\infty} A_i B^2_{i-2}(t_j) = \frac{1}{t_{j+1} - t_{j-1}} \left[A_j(t_{j+1} - t_j) + A_{j+1}(t_j - t_{j-1}) \right] \tag{14}$$

Its justification is left to Exercise 6.3.26. If the interpolation conditions (13) are now imposed, we obtain the following system of equations, which gives the necessary and sufficient conditions on the coefficients:

$$A_j(t_{j+1} - t_j) + A_{j+1}(t_j - t_{j-1}) = y_j(t_{j+1} - t_{j-1}) \quad (0 \leqq j \leqq n) \tag{15}$$

This is a system of $n + 1$ linear equations in $n + 2$ unknowns $A_0, A_1, \ldots, A_{n+1}$.

One way to solve Equation (15) is to assign any value to A_0 and then use Equation (15) to compute for $A_1, A_2, \ldots, A_{n+1}$, recursively. For this purpose, the equations could be rewritten as

$$A_{j+1} = \alpha_j + \beta_j A_j \quad (0 \leqq j \leqq n) \tag{16}$$

where these abbreviations have been used:

$$\begin{cases} \alpha_j = y_j \left(\dfrac{t_{j+1} - t_{j-1}}{t_j - t_{j-1}} \right) \\ \beta_j = \dfrac{t_j - t_{j+1}}{t_j - t_{j-1}} \end{cases} \quad (0 \leqq j \leqq n)$$

To keep the coefficients small in magnitude, we recommend selecting A_0 such that the expression

$$\Phi = \sum_{i=0}^{n+1} A_i^2$$

will be a minimum. To determine this value of A_0, we proceed as follows: By successive substitution using Equation (16), we can show that

$$A_{j+1} = \gamma_j + \delta_j A_0 \quad (0 \leqq j \leqq n) \tag{17}$$

where the coefficients γ_j and δ_j are obtained recursively by this algorithm:

$$\begin{cases} \gamma_0 = \alpha_0, & \delta_0 = \beta_0 \\ \gamma_j = \alpha_j + \beta_j \gamma_{j-1}, & \delta_j = \beta_j \delta_{j-1} \quad (1 \leqq j \leqq n) \end{cases} \tag{18}$$

Then Φ is a quadratic function of A_0 as follows:

$$\begin{aligned}\Phi &= A_0^2 + A_1^2 + \cdots + A_{n+1}^2\\ &= A_0^2 + (\gamma_0 + \delta_0 A_0)^2 + (\gamma_1 + \delta_1 A_0)^2 + \cdots + (\gamma_n + \delta_n A_0)^2\end{aligned}$$

To find the minimum of Φ, we take its derivative with respect to A_0 and set it equal to zero:

$$\frac{d\Phi}{dA_0} = 2A_0 + 2(\gamma_0 + \delta_0 A_0)\delta_0 + 2(\gamma_1 + \delta_1 A_0)\delta_1 + \cdots + 2(\gamma_n + \delta_n A_0)\delta_n = 0$$

This is equivalent to $qA_0 + p = 0$, where

$$\begin{cases} q = 1 + \delta_0^2 + \delta_1^2 + \cdots + \delta_n^2 \\ p = \gamma_0\delta_0 + \gamma_1\delta_1 + \cdots + \gamma_n\delta_n \end{cases}$$

Pseudocode and a Curve-Fitting Example

A procedure that computes coefficients $A_0, A_1, \ldots, A_{n+1}$ in the manner previously outlined is given now. In its calling sequence, $(t_i)_{0:n}$ is the knot array, $(y_i)_{0:n}$ is the array of abscissa points, $(a_i)_{0:n+1}$ is the array of A_i coefficients, and $(h_i)_{0:n+1}$ is an array that contains $h_i = t_i - t_{i-1}$. Only n, (t_i), and (y_i) are input values. They are available unchanged when the routine is finished. Arrays (a_i) and (h_i) are computed and available as output.

***BSpline 2_Coef* Pseudocode**

```
procedure BSpline2_Coef(n, (t_i), (y_i), (a_i), (h_i))
integer i, n;   real δ, γ, p, q
real array (a_i)_{0:n+1}, (h_i)_{0:n+1}, (t_i)_{0:n}, (y_i)_{0:n}
for i = 1 to n
    h_i ← t_i − t_{i−1}
end for
h_0 ← h_1
h_{n+1} ← h_n
δ ← −1
γ ← 2y_0
p ← δγ
q ← 2
for i = 1 to n
    r ← h_{i+1}/h_i
    δ ← −rδ
    γ ← −rγ + (r + 1)y_i
    p ← p + γδ
    q ← q + δ^2
end for
a_0 ← −p/q
for i = 1 to n + 1
    a_i ← [(h_{i−1} + h_i)y_{i−1} − h_i a_{i−1}]/h_{i−1}
end for
end procedure BSpline2_Coef
```

Next we give a procedure function *BSpline2_Eval* for computing values of the quadratic spline given by $S(x) = \sum_{i=0}^{n+1} A_i B_{i-2}^2(x)$. Its calling sequence has some of the same variables as in the preceding pseudocode. The input variable x is a single real number that should lie between t_0 and t_n. The result of Exercise 6.3.26 is used.

BSpline2_Eval
Pseudocode

```
real function BSpline2_Eval(n, (t_i), (a_i), (h_i), x)
integer i, n;   real d, e, x;   real array (a_i)_{0:n+1}, (h_i)_{0:n+1}, (t_i)_{0:n}
for i = n − 1 to 0 step −1
    if x − t_i ≧ 0 then exit loop
end for
i ← i + 1
d ← [a_{i+1}(x − t_{i−1}) + a_i(t_i − x + h_{i+1})]/(h_i + h_{i+1})
e ← [a_i(x − t_{i−1} + h_{i−1}) + a_{i−1}(t_{i−1} − x + h_i)]/(h_{i−1} + h_i)
BSpline2_Eval ← [d(x − t_{i−1}) + e(t_i − x)]/h_i
end function BSpline2_Eval
```

Using the table of 20 points from Section 6.2, we can compare the resulting natural cubic spline curve with the quadratic spline produced by the procedures *BSpline2_Coef* and *BSpline2_Eval*. The first of these curves is shown in Figure 6.8 (p. 273), and the second is in Figure 6.16. The latter is reasonable, but perhaps not as pleasing as the former. These curves show once again that cubic natural splines are simple and elegant functions for curve fitting.

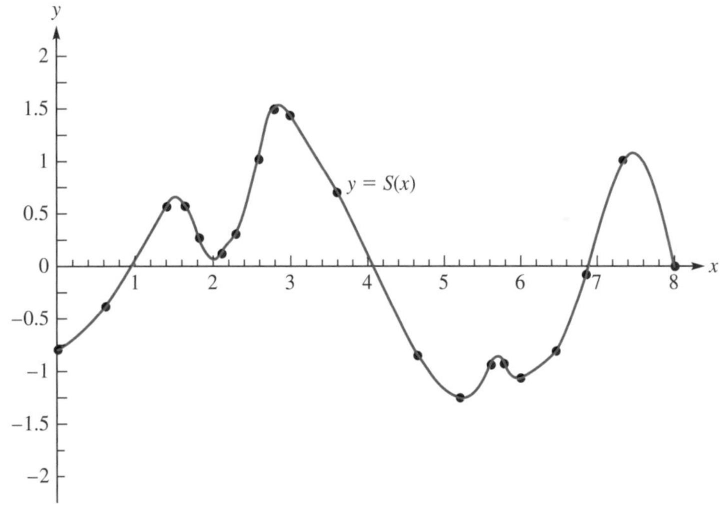

FIGURE 6.16 Quadratic interpolating spline

Schoenberg's Process

An efficient process due to Schoenberg [1967] can also be used to obtain B spline approximations to a given function. Its quadratic version is defined by

Quadratic Case

$$S(x) = \sum_{i=-\infty}^{\infty} f(\tau_i)B_i^2(x) \quad \text{where} \quad \tau_i = \frac{1}{2}(t_{i+1} + t_{i+2}) \tag{19}$$

Here, of course, the knots are $\{t_i\}_{i=-\infty}^{\infty}$, and the points where f must be evaluated are midpoints between the knots.

Equation (19) is useful in producing a quadratic spline function that approximates f. The salient properties of this process are as follows:

■ **Properties**

Schoenberg's Process

1. If $f(x) = ax + b$, then $S(x) = f(x)$.
2. If $f(x) \geqq 0$ everywhere, then $S(x) \geqq 0$ everywhere.
3. $\max_x |S(x)| \leqq \max_x |f(x)|$.
4. If f is continuous on $[a, b]$, if $\delta = \max_i |t_{i+1} - t_i|$, and if $\delta < b - a$, then for x in $[a, b]$,
$$|S(x) - f(x)| \leqq \frac{3}{2} \max_{a \leqq u \leqq v \leqq u+\delta \leqq b} |f(u) - f(v)|$$
5. The graph of S does not cross any line in the plane a greater number of times than does the graph of f.

Some of these properties are elementary; others are more abstruse. Property 1 is outlined in Exercise 6.3.29. Property 2 is obvious because $B_i^2(x) \geqq 0$ for all x. Property 3 follows easily from Equation (19) because if $|f(x)| \leqq M$, then

$$|S(x)| \leqq \left| \sum_{i=-\infty}^{\infty} f(\tau_i) B_i^2(x) \right| \leqq \sum_{i=-\infty}^{\infty} |f(\tau_i)| B_i^2(x) \leqq M \sum_{i=-\infty}^{\infty} B_i^2(x) = M$$

Properties 4 and 5 are accepted without proof. Their significance, however, should not be overlooked. By Property 4, we can make the function S close to a continuous function f simply by making the *mesh size* δ small. This is because $f(u) - f(v)$ can be made as small as we wish simply by imposing the inequality $|u - v| \leqq \delta$ (uniform continuity property). Property 5 can be interpreted as a shape-preserving attribute of the approximation process. In a crude interpretation, S should not exhibit more undulations than f.

Pseudocode: Schoenberg's Process

A pseudocode to obtain a spline approximation by means of Schoenberg's process is developed here. Suppose that f is defined on an interval $[a, b]$ and that the spline approximation of Equation (19) is wanted on the same interval. We define **nodes** $\tau_i = a + ih$, where $h = (b - a)/n$. Here, i can be any integer, but the nodes in $[a, b]$ are only $\tau_0, \tau_1, \dots, \tau_n$. To have $\tau_i = \frac{1}{2}(t_{i+1} + t_{i+2})$, we define the **knots** $t_i = a + (i - \frac{3}{2})h$. In Equation (19), the only B splines B_i^2 that are *active* on $[a, b]$ are $B_{-1}^2, B_0^2, \dots, B_{n+1}^2$. Hence, for our purposes, Equation (19) becomes

$$S(x) = \sum_{i=-1}^{n+1} f(\tau_i) B_i^2(x) \tag{20}$$

Thus, we require the values of f at $\tau_{-1}, \tau_0, \dots, \tau_{n+1}$. Two of these nodes are outside the interval $[a, b]$. Therefore, we furnish linearly extrapolated values in the code by defining

Extrapolation Values

$$\begin{cases} f(\tau_{-1}) = 2f(\tau_0) - f(\tau_1) \\ f(\tau_{n+1}) = 2f(\tau_n) - f(\tau_{n-1}) \end{cases}$$

To use the formulas in Exercise 6.3.26, we write

$$S(x) = \sum_{i=1}^{n+3} D_i B_{i-2}^2(x) \quad [D_i = f(\tau_{i-2})]$$

A pseudocode to compute $D_1, D_2, \ldots, D_{n+3}$ is given now. In the calling sequence for procedure *Schoenberg_Coef*, f is an external function. After execution, the $n + 3$ desired coefficients are in the (d_i) array.

Schoenberg_Coef **Pseudocode**

```
procedure Schoenberg_Coef(f, a, b, n, (d_i))
integer i;   real a, b, h;   real array (d_i)_{1:n+3}
external function f
h ← (b − a)/n
for i = 2 to n + 2
    d_i ← f(a + (i − 2)h)
end for
d_1 ← 2d_2 − d_3
d_{n+3} ← 2d_{n+2} − d_{n+1}
end procedure Schoenberg_Coef
```

After the coefficients D_i have been obtained by the procedure just given, we can recover values of the spline $S(x)$ in Equation (20). Here, we use the algorithm of Exercise 6.3.26. Given an x, we first need to know where it is relative to the knots. To determine k such that $t_{k-1} \leqq x \leqq t_k$, we notice that k should be the largest integer such that $t_{k-1} \leqq x$. This inequality is equivalent to the inequality $k \leqq \frac{5}{2} + (x - a)/h$, as is easily verified. This explains the calculations of k in the pseudocode. The location of x is indicated in Figure 6.17. In the calling sequence for function *Schoenberg_Eval*, a and b are the ends of the interval, and x is a point where the value of $S(x)$ is desired. The procedure determines knots t_i in such a way that the equally spaced points τ_i in the preceding procedure satisfy $\tau_i = \frac{1}{2}(t_{i+1} + t_{i+2})$.

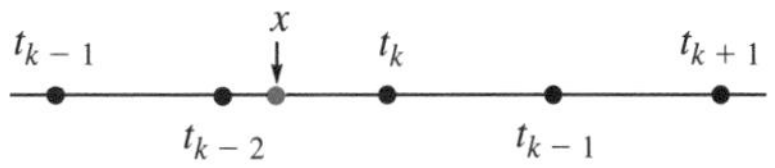

FIGURE 6.17
Location of x

Schoenberg_Eval **Pseudocode**

```
real function Schoenberg_Eval(a, b, n, (d_i), x)
integer k:   real c, h, p, w;   real array (d_i)_{1:n+3}
h ← (b − a)/n
k ← integer[(x − a)/h + 5/2]
p ← x − a − (k − 5/2)h
c ← [d_{k+1}p + d_k(2h − p)]/(2h)
e ← [d_k(p + h) + d_{k−1}(h − p)]/(2h)
Schoenberg_Eval ← [cp + e(h − p)]/h
end function Schoenberg_Eval
```

Bézier Curves

Control Points

In computer-aided design, it is useful to have a procedure for producing a curve that goes through (or near to) some **control points**, or a curve that can be easily manipulated to give a desired shape. High-degree polynomial interpolation is generally not suitable for this sort of task, as one might guess from the negative remarks previously made about them. Experience shows that if one specifies a number of control points through which the polynomial must pass, the overall shape of the resulting curve may be severely disappointing!

Polynomials can be used in a different way, however, leading to **Bézier curves**. Bézier curves use as a basis for the space Π_n (all polynomials of degree not exceeding n) a special set of polynomials that lend themselves to the task at hand. We standardize to the interval [0, 1] and fix a value of n. Next, we define basic polynomial functions

Basic Polynomial Functions

$$\varphi_{ni}(x) = \binom{n}{i} x^i (1-x)^{n-i} \quad (0 \leqq i \leqq n)$$

The polynomials φ_{ni} are the constituents of the **Bernstein polynomials**. For a continuous function f defined on [0, 1], Bernstein, in 1912, proved that the sequence of polynomials

Bernstein Polynomials

$$p_n(x) = \sum_{i=0}^{n} f\left(\frac{i}{n}\right) \varphi_{ni}(x) \quad (n \geqq 1)$$

converges uniformly to f, which provides a very attractive proof of the Weierstrass Approximation Theorem.

The graphs of a few polynomials φ_{ni} are shown in Figure 6.18, where we used $n = 7$ and $i = 0, 1, 5$. The Bernstein basic polynomials are found in mathematical software systems such as Maple, Mathematica, or MATLAB, for example.

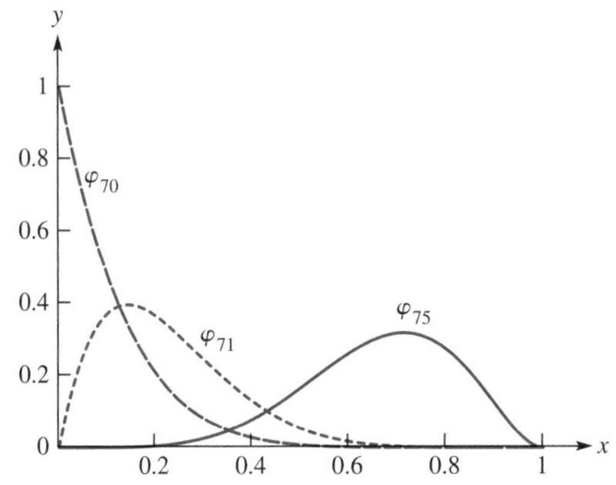

FIGURE 6.18 First few Bernstein basis polynomials

Bernstein polynomials have two salient properties.

■ **Properties**

Bernstein Polynomials

For all x satisfying $0 \leqq x \leqq 1$,

1. $\varphi_{ni}(x) \geqq 0$
2. $\sum_{i=0}^{n} \varphi_{ni}(x) = 1$

Any set of functions having these two properties is called a **partition of unity** on the interval [0, 1]. Notice that the second equation shown is actually valid for all real x. The set $\{\varphi_{n0}, \varphi_{n1}, \dots, \varphi_{nn}\}$ is a basis for the space Π_n. Consequently, every polynomial of degree at most n has a representation

$$\sum_{i=0}^{n} a_i \varphi_{ni}(x)$$

If we want to create a polynomial that comes close to interpolating values $(i/n, y_i)$ for $0 \leqq i \leqq n$, we can use $\sum_{i=0}^{n} y_i \varphi_{ni}$ to start and then, after examining the resulting curve, adjust the coefficients to change the shape of the curve. This is one procedure that can be used in computer-aided design. Changing the value of y_i will change the curve principally in the vicinity of i/n because of the local nature of the basic polynomials φ_{ni}.

Another way in which these polynomials can be used is in creating curves that are not simply graphs of a function f. Here, we turn to a vector form of the suggested procedure. If $n+1$ vectors $v_0, v_1, \ldots, v_n$ are prescribed, say, in $\mathbb{R}^2$ or $\mathbb{R}^3$, the expression

$$u(t) = \sum_{i=0}^{n} \varphi_{ni}(t) v_i \qquad (0 \leqq t \leqq 1)$$

makes sense, since the right-hand side is (for each t) a linear combination of the vectors v_i. As t runs over the interval $[0, 1]$, the vector $u(t)$ describes a curve in the space where the vectors v_i are situated. This curve lies in the **convex hull** of the vectors v_i, because $u(t)$ is a *convex* linear combination of the v_i. This requires the two properties mentioned of φ_{ni}.

To illustrate this procedure, we have selected seven points in the plane and have drawn the closed curve generated by the given equation; that is, by the vector $u(t)$. Figure 6.19 shows the resulting curve as well as the **control points**, which are the vertices of the polygon. Mathematical software systems such as Maple, Mathematica, or MATLAB can be used to do this.

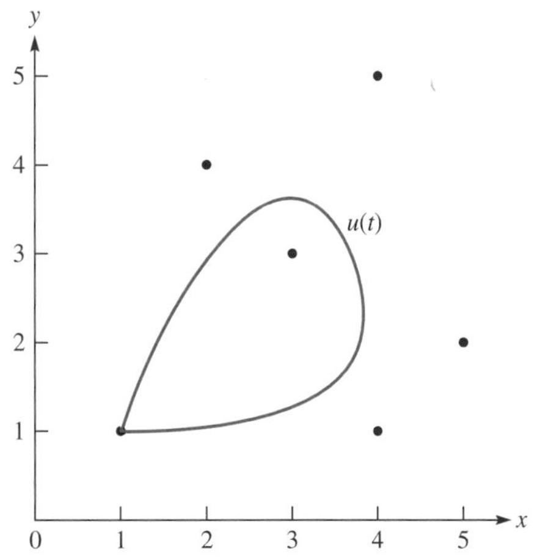

FIGURE 6.19 Curve using control points

A glance at Figure 6.18 suggests that perhaps B splines can be used in the role of the Bernstein functions φ_{ni}. Indeed, that is the case, and B splines have taken over in most programs for computer-aided design. Thus, to obtain a curve that comes close to a set of points (t_i, y_i), we can set up a system of B splines (for example, cubic B splines) having knots t_i. Then the linear combination $\sum_{i=0}^{n} y_i B_i^3$ can be examined to see whether it has the desired shape. Here, of course, B_i^3 denotes a cubic B spline whose support is the interval (t_i, t_{i+4}).

The vector case is like that described previously, except that the functions φ_{ni} are replaced by B_i^3. Also, it is easier to take the knots as integers and let t run from 0 to n. The Properties 1 and 2 of the φ_{ni} displayed are also shared by the B splines.

Summary 6.3

- The **B spline of degree 0** is

$$B_i^0(x) = \begin{cases} 1 & (t_i \leqq x < t_{i+1}) \\ 0 & (\text{otherwise}) \end{cases}$$

Higher-degree B splines are defined recursively:

$$B_i^k(x) = \left(\frac{x - t_i}{t_{i+k} - t_i}\right) B_i^{k-1}(x) + \left(\frac{t_{i+k+1} - x}{t_{i+k+1} - t_{i+1}}\right) B_{i+1}^{k-1}(x)$$

where $k = 1, 2, \ldots$ and $i = 0, \pm 1, \pm 2, \ldots$.

- Some properties are

$$\begin{cases} B_i^k(x) = 0 & x \notin [t_i, t_{i+k+1}) \\ B_i^k(x) > 0 & x \in (t_i, t_{i+k+1}) \end{cases}$$

An efficient method to evaluate a function of the form

$$f(x) = \sum_{i=-\infty}^{\infty} C_i^k B_i^k(x)$$

is to use

$$C_i^{j-1} = C_i^j \left(\frac{x - t_i}{t_{i+j} - t_i}\right) + C_{i-1}^j \left(\frac{t_{i+j} - x}{t_{i+j} - t_i}\right)$$

- The **derivative of B splines** is

$$\frac{d}{dx} B_i^k(x) = \left(\frac{k}{t_{i+k} - t_i}\right) B_i^{k-1}(x) - \left(\frac{k}{t_{i+k+1} - t_{i+1}}\right) B_{i+1}^{k-1}(x)$$

A useful formula is

$$\frac{d}{dx} \sum_{i=-\infty}^{\infty} c_i B_i^k(x) = \sum_{i=-\infty}^{\infty} d_i B_i^{k-1}(x)$$

where $d_i = k(c_i - c_{i-1})/(t_{i+k} - t_i)$. A basic result needed for integration is

$$\int_{-\infty}^{x} B_i^k(s)\, ds = \left(\frac{t_{i+k+1} - t_i}{k+1}\right) \sum_{j=i}^{\infty} B_j^{k+1}(x)$$

A resulting useful formula is

$$\int_{-\infty}^{x} \sum_{i=-\infty}^{\infty} c_i B_i^k(s)\, ds = \sum_{i=-\infty}^{\infty} e_i B_i^{k+1}(x)$$

where $e_i = 1/(k+1) \sum_{j=-\infty}^{i} c_j (t_{j+k+1} - t_j)$.

- To determine the coefficients in the expression

$$S(x) = \sum_{i=-\infty}^{\infty} A_i B_{i-k}^2(x)$$

so that the resulting spline function interpolates a prescribed table, we use the condition

$$A_j(t_{j+1} - t_j) + A_{j+1}(t_j - t_{j-1}) = y_j(t_{j+1} - t_{j-1}) \quad (0 \leqq j \leqq n)$$

This is a system of $n + 1$ linear equations in $n + 2$ unknowns $A_0, A_1, \ldots, A_{n+1}$ that can be solved recursively.

- **Schoenberg's process** is an efficient process to obtain B spline approximations to a given function. For example, its quadratic version is defined by

$$S(x) = \sum_{i=-\infty}^{\infty} f(\tau_i) B_i^2(x)$$

where $\tau_i = \frac{1}{2}(t_{i+1} + t_{i+2})$ and the knots are $\{t_i\}_{i=-\infty}^{\infty}$. The points τ_i where f must be evaluated are midpoints between the knots.

- **Bézier curves** are used in computer-aided design for producing a curve that goes through (or near to) *control points*, or a curve that can be manipulated easily to give a desired shape. Bézier curves use **Bernstein polynomials**. For a continuous function f defined on $[0, 1]$, the sequence of Bernstein polynomials

$$p_n(x) = \sum_{i=0}^{n} f\left(\frac{i}{n}\right)\varphi_{ni}(x) \qquad (n \geqq 1)$$

converges uniformly to f. The polynomials φ_{ni} are

$$\varphi_{ni}(x) = \binom{n}{i} x^i(1-x)^{n-i} \qquad (0 \leqq i \leqq n)$$

Exercises 6.3

1. Show that the functions $f_n(x) = \cos nx$ are generated by this recursive definition:
$$\begin{cases} f_0(x) = 1, \quad f_1(x) = \cos x \\ f_{n+1}(x) = 2f_1(x)f_n(x) - f_{n-1}(x) \quad (n \geqq 1) \end{cases}$$

[a]2. What functions are generated by the following recursive definition?
$$\begin{cases} f_0(x) = 1, \quad f_1(x) = x \\ f_{n+1}(x) = 2xf_n(x) - f_{n-1}(x) \quad (n \geqq 1) \end{cases}$$

[a]3. Find an expression for $B_i^2(x)$ and verify that it is piecewise quadratic. Show that $B_i^2(x)$ is zero at every knot except
$$B_i^2(t_{i+1}) = \frac{t_{i+1} - t_i}{t_{i+2} - t_i}, \qquad B_i^2(t_{i+2}) = \frac{t_{i+3} - t_{i+2}}{t_{i+3} - t_{i+1}}$$

4. Verify Equation (5).

[a]5. Establish that $\sum_{i=-\infty}^{\infty} f(t_i)B_{i-1}^1(x)$ is a first-degree spline that interpolates f at every knot. What is the zero-degree spline that does so?

6. Show that if $t_m \leqq x < t_{m+1}$, then
$$\sum_{i=-\infty}^{\infty} c_i B_i^k(x) = \sum_{i=m-k}^{m} c_i B_i^k(x)$$

7. Let $h_i = t_{i+1} - t_i$. Show that if
$$S(x) = \sum_{i=-\infty}^{\infty} c_i B_i^2(x)$$
$$c_{i-1}h_{i-1} + c_{i-2}h_i = y_i(h_i + h_{i-1})$$
for all i, then $S(t_m) = y_m$ for all m.
Hint: Use Exercise 6.3.3.

8. Show that the coefficients C_i^{j-1} generated by Equation (7) satisfy the condition $\min_i C_i^{j-1} \leqq f(x) \leqq \max_i C_i^{j-1}$.

9. For equally spaced knots, show that $k(k+1)^{-1}B_i^k(x)$ lies in the interval with endpoints $B_i^{k-1}(x)$ and $B_{i+1}^{k-1}(x)$.

10. Show that $B_i^k(x) = B_0^k(x - t_i)$ if the knots are the integers on the real line ($t_i = i$).

11. Show that
$$\int_{-\infty}^{\infty} B_i^k(x)\,dx = \frac{t_{i+k+1} - t_i}{k+1}$$

12. Show that the class of all spline functions of degree m that have knots $x_0, x_1, \ldots, x_n$ *includes* the class of polynomials of degree m.

13. Establish Equation (8) by induction.

[a]14. Which B splines B_i^k have a nonzero value on the interval (t_n, t_m)? Explain.

[a]15. Show that on $[t_i, t_{i+1}]$ we have
$$B_i^k(x) = \frac{(x - t_i)^k}{(t_{i+1} - t_i)(t_{i+2} - t_i)\cdots(t_{i+k} - t_i)}$$

[a]16. Is a spline of the form $S(x) = \sum_{i=-\infty}^{\infty} c_i B_i^k(x)$ *uniquely* determined by a finite set of interpolation conditions $S(t_i) = y_i$ $(0 \leqq i \leqq n)$? Why or why not?

[a]17. If the spline function $S(x) = \sum_{i=-\infty}^{\infty} c_i B_i^k(x)$ vanishes at each knot, must it be identically zero? Why or why not?

18. What is the necessary and sufficient condition on the coefficients in order that $\sum_{i=-\infty}^{\infty} c_i B_i^k = 0$? State and prove.

[a]19. Expand the function $f(x) = x$ in an infinite series $\sum_{i=-\infty}^{\infty} c_i B_i^1$.

[a]**20.** Establish that $\sum_{i=-\infty}^{\infty} B_i^k$ is a constant function by means of Equation (9).

21. Show that if $k \geqq 2$, then

$$\frac{d^2}{dx^2} \sum_{i=-\infty}^{\infty} c_i B_i^k = k(k-1) \sum_{i=-\infty}^{\infty} \left[\frac{c_i - c_{i-1}}{(t_{i+k} - t_i)(t_{i+k-1} - t_i)} - \left[\frac{c_{i-1} - c_{i-2}}{(t_{i+k-1} - t_{i-1})(t_{i+k-1} - t_i)} \right] B_i^{k-2} \right]$$

22. Prove that if the knots are taken to be the integers, then

$$B_{-1}^1(x) = \max\{0, 1 - |x|\}.$$

23. Letting the knots be the integers, show that

$$B_0^2(x) = \begin{cases} 0 & (x < 0) \\ \frac{1}{2}x^2 & (0 \leqq x < 1) \\ \frac{1}{2}(6x - 3 - 2x^2) & (1 \leqq x < 2) \\ \frac{1}{2}(3 - x)^2 & (2 \leqq x < 3) \\ 0 & (x \geqq 3) \end{cases}$$

[a]**24.** Establish formulas

$$B_{i-1}^2(t_i) = \frac{t_i - t_{i-1}}{t_{i+1} - t_{i-1}} = \frac{h_{i-1}}{h_i + h_{i-1}}$$

$$B_{i-2}^2(t_i) = \frac{t_{i+1} - t_i}{t_{i+1} - t_{i-1}} = \frac{h_i}{h_i + h_{i-1}}$$

where $h_i = t_{i+1} - t_i$.

25. Show by induction that if

$$A_j = \frac{1}{t_{j-1} - t_{j-2}} \left[y_{j-1}(t_j - t_{j-2}) - A_{j-1}(t_j - t_{j-1}) \right]$$

for $j = 2, 3, \ldots, n+1$, then

$$\sum_{i=0}^{n+1} A_i B_{i-2}^2(t_j) = y_j \quad (0 \leqq j \leqq n)$$

26. Show that if we set $S(x) = \sum_{i=-\infty}^{\infty} A_i B_{i-2}^2(x)$ and $t_{j-1} \leqq x \leqq t_j$, then

$$S(x) = \frac{1}{t_j - t_{j-1}} [d(x - t_{j-1}) + e(t_j - x)]$$

with

$$d = \frac{1}{t_{j+1} - t_{j-1}} [A_{j+1}(x - t_{j-1}) + A_j(t_{j+1} - x)]$$

and

$$e = \frac{1}{t_j - t_{j-2}} [A_j(x - t_{j-2}) + A_{j-1}(t_j - x)]$$

27. Verify Equations (17) and (18) by induction, using Equation (16).

[a]**28.** If points $\tau_0 < \tau_1 < \cdots < \tau_n$ are given, can we always determine points t_i such that $t_i < t_{i+1}$ and $\tau_i = \frac{1}{2}(t_{i+1} + t_{i+2})$? Why or why not?

29. Show that if $f(x) = x$, then Schoenberg's process produces $S(x) = x$.

[a]**30.** Show that $x^2 = \sum_{i=-\infty}^{\infty} t_{i+1} t_{i+2} B_i^2(x)$.

31. Let $f(x) = x^2$. Assume that $t_{i+1} - t_i \leqq \delta$ for all i. Show that the quadratic spline approximation to f given by Equation (19) differs from f by no more than $\delta^2/4$. *Hint*: Use the preceding problem and the fact that $\sum_{i=-\infty}^{\infty} B_i^2 \equiv 1$.

[a]**32.** Verify (for $k > 0$) that $B_i^k(t_j) = 0$ if and only if $j \leqq i$ or $j \geqq i + k + 1$.

[a]**33.** What is the maximum value of B_i^2 and where does it occur?

34. Let the knots be the integers, and prove that

$$B_0^3(x) = \begin{cases} 0 & (x < 0) \\ \frac{1}{6}x^3 & (0 \leqq x < 1) \\ \frac{1}{6}(4 - 3x(x-2)^2) & (1 \leqq x < 2) \\ \frac{1}{6}(4 + 3(x-4)(x-2)^2) & (2 \leqq x < 3) \\ \frac{1}{6}(4 - x)^3 & (3 \leqq x < 4) \\ 0 & (x \geqq 4) \end{cases}$$

35. In the theory of Bézier curves, using the Bernstein basic polynomials, show that the curve passes through the first point, v_0.

36. Show that a **linear B spline** with integer knots can be written in matrix form as

$$S(x) = [x \quad 1] \begin{bmatrix} -1 & 1 \\ 2 & 0 \end{bmatrix} \begin{bmatrix} c_1 \\ c_0 \end{bmatrix} = b_{10}c_0 + b_{11}c_1$$

where

$$B_0^1(x) = \begin{cases} b_{10} = x & (0 \leqq x < 1) \\ b_{11} = 2 - x & (1 \leqq x < 2) \\ 0 & \text{(otherwise)} \end{cases}$$

37. Show that the **quadratic B spline** with integer knots can be written in matrix form as

$$S(x) = \tfrac{1}{2}[x^2 \quad x \quad 1] \begin{bmatrix} 1 & -2 & 1 \\ -6 & 6 & 0 \\ 9 & -3 & 0 \end{bmatrix} \begin{bmatrix} c_2 \\ c_1 \\ c_0 \end{bmatrix}$$

$$= b_{20}c_0 + b_{21}c_1 + b_{22}c_2$$

where

$$B_0^2(x) = \begin{cases} b_{20} & (0 \leqq x < 1) \\ b_{21} & (1 \leqq x < 2) \\ b_{22} & (2 \leqq x < 3) \\ 0 & \text{(otherwise)} \end{cases}$$

Hint: See Exercise 6.3.23.

38. Show that the **cubic B spline** with integer knots can be written as

$$S(x) = \frac{1}{6}[x^3 \quad x^2 \quad x \quad 1] \begin{bmatrix} -1 & 3 & -3 & 1 \\ 12 & -24 & 12 & 0 \\ -48 & 60 & -12 & 0 \\ 64 & -44 & 4 & 0 \end{bmatrix} \begin{bmatrix} c_3 \\ c_2 \\ c_1 \\ c_0 \end{bmatrix}$$
$$= b_{30}c_0 + b_{31}c_1 + b_{32}c_2 + b_{33}c_3$$

where

$$B_0^3(x) = \begin{cases} b_{30} & (0 \leqq x < 1) \\ b_{31} & (1 \leqq x < 2) \\ b_{32} & (2 \leqq x < 3) \\ b_{33} & (3 \leqq x < 4) \\ 0 & \text{(otherwise)} \end{cases}$$

Hint: See Exercise 6.3.34.

Computer Exercises 6.3

1. Using an automatic plotter, graph B_0^k for $k = 0, 1, 2, 3, 4$. Use integer knots $t_i = i$ over the interval $[0, 5]$.

2. Let $t_i = i$ (so the knots are the integer points on the real line). Print a table of 100 values of the function $3B_7^1 + 6B_8^1 - 4B_9^1 + 2B_{10}^1$ on the interval $[6, 14]$. Using a plotter, construct the graph of this function on the given interval.

3. (Continuation) Repeat for the function $3B_7^2 + 6B_8^2 - 4B_9^2 + 2B_{10}^2$.

4. Assuming that $S(x) = \sum_{i=0}^{n} c_i B_i^k(x)$, write a procedure to evaluate $S'(x)$ at a specified x. Input is $n, k, x, t_0, \dots, t_{n+k+1}$ and $c_0, c_1, \dots, c_n$.

5. Write a procedure to evaluate $\int_a^b S(x)\,dx$, using the assumption that $S(x) = \sum_{i=0}^{n} c_i B_i^k(x)$. Input will be $n, k, a, b, c_0, c_1, \dots, c_n, t_0, \dots, t_{n+k+1}$.

6. (**March of the B splines**) Produce graphs of several B splines of the same degree *marching* across the x-axis. Use mathematical software such as MATLAB, Maple, or Mathematica.

[a]**7.** Historians have estimated the size of the Spanish Army of Flanders in the Spanish Netherlands as follows:

Date	Sept. 1572	Dec. 1573	Mar. 1574
Number	67, 259	62, 280	62, 350

Jan. 1575	May 1576	Feb. 1578	Sept. 1580
59, 250	51, 457	27, 603	45, 435

Oct. 1582	Apr. 1588	Nov. 1591	Mar. 1607
61, 162	63, 455	62, 164	41, 471

Fit the table with a quadratic B spline, and use it to find the average size of the army during the period given. (The *average* is defined by an integral.)

8. Rewrite procedures *BSpline2_Coef* and *BSpline_Eval* so that the array (h_i) is not used.

9. Rewrite procedures *BSpline2_Coef* and *BSpline2_Eval* for the special case of equally spaced knots, simplifying the code where possible.

10. Write a procedure to produce a spline approximation to $F(x) = \int_a^x f(t)\,dt$. Assume that $a \leqq x \leqq b$. Begin by finding a quadratic spline interpolant to f at the n points $t_i = a + i(b-a)/n$. Test your program on the following:

a. $f(x) = \sin x \quad (0 \leqq x \leqq \pi)$
b. $f(x) = e^x \quad (0 \leqq x \leqq 4)$
c. $f(x) = (x^2 + 1)^{-1} \quad (0 \leqq x \leqq 2)$

11. Write a procedure to produce a spline function that approximates $f'(x)$ for a given f on a given interval $[a, b]$. Begin by finding a quadratic spline interpolant to f at $n + 1$ points evenly spaced in $[a, b]$, including endpoints. Test your procedure on the functions suggested in the preceding computer exercise.

12. Define f on $[0, 6]$ to be a polygonal line that joins points $(0, 0)$, $(1, 2)$, $(3, 3)$, $(5, 3)$, and $(6, 0)$. Determine spline approximations to f, using Schoenberg's process and taking 7, 13, 19, 25, and 31 knots.

13. Write suitable code to calculate $\sum_{i=-\infty}^{\infty} f(s_i)B_i^2(x)$ with $s_i = \frac{1}{2}(t_{i+1} + t_{i+2})$. Assume that f is defined on $[a, b]$ and that x will lie in $[a, b]$. Assume also that $t_1 < a < t_2$ and $t_{n+1} < b < t_{n+2}$. (Make no assumption about the spacing of knots.)

14. Write a procedure to carry out this approximation scheme:

$$S(x) = \sum_{i=-\infty}^{\infty} f(\tau_i)B_i^3(x),$$

where

$$\tau_i = \frac{1}{3}(t_{i+1} + t_{i+2} + t_{i+3})$$

Assume that f is defined on $[a, b]$ and that $\tau_i = a + ih$ for $0 \leq i \leq n$, where $h = (b - a)/n$.

15. Using a mathematical software system such as MATLAB, Maple, or Mathematica with B spline routines, compute and plot the spline curve in Figure 6.16 (p. 273) based on the 20 data points from Section 6.2. Vary the degree of the B splines from 0, 1, 2, 3, through 4 and observe the resulting curves.

16. Using B splines, write a program to perform a natural cubic spline interpolation at knots $t_0 < t_1 < \cdots < t_n$.

17. The documentation preparation system LaTeX is widely available and contains facilities for drawing some simple curves such as Bézier curves. Use this system to reproduce the following figure.

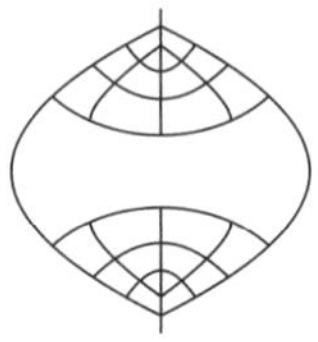

18. Show how to use mathematical software such as found in MATLAB, Maple, or Mathematica to plot the functions corresponding to

a. Figure 6.14. **b.** Figure 6.15.

c. Figure 6.18. **d.** Figure 6.19.

19. (**Computer-Aided Geometric Design**) Use mathematical software for drawing two-dimensional Bézier spline curves, and graph the script number five shown, using spline points and control points. See Farin [1990], Sauer [2012], and Yamaguchi [1988] for additional details.

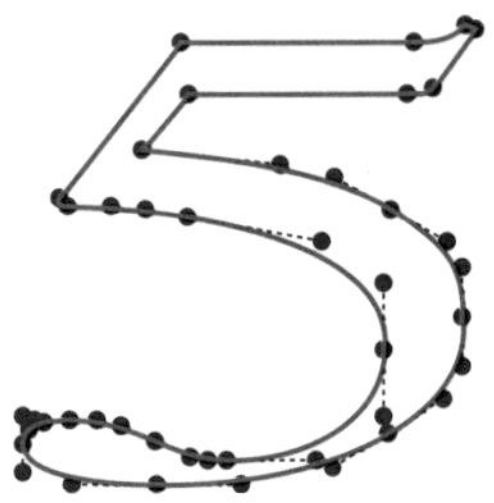

7 Initial Values Problems

In a simple electrical circuit, the current $\mathcal{I}$ in amperes is a function of time: $\mathcal{I}(t)$. The function $\mathcal{I}(t)$ satisfies an ordinary differential equation of the form

$$\frac{d\mathcal{I}}{dt} = f(t, \mathcal{I})$$

Here, the right-hand side is a function of t and $\mathcal{I}$ that depends on the circuit and on the nature of the electromotive force supplied to the circuit.

A model to account for the way in which two different animal species sometimes interact is the **predator-prey model**. If $u(t)$ is the number of individuals in the predator species and $v(t)$ the number of individuals in the prey species, then under suitable simplifying assumptions and with appropriate constants a, b, c, and d,

$$\begin{cases} \dfrac{du}{dt} = a(v+b)u \\ \dfrac{dv}{dt} = c(u+d)v \end{cases}$$

This is a pair of nonlinear ordinary differential equations (ODEs) that govern the populations of the two species (as functions of time t).

Numerical procedures are developed for solving such problems.

7.1 Taylor Series Methods

First, we present a general discussion of ordinary differential equations and their solutions.

Initial-Value Problem: Analytical versus Numerical Solution

An **ordinary differential equation** (ODE) is an equation that involves one or more derivatives of an unknown function. A **solution** of a differential equation is a specific function that satisfies the equation. Here are some examples of differential equations with their solutions. In each case, t is the independent variable and x is the dependent variable. Thus, x is the name of the unknown function of the independent variable t:

ODE Examples

Equation	Solution
$x' - x = e^t$	$x(t) = te^t + ce^t$
$x'' + 9x = e^t$	$x(t) = c_1 \sin 3t + c_2 \cos 3t$
$x' + (1/2)x = 0$	$x(t) = \sqrt{c - t}$

In these three examples, the letter c denotes an arbitrary constant. The fact that such constants appear in the solutions is an indication that a differential equation does not, in general, determine a unique solution function. When occurring in a scientific problem, a differential equation is usually accompanied by auxiliary conditions that (together with the differential equation) specify the unknown function precisely.

In this chapter, we concentrate on one type of differential equation and one type of auxiliary condition: the **initial-value problem** for a first-order differential equation. The standard form that has been adopted is

IVP Standard Form

$$\begin{cases} x' = f(t, x) \\ x(a) \text{ is given} \end{cases} \tag{1}$$

It is understood that x is a function of t, so the differential equation written in more detail looks like this:

$$\frac{dx(t)}{dt} = f(t, x(t))$$

Problem (1) is termed an initial-value problem because t can be interpreted as time and $t = a$ can be thought of as the initial instant in time. We want to be able to determine the value of x at any time t before or after a.

Here are some examples of initial-value problems, together with their solutions:

IVP Examples

Equation	**Initial Value**	**Solution**
$x' = x + 1$	$x(0) = 0$	$x = e^t - 1$
$x' = 6t - 1$	$x(1) = 6$	$x = 3t^2 - t + 4$
$x' = t/(x + 1)$	$x(0) = 0$	$x = \sqrt{t^2 + 1} - 1$

Although many methods exist for obtaining analytical solutions of differential equations, they are primarily limited to special differential equations. When applicable, they produce a solution in the form of a formula, such as shown in the preceding examples. Frequently, however, in practical problems, a differential equation is not amenable to solution by special methods, and a numerical solution must be sought. Even when a formal solution can be obtained, a numerical solution may be preferable, especially if the formal solution is very complicated. A numerical solution of a differential equation is usually obtained in the form of a table; the functional form of the solution remains unknown insofar as a specific formula is concerned.

The form of the differential equation adopted here permits the function f to depend on t and x. If f does not involve x, as in the second preceding example, then the differential equation can be solved by a direct process of indefinite integration. To illustrate, consider the initial-value problem

Direct Approach of Indefinite Integration

$$\begin{cases} x' = 3t^2 - 4t^{-1} + (1 + t^2)^{-1} \\ x(5) = 17 \end{cases} \tag{2}$$

The differential equation can be integrated to produce

$$x(t) = t^3 - 4 \ln t + \arctan t + C$$

The constant C can then be chosen so that $x(5) = 17$. We can use a mathematical software system such as MATLAB, Maple, or Mathematica to solve this differential equation explicitly and thereby find the value of this constant as $C = 4 \ln(5) - \arctan(5) - 108$.

We often want a numerical solution to a differential equation because (a) the *closed-form* solution may be very complicated and difficult to evaluate or (b) there is no other choice; that is, no *closed-form* solution can be found. Consider, for instance, the differential equation

No Closed-Form Solution

$$x' = e^{-\sqrt{t^2 - \sin t}} + \ln|\sin t + \tanh t^3| \tag{3}$$

The solution is obtained by taking the integral or antiderivative of the right-hand side. It can be done in principle, but not in practice. In other words, a function x exists for which dx/dt is the right-hand member of Equation (3), but it is not possible to write $x(t)$ in terms of familiar functions.

Solving ordinary differential equations on a computer may require a large number of steps with small step size, so a significant amount of roundoff error can accumulate. Consequently, multiple-precision computations may be necessary on small-word-length computers.

An Example of a Practical Problem

Many practical problems in dynamics involve Newton's three **Laws of Motion**, particularly the Second Law. It states symbolically that $F = ma$, where F is the force acting on a body of mass m and a is the resulting acceleration of that body. This law is a differential equation in disguise because a, the acceleration, is the derivative of velocity and velocity is, in turn, the derivative of the position. We illustrate with a simplified model of a rocket being fired at time $t = 0$. Its motion is to be vertically upward, and we measure its height with the variable x. The propulsive force is a constant value, namely, 5370. (Units are chosen to be consistent with each other.) There is a negative force due to air resistance whose magnitude is $v^{3/2}/\ln(2 + v)$, where v is the velocity of the rocket. The mass is decreasing at a steady rate due to the burning of fuel and is taken to be $321 - 24t$. The independent variable is time, t. The fuel is completely consumed by the time $t = 10$. There is a downward force, due to gravity, of magnitude 981. Putting all these terms into the equation $F = ma$, we have

Simplified Model of a Rocket

$$5370 - 981 - v^{3/2}/\ln(2 + v) = (321 - 24t)v' \tag{4}$$

The initial condition is $v = 0$ at $t = 0$.

We shall develop methods to solve such differential equations in the succeeding sections. Moreover, one can also invoke a mathematical software system to solve this problem.

Using an ODE Computer Code

A computer code for solving ordinary differential equations produces a table of discrete values, whereas the mathematical solution is a continuous function. One may need additional values within an interval for various purposes, such as plotting. Interpolation procedures can be used to obtain all values of the approximate numerical solution within a given interval. For example, a piecewise polynomial interpolation scheme may yield a numerical solution that is continuous and has a continuous first derivative matching the derivative of the solution. In using any ODE solver, an approximation to $x'(t)$ is available from the fact that

$$x'(t) = f(t, x)$$

Mathematical packages for solving ODEs may include automatic plotting capabilities because the best way to make sense out of the large amount of data that may be returned as the *solution* is to display the solution curves on a graphical monitor or plot them on paper.

Solving Differential Equations and Integration

There is a close connection between solving differential equations and integration. Consider the differential equation

$$\begin{cases} \dfrac{dx}{dr} = f(r, x) \\ x(a) = s \end{cases}$$

Integrating from t to $t + h$, we have

$$\int_t^{t+h} dx = \int_t^{t+h} f(r, x(r))\, dr$$

Hence, we obtain

$$x(t + h) = x(t) + \int_t^{t+h} f(r, x(r))\, dr$$

Integration Rules Yield ODE Methods

Replacing the integral with one of the numerical integration rules from Chapter 5, we obtain a formula for solving the differential equation. For example, Euler's method, Equation (6), is obtained from the left rectangle approximation. (See Exercise 5.2.33):

$$\int_t^{t+h} f(r, x(r))\, dr \approx hf(t, x(t))$$

The trapezoid rule

$$\int_t^{t+h} f(r, x(r))\, dr \approx \frac{h}{2}[f(t, x(t)) + f(t + h, x(t + h))]$$

gives the formula

$$x(t + h) = x(t) + \frac{h}{2}[f(t, x(t)) + f(t + h, x(t + h))]$$

Since $x(t + h)$ appears on both sides of this equation, it is called an **implicit formula**. If Euler's method

$$x(t + h) = x(t) + hf(t, x(t))$$

is used for the $x(t + h)$ on the right-hand side, then we obtain the Runge-Kutta formula of order 2—namely, Equation (10) in Section 7.2.

Using the Fundamental Theorem of Calculus, we can easily show that an approximate numerical value for the integral

Integration versus Solving IVP

$$\int_a^b f(r, x(r))\, dr$$

can be computed by solving the following initial-value problem for $x(b)$:

$$\begin{cases} \dfrac{dx}{dr} = f(r, x) \\ x(a) = 0 \end{cases}$$

Vector Fields

Consider a generic first-order differential equation with prescribed initial condition:

Generic First-Order ODE

$$\begin{cases} x'(t) = f(t, x(t)) \\ x(a) = b \end{cases}$$

Before addressing the question of solving such an initial-value problem numerically, it is helpful to think about the intuitive meaning of the equation. The function f provides the slope of the solution function in the tx-plane. At every point where $f(t, x)$ is defined, we can imagine a short line segment being drawn through that point and having the prescribed slope. We cannot graph *all* of these short segments, but we can draw as many as we wish, in the hope of understanding how the solution function $x(t)$ traces its way through this forest of line segments while keeping its slope at every point equal to the slope of the line segment drawn at that point. The diagram of line segments illustrates discretely the so-called **vector field** of the differential equation.

For example, let us consider the equation

$$x' = \sin(x + t^2)$$

First Vector Field Example

with initial value $x(0) = 0$. In the rectangle described by the inequalities $-4 \leqq x \leqq 4$ and $-4 \leqq t \leqq 4$, we can direct mathematical software, such as MATLAB, to furnish a picture of the vector field engendered by our differential equation. Using commands in an environment of windows, we bring up a window with the differential equation shown in a rectangle. Behind the scenes, the mathematical software carries out immense calculations to provide the vector field for this differential equation, and displays it, correctly labeled. To see the solution going through any point in the diagram, it is necessary only to use the mouse to position the pointer on such a point. By clicking the left mouse button, the software displays the solution sought. By using such a software tool, one can see immediately the effect of changing initial conditions. For the problem under consideration, several solution curves (corresponding to different initial values) are shown in Figure 7.1.

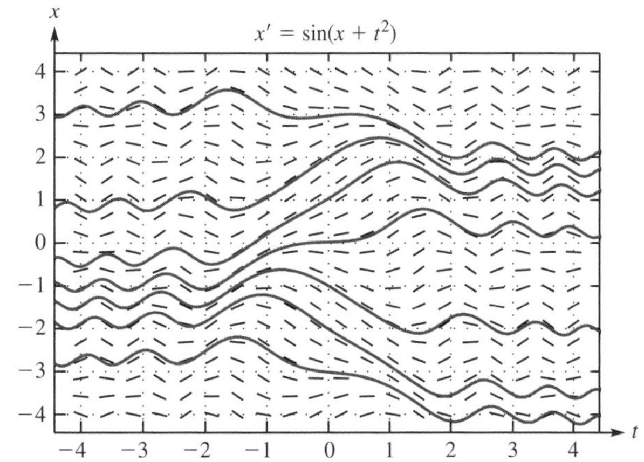

FIGURE 7.1 Vector field and some solution curves for $x' = \sin(x + t^2)$

Second-Order Field Example

Another example, treated in the same way, is the differential equation

$$x' = x^2 - t$$

Figure 7.2 (p. 304) shows a vector field for this equation and some of its solutions. Notice the phenomenon of many quite different curves all seeming to arise from the same initial

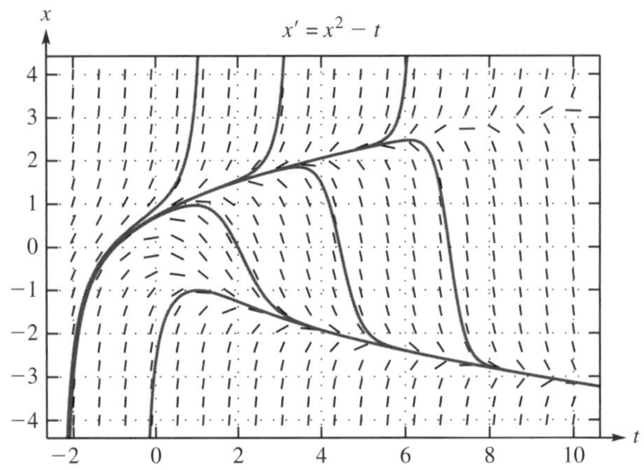

FIGURE 7.2
Vector field and some solution curves for $x' = x^2 - t$

condition. *What is happening here?* This is an extreme example of a differential equation whose solutions are exceedingly sensitive to the initial condition! Expect trouble in solving this differential equation with an initial value prescribed at $t = -2$.

How do we know that the differential equation $x' = x^2 - t$*, together with an initial value,* $x(t_0) = x_0$*, has a unique solution?* There are many theorems in the subject of differential equations that concern such existence and uniqueness questions. One of the easiest to use is as follows.

Theorem 1 **Uniqueness of Initial-Value Problems**

If f and $\partial f/\partial y$ are continuous in the rectangle defined by $|t - t_0| < \alpha$ and $|x - x_0| < \beta$, then the initial-value problem $x' = f(t, x)$, $x(t_0) = x_0$ has a unique continuous solution in some interval $|t - t_0| < \epsilon$.

From the theorem just quoted, we cannot conclude that the solution in question is defined for $|t - t_0| < \beta$. However, the value of ϵ in the theorem is at least β/M, where M is an upper bound for $|f(t, x)|$ in the original rectangle.

Taylor Series Methods

The numerical method described in this section does not have the utmost generality, but it is natural and capable of high precision. Its principle is to represent the solution of a differential equation locally by a few terms of its Taylor series.

In what follows, we shall assume that our solution function x is represented by its Taylor series*

Taylor Series

$$x(t + h) = x(t) + hx'(t) + \frac{1}{2!}h^2x''(t) + \frac{1}{3!}h^3x'''(t) + \frac{1}{4!}h^4x^{(iv)}(t) + \frac{1}{5!}h^5x^{(v)}(t) + \cdots + \frac{1}{m!}h^m x^{(m)}(t) + \cdots \tag{5}$$

*Remember that some functions such as e^{-1/x^2} are smooth, but *not* represented by a Taylor series at 0.

For numerical purposes, the Taylor series truncated after $m+1$ terms enables us to compute $x(t+h)$ rather accurately if h is small and if $x(t), x'(t), x''(t), \ldots, x^{(m)}(t)$ are known. When only terms through $h^m x^{(m)}(t)/m!$ are included in the Taylor series, the method that results is called the **Taylor series method of order m**. We begin with the case $m = 1$.

Euler's Method and Pseudocode

The Taylor series method of order 1 is known as **Euler's method**. To find approximate values of the solutions to the initial-value problem

$$\begin{cases} x' = f(t, x(t)) \\ x(a) = x_a \end{cases}$$

over the interval $[a, b]$, the first two terms in the Taylor series (5) are used:

$$x(t+h) \approx x(t) + hx'(t)$$

Hence, the formula

Euler's Method

$$x(t+h) = x(t) + hf(t, x(t)) \tag{6}$$

can be used to step from $t = a$ to $t = b$ with n steps of size $h = (b-a)/n$.

The pseudocode for Euler's method can be written as follows, where some prescribed values for n, a, b, and x_a are used:

***Euler* Pseudocode**

```
program Euler
integer k;   real h, t;   integer n ← 100
external function f
real a ← 1, b ← 2, x ← −4
h ← (b − a)/n
t ← a
output 0, t, x
for k = 1 to n
    x ← x + hf(t, x)
    t ← t + h
    output k, t, x
end for
end program Euler
```

To use this program, a code for $f(t, x)$ is needed, as shown in Example 1.

EXAMPLE 1 Using Euler's method, compute an approximate value for $x(2)$ for the differential equation $x' = 1 + x^2 + t^3$ with the initial value $x(1) = -4$ using 100 steps.

Solution Use the pseudocode above with the initial values given and combine with the following function:

```
real function f(t, x)
real t, x
f ← 1 + x² + t³
end function
```

The computed value is $x(2) \approx 4.23585$. ■

We can write a computer program to execute Euler's method on this very simple problem:

$$\begin{cases} x'(t) = x \\ x(0) = 1 \end{cases}$$

We obtain the results $x(2) \approx 7.3891$. The plot produced by the code is shown in Figure 7.3. The solution, $x(t) = e^t$, is the solid curve, and the points produced by Euler's method are shown by dots. *Can you understand why the dots are always below the curve?*

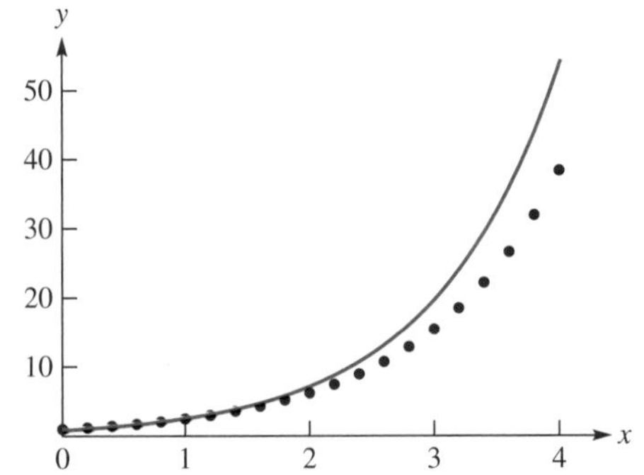

FIGURE 7.3
Euler's method curves

Important Questions

Before accepting these results and continuing, one should raise some questions such as: *How accurate are the answers? Are higher-order Taylor series methods ever needed?* Unfortunately, Euler's method is *not* very accurate because only two terms in the Taylor series (5) are used. Consequently, the truncation error is $\mathcal{O}(h^2)$.

Taylor Series Method of Higher Order

Example 1 can be used to explain the Taylor series method of higher order. Consider again the initial-value problem

Sample IVP

$$\begin{cases} x' = 1 + x^2 + t^3 \\ x(1) = -4 \end{cases} \tag{7}$$

If the functions in the differential equation are differentiated several times with respect to t, the results are as follows. (Remember that a function of x must be differentiated with respect to t by using the *chain rule*.)

$$\begin{aligned} x' &= 1 + x^2 + t^3 \\ x'' &= 2xx' + 3t^2 \\ x''' &= 2xx'' + 2x'x' + 6t \\ x^{(iv)} &= 2xx''' + 6x'x'' + 6 \end{aligned} \tag{8}$$

If numerical values of t and $x(t)$ are known, these four formulas, applied in order, yield $x'(t), x''(t), x'''(t)$, and $x^{(iv)}(t)$. Thus, it is possible from this work to use the first five terms in the Taylor series, Equation (5). Since $x(1) = -4$, we have a suitable starting point, and we select $n = 100$, which determines h. Next, we can compute an approximation to $x(a+h)$ from Formulas (5) and (8). The same process can be repeated to compute $x(a + 2h)$ using $x(a + h), x'(a + h), \ldots, x^{(iv)}(a + h)$.

Pseudocode and Results

Here is the pseudocode:

Taylor Pseudocode

```
program Taylor
integer k;   real h, t, x, x′, x″, x‴, x^(iv)
integer n ← 100
real a ← 1, b ← 2, x ← −4
h ← (b − a)/n
t ← a
output 0, t, x
for k = 1 to n
    x′ ← 1 + x^2 + t^3
    x″ ← 2xx′ + 3t^2
    x‴ ← 2xx″ + 2(x′)^2 + 6t
    x^(iv) ← 2xx‴ + 6x′x″ + 6
    x ← x + h [x′ + (1/2)h [x″ + (1/3)h [x‴ + (1/4)h [x^(iv)]]]]
    t ← a + kh
    output k, t, x
end for
end program Taylor
```

A few words of explanation may be helpful here. In this example, we compute the solution of the differential equation over the interval $a = 1 \leqq t \leqq 2 = b$ using 100 steps. In each step, the current value of t is an integer multiple of the step size h. The assignment statements that define x', x'', x''', and $x^{(iv)}$ are simply carrying out calculations of the derivatives according to Equation (8). The final calculation carries out the evaluation of the Taylor series in Equation (5) using five terms. Since this equation is a polynomial in h, it is evaluated most efficiently by using nested multiplication, which explains the formula for x in the pseudocode. The computation $t \leftarrow t + h$ may cause a small amount of roundoff error to accumulate in the value of t. This is minimized by using $t \leftarrow a + kh$.

As one might expect, the results of using only two terms in the Taylor series (Euler's method) are *not* as accurate as when five terms are used:

Computer Results

Euler's Method	**Taylor Series Method (Order 4)**
$x(2) \approx 4.23585\,41$	$x(2) \approx 4.37120\,96$

By further analysis, one can prove that the correct value to more significant figures is $x(2) \approx 4.37122\,1866$. Here, the computations were done with more precision just to show that lack of precision was not a contributing factor.

Types of Errors

When the pseudocode described above is programmed and run on a computer, what sort of accuracy can we expect? Are all the digits printed by the machine for the variable x accurate? Of course not! On the other hand, it is not easy to say how many digits *are* reliable. Here is a coarse assessment. Since terms up to $\frac{1}{24}h^4x^{(iv)}(t)$ are included, the first term *not* included in the Taylor series is $\frac{1}{120}h^5x^{(v)}(t)$. The error may be larger than this, but the factor $h^5 = (10^{-2})^5 \approx 10^{-10}$ is affecting only the tenth decimal place. The printed solution is perhaps accurate to eight decimal places. Bridges or airplanes should *not* be built on such *shoddy analysis*, but for now, our attention is focused on the general form of the procedure.

Local Truncation Error

Actually, there are two types of errors to consider. At each step, if $x(t)$ is known and $x(t+h)$ is computed from the first few terms of the Taylor series, an error occurs because we have truncated the Taylor series. This error, then, is called the **truncation error** or, to be more precise, the **local truncation error**. In the preceding example, it is roughly $\frac{1}{120}h^5x^{(v)}(\xi)$. In this situation, we say that the local truncation error is *of order* h^5, abbreviated by $\mathcal{O}(h^5)$.

The second type of error obviously present is due to the accumulated effects of all local truncation errors. Indeed, the calculated value of $x(t+h)$ is in error because $x(t)$ is already wrong (because of previous truncation errors) and because another local truncation error occurs in the computation of $x(t+h)$ by means of the Taylor series.

Roundoff Error

Additional sources of errors must be considered in a complete theory. One is **roundoff error**. Although not serious in any one step of the solution procedure, after hundreds or thousands of steps, it may accumulate and contaminate the calculated solution seriously! Remember that an error that is made at a certain step is carried forward into all succeeding steps. Depending on the differential equation and the method that is used to solve it, such errors may be magnified by succeeding steps.

Taylor Series Method Using Symbolic Computations

Various routine mathematical calculations of both a nonnumerical and a numerical type, including differentiation and integration of even rather complicated expressions, can now be turned over to the computer. Of course, this applies only to a restricted class of functions, but this class is broad enough to include all the functions that one encounters in the typical calculus textbook. With the use of such a program for symbolic computations, the Taylor series method of high order can be carried out without difficulty. Using the algebraic manipulation potentialities in mathematical software such as MATLAB, Maple or Mathematica, we can write code to solve the initial value problem (7). The final result is $x(2) \approx 4.37121\,00522\,49692\,27234\,569$.

Summary 7.1

- We wish to solve the first-order initial-value problem

$$\begin{cases} x'(t) = f(t, x(t)) \\ x(a) = x_a \end{cases}$$

  over the interval $[a, b]$ with step size $h = (b - a)/n$.

- The **Taylor series method of order *m*** is

$$x(t+h) = x(t) + hx'(t) + \frac{1}{2!}h^2x''(t) + \frac{1}{3!}h^3x'''(t) + \frac{1}{4!}h^4x^{(iv)}(t) + \frac{1}{5!}h^5x^{(v)}(t) + \cdots + \frac{1}{m!}h^mx^{(m)}(t)$$

  where all of the derivatives $x'', x''', \ldots, x^{(m)}$ have been determined analytically.

- **Euler's method** is the Taylor series method of order 1 and can be written as

$$x(t+h) = x(t) + hf(t, x(t))$$

Because only two terms in the Taylor series are used, the truncation error is large, and the results cannot be computed with much accuracy. Consequently, higher-order Taylor series methods are used most often. Of course, they require determining more derivatives, with more chances for mathematical errors.

Exercises 7.1

1. Give the solutions of these differential equations:

[a]**a.** $x' = t^3 + 7t^2 - t^{1/2}$ [a]**b.** $x' = x$

c. $x' = -x$ **d.** $x'' = -x$

[a]**e.** $x'' = x$ **f.** $x'' + x' - 2x = 0$
Hint: Try $x = e^{at}$.

2. Give the solutions of these initial-value problems:

[a]**a.** $x' = t^2 + t^{1/3}$ $x(0) = 7$

b. $x' = 2x$ $x(0) = 15$

c. $x'' = -x$ $x(\pi) = 0$ $x'(\pi) = 3$

3. Solve the following differential equations:

a. $x' = 1 + x^2$ *Hint*: $1 + \tan^2 t = \sec^2 t$

b. $x' = \sqrt{1 - x^2}$ *Hint*: $\sin^2 t + \cos^2 t = 1$

[a]**c.** $x' = t^{-1} \sin t$
Hint: See Computer Exercise 5.1.2.

[a]**d.** $x' + tx = t^2$
Hint: Multiply by $f(t) = \exp(t^2/2)$ so left-hand side becomes $(xf)'$.

[a]**4.** Solve Exercise 7.1.3b by substituting a power series $x(t) = \sum_{n=0}^{\infty} a_n t^n$ and then determining appropriate values of the coefficients.

5. Determine x'' when $x' = xt^2 + x^3 + e^x t$.

[a]**6.** Find a polynomial p with the property $p - p' = t^3 + t^2 - 2t$.

7. The general first-order linear differential equation is $x' + px + q = 0$, where p and q are functions of t. Show that the solution is $x = -y^{-1}(z + c)$, where y and z are functions obtained as follows: Let u be an antiderivative of p. Put $y = e^u$, and let z be an antiderivative of yq.

8. Here is an initial-value problem that has two solutions: $x' = x^{1/3}$, $x(0) = 0$. Verify that the two solutions are $x_1(t) = 0$ and $x_2(t) = \left(\frac{2}{3}t\right)^{3/2}$ for $t \geq 0$. If the Taylor series method is applied, what happens?

[a]**9.** Consider the problem $x' = x$. If the initial condition is $x(0) = c$, then the solution is $x(t) = ce^t$. If a roundoff error of ε occurs in reading the value of c into the computer, what effect is there on the solution at the point $t = 10$? At $t = 20$? Do the same for $x' = -x$.

[a]**10.** If the Taylor series method is used on the initial-value problem $x' = t^2 + x^3$, $x(0) = 0$, and if we intend to use the derivatives of x up to and including $x^{(iv)}$, what are the five main equations that must be programmed?

11. In solving the following differential equations by the Taylor series method of order n, what are the main equations in the algorithm?

[a]**a.** $x' = x + e^x, \quad n = 4$

b. $x' = x^2 - \cos x, \quad n = 5$

[a]**12.** Calculate an approximate value for $x(0.1)$ using one step of the Taylor series method of order 3 on the ordinary differential equation

$$\begin{cases} x'' = x^2 e^t + x' \\ x(0) = 1, \quad x'(0) = 2 \end{cases}$$

13. Suppose that a differential equation is solved numerically on an interval $[a, b]$ and that the local truncation error is ch^p. Show that if all truncation errors have the same sign (the worst possible case), then the total truncation error is $(b - a)ch^{p-1}$, where $h = (b - a)/n$.

[a]**14.** If we plan to use the Taylor series method with terms up to h^{20}, how should the computation $\sum_{n=0}^{20} x^{(n)}(t)h^n/n!$ be carried out? Assume that $x(t)$, $x^{(1)}(t)$, $x^{(2)}(t), \ldots,$ and $x^{(20)}(t)$ are available.
Hint: Only a few statements suffice.

15. Explain how to use the ODE method that is based on the trapezoid rule:

$$\widehat{x}(t+h) = x(t) + hf(t, x(t))$$

$$x(t+h) = x(t) + \frac{h}{2}[f(t, x(t)) + f(t+h, \widehat{x}(t+h))]$$

This is called the **improved Euler's method** or **Heun's method**. Here, $\widehat{x}(t+h)$ is computed by using Euler's method.

16. (Continuation) Use the improved Euler's method to solve the following differential equation over the interval [0, 1] with step size $h = 0.1$:

$$\begin{cases} x' = -x + t + \frac{1}{2} \\ x(0) = 1 \end{cases}$$

17. Consider the initial-value problem

$$\begin{cases} x' = -100x^2 \\ x(0) = 1 \end{cases}$$

In the improved Euler's method, replace $\widehat{x}(t+h)$ with $x(t+h)$ and try to solve with one step of size $h = 0.1$. Explain what happens. Find the closed-form solution by substituting $x = (a + bt)^c$ and determining a, b, c.

Computer Exercises 7.1

[a]**1.** Write and test a program for applying the Taylor series method to the initial-value problem

$$\begin{cases} x' = x + x^2 \\ x(1) = \dfrac{e}{16 - e} = 0.20466\,34172\,89155\,26943 \end{cases}$$

Generate the solution in the interval [1, 2.77]. Use derivatives up to $x^{(v)}$ in the Taylor series. Use $h = 1/100$. Print out for comparison the values of the exact solution $x(t) = e^t/(16 - e^t)$. Verify that it is the exact solution.

2. Write a program to solve each problem on the indicated intervals. Use the Taylor series method with $h = 1/100$, and include terms to h^3. Account for any difficulties.

a. $\begin{cases} x' = t + x^2 \quad \text{on } [0, 0.9] \\ x(0) = 1 \end{cases}$

[a]**b.** $\begin{cases} x' = x - t \quad \text{on } [1, 1.75] \\ x(1) = 1 \end{cases}$

[a]**c.** $\begin{cases} x' = tx + t^2x^2 \quad \text{on } [2, 5] \\ x(2) = -0.63966\,25333 \end{cases}$

[a]**3.** Solve the differential equation $x' = x$ with initial value $x(0) = 1$ by the Taylor series method on the interval [0, 10]. Compare the result with the exact solution $x(t) = e^t$. Use derivatives up to and including the tenth. Use step size $h = 1/100$.

4. Solve for $x(1)$:

[a]**a.** $x' = 1 + x^2, \quad x(0) = 0$

b. $x' = (1 + t)^{-1}x, \quad x(0) = 1$

Use the Taylor series method of order 5 with $h = 1/100$, and compare with the exact solutions, which are $\tan t$ and $1 + t$, respectively.

[a]**5.** Solve the initial-value problem $x' = t + x + x^2$ on the interval [0, 1] with initial condition $x(1) = 1$. Use the Taylor series method of order 5.

6. Solve the initial-value problem $x' = (x+t)^2$ with $x(0) = -1$ on the interval [0, 1] using the Taylor series method with derivatives up to and including the fourth. Compare this to Taylor series methods of orders 1, 2, and 3.

[a]**7.** Write a program to solve on the interval [0, 1] the initial-value problem

$$\begin{cases} x' = tx \\ x(0) = 1 \end{cases}$$

using the Taylor series method of order 20; that is, include terms in the Taylor series up to and including h^{20}. Observe that a simple recursive formula can be used to obtain $x^{(n)}$ for $n = 1, 2, \ldots, 20$.

8. Write a program to solve the initial-value problem $x' = \sin x + \cos t$, using the Taylor series method. Continue the solution from $t = 2$ to $t = 5$, starting with $x(2) = 0.32$. Include terms up to and including h^3.

[a]**9.** Write a program to solve the initial-value problem $x' = e^t x$ with $x(2) = 1$ on the interval $0 \leqq t \leqq 2$ using the Taylor series method. Include terms up to h^4.

[a]**10.** Write a program to solve $x' = tx + t^4$ on the interval $0 \leqq t \leqq 5$ with $x(5) = 3$. Use the Taylor series method with terms to h^4.

11. Write a program to solve the initial-value problem of the example in this section over the interval [1, 3]. Explain.

12. Compute a table, at 101 equally spaced points in the interval [0, 2], of the **Dawson integral**

$$f(x) = \exp\left(-x^2\right) \int_0^x \exp\left(t^2\right) dt$$

by numerically solving, with the Taylor series method of suitable order, an initial-value problem of which f is the solution. Make the table accurate to eight decimal places, and print only eight decimal places.
Hint: Find the relationship between $f'(x)$ and $xf(x)$. The Fundamental Theorem of Calculus is useful.

Check values: $f(1) = 0.5380795069$ and $f(2) = 0.3013403889$.

13. Solve the initial-value problem $x' = t^3 + e^x$ with $x(3) = 7.4$ on the interval $0 \leq t \leq 3$ by means of the fourth-order Taylor series method.

14. Use a symbolic manipulation package such as Maple to solve the differential equations of Example 1 by the fourth-order Taylor series method to high accuracy, carrying 24 decimal digits.

15. Program the pseudocodes Euler and Taylor and compare the numerical results to those given in the text.

16. (Continuation) Repeat by calling directly an ordinary differential equation solver routine within a mathematical software system such as MATLAB, Maple, or Mathematica.

17. Use mathematical software such as MATLAB, Maple, or Mathematica to find analytical or numerical solutions for these ordinary differential equations:

    a. ODE (2) **b.** ODE (3) **c.** ODE (4)

18. Write computer programs to reproduce these figures:

    a. Figure 7.1 **b.** Figure 7.2 **c.** Figure 7.3

7.2 Runge-Kutta Methods

Introduction

The methods named after Carl Runge and Wilhelm Kutta are designed to imitate the Taylor series method without requiring analytic differentiation of the original differential equation. Recall that in using the Taylor series method on the initial-value problem

$$\begin{cases} x' = f(t, x) \\ x(a) = x_a \end{cases} \tag{1}$$

Solving ODEs Without Differentiation

we need to obtain $x'', x''', \ldots$ by differentiating the function f. This requirement can be a serious obstacle to using the method. The user of this method must do some preliminary analytical work before writing a computer program. Ideally, a method for solving Equation (1) should involve nothing more than writing a code to evaluate f. The Runge-Kutta methods accomplish this.

For purposes of exposition, the Runge-Kutta method of order 2 is presented, although its low precision usually precludes its use in actual scientific calculations. Later, the Runge-Kutta method of order 4 is given *without* a derivation. It is in common use. The order-2 Runge-Kutta procedure does find application in real-time calculations on small computers. For example, it is used in some aircraft by the on-board mini-computer.

At the heart of any method for solving an initial-value problem is a procedure for advancing the solution function one step at a time; that is, a formula must be given for $x(t + h)$ in terms of known quantities. As examples of known quantities, we can cite $x(t), x(t - h), x(t - 2h), \ldots$ if the solution process has gone through a number of steps. At the beginning, only $x(a)$ is known. Of course, we assume that $f(t, x)$ can be computed for any point (t, x).

Taylor Series for $f(x, y)$

Before explaining the Runge-Kutta method of order 2, let us present the **Taylor series in two variables**. The infinite series is

Taylor Series in Two Variables

$$f(x + h, y + k) = \sum_{i=0}^{\infty} \frac{1}{i!} \left(h \frac{\partial}{\partial x} + k \frac{\partial}{\partial y} \right)^i f(x, y) \tag{2}$$

This series is analogous to the Taylor series in one variable given by Equation (11) in Section 1.2. The mysterious-looking terms in Equation (2) are interpreted as follows:

$$\left(h\frac{\partial}{\partial x}+k\frac{\partial}{\partial y}\right)^0 f(x,y) = f$$

$$\left(h\frac{\partial}{\partial x}+k\frac{\partial}{\partial y}\right)^1 f(x,y) = h\frac{\partial f}{\partial x}+k\frac{\partial f}{\partial y}$$

$$\left(h\frac{\partial}{\partial x}+k\frac{\partial}{\partial y}\right)^2 f(x,y) = h^2\frac{\partial^2 f}{\partial x^2}+2hk\frac{\partial^2 f}{\partial x\partial y}+k^2\frac{\partial^2 f}{\partial y^2}$$

$$\vdots$$

where f and all partial derivatives are evaluated at (x, y). As in the one-variable case, if the Taylor series is truncated, an error term or remainder term is needed to restore the equality. Here is the appropriate equation:

TS with Error Term

$$f(x+h, y+k) = \sum_{i=0}^{n-1}\frac{1}{i!}\left(h\frac{\partial}{\partial x}+k\frac{\partial}{\partial y}\right)^i f(x,y) + \frac{1}{n!}\left(h\frac{\partial}{\partial x}+k\frac{\partial}{\partial y}\right)^n f(\overline{x},\overline{y}) \tag{3}$$

The point $(\overline{x}, \overline{y})$ lies on the line segment that joins (x, y) to $(x+h, y+k)$ in the plane.

In applying Taylor series, we use subscripts to denote partial derivatives. So, for instance, we define

$$f_x = \frac{\partial f}{\partial x} \qquad f_t = \frac{\partial f}{\partial t}, \qquad f_{xx} = \frac{\partial^2 f}{\partial x^2}, \qquad f_{xt} = \frac{\partial^2 f}{\partial t\,\partial x} \tag{4}$$

We are dealing with functions for which the order of these subscripts is immaterial; for example, $f_{xt} = f_{tx}$. Thus, we have

TS with First Few Terms

$$\begin{aligned} f(x+h, y+k) = f &+ (hf_x + kf_y) \\ &+ \frac{1}{2!}\left(h^2 f_{xx} + 2hkf_{xy} + k^2 f_{yy}\right) \\ &+ \frac{1}{3!}\left(h^3 f_{xxx} + 3h^2 k f_{xxy} + 3hk^2 f_{xyy} + k^3 f_{yyy}\right) \\ &+ \cdots \end{aligned}$$

As special cases, we notice that

Special Cases

$$f(x+h, y) = f + hf_x + \frac{h^2}{2!}f_{xx} + \frac{h^3}{3!}f_{xxx} + \cdots$$

$$f(x, y+k) = f + kf_y + \frac{k^2}{2!}f_{yy} + \frac{k^3}{3!}f_{yyy} + \cdots$$

Runge-Kutta Method of Order 2

In the Runge-Kutta method of order 2, a formula is adopted that has two function evaluations of the special form

$$\begin{cases} K_1 = hf(t, x) \\ K_2 = hf(t+\alpha h, x+\beta K_1) \end{cases}$$

and a linear combination of these is added to the value of x at t to obtain the value at $t+h$:

$$x(t+h) = x(t) + w_1 K_1 + w_2 K_2$$

or, equivalently,

$$x(t+h) = x(t) + w_1 h f(t,x) + w_2 h f(t+\alpha h, x + \beta h f(t,x)) \tag{5}$$

The objective is to determine constants w_1, w_2, α, and β so that Equation (5) is as accurate as possible. Explicitly, we want to reproduce as many terms as possible in the Taylor series

$$x(t+h) = x(t) + hx'(t) + \frac{1}{2!}h^2 x''(t) + \frac{1}{3!}h^3 x'''(t) + \frac{1}{4!}h^4 x^{(iv)}(t) + \frac{1}{5!}h^5 x^{(v)}(t) + \cdots \tag{6}$$

Deriving Runge-Kutta Methods of Order 2

Now compare Equation (5) with Equation (6). One way to force them to agree up through the term in h is to set $w_1 = 1$ and $w_2 = 0$ because $x' = f$. However, this simply reproduces Euler's method (described in the preceding section), and its order of precision is only 1. Agreement up through the h^2 term is possible by a more adroit choice of parameters. To see how, apply the two-variable form of the Taylor series to the final term in Equation (5). We use $n = 2$ in the two-variable Taylor series given by Formula (3), with t, αh, x, and $\beta h f$ playing the role of x, h, y, and k, respectively:

$$f(t+\alpha h, x+\beta h f) = f + \alpha h f_t + \beta h f f_x + \frac{1}{2}\left(\alpha h \frac{\partial}{\partial t} + \beta h f \frac{\partial}{\partial x}\right)^2 f(\overline{x}, \overline{y})$$

Using the above equation results in a new form for Equation (5). We have

$$x(t+h) = x(t) + (w_1 + w_2)hf + \alpha w_2 h^2 f_t + \beta w_2 h^2 f f_x + \mathcal{O}(h^3) \tag{7}$$

Equation (6) is also given a new form by using differential Equation (1). Since $x' = f$, we have

$$x'' = \frac{dx'}{dt} = \frac{df(t,x)}{dt} = \left(\frac{\partial f}{\partial t}\right)\left(\frac{dt}{dt}\right) + \left(\frac{\partial f}{\partial x}\right)\left(\frac{dx}{dt}\right) = f_t + f_x f$$

So Equation (6) implies that

$$x(t+h) = x + hf + \frac{1}{2}h^2 f_t + \frac{1}{2}h^2 f f_x + \mathcal{O}(h^3) \tag{8}$$

Agreement between Equations (7) and (8) is achieved by stipulating that

$$w_1 + w_2 = 1, \qquad \alpha w_2 = \frac{1}{2}, \qquad \beta w_2 = \frac{1}{2} \tag{9}$$

A convenient solution to Equation (9) is

$$\alpha = 1, \qquad \beta = 1, \qquad w_1 = \frac{1}{2}, \qquad w_2 = \frac{1}{2}$$

The resulting **second-order Runge-Kutta method** is then, from Equation (5),

Runge-Kutta Method of Order 2

$$x(t+h) = x(t) + \frac{h}{2}f(t,x) + \frac{h}{2}f(t+h, x+hf(t,x))$$

or, equivalently,

$$x(t+h) = x(t) + \frac{1}{2}(K_1 + K_2) \tag{10}$$

where

$$\begin{cases} K_1 = hf(t, x) \\ K_2 = hf(t + h, x + K_1) \end{cases}$$

Formula (10) shows that the solution function at $t + h$ is computed at the expense of two evaluations of the function f.

Notice that other solutions for the nonlinear System (9) are possible. For example, α can be arbitrary, and then

$$\beta = \alpha, \qquad w_1 = 1 - \frac{1}{2\alpha}, \qquad w_2 = \frac{1}{2\alpha}$$

One can show (see Exercise 7.2.10) that the error term for Runge-Kutta methods of order 2 is

Error Term

$$\frac{h^3}{4}\left(\frac{2}{3} - \alpha\right)\left(\frac{\partial}{\partial t} + f\frac{\partial}{\partial x}\right)^2 f + \frac{h^3}{6} f_x \left(\frac{\partial}{\partial t} + f\frac{\partial}{\partial x}\right) f \tag{11}$$

Notice that the method with $\alpha = \frac{2}{3}$ is especially interesting. However, none of the second-order Runge-Kutta methods is widely used on large computers because the error is only $\mathcal{O}(h^3)$.

Runge-Kutta Method of Order 4

One algorithm in common use for the initial-value Problem (1) is the classical **fourth-order Runge-Kutta method**. Its formulas are as follows:

$$x(t + h) = x(t) + \frac{1}{6}(K_1 + 2K_2 + 2K_3 + K_4) \tag{12}$$

where

Runge-Kutta Method of Order 4

$$\begin{cases} K_1 = hf(t, x) \\ K_2 = hf\left(t + \frac{1}{2}h, x + \frac{1}{2}K_1\right) \\ K_3 = hf\left(t + \frac{1}{2}h, x + \frac{1}{2}K_2\right) \\ K_4 = hf(t + h, x + K_3) \end{cases}$$

The derivation of the Runge-Kutta formulas of order 4 is tedious. Very few textbooks give the details. Two exceptions are the books of Henrici [1962] and Ralston [1965]. There exist higher-order Runge-Kutta formulas, and they are still more laborious to derive. However, symbolic manipulation software packages such as in MATLAB, Maple, or Mathematica can be used to develop the formulas.

As shown, the solution at $x(t + h)$ is obtained at the expense of evaluating the function f four times. The final formula agrees with the Taylor expansion up to and including the term in h^4. The error therefore contains h^5, but no lower powers of h. Without knowing the coefficient of h^5 in the error, we cannot be precise about the local truncation error. In treatises devoted to this subject, these matters are explored further. See, for example, Butcher [1987] or Gear [1971].

Pseudocode

Here is a pseudocode to implement the classical Runge-Kutta method of order 4:

RK4 **Pseudocode**

```
procedure RK4(f, t, x, h, n)
integer j, n;      real K_1, K_2, K_3, K_4, h, t, t_a, x
external function f
output 0, t, x
t_a ← t
for j = 1 to n
    K_1 ← hf(t, x)
    K_2 ← hf(t + (1/2)h, x + (1/2)K_1)
    K_3 ← hf(t + (1/2)h, x + (1/2)K_2)
    K_4 ← hf(t + h, x + K_3)
    x ← x + (1/6)(K_1 + 2K_2 + 2K_3 + K_4)
    t ← t_a + jh
    output j, t, x
end for
end procedure RK4
```

To illustrate the use of the preceding pseudocode, consider the initial-value problem

$$\begin{cases} x' = 2 + (x - t - 1)^2 \\ x(1) = 2 \end{cases} \tag{13}$$

whose exact solution is $x(t) = 1 + t + \tan(t - 1)$. A pseudocode to solve this problem on the interval $[1, 1.5625]$ by the Runge-Kutta procedure follows. The step size needed is calculated by dividing the length of the interval by the number of steps, say, $n = 72$.

Test_RK4 **Pseudocode**

```
program Test_RK4
real h, t;      external function f
integer n ← 72
real a ← 1, b ← 1.5625, x ← 2
h ← (b − a)/n
t ← a
call RK4(f, t, x, h, n)
end program Test_RK4

real function f(t, x)
real t, x
f ← 2 + (x − t − 1)^2
end function f
```

We include an external-function statement both in the main program and in procedure *RK4* because the procedure f is passed in the argument list of *RK4*. The final value of the computed numerical solution is $x(1.5625) = 3.19293\,7699$.

General-purpose routines incorporating the Runge-Kutta algorithm usually include additional programming to monitor the truncation error and make necessary adjustments in the step size as the solution progresses. In general terms, the step size can be large when the solution is slowly varying, but should be small when it is rapidly varying. Such a program is presented in Section 7.3.

Summary 7.2

- The **second-order Runge-Kutta method** is

$$x(t+h) = x(t) + \frac{1}{2}(K_1 + K_2)$$

where

$$\begin{cases} K_1 = hf(t, x) \\ K_2 = hf(t+h, x+K_1) \end{cases}$$

This method requires two evaluations of the function f per step. It is equivalent to a Taylor series method of order 2.

- One of the most popular single-step methods for solving ODEs is the **fourth-order Runge-Kutta method**

$$x(t+h) = x(t) + \frac{1}{6}(K_1 + 2K_2 + 2K_3 + K_4)$$

where

$$\begin{cases} K_1 = hf(t, x) \\ K_2 = hf\left(t + \frac{1}{2}h, x + \frac{1}{2}K_1\right) \\ K_3 = hf\left(t + \frac{1}{2}h, x + \frac{1}{2}K_2\right) \\ K_4 = hf(t+h, x+K_3) \end{cases}$$

It needs four evaluations of the function f per step. Since it is equivalent to a Taylor series method of order 4, it has truncation error of order $\mathcal{O}(h^5)$. The small number of function evaluations and high-order truncation error account for its popularity.

Exercises 7.2

1. Derive the equations needed to apply the fourth-order Taylor series method to the differential equation $x' = tx^2 + x - 2t$. Compare them in complexity with the equations required for the fourth-order Runge-Kutta method.

2. Put these differential equations into a form suitable for numerical solution by the Runge-Kutta method.

   a. $x + 2xx' - x' = 0$ **b.** $\log x' = t^2 - x^2$

   [a]**c.** $(x')^2(1 - t^2) = x$

[a]**3.** Solve the differential equation

$$\begin{cases} \dfrac{dx}{dt} = -tx^2 \\ x(0) = 2 \end{cases}$$

at $t = -0.2$, correct to two decimal places, using one step of the Taylor series method of order 2 and one step of the Runge-Kutta method of order 2.

4. Consider the ordinary differential equation

$$\begin{cases} x' = (tx)^3 - (x/t)^2 \\ x(1) = 1 \end{cases}$$

Take one step of the Taylor series method of order 2 with $h = 0.1$ and then use the Runge-Kutta method of order 2 to recompute $x(1.1)$. Compare answers.

5. In solving the following differential equations by using a Runge-Kutta procedure, it is necessary to write code for a function $f(t, x)$. Do so for each of the following:

   [a]**a.** $x' = t^2 + tx' - 2xx'$ **b.** $x' = e^t + x' \cos x + t^2$

6. Consider the ordinary differential equation $x' = t^3x^2 - 2x^3/t^2$ with $x(1) = 1$. Determine the equations that would be used in applying the Taylor series method of order 3 and the Runge-Kutta method of order 4.

7. Consider the **third-order Runge-Kutta method**:

$$x(t+h) = x(t) + \frac{1}{9}(2K_1 + 3K_2 + 4K_3)$$

where

$$\begin{cases} K_1 = hf(t, x) \\ K_2 = hf\left(t + \frac{1}{2}h, x + \frac{1}{2}K_1\right) \\ K_3 = hf\left(t + \frac{3}{4}h, x + \frac{3}{4}K_2\right) \end{cases}$$

a. Show that it agrees with the Taylor series method of the same order for the differential equation $x' = x + t$.

b. Prove that this third-order Runge-Kutta method reproduces the Taylor series of the solution up to and including terms in h^3 for *any* differential equation.

[a]**8.** Describe how the fourth-order Runge-Kutta method can be used to produce a table of values for the function

$$f(x) = \int_0^x e^{-t^2}\,dt$$

at 100 equally spaced points in the unit interval.
Hint: Find an appropriate initial-value problem whose solution is f.

9. Show that the fourth-order Runge-Kutta formula reduces to a simple form when applied to an ordinary differential equation of the form

$$x' = f(t)$$

[a]**10.** Establish the error term (11) for Runge-Kutta methods of order 2.

[a]**11.** On a certain computer, it was found that when the fourth-order Runge-Kutta method was used over an interval $[a, b]$ with $h = (b - a)/n$, the total error due to roundoff was about $36n2^{-50}$ and the total truncation error was $9nh^5$, where n is the number of steps and h is the step size. What is an optimum value of h?
Hint: Minimize the total error: roundoff error plus truncation error.

[a]**12.** How would you solve the initial-value problem

$$\begin{cases} x' = \sin x + \sin t \\ x(0) = 0 \end{cases}$$

on the interval $[0, 1]$ if ten decimal places of accuracy, 10^{-10}, are required? Assume that you use a computer with adequate precision, and assume that the fourth-order Runge-Kutta method involves truncation error of magnitude $100h^5$.

13. An important theorem of calculus states that the equation $f_{tx} = f_{xt}$ is true, provided that at least one of these two partial derivatives exists and is continuous. Test this equation on some functions, such as $f(t, x) = xt^2 + x^2t + x^3t^4$, $\log(x - t^{-1})$, and $e^x \sinh(t + x) + \cos(2x - 3t)$.

14. **a.** If $x' = f(t, x)$, then

$$x'' = Df, \quad x''' = D^2 f + f_x Dff$$

where

$$D = \frac{\partial}{\partial t} + f\frac{\partial}{\partial x}, \quad D^2 = \frac{\partial^2}{\partial t^2} + 2f\frac{\partial^2}{\partial x\,\partial t} + f^2\frac{\partial^2}{\partial x^2}$$

Verify these equations.

[a]**b.** Determine $x^{(iv)}$ in a similar form.

[a]**15.** Derive the two-variable form of the Taylor series from the one-variable form by considering the function of one variable $\phi(t) = f(x + th, y + tk)$ and expanding it by Taylor's Theorem.

16. The Taylor series expansion about point (a, b) in terms of two variables x and y is given by

$$f(x, y) = \sum_{i=0}^{\infty} \frac{1}{i!}\left((x - a)\frac{\partial}{\partial x} + (y - b)\frac{\partial}{\partial y}\right)^i f(a, b)$$

Show that Formula (2) can be obtained from this form by a change of variables.

[a]**17.** (Continuation) Using the form given in the preceding problem, determine the first four nonzero terms in the Taylor series for $f(x, y) = \sin x + \cos y$ about the point $(0, 0)$. Compare the result to the known series for $\sin x$ and $\cos y$. Make a conjecture about the Taylor series for functions that have the special form $f(x, y) = g(x) + h(y)$.

[a]**18.** For the function $f(x, y) = y^2 - 3\ln x$, write the first six terms in the Taylor series of $f(1 + h, 0 + k)$.

[a]**19.** Using the truncated Taylor series about $(1, 1)$, give a three-term approximation to $e^{(1-xy)}$.
Hint: Use Exercise 7.2.16.

[a]**20.** The function $f(x, y) = xe^y$ can be approximated by the Taylor series in two variables by $f(x + h, y + k) \approx (Ax + B)e^y$. Determine A and B when terms through the second partial derivatives are used in the series.

[a]**21.** For $f(x, y) = (y - x)^{-1}$, the Taylor series can be written as

$$f(x + h, y + k) = Af + Bf^2 + Cf^3 + \cdots$$

where $f = f(x, y)$. Determine the coefficients A, B, and C.

[a]**22.** Consider the function e^{x^2+y}. Determine its Taylor series about the point $(0, 1)$ through second-partial-derivative terms. Use this result to obtain an approximate value for $f(0.001, 0.998)$.

23. Show that the improved Euler's method is a Runge-Kutta method of order 2.

Computer Exercises 7.2

1. Run the sample pseudocode given in the text for differential Equation (13) to illustrate the Runge-Kutta method.

[a]**2.** Solve the initial-value problem $x' = x/t + t\sec(x/t)$ with $x(0) = 0$ by the fourth-order Runge-Kutta method. Continue the solution to $t = 1$ using step size $h = 2^{-7}$. Compare the numerical solution with the exact solution, which is $x(t) = t\arcsin t$. Define $f(0, 0) = 0$, where $f(t, x) = x/t + t\sec(x/t)$.

3. Select one of the following initial-value problems, and compare the numerical solutions obtained with fourth-order Runge-Kutta formulas and fourth-order Taylor series. Use different values of $h = 2^{-n}$, for $n = 2, 3, \ldots, 7$, to compute the solution on the interval $[1, 2]$.

a. $x' = 1 + x/t, \quad x(1) = 1$

[a]**b.** $x' = 1/x^2 - xt, \quad x(1) = 1$

[a]**c.** $x' = 1/t^2 - x/t - x^2, \quad x(1) = -1$

[a]**4.** Select a Runge-Kutta routine from a program library, and test it on the initial-value problem $x' = (2 - t)x$ with $x(2) = 1$. Compare with the exact solution, $x = \exp\left[-\left(\frac{1}{2}\right)(t-2)^2\right]$.

[a]**5.** **(Ill-Conditioned ODE)** Solve the ordinary differential equation $x' = 10x + 11t - 5t^2 - 1$ with initial value $x(0) = 0$. Continue the solution from $t = 0$ to $t = 3$, using the fourth-order Runge-Kutta method with $h = 2^{-8}$. Print the numerical solution and the exact solution $(t^2/2 - t)$ at every tenth step, and draw a graph of the two solutions. Verify that the solution of the same differential equation with initial value $x(0) = \varepsilon$ is $\varepsilon e^{10t} + t^2/2 - t$ and thus account for the discrepancy between the numerical and exact solutions of the original problem.

[a]**6.** Solve the initial-value problem $x' = x\sqrt{x^2 - 1}$ with $x(0) = 1$ by the Runge-Kutta method on the interval $0 \leqq t \leqq 1.6$, and account for any difficulties. Then, using negative h, solve the same differential equation on the same interval with initial value $x(1.6) = 1.0$.

7. The following pathological example has been given by Dahlquist and Björck [1974]. Consider the differential equation $x' = 100(\sin t - x)$ with initial value $x(0) = 0$. Integrate it with the fourth-order Runge-Kutta method on the interval $[0, 3]$, using step sizes $h = 0.015, 0.020, 0.025, 0.030$. Observe the numerical instability!

[a]**8.** Consider the differential equation

$$\begin{cases} x' = \begin{cases} x + t, & -1 \leqq t \leqq 0 \\ x - t, & 0 \leqq t \leqq 1 \end{cases} \\ x(-1) = 1 \end{cases}$$

Using the Runge-Kutta procedure *RK4* with step size $h = 0.1$, solve this problem over the interval $[-1, 1]$. Now solve by using $h = 0.09$. Which numerical solution is more accurate and why?
Hint: The true solution is given by $x = e^{(t+1)} - (t + 1)$ if $t \leqq 0$ and $x = e^{(t+1)} - 2e^t + (t + 1)$ if $t \geqq 0$.

[a]**9.** Solve $t - x' + 2xt = 0$ with $x(0) = 0$ on the interval $[0, 10]$ using the Runge-Kutta formulas with $h = 0.1$. Compare with the true solution: $\frac{1}{2}(e^{t^2} - 1)$. Draw a graph or have one created by an automatic plotter. Then graph the logarithm of the solution.

10. Write a program to solve $x' = \sin(xt) + \arctan t$ on $1 \leqq t \leqq 7$ with $x(2) = 4$ using the Runge-Kutta procedure *RK4*.

11. The general form of Runge-Kutta methods of order 2 is given by Equation (5). Write and test procedure *RK2*(f, t, x, h, α, n) for carrying out n steps with step size h and initial conditions t and x for several given α values.

12. We want to solve

$$\begin{cases} x' = e^t x^2 + e^3 \\ x(2) = 4 \end{cases}$$

at $x(5)$ with step size 0.5. Solve it in the following two ways.

a. Code the function $f(t, x)$ that is needed and use procedure *RK4*.

b. Write a short program that uses the Taylor series method including terms up to h^4.

13. Plot the solution for differential equation (13).

14. Select a differential equation with a known solution and compare the classical fourth-order Runge-Kutta method with one or both of the following ones. Print the errors at each step. Is the ratio of the two errors a constant at each step? What are the advantages and disadvantages of each method?

a. A **fourth-order Runge-Kutta method** similar to the classical one is given by

$$x(t+h) = x(t) + \frac{1}{6}(K_1 + 4K_3 + K_4)$$

where

$$\begin{cases} K_1 = hf(t, x) \\ K_2 = hf\left(t + \frac{1}{2}h,\ x + \frac{1}{2}K_1\right) \\ K_3 = hf\left(t + \frac{1}{2}h,\ x + \frac{1}{4}K_1 + \frac{1}{4}K_2\right) \\ K_4 = hf(t + h,\ x - K_2 + 2K_3) \end{cases}$$

See England [1969] or Shampine, Allen, and Pruess [1997].

b. Another **fourth-order Runge-Kutta method** is given by

$$x(t+h) = x(t) + w_1K_1 + w_2K_2 + w_3K_3 + w_4K_4$$

where

$$\begin{cases} K_1 = hf(t, x) \\ K_2 = hf\left(t + \frac{2}{5}h,\ x + \frac{2}{5}K_1\right) \\ K_3 = hf\left(t + \frac{1}{16}\left(14 - 3\sqrt{5}\right)h,\ x + c_{31}K_1 + c_{32}K_2\right) \\ K_4 = hf(t + h,\ x + c_{41}K_1 + c_{42}K_2 + c_{43}K_3) \end{cases}$$

Here the appropriate constants are

$$c_{31} = \frac{3\left(-963 + 476\sqrt{5}\right)}{1024}, \quad c_{32} = \frac{5\left(757 - 324\sqrt{5}\right)}{1024}$$

$$c_{41} = \frac{-3365 + 2094\sqrt{5}}{6040}, \quad c_{42} = \frac{-975 - 3046\sqrt{5}}{2552}$$

$$c_{43} = \frac{32\left(14595 + 6374\sqrt{5}\right)}{2\,40845}$$

$$w_1 = \frac{263 + 24\sqrt{5}}{1812}, \quad w_2 = \frac{125\left(1 - 8\sqrt{5}\right)}{3828}$$

$$w_3 = \frac{1024\left(3346 + 1623\sqrt{5}\right)}{59\,24787}, \quad w_4 = \frac{2\left(15 - 2\sqrt{5}\right)}{123}$$

Note: There are any number of Runge-Kutta methods of any order. The higher the order, the more complicated are the formulas. Since the one given by Equation (12) has error $\mathcal{O}(h^5)$ and is rather simple, it is the most popular fourth-order Runge-Kutta method. The error term for the method of part **b** of this problem is also $\mathcal{O}(h^5)$, and it is optimum in a certain sense.
(See Ralston [1965] for details.)

15. A **fifth-order Runge-Kutta method** is given by

$$x(t+h) = x(t) + \frac{1}{24}K_1 + \frac{5}{48}K_4 + \frac{27}{56}K_5 + \frac{125}{336}K_6$$

where

$$\begin{cases} K_1 = hf(t, x) \\ K_2 = hf\left(t + \frac{1}{2}h,\ x + \frac{1}{2}K_1\right) \\ K_3 = hf\left(t + \frac{1}{2}h,\ x + \frac{1}{4}K_1 + \frac{1}{4}K_2\right) \\ K_4 = hf(t + h,\ x - K_2 + 2K_3) \\ K_5 = hf\left(t + \frac{2}{3}h,\ x + \frac{7}{27}K_1 + \frac{10}{27}K_2 + \frac{1}{27}K_4\right) \\ K_6 = hf\left(t + \frac{1}{5}h,\ x + \frac{28}{625}K_1 - \frac{1}{5}K_2 + \frac{546}{625}K_3 \right. \\ \qquad\qquad \left. + \frac{54}{625}K_4 - \frac{378}{625}K_5\right) \end{cases}$$

Write and test a procedure that uses this formula.

16. **a.** Use a symbol manipulation package such as Maple or Mathematica to find the general Runge-Kunge method of order 2.

b. Repeat for order 3.

17. (**Delay Ordinary Differential Equation**) Investigate procedures for determining the numerical solution of an ordinary differential equation with a constant delay such as

$$x'(t) = -x(t) + x(t - 20) + \frac{1}{20}\cos\left(\frac{t}{20}\right) + \sin\left(\frac{t}{20}\right) - \sin\left(\frac{t}{20} - 1\right)$$

on the interval $0 \leqq t \leqq 1000$, where $x(t) = \sin\left(t/20\right)$ for $t \leqq 0$. Use a step size less than or equal to 20 so that no overlapping occurs. Compare to the exact solution $x(t) = \sin(t/20)$.

18. Write a software for program *Test_RK4* and routine *RK4*, and verify the numerical results given in the text.

7.3 Adaptive Runge-Kutta and Multistep Methods

An Adaptive Runge-Kutta-Fehlberg Method

In realistic situations involving the numerical solution of initial-value problems, there is always a need to estimate the precision attained in the computation. Usually, an error tolerance is prescribed, and the numerical solution must not deviate from the true solution beyond this tolerance. Once a method has been selected, the error tolerance dictates the largest allowable step size. Even if we consider only the local truncation error, determining an appropriate step size may be difficult. Moreover, often a small step size is needed on one portion of the solution curve, whereas a larger one may suffice elsewhere.

For the reasons given, various methods have been developed for *automatically* adjusting the step size in algorithms for the initial-value problem. One simple procedure is now described. Consider the classical fourth-order Runge-Kutta method discussed in Section 7.2. To advance the solution curve from t to $t+h$, we can take one step of size h using the Runge-Kutta formulas. But we can also take *two* steps of size $h/2$ to arrive at $t+h$. If there were no truncation error, the value of the numerical solution $x(t+h)$ would be the same for both procedures. The difference in the numerical results can be taken as an estimate of the local truncation error. So, in practice, if this difference is within the prescribed tolerance, the current step size h is satisfactory. If this difference exceeds the tolerance, the step size is halved. If the difference is very much less than the tolerance, the step size is doubled.

Schemes for Automatic Step-Size Adjustment

The procedure just outlined is easily programmed but rather wasteful of computing time and is not recommended. A more sophisticated method was developed by Fehlberg [1969]. The **Runge-Kutta-Fehlberg method of order 4** is

$$x(t+h) = x(t) + \frac{25}{216}K_1 + \frac{1408}{2565}K_3 + \frac{2197}{4104}K_4 - \frac{1}{5}K_5 \tag{1}$$

where

Runge-Kutta-Fehlberg Method of Order 4

$$\begin{cases} K_1 = hf(t, x) \\ K_2 = hf\left(t + \frac{1}{4}h,\ x + \frac{1}{4}K_1\right) \\ K_3 = hf\left(t + \frac{3}{8}h,\ x + \frac{3}{32}K_1 + \frac{9}{32}K_2\right) \\ K_4 = hf\left(t + \frac{12}{13}h,\ x + \frac{1932}{2197}K_1 - \frac{7200}{2197}K_2 + \frac{7296}{2197}K_3\right) \\ K_5 = hf\left(t + h,\ x + \frac{439}{216}K_1 - 8K_2 + \frac{3680}{513}K_3 - \frac{845}{4104}K_4\right) \end{cases}$$

Since this scheme requires one more function evaluation than the classical Runge-Kutta method of order 4, it is of questionable value alone. However, with an additional function evaluation

$$K_6 = hf\left(t + \frac{1}{2}h,\ x - \frac{8}{27}K_1 + 2K_2 - \frac{3544}{2565}K_3 + \frac{1859}{4104}K_4 - \frac{11}{40}K_5\right) \tag{2}$$

we can obtain a **fifth-order Runge-Kutta method**, namely,

Runge-Kutta Method of Order 5

$$x(t+h) = x(t) + \frac{16}{135}K_1 + \frac{6656}{12825}K_3 + \frac{28561}{56430}K_4 - \frac{9}{50}K_5 + \frac{2}{55}K_6 \tag{3}$$

The difference between the values of $x(t+h)$ obtained from the fourth- and fifth-order procedures is an estimate of the local truncation error in the fourth-order procedure. So six function evaluations give a fifth-order approximation, together with an error estimate!

Pseudocode

A pseudocode for the **Runge-Kutta-Fehlberg method** is given in procedure *RK45*:

***RK45* Pseudocode**

```
procedure RK45(f, t, x, h, ε)
real ε, K1, K2, K3, K4, K5, K6, h, t, x, x4
external function f
real c20 ← 0.25, c21 ← 0.25
real c30 ← 0.375, c31 ← 0.09375, c32 ← 0.28125
real c40 ← 12./13., c41 ← 1932./2197.
real c42 ← −7200./2197., c43 ← 7296./2197.
real c51 ← 439./216., c52 ← −8.
real c53 ← 3680./513., c54 ← −845./4104.
real c60 ← 0.5, c61 ← −8./27., c62 ← 2.
real c63 ← −3544./2565., c64 ← 1859./4104.
real c65 ← −0.275
real a1 ← 25./216., a2 ← 0., a3 ← 1408./2565.
real a4 ← 2197./4104., a5 ← −0.2
real b1 ← 16./135., b2 ← 0., b3 ← 6656./12825.
real b4 ← 28561./56430., b5 ← −0.18
real b6 ← 2./55.
K1 ← hf(t, x)
K2 ← hf(t + c20h, x + c21K1)
K3 ← hf(t + c30h, x + c31K1 + c32K2)
K4 ← hf(t + c40h, x + c41K1 + c42K2 + c43K3)
K5 ← hf(t + h, x + c51K1 + c52K2 + c53K3 + c54K4)
K6 ← hf(t + c60h, x + c61K1 + c62K2 + c63K3 + c64K4 + c65K5)
x4 ← x + a1K1 + a3K3 + a4K4 + a5K5
x ← x + b1K1 + b3K3 + b4K4 + b5K5 + b6K6
t ← t + h
ε ← |x − x4|
end procedure RK45
```

Of course, the programmer may wish to consider various optimization techniques such as assigning numerical values to the coefficients with decimal expansions corresponding to the precision of the computer being used so that the fractions do not need to be recomputed at each call to the procedure.

We can use the *RK45* procedure in a nonadaptive fashion such as in the following test program:

```
program Test_RK45
integer k;   real t, h, ε;   external function f
integer n ← 72
real a ← 1.0, b ← 1.5625, x ← 2.0
```

(Continued)

Test_RK45 Pseudocode

```
h ← (b − a)/n
t ← a
output 0, t, x
for k = 1 to n
    call RK45(f, t, x, h, ε)
    output k, t, x, ε
end for
end program Test_RK45

real function f(t, x)
real t, x
f ← 2.0 + (x − t − 1.0)^2
end function f
```

Here, we print the error estimation at each step. However, we can use it in an adaptive procedure, since the error estimate ε can tell us when to adjust the step size to control the *single-step error*.

A Simple Adaptive Procedure

We now describe a simple adaptive procedure. In the *RK45* procedure, the fourth- and fifth-order approximations for $x(t+h)$, say, x_4 and x_5, are computed from six function evaluations, and the error estimate $\varepsilon = |x_4 - x_5|$ is known. From user-specified bounds on the allowable error estimate ($\varepsilon_{\min} \leqq \varepsilon \leqq \varepsilon_{\max}$), the step size h is doubled or halved as needed to keep ε within these bounds. A range for the allowable step size h is also specified by the user ($h_{\min} \leqq |h| \leqq h_{\max}$). Clearly, the user must set the bounds ($\varepsilon_{\min}, \varepsilon_{\max}, h_{\min}, h_{\max}$) carefully so that the adaptive procedure does not get caught in a loop, trying repeatedly to halve and double the step size from the same point to meet error bounds that are too restrictive for the given differential equation.

Basically, our adaptive process is as follows:

■ Algorithm

Overview of Adaptive Process

1. Given a step size h and an initial value $x(t)$, the *RK45* routine computes the value $x(t+h)$ and an error estimate ε.
2. If $\varepsilon_{\min} \leqq \varepsilon \leqq \varepsilon_{\max}$, then the step size h is not changed and the next step is taken by repeating step 1 with initial value $x(t+h)$.
3. If $\varepsilon < \varepsilon_{\min}$, then h is replaced by $2h$, provided that $|2h| \leqq h_{\max}$.
4. If $\varepsilon > \varepsilon_{\max}$, then h is replaced by $h/2$, provided that $|h/2| \geqq h_{\min}$.
5. If $h_{\min} \leqq |h| \leqq h_{\max}$, then the step is repeated by returning to step 1 with $x(t)$ and the new h value.

The procedure for this adaptive scheme is *RK45_Adaptive*. In the parameter list of the pseudocode, f is the function $f(t, x)$ for the differential equation, t and x contain the initial values, h is the initial step size, t_b is the final value for t, *itmax* is the maximum number of steps to be taken in going from $a = t_a$ to $b = t_b$, $\varepsilon_{\min}$ and $\varepsilon_{\max}$ are lower and upper bounds on the allowable error estimate ε, $h_{\min}$ and $h_{\max}$ are bounds on the step size h, and *iflag* is an error flag that returns one of the following values:

iflag	Meaning
0	Successful march from t_a to t_b
1	Maximum number of iterations reached

On return, t and x are the exit values, and h is the final step size value considered or used:

_RK45_Adaptive_ Pseudocode

```
procedure RK45_Adaptive(f, t, x, h, t_b, itmax, ε_max, ε_min, h_min, h_max, iflag)
integer iflag, itmax, n;   external function f
real ε, ε_max, ε_min, d, h, h_min, h_max, t, t_b, x, x_save, t_save
real δ ← (1/2) × 10^(−5)
output 0, h, t, x
iflag ← 1
k ← 0
while k ≦ itmax
    k ← k + 1
    if |h| < h_min then h ← sign(h)h_min
    if |h| > h_max then h ← sign(h)h_max
    d ← |t_b − t|
    if d ≦ |h| then
        iflag ← 0
        if d ≦ δ · max{|t_b|, |t|} then exit loop
        h ← sign(h)d
    end if
    x_save ← x
    t_save ← t
    call RK45(f, t, x, h, ε)
    output n, h, t, x, ε
    if iflag = 0 then exit loop
    if ε < ε_min then h ← 2h
    if ε > ε_max then
        h ← h/2
        x ← x_save
        t ← t_save
        k ← k − 1
    end if
end while
end procedure RK45_Adaptive
```

In the pseudocode, notice that several conditions must be checked to determine the size of the final step, since floating-point arithmetic is involved and the step size varies.

Repeat the computer example in the previous section using *RK45_Adaptive*, which allows variable step size, instead of *RK4*. Compare the accuracy of these two computed solutions. (See Computer Exercise 7.4.22.)

An Industrial Example

A first-order differential equation that arose in the modeling of an industrial chemical process is as follows:

$$\begin{cases} x' = a + b\sin t + cx \\ x(0) = 0 \end{cases} \tag{4}$$

in which $a = 3$, $b = 5$, and $c = 0.2$ are constants. This equation is amenable to the solution techniques of calculus, in particular the use of an integrating factor. However, the analytic solution is complicated, and a numerical solution may be preferable.

To solve this problem numerically using the adaptive Runge-Kutta formulas, identify (and program) the function f that appears in the general description. In this problem, it is $f(t, x) = 3 + 5\sin t + 0.2x$. Here is a brief pseudocode for solving the equation on the interval [0, 10] with particular values assigned to the parameters in the routine *RK45_Adaptive*:

***Test_RK45* Pseudocode**

```
program Test_RK45_Adaptive
integer iflag;      real t, x, h, t_b;   external function f
integer itmax ← 1000
real ε_max ← 10^-5, ε_min ← 10^-8, h_min ← 10^-6, h_max ← 1.0
t ← 0.0;   x ← 0.0;   h ← 0.01;   t_b ← 10.0
call RK45_Adaptive(f, t, x, h, t_b, itmax, ε_max, ε_min, h_min, h_max, iflag)
output itmax, iflag
end program Test_RK45_Adaptive

real function f(t, x)
real t, x
f ← 3 + 5 sin(t) + 0.2x
end function f
```

We obtain the approximation $x(10) \approx 135.917$. The output from the code is a table of values that can be sent to a plotting routine. The resulting graph helps the user to visualize the solution curve.

Adams-Bashforth-Moulton Formulas

We now introduce a strategy in which numerical quadrature formulas are used to solve a single first-order ordinary differential equation. The model equation is

$$x'(t) = f(t, x(t))$$

and we suppose that the values of the unknown function have been computed at several points to the left of t, namely, $t, t - h, t - 2h, \dots, t - (n - 1)h$. We want to compute $x(t + h)$. By the theorems of calculus, we can write

$$\begin{aligned} x(t + h) &= x(t) + \int_t^{t+h} x'(s)\, ds \\ &= x(t) + \int_t^{t+h} f(s, x(s))\, ds \\ &\approx x(t) + \sum_{j=1}^{n} c_j f_j \end{aligned}$$

where the abbreviation $f_j = f(t - (j - 1)h, x(t - (j - 1)h))$ has been used. In the last line of the above equation, we have brought in a suitable numerical integration rule. The simplest case of such a formula over interval [0, 1] uses values of the integrand at points

$0, -1, -2, \dots, 1 - n$ in the case of an **Adams-Bashforth formula**. Once we have such a basic rule, a change of variable produces the rule for any other interval with any other uniform spacing.

Let's find a rule of the form

$$\int_0^1 F(r)\,dr \approx c_1 F(0) + c_2 F(-1) + \cdots + c_n F(1-n)$$

There are n coefficients c_j at our disposal. We know from interpolation theory that the rule can be made exact for all polynomials of degree $n-1$. It suffices that we insist on integrating each function $1, r, r^2, \dots, r^{n-1}$ exactly. Hence, we write down the appropriate equation:

$$\int_0^1 r^{i-1}\,dt = \sum_{j=1}^{n} c_j (1-j)^{i-1} \qquad (1 \leqq i \leqq n)$$

This is a linear system $\boldsymbol{Au} = \boldsymbol{b}$ of n equations in n unknowns. The elements of the matrix $\boldsymbol{A}$ are $A_{ij} = (1-j)^{i-1}$, and the right-hand side is $b_i = 1/i$.

When this program is run, the output is the vector of coefficients $\left(55/24,\ -59/24,\ 37/24,\ -3/8\right)$. Of course, higher-order formulas are obtained by changing the value of n in the code. To get the **Adams-Moulton formulas**, we start with a quadrature rule of the form

$$\int_0^1 G(r)\,dr \approx \sum_{j=1}^{n} C_j G(2-j)$$

A program similar to the one above yields the coefficients $\left(9/24, 19/24, -5/24, 1/24\right)$. The distinction between the two quadrature rules is that one involves the value of the integrand at 1 and the other does not.

How do we arrive at formulas for $\int_t^{t+h} g(s)\,ds$ from the work already done? Use the change of variable from s to σ given by $s = h\sigma - t$. In these considerations, think of t as a constant. The new integral is $h \int_0^1 g(h\sigma + t)\,d\sigma$, which can be treated with either of the two formulas already designed for the interval [0, 1]. For example, we have

Adams-Bashforth Quadrature Rules

$$\begin{aligned}
\int_t^{t+h} F(r)\,dr &\approx \frac{h}{24}\left[55F(t) - 59F(t-h) + 37F(t-2h) - 9F(t-3h)\right] \\
\int_t^{t+h} G(r)\,dr &\approx \frac{h}{24}\left[9G(t+h) + 19G(t) - 5G(t-h) + G(t-2h)\right]
\end{aligned} \tag{5}$$

The method of undetermined coefficients used here to obtain the quadrature rules does not, by itself, provide the error terms that we would like to have. An assessment of the error can be made from interpolation theory, because the methods considered here come from integrating an interpolating polynomial. Details can be found in more advanced books. You can experiment with some of the Adams-Bashforth-Moulton formulas in Computer Exercises 7.3.2 and 7.3.4. These methods are taken up again in Section 7.5.

Stability Analysis

Let us now resume the discussion of errors that inevitably occur in the numerical solution of an initial-value problem

$$\begin{cases} x' = f(t, x) \\ x(a) = s \end{cases} \tag{6}$$

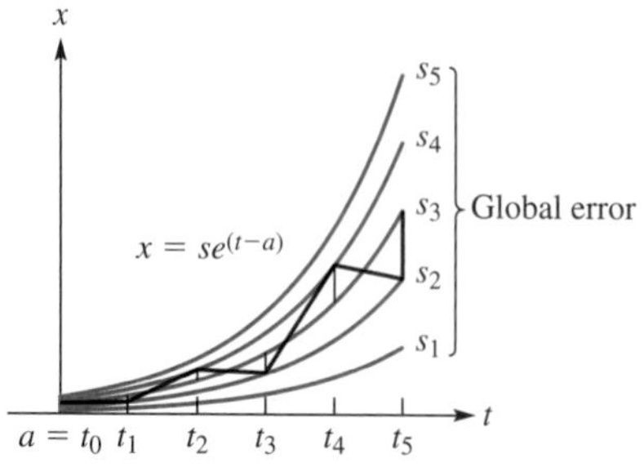

FIGURE 7.4
Solution curves to $x' = x$ with $x(a) = s$

The exact solution is a function $x(t)$. It depends on the initial value s, and to show this, we write $x(t, s)$. The differential equation therefore gives rise to a family of solution curves, each corresponding to one value of the parameter s. For example, the differential equation

Example of Divergent Solution Curves

$$\begin{cases} x' = x \\ x(a) = s \end{cases}$$

gives rise to the family of solution curves $x = se^{(t-a)}$ that differ in their initial values $x(a) = s$. A few such curves are shown in Figure 7.4. The fact that the curves there diverge from one another as t increases has important numerical significance. Suppose, for instance, that initial value s is read into the computer with some roundoff error. Then even if all subsequent calculations are precise and *no truncation errors* occur, the computed solution is still wrong! An error made at the beginning has the effect of selecting the wrong *curve* from the family of all solution curves. Since these curves diverge from one another, any minute error made at the beginning is responsible for an eventual complete loss of accuracy. This phenomenon is not restricted to errors made in the first step, because each point in the numerical solution can be interpreted as the initial value for succeeding points.

For an example in which this difficulty does *not* arise, consider

Example of Convergent Solution Curves

$$\begin{cases} x' = -x \\ x(a) = s \end{cases} \tag{7}$$

Its solutions are $x = se^{-(t-a)}$. As t increases, these curves come closer together, as in Figure 7.5. Thus, errors made in the numerical solution still result in selecting the wrong curve, but the effect is not as serious because the curves coalesce.

At a given step, the global error of an approximate solution to an ordinary differential equation contains both the local error at that step and the accumulative effect of all the local

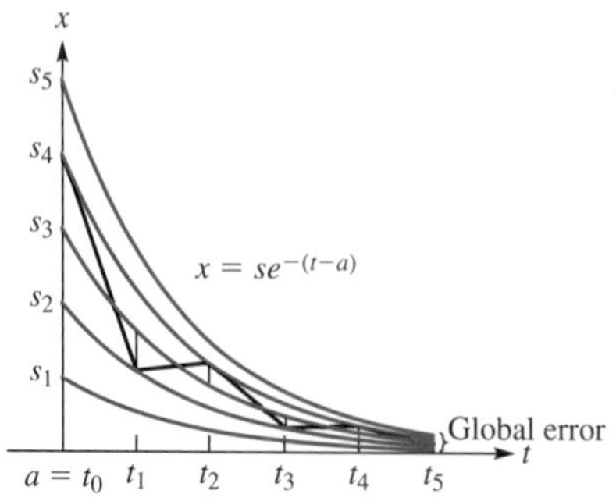

FIGURE 7.5
Solution curves to $x' = -x$ with $x(a) = s$

errors at all previous steps. For divergent solution curves, the local errors at each step are magnified over time, and the global error may be greater than the sum of all the local errors. In Figures 7.4 and 7.5, the steps in the numerical solution are indicated by dots connected by dark lines. Also, the local errors are indicated by small vertical bars and the global error by a vertical bar at the right end of the curves.

Conditions for Curves to Diverge or Converge

For convergent solution curves, the local errors at each step are reduced over time, and the global error may be less than the sum of all the local errors. *For the general differential Equation (6), how can the two modes of behavior just discussed be distinguished?* It is simple. If $f_x > \delta$ for some positive δ, the curves **diverge**. However, if $f_x < -\delta$, they **converge**. To see why, consider two nearby solution curves that correspond to initial values s and $s+h$. By Taylor series, we have

$$x(t, s+h) = x(t, s) + h\frac{\partial}{\partial s}x(t, s) + \frac{1}{2}h^2\frac{\partial^2}{\partial s^2}x(t, s) + \cdots$$

whence

$$x(t, s+h) - x(t, s) \approx h\frac{\partial}{\partial s}x(t, s)$$

Thus, the divergence of the curves means that

$$\lim_{t\to\infty} |x(t, s+h) - x(t, s)| = \infty$$

and can be written as

$$\lim_{t\to\infty} \left|\frac{\partial}{\partial s}x(t, s)\right| = \infty$$

To calculate this partial derivative, start with the differential equation satisfied by $x(t, s)$:

$$\frac{\partial}{\partial t}x(t, s) = f(t, x(t, s))$$

and differentiate partially with respect to s:

$$\frac{\partial}{\partial s}\frac{\partial}{\partial t}x(t, s) = \frac{\partial}{\partial s}f(t, x(t, s))$$

Hence, we obtain

$$\frac{\partial}{\partial t}\frac{\partial}{\partial s}x(t, s) = f_x(t, x(t, s))\frac{\partial}{\partial s}x(t, s) + f_t(t, x(t, s))\frac{\partial t}{\partial s} \tag{8}$$

But s and t are independent variables (a change in s produces no change in t), so $\partial t/\partial s = 0$. If s is now fixed and if we put $u(t) = (\partial/\partial s)x(t, s)$ and $q(t) = f_x(t, x(t, s))$, then Equation (8) becomes

$$u' = qu \tag{9}$$

This is a linear differential equation with solution $u(t) = ce^{Q(t)}$, where Q is the indefinite integral (antiderivative) of q. The condition $\lim_{t\to\infty} |u(t)| = \infty$ is met if $\lim_{t\to\infty} Q(t) = \infty$. This situation, in turn, occurs if $q(t)$ is positive and bounded away from zero because then

$$Q(t) = \int_a^t q(\theta)\, d\theta > \int_a^t \delta\, d\theta = \delta(t-a) \to \infty$$

as $t \to \infty$ if $f_x = q > \delta > 0$.

To illustrate, consider the differential equation $x' = t + \tan x$. The solution curves diverge from one another as $t \to \infty$ because $f_x(t, x) = \sec^2 x > 1$.

Summary 7.3

- The **Runge-Kutta-Fehlberg method** is

$$\widetilde{x}(t) = x(t) + \frac{25}{216}K_1 + \frac{1408}{2565}K_3 + \frac{2197}{4104}K_4 - \frac{1}{5}K_5$$

$$x(t+h) = x(t) + \frac{16}{135}K_1 + \frac{6656}{12825}K_3 + \frac{28561}{56430}K_4 - \frac{9}{50}K_5 + \frac{2}{55}K_6$$

where

$$\begin{cases} K_1 = hf(t, x) \\ K_2 = hf\left(t + \frac{1}{4}h,\ x + \frac{1}{4}K_1\right) \\ K_3 = hf\left(t + \frac{3}{8}h,\ x + \frac{3}{32}K_1 + \frac{9}{32}K_2\right) \\ K_4 = hf\left(t + \frac{12}{13}h,\ x + \frac{1932}{2197}K_1 - \frac{7200}{2197}K_2 + \frac{7296}{2197}K_3\right) \\ K_5 = hf\left(t + h,\ x + \frac{439}{216}K_1 - 8K_2 + \frac{3680}{513}K_3 - \frac{845}{4104}K_4\right) \\ K_6 = hf\left(t + \frac{1}{2}h,\ x - \frac{8}{27}K_1 + 2K_2 - \frac{3544}{2565}K_3 + \frac{1859}{4104}K_4 - \frac{11}{40}K_5\right) \end{cases}$$

The quantity $\varepsilon = |x(t+h) - \widetilde{x}|$ can be used in an adaptive step-size procedure.

- A fourth-order multistep method is the **Adams-Bashforth-Moulton method**:

$$\begin{aligned} \widetilde{x}(t+h) = x(t) + \frac{h}{24}\big[&55f(t, x(t)) - 59f(t-h, x(t-h)) \\ &+ 37f(t-2h, x(t-2h)) - 9f(t-3h, x(t-3h))\big] \\ x(t+h) = x(t) + \frac{h}{24}[&9f(t+h, \widetilde{x}(t+h)) + 19f(t, x)(t)) \\ &- 5f(t-h, x(t-h)) + f(t-2h, x(t-2h))] \end{aligned}$$

The value $\widetilde{x}(t+h)$ is the **predicted value**, and $x(t+h)$ is the **corrected value**. The truncation errors for these two formulas are $\mathcal{O}(h^5)$. Since the value of $x(a)$ is given, the values for $x(a+h), x(a+2h), x(a+3h), x(a+4h)$ are computed by some single-step method such as the fourth-order Runge-Kutta method.

Exercises 7.3

[a]**1.** Solve the problem

$$\begin{cases} x' = -x \\ x(0) = 1 \end{cases}$$

by using the trapezoid rule, as discussed at the beginning of this chapter. Compare the true solution at $t = 1$ to the approximate solution obtained with n steps. Show, for example, that for $n = 5$, the error is 0.00123.

[a]**2.** Derive an implicit multistep formula based on Simpson's rule, involving uniformly spaced points $x(t-h)$, $x(t)$, and $x(t+h)$, for numerically solving the ordinary differential equation $x' = f$.

3. An alert student noticed that the coefficients in the Adams-Bashforth formula add up to 1. Why is that so?

[a]**4.** Derive a formula of the form

$$x(t+h) = ax(t) + bx(t-h) + h[cx'(t+h) + dx''(t) + ex'''(t-h)]$$

that is accurate for polynomials of as high a degree as possible.
Hint: Use polynomials $1, t, t^2$, and so on.

[a]**5.** Determine the coefficients of an implicit, one-step, ordinary differential equation method of the form

$$x(t+h) = ax(t) + bx'(t) + cx'(t+h)$$

so that it is exact for polynomials of as high a degree as possible. What is the order of the error term?

6. The differential equation that is used to illustrate the adaptive Runge-Kutta program can be solved with an integrating factor. Do so.

7. Establish Equation (9).

[a]**8.** The initial-value problem $x' = (1+t^2)x$ with $x(0) = 1$ is to be solved on the interval $[0, 9]$. How sensitive is $x(9)$ to perturbations in the initial value $x(0)$?

9. For each differential equation, determine regions in which the solution curves tend to diverge from one another as t increases:

[a]**a.** $x' = \sin t + e^x$ **b.** $x' = x + te^{-t}$

[a]**c.** $x' = xt$ **d.** $x' = x^3(t^2+1)$

[a]**e.** $x' = \cos t - e^x$ **f.** $x' = (1-x^3)(1+t^2)$

10. For the differential equation $x' = t(x^3 - 6x^2 + 15x)$, determine whether the solution curves diverge from one another as $t \to \infty$.

[a]**11.** Determine whether the solution curves of $x' = (1+t^2)^{-1}x$ diverge from one another as $t \to \infty$.

Computer Exercises 7.3

1. Use mathematical software to solve systems of linear equations whose solutions are

a. Adams-Bahforth coefficients

b. Adams-Moulton coefficients

2. The **second-order Adams-Bashforth-Moulton method** is given by

$$\widetilde{x}(t+h) = x(t) + \frac{h}{2}[3f(t, x(t)) - f(t-h, x(t-h))]$$

$$x(t+h) = x(t) + \frac{h}{2}[f(t+h, \widetilde{x}(t+h)) + f(t, x(t))]$$

The approximate single-step error is $\varepsilon \equiv K|x(t+h) - \widetilde{x}(t+h)|$, where $K = \frac{1}{6}$. Using ε to monitor the convergence, write and test an adaptive procedure for solving an ODE of your choice using these formulas.

3. (Continuation) Carry out the instructions of the previous computer problem for the **third-order Adams-Bashforth-Moulton method**:

$$\widetilde{x}(t+h) = x(t) + \frac{h}{12}[23f(t, x(t)) - 16f(t-h, x(t-h)) + 5f(t-2h, x(t-2h))]$$

$$x(t+h) = x(t) + \frac{h}{12}[5f(t+h, \widetilde{x}(t+h))8f(t, x(t)) - f(t-h, x(t-h))]$$

where $K = \frac{1}{10}$ in the expression for the approximate single-step error.

4. **(Predictor-Corrector Scheme)** Using the **fourth-order Adams-Bashforth-Moulton method**, derive the predictor-corrector scheme given by the following equations:

$$\widetilde{x}(t+h) = x(t) + \frac{h}{24}[55f(t, x(t)) - 59f(t-h, x(t-h)) + 37f(t-2h, x(t-2h)) - 9f(t-3h, x(t-3h))]$$

$$x(t+h) = x(t) + \frac{h}{24}[9f(t+h, \widetilde{x}(t+h)) + 19f(t, x(t)) - 5f(t-h, x(t-h)) + f(t-2h, x(t-2h))]$$

Write and test a procedure for the Adams-Bashforth-Moulton method.
Note: This is a multistep process because values of x at t, $t-h$, $t-2h$, and $t-3h$ are used to determine the *predicted* value $\widetilde{x}(t+h)$, which, in turn, is used with values of x at t, $t-h$, and $t-2h$ to obtain the *corrected* value $x(t+h)$. The error terms for these formulas are $(251/720)h^5 f^{(iv)}(\xi)$ and $-(19/720)h^5 f^{(iv)}(\eta)$, respectively. (See Section 7.5 for additional discussion of these methods.)

[a]**5.** Solve

$$\begin{cases} x' = \dfrac{3x}{t} + \dfrac{9}{2}t - 13 \\ x(3) = 6 \end{cases}$$

at $x\left(\frac{1}{2}\right)$ using procedure *RK45_Adaptive* to obtain the desired solution to nine decimal places. Compare with the true solution:

$$x = t^3 - \frac{9}{2}t^2 + \frac{13}{2}t$$

[a]**6.** (Continuation) Repeat the previous problem for $x\left(-\frac{1}{2}\right)$.

7. It is known that the fourth-order Runge-Kutta method described in Equation (12) of Section 7.2 has a local truncation error that is $\mathcal{O}(h^5)$. Devise and carry out a numerical experiment to test this.
Suggestions: Take just one step in the numerical solution of a nontrivial differential equation whose solution is known beforehand. However, use a variety of values for h, such as 2^{-n}, where $1 \leqq n \leqq 24$. Test whether the ratio of errors to h^5 remains bounded as $h \to 0$. A multiple-precision calculation may be needed. Print the indicated ratios.

8. Compute the numerical solution of

$$\begin{cases} x' = -x \\ x(0) = 1 \end{cases}$$

using the **midpoint method**

$$x_{n+1} = x_{n-1} + 2hx'_n$$

with $x_0 = 1$ and $x_1 = -h + \sqrt{1+h^2}$. Are there any difficulties in using this method for this problem? Carry out an analysis of the stability of this method.
Hint: Consider fixed h and assume $x_n = \lambda^n$.

[a]**9.** Tabulate and graph the function $[1 - \ln v(x)]v(x)$ on $[0, e]$, where $v(x)$ is the solution of the initial-value problem $(dv/dx)[\ln v(x)] = 2x$, $v(0) = 1$.
Check value: $v(1) = e$.

10. Determine the numerical value of

$$2\pi \int_4^5 \frac{e^s}{s}\, ds$$

in three ways: solving the integral, an ordinary differential equation, and using the exact formula.

11. Compute and print a table of the function

$$f(\phi) = \int_0^\phi \sqrt{1 - \tfrac{1}{4}\sin^2\theta}\, d\theta$$

by solving an appropriate initial-value problem. Cover the interval $[0, 90°]$ with steps of $1°$ and use the Runge-Kutta method of order 4.
Check values: Use $f(30°) = 0.51788\,193$ and $f(90°) = 1.46746\,221$.

Note: This is an example of an elliptic integral of the second kind. It arises in finding an arc length on an ellipse and in many engineering problems.

[a]**12.** By solving an appropriate initial-value problem, make a table of the function

$$f(x) = \int_{1/x}^\infty \frac{dt}{te^t}$$

on the interval $[0, 1]$. Determine how well f is approximated by $xe^{-1/x}$.
Hint: Let $t = -\ln s$.

[a]**13.** By solving an appropriate initial-value problem, make a table of the function

$$f(x) = \frac{2}{\sqrt{\pi}} \int_0^x e^{-t^2}\, dt$$

on the interval $0 \leqq x \leqq 2$. Determine how accurately $f(x)$ is approximated on this interval by the function

$$g(x) = 1 - \left(ay + by^2 + cy^3\right)\frac{2}{\sqrt{\pi}}e^{-x^2}$$

where

$$\begin{cases} a = 0.30842\,84, & b = -0.08497\,13 \\ c = 0.66276\,98, & y = (1 + 0.47047x)^{-1} \end{cases}$$

14. Use the Runge-Kutta method to compute

$$\int_0^1 \sqrt{1+s^3}\, ds$$

[a]**15.** Write and run a program to print an accurate table of the **sine integral**

$$\mathrm{Si}(x) = \int_0^x \frac{\sin r}{r}\, dr$$

The table should cover the interval $0 \leqq x \leqq 1$ in steps of size 0.01.
(Use $\sin(0)/0 = 1$. See Computer Exercise 5.1.9.)

16. Compute a table of the function

$$\mathrm{Shi}(x) = \int_0^x \frac{\sinh t}{t}\, dt$$

by finding an initial-value problem that it satisfies and then solving the initial-value problem. Your table should be accurate to nearly machine precision.
(Use $\sinh(0)/\,0 = 1$.)

17. Design and carry out a numerical experiment to verify that a slight perturbation in an initial-value problem can cause catastrophic errors in the numerical solution.
Note: An **initial-value problem** is an ordinary differential equation with conditions specified only at the initial point. (Compare this with a *boundary value problem* as given in Chapter 11.)

18. Run example programs for solving the industrial example in Equation (4), compare the solutions, and produce the plots.

19. Another adaptive Runge-Kutta method was developed by England [1969]. The **Runge-Kutta-England method** is similar to the Runge-Kutta-Fehlberg method in that it combines a fourth-order Runge-Kutta formula and a companion fifth-order one. To reduce the number of function evaluations, the formulas are derived so that some of the same function evaluations are used in each pair of formulas. (A fourth-order Runge-Kutta formula requires at least four function evaluations, and a fifth-order one requires at least six.) The Runge-Kutta-England method uses the fourth-order Runge-Kutta methods in Computer Exercise 7.2.14a and takes two half-steps as follows:

$$x\left(t + \frac{1}{2}h\right) = x(t) + \frac{1}{6}(K_1 + 4K_3 + K_4)$$

where

$$\begin{cases} K_1 = \frac{1}{2}hf(t, x(t)) \\ K_2 = \dfrac{1}{2}hf\left(t + \frac{1}{4}h,\ x(t) + \frac{1}{2}K_1\right) \\ K_3 = \dfrac{1}{2}hf\left(t + \frac{1}{4}h,\ x(t) + \frac{1}{4}K_1 + \frac{1}{4}K_2\right) \\ K_4 = \dfrac{1}{2}hf\left(t + \frac{1}{2}h,\ x(t) - K_2 + 2K_3\right) \end{cases}$$

and

$$x(t+h) = x\left(t + \tfrac{1}{2}h\right) + \frac{1}{6}\left(K_5 + 4K_7 + K_8\right)$$

where

$$\begin{cases} K_5 = \dfrac{1}{2}hf\left(t + \frac{1}{2}h,\ x\left(t + \frac{1}{2}h\right)\right) \\ K_6 = \dfrac{1}{2}hf\left(t + \frac{3}{4}h,\ x\left(t + \frac{1}{2}h\right) + \frac{1}{2}K_5\right) \\ K_7 = \dfrac{1}{2}hf\left(t + \frac{3}{4}h,\ x\left(t + \frac{1}{2}h\right) + \frac{1}{4}K_5 + \frac{1}{4}K_6\right) \\ K_8 = \dfrac{1}{2}hf\left(t + h,\ x\left(t + \frac{1}{2}h\right) - K_6 + 2K_7\right) \end{cases}$$

With these two half-steps, there are enough function evaluations so that only one more

$$K_9 = \frac{1}{2}hf\Big(t + h, x(t) - \tfrac{1}{12}(K_1 + 96K_2 - 92K_3 + 121K_4 - 144K_5 - 6K_6 + 12K_7\Big)$$

is needed to obtain a fifth-order Runge-Kutta method:

$$\widehat{x}(t+h) = x(t) + \frac{1}{90}\left(14K_1 + 64K_3 + 32K_3 - 8K_5 + 64K_7 + 15K_8 - K_9\right)$$

An adaptive procedure can be developed by using an error estimation based on the two values $x(t+h)$ and $\widehat{x}(t+h)$. Program and test such a procedure. (See, for example, Shampine, Allen, and Pruess [1997].)

20. Investigate the numerical solution of the initial-value problem

$$\begin{cases} x' = -\sqrt{1 - x^2} \\ x(0) = 1 \end{cases}$$

This problem is ill-conditioned, since $x(t) = \cos t$ is a solution and $x(t) = 1$ is also. For more information on this and other test problems, see Cash [2003] or `www2.imperial.ac.uk/~jcash/`.

21. (**Student Research Project**) Learn and write a report about algebraic differential equations.

22. Write software to implement the following pseudocodes and verify the numerical results given in the text for IVP (13) in Section 7.2 and for IVP (4) in this section.

    a. *Test_RK4* and *RK4*

    b. *Test_RK45* and *RK45*

    c. *Test_RK45_Adaptive* and *RK45_Adaptive*

7.4 Methods for First and Higher-Order Systems

So far in our study of the numerical solutions of ordinary differential equations, we have restricted our attention to a single differential equation of the first order with an accompanying auxiliary condition. Scientific and technological problems often lead to more complicated situations, however. The next degree of complication occurs with *systems* of several first-order equations.

Uncoupled and Coupled Systems

The sun and the nine planets form a system of *particles* moving under the jurisdiction of Newton's law of gravitation. The position vectors of the planets constitute a system of 27 functions, and the Newtonian laws of motion can be written, then, as a system of 54 first-order ordinary differential equations. In principle, the past and future positions of the planets can be obtained by solving these equations numerically.

Taking an example of more modest scope, we consider two equations with two auxiliary conditions. Let x and y be two functions of t subject to the system

Sample ODE Coupled System

$$\begin{cases} x'(t) = x(t) - y(t) + 2t - t^2 - t^3 \\ y'(t) = x(t) + y(t) - 4t^2 + t^3 \end{cases} \tag{1}$$

with initial conditions

$$\begin{cases} x(0) = 1 \\ y(0) = 0 \end{cases}$$

This is an example of an initial-value problem that involves a system of two first-order differential equations. Note that in the example given, it is not possible to solve either of the two differential equations by itself because the first equation governing x' involves the unknown function y, and the second equation governing y' involves the unknown function x. In this situation, we say that the two differential equations are **coupled**.

The reader is invited to verify that the analytic solution is

Analytic Solution

$$\begin{cases} x(t) = e^t \cos(t) + t^2 = \cos(t)[\cosh(t) + \sinh(t)] + t^2 \\ y(t) = e^t \sin(t) - t^3 = \sin(t)[\cosh(t) + \sinh(t)] - t^3 \end{cases}$$

Let us look at another example that is superficially similar to the first but is actually simpler:

Sample ODE Uncoupled System

$$\begin{cases} x'(t) = x(t) + 2t - t^2 - t^3 \\ y'(t) = y(t) - 4t^2 + t^3 \end{cases} \tag{2}$$

with initial conditions

$$\begin{cases} x(0) = 1 \\ y(0) = 0 \end{cases}$$

These two equations are *not* coupled and can be solved separately as two unrelated initial-value problems (using, for instance, the methods of Sections 7.1–3). Naturally, our concern here is with systems that are coupled, although methods that solve coupled systems also solve those that are not. The procedures extend to systems, whether coupled or uncoupled.

Taylor Series Method

We illustrate the Taylor series method for System (1) and begin by differentiating the equations constituting it:

$$\begin{cases} x' = x - y + 2t - t^2 - t^3 \\ y' = x + y - 4t^2 + t^3 \end{cases}$$

$$\begin{cases} x'' = x' - y' + 2 - 2t - 3t^2 \\ y'' = x' + y' - 8t + 3t^2 \end{cases}$$

$$\begin{cases} x''' = x'' - y'' - 2 - 6t \\ y''' = x'' + y'' - 8 + 6t \end{cases}$$

$$\begin{cases} x^{(iv)} = x''' - y''' - 6 \\ y^{(iv)} = x''' + y''' + 6 \end{cases}$$

$$\begin{cases} x^{(v)} = x^{(iv)} - y^{(iv)} \\ y^{(v)} = x^{(iv)} + y^{(iv)} \end{cases}$$

etc.

A program to proceed from $x(t)$ to $x(t+h)$ and from $y(t)$ to $y(t+h)$ is easily written by using a few terms of the Taylor series:

Taylor Series for $x(t+h)$ and $y(t+h)$

$$x(t+h) = x + hx' + \frac{h^2}{2}x'' + \frac{h^3}{6}x''' + \frac{h^4}{24}x^{(iv)} + \frac{h^5}{120}x^{(v)} + \cdots$$

$$y(t+h) = y + hy' + \frac{h^2}{2}y'' + \frac{h^3}{6}y''' + \frac{h^4}{24}y^{(iv)} + \frac{h^5}{120}y^{(v)} + + \cdots$$

together with equations for the various derivatives. Here, x and y and all their derivatives are functions of t; that is, $x = x(t)$, $y = y(t)$, $x' = x'(t)$, $y'' = y''(t)$, and so on.

A pseudocode program that generates and prints a numerical solution from 0 to 1 in 100 steps is as follows. Terms up to h^4 have been used in the Taylor series.

***Taylor_System1* Pseudocode**

```
program Taylor_System1
integer k;   real h, t, x, y, x', y', x'', y'', x''', y''', x^(iv), y^(iv)
integer nsteps ← 100;   real a ← 0, b ← 1
x ← 1;   y ← 0;   t ← a
output 0, t, x, y
h ← (b − a)/nsteps
for k = 1 to nsteps
    x' ← x − y + t(2 − t(1 + t))
    y' ← x + y + t^2(−4 + t)
    x'' ← x' − y' + 2 − t(2 + 3t)
    y'' ← x' + y' + t(−8 + 3t)
    x''' ← x'' − y'' − 2 − 6t
    y''' ← x'' + y'' − 8 + 6t
    x^(iv) ← x''' − y''' − 6
    y^(iv) ← x''' + y''' + 6
    x^(v) ← x^(iv) − y^(iv)
    y^(v) ← x^(iv) + y^(iv)
    x ← x + h[x' + (1/2)h[x'' + (1/3)h[x''' + (1/4)h[x^(iv)]]]]
    y ← y + h[y' + (1/2)h[y'' + (1/3)h[y''' + (1/4)h[y^(iv)]]]]
    t ← t + h
    output k, t, x, y
end for
end program Taylor_System1
```

Vector Notation

Observe that System (1) can be written in vector notation as

ODE System1 in Vector Form

$$\begin{bmatrix} x' \\ y' \end{bmatrix} = \begin{bmatrix} x - y + 2t - t^2 - t^3 \\ x + y - 4t^2 + t^3 \end{bmatrix} \tag{3}$$

with initial conditions

$$\begin{bmatrix} x(0) \\ y(0) \end{bmatrix} = \begin{bmatrix} 1 \\ 0 \end{bmatrix}$$

This is a special case of a more general problem that can be written as

$$\begin{cases} \boldsymbol{X}' = \boldsymbol{F}(t, \boldsymbol{X}) \\ \boldsymbol{X}(a) = \boldsymbol{S} \quad \text{(given)} \end{cases} \tag{4}$$

where

$$\boldsymbol{X} = \begin{bmatrix} x \\ y \end{bmatrix}, \qquad \boldsymbol{X}' = \begin{bmatrix} x' \\ y' \end{bmatrix}$$

and $\boldsymbol{F}$ is the vector whose two components are given by the right-hand sides in Equation (1). Since $\boldsymbol{F}$ depends on t and $\boldsymbol{X}$, we write $\boldsymbol{F}(t, \boldsymbol{X})$.

Systems of ODEs

We can continue this idea in order to handle a system of n first-order differential equations. First, we write them as

First-Order ODEs

$$\begin{cases} x_1' = f_1(t, x_1, x_2, \ldots, x_n) \\ x_2' = f_2(t, x_1, x_2, \ldots, x_n) \\ \quad\vdots \\ x_n' = f_n(t, x_1, x_2, \ldots, x_n) \\ x_1(a) = s_1, x_2(a) = s_2, \ldots, \; x_n(a) = s_n \quad \text{(all given)} \end{cases}$$

Then we let

$$\boldsymbol{X} = \begin{bmatrix} x_1 \\ x_2 \\ \vdots \\ x_n \end{bmatrix}, \qquad \boldsymbol{X}' = \begin{bmatrix} x_1' \\ x_2' \\ \vdots \\ x_n' \end{bmatrix}, \qquad \boldsymbol{F} = \begin{bmatrix} f_1 \\ f_2 \\ \vdots \\ f_n \end{bmatrix}, \qquad \boldsymbol{S} = \begin{bmatrix} s_1 \\ s_2 \\ \vdots \\ s_n \end{bmatrix}$$

and we obtain Equation (4), which is an ordinary differential equation written in vector notation.

Taylor Series Method: Vector Notation

The **m-order Taylor series method** would be written as

Taylor Series Method mth Order

$$\boldsymbol{X}(t + h) = \boldsymbol{X} + h\boldsymbol{X}' + \frac{h^2}{2}\boldsymbol{X}'' + \frac{h^3}{3!}\boldsymbol{X}''' + \frac{h^4}{4!}\boldsymbol{X}^{(iv)} + \frac{h^5}{5!}\boldsymbol{X}^{(v)} + \cdots + \frac{h^m}{m!}\boldsymbol{X}^{(m)} \tag{5}$$

where $\boldsymbol{X} = \boldsymbol{X}(t)$, $\boldsymbol{X}' = \boldsymbol{X}'(t)$, $\boldsymbol{X}'' = \boldsymbol{X}''(t)$, $\boldsymbol{X}''' = \boldsymbol{X}'''(t)$, and so on.

A pseudocode for the Taylor series method of order 4 applied to the preceding problem can be easily rewritten by a simple change of variables and the introduction of an array and an inner loop.

Taylor_System2 Pseudocode

```
program Taylor_System2
integer i, k;   real h, t;   real array (x_i)_{1:n}, (d_ij)_{1:n×1:4}
integer n ← 2, nsteps ← 100
real a ← 0, b ← 1
t ← 0;   (x_i) ← (1, 0)
output 0, t, (x_i)
h ← (b − a)/nsteps
for k = 1 to nsteps
    d_11 ← x_1 − x_2 + t(2 − t(1 + t))
    d_21 ← x_1 + x_2 + t^2(−4 + t)
    d_12 ← d_11 − d_21 + 2 − t(2 + 3t)
    d_22 ← d_11 + d_21 + t(−8 + 3t)
    d_13 ← d_12 − d_22 − 2 − 6t
    d_23 ← d_12 + d_22 − 8 + 6t
    d_14 ← d_13 − d_23 − 6
    d_24 ← d_13 + d_23 + 6
    for i = 1 to n
        x_i ← x_i + h [d_i1 + (1/2)h [d_i2 + (1/3)h [d_i3 + (1/4)h [d_i4]]]]
    end for
    t ← t + h
    output k, t, (x_i)
end for
end program Taylor_System2
```

Here, a two-dimensional array is used instead of all the different derivative variables; that is, $d_{ij} \leftrightarrow x_i^{(j)}$. In fact, this and other methods in this chapter become particularly easy to program if the computer language supports vector operations.

Runge-Kutta Method

The Runge-Kutta methods also extend to systems of differential equations. The classical **fourth-order Runge-Kutta method for System** (4) uses these formulas:

Runge-Kutta Method Fourth-Order in Vector Form

$$X(t+h) = X + \frac{h}{6}(K_1 + 2K_2 + 2K_3 + K_4) \tag{6}$$

where

$$\begin{cases} K_1 = F(t, X) \\ K_2 = F\left(t + \frac{1}{2}h, x + \frac{1}{2}hk_1\right) \\ K_3 = F\left(t + \frac{1}{2}h, x + \frac{1}{2}hk_2\right) \\ K_4 = F(t + h, X + hK_3) \end{cases}$$

Here, $X = X(t)$, and all quantities are vectors with n components except variables t and h.

Pseudocode

A procedure for carrying out the Runge-Kutta procedure is given next. It is assumed that the system to be solved is in the form of Equation (4) and that there are n equations in the system. The user furnishes the initial value of t, the initial value of $\boldsymbol{X}$, the step size h, and the number of steps to be taken, *nsteps*. Furthermore, procedure *XP_System*$(n, t, (x_i), (f_i))$ is needed, which evaluates the right-hand side of Equation (4) for a given value of array (x_i) and stores the result in array (f_i). (The name *XP_System2* is chosen as an abbreviation of $\boldsymbol{X}'$ for a system.)

RK4_System1 Pseudocode

```
procedure RK4_System1(n, h, t, (x_i), nsteps)
integer i, j, n;   real h, t;   real array (x_i)_{1:n}
allocate real array (y_i)_{1:n}, (K_{i,j})_{1:n×1:4}
output 0, t, (x_i)
for j = 1 to nsteps
    call XP_System(n, t, (x_i), (K_{i,1}))
    for i = 1 to n
        y_i ← x_i + (1/2)hK_{i,1}
    end for
    call XP_System(n, t + h/2, (y_i), (K_{i,2}))
    for i = 1 to n
        y_i ← x_i + (1/2)hK_{i,2}
    end for
    call XP_System(n, t + h/2, (y_i), (K_{i,3}))
    for i = 1 to n
        y_i ← x_i + hK_{i,3}
    end for
    call XP_System(n, t + h, (y)_i, (K_{i,4}))
    for i = 1 to n
        x_i ← x_i + (1/6)h[K_{i,1} + 2K_{i,2} + 2K_{i,3} + K_{i,4}]
    end for
    t ← t + h
    output j, t, (x_i)
end for
deallocate array (y_i), (K_{i,j})
end procedure RK4_System1
```

To illustrate the use of this procedure, we again use System (1) for our example. Of course, it must be rewritten in the form of Equation (4). A suitable main program and a procedure for computing the right-hand side of Equation (4) follow:

Test_RK4_System1 Pseudocode

```
program Test_RK4_System1
integer n ← 2, nsteps ← 100
real a ← 0, b ← 1
real h, t;   real array (x_i)_{1:n}
t ← 0
(x_i) ← (1, 0)
h ← (b − a)/nsteps
```

(Continued)

```
call RK4_System1(n, h, t, (x_i), nsteps)
end program Test_RK4_System1

procedure XP_System(n, t, (x_i), (f_i))
real array (x_i)_1:n, (f_i)_1:n
integer n
real t
f_1 ← x_1 − x_2 + t(2 − t(1 + t))
f_2 ← x_1 + x_2 − t^2(4 − t)
end procedure XP_System
```

We use a numerical experiment to compare the results of the Taylor series method and the Runge-Kutta method with the analytic solution of System (1). At the point $t = 1.0$, the results are as follows:

Computer Results

	Taylor Series	**Runge-Kutta**	**Analytic Solution**
$x(1.0) \approx$	2.46869 40	2.46869 42	2.46869 39399
$y(1.0) \approx$	1.28735 46	1.28735 61	1.28735 52872

We can use mathematical software routines found in MATLAB, Maple, or Mathematica to obtain the numerical solution of the system of ordinary differential equations (1). For t over the interval $[0, 1]$, we invoke an ODE procedure to march from $t = 0$ at which $x(0) = 1$ and $y(0) = 0$ to $t = 1$ at which $x(1) = 2.468693912$ and $y(1) = 1.287355325$.

To obtain the numerical solution of the ordinary differential equation defined for t over the interval $[1, 1.5]$, invoke an ordinary differential equation solving procedure to march from $t = 0$ at which $x(1) = 2$ and $y(1) = -2$ to $t = 1.5$ at which $x(1.5) \approx 15.5028$ and $y(1.5) \approx 6.15486$.

Autonomous ODE

When we wrote the system of differential equations in vector form

$$\boldsymbol{X}' = \boldsymbol{F}(t, \boldsymbol{X})$$

we assumed that the variable t was explicitly separated from the other variables and treated differently. It is not necessary to do this. Indeed, we can introduce a new variable x_0 that is t in disguise and add a new differential equation $x_0' = 1$. A new initial condition must also be provided, $x_0(a) = a$. In this way, we increase the number of differential equations from n to $n + 1$ and obtain a system written in the more elegant vector form

$$\begin{cases} \boldsymbol{X}' = \boldsymbol{F}(\boldsymbol{X}) \\ \boldsymbol{X}(a) = \boldsymbol{S} \quad \text{(given)} \end{cases}$$

Consider the system of two equations given by Equation (1). We write it as a system with three variables by letting

$$x_0 = t, \qquad x_1 = x, \qquad x_2 = y$$

Thus, we have

$$\begin{bmatrix} x_0' \\ x_1' \\ x_2' \end{bmatrix} = \begin{bmatrix} 1 \\ x_1 - x_2 + 2x_0 - x_0^2 - x_0^3 \\ x_1 + x_2 - 4x_0^2 + x_0^3 \end{bmatrix}$$

The auxiliary condition for the vector $\boldsymbol{X}$ is $\boldsymbol{X}(0) = [0, 1, 0]^T$.

As a result of the preceding remarks, we sacrifice no generality in considering a system of $n + 1$ first-order differential equations written as

$$\begin{cases} x_0' = f_0(x_0, x_1, x_2, \dots, x_n) \\ x_1' = f_1(x_0, x_1, x_2, \dots, x_n) \\ x_2' = f_2(x_0, x_1, x_2, \dots, x_n) \\ \quad \vdots \\ x_n' = f_n(x_0, x_1, x_2, \dots, x_n) \\ x_0(a) = s_0, x_1(a) = s_1, x_2(a) = s_2, \dots,\ x_n(a) = s_n \quad \text{(all given)} \end{cases}$$

We can write this system in general vector notation as

IVP Autonomous Vector Form

$$\begin{cases} \boldsymbol{X}' = \boldsymbol{F}(\boldsymbol{X}) \\ \boldsymbol{X}(a) = \boldsymbol{S} \quad \text{(given)} \end{cases} \tag{7}$$

where

$$\boldsymbol{X} = \begin{bmatrix} x_0 \\ x_1 \\ x_2 \\ \vdots \\ x_n \end{bmatrix}, \quad \boldsymbol{X}' = \begin{bmatrix} x_0' \\ x_1' \\ x_2' \\ \vdots \\ x_n' \end{bmatrix}, \quad \boldsymbol{F} = \begin{bmatrix} f_0 \\ f_1 \\ f_2 \\ \vdots \\ f_n \end{bmatrix}, \quad \boldsymbol{S} = \begin{bmatrix} s_0 \\ s_1 \\ s_2 \\ \vdots \\ s_n \end{bmatrix}$$

A system of differential equations without the t variable explicitly present is said to be **autonomous**. The numerical methods that we discuss do not require that $x_0 = t$ or $f_0 = 1$ or $s_0 = a$.

For an autonomous system, the classical **fourth-order Runge-Kutta method** for System (6) uses these formulas:

$$\boldsymbol{X}(t + h) = \boldsymbol{X} + \frac{h}{6}(\boldsymbol{K}_1 + 2\boldsymbol{K}_2 + 2\boldsymbol{K}_3 + \boldsymbol{K}_4) \tag{8}$$

where

Runge-Kutta Method Fourth Order

$$\begin{cases} \boldsymbol{K}_1 = \boldsymbol{F}(\boldsymbol{X}) \\ \boldsymbol{K}_2 = \boldsymbol{F}\big(\boldsymbol{X} + \tfrac{1}{2}h\boldsymbol{K}_1\big) \\ \boldsymbol{K}_3 = \boldsymbol{F}\big(\boldsymbol{X} + \tfrac{1}{2}h\boldsymbol{K}_2\big) \\ \boldsymbol{K}_4 = \boldsymbol{F}(\boldsymbol{X} + h\boldsymbol{K}_3) \end{cases}$$

Here, $\boldsymbol{X} = \boldsymbol{X}(t)$, and all quantities are vectors with $n+1$ components except the variables h.

In the previous example, the procedure *RK4_System1* would need to be modified by beginning the arrays with 0 rather than 1 and omitting the variable t. (We call it *RK4_System2* and leave it as Computer Exercise 7.4.4.) Then the calling programs would be as follows:

```
program Test_RK4_System2
real h, t;   real array (x_i)_{0:n}
integer n ← 2, nsteps ← 100
```

(Continued)

Test_RK4_System2 Pseudocode

real $a \leftarrow 0,\ b \leftarrow 1$
$(x_i) \leftarrow (0, 1, 0)$
$h \leftarrow (b - a)/nsteps$
call *RK4_System2*$(n, h, (x_i), nsteps)$
end program *Test_RK4_System2*

procedure *XP_System*$(n, (x_i), (f_i))$
real array $(x_i)_{0:n}, (f_i)_{0:n}$
integer n
$f_0 \leftarrow 1$
$f_1 \leftarrow x_1 - x_2 + x_0(2 - x_0(1 + x_0))$
$f_2 \leftarrow x_1 + x_2 - x_0^2(4 - x_0)$
end procedure *XP_System*

It is typical in ordinary differential equation solvers, such as those found in mathematical software libraries, for the user to interface with them by writing a subprogram in a nonautonomous format. In other words, the ordinary differential equation solver takes as input both the independent variable and the dependent variable and returns values for the right-hand side to the ordinary differential equation. Consequently, the nonautonomous programming convention may seem more natural to those who are using these software packages.

It is a useful exercise to find a physical application in your field of study or profession involving the solution of an ordinary differential equation. It is instructive to analyze and solve the physical problem by determining the appropriate numerical method and translating the problem into the format that is compatible with the available software.

Higher-Order Differential Equations and Systems

Consider the initial-value problem for ordinary differential equations of order higher than 1. A differential equation of order n is normally accompanied by n auxiliary conditions. This many initial conditions are needed to specify the solution of the differential equation precisely (assuming certain smoothness conditions are present). Take, for example, a particular second-order initial-value problem

Sample IVP Second Order

$$\begin{cases} x''(t) = -3\cos^2(t) + 2 \\ x(0) = 0, \quad x'(0) = 0 \end{cases} \tag{9}$$

Without the auxiliary conditions, the general analytic solution is

$$x(t) = \frac{1}{4}t^2 + \frac{3}{8}\cos(2t) + c_1 t + c_2$$

where c_1 and c_2 are arbitrary constants. To select one specific solution, c_1 and c_2 must be fixed, and two initial conditions allow this to be done. In fact, $x(0) = 0$ yields $c_2 = -\frac{3}{8}$, and $x'(0) = 0$ forces $c_1 = 0$.

Higher-Order Differential Equations

In general, higher-order problems can be much more complicated than this simple example because System (9) has the special property that the function on the right-hand side of the differential equation does not involve x. The most general form of an ordinary differential equation with initial conditions that we shall consider is

IVP

$$\begin{cases} x^{(n)} = f(t, x, x', x'', \dots, x^{(n-1)}) \\ x(a), x'(a), x''(a), \dots, x^{(n-1)}(a) \quad \text{(all given)} \end{cases} \tag{10}$$

This can be solved numerically by turning it into a system of *first-order* differential equations. To do so, we define new variables $x_1, x_2, \dots, x_n$ as follows:

$$x_1 = x, \qquad x_2 = x', \qquad x_3 = x'', \qquad \dots, \qquad x_{n-1} = x^{(n-2)}, \qquad x_n = x^{(n-1)}$$

Consequently, the original Initial-Value Problem (10) is equivalent to

$$\begin{cases} x_1' = x_2 \\ x_2' = x_3 \\ \quad\vdots \\ x_{n-1}' = x_n \\ x_n' = f(t, x_1, x_2, \dots, x_n) \\ x_1(a), x_2(a), \dots, x_n(a) \quad \text{(all given)} \end{cases}$$

or, in vector notation,

IVP Vector Form

$$\begin{cases} \boldsymbol{X}' = \boldsymbol{F}(t, \boldsymbol{X}) \\ \boldsymbol{X}(a) = \boldsymbol{S} \quad \text{(given)} \end{cases} \tag{11}$$

where

$$\begin{aligned} \boldsymbol{X} &= [x_1, x_2, \dots, x_n]^T \\ \boldsymbol{X}' &= [x_1', x_2', \dots, x_n']^T \\ \boldsymbol{F} &= [x_2, x_3, x_4, \dots, x_n, f]^T \end{aligned}$$

and

$$\boldsymbol{X}(a) = [x_1(a), x_2(a), \dots, x_n(a)]$$

Whenever a problem must be transformed by introducing new variables, it is recommended that a **dictionary** be provided to show the relationship between the new and the old variables. At the same time, this information, together with the differential equations and the initial values, can be displayed in a chart. Such systematic bookkeeping can be helpful in a complicated situation.

To illustrate, let us transform the initial-value problem

Sample IVP Third Order

$$\begin{cases} x''' = \cos x + \sin x' - e^{x''} + t^2 \\ x(0) = 3, \quad x'(0) = 7, \quad x''(0) = 13 \end{cases} \tag{12}$$

into a form suitable for solution by the Runge-Kutta procedure. A chart summarizing the transformed problem is as follows:

Old Variable	New Variable	Initial Value	Differential Equation
x	x_1	3	$x_1' = x_2$
x'	x_2	7	$x_2' = x_3$
x''	x_3	13	$x_3' = \cos x_1 + \sin x_2 - e^{x_3} + t^2$

So the corresponding first-order system is

$$X' = \begin{bmatrix} x_2 \\ x_3 \\ \cos x_1 + \sin x_2 - e^{x_3} + t^2 \end{bmatrix}$$

and $X(0) = [3, 7, 13]^T$.

Systems of Higher-Order Differential Equations

By systematically introducing new variables, we can transform a system of differential equations of various orders into a larger system of first-order equations. For instance, the system

Sample IVP Fifth Order

$$\begin{cases} x'' = x - y - (3x')^2 + (y')^3 + 6y'' + 2t \\ y''' = y'' - x' + e^x - t \\ x(1) = 2, \quad x'(1) = -4, \quad y(1) = -2, \quad y'(1) = 7, \quad y''(1) = 6 \end{cases} \tag{13}$$

can be solved by the Runge-Kutta procedure if we first transform it according to the following chart:

Old Variable	New Variable	Initial Value	Differential Equation
x	x_1	2	$x_1' = x_2$
x'	x_2	-4	$x_2' = x_1 - x_3 - 9x_2^2 + x_4^3 + 6x_5 + 2t$
y	x_3	-2	$x_3' = x_4$
y'	x_4	7	$x_4' = x_5$
y''	x_5	6	$x_5' = x_5 - x_2 + e^{x_1} - t$

Hence, we have

$$X' = \begin{bmatrix} x_2 \\ x_1 - x_3 - 9x_2^2 + x_4^3 + 6x_5 + 2t \\ x_4 \\ x_5 \\ x_5 - x_2 + e^{x_1} - t \end{bmatrix}$$

and $X(1) = [2, -4, -2, 7, 6]^T$.

Autonomous ODE Systems

We notice that t is present on the right-hand side of Equation (11) and that therefore the equations $x_0 = t$ and $x_0' = 1$ can be introduced to form an autonomous system of ordinary differential equations in vector notation. It is easy to show that a higher-order system of differential equations having the form in Equation (10) can be written in vector notation as

IVP Vector Form

$$\begin{cases} X' = F(X) \\ X(a) = S \quad \text{(given)} \end{cases}$$

where

$$X = [x_0, x_1, x_2, \ldots, x_n]^T$$
$$X' = [x_0', x_1', x_2', \ldots, x_n']^T$$
$$F = [1, x_2, x_3, x_4, \ldots, x_n, f]^T$$

and

$$X(a) = [a, x_1(a), x_2(a), \ldots, x_n(a)]$$

As an example, the ordinary differential equation system in Equation (12) can be written in autonomous form as

Sample IVP Autonomous Form

$$X' = \begin{bmatrix} 1 \\ x_2 \\ x_1 - x_3 - 9x_2^2 + x_4^3 + 6x_5 + 2x_0 \\ x_4 \\ x_5 \\ x_5 - x_2 + e^{x_1} - x_0 \end{bmatrix}$$

and $X(1) = [1, 2, -4, -2, 7, 6]^T$.

Summary 7.4

- A **system of ordinary differential equations**

$$\begin{cases} x_1' = f_1(t, x_1, x_2, \ldots, x_n) \\ x_2' = f_2(t, x_1, x_2, \ldots, x_n) \\ \quad \vdots \\ x_n' = f_n(t, x_1, x_2, \ldots, x_n) \\ x_1(a) = s_1, x_2(a) = s_2, \ldots, x_n(a) = s_n, \quad \text{(all given)} \end{cases}$$

can be written in vector notation as

$$\begin{cases} X' = F(t, X) \\ X(a) = S \quad \text{(given)} \end{cases}$$

where we define the following n component vectors

$$\begin{cases} X = [x_1, x_2, \ldots, x_n]^T \\ X' = [x_1', x_2', \ldots, x_n']^T \\ F = [f_1, f_2, \ldots, f_n]^T \\ X(a) = [x_1(a), x_2(a), \ldots, x_n(a)]^T \end{cases}$$

- The **Taylor series method of order m** is

$$X(t+h) = X + hX' + \frac{h^2}{2}X'' + \frac{h^3}{3!}X''' + \cdots + \frac{h^m}{m!}X^{(m)}$$

where $X = X(t)$, $X' = X'(t)$, $X'' = X''(t)$, $X''' = X'''(t)$, and so on.

- The **Runge-Kutta method of order 4** is

$$X(t+h) = X + \frac{h}{6}(K_1 + 2K_2 + 2K_3 + K_4)$$

where

$$\begin{cases} K_1 = F(t, X) \\ K_2 = F\left(t + \frac{1}{2}h, X + \frac{1}{2}hK_1\right) \\ K_3 = F\left(t + \frac{1}{2}h, X + \frac{1}{2}hK_2\right) \\ K_4 = F(t + h, X + hK_3) \end{cases}$$

Here, $X = X(t)$, and all quantities are vectors with n components except variables t and h.

- We can absorb the t variable into the vector by letting $x_0 = t$ and then writing the autonomous form for the **system of ordinary differential equations in vector notation** as

$$\begin{cases} X' = F(X) \\ X(a) = S \quad \text{(given)} \end{cases}$$

where vectors are defined to have $n + 1$ components. Then

$$\begin{cases} X = [x_0, x_1, x_2, \dots, x_n]^T \\ X' = [x_0', x_1', x_2', \dots, x_n']^T \\ F = [1, f_1, f_2, \dots, f_n]^T \\ X(a) = [a, x_1(a), x_2(a), \dots, x_n(a)]^T \end{cases}$$

- The **Runge-Kutta method of order 4** for the system of ordinary differential equations in autonomous form is

$$X(t + h) = X + \frac{h}{6}(K_1 + 2K_2 + 2K_3 + K_4)$$

where

$$\begin{cases} K_1 = F(X) \\ K_2 = F\left(X + \frac{1}{2}hK_1\right) \\ K_3 = F\left(X + \frac{1}{2}hK_2\right) \\ K_4 = F(X + hK_3) \end{cases}$$

Here, $X = X(t)$, and all quantities F and K_i are vectors with $n+1$ components except the variables t and h.

- A **single nth-order ordinary differential equation with initial values** has the form

$$\begin{cases} x^{(n)} = f(t, x, x', x'', \dots, x^{(n-1)}) \\ x(a), x'(a), x''(a), \dots, x^{(n-1)}(a) \quad \text{(all given)} \end{cases}$$

It can be turned into a system of first-order equations of the form

$$\begin{cases} X' = F(t, X) \\ X(a) = S \quad \text{(given)} \end{cases}$$

where

$$\begin{cases} X = [x_1, x_2, \ldots, x_n]^T \\ X' = [x_1', x_2', \ldots, x_n']^T \\ F = [x_2, x_3, x_4, \ldots, x_n, f]^T \\ X(a) = [x_1(a), x_2(a), \ldots, x_n(a)]^T \end{cases}$$

- We can absorb the variable t into the vector notation by letting $x_0 = t$ and extending the vectors to length $n + 1$. Thus, a **single nth-order ordinary differential equation** can be written as

$$\begin{cases} X' = F(X) \\ X(a) = S \quad \text{(given)} \end{cases}$$

where

$$\begin{cases} X = [x_0, x_1, x_2, \ldots, x_n]^T \\ X' = [x_0', x_1', x_2', \ldots, x_n']^T \\ F = [1, x_2, x_3, x_4, \ldots, x_n, f]^T \\ X(a) = [a, x_1(a), x_2(a), \ldots, x_n(a)] \end{cases}$$

Exercises 7.4

[a]**1.** Consider

$$\begin{cases} x' = y, & x(0) = -1 \\ y' = x, & y(0) = 0 \end{cases}$$

Write down the equations, without derivatives, to be used in the Taylor series method of order 5.

[a]**2.** How would you solve this system of differential equations numerically?

$$\begin{cases} x_1' = x_1^2 + e^t - t^2 \\ x_2' = x_2 - \cos t \\ x_1(0) = 0, \quad x_2(1) = 0 \end{cases}$$

[a]**3.** How would you solve the initial-value problem

$$\begin{cases} x_1'(t) = x_1(t)e^t + \sin t - t^2 \\ x_2'(t) = [x_2(t)]^2 - e^t + x_2(t) \\ x_1(1) = 2, \quad x_2(1) = 4 \end{cases}$$

if a computer program were available to solve an initial-value problem of the form $x' = f(t, x)$ involving a single unknown function $x = x(t)$?

[a]**4.** Write an equivalent system of first-order differential equations without t appearing on the right-hand side:

$$\begin{cases} x' = x^2 + \log(y) + t^2 \\ y' = e^y - \cos(x) + \sin(tx) - (xy)^7 \\ x(0) = 1, \quad y(0) = 3 \end{cases}$$

[a]**5.** Turn this differential equation into a system of first-order equations suitable for applying the Runge-Kutta method:

$$\begin{cases} x''' = 2x' + \log(x'') + \cos(x) \\ x(0) = 1, \quad x'(0) = -3, \quad x''(0) = 5 \end{cases}$$

6. a. Assuming that a program is available for solving initial-value problems of the form in Equation (11), how can it be used to solve the following differential equation?

$$\begin{cases} x''' = t + x + 2x' + 3x'' \\ x(1) = 3, \quad x'(1) = -7, \quad x''(1) = 4 \end{cases}$$

b. How would this problem be solved if the initial conditions were $x(1) = 3$, $x'(1) = -7$, and $x'''(1) = 0$?

[a]7. How would you solve this differential equation problem numerically?

$$\begin{cases} x_1'' = x_1' + x_1^2 - \sin t \\ x_2'' = x_2 - (x_2')^{1/2} + t \end{cases}$$

with initial conditins

$$x_1(0) = 1, \quad x_2(1) = 3, \quad x_1'(0) = 0, \quad x_2'(1) = -2$$

[a]8. Convert to a first-order system the orbital equations

$$\begin{cases} x'' + x(x^2 + y^2)^{-3/2} = 0 \\ y'' + y(x^2 + y^2)^{-3/2} = 0 \end{cases}$$

with initial conditions

$$x(0) = 0.5 \quad x'(0) = 0.75 \quad y(0) = 0.25 \quad y'(0) = 1.0$$

[a]9. Rewrite the following equation as a system of first-order differential equations without t appearing on the right-hand side:

$$\begin{cases} x^{(iv)} = (x''')^2 + \cos(x'x'') - \sin(tx) + \log\left(\dfrac{x}{t}\right) \\ x(0) = 1, \quad x'(0) = 3, \quad x''(0) = 4, \quad x'''(0) = 5 \end{cases}$$

[a]10. Express the system of ordinary differential equations

$$\begin{cases} \dfrac{d^2z}{dt^2} - 2t\dfrac{dz}{dt} = 2te^{xz} \\ \dfrac{d^2x}{dt^2} - 2xz\dfrac{dx}{dt} = 3x^2yt^2 \\ \dfrac{d^2y}{dt^2} - e^y\dfrac{dy}{dt} = 4xt^2z \\ z(1) = x''(1) = y'(1) = 2 \\ z'(1) = x(1) = y(1) = 3 \end{cases}$$

as a system of first-order ordinary differential equations.

11. Determine a system of first-order equations equivalent to each of the following:

[a]**a.** $x''' + x'' \sin x + tx' + x = 0$

b. $x^{(iv)} + x'' \cos x' + txx' = 0$

c. $\begin{cases} x'' = 3x^2 - 7y^2 + \sin t + \cos(x'y') \\ y''' = y + x^2 - \cos t - \sin(xy'') \end{cases}$

[a]12. Consider

$$\begin{cases} x'' = x' - x \\ x(0) = 0, \quad x'(0) = 1 \end{cases}$$

Determine the associated first-order system and its auxiliary initial conditions.

[a]13. The problem

$$\begin{cases} x''(t) = x + y - 2x' + 3y' + \log t \\ y''(t) = 2x - 3y + 5x' + ty' - \sin t \\ x(0) = 1, \quad x'(0) = 2 \\ y(0) = 3, \quad y'(0) = 4 \end{cases}$$

is to be put into the form of an autonomous system of five first-order equations. Give the resulting system and the appropriate initial values.

14. Write procedure *XP_System* for use with the fourth-order Runge-Kutta routine *RK4_System1* for the following differential equation:

$$\begin{cases} x''' = 10e^{x''} - x''' \sin(x'x) - (xt)^{10} \\ x(2) = 6.5, \quad x'(2) = 4.1, \quad x''(2) = 3.2 \end{cases}$$

15. If we are going to solve the initial-value problem

$$\begin{cases} x''' = x' - tx'' + x + \ln t \\ x(1) = x'(1) = x''(1) = 1 \end{cases}$$

using Runge-Kutta formulas, how should the problem be transformed?

16. Convert this problem involving differential equations into an autonomous system of first-order equations (with initial values):

$$\begin{cases} 3x' + \tan x'' - x^2 = \sqrt{t^2 + 1} + y^2 + (y')^2 \\ -3y' + \cot y'' + y^2 = t^2 + (x+1)^{1/2} + 4x' \\ x(1) = 2, \quad x'(1) = -2, \quad y(1) = 7, \quad y'(1) = 3 \end{cases}$$

17. Follow the instructions in the preceding problem on this example:

$$\begin{cases} txyz + x'y'/t = tx^2 + x/y'' + z \\ t^2x/z + y'z't = y^2 - (z'')^2x + x'y' \\ tyz - x'z'y' = z^2 - zx'' - (yz)' \\ x(3) = 1, \quad y(3) = 2, \quad z(3) = 4 \\ x'(3) = 5, \; y'(3) = 6, \; z'(3) = 7 \end{cases}$$

18. Turn this pair of differential equations into a second-order differential equation involving x alone:

$$\begin{cases} x' = -x + axy \\ y' = 3y - xy \end{cases}$$

Computer Exercises 7.4

[a]**1.** Solve the system of differential equations (1) by using two different methods given in this section and compare the results with the analytic solution.

[a]**2.** Solve the initial-value problem

$$\begin{cases} x' = t + x^2 - y \\ y' = t^2 - x + y^2 \\ x(0) = 3, \quad y(0) = 2 \end{cases}$$

by means of the Taylor series method using $h = 1/128$ on the interval $[0, 0.38]$. Include terms involving three derivatives in x and y. How accurate are the computed function values?

3. Write the Runge-Kutta procedure to solve

$$\begin{cases} x_1' = -3x_2 \\ x_2' = \frac{1}{3}x_1 \\ x_1(0) = 0, \quad x_2(0) = 1 \end{cases}$$

on the interval $0 \leqq t \leqq 4$. Plot the solution.

[a]**4.** Write procedure *RK4_System2* and a driver program for solving the ordinary differential equation system given by Equation (2). Use $h = -10^{-2}$, and print out the values of x_0, x_1, and x_2, together with the true solution on the interval $[-1, 0]$. Verify that the true solution is $x(t) = e^t + 6 + 6t + 4t^2 + t^3$ and $y(t) = e^t - t^3 + t^2 + 2t + 2$.

[a]**5.** Using the Runge-Kutta procedure, solve the following initial-value problem on the interval $0 \leqq t \leqq 2\pi$. Plot the resulting curves $(x_1(t), x_2(t))$ and $(x_3(t), x_4(t))$. They should be circles.

$$\begin{cases} \mathbf{X}' = \begin{bmatrix} x_3 \\ x_4 \\ -x_1\left(x_1^2 + x_2^2\right)^{-3/2} \\ -x_2\left(x_1^2 + x_2^2\right)^{-3/2} \end{bmatrix} \\ \mathbf{X}(0) = [1, 0, 0, 1]^T \end{cases}$$

6. Solve the problem

$$\begin{cases} x_0' = 1 \\ x_1' = -x_2 + \cos x_0 \\ x_2' = \quad x_1 + \sin x_0 \\ x_0(1) = 1, \quad x_1(1) = 0, \quad x_2(1) = -1 \end{cases}$$

Use the Runge-Kutta method and the interval $-1 \leqq t \leqq 2$.

[a]**7.** Write and test a program, using the Taylor series method of order 5, to solve the system

$$\begin{cases} x' = tx - y^2 + 3t \\ y' = x^2 - ty - t^2 \\ x(5) = 2, \quad y(5) = 3 \end{cases}$$

on the interval $[5, 6]$ using $h = 10^{-3}$. Print values of x and y at steps of 0.1.

8. Print a table of $\sin t$ and $\cos t$ on the interval $[0, \pi/2]$ by numerically solving the system

$$\begin{cases} x' = y \\ y' = -x \\ x(0) = 0, \quad y(0) = 1 \end{cases}$$

9. Write a program for using the Taylor series method of order 3 to solve the system

$$\begin{cases} x' = tx + y' - t^2 \\ y' = ty + 3t \\ z' = tz - y' + 6t^3 \\ x(0) = 1, \quad y(0) = 2 \quad z(0) = 3 \end{cases}$$

on the interval $[0, 0.75]$ using $h = 0.01$.

10. Write and test a short program for solving the system of differential equations

$$\begin{cases} y' = x^3 - t^2y - t^2 \\ x' = tx^2 - y^4 + 3t \\ y(2) = 5, \quad x(2) = 3 \end{cases}$$

over the interval $[2, 5]$ with $h = 0.25$. Use the Taylor series method of order 4.

11. Recode and test procedure *RK4_System2* using a computer language that supports vector operations.

12. Verify the numerical results given in the text for the system of differential equations (1) from programs *Test_RK4_System1* and *RK4_System2*.

13. (Continuation) Use mathematical software such as MATLAB, Maple, or Mathematica containing symbolic manipulation capabilities to verify the analytic solution for the system of differential equations (1).

14. (Continuation) Use mathematical software routines such as in MATLAB, Maple, or Mathematica to verify the numerical solutions given in the text. Plot the resulting solution curve. Compare with the results from programs *Test_RK4_System1* and *Test_RK4_System2*.

15. Use *RK4_System1* to solve each of the following for $0 \leqq t \leqq 1$. Use $h = 2^{-k}$ with $k = 5, 6,$ and 7, and compare results.

a. $$\begin{cases} x'' = 2(e^{2t} - x^2)^{1/2} \\ x(0) = 0, \quad x'(0) = 1 \end{cases}$$

b. $$\begin{cases} x'' = x^2 - y + e^t \\ y'' = x - y^2 - e^t \\ x(0) = 0, \quad x'(0) = 0 \\ y(0) = 1, \quad y'(0) = -2 \end{cases}$$

16. Solve the **Airy differential equation**

$$\begin{cases} x'' = tx \\ x(0) = 0.35502\,80538\,87817 \\ x'(0) = -0.25881\,94037\,92807 \end{cases}$$

on the interval [0, 4.5] using the Runge-Kutta method. **Check value**: $x(4.5) = 0.00033\,02503$ is correct.

17. Solve

$$\begin{cases} x'' + x' + x^2 - 2t = 0 \\ x(0) = 0, \quad x'(0) = 0.1 \end{cases}$$

on [0, 3] by any convenient method. If a plotter is available, graph the solution.

18. Solve

$$\begin{cases} x'' = 2x' - 5x \\ x(0) = 0, \quad x'(0) = 0.4 \end{cases}$$

on the interval [−2, 0].

19. Write computer programs based on the pseudocode in the text to find the numerical solution of these ordinary differential equation systems:

a. IVP (9) **b.** IVP (12) **c.** IVP (13)

20. (Continuation) Use mathematical software such as MATLAB, Maple, or Mathematica with symbolic manipulation capabilities to find their analytical solutions.

21. (Continuation) Use mathematical software routines such as in MATLAB, Maple, or Mathematica to verify the numerical solutions for these ordinary differential equation systems. Plot the resulting solution curves.

7.5 Adams-Bashforth-Moulton Methods

A Predictor-Corrector Scheme

The procedures explained so far have solved the initial-value problem

$$\begin{cases} X' = F(X) \\ X(a) = S \quad \text{(given)} \end{cases} \tag{1}$$

by means of **single-step** numerical methods. In other words, if the solution $X(t)$ is known at a particular point t, then $X(t + h)$ can be computed with no knowledge of the solution at points earlier than t. The Runge-Kutta and Taylor series methods compute $X(t + h)$ in terms of $X(t)$ and various values of F.

Single-Step versus Multi-Step Methods

More efficient methods can be devised if several values $X(t), X(t - h), X(t - 2h), \ldots$ are used in computing $X(t + h)$. Such methods are called **multi-step** methods. They have the obvious drawback that at the beginning of the numerical solution, no prior values of X are available. So it is usual to start a numerical solution with a single-step method, such as the Runge-Kutta procedure, and transfer to a multistep procedure for efficiency as soon as enough starting values have been computed.

An example of a multistep formula is known as the **Adams-Bashforth formula** (see Section 7.3 (p. 325) and the related Computer Exercises 7.3.2–4). It is

Adams-Bashforth Predictor Step

$$\widetilde{X}(t + h) = X(t) + \frac{h}{24}\{55F[X(t)] - 59F[X(t - h)] + 37F[X(t - 2h)] - 9F[X(t - 3h)]\} \tag{2}$$

Here, $\widetilde{X}(t+h)$ is the predicted value of $X(t+h)$ computed by using Formula (2). If the solution X has been computed at the four points t, $t-h$, $t-2h$, and $t-3h$, then Formula (2) can be used to compute $\widetilde{X}(t+h)$. If this is done systematically, then only *one* evaluation of F is required for each step. This represents a considerable savings over the fourth-order Runge-Kutta procedure; the latter requires *four* evaluations of F per step. (Of course, a consideration of truncation error and stability might permit a larger step size in the Runge-Kutta method and make it much more competitive.)

In practice, Formula (2) is never used by itself. Instead, it is used as a *predictor*, and then another formula is used as a *corrector*. The corrector that is usually used with Formula (2) is the **Adams-Moulton formula**:

Adams-Moulton Corrector Step

$$X(t+h) = X(t) + \frac{h}{24}\{9F[\widetilde{X}(t+h] + 19F[X(t)] - 5F[X(t-h)] + F[X(t-2h)]\} \tag{3}$$

Thus, Formula (2) predicts a tentative value of $X(t+h)$, and Formula (3) computes this X value more accurately. The combination of the two formulas results in a **predictor-corrector scheme**.

Adams-Moulton (Predictor-Corrector) Method

With initial values of X specified at a, three steps of a Runge-Kutta method can be performed to determine enough X values that the Adams-Bashforth-Moulton procedure can begin. The fourth-order Adams-Bashforth and Adams-Moulton formulas, started with the fourth-order Runge-Kutta method, are referred to as the **Adams-Moulton method**. Predictor and corrector formulas of the same order are used so that only one application of the corrector formula is needed. Some suggest iterating the corrector formula, but experience has demonstrated that the best overall approach is only *one* application per step.

Pseudocode

How the Scheme Works

Storage of the approximate solution at previous steps in the Adams-Moulton method is usually handled either by storing in an array of dimension larger than the total number of steps to be taken or by physically shifting data after each step (discarding the oldest data and storing the newest in their place). If an adaptive process is used, the total number of steps to be taken cannot be determined beforehand. Physical shifting of data can be eliminated by cycling the indices of a storage array of fixed dimension. For the Adams-Moulton method, the x_i data for $X(t)$ are stored in a two-dimensional array with entries z_{im} in locations $m = 1, 2, 3, 4, 5, 1, 2, \ldots$ for $t = a, a+h, a+2h, a+3h, a+4h, a+5h, a+6h, \ldots$, respectively. The sketch in Figure 7.6 shows the first several t values with corresponding m values and abbreviations for the formulas used.

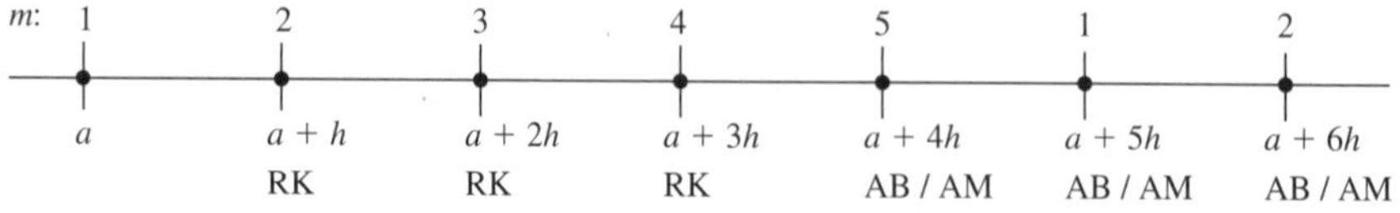

FIGURE 7.6 Starting values for applications of RK and AB/AM methods

An error analysis can be conducted after each step of the Adams-Moulton method. If $x_i^{(p)}$ is the numerical approximation of the ith equation in System (1) at $t+h$ obtained by predictor Formula (2) and x_i is that from corrector Formula (3) at $t+h$, then it can be shown that the single-step error for the ith component at $t + h$ is given approximately by

$$\varepsilon_i = \frac{19}{270}\frac{\left|x_i - x_i^{(p)}\right|}{|x_i|}$$

So we compute

$$\text{est} = \max_{1 \leqq i \leqq n} |\varepsilon_i|$$

in the Adams-Moulton procedure *AM_System* to obtain an estimate of the maximum single-step error at $t + h$.

A control procedure is needed that calls the Runge-Kutta procedure three times and then calls the Adams-Moulton predictor-corrector scheme to compute the remaining steps. Such a procedure for doing *nsteps* steps with a fixed step size h follows:

***AMRK* Pseudocode**

```
procedure AMRK(n, h, (x_i), nsteps)
integer i, k, m, n;   real est, h;   real array (x_i)_{0:n}
allocate real array (f_ij)_{0:n×0:4}, (z_ij)_{0:n×0:4}
m ← 0
output h
output 0, (x_i)
for i = 0 to n
    z_im ← x_i
end for
for k = 1 to 3
    call RK_System(m, n, h, (z_ij), (f_ij))
    output k, (z_im)
end for
for k = 4 to nsteps
    call AM_System(m, n, h, est, (z_ij), (f_ij), )
    output k, (z_im)
    output est
end for
for i = 0 to n
    x_i ← z_im
end for
deallocate array (f, z)
end procedure AMRK
```

The Adams-Moulton method for a system and the computation of the single-step error are accomplished in the following pseudocode:

```
procedure AM_System(m, n, h, est, (z_ij), (f_ij))
integer i, j, k, m, mp1;   real d, d_max, est, h
real array (z_ij)_{0:n×0:4}, (f_ij)_{0:n×0:4}
allocate real array (s_i)_{0:n}, (y_i)_{0:n}
real array (a_i)_{1:4} ← (55, −59, 37, −9)
real array (b_i)_{1:4} ← (9, 19, −5, 1)
mp1 ← (1 + m)mod5
```

(Continued)

AM_System Pseudocode

```
call XP_System(n, (z_im), (f_im))
for i = 0 to n
    s_i ← 0
end for
for k = 1 to 4
    j ← (m − k + 6)mod5
    for i = 0 to n
        s_i ← s_i + a_k f_ij
    end for
end for
for i = 0 to n
    y_i ← z_im + hs_i/24
end for
call XP_System(n, (y_i), (f_i,mp1))
for i = 0 to n
    s_i ← 0
end for
for k = 1 to 4
    j ← (mp1 − k + 6)mod5
    for i = 0 to n
        s_i ← s_i + b_k f_ij
    end for
end for
for i = 0 to n
    z_i,mp1 ← z_im + hs_i/24
end for
m ← mp1
d_max ← 0
for i = 0 to n
    d ← |z_im − y_i|/|z_im|
    if d > d_max then
        d_max ← d
        j ← i
    end if
end for
est ← 19d_max/270
deallocate array (s, y)
end procedure AM_System
```

Here, the function evaluations are stored cyclically in f_{im} for use by Formulas (2) and (3). Various optimization techniques are possible in this pseudocode. For example, the programmer may wish to move the computation of $\frac{1}{24}h$ outside of the loops.

A companion Runge-Kutta procedure is needed, which is a modification of procedure *RK4_System2* from Section 7.4:

```
procedure RK_System(m, n, h, (z_ij), (f_ij))
integer i, m, mp1, n;   real h;   real array (z_ij)_0:n×0:4, (f_ij)_0:n×0:4
allocate real array (g_ij)_0:n×0:3, (y_i)_0:n
```

(Continued)

***RK_System* Pseudocode**

```
mp1 ← (1 + m)mod5
call XP_System(n, (z_im), (f_im))
for i = 0 to n
    y_i ← z_im + (1/2)h f_im
end for
call XP_System(n, (y_i), (g_i,1))
for i = 0 to n
    y_i ← z_im + (1/2)h g_i,1
end for
call XP_System(n, (y_i), (g_i,2))
for i = 0 to n
    y_i ← z_im + h g_i,2
end for
call XP_System(n, (y_i), (g_i,3))
for i = 0 to n
    z_i,mp1 ← z_im + h[f_im + 2g_i,1 + 2g_i,2 + g_i,3]/6
end for
m ← mp1
deallocate array (g_ij), (y_i)
end procedure RK_System
```

As before, the programmer may wish to move $\frac{1}{6}h$ out of the loop.

To use the Adams-Moulton pseudocode, we supply the procedure *XP_System* that defines the system of ordinary differential equations and write a driver program with a call to procedure *AMRK*. The complete program then consists of the following five parts: the main program and procedures *XP_System*, *AMRK*, *RK_System*, and *AM_System*.

As an illustration, the pseudocode for IVP (12) in Section 7.4 is as follows:

***Test_AMRK* Pseudocode**

```
program Test_AMRK
real h;   real array (x_i)_0:n
integer n ← 5, nsteps ← 100
real a ← 0, b ← 1
(x_i) ← (1, 2, −4, −2, 7, 6)
h ← (b − a)/nsteps
call AMRK(n, h, (x_i), nsteps)
end program Test_AMRK

procedure XP_System(n, (x_i), (f_i))
integer n;   real array (x_i)_0:n, (f_i)_0:n
f_0 ← 1
f_1 ← x_2
f_2 ← x_1 − x_3 − 9x_2^2 + x_4^3 + 6x_5 + 2x_0
f_3 ← x_4
f_4 ← x_5
f_5 ← x_5 − x_2 + e^(x_1) − x_0
end procedure XP_System
```

Here, we have programmed this procedure for an autonomous system of ordinary differential equations.

An Adaptive Scheme

Since an estimate of the error is available from the Adams-Moulton method, it is natural to replace procedure *AMRK* with one that employs an adaptive scheme—that is, one that changes the step size. A procedure similar to the one used in Section 7.3 is outlined here. The Runge-Kutta method is used to compute the first three steps, and then the Adams-Moulton method is used. If the error test determines that halving or doubling of the step size is necessary in the first step using the Adams-Moulton method, then the step size is halved or doubled, and the whole process starts again with the initial values—so at least one step of the Adams-Moulton method must take place. If during this process the error test indicates that halving is required at some point within the interval $[a, b]$, then the step size is halved. A retreat is made back to a previously computed value, and after three Runge-Kutta steps have been computed, the process continues, using the Adams-Moulton method again but with the new step size. In other words, the point at which the error was too large should be computed by the Adams-Moulton method, not the Runge-Kutta method. Doubling the step size is handled in an analogous manner. Doubling the step size requires only saving an appropriate number of previous values; however, one can simplify this process (whether halving or doubling the step size) by always backing up two steps with the *old* step size and then using this as the beginning point of a *new* initial-value problem with the *new* step size. Other, more complicated procedures can be designed and can be the subject of numerical experimentation. (See Computer Exercise 7.5.3.)

How Adaptive Scheme Works

An Engineering Example

In chemical engineering, a complicated production activity may involve several reactors connected with inflow and outflow pipes. The concentration of a certain chemical in the ith reactor is an unknown quantity, x_i. Each x_i is a function of time. If there are n reactors, the whole process is governed by a system of n differential equations of the form

$$\begin{cases} \boldsymbol{X}' = \boldsymbol{A}\boldsymbol{X} + \boldsymbol{V} \\ \boldsymbol{X}(0) = \boldsymbol{S} \quad \text{(given)} \end{cases}$$

where $\boldsymbol{X}$ is the vector containing the unknown quantities x_i, $\boldsymbol{A}$ is an $n \times n$ matrix, and $\boldsymbol{V}$ is a constant vector. The entries in $\boldsymbol{A}$ depend on the flow rates permitted between different reactors of the system.

There are several approaches to solving this problem. One is to diagonalize the matrix $\boldsymbol{A}$ by finding a nonsingular matrix $\boldsymbol{P}$ for which is $\boldsymbol{P}^{-1}\boldsymbol{A}\boldsymbol{P}$ is diagonal and then using the matrix exponential function to solve the system in an analytic form. This is a task that mathematical software can handle. On the other hand, we can simply turn the problem over to an *ODE solver* and get the numerical solution. One piece of information that is always wanted in such a problem is a description of the **steady state** of the system. That means the values of all variables at $t = \infty$. Each function x_i should be a linear combination of exponential functions of the form $t \mapsto e^{\lambda t}$, in which $\lambda < 0$.

Here is a simple example that can illustrate all of this:

Sample ODE 1

$$\begin{bmatrix} x_1' \\ x_2' \\ x_3' \end{bmatrix} = \begin{bmatrix} -8/3 & -4/3 & 1 \\ -17/3 & -4/3 & 1 \\ -35/3 & 14/3 & -2 \end{bmatrix} \begin{bmatrix} x_1 \\ x_2 \\ x_3 \end{bmatrix} + \begin{bmatrix} 12 \\ 29 \\ 48 \end{bmatrix} \tag{4}$$

Using mathematical software such as MATLAB, Maple, or Mathematica, we can obtain a closed-form solution:

$$\begin{cases} x(t) = \frac{1}{6}e^{-3t}(6 - 50e^t + 10e^{2t} + 34e^{3t}) \\ y(t) = \frac{1}{6}e^{-3t}(12 - 125e^t + 40e^{2t} + 73e^{3t}) \\ z(t) = \frac{1}{6}e^{-3t}(14 - 200e^t + 70e^{2t} + 116e^{3t}) \end{cases}$$

For a system of ordinary differential equations with a large number of variables, it may be more convenient to represent them in a computer program with an array such as `x(i,t)` rather than by separate variables names. To see the numerical value of the analytic solution at a single point, say, $t = 2.5$, we obtain $x(2.5) \approx 5.74788$, $y(2.5) \approx 12.5746$, $z(2.5) \approx 20.0677$. Also, we can produce a graphing of the analytic solution to the problem.

Finally, the programs presented in this section can be used to generate a numerical solution on a prescribed interval with a prescribed number of points.

Stiff ODEs and an Example

In many applications of differential equations there are several functions to be *tracked* together as functions of time. A system of ordinary differential equations may be used to model the physical phenomena. In such a situation, it can happen that different solution functions (or different components of a single solution) have quite disparate behavior that makes the selection of the step size in the numerical solution problematic. For example, one component of a function may require a small step in the numerical solution because it is varying rapidly, whereas another component may vary slowly and not require a small size for its computation. Such a system is said to be **stiff**. Figure 7.7 illustrates a slowly varying solution surrounded by other solutions with rapidly decaying transients.

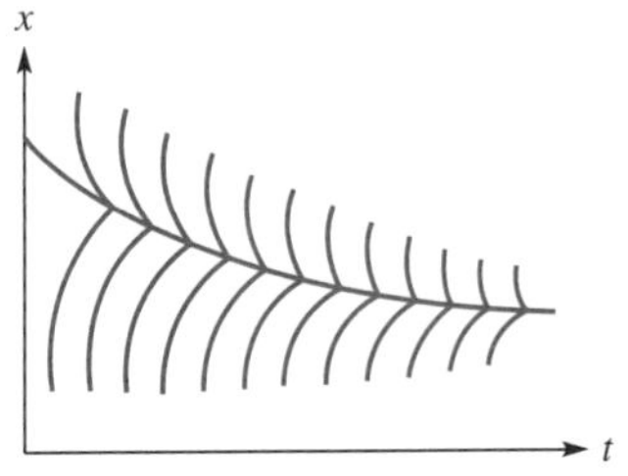

FIGURE 7.7 Solution curves for a stiff ODE

An example illustrates this possibility. Consider a system of two differential equations with initial conditions:

Sample ODE 2

$$\begin{cases} x' = -20x - 19y, & x(0) = 2 \\ y' = -19x - 20y, & y(0) = 0 \end{cases} \tag{5}$$

The solution is easily seen to be

True Solution

$$\begin{cases} x(t) = e^{-39t} + e^{-t} \\ y(t) = e^{-39t} - e^{-t} \end{cases}$$

The component e^{-39t} quickly becomes negligible as t increases, starting at 0. The solution is then approximately given by $x(t) = -y(t) = e^{-t}$, and this function is smooth and decreasing to 0. It would seem that in almost any numerical solution, a large step size could be used. However, let's examine the simplest of numerical procedures: **Euler's method**. It generates the solution by using the following equations:

Difference Equations

$$\begin{cases} x_{n+1} = x_n + h(-20x_n - 19y_n), & x_0 = 2 \\ y_{n+1} = y_n + h(-19x_n - 20y_n), & y_0 = 0 \end{cases}$$

These difference equations can be solved in closed form, and we have

Closed Form

$$\begin{cases} x_n = (1 - 39h)^n + (1 - h)^n \\ y_n = (1 - 39h)^n - (1 - h)^n \end{cases}$$

For the numerical solution to converge to 0 (and thus imitate the actual solution), it is necessary that $h < \frac{2}{39}$. If we were solving only the differential equation $x' = -x$ to get the solution $x(t) = e^{-t}$, the step size could be as large as $h = 2$ to get the correct behavior as t increased. (See Exercise 7.5.2.)

To see that numerical success (in the sense of being able to use a reasonable step size) depends on the method used, let us consider the implicit Euler method. For a single differential equation, this employs the formula

$$x_{n+1} = x_n + hf(t_{n+1}, x_{n+1})$$

Since x_{n+1} appears on both sides of this equation, the equation must be solved for x_{n+1}. In the example being considered, the Euler equations are

Euler's Equations

$$\begin{cases} x_{n+1} = x_n + h(-20x_{n+1} - 19y_{n+1}) \\ y_{n+1} = y_n + h(-19x_{n+1} - 20y_{n+1}) \end{cases}$$

This pair of equations has the form $\boldsymbol{X}_{n+1} = \boldsymbol{X}_n + \boldsymbol{A}\boldsymbol{X}_{n+1}$, where $\boldsymbol{A}$ is the 2×2 matrix in the previous pair of equations and $\boldsymbol{X}_n$ is the vector having components x_n and y_n. This matrix equation can be written $(\boldsymbol{I} - \boldsymbol{A})\boldsymbol{X}_{n+1} = \boldsymbol{X}_n$ or $\boldsymbol{X}_{n+1} = (\boldsymbol{I} - \boldsymbol{A})^{-1}\boldsymbol{X}_n$. A consequence is that the explicit solution is $\boldsymbol{X}_n = (\boldsymbol{I} - \boldsymbol{A})^{-n}\boldsymbol{X}_0$. At this point, it is necessary to appeal to a result concerning such iterative processes. For $\boldsymbol{X}_n$ to converge to 0 for any choice of initial vector $\boldsymbol{X}_0$, it is necessary and sufficient that all eigenvalues of $(\boldsymbol{I} - \boldsymbol{A})^{-1}$ be less than one in modulus (see Kincaid and Cheney [2002]). Equivalently, the eigenvalues of $\boldsymbol{I} - \boldsymbol{A}$ should be greater than 1 in modulus. An easy calculation shows that for positive h this condition is met, without further hypotheses. Thus, the implicit Euler method can be used with any *reasonable* step size on this problem. In the literature on stiff equations, much more information can be found, and there are books that address this topic thoroughly. Some essential references are Dekker and Verwer [1984], Gear [1971], Miranker [1981], and Shampine and Gordon [1975].

In general, stiff ordinary differential equations are rather difficult to solve. This is compounded by the fact that in most cases, you do not know beforehand whether an ordinary differential equation that you're trying to solve numerically is stiff. Software packages usually have ordinary differential equation solvers specifically designed to handle stiff ordinary differential equations. Some of these procedures may vary both the step size and the order of the method. In such algorithms, the Jacobian matrix $\partial \boldsymbol{F}/\partial Xy$ may play a role. Solving an associated linear system involving the Jacobian matrix is critical to the reliability and efficiency of the code. The Jacobian matrix may be sparse, an indication that the function F does not depend on some of the variables in the problem.

History of ODE Numerical Methods

Additional Reading

For readers interested in the history of numerical analysis, we recommend the book by Goldstine [1977]. The textbook on differential equations by Moulton [1930] gives some insight into the numerical methods used prior to the advent of high-speed computing machines. Also Moulton gives some of the history, going back to Newton! The calculation of orbits in celestial mechanics has always been a stimulus for the invention of numerical methods; the needs of ballistic science have been also. Moulton mentions that the retardation of a projectile by air resistance is a very complicated function of velocity that necessitates numerical solution of the otherwise simple equations of ballistics.

Summary 7.5

- For the autonomous form for a system of ordinary differential equations in vector notation

$$\begin{cases} \boldsymbol{X}' = \boldsymbol{F}(\boldsymbol{X}) \\ \boldsymbol{X}(a) = \boldsymbol{S} \quad \text{(given)} \end{cases}$$

the **Adams-Bashforth-Moulton method of fourth order** is

$$\begin{aligned}\widetilde{\boldsymbol{X}}(t+h) = \boldsymbol{X}(t) + \frac{h}{24}\{55\boldsymbol{F}[\boldsymbol{X}(t)] - 59\boldsymbol{F}[\boldsymbol{X}(t-h)] + 37\boldsymbol{F}[\boldsymbol{X}(t-2h)] \\ - 9\boldsymbol{F}[\boldsymbol{X}(t-3h)]\}\end{aligned}$$

$$\begin{aligned}\boldsymbol{X}(t+h) = \boldsymbol{X}(t) + \frac{h}{24}\{9\boldsymbol{F}[\widetilde{\boldsymbol{X}}(t+h)] + 19\boldsymbol{F}[\boldsymbol{X}(t)] - 5\boldsymbol{F}[\boldsymbol{X}(t-h)] \\ + \boldsymbol{F}[\boldsymbol{X}(t-2h)]\}\end{aligned}$$

Here, $\widetilde{\boldsymbol{X}}(t+h)$ is the **predictor**, and $\boldsymbol{X}(t+h)$ is the **corrector**. The Adams-Bashforth-Moulton method needs *five* evaluations of $\boldsymbol{F}$ per step. With the initial vector $\boldsymbol{X}(a)$ given, the values for $\boldsymbol{X}(a+h)$, $\boldsymbol{X}(a+2h)$, $\boldsymbol{X}(a+3h)$ are computed by the Runge-Kutta method of fourth order. Then the Adams-Bashforth-Moulton method can be used repeatedly. The predicted value $\widetilde{\boldsymbol{X}}$ is computed from the four $\boldsymbol{X}$ values at $t, t-h, t-2h$, and $t-3h$, and then the corrected value $\boldsymbol{X}(t+h)$ can be computed by using the predictor value $\widetilde{\boldsymbol{X}}(t+h)$ and previously evaluated values of $\boldsymbol{F}$ at t, $t-h$, and $t-2h$.

Exercises 7.5

[a]**1.** Find the general solution of this system by turning it into a first-order system of four equations:

$$\begin{cases} x'' = \alpha y \\ y'' = \beta x \end{cases}$$

2. Verify the assertions made about the step size h in the discussion of stiff Equation (5).

3. Write autonomous systems of first-order differential equations for each of these:

a. $\begin{cases} y'' + yz = 0 \\ z' + 2yz = 4 \\ y(0) = 1, \; y'(0) = 0, \; z(0) = 3 \end{cases}$

b. $\begin{cases} x''' - \sin(x'') + e^t x' + 2t \cos x = 25 \\ x(0) = 5, \; x'(0) = 3, \; x''(0) = 7 \end{cases}$

c. $\begin{cases} x''' - [\sin x'' + e^t x']^2 + \cos x = 0 \\ x(0) = 3, \; x'(0) = 4, \; x''(0) = 5 \end{cases}$

4. Correct the following systems of higher-order differential equations into a system of first-order equations in which t does not appear explicitly:

$$\begin{cases} x''' - 5tx''y'' - \ln(x')z = 0 \\ y'' - \sin(ty) + 7tx'' = 0 \\ z' + 16ty' - e^t zx' = 0 \end{cases}$$

Computer Exercises 7.5

1. Test the procedure *AMRK* on the system given in Computer Exercise 7.4.2.

2. The single-step error is closely controlled by using fourth-order formulas; however, the roundoff error in performing the computations in Equations (3) and (4) can be large. It is logical to carry these out in what is known as **partial double-precision** arithmetic. The function $\boldsymbol{F}$ would be evaluated in single precision at the desired points $\boldsymbol{X}(t+ih)$, but the linear combination $\sum_i c_i \boldsymbol{F}(\boldsymbol{X}(t+ih))$ would be accumulated in double precision. Also, the addition of $\boldsymbol{X}(t)$ to this result is done in double precision. Recode the Adams-Moulton method so that partial double-precision arithmetic is used. Compare this code with that in the text for a system with a known solution. How do they compare with regard to roundoff error at each step?

3. Write and test an adaptive process similar to *RK45_Adaptive* in Section 7.3 with calling sequence

procedure *AMRK_Adaptive*$(n, h, t_a, t_b, (x_i),$
$itmax, \varepsilon_{\min}, \varepsilon_{\max}, h_{\min}, h_{\max}, iflag)$

This routine should carry out the adaptive procedure outlined in this section and be used in place of the *AMRK* procedure.

4. Solve the predator-prey problem in the example at the beginning of this chapter with $a =- 10^{-2}, b = -\frac{1}{4}\times 10^2$, $c = 10^{-2}$ and $d =- 10^2$ and with initial values $u(0) = 80$, $v(0) = 30$. Plot u (the prey) and v (the predator) as functions of time t.

5. Solve and plot the numerical solution of the system of ordinary differential equations given by Equation (4) using mathematical software such as MATLAB, Maple, or Mathematica.

6. (Continuation) Repeat for Equation (5) using a routine specifically designed to handle stiff ordinary differential equations.

7a. Solve the following test problem and plot the solution curves. This problem corresponds to a recently discovered stable orbit that arises in the restricted three-body problem in which the orbits are co-planar. The two spatial coordinates of the jth body are x_{1j} and x_{2j} for $j = 1, 2, 3$. Each of the six coordinates satisfies a second-order differential equation:

$$x''_{ij} = \sum_{\substack{k=1\\k\neq j}}^{3} m_k\left(x_{ik} - x_{ij}\right)/d^3_{jk}$$

where $d^2_{jk} = \sum_{i=1}^{2}(x_{ij} - x_{ik})^2$ for $k, j = 1, 2, 3$. Assume that the bodies have equal mass, say, $m_1 = m_2 = m_3 = 1$, and with the appropriate starting conditions, they will follow the same figure-eight orbit as a periodic steady-state solution. When the system is rewritten as a first-order system, the dimension of the problem is twelve, and the initial conditions at $t = 0$ are given by

$$\begin{cases} x_{11} = -0.97000436, & x'_{11} = 0.466203685\\ x_{21} = 0.24308753, & x'_{21} = 0.43236573\\ x_{12} = 0.0, & x'_{12} = -0.93240737\\ x_{22} = 0.0, & x'_{22} = -0.86473146\\ x_{13} = 0.97000436, & x'_{13} = 0.466203685\\ x_{23} = -0.24308753, & x'_{23} = 0.43236573 \end{cases}$$

Solve the problem for $t \in [0, 20]$.

7b. Solve the following test problem and plot the solution curves. The **Lorenz problem** is well known, and it arises in the study of dynamical systems:

$$\begin{cases} vx'_1 = 10(x_2 - x_1)\\ x'_2 = x_1(28 - x_3) - x_2\\ x'_3 = x_1x_2 - \frac{8}{3}x_3\\ x_1(0) = 15, \quad x_2(0) = 15, \quad x_3(0) = 36 \end{cases}$$

Solve the problem for $t \in [0, 20]$. It is known to have solutions that are potentially poorly conditioned. *Note*: For additional details on these problems, see Enright [2006].

8. Write a computer program based on pseudocode *Test_AMRK* to find the numerical solution to the ordinary differential equation systems, and compare the results with that by using a built-in routine such as in MATLAB, Maple, or Mathematica. Plot the resulting solution curves.

9. (**Tacoma Narrows Bridge**) In 1940, the third-longest suspension bridge in the world collapsed in a high wind.

FIGURE 7.8 Tacoma Narrows Bridge collapsing in 1940.

The following system of differential equations is a mathematical model that attempts to explain how twisting oscillations can be magnified and cause such a calamity:

$$\begin{cases} y'' = -y'd - [K/(ma)]\left[e^{a(y-\ell\sin\theta)} \right. \\ \qquad - 1 + e^{a(y+\ell\sin\theta)} \\ \qquad \left. - 1\right] + 0.2W\sin\omega t \\ \theta'' = -\theta y'd + (3\cos\theta/\ell)[K/(ma)]\left[e^{a(y-\ell\sin\theta)} \right. \\ \qquad \left. - e^{a(y+\ell\sin\theta)}\right] \end{cases}$$

The last term in the y equation is the forcing term for the wind W, which adds a strictly vertical oscillation to the bridge. Here, the roadway has width 2ℓ hanging between two suspended cables, y is the current distance from the center of the roadway as it hangs below its equilibrium point, and θ is the angle the roadway makes with the horizontal. Also, Newton's Law $F = ma$ is used and Hooke's constant K. Explore how ODE solvers are used to generate numerical trajectories for various parameter settings. Illustrate different types of phenomena that are available in this model. For additional details, see McKenna and Tuama [2001] and Sauer [2012].

8 More on Linear Systems

In applications that involve partial differential equations, large linear systems arise with sparse coefficient matrices such as

$$A = \begin{bmatrix} 4 & -1 & 0 & -1 & 0 & 0 & 0 & 0 & 0 \\ -1 & 4 & -1 & 0 & -1 & 0 & 0 & 0 & 0 \\ 0 & -1 & 4 & 0 & 0 & -1 & 0 & 0 & 0 \\ -1 & 0 & 0 & 4 & -1 & 0 & -1 & 0 & 0 \\ 0 & -1 & 0 & -1 & 4 & -1 & 0 & -1 & 0 \\ 0 & 0 & -1 & 0 & -1 & 4 & 0 & 0 & -1 \\ 0 & 0 & 0 & -1 & 0 & 0 & 4 & -1 & 0 \\ 0 & 0 & 0 & 0 & -1 & 0 & -1 & 4 & -1 \\ 0 & 0 & 0 & 0 & 0 & -1 & 0 & -1 & 4 \end{bmatrix}$$

Gaussian elimination may cause *fill-in* of the zero entries by nonzero values. On the other hand, iterative methods preserve its sparse structure.

8.1 Matrix Factorizations

LU Factorization

An $n \times n$ system of linear equations can be written in matrix form

$$\boldsymbol{Ax} = \boldsymbol{b} \tag{1}$$

where the coefficient matrix $\boldsymbol{A}$ has the form

$$\boldsymbol{A} = \begin{bmatrix} a_{11} & a_{12} & a_{13} & \cdots & a_{1n} \\ a_{21} & a_{22} & a_{23} & \cdots & a_{2n} \\ a_{31} & a_{32} & a_{33} & \cdots & a_{3n} \\ \vdots & \vdots & \vdots & \ddots & \vdots \\ a_{n1} & a_{n2} & a_{n3} & \cdots & a_{nn} \end{bmatrix}$$

Our main objective is to show that the naive Gaussian algorithm applied to $\boldsymbol{A}$ yields a factorization of $\boldsymbol{A}$ into a product of two simple matrices, one unit *lower triangular*:

$$\boldsymbol{L} = \begin{bmatrix} 1 & & & & \\ \ell_{21} & 1 & & & \\ \ell_{31} & \ell_{32} & 1 & & \\ \vdots & \vdots & \vdots & \ddots & \\ \ell_{n1} & \ell_{n2} & \ell_{n3} & \cdots & 1 \end{bmatrix}$$

and the other *upper triangular*:

$$U = \begin{bmatrix} u_{11} & u_{12} & u_{13} & \cdots & u_{1n} \\ & u_{22} & u_{23} & \cdots & u_{2n} \\ & & u_{33} & \cdots & u_{3n} \\ & & & \ddots & \vdots \\ & & & & u_{nn} \end{bmatrix}$$

In short, we refer to this as an $\boldsymbol{LU}$ **factorization** of A; that is, $A = LU$.

Numerical Example

The system of Equations (2) of Section 2.1 can be written succinctly in matrix form:

Sample 4 × 4 Linear System

$$\begin{bmatrix} 6 & -2 & 2 & 4 \\ 12 & -8 & 6 & 10 \\ 3 & -13 & 9 & 3 \\ -6 & 4 & 1 & -18 \end{bmatrix} \begin{bmatrix} x_1 \\ x_2 \\ x_3 \\ x_4 \end{bmatrix} = \begin{bmatrix} 16 \\ 26 \\ -19 \\ -34 \end{bmatrix} \tag{2}$$

Furthermore, the operations that led from this system to Equation (5) of Section 2.1, that is, the system

After Forward Elimination

$$\begin{bmatrix} 6 & -2 & 2 & 4 \\ 0 & -4 & 2 & 2 \\ 0 & 0 & 2 & -5 \\ 0 & 0 & 0 & -3 \end{bmatrix} \begin{bmatrix} x_1 \\ x_2 \\ x_3 \\ x_4 \end{bmatrix} = \begin{bmatrix} 16 \\ -6 \\ -9 \\ -3 \end{bmatrix} \tag{3}$$

could be effected by an appropriate matrix multiplication. The forward elimination phase can be interpreted as starting from Equation (1) and proceeding to

$$\boldsymbol{MAx} = \boldsymbol{Mb} \tag{4}$$

where M is a matrix chosen so that MA is the coefficient matrix for System (3). Hence, we have

$$MA = \begin{bmatrix} 6 & -2 & 2 & 4 \\ 0 & -4 & 2 & 2 \\ 0 & 0 & 2 & -5 \\ 0 & 0 & 0 & -3 \end{bmatrix} \equiv U$$

which is an upper triangular matrix.

Step 1 of naive Gaussian elimination results in Equation (3) of Section 2.1 or the system

Step 1

$$\begin{bmatrix} 6 & -2 & 2 & 4 \\ 0 & -4 & 2 & 2 \\ 0 & -12 & 8 & 1 \\ 0 & 2 & 3 & -14 \end{bmatrix} \begin{bmatrix} x_1 \\ x_2 \\ x_3 \\ x_4 \end{bmatrix} = \begin{bmatrix} 16 \\ -6 \\ -27 \\ -18 \end{bmatrix}$$

This step can be accomplished by multiplying (1) by a lower triangular matrix M_1:

$$\boldsymbol{M}_1\boldsymbol{Ax} = \boldsymbol{M}_1\boldsymbol{b}$$

where

$$M_1 = \begin{bmatrix} 1 & 0 & 0 & 0 \\ -2 & 1 & 0 & 0 \\ -\frac{1}{2} & 0 & 1 & 0 \\ 1 & 0 & 0 & 1 \end{bmatrix}$$

Notice the special form of M_1. The diagonal elements are all 1's, and the only other nonzero elements are in the first column. These numbers are the *negatives of the multipliers* located in the positions where they created 0's as coefficients in Step 1 of the forward elimination phase. To continue, Step 2 resulted in Equation (4) of Section 2.1 or the system

Step 2

$$\begin{bmatrix} 6 & -2 & 2 & 4 \\ 0 & -4 & 2 & 2 \\ 0 & 0 & 2 & -5 \\ 0 & 0 & 4 & -13 \end{bmatrix} \begin{bmatrix} x_1 \\ x_2 \\ x_3 \\ x_4 \end{bmatrix} = \begin{bmatrix} 16 \\ -6 \\ -9 \\ -21 \end{bmatrix}$$

which is equivalent to

$$M_2 M_1 A x = M_2 M_1 b$$

where

$$M_2 = \begin{bmatrix} 1 & 0 & 0 & 0 \\ 0 & 1 & 0 & 0 \\ 0 & -3 & 1 & 0 \\ 0 & \frac{1}{2} & 0 & 1 \end{bmatrix}$$

Again, M_2 differs from an identity matrix by the presence of the negatives of the multipliers in the second column from the diagonal down. Finally, Step 3 gives System (3), which is equivalent to

Step 3

$$M_3 M_2 M_1 A x = M_3 M_2 M_1 b$$

where

$$M_3 = \begin{bmatrix} 1 & 0 & 0 & 0 \\ 0 & 1 & 0 & 0 \\ 0 & 0 & 1 & 0 \\ 0 & 0 & -2 & 1 \end{bmatrix}$$

Now the forward elimination phase is complete, and with

$$M = M_3 M_2 M_1 \tag{5}$$

we have the upper triangular coefficient System (3).

Using Equations (4) and (5), we can give a different interpretation of the forward elimination phase of naive Gaussian elimination. Now we see that

$$\begin{aligned} A &= M^{-1} U \\ &= M_1^{-1} M_2^{-1} M_3^{-1} U \\ &= LU \end{aligned}$$

Since each $\boldsymbol{M}_k$ has such a special form, its inverse is obtained by simply changing the signs of the negative multiplier entries! Hence, we have

Unit Lower Triangular

$$\boldsymbol{L} = \begin{bmatrix} 1 & 0 & 0 & 0 \\ 2 & 1 & 0 & 0 \\ \frac{1}{2} & 0 & 1 & 0 \\ -1 & 0 & 0 & 1 \end{bmatrix} \begin{bmatrix} 1 & 0 & 0 & 0 \\ 0 & 1 & 0 & 0 \\ 0 & 3 & 1 & 0 \\ 0 & -\frac{1}{2} & 0 & 1 \end{bmatrix} \begin{bmatrix} 1 & 0 & 0 & 0 \\ 0 & 1 & 0 & 0 \\ 0 & 0 & 1 & 0 \\ 0 & 0 & 2 & 1 \end{bmatrix}$$

$$= \begin{bmatrix} 1 & 0 & 0 & 0 \\ 2 & 1 & 0 & 0 \\ \frac{1}{2} & 3 & 1 & 0 \\ -1 & -\frac{1}{2} & 2 & 1 \end{bmatrix}$$

It is somewhat amazing that $\boldsymbol{L}$ is a unit lower triangular matrix composed of the multipliers. Notice that in forming $\boldsymbol{L}$, we did not determine $\boldsymbol{M}$ first and then compute $\boldsymbol{M}^{-1} = \boldsymbol{L}$. (Why?)

It is easy to verify that

***LU* Factorization**

$$\boldsymbol{LU} = \begin{bmatrix} 1 & 0 & 0 & 0 \\ 2 & 1 & 0 & 0 \\ \frac{1}{2} & 3 & 1 & 0 \\ -1 & -\frac{1}{2} & 2 & 1 \end{bmatrix} \begin{bmatrix} 6 & -2 & 2 & 4 \\ 0 & -4 & 2 & 2 \\ 0 & 0 & 2 & -5 \\ 0 & 0 & 0 & -3 \end{bmatrix}$$

$$= \begin{bmatrix} 6 & -2 & 2 & 4 \\ 12 & -8 & 6 & 10 \\ 3 & -13 & 9 & 3 \\ -6 & 4 & 1 & -18 \end{bmatrix} = \boldsymbol{A}$$

We see that $\boldsymbol{A}$ is **factored** or **decomposed** into a unit lower triangular matrix $\boldsymbol{L}$ and an upper triangular matrix $\boldsymbol{U}$. The matrix $\boldsymbol{L}$ consists of the multipliers located in the positions of the elements they annihilated from $\boldsymbol{A}$, of unit diagonal elements, and of 0 upper triangular elements. In fact, we now know the general form of $\boldsymbol{L}$ and can just write it down directly using the multipliers *without* forming the $\boldsymbol{M}_k$'s and the $\boldsymbol{M}_k^{-1}$'s. The matrix $\boldsymbol{U}$ is upper triangular (not generally having unit diagonal) and is the final coefficient matrix after the forward elimination phase is completed.

It should be noted that the pseudocode *Naive_Gauss* of Section 2.1 replaces the original coefficient matrix with its $\boldsymbol{LU}$ factorization. The elements of $\boldsymbol{U}$ are in the upper triangular part of the (a_{ij})-array including the diagonal. The entries below the main diagonal in $\boldsymbol{L}$ (that is, the multipliers) are found below the main diagonal in the (a_{ij})-array. Since it is known that $\boldsymbol{L}$ has a unit diagonal, nothing is lost by not storing the 1's. (In fact, we have run out of room in the (a_{ij})-array anyway!)

Formal Derivation

To see formally how the Gaussian elimination (in naive form) leads to an $\boldsymbol{LU}$-factorization, it is necessary to show that each row operation used in the algorithm can be effected by multiplying $\boldsymbol{A}$ on the left by an elementary matrix. Specifically, if we wish to subtract λ times row p from row q, we first apply this operation to the $n \times n$ identity matrix to create an elementary matrix $\boldsymbol{M}_{qp}$. Then we form the matrix product $\boldsymbol{M}_{qp}\boldsymbol{A}$.

Before proceeding, let us verify that $\boldsymbol{M}_{qp}\boldsymbol{A}$ is obtained by subtracting λ times row p from row q in matrix $\boldsymbol{A}$. Assume that $p < q$ (for in the naive algorithm, this is always true). Then the elements of $\boldsymbol{M}_{qp} = (m_{ij})$ are

$$m_{ij} = \begin{cases} 1, & \text{if } i = j \\ -\lambda, & \text{if } i = q \text{ and } j = p \\ 0, & \text{otherwise} \end{cases}$$

Therefore, the elements of $\boldsymbol{M}_{qp}\boldsymbol{A}$ are given by

$$(\boldsymbol{M}_{qp}\boldsymbol{A})_{ij} = \sum_{s=1}^{n} m_{is}a_{sj} = \begin{cases} a_{ij} & \text{if } i \neq q \\ a_{qj} - \lambda a_{pj} & \text{if } i = q \end{cases}$$

The qth row of $\boldsymbol{M}_{qp}\boldsymbol{A}$ is the sum of the qth row of $\boldsymbol{A}$ and $-\lambda$ times the pth row of $\boldsymbol{A}$, as was to be proved.

Step k of Gaussian elimination corresponds to the matrix $\boldsymbol{M}_k$, which is the product of $n-k$ elementary matrices:

$$\boldsymbol{M}_k = \boldsymbol{M}_{nk}\boldsymbol{M}_{n-1,k}\cdots\boldsymbol{M}_{k+1,k}$$

Notice that each elementary matrix $\boldsymbol{M}_{ik}$ here is lower triangular because $i > k$, and therefore $\boldsymbol{M}_k$ is also lower triangular. If we carry out the Gaussian forward elimination process on $\boldsymbol{A}$, the result will be an upper triangular matrix $\boldsymbol{U}$. On the other hand, the result is obtained by applying a succession of factors such as $\boldsymbol{M}_k$ to the left of $\boldsymbol{A}$. Hence, the entire process is summarized by writing

$$\boldsymbol{M}_{n-1}\cdots\boldsymbol{M}_2\boldsymbol{M}_1\boldsymbol{A} = \boldsymbol{U}$$

Since each $\boldsymbol{M}_k$ is invertible, we have

$$\boldsymbol{A} = \boldsymbol{M}_1^{-1}\boldsymbol{M}_2^{-1}\cdots\boldsymbol{M}_{n-1}^{-1}\boldsymbol{U}$$

Each $\boldsymbol{M}_k$ is lower triangular having 1's on its main diagonal (unit lower triangular). Each inverse $\boldsymbol{M}_k^{-1}$ has the same property, and the same is true of their product. Hence, the matrix

$$\boldsymbol{L} = \boldsymbol{M}_1^{-1}\boldsymbol{M}_2^{-1}\cdots\boldsymbol{M}_{n-1}^{-1} \tag{6}$$

is unit lower triangular, and we have

$$\boldsymbol{A} = \boldsymbol{L}\boldsymbol{U}$$

This is the so-called $\boldsymbol{LU}$ **factorization** of $\boldsymbol{A}$. Our construction of it depends upon *not* encountering any 0 divisors in the algorithm. It is easy to give examples of matrices that have *no* $\boldsymbol{LU}$ factorization; one of the simplest is

$$\boldsymbol{A} = \begin{bmatrix} 0 & 1 \\ 1 & 1 \end{bmatrix}$$

(Also, see Exercise 8.1.4.)

■ **Theorem 1** ***LU* Factorization Theorem**

Let $\boldsymbol{A} = (a_{ij})$ be an $n \times n$ matrix. Assume that the forward elimination phase of the naive Gaussian algorithm is applied to $\boldsymbol{A}$ without encountering any 0 divisors. Let the resulting matrix be denoted by $\widetilde{A} = (\widetilde{a}_{ij})$. If

$$\boldsymbol{L} = \begin{bmatrix} 1 & 0 & 0 & \cdots & 0 \\ \widetilde{a}_{21} & 1 & 0 & \cdots & 0 \\ \widetilde{a}_{31} & \widetilde{a}_{32} & 1 & \cdots & 0 \\ \vdots & \vdots & \ddots & \ddots & \vdots \\ \widetilde{a}_{n1} & \widetilde{a}_{n2} & \cdots & \widetilde{a}_{n,n-1} & 1 \end{bmatrix}$$

and

$$\boldsymbol{U} = \begin{bmatrix} \widetilde{a}_{11} & \widetilde{a}_{12} & \widetilde{a}_{13} & \cdots & \widetilde{a}_{1n} \\ 0 & \widetilde{a}_{22} & \widetilde{a}_{23} & \cdots & \widetilde{a}_{2n} \\ 0 & 0 & \widetilde{a}_{33} & \cdots & \widetilde{a}_{3n} \\ \vdots & \vdots & \ddots & \ddots & \vdots \\ 0 & 0 & \cdots & 0 & \widetilde{a}_{nn} \end{bmatrix}$$

then

$$\boldsymbol{A} = \boldsymbol{LU}$$

Proof We define the Gaussian algorithm formally as follows. Let $\boldsymbol{A}^{(1)} = \boldsymbol{A}$. Then we compute $\boldsymbol{A}^{(2)}, \boldsymbol{A}^{(3)}, \dots, \boldsymbol{A}^{(n)}$ recursively by the naive Gaussian algorithm, following these equations:

$$a_{ij}^{(k+1)} = a_{ij}^{(k)} \qquad \text{(if } i \leqq k \text{ or } j < k) \tag{7}$$

$$a_{ij}^{(k+1)} = \frac{a_{ik}^{(k)}}{a_{kk}^{(k)}} \qquad \text{(if } i > k \text{ and } j = k) \tag{8}$$

$$a_{ij}^{(k+1)} = a_{ij}^{(k)} - \left(\frac{a_{ik}^{(k)}}{a_{kk}^{(k)}}\right) a_{kj}^{(k)} \qquad \text{(if } i > k \text{ and } j > k) \tag{9}$$

These equations describe in a precise form the forward elimination phase of the naive Gaussian elimination algorithm.

For example, Equation (7) states that in proceeding from $\boldsymbol{A}^{(k)}$ to $\boldsymbol{A}^{(k+1)}$, we do not alter rows $1, 2, \dots, k$ or columns $1, 2, \dots, k-1$. Equation (8) shows how the multipliers are computed and stored in passing from $\boldsymbol{A}^{(k)}$ to $\boldsymbol{A}^{(k+1)}$. Finally, Equation (9) shows how multiples of row k are subtracted from rows $k+1, k+2, \dots, n$ to produce $\boldsymbol{A}^{(k+1)}$ from $\boldsymbol{A}^{(k)}$.

Notice that $\boldsymbol{A}^{(n)}$ is the final result of the process. (It was referred to as $\widetilde{A}$ in the statement of the theorem.) The formal definitions of $\boldsymbol{L} = (\ell_{ik})$ and $\boldsymbol{U} = (u_{kj})$ are therefore

$$\begin{cases} \ell_{ik} = 1 & (i = k) \quad (10) \\ \ell_{ik} = a_{ik}^{(n)} & (k < i) \quad (11) \\ \ell_{ik} = 0 & (k > i) \quad (12) \end{cases}$$

$$\begin{cases} u_{kj} = a_{kj}^{(n)} & (j \geqq k) \quad (13) \\ u_{kj} = 0 & (j < k) \quad (14) \end{cases}$$

Now we draw some consequences of these equations.

First, it follows immediately from Equation (7) that

$$a_{ij}^{(i)} = a_{ij}^{(i+1)} = \cdots = a_{ij}^{(n)} \tag{15}$$

Likewise, we have, from Equation (7),

$$a_{ij}^{(j+1)} = a_{ij}^{(j+2)} = \cdots = a_{ij}^{(n)} \qquad (j < n) \tag{16}$$

From Equations (16) and (8), we now have

$$a_{ij}^{(n)} = a_{ij}^{(j+1)} = \frac{a_{ij}^{(j)}}{a_{jj}^{(j)}} \qquad (j < n) \tag{17}$$

From Equations (17) and (11), it follows that

$$\ell_{ik} = a_{ik}^{(n)} = \frac{a_{ik}^{(k)}}{a_{kk}^{(k)}} \qquad (k < i) \tag{18}$$

From Equations (13) and (15), we have

$$u_{kj} = a_{kj}^{(n)} = a_{kj}^{(k)} \qquad (k \leqq j) \tag{19}$$

With the aid of all these equations, we can now prove that $\boldsymbol{LU} = \boldsymbol{A}$. First, consider the case $i \leqq j$. Then we obtain

$$\begin{aligned}
(\boldsymbol{LU})_{ij} &= \sum_{k=1}^{n} \ell_{ik} u_{kj} && \text{[definition of multiplication]} \\
&= \sum_{k=1}^{i} \ell_{ik} u_{kj} && \text{[by Equation (12)]} \\
&= \sum_{k=1}^{i-1} \ell_{ik} u_{kj} + u_{ij} && \text{[by Equation (10)]} \\
&= \sum_{k=1}^{i-1} \left[\frac{a_{ik}^{(k)}}{a_{kk}^{(k)}} \right] a_{kj}^{(k)} + a_{ij}^{(i)} && \text{[by Equations (18) and (19)]} \\
&= \sum_{k=1}^{i-1} \left[a_{ij}^{(k)} - a_{ij}^{(k+1)} \right] + a_{ij}^{(i)} && \text{[by Equation (9)]} \\
&= a_{ij}^{(1)} = a_{ij}
\end{aligned}$$

In the remaining case, $i > j$, we have

$$\begin{aligned}
(\boldsymbol{LU})_{ij} &= \sum_{k=1}^{n} \ell_{ik} u_{kj} && \text{[definition of multiplication]} \\
&= \sum_{k=1}^{j} \ell_{ik} u_{kj} && \text{[by Equation (14)]} \\
&= \sum_{k=1}^{j} \left[\frac{a_{ik}^{(k)}}{a_{kk}^{(k)}} \right] a_{kj}^{(k)} && \text{[by Equations (18) and (19)]} \\
&= \sum_{k=1}^{j-1} \left[\frac{a_{ik}^{(k)}}{a_{kk}^{(k)}} \right] a_{kj}^{(k)} + a_{ij}^{(j)} \\
&= \sum_{k=1}^{j-1} \left[a_{ij}^{(k)} - a_{ij}^{(k+1)} \right] + a_{ij}^{(j)} && \text{[by Equation (9)]} \\
&= a_{ij}^{(1)} = a_{ij}
\end{aligned}$$

■

Pseudocode

The following is the pseudocode for carrying out the $\boldsymbol{LU}$ factorization, which is sometimes called the **Doolittle factorization**:

Doolittle Factorization Pseudocode

```
integer i, k, n;   real array (a_ij)_{1:n×1:n}, (ℓ_ij)_{1:n×1:n}, (u_ij)_{1:n×1:n}
for k = 1 to n
    ℓ_kk ← 1
    for j = k to n
        u_kj ← a_kj − Σ_{s=1}^{k−1} ℓ_ks u_sj
    end do
    for i = k + 1 to n
        ℓ_ik ← (a_ik − Σ_{s=1}^{k−1} ℓ_is u_sk) / u_kk
    end do
end do
```

Solving Linear Systems Using *LU* Factorization

Once the $\boldsymbol{LU}$-factorization of $\boldsymbol{A}$ is available, we can solve the system

$$\boldsymbol{Ax} = \boldsymbol{b}$$

by writing

$$\boldsymbol{LUx} = \boldsymbol{b}$$

Then we solve two triangular systems:

Unit Lower Triangular System

$$\boldsymbol{Lz} = \boldsymbol{b} \tag{20}$$

for $\boldsymbol{z}$ and

Upper Triangular System

$$\boldsymbol{Ux} = \boldsymbol{z} \tag{21}$$

for $\boldsymbol{x}$. This is particularly useful for problems that involve the same coefficient matrix $\boldsymbol{A}$ and many different right-hand vectors $\boldsymbol{b}$.

Since $\boldsymbol{L}$ is unit lower triangular, $\boldsymbol{z}$ is obtained by the pseudocode

Forward Substitution Pseudocode

```
integer i, n;   real array (b_i)_{1:n}, (ℓ_ij)_{1:n×1:n}, (z_i)_{1:n}
z_1 ← b_1
for i = 2 to n
    z_i ← b_i − Σ_{j=1}^{i−1} ℓ_ij z_j
end for
```

Likewise, x is obtained by the pseudocode

Backward Substitution Pseudocode

```
integer i, n;   real array (u_ij)_{1:n×1:n}, (x_i)_{1:n}, (z_i)_{1:n}
x_n ← z_n/u_nn
for i = n − 1 to 1 step −1
    x_i ← (z_i − Σ_{j=i+1}^{n} u_ij x_j) / u_ii
end for
```

The first of these two algorithms applies the forward phase of Gaussian elimination to the right-hand-side vector $\boldsymbol{b}$. (Recall that the ℓ_{ij}'s are the *multipliers* that have been stored in the array (a_{ij}).)

The easiest way to verify this assertion is to use Equation (6) and to rewrite the equation

$$\boldsymbol{L}z = \boldsymbol{b}$$

in the form

$$\boldsymbol{M}_1^{-1}\boldsymbol{M}_2^{-1}\cdots\boldsymbol{M}_{n-1}^{-1}z = \boldsymbol{b}$$

From this, we get immediately

$$z = \boldsymbol{M}_{n-1}\cdots\boldsymbol{M}_2\boldsymbol{M}_1\boldsymbol{b}$$

Thus, the same operations used to reduce $\boldsymbol{A}$ to $\boldsymbol{U}$ are to be used on $\boldsymbol{b}$ to produce z.

Another way to solve Equation (20) is to note that what must be done is to form

$$\boldsymbol{M}_{n-1}\boldsymbol{M}_{n-2}\cdots\boldsymbol{M}_2\boldsymbol{M}_1\boldsymbol{b}$$

This can be accomplished by using only the array (b_i) by putting the results back into $\boldsymbol{b}$; that is,

$$\boldsymbol{b} \leftarrow \boldsymbol{M}_k\boldsymbol{b}$$

We know what $\boldsymbol{M}_k$ looks like because it is made up of negative multipliers that have been saved in the array (a_{ij}). Consequently, we have

$$\boldsymbol{M}_k\boldsymbol{b} = \begin{bmatrix} 1 & & & & & & \\ & \ddots & & & & & \\ & & 1 & & & & \\ & & -a_{k+1,k} & 1 & & & \\ & & \vdots & & \ddots & & \\ & & -a_{ik} & & & 1 & \\ & & \vdots & & & & \ddots & \\ & & -a_{nk} & & & & & 1 \end{bmatrix} \begin{bmatrix} b_1 \\ \vdots \\ b_k \\ b_{k+1} \\ \vdots \\ b_i \\ \vdots \\ b_n \end{bmatrix}$$

The entries b_1 to b_k are not changed by this multiplication, while b_i (for $i \geqq k+1$) is replaced by $-a_{ik}b_k + b_i$. Hence, the following pseudocode updates the array (b_i) based on the stored multipliers in the array a:

Update rhs b Pseudocode

```
integer i, k, n;   real array (a_ij)_{1:n×1:n}, (b_i)_{1:n}
for k = 1 to n − 1
    for i = k + 1 to n
        b_i ← b_i − a_ik b_k
    end for
end for
```

This pseudocode should be familiar. It is the process for updating $\boldsymbol{b}$ from Section 2.2 (p. 91).

The algorithm for solving Equation (21) is the back substitution phase of the naive Gaussian elimination process.

LDL^T Factorization

In the $\boldsymbol{LDL}^T$ factorization, $\boldsymbol{L}$ is unit lower triangular, and $\boldsymbol{D}$ is a diagonal matrix. This factorization can be carried out if $\boldsymbol{A}$ is symmetric and has an ordinary $\boldsymbol{LU}$ factorization, with $\boldsymbol{L}$ unit lower triangular. To see this, we start with

$$\boldsymbol{LU} = \boldsymbol{A} = \boldsymbol{A}^T = (\boldsymbol{LU})^T = \boldsymbol{U}^T\boldsymbol{L}^T$$

Since $\boldsymbol{L}$ is unit lower triangular, it is invertible, and we can write $\boldsymbol{U} = \boldsymbol{L}^{-1}\boldsymbol{U}^T\boldsymbol{L}^T$. Then $\boldsymbol{U}(\boldsymbol{L}^T)^{-1} = \boldsymbol{L}^{-1}\boldsymbol{U}^T$. Since the right side of this equation is lower triangular and the left side is upper triangular, both sides are diagonal, say, $\boldsymbol{D}$. From the equation $\boldsymbol{U}(\boldsymbol{L}^T)^{-1} = \boldsymbol{D}$, we have $\boldsymbol{U} = \boldsymbol{DL}^T$ and $\boldsymbol{A} = \boldsymbol{LU} = \boldsymbol{LDL}^T$.

We now derive the pseudocode for obtaining the $\boldsymbol{LDL}^T$ factorization of a symmetric matrix $\boldsymbol{A}$ in which $\boldsymbol{L}$ is unit lower triangular and $\boldsymbol{D}$ is diagonal. In our analysis, we write a_{ij} as generic elements of $\boldsymbol{A}$ and ℓ_{ij} as generic elements of $\boldsymbol{L}$. The diagonal of $\boldsymbol{D}$ has elements d_{ii}, or d_i. From the equation $\boldsymbol{A} = \boldsymbol{LDL}^T$, we have

$$\begin{aligned} a_{ij} &= \sum_{\nu=1}^{n}\sum_{\mu=1}^{n} \ell_{i\nu} d_{\nu\mu} \ell^T_{\mu j} \\ &= \sum_{\nu=1}^{n}\sum_{\mu=1}^{n} \ell_{i\nu} d_{\nu} \delta_{\nu\mu} \ell_{j\mu} \\ &= \sum_{\nu=1}^{n} \ell_{i\nu} d_{\nu} \ell_{j\nu} \qquad (1 \leqq i, j \leqq n) \end{aligned}$$

Use the fact that $\ell_{ij} = 0$ when $j > i$ and $\ell_{ii} = 1$ to continue the argument

$$a_{ij} = \sum_{\nu=1}^{\min(i,j)} \ell_{i\nu} d_{\nu} \ell_{j\nu} \qquad (1 \leqq i, j \leqq n)$$

Assume now that $j \leqq i$. Then

$$\begin{aligned} a_{ij} &= \sum_{\nu=1}^{j} \ell_{i\nu} d_{\nu} \ell_{j\nu} \\ &= \sum_{\nu=1}^{j-1} \ell_{i\nu} d_{\nu} \ell_{j\nu} + \ell_{ij} d_j \ell_{jj} \\ &= \sum_{\nu=1}^{j-1} \ell_{i\nu} d_{\nu} \ell_{j\nu} + \ell_{ij} d_j \qquad (1 \leqq j \leqq i \leqq n) \end{aligned}$$

In particular, let $j = i$. We get

$$a_{ii} = \sum_{\nu=1}^{i-1} \ell_{i\nu} d_\nu \ell_{i\nu} + d_i \qquad (1 \leqq i \leqq n)$$

Equivalently, we have

$$d_i = a_{ii} - \sum_{\nu=1}^{i-1} d_\nu \ell_{i\nu}^2 \qquad (1 \leqq i \leqq n)$$

Particular cases of this are

$$\begin{cases} d_1 = a_{11} \\ d_2 = a_{22} - d_1 \ell_{21}^2 \\ d_3 = a_{33} - d_1 \ell_{31}^2 - d_2 \ell_{32}^2 \\ \quad \text{etc.} \end{cases}$$

Now we can limit our attention to the cases $1 \leqq j < i \leqq n$, where we have

$$a_{ij} = \sum_{\nu=1}^{j-1} \ell_{i\nu} d_\nu \ell_{j\nu} + \ell_{ij} d_j \qquad (1 \leqq j < i \leqq n)$$

Solving for ℓ_{ij}, we obtain

$$\ell_{ij} = \left[a_{ij} - \sum_{\nu=1}^{j-1} \ell_{i\nu} d_\nu \ell_{j\nu} \right] \Big/ d_j \qquad (1 \leqq j < i \leqq n)$$

Let's do some checking. Taking $j = 1$, we have

$$\ell_{i1} = a_{i1}/d_1 \qquad (2 \leqq i \leqq n)$$

This formula produces column one in $\boldsymbol{L}$. Taking $j = 2$, we have

$$\ell_{i2} = (a_{i2} - \ell_{i1} d_1 \ell_{21})/d_2 \qquad (3 \leqq i \leqq n)$$

This formula produces column two in $\boldsymbol{L}$.

The formal algorithm for the $\boldsymbol{LDL}^T$ factorization is as follows:

LDL^T Factorization Pseudocode

integer i, j, n, ν; **real array** $(a_{ij})_{1:n\times 1:n}, (\ell_{ij})_{1:n\times 1:n}, (d_i)_{1:n}$
for $j = 1$ **to** n
  $\ell_{jj} = 1$
  $d_j = a_{jj} - \sum_{\nu=1}^{j-1} d_\nu \ell_{j\nu}^2$
  for $i = j + 1$ **to** n
    $\ell_{ji} = 0$
    $\ell_{ij} = \left(a_{ij} - \sum_{\nu=1}^{j-1} \ell_{i\nu} d_\nu \ell_{j\nu} \right) \Big/ d_j$
  end for
end for

EXAMPLE 1 Determine the LDL^T factorization of the matrix

$$A = \begin{bmatrix} 4 & 3 & 2 & 1 \\ 3 & 3 & 2 & 1 \\ 2 & 2 & 2 & 1 \\ 1 & 1 & 1 & 1 \end{bmatrix}$$

Solution First, we determine the LU factorization:

$$A = \begin{bmatrix} 1 & 0 & 0 & 0 \\ \frac{3}{4} & 1 & 0 & 0 \\ \frac{1}{2} & \frac{2}{3} & 1 & 0 \\ \frac{1}{4} & \frac{1}{3} & \frac{1}{2} & 1 \end{bmatrix} \begin{bmatrix} 4 & 3 & 2 & 1 \\ 0 & \frac{3}{4} & \frac{1}{2} & \frac{1}{4} \\ 0 & 0 & \frac{2}{3} & \frac{1}{3} \\ 0 & 0 & 0 & \frac{1}{2} \end{bmatrix} = LU$$

Then extract the diagonal elements from U and place them into a diagonal matrix D, writing

$$U = \begin{bmatrix} 4 & 0 & 0 & 0 \\ 0 & \frac{3}{4} & 0 & 0 \\ 0 & 0 & \frac{2}{3} & 0 \\ 0 & 0 & 0 & \frac{1}{2} \end{bmatrix} \begin{bmatrix} 1 & \frac{3}{4} & \frac{1}{2} & \frac{1}{4} \\ 0 & 1 & \frac{2}{3} & \frac{1}{3} \\ 0 & 0 & 1 & \frac{1}{2} \\ 0 & 0 & 0 & 1 \end{bmatrix} = DL^T$$

Clearly, we have $A = LDL^T$. ■

Cholesky Factorization

Any symmetric matrix that has an LU factorization in which L is unit lower triangular has an LDL^T factorization. The Cholesky factorization $A = LL^T$ is a simple consequence of it for the case in which A is symmetric and positive definite.

Suppose in the factorization

$$A = LU$$

Factorizations

the matrix L is lower triangular and the matrix U is upper triangular. When L is *unit* lower triangular, it is called the **Doolittle factorization**. When U is *unit* upper triangular, it goes by the name **Crout factorization**. In the case in which A is symmetric positive definite and $U = L^T$, it is called the **Cholesky factorization**. The mathematician André Louis Cholesky proved the following result.

■ **Theorem 2** **Cholesky Theorem on LL^T Factorization**

If A is a real, symmetric, and positive definite matrix, then it has a unique factorization

$$A = LL^T$$

in which L is lower triangular with a positive diagonal.

Recall that a matrix A is **symmetric and positive definite** if $A = A^T$ and $x^TAx > 0$ for every nonzero vector x. It follows at once that A is nonsingular because A obviously cannot map any nonzero vector into 0. Moreover, by considering special vectors of the form $x = (x_1, x_2, \ldots, x_k, 0, 0, \ldots, 0)^T$, we see that the leading principal minors of A are also positive definite. Theorem 1 implies that A has an LU decomposition. By the symmetry of

$\boldsymbol{A}$, we then have, from the previous discussion

$$\boldsymbol{A} = \boldsymbol{L}\boldsymbol{D}\boldsymbol{L}^T$$

It can be shown that $\boldsymbol{D}$ is positive definite, and thus its elements d_{ii} are positive. Denoting by $\boldsymbol{D}^{1/2}$ the diagonal matrix whose diagonal elements are $\sqrt{d_{ii}}$, we have

$$\boldsymbol{A} = \widetilde{\boldsymbol{L}}\widetilde{\boldsymbol{L}}^T$$

where $\widetilde{\boldsymbol{L}} \equiv \boldsymbol{L}\boldsymbol{D}^{1/2}$, which is the Cholesky factorization. We leave the proof of uniqueness to the reader.

The algorithm for the Cholesky factorization is a special case of the general $\boldsymbol{LU}$ factorization algorithm. If $\boldsymbol{A}$ is real, symmetric, and positive definite, then by Theorem 2, it has a unique factorization of the form

$$\boldsymbol{A} = \boldsymbol{L}\boldsymbol{L}^T$$

in which $\boldsymbol{L}$ is lower triangular and has positive diagonal. Thus, in the equation $\boldsymbol{A} = \boldsymbol{LU}$, we must have $\boldsymbol{U} = \boldsymbol{L}^T$. In the kth step of the general algorithm, the diagonal entry is computed by

$$\ell_{kk} = \left(a_{kk} - \sum_{s=1}^{k-1} \ell_{ks}^2 \right)^{1/2} \tag{22}$$

The algorithm for the **Cholesky factorization** is as follows:

Cholesky Factorization Pseudocode

```
integer i, k, n, s;   real array (a_ij)_{1:n×1:n}, (ℓ_ij)_{1:n×1:n}
for k = 1 to n
    ℓ_kk ← ( a_kk − Σ_{s=1}^{k−1} ℓ_ks² )^{1/2}
    for i = k + 1 to n
        ℓ_ik ← ( a_ik − Σ_{s=1}^{k−1} ℓ_is ℓ_ks ) / ℓ_kk
    end do
end do
```

Theorem 2 guarantees that $\ell_{kk} > 0$. Observe that Equation (22) gives us the following bound:

$$a_{kk} = \sum_{s=1}^{k} \ell_{ks}^2 \geqq \ell_{kj}^2 \qquad (j \leqq k)$$

from which we conclude that

$$|\ell_{kj}| \leqq \sqrt{a_{kk}} \qquad (1 \leqq j \leqq k)$$

Hence, any element of $\boldsymbol{L}$ is bounded by the square root of a corresponding diagonal element in $\boldsymbol{A}$. This implies that the elements of $\boldsymbol{L}$ do not become large relative to $\boldsymbol{A}$ even without any pivoting. In the Cholesky algorithm (and the Doolittle algorithms), the dot products of vectors should be computed in double precision to avoid a buildup of roundoff errors.

EXAMPLE 2 Determine the Cholesky factorization of the matrix in Example 1.

Solution Using the results from Example 1, we write

$$A = LDL^T = (LD^{1/2})(D^{1/2}L^T) = \widetilde{L}\widetilde{L}^T$$

where

$$\widetilde{L} = LD^{1/2}$$
$$= \begin{bmatrix} 1 & 0 & 0 & 0 \\ \frac{3}{4} & 1 & 0 & 0 \\ \frac{1}{2} & \frac{2}{3} & 1 & 0 \\ \frac{1}{4} & \frac{1}{3} & \frac{1}{2} & 1 \end{bmatrix} \begin{bmatrix} 2 & 0 & 0 & 0 \\ 0 & \frac{1}{2}\sqrt{3} & 0 & 0 \\ 0 & 0 & \sqrt{\frac{2}{3}} & 0 \\ 0 & 0 & 0 & \frac{1}{\sqrt{2}} \end{bmatrix}$$
$$= \begin{bmatrix} 2 & 0 & 0 & 0 \\ \frac{3}{2} & \frac{1}{2}\sqrt{3} & 0 & 0 \\ 1 & \frac{1}{3}\sqrt{3} & \sqrt{\frac{2}{3}} & 0 \\ \frac{1}{2} & \frac{1}{6}\sqrt{3} & \frac{1}{2}\sqrt{\frac{2}{3}} & \frac{1}{\sqrt{2}} \end{bmatrix} = \begin{bmatrix} 2.0000 & 0 & 0 & 0 \\ 1.5000 & 0.8660 & 0 & 0 \\ 1.0000 & 0.5774 & 0.8165 & 0 \\ 0.5000 & 0.2887 & 0.4082 & 0.7071 \end{bmatrix}$$

Clearly, $\widetilde{L}$ is the lower triangular matrix in the Cholesky factorization

$$A = \widetilde{L}\widetilde{L}^T$$

■

Multiple Right-Hand Sides

Many software packages for solving linear systems allow the input of multiple right-hand sides. Suppose an $n \times m$ matrix B is

Solving $Ax = B$

$$B = [b^{(1)}, b^{(2)}, \dots, b^{(m)}]$$

in which each column corresponds to a right-hand side of the m linear systems

$$Ax^{(j)} = b^{(j)}$$

for $1 \leqq j \leqq m$. Thus, we can write

$$A[x^{(1)}, x^{(2)}, \dots, x^{(m)}] = [b^{(1)}, b^{(2)}, \dots, b^{(m)}]$$

or

$$AX = B$$

From Section 2.2, procedure *Gauss* can be used once to produce a factorization of A, and procedure *Solve* can be used m times with right-hand side vectors $b^{(j)}$ to find the m solution vectors $x^{(j)}$ for $1 \leqq j \leqq m$.

Since the factorization phase can be done in $\frac{1}{3}n^3$ long operations while each of the back substitution phases requires n^2 long operations, this entire process can be done in $\frac{1}{3}n^3 + mn^2$ long operations. This is much less than $m\left(\frac{1}{3}n^3 + n^2\right)$, which is what it would take if each of the m linear systems were solved separately.

Operation Count

Computing A^{-1}

In some applications, such as in statistics, it may be necessary to compute the inverse of a matrix A and explicitly display it as A^{-1}. If an $n \times n$ matrix A has an inverse, it is an $n \times n$

matrix $\boldsymbol{X}$ with the property that

Computing A Inverse

$$\boldsymbol{A}\boldsymbol{X} = \boldsymbol{I} \tag{23}$$

where $\boldsymbol{I}$ is the identity matrix. If $\boldsymbol{x}^{(j)}$ denotes the jth column of $\boldsymbol{X}$ and $\boldsymbol{I}^{(j)}$ denotes the jth column of $\boldsymbol{I}$, then matrix Equation (23) can be written as

$$\boldsymbol{A}[\boldsymbol{x}^{(1)}, \boldsymbol{x}^{(2)}, \dots, \boldsymbol{x}^{(n)}] = [\boldsymbol{I}^{(1)}, \boldsymbol{I}^{(2)}, \dots, \boldsymbol{I}^{(n)}]$$

This can be written as n linear systems of equations of the form

$$\boldsymbol{A}\boldsymbol{x}^{(j)} = \boldsymbol{I}^{(j)} \qquad (1 \leqq j \leqq n)$$

This can be done by using procedures *Gauss* and *Solve* from Section 2.2. Use procedure *Gauss* once to produce a factorization of $\boldsymbol{A}$, and use procedure *Solve* n times with the right-hand side vectors $\boldsymbol{I}^{(j)}$ for $1 \leqq j \leqq n$. This is equivalent to solving, one at a time, for the columns of $\boldsymbol{A}^{-1}$, which are $\boldsymbol{x}^{(j)}$. Hence, we obtain

$$\boldsymbol{A}^{-1} = [\boldsymbol{x}^{(1)}, \boldsymbol{x}^{(2)}, \dots, \boldsymbol{x}^{(n)}]$$

Caution

A word of caution on computing the inverse of a matrix: In solving a linear system $\boldsymbol{A}\boldsymbol{x} = \boldsymbol{b}$, it is not advisable to determine $\boldsymbol{A}^{-1}$ and then compute the matrix-vector product $\boldsymbol{x} = \boldsymbol{A}^{-1}\boldsymbol{b}$ because this requires many unnecessary calculations, compared to directly solving $\boldsymbol{A}\boldsymbol{x} = \boldsymbol{b}$ for $\boldsymbol{x}$.

Example Using Software Packages

A **permutation matrix** is an $n \times n$ matrix $\boldsymbol{P}$ that arises from the identity matrix by permuting its rows. It then turns out that permuting the rows of any $n \times n$ matrix $\boldsymbol{A}$ can be accomplished by multiplying $\boldsymbol{A}$ on the left by $\boldsymbol{P}$. Every permutation matrix is invertible, since the rows still form a basis for $\mathbb{R}^n$. When Gaussian elimination with row pivoting is performed on a matrix $\boldsymbol{A}$, the result is expressible as

Solving $Ax = b$ Using $PA = LU$

$$\boldsymbol{P}\boldsymbol{A} = \boldsymbol{L}\boldsymbol{U}$$

where $\boldsymbol{L}$ is lower triangular and $\boldsymbol{U}$ is upper triangular. The matrix $\boldsymbol{P}\boldsymbol{A}$ is $\boldsymbol{A}$ with its rows rearranged.

If we have the $\boldsymbol{L}\boldsymbol{U}$ *factorization of* $\boldsymbol{P}\boldsymbol{A}$*, how do we solve the system* $\boldsymbol{A}\boldsymbol{x} = \boldsymbol{b}$*?* First, write it as

$$\boldsymbol{P}\boldsymbol{A}\boldsymbol{x} = \boldsymbol{P}\boldsymbol{b}$$

then $\boldsymbol{L}\boldsymbol{U}\boldsymbol{x} = \boldsymbol{P}\boldsymbol{b}$. Let $\boldsymbol{y} = \boldsymbol{U}\boldsymbol{x}$, so that our problem is now

$$\begin{cases} \boldsymbol{L}\boldsymbol{y} = \boldsymbol{P}\boldsymbol{b} \\ \boldsymbol{U}\boldsymbol{x} = \boldsymbol{y} \end{cases}$$

The first equation is easily solved for $\boldsymbol{y}$, and then the second equation is easily solved for $\boldsymbol{x}$. Mathematical software systems such as MATLAB, Maple, and Mathematica produce factorizations of the form $\boldsymbol{P}\boldsymbol{A} = \boldsymbol{L}\boldsymbol{U}$ upon command.

EXAMPLE 3 Use the mathematical software systems in Maple, MATLAB, and Mathematica to find the $\boldsymbol{LU}$ factorization of this matrix:

$$\boldsymbol{A} = \begin{bmatrix} 6 & -2 & 2 & 4 \\ 12 & -8 & 6 & 10 \\ 3 & -13 & 9 & 3 \\ -6 & 4 & 1 & -18 \end{bmatrix} \tag{24}$$

Solution First, we use Maple and find this factorization:

Maple

$$\boldsymbol{A} = \boldsymbol{LU} = \begin{bmatrix} 1 & 0 & 0 & 0 \\ 2 & 1 & 0 & 0 \\ \frac{1}{2} & 3 & 1 & 0 \\ -1 & -\frac{1}{2} & 2 & 1 \end{bmatrix} \begin{bmatrix} 6 & -2 & 2 & 4 \\ 0 & -4 & 2 & 2 \\ 0 & 0 & 2 & -5 \\ 0 & 0 & 0 & -3 \end{bmatrix}$$

Next, we use MATLAB and find a different factorization:

$$\boldsymbol{PA} = \widehat{\boldsymbol{L}}\widehat{\boldsymbol{U}}$$

MATLAB

$$\widehat{\boldsymbol{L}} = \begin{bmatrix} 1.0000 & 0 & 0 & 0 \\ 0.2500 & 1.0000 & 0 & 0 \\ -0.5000 & 0 & 1.0000 & 0 \\ 0.5000 & -0.1818 & 0.0909 & 1.0000 \end{bmatrix}$$

$$\widehat{\boldsymbol{U}} = \begin{bmatrix} 12.0000 & -8.0000 & 6.0000 & 10.0000 \\ 0 & -11.0000 & 7.5000 & 0.5000 \\ 0 & 0 & 4.0000 & -13.0000 \\ 0 & 0 & 0 & 0.2727 \end{bmatrix}$$

$$\boldsymbol{P} = \begin{bmatrix} 0 & 1 & 0 & 0 \\ 0 & 0 & 1 & 0 \\ 0 & 0 & 0 & 1 \\ 1 & 0 & 0 & 0 \end{bmatrix}$$

where $\boldsymbol{P}$ is a permutation matrix corresponding to the pivoting strategy used. Finally, we use Mathematica to create this $\boldsymbol{LU}$ decomposition:

Mathematica

$$\begin{bmatrix} 3 & -13 & 9 & 3 \\ -2 & -22 & 19 & -12 \\ 2 & -\frac{12}{11} & \frac{52}{11} & -\frac{166}{11} \\ 4 & -2 & \frac{22}{13} & -\frac{6}{13} \end{bmatrix}$$

The output is in a compact store scheme that contains both the lower triangular matrix and the upper triangular matrix in a single matrix. However, the storage arrangement may be complicated because the rows are usually permuted during the factorization in an effort to make the solution process numerically stable. Verify that this factorization corresponds to the permutation of rows of matrix $\boldsymbol{A}$ in the order 3, 4, 1, 2. ■

Summary 8.1

- If $\boldsymbol{A} = (a_{ij})$ is an $n \times n$ matrix such that the forward elimination phase of the naive Gaussian algorithm can be applied to $\boldsymbol{A}$ without encountering any zero divisors, then the resulting matrix can be denoted by $\widetilde{A} = (\widetilde{a}_{ij})$, where

$$\boldsymbol{L} = \begin{bmatrix} 1 & 0 & 0 & \cdots & 0 \\ \widetilde{a}_{21} & 1 & 0 & \cdots & 0 \\ \widetilde{a}_{31} & \widetilde{a}_{32} & 1 & \cdots & 0 \\ \vdots & \vdots & \ddots & \ddots & \vdots \\ \widetilde{a}_{n1} & \widetilde{a}_{n2} & \cdots & \widetilde{a}_{n,n-1} & 1 \end{bmatrix}$$

and

$$\boldsymbol{U} = \begin{bmatrix} \widetilde{a}_{11} & \widetilde{a}_{12} & \widetilde{a}_{13} & \cdots & \widetilde{a}_{1n} \\ 0 & \widetilde{a}_{22} & \widetilde{a}_{23} & \cdots & \widetilde{a}_{2n} \\ 0 & 0 & \widetilde{a}_{33} & \cdots & \widetilde{a}_{3n} \\ \vdots & \vdots & \ddots & \ddots & \vdots \\ 0 & 0 & \cdots & 0 & \widetilde{a}_{nn} \end{bmatrix}$$

This is the $\boldsymbol{LU}$ **factorization** of $\boldsymbol{A}$, so $\boldsymbol{A} = \boldsymbol{LU}$, where $\boldsymbol{L}$ is a unit lower triangular and $\boldsymbol{U}$ is upper triangular. When we carry out the Gaussian forward elimination process on $\boldsymbol{A}$, the result is the upper triangular matrix $\boldsymbol{U}$. The matrix $\boldsymbol{L}$ is the unit lower triangular matrix whose entries are negatives of the multipliers in the locations of the elements they zero out.

- We can also give a formal description as follows. The matrix $\boldsymbol{U}$ can be obtained by applying a succession of matrices $\boldsymbol{M}_k$ to the left of $\boldsymbol{A}$. The kth step of Gaussian elimination corresponds to a unit lower triangular matrix $\boldsymbol{M}_k$, which is the product of $n-k$ elementary matrices

$$\boldsymbol{M}_k = \boldsymbol{M}_{nk}\,\boldsymbol{M}_{n-1,k}\cdots\boldsymbol{M}_{k+1,k}$$

where each **elementary matrix** $\boldsymbol{M}_{ik}$ is unit lower triangular. If $\boldsymbol{M}_{qp}\boldsymbol{A}$ is obtained by subtracting λ times row p from row q in matrix $\boldsymbol{A}$ with $p < q$, then the elements of $\boldsymbol{M}_{qp} = (m_{ij})$ are

$$m_{ij} = \begin{cases} 1, & \text{if } i = j \\ -\lambda, & \text{if } i = q \text{ and } j = p \\ 0, & \text{otherwise} \end{cases}$$

The entire Gaussian elimination process is summarized by writing

$$\boldsymbol{M}_{n-1}\cdots\boldsymbol{M}_2\boldsymbol{M}_1\boldsymbol{A} = \boldsymbol{U}$$

Since each $\boldsymbol{M}_k$ is invertible, we have

$$\boldsymbol{A} = \boldsymbol{M}_1^{-1}\boldsymbol{M}_2^{-1}\cdots\boldsymbol{M}_{n-1}^{-1}\boldsymbol{U}$$

Each $\boldsymbol{M}_k$ is a unit lower triangular matrix, and the same is true of each inverse $\boldsymbol{M}_k^{-1}$, as well as their products. Hence, the matrix

$$\boldsymbol{L} = \boldsymbol{M}_1^{-1}\boldsymbol{M}_2^{-1}\cdots\boldsymbol{M}_{n-1}^{-1}$$

is unit lower triangular.

- For symmetric matrices, we have the $\boldsymbol{LDL}^T$ factorization, and for symmetric positive definite matrices, we have the $\boldsymbol{LL}^T$ factorization, which is also known as Cholesky factorization.

- If the $\boldsymbol{LU}$ factorization of $\boldsymbol{A}$ is available, we can solve the system

$$\boldsymbol{Ax} = \boldsymbol{b}$$

by solving two triangular systems:

$$\begin{cases} Ly = b & \text{for } y \\ Ux = y & \text{for } x \end{cases}$$

This is useful for problems that involve the same coefficient matrix A and many different right-hand vectors b. For example, let B be an $n \times m$ matrix of the form

$$B = [b^{(1)}, b^{(2)}, \dots, b^{(m)}]$$

where each column corresponds to a right-hand side of the m linear systems

$$Ax^{(j)} = b^{(j)} \qquad (1 \leqq j \leqq m)$$

Thus, we can write

$$A\left[x^{(1)}, x^{(2)}, \dots, x^{(m)}\right] = \left[b^{(1)}, b^{(2)}, \dots, b^{(m)}\right]$$

or

$$AX = B$$

A special case of this is to compute the inverse of an $n \times n$ invertible matrix A. We write

$$AX = I$$

where I is the identity matrix. If $x^{(j)}$ denotes the jth column of X and $I^{(j)}$ denotes the jth column of I, this can be written as

$$A\left[x^{(1)}, x^{(2)}, \dots, x^{(n)}\right] = \left[I^{(1)}, I^{(2)}, \dots, I^{(n)}\right]$$

or as n linear systems of equations of the form

$$Ax^{(j)} = I^{(j)} \qquad (1 \leqq j \leqq n)$$

We can use LU factorization to solve these n systems efficiently, obtaining

$$A^{-1} = \left[x^{(1)}, x^{(2)}, \dots, x^{(n)}\right]$$

- When Gaussian elimination with row pivoting is performed on a matrix A, the result is expressible as

$$PA = LU$$

where P is a **permutation matrix**, L is unit lower triangular, and U is upper triangular. Here, the matrix PA is A with its rows interchanged. We can solve the system $Ax = b$ by solving

$$\begin{cases} Ly = Pb & \text{for } y \\ Ux = y & \text{for } x \end{cases}$$

Exercises 8.1

1. Using naive Gaussian elimination, factor the following matrices in the form $A = LU$, where L is a unit lower triangular matrix and U is an upper triangular matrix.

[a]**a.** $A = \begin{bmatrix} 3 & 0 & 3 \\ 0 & -1 & 3 \\ 1 & 3 & 0 \end{bmatrix}$

b. $A = \begin{bmatrix} 1 & 0 & \frac{1}{3} & 0 \\ 0 & 1 & 3 & -1 \\ 3 & -3 & 0 & 6 \\ 0 & 2 & 4 & -6 \end{bmatrix}$

c. $A = \begin{bmatrix} -20 & -15 & -10 & -5 \\ 1 & 0 & 0 & 0 \\ 0 & 1 & 0 & 0 \\ 0 & 0 & 1 & 0 \end{bmatrix}$

2. Consider the matrix

$$A = \begin{bmatrix} 1 & 0 & 0 & 2 \\ 0 & 3 & 0 & 0 \\ 0 & 9 & 4 & 0 \\ 5 & 0 & 8 & 10 \end{bmatrix}$$

[a]**a.** Determine a unit lower triangular matrix $\boldsymbol{M}$ and an upper triangular matrix $\boldsymbol{U}$ such that $\boldsymbol{MA} = \boldsymbol{U}$.

b. Determine a unit lower triangular matrix $\boldsymbol{L}$ and an upper triangular matrix $\boldsymbol{U}$ such that $\boldsymbol{A} = \boldsymbol{LU}$. Show that $\boldsymbol{ML} = \boldsymbol{I}$ so that $\boldsymbol{L} = \boldsymbol{M}^{-1}$.

3. Consider the matrix

$$A = \begin{bmatrix} 25 & 0 & 0 & 0 & 1 \\ 0 & 27 & 4 & 3 & 2 \\ 0 & 54 & 58 & 0 & 0 \\ 0 & 108 & 116 & 0 & 0 \\ 100 & 0 & 0 & 0 & 24 \end{bmatrix}$$

[a]**a.** Determine the unit lower triangular matrix $\boldsymbol{M}$ and the upper triangular matrix $\boldsymbol{U}$ such that $\boldsymbol{MA} = \boldsymbol{U}$.

b. Determine $\boldsymbol{M}^{-1} = \boldsymbol{L}$ such that $\boldsymbol{A} = \boldsymbol{LU}$.

4. Consider the matrix

$$A = \begin{bmatrix} 2 & 2 & 1 \\ 1 & 1 & 1 \\ 3 & 2 & 1 \end{bmatrix}$$

a. Show that $\boldsymbol{A}$ *cannot* be factored into the product of a unit lower triangular matrix and an upper triangular matrix.

[a]**b.** Interchange the rows of $\boldsymbol{A}$ so that this can be done.

5. Consider the matrix

$$A = \begin{bmatrix} a & 0 & 0 & z \\ 0 & b & 0 & 0 \\ 0 & x & c & 0 \\ w & 0 & y & d \end{bmatrix}$$

[a]**a.** Determine a unit lower triangular matrix $\boldsymbol{M}$ and an upper triangular matrix $\boldsymbol{U}$ such that $\boldsymbol{MA} = \boldsymbol{U}$.

[a]**b.** Determine a lower triangular matrix $\boldsymbol{L}'$ and a unit upper triangular matrix $\boldsymbol{U}'$ such that $\boldsymbol{A} = \boldsymbol{L}'\boldsymbol{U}'$.

6. Consider the matrix

$$A = \begin{bmatrix} 4 & -1 & -1 & 0 \\ -1 & 4 & 0 & -1 \\ -1 & 0 & 4 & -1 \\ 0 & -1 & -1 & 4 \end{bmatrix}$$

Factor $\boldsymbol{A}$ in the following ways:

[a]**a.** $\boldsymbol{A} = \boldsymbol{LU}$, where $\boldsymbol{L}$ is unit lower triangular and $\boldsymbol{U}$ is upper triangular.

[a]**b.** $\boldsymbol{A} = \boldsymbol{LDU}'$, where $\boldsymbol{L}$ is unit lower triangular, $\boldsymbol{D}$ is diagonal, and $\boldsymbol{U}'$ is unit upper triangular.

[a]**c.** $\boldsymbol{A} = \boldsymbol{L}'\boldsymbol{U}'$, where $\boldsymbol{L}'$ is lower triangular and $\boldsymbol{U}'$ is unit upper triangular.

[a]**d.** $\boldsymbol{A} = (\boldsymbol{L}'')(\boldsymbol{L}'')^T$, where $\boldsymbol{L}''$ is lower triangular.

[a]**e.** Evaluate the determinant of $\boldsymbol{A}$.
Hint: $\det(\boldsymbol{A}) = \det(\boldsymbol{L})\det(\boldsymbol{D})\det(\boldsymbol{U}') = \det(\boldsymbol{D})$.

7. Consider the 3×3 Hilbert matrix

$$A = \begin{bmatrix} 1 & \frac{1}{2} & \frac{1}{3} \\ \frac{1}{2} & \frac{1}{3} & \frac{1}{4} \\ \frac{1}{3} & \frac{1}{4} & \frac{1}{5} \end{bmatrix}$$

Repeat the preceding problem using this matrix.

[a]**8.** Find the $\boldsymbol{LU}$ decomposition, where $\boldsymbol{L}$ is unit lower triangular, for

$$A = \begin{bmatrix} 1 & 0 & 0 & 1 \\ 1 & 1 & 0 & -1 \\ -1 & 1 & 1 & 1 \\ 1 & -1 & 1 & -1 \end{bmatrix}$$

9. Consider

$$A = \begin{bmatrix} 2 & -1 & 2 \\ 2 & -3 & 3 \\ 6 & -1 & 8 \end{bmatrix}$$

[a]**a.** Find the matrix factorization $\boldsymbol{A} = \boldsymbol{LDU}'$, where $\boldsymbol{L}$ is unit lower triangular, $\boldsymbol{D}$ is diagonal, and $\boldsymbol{U}'$ is unit upper triangular.

[a]**b.** Use this decomposition of $\boldsymbol{A}$ to solve $\boldsymbol{Ax} = \boldsymbol{b}$, where $\boldsymbol{b} = [-2, -5, 0]^T$.

[a]**10.** Repeat the preceding problem for

$$A = \begin{bmatrix} -2 & 1 & -2 \\ -4 & 3 & -3 \\ 2 & 2 & 4 \end{bmatrix}, \qquad b = \begin{bmatrix} 1 \\ 4 \\ 4 \end{bmatrix}$$

11. Consider the system of equations

$$\begin{cases} 6x_1 = 12 \\ 6x_2 + 3x_1 = -12 \\ 7x_3 - 2x_2 + 4x_1 = 14 \\ 21x_4 + 9x_3 - 3x_2 + 5x_1 = -2 \end{cases}$$

a. Solve for x_1, x_2, x_3, and x_4 (in order) by forward substitution.

b. Write this system in matrix notation $\boldsymbol{Ax} = \boldsymbol{b}$, where $\boldsymbol{x} = [x_1, x_2, x_3, x_4]^T$. Determine the $\boldsymbol{LU}$ factorization $\boldsymbol{A} = \boldsymbol{LU}$, where $\boldsymbol{L}$ is unit lower triangular and $\boldsymbol{U}$ is upper triangular.

[a]**12.** Given

$$A = \begin{bmatrix} 3 & 2 & -1 \\ 5 & 3 & 2 \\ -1 & 1 & -3 \end{bmatrix}$$

$$L^{-1} = \begin{bmatrix} 1 & 0 & 0 \\ -\frac{5}{3} & 1 & 0 \\ -8 & 5 & 1 \end{bmatrix}, \quad U = \begin{bmatrix} 3 & 2 & -1 \\ 0 & -\frac{1}{3} & \frac{11}{3} \\ 0 & 0 & 15 \end{bmatrix}$$

obtain the inverse of A by solving $UX^{(j)} = L^{-1}I^{(j)}$ for $j = 1, 2, 3$.

13. Using the system of Equation (2), form $M = M_3M_2M_1$ and determine M^{-1}. Verify that $M^{-1} = L$. Why is this, in general, not a good idea?

14. Consider the matrix $A = \text{Tridiagonal}\,(a_{i,i-1}, a_{ii}, a_{i,i+1})$, where $a_{ii} \neq 0$. Here is the 4×4 case.

[a]**a.** Establish the algorithm

```
integer i
real array (a_ij)1:n×1:n, (ℓ_ij)1:n×1:n, (u_ij)1:n×1:n
ℓ_11 ← a_11
for i = 2 to 4
    ℓ_i,i−1 ← a_i,i−1
    u_i−1,i ← a_i−1,i / ℓ_i−1,i−1
    ℓ_i,i ← a_i,i − ℓ_i,i−1 u_i−1,i
end for
```

for determining the elements of a lower bidiagonal matrix $L = (\ell_{ij})$ and a *unit* upper bidiagonal matrix $U = (u_{ij})$ such that $A = LU$.

b. Establish the algorithm

```
integer i;
real array (a_ij)1:n×1:n, (ℓ_i,j)1:n×1:n, (u_i,j)1:n×1:n
u_11 ← a_11
for i = 2 to 4
    u_i−1,i ← a_i−1,i
    ℓ_i,i−1 ← a_i,i−1 / u_i−1,i−1
    u_i,i ← a_i,i − ℓ_i,i−1 u_i−1,i
end for
```

for determining the elements of a *unit* lower bidiagonal matrix $L = (\ell_{ij})$ and an upper bidiagonal matrix $U = (u_{ij})$ such that $A = LU$.

By extending the loops, we can generalize these algorithms to $n \times n$ tridiagonal matrices.

15. Show that the equation $Ax = B$ can be solved by Gaussian elimination with scaled partial pivoting in $(n^3/3) + mn^2 + \mathcal{O}(n^2)$ multiplications and divisions, where A, X, and B are matrices of order $n \times n$, $n \times m$, and $n \times m$, respectively. Thus, if B is $n \times n$, then the $n \times n$ solution matrix X can be found by Gaussian elimination with scaled partial pivoting in $\frac{4}{3}n^3 + \mathcal{O}(n^2)$ multiplications and divisions.

Hint: If $X^{(j)}$ and $B^{(j)}$ are the jth columns of X and B, respectively, then $AX^{(j)} = B^{(j)}$.

16. Let $\mathcal{X}$ be a square matrix that has the form

$$\mathcal{X} = \begin{bmatrix} A & B \\ C & D \end{bmatrix}$$

where A and D are square matrices and A^{-1} exists. It is known that $\mathcal{X}^{-1}$ exists if and only if $(D - CA^{-1}B)^{-1}$ exists. Verify that $\mathcal{X}^{-1}$ is given by

$$\mathcal{X} = \begin{bmatrix} I & -A^{-1}B \\ 0 & I \end{bmatrix} \begin{bmatrix} A^{-1} & 0 \\ 0 & (D - CA^{-1}B)^{-1} \end{bmatrix} \times \begin{bmatrix} I & 0 \\ -CA^{-1} & I \end{bmatrix}$$

As an application, compute the inverse of the following:

[a]**a.** $\mathcal{X} = \left[\begin{array}{cc|cc} 1 & 0 & 0 & 1 \\ 0 & 1 & 1 & 0 \\ \hline 1 & 0 & 1 & 2 \\ 0 & 0 & 0 & 1 \end{array}\right]$

[a]**b.** $\mathcal{X} = \left[\begin{array}{ccc|c} 1 & 0 & 0 & 1 \\ 0 & 1 & 0 & 1 \\ 0 & 0 & 1 & 1 \\ \hline 1 & 1 & 1 & 2 \end{array}\right]$

17. Let A be an $n \times n$ complex matrix such that A^{-1} exists. Verify that

$$\begin{bmatrix} A & \overline{A} \\ -Ai & -\overline{A}i \end{bmatrix}^{-1} = \frac{1}{2}\begin{bmatrix} A^{-1} & A^{-1}i \\ \overline{A}^{-1} & -\overline{A}^{-1}i \end{bmatrix}$$

where $\overline{A}$ denotes the complex conjugate of A; if $A = (a_{ij})$, then $\overline{A} = (\overline{a}_{ij})$. Recall that for a complex number $z = a + bi$, where a and b are real, and $\overline{z} = a - bi$.

18. Find the LU factorization of this matrix:

$$A = \begin{bmatrix} 2 & 2 & 1 \\ 4 & 7 & 2 \\ 2 & 11 & 5 \end{bmatrix}$$

19. **a.** Prove that the product of two lower triangular matrices is lower triangular.

b. Prove that the product of two unit lower triangular matrices is unit lower triangular.

c. Prove that the inverse of a unit lower triangular matrix is unit lower triangular.

d. By using the transpose operation, prove that all of the preceding results are true for upper triangular matrices.

20. Let $\boldsymbol{L}$ be lower triangular, $\boldsymbol{U}$ be upper triangular, and $\boldsymbol{D}$ be diagonal.

a. If $\boldsymbol{L}$ and $\boldsymbol{U}$ are both unit triangular and $\boldsymbol{LDU}$ is diagonal, does it follow that $\boldsymbol{L}$ and $\boldsymbol{U}$ are diagonal?

b. If $\boldsymbol{LDU}$ is nonsingular and diagonal, does it follow that $\boldsymbol{L}$ and $\boldsymbol{U}$ are diagonal?

c. If $\boldsymbol{L}$ and $\boldsymbol{U}$ are both unit triangular and if $\boldsymbol{LDU}$ is diagonal, does it follow that $\boldsymbol{L} = \boldsymbol{U} = \boldsymbol{I}$?

21. Determine the $\boldsymbol{LDL}^T$ factorization for the following matrix:

$$A = \begin{bmatrix} 1 & 2 & -1 & 1 \\ 2 & 3 & -4 & 3 \\ -1 & -4 & -1 & 3 \\ 1 & 3 & 3 & 0 \end{bmatrix}$$

22. Find the Cholesky factorization of

$$A = \begin{bmatrix} 4 & 6 & 10 \\ 6 & 25 & 19 \\ 10 & 19 & 62 \end{bmatrix}$$

23. Consider the system

$$\begin{bmatrix} A & \mathbf{0} \\ B & C \end{bmatrix} \begin{bmatrix} x \\ y \end{bmatrix} = \begin{bmatrix} b \\ d \end{bmatrix}$$

Show how to solve the system more cheaply using the submatrices rather than the overall system. Give an estimate of the computational cost of both the new and old approaches. This problem illustrates solving a block linear system with a special structure.

24. Determine the $\boldsymbol{LDL}^T$ factorization of the matrix

$$A = \begin{bmatrix} 5 & 35 & -20 & 65 \\ 35 & 244 & -143 & 461 \\ -20 & -143 & 73 & -232 \\ 65 & 461 & -232 & 856 \end{bmatrix}$$

Can you find the Cholesky factorization?

25. (**Sparse Factorizations**) Consider the following sparse symmetric matrices with the nonzero pattern shown where nonzero entries in the matrix are indicated by the $\times$ symbol and zero entries are a blank. Show the nonzero pattern in the matrix $\boldsymbol{L}$ for the Cholesky factorization by using the symbol $+$ for the fill-in of a zero entry by a nonzero entry.

a. $$A = \left[\begin{array}{ccc|ccc|ccc} \times & \times & & \times & & & & & \\ \times & \times & \times & & \times & & & & \\ & \times & \times & & & \times & & & \\ \hline \times & & & \times & \times & & \times & & \\ & \times & & \times & \times & \times & & \times & \\ & & \times & & \times & \times & & & \times \\ \hline & & & \times & & & \times & & \\ & & & & \times & & \times & \times & \times \\ & & & & & \times & & \times & \times \end{array}\right]$$

b. $$A = \left[\begin{array}{ccc|ccc|ccc} \times & & & & \times & & \times & & \\ & \times & & & \times & & & \times & \\ & & \times & & & \times & \times & & \\ \hline & & & \times & & \times & & \times & \\ \times & \times & & & \times & & & & \times \\ & & \times & \times & & \times & & & \times \\ \hline \times & & \times & & & & \times & & \times \\ & \times & & \times & & & & \times & \times \\ & & & & \times & \times & \times & \times & \times \end{array}\right]$$

c. $$A = \left[\begin{array}{ccc|ccc|ccc} \times & & \times & & & & \times & & \\ & \times & \times & & & & & & \times \\ \times & \times & \times & & & & & \times & \\ \hline & & & \times & & \times & \times & & \\ & & & & \times & \times & & & \times \\ & & & \times & \times & \times & & \times & \\ \hline \times & & & \times & & & \times & \times & \\ & & \times & & & \times & \times & \times & \times \\ & \times & & & \times & & & \times & \times \end{array}\right]$$

Computer Exercises 8.1

1. Write and test a procedure for implementing the algorithms of Exercise 8.1.14.

2. The $n \times n$ factorization $\boldsymbol{A} = \boldsymbol{LU}$, where $\boldsymbol{L} = (\ell_{ij})$ is lower triangular and $\boldsymbol{U} = (u_{ij})$ is upper triangular, can be computed directly by the following algorithm (provided zero divisions are not encountered): Specify either ℓ_{11} or u_{11} and compute the other such that $\ell_{11}u_{11} = a_{11}$. Compute the first column in $\boldsymbol{L}$ by

$$\ell_{i1} = \frac{a_{i1}}{u_{11}} \qquad (1 \leqq i \leqq n)$$

and compute the first row in $\boldsymbol{U}$ by

$$u_{1j} = \frac{a_{1j}}{\ell_{11}} \qquad (1 \leqq j \leqq n)$$

Now suppose that columns $1, 2, \dots, k-1$ have been computed in $\boldsymbol{L}$ and that rows $1, 2, \dots, k-1$ have been computed in $\boldsymbol{U}$. At the kth step, specify either ℓ_{kk} or u_{kk}, and compute the other such that

$$\ell_{kk}u_{kk} = a_{kk} - \sum_{m=1}^{k-1} \ell_{km}u_{mk}$$

Compute the kth column in $\boldsymbol{L}$ by

$$\ell_{ik} = \frac{1}{u_{kk}}\left(a_{ik} - \sum_{m=1}^{k-1} \ell_{im}u_{mk}\right) \qquad (k \leqq i \leqq n)$$

and compute the kth row in $\boldsymbol{U}$ by

$$u_{kj} = \frac{1}{\ell_{kk}}\left(a_{kj} - \sum_{m=1}^{k-1} \ell_{km}u_{mj}\right), \qquad (k \leqq j \leqq n)$$

This algorithm is continued until all elements of $\boldsymbol{U}$ and $\boldsymbol{L}$ are completely determined. When $\ell_{ii} = 1\ (1 \leqq i \leqq n)$, this procedure is called the **Doolittle factorization**, and when $u_{jj} = 1\ (1 \leqq j \leqq n)$, it is known as the **Crout factorization**.

Define the test matrix

$$\boldsymbol{A} = \begin{bmatrix} 5 & 7 & 6 & 5 \\ 7 & 10 & 8 & 7 \\ 6 & 8 & 10 & 9 \\ 5 & 7 & 9 & 10 \end{bmatrix}$$

Using the algorithm above, compute and print factorizations so that the diagonal entries of $\boldsymbol{L}$ and $\boldsymbol{U}$ are of the following forms:

diag($\boldsymbol{L}$)	diag($\boldsymbol{U}$)	
[1, 1, 1, 1]	[?, ?, ?, ?]	Doolittle
[?, ?, ?, ?]	[1, 1, 1, 1]	Crout
[1, ?, 1, ?]	[?, 1, ?, 1]	
[?, 1, ?, 1]	[1, ?, 1, ?]	
[?, ?, 7, 9]	[3, 5, ?, ?]	

Here the question mark means that the entry is to be computed. Write code to check the results by multiplying $\boldsymbol{L}$ and $\boldsymbol{U}$ together.

3. Write

procedure *Poly*$(n, (a_{ij}), (c_i), k, (y_{ij}))$

for computing the $n \times n$ matrix $p_k(\boldsymbol{A})$ stored in array (y_{ij}):

$$y_k = p_k(\boldsymbol{A}) = c_0\boldsymbol{I} + c_1\boldsymbol{A} + c_2\boldsymbol{A}^2 + \cdots + c_k\boldsymbol{A}^k$$

where $\boldsymbol{A}$ is an $n \times n$ matrix and p_k is a kth-degree polynomial. Here (c_i) are real constants for $0 \leqq i \leqq k$. Use nested multiplication and write efficient code. Test procedure *Poly* on the following data:

Case 1.

$\boldsymbol{A} = \boldsymbol{I}_5, \quad p_3(x) = 1 - 5x + 10x^3$

Case 2.

$$\boldsymbol{A} = \begin{bmatrix} 1 & 2 \\ 3 & 4 \end{bmatrix}, \quad p_2(x) = 1 - 2x + x^2$$

Case 3.

$$\boldsymbol{A} = \begin{bmatrix} 0 & 2 & 4 \\ 0 & 0 & 8 \\ 0 & 0 & 0 \end{bmatrix}, \quad p_3(x) = 1 + 3x - 3x^2 + x^3$$

[a] **Case 4.**

$$\boldsymbol{A} = \begin{bmatrix} 2 & -1 & 0 & 0 \\ -1 & 2 & -1 & 0 \\ 0 & -1 & 2 & -1 \\ 0 & 0 & -1 & 2 \end{bmatrix}$$

$p_5(x) = 10 + x - 2x^2 + 3x^3 - 4x^4 + 5x^5$

Case 5.

$$\boldsymbol{A} = \begin{bmatrix} -20 & -15 & -10 & -5 \\ 1 & 0 & 0 & 0 \\ 0 & 1 & 0 & 0 \\ 0 & 0 & 1 & 0 \end{bmatrix}$$

$p_4(x) = 5 + 10x + 15x^2 + 20x^3 + x^4$

Case 6.

$$\boldsymbol{A} = \begin{bmatrix} 5 & 7 & 6 & 5 \\ 7 & 10 & 8 & 7 \\ 6 & 8 & 10 & 9 \\ 5 & 7 & 9 & 10 \end{bmatrix},$$

$p_4(x) = 1 - 100x + 146x^2 - 35x^3 + x^4$

4. Write and test a procedure for determining $\boldsymbol{A}^{-1}$ for a given square matrix $\boldsymbol{A}$ of order n. Your procedure should use procedures *Gauss* and *Solve*.

5. Write and test a procedure to solve the system $\boldsymbol{AX} = \boldsymbol{B}$ in which $\boldsymbol{A}$, $\boldsymbol{X}$, and $\boldsymbol{B}$ are matrices of order $n \times n$, $n \times m$, and $n \times m$, respectively. Verify that the procedure works on several test cases, one of which has $\boldsymbol{B} = \boldsymbol{I}$ so that the solution $\boldsymbol{X}$ is the inverse of $\boldsymbol{A}$.
Hint: See Exercise 8.1.15.

6. Write and test a procedure for directly computing the inverse of a tridiagonal matrix. Assume that pivoting is not necessary.

7. (Continuation) Test the procedure of the preceding computer problem on the symmetric tridiagonal matrix $\boldsymbol{A}$ of order 10:

$$\boldsymbol{A} = \begin{bmatrix} -2 & 1 & & & & \\ 1 & -2 & 1 & & & \\ & 1 & -2 & 1 & & \\ & & \ddots & \ddots & \ddots & \\ & & & 1 & -2 & 1 \\ & & & & 1 & -2 \end{bmatrix}$$

The inverse of this matrix is known to be

$$(\boldsymbol{A}^{-1})_{ij} = (\boldsymbol{A}^{-1})_{ji} = \frac{-i(n+1-j)}{(n+1)} \qquad (i \leqq j)$$

8. Investigate the numerical difficulties in inverting the following matrix:

$$A = \begin{bmatrix} -0.0001 & 5.096 & 5.101 & 1.853 \\ 0. & 3.737 & 3.740 & 3.392 \\ 0. & 0. & 0.006 & 5.254 \\ 0. & 0. & 0. & 4.567 \end{bmatrix}$$

9. Consider the following two test matrices:

$$A = \begin{bmatrix} 4 & 6 & 10 \\ 6 & 25 & 19 \\ 10 & 19 & 62 \end{bmatrix}, \qquad B = \begin{bmatrix} 4 & 6 & 10 \\ 6 & 13 & 19 \\ 10 & 19 & 62 \end{bmatrix}$$

Show that the first Cholesky factorization has all integers in the solution, while the second one is all integers until the last step, where there is a square root.

a. Program the Cholesky algorithm.

b. Use MATLAB, Maple, or Mathematica to find the Cholesky factorizations.

10. Let A be real, symmetric, and positive definite. Is the same true for the matrix obtained by removing the first row and column of A?

11. Devise a code for inverting a unit lower triangular matrix. Test it on the following matrix:

$$\begin{bmatrix} 1 & 0 & 0 & 0 \\ 3 & 1 & 0 & 0 \\ 5 & 2 & 1 & 0 \\ 7 & 4 & -3 & 1 \end{bmatrix}$$

12. Verify Example 1 using MATLAB, Maple, or Mathematica.

13. In Example 3, verify the factorizations of matrix A using MATLAB, Maple, and Mathematica.

14. Find the $PA = LU$ factorization of this matrix:

$$A = \begin{bmatrix} -0.05811 & -0.11696 & 0.51004 & -0.31330 \\ -0.04291 & 0.56850 & 0.07041 & 0.68747 \\ -0.01652 & 0.38953 & 0.01203 & -0.52927 \\ -0.06140 & 0.32179 & -0.22094 & 0.42448 \end{bmatrix}$$

which was studied by Wilkinson [1965, p. 640].

8.2 Eigenvalues and Eigenvectors

Av Scalar Multiple of *v*

Let A be an $n \times n$ matrix. We ask the following natural question about A: *Are there nonzero vectors* v *for which* Av *is a scalar multiple of* v? Although we pose this question in the spirit of pure curiosity, there are many situations in scientific computation in which this question arises.

The answer to our question is a qualified *Yes*! We must be willing to consider complex scalars, as well as vectors with complex components. With that broadening of our viewpoint, such vectors always exist. Here are two examples. In the first, we need not bring in complex numbers to illustrate the situation, whereas in the second, the vectors and scalar factors must be complex.

EXAMPLE 1 Let $A = \begin{bmatrix} 3 & 2 \\ 7 & -2 \end{bmatrix}$. Find a nonzero vector v for which Av is a multiple of v.

Solution You can easily verify that

$$A\begin{bmatrix} 1 \\ 1 \end{bmatrix} = \begin{bmatrix} 5 \\ 5 \end{bmatrix} = 5\begin{bmatrix} 1 \\ 1 \end{bmatrix}$$

$$A\begin{bmatrix} 2 \\ -7 \end{bmatrix} = \begin{bmatrix} -8 \\ 28 \end{bmatrix} = -4\begin{bmatrix} 2 \\ -7 \end{bmatrix}$$

We have two different answers (but we have not revealed how to find them). ■

EXAMPLE 2 Repeat the preceding example with the matrix $A = \begin{bmatrix} 1 & 1 \\ -2 & 3 \end{bmatrix}$.

Solution As in Example 1, it can be verified that

$$A \begin{bmatrix} 1 \\ 1+i \end{bmatrix} = (2+i) \begin{bmatrix} 1 \\ 1+i \end{bmatrix}$$
$$A \begin{bmatrix} 1 \\ 1-i \end{bmatrix} = (2-i) \begin{bmatrix} 1 \\ 1-i \end{bmatrix}$$

In these equations, $i = \sqrt{-1}$. Surprisingly, we find answers involving complex numbers even though the matrix does not contain any complex entries! ■

Eigenvalues
Eigenvectors

When the equation $Ax = \lambda x$ is valid and x is not zero, we say that λ is an **eigenvalue** of A and x is an accompanying **eigenvector**. Thus, in Example 1, the matrix has 5 as an eigenvalue with accompanying eigenvector $[1, 1]^T$, and -4 is another eigenvalue with accompanying eigenvector $[2, -7]^T$. Example 2 emphasizes that a real matrix may have complex eigenvalues and complex eigenvectors. Notice that an equation $A\mathbf{0} = \lambda\mathbf{0}$ and an equation $A\mathbf{0} = \mathbf{0}x$ say nothing useful about eigenvalues and eigenvectors of A.

Many problems in science lead to eigenvalue problems in which the principal question usually is: *What are the eigenvalues of a given matrix, and what are the accompanying eigenvectors*? An outstanding application of eigenvalues and eigenvectors is to systems of linear differential equations, which we discuss later.

Eigenspace

Notice that if $Ax = \lambda x$ and $x \neq 0$, then every nonzero multiple of x is an eigenvector (with the same eigenvalue). If λ is an eigenvalue of an $n \times n$ matrix A, then the set $\{x\colon Ax = \lambda x\}$ is a subspace of $\mathbb{R}^n$ called an **eigenspace**. It is necessarily of dimension at least 1.

Calculating Eigenvalues and Eigenvectors

Given a square matrix A, how does one discover its eigenvalues? Begin by observing that the equation $Ax = \lambda x$ is equivalent to $(A - \lambda I)x = \mathbf{0}$. Since we are interested in nonzero solutions to this equation, the matrix $A - \lambda I$ must be noninvertible, and therefore $\text{Det}(A - \lambda I) = 0$. This is how (in principle) we can find all the eigenvalues of A. Specifically, form the function p defined by

Characteristic Polynomial

$$p(\lambda) = \text{Det}(A - \lambda I)$$

and find the zeros of p. It turns out that p is a polynomial of degree n and must have n zeros, provided that we allow complex zeros and count each zero a number of times equal to its multiplicity. Even if the matrix A is real, we must be prepared for complex eigenvalues. The polynomial just described is called the **characteristic polynomial** of the matrix A. If this polynomial has a repeated factor, such as $(\lambda - 3)^k$, then we say that 3 is a root of **multiplicity** k. Such roots are still eigenvalues, but they can be troublesome when $k > 1$.

To illustrate the calculation of eigenvalues, let us use the matrix in Example 1, namely,

Eigenvalues Calculation

$$A = \begin{bmatrix} 3 & 2 \\ 7 & -2 \end{bmatrix}$$

The characteristic polynomial is

$$p(\lambda) = \text{Det}(A - \lambda I) = \text{Det} \begin{bmatrix} 3-\lambda & 2 \\ 7 & -2-\lambda \end{bmatrix} = (3-\lambda)(-2-\lambda) - 14$$
$$= \lambda^2 - \lambda - 20 = (\lambda - 5)(\lambda + 4)$$

The eigenvalues are 5 and -4.

Mathematical Software

We can carry out this calculation with one or two commands in MATLAB, Maple, or Mathematica. We can determine the characteristic polynomial and subsequently compute its zeros. This gives us the two roots of the characteristic polynomial, which are the eigenvalues 5 and -4. These mathematical software systems also have single commands to produce a list of eigenvalues, computed in the best possible way, which is usually *not* to determine the characteristic polynomial and subsequently compute its zeros!

In general, an $n \times n$ matrix has a characteristic polynomial of degree n, and its roots are the eigenvalues of $\boldsymbol{A}$. Since the calculation of zeros of a polynomial is numerically challenging if not unstable, this straightforward procedure is not recommended. (See Computer Exercise 8.2.2 for an experiment pertaining to this situation.) For small values of n, it may be quite satisfactory, however. It is called the **direct method** for computing eigenvalues.

Once an eigenvalue λ has been determined for a matrix $\boldsymbol{A}$, an eigenvector can be computed by solving the system $(\boldsymbol{A} - \lambda \boldsymbol{I})\boldsymbol{x} = \boldsymbol{0}$. Thus, in Example 1, we must solve $(\boldsymbol{A} - 5\boldsymbol{I})\boldsymbol{x} = \boldsymbol{0}$, or

Eigenvector Calculation

$$\begin{bmatrix} -2 & 2 \\ 7 & -7 \end{bmatrix} \begin{bmatrix} x_1 \\ x_2 \end{bmatrix} = \begin{bmatrix} 0 \\ 0 \end{bmatrix}$$

Of course, this matrix is singular, and the homogeneous equation has nontrivial solutions, such as $[1, 1]^T$. The other eigenvalue is treated in the same way, leading to an eigenvector $[2, -7]^T$. Any scalar multiple of an eigenvector is also an eigenvector.

Using Software

This work can be done by using mathematical software to find an eigenvector for each eigenvalue λ via the null space of the matrix $\boldsymbol{A} - \lambda \boldsymbol{I}$. Also, we can use a single command to compute all the eigenvalues directly or request the calculation of all the eigenvalues and eigenvectors at once. The MATLAB command `[V,D] = eig(A)` produces two arrays, `V` and `D`. The array `V` has eigenvectors of $\boldsymbol{A}$ as its columns, and the array `D` contains all the eigenvalues of $\boldsymbol{A}$ on its diagonal. The program returns a vector of unit length such as $[0.7071, 0.7071]^T$. That vector by itself provides a basis for the null space of $\boldsymbol{A} - 5\boldsymbol{I}$. (Maple and Mathematica have commands for computing eigenvalues and eigenvectors.)

Notice that the eigenvalue-eigenvector problem is nonlinear. The equation

$$\boldsymbol{A}\boldsymbol{x} = \lambda \boldsymbol{x}$$

has two unknowns, λ and $\boldsymbol{x}$. They appear in the equation multiplied together. If either $\boldsymbol{x}$ or λ were known, finding the other would be a linear problem and very easy.

Mathematical Software

A typical, mundane use of mathematical software such as MATLAB might be to compute the eigenvalues and eigenvalues of this matrix

Using MATLAB

$$\boldsymbol{A} = \begin{bmatrix} 1 & 3 & -7 \\ -3 & 4 & 1 \\ 2 & -5 & 3 \end{bmatrix} \tag{1}$$

with a command such as

```
[V,D] = eig(A)
```

MATLAB responds instantly with the eigenvectors in the array `V` and the eigenvalues in the diagonal array `D`. The real eigenvalue is 0.0214, and the complex pair of eigenvalues are

$3.9893 \pm 5.5601i$. Behind the scenes, much complicated computing may be taking place! The general procedure has these components: First, by means of similarity transformations, $\boldsymbol{A}$ is put into lower Hessenberg form. This means that all elements below the first subdiagonal are zero. Thus, the new matrix $\boldsymbol{A} = (a_{ij})$ satisfies $a_{ij} = 0$ when $i > j + 1$. Similarity transformations ensure that the eigenvalues are not disturbed. If $\boldsymbol{A}$ is real, further similarity transformations put $\boldsymbol{A}$ into a near-diagonal form in which each diagonal element is either a single real number or a 2×2 real matrix whose eigenvalues are a pair of conjugate complex numbers. Creating the additional zeros just below the diagonal requires some iterative process, because after all, we are in effect computing the zeros of a polynomial. The iterative process is reminiscent of the power method that is described in Section 8.3.

Maple can be used to compute the eigenvalues and eigenvectors. The quantities are computed in exact arithmetic and then converted to floating-point. In some versions of MATLAB, symbolic computations are available. In Mathematica, we can use either numerical or symbolical commands to obtain similar results.

LAPACK

The best advice for anyone who is confronted with challenging eigenvalue problems is to use the software in the package LAPACK. Special eigenvalue algorithms for various types of matrices are available there. For example, if the matrix in question is real and symmetric, use an algorithm tailored for that case. There are about a dozen categories available to choose from in LAPACK. MATLAB itself employs some of the routines in LAPACK.

Properties of Eigenvalues

A theorem that summarizes the special properties of a matrix that impinge on the computing of its eigenvalues follows.

■ Theorem 1 **Matrix Eigenvalue Properties**

The following statements are true for any square matrix $\boldsymbol{A}$:

1. If λ is an eigenvalue of $\boldsymbol{A}$, then $p(\lambda)$ is an eigenvalue of $p(\boldsymbol{A})$, for any polynomial p. In particular, λ^k is an eigenvalue of $\boldsymbol{A}^k$.
2. If $\boldsymbol{A}$ is invertible and λ is an eigenvalue of $\boldsymbol{A}$, then $p(1/\lambda)$ is an eigenvalue of $p(\boldsymbol{A}^{-1})$, for any polynomial p. In particular, λ^{-1} is an eigenvalue of $\boldsymbol{A}^{-1}$.
3. If $\boldsymbol{A}$ is real and symmetric, then its eigenvalues are real.
4. If $\boldsymbol{A}$ is complex and Hermitian, then its eigenvalues are real.
5. If $\boldsymbol{A}$ is Hermitian and positive definite, then its eigenvalues are positive.
6. If $\boldsymbol{P}$ is invertible, then $\boldsymbol{A}$ and $\boldsymbol{PAP}^{-1}$ have the same characteristic polynomial (and the same eigenvalues).

Symmetric
Hermitian
Conjugate Transpose
Positive Definite

Recall that a matrix $\boldsymbol{A}$ is **symmetric** if $\boldsymbol{A} = \boldsymbol{A}^T$, where $\boldsymbol{A}^T = (a_{ji})$ is the **transpose** of $\boldsymbol{A} = (a_{ij})$. On the other hand, a complex matrix $\boldsymbol{A}$ is **Hermitian** if $\boldsymbol{A} = \boldsymbol{A}^*$, where $\boldsymbol{A}^* = \overline{\boldsymbol{A}}^T = (\overline{a}_{ji})$. Here $\boldsymbol{A}^*$ is the **conjugate transpose** of the matrix $\boldsymbol{A}$. Using the syntax of programming, we can write $\boldsymbol{A}^T(i, j) = \boldsymbol{A}(j, i)$ and $\boldsymbol{A}^*(i, j) = \overline{\boldsymbol{A}}(j, i)$. Recall also that $\boldsymbol{A}$ is **positive definite** if $\boldsymbol{x}^T\boldsymbol{A}\boldsymbol{x} > 0$ for all nonzero vectors $\boldsymbol{x}$.

Similar Matrices

Two matrices $\boldsymbol{A}$ and $\boldsymbol{B}$ are **similar** to each other if there exists an invertible matrix $\boldsymbol{P}$ such that $\boldsymbol{B} = \boldsymbol{P}\boldsymbol{A}\boldsymbol{P}^{-1}$. Similar matrices have the same characteristic polynomial

$$\begin{aligned}
\text{Det}(\boldsymbol{B} - \lambda \boldsymbol{I}) &= \text{Det}(\boldsymbol{P}\boldsymbol{A}\boldsymbol{P}^{-1} - \lambda \boldsymbol{I}) \\
&= \text{Det}(\boldsymbol{P}(\boldsymbol{A} - \lambda \boldsymbol{I})\boldsymbol{P}^{-1}) \\
&= \text{Det}(\boldsymbol{P}) \cdot \text{Det}(\boldsymbol{A} - \lambda \boldsymbol{I}) \cdot \text{Det}(\boldsymbol{P}^{-1}) \\
&= \text{Det}(\boldsymbol{A} - \lambda \boldsymbol{I})
\end{aligned}$$

Thus, we have this important theorem.

Theorem 2 — Eigenvalues of Similar Matrices

Similar matrices have the same eigenvalues.

This theorem suggests a strategy for finding eigenvalues of $\boldsymbol{A}$. Transform the matrix $\boldsymbol{A}$ to a matrix $\boldsymbol{B}$ using a similarity transformation

$$\boldsymbol{B} = \boldsymbol{P}\boldsymbol{A}\boldsymbol{P}^{-1}$$

in which $\boldsymbol{B}$ has a special structure, and then find the eigenvalues of matrix $\boldsymbol{B}$. Specifically, if $\boldsymbol{B}$ is *triangular* or *diagonal*, the eigenvalues of $\boldsymbol{B}$ (and those of $\boldsymbol{A}$) are simply the diagonal elements of $\boldsymbol{B}$.

Unitarily Similar

Unitary

Matrices $\boldsymbol{A}$ and $\boldsymbol{B}$ are said to be **unitarily similar** to each other if $\boldsymbol{B} = \boldsymbol{U}^*\boldsymbol{A}\boldsymbol{U}$ for some unitary matrix $\boldsymbol{U}$. Recall that a matrix $\boldsymbol{U}$ is **unitary** if $\boldsymbol{U}\boldsymbol{U}^* = \boldsymbol{I}$. This brings us naturally to another important theorem and two corollaries.

Theorem 3 — Schur's Theorem

Every square matrix is unitarily similar to a triangular matrix.

In this theorem, an arbitrary complex $n \times n$ matrix $\boldsymbol{A}$ is given, and the assertion made is that a unitary matrix $\boldsymbol{U}$ exists such that:

$$\boldsymbol{U}\boldsymbol{A}\boldsymbol{U}^* = \boldsymbol{T} \tag{2}$$

where $\boldsymbol{U}\boldsymbol{U}^* = \boldsymbol{I}$ and $\boldsymbol{T}$ is a triangular matrix.

The proof of Schur's Theorem can be found in Kincaid and Cheney [2002] and Golub and Van Loan [1996].

Corollary 1 — Matrix Similar to a Triangular Matrix

Every square real matrix is similar to a triangular matrix.

Thus, the factorization

$$\boldsymbol{P}\boldsymbol{A}\boldsymbol{P}^{-1} = \boldsymbol{T}$$

is possible, where $\boldsymbol{T}$ is triangular, $\boldsymbol{P}$ is invertible, and $\boldsymbol{A}$ is real.

EXAMPLE 3 We illustrate Schur's Theorem by finding the decomposition of this 2×2 matrix:

$$A = \begin{bmatrix} 3 & -2 \\ 8 & 3 \end{bmatrix}$$

Solution From the characteristic equation

$$\text{Det}(A - \lambda I) = \lambda^2 - 6\lambda + 25 = 0$$

the eigenvalues are $3 \pm 4i$. By solving $A - \lambda I = 0$ with each of these eigenvalues, the corresponding eigenvectors are $v_1 = [i, 2]^T$ and $v_2 = [-i, 2]^T$. Using the Gram-Schmidt orthogonalization process, we obtain $u_1 = v_1$ and $u_2 = v_2 - [v_2^* u_1 / u_1^* u_1] u_1 = [-2, -i]^T$. After normalizing these vectors, we obtain the unitary matrix

$$U = \frac{1}{\sqrt{5}} \begin{bmatrix} i & -2 \\ 2 & -i \end{bmatrix}$$

which satisfies the property $UU^* = I$, Finally, we obtain the Schur form

$$UAU^* = \begin{bmatrix} 3+4i & -6 \\ 0 & 3-4i \end{bmatrix}$$

which is an upper triangular matrix with the eigenvalues on the diagonal. ■

■ **Corollary 2** **Hermitian Matrix Unitarily Similar to a Diagonal Matrix**

Every square Hermitian matrix is unitarily similar to a diagonal matrix.

Corollary 2 says that a Hermitian matrix $A = A^*$ can be factored as

$$UAU^* = D$$

where D is diagonal and U is unitary.

This follows from Corollary 1 since a Hermitian matrix ($A = A^*$) is unitarily similar to a triangular metrix T

$$UAU^* = T$$

where $UU^* = I$. Furthermore, we have

$$U^* A^* U = T^*$$

Since $A = A^*$, we obtain $T = T^*$, which must be a diagonal matrix.

Deflation Process

Most numerical methods for finding eigenvalues of an $n \times n$ matrix A proceed by determining such similarity transformations. Then one eigenvalue at a time, say, λ, is computed, and a **deflation process** is used to produce an $(n-1) \times (n-1)$ matrix $\widetilde{A}$ whose eigenvalues are the same as those of A, except for λ. Any such procedure can be repeated with the matrix $\widetilde{A}$ to find as many eigenvalues of the matrix A as desired. In practice, this strategy must be used cautiously because the successive eigenvalues may be infected with roundoff error!

Gershgorin's Theorem

Sometimes it is necessary to determine in a coarse manner where the eigenvalues of a matrix are situated in the complex plane $\mathbb{C}$. The most famous of these so-called **localization theorems** is the following.

Theorem 4 **Gershgorin's Theorem**

All eigenvalues of an $n \times n$ matrix $\boldsymbol{A} = (a_{ij})$ are contained in the union of the n discs $C_i = C_i(a_{ii}, r_i)$ in the complex plane with center a_{ii} and radii r_i given by the sum of the magnitudes of the off-diagonal entries in the ith row.

The matrix $\boldsymbol{A}$ can have either real or complex entires. The region containing the eigenvalues of $\boldsymbol{A}$ can be written

$$\bigcup_{i=1}^{n} C_i = \bigcup_{i=1}^{n} \{z \in \mathbb{C} : |z - a_{ii}| \leqq r_i\} \tag{3}$$

where the radii are $r_i = \sum_{\substack{j=1 \\ j \neq i}}^{n} |a_{ij}|$.

The eigenvalues of $\boldsymbol{A}$ and $\boldsymbol{A}^T$ are the same because the characteristic equation involves the determinant, which is the same for a matrix and its transpose. Therefore, we can apply the Gershgorin Theorem to $\boldsymbol{A}^T$ and obtain the following useful result.

Corollary 3 **More Gershgorin Discs**

All eigenvalues of an $n \times n$ matrix $\boldsymbol{A} = (a_{ij})$ are contained in the union of the n discs $D_i = D_i(a_{ii}, s_i)$ in the complex plane having center at a_{ii} and radii s_i given by the sum of the magnitudes of the columns of $\boldsymbol{A}$.

Consequently, the region containing the eigenvalues of $\boldsymbol{A}$ can be written as

$$\bigcup_{i=1}^{n} D_i = \bigcup_{i=1}^{n} \{z \in \mathbb{C} : |z - a_{ii}| \leqq s_i\} \tag{4}$$

where the radii are $s_i = \sum_{\substack{i=1 \\ i \neq j}}^{n} |a_{ij}|$. Finally, the region containing the eigenvalues of $\boldsymbol{A}$ is

$$\left(\bigcup_{i=1}^{n} C_i\right) \bigcap \left(\bigcup_{i=1}^{n} D_i\right) \tag{5}$$

This may result in tighter bounds on the eigenvalues in some case. Also, a useful localization result is

Corollary 4 **Localization**

For a matrix $\boldsymbol{A}$, the union of any k Gershgorin discs that do not intersect the remaining $n - k$ circles contains exactly k (counting multiplicities) of the eigenvalues of $\boldsymbol{A}$.

For a strictly diagonally dominant matrix, zero cannot lie in any of its Gershgorin discs, so it must be invertible. Consequently, we obtain the following results.

Corollary 5 **Strictly Diagonally Dominant Matrix**

Every strictly diagonally dominant matrix is invertible.

EXAMPLE 4 Consider the matrix

$$A = \begin{bmatrix} 4-i & 2 & i \\ -1 & 2i & 2 \\ 1 & -1 & -5 \end{bmatrix}$$

Draw the Gershgorin discs.

Solution Using the rows of A, we find that the Gershgorin discs are $C_1(4-i, 3)$, $C_2(2i, 3)$, and $C_3(-5, 2)$. By using the columns of A, we obtain more Gershgorin discs: $D_1(4-i, 2)$, $D_2(2i, 3)$, and $D_3(-5, 3)$. Consequently, all the eigenvalues of A are in the three discs D_1, C_2, and C_3, as shown in Figure 8.1. By other means, we compute the eigenvalues of A as $\lambda_1 = 3.7208 - 1.05461i$, $\lambda_2 = 4.5602 + -0.2849i$, and $\lambda_3 = -0.1605 + 2.3395i$. In Figure 8.1, the center of the discs are designated by dots • and the eigenvalues by $*$. ■

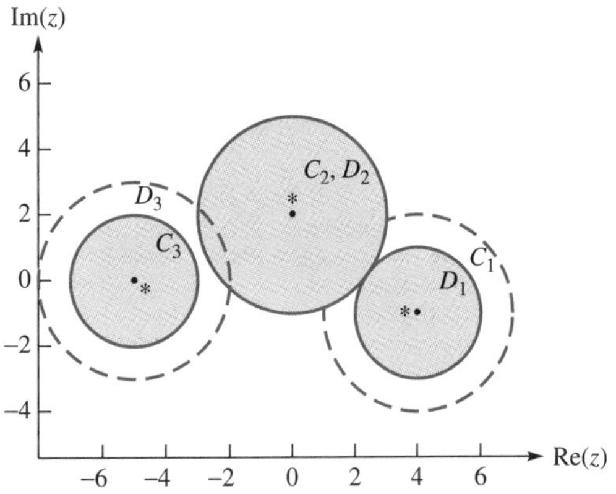

FIGURE 8.1 Gershorgin discs

Singular Value Decomposition

This subsection requires some further knowledge of linear algebra, in particular the diagonalization of symmetric matrices, eigenvalues, eigenvectors, rank, column space, and norms. See online, at the textbook Web site, Appendix D2 for a brief reveiw of these topics. (In the discussion following, we assume that the Euclidean norm is being used.)

The singular value decomposition is a general-purpose tool that has many uses, particularly in least-squares problems (Chapter 9). It can be applied to any matrix, whether square or not. We begin by stating that the **singular values** of a matrix A are the nonnegative square roots of the eigenvalues of $A^T A$.

■ **Theorem 5** **Matrix Spectral Theorem**

Let A be $m \times n$. Then $A^T A$ is an $n \times n$ symmetric matrix, and it can be diagonalized by an orthogonal matrix, say, Q:

$$A^T A = Q D Q^{-1}$$

where $Q Q^T = Q^T Q = I$ and D is a diagonal $n \times n$ matrix.

Furthermore, the diagonal matrix D contains the eigenvalues of $A^T A$ on its diagonal. This follows from the fact that $A^T A Q = Q D$, so the columns of Q are eigenvectors of $A^T A$.

If λ is an eigenvalue of A^TA and if x is a corresponding eigenvector, then $A^TAx = \lambda x$, whence

$$||Ax||^2 = (Ax)^T(Ax) = x^TA^TAx = x^T\lambda x = \lambda||x||^2$$

This equation shows that λ is real and nonnegative. We can order the eigenvalues as $\lambda_1 \geqq \lambda_2 \geqq \cdots \geqq \lambda_n \geqq 0$. (Reordering the eigenvalues requires reordering the columns of Q.)

Singular Values

The numbers $\sigma_j = +\sqrt{\lambda_j}$ are the **singular values** of A.

Since Q is an orthogonal matrix, its columns form an orthonormal base for $\mathbb{R}^n$. They are unit eigenvectors of A^TA, so if v_j is the jth column of Q, then $A^TAv_j = \lambda_j v_j$. Some of the eigenvalues of A^TA can be zero. Define r by the condition

$$\lambda_1 \geqq \lambda_2 \geqq \cdots \geqq \lambda_r > 0 = \lambda_{r+1} = \cdots = \lambda_n$$

For a review of concepts such as rank, orthogonal basis, orthonormal basis, column space, null space, and so on, see Appendix D.2 on the textbook Web site.

■ **Theorem 6** **Orthogonal Basis Theorem**

If the rank of A is r, then an orthogonal basis for the column space of A is $\{Av_j : 1 \leqq j \leqq r\}$.

Proof Observe that

$$(Av_k)^T(Av_j) = v_k^TA^TAv_j = v_k^T\lambda_j v_j = \lambda_j\delta_{kj}$$

This establishes the orthogonality of the set $\{Av_j : 1 \leqq j \leqq n\}$. By letting $k = j$, we get $||Av_j||^2 = \lambda_j$. Hence, $Av_j \neq 0$ if and only if $1 \leqq j \leqq r$. If w is any vector in the column space of A, then $w = Ax$ for some x in $\mathbb{R}^n$. Putting $x = \sum_{j=1}^n c_jv_j$, we get

$$w = Ax = \sum_{j=1}^{n} c_jAv_j = \sum_{j=1}^{r} c_jAv_j$$

and therefore, w is in the span of $\{Av_1, Av_2, \dots, Av_r\}$. ■

The preceding theorem gives a reasonable way of computing the rank of a numerical matrix. First, compute its singular values. Any that are very small can be assumed to be zero. The remaining ones are strongly positive, and if there are r of them, we take r to be

Rank

the numerically computed **rank** of A.

A **singular value decomposition** of an $m \times n$ matrix A is any representation of A in the form

Singular Value Decomposition

$$A = UDV^T \tag{6}$$

where U and V are orthogonal matrices and D is an $m \times n$ *diagonal* matrix having nonnegative diagonal entries that are ordered $d_{11} \geqq d_{22} \geqq \cdots \geqq 0$. Then from Exercise 8.2.4, it follows that the diagonal elements d_{ii} are necessarily the singular values of A. Note that the matrix U is $m \times m$ and V is $n \times n$. A nonsquare matrix D is **diagonal** if the only elements that are not zero are among those whose two indices are equal.

Now a singular value decomposition of A (there are many of them) can be described. Start with the vectors $v_1, v_2, \dots, v_r$. Normalize the vectors Av_j to get vectors u_j. Thus, we have

$$u_j = Av_j/||Av_j|| \qquad (1 \leqq j \leqq r)$$

Extend this set to an orthonormal base for $\mathbb{R}^m$. Let $\boldsymbol{U}$ be the $m \times m$ matrix whose columns are $\boldsymbol{u}_1, \boldsymbol{u}_2, \ldots, \boldsymbol{u}_m$. Define $\boldsymbol{D}$ to be the $m \times n$ matrix consisting of zeros except for $\sigma_1, \sigma_2, \ldots, \sigma_r$ on its diagonal. Let $\boldsymbol{V} = \boldsymbol{Q}$, where $\boldsymbol{Q}$ is as discussed.

To verify the equation $\boldsymbol{A} = \boldsymbol{U}\boldsymbol{D}\boldsymbol{V}^T$, first note that $\sigma_j = ||\boldsymbol{A}\boldsymbol{v}_j||_2$ and that $\sigma_j \boldsymbol{u}_j = \boldsymbol{A}\boldsymbol{v}_j$. Then compute $\boldsymbol{U}\boldsymbol{D}$. Since $\boldsymbol{D}$ is diagonal, this is easy. We get

$$\begin{aligned}\boldsymbol{U}\boldsymbol{D} &= [\boldsymbol{u}_1, \boldsymbol{u}_2, \ldots, \boldsymbol{u}_m]\boldsymbol{D} = [\sigma_1\boldsymbol{u}_1, \sigma_2\boldsymbol{u}_2, \ldots, \sigma_r\boldsymbol{u}_r, 0, \ldots, 0]\\ &= [\boldsymbol{A}\boldsymbol{v}_1, \boldsymbol{A}\boldsymbol{v}_2, \ldots, \boldsymbol{A}\boldsymbol{v}_r, \ldots, \boldsymbol{A}\boldsymbol{v}_n] = \boldsymbol{A}\boldsymbol{Q} = \boldsymbol{A}\boldsymbol{V}\end{aligned}$$

This implies that

Verifying $A = UDV^T$

$$\boldsymbol{A} = \boldsymbol{U}\boldsymbol{D}\boldsymbol{V}^T$$

The **condition number** of a matrix can be expressed in terms of its singular values

Condition Number

$$\kappa(\boldsymbol{A}) = \sqrt{\frac{\sigma_{\max}}{\sigma_{\min}}} \tag{7}$$

since $||\boldsymbol{A}||_2^2 = \rho(\boldsymbol{A}^T\boldsymbol{A}) = \sigma_{\max}(\boldsymbol{A})$ and $||\boldsymbol{A}^{-1}||_2^2 = \rho(\boldsymbol{A}^{-T}\boldsymbol{A}^{-1}|| = \sigma_{\min}(\boldsymbol{A})$.

Numerical Examples of Singular Value Decomposition

The numerical determination of a singular value decomposition is best left to mathematical software such as MATLAB, Maple, Mathematica, LAPACK, and other software packages. Usually, they do *not* form $\boldsymbol{A}^T\boldsymbol{A}$ in the numerical work because its **condition number** may be much worse than that of $\boldsymbol{A}$. This phenomenon is easily illustrated by the matrices

$$\boldsymbol{A} = \begin{bmatrix} 1 & 1 & 1\\ \varepsilon & 0 & 0\\ 0 & \varepsilon & 0\\ 0 & 0 & \varepsilon\end{bmatrix}, \qquad \boldsymbol{A}^T\boldsymbol{A} = \begin{bmatrix} 1+\varepsilon^2 & 1 & 1\\ 1 & 1+\varepsilon^2 & 1\\ 1 & 1 & 1+\varepsilon^2\end{bmatrix}$$

There are small values of ε for which $\boldsymbol{A}$ has rank 3 and $\boldsymbol{A}^T\boldsymbol{A}$ has rank 1 (in the computer).

EXAMPLE 5 In Example 2 of Section 1.1 (p. 4), we encountered this matrix:

$$\boldsymbol{A} = \begin{bmatrix} 0.1036 & 0.2122\\ 0.2081 & 0.4247\end{bmatrix}$$

Determine its eigenvalues, singular values, and condition number.

Solution By using mathematical software, it is easy to find the eigenvalues $\lambda_1(\boldsymbol{A}) \approx -0.0003$ and $\lambda_2(\boldsymbol{A}) \approx 0.5286$. We can form the matrix

$$\boldsymbol{A}^T\boldsymbol{A} = \begin{bmatrix} 0.0540 & 01104\\ 0.1104 & 0.2254\end{bmatrix}$$

and find its eigenvalues $\lambda_1(\boldsymbol{A}^T\boldsymbol{A}) \approx 0.3025 \times 10^{-4}$ and $\lambda_2(\boldsymbol{A}^T\boldsymbol{A}) \approx 0.2794$. Therefore, the singular values are

$$\sigma_1(\boldsymbol{A}) = \sqrt{|\lambda_1(\boldsymbol{A}^T\boldsymbol{A})|} \approx 0.0003, \qquad \sigma_2(\boldsymbol{A}) = \sqrt{|\lambda_2(\boldsymbol{A}^T\boldsymbol{A})|} \approx 0.5286$$

Also, we can obtain the singular values directly as $\sigma_1 \approx 0.0003$ and $\sigma_2 \approx 0.5286$ using mathematical software. Consequently, the condition number is $\kappa(\boldsymbol{A}) = \sigma_2/\sigma_1 \approx 1747.6$. Because of this large condition number, we now understand why there was difficultly in solving the linear system of equations with this coefficient matrix! ■

EXAMPLE 6 Calculate the singular value decomposition of the matrix

$$A = \begin{bmatrix} 1 & 1 \\ 0 & 1 \\ 1 & 0 \end{bmatrix} \tag{8}$$

Solution Here, the matrix A is $m \times n$ with $m = 3$ and $n = 2$. First, we find that the eigenvalues of the matrix

$$A^T A = \begin{bmatrix} 2 & 1 \\ 1 & 2 \end{bmatrix}$$

arranged in descending order are $\lambda_1 = 3$ and $\lambda_2 = 1$. So there are 2 nonzero eigenvalues of the matrix $A^T A$. Next, we determine that the eigenvectors of the matrix $A^T A$ are $[1, 1]^T$ for $\lambda_1 = 3$ and $[1, -1]^T$ for $\lambda_2 = 1$. Consequently, the orthonormal set of eigenvectors of $A^T A$ are $\left[\frac{1}{2}\sqrt{2}, \frac{1}{2}\sqrt{2}\right]^T$ for $\lambda_1 = 3$ and $\left[\frac{1}{2}\sqrt{2}, -\frac{1}{2}\sqrt{2}\right]^T$ for $\lambda_2 = 1$. Then we arrange them in the same order as the eigenvalues to form the column vectors of the $n \times n$ matrix V:

$$V = \begin{bmatrix} v_1 & v_2 \end{bmatrix} = \begin{bmatrix} \frac{1}{2}\sqrt{2} & \frac{1}{2}\sqrt{2} \\ \frac{1}{2}\sqrt{2} & -\frac{1}{2}\sqrt{2} \end{bmatrix}$$

Now we form the $m \times n$ singular value matrix D by placing $\sigma_1 = \sqrt{3}$ and $\sigma_2 = 1$ on the leading diagonal

$$D = \begin{bmatrix} \sqrt{3} & 0 \\ 0 & \sqrt{1} \\ 0 & 0 \end{bmatrix}$$

Here, on the leading diagonal are the square roots of the eigenvalues of $A^T A$ in descending order, and the rest of the entries of the matrix D are zeros. Next, we compute vectors $u_i = \sigma_i^{-1} A v_i$ for $i = 1$ and 2, which form the column vectors of the $m \times m$ matrix U. In this case, we find

$$u_1 = \sigma_1^{-1} A v_1 = \frac{1}{3}\sqrt{3} \begin{bmatrix} 1 & 1 \\ 0 & 1 \\ 1 & 0 \end{bmatrix} \begin{bmatrix} \frac{1}{2}\sqrt{2} \\ \frac{1}{2}\sqrt{2} \end{bmatrix} = \begin{bmatrix} \frac{1}{3}\sqrt{6} \\ \frac{1}{6}\sqrt{6} \\ \frac{1}{6}\sqrt{6} \end{bmatrix}$$

and

$$u_2 = \sigma_2^{-1} A v_2 = \begin{bmatrix} 1 & 1 \\ 0 & 1 \\ 1 & 0 \end{bmatrix} \begin{bmatrix} \frac{1}{2}\sqrt{2} \\ -\frac{1}{2}\sqrt{2} \end{bmatrix} = \begin{bmatrix} 0 \\ -\frac{1}{2}\sqrt{2} \\ \frac{1}{2}\sqrt{2} \end{bmatrix}$$

Finally, we add to the matrix U the rest of the $m - r$ vectors using the Gram-Schmidt orthogonalization process in Section 9.2 (p. 444). So we make the vector u_3 perpendicular to u_1 and u_2:

$$u_3 = e_1 - \left(u_1^T e_1\right) u_1 - \left(u_1^T e_2\right) u_2 = \begin{bmatrix} \frac{1}{3} \\ -\frac{1}{3} \\ -\frac{1}{3} \end{bmatrix}$$

Normalizing the vector $\boldsymbol{u}_3$, we get

$$\widetilde{\boldsymbol{u}}_3 = \begin{bmatrix} \frac{1}{3}\sqrt{3} \\ -\frac{1}{3}\sqrt{3} \\ -\frac{1}{3}\sqrt{3} \end{bmatrix}$$

So we have the matrix

$$\boldsymbol{U} = \begin{bmatrix} \boldsymbol{u}_1 & \boldsymbol{u}_2 & \widetilde{\boldsymbol{u}}_3 \end{bmatrix} = \begin{bmatrix} \frac{1}{3}\sqrt{6} & 0 & \frac{1}{3}\sqrt{3} \\ \frac{1}{6}\sqrt{6} & \frac{1}{2}\sqrt{2} & -\frac{1}{3}\sqrt{3} \\ \frac{1}{6}\sqrt{6} & -\frac{1}{2}\sqrt{2} & -\frac{1}{3}\sqrt{3} \end{bmatrix}$$

The singular value decomposition of the matrix $\boldsymbol{A}$ is

$$\boldsymbol{A} = \boldsymbol{U}\boldsymbol{D}\boldsymbol{V}^T$$

$$\begin{bmatrix} 1 & 1 \\ 0 & 1 \\ 1 & 0 \end{bmatrix} = \begin{bmatrix} \frac{1}{3}\sqrt{6} & 0 & \frac{1}{3}\sqrt{3} \\ \frac{1}{6}\sqrt{6} & \frac{1}{2}\sqrt{2} & -\frac{1}{3}\sqrt{3} \\ \frac{1}{6}\sqrt{6} & -\frac{1}{2}\sqrt{2} & -\frac{1}{3}\sqrt{3} \end{bmatrix} \begin{bmatrix} \sqrt{3} & 0 \\ 0 & \sqrt{1} \\ 0 & 0 \end{bmatrix} \begin{bmatrix} \frac{1}{2}\sqrt{2} & \frac{1}{2}\sqrt{2} \\ \frac{1}{2}\sqrt{2} & -\frac{1}{2}\sqrt{2} \end{bmatrix}$$

So there we have it! Fortunately, there is mathematical software for doing all of this instantly! We can verify the results by computing the matrix $\boldsymbol{A}$ from the factorization on the right-hand side. ■

See Chapters 9 and 13 for some important applications of the singular value decomposition. Further examples are given there and in the problems of those chapters.

Application: Linear Differential Equations

The application of eigenvalue theory to systems of linear differential equations is briefly explained here. Let us start with a single linear differential equation with one *dependent* variable x. The *independent* variable is t and often represents time. We write $x' = ax$, or in more detail $(d/dt)x(t) = ax(t)$. There is a family of solutions, namely, $x(t) = ce^{at}$, where c is an arbitrary real parameter. If an initial value $x(0)$ is prescribed, we shall need parameter c to get the initial value right.

A pair of **linear differential equations** with two dependent variables, x_1 and x_2, looks like this:

$$\begin{cases} x_1' = a_{11}x_1 + a_{12}x_2 \\ x_2' = a_{21}x_1 + a_{22}x_2 \end{cases} \tag{9}$$

The general form of a system of n linear first-order differential equations, with constant coefficients, is simply

$$\boldsymbol{x}' = \boldsymbol{A}\boldsymbol{x}$$

Here, $\boldsymbol{A}$ is an $n \times n$ numerical matrix, and the vector $\boldsymbol{x}$ has n components, x_j, each being a function of t. Differentiation is with respect to t. To solve this, we are guided by the easy case of $n = 1$, discussed above. Here, we try

$$\boldsymbol{x}(t) = e^{\lambda t}\boldsymbol{v}$$

where $\boldsymbol{v}$ is a constant vector. Taking the derivative of $\boldsymbol{x}$, we have $\boldsymbol{x}' = \lambda e^{\lambda t}\boldsymbol{v}$. Now the system of equations has become $\lambda e^{\lambda t}\boldsymbol{v} = \boldsymbol{A}e^{\lambda t}v$, or $\lambda \boldsymbol{v} = \boldsymbol{A}\boldsymbol{v}$. This is how eigenvalues come into the process. We have proved the following result.

Theorem 7

Linear Differential Equations

If λ is an eigenvalue of the matrix $\boldsymbol{A}$ and if $\boldsymbol{v}$ is an accompanying eigenvector, then one solution of the differential equation $\boldsymbol{x}' = \boldsymbol{A}\boldsymbol{x}$ is $\boldsymbol{x}(t) = e^{\lambda t}\boldsymbol{v}$.

Application: A Vibration Problem

Eigenvalue-eigenvector analysis can be utilized for a variety of differential equations. Consider the system of two masses and three springs shown in Figure 8.2. Here, the masses are constrained to move only in the horizontal direction.

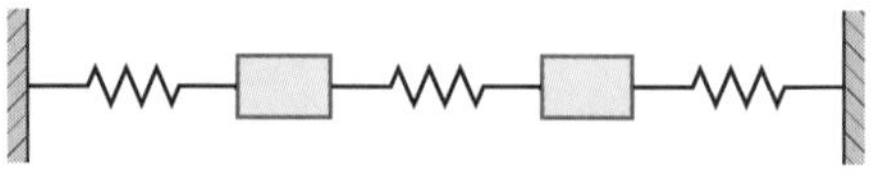

FIGURE 8.2 Two-mass vibration problem

From this situation, we write the equations of motion in matrix-vector form:

$$\begin{bmatrix} x_1'' \\ x_2'' \end{bmatrix} = \begin{bmatrix} -\beta & \alpha \\ \alpha & -\beta \end{bmatrix} \begin{bmatrix} x_1 \\ x_2 \end{bmatrix}, \qquad \boldsymbol{x}'' = \boldsymbol{A}\boldsymbol{x} \tag{10}$$

By assuming that the solution is purely oscillatory (no damping), we have

$$\boldsymbol{x} = \boldsymbol{v}e^{i\omega t}$$

In matrix form, we get

$$\begin{bmatrix} x_1 \\ x_2 \end{bmatrix} = \begin{bmatrix} v_1 \\ v_2 \end{bmatrix} e^{i\omega t}$$

By differentiation, we obtain

$$\boldsymbol{x}'' = -\omega^2 \boldsymbol{v}e^{i\omega t} = -\omega^2 \boldsymbol{x}$$

and

$$\begin{bmatrix} -\beta & \alpha \\ \alpha & -\beta \end{bmatrix} \boldsymbol{x} = -\omega^2 \boldsymbol{x}$$

This is the eigenvalue problem

$$\boldsymbol{A}\boldsymbol{x} = \lambda \boldsymbol{x}$$

where $\lambda = -\omega^2$. Eigenvalues can be found from the characteristic equation:

$$\text{Det}(\boldsymbol{A} + \omega^2 \boldsymbol{I}) = \det \begin{bmatrix} \omega^2 - \beta & \alpha \\ \alpha & \omega^2 - \beta \end{bmatrix} = 0$$

We find

$$(\omega^2 - \beta)^2 - \alpha^2 = \omega^4 - 2\beta\omega^2 + (\beta^2 - \alpha^2) = 0$$

$$\omega^2 = \frac{1}{2}\left[2\beta \pm \sqrt{4\beta^2 - 4(\beta^2 - \alpha^2)}\right] = \beta \pm \alpha$$

For simplicity, we now assume unit masses and unit springs so that $\beta = 2$ and $\alpha = 1$. Then we obtain

$$A = \begin{bmatrix} -2 & 1 \\ 1 & -2 \end{bmatrix}$$

Then the roots of the characteristic equations are $\omega_1^2 = \beta + \alpha = 3$ and $\omega_2^2 = \beta - \alpha = 1$. Next, we can find the eigenvectors. For the first eigenvalue, we obtain

$$(A + \omega_1^2 I)v_1 = 0 \quad \Rightarrow \quad \begin{bmatrix} 1 & 1 \\ 1 & 1 \end{bmatrix} \begin{bmatrix} v_{11} \\ v_{12} \end{bmatrix} = 0$$

Since $v_{11} = -v_{12}$, we obtain the eigenvector

$$v_1 = \begin{bmatrix} 1 \\ -1 \end{bmatrix}$$

For the second eigenvector, we have

$$(A + \omega_2^2 I)v_2 = 0 \quad \Rightarrow \quad \begin{bmatrix} -1 & 1 \\ 1 & -1 \end{bmatrix} \begin{bmatrix} v_{21} \\ v_{22} \end{bmatrix} = 0$$

Since $v_{21} = -v_{22}$, we obtain the eigenvector

$$v_2 = \begin{bmatrix} 1 \\ 1 \end{bmatrix}$$

The general solution for the equations of motion for the two-mass system is

$$x(t) = c_1 v_1 e^{i\omega_1 t} + c_2 v_1 e^{-i\omega_1 t} + c_3 v_2 e^{i\omega_2 t} + c_4 v_2 e^{-i\omega_2 t}$$

Because the solution was for the square of the frequency, each frequency is used twice (one positive and one negative). We can use initial conditions to solve for the unknown coefficients.

Summary 8.2

- An eigenvalue λ and eigenvector x satisfy the equation $Ax = \lambda x$. The direct method to compute the eigenvalues is to find the roots of the characteristic equation $p(\lambda) = \det(A - \lambda I) = 0$. Then, for each eigenvalue λ, the eigenvectors can be found by solving the homogeneous system $(A - \lambda I)x = 0$. There are software packages for finding the eigenvalue-eigenvector pairs using more-sophisticated methods.
- There are many useful properties for matrices that influence their eigenvalues. For example, the eigenvalues are real when A is symmetric or Hermitian. The eigenvalues are positive when A is symmetric or Hermitian positive definite.
- Many eigenvalue procedures involve similarity or unitary transformations to produce triangular or diagonal matrices.
- Gershgorin's discs can be used to localize the eigenvalues by finding coarse estimates of them.
- The **singular value decomposition** of an $m \times n$ matrix A is

$$A = UDV^T$$

where D is an $m \times n$ diagonal matrix whose diagonal entries are the singular values, U is an $m \times m$ orthogonal matrix, and V is an $n \times n$ orthogonal matrix. The singular values of A are the nonnegative square roots of the eigenvalues of $A^T A$.

Exercises 8.2

[a]**1.** Are $[i, -1+i]^T$ and $[-i, -1-i]^T$ eigenvectors of the matrix in Example 2? Show why or why not.

2. Prove that if λ is an eigenvalue of a real matrix with eigenvector $\boldsymbol{x}$, then $\overline{\lambda}$ is also an eigenvalue with eigenvector $\overline{\boldsymbol{x}}$. (For a complex number $z = x + iy$, the conjugate is defined by $\overline{z} = x - iy$.)

[a]**3.** Let

$$A = \begin{bmatrix} \cos\theta & -\sin\theta \\ \sin\theta & \cos\theta \end{bmatrix}$$

Account for the fact that the matrix A has the effect of rotating vectors counterclockwise through an angle θ and thus cannot map any vector into a multiple of itself.

4. Let $\boldsymbol{A}$ be an $m \times n$ matrix such that $\boldsymbol{A} = \boldsymbol{U}\boldsymbol{D}\boldsymbol{V}^T$, where $\boldsymbol{U}$ and $\boldsymbol{V}$ are orthogonal and $\boldsymbol{D}$ is diagonal and nonnegative. Prove that the diagonal elements of $\boldsymbol{D}$ are the singular values of $\boldsymbol{A}$.

5. Let $\boldsymbol{A}$, $\boldsymbol{U}$, $\boldsymbol{D}$, and $\boldsymbol{V}$ be as in the singular value decomposition: $\boldsymbol{A} = \boldsymbol{U}\boldsymbol{D}\boldsymbol{V}^T$. Let r be as described in the text. Define $\boldsymbol{U}_r$ to consist of the first r columns of $\boldsymbol{U}$. Let $\boldsymbol{V}_r$ consist of the first r columns of $\boldsymbol{V}$, and let $\boldsymbol{D}_r$ be the $r \times r$ matrix having the same diagonal as $\boldsymbol{D}$. Prove that $\boldsymbol{A} = \boldsymbol{U}_r\boldsymbol{D}_r\boldsymbol{V}_r^T$. (This factorization is called the **economical version** of the singular value decomposition.)

6. A linear map P is a **projection** if $P^2 = P$. We can use the same terminology for an $n \times n$ matrix: $\boldsymbol{A}^2 = \boldsymbol{A}$ is the projection property. Use the **Pierce decomposition**, $I = A + (I - A)$, to show that every point in $\mathbb{R}^n$ is the sum of a vector in the range of $\boldsymbol{A}$ and a vector in the null space of $\boldsymbol{A}$. What are the eigenvalues of a projection?

7. Find all of the Gershgorin discs for the following matrices. Indicate the smallest region(s) containing all of the eigenvalues:

[a]**a.** $\begin{bmatrix} 3 & -1 & 1 \\ 2 & 4 & -2 \\ 3 & -1 & 9 \end{bmatrix}$ **b.** $\begin{bmatrix} 3 & 1 & 2 \\ -1 & 4 & -1 \\ 1 & -2 & 9 \end{bmatrix}$

[a]**c.** $\begin{bmatrix} 1-i & 1 & i \\ 0 & 2i & 2 \\ 1 & 0 & 2 \end{bmatrix}$

8. (**Multiple Choice**) Let $\boldsymbol{A}$ be an $n \times n$ invertible (nonsingular) matrix. Let $\boldsymbol{x}$ be a nonzero vector. Suppose that $\boldsymbol{A}\boldsymbol{x} = \lambda\boldsymbol{x}$. Which equation does *not* follow from these hypotheses?

a. $\boldsymbol{A}^k\boldsymbol{x} = \lambda^k\boldsymbol{x}$ **b.** $\lambda^{-k}\boldsymbol{x} = (\boldsymbol{A}^{-1})^k\boldsymbol{x}$ for $k \geq 0$

c. $p(\boldsymbol{A})\boldsymbol{x} = p(\lambda)\boldsymbol{x}$ for any polynomial p

d. $\boldsymbol{A}^k\boldsymbol{x} = (1-\lambda)^k\boldsymbol{x}$

[a]**9.** (**Multiple Choice**) For what values of s will the matrix $\boldsymbol{I} - s\boldsymbol{v}\boldsymbol{v}^*$ be unitary, where $\boldsymbol{v}$ is a column vector of unit length?

a. 0,1 **b.** 0,2 **c.** 1,2

d. $0, \sqrt{2}$ **e.** None of these.

10. (**Multiple Choice**) Let $\boldsymbol{U}$ and $\boldsymbol{V}$ be unitary $n \times n$ matrices, possibly complex. Which conclusion is *not* justified?

a. $\boldsymbol{U} + \boldsymbol{V}$ is unitary. **b.** $\boldsymbol{U}^*$ is unitary.

c. $\boldsymbol{U}\boldsymbol{V}$ is unitary.

d. $\boldsymbol{U} - \boldsymbol{v}\boldsymbol{v}^*$ is unitary when $||\boldsymbol{v}|| = \sqrt{2}$ and $\boldsymbol{v}$ is a column vector.

e. None of these.

[a]**11.** (**Multiple Choice**) Which assertion is true?

a. Every $n \times n$ matrix has n distinct (different) eigenvalues.

b. The eigenvalues of a real matrix are real.

c. If $\boldsymbol{U}$ is a unitary matrix, then $\boldsymbol{U}^* = \boldsymbol{U}^T$.

d. A square matrix and its transpose have the same eigenvalues.

e. None of these.

12. (**Multiple Choice**) Consider the symmetric matrix

$$A = \begin{bmatrix} 1 & 3 & 4 & -1 \\ 3 & 7 & -6 & 1 \\ 4 & -6 & 3 & 0 \\ -1 & 1 & 0 & 5 \end{bmatrix}$$

What is the smallest interval derived from Gershgorin's Theorem such that all eigenvalues of the matrix $\boldsymbol{A}$ lie in that interval?

a. $[-7, 9]$ **b.** $[-7, 13]$ **c.** $[3, 7]$

d. $[-3, 17]$ **e.** None of these.

[a]**13.** (**True or False**) Gershgorin's Theorem asserts that every eigenvalue λ of an $n \times n$ matrix $\boldsymbol{A}$ must satisfy one of these inequalities:

$$|\lambda - a_{ii}| \leq \sum_{\substack{j=1 \\ j \neq i}}^{n} |a_{ij}| \quad \text{for} \quad 1 \leq i \leq n.$$

14. (**True or False**) A consequence of Schur's Theorem is that every square matrix $\boldsymbol{A}$ can be factored as $\boldsymbol{A} = \boldsymbol{P}\boldsymbol{T}\boldsymbol{P}^{-1}$, where $\boldsymbol{P}$ is a nonsingular matrix and $\boldsymbol{T}$ is upper triangular.

[a]15. (**True or False**) A consequence of Schur's Theorem is that every (real) symmetric matrix $\boldsymbol{A}$ can be factored in the form $\boldsymbol{A} = \boldsymbol{P}\boldsymbol{D}\boldsymbol{P}^{-1}$, where $\boldsymbol{P}$ is unitary and $\boldsymbol{D}$ is diagonal.

16. Explain why $||\boldsymbol{U}\boldsymbol{B}||_2 = ||\boldsymbol{B}||_2$ for any matrix $\boldsymbol{B}$ when $\boldsymbol{U}^T\boldsymbol{U} = \boldsymbol{I}$.

17. Consider the matrix $\boldsymbol{A} = \begin{bmatrix} 4 & -\frac{1}{2} & 0 \\ \frac{3}{5} & 5 & -\frac{3}{5} \\ 0 & \frac{1}{2} & 3 \end{bmatrix}$.

Plot the Gershgorin discs in the complex plane for $\boldsymbol{A}$ and $\boldsymbol{A}^T$ as well as indicate the locations of the eigenvalues.

18. (Continuation) Let $\boldsymbol{B}$ be the matrix obtained by changing the negative entries in $\boldsymbol{A}$ to positive numbers. Repeat the process for $\boldsymbol{B}$.

19. (Continuation) Repeat for $\boldsymbol{C} = \begin{bmatrix} 4 & 0 & -2 \\ 1 & 2 & 0 \\ 1 & 1 & 9 \end{bmatrix}$.

20. Find the Schur decomposition of

$$\boldsymbol{A} = \begin{bmatrix} 5 & 7 \\ -2 & -4 \end{bmatrix}.$$

Computer Exercises 8.2

1. Use MATLAB, Maple, Mathematica, or other computer programs available to compute the eigenvalues and eigenvectors of these matrices:

   [a]**a.** $\boldsymbol{A} = \begin{bmatrix} 1 & 7 \\ 2 & -5 \end{bmatrix}$

   b. $\begin{bmatrix} 4 & -7 & 3 & 2 & 3 \\ 1 & 6 & 11 & -1 & 2 \\ 5 & -5 & -2 & -4 & 1 \\ 9 & -3 & 1 & 6 & 5 \\ 3 & 2 & 5 & -5 & 1 \end{bmatrix}$

   c. Let $n = 12$, $a_{ij} = i/j$ when $i \leqq j$, and $a_{ij} = j/i$ when $i > j$. Find the eigenvalues.

   [a]**d.** Create an $n \times n$ matrix with a tridiagonal structure and nonzero elements $(-1, 2, -1)$ in each row. For $n = 5$ and 20, find all of the eigenvalues, and verify that they are $2 - 2\cos(j\pi/(n+1))$.

   e. For any positive integer n, form the symmetric matrix $\boldsymbol{A}$ whose upper triangular part is given by

$$\begin{bmatrix} n & n-1 & n-2 & n-3 & \cdots & 2 & 1 \\ & n-1 & n-2 & n-3 & \cdots & 2 & 1 \\ & & n-2 & n-3 & \cdots & 2 & 1 \\ & & & \ddots & \cdots & \vdots & \vdots \\ & & & & \ddots & 2 & 1 \\ & & & & & 2 & 1 \\ & & & & & & 1 \end{bmatrix}$$

   The eigenvalues of $\boldsymbol{A}$ are $1/\{2-2\cos[(2i-1)\pi/(2n+1)]\}$. (See Frank [1958] and Gregory and Karney [1969].) Numerically verify this result for $n = 30$.

2. Use MATLAB to compute the eigenvalues of a random 100×100 matrix by direct use of the command `eig` and by use of the commands `poly` and `roots`. Use the timing functions to determine the CPU time for each.

3. Let p be the polynomial of degree 20 whose roots are the integers $1, 2, \dots, 20$. Find the usual power form of this polynomial so that

$$p(t) = t^{20} + a_{19}t^{19} + a_{18}t^{18} + \cdots + a_0$$

   Next, form the so-called **companion matrix**, which is 20×20 and has zeros in all positions except all 1's on the superdiagonal and the coefficients $-a_0, -a_1, \dots, -a_{19}$ as its bottom row. Find the eigenvalues of this matrix, and account for any difficulties encountered.

4. (**Student Research Project**) Investigate some modern methods for computing eigenvalues and eigenvectors. For the symmetric case, see the book by Parlett [1997]. Also, read the LAPACK User's Guide. (See Anderson, et al. [1999].)

5. (**Student Research Project**) Experiment with the Cayley-Hamilton Theorem, which asserts that every square matrix satisfies its own characteristic equation. Check this numerically by using MATLAB or some other mathematical software system. Use matrices of size 3, 6, 9, 12, and 15, and account for any surprises. If you can use higher-precision arithmetic do so—MATLAB works with 15 digits of precision.

6. (**Student Research Project**) Experiment with the QR algorithm and the singular value decomposition of matrices—for example, using MATLAB. Try examples with four types of equations $\boldsymbol{A}\boldsymbol{x} = \boldsymbol{b}$—namely, (a) the

system has a unique solution; (b) the system has many solutions; (c) the system is inconsistent but has a unique least-squares solution; (d) the system is inconsistent and has many least-squares solutions.

7. Using mathematical software such as MATLAB, Maple, or Mathematica on each of the following matrices, compute the eigenvalues via the characteristic polynomial, compute the eigenvectors via the null space of the matrix, and compute the eigenvalues and eigenvectors directly:

[a]a. $\begin{bmatrix} 3 & 2 \\ 7 & -1 \end{bmatrix}$ [a]b. $\begin{bmatrix} 1 & 3 & -7 \\ -3 & 4 & 1 \\ 2 & -5 & 3 \end{bmatrix}$

8. Using mathematical software such as MATLAB, Maple, or Mathematica, determine the execution time for computing all eigenvalues of a 1000×1000 matrix with random entries.

9. Using mathematical software such as MATLAB, Maple, or Mathematica, compute the Schur factorization of these complex matrices, and verify the results according to Schur's Theorem and its corollaries:

a. $\begin{bmatrix} 3-i & 2-i \\ 2+i & 3+i \end{bmatrix}$ b. $\begin{bmatrix} 2+i & 3+i \\ 3-i & 2-i \end{bmatrix}$

c. $\begin{bmatrix} 2-i & 2+i \\ 3-i & 3+i \end{bmatrix}$

10. Using mathematical software such as MATLAB, Maple, or Mathematica, compute the singular value decomposition of these matrices, and verify that each result satisfies the equation $A = UDV^T$:

a. $\begin{bmatrix} 1 & 1 \\ 0 & 1 \\ 1 & 0 \end{bmatrix}$ b. $\begin{bmatrix} 1 & 3 & -2 \\ 2 & 7 & 5 \\ -2 & -3 & 4 \\ 5 & -3 & -2 \end{bmatrix}$

Create the diagonal matrix $D = U^T AV$ to check the results (always recommended). One can see the effects of roundoff errors in these calculations, for the off-diagonal elements in D are theoretically zero.

[a]11. Consider $A = \begin{bmatrix} 5 & 4 & 1 & 1 \\ 4 & 5 & 1 & 1 \\ 1 & 1 & 4 & 2 \\ 1 & 1 & 2 & 4 \end{bmatrix}$

Find the eigenvalues and accompanying eigenvectors of this matrix, from Gregory and Karney [1969], without using software.
Hint: The answers can be integers.

12. Find the singular value decomposition of these matrices:

a. $\begin{bmatrix} 2 & 1 & -2 \end{bmatrix}$ b. $\begin{bmatrix} 3 \\ 4 \end{bmatrix}$

c. $\begin{bmatrix} -\frac{5}{2} + 3\sqrt{3} & \frac{5}{2}\sqrt{3} + 3 \end{bmatrix}$

d. $\begin{bmatrix} 2 & 2 & 2 & 2 \\ \frac{17}{10} & \frac{1}{10} & -\frac{17}{10} & -\frac{1}{10} \\ \frac{3}{5} & \frac{9}{5} & -\frac{3}{5} & -\frac{9}{5} \end{bmatrix}$

e. $\begin{bmatrix} \frac{7}{2} - \frac{13}{6}\sqrt{6} & \frac{7}{2} + \frac{13}{6}\sqrt{6} \\ -\frac{7}{2} - \frac{13}{6}\sqrt{6} & -\frac{7}{2} + \frac{13}{6}\sqrt{6} \\ -\frac{13}{6}\sqrt{6} & \frac{13}{6}\sqrt{6} \end{bmatrix}$

13. Consider $B = \begin{bmatrix} -149 & -50 & -154 \\ 537 & 180 & 546 \\ -27 & -9 & -25 \end{bmatrix}$.

Find the eigenvalues, singular values, and condition number of the matrix B.

8.3 Power Method

Mathematical Derivation

A procedure called the **power method** can be employed to compute eigenvalues of a given matrix. It is an example of an iterative process that, under the right circumstances, produces a sequence converging to an eigenvalue of a given matrix.

Suppose that A is an $n \times n$ matrix, and that its eigenvalues (which we do not know) have the following property:

Eigenvalues

$$|\lambda_1| > |\lambda_2| \geqq |\lambda_3| \geqq \cdots \geqq |\lambda_n|$$

Notice the strict inequality in this hypothesis. Except for that, we are simply ordering the eigenvalues according to decreasing absolute value. (This is only a matter of notation.) Each eigenvalue has a nonzero eigenvector $\boldsymbol{u}^{(i)}$ and

Eigenvectors

$$\boldsymbol{A}\boldsymbol{u}^{(i)} = \lambda_i \boldsymbol{u}^{(i)} \qquad (i = 1, 2, \ldots, n) \tag{1}$$

We assume that there is a linearly independent set of n eigenvectors $\{\boldsymbol{u}^{(1)}, \boldsymbol{u}^{(2)}, \ldots, \boldsymbol{u}^{(n)}\}$. It is necessarily a basis for $\mathbb{C}^n$.

We want to compute the *single* eigenvalue of maximum modulus (the **dominant eigenvalue**) and an associated eigenvector. We select an arbitrary starting vector, $\boldsymbol{x}^{(0)} \in \mathbb{C}^n$, and express it as a linear combination of the eigenvectors $\boldsymbol{u}^{(1)}, \boldsymbol{u}^{(2)}, \ldots, \boldsymbol{u}^{(n)}$:

Linear Combination

$$\boldsymbol{x}^{(0)} = c_1\boldsymbol{u}^{(1)} + c_2\boldsymbol{u}^{(2)} + \cdots + c_n\boldsymbol{u}^{(n)}$$

In this equation, we must assume that $c_1 \neq 0$. Since the coefficients can be absorbed into the vectors $\boldsymbol{u}^{(i)}$, there is no loss of generality in assuming that

$$\boldsymbol{x}^{(0)} = \boldsymbol{u}^{(1)} + \boldsymbol{u}^{(2)} + \cdots + \boldsymbol{u}^{(n)} \tag{2}$$

Then we repeatedly carry out matrix-vector multiplication, using the matrix $\boldsymbol{A}$ to produce a sequence of vectors. Specifically, we have

Sequence of Vectors

$$\begin{cases} \boldsymbol{x}^{(1)} = \boldsymbol{A}\boldsymbol{x}^{(0)} \\ \boldsymbol{x}^{(2)} = \boldsymbol{A}\boldsymbol{x}^{(1)} = \boldsymbol{A}^2\boldsymbol{x}^{(0)} \\ \boldsymbol{x}^{(3)} = \boldsymbol{A}\boldsymbol{x}^{(2)} = \boldsymbol{A}^3\boldsymbol{x}^{(0)} \\ \quad\vdots \\ \boldsymbol{x}^{(k)} = \boldsymbol{A}\boldsymbol{x}^{(k-1)} = \boldsymbol{A}^k\boldsymbol{x}^{(0)} \\ \quad\vdots \end{cases}$$

In general, we have

$$\boldsymbol{x}^{(k)} = \boldsymbol{A}^k\boldsymbol{x}^{(0)} \qquad (k = 1, 2, 3, \ldots)$$

Substituting $\boldsymbol{x}^{(0)}$ in Equation (2), we obtain

$$\begin{aligned} \boldsymbol{x}^{(k)} &= \boldsymbol{A}^k\boldsymbol{x}^{(0)} \\ &= \boldsymbol{A}^k\boldsymbol{u}^{(1)} + \boldsymbol{A}^k\boldsymbol{u}^{(2)} + \boldsymbol{A}^k\boldsymbol{u}^{(3)} + \cdots + \boldsymbol{A}^k\boldsymbol{u}^{(n)} \\ &= \lambda_1^k\boldsymbol{u}^{(1)} + \lambda_2^k\boldsymbol{u}^{(2)} + \lambda_3^k\boldsymbol{u}^{(3)} + \cdots + \lambda_n^k\boldsymbol{u}^{(n)} \end{aligned}$$

by using Equation (1). This can be written in the form

$$\boldsymbol{x}^{(k)} = \lambda_1^k\left[\boldsymbol{u}^{(1)} + \left(\frac{\lambda_2}{\lambda_1}\right)^k\boldsymbol{u}^{(2)} + \left(\frac{\lambda_3}{\lambda_1}\right)^k\boldsymbol{u}^{(3)} + \cdots + \left(\frac{\lambda_n}{\lambda_1}\right)^k\boldsymbol{u}^{(n)}\right]$$

Since $|\lambda_1| > |\lambda_j|$ for $j > 1$, we have $|\lambda_j/\lambda_1| < 1$ and $(\lambda_j/\lambda_1)^k \to 0$ as $k \to \infty$. To simplify the notation, we write the above equation in the form

$$\boldsymbol{x}^{(k)} = \lambda_1^k\left[\boldsymbol{u}^{(1)} + \boldsymbol{\varepsilon}^{(k)}\right] \tag{3}$$

where $\boldsymbol{\varepsilon}^{(k)} \to 0$ as $k \to \infty$. We let φ be any complex-valued *linear functional* on $\mathbb{C}^n$ such that $\varphi(\boldsymbol{u}^{(1)}) \neq 0$. Recall that φ is a **linear functional** if $\varphi(a\boldsymbol{x} + b\boldsymbol{y}) = a\varphi(\boldsymbol{x}) + b\varphi(\boldsymbol{y})$ for scalars a and b and vectors $\boldsymbol{x}$ and $\boldsymbol{y}$. For example, $\varphi(\boldsymbol{x}) = x_j$ for some fixed j $(1 \leqq j \leqq n)$, is a linear functional. Now, looking back at Equation (3), we apply φ to it:

Linear Functional

$$\varphi\left(\boldsymbol{x}^{(k)}\right) = \lambda_1^k\left[\varphi\left(\boldsymbol{u}^{(1)}\right) + \varphi\left(\boldsymbol{\varepsilon}^{(k)}\right)\right]$$

Next, we form ratios $r_1, r_2, \ldots$ as follows:

Ratio Converges to Dominant Eigenvalue

$$r_k \equiv \frac{\varphi\left(\boldsymbol{x}^{(k+1)}\right)}{\varphi\left(\boldsymbol{x}^{(k)}\right)} = \lambda_1 \left[\frac{\varphi\left(\boldsymbol{u}^{(1)}\right) + \varphi\left(\boldsymbol{\varepsilon}^{(k+1)}\right)}{\varphi\left(\boldsymbol{u}^{(1)}\right) + \varphi\left(\boldsymbol{\varepsilon}^{(k)}\right)}\right] \to \lambda_1 \qquad \text{as} \quad k \to \infty$$

Hence, we are able to compute the dominant eigenvalue λ_1 as the limit of the sequence $\{r_k\}$. With a little more care, we can get an accompanying eigenvector. In the definition of the vectors $\boldsymbol{x}^{(k)}$ in Equation (2), we see nothing to prevent the vectors from growing or converging to zero. Normalization cures this problem, as in one of the following pseudocodes.

Power Method Pseudocode

Here we present pseudocode for calculating the dominant eigenvalue and an associated eigenvector for a prescribed matrix $\boldsymbol{A}$. In each algorithm, φ is a linear functional chosen by the user. For example, one can use $\varphi(\boldsymbol{x}) = x_1$ (the first component of the vector).

Power Method Pseudocode

Power Method Pseudocode

```
integer k, kmax, n;    real r
real array (A)1:n×1:n, (x)1:n, (y)1:n
external function φ
output 0, x
for k = 1 to kmax
    y ← Ax
    r ← φ(y)/φ(x)
    x ← y
    output k, x, r
end do
```

We use a simple 2×2 matrix such as

$$\boldsymbol{A} = \begin{bmatrix} 3 & 1 \\ 1 & 3 \end{bmatrix}$$

to give a geometric illustration of the power method as shown in Figure 8.3. Clearly, the eigenvalues are $\lambda_1 = 2$ and $\lambda_2 = 4$ with eigenvectors $\boldsymbol{v}^{(1)} = [-1, 1]^T$ and $\boldsymbol{v}^{(2)} = [1, 1]^T$, respectively. Starting with $\boldsymbol{x}^{(0)} = [0, 1]^T$, the power method repeatedly multiplies the matrix $\boldsymbol{A}$ by a vector. It produces a sequence of vectors $\boldsymbol{x}^{(1)}, \boldsymbol{x}^{(2)}$, and so on that move in the direction of the eigenvector $\boldsymbol{v}^{(2)}$, which corresponds to the dominant eigenvalue $\lambda_2 = 4$.

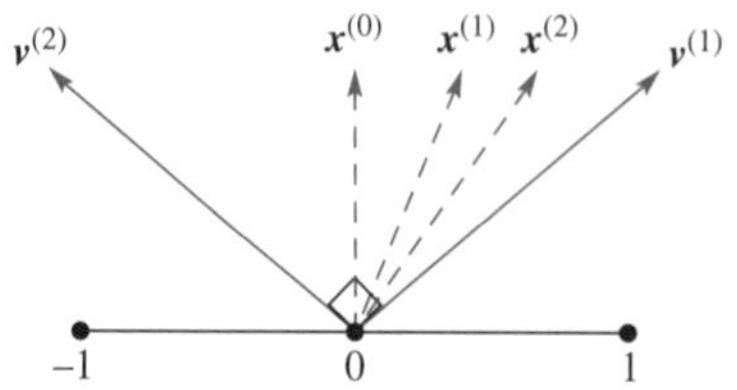

FIGURE 8.3
In 2D, power method illustration

We can easily modify this algorithm to produce normalized eigenvectors by using the infinity vector norm $||\boldsymbol{x}||_\infty = \max_{1 \leqq j \leqq n} |x_j|$, as in the following code:

Modified Power Method Algorithm with Normalization

Normalized Power Method Pseudocode

```
integer k, kmax, n;   real r
real array (A)_{1:n×1:n}, (x)_{1:n}, (y)_{1:n}
external function φ
output 0, x
for k = 1 to kmax
    y ← Ax
    r ← φ(y)/φ(x)
    x ← y/||y||_∞
    output k, x, r
end do
```

Aitken Acceleration

From a given sequence $\{r_k\}$, we can construct another sequence $\{s_k\}$ by means of the **Aitken acceleration** formula

Aitken Acceleration

$$s_k = r_k - \frac{(r_k - r_{k-1})^2}{r_k - 2r_{k-1} + r_{k-2}} \qquad (k \geqq 3)$$

If the original sequence $\{r_k\}$ converges to r and if certain other conditions are satisfied, then the new sequence $\{s_k\}$ converges to r more rapidly than the original one. (For details, see Kincaid and Cheney [2002].) Because subtractive cancellation may eventually spoil the results, the Aitken acceleration process should be stopped soon after the values become apparently stationary.

EXAMPLE 1 Use the modified power method algorithm and Aitken acceleration to find the dominant eigenvalue and an eigenvector of the given matrix $\boldsymbol{A}$, with vector $\boldsymbol{x}^{(0)}$ and $\varphi(\boldsymbol{x})$ given as follows:

$$\boldsymbol{A} = \begin{bmatrix} 6 & 5 & -5 \\ 2 & 6 & -2 \\ 2 & 5 & -1 \end{bmatrix}, \qquad \boldsymbol{x}^{(0)} = \begin{bmatrix} -1 \\ 1 \\ 1 \end{bmatrix}, \qquad \varphi(\boldsymbol{x}) = x_2$$

Solution After coding and running the modified power method algorithm with Aitken acceleration, we obtain the following results:

Computer Results

$$\begin{array}{lll}
\boldsymbol{x}^{(0)} = [-1.0000, 1.0000, 1.0000]^T & & \\
\boldsymbol{x}^{(1)} = [-1.0000, 0.3333, 0.3333]^T & r_0 = 2.0000 & \\
\boldsymbol{x}^{(2)} = [-1.0000, -0.1111, -0.1111]^T & r_1 = -2.0000 & \\
\boldsymbol{x}^{(3)} = [-1.0000, -0.4074, -0.4074]^T & r_2 = 22.0000 & \\
\boldsymbol{x}^{(4)} = [-1.0000, -0.6049, -0.6049]^T & r_3 = 8.9091 & s_3 = 13.5294 \\
\boldsymbol{x}^{(5)} = [-1.0000, -0.7366, -0.7366]^T & r_4 = 7.3061 & s_4 = 7.0825 \\
\boldsymbol{x}^{(6)} = [-1.0000, -0.8244, -0.8244]^T & r_5 = 6.7151 & s_5 = 6.3699 \\
\vdots & \vdots & \vdots \\
\boldsymbol{x}^{(14)} = [-1.0000, -0.9931, -0.9931]^T & r_{13} = 6.0208 & s_{13} = 6.0005
\end{array}$$

The Aitken-accelerated sequence, s_k, converges noticeably faster than the sequence $\{r_k\}$. The actual dominant eigenvalue and an associated eigenvector are

$$\lambda_1 = 6, \qquad \boldsymbol{u}^{(1)} = [1, 1, 1]^T$$

The coding of the modified power method is very simple, and we leave the actual implementation as an exercise. We also use the simple *infinity-norm* for normalizing the vectors. The final vectors and estimates of the eigenvalue are displayed with 15 decimals digits. ■

In such a problem, one should always seek an independent verification of the purported answer. Here, we simply compute $\boldsymbol{Ax}$ to see whether it coincides with $s_{14}\boldsymbol{x}$. The last few commands in the code are doing this rough checking, taking s_{14} as probably the best estimate of the eigenvalue and the last $\boldsymbol{x}$-vector as the best estimate of an eigenvector. The results after 14 steps are *not* very accurate. For better accuracy, take 80 steps!

Inverse Power Method

It is possible to compute other eigenvalues of a matrix by using modifications of the power method. For example, if $\boldsymbol{A}$ is invertible, we can compute its eigenvalue of smallest magnitude by noting this logical equivalence:

$$\boldsymbol{Ax} = \lambda \boldsymbol{x} \iff \boldsymbol{x} = \boldsymbol{A}^{-1}(\lambda \boldsymbol{x}) \iff \boldsymbol{A}^{-1}\boldsymbol{x} = \frac{1}{\lambda}\boldsymbol{x}$$

Thus, the smallest eigenvalue of $\boldsymbol{A}$ in magnitude is the reciprocal of the largest eigenvalue of $\boldsymbol{A}^{-1}$. We compute it by applying the power method to $\boldsymbol{A}^{-1}$ and taking the reciprocal of the result.

Computing Smallest Eigenvalue by Inverse Power Method

Suppose that there is a single smallest eigenvalue of $\boldsymbol{A}$, which is λ_n with our usual ordering:

$$|\lambda_1| \geqq |\lambda_2| \geqq |\lambda_3| \geqq \cdots \geqq |\lambda_{n-1}| > |\lambda_n| > 0$$

It follows that $\boldsymbol{A}$ is invertible. (Why?) The eigenvalues of $\boldsymbol{A}^{-1}$ are λ_j^{-1} for $1 \leqq j \leqq n$. Therefore, we have

$$|\lambda_n^{-1}| > |\lambda_{n-1}^{-1}| \geqq \cdots \geqq |\lambda_1^{-1}| > 0$$

We can use the power method on the matrix $\boldsymbol{A}^{-1}$ to compute its dominant eigenvalue λ_n^{-1}. The reciprocal of this is the eigenvalue of $\boldsymbol{A}$ that we sought. Notice that we need not compute $\boldsymbol{A}^{-1}$ because the equation

$$\boldsymbol{x}^{(k+1)} = \boldsymbol{A}^{-1}\boldsymbol{x}^{(k)}$$

is equivalent to the equation

$$\boldsymbol{A}\boldsymbol{x}^{(k+1)} = \boldsymbol{x}^{(k)}$$

so the vector $\boldsymbol{x}^{(k+1)}$ can be more easily computed by solving this last linear system. To do this, we first find the $\boldsymbol{LU}$ factorization of $\boldsymbol{A}$, namely, $\boldsymbol{A} = \boldsymbol{LU}$. Then we repeatedly update the right-hand side and back solve:

$$\boldsymbol{U}\boldsymbol{x}^{(k+1)} = \boldsymbol{L}^{-1}\boldsymbol{x}^{(k)}$$

to obtain $\boldsymbol{x}^{(1)}, \boldsymbol{x}^{(2)}, \ldots$.

EXAMPLE 2 Compute the smallest eigenvalue and an associated eigenvector of this matrix

$$\boldsymbol{A} = \frac{1}{3}\begin{bmatrix} -154 & 528 & 407 \\ 55 & -144 & -121 \\ -132 & 396 & 318 \end{bmatrix}$$

using the following initial vector and linear function:

$$\boldsymbol{x}^{(0)} = [1, 2, 3]^T, \qquad \varphi(\boldsymbol{x}) = x_2$$

Solution We decide to take the easy route and use the inverse of **A** for producing the successive $\boldsymbol{x}$ vectors. We leave the actual implementation as an exercise. The ratios r_k are saved, and once they are complete, the Aitken accelerated values, s_k, are computed. Notice that at the end, we want the reciprocal of the limiting ratio. Hence, it is easier to use reciprocals at every step in the code. Thus, you see $r_k = x_2/y_2$ rather than y_2/x_2, and these ratios should converge to the smallest eigenvalue of $\boldsymbol{A}$. The final results after 80 steps are these:

$$\boldsymbol{x} = [0.26726101285547, -0.53452256017715, 0.80178375118802]^T$$
$$s_{80} = 3.33333333343344$$

We can divide each entry in $\boldsymbol{x}$ by the first component and arrive at

$$\boldsymbol{x} = [1.0, -2.00000199979120, 3.00000266638827]^T$$

The eigenvalue is actually $\frac{10}{3}$, and the eigenvector is $[1, -2, 3]^T$. The discrepancy between $\boldsymbol{Ax}$ and $s_{80}\boldsymbol{x}$ is about 2.6×10^{-6}. ■

Example: Inverse Power Method

Using mathematical software on a small example,

$$\boldsymbol{A} = \begin{bmatrix} 6 & 5 & -5 \\ 2 & 6 & 2 \\ 2 & 5 & -1 \end{bmatrix} \tag{4}$$

we can first get $\boldsymbol{A}^{-1}$ and then use the power method. (We have changed one entry in the matrix $\boldsymbol{A}$ from Example 1 to solve a different problem.) We leave the implementation of the code as an exercise. In the code, r is the reciprocal of the quantity r in the original power method. Thus, at the end of the computation, r should be the eigenvalue of $\boldsymbol{A}$ that has the smallest absolute value. After the prescribed 30 steps, we find that $r = 0.214$ and $\boldsymbol{x} = [0.7916, 0.5137, 0.3308]^T$. As usual, we can verify the result independently by computing $\boldsymbol{Ax}$ and $r\boldsymbol{x}$, which should be equal. The method just illustrated is called the **inverse power method**. On larger examples, the successive vectors should be computed not via $\boldsymbol{A}^{-1}$ but rather by solving the equation $\boldsymbol{Ay} = \boldsymbol{x}$ for $\boldsymbol{y}$. In mathematical software systems such as MATLAB, Maple, and Mathematica, this can be done with a single command. Alternatively, one can get the $\boldsymbol{LU}$ factorization of $\boldsymbol{A}$ and solve $\boldsymbol{Lz} = \boldsymbol{x}$ and $\boldsymbol{Uy} = \boldsymbol{z}$.

In this example, two eigenvalues are complex. Since the matrix is real, they must be conjugate pairs of the form $\alpha + \beta i$ and $\alpha - \beta i$. They have the same magnitude; thus, the hypothesis $|\lambda_1| > |\lambda_2|$ needed in the convergence proof of the power method is violated. *What happens when the power method is applied to* $\boldsymbol{A}$? The values of r for $k = 26$ to 30 are 0.76, −53.27, 8.86, 2.69, and −9.42. We leave the implementation of the code as a computer exercise.

Shifted (Inverse) Power Method

Other eigenvalues of a matrix (besides the largest and smallest) can be computed by exploiting the following logical equivalences:

$$\boldsymbol{Ax} = \lambda\boldsymbol{x} \iff (\boldsymbol{A} - \mu\boldsymbol{I})\boldsymbol{x} = (\lambda - \mu)\boldsymbol{x} \iff (\boldsymbol{A} - \mu\boldsymbol{I})^{-1}\boldsymbol{x} = \frac{1}{\lambda - \mu}\boldsymbol{x}$$

If we want to compute an eigenvalue of $\boldsymbol{A}$ that is close to a given number μ, we can apply the inverse power method to $\boldsymbol{A} - \mu \boldsymbol{I}$ and take the reciprocal of the limiting value of r. This should be $\lambda - \mu$.

We can also compute an eigenvalue of $\boldsymbol{A}$ that is *farthest* from a given number μ. Suppose that for some eigenvalue λ_j of matrix $\boldsymbol{A}$, we have

$$|\lambda_j - \mu| > \varepsilon \qquad \text{and} \qquad 0 < |\lambda_i - \mu| < \varepsilon \qquad \text{for all } i \neq j$$

Consider the *shifted* matrix $\boldsymbol{A} - \mu \boldsymbol{I}$. Applying the power method to the shifted matrix $\boldsymbol{A} - \mu \boldsymbol{I}$, we compute ratios r_k that converge to $\lambda_j - \mu$. This procedure is called the **shifted power method**.

If we want to compute the eigenvalue of $\boldsymbol{A}$ that is *closest* to a given number μ, a variant of the procedure is needed. Suppose that λ_j is an eigenvalue of $\boldsymbol{A}$ such that

$$0 < |\lambda_j - \mu| < \varepsilon \qquad \text{and} \qquad |\lambda_i - \mu| > \varepsilon \qquad \text{for all } i \neq j$$

Consider the *shifted* matrix $\boldsymbol{A} - \mu \boldsymbol{I}$. The eigenvalues of this matrix are $\lambda_i - \mu$. Applying the inverse power method to $\boldsymbol{A} - \mu \boldsymbol{I}$ gives an approximate value for $(\lambda_j - \mu)^{-1}$. We can use the explicit inverse of $\boldsymbol{A} - \mu \boldsymbol{I}$ or the $\boldsymbol{LU}$ factorization $\boldsymbol{A} - \mu \boldsymbol{I} = \boldsymbol{LU}$. Now we repeatedly solve the equations

$$(\boldsymbol{A} - \mu \boldsymbol{I})\boldsymbol{x}^{(k+1)} = \boldsymbol{x}^{(k)}$$

by solving instead $\boldsymbol{U}\boldsymbol{x}^{(k+1)} = \boldsymbol{L}^{-1}\boldsymbol{x}^{(k)}$. Since the ratios r_k converge to $(\lambda_j - \mu)^{-1}$, we have

$$\lambda_j = \mu + \left(\lim_{k\to\infty} r_k\right)^{-1} = \mu + \lim_{k\to\infty} \frac{1}{r_k}$$

This algorithm is called the **shifted inverse power method**.

Example: Shifted Inverse Power Method

To illustrate the shifted inverse power method, we consider this matrix

$$\boldsymbol{A} = \begin{bmatrix} 1 & 3 & 7 \\ 2 & -4 & 5 \\ 3 & 4 & -6 \end{bmatrix} \tag{5}$$

and we write mathematical software to compute the eigenvalue closest to -6. The code we use takes ratios of y_2/x_2, and we are therefore expecting convergence of these ratios to $\lambda + 6$. After eight steps, we have $r = 0.9590$ and $\boldsymbol{x} = [-0.7081, 0.6145, 0.3478]^T$. Hence, the eigenvalue should be $\lambda = 0.9590 - 6 = -5.0410$. We can ask MATLAB to confirm the eigenvalue and eigenvector by computing both $\boldsymbol{Ax}$ and $\lambda\boldsymbol{x}$ to be approximately $[3.57, -3.10, -1.75]^T$.

Summary 8.3

- We have considered the following methods for computing eigenvalues of a matrix. In the **power method**, we approximate the largest eigenvalue λ_1 by generating a sequence of points using the formula

$$\boldsymbol{x}^{(k+1)} = \boldsymbol{A}\boldsymbol{x}^{(k)}$$

and then forming a sequence $r_k = \varphi(\boldsymbol{x}^{(k+1)})/\varphi(\boldsymbol{x}^{(k)})$, where φ is a linear functional. Under the right circumstances, this sequence, r_k, will converge to the largest eigenvalue of $\boldsymbol{A}$.

- In the **inverse power method**, we find the smallest eigenvalue λ_n by using the preceding process on the inverse of the matrix. The reciprocal of the largest eigenvalue of A^{-1} is the smallest eigenvalue of A. We can also describe this process as one of computing the sequence so that

$$Ax^{(k+1)} = x^{(k)}$$

- In the **shifted power method**, we find the eigenvalue that is farthest from a given number μ by seeking the largest eigenvalue of $A - \mu I$. This involves an iteration to produce a sequence

$$x^{(k+1)} = (A - \mu I)x^{(k)}$$

- In the **shifted inverse power method**, we find the eigenvalue that is closest to μ by applying the inverse power method to $A - \mu I$. This requires solving the equation

$$(A - \mu I)x^{(k+1)} = x^{(k)} \qquad (A - \mu I = LU)$$

Exercises 8.3

[a]1. (**Multiple Choice**) Let $A = \begin{bmatrix} 5 & 2 \\ 4 & 7 \end{bmatrix}$.

The power method has been applied to the matrix A. The result is a long list of vectors that seem to settle down to a vector of the form $[h, 1]^T$, where $|h| < 1$. What is the largest eigenvalue, approximately, in terms of that number h?

a. $4h + 7$ **b.** $5h + 2$ **c.** $1/h$

d. $5h + 4$ **e.** None of these.

[a]2. (**Multiple Choice**) What is the expected final output of the following pseudocode? Here y_1 and x_1 are the first components of y and x, respectively.

```
integer n, kmax;   real r
real array (A^-1)1:n×1:n, (x)1:n, (y)1:n
for k = 1 to 30
      y ← A^-1 x
      r ← y1/x1
      x ← y/||y||
      output r, x
end do
```

a. r is the eigenvalue of A largest in magnitude, and x is an accompanying eigenvector.

b. $r = 1/\lambda$, where λ is the smallest eigenvalue of A, and x is such that $Ax = \lambda x$.

c. A vector x such that $Ax = rx$, where r is the eigenvalue of A having the smallest magnitude.

d. r is the largest (in magnitude) eigenvalue of A and x is a corresponding eigenvector of A.

e. None of these.

3. Briefly describe how to compute the following:

   a. The dominant eigenvalue and associate eigenvector.

   b. The next dominant eigenvalue and associated eigenvector.

   c. The least dominant eigenvalue and associated eigenvector.

   d. An eigenvalue other than the dominant or least dominant eigenvalue and associated eigenvectors.

4. Let $A = \begin{bmatrix} 2 & -1 & 0 \\ -1 & 2 & -1 \\ 0 & -1 & 2 \end{bmatrix}$

   Carry out several iterations of the power method, starting with $x^{(0)} = (1, 1, 1)$. What is the purpose of this procedure?

5. Let $B = A - 4I = \begin{bmatrix} -2 & -1 & 0 \\ -1 & -2 & -1 \\ 0 & -1 & -2 \end{bmatrix}$.

   Carry out some iterations of the power method applied to B, starting with $x^{(0)} = (1, 1, 1)$. What is the purpose of this procedure?

6. Let $C = A^{-1} = \frac{1}{4}\begin{bmatrix} 3 & 2 & 1 \\ 2 & 4 & 2 \\ 1 & 2 & 3 \end{bmatrix}$.

   Carry out a few iterations of the power method applied to C, starting with $x^{(0)} = (1, 1, 1)$. What is the purpose of this procedure?

7. The **Rayleigh quotient** is the expression $\langle x, x\rangle_A / \langle x, x\rangle = x^T Ax / x^T x$. How can the Rayleigh quotient be used when $Ax = \lambda x$?

Computer Exercises 8.3

1. Use the power method, the inverse power method, and their shifted forms as well as Aitken's acceleration to find some or all of the eigenvalues of the following matrices:

a. $\begin{bmatrix} 5 & 4 & 1 & 1 \\ 4 & 5 & 1 & 1 \\ 1 & 1 & 4 & 2 \\ 1 & 1 & 2 & 4 \end{bmatrix}$ **b.** $\begin{bmatrix} 2 & 3 & 4 \\ 7 & -1 & 3 \\ 1 & -1 & 5 \end{bmatrix}$

c. $\begin{bmatrix} -2 & 1 & 0 & 0 & 0 \\ 1 & -2 & 1 & 0 & 0 \\ 0 & 1 & -2 & 1 & 0 \\ 0 & 0 & 1 & -2 & 1 \\ 0 & 0 & 0 & 1 & -2 \end{bmatrix}$

2. Redo the examples in this section, using either MATLAB, Maple, or Mathematica.

3. Modify and test the pseudocode for the power method to normalize the vector so that the largest component is always 1 in the infinity-norm. This procedure gives the eigenvector and eigenvalue without having to compute a linear functional.

[a]4. Let $A = \begin{bmatrix} -57 & 192 & 148 \\ 20 & -53 & -44 \\ -48 & 144 & 115 \end{bmatrix}$

Find the eigenvalues that are close to −4, 2, and 8 by using the inverse power method.

5. Using mathematical software such as MATLAB, Maple, or Mathematica, write and execute code for implementing the methods in Section 8.3. Verify that the results are consistent with those described in the text.

a. Example 1 using the modified power method.

b. Example 2 using the inverse power method with Aitken acceleration.

c. Matrix (4) using the inverse power method.

d. Matrix (5) using the shifted power method.

6. Consider the matrix $A = \begin{bmatrix} 1 & 1 & \frac{1}{2} \\ 1 & 1 & \frac{1}{4} \\ \frac{1}{2} & \frac{1}{4} & 2 \end{bmatrix}$

a. Use the normalized power method starting with $x^{(0)} = [1, 1, 1]^T$, and find the dominant eigenvalue and eigenvector of the matrix A.

b. Repeat, starting with the initial value $x^{(0)} = [-0.64966116, 0.74822116, 0]^T$. Explain the results. See Ralston [1965, p. 475–476].

7. Let $A = \begin{bmatrix} -4 & 14 & 0 \\ -5 & 13 & 0 \\ -1 & 0 & 2 \end{bmatrix}$.

Code and apply each of the following:

a. The modified power algorithm starting with $x^{(0)} = [1, 1, 1]^T$ as well as the Aitken's acceleration process.

b. The inverse power algorithm.

c. The shifted power algorithm.

d. The shifted inverse power algorithm.

8. (Continuation) Let $B = \begin{bmatrix} 4 & -1 & 1 \\ -1 & 3 & -2 \\ 1 & -2 & 3 \end{bmatrix}$.

Repeat the previous exercise starting with $x^{(0)} = [1, 0, 0]^T$.

9. (Continuation) Let $C = \begin{bmatrix} -8 & -5 & 8 \\ 6 & 3 & -8 \\ -3 & 1 & 9 \end{bmatrix}$.

Use $x^{(0)} = [1, 1, 1]^T$. Repeat the previous exercise starting with $x^{(0)} = [1, 0, 0]^T$.

10. By means of the power method, find an eigenvalue and associated eigenvector of these matrices from the historical books by Fox [1957] and Wilkinson [1965]. Use the given starting values and carry out the procedure with and without normalization. Verify your results by using mathematical software such as MATLAB, Maple, or Mathematica.

a. $\begin{bmatrix} 0.9901 & 0.002 \\ -0.0001 & 0.9904 \end{bmatrix}, \quad x^{(0)} = [1, 0.9]^T$

b. $\begin{bmatrix} 8 & -1 & -5 \\ -4 & 4 & -2 \\ 18 & -5 & -7 \end{bmatrix}, \quad x^{(0)} = [1, 0.8, 1]^T$

c. $\begin{bmatrix} 1 & 1 & 3 \\ 1 & -2 & 1 \\ 3 & 1 & 3 \end{bmatrix}, \quad x^{(0)} = [1, 1, 1]^T$

d. $\begin{bmatrix} -2 & -1 & 4 \\ 2 & 1 & -2 \\ -1 & -1 & 3 \end{bmatrix}, \quad x^{(0)} = [3, 1, 2]^T$

11. Find all of the eigenvalues and associated eigenvectors of these matrices from Fox [1957] and Wilkinson [1965] by means of the power method and variations of it.

Verify your results by using mathematical software such as MATLAB, Maple, or Mathematica.

a. $\begin{bmatrix} 2 & 1 \\ 4 & 2 \end{bmatrix}$

b. $\begin{bmatrix} 0.4812 & 0.0023 \\ -0.0024 & 0.4810 \end{bmatrix}$

c. $\begin{bmatrix} 1 & 1 & 0 \\ -1+10^{-8} & 3 & 0 \\ 0 & 1 & 1 \end{bmatrix}$

d. $\begin{bmatrix} 5 & -1 & -2 \\ -1 & 3 & -2 \\ -2 & -2 & 5 \end{bmatrix}$

e. $\begin{bmatrix} 0.987 & 0.400 & -0.487 \\ -0.079 & 0.500 & -0.479 \\ 0.082 & 0.400 & 0.418 \end{bmatrix}$

8.4 Iterative Solutions of Linear Systems

In this section, we explore a completely different strategy for solving a nonsingular linear system

$$\boldsymbol{A}\boldsymbol{x} = \boldsymbol{b} \tag{1}$$

This alternative approach is often used on enormous problems that arise in solving partial differential equations numerically. In that subject, systems having hundreds of thousands of equations arise routinely. (See Section 12.3.)

Vector and Matrix Norms

We first present a brief overview of vector and matrix norms because they are useful in the discussion of errors and in the stopping criteria for iterative methods. Norms can be defined on any vector space, but we usually use $\mathbb{R}^n$ or $\mathbb{C}^n$. A vector norm $||\boldsymbol{x}||$ can be thought of as the **length** or **magnitude** of a vector $\boldsymbol{x} \in \mathbb{R}^n$. A **vector norm** is any mapping from $\mathbb{R}^n$ to $\mathbb{R}$ that obeys these three properties:

Vector Norm

$$\begin{aligned} ||x|| &> 0 \text{ if } \boldsymbol{x} \neq \boldsymbol{0} \\ ||\alpha \boldsymbol{x}|| &= |\alpha|\,||\boldsymbol{x}|| \\ ||\boldsymbol{x}+\boldsymbol{y}|| &\leqq ||\boldsymbol{x}|| + ||\boldsymbol{y}|| \qquad \textbf{(triangle inequality)} \end{aligned}$$

for vectors $\boldsymbol{x}, \boldsymbol{y} \in \mathbb{R}^n$ and scalars $\alpha \in \mathbb{R}$. Examples of vector norms for the vector $\boldsymbol{x} = (x_1, x_2, \dots, x_n)^T \in \mathbb{R}^n$ are

$\ell_1, \ell_2, \ell_\infty$-vector Norm

$$\begin{aligned} ||\boldsymbol{x}||_1 &= \sum_{i=1}^{n} |x_i| && \ell_1\textbf{-vector norm} \\ ||\boldsymbol{x}||_2 &= \left(\sum_{i=1}^{n} x_i^2 \right)^{1/2} && \textbf{Euclidean}/\ell_2\textbf{-vector norm} \\ ||\boldsymbol{x}||_\infty &= \max_{1 \leqq i \leqq n} |x_i| && \ell_\infty\textbf{-vector norm} \end{aligned}$$

For $n \times n$ matrices, we can also have **matrix norms**, subject to the same requirements:

Matrix Norm

$$\begin{aligned} &||\boldsymbol{A}|| > 0 \text{ if } \boldsymbol{A} \neq 0 \\ &||\alpha \boldsymbol{A}|| = |\alpha|\,||\boldsymbol{A}|| \\ &||\boldsymbol{A}+\boldsymbol{B}|| \leqq ||\boldsymbol{A}|| + ||\boldsymbol{B}|| \qquad \textbf{(triangular inequality)} \end{aligned}$$

for matrices $\boldsymbol{A}, \boldsymbol{B}$, and scalars α.

We usually prefer matrix norms that are related to a vector norm. For a vector norm $||\cdot||$, the **subordinate matrix norm** is defined by

Subordinate Matrix Norm

$$||A|| \equiv \sup\{||Ax|| : x \in \mathbb{R}^n \text{ and } ||x|| = 1\}$$

Here, A is an $n \times n$ matrix. For a subordinate matrix norm, some additional properties are

Matrix Norm Properties

$$\begin{aligned} ||I|| &= 1 \\ ||Ax|| &\leqq ||A||\,||x|| \\ ||AB|| &\leqq ||A||\,||B|| \end{aligned}$$

There are two meanings associated with the notation $||\cdot||_p$, one for vectors and another for matrices. The context determines which one is intended. Examples of subordinate matrix norms for an $n \times n$ matrix A are

$\ell_1, \ell_2, \ell_\infty$-matrix Norms

$$||A||_1 = \max_{1 \leqq j \leqq n} \sum_{i=1}^{n} |a_{ij}| \qquad \ell_1\text{-matrix norm}$$

$$||A||_2 = \max_{1 \leqq i \leqq n} \sqrt{|\sigma_{\max}|} \qquad \text{spectral }/\ell_2\text{-matrix norm}$$

$$||A||_\infty = \max_{1 \leqq i \leqq n} \sum_{j=1}^{n} |a_{ij}| \qquad \ell_\infty\text{-matrix norm}$$

Singular Value/ Spectral Radius

Here, σ_i are the eigenvalues of $A^T A$, which are called the **singular values** of A. The largest $\sigma_{\max}$ in absolute value is termed the **spectral radius** of A. (See Sections 7.2 and 9.3 for a discussion of singular values.)

Condition Number and Ill-Conditioning

An important quantity that has some influence in the numerical solution of a linear system

Condition Number

$$Ax = b$$

is the **condition number**, which is defined as

$$\kappa(A) = \|A\|_2\,\|A^{-1}\|_2$$

It turns out that it is not necessary to compute the inverse of A to obtain an estimate of the condition number. Also, it can be shown that the condition number $\kappa(A)$ gauges the transfer of error from the matrix A and the vector b to the solution x.

■ Rule

Rule of Thumb

If $\kappa(A) = 10^k$, then one can expect to lose at least k digits of precision in solving the system $Ax = b$.

Ill-Conditioned Matrix

If the linear system is sensitive to perturbations in the elements of A, or to perturbations of the components of b, then this fact is reflected in A having a large condition number. In such a case, the matrix A is said to be **ill-conditioned**. Briefly, the larger the condition number, the more ill-conditioned the system.

Suppose we want to solve an invertible linear system of equations

$$Ax = b$$

for a given coefficient matrix A and right-hand side b, but there may have been perturbations of the data owing to uncertainty in the measurements and roundoff errors in the calculations. Suppose that the right-hand side is perturbed by an amount assigned the symbol $\delta \boldsymbol{b}$ and the corresponding solution is perturbed an amount denoted by the symbol $\delta \boldsymbol{x}$. Then we have

$$\boldsymbol{A}(\boldsymbol{x} + \delta \boldsymbol{x}) = \boldsymbol{A}\boldsymbol{x} + \boldsymbol{A}\delta \boldsymbol{x} = \boldsymbol{b} + \delta \boldsymbol{b}$$

where

$$\boldsymbol{A}\delta \boldsymbol{x} = \delta \boldsymbol{b}$$

From the original linear system $\boldsymbol{A}\boldsymbol{x} = \boldsymbol{x}$ and norms, we have

$$||\boldsymbol{b}|| = ||\boldsymbol{A}\boldsymbol{x}|| \leqq ||\boldsymbol{A}|| \, ||\boldsymbol{x}||$$

which gives us

$$\frac{1}{||\boldsymbol{x}||} \leqq \frac{||\boldsymbol{A}||}{||\boldsymbol{b}||}$$

From the perturbed linear system $\boldsymbol{A}\delta \boldsymbol{x} = \delta \boldsymbol{b}$, we obtain $\delta \boldsymbol{x} = \boldsymbol{A}^{-1}\delta \boldsymbol{b}$ and

$$||\delta \boldsymbol{x}|| \leqq ||\boldsymbol{A}^{-1}|| \, ||\delta \boldsymbol{b}||$$

Combining the two inequalities above, we obtain

Condition Number Bound

$$\frac{||\delta \boldsymbol{x}||}{||\boldsymbol{x}||} \leqq \kappa(\boldsymbol{A}) \frac{||\delta \boldsymbol{b}||}{||\boldsymbol{b}||}$$

which contains the condition number of the original matrix $\boldsymbol{A}$.

As an example of an ill-conditioned matrix consider the Hilbert matrix

3 × 3 Hilbert Matrix

$$H_3 = \begin{bmatrix} 1 & \frac{1}{2} & \frac{1}{3} \\ \frac{1}{2} & \frac{1}{3} & \frac{1}{4} \\ \frac{1}{3} & \frac{1}{4} & \frac{1}{5} \end{bmatrix}$$

We can use the MATLAB commands to generate the matrix and then to compute both the condition number using the 2-norm and the determinant of the matrix. We find the condition number is 524.0568 and the determinant is 4.6296×10^{-4}. In solving linear systems, the condition number of the coefficient matrix measures the sensitivity of the system to errors in the data. When the condition number is large, the computed solution of the system may be dangerously in error! Further checks should be made before accepting the solution as being accurate. Values of the condition number near 1 indicate a **well-conditioned matrix** whereas large values indicate an **ill-conditioned matrix**. Using the determinant to check for singularity is appropriate only for matrices of modest size. Using mathematical software, one can compute the condition number to check for singular or near-singular matrices.

Well/Ill-Conditioned Matrices

A goal in the study of numerical methods is to acquire an awareness of whether a numerical result can be trusted or whether it may be suspect (and therefore in need of further analysis). The condition number provides some evidence regarding this question. With the advent of sophisticated mathematical software systems, an estimate of the condition number is often returned, along with an approximate solution so that one can judge the trustworthiness of the results. In fact, some solution procedures involve advanced features that depend on an estimated condition number and may switch solution techniques based on it. For example, this criterion may result in a switch of the solution technique from a variant of Gaussian elimination to a least-squares solution for an ill-conditioned system.

Unsuspecting users may *not* realize that this has happened unless they look at all of the results, including the estimate of the condition number. (Condition numbers can also be associated with other numerical problems, such as locating roots of equations.)

Basic Iterative Methods

The iterative-method strategy produces a sequence of approximate solution vectors $\boldsymbol{x}^{(0)}$, $\boldsymbol{x}^{(1)}, \boldsymbol{x}^{(2)}, \dots$ for system $\boldsymbol{Ax} = \boldsymbol{b}$. The numerical procedure is designed so that, in principle, the sequence of vectors converges to the actual solution. The process can be stopped when sufficient precision has been attained. This stands in contrast to the Gaussian elimination algorithm, which has no provision for stopping midway and offering up an approximate solution. A general iterative algorithm for solving System (1) goes as follows: Select a nonsingular matrix $\boldsymbol{Q}$, and having chosen an arbitrary starting vector $\boldsymbol{x}^{(0)}$, generate vectors $\boldsymbol{x}^{(1)}, \boldsymbol{x}^{(2)}, \dots$ recursively from the equation

General Iteration

$$\boldsymbol{Qx}^{(k)} = (\boldsymbol{Q} - \boldsymbol{A})\boldsymbol{x}^{(k-1)} + \boldsymbol{b} \qquad (k = 1, 2, \dots) \tag{2}$$

To see that this is sensible, suppose that the sequence $\boldsymbol{x}^{(k)}$ does converge, to a vector $\boldsymbol{x}^*$, say. Then by taking the limit as $k \to \infty$ in System (2), we get

$$\boldsymbol{Qx}^* = (\boldsymbol{Q} - \boldsymbol{A})\boldsymbol{x}^* + \boldsymbol{b}$$

Richardson Iteration

This leads to $\boldsymbol{Ax}^* = \boldsymbol{b}$. Thus, if the sequence converges, its limit is a solution to the original System (1). For example, the **Richardson iteration** uses $\boldsymbol{Q} = \boldsymbol{I}$.

An outline of the pseudocode for carrying out the general iterative procedure (2) follows:

General Iterative Pseudocode

```
integer k, kmax
real array (x^(0))_1:n, (b)_1:n, (c)_1:n, (x)_1:n, (y)_1:n, (A)_1:n×1:n, (Q)_1:n×1:n
x ← x^(0)
for k = 1 to kmax
    y ← x
    c ← (Q − A)x + b
    solve Qx = c
    output k, x
    if ‖x − y‖ < ε then
        output "convergence"
        stop
    end if
end for
output "maximum iteration reached"
```

In choosing the invertible matrix $\boldsymbol{Q}$, we are influenced by the following.

- System (2) should be *easy* to solve for $\boldsymbol{x}^{(k)}$, when the right-hand side is known.
- Matrix $\boldsymbol{Q}$ should be chosen to ensure that the sequence $\boldsymbol{x}^{(k)}$ converges, no matter what initial vector is used. Ideally, this convergence will be rapid.

One should not believe that it is necessary to compute the inverse of $\boldsymbol{Q}$ to carry out an iterative procedure. For small systems, we can easily compute the inverse of $\boldsymbol{Q}$, but in general, this is definitely *not* to be done! We want to solve a linear system in which $\boldsymbol{Q}$ is the coefficient matrix. As was mentioned previously, we want to select $\boldsymbol{Q}$ so that a linear system with $\boldsymbol{Q}$ as the coefficient matrix is easy to solve. Examples of such matrices are diagonal, tridiagonal, banded, lower triangular, and upper triangular.

Now, let's view System (1) in its detailed form

$$\sum_{j=1}^{n} a_{ij}x_j = b_i \qquad (1 \leqq i \leqq n) \tag{3}$$

Solving the ith equation for the ith unknown term, we obtain an equation that describes the **Jacobi method**:

Jacobi Method

$$x_i^{(k)} = \left[-\sum_{\substack{j=1 \\ j\neq i}}^{n} (a_{ij}/a_{ii})x_j^{(k-1)} + (b_i/a_{ii}) \right] \qquad (1 \leqq i \leqq n) \tag{4}$$

Here, we assume that all diagonal elements are nonzero. (If this is not the case, we can usually rearrange the equations so that it is.)

In the Jacobi method above, the equations are solved in order. The components $x_j^{(k-1)}$ and the corresponding new values $x_j^{(k)}$ can be used immediately in their place. If this is done, we have the **Gauss-Seidel method**:

Gauss-Seidel Method

$$x_i^{(k)} = \left[-\sum_{\substack{j=1 \\ j<i}}^{n} (a_{ij}/a_{ii})x_j^{(k)} - \sum_{\substack{j=1 \\ j>i}}^{n} (a_{ij}/a_{ii})x_j^{(k-1)} + (b_i/a_{ii}) \right] \tag{5}$$

If $x^{(k-1)}$ is not saved, then we can dispense with the superscripts in the pseudocode as follows:

Gauss-Seidel Pseudocode

> **integer** $i, j, k, kmax, n$; **real array** $(a_{ij})_{1:n\times 1:n}, (b_i)_{1:n}, (x_i)_{1:n}$
> **for** $k = 1$ **to** $kmax$
>   **for** $i = 1$ **to** n
> $$x_i \leftarrow \left[b_i - \sum_{\substack{j=1 \\ j\neq i}}^{n} a_{ij}x_j \right] \Big/ a_{ii}$$
>   **end for**
> **end for**

An acceleration of the Gauss-Seidel method is possible by the introduction of a relaxation factor ω, resulting in the **successive overrelaxation (SOR) method**:

SOR Method

$$x_i^{(k)} = \omega \left\{ \left[-\sum_{\substack{j=1 \\ j<i}}^{n} (a_{ij}/a_{ii})x_j^{(k)} - \sum_{\substack{j=1 \\ j>i}}^{n} (a_{ij}/a_{ii})x_j^{(k-1)} + (b_i/a_{ii}) \right] \right\} + (1-\omega)x_i^{(k-1)} \tag{6}$$

The SOR method with $\omega = 1$ reduces to the Gauss-Seidel method.

We now consider numerical examples using iterative methods associated with the names Jacobi, Gauss-Seidel, and successive overrelaxation.

EXAMPLE 1 **(Jacobi iteration)** Let

$$A = \begin{bmatrix} 2 & -1 & 0 \\ -1 & 3 & -1 \\ 0 & -1 & 2 \end{bmatrix}, \qquad b = \begin{bmatrix} 1 \\ 8 \\ -5 \end{bmatrix}$$

Carry out a number of iterations of the Jacobi method, starting with the zero initial vector.

Solution Rewriting the equations, we have the Jacobi method:

$$\begin{aligned} x_1^{(k)} &= \frac{1}{2}x_2^{(k-1)} + \frac{1}{2} \\ x_2^{(k)} &= \frac{1}{3}x_1^{(k-1)} + \frac{1}{3}x_3^{(k-1)} + \frac{8}{3} \\ x_3^{(k)} &= \frac{1}{2}x_2^{(k-1)} - \frac{5}{2} \end{aligned}$$

Taking the initial vector to be $\boldsymbol{x}^{(0)} = [0, 0, 0]^T$, we find (with the aid of a computer program or a programmable calculator) that

$$\begin{aligned} \boldsymbol{x}^{(0)} &= [0, 0, 0]^T \\ \boldsymbol{x}^{(1)} &= [0.5000, 2.6667, -2.5000]^T \\ \boldsymbol{x}^{(2)} &= [1.8333, 2.0000, -1.1667]^T \\ &\vdots \\ \boldsymbol{x}^{(21)} &= [2.0000, 3.0000, -1.0000]^T \end{aligned}$$

The actual solution (to four decimal places rounded) is obtained. ■

In the Jacobi method, $\boldsymbol{Q}$ is taken to be the diagonal of $\boldsymbol{A}$:

$$\boldsymbol{Q} = \begin{bmatrix} 2 & 0 & 0 \\ 0 & 3 & 0 \\ 0 & 0 & 2 \end{bmatrix}$$

Now

$$\boldsymbol{Q}^{-1} = \begin{bmatrix} \frac{1}{2} & 0 & 0 \\ 0 & \frac{1}{3} & 0 \\ 0 & 0 & \frac{1}{2} \end{bmatrix}, \qquad \boldsymbol{Q}^{-1}\boldsymbol{A} = \begin{bmatrix} 1 & -\frac{1}{2} & 0 \\ -\frac{1}{3} & 1 & -\frac{1}{3} \\ 0 & -\frac{1}{2} & 1 \end{bmatrix}$$

The Jacobi iteration matrix and constant vector are

Jacobi Iteration Matrix

$$\boldsymbol{B} = \boldsymbol{I} - \boldsymbol{Q}^{-1}\boldsymbol{A} = \begin{bmatrix} 0 & \frac{1}{2} & 0 \\ \frac{1}{3} & 0 & \frac{1}{3} \\ 0 & \frac{1}{2} & 0 \end{bmatrix}, \qquad \boldsymbol{h} = \boldsymbol{Q}^{-1}\boldsymbol{b} = \begin{bmatrix} \frac{1}{2} \\ \frac{8}{3} \\ -\frac{5}{2} \end{bmatrix}$$

One can see that $\boldsymbol{Q}$ is *close* to $\boldsymbol{A}$, $\boldsymbol{Q}^{-1}\boldsymbol{A}$ is *close* to $\boldsymbol{I}$, and $\boldsymbol{I} - \boldsymbol{Q}^{-1}\boldsymbol{A}$ is *small*. We write the Jacobi method as

$$\boldsymbol{x}^{(k)} = \boldsymbol{B}\boldsymbol{x}^{(k-1)} + \boldsymbol{h}$$

EXAMPLE 2 **(Gauss-Seidel iteration)** Repeat the preceding example using the Gauss-Seidel method.

Solution The idea of the Gauss-Seidel method is simply to accelerate the convergence by incorporating each vector as soon as it has been computed. Obviously, it would be more efficient in the Jacobi method to use the updated value $x_1^{(k)}$ in the second equation instead of the old value $x_1^{(k-1)}$. Similarly, $x_2^{(k)}$ could be used in the third equation in place of $x_2^{(k-1)}$. Using the new iterates as soon as they become available, we have the Gauss-Seidel method:

$$\begin{aligned} x_1^{(k)} &= \frac{1}{2}x_2^{(k-1)} + \frac{1}{2} \\ x_2^{(k)} &= \frac{1}{3}x_1^{(k)} + \frac{1}{3}x_3^{(k-1)} + \frac{8}{3} \\ x_3^{(k)} &= \frac{1}{2}x_2^{(k)} - \frac{5}{2} \end{aligned}$$

Starting with the initial vector zero, some of the iterates are

$$\begin{aligned} \boldsymbol{x}^{(0)} &= [0, 0, 0]^T \\ \boldsymbol{x}^{(1)} &= [0.5000, 2.8333, -1.0833]^T \\ \boldsymbol{x}^{(2)} &= [1.9167, 2.9444, -1.0278]^T \\ &\vdots \\ \boldsymbol{x}^{(9)} &= [2.0000, 3.0000, -1.0000]^T \end{aligned}$$

In this example, the convergence of the Gauss-Seidel method is approximately twice as fast as that of the Jacobi method. ■

In the iterative algorithm that goes by the name Gauss-Seidel, $\boldsymbol{Q}$ is chosen as the lower triangular part of $\boldsymbol{A}$, including the diagonal. Using the data from the previous example, we now find that

$$\boldsymbol{Q} = \begin{bmatrix} 2 & 0 & 0 \\ -1 & 3 & 0 \\ 0 & -1 & 2 \end{bmatrix}$$

The usual row operations give us

$$\boldsymbol{Q}^{-1} = \begin{bmatrix} \frac{1}{2} & 0 & 0 \\ \frac{1}{6} & \frac{1}{3} & 0 \\ \frac{1}{12} & \frac{1}{6} & \frac{1}{2} \end{bmatrix}, \qquad \boldsymbol{Q}^{-1}\boldsymbol{A} = \begin{bmatrix} 1 & -\frac{1}{2} & 0 \\ 0 & \frac{5}{6} & -\frac{1}{3} \\ 0 & -\frac{1}{12} & \frac{5}{6} \end{bmatrix}$$

Again, we emphasize that in a practical problem we would not compute $\boldsymbol{Q}^{-1}$. The Gauss-Seidel iterative matrix and constant vector are

Gauss-Seidel Iteration Matrix

$$\mathcal{L} = \boldsymbol{I} - \boldsymbol{Q}^{-1}\boldsymbol{A} = \begin{bmatrix} 0 & \frac{1}{2} & 0 \\ 0 & \frac{1}{6} & \frac{1}{3} \\ 0 & \frac{1}{12} & \frac{1}{6} \end{bmatrix}, \qquad \boldsymbol{h} = \boldsymbol{Q}^{-1}\boldsymbol{b} = \begin{bmatrix} \frac{1}{2} \\ \frac{17}{6} \\ -\frac{13}{12} \end{bmatrix}$$

We write the Gauss-Seidel method as

$$\boldsymbol{x}^{(k)} = \mathcal{L}\boldsymbol{x}^{(k-1)} + \boldsymbol{h}$$

EXAMPLE 3 **(SOR iteration)** Repeat the preceding example using the SOR iteration with $\omega = 1.1$.

Solution Introducing a relaxation factor ω into the Gauss-Seidel method, we have the SOR method:

$$x_1^{(k)} = \omega\left[\frac{1}{2}x_2^{(k-1)} + \frac{1}{2}\right] + (1-\omega)x_1^{(k-1)}$$

$$x_2^{(k)} = \omega\left[\frac{1}{3}x_1^{(k)} + \frac{1}{3}x_3^{(k-1)} + \frac{8}{3}\right] + (1-\omega)x_2^{(k-1)}$$

$$x_3^{(k)} = \omega\left[\frac{1}{2}x_2^{(k)} - \frac{5}{2}\right] + (1-\omega)x_3^{(k-1)}$$

Starting with the initial vector of zeros and with $\omega = 1.1$, some of the iterates are

$$\begin{aligned}
\boldsymbol{x}^{(0)} &= [0, 0, 0]^T \\
\boldsymbol{x}^{(1)} &= [0.5500, 3.1350, -1.0257]^T \\
\boldsymbol{x}^{(2)} &= [2.2193, 3.0574, -0.9658]^T \\
&\ \vdots \\
\boldsymbol{x}^{(7)} &= [2.0000, 3.0000, -1.0000]^T
\end{aligned}$$

In this example, the convergence of the SOR method is faster than that of the Gauss-Seidel method. ■

In the iterative algorithm that goes by the name successive overrelaxation (SOR), $\boldsymbol{Q}$ is chosen as the lower triangular part of $\boldsymbol{A}$ including the diagonal, but each diagonal element a_{ij} is replaced by a_{ij}/ω, where ω is the **relaxation factor**. (Initial work on the SOR method was done by Southwell [1946] and Young [1950].)

From the previous example, this means that

$$\boldsymbol{Q} = \begin{bmatrix} \frac{20}{11} & 0 & 0 \\ -1 & \frac{30}{11} & 0 \\ 0 & -1 & \frac{20}{11} \end{bmatrix}$$

Now

$$\boldsymbol{Q}^{-1} = \begin{bmatrix} \frac{11}{20} & 0 & 0 \\ \frac{121}{600} & \frac{11}{30} & 0 \\ \frac{1331}{12000} & \frac{121}{600} & \frac{11}{20} \end{bmatrix}, \qquad \boldsymbol{Q}^{-1}\boldsymbol{A} = \begin{bmatrix} \frac{11}{10} & -\frac{11}{20} & 0 \\ \frac{11}{300} & \frac{539}{600} & -\frac{11}{30} \\ \frac{121}{6000} & \frac{671}{12000} & \frac{539}{600} \end{bmatrix}$$

The SOR iteration matrix and constant vector are

SOR Iteration Matrix

$$\mathcal{L}_\omega = \boldsymbol{I} - \boldsymbol{Q}^{-1}\boldsymbol{A} = \begin{bmatrix} -\frac{1}{10} & \frac{11}{20} & 0 \\ -\frac{11}{300} & \frac{61}{600} & \frac{11}{30} \\ -\frac{121}{6000} & -\frac{671}{12000} & \frac{61}{600} \end{bmatrix}, \qquad \boldsymbol{h} = \boldsymbol{Q}^{-1}\boldsymbol{b} = \begin{bmatrix} \frac{11}{20} \\ \frac{627}{200} \\ -\frac{4103}{4000} \end{bmatrix}$$

We write the SOR method as

SOR Method

$$\boldsymbol{x}^{(k)} = \mathcal{L}_\omega \boldsymbol{x}^{(k-1)} + \boldsymbol{h}$$

Pseudocode

We can write pseudocode for the Jacobi, Gauss-Seidel, and SOR methods assuming that the linear System (1) is stored in matrix-vector form:

Jacobi **Pseudocode**

```
procedure Jacobi(A, b, x)
real kmax ← 100, δ ← 10^{-10}, ε ← ½ × 10^{-4}
integer i, j, k, kmax, n;   real diag, sum
real array (A)_{1:n×1:n}, (b)_{1:n}, (x)_{1:n}, (y)_{1:n}
n ← size(A)
for k = 1 to kmax
    y ← x
    for i = 1 to n
        sum ← b_i
        diag ← a_{ii}
        if |diag| < δ then
            output "diagonal element too small"
            return
        end if
        for j = 1 to n
            if j ≠ i then
                sum ← sum − a_{ij} y_j
            end if
        end for
        x_i ← sum/diag
    end for
    output k, x
    if ||x − y|| < ε then
        output k, x
        return
    end if
end for
output "maximum iterations reached"
return
end Jacobi
```

Here, the vector $\boldsymbol{y}$ contains the old iterate values, and the vector $\boldsymbol{x}$ contains the updated ones. The values of $kmax$, δ, and ε are set either in a parameter statement or as global variables.

The pseudocode for the procedure $Gauss\_Seidel(\boldsymbol{A}, \boldsymbol{b}, \boldsymbol{x})$ is the same as that for the shown Jacobi pseudocode except that the innermost j-loop is replaced by the following:

Gauss-Seidel Pseudocode

```
for j = 1 to i − 1
    sum ← sum − a_{ij} x_j
end for
for j = i + 1 to n
    sum ← sum − a_{ij} x_j
end for
```

The pseudocode for procedure $SOR(\boldsymbol{A}, \boldsymbol{b}, \boldsymbol{x}, \omega)$ is the same as that for the Gauss-Seidel pseudocode with the statement following the j-loop replaced by the following:

SOR Pseudocode

```
x_i ← sum/diag
x_i ← ωx_i + (1 − ω)y_i
```

In the solution of partial differential equations, iterative methods are frequently used to solve large sparse linear systems, which often have special structures. The partial derivatives are approximated by stencils composed of relatively few points, such as 5, 7, or 9. This leads to only a few nonzero entries per row in the linear system. In such systems, the coefficient matrix $\boldsymbol{A}$ is usually *not* stored because the matrix-vector product can be written directly in the code. See Section 12.3 for additional details on this and how it is related to solving elliptic partial differential equations.

Large Sparse Linear Systems

Convergence Theorems

For the analysis of the method described by System (2), we write

$$\boldsymbol{x}^{(k)} = \boldsymbol{Q}^{-1}\left[(\boldsymbol{Q} - \boldsymbol{A})\boldsymbol{x}^{(k-1)} + \boldsymbol{b}\right]$$

or

$$\boldsymbol{x}^{(k)} = \mathcal{G}\boldsymbol{x}^{(k-1)} + \boldsymbol{h} \tag{7}$$

where the iteration matrix and vector are

Iteration Matrix and Vector

$$\mathcal{G} = \boldsymbol{I} - \boldsymbol{Q}^{-1}\boldsymbol{A}, \qquad \boldsymbol{h} = \boldsymbol{Q}^{-1}\boldsymbol{b}$$

Notice that in the pseudocode, we do *not* compute $\boldsymbol{Q}^{-1}$. The matrix $\boldsymbol{Q}^{-1}$ is used to facilitate the analysis. Now let $\boldsymbol{x}$ be the solution of System (1). Since $\boldsymbol{A}$ is invertible, $\boldsymbol{x}$ exists and is unique. We have, from Equation (7),

$$\begin{aligned}\boldsymbol{x}^{(k)} - \boldsymbol{x} &= (\boldsymbol{I} - \boldsymbol{Q}^{-1}\boldsymbol{A})\boldsymbol{x}^{(k-1)} - \boldsymbol{x} + \boldsymbol{Q}^{-1}\boldsymbol{b}\\ &= (\boldsymbol{I} - \boldsymbol{Q}^{-1}\boldsymbol{A})\boldsymbol{x}^{(k-1)} - (\boldsymbol{I} - \boldsymbol{Q}^{-1}\boldsymbol{A})\boldsymbol{x}\\ &= (\boldsymbol{I} - \boldsymbol{Q}^{-1}\boldsymbol{A})(\boldsymbol{x}^{(k-1)} - \boldsymbol{x})\end{aligned}$$

One can interpret $\boldsymbol{e}^{(k)} \equiv \boldsymbol{x}^{(k)} - \boldsymbol{x}$ as the current **error vector**. Thus, we have

Error Vector

$$\boldsymbol{e}^{(k)} = (\boldsymbol{I} - \boldsymbol{Q}^{-1}\boldsymbol{A})\boldsymbol{e}^{(k-1)} \tag{8}$$

We want $\boldsymbol{e}^{(k)}$ to become *smaller* as k increases. Equation (8) shows that $\boldsymbol{e}^{(k)}$ is *smaller* than $\boldsymbol{e}^{(k-1)}$ if $\boldsymbol{I} - \boldsymbol{Q}^{-1}\boldsymbol{A}$ is *small*, in some sense. In turn, $\boldsymbol{Q}^{-1}\boldsymbol{A}$ should be *close* to $\boldsymbol{I}$ and $\boldsymbol{Q}$ should be *close* to $\boldsymbol{A}$. (Norms can be used to make *small* and *close* precise.)

■ Theorem 1 **Spectral Radius Theorem**

In order that the sequence generated by

$$\boldsymbol{Q}\boldsymbol{x}^{(k)} = (\boldsymbol{Q} - \boldsymbol{A})\boldsymbol{x}^{(k-1)} + \boldsymbol{b}$$

to converge, no matter what starting point $\boldsymbol{x}^{(0)}$ is selected, it is necessary and sufficient that all eigenvalues of $\boldsymbol{I} - \boldsymbol{Q}^{-1}\boldsymbol{A}$ lie in the open unit disc, $|z| < 1$, in the complex plane.

The conclusion of this theorem can also be written as

Spectral Radius

$$\rho(\boldsymbol{I} - \boldsymbol{Q}^{-1}\boldsymbol{A}) < 1$$

where ρ is the **spectral radius** function: For any $n \times n$ matrix $\boldsymbol{G}$, having eigenvalues λ_i, we define

$$\rho(\boldsymbol{G}) = \max_{1 \leqq i \leqq n} |\lambda_i|$$

EXAMPLE 4 Determine whether the Jacobi, Gauss-Seidel, and SOR methods (with $\omega = 1.1$) of Example 3 converge for all initial iterates.

Solution For the Jacobi method, we can easily compute the eigenvalues of the relevant matrix $\boldsymbol{B}$. The steps are

$$\det(\boldsymbol{B} - \lambda \boldsymbol{I}) = \det \begin{bmatrix} -\lambda & \frac{1}{2} & 0 \\ \frac{1}{3} & -\lambda & \frac{1}{3} \\ 0 & \frac{1}{2} & -\lambda \end{bmatrix} = -\lambda^3 + \frac{1}{6}\lambda + \frac{1}{6}\lambda = 0$$

The eigenvalues are $\lambda = 0, \pm\sqrt{1/3} \approx \pm 0.5774$. Thus, by Spectral Radius Theorem, the Jacobi iteration succeeds for any starting vector in this example.

For the Gauss-Seidel method, the eigenvalues of the iteration matrix $\mathcal{L}$ are determined from

$$\det(\mathcal{L} - \lambda \boldsymbol{I}) = \det \begin{bmatrix} -\lambda & \frac{11}{20} & 0 \\ 0 & \frac{1}{6} - \lambda & \frac{1}{3} \\ 0 & \frac{1}{12} & \frac{1}{6} - \lambda \end{bmatrix} = -\lambda\left(\frac{1}{6} - \lambda\right)^2 + \frac{1}{36}\lambda = 0$$

The eigenvalues are $\lambda = 0, 0, \frac{1}{3} \approx 0.333$. Hence, the Gauss-Seidel iteration converges for any initial vector in this example.

For the SOR method with $\omega = 1.1$, the eigenvalues of the iteration matrix $\mathcal{L}_\omega$ are determined from

$$\det(\mathcal{L}_\omega - \lambda \boldsymbol{I}) = \det \begin{bmatrix} -\frac{1}{10} - \lambda & \frac{11}{20} & 0 \\ -\frac{11}{300} & \frac{61}{600} - \lambda & \frac{11}{30} \\ -\frac{121}{6000} & \frac{671}{12000} & \frac{61}{600} - \lambda \end{bmatrix}$$

$$= \left(-\frac{1}{10} - \lambda\right)\left(\frac{61}{600} - \lambda\right)^2 - \frac{121}{6000} \cdot \frac{11}{30} \cdot \frac{11}{20}$$

$$+ \frac{11}{20} \cdot \frac{11}{300}\left(\frac{61}{600} - \lambda\right) - \left(-\frac{1}{10} - \lambda\right)\frac{671}{12000} \cdot \frac{11}{30}$$

$$= -\frac{1}{1000} + \frac{31}{3000}\lambda + \frac{31}{3000}\lambda^2 - \lambda^3 = 0$$

The eigenvalues are $\lambda \approx 0.1200, 0.0833, -0.1000$. Hence, the SOR iteration converges for any initial vector in this example. ■

A condition that is easier to verify than the inequality $\rho(\boldsymbol{I} - \boldsymbol{Q}^{-1}\boldsymbol{A}) < 1$ is the dominance of the diagonal elements over the other elements in the same row. As defined in Section 2.3, we can use the property of **diagonal dominance**

Diagonal Dominance

$$|a_{ii}| > \sum_{\substack{j=1 \\ j \neq i}}^{n} |a_{ij}|$$

to determine whether the Jacobi and Gauss-Seidel methods converge.

■ **Theorem 2** **Jacobi and Gauss-Seidel Convergence Theorem**

If $\boldsymbol{A}$ is diagonally dominant, then the Jacobi and Gauss-Seidel methods converge for any starting vector $\boldsymbol{x}^{(0)}$.

Notice that this is a sufficient, but not a necessary condition. Indeed, there are matrices that are *not* diagonally dominant for which these methods converge.

Another important property follows:

■ **Definition 1** **Symmetric Positive Definite**

Matrix $\boldsymbol{A}$ is **symmetric positive definite (SPD)** if $\boldsymbol{A} = \boldsymbol{A}^T$ and $\boldsymbol{x}^T\boldsymbol{A}\boldsymbol{x} > 0$ for all nonzero real vectors $\boldsymbol{x}$.

For a matrix $\boldsymbol{A}$ to be SPD, it is necessary and sufficient that $\boldsymbol{A} = \boldsymbol{A}^T$ and that all eigenvalues of $\boldsymbol{A}$ are positive.

■ **Theorem 3** **SOR Convergence Theorem**

Suppose that the matrix $\boldsymbol{A}$ has positive diagonal elements and that $0 < \omega < 2$. The SOR method converges for any starting vector $\boldsymbol{x}^{(0)}$ if and only if $\boldsymbol{A}$ is symmetric and positive definite.

Matrix Formulation of Iterative Methods

For the formal theory of iterative methods, we split the matrix $\boldsymbol{A}$ into the sum of a nonzero diagonal matrix $\boldsymbol{D}$, a strictly lower triangular matrix $\boldsymbol{C}_L$, and a strictly upper triangular matrix $\boldsymbol{C}_U$ such that

$$\boldsymbol{A} = \boldsymbol{D} - \boldsymbol{C}_L - \boldsymbol{C}_U$$

Here, $\boldsymbol{D} = \operatorname{diag}(\boldsymbol{A})$, $\boldsymbol{C}_L = (-a_{ij})_{i>j}$, and $\boldsymbol{C}_U = (-a_{ij})_{i<j}$. The linear System (3) can be written as

$$(\boldsymbol{D} - \boldsymbol{C}_L - \boldsymbol{C}_U)\boldsymbol{x} = \boldsymbol{b}$$

From Equation (4), the **Jacobi method** in matrix-vector form is

Jacobi Method

$$\boldsymbol{D}\boldsymbol{x}^{(k)} = (\boldsymbol{C}_L + \boldsymbol{C}_U)\boldsymbol{x}^{(k-1)} + \boldsymbol{b}$$

This corresponds to Equation (2) with $\boldsymbol{Q} = \operatorname{diag}(\boldsymbol{A}) = \boldsymbol{D}$. From Equation (5), the **Gauss-Seidel method** becomes

Gauss-Seidel Method

$$(\boldsymbol{D} - \boldsymbol{C}_L)\boldsymbol{x}^{(k)} = \boldsymbol{C}_U\boldsymbol{x}^{(k-1)} + \boldsymbol{b}$$

This corresponds to Equation (2) with $\boldsymbol{Q} = \operatorname{diag}(\boldsymbol{A}) + \text{lower triangular}(\boldsymbol{A}) = \boldsymbol{D} - \boldsymbol{C}_L$. From Equation (6), the **SOR method** can be written as

SOR Method

$$(\boldsymbol{D} - \omega\boldsymbol{C}_L)\boldsymbol{x}^{(k)} = [\omega\boldsymbol{C}_U + (1-\omega)\boldsymbol{D}]\boldsymbol{x}^{(k-1)} + \omega\boldsymbol{b}$$

This corresponds to Equation (2) with $\boldsymbol{Q} = (1/\omega)\operatorname{diag}(\boldsymbol{A}) + \text{lower triangular}(\boldsymbol{A}) = (1/\omega)(\boldsymbol{D} - \omega\boldsymbol{C}_L)$.

In summary, the iteration matrix and constant vector for the basic three iterative methods (Jacobi, Gauss-Seidel, and SOR) can be written in terms of this splitting. For the **Jacobi method**, we have $\boldsymbol{Q} = \boldsymbol{D}$ and

Jacobi Matrix

$$\begin{aligned} \boldsymbol{B} &= \boldsymbol{I} - \boldsymbol{Q}^{-1}\boldsymbol{A} = \boldsymbol{D}^{-1}(\boldsymbol{C}_L + \boldsymbol{C}_U) \\ \boldsymbol{h} &= \boldsymbol{Q}^{-1}\boldsymbol{b} = \boldsymbol{D}^{-1}\boldsymbol{b} \end{aligned}$$

For the **Gauss-Seidel method**, we have $\boldsymbol{Q} = \boldsymbol{D} - \boldsymbol{C}_L$ and

Gauss-Seidel Matrix

$$\begin{aligned} \mathcal{L} &= \boldsymbol{I} - \boldsymbol{Q}^{-1}\boldsymbol{A} = (\boldsymbol{D} - \boldsymbol{C}_L)^{-1}\boldsymbol{C}_U \\ \boldsymbol{h} &= \boldsymbol{Q}^{-1}\boldsymbol{b} = (\boldsymbol{D} - \boldsymbol{C}_L)^{-1}\boldsymbol{b} \end{aligned}$$

For the **SOR method**, we have $\boldsymbol{Q} = 1/\omega(\boldsymbol{D} - \omega\boldsymbol{C}_L)$ and

SOR Matrix

$$\begin{aligned} \mathcal{L}_\omega &= \boldsymbol{I} - \boldsymbol{Q}^{-1}\boldsymbol{A} = (\boldsymbol{D} - \omega\boldsymbol{C}_L)^{-1}[\omega\boldsymbol{C}_U + (1-\omega)\boldsymbol{D}] \\ \boldsymbol{h} &= \boldsymbol{Q}^{-1}\boldsymbol{b} = \omega(\boldsymbol{D} - \omega\boldsymbol{C}_L)^{-1}\boldsymbol{b} \end{aligned}$$

Another View of Overrelaxation

In some cases, the rate of convergence of the basic iterative scheme (2) can be improved by the introduction of an auxiliary vector and an *acceleration parameter* ω as follows:

$$\begin{aligned} \boldsymbol{Q}\boldsymbol{z}^{(k)} &= (\boldsymbol{Q} - \boldsymbol{A})\boldsymbol{x}^{(k-1)} + \boldsymbol{b} \\ \boldsymbol{x}^{(k)} &= \omega\boldsymbol{z}^{(k)} + (1-\omega)\boldsymbol{x}^{(k-1)} \end{aligned}$$

or

Overrelaxation

$$\boldsymbol{x}^{(k)} = \omega\left\{(\boldsymbol{I} - \boldsymbol{Q}^{-1}\boldsymbol{A})\boldsymbol{x}^{(k-1)} + \boldsymbol{Q}^{-1}\boldsymbol{b}\right\} + (1-\omega)\boldsymbol{x}^{(k-1)}$$

The parameter ω gives a weighting in favor of the updated values. When $\omega = 1$, this procedure reduces to the basic iterative method, and when $1 < \omega < 2$, the rate of convergence may be improved, which is called **overrelaxation**. When $\boldsymbol{Q} = \boldsymbol{D}$, we have the **Jacobi overrelaxation (JOR) method**:

JOR Method

$$\boldsymbol{x}^{(k)} = \omega\left\{\boldsymbol{B}\boldsymbol{x}^{(k-1)} + \boldsymbol{h}\right\} + (1-\omega)\boldsymbol{x}^{(k-1)}$$

Overrelaxation has particular advantages when used with the Gauss-Seidel method in a slightly different way:

$$\begin{aligned} \boldsymbol{D}\boldsymbol{z}^{(k)} &= \boldsymbol{C}_L\boldsymbol{x}^{(k)} + \boldsymbol{C}_U\boldsymbol{x}^{(k-1)} + \boldsymbol{b} \\ \boldsymbol{x}^{(k)} &= \omega\boldsymbol{z}^{(k)} + (1-\omega)\boldsymbol{x}^{(k-1)} \end{aligned}$$

and we have the **SOR method**:

SOR Method

$$\boldsymbol{x}^{(k)} = \mathcal{L}_\omega\boldsymbol{x}^{(k-1)} + \boldsymbol{h}$$

Conjugate Gradient Method

The conjugate gradient method is one of the most popular iterative methods for solving sparse systems of linear equations. This is particularly true for systems that arise in the numerical solutions of partial differential equations. (See Section 12.1.)

We begin with a brief presentation of definitions and associated notation. Assume that the real $n \times n$ matrix $\boldsymbol{A}$ is **symmetric**, meaning that $\boldsymbol{A}^T = \boldsymbol{A}$. The **inner product** of two vectors $\boldsymbol{u} = (u_1, u_2, \ldots, u_n)$ and $\boldsymbol{v} = (v_1, v_2, \ldots, v_n)$ can be written as $\langle \boldsymbol{u}, \boldsymbol{v} \rangle = \boldsymbol{u}^T\boldsymbol{v} = \sum_{i=1}^{n} u_i v_i$, which is the scalar sum. Note that $\langle \boldsymbol{u}, \boldsymbol{v} \rangle = \langle \boldsymbol{v}, \boldsymbol{u} \rangle$. If $\boldsymbol{u}$ and $\boldsymbol{v}$ are mutually orthogonal, then $\langle \boldsymbol{u}, \boldsymbol{v} \rangle = 0$. An $\boldsymbol{A}$-**inner product** of two vectors $\boldsymbol{u}$ and $\boldsymbol{v}$ is defined as

A-inner Product

$$\langle \boldsymbol{u}, \boldsymbol{v} \rangle_A = \langle \boldsymbol{A}\boldsymbol{u}, \boldsymbol{v} \rangle = \boldsymbol{u}^T \boldsymbol{A}^T \boldsymbol{v}$$

Two nonzero vectors $\boldsymbol{u}$ and $\boldsymbol{v}$ are $\boldsymbol{A}$-**conjugate** if $\langle \boldsymbol{u}, \boldsymbol{v} \rangle_A = 0$. An $n \times n$ matrix $\boldsymbol{A}$ is **positive definite** if

Positive Definite

$$\langle \boldsymbol{x}, \boldsymbol{x} \rangle_A > 0$$

for all nonzero vectors $\boldsymbol{x} \in \mathbb{R}^n$. In general, expressions such as $\langle \boldsymbol{u}, \boldsymbol{v} \rangle$ and $\langle \boldsymbol{u}, \boldsymbol{v} \rangle_A$ reduce to 1×1 matrices and are treated as scalar values. A **quadratic form** is a scalar quadratic function of a vector of the form

Quadratic Form

$$f(\boldsymbol{x}) = \frac{1}{2}\langle \boldsymbol{x}, \boldsymbol{x} \rangle_A - \langle \boldsymbol{b}, \boldsymbol{x} \rangle + c$$

Here, $\boldsymbol{A}$ is a matrix, $\boldsymbol{x}$ and $\boldsymbol{b}$ are vectors, and c is a scalar constant. The **gradient** of a quadratic form is

Gradient

$$f'(\boldsymbol{x}) = \left[\partial f(\boldsymbol{x})/\partial x_1, \quad \partial f(\boldsymbol{x})/\partial x_2, \quad \cdots, \quad \partial f(\boldsymbol{x})/\partial x_n\right]^T$$

We can derive the following:

$$f'(\boldsymbol{x}) = \frac{1}{2}\boldsymbol{A}^T\boldsymbol{x} + \frac{1}{2}\boldsymbol{A}\boldsymbol{x} - \boldsymbol{b}$$

If $\boldsymbol{A}$ is symmetric, this reduces to

$$f'(\boldsymbol{x}) = \boldsymbol{A}\boldsymbol{x} - \boldsymbol{b}$$

Setting the gradient to zero, we obtain the linear system to be solved, $\boldsymbol{A}\boldsymbol{x} = \boldsymbol{b}$. Therefore, the solution of $\boldsymbol{A}\boldsymbol{x} = \boldsymbol{b}$ is a critical point of $f(\boldsymbol{x})$. If $\boldsymbol{A}$ is symmetric and positive definite, then $f(\boldsymbol{x})$ is minimized by the solution of $\boldsymbol{A}\boldsymbol{x} = \boldsymbol{b}$. So an alternative way of solving the linear system $\boldsymbol{A}\boldsymbol{x} = \boldsymbol{b}$ is by finding an $\boldsymbol{x}$ that minimizes $f(\boldsymbol{x})$.

We want to solve the linear system

$$\boldsymbol{A}\boldsymbol{x} = \boldsymbol{b}$$

where the $n \times n$ matrix $\boldsymbol{A}$ is symmetric and positive definite.

Suppose that $\{\boldsymbol{p}^{(1)}, \boldsymbol{p}^{(2)}, \ldots, \boldsymbol{p}^{(k)}, \ldots, \boldsymbol{p}^{(n)}\}$ is a set containing a sequence of n mutually conjugate **direction vectors**. Then they form a basis for the space $\mathbb{R}^n$. Hence, we can expand the true solution vector $\boldsymbol{x}^*$ of $\boldsymbol{A}\boldsymbol{x} = \boldsymbol{b}$ into a linear combination of these basis vectors:

Direction Vectors

$$\boldsymbol{x}^* = \alpha_1 \boldsymbol{p}^{(1)} + \alpha_2 \boldsymbol{p}^{(2)} + \cdots + \alpha^{(k)} \boldsymbol{p}^{(k)} + \cdots + \alpha_n \boldsymbol{p}^{(n)}$$

where the coefficients are given by

$$\alpha_k = \langle \boldsymbol{p}^{(k)}, \boldsymbol{b} \rangle / \langle \boldsymbol{p}^{(k)}, \boldsymbol{p}^{(k)} \rangle_A$$

This can be viewed as a direct method for solving the linear system $\boldsymbol{A}\boldsymbol{x} = \boldsymbol{b}$: First find the sequence of n conjugate direction vectors $\boldsymbol{p}^{(k)}$, and then compute the coefficients α_k. However, in practice, this approach is impractical because it takes too much computer time and storage.

On the other hand, if we view the conjugate gradient method as an iterative method, then we are able to solve large sparse linear systems in a reasonable amount of time and storage. The key is carefully choosing a small set of the conjugate direction vectors $\boldsymbol{p}^{(k)}$ so that we do not need them all to obtain a good approximation to the true solution vector.

Start with an initial guess $\boldsymbol{x}^{(0)}$ to the true solution $\boldsymbol{x}^*$. We can assume without loss of generality that $\boldsymbol{x}^{(0)}$ is the zero vector. The true solution $\boldsymbol{x}^*$ is also the unique minimizer of

Discussion of Conjugate Gradient Method

$$f(\boldsymbol{x}) = \frac{1}{2}\langle \boldsymbol{x}, \boldsymbol{x}\rangle_A - \langle \boldsymbol{x}, \boldsymbol{x}\rangle = \frac{1}{2}\boldsymbol{x}^T A\boldsymbol{x} - \boldsymbol{x}^T\boldsymbol{x}$$

for $\boldsymbol{x} \in \mathbb{R}^n$. This suggests taking the first basis vector $\boldsymbol{p}^{(1)}$ to be the gradient of f at $\boldsymbol{x} = \boldsymbol{x}^{(0)}$, which equals $-\boldsymbol{b}$. The other vectors in the basis are now conjugate to the gradient—hence the name *conjugate gradient method*. The kth residual vector is

$$\boldsymbol{r}^{(k)} = \boldsymbol{b} - A\boldsymbol{x}^{(k)}$$

The gradient descent method moves in the direction $\boldsymbol{r}^{(k)}$. Take the direction closest to the gradient vector $\boldsymbol{r}^{(k)}$ by insisting that the direction vectors $\boldsymbol{p}^{(k)}$ be conjugate to each other. Putting all this together, we obtain the expression

$$\boldsymbol{p}^{(k+1)} = \boldsymbol{r}^{(k)} - \left[\left\langle \boldsymbol{p}^{(k)}, \boldsymbol{r}^{(k)}\right\rangle_A / \left\langle \boldsymbol{p}^{(k)}, \boldsymbol{p}^{(k)}\right\rangle_A\right] \boldsymbol{p}_k$$

After some simplifications, the algorithm is obtained for solving the linear system $A\boldsymbol{x} = \boldsymbol{b}$, where the coefficient matrix A is real, symmetric, and positive definite. The input vector $\boldsymbol{x}^{(0)}$ is an initial approximation to the solution or the zero vector.

In theory, the conjugate gradient iterative method solves a system of n linear equations in at most n steps, if the matrix A is symmetric and positive definite. Moreover, the nth iterative vector $\boldsymbol{x}^{(n)}$ is the unique minimizer of the quadratic function

Quadratic Function

$$q(\boldsymbol{x}) = \frac{1}{2}\boldsymbol{x}^T A\boldsymbol{x} - \boldsymbol{x}^T\boldsymbol{b}$$

When the conjugate gradient method was introduced by Hestenes and Stiefel [1952], the initial interest in it waned once it was discovered that this finite-termination property was *not* obtained in practice. But two decades later, there was renewed interest in this method when it was viewed as an iterative process by Reid [1971] and others. In practice, the solution of a system of linear equations can often be found with satisfactory precision in a number of steps considerably less than the order of the system.

CG Pseudocode and Features

Here is a pseudocode for the **conjugate gradient algorithm**:

Conjugate Gradient Pseudocode

```
k ← 0; x ← 0; r ← b − Ax; δ ← ⟨r, r⟩
while (√δ > ε√⟨b, b⟩) and (k < k_max)
    k ← k + 1
    if k = 1 then
        p ← r
    else
        β ← δ/δ_old
        p ← r + βp
    end if
    w ← Ap
    α ← δ/⟨p, w⟩
    x ← x + αp
    r ← r − αw
    δ_old ← δ
    δ ← ⟨r, r⟩
end while
```

Here, ε is a parameter used in the convergence criterion (such as $\varepsilon = 10^{-5}$), and k_{max} is the maximum number of iterations allowed. Usually, the number of iterations needed is much less than the size of the linear system. We save the previous value of δ in the variable δ_{old}. If a good guess for the solution vector $\boldsymbol{x}$ is known, then it should be used as an initial vector instead of the zero vector. The variable ε is the desired convergence tolerance. The algorithm produces not only a sequence of vectors $\boldsymbol{x}^{(i)}$ that converges to the solution, but an orthogonal sequence of residual vectors $\boldsymbol{r}^{(i)} = \boldsymbol{b} - \boldsymbol{A}\boldsymbol{x}^{(i)}$ and an $\boldsymbol{A}$-orthogonal sequence of search direction vectors $\boldsymbol{p}^{(i)}$, namely, $\langle \boldsymbol{r}^{(i)}, \boldsymbol{r}^{(j)} \rangle = 0$ if $i \neq j$ and $\langle \boldsymbol{p}^{(i)}, \boldsymbol{A}\boldsymbol{p}^{(j)} \rangle = 0$ if $i \neq j$.

Features of Conjugate Gradient

The main computational features of the conjugate gradient algorithm are complicated to derive, but the final conclusion is that in each step, only *one* matrix-vector multiplication is required and only a few dot-products are computed. These are extremely desirable attributes in solving large and sparse linear systems. Also, unlike Gaussian elimination, there is no fill-in, so only the nonzero entries in $\boldsymbol{A}$ need to be stored in the computer memory. For some partial differential equation problems, the equations in the linear system can be represented by stencils that describe the nonzero structure within the coefficient matrix. Sometimes these stencils are used in a computer program rather than storing the nonzero entries in the coefficient matrix.

EXAMPLE 5 Use the conjugate gradient method to solve this linear system:

$$\begin{bmatrix} 2 & -1 & 0 \\ -1 & 3 & -1 \\ 0 & -1 & 2 \end{bmatrix} \begin{bmatrix} x_1 \\ x_2 \\ x_3 \end{bmatrix} = \begin{bmatrix} 1 \\ 8 \\ -5 \end{bmatrix}$$

Solution Programming the pseudocode, we obtain the iterates

$$\boldsymbol{x}^{(0)} = [0.00000,\ 0.00000,\ 0.00000]^T$$
$$\boldsymbol{x}^{(1)} = [0.29221,\ 2.33766,\ -1.46108]^T$$
$$\boldsymbol{x}^{(2)} = [1.82254,\ 2.60772,\ -1.55106]^T$$
$$\boldsymbol{x}^{(3)} = [2.00000,\ 3.00000,\ -1.00000]^T$$

In only three iterations, we have the answer accurate to full machine precision! This illustrates the finite termination property of this method. The matrix $\boldsymbol{A}$ is symmetric positive definite and the eigenvalues of $\boldsymbol{A}$ are 1, 2, 4. This simple example may be a bit misleading because one cannot expect such rapid convergence in realistic applications. (The rate of convergence depends on various properties of the linear system.) In fact, this example is too small to illustrate the power of advanced iterative methods on very large and sparse systems. ■

Preconditioning

The conjugate gradient method may converge slowly when the matrix $\boldsymbol{A}$ is ill-conditioned; however, the convergence can be accelerated by a technique called **preconditioning**. This involves a matrix $\boldsymbol{M}^{-1}$ that approximates $\boldsymbol{A}$ so that $\boldsymbol{M}^{-1}\boldsymbol{A}$ is well-conditioned and $\boldsymbol{M}\boldsymbol{x} = \boldsymbol{y}$ is easily solved. For many very large and sparse linear systems, preconditioned conjugate gradient methods have now become the iterative methods of choice! For additional details, see Golub and Van Loan [1996], as well as many other standard textbooks and references.

Summary 8.4

- For the linear system

$$\boldsymbol{A}\boldsymbol{x} = \boldsymbol{b}$$

the general form of an iterative method is

$$\boldsymbol{x}^{(k)} = \mathcal{G}\boldsymbol{x}^{(k-1)} + \boldsymbol{h}$$

where the **iteration matrix** and vector are

$$\mathcal{G} = \boldsymbol{I} - \boldsymbol{Q}^{-1}\boldsymbol{A} \qquad \boldsymbol{h} = \boldsymbol{Q}^{-1}\boldsymbol{b}$$

The **error vector** is

$$\boldsymbol{e}^{(k)} = (\boldsymbol{I} - \boldsymbol{Q}^{-1}\boldsymbol{A})\boldsymbol{e}^{(k-1)}$$

- In detail, we consider the linear system in the form

$$\sum_{j=1}^{n} a_{ij}x_j = b_i \qquad (1 \leqq i \leqq n)$$

assuming that $a_{ii} \neq 0$.

- The **Jacobi method** is

$$x_i^{(k)} = \sum_{\substack{j=1 \\ j\neq i}}^{n} (-a_{ij}/a_{ii})x_j^{(k-1)} - (b_i/a_{ii}) \qquad (1 \leqq i \leqq n)$$

- The **Gauss-Seidel method** is

$$x_i^{(k)} = \sum_{\substack{j=1 \\ j<i}}^{n} (-a_{ij}/a_{ii})x_i^{(k)} + \sum_{\substack{j=1 \\ j>i}}^{n} (-a_{ij}/a_{ii})x_j^{(k-1)} - (b_i/a_{ii})$$

- The **SOR method** is

$$x_i^{(k)} = \omega\left\{ \sum_{\substack{j=1 \\ j<i}}^{n} (-a_{ij}/a_{ii})x_i^{(k)} + \sum_{\substack{j=1 \\ j>i}}^{n} (-a_{ij}/a_{ii})x_j^{(k-1)} - (b_i/a_{ii}) \right\} + (1-\omega)x_i^{(k-1)}$$

The SOR method reduces to the Gauss-Seidel method when $\omega = 1$.

- For a matrix formulation, we split the matrix $\boldsymbol{A} = (a_{ij})$:

$$\boldsymbol{A} = \boldsymbol{D} - \boldsymbol{C}_L - \boldsymbol{C}_U$$

where $\boldsymbol{D} = \text{diag}(\boldsymbol{A})$ is a nonzero diagonal matrix, $\boldsymbol{C}_L = (-a_{ij})_{i>j}$ is a strictly lower triangular matrix, and $\boldsymbol{C}_U = (-a_{ij})_{i<j}$ is a strictly upper triangular matrix.

- The **Jacobi method** in matrix-vector form is

$$\boldsymbol{D}\boldsymbol{x}^{(k)} = (\boldsymbol{C}_L + \boldsymbol{C}_U)\boldsymbol{x}^{(k-1)} + \boldsymbol{b}$$

- The **Gauss-Seidel method** is

$$(\boldsymbol{D} - \boldsymbol{C}_L)\boldsymbol{x}^{(k)} = \boldsymbol{C}_U\boldsymbol{x}^{(k-1)} + \boldsymbol{b}$$

- The **SOR method** is

$$(\boldsymbol{D} - \omega\boldsymbol{C}_L)\boldsymbol{x}^{(k)} = [\omega\boldsymbol{C}_U + (1-\omega)\boldsymbol{D}]\boldsymbol{x}^{(k-1)} + \omega\boldsymbol{b}$$

The splitting matrices, iteration matrices, and constant vectors are as follows:

- For the **Jacobi method**, we have

$$Q = D$$
$$B = D^{-1}(C_L + C_U)$$
$$h = D^{-1}b$$

- For the **Gauss-Seidel method**, we have

$$Q = D - C_L$$
$$\mathcal{L} = (D - C_L)^{-1}C_U$$
$$h = (D - C_L)^{-1}b$$

- For the **SOR method**, we have

$$Q = \frac{1}{\omega}(D - \omega C_L)$$
$$\mathcal{L}_\omega = (D - \omega C_L)^{-1}[\omega C_U + (1-\omega)D]$$
$$h = \omega(D - \omega C_L)^{-1}b$$

- An iterative method converges for a specific matrix A if and only if

$$\rho(I - Q^{-1}A) < 1$$

If A is diagonally dominant, then the Jacobi and Gauss-Seidel methods converge for any $x^{(0)}$. The SOR method converges, for $0 < \omega < 2$ and any $x^{(0)}$, if and only if A is symmetric and positive definite with positive diagonal elements.

Exercises 8.4

[a]**1.** Give an alternative solution to Example 4.

2. Write the matrix formula for the Gauss-Seidel overrelaxation method.

[a]**3.** (**Multiple Choice**) In solving a system of equations $Ax = b$, it is often convenient to use an iterative method, which generates a sequence of $x^{(k)}$ vectors that should converge to a solution. The process is stopped when sufficient accuracy has been attained. A general procedure is to obtain $x^{(k)}$ by solving $Qx^{(k)} = (Q - A)x^{(k-1)} + b$. Here, Q is a certain matrix that is usually connected somehow to A. The process is repeated, starting with any available guess, $x^{(0)}$. What hypothesis guarantees that the method works, no matter what starting point is selected?

a. $||Q|| < 1$ **b.** $||QA|| < 1$

c. $||I - QA|| < 1$ **d.** $||I - Q^{-1}A|| < 1$

e. None of these.

Hint: The spectral radius is less than or equal to the norm.

4. (**Multiple Choice**) From a vector norm, we can create a subordinate matrix norm. Which relation is satisfied by every subordinate matrix norm?

a. $||Ax|| \geqq ||A||\,||x||$ **b.** $||I|| = 1$

c. $||AB|| \geqq ||A||\,||B||$ **d.** $||A + B|| \geqq ||A|| + ||B||$

e. None of these.

[a]**5.** (**Multiple Choice**) The condition for diagonal dominance of a matrix A is:

a. $|a_{ii}| < \sum_{j=1, j\neq i}^{n} |a_{ij}|$ **b.** $|a_{ii}| \geqq \sum_{j=1, j\neq i}^{n} |a_{ij}|$

c. $|a_{ii}| < \sum_{j=1}^{n} |a_{ij}|$ **d.** $|a_{ii}| > \sum_{j=1}^{n} |a_{ij}|$

e. None of these.

6. (**Multiple Choice**) A necessary and sufficient condition for the standard iteration formula $x^{(k)} = \mathcal{G}x^{(k-1)} + h$ to produce a sequence $x^{(k)}$ that converges to a solution of the equation $(I - \mathcal{G})x = h$ is that:

a. The spectral radius of $\mathcal{G}$ is greater than 1.

b. The matrix $\mathcal{G}$ is diagonally dominant.

c. The spectral radius of $\mathcal{G}$ is less than 1.

d. $\mathcal{G}$ is nonsingular. **e.** None of these.

7. (**Multiple Choice**) A sufficient condition for the Jacobi method to converge for the linear system $\boldsymbol{Ax} = \boldsymbol{b}$.

a. $\boldsymbol{A} - \boldsymbol{I}$ is diagonally dominant.

b. $\boldsymbol{A}$ is diagonally dominant.

c. $\mathcal{G}$ is nonsingular.

d. The spectral radius of $\mathcal{G}$ is less than 1.

e. None of these.

8. (**Multiple Choice**) A sufficient condition for the Gauss-Seidel method to work on the linear system $\boldsymbol{Ax} = \boldsymbol{b}$.

a. $\boldsymbol{A}$ is diagonally dominant.

b. $\boldsymbol{A} - \boldsymbol{I}$ is diagonally dominant.

c. The spectral radius of $\boldsymbol{A}$ is less than 1.

d. $\mathcal{G}$ is nonsingular. **e.** None of these.

[a]**9.** (**Multiple Choice**) Necessary and sufficient conditions for the SOR method, where $0 < \omega < 2$, to work on the linear system $\boldsymbol{Ax} = \boldsymbol{b}$.

a. $\boldsymbol{A}$ is diagonally dominant. **b.** $\rho(\boldsymbol{A}) < 1$.

c. $\boldsymbol{A}$ is symmetric positive definite.

d. $\boldsymbol{x}^{(0)} = 0$. **e.** None of these.

10. The **Frobenius norm**

$$||A||_F = \sqrt{\sum_{i=1}^{n}\sum_{j=1}^{n} |a_{ij}|^2}$$

is frequently used because it is so easy to compute. Find the value of this norm for these matrices:

a. $\begin{bmatrix} 1 & 2 & 3 \\ 0 & 5 & 4 \\ 2 & 1 & 3 \end{bmatrix}$ **b.** $\begin{bmatrix} 0 & 0 & 1 & 2 \\ 3 & 0 & 5 & 4 \\ 1 & 1 & 1 & 2 \\ 1 & 3 & 2 & 2 \end{bmatrix}$

c. $\begin{bmatrix} 1 & 1 & 1 & 1 & 1 \\ 2 & 3 & 4 & 5 & 6 \\ 0 & 1 & 0 & 1 & 0 \\ 3 & 4 & 3 & 4 & 3 \\ 5 & 5 & 5 & 5 & 5 \end{bmatrix}$ **d.** $\begin{bmatrix} -3 & -1 & 3 & -1 \\ 1 & 3 & 1 & -3 \\ 3 & -1 & 3 & 1 \\ 0 & 3 & 0 & 1 \end{bmatrix}$

11. Determine the condition numbers $\kappa(\boldsymbol{A})$ of these matrices:

a. $\begin{bmatrix} -2 & 1 & 0 \\ 1 & -2 & 1 \\ 0 & 1 & -2 \end{bmatrix}$ **b.** $\begin{bmatrix} 0 & 0 & 1 \\ 0 & 1 & 0 \\ 1 & 1 & 1 \end{bmatrix}$

c. $\begin{bmatrix} 3 & 0 & 0 \\ 0 & 2 & 0 \\ 0 & 0 & 1 \end{bmatrix}$ **d.** $\begin{bmatrix} -2 & -1 & 2 & -1 \\ 1 & 2 & 1 & -2 \\ 2 & -1 & 2 & 1 \\ 0 & 2 & 0 & 1 \end{bmatrix}$

Computer Exercises 8.4

1. Redo several or all of Examples 1–5 using the linear system involving one of the following coefficient matrix and right-hand side vector pairs:

a. $\boldsymbol{A} = \begin{bmatrix} 5 & -1 \\ -1 & 3 \end{bmatrix}, \qquad \boldsymbol{b} = \begin{bmatrix} 7 \\ 4 \end{bmatrix}$

b. $\boldsymbol{A} = \begin{bmatrix} 5 & -1 & 0 \\ -1 & 3 & -1 \\ 0 & -1 & 2 \end{bmatrix}, \qquad \boldsymbol{b} = \begin{bmatrix} 7 \\ 4 \\ 5 \end{bmatrix}$

c. $\boldsymbol{A} = \begin{bmatrix} 2 & -1 & 0 \\ -1 & 6 & -2 \\ 4 & -3 & 8 \end{bmatrix}, \qquad \boldsymbol{b} = \begin{bmatrix} 1 \\ 3 \\ 9 \end{bmatrix}$

d. $\boldsymbol{A} = \begin{bmatrix} 7 & 3 & -1 \\ 3 & 8 & 1 \\ -1 & 1 & 4 \end{bmatrix}, \qquad \boldsymbol{b} = \begin{bmatrix} 3 \\ -4 \\ 2 \end{bmatrix}$

2. Using the Jacobi, Gauss-Seidel, and SOR ($\omega = 1.1$) iterative methods, write and execute a computer program to solve the following linear system to four decimal places (rounded) of accuracy:

$$\begin{bmatrix} 7 & 1 & -1 & 2 \\ 1 & 8 & 0 & -2 \\ -1 & 0 & 4 & -1 \\ 2 & -2 & -1 & 6 \end{bmatrix} \begin{bmatrix} x_1 \\ x_2 \\ x_3 \\ x_4 \end{bmatrix} = \begin{bmatrix} 3 \\ -5 \\ 4 \\ -3 \end{bmatrix}$$

Compare the number of iterations needed in each case. *Hint*: The exact solution is $\boldsymbol{x} = (1, -1, 1, -1)^T$.

[a]**3.** Using the Jacobi, Gauss-Seidel, and the SOR ($\omega = 1.4$) iterative methods, write and run code to solve the following linear system to four decimal places of accuracy:

$$\begin{bmatrix} 7 & 3 & -1 & 2 \\ 3 & 8 & 1 & -4 \\ -1 & 1 & 4 & -1 \\ 2 & -4 & -1 & 6 \end{bmatrix} \begin{bmatrix} x_1 \\ x_2 \\ x_3 \\ x_4 \end{bmatrix} = \begin{bmatrix} -1 \\ 0 \\ -3 \\ 1 \end{bmatrix}$$

Compare the number of iterations in each case. *Hint*: Here, the exact solution is $\boldsymbol{x} = (-1, 1, -1, 1)^T$.

4. (Continuation) Solve the system using the SOR iterative method with values of $\omega = 1(0.1)2$. Plot the number of iterations for convergence versus the values of ω. Which value of ω results in the fastest convergence?

5. Program and run the Jacobi, Gauss-Seidel, and SOR methods for the system of Example 1 as follows:

a. Use equations involving the splitting matrix Q.

b. Use matrix-vector multiplication.

c. Use the equation formulations in Examples 1-3.

6. (Continuation) Select one or more of the systems in Computer Exercise 8.4.1, and rerun these programs.

[a]**7.** Consider the linear system

$$\begin{bmatrix} 9 & -3 \\ -2 & 8 \end{bmatrix} \begin{bmatrix} x_1 \\ x_2 \end{bmatrix} = \begin{bmatrix} 6 \\ -4 \end{bmatrix}$$

Using mathematical software, compare solving it by using the Jacobi method and the Gauss-Seidel method starting with $\boldsymbol{x}^{(0)} = (0, 0)^T$.

8. (Continuation)

a. Change the $(1, 1)$ entry from 9 to 1 so that the coefficient matrix is no longer diagonally dominant and see whether the Gauss-Seidel method still works. Explain why or why not.

b. Then change the $(2, 2)$ entry from 8 to 1 as well and test. Again explain the results.

9. Use the conjugate gradient method to solve this linear system:

$$\begin{bmatrix} 2.0 & -0.3 & -0.2 \\ -0.3 & 2.0 & -0.1 \\ -0.2 & -0.1 & 2.0 \end{bmatrix} \begin{bmatrix} x_1 \\ x_2 \\ x_3 \end{bmatrix} = \begin{bmatrix} 7 \\ 5 \\ 3 \end{bmatrix}$$

10. (**Euler-Bernoulli Beam**) A simple model for a bending beam under stress involves the Euler-Bernoulli differential equation. A finite difference discretization converts it into a system of linear equations. As the size of the discretization decreases, the linear system becomes larger and more ill-conditioned.

a. For a beam pinned at both ends, we obtain the following banded system of linear equations with a bandwidth of five:

$$\begin{bmatrix}
12 & -6 & \frac{4}{3} & & & & & & \\
-4 & 6 & -4 & 1 & & & & & \\
1 & -4 & 6 & -4 & 1 & & & & \\
 & 1 & -4 & 6 & -4 & 1 & & & \\
 & & \ddots & \ddots & \ddots & \ddots & \ddots & \ddots & \\
 & & & 1 & -4 & 6 & -4 & 1 & \\
 & & & & 1 & -4 & 6 & -4 & 1 \\
 & & & & & 1 & -4 & 6 & -4 \\
 & & & & & & \frac{4}{3} & -6 & 12
\end{bmatrix}
\times
\begin{bmatrix} y_1 \\ y_2 \\ y_3 \\ y_4 \\ \vdots \\ y_{n-3} \\ y_{n-2} \\ y_{n-1} \\ y_n \end{bmatrix}
=
\begin{bmatrix} b_1 \\ b_2 \\ b_3 \\ b_4 \\ \vdots \\ b_{n-3} \\ b_{n-2} \\ b_{n-1} \\ b_n \end{bmatrix}$$

The right-hand side represents forces on the beam. Set the right-hand side so that there is a known solution, such as a sag in the middle of the beam. Using an iterative method, repeatedly solve the system by allowing n to increase. Does the error in the solution increase when n increases? Use mathematical software that computes the condition number of the coefficient matrix to explain what is happening.

b. The linear system of equations for a **cantilever beam** with a free boundary condition at only one end is

$$\begin{bmatrix}
12 & -6 & \frac{4}{3} & & & & & & \\
-4 & 6 & -4 & 1 & & & & & \\
1 & -4 & 6 & -4 & 1 & & & & \\
 & 1 & -4 & 6 & -4 & 1 & & & \\
 & & \ddots & \ddots & \ddots & \ddots & \ddots & \ddots & \\
 & & & 1 & -4 & 6 & -4 & 1 & \\
 & & & & 1 & -4 & 6 & -4 & 1 \\
 & & & & & 1 & -\frac{93}{25} & \frac{111}{25} & -\frac{43}{25} \\
 & & & & & & \frac{12}{25} & \frac{24}{25} & \frac{12}{25}
\end{bmatrix}
\times
\begin{bmatrix} y_1 \\ y_2 \\ y_3 \\ y_4 \\ \vdots \\ y_{n-3} \\ y_{n-2} \\ y_{n-1} \\ y_n \end{bmatrix}
=
\begin{bmatrix} b_1 \\ b_2 \\ b_3 \\ b_4 \\ \vdots \\ b_{n-3} \\ b_{n-2} \\ b_{n-1} \\ b_n \end{bmatrix}$$

Repeat the numerical experiment for this system. See Sauer [2012] for additional details.

11. Consider this sparse linear system:

$$\left[\begin{array}{cccccccccc}
3 & -1 & & & & & & & & \frac{1}{2} \\
-1 & 3 & -1 & & & & & & \frac{1}{2} & \\
 & -1 & 3 & -1 & & & & \frac{1}{2} & & \\
 & & \ddots & \ddots & \ddots & \cdot^{\cdot^{\cdot}} & & & & \\
 & & & -1 & 3 & -1 & & & & \\
 & & & \cdot^{\cdot^{\cdot}} & \ddots & \ddots & \ddots & & & \\
 & & \frac{1}{2} & & & -1 & 3 & -1 & & \\
 & \frac{1}{2} & & & & & -1 & 3 & -1 & \\
\frac{1}{2} & & & & & & & -1 & 3 &
\end{array}\right]
\times \begin{bmatrix} x_1 \\ x_2 \\ x_3 \\ \vdots \\ \vdots \\ \vdots \\ x_{n-2} \\ x_{n-1} \\ x_n \end{bmatrix}
= \begin{bmatrix} 2.5 \\ 1.5 \\ 1.5 \\ \vdots \\ 1.0 \\ \vdots \\ 1.5 \\ 1.5 \\ 2.5 \end{bmatrix}$$

The true solution is $x = [1, 1, 1, \ldots, 1, 1, 1]^T$. Use an iterative method to solve this system for increasing values of n.

12. Consider the sample two-dimensional linear system

$$A x \begin{bmatrix} 3 & 2 \\ 2 & 6 \end{bmatrix} \begin{bmatrix} x_1 \\ x_1 \end{bmatrix} = \begin{bmatrix} 2 \\ -8 \end{bmatrix} = \boldsymbol{b}$$

and $\boldsymbol{c} = 0$. Plot graphs to show the following:

a. The solution lies at the intersection of two lines.

b. Graph of the quadratic form

$$\boldsymbol{F}(\boldsymbol{x}) = c + b^T x + \frac{1}{2}\boldsymbol{x}^T \boldsymbol{A}\boldsymbol{x}$$

showing that the minimum point of this surface is the solution of $\boldsymbol{A}\boldsymbol{x} = \boldsymbol{b}$.

c. Contours of the quadratic form so that each ellipsoidal curve has a constant value.

d. Gradient $F'(\boldsymbol{x})$ of the quadratic form. Show that for every $\boldsymbol{x}$, the gradient points in the direction of the steepest increase of $F(\boldsymbol{x})$ and is orthogonal to the contour lines. (See Section 13.2.)

9 Least Squares Methods and Fourier Series

Surface tension S in a liquid is known to be a linear function of temperature T. For a particular liquid, measurements have been made of the surface tension at certain temperatures. The results were as follows:

T	0	10	20	30	40	80	90	95
S	68.0	67.1	66.4	65.6	64.6	61.8	61.0	60.0

How can the most probable values of the constants in the equation

$$S = aT + b$$

be determined? Methods for solving such problems are developed in this chapter.

9.1 Method of Least Squares

Linear Least Squares

In experimental, social, and behavioral sciences, an experiment or survey often produces a mass of data. To interpret the data, the investigator may resort to graphical methods. For instance, an experiment in physics might produce a numerical table of the form

$m + 1$ Data Points (x_i, y_i)

x	x_0	x_1	$\cdots$	x_m
y	y_0	y_1	$\cdots$	y_m

(1)

and from it, $m + 1$ points on a graph could be plotted. Suppose that the resulting graph looks like Figure 9.1. A reasonable tentative conclusion is that the underlying function is *linear* and that the failure of the points to fall *precisely* on a straight line is due to experimental error. Proceeding on this assumption—or if theoretical reasons exist for believing that the function is indeed **linear**—the next step is to determine the correct function. Assuming that

Linear Case

$$y = ax + b$$

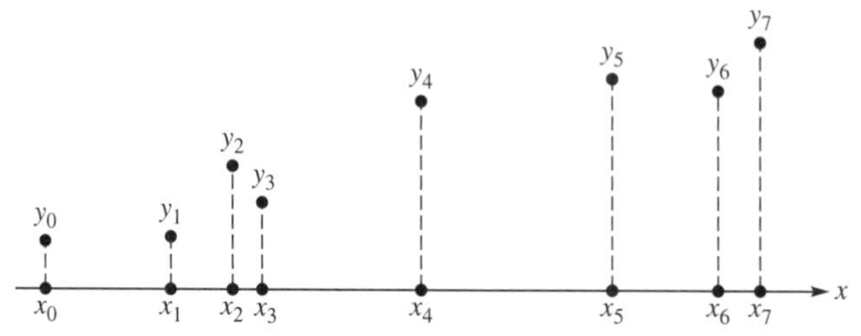

FIGURE 9.1 Experimental data

we want to find the coefficients a and b. Thinking geometrically, we ask:

What line most nearly passes through the eight points plotted?

To answer this question, suppose that a guess is made about the correct values of a and b. This is equivalent to deciding on a specific line to represent the data. In general, the data points may not fall on the line $y = ax + b$. If by chance the kth datum falls on the line, then

$$ax_k + b - y_k = 0$$

If it does not, then there is a **discrepancy** or **error** of magnitude

$$|\, ax_k + b - y_k \,|$$

The total absolute error for all $m + 1$ points is therefore

$$\sum_{k=0}^{m} |\, ax_k + b - y_k \,|$$

ℓ_1 Approximation

This is a function of a and b, and it would be reasonable to choose a and b so that the function assumes its minimum value. This problem is an example of **ℓ_1 approximation** and can be solved by the techniques of **linear programming**, a subject dealt with in Chapter 14. (The methods of calculus do not work on this function because it is not generally differentiable.)

In practice, it is common to minimize a different error function of a and b:

Minimize $\varphi(a, b)$

$$\varphi(a, b) = \sum_{k=0}^{m} (ax_k + b - y_k)^2 \tag{2}$$

ℓ_2 Approximation

This function is suitable because of statistical considerations. Explicitly, if the errors follow a *normal probability distribution*, then the minimization of φ produces a best estimate of a and b. This is called an **ℓ_2 approximation**. Another advantage is that the methods of calculus can be used on Equation (2).

The ℓ_1 and ℓ_2 approximations are related to specific cases of the **ℓ_p norm** defined by

ℓ_p Norm

$$\|x\|_p = \left\{ \sum_{i=1}^{n} |x_i|^p \right\}^{1/p}, \qquad (1 \leqq p < \infty)$$

for the vector $x = [x_1, x_2, \ldots, x_n]^T$.

Let us try to make $\varphi(a, b)$ a minimum. By calculus, the conditions

$$\frac{\partial \varphi}{\partial a} = 0 \qquad \frac{\partial \varphi}{\partial b} = 0$$

(partial derivatives of φ with respect to a and b, respectively) are *necessary* at the minimum. Taking derivatives in Equation (2), we obtain

$$\begin{cases} \displaystyle\sum_{k=0}^{m} 2(ax_k + b - y_k)x_k = 0 \\ \displaystyle\sum_{k=0}^{m} 2(ax_k + b - y_k) = 0 \end{cases}$$

This is a pair of simultaneous linear equations in the unknowns a and b. They are called the **normal equations** and can be written as

Normal Equations

$$\begin{cases} \left(\sum_{k=0}^{m} x_k^2\right) a + \left(\sum_{k=0}^{m} x_k\right) b = \sum_{k=0}^{m} y_k x_k \\ \left(\sum_{k=0}^{m} x_k\right) a + \quad (m+1) b = \sum_{k=0}^{m} y_k \end{cases} \tag{3}$$

Here, of course, $\sum_{k=0}^{m} 1 = m + 1$, which is the number of data points. To simplify the notation, we set

$$p = \sum_{k=0}^{n} x_k, \qquad q = \sum_{k=0}^{n} y_k, \qquad r = \sum_{k=0}^{n} x_k y_k, \qquad s = \sum_{k=0}^{n} x_k^2$$

The system of Equations (3) is now

2 × 2 Linear System

$$\begin{bmatrix} s & p \\ p & m+1 \end{bmatrix} \begin{bmatrix} a \\ b \end{bmatrix} = \begin{bmatrix} r \\ q \end{bmatrix}$$

We solve this pair of equations by Gaussian elimination and obtain the following algorithm. Alternatively, since this is a 2 × 2 linear system, we can use Cramer's Rule* to solve it. The determinant of the coefficient matrix is

Using Cramer's Rule

$$d = \text{Det} \begin{bmatrix} s & p \\ p & m+1 \end{bmatrix} = (m+1)s - p^2$$

Moreover, we obtain

$$a = \frac{1}{d} \text{Det} \begin{bmatrix} r & p \\ q & m+1 \end{bmatrix} = \frac{1}{d} [(m+1)r - pq]$$

$$b = \frac{1}{d} \text{Det} \begin{bmatrix} s & r \\ p & q \end{bmatrix} = \frac{1}{d} [sq - pr]$$

We can write this as an algorithm:

■ Algorithm

Linear Least Squares

The coefficients in the least-squares line $y = ax + b$ through the set of $m + 1$ data points (x_k, y_k) for $k = 0, 1, 2, \ldots, m$ are computed (in order) as follows:

1. $p = \sum_{k=0}^{m} x_k$
2. $q = \sum_{k=0}^{m} y_k$
3. $r = \sum_{k=0}^{m} x_k y_k$
4. $s = \sum_{k=0}^{m} x_k^2$
5. $d = (m+1)s - p^2$
6. $a = [(m+1)r - pq] / d$
7. $b = [sq - pr] / d$

*Cramer's Rule is given in Appendix D.

Another form of this result is

Coefficients a and b in Detail

$$a = \frac{1}{d}\left[(m+1)\left(\sum_{k=0}^{m} x_k y_k\right) - \left(\sum_{k=0}^{m} x_k\right)\left(\sum_{k=0}^{m} y_k\right)\right]$$
$$b = \frac{1}{d}\left[\left(\sum_{k=0}^{m} x_k^2\right)\left(\sum_{k=0}^{m} y_k\right) - \left(\sum_{k=0}^{m} x_k\right)\left(\sum_{k=0}^{m} x_k y_k\right)\right] \tag{4}$$

where

$$d = (m+1)\left(\sum_{k=0}^{m} x_k^2\right) - \left(\sum_{k=0}^{m} x_k\right)^2$$

Linear Example

The preceding analysis illustrates the **least-squares** procedure in the simple linear case.

EXAMPLE 1 As a concrete example, find the linear least-squares solution for the following table of values:

x	4	7	11	13	17
y	2	0	2	6	7

Plot the original data points and the line using a finer set of grid points.

Solution The equations in Algorithm 1 lead to this system of two equations:

$$\begin{cases} 644a + 52b = 227 \\ 52a + 5b = 17 \end{cases}$$

whose solution is $a = 0.4864$ and $b = -1.6589$. By Equation (3), we obtain the value $\varphi(a, b) = 10.7810$. Figure 9.2 is a plot of the given data and the linear least-squares straight line.

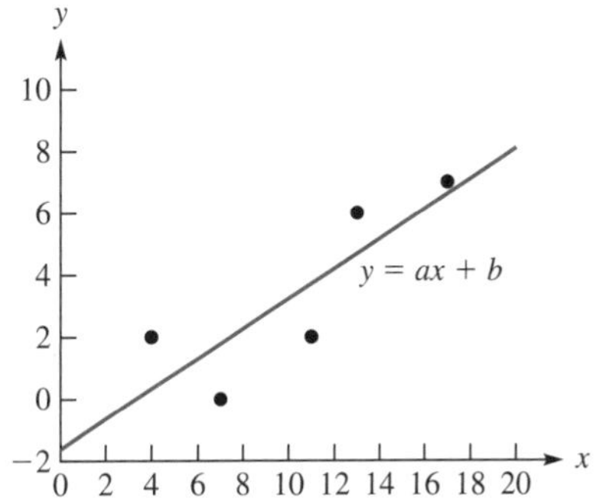

FIGURE 9.2
Linear least squares

■

We can use mathematical software such as MATLAB, Maple, or Mathematica to fit a linear least-squares polynomial to the data and verify the value of φ. (See Computer Exercise 9.1.5.)

Four Data Points

To understand what is going on here, we want to determine the equation of a line of the form $y = ax + b$ that fits the data best in the least-squares sense. With four data points (x_i, y_i), we have four equations $y_i = ax_i + b$ for $i = 1, 2, 3, 4$ that can be written as

$$\boldsymbol{Ax} = \boldsymbol{y}$$

where

4 × 2 Linear System

$$\begin{bmatrix} x_1 & 1 \\ x_2 & 1 \\ x_3 & 1 \\ x_4 & 1 \end{bmatrix} \begin{bmatrix} a \\ b \end{bmatrix} = \begin{bmatrix} y_1 \\ y_2 \\ y_3 \\ y_4 \end{bmatrix}$$

In general, we want to solve a linear system

$m \times n$ Linear System

$$\boldsymbol{Ax} = \boldsymbol{b}$$

where $\boldsymbol{A}$ is an $m \times n$ matrix and $m > n$. The solution coincides with the solution of the **normal equations**

Normal Equations

$$\boldsymbol{A}^T\boldsymbol{Ax} = \boldsymbol{A}^T\boldsymbol{b}$$

This corresponds to minimizing $||\boldsymbol{Ax} - \boldsymbol{b}||_2^2$.

Nonpolynomial Example

The method of least squares is not restricted to linear (first-degree) polynomials or to any specific functional form. Suppose, for instance, that we want to fit a table of values (x_k, y_k), where $k = 0, 1, \ldots, m$, by a function of the form

Nonpolynomial Case

$$y = a \ln x + b \cos x + ce^x$$

in the least-squares sense. The unknowns in this problem are the three coefficients a, b, and c. We consider the function

$$\varphi(a, b, c) = \sum_{k=0}^{m} (a \ln x_k + b \cos x_k + ce^{x_k} - y_k)^2$$

and set $\partial\varphi/\partial a = 0$, $\partial\varphi/\partial b = 0$, and $\partial\varphi/\partial c = 0$. This results in the following three normal equations:

3 × 3 Normal Equations

$$\begin{cases} a\displaystyle\sum_{k=0}^{m}(\ln x_k)^2 + b\sum_{k=0}^{m}(\ln x_k)(\cos x_k) + c\sum_{k=0}^{m}(\ln x_k)e^{x_k} = \sum_{k=0}^{m} y_k \ln x_k \\ a\displaystyle\sum_{k=0}^{m}(\ln x_k)(\cos x_k) + b\sum_{k=0}^{m}(\cos x_k)^2 + c\sum_{k=0}^{m}(\cos x_k)e^{x_k} = \sum_{k=0}^{m} y_k \cos x_k \\ a\displaystyle\sum_{k=0}^{m}(\ln x_k)e^{x_k} + b\sum_{k=0}^{m}(\cos x_k)e^{x_k} + c\sum_{k=0}^{m}(e^{x_k})^2 = \sum_{k=0}^{m} y_k e^{x_k} \end{cases}$$

EXAMPLE 2 Fit a function of the form $y = a \ln x + b \cos x + ce^x$ to the following table values:

x	0.24	0.65	0.95	1.24	1.73	2.01	2.23	2.52	2.77	2.99
y	0.23	−0.26	−1.10	−0.45	0.27	0.10	−0.29	0.24	0.56	1.00

Solution Using the table and the equations above, we obtain the 3 × 3 system

$$\begin{cases} 6.79410a - 5.34749b + 63.25889c = 1.61627 \\ -5.34749a + 5.10842b - 49.00859c = -2.38271 \\ 63.25889a - 49.00859b + 1002.50650c = 26.77277 \end{cases}$$

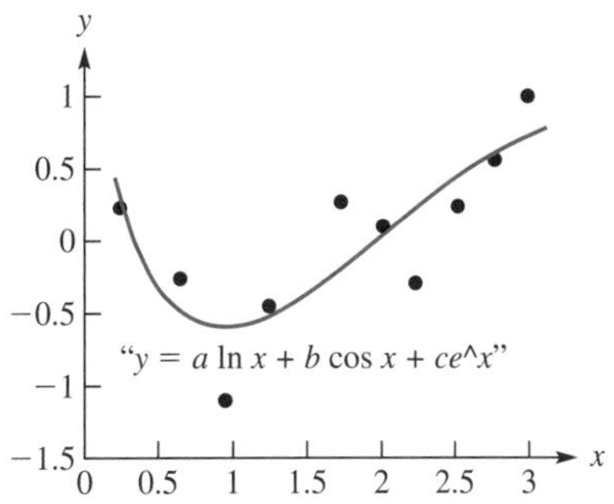

FIGURE 9.3 Nonpolynomial least squares

It has the solution $a = -1.04103$, $b = -1.26132$, and $c = 0.03073$. So the curve

$$y = -1.04103 \ln x - 1.26132 \cos x + 0.03073 e^x$$

has the required form and fits the table in the least-squares sense. The value of $\varphi(a, b, c)$ is 0.92557. Figure 9.3 is a plot of the given data and the nonpolynomial least-squares curve. ■

We can use mathematical software such as MATLAB, Maple, or Mathematica to verify these results and to plot the solution curve. (See Computer Exercise 9.1.6.)

Basis Functions $\{g_0, g_1, \ldots, g_n\}$

The principle of least squares, illustrated in these two simple cases, can be extended to general linear families of functions without involving any new ideas. Suppose that the data in Equation (1) are thought to conform to a relationship such as

More General Case

$$y = \sum_{j=0}^{n} c_j g_j(x) \tag{5}$$

in which the functions $g_0, g_1, \ldots, g_n$ (called **basis functions**) are known and held fixed. The coefficients $c_0, c_1, \ldots, c_n$ are to be determined according to the principle of least squares. In other words, we define the expression

Basis Functions

$$\varphi(c_0, c_1, \ldots, c_n) = \sum_{k=0}^{m} \left[\sum_{j=0}^{n} c_j g_j(x_k) - y_k \right]^2 \tag{6}$$

and select the coefficients to make it as small as possible. Of course, the expression $\varphi(c_0, c_1, \ldots, c_n)$ is the sum of the squares of the errors associated with each entry (x_k, y_k) in the given table.

Proceeding as before, we write down as necessary conditions for the minimum n equations

$$\frac{\partial \varphi}{\partial c_i} = 0 \qquad (0 \leqq i \leqq n)$$

These partial derivatives are obtained from Equation (7). Indeed, we have

$$\frac{\partial \varphi}{\partial c_i} = \sum_{k=0}^{m} 2 \left[\sum_{j=0}^{n} c_j g_j(x_k) - y_k \right] g_i(x_k) \qquad (0 \leqq i \leqq n)$$

When set equal to zero, the resulting equations can be rearranged as

Normal Equations

$$\sum_{j=0}^{n}\left[\sum_{k=0}^{m} g_i(x_k)g_j(x_k)\right]c_j = \sum_{k=0}^{m} y_k g_i(x_k) \qquad (0 \leqq i \leqq n) \tag{7}$$

These are the **normal equations** in this situation and serve to determine the best values of the parameters $c_0, c_1, \ldots, c_n$. The normal equations are linear in c_i; thus, in principle, they can be solved by the method of Gaussian elimination (see Chapter 2).

Necessary Conditions

In practice, the normal equations may be difficult to solve if care is not taken in choosing the basis functions $g_0, g_1, \ldots, g_n$. First, the set $\{g_0, g_1, \ldots, g_n\}$ should be **linearly independent**. This means that no linear combination $\sum_{i=0}^{n} c_i g_i$ can be the zero function (except in the trivial case when $c_0 = c_1 = \cdots = c_n = 0$). Second, the functions $g_0, g_1, \ldots, g_n$ should be *appropriate* to the problem at hand. Finally, one should choose a set of basis functions that is *well conditioned* for numerical work. We elaborate on this aspect of the problem in Section 9.2.

Summary 9.1

- We wish to find a line $y = ax + b$ that most nearly passes through the $m + 1$ pairs of points (x_i, y_i) for $0 \leqq i \leqq m$. An example of **ℓ_1 approximation** is to choose a and b so that the total absolute error for all these points is minimized:

$$\sum_{k=0}^{m} |\, ax_k + b - y_k \,|$$

This can be solved by the techniques of linear programming.

- An **ℓ_2 approximation** will minimize a different error function of a and b:

$$\varphi(a, b) = \sum_{k=0}^{m} (ax_k + b - y_k)^2$$

The minimization of φ produces a best estimate of a and b in the least-squares sense. One solves the **normal equations**

$$\begin{cases} \left(\sum_{k=0}^{m} x_k^2\right) a + \left(\sum_{k=0}^{m} x_k\right) b = \sum_{k=0}^{m} y_k x_k \\ \left(\sum_{k=0}^{m} x_k\right) a + \quad (m+1) b = \sum_{k=0}^{m} y_k \end{cases}$$

- In a more general case, the data points conform to a relationship such as

$$y = \sum_{j=0}^{n} c_j g_j(x)$$

in which the **basis functions** $g_0, g_1, \ldots, g_n$ are known and held fixed. The coefficients $c_0, c_1, \ldots, c_n$ are to be determined according to the principle of least squares. The **normal equations** in this situation are

$$\sum_{j=0}^{n}\left[\sum_{k=0}^{m} g_i(x_k)g_j(x_k)\right]c_j = \sum_{k=0}^{m} y_k g_i(x_k) \qquad (0 \leqq i \leqq n)$$

and can be solved, in principle, by the method of Gaussian elimination to determine the best values of the parameters $c_0, c_1, \ldots, c_n$.

Exercises 9.1

[a] **1.** Using the method of least squares, find the constant function that best fits the following data:

x	-1	2	3
y	$\frac{5}{4}$	$\frac{4}{3}$	$\frac{5}{12}$

[a] **2.** Determine the *constant* function c that is produced by the least-squares theory applied to a table of $m+1$ data points. Does the resulting formula involve the points x_k in any way? Apply your general formula to the preceding exercise.

[a] **3.** Find an equation of the form $y = ae^{x^2} + bx^3$ that best fits the points $(-1, 0)$, $(0, 1)$, and $(1, 2)$ in the least-squares sense.

4. Suppose that the x points in a table of $m+1$ data points are situated symmetrically about 0 on the x-axis. In this case, there is an especially simple formula for the line that best fits the points. Find it.

[a] **5.** Find the equation of a parabola of form $y = ax^2 + b$ that best represents the following data. Use the method of least squares.

x	-1	0	1
y	3.1	0.9	2.9

6. Suppose that a table of $m+1$ data points is known to conform to a function like $y = x^2 - x + c$. What value of c is obtained by the least-squares theory?

[a] **7.** Suppose that a table of $m+1$ data points is thought to be represented by a function $y = c \log x$. If so, what value for c emerges from the least-squares theory?

8. Show that Equation (4) is the solution of Equation (3).

9. (Continuation) How do we know that divisor d is not zero? In fact, show that d is positive for $m \geq 1$.
Hint: Show that

$$d = \sum_{k=0}^{m} \sum_{l=0}^{k-1} (x_k - x_l)^2$$

by induction on m. The Cauchy-Schwarz inequality can also be used to prove that $d > 0$.

10. (Continuation) Show that a and b can also be computed as follows:

$$\widehat{x} = \frac{1}{m+1} \sum_{k=0}^{m} x_k, \quad \widehat{y} = \frac{1}{m+1} \sum_{k=0}^{m} y_k$$

$$c = \sum_{k=0}^{m} (x_k - \widehat{x})^2,$$

$$a = \frac{1}{c} \sum_{k=0}^{m} (x_k - \widehat{x})(y_k - \widehat{y}), \quad b = \widehat{y} - a\widehat{x}$$

Hint: Show that $d = (m+1)c$.

[a] **11.** How do we know that the coefficients $c_0, c_1, \ldots, c_n$ that satisfy the normal Equations (7) do not lead to a maximum in the function defined by Equation (6)?

[a] **12.** If a table of $m+1$ data points is thought to conform to a relationship $y = \log(cx)$, what is the value of c obtained by the method of least squares?

[a] **13.** What straight line best fits the following data

x	1	2	3	4
y	0	1	1	2

in the least-squares sense?

14. In analytic geometry, we learn that the distance from a point (x_0, y_0) to a line represented by the equation $ax + by = c$ is $(ax_0 + by_0 - c)(a^2 + b^2)^{-1/2}$. Determine a straight line that fits a table of data points (x_i, y_i), for $0 \leq i \leq m$, in such a way that the sum of the squares of the distances from the points to the line is minimized.

15. Show that if a straight line is fitted to a table (x_i, y_i) by the method of least squares, then the line will pass through the point (x^*, y^*), where x^* and y^* are the averages of the x_i's and y_i's, respectively.

[a] **16.** The viscosity V of a liquid is known to vary with temperature according to a quadratic law $V = a + bT + cT^2$. Find the best values of a, b, and c for the following table:

T	1	2	3	4	5	6	7
V	2.31	2.01	1.80	1.66	1.55	1.47	1.41

17. An experiment involves two independent variables x and y and one dependent variable z. How can a function $z = a + bx + cy$ be fitted to the table of points (x_k, y_k, z_k)? Give the normal equations.

[a] **18.** Find the best function (in the least-squares sense) that fits the following data points and is of the form $f(x) = a \sin \pi x + b \cos \pi x$:

x	-1	$-\frac{1}{2}$	0	$\frac{1}{2}$	1
y	-1	0	1	2	1

[a]**19.** Find the quadratic polynomial that best fits the following data in the sense of least squares:

x	-2	-1	0	1	2
y	2	1	1	1	2

[a]**20.** What line best represents the following data in the least-squares sense?

x	0	1	2
y	5	-6	7

[a]**21.** What constant c makes the expression

$$\sum_{k=0}^{m} [f(x_k) - ce^{x_k}]^2$$

as small as possible?

22. Show that the formula for the best line to fit data (k, y_k) at the integers k for $1 \leq k \leq n$ is

$$y = ax + b$$

where

$$a = \frac{6}{n(n^2-1)}\left[2\sum_{k=1}^{n} ky_k - (n+1)\sum_{k=1}^{n} y_k\right]$$

$$b = \frac{2}{n(n-1)}\left[(2n+1)\sum_{k=1}^{n} y_k - 3\sum_{k=1}^{n} ky_k\right]$$

23. Establish the normal equations and verify the results in Example 1.

24. A vector $\boldsymbol{v}$ is asserted to be the least-squares solution of an inconsistent system $\boldsymbol{Ax} = \boldsymbol{b}$. How can we test $\boldsymbol{v}$ without going through the entire least-squares procedure?

25. Find the normal equations for the following data points:

x	1.0	2.0	2.5	3.0
y	3.7	4.1	4.3	5.0

Determine the straight line that best fits the data in the least-squares sense. Plot the data point and the least-squares line.

26. For the case $n = 4$, show directly that by forming the normal equations from the data points (x_i, y_i), we obtain the results in Theorem 1.

Computer Exercises 9.1

1. Write a procedure that sets up the normal system of Equations (7). Using that procedure and other routines, such as *Gauss* and *Solve* from Section 2.2, verify the solution given for the problem involving $\ln x$, $\cos x$, and e^x in the subsection entitled "Nonpolynomial Example."

2. Write a procedure that fits a straight line to Table (1). Use this procedure to find the constants in the equation $S = aT + b$ for the table in the example that begins this chapter (p. 426). Also, verify the results obtained for Example 1 (p. 429).

3. Write and test a program that takes $m + 1$ points in the plane (x_i, y_i), where $0 \leq i \leq m$, with $x_0 < x_1 < \cdots < x_m$, and computes the best linear fit by the method of least squares. Then the program should create a plot of the points and the best line determined by the least-squares method.

4. The U.S. Federal Government Internal Revenue Service publishes the following table of values having to do with minimal distributions of pension plans:

x	1	2	3	4	5	6
y	29.9	29.0	28.1	27.1	26.2	25.3

7	8	9	10	11	12
24.4	23.6	22.7	21.8	21.0	20.1

13	14	15	16
19.3	18.5	17.7	16.9

What simple function represents the data? Use Equation (5), and plot the data and the results using either plotting software such as gnuplot or some mathematics software system such as Maple, MATLAB, or Mathematica.

5. Using mathematical software such as MATLAB, Maple, or Mathematica, fit a linear least-squares polynomial to the data in Example 1. Then plot the original data and the polynomial using a fine set of grid points.

6. (Continuation) Verify the results in Example 2 and plot the curve.

9.2 Orthogonal Systems and Chebyshev Polynomials

Orthonormal Basis Functions $\{g_0, g_1, \dots, g_n\}$

Once the functions $g_0, g_1, \dots g_n$ of Equation (5) in Section 9.1 have been chosen, the least-squares problem can be interpreted as follows: The set of all functions g that can be expressed as linear combinations of $g_0, g_1, \dots, g_n$ is a vector space $\mathcal{G}$. (Familiarity with vector spaces is not essential to understanding the discussion here.) In symbols, we have

$$\mathcal{G} = \left\{ g\colon \text{there exist } c_0, c_1, \dots, c_n, \text{ such that } g(x) = \sum_{j=0}^{n} c_j g_j(x) \right\}$$

The function that is being sought in the least-squares problem is thus an element of the vector space $\mathcal{G}$. Since the functions $g_0, g_1, \dots, g_n$ form a **basis** for $\mathcal{G}$, the set is not linearly dependent. However, a given vector space has many different bases, and they can differ drastically in their numerical properties.

Let us turn our attention away from the given basis $\{g_0, g_1, \dots, g_n\}$ to the vector space $\mathcal{G}$ generated by that basis. Without changing $\mathcal{G}$, we ask:

What basis for $\mathcal{G}$ should be chosen for numerical work?

In the present problem, the principal numerical task is to solve the normal equations—that is, Equation (7) in Section 9.1:

Normal Equations

$$\sum_{j=0}^{n} \left[\sum_{k=0}^{m} g_i(x_k) g_j(x_k) \right] c_j = \sum_{k=0}^{m} y_k g_i(x_k) \qquad (0 \leqq i \leqq n) \tag{1}$$

The nature of this system obviously depends on the basis $\{g_0, g_1, \dots, g_n\}$. We want these equations to be *easily* solved or to be capable of being *accurately* solved. The ideal situation occurs when the coefficient matrix in Equation (1) is the identity matrix. This happens if the basis $\{g_0, g_1, \dots, g_n\}$ has the property of **orthonormality**:

Orthonormality Property

$$\sum_{k=0}^{m} g_i(x_k) g_j(x_k) = \delta_{ij} = \begin{cases} 1 & i = j \\ 0 & i \neq j \end{cases}$$

In the presence of this property, Equation (1) simplifies dramatically to

$$c_j = \sum_{k=0}^{m} y_k g_j(x_k) \qquad (0 \leqq j \leqq n)$$

which is no longer a system of equations to be solved but rather an explicit formula for the coefficients c_j.

Under rather general conditions, the space $\mathcal{G}$ has a basis that is orthonormal in the sense just described. A procedure known as the **Gram-Schmidt process** can be used to obtain such a basis. There are some situations in which the effort of obtaining an orthonormal basis is justified, but simpler procedures often suffice. We now describe one such procedure.

Remember that our goal is to make Equation (1) well disposed for numerical solution. We want to avoid any matrix of coefficients that involves the difficulties encountered in connection with the **Hilbert matrix**. (See Computer Exercise 2.2.4.) This objective can be met if the basis for the space $\mathcal{G}$ is well chosen.

We now consider the space $\mathcal{G}$ that consists of all polynomials of degree $\leqq n$, which is an important example of the least-squares theory. It may seem natural to use the following $n + 1$ functions as a basis for $\mathcal{G}$:

Basis Functions

$$g_0(x) = 1, \qquad g_1(x) = x, \qquad g_2(x) = x^2, \qquad \dots, \qquad g_n(x) = x^n$$

Using this basis, we write a typical element of the space $\mathcal{G}$ in the form

$$g(x) = \sum_{j=0}^{n} c_j g_j(x) = \sum_{j=0}^{n} c_j x^j = c_0 + c_1 x + c_2 x^2 + \cdots + c_n x^n$$

This basis, however natural, is almost always a *poor* choice for numerical work. For many purposes, the Chebyshev polynomials (suitably defined for the interval involved) do form a *good* basis.

Figure 9.4 gives an indication of why the monomial x^j does not form a good basis for numerical work: These functions are too much alike! If a function g is given and we wish to express it as a linear combination of the monomials, $g(x) = \sum_{j=0}^{n} c_j x^j$, it is difficult to determine the coefficients c_j precisely. Figure 9.4 also shows a few of the Chebyshev polynomials; they are quite different from one another.

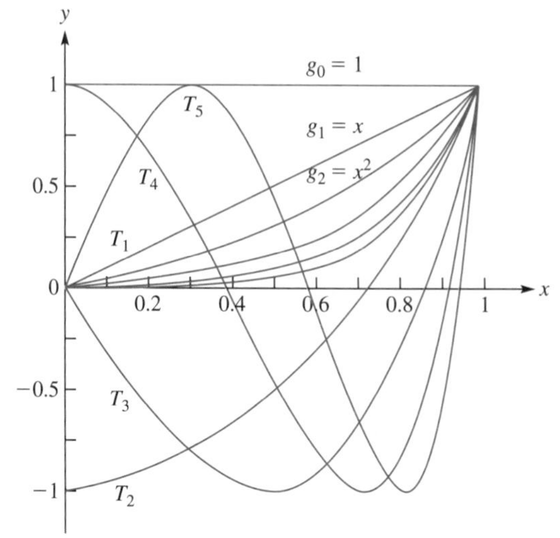

FIGURE 9.4 Polynomials x^k and Chebyshev polynomials T_k

For simplicity, assume that the points in our least-squares problem have the property

$$-1 = x_0 < x_1 < \cdots < x_m = 1$$

Then the **Chebyshev polynomials** for the interval $[-1, 1]$ can be used. The traditional notation is

Chebyshev Polynomials

$$\begin{cases} T_0(x) = 1, & T_1(x) = x, & T_2(x) = 2x^2 - 1 \\ T_3(x) = 4x^3 - 3x, & T_4(x) = 8x^4 - 8x^2 + 1 & \text{etc.} \end{cases}$$

A recursive formula for these polynomials is

Recurrence Relation

$$T_j(x) = 2xT_{j-1}(x) - T_{j-2}(x) \qquad (j \geqq 2) \tag{2}$$

This formula, together with the equations $T_0(x) = 1$ and $T_1(x) = x$, provides a formal definition of the Chebyshev polynomials. Alternatively, we can write $T_k(x) = \cos(k \arccos x)$.

Chebyshev Polynomial Case

Linear combinations of Chebyshev polynomials are easy to evaluate because a special nested multiplication algorithm applies. To describe this procedure, consider an arbitrary linear combination of $T_0, T_1, T_2, \ldots, T_n$:

$$g(x) = \sum_{j=0}^{n} c_j T_j(x)$$

An algorithm to compute $g(x)$ for any given x goes as follows:

Computing $g(x)$

$$\begin{cases} w_{n+2} = w_{n+1} = 0 \\ w_j = c_j + 2xw_{j+1} - w_{j+2} \qquad (j = n, n-1, \ldots, 0) \\ g(x) = w_0 - xw_1 \end{cases} \tag{3}$$

To see that this algorithm actually produces $g(x)$, we write down the series for g, shift some indices, and use Formulas (2) and (3):

$$\begin{aligned}
g(x) &= \sum_{j=0}^{n} c_j T_j(x) \\
&= \sum_{j=0}^{n} (w_j - 2xw_{j+1} + w_{j+2}) T_j \\
&= \sum_{j=0}^{n} w_j T_j - 2x \sum_{j=0}^{n} w_{j+1} T_j + \sum_{j=0}^{n} w_{j+2} T_j \\
&= \sum_{j=0}^{n} w_j T_j - 2x \sum_{j=1}^{n+1} w_j T_{j-1} + \sum_{j=2}^{n+2} w_j T_{j-2} \\
&= \sum_{j=0}^{n} w_j T_j - 2x \sum_{j=1}^{n} w_j T_{j-1} + \sum_{j=2}^{n} w_j T_{j-2} \\
&= w_0 T_0 + w_1 T_1 + \sum_{j=2}^{n} w_j T_j - 2xw_1 T_0 - 2x \sum_{j=2}^{n} w_j T_{j-1} + \sum_{j=2}^{n} w_j T_{j-2} \\
&= w_0 + xw_1 - 2xw_1 + \sum_{j=2}^{n} w_j (T_j - 2xT_{j-1} + T_{j-2}) \\
&= w_0 - xw_1
\end{aligned}$$

In general, it is best to arrange the data so that all the abscissas $\{x_i\}$ lie in the interval $[-1, 1]$. Then, if the first few Chebyshev polynomials are used as a basis for the polynomials, the normal equations should be reasonably well conditioned. We have not given a technical definition of this term; it can be interpreted informally to mean that Gaussian elimination with pivoting produces an accurate solution to the normal equations.

If the original data do not satisfy $\min\{x_k\} = -1$ and $\max\{x_k\} = 1$ but lie instead in another interval $[a, b]$, then the change of variable

Change of Variables

$$x = \frac{1}{2}(b-a)z + \frac{1}{2}(a+b)$$

produces a variable z that traverses $[-1, 1]$ as x traverses $[a, b]$.

Outline of Algorithm

Here is an outline of a procedure, based on the preceding discussion, that produces a polynomial of degree $\leqq (n+1)$ that best fits a given table of values $(x_k, y_k)(0 \leqq k \leqq m)$. Here, m is usually much greater than n.

■ **Algorithm**

Polynomial Fitting

1. Find the smallest interval $[a, b]$ that contains all the x_k. Thus, let $a = \min\{x_k\}$ and $b = \max\{x_k\}$.
2. Make a transformation to the interval $[-1, 1]$ by defining
$$z_k = \frac{2x_k - a - b}{b - a} \qquad (0 \leqq k \leqq m)$$
3. Decide on the value of n to be used. In this situation, 8 or 10 would be a large value for n.
4. Using Chebyshev polynomials as a basis, generate the $(n+1) \times (m+1)$ normal equations

$(n+1) \times (m+1)$ Normal Equations

$$\sum_{j=0}^{n} \left[\sum_{k=0}^{m} T_i(z_k) T_j(z_k) \right] c_j = \sum_{k=0}^{m} y_k T_i(z_k) \qquad (0 \leqq i \leqq m) \tag{4}$$

5. Use an equation-solving routine to solve the normal equations for coefficients $c_0, c_1, \dots, c_n$ in the function
$$g(x) = \sum_{j=0}^{n} c_j T_j(x)$$
6. The polynomial that is being sought is
$$g\left(\frac{2x - a - b}{b - a}\right)$$

The details of Step 4 are as follows: Begin by introducing a double-subscripted variable:

$$t_{jk} = T_j(z_k) \qquad 0 \leqq k \leqq m, \; 0 \leqq j \leqq n$$

The $(n+1) \times (m+1)$ matrix $\boldsymbol{T} = (t_{jk})$ can be computed efficiently by using the recursive definition of the Chebyshev polynomials, Equation (2), as in the following segment of pseudocode:

Pseudocode to Compute *T* Matrix

```
integer j, k, m;   real array (t_ij)_{0:n×0:m}, (z_i)_{0:n}
for k = 0 to m
    t_0k ← 1
    t_1k ← z_k
    for j = 2 to n
        t_jk ← 2 z_k t_{j−1,k} − t_{j−2,k}
    end for
end for
```

The normal equations have a coefficient matrix $\boldsymbol{A} = (a_{ij})_{0:n\times 0:n}$ and a right-hand side $\boldsymbol{b} = (b_i)_{0:n}$ given by

Normal Equations $\boldsymbol{A} = (a_{ij}), \boldsymbol{b} = (b_i)$

$$\begin{aligned} a_{ij} &= \sum_{k=0}^{m} T_i(z_k)T_j(z_k) = \sum_{k=0}^{m} t_{ik}t_{jk} \qquad (0 \leqq i,\ j \leqq n) \\ b_i &= \sum_{k=0}^{m} y_k T_i(z_k) = \sum_{k=0}^{m} y_k t_{ik} \qquad (0 \leqq i \leqq n) \end{aligned} \tag{5}$$

The pseudocode to calculate $\boldsymbol{A}$ and $\boldsymbol{b}$ follows:

Pseudocode to Compute $\boldsymbol{A}$ and $\boldsymbol{b}$

```
real array (a_ij)_{0:n×0:n}, (b_i)_{0:n}, (t_ij)_{0:n×0:m}, (y_i)_{0:n}
integer i, j, m, n;   real s
for i = 0 to n
    s ← 0
    for k = 0 to m
        s ← s + y_k t_ik
    end for
    b_i ← s
    for j = i to n
        s ← 0
        for k = 0 to m
            s ← s + t_ik t_jk
        end for
        a_ij ← s
        a_ji ← s
    end for
end for
```

To fit data with polynomials, other methods exist that employ systems of polynomials tailor-made for a given set of abscissas. The method outlined above is, however, simple and direct.

Smoothing Data: Polynomial Regression

One of the important applications of the least-squares procedure is in the smoothing of data. In this context, **smoothing** refers to the fitting of a "smooth" curve to a set of "noisy" values (that is, the values contain experimental errors). If one knows the type of function to which the data should conform, then the least-squares procedure can be used to compute any unknown parameters in the function. This has been amply illustrated in the examples given previously. However, if one simply wishes to smooth the data by fitting them with any convenient function, then polynomials of increasing degree can be used until a reasonable balance between good fit and smoothness is obtained.

This idea is illustrated by the experimental data depicted in the following table, which shows 20 points (x_i, y_i):

20 Data Points (x_i, y_i)

x	−1.0	−0.92	−0.84	−0.8	−0.72	−0.64	−0.56	−0.48	−0.36
y	4.0	1.0	5.0	7.0	6.0	3.0	2.0	2.0	5.0

−0.24	−0.12	0.0	0.12	0.2	0.32	0.4	0.52	0.64	0.76	0.92
12.0	13.0	10.0	7.0	4.0	−2.0	−6.0	−8.0	−2.0	4.0	9.0

Of course, a polynomial of degree 19 can be determined that passes through these points *exactly*. But if the points are contaminated by experimental errors, our purposes are better served by some lower-degree polynomial that fits the data *approximately* in the least-squares sense. In statistical parlance, this is the problem of **curvilinear regression**. A good software library will contain code for the polynomial fitting of empirical data using a least-squares criterion. Such programs will determine the fitting polynomials of degrees 0, 1, 2, ... with a minimum of computing effort and with high precision. One can, of course, use the techniques illustrated already in this chapter, although they are not at all streamlined. Thus, with the Chebyshev polynomials as a basis, we can set up and solve the normal equations for $n = 0, 1, 2, \ldots$ and plot the resulting functions. Some of the polynomials obtained in this way for the data of the table are shown in Figure 9.5.

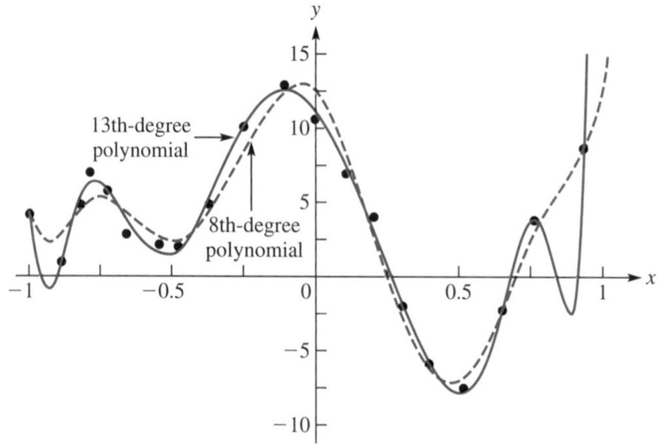

FIGURE 9.5 Polynomial of degree 8 (dashed line) and polynomial of degree 13 (solid line)

A Procedure for Polynomial Regression

An efficient procedure for polynomial regression, given by Forsythe [1957], is now explained. This procedure uses a system of orthogonal polynomials that are tailor-made for the problem at hand. We begin with a table of experimental values:

Experimental Values

x	x_0	x_1	$\cdots$	x_m
y	y_0	y_1	$\cdots$	y_m

The ultimate objective is to replace this table by a suitable polynomial of modest degree, with the experimental errors of the table somehow suppressed. We do *not* know what degree of polynomial should be used.

For statistical purposes, a reasonable hypothesis is that there is a polynomial

Polynomial Regression

$$p_N(x) = \sum_{i=0}^{N} a_i x^i$$

that represents the trend of the table and that the given tabular values obey the equation

$$y_i = p_N(x_i) + \varepsilon_i \qquad (0 \leqq i \leqq m)$$

In this equation, ε_i represents an observational error that is present in y_i. A further reasonable hypothesis is that these errors are independent random variables that are normally distributed.

For a fixed value of n, we have already discussed a method of determining p_n by the method of least squares. Thus, a system of normal equations can be set up to determine the coefficients of p_n. Once these are known, a quantity called the **variance** can be computed from the formula

Variance

$$\sigma_n^2 = \frac{1}{m-n}\sum_{i=0}^{m}[y_i - p_n(x_i)]^2 \qquad (m > n) \tag{6}$$

Statistical theory tells us that if the trend of the table is truly a polynomial of degree N (but infected by noise), then

$$\sigma_0^2 > \sigma_1^2 > \cdots > \sigma_N^2 = \sigma_{N+1}^2 = \sigma_{N+2}^2 = \cdots = \sigma_{m-1}^2$$

This fact suggests the following strategy for dealing with the case in which N is not known: Compute $\sigma_0^2, \sigma_1^2, \ldots$ in succession. As long as these are decreasing significantly, continue the calculation. When an integer N is reached for which $\sigma_N^2 \approx \sigma_{N+1}^2 \approx \sigma_{N+2}^2 \approx \cdots$, stop and declare p_N to be the polynomial sought.

If $\sigma_0^2, \sigma_1^2, \ldots$ are to be computed directly from the definition in Equation (6), then each of the polynomials $p_0, p_1, \ldots$ will have to be determined. The procedure described next can avoid the determination of all but the one desired polynomial.

In the remainder of the discussion, the abscissas x_i are to be held fixed. These points are assumed to be distinct, although the theory can be extended to include cases in which some points repeat. If f and g are two functions whose domains include the points $\{x_0, x_1, \ldots, x_m\}$, then the following notation is used:

Inner Product

$$\langle f, g\rangle = \sum_{i=0}^{m} f(x_i)g(x_i) \tag{7}$$

This quantity is called the *inner product* of f and g. Much of our discussion does not depend on the exact form of the inner product but only on certain of its properties. An **inner product** $\langle\cdot\,,\,\cdot\rangle$ has the following properties:

■ **Properties**

Defining Properties of an Inner Product

1. $\langle f, g\rangle = \langle g, f\rangle$
2. $\langle f, f\rangle > 0$ unless $f(x_i) = 0$ for all i
3. $\langle af, g\rangle = a\langle f, g\rangle$ where $a \in \mathbb{R}$
4. $\langle f, g+h\rangle = \langle f, g\rangle + \langle f, h\rangle$

The reader should verify that the inner product defined in Equation (7) has the properties listed.

Orthogonal Functions

A set of functions is now said to be **orthogonal** if $\langle f, g\rangle = 0$ for any two different functions f and g in that set. An orthogonal set of polynomials can be generated recursively by the following formulas:

Recurrence Relation

$$\begin{cases} q_0(x) = 1 \\ q_1(x) = x - \alpha_0 \\ q_{n+1}(x) = xq_n(x) - \alpha_n q_n(x) - \beta_n q_{n-1}(x) \qquad (n \geqq 1) \end{cases}$$

where

$$\alpha_n = \frac{\langle xq_n, q_n\rangle}{\langle q_n, q_n\rangle}, \qquad \beta_n = \frac{\langle xq_n, q_{n-1}\rangle}{\langle q_{n-1}, q_{n-1}\rangle}$$

In these formulas, a slight abuse of notation occurs where "xq_n" is used to denote the function whose value at x is $xq_n(x)$.

To understand how this definition leads to an orthogonal system, let's examine a few cases. First, we have

First Few Cases

$$\langle q_1, q_0\rangle = \langle x - \alpha_0, q_0\rangle = \langle xq_0 - \alpha_0 q_0, q_0\rangle = \langle xq_0, q_0\rangle - \alpha_0\langle q_0, q_0\rangle = 0$$

Notice that several properties of an inner product listed previously have been used here. Also, the definition of α_0 was used. Another of the first few cases is this:

$$\begin{aligned}\langle q_2, q_1\rangle &= \langle xq_1 - \alpha_1 q_1 - \beta_1 q_0, q_1\rangle \\ &= \langle xq_1, q_1\rangle - \alpha_1\langle q_1, q_1\rangle - \beta_1\langle q_0, q_1\rangle = 0\end{aligned}$$

Here, the definition of α_1 has been used, as well as the fact (established above) that $\langle q_1, q_0\rangle = 0$. The next step in a formal proof is to verify that $\langle q_2, q_0\rangle = 0$. Then an inductive proof completes the argument.

One part of this proof consists in showing that the coefficients α_n and β_n are well defined. This means that the denominators $\langle q_n, q_n\rangle$ are not zero. To verify that this is the case, suppose that $\langle q_n, q_n\rangle = 0$. Then $\sum_{i=0}^{m}[q_n(x_i)]^2 = 0$, and consequently, $q_n(x_i) = 0$ for *each* value of i. This means that the polynomial q_n has $m+1$ roots, $x_0, x_1, \dots, x_m$. Since the degree n is less than m, we conclude that q_n is the zero polynomial. However, this is *not* possible because obviously

$$\begin{cases} q_0(x) = 1 \\ q_1(x) = x - \alpha_0 \\ q_2(x) = x^2 + \text{(lower-order terms)} \end{cases}$$

and so on. Observe that this argument requires $n < m$.

The system of orthogonal polynomials $\{q_0, q_1, \dots, q_{m-1}\}$ generated by the above algorithm is a basis for the vector space $\prod_{m-1}$ of all polynomials of degree at most $m-1$. It is clear from the algorithm that each q_n starts with the highest term x^n. If it is desired to express a given polynomial p of degree n ($n \leqq m-1$) as a linear combination of $q_0, q_1, \dots, q_n$, this can be done as follows: Set

$$p = \sum_{i=0}^{n} a_i q_i \tag{8}$$

On the right-hand side, only one summand contains x^n. It is the term $a_n q_n$. On the left-hand side, there is also a term in x^n. One chooses a_n so that $a_n x^n$ on the right is equal to the corresponding term in p. Now write

Using Inner Products

$$p - a_n q_n = \sum_{i=0}^{n-1} a_i q_i$$

On both sides of this equation, there are polynomials of degree at most $n-1$ (because of the choice of a_n). Hence, we can now choose a_{n-1} in the way we chose a_n; that is, choose a_{n-1} so that the terms in x^{n-1} are the same on both sides. By continuing in this way, we discover the unique values that the coefficients a_i must have. This establishes that $\{q_0, q_1, \dots, q_n\}$ is a basis for $\prod_n$, for $n = 0, 1, \dots, m-1$.

Another Approach

Another way of determining the coefficients a_i (once we know that they exist!) is to take the inner product of both sides of Equation (8) with q_j. The result is

$$\langle p, q_j \rangle = \sum_{i=0}^{n} a_i \langle q_i, q_j \rangle \qquad (0 \leqq j \leqq n)$$

Since the set $q_0, q_1, \ldots, q_n$ is orthogonal, $\langle q_i, q_j \rangle = 0$ for each i different from j. Hence, we obtain

$$\langle p, q_j \rangle = a_j \langle q_j, q_j \rangle$$

This gives a_j as a quotient of two inner products.

Least-Squares Problem Revisited

Now we return to the least-squares problem. Let F be a function that we wish to fit by a polynomial p_n of degree n. We shall find the polynomial that minimizes the expression

Least-Squares Problem

$$\sum_{i=0}^{m} [F(x_i) - p_n(x_i)]^2$$

The solution is given by the formulas

$$p_n = \sum_{i=0}^{n} c_i q_i, \qquad c_i = \frac{\langle F, q_i \rangle}{\langle q_i, q_i \rangle} \tag{9}$$

It is especially noteworthy that c_i does *not* depend on n. This implies that the various polynomials $p_0, p_1, \ldots$ that we are seeking can all be obtained by simply truncating *one* series—namely, $\sum_{i=0}^{m-1} c_i q_i$. To prove that p_n, as given in Equation (9), solves our problem, we return to the normal equation, Equation (1). The basic functions now being used are $q_0, q_1, \ldots, q_n$. Thus, the normal equations are

Normal Equations

$$\sum_{j=0}^{n} \left[\sum_{k=0}^{m} q_i(x_k) q_j(x_k) \right] c_j = \sum_{k=0}^{m} y_k q_i(x_k) \qquad (0 \leqq i \leqq n)$$

Using the inner product notation, we get

$$\sum_{j=0}^{n} \langle q_i, q_j \rangle c_j = \langle F, q_i \rangle \qquad (0 \leqq i \leqq n)$$

where F is some function such that $F(x_k) = y_k$ for $0 \leqq k \leqq m$. Next, apply the orthogonality property $\langle q_i, q_j \rangle = 0$ when $i \neq j$. The result is

Results $c = (c_i)$

$$\langle q_i, q_i \rangle c_i = \langle F, q_i \rangle \qquad (0 \leqq i \leqq n) \tag{10}$$

Now we return to the variance numbers $\sigma_0^2, \sigma_1^2, \ldots$ and show how they can be easily computed. First, an important observation: The set $\{q_0, q_1, \ldots, q_n, F - p_n\}$ is orthogonal! The only new fact here is that $\langle F - p_n, q_i \rangle = 0$ for $0 \leqq i \leqq n$.

To check this, write

Check Results

$$\begin{aligned} \langle F - p_n, q_i \rangle &= \langle F, q_i \rangle - \langle p_n, q_i \rangle \\ &= \langle F, q_i \rangle - \left\langle \sum_{j=0}^{n} c_j q_j, q_i \right\rangle \end{aligned}$$

$$
\begin{aligned}
&= \langle F, q_i \rangle - \sum_{j=0}^{n} c_j \langle q_j, q_i \rangle \\
&= \langle F, q_i \rangle - c_i \langle q_i, q_i \rangle = 0
\end{aligned}
$$

In this computation, we used Equations (9) and (10). Since p_n is a linear combination of $q_0, q_1, \ldots, q_n$, it follows easily that

$$\langle F - p_n, p_n \rangle = 0$$

Now recall that the variance σ_n^2 was defined by

$$\sigma_n^2 = \frac{\rho_n}{m-n}, \qquad \rho_n = \sum_{i=0}^{m} [y_i - p_n(x_i)]^2$$

The quantities ρ_n can be written in another way:

$$
\begin{aligned}
\rho_n &= \langle F - p_n, F - p_n \rangle \\
&= \langle F - p_n, F \rangle \\
&= \langle F, F \rangle - \langle F, p_n \rangle \\
&= \langle F, F \rangle - \sum_{i=0}^{n} c_i \langle F, q_i \rangle \\
&= \langle F, F \rangle - \sum_{i=0}^{n} \frac{\langle F, q_i \rangle^2}{\langle q_i, q_i \rangle}
\end{aligned}
$$

Thus, the numbers $\rho_0, \rho_1, \ldots$ can be generated recursively by the algorithm

Recurrence Relations for ρ_n

$$
\begin{cases}
\rho_0 = \langle F, F \rangle - \dfrac{\langle F, q_0 \rangle^2}{\langle q_0, q_0 \rangle} \\[2ex]
\rho_n = \rho_{n-1} - \dfrac{\langle F, q_n \rangle^2}{\langle q_n, q_n \rangle} \qquad (n \geq 1)
\end{cases}
$$

Gram-Schmidt Process

The **projection operator** is defined as

$$\text{proj}_{\boldsymbol{y}}\, \boldsymbol{x} = \frac{\langle \boldsymbol{x}, \boldsymbol{y} \rangle}{\langle \boldsymbol{y}, \boldsymbol{y} \rangle}\, \boldsymbol{y}$$

that projects the vector $\boldsymbol{x}$ orthogonally onto the vector $\boldsymbol{y}$. The Gram-Schmidt process can be written as

$$
\begin{aligned}
z_1 &= \boldsymbol{v}_1, & q_1 &= \frac{z_1}{||z_1||} \\
z_2 &= \boldsymbol{v}_2 - \text{proj}_{\boldsymbol{z}_1} \boldsymbol{v}_2, & q_2 &= \frac{z_2}{||z_2||} \\
z_3 &= \boldsymbol{v}_3 - \text{proj}_{\boldsymbol{z}_1} \boldsymbol{v}_3 - \text{proj}_{\boldsymbol{z}_2} \boldsymbol{v}_3, & q_3 &= \frac{z_3}{||z_3||}
\end{aligned}
$$

In general, the k step is

$$z_k = \boldsymbol{v}_k - \sum_{j=1}^{k-1} \text{proj}_{\boldsymbol{v}_j} \boldsymbol{v}_k, \qquad q_k = \frac{z_k}{||z_k||}$$

Here $\{z_1, z_2, z_3, \dots, z_k\}$ is an orthogonal set and $\{\boldsymbol{q}_1, \boldsymbol{q}_2, \boldsymbol{q}_3, \dots, \boldsymbol{q}_k\}$ is an orthonormal set. When implemented on a computer, the Gram-Schmidt process is numerically unstable because the vectors z_k may not be exactly orthogonal due to roundoff errors. By a minor modification, the Gram-Schmidt process can be stabilized. Instead of computing the vectors $\boldsymbol{u}_k$ as above, it can be computed a term at a time. A computer algorithm for the **modified Gram-Schmidt process**:

```
for j = 1 to k
    for i = 1 to j − 1
        s ← ⟨v_j, v_i⟩
        v_j ← v_j − s v_i
    end for
v_i ← v_j/||v_j||
end for
```

Here the vectors $\boldsymbol{v}_1, \boldsymbol{v}_2, \dots, \boldsymbol{v}_k$ are replaced with orthonormal vectors that span the same subspace. The i-loop removes components in the $\boldsymbol{v}_i$ direction followed by normalization of the vector. In exact arithmetic, this computation gives the same results as the original form above. However, it produces smaller errors in finite-precision computer arithmetic.

EXAMPLE 1 Consider the vectors $\boldsymbol{v}_1 = (1, \varepsilon, 0, 0)$, $\boldsymbol{v}_1 = (1, 0, \varepsilon, 0)$, and $\boldsymbol{v}_1 = (1, 0, 0, \varepsilon)$. Assume ε is a small number. Carry out the standard Gram-Schmidt procedure and the modified Gram-Schmidt procedure. Check the orthogonality conditions of the resulting vectors.

Solution Using the classical Gram-Schmidt process, we obtain $\boldsymbol{u}_1 = (1, \varepsilon, 0, 0)$, $\boldsymbol{u}_2 = (0, -1, 1, 0)/\sqrt{2}$, and $\boldsymbol{u}_3 = (0, -1, 0, 1)/\sqrt{2}$. Using the modified Gram-Schmidt process, we find $z_1 = (1, \varepsilon, 0, 0)$, $z_2 = (0, -1, 1, 0)/\sqrt{2}$, and $z_3 = (0, -1, -1, 2)/\sqrt{6}$. Checking orthogonality, we find $\langle \boldsymbol{u}_2, \boldsymbol{u}_3\rangle = \frac{1}{2}$ and $\langle z_2, z_3\rangle = 0$. ■

Summary 9.2

- We use Chebyshev polynomials $\{T_j\}$ as an orthogonal basis that can be generated recursively by

$$T_j(x) = 2xT_{j-1}(x) - T_{j-2}(x) \qquad (j \geqq 2)$$

with $T_0(x) = 1$ and $T_1(x) = x$. The coefficient matrix $A = (a_{ij})_{0:n\times 0:n}$ and the right-hand side $\boldsymbol{b} = (b_i)_{0:n}$ of the normal equations are

$$a_{ij} = \sum_{k=0}^{m} T_i(z_k)T_j(z_k) \qquad (0 \leqq i,\ j \leqq n)$$

$$b_i = \sum_{k=0}^{m} y_k T_i(z_k) \qquad (0 \leqq i \leqq n)$$

A linear combination of Chebyshev polynomials

$$g(x) = \sum_{j=0}^{n} c_j T_j(x)$$

can be evaluated recursively:

$$\begin{cases} w_{n+2} = w_{n+1} = 0 \\ w_j = c_j + 2xw_{j+1} - w_{j+2} & (j = n, n-1, \ldots, 0) \\ g(x) = w_0 - xw_1 \end{cases}$$

- We discuss smoothing of data by polynomial regression.

Exercises 9.2

1. Let $g_0, g_1, \ldots, g_n$ be a set of functions such that $\sum_{k=0}^{m} g_i(x_k)g_j(x_k) = 0$ if $i \neq j$. What linear combination of these functions best fits the data at the beginning of Section 9.1 (p. 426)?

[a]2. Consider polynomials $g_0, g_1, \ldots, g_n$ defined by $g_0(x) = 1$, $g_1(x) = x - 1$, and $g_j(x) = 3xg_{j-1}(x) + 2g_{j-2}(x)$. Develop an efficient algorithm for computing values of the function $f(x) = \sum_{j=0}^{n} c_j g_j(x)$.

[a]3. Show that $\cos n\theta = 2\cos\theta\cos(n-1)\theta - \cos(n-2)\theta$. *Hint*: Use the familiar identity $\cos(A \mp B) = \cos A \cos B \pm \sin A \sin B$.

4. (Continuation) Show that if $f_n(x) = \cos(n \arccos x)$, then $f_0(x) = 1$, $f_1(x) = x$, and $f_n(x) = 2xf_{n-1}(x) - f_{n-2}(x)$.

[a]5. (Continuation) Show that an alternate definition of Chebyshev polynomials is $T_n(x) = \cos(n \arccos x)$ for $-1 \leqq x \leqq 1$.

[a]6. (Continuation) Give a one-line proof that $T_n(T_m(x)) = T_{nm}(x)$.

[a]7. (Continuation) Show that $|T_n(x)| \leqq 1$ for x in the interval $[-1, 1]$.

[a]8. Define $g_k(x) = T_k\left(\frac{1}{2}x + \frac{1}{2}\right)$. What recursive relation do these functions satisfy?

9. Show that $T_0, T_2, T_4, \ldots$ are even and that $T_1, T_3, \ldots$ are odd functions. Recall that an even function satisfies the equation $f(x) = f(-x)$; an odd function satisfies the equation $f(x) = -f(-x)$.

[a]10. Count the number of operations involved in the algorithm used to compute $g(x) = \sum_{j=0}^{n} c_j T_j(x)$.

11. Show that the algorithm for computing $g(x) = \sum_{j=0}^{n} c_j T_j(x)$ can be modified to read

$$\begin{cases} w_{n-1} = c_{n-1} + 2xc_n \\ w_k = c_k + 2xw_{k+1} - w_{k-2} & (n-2 \geqq k \geqq 1) \\ g(x) = c_0 + xw_1 - w_2 \end{cases}$$

thus making w_{n+2}, w_{n+1}, and w_0 unnecessary.

[a]12. (Continuation) Count the operations for the algorithm in the preceding problem.

[a]13. Determine $T_6(x)$ as a polynomial in x.

14. Verify the four properties of an inner product that were listed in the text, using Definition (7).

15. Verify these formulas:

$$p_0(x) = \frac{1}{m+1}\sum_{i=0}^{m} y_i,$$

$$\beta_n = \frac{\langle q_n, q_n \rangle}{\langle q_{n-1}, q_{n-1} \rangle}, \quad c_n = \frac{\rho_{n-1} - \rho_n}{\langle F, q_n \rangle}$$

16. Complete the proof that the algorithm for generating the orthogonal system of polynomials works.

[a]17. There is a function f of the form

$$f(x) = \alpha x^{12} + \beta x^{13}$$

for which $f(0.1) = 6 \times 10^{-13}$ and $f(0.9) = 3 \times 10^{-2}$. What is it? Are α and β sensitive to perturbations in the two given values of $f(x)$?

18. (**Multiple Choice**) Let $\boldsymbol{x}_1 = [2, 2, 1]^T$, $\boldsymbol{x}_2 = [1, 1, 5]^T$, and $\boldsymbol{x}_3 = [-3, 2, 1]^T$. If the Gram-Schmidt process is applied to this ordered set of vectors to produce an orthonormal set $\{\boldsymbol{u}_1, \boldsymbol{u}_2, \boldsymbol{u}_3\}$, what is $\boldsymbol{u}_1$?

a. $\left[\frac{2}{3}, \frac{2}{3}, \frac{1}{3}\right]^T$ **b.** $[2, 2, 1]^T$

c. $\left[\frac{2}{5}, \frac{2}{5}, \frac{1}{5}\right]^T$ **d.** $[1, 0, 0]^T$

e. None of these.

19. (**Multiple Choice**, continued) What is $\boldsymbol{u}_2$?

a. $\frac{1}{\sqrt{27}}[1, 1, 5]^T$ **b.** $\frac{1}{\sqrt{18}}[-1, -1, 4]^T$

c. $[2, 2, 1]^T$ **d.** $[1, 1, -4]^T$

e. None of these.

Computer Exercises 9.2

1. Carry out an experiment in data smoothing as follows: Start with a polynomial of modest degree, say, 7. Compute 100 values of this polynomial at random points in the interval $[-1, 1]$. Perturb these values by adding random numbers chosen from a small interval, say, $\left[-\frac{1}{8}, \frac{1}{8}\right]$. Try to recover the polynomial from these perturbed values by using the method of least squares.

2. Write **real function** *Cheb*(n, x) for evaluating $T_n(x)$. Use the recursive formula satisfied by Chebyshev polynomials. Do not use a subscripted variable. Test the program on these 15 cases: $n = 0, 1, 3, 6, 12$ and $x = 0, -1, 0.5$.

3. Write **real function** *Cheb*$(n, x, (y_i))$ to calculate $T_0(x), T_1(x), \dots, T_n(x)$, and store these numbers in the array (y_i). Use your routine, together with suitable plotting routines, to obtain graphs of $T_0, T_1, T_2, \dots, T_8$ on $[-1, 1]$.

4. Write **real function** $F(n, (c_i), x)$ for evaluating $f(x) = \sum_{j=0}^{n} c_j T_j(x)$. Test your routine by means of the formula $\sum_{k=0}^{\infty} t^k T_k(x) = (1 - tx)/(1 - 2tx + t^2)$, valid for $|t| < 1$. If $|t| \leqq \frac{1}{2}$, then only a few terms of the series are needed to give full machine precision. Add terms in ascending order of magnitude.

5. Obtain a graph of T_n for some reasonable value of n by means of the following idea: Generate 100 equally spaced angles θ_i in the interval $[0, \pi]$. Define $x_i \cos\theta_i$ and $y_i = T_n(x_i) = \cos(n \arccos x_i) = \cos n\theta_i$. Send the points (x_i, y_i) to a suitable plotting routine.

6. Write suitable code to carry out the procedure outlined in the text for fitting a table with a linear combination of Chebyshev polynomials. Test it in the manner of Computer Exercise 9.2.1, first by using an unperturbed polynomial. Find out experimentally how large n can be in this process before roundoff errors become serious.

[a]7. Define $x_k = \cos[(2k - 1)\pi/(2m)]$. Select modest values of n and $m > 2n$. Compute and print the matrix $\boldsymbol{A}$ whose elements are

$$a_{ij} = \sum_{k=0}^{m} T_i(x_k)T_j(x_k) \qquad (0 \leqq i,\ j \leqq n)$$

Interpret the results in terms of the least-squares polynomial-fitting problem.

8. Program the algorithm for finding $\sigma_0^2, \sigma_1^2, \dots$ in the polynomial regression problem.

9. Program the complete polynomial regression algorithm. The output should be α_n, β_n, σ_n^2, and c_n for $0 \leqq n \leqq N$, where N is determined by the condition $\sigma_{N-1}^2 > \sigma_N^2 \approx \sigma_{N+1}^2$.

10. Using orthogonal polynomials, find the quadratic polynomial that fits the following data in the sense of least squares:

a.

x	-1	$-\frac{1}{2}$	0	$\frac{1}{2}$	1
y	-1	0	1	2	1

b.

x	-2	-1	0	1	2
y	2	1	1	1	2

9.3 Examples of the Least-Squares Principle

Inconsistent Systems

The principle of least squares is also used in other situations. In one of these, we attempt to *solve* an inconsistent system of linear equations of the form

Case: Inconsistent System

$$\sum_{j=0}^{n} a_{kj} x_j = b_k \qquad (0 \leqq k \leqq m) \tag{1}$$

in which $m > n$. Here, there are $m + 1$ equations, but only $n + 1$ unknowns. If a given $n + 1$-tuple $(x_0, x_1, \dots, x_n)$ is substituted on the left, the discrepancy between the two sides of the kth equation is termed the ***k*th residual**. Ideally, of course, all residuals should be zero. If it is not possible to select $(x_0, x_1, \dots, x_n)$ so as to make all residuals zero, System (1) is said to be **inconsistent** or **incompatible**. In this case, an alternative is to minimize the sum of the squares of the residuals. So we are led to minimize the expression

$$\varphi(x_0, x_1, \ldots, x_n) = \sum_{k=0}^{m} \left(\sum_{j=0}^{n} a_{kj} x_j - b_k \right)^2 \tag{2}$$

by making an appropriate choice of $(x_0, x_1, \ldots, x_n)$. Proceeding as before, we take partial derivatives with respect to x_i and set them equal to zero, thereby arriving at the normal equations

Normal Equations

$$\sum_{j=0}^{n} \left(\sum_{k=0}^{m} a_{ki} a_{kj} \right) x_j = \sum_{k=0}^{m} b_k a_{ki} \qquad (0 \leqq i \leqq n) \tag{3}$$

This is a linear system of just $n + 1$ equations involving unknowns $x_0, x_1, \ldots, x_n$. It can be shown that this system is consistent, provided that the column vectors in the original coefficient array are linearly independent. System (3) can be solved, for instance, by Gaussian elimination. The solution of System (3) is then a best approximate solution of Equation (1) in the least-squares sense.

Modified Gram-Schmidt Process

Special methods have been devised for the problem just discussed. Generally, they gain in precision over the simple approach outlined above. One such algorithm for solving System (1),

$$\boldsymbol{Ax} = \boldsymbol{b}$$

begins by factoring

***QR* Factorization**

$$\boldsymbol{A} = \boldsymbol{QR}$$

where matrix $\boldsymbol{Q}$ is $(m + 1) \times (n + 1)$ satisfying

$$\boldsymbol{Q}^T \boldsymbol{Q} = \boldsymbol{I}$$

and matrix $\boldsymbol{R}$ is $(n + 1) \times (n + 1)$ satisfying $r_{ii} > 0$ and $r_{ij} = 0$ for $j < i$. Then the least-squares solution is obtained by an algorithm called the **modified Gram-Schmidt process**.

A more elaborate (and more versatile) algorithm depends on the **singular value decomposition** of the matrix $\boldsymbol{A}$. This is a factoring,

$$\boldsymbol{A} = \boldsymbol{U \Sigma V}^T$$

in which $\boldsymbol{U}^T\boldsymbol{U} = \boldsymbol{I}_{m+1}$, $\boldsymbol{V}^T\boldsymbol{V} = \boldsymbol{I}_{n+1}$, and $\boldsymbol{\Sigma}$ is an $(m + 1) \times (n + 1)$ diagonal matrix that has nonnegative entries. For these more reliable procedures, the reader is referred to material at the end of this section and to Stewart [1973] and Lawson and Hanson [1995].

Use of a Weight Function $w(x)$

Another important example of the principle of least squares occurs in fitting or approximating functions on *intervals* rather than discrete sets. For example, a given function f defined on an interval $[a, b]$ may have to be approximated by a function such as

Case: Functions on Intervals

$$g(x) = \sum_{j=0}^{n} c_j g_j(x)$$

It is natural, then, to attempt to minimize the expression

Minimize φ

$$\varphi(c_0, c_1, \dots, c_n) = \int_a^b [g(x) - f(x)]^2\,dx \tag{4}$$

by choosing coefficients appropriately. In some applications, it is desirable to force functions g and f into better agreement in certain parts of the interval. For this purpose, we can modify Equation (4) by including a positive **weight function** $w(x)$, which can, of course, be $w(x) \equiv 1$ if all parts of the interval are to be treated the same. The result is

$$\varphi(c_0, c_1, \dots, c_n) = \int_a^b [g(x) - f(x)]^2 w(x)\,dx$$

The minimum of φ is again sought by differentiating with respect to each c_i and setting the partial derivatives equal to zero. The result is a system of normal equations:

Normal Equations

$$\sum_{j=0}^{n} \left[\int_a^b g_i(x) g_j(x) w(x)\,dx \right] c_j = \int_a^b f(x) g_i(x) w(x)\,dx \qquad (0 \leqq i \leqq n) \tag{5}$$

This is a system of $n+1$ linear equations in $n+1$ unknowns $c_0, c_1, \dots, c_n$ and can be solved by Gaussian elimination. Earlier remarks about choosing a good basis apply here also. The ideal situation is to have functions $g_0, g_1, \dots, g_n$ that have the orthogonality property:

Orthogonality Property

$$\int_a^b g_i(x) g_j(x) w(x)\,dx = 0 \qquad (i \neq j) \tag{6}$$

Many such orthogonal systems have been developed over the years.

For example, **Chebyshev polynomials** form one such system, namely,

Chebyshev Polynomials

$$\int_{-1}^{1} T_i(x) T_j(x) (1 - x^2)^{-1/2}\,dx = \begin{cases} 0, & i \neq j \\ \dfrac{\pi}{2}, & i = j > 0 \\ \pi, & i = j = 0 \end{cases}$$

The weight function $(1 - x^2)^{-1/2}$ assigns heavy weight to the ends of the interval $[-1, 1]$.

If a sequence of nonzero functions $g_0, g_1, \dots, g_n$ is orthogonal according to Equation (6), then the sequence $\lambda_0 g_0, \lambda_1 g_1, \dots, \lambda_n g_n$ is orthonormal for appropriate positive real numbers λ_j, namely,

$$\lambda_j = \left\{ \int_a^b [g_j(x)]^2 w(x)\,dx \right\}^{-1/2}$$

Nonlinear Example

As another example of the least-squares principle, here is a nonlinear problem. Suppose that a table of points (x_k, y_k) is to be fitted by a function of the form

Case: $y = e^{cx}$

$$y = e^{cx}$$

Proceeding as before leads to the problem of minimizing the function

Minimize φ

$$\varphi(c) = \sum_{k=0}^{m} (e^{cx_k} - y_k)^2$$

The minimum occurs for a value of c such that

$$0 = \frac{\partial \varphi}{\partial c} = \sum_{k=0}^{m} 2(e^{cx_k} - y_k)e^{cx_k}x_k$$

This equation is nonlinear in c. One could contemplate solving it by Newton's method or the secant method. On the other hand, the problem of minimizing $\varphi(c)$ could be attacked directly. Since there can be multiple roots in the normal equation and local minima in φ itself, a direct minimization of φ would be safer. This type of difficulty is typical of **nonlinear least-squares problems**. Consequently, other methods of curve fitting are often preferred if the unknown parameters do not occur linearly in the problem.

Nonlinear Least-Squares Problems

Alternatively, this particular example can be linearized by a change of variables $z = \ln y$ and by considering

Linearized

$$z = cx$$

The problem of minimizing the function

Minimizing φ

$$\varphi(c) = \sum_{k=0}^{m} (cx_k - z_k)^2 \qquad z_k = \ln y_k$$

is easy and leads to

$$c = \sum_{k=0}^{m} z_k x_k \Bigg/ \sum_{k=0}^{m} x_k^2$$

This value of c is *not* the solution of the original problem, but may be satisfactory in some applications.

Linear and Nonlinear Example

The final example contains elements of linear and nonlinear theory. Suppose that an (x_k, y_k) table is given with $m + 1$ entries and that a functional relationship such as

Case: $y = a\sin(bx)$

$$y = a\sin(bx)$$

is suspected.

Can the least-squares principle be used to obtain the appropriate values of the parameters a and b?

Notice that parameter b enters this function in a nonlinear way, creating some difficulty, as will be seen. According to the principle of least squares, the parameters should be chosen such that the expression

$$\sum_{k=0}^{m} [a\sin(bx_k) - y_k]^2$$

has a minimum value. The minimum value is sought by differentiating this expression with respect to a and b and setting these partial derivatives equal to zero. The results are

$$\begin{cases} \displaystyle\sum_{k=0}^{m} 2[a\sin(bx_k) - y_k]\sin(bx_k) = 0 \\ \displaystyle\sum_{k=0}^{m} 2[a\sin(bx_k) - y_k]ax_k\cos(bx_k) = 0 \end{cases}$$

If b were known, a could be obtained from either equation. The correct value of b is the one for which these corresponding two a values are identical. So each of the preceding equations should be solved for a, and the results set equal to each other. This process leads to the equation

$$\frac{\sum_{k=0}^{m} y_k \sin bx_k}{\sum_{k=0}^{m} (\sin bx_k)^2} = \frac{\sum_{k=0}^{m} x_k y_k \cos bx_k}{\sum_{k=0}^{m} x_k \sin bx_k \cos bx_k}$$

which can now be solved for parameter b, using, for example, the bisection method or the secant method. Then either side of this equation can be evaluated as the value of a.

Additional Details on SVD

The singular value decomposition (SVD) of a matrix is a factorization that can reveal important properties of the matrix that otherwise could escape detection. For example, from the SVD decomposition of a square matrix one could be alerted to the near-singularity of the matrix. Or from the SVD factorization of a nonsquare matrix an unexpected loss of rank could be revealed. Since the SVD factorization of a matrix yields a complete orthogonal decomposition, it provides a technique for computing the least-squares solution of a system of equations and at the same time producing the norm of the error vector.

Suppose that a given $m \times n$ matrix has the factorization

SVD Factorization

$$\boldsymbol{A} = \boldsymbol{U}\boldsymbol{D}\boldsymbol{V}^T$$

where $\boldsymbol{U} = [\boldsymbol{u}_1, \boldsymbol{u}_2, \dots, \boldsymbol{u}_m]$ is an $m \times m$ orthogonal matrix, $\boldsymbol{V} = [\boldsymbol{v}_1, \boldsymbol{v}_2, \dots, \boldsymbol{v}_n]$ is an $n \times n$ orthogonal matrix, and the $m \times n$ diagonal matrix $\boldsymbol{D}$ contains the singular values of $\boldsymbol{A}$ on its diagonal, listed in decreasing order. The singular values of a matrix $\boldsymbol{A}$ are the positive square roots of the eigenvalues of $\boldsymbol{A}^T\boldsymbol{A}$. These are denoted by $\sigma_1 \geqq \sigma_2 \geqq \cdots \geqq \sigma_r \geqq 0$. In detail, we have

$$\boldsymbol{U}^T\boldsymbol{A}\boldsymbol{V} = \boldsymbol{D} = \begin{bmatrix} \sigma_1 & & & & & & \\ & \sigma_2 & & & & & \\ & & \ddots & & & & \\ & & & \sigma_r & & & \\ & & & & 0 & & \\ & & & & & \ddots & \\ & & & & & & 0 \\ & & & & & & \\ & & & & & & \end{bmatrix}_{m\times n}$$

where $\boldsymbol{U}^T\boldsymbol{U} = \boldsymbol{I}_m$ and $\boldsymbol{V}^T\boldsymbol{V} = \boldsymbol{I}_n$. (In the above matrix, blank space corresponds to zero entries.) Moreover, we have $\boldsymbol{A}\boldsymbol{v}_i = \sigma_i\boldsymbol{u}_i$ and $\sigma_i = ||\boldsymbol{A}\boldsymbol{v}_i||_2$ where $\boldsymbol{v}_i$ is column i in $\boldsymbol{V}$ and $\boldsymbol{u}_i$ is column i in $\boldsymbol{U}$. Since $\boldsymbol{U}$ is orthogonal, we obtain

$$\begin{aligned} \left|\left|\boldsymbol{A}\boldsymbol{x} - \boldsymbol{b}\right|\right|_2^2 &= \left|\left|\boldsymbol{U}^T(\boldsymbol{A}\boldsymbol{x} - \boldsymbol{b})\right|\right|_2^2 = \left|\left|\boldsymbol{U}^T\boldsymbol{A}\boldsymbol{x} - \boldsymbol{U}^T\boldsymbol{b}\right|\right|_2^2 \\ &= \left|\left|\boldsymbol{U}^T\boldsymbol{A}(\boldsymbol{V}\boldsymbol{V}^T)\boldsymbol{x} - \boldsymbol{U}^T\boldsymbol{b}\right|\right|_2^2 \end{aligned}$$

$$= \left|\left|(U^T A V)(V^T x) - U^T b\right|\right|_2^2$$
$$= \left|\left|D V^T x - U^T b\right|\right|_2^2 = \left|\left|D y - c\right|\right|_2^2$$
$$= \sum_{i=1}^{r} (\sigma_i y_i - c_i)^2 + \sum_{i=r+1}^{m} c_i^2$$

where $y = V^T x$ and $c = U^T b$. Here, y is defined by $y_i = c_i/\sigma_j$ and x by $x = V y$. Since $c_i = u_i^T b$ and $x = V y$, if $y_i = \sigma_i^{-1} c_i$ for $1 \leq i \leq r$ then the least-squares solution is

$$x_{LS} = \sum_{i=1}^{n} y_i v_i = \sum_{i=1}^{r} \sigma_i^{-1} c_i v_i = \sum_{i=1}^{r} \sigma_i^{-1} \left(u_i^T b\right) v_i$$

and

$$\left|\left|A x_{LS} - b\right|\right|_2^2 = \sum_{i=r+1}^{m} c_i^2 = \sum_{i=r+1}^{m} \left(u_i^T b\right)^2$$

which is the smallest of all two-norm minimizers. For additional, details see Golub and Van Loan [1996].

In conclusion, we obtain the following theorem.

Theorem 1 **SVD Least-Squares Theorem**

Let A be an $m \times n$ matrix of rank r. Let the SVD factorization be

$$A = U D V^T$$

The least-squares solution of the system $Ax = b$ is $x_{LS} = \sum_{i=1}^{n} (\sigma_i^{-1} c_i) v_i$, where $c_i = u_i^T b$. If there exist many least-squares solutions to the given system, then the one of least 2-norm is x as described above.

EXAMPLE 1 Find the least-squares solution of this nonsquare system

$$\begin{bmatrix} 1 & 1 \\ 0 & 1 \\ 1 & 0 \end{bmatrix} \begin{bmatrix} x \\ y \\ z \end{bmatrix} = \begin{bmatrix} 1 \\ -1 \\ 1 \end{bmatrix}$$

using the singular value decomposition:

$$\begin{bmatrix} 1 & 1 \\ 0 & 1 \\ 1 & 0 \end{bmatrix} = \begin{bmatrix} \frac{1}{3}\sqrt{6} & 0 & \frac{1}{3}\sqrt{3} \\ \frac{1}{6}\sqrt{6} & \frac{1}{2}\sqrt{2} & -\frac{1}{3}\sqrt{3} \\ \frac{1}{6}\sqrt{6} & -\frac{1}{2}\sqrt{2} & -\frac{1}{3}\sqrt{3} \end{bmatrix} \begin{bmatrix} \sqrt{3} & 0 \\ 0 & 1 \\ 0 & 0 \end{bmatrix} \begin{bmatrix} \frac{1}{2}\sqrt{2} & \frac{1}{2}\sqrt{2} \\ \frac{1}{2}\sqrt{2} & -\frac{1}{2}\sqrt{2} \end{bmatrix}$$

Solution We have $r = \text{rank}(A) = 2$ and the singular values $\sigma_1 = \sqrt{3}$ and $\sigma_2 = 1$. This leads to

$$c_1 = u_1^T b = \begin{bmatrix} \frac{1}{3}\sqrt{6} & \frac{1}{6}\sqrt{6} & \frac{1}{6}\sqrt{6} \end{bmatrix} \begin{bmatrix} 1 \\ -1 \\ 1 \end{bmatrix} = \frac{1}{3}\sqrt{6}$$

and

$$c_2 = \boldsymbol{u}_2^T \boldsymbol{b} = \begin{bmatrix} 0 & -\frac{1}{2}\sqrt{2} & \frac{1}{2}\sqrt{2} \end{bmatrix} \begin{bmatrix} 1 \\ -1 \\ 1 \end{bmatrix} = \sqrt{2}$$

and

$$x_{LS} = \left(\sigma_1^{-1} c_1\right) \boldsymbol{v}_1 + \left(\sigma_2^{-1} c_2\right) \boldsymbol{v}_2 = \frac{1}{\sqrt{3}} \left(\frac{1}{3}\sqrt{6}\right) \begin{bmatrix} \frac{1}{2}\sqrt{2} \\ \frac{1}{2}\sqrt{2} \end{bmatrix} + \sqrt{2} \begin{bmatrix} \frac{1}{2}\sqrt{2} \\ -\frac{1}{2}\sqrt{2} \end{bmatrix}$$

$$= \begin{bmatrix} \frac{1}{3} \\ \frac{1}{3} \end{bmatrix} + \begin{bmatrix} 1 \\ 1 \end{bmatrix} = \begin{bmatrix} \frac{4}{3} \\ -\frac{2}{3} \end{bmatrix}$$

This solution is the same as that from the normal equations. ■

Using the Singular Value Decomposition

This material requires the theory of the singular value decomposition discussed in Section 8.2.

An important application of the singular value decomposition is in the matrix least-squares problem, to which we now return. For any system of linear equations

$$\boldsymbol{A}\boldsymbol{x} = \boldsymbol{b}$$

we want to define a unique **minimal solution**. This is described as follows. Let $\boldsymbol{A}$ be $m \times n$, and define

Minimal Solution

$$\rho = \inf\{||\boldsymbol{A}\boldsymbol{x} - \boldsymbol{b}||_2 : \boldsymbol{x} \in \mathbb{R}^n\}$$

The minimal solution of our system is taken to be the point of smallest norm in the set $\{\boldsymbol{x}: ||\boldsymbol{A}\boldsymbol{x} - \boldsymbol{b}||_2 = \rho\}$. If the system is *consistent*, then $\rho = 0$, and we are simply asking for the point of least norm among all solutions. If the system is *inconsistent*, we want $\boldsymbol{A}\boldsymbol{x}$ to be as close as possible to $\boldsymbol{b}$; that is, $||\boldsymbol{A}\boldsymbol{x} - \boldsymbol{b}||_2 = \rho$. If there are many such points, we choose the one closest to the origin.

The minimal solution is produced by using the *pseudo-inverse* of $\boldsymbol{A}$, and this object, in turn, can be computed from the singular value decomposition of $\boldsymbol{A}$ as discussed in Section 8.2. First, consider a diagonal $m \times n$ matrix of the following form, where the σ_j are positive numbers:

$$\boldsymbol{D} = \begin{bmatrix} \sigma_1 & & & & & & \\ & \sigma_2 & & & & & \\ & & \ddots & & & & \\ & & & \sigma_r & & & \\ & & & & 0 & & \\ & & & & & \ddots & \\ & & & & & & 0 \\ & & & & & & \end{bmatrix}_{m \times n}$$

Its pseudo-inverse $\boldsymbol{D}^+$ is defined to be of the same form, except that it is to be $n \times m$ and it has $1/\sigma_j$ on its diagonal. For example, we have

$$\boldsymbol{D} = \begin{bmatrix} 5 & 0 & 0 \\ 0 & 2 & 0 \end{bmatrix}, \qquad \boldsymbol{D}^+ = \begin{bmatrix} \frac{1}{5} & 0 \\ 0 & \frac{1}{2} \\ 0 & 0 \end{bmatrix}$$

If $\boldsymbol{A}$ is any $m \times n$ matrix and if $\boldsymbol{U}\boldsymbol{D}\boldsymbol{V}^T$ is one of its **singular value decompositions**, we define the **pseudo-inverse** of $\boldsymbol{A}$ to be

Pseudo-Inverse

$$\boldsymbol{A}^+ = \boldsymbol{V}\boldsymbol{D}^+\boldsymbol{U}^T$$

We do not stop to prove that the pseudo-inverse of $\boldsymbol{A}$ is unique if we impose the order $\sigma_1 \geqq \sigma_2 \geqq \cdots$.

■ **Theorem 2** **Minimal Solution Theorem**

Consider a system of linear equations $\boldsymbol{A}\boldsymbol{x} = \boldsymbol{b}$, in which $\boldsymbol{A}$ is an $m \times n$ matrix. The minimal solution of the system is $\boldsymbol{A}^+\boldsymbol{b}$.

Proof Use the notation established above, and let $\boldsymbol{x}$ be any point in $\mathbb{R}^n$. Define $\boldsymbol{y} = \boldsymbol{V}^T\boldsymbol{x}$ and $\boldsymbol{c} = \boldsymbol{U}^T\boldsymbol{b}$. Using the properties of $\boldsymbol{V}$ and $\boldsymbol{U}$, we obtain

$$\begin{aligned}
\rho &= \inf_{\boldsymbol{x}} ||\boldsymbol{A}\boldsymbol{x} - \boldsymbol{b}||_2 \\
&= \inf_{\boldsymbol{x}} ||\boldsymbol{U}\boldsymbol{D}\boldsymbol{V}^T\boldsymbol{x} - \boldsymbol{b}||_2 \\
&= \inf_{\boldsymbol{x}} ||\boldsymbol{U}^T(\boldsymbol{U}\boldsymbol{D}\boldsymbol{V}^T\boldsymbol{x} - \boldsymbol{b})||_2 \\
&= \inf_{\boldsymbol{x}} ||\boldsymbol{D}\boldsymbol{V}^T\boldsymbol{x} - \boldsymbol{U}^T\boldsymbol{b}||_2 \\
&= \inf_{\boldsymbol{y}} ||\boldsymbol{D}\boldsymbol{y} - \boldsymbol{c}||_2
\end{aligned}$$

Exploiting the special nature of $\boldsymbol{D}$, we have

$$||\boldsymbol{D}\boldsymbol{y} - \boldsymbol{c}||_2^2 = \sum_{i=1}^{r} (\sigma_i y_i - c_i)^2 + \sum_{i=r+1}^{m} c_i^2$$

To minimize this last expression, we define $y_i = c_i/\sigma_i$ for $1 \leqq i \leqq r$. The other components can remain unspecified. But to get the $\boldsymbol{y}$ of least norm, we must set $y_i = 0$ for $r+1 \leqq i \leqq m$. This construction is carried out by the pseudo-inverse $\boldsymbol{D}^+$, so $\boldsymbol{y} = \boldsymbol{D}^+\boldsymbol{c}$. Hence, we obtain

$$\boldsymbol{x} = \boldsymbol{V}\boldsymbol{y} = \boldsymbol{V}\boldsymbol{D}^+\boldsymbol{c} = \boldsymbol{V}\boldsymbol{D}^+\boldsymbol{U}^T\boldsymbol{b} = \boldsymbol{A}^+\boldsymbol{b}$$

Let us express the minimal solution in another form, taking advantage of the zero components in the vector $\boldsymbol{y}$. Since $y_i = 0$ for $i > r$, we require only the first r components of $\boldsymbol{y}$. These are given by $y_i = c_i/\sigma_i$. Now it is evident that only the first r components of $\boldsymbol{c}$ are needed. Since $\boldsymbol{c} = \boldsymbol{U}^T\boldsymbol{b}$, c_i is the inner product of row i in $\boldsymbol{U}^T$ with the vector $\boldsymbol{b}$. That is the same as the inner product of the ith column of $\boldsymbol{U}$ with $\boldsymbol{b}$. Thus,

$$y_i = \boldsymbol{u}_i^T \boldsymbol{b}/\sigma_i \qquad (1 \leqq i \leqq r)$$

The minimal solution, which we may denote by $\boldsymbol{x}^*$, is then

$$\boldsymbol{x}^* = \boldsymbol{V}\boldsymbol{y} = \sum_{i=1}^{r} y_i \boldsymbol{v}_i$$

■

Examples

Mathematical Software

An example of this procedure can be carried out in mathematical software such as MATLAB, Maple, or Mathematica. We can generate a system of 20 equations with three unknowns by a random process. This technique is often used in testing software, especially in benchmarking studies, in which a large number of examples is run with careful timing. The software has a provision for generating random matrices. When executed, the computer program first exhibits the random input. The three singular values of matrix $\boldsymbol{A}$ are displayed. Then the diagonal 20×3 matrix $\boldsymbol{D}$ is displayed. A check on the numerical work is made by computing $\boldsymbol{U}\boldsymbol{D}\boldsymbol{V}^T$, which should equal $\boldsymbol{A}$. Then the pseudo-inverse of $\boldsymbol{D}^+$ is computed. Next, the pseudo-inverse $\boldsymbol{A}^+$ is computed. The minimal solution, $\boldsymbol{x} = \boldsymbol{A}^+\boldsymbol{b}$, is computed, as well as the residual vector, $\boldsymbol{r} = \boldsymbol{A}^+\boldsymbol{b} = \boldsymbol{b}$. Then the orthogonality condition $\boldsymbol{A}^T\boldsymbol{r} = \boldsymbol{0}$ is checked. This program is therefore carrying out all the steps described above for obtaining the minimal solution of a system of equations. Another example is given below to show what happens in the case of a loss in rank. (See Computer Exercise 9.3.10.)

Pseudo-Inverse

In problems of this type, the user must examine the singular values and decide whether any are small enough to warrant being set equal to zero. The necessity of this step becomes clear when we look at the definition of $\boldsymbol{D}^+$. The reciprocals of the singular values are the principal constituents of this matrix. Any very small singular value that is *not* set equal to zero will therefore have a disruptive effect on the subsequent calculations. A rule of thumb that has been recommended is to drop any singular value whose magnitude is less than σ_1 times the inherent accuracy of the coefficient matrix. Thus, if the data are accurate to three decimal places and if $\sigma_1 = 5$, then any σ_i less than 0.005 should be set equal to zero.

An example of a small matrix having a **near-deficiency in rank** is given next. In the Maple program, certain singular values are set equal to zero if they fail to meet the relative size criterion mentioned in the previous paragraph. Also, we have added, as a check on the calculations, a verification of the following four **Penrose properties** for a pseudo-matrix.

■ **Theorem 3** **Penrose Properties of the Pseudo-Inverse**

Penrose Properties

The pseudo-inverse $\boldsymbol{A}^+$ for the matrix $\boldsymbol{A}$ has these four properties:

$$\boldsymbol{A} = \boldsymbol{A}\boldsymbol{A}^+\boldsymbol{A} \qquad \boldsymbol{A}^+ = \boldsymbol{A}^+\boldsymbol{A}\boldsymbol{A}^+$$
$$\boldsymbol{A}\boldsymbol{A}^+ = (\boldsymbol{A}\boldsymbol{A}^+)^T \qquad \boldsymbol{A}^+\boldsymbol{A} = (\boldsymbol{A}^+\boldsymbol{A})^T$$

We can use mathematical software such as MATLAB, Maple, or Mathematica for finding the pseudo-inverse of a matrix that has a deficiency in rank. For example, consider this 5×3 matrix:

$$A = \begin{bmatrix} -85 & -55 & -115 \\ -35 & 97 & -167 \\ 79 & 56 & 102 \\ 63 & 57 & 69 \\ 45 & -8 & 97.5 \end{bmatrix} \tag{7}$$

A tolerance value is set so that in the evaluation of singular values any value whose magnitude is less than the tolerance is treated as zero. We can verify the Penrose properties for this matrix. (See Computer Exercise 9.3.11.)

Orthogonal Matrices and Spectral Theorem

A matrix $\boldsymbol{Q}$ is said to be **orthogonal** if

$$\boldsymbol{Q}\boldsymbol{Q}^T = \boldsymbol{Q}^T\boldsymbol{Q} = \boldsymbol{I}$$

This forces $\boldsymbol{Q}$ to be square and nonsingular. Furthermore,

$$\boldsymbol{Q}^{-1} = \boldsymbol{Q}^T$$

With this concept available, we can state one of the principal theorems of linear algebra: the spectral theorem for symmetric matrices.

Theorem 4 **Spectral Theorem for Symmetric Matrices**

If $\boldsymbol{A}$ is a symmetric real matrix, then there exists an orthogonal matrix $\boldsymbol{Q}$ such that $\boldsymbol{Q}^T\boldsymbol{A}\boldsymbol{Q}$ is a diagonal matrix.

The equation

$$\boldsymbol{Q}^T\boldsymbol{A}\boldsymbol{Q} = \boldsymbol{D}$$

is equivalent to

$$\boldsymbol{A}\boldsymbol{Q} = \boldsymbol{Q}\boldsymbol{D}$$

If $\boldsymbol{D}$ is diagonal, the columns $\boldsymbol{v}_i$ of $\boldsymbol{Q}$ obey the equation

$$\boldsymbol{A}\boldsymbol{v}_i = d_{ii}\boldsymbol{v}_i$$

In other words, the columns of $\boldsymbol{Q}$ form an orthonormal system of eigenvectors of $\boldsymbol{A}$, and the diagonal elements of $\boldsymbol{D}$ are the eigenvalues of $\boldsymbol{A}$.

Summary 9.3

- We attempt to solve an **inconsistent system**

$$\sum_{j=0}^{n} a_{kj}x_j = b_k \qquad (0 \leqq k \leqq m)$$

in which there are $m+1$ equations but only $n+1$ unknowns with $m > n$. We minimize the sum of the squares of the residuals and are led to minimize the expression

$$\varphi(x_0, x_1, \ldots, x_n) = \sum_{k=0}^{m} \left(\sum_{j=0}^{n} a_{kj} x_j - b_k \right)^2$$

We solve the $(n+1) \times (n+1)$ system of **normal equations**

$$\sum_{j=0}^{n} \left(\sum_{k=0}^{m} a_{ki} a_{kj} \right) x_j = \sum_{k=0}^{m} b_k a_{ki} \qquad (0 \leq i \leq n)$$

by Gaussian elimination, and the solution is a best approximate solution of the original system in the least-squares sense.

Exercises 9.3

1. Analyze the least-squares problem of fitting data by a function of the form $y = x^c$.

[a]**2.** Show that the **Hilbert matrix** (Computer Exercise 2.2.4) arises in the normal equations when we minimize

$$\int_0^1 \left[\sum_{j=0}^{n} c_j x^j - f(x) \right]^2 dx$$

[a]**3.** Find a function of the form $y = e^{cx}$ that best fits this table:

x	0	1
y	$\frac{1}{2}$	1

[a]**4.** (Continuation) Repeat the preceding problem for the following table:

x	0	1
y	a	b

5. (Continuation) Repeat the preceding problem under the supposition that b is negative.

[a]**6.** Show that the normal equation for the problem of fitting $y = e^{cx}$ to points $(1, -12)$ and $(2, 7.5)$ has two real roots: $c = \ln 2$ and $c = 0$. Which value is correct for the fitting problem?

7. Consider the inconsistent System (1). Suppose that each equation has associated with it a positive number w_i indicating its relative importance or reliability. How should Equations (2) and (3) be modified to reflect this?

[a]**8.** Determine the best approximate solution of the inconsistent system of linear equations

$$\begin{cases} 2x + 3y = 1 \\ x - 4y = -9 \\ 2x - y = -1 \end{cases}$$

in the least-squares sense.

9.[a]**a.** Find the constant c for which cx is the best approximation in the sense of least squares to the function $\sin x$ on the interval $[0, \pi/2]$.

[a]**b.** Do the same for e^x on $[0, 1]$.

10. Analyze the problem of fitting a function $y = (c - x)^{-1}$ to a table of $m + 1$ points.

11. Show that the normal equations for the least-squares solution of $\boldsymbol{Ax} = \boldsymbol{b}$ can be written $(\boldsymbol{A}^T \boldsymbol{A})\boldsymbol{x} = \boldsymbol{A}^T \boldsymbol{b}$.

12. Derive the normal equations given by System (5).

13. A table of values (x_k, y_k), where $k = 0, 1, \ldots, m$, is obtained from an experiment. When plotted on semilogarithmic graph paper, the points lie nearly on a straight line, implying that $y \approx e^{ax+b}$. Suggest a simple procedure for obtaining parameters a and b.

[a]**14.** In fitting a table of values to a function of the form $a + bx^{-1} + cx^{-2}$, we try to make each point lie on the curve. This leads to $a + bx_k^{-1} + cx_k^{-2} = y_k$ for $0 \leq k \leq m$. An equivalent equation is $ax_k^2 + bx_k + c = y_k x_k^2$ for $0 \leq k \leq m$. Are the least-squares problems for these systems of equations equivalent?

[a]**15.** A table of points (x_k, y_k) is plotted and appears to lie on a hyperbola of the form $y = (a + bx)^{-1}$. How can

the *linear* theory of least squares be used to obtain good estimates of a and b?

[a]**16.** Consider $f(x) = e^{2x}$ over $[0, \pi]$. We wish to approximate the function by a trigonometric polynomial of the form $p(x) = a + b\cos(x) + c\sin(x)$. Determine the linear system to be solved for determining the least-squares fit of p to f.

[a]**17.** Find the constant c that makes the expression

$$\int_0^1 (e^x - cx)^2\, dx$$

a minimum.

18. Show that in every least-squares matrix problem, the normal equations have a symmetric coefficient matrix.

19. Verify that the following steps produce the least-squares solution of $\boldsymbol{Ax} = \boldsymbol{b}$.

a. Factor $\boldsymbol{A} = \boldsymbol{QR}$, where $\boldsymbol{Q}$ and $\boldsymbol{R}$ have the properties described in the text.

b. Define $\boldsymbol{y} = \boldsymbol{Q}^T\boldsymbol{b}$.

c. Solve the lower triangular system $\boldsymbol{Rx} = \boldsymbol{y}$.

[a]**20.** What value of c should be used if a table of experimental data (x_i, y_i) for $0 \leqq i \leqq m$ is to be represented by the formula $y = c\sin x$? An explicit usable formula for c is required. Use the principle of least squares.

21. Refer to the formulas leading to the minimal solution of the system $\boldsymbol{Ax} = \boldsymbol{b}$. Prove that the $\boldsymbol{y}$-vector is given by the formula $y_i = \sigma_i^{-2}\boldsymbol{b}^T\boldsymbol{A}\boldsymbol{v}_i$ for $1 \leqq i \leqq r$.

22. Prove that the pseudo-inverse satisfies the four Penrose equations.

23. Use the four Penrose properties to find the pseudo-inverse of the matrix $[a, 0]^T$, where $a > 0$. Prove that the pseudo-inverse is a discontinuous function of a.

24. Use the technique suggested in the preceding problem to find the pseudo-inverse of the $m \times n$ matrix consisting solely of 1's.

25. Use the Penrose equations to find the pseudo-inverse of any $1 \times n$ matrix and any $m \times 1$ matrix.

26. (**Multiple Choice**) Let $\boldsymbol{A} = \boldsymbol{PDQ}$, where $\boldsymbol{A}$ is an $m \times n$ matrix, $\boldsymbol{P}$ is an $m \times m$ unitary matrix, $\boldsymbol{D}$ is an $m \times n$ diagonal matrix, and $\boldsymbol{Q}$ is an $n \times n$ unitary matrix. Which equation can be deduced from those hypotheses?

a. $\boldsymbol{A}^* = \boldsymbol{P}^*\boldsymbol{D}^*\boldsymbol{Q}^*$ **b.** $\boldsymbol{A}^{-1} = \boldsymbol{Q}^*\boldsymbol{D}^{-1}\boldsymbol{P}^*$
c. $\boldsymbol{D} = \boldsymbol{PAQ}$ **d.** $\boldsymbol{A}^*\boldsymbol{A} = \boldsymbol{Q}^*\boldsymbol{D}^*\boldsymbol{DQ}$
e. None of these.

27. (**Multiple Choice**, continued) Assume the hypotheses of the preceding problem. Use the notation $^+$ to indicate a pseudo-inverse. Which equation is correct?

a. $\boldsymbol{A}^+ = \boldsymbol{PD}^+\boldsymbol{Q}$ **b.** $\boldsymbol{A}^* = \boldsymbol{Q}^*\boldsymbol{D}^{-1}\boldsymbol{P}^*$
c. $\boldsymbol{A}^+ = \boldsymbol{Q}^*\boldsymbol{D}^+\boldsymbol{P}^*$ **d.** $\boldsymbol{A}^{-1} = \boldsymbol{Q}^*\boldsymbol{D}^+\boldsymbol{P}^*$
e. None of these.

28. (**Multiple Choice**) Let $\boldsymbol{D}$ be an $m \times n$ diagonal matrix with diagonal elements $p_1, p_2, \ldots, p_r, 0, 0, \ldots, 0$. Here all the numbers p_i, for $1 \leqq i \leqq r$, are positive. Which assertion is *not* valid?

a. $\boldsymbol{D}^+$ is the $m \times n$ diagonal matrix with diagonal elements $(1/p_1, 1/p_2, \ldots, 1/p_r, 0, 0, \ldots, 0)$

b. $\boldsymbol{D}^+$ is the $n \times m$ diagonal matrix with diagonal elements $(1/p_1, 1/p_2, \ldots, 1/p_r, 0, 0, \ldots, 0)$

c. $(\boldsymbol{D}^+)^* = (\boldsymbol{D}^*)^+$

d. $\boldsymbol{D}^{++} = \boldsymbol{D}$

e. None of these.

29. (**Multiple Choice**) Consider an inconsistent system of equations $\boldsymbol{Ax} = \boldsymbol{b}$. Let $\boldsymbol{U}$ be a unitary matrix and let $\boldsymbol{E} = \boldsymbol{U}^*\boldsymbol{A}$. Let $\boldsymbol{v}$, $\boldsymbol{w}$, and $\boldsymbol{z}$ be vectors such that $\boldsymbol{Uv} = \boldsymbol{Eb}$, $\boldsymbol{Uw} = \boldsymbol{E}^*\boldsymbol{b}$, $\boldsymbol{Ey} = \boldsymbol{U}^*\boldsymbol{b}$, and $\boldsymbol{Ex} = \boldsymbol{Ub}$. A vector that solves the least-squares problem for the original system $\boldsymbol{Ax} = \boldsymbol{b}$ is:

a. $\boldsymbol{v}$ **b.** $\boldsymbol{w}$ **c.** $\boldsymbol{y}$ **d.** $\boldsymbol{z}$
b. None of these.

Computer Exercises 9.3

[a]**1.** Using the method suggested in the text, fit the data in the table

x	0.1	0.2	0.3	0.4	0.5	0.6	0.7	0.8
y	0.6	1.1	1.6	1.8	2.0	1.9	1.7	1.3

by a function $y = a\sin bx$.

2. (**Prony's Method, $n = 1$**) To fit a table of the form

x	1	2	$\cdots$	m
y	y_1	y_2	$\cdots$	y_m

by the function $y = ab^x$, we can proceed as follows: If y is actually ab^x, then $y_k = ab^k$ and $y_{k+1} = by_k$ for

$k = 1, 2, \ldots, m-1$. So we determine b by solving this system of equations using the least-squares method. Having found b, we find a by solving the equations $y_k = ab^k$ in the least-squares sense. Write a program to carry out this procedure, and test it on an artificial example.

3. (Continuation) Modify the procedure of the preceding computer problem to handle any case of equally spaced points.

4. A quick way of fitting a function of the form

$$f(x) \approx \frac{a + bx}{1 + cx}$$

is to apply the least-squares method to the problem $(1 + cx)f(x) \approx a + bx$. Use this technique to fit the **world population** data given here (in billions):

Year	Population
1000	0.340
1650	0.545
1800	0.907
1900	1.61
1950	2.56
1960	3.15
1970	3.65
1980	4.20
1990	5.30
2000	6.12
2010	6.98

Determine when the world population becomes infinite!

5. (**Student Research Project**) Explore the question of whether the least-squares method should be used to predict. For example, study the variances in the preceding problem to determine whether a polynomial of any degree would be satisfactory.

6. Write a procedure that takes as input an $(m+1) \times (n+1)$ matrix $\boldsymbol{A}$ and an $m+1$ vector $\boldsymbol{b}$ and returns the least-squares solution of the system $\boldsymbol{Ax} = \boldsymbol{b}$.

7. Write a Maple program to find the minimal solution of any system of equations, $\boldsymbol{Ax} = \boldsymbol{b}$.

8. (Continuation) Write a MATLAB program for the task in the preceding problem.

9. Investigate some of the newer methods for solving inconsistent linear equations $\boldsymbol{Ax} = \boldsymbol{b}$, when the criterion is to make $\boldsymbol{Ax}$ close to $\boldsymbol{b}$ in one of the other useful norms, namely, the maximum norm $||\boldsymbol{x}||_\infty = \max_{1 \leqq i \leqq n} |x_i|$ or the ℓ_1 norm $||\boldsymbol{x}||_1 = \sum_{i=1}^{n} |x_i|$. Use some of the available software.

10. Using mathematical software such as MATLAB, Maple, or Mathematica, generate a system of 20 equations with 3 unknowns by a random-number generator. Form the pseudo-inverse matrix and verify the properties in Theorem 2.

11. (Continuation.) Repeat using Matrix (7).

12. Write a computer program for carrying out the least-squares curve fit using Chebyshev polynomials. Test the code on a suitable data set and plot the results.

9.4 Fourier Series

Introduction

Introduced by Jean Baptiste Joseph Fourier (1763–1830), Fourier series decompose a periodic function into a linear combinations of sines and cosines.* Since then, this idea has been expanded into an entire area of study known as **Fourier analysis**, which has applications in acoustics, optics, vibrations, quantum mechanics, signal processing, and many other fields.

Least-Squares Approximation

First, we establish the least-squares approximation of a continuous function f (which is periodic over the interval $[-\pi, \pi]$, in the space of trigonometric polynomials, denoted $T[-\pi, \pi]$) and show that it can be spanned by the orthogonal set

$$W = \{1, \cos x, \sin x, \cos 2x, \sin 2x, \ldots, \cos Nx, \sin Nx\}$$

*In 1798, Joseph Fourier was Napoleon's scientific advisor during France's expedition to Egypt!

where N is a positive integer. The space is geometrically defined using the **inner product**

Inner Products

$$\langle f, g\rangle = \int_{-\pi}^{\pi} f(x)g(x)\,dx \tag{1}$$

Any continuous function f on the interval $[-\pi, \pi]$ can be approximated by linear combinations of elements from the set W, as closely as needed with a sufficiently large value of N. Throughout, we assume n, m, j, k, N, etc. are integers.

The vectors in W are mutually orthogonal

$$|| 1 ||_2^2 = \langle 1, 1\rangle = 2\pi$$

$$|| \cos nx ||_2^2 = \langle \cos nx, \cos nx\rangle = \pi$$

$$|| \sin nx ||_2^2 = \langle \sin nx, \sin nx\rangle = \pi$$

where integer $n \neq 0$. (See Example 1, p. 461.) Moreover, an orthonormal basis is

Orthonormal Basis

$$\begin{aligned} U &= \{g_0, g_1, g_2, \ldots, g_{2N-1}, g_{2N}\} \\ &= \left\{\frac{1}{\sqrt{2\pi}}, \frac{1}{\sqrt{\pi}}\cos x, \frac{1}{\sqrt{\pi}}\sin x, \frac{1}{\sqrt{\pi}}\cos 2x, \right. \\ &\qquad \left. \frac{1}{\sqrt{\pi}}\sin 2x, \ldots, \frac{1}{\sqrt{\pi}}\cos Nx, \frac{1}{\sqrt{\pi}}\sin Nx\right\} \end{aligned}$$

We use U and the formula

$$g(x) = \text{Proj}_T f = \langle f, g_0\rangle g_0 + \langle f, g_1\rangle g_1 + \langle f, g_2\rangle g_2 + \cdots + \langle f, g_{2N-1}\rangle g_{2N-1} + \langle f, g_{2N}\rangle g_{2N}$$

to find the least-squares approximation of g in terms of f in the general form

$$\begin{aligned} g(x) = \left\langle f, \frac{1}{\sqrt{2\pi}}\right\rangle \frac{1}{\sqrt{2\pi}} &+ \left\langle f, \frac{1}{\sqrt{\pi}}\cos x\right\rangle \frac{1}{\sqrt{\pi}}\cos x + \left\langle f, \frac{1}{\sqrt{\pi}}\sin x\right\rangle \frac{1}{\sqrt{\pi}}\sin x \\ &+ \cdots + \left\langle f, \frac{1}{\sqrt{\pi}}\cos Nx\right\rangle \frac{1}{\sqrt{\pi}}\cos Nx + \left\langle f, \frac{1}{\sqrt{\pi}}\sin Nx\right\rangle \frac{1}{\sqrt{\pi}}\sin Nx \end{aligned}$$

We introduce this convenient notation

Coefficients a_0, a_n, b_n

$$\frac{1}{2}a_0 = \left\langle f, \frac{1}{\sqrt{2\pi}}\right\rangle \frac{1}{\sqrt{2\pi}} = \frac{1}{2\pi}\int_{-\pi}^{\pi} f(x)\,dx$$

$$a_n = \left\langle f, \frac{1}{\sqrt{\pi}}\cos nx\right\rangle \frac{1}{\sqrt{\pi}} = \frac{1}{\pi}\int_{-\pi}^{\pi} f(x)\cos nx\,dx \qquad (1 \leqq n \leqq N)$$

$$b_n = \left\langle f, \frac{1}{\sqrt{\pi}}\sin nx\right\rangle \frac{1}{\sqrt{\pi}} = \frac{1}{\pi}\int_{-\pi}^{\pi} f(x)\sin nx\,dx \qquad (1 \leqq n \leqq N)$$

Orthogonality Properties

The following trigonometric identities are particularly useful in Fourier series because they express products of sines and cosines as sums.

Trigonometric Identities

$$\cos mx \cos nx = \cos[(m+n)x] + \cos[(m-n)x]$$

$$\sin mx \sin nx = \sin[(m+n)x] + \sin[(m-n)x]$$

$$\sin x \cos y = \frac{1}{2}\left[\sin(x-y) + \sin(x+y)\right]$$

Establish these **orthogonality properties**:

Orthogonality Properties

$$\langle \cos mx, \cos nx \rangle = \pi \delta_{mn} = \begin{cases} 0, & (m \neq n) \\ \pi, & (m = n \neq 0) \end{cases}$$

$$\langle \sin mx, \sin nx \rangle = \pi \delta_{mn} = \begin{cases} 0, & (m \neq n) \\ \pi, & (m = n \neq 0) \end{cases}$$

$$\langle \cos mx, \sin nx \rangle = 0$$

Here δ_{mn} is the **Kronecker delta**. These properties can be used to establish that W is an orthogonal set and U is an orthonormal set.

EXAMPLE 1 Establish the second property above, when $m \neq n$. (We leave the others as exercises.)

Solution

$$\begin{aligned} \langle \sin mx, \sin nx \rangle &= \int_{-\pi}^{\pi} \sin mx \, \sin nx \, dx \\ &= \frac{1}{2} \int_{-\pi}^{\pi} \big[\sin[(m+n)x] + \sin[(m-n)x] \big] \, dx \\ &= -\frac{1}{2} \left[\frac{1}{m+n} \cos[(m+n)x] + \frac{1}{m-n} \cos[(m-n)x] \right]_{-\pi}^{\pi} = 0 \end{aligned}$$

■

Standard Integrals

The computation of the Fourier series is based on these standard integrations formulas

Integration Formulas

$$\int_{-\pi}^{\pi} \cos mx \, dx = 0$$

$$\int_{-\pi}^{\pi} \sin mx \, dx = 0$$

$$\int_{-\pi}^{\pi} \sin mx \, \cos nx \, dx = 0$$

$$\int_{-\pi}^{\pi} \cos mx \, \cos nx \, dx = \pi \delta_{mn} = \begin{cases} 0, & m \neq n \\ \pi, & m = n \end{cases}$$

$$\int_{-\pi}^{\pi} \sin mx \, \sin nx \, dx = \pi \delta_{mn} = \begin{cases} 0, & m \neq n \\ \pi, & m = n \end{cases}$$

for integers $m \neq 0$ and $n \neq 0$.

Some other useful standard integration formulas are

Useful Integrals

$$\int x \cos mx \, dx = \frac{1}{m} x \sin mx + \frac{1}{m^2} \cos mx + C \tag{2}$$

$$\int x \sin mx \, dx = -\frac{1}{m} x \cos mx + \frac{1}{m^2} \sin mx + C \tag{3}$$

which are obtained by **integration by parts**

Integration by Parts

$$\int u \, dv = uv - \int v \, du$$

For example, let $u = x$ and $dv = \cos mx \, dx$ in Formula (2).

Trigonometric Polynomial

Consider the **trigonometric polynomial** of degree N of the form

$$\begin{aligned} p_N(x) &= \frac{1}{2}a_0 + a_1 \cos x + b_1 \sin x + \cdots + a_N \cos Nx + b_N \sin Nx \\ &= \frac{1}{2}a_0 + \sum_{n=1}^{N} \left[a_n \cos nx + b_n \sin nx\right] \end{aligned} \tag{4}$$

where $a_N \neq 0$ or $b_N \neq 0$. It is sometimes more advantageous to expand a function into a trigonometric series rather than as a power series. For example, it makes sense to express some phenomena in nature as periodic functions. Recall that a Taylor series of a function f we can be written as an **N-th degree polynomial approximation**

$$f(x) \approx \sum_{n=1}^{N} c_n (x-c)^k$$

where $c_n = f^{(n)}(c)/n!$. For Fourier series, we find formulas for the coefficients in terms of integrals of f rather than derivatives of f.

Consider a subspace of continuous functions over $[-\pi, \pi]$ spanned by the elements of the set W. The best approximation to f by functions in W is the **N-th order Fourier approximation** to f over $[-\pi, \pi]$. It is given by the orthogonal projection onto W since the functions in the set are orthogonal. The standard formulas for the orthogonal projections are

$$a_n = \frac{\langle f, \cos nx\rangle}{\langle \cos nx, \cos nx\rangle} \qquad (n \geqq 0)$$

$$b_n = \frac{\langle f, \sin nx\rangle}{\langle \sin nx, \sin nx\rangle} \qquad (n \geqq 1)$$

Using $\langle \sin nx, \sin nx\rangle = \pi$, $\langle \cos nx, \cos nx\rangle = \pi$, and the inner product Formula (1), we obtain

Fourier Coefficients

$$a_n = \frac{1}{\pi}\int_{-\pi}^{\pi} f(x)\cos nx\, dx \qquad (n \geqq 0)$$

$$b_n = \frac{1}{\pi}\int_{-\pi}^{\pi} f(x)\sin nx\, dx \qquad (n \geqq 1)$$

We can verify a_0 using the constant function $f(x) = 1$ and the orthogonal projection

$$\frac{\langle f, 1\rangle}{\langle 1, 1\rangle} = \left[\int_{-\pi}^{\pi} f(x)\cdot 1\, dx\right] \Big/ \left[\int_{-\pi}^{\pi} 1 \cdot 1\, dx\right] = \frac{1}{2\pi}\int_{-\pi}^{\pi} f(x)\, dx = \frac{1}{2}a_0$$

Solving for a_0, we have

$$a_0 = \frac{1}{\pi}\int_{-\pi}^{\pi} f(x)\, dx$$

Clearly, formula for a_n holds when $n = 0$ since $\cos 0 = 1$. For clarity, we often write the formula for a_0 separately from that for a_n.

The **N-th order Fourier approximation** of $f(x)$ using the **Fourier coefficients** a_0, a_n, and b_n can be written as Summation (4). This approximation becomes increasingly better as N increases in the sense that $||f - p_N||$ becomes smaller, which is called the **means square error** in the approximation. The **Fourier series** of f on the interval $[-\pi, \pi]$ is the infinite sum

Fourier Series of f

$$f(x) = \frac{1}{2}a_0 + \sum_{n=1}^{\infty} \left[a_n \cos nx + b_n \sin nx\right] \tag{5}$$

Fourier series can be shown to hold for a wider class of functions such as **piecewise continuous functions**; that is, when $f(x)$ is continuous except perhaps for a finite number of removable or jump discontinuities. Here is a typical theorem with regard to Fourier series for continuous and piecewise continuous functions.

Theorem 1 **Fourier Convergence Theorem**

The Fourier Series (5) is convergent if f is a continuous periodic function on $[-\pi, \pi]$ with period 2π or if f and f' are piecewise continuous. Where f is continuous, the sum of the Fourier series equals $f(c)$ at all numbers c. When f is discontinuous at a point c, the sum of the Fourier series is the average of the left and right limits; namely, $\frac{1}{2}[f(c^+) + f(c^-)]$, where $f(c^+) = \lim_{x\to c^+} f(x)$ and $f(c^-) = \lim_{x\to c^-} f(x)$.

The **Dirichlet conditions** are sufficient conditions for a real-valued periodic function $f(x)$ to be the same as the Fourier series at points of continuity, and the behavior of the Fourier series at points of discontinuity are prescribed as above. The **Gibbs phenomenon** describes the peculiar behavior of the Fourier series approximations at simple discontinuities. Although there may be large oscillations near the jumps, these overshoots (**ringing**) do not die out as the frequency increases, but they do approach a finite limit.

In summary, the Fourier series for the continuous function $f(x)$, periodic on $[-\pi, \pi]$, is

Fourier Series of f

$$f(x) = \frac{1}{2}a_0 + \sum_{n=1}^{\infty} \left[a_n \cos nx + b_n \sin nx\right] \tag{6}$$

where

Fourier Coefficients

$$a_0 = \frac{1}{\pi}\int_{-\pi}^{\pi} f(x)\,dx \tag{7}$$

$$a_n = \frac{1}{\pi}\int_{-\pi}^{\pi} f(x)\cos nx\,dx \qquad (n \geqq 1) \tag{8}$$

$$b_n = \frac{1}{\pi}\int_{-\pi}^{\pi} f(x)\sin nx\,dx \qquad (n \geqq 1) \tag{9}$$

If f is continuous at x, the Fourier Series (6) converges to $f(x)$ and the equality symbol $=$ is used. On the other hand, if f is discontinuous at x, the Fourier series (6) may *not* converge to $f(x)$ and the approximation symbol $\approx$ is used. In the latter case, the Fourier series converges to the **midpoint of the jump**.

Cosine Series and Sine Series

Symmetry in a function $f(x)$ can be exploited to reduce the computation effort of finding the Fourier coefficients and series when the symmetry exists either about the vertical axis or the origin.

An **even function** $f_e(x)$ is symmetric with respect to the vertical axis ($x = 0$) and satisfies

Even Function

$$f_e(-x) = f_e(x) \tag{10}$$

An **odd function** $f_o(z)$ is symmetric with respect to the origin and satisfies

Odd Function

$$f_o(-x) = -f_o(x) \tag{11}$$

Based on these results, we obtain the following:

If $f_e(x)$ is an **even function**, then $f_e(x)\sin nx$ is odd and the Fourier coefficients b_n are

$$b_n = 0 \qquad (n \geqq 1)$$

Hence, we obtain a **cosine series**

Cosine Series

$$f_e(x) = \frac{1}{2}a_0 + \sum_{n=1}^{\infty} a_n \cos nx \tag{12}$$

where

$$a_0 = \frac{2}{\pi}\int_0^{\pi} f_e(x)\,dx$$

$$a_n = \frac{2}{\pi}\int_0^{\pi} f_e(x)\cos nx\,dx \qquad (n \geqq 1)$$

If $f_o(x)$ is an **odd function**, then $f_o(x)\cos nx$ is an odd function and the Fourier coefficients a_n are

$$a_n = 0 \qquad (n \geqq 0)$$

Hence, we obtain a **sine series**

Sine Series

$$f_o(x) = \sum_{n=1}^{\infty} b_n \sin nx \tag{13}$$

where

$$b_n = \frac{2}{\pi}\int_0^{\pi} f_o(x)\sin nx\,dx \qquad (n \geqq 1)$$

Sawtooth Wave Function on $[-\pi, \pi]$

EXAMPLE 2 Determine the Nth degree polynomial approximation to the Fourier series for the periodic **sawtooth wave function**

Sawtooth Wave Function on $[-\pi, \pi]$

$$f(x) = x \quad \text{on} \quad (-\pi, \pi) \quad \text{with} \quad f(-\pi) = 0 = f(\pi)$$

Solution Since $f(x) = x$ is an odd function on $(-\pi, \pi)$, we obtain immediately

$$a_n = 0 \qquad (n \geqq 0)$$

We can verify this for the a_0 coefficient

$$a_0 = \frac{1}{\pi}\int_{-\pi}^{\pi} x\,dx = \frac{1}{\pi}\left.\frac{x^2}{2}\right|_{-\pi}^{\pi} = 0$$

Consequently, we need only determine the b_n coefficients

$$\begin{aligned} b_n &= \frac{1}{\pi}\int_{-\pi}^{\pi} x \sin nx\,dx = \frac{2}{\pi}\int_{0}^{\pi} x \sin nx\,dx \\ &= \frac{2}{\pi}\left[-\frac{1}{n}x\cos nx + \frac{1}{n^2}\sin nx\right]_0^{\pi} \\ &= -\frac{2}{n}\cos n\pi + \frac{2}{\pi n^2}\sin n\pi \\ &= 2(-1)^{n+1}\frac{1}{n} \qquad (n \geqq 1) \end{aligned}$$

Here we use integration Formula (3) as well as $\cos n\pi = (-1)^n$ and $\sin n\pi = 0$. Therefore, we have

$$f(x) \approx p_N(x) = 2\sum_{n=1}^{N}(-1)^{n+1}\frac{1}{n}\sin nx \tag{14}$$

In detail, we obtain

Sawtooth Polynomial p_N

$$p_N(x) = 2\left[\sin x - \frac{1}{2}\sin 2x + \frac{1}{3}\sin 3x - \frac{1}{4}\sin 4x + \cdots + (-1)^{N+1}\frac{1}{N}\sin Nx\right]$$

As expected for an odd function, this is a sine series. Also, $f(0) = 0$ and $f(n\pi) = 0$ for all integers n. So at the discontinuous points $x = n\pi$, the Fourier series converges to 0, which is the midpoint of the jump. See Figure 9.6.

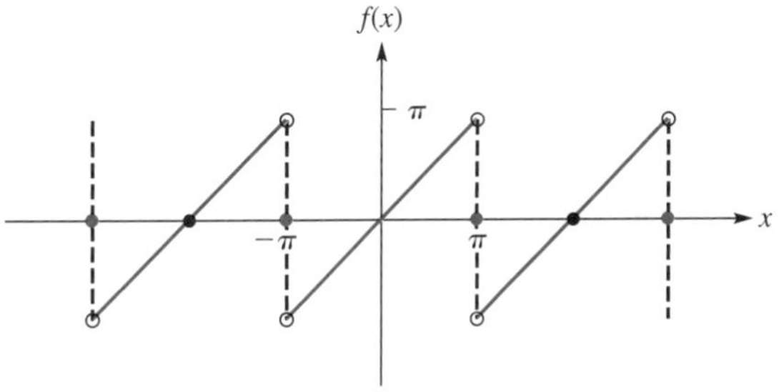

FIGURE 9.6 Sawtooth wave on $[-\pi, \pi]$

■

Infinite Fourier Series

Assume that a continuous periodic function $f(x)$ on $[-\pi, \pi]$ can be expressed as a convergent infinite series of sines and cosines such as (6). It turns out that the formulas for the coefficients a_0, a_n and b_n are the same as those given in formulas (7)–(9), respectively.

Here we have assumed that f is a continuous function over the interval $[-\pi, \pi]$ such that the infinite trigonometric series of sines and cosines exists and can be integrated term-by-term.

$$\begin{aligned} \int_{-\pi}^{\pi} f(x)\,dx &= \frac{1}{2}\int_{-\pi}^{\pi} a_0\,dx + \int_{-\pi}^{\pi}\sum_{n=1}^{\infty}\left[a_n\cos nx + b_n \sin nx\right]dx \\ &= \frac{1}{2}(2\pi)a_0 + \sum_{n=1}^{\infty} a_n\int_{-\pi}^{\pi}\cos nx\,dx + \sum_{n=1}^{\infty} b_n\int_{-\pi}^{\pi}\sin nx\,dx \end{aligned}$$

Using standard integration formulas, both of the integrals above have the value 0 and we obtain Formula (7) for a_0. Multiplying infinite Series (6) for $f(x)$ by $\cos nx$ and integrating term-by-term, we obtain Formula (8) for a_n. Similarly, we can determine Formula (9) for b_n. We leave verifying a_n and b_n as exercises.

Periodic on $[-P/2, P/2]$

We can change the interval from $[-\pi, \pi]$ to $[-P/2, P/2]$ by using the change of variables $x_{\text{new}} = [P/2\pi]x_{\text{old}}$. Consequently, for a function $f(x)$ periodic on $[-P/2, P/2]$ with period P, we can write the Fourier series as

$$f(x) = \frac{1}{2}a_0 + \sum_{n=1}^{\infty}\left[a_n \cos\left(\frac{2\pi n}{P}x\right) + b_n \sin\left(\frac{2\pi nx}{P}\right)\right] \tag{15}$$

where

$$a_0 = \frac{2}{P}\int_{-P/2}^{P/2} f(x)\,dx$$

$$a_n = \frac{2}{P}\int_{-P/2}^{P/2} f(x)\cos\left(\frac{2\pi n}{P}x\right)dx \qquad (n \geqq 1)$$

$$b_n = \frac{2}{P}\int_{-P/2}^{P/2} f(x)\sin\left(\frac{2\pi n}{P}x\right)dx \qquad (n \geqq 1)$$

Clearly, $P = 2\pi$ gives the special case $[-\pi, \pi]$ and the Fourier series given by Formulas (6)–(9).

Periodic on $[-L, L]$

Next, we find that the Fourier series for a function $f(x)$ periodic on $[-L, L]$, with period $P = 2L$, is given by

$$f(x) = \frac{1}{2}a_0 + \sum_{n=1}^{\infty}\left[a_n \cos\left(\frac{n\pi}{L}x\right) + b_n \sin\left(\frac{n\pi}{L}x\right)\right] \tag{16}$$

where

$$a_0 = \frac{1}{L}\int_{-L}^{L} f(x)\,dx$$

$$a_n = \frac{1}{L}\int_{-L}^{L} f(x)\cos\left(\frac{n\pi}{L}x\right)dx \qquad (n \geqq 1)$$

$$b_n = \frac{1}{L}\int_{-L}^{L} f(x)\sin\left(\frac{n\pi}{L}x\right)dx \qquad (n \geqq 1)$$

Again, $[-\pi, \pi]$ is a special case with $L = \pi$.

Periodic on $[0, 2L]$

Similarly, if a periodic function $f(x)$ is defined over $[0, 2L]$, with period $P = 2L$, the Fourier series becomes

$$f(x) = \frac{1}{2}a_0 + \sum_{n=1}^{\infty}\left[a_n \cos\left(\frac{n\pi}{L}x\right) + b_n \sin\left(\frac{n\pi}{L}x\right)\right] \tag{17}$$

where

$$a_0 = \frac{1}{L}\int_0^{2L} f(x)\,dx$$

$$a_n = \frac{1}{L}\int_0^{2L} f(x)\cos\left(\frac{n\pi}{L}x\right)dx \qquad (n \geqq 1)$$

$$b_n = \frac{1}{L}\int_0^{2L} f(x)\sin\left(\frac{n\pi}{L}x\right)dx \qquad (n \geqq 1)$$

Clearly, $[0, 2\pi]$ is a special case with $L = \pi$. Moreover, any interval $[x_0, x_0 + 2L]$ can be used for a periodic function with period $P = 2L$.

Fourier Series and Music

Fourier series can be use to analysis the sounds from musical instruments or to synthesis them as well as other sounds. The difference between wavefronts can be expressed as a Fourier series such as

$$p_N(x) = a_0 + \sum_{n=1}^{N}\left[a_n \cos\left(\frac{\pi}{L}x\right) + b_n \sin\left(\frac{\pi}{L}x\right)\right]$$

Since a sound can be expressed as a sum of simple pure sounds, the difference between two instruments can be attributed to the relative sizes of the Fourier coefficients from these respective wavefronts. The N-th term of a Fourier series is called the **N-th harmonic** of p_N with **amplitude** $A_N = \sqrt{a_N^2 + b_N^2}$, and **energy** A_N^2.

Sawtooth Wave Function on [0, 2L]

EXAMPLE 3 **Sawtooth Wave on [0, 2L]**

Now we find the N-th order Fourier series approximation to the **sawtooth wave function**

$$f(x) = \frac{1}{2L}x \quad \text{on} \quad (0, 2L) \quad \text{with} \quad f(0) = \frac{1}{2} = f(2L)$$

and then we plot the function f as well as some of the Fourier series approximations.

Solution The sawtooth wave resembles the teeth on a saw blade. This waveform is one of the best ones for synthesizing string musical instruments such as violins and cellos.

For example, consider a string of length $2L$ plucked at the right end and fixed at the left end. First, we compute

$$\begin{aligned} a_0 &= \frac{1}{L}\int_0^{2L} \frac{x}{2L}\,dx \\ &= \frac{1}{4L^2}x^2\Big|_0^{2L} = 1 \end{aligned}$$

Next, we have

$$\begin{aligned} a_n &= \frac{1}{L}\int_0^{2L} \frac{1}{2L}x\cos\left(\frac{n\pi}{L}x\right)dx \\ &= \frac{1}{2L^2}\left[\frac{L}{n\pi}x\sin\left(\frac{n\pi}{L}x\right) + \frac{L^2}{n^2\pi^2}\cos\left(\frac{n\pi}{L}x\right)\right]_0^{2L} \\ &= \frac{1}{2L^2}\left[\frac{2L^2}{n\pi}\sin 2n\pi + \frac{L^2}{n^2\pi^2}\cos 2n\pi - \frac{L^2}{n^2\pi^2}\right] = 0 \qquad (n \geqq 1) \end{aligned}$$

We have used integration-by-parts Formula (2), as well as $\sin 2n\pi = 0$ and $\cos 2n\pi = -1$. Finally, we have

$$\begin{aligned} b_n &= \frac{1}{L}\int_0^{2L} \frac{1}{2L} x \sin\left(\frac{n\pi}{L}x\right) dx \\ &= \frac{1}{2L^2}\left[-\frac{L}{n\pi} x \cos\left(\frac{n\pi}{L}x\right) + \frac{L^2}{n^2\pi^2}\sin\left(\frac{n\pi}{L}x\right)\right]_0^{2L} \\ &= \frac{1}{2L^2}\left[-\frac{2L^2}{n\pi}\cos 2n\pi + \frac{L}{n^2\pi^2}\sin 2n\pi\right] \\ &= -\frac{1}{n\pi} \qquad (n \geqq 1) \end{aligned}$$

We have used integration Formula (3), as well as $\cos 2n\pi = 1$ and $\sin 2n\pi = 0$.

The N-th order Fourier series approximation of the sawtooth function f is

$$f(x) \approx p_N(x) = \frac{1}{2} - \frac{1}{\pi}\sum_{n=1}^{N}\frac{1}{n}\sin\left(\frac{n\pi}{L}x\right) \tag{18}$$

When $L = \pi$, the approximations over $[0, 2\pi]$ is

$$p_N(x) = \frac{1}{2} - \frac{1}{\pi}\left[\sin x + \frac{1}{2}\sin 2x + \frac{1}{3}\sin 3x + \cdots + \frac{1}{N}\sin Nx\right]$$

■

Fourier Series Examples

Here are some common Fourier series, which are periodic on $[0, 2L]$ with $P = 2L$. (See Figures 9.7–9.9 over $[0, 2\pi]$ with approximations p_2, p_4, and p_{10}.)

$$\textbf{Sawtooth Wave} \quad ST(x) = \frac{1}{2} - \frac{1}{\pi}\sum_{n=1}^{\infty}\frac{1}{n}\sin\left(\frac{n\pi}{L}x\right) \tag{19}$$

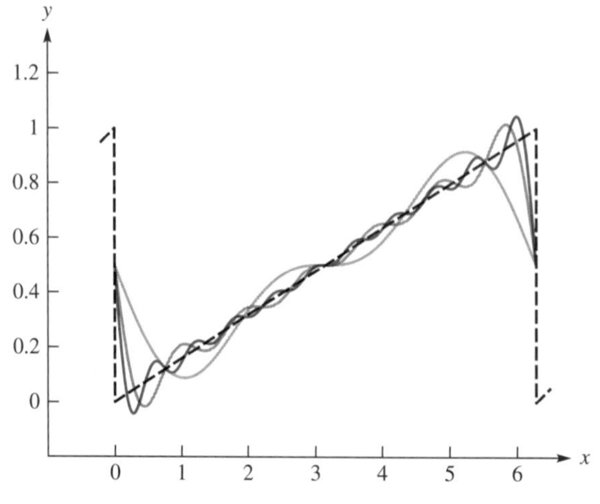

FIGURE 9.7 Sawtooth Wave on $[0, 2\pi]$ with approximations p_2, p_6, and p_{10}

Square Wave $$SW(x) = \frac{4}{\pi} \sum_{n=1,3,5,\ldots}^{\infty} \frac{1}{n} \sin\left(\frac{n\pi}{L}x\right) \tag{20}$$

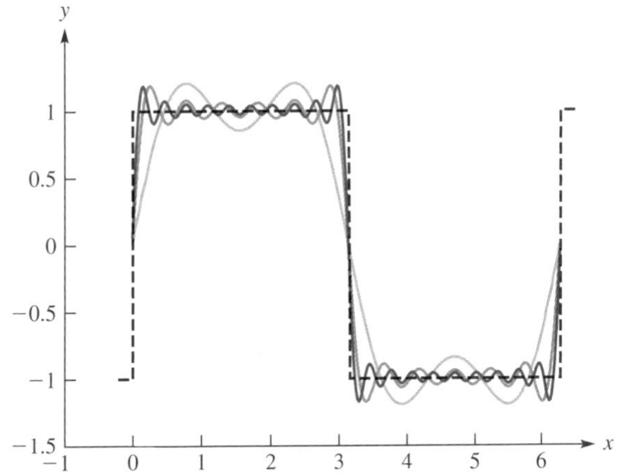

FIGURE 9.8 Square Wave on $[0, 2\pi]$ with approximations p_2, p_6, and p_{10}

Triangle Wave $$TW(x) = \frac{8}{\pi^2} \sum_{n=1,3,5,\ldots}^{\infty} (-1)^{(n-1)/2} \frac{1}{n^2} \sin\left(\frac{n\pi}{L}x\right) \tag{21}$$

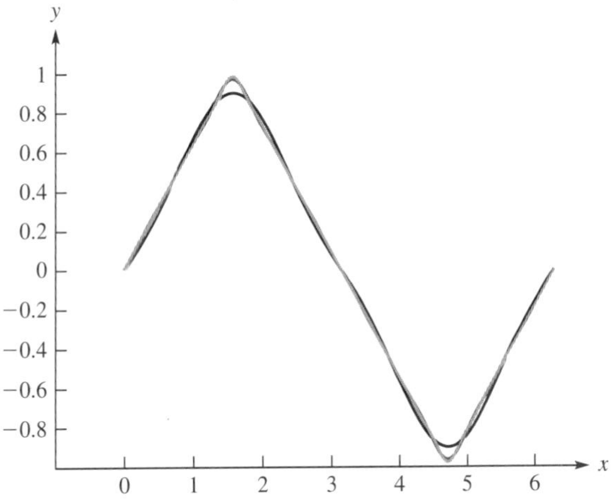

FIGURE 9.9 Triangle Wave on $[0, 2\pi]$ with approximations p_2, p_6, and p_{10}

Euler's and de Moivre Formulas

In the complex plane $\mathbb{C}^n$, $i = \sqrt{-1}$ and we can use **Euler's equations**

Euler's Equation

$$e^{i\theta} = \cos\theta + i\sin\theta$$
$$e^{-i\theta} = \cos\theta - i\sin\theta$$

to obtain

$$\cos\theta = \frac{1}{2}\left(e^{i\theta} + e^{-i\theta}\right)$$
$$\sin\theta = \frac{1}{2i}\left(e^{i\theta} - e^{-i\theta}\right)$$

Moreover, the **de Moivre's formula** is

de Moivre's Formula

$$(\cos\theta + i\sin\theta)^k = \cos k\theta + i\sin k\theta$$

Also, we have these famous identity

$$\begin{aligned} e^{i2\pi} &= \cos 2\pi + i\sin 2\pi = 1 \\ e^{i2\pi k} &= \cos 2\pi k + i\sin 2\pi k = 1 \end{aligned}$$

We can write a complex number x in polar form (r, θ), with a **magnitude** r and an **angle** θ, as

$$x = re^{i\theta} = r(\cos\theta + i\sin\theta)$$

So we obtain

$$x^k = r^k(\cos k\theta + i\sin k\theta)$$

Complex Fourier Series

Suppose the function $f(x)$ has period P on the interval $[-P/2, P/2]$. Often it is convenient to write the Fourier series, containing real functions (sines and cosines), as a sum of exponential functions in the form

$$f(x) = \sum_{n=-\infty}^{\infty} \alpha_n e^{in\omega_0 x} \tag{22}$$

where $\omega_0 = 2\pi/P$ is the **frequency** (for example, radians per second). Here the α_n are the complex Fourier coefficients.

Recall Equation (15), which is the Fourier series for the periodic function $f(x)$ of period P on the interval $[-P/2, P/2]$. By substituting these identities

$$\begin{aligned} \cos(n\omega_0 x) &= \frac{1}{2}\left[e^{in\omega_0 x} + e^{-in\omega_0 x}\right] \\ \sin(n\omega_0 x) &= \frac{1}{2i}\left[e^{in\omega_0 x} - e^{-in\omega_0 x}\right] \end{aligned}$$

we find the relationship between the trigonometric and exponential forms of the series

$$\begin{cases} \alpha_n = \frac{1}{2}(a_n + ib_n) & n < 0 \\ \alpha_0 = \frac{1}{2}a_0 & n = 0 \\ \alpha_n = \frac{1}{2}(a_n - b_n) & n > 0 \end{cases} \tag{23}$$

Thus, we have

$$f(x) = \alpha_0 + \sum_{n=1}^{\infty}\left[\alpha_n e^{in\omega_0 x} + \alpha_{-n} e^{-in\omega_0 x}\right]$$

Note that α_{-n} is the complex conjugate of α_n. (See Exercise 9.4.16.) To find the coefficients α_n, multiply each side of Series (22) by $e^{-im\omega_0 x}$ and integrate over $[-P/2, P/2]$ yielding

$$\int_{-P/2}^{P/2} f(x)e^{-im\omega_0 x}\,dx = \sum_{n=-\infty}^{\infty}\int_{-P/2}^{P/2} \alpha_n e^{i(n-m)\omega_0 x}\,dx$$

In the integral on the right-hand side, all the terms are zero except those for which $m = n$ because the terms with different exponents are orthogonal. Consequently, we obtain

$$\int_{-P/2}^{P/2} f(x)e^{-im\omega_0 x}\,dx = \alpha_n \int_{-P/2}^{P/2} dx = \alpha_n P$$

So we find that

$$\alpha_n = \frac{1}{P}\int_{-P/2}^{P/2} f(x)e^{-in\omega_0 x}\,dx$$

for all integers n.

Roots of Unity

Roots of unity arise when finding the complex roots of the polynomial

$$x^n = 1$$

(Remember that a n-th degree polynomial has n complex roots.) Let $x = re^{i\theta}$ in polar form with $r = 1$. Then $1 = x^n = r^n e^{in\theta} = e^{in\theta}$. Since $1 = e^{i2\pi k}$ for $k \geqq 0$, this leaves us with $e^{in\theta} = e^{i2\pi k}$. Taking the natural logarithm of both sides, we obtain $in\theta = i2\pi k$ and $\theta = 2\pi(k/n)$. So the solutions to $x^n = 1$ are given by $x = e^{i2\pi(k/n)}$ for $k = 0, 1, 2, \ldots, n-1$. Each $k = 0, 1, 2, \ldots, n-1$ gives a distinct value of θ, but once we get to $k \geqq n$ the values begin to repeat. Hence, the solutions to $x^n = 1$ are given by

$$x = e^{i2\pi(k/n)} = \cos\left(2\pi\frac{k}{n}\right) + i\sin\left(2\pi\frac{k}{n}\right) \qquad (k = 0, 1, 2, \ldots, n-1)$$

All roots of unity lie on the unit circle in the complex plane. The roots are the vertices of a regular n polygon in the complex plane. For $n \geqq 1$, the sum of the n-th roots of unity is zero.

For a given positive integer n, the roots of the **cyclotomic equation** $x^n = 1$ are the n solutions

$$x = \sqrt[n]{1} = e^{i2\pi(k/n)} \qquad (0 \leqq k \leqq n-1)$$

This is called the ***n*-th roots of unity** or **de Moivre numbers**. Moreover, the n-th roots of unity can be expressed as

n-th Roots of Unity

$$\omega_n^k = e^{i2\pi(k/n)} = \begin{cases} \cos 0 + i\sin 0 = 1 & (k = 0) \\ \cos\left(2\pi\frac{k}{n}\right) + i\sin\left(2\pi\frac{k}{n}\right) \neq 1 & (1 \leqq k \leqq n-1) \end{cases}$$

The first of these ($k = 0$) is called the **primitive *n*-th root of unity**.

Sample ω_n^k

Examples of the first few of the n-th roots of unity are

n	$k = 0$	$k = 1$	$k = 2$	$k = 3$
1	$\omega_1^0 = e^{i2\pi(0/1)} = 1$			
2	$\omega_2^0 = e^{i2\pi(0/2)} = 1$	$\omega_2^1 = e^{i2\pi(1/2)} = -1$		
3	$\omega_3^0 = e^{i2\pi(0/3)} = 1$	$\omega_3^1 = e^{i2\pi(1/3)} = -\frac{1}{2} + i\frac{1}{2}\sqrt{3}$	$\omega_3^2 = e^{i2\pi(2/3)} = -\frac{1}{2} - i\frac{1}{2}\sqrt{3}$	
4	$\omega_4^0 = e^{i2\pi(0/4)} = 1$	$\omega_4^1 = e^{i2\pi(1/4)} = i$	$\omega_4^2 = e^{i2\pi(2/4)} = -1$	$\omega_4^3 = e^{i2\pi(3/4)} = -i$

Three cases are shown in Figure 9.10.

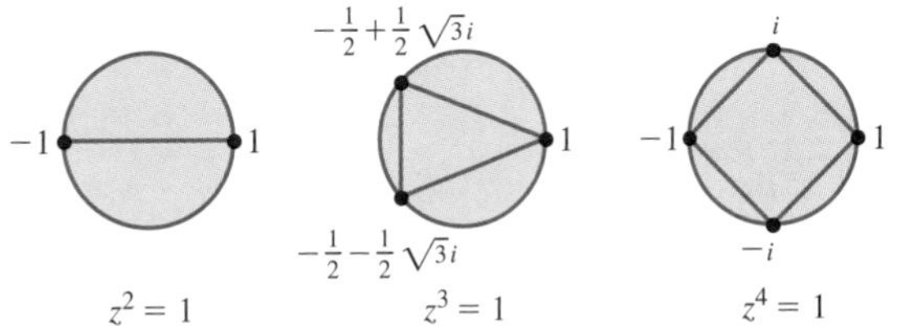

FIGURE 9.10
n-th roots of unity

Discrete Fourier Transform

The **Discrete Fourier Transform (DFT)** of the vector $\boldsymbol{x} = [x_0, x_1, x_2, \ldots, x_{n-1}]^T$ to the vector $\boldsymbol{y} = [y_0, y_1, y_2, \ldots, y_{n-1}]^T$ is given by the sequence

$$y_j = \sum_{k=0}^{n-1} \omega_n^{jk} x_k \qquad (0 \le j \le n-1)$$

In matrix notation, the **DFT** is

DFT Matrix Form

$$\boldsymbol{y} = \boldsymbol{F}_n \boldsymbol{x}$$

Here $\boldsymbol{x}$ is the original input signal and $\boldsymbol{y}$ is the DFT of the signal. Also, DFTs are used for trigonometric interpolation among other things. The transformation matrix is defined by

Vandermonde Matrix

$$\boldsymbol{F}_n = (F_{jk})_{n\times n} = \left(\omega_n^{jk}\right)_{0 \leqq j \leqq n-1, 0 \leqq k \leqq n-1}$$

This is known as the **Vandermonde matrix** of the roots of unity ω_n^k. To make the matrix unitary, multiply $\boldsymbol{F}_n$ by the normalization factor $1/\sqrt{n}$.

Notice this pattern

$$\boldsymbol{F}_1 = \omega_1^0 = 1, \qquad \boldsymbol{F}_2 = \begin{bmatrix} \omega_2^0 & \omega_2^0 \\ \omega_2^0 & \omega_2^1 \end{bmatrix} = \begin{bmatrix} 1 & 1 \\ 1 & -1 \end{bmatrix}$$

$$\boldsymbol{F}_4 = \begin{bmatrix} \omega_4^0 & \omega_4^0 & \omega_4^0 & \omega_4^0 \\ \omega_4^0 & \omega_4^1 & \omega_4^2 & \omega_4^3 \\ \omega_4^0 & \omega_4^2 & \omega_4^4 & \omega_4^6 \\ \omega_4^0 & \omega_4^3 & \omega_4^6 & \omega_4^9 \end{bmatrix} = \begin{bmatrix} 1 & 1 & 1 & 1 \\ 1 & -i & -1 & i \\ 1 & -1 & 1 & -1 \\ 1 & i & -1 & -i \end{bmatrix}$$

For $n = 4$, we have

$$\boldsymbol{y} = \begin{bmatrix} y_0 \\ y_1 \\ y_2 \\ y_3 \end{bmatrix} = \begin{bmatrix} 1 & 1 & 1 & 1 \\ 1 & \omega_4^1 & \omega_4^2 & \omega_4^3 \\ 1 & \omega_4^2 & \omega_4^4 & \omega_4^6 \\ 1 & \omega_4^3 & \omega_4^6 & \omega_4^9 \end{bmatrix} \begin{bmatrix} x_0 \\ x_1 \\ x_2 \\ x_3 \end{bmatrix} = \boldsymbol{F}_4 \boldsymbol{x}$$

where

$$\boldsymbol{F}_4 = \left[\begin{array}{cc|cc} 1 & 1 & 1 & 1 \\ 1 & -i & -1 & i \\ \hline 1 & -1 & 1 & -1 \\ 1 & i & -1 & -i \end{array}\right]$$

The **Inverse DFT** is the sequence

$$x_k = \frac{1}{n} \sum_{j=0}^{n-1} \omega_n^{-jk} y_j \qquad (0 \le k \le n-1)$$

In matrix notation, the inverse of the DFT is

Inverse DFT Matrix Form

$$\boldsymbol{x} = \boldsymbol{F}_n^{-1} \boldsymbol{y}$$

where

$$\boldsymbol{F}_n^{-1} = \frac{1}{n} \boldsymbol{F}_n^*$$

where F^* is the conjugate transpose of F. For example with $n = 4$, we have

$$F_4^{-1} = \frac{1}{4}\begin{bmatrix} 1 & 1 & 1 & 1 \\ 1 & \omega_4^{-1} & \omega_4^{-2} & \omega_4^{-3} \\ 1 & \omega_4^{-2} & \omega_4^{-4} & \omega_4^{-6} \\ 1 & \omega_4^{-3} & \omega_4^{-6} & \omega_4^{-9} \end{bmatrix} = \frac{1}{4}\left[\begin{array}{cc|cc} 1 & 1 & 1 & 1 \\ 1 & i & -1 & -i \\ \hline 1 & -1 & 1 & -1 \\ 1 & -i & -1 & i \end{array}\right]$$

Moreover, we can show that

$$F_4 P_4 = \left[\begin{array}{cc|cc} 1 & 1 & 1 & 1 \\ 1 & -1 & -i & i \\ \hline 1 & 1 & -1 & -1 \\ 1 & -1 & i & -i \end{array}\right] = \begin{bmatrix} F_2 & D_2 F_2 \\ F_2 & -D_2 F_2 \end{bmatrix}$$

Thus, F_4 can be rearranged into diagonally scaled blocks of F_2, which holds for any even n. Here we use the permutation matrix

$$P_4 = \begin{bmatrix} e_1 & e_3 & e_2 & e_4 \end{bmatrix} = \begin{bmatrix} 1 & 0 & 0 & 0 \\ 0 & 0 & 1 & 0 \\ 0 & 1 & 0 & 0 \\ 0 & 0 & 0 & 1 \end{bmatrix}$$

and the diagonal matrix

$$D_2 = \begin{bmatrix} \omega_4^0 & 0 \\ 0 & \omega_4^1 \end{bmatrix} = \begin{bmatrix} 1 & 0 \\ 0 & -i \end{bmatrix}$$

In general, P_n is a permutation matrix that groups even-numbered columns and odd-numbered columns and the diagonal matrix is

$$D_{n/2} = \text{Diag}(1, \omega_n, \omega_n^2, \ldots, \omega_n^{(n/2)-1})$$

We can find F_n by applying $F_{n/2}$ to the even and to the odd subsequences. Then scaling the results by $\pm D_{n/2}$, where necessary. This recursive DFT is called the **Fast Fourier Transform (FFT)**.

Computing the DFT of the four-point problem reduces to computing the DFT of the two-point even and odd subsequences.

Theorem 2 **Recursive Computation of DFT**

The Discrete Fourier Transformation of an n-point sequence can be computed by two $n/2$-point Discrete Fourier Transformations (n even).

Fast Fourier Transforms

Direct computation of DFT requires $O(n^2)$ operations, but this can be reduced to $O(n \log_2 n)$ by exploiting the efficiencies of the Fast Fourier Transform (FFT). The case used most often is when n is a power of two; namely, $n = 2^r$. To reduce the total effort required in finding the FFT, we make use of the periodicity of the complex exponential function and clever reordering of the computations.

Both of the indices on the components of the transform and on the summation run from 0 to $n-1$. Exact values of j can be written in binary form as $j = 2^{r-1} j_r + \cdots + 2^2 j_3 + 2 j_2 + j_1$, where each of the numbers $j_1, j_2, \ldots, j_r$ is either 0 or 1. This reordering of the k values as

Bit Reversal

they related to the j values is know as **bit reversal**. This is one of the cleverness in the FFT computations!

The **Cooley-Tukey Fast Fourier (FFT) algorithm** rearranges the input values in bit-reversal order and then builds the output transformation. The basic idea is breaking a transform of length N into two transforms of length $N/2$ using the identity

Danielson-Lanczo Identity

$$\begin{aligned}\sum_{n=0}^{N-1} a_n e^{-i2\pi nk/N} &= \sum_{n=0}^{N/2-1} a_{2n} e^{-i2\pi(2n)k/N} + \sum_{n=0}^{N/2-1} a_{2n+1} e^{-i2\pi(2n+1)k/N} \\ &= \sum_{n=0}^{N/2-1} a_n^{\text{even}}\, e^{-i2\pi nk/(N/2)} + e^{-i2\pi k/N} \sum_{n=0}^{N/2-1} a_n^{\text{odd}}\, e^{-i2\pi nk/(N/2)}\end{aligned}$$

This is also called the **Danielson-Lanczo Lemma**. It can be visualized via the **Fourier matrix**

Fourier Matrix

$$\boldsymbol{F}_n = (F_{jk})_{n\times n}$$

with entries

$$F_{jk} = e^{i2\pi(jk)/n} \equiv \omega_n^{jk} \qquad (0 \leqq j \leqq n-1,\, 0 \leqq k \leqq n-1)$$

where $\omega_n = e^{(i2\pi)/n}$. Multiplying by $1/\sqrt{n}$ makes the matrix unitary.

For example, we can show that

$$\boldsymbol{F}_4 = \begin{bmatrix} 1 & 1 & 1 & 1 \\ 1 & i & -1 & -i \\ 1 & -1 & 1 & -1 \\ 1 & -i & -1 & -i \end{bmatrix} = \left[\begin{array}{cc|cc} 1 & 0 & 1 & 0 \\ 0 & 1 & 0 & i \\ \hline 1 & 0 & -1 & 0 \\ 0 & 1 & 0 & -i \end{array}\right] \left[\begin{array}{cc|cc} 1 & 1 & 0 & 0 \\ 1 & i^2 & 0 & 0 \\ \hline 0 & 0 & 1 & 1 \\ 0 & 0 & 1 & i^2 \end{array}\right] \begin{bmatrix} 1 & 0 & 0 & 0 \\ 0 & 0 & 1 & 0 \\ 0 & 1 & 0 & 0 \\ 0 & 0 & 0 & 1 \end{bmatrix}$$

In matrix form, we can write

$$\boldsymbol{F}_4 = \begin{bmatrix} \boldsymbol{I}_2 & \boldsymbol{D}_2 \\ \boldsymbol{I}_2 & -\boldsymbol{D}_2 \end{bmatrix} \begin{bmatrix} \boldsymbol{F}_2 & \boldsymbol{0} \\ \boldsymbol{0} & \boldsymbol{F}_2 \end{bmatrix} \begin{bmatrix} \text{even-odd} \\ \text{shuffle} \end{bmatrix}$$

where

$$\boldsymbol{F}_2 = \begin{bmatrix} 1 & 1 \\ 1 & i^2 \end{bmatrix} = \begin{bmatrix} 1 & 1 \\ 1 & -1 \end{bmatrix}$$

So in general, we have

$$\boldsymbol{F}_{2n} = \begin{bmatrix} \boldsymbol{I}_n & \boldsymbol{D}_n \\ \boldsymbol{I}_n & -\boldsymbol{D}_n \end{bmatrix} \begin{bmatrix} \boldsymbol{F}_n & \boldsymbol{0} \\ \boldsymbol{0} & \boldsymbol{F}_n \end{bmatrix} \begin{bmatrix} \text{even-odd} \\ \text{shuffle} \end{bmatrix}$$

and repeating we obtain

$$\begin{bmatrix} \boldsymbol{F}_n & \boldsymbol{0} \\ \boldsymbol{0} & \boldsymbol{F}_n \end{bmatrix} = \left[\begin{array}{cc|cc} \boldsymbol{I}_{n/2} & \boldsymbol{D}_{n/2} & \boldsymbol{0} & \boldsymbol{0} \\ \boldsymbol{I}_{n/2} & -\boldsymbol{D}_{n/2} & \boldsymbol{0} & \boldsymbol{0} \\ \hline \boldsymbol{0} & \boldsymbol{0} & \boldsymbol{I}_{n/2} & \boldsymbol{D}_{n/2} \\ \boldsymbol{0} & \boldsymbol{0} & \boldsymbol{I}_{n/2} & -\boldsymbol{D}_{n/2} \end{array}\right] \left[\begin{array}{cc|cc} \boldsymbol{F}_{n/2} & \boldsymbol{0} & \boldsymbol{0} & \boldsymbol{0} \\ \boldsymbol{0} & \boldsymbol{F}_{n/2} & \boldsymbol{0} & \boldsymbol{0} \\ \hline \boldsymbol{0} & \boldsymbol{0} & \boldsymbol{F}_{n/2} & \boldsymbol{0} \\ \boldsymbol{0} & \boldsymbol{0} & \boldsymbol{0} & \boldsymbol{F}_{n/2} \end{array}\right] \left[\begin{array}{c} \text{Shuffle:} \\ \text{even-odd} \\ 0, 2 \,(\text{mod } 4) \\ \hline \text{even-odd} \\ 1, 3 \,(\text{mod } 4) \end{array}\right]$$

Mathematical Software

Mathematical software systems such as MATLAB, Maple, and Mathematica have routines for computing Fourier series and plotting them. Web pages associated with these packages are particularly useful. Moreover, there are software packages specifically designed for handling Fourier series.

Summary 9.4

- The formulas for the **orthogonal projections** of functions f and g are

$$a_n = \frac{\langle f, \cos nx\rangle}{\langle \cos nx, \cos nx\rangle} \qquad (n \geqq 0)$$

$$b_n = \frac{\langle f, \sin nx\rangle}{\langle \sin nx, \sin nx\rangle} \qquad (n \geqq 1)$$

where the **inner product** of f and g is

$$\langle f, g\rangle = \int_{\pi}^{\pi} f(x)g(x)\,dx$$

- The **N-th order Fourier approximation** for a function $f(x)$ periodic over $[-\pi, \pi]$ is

$$p_N(x) = \frac{1}{2}a_0 + \sum_{n=1}^{N} \left[a_n \cos nx + b_n \sin nx\right]$$

where the **Fourier coefficients** are

$$a_n = \frac{1}{\pi}\int_{-\pi}^{\pi} f(x)\cos nx\,dx \qquad (0 \leqq n \leqq N)$$

$$b_n = \frac{1}{\pi}\int_{-\pi}^{\pi} f(x)\sin nx\,dx \qquad (1 \leqq n \leqq N)$$

- The Fourier series for $f(x)$ periodic on $[-\pi, \pi]$ is

$$f(x) = \frac{1}{2}a_0 + \sum_{n=1}^{\infty} \left[a_n \cos nx + b_n \sin nx\right]$$

where

$$a_n = \frac{1}{\pi}\int_{-\pi}^{\pi} f(x)\cos nx\,dx \qquad (n \geqq 0)$$

$$b_n = \frac{1}{\pi}\int_{-\pi}^{\pi} f(x)\sin nx\,dx \qquad (n \geqq 1)$$

- An **even function** f is symmetric with respect to the vertical axis ($x = 0$) and satisfies

$$f(-x) = f(x)$$

- An **odd function** f is symmetric with respect to the origin and satisfies

$$f(-x) = -f(x)$$

- If $f_e(x)$ is an **even function**, we obtain a **cosine series**

$$f_e(x) = \frac{1}{2}a_0 + \sum_{n=1}^{\infty} a_n \cos nx$$

where

$$a_n = \frac{2}{\pi}\int_{0}^{\pi} f_e(x)\cos nx\,dx \qquad (n \geqq 0)$$

- If $f_o(x)$ is an **odd function**, we obtain a **sine series**

$$f_o(x) = \sum_{n=1}^{\infty} b_n \sin nx$$

where

$$b_n = \frac{2}{\pi}\int_0^{\pi} f_o(x)\sin nx\,dx \qquad (n \geqq 1)$$

- The **sawtooth wave** Fourier series over $[-\pi, \pi]$ is

$$f(x) \approx p_N(x) = 2\sum_{n=1}^{N}(-1)^{n+1}\frac{1}{n}\sin nx$$

- The Fourier series for a periodic function $f(x)$ with period P over $[-P/2, P/2]$ is

$$f(x) = \frac{1}{2}a_0 + \sum_{n=1}^{\infty}\left[a_n\cos\left(\frac{2n\pi}{P}x\right) + b_n\sin\left(\frac{2n\pi}{P}x\right)\right]$$

where the Fourier coefficients are

$$a_n = \frac{1}{P}\int_{-P/2}^{P/2} f(x)\cos\left(\frac{2n\pi}{P}x\right)dx \qquad (n \geqq 0)$$

$$b_n = \frac{1}{P}\int_{-P/2}^{P/2} f(x)\sin\left(\frac{2n\pi}{P}x\right)dx \qquad (n \geqq 1)$$

- For $f(x)$ periodic on $[-L, L]$, the Fourier series is

$$f(x) = \frac{1}{2}a_0 + \sum_{n=1}^{\infty}\left[a_n\cos\left(\frac{n\pi}{L}x\right) + b_n\sin\left(\frac{n\pi x}{L}\right)\right]$$

where

$$a_n = \frac{1}{L}\int_{-L}^{L} f(x)\cos\left(\frac{n\pi}{L}x\right)dx \qquad (n \geqq 0)$$

$$b_n = \frac{1}{L}\int_{-L}^{L} f(x)\sin\left(\frac{n\pi}{L}x\right)dx \qquad (n \geqq 1)$$

- A function $f(x)$ periodic on $[0, 2L]$ can be approximated by a Fourier series of the form

$$f(x) = \frac{1}{2}a_0 + \sum_{n=1}^{\infty}\left[a_n\cos\left(\frac{n\pi}{L}x\right) + b_n\sin\left(\frac{n\pi}{L}x\right)\right]$$

where

$$a_n = \frac{1}{L}\int_0^{2L} f(x)\cos\left(\frac{n\pi}{L}x\right)dx \qquad (n \geqq 0)$$

$$b_n = \frac{1}{L}\int_0^{2L} f(x)\sin\left(\frac{n\pi}{L}x\right)dx \qquad (n \geqq 1)$$

- Fourier series for some common periodic functions on $[0, 2L]$ are

Sawtooth Wave $\quad ST(x) = \frac{1}{2} - \frac{1}{\pi}\sum_{n=1}^{\infty}\frac{1}{n}\sin\left(\frac{n\pi}{L}x\right)$

$$\textbf{Square Wave} \qquad SW(x) = \frac{4}{\pi} \sum_{n=1,3,5,\ldots}^{\infty} \frac{1}{n} \sin\left(\frac{n\pi}{L}x\right)$$

$$\textbf{Triangle Wave} \qquad TW(x) = \frac{8}{\pi^2} \sum_{n=1,3,5,\ldots}^{\infty} (-1)^{(n-1)/2} \frac{1}{n^2} \sin\left(\frac{n\pi}{L}x\right)$$

- The **complex Fourier series** for a $f(x)$ of period P over $[-P/2, P/2]$ is

$$f(x) = \sum_{n=-\infty}^{\infty} \alpha_n e^{in\omega_0 x}$$

where

$$\alpha_n = \int_{-P/2}^{P/2} f(x) e^{-in\omega_0 x}\, dx$$

- **Euler's formula** is

$$e^{\pm i\theta} = \cos\theta \pm i \sin\theta$$

- **de Moivre's formula** is

$$(\cos\theta + i\sin\theta)^k = \cos k\theta + i \sin k\theta$$

- The k solutions of the ***n*-th roots of unity**, or **de Moivre numbers**, are

$$x = \sqrt[n]{1} = e^{i2\pi(k/n)} \qquad (0 \leqq k \leqq n-1)$$

- The n-th roots of unity are given by

$$\omega_n^k = e^{i2\pi(k/n)} = \begin{cases} \cos 0 + i \sin 0 = 1 & (k = 0) \\ \cos\left(2\pi\frac{k}{n}\right) + i\sin\left(2\pi\frac{k}{n}\right) \neq 1 & (1 \leqq k \leqq n-1) \end{cases}$$

- The **Discrete Fourier Transformation (DFT)** of $\boldsymbol{x} = [x_0, x_1, x_2, \ldots, x_{n-1}]^T$ is $\boldsymbol{y} = [y_0, y_1, y_2, \ldots, y_{n-1}]^T$ is given in matrix-vector form as

$$\boldsymbol{y} = \boldsymbol{F}_n \boldsymbol{x}$$

where

$$\boldsymbol{F}_n = \left(F_{jk}\right)_{n\times n} = \left(\omega_n^{jk}\right)_{0\leqq j\leqq n-1, 0\leqq k\leqq n-1}$$

which is the **Vandermonde matrix** of the roots of unity.

- The **inverse of the DFT** in matrix vector form is

$$\boldsymbol{x} = \boldsymbol{F}_n^{-1} \boldsymbol{y}$$

where

$$\boldsymbol{F}_n^{-1} = \frac{1}{n} \boldsymbol{F}_n^*$$

- In the **Fast Fourier Transformation**, we have

$$\boldsymbol{F}_{2n} = \begin{bmatrix} \boldsymbol{I}_n & \boldsymbol{D}_n \\ \boldsymbol{I}_n & -\boldsymbol{D}_n \end{bmatrix} \begin{bmatrix} \boldsymbol{F}_n & \boldsymbol{0} \\ \boldsymbol{0} & \boldsymbol{F}_n \end{bmatrix} \begin{bmatrix} \text{even-odd} \\ \text{shuffle} \end{bmatrix}$$

Exercises 9.4

1. Establish the orthogonality properties (p. 461).

2. Establish these properties:

   a. Product of two even functions is even.

   b. Product of two odd functions is even.

   c. Product of an even and an odd function is odd.

3. For the sawtooth wave, show how the results of Examples 2 and 3 are related; that is, Equations (14) and (18).

4. For a continuous function $f(x)$ periodic over the interval $[-\pi, \pi]$ determine Formulas (7), (8), and (9) for the coefficients a_0, a_n, and b_n of the Fourier Series (6) using term-by-term integration.

[a]5. For $f(x) = x^2$ on the interval $[-1, 1]$ with period 2, determine the Fourier series. Show that $\dfrac{\pi^2}{6} = \displaystyle\sum_{n=1}^{\infty} \frac{1}{n^2}$.

6. Starting with the Fourier series for a period function f over $[-\pi, \pi]$ and Formulas (6)–(9), write out the details for using a change of variables and the substitution rule to convert to the formulas for the Fourier series over these intervals:

   a. $[-L, L]$

   b. $[-P/2, P/2]$

   c. $[0, 2L]$

   d. $[x_0, x_0 + 2L]$

7. Find the Fourier series of these periodic functions given over the given intervals:

   a. $f(x) = \begin{cases} \dfrac{x}{2}, & 0 \leqq x < L \\ 2 - \dfrac{x}{L}, & L \leqq x < 2L \end{cases}$

   b. $f(x) = \begin{cases} 2 + x, & 0 \leqq x < L \\ 2 - \dfrac{x}{L}, & L \leqq x < 2L \end{cases}$

   c. $f(x) = \begin{cases} 0, & -\pi \leqq x \leqq 0 \\ \sin x, & 0 < x \leqq \pi \end{cases}$

   [a]d. $f(x) = \begin{cases} -1, & -\pi < x < -\dfrac{\pi}{2} \\ 0, & -\dfrac{\pi}{2} \leqq x \leqq \dfrac{\pi}{2} \\ 1, & \dfrac{\pi}{2} < x < \pi \end{cases}$

8. Write out the Fourier series for a periodic function $f(x)$ over $[-L, L]$ with period $P = 2L$ under these conditions.

   a. $f(x)$ is an even function.

   b. $f(x)$ is an odd function.

9. Explain why it is possible for $a_0 \neq 0$ in Example 3, but $f(-x) = -f(x)$.

10. Suppose we are given a periodic function with period $P = 2L$, but the function $f(x)$ is defined only on the interval $[0, L]$. Show how it can be extended to the interval $[-L, L]$ under the following conditions.

    a. $f(x)$ is an even function.

    b. $f(x)$ is an odd function.

11. [a]a. Suppose a complex number z has n different n-th roots in the complex plane. So $z = r(\cos\theta + i\sin\theta)$ satisfies the equation

    $$z^n = a$$

    for a nonzero number a. Show how to determine the n-th roots of a.

    [a]b. What are the four different fourth roots of two?

12. Show that $+1$ is always an n-th root of unity, but -1 is such a root only if n is even.

13. Establish these beautiful identities

    a. $e^{2\pi i} = 1$

    b. $ix = \ln(\cos x + i \sin x)$

14. Derive these Fourier series for the periodic function given over $[-\pi, \pi]$:

    a. $\pi^2 - x^2 = \dfrac{2}{3}\pi^2 + \displaystyle\sum_{n=1}^{\infty} -\frac{4}{n^2}(-1)^n \cos nx$

    b. $3x = \displaystyle\sum_{n=1}^{\infty} -\frac{6}{n}(-1)^n \sin nx$

    c. $x^3 = \displaystyle\sum_{n=1}^{\infty} 2(-1)^n \left(\frac{6}{n^3} - \frac{\pi^2}{n}\right) \sin nx$

15. Establish that the complex conjugate of ω_n^k is ω_n^{-k}.

16. Establish (23), which is the relationships between the Fourier coefficients α_n in the complex representation and the coefficients a_n and b_n of the real sine and cosine representation.

[a]**17.** Show how to take advantage of symmetries for the efficient computation of a DFT when $n = 4$.

18. Consider the discontinuous periodic sawtooth function with period 2π given by

$$f(x) = \begin{cases} 0, & -\pi \leqq x < 0 \\ x, & 0 \leqq x < \pi \end{cases}$$

a. Determine the Fourier series coefficients $\frac{1}{2}a_0$, a_n, b_n using the identity $\cos n\pi = (-1)^n$ to simplify.

b. Write out a few terms of the series for $f(x)$ where it is continuous and the equality holds.

c. Find the value of the series at the points of discontinuity; namely, $x = n\pi$. Explain.

d. Plot the function and Fourier series approximations p_N with $N = 5, 10, 15$ and 20.

Computer Exercises 9.4

1. Use mathematical software to solve the following periodic functions $f(x)$ over the given intervals. First, compute the Fourier coefficients of $f(x)$. Second, compute the Fourier series of $f(x)$. Next, plot $f(x)$ and some of the approximations such as p_2, p_4, and p_6. Briefly discuss the results.

a. $f(x) = \begin{cases} 0, & -\pi \leqq x < 0 \\ 1, & 0 \leqq x < \pi \end{cases}$

with $f(x + 2\pi) = f(x)$.

b. (Continuation) Use the results to show that

$$\frac{\pi}{4} = 1 - \frac{1}{3} + \frac{1}{5} - \frac{1}{7} + \cdots$$

c. $f(x) = |x|$ on $-1 \leqq x \leqq 1$ with $f(x + 2) = f(x)$.

d. (Continuation) Use the results to show that

$$\frac{\pi^2}{8} = 1 + \frac{1}{3^2} + \frac{1}{5^2} + \frac{1}{7^2} + \cdots$$

2. For the equation $x^n = 1$, consider the pseudocode for the n-th roots of unity:

```
for k = 1 to n
    omega(k) = exp(i * 2 * pi * (k/n))
end for
```

a. Compute these roots using MATLAB or some other mathematical software system or programming language.

b. Check that

```
omega(k)* * n = 1
```

c. Determine the results of these element-by-element computations:

```
sum(omega);  sum(omega.* * 2);  prod(omega)
```

d. For various values of integer n, compute the determinant of this matrix

$$\begin{bmatrix} 1 & 1 & 1 & \cdots & 1 \\ 1 & x & x^2 & \cdots & x^{n-1} \\ 1 & x^2 & x^4 & \cdots & x^{2(n-1)} \\ 1 & x^3 & x^6 & \cdots & x^{3(n-1)} \\ \vdots & \vdots & \vdots & \cdots & \vdots \\ 1 & x^{n-1} & x^{2(n-1)} & \cdots & x^{(n-1)^2} \end{bmatrix}$$

where $x = \omega_k$ where $k = 0, 1, 2, \ldots, n-1$.

e. (Continuation) Determine a formula for the determinant of this matrix.

3. Derive the following matrix

$$\boldsymbol{F}_4 = \begin{bmatrix} 1 & 1 & 1 & 1 \\ 1 & i & i^2 & i^3 \\ 1 & i^2 & i^4 & i^6 \\ 1 & i^3 & i^6 & i^9 \end{bmatrix}$$

Use mathematical software to verify it.

4. Use mathematical software to directly compute

a. $\boldsymbol{F}_4$, $\boldsymbol{F}_4^{-1}$, $\boldsymbol{F}_8$, and $\boldsymbol{F}_8^{-1}$.

b. $\boldsymbol{F}_{2n} = \begin{bmatrix} \boldsymbol{I}_n & \boldsymbol{D}_n \\ \boldsymbol{I}_n & -\boldsymbol{D}_n \end{bmatrix} \begin{bmatrix} \boldsymbol{F}_n & \boldsymbol{0} \\ \boldsymbol{0} & \boldsymbol{F}_n \end{bmatrix} \begin{bmatrix} \text{even-odd} \\ \text{shuffle} \end{bmatrix}$

for $\boldsymbol{F}_4$ and $\boldsymbol{F}_8$.

5. Use Mathematical software to plot the following functions and several of their Fourier series approximations

a. Sawtooth Wave, Equation (19)

b. Square Wave, Equation (20)

c. Triangle Wave, Equation (21)

[a]**6.** The **square wave** of amplitude $A = 10$ and period $P = 10$ is the periodic function

$$f(x) = \begin{cases} -A, & -P/2 < x < 0 \\ A, & 0 < x < P/2 \end{cases}$$

with $f(x) = f(x+P)$. Compute the Fourier series coefficients and the trigonometric polynomial in a simplified form. Plot the function and the approximations for the Fourier series using 5, 10, 15, and 20 terms.

7. (Continuation) Re-do the previous exercise using the complex Fourier series, which may have an advantage when the magnitude of the coefficients are of interest.

8. Consider the periodic function defined by

[a]**a.** $f(x) = 2 + x$ on $-2 < x < 2$

b. $f(x) = x^2$ on $-1 < x < 1$

Use mathematical software to compute the coefficients of the first 10 terms in the Fourier series. Then plot the function and these approximations.

9. Consider this periodic function:

$$f(x) = \begin{cases} 0, & -\pi < x < 0 \\ x, & 0 < x < \pi \end{cases}$$

with period 2π. Compute the trigonometric polynomial, in simplified form, for the Fourier coefficients and series. Plot the function and the approximations for the Fourier series with 5, 10, 15, and 20 terms.
Hint: Use the identity $\cos n\pi = (-1)^n$ and the fact that the series converges to $\pi/2$ at points of discontinuity $x = n\pi$.

10 Monte Carlo Methods and Simulation

A highway engineer wishes to simulate the flow of traffic for a proposed design of a major freeway intersection. The information that is obtained will then be used to determine the capacity of *storage lanes* (in which cars must slow down to yield the right of way). The intersection has the form shown in Figure 10.1, and various flows (cars per minute) are postulated at the points where arrows are drawn. By writing and running a simulation program, the engineer can study the effect of different speed limits, determine which flows lead to saturation (bottlenecks), and so on. Some techniques for constructing such programs are developed in this chapter.

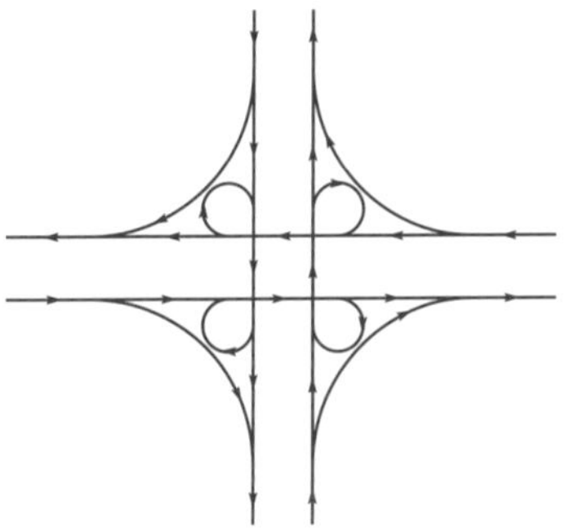

FIGURE 10.1
Traffic flow

10.1 Random Numbers

This chapter differs from most of the others in its point of view. Instead of addressing clear-cut mathematical problems, it attempts to develop methods for simulating complicated processes or phenomena. If the computer can be made to imitate an experiment or a process, then by repeating the computer simulation with different data, we can draw statistical conclusions. In such an approach, the conclusions may lack a high degree of mathematical precision but still be sufficiently accurate to enable us to understand the process being simulated.

Particular emphasis is given to problems in which the computer simulation involves an element of *chance*. The whimsical name of *Monte Carlo methods* was applied some years ago by Stanislaw M. Ulam (1909–1984) to this way of imitating reality by a computer. Since chance or randomness is part of the method, we begin with the elusive concept of **random numbers**.

Random Numbers

Consider a sequence of real numbers $x_1, x_2, \ldots$ all lying in the unit interval (0, 1). Expressed informally, the sequence is **random** if the numbers seem to be distributed haphazardly throughout the interval and if there seems to be no pattern in the progression $x_1, x_2, \ldots$ For example, if all the numbers in decimal form begin with the digit 3, then the numbers are clustered in the subinterval $0.3 \leq x < 0.4$ and are not randomly distributed in (0, 1). If the numbers are monotonically increasing, they are not random. If each x_i is obtained from its predecessor by a simple continuous function, say, $x_i = f(x_{i-1})$, then the sequence is not random (although it might appear to be so). A precise definition of **randomness** is quite difficult to formulate, and the interested reader may wish to consult an article by Chaitlin [1975], in which randomness is related to the complexity of computer algorithms! Thus, it seems best, at least in introductory material, to accept intuitively the notion of a random sequence of numbers in an interval and to accept certain algorithms for generating sequences that are more or less random.

Randomness

A recommended reference is the book of Niederreiter [1992].

Random-Number Algorithms and Generators

Random-Number Generators

Most computer systems have **random-number generators**, which are procedures that produce either a single random number or an entire array of random numbers with each call. In this chapter, we call such a procedure *Random*. The reader can use a random-number generator available on his or her own computing system, one available within the computer language being used, or one of the generators described next. For example, random-number generators are contained in mathematical software systems such as MATLAB, Maple, and Mathematica as well as many computer programming languages. These random-number procedures return one or an array of uniformly distributed pseudo-random numbers in the unit interval (0, 1) depending on whether the argument is a scalar variable or an array. A random seed procedure restarts or queries the pseudo-random-number generator. The random number generator can produce hundreds of thousands of pseudo-random numbers before repeating itself, at least theoretically.

Uniformly Distributed

For the problems in this chapter, select a routine to provide random numbers uniformly distributed in the interval (0, 1). A sequence of numbers is **uniformly distributed** in the interval (0, 1) if no subset of the interval contains more than its share of the numbers. In particular, the probability that an element x drawn from the sequence falls within the subinterval $[a, a + h]$ should be h and hence independent of the number a. Similarly, if $p_i = (x_i, y_i)$ are random points in the plane uniformly distributed in some rectangle, then the number of these points that fall inside a small square of area k should depend only on k and not on where the square is located inside the rectangle.

Pseudo-Random Numbers

Random numbers produced by a computer code cannot be truly random because the manner in which they are produced is completely *deterministic*; that is, no element of chance is actually present. But the sequences that are produced by these routines appear to be random, and they do pass certain tests for randomness. Some authors prefer to emphasize this point by calling such computer-generated sequences **pseudo-random numbers**.

If the reader wishes to program a random-number generator, the following one should be satisfactory on a computer that has 32-bit word length. This algorithm generates n random numbers $x_1, x_2, \ldots, x_n$ uniformly distributed in the open interval (0, 1) by means of the following recursive pseudocode:

Random-Number Recursive Pseudocode

integer array $(\ell_i)_{0:n}$; **real array** $(x_i)_{1:n}$
$\ell_0 \leftarrow$ any integer such that $1 < \ell_0 < 2^{31} - 1$
for $i = 1$ **to** n
  $\ell_i \leftarrow$ remainder when $7^5 \ell_{i-1}$ is divided by $2^{31} - 1$
  $x_i \leftarrow \ell_i / (2^{31} - 1)$
end for

Here, all ℓ_i's are integers in the range $1 < \ell_i < 2^{31} - 1$. The initial integer ℓ_0 is called the **seed** for the sequence and is selected as any integer between 1 and the Mersenne prime number $2^{31} - 1 = 2147483647$.

For information on portable random-number generators, the reader should consult the article by Schrage [1979]. A fast *normal* random-number generator can be written in only a few lines of code as presented in Leva [1992]. It is based on the ratio of uniform deviates method of Kinderman and Monahan [1977].

An external function procedure to generate a new array of pseudo-random numbers per call could be based on the following pseudocode:

***Random* Pseudocode**

real procedure $Random((x_i))$
integer $seed, i, n$; **real array** $(x_i)_{1:n}$
integer $k \leftarrow 16807,\ j \leftarrow 2147483647$
$seed \leftarrow$ select initial value for $seed$
$n \leftarrow size((x_i))$
for $i = 1$ **to** n
  $seed \leftarrow \text{mod}(k \cdot seed,\ j)$
  $x_i \leftarrow \text{real}(seed)/real(j)$
end for
end procedure *Random*

To allow adequate representation of the numbers involved in procedure *Random*, it must be written by using double or extended precision for use on a 32-bit computer; otherwise, it will produce nonrandom numbers.

Recall that here and elsewhere, $\text{mod}(n, m)$ is the remainder when n is divided by m; that is, it results in $n - [\text{integer}(n/m)]m$, where $\text{integer}(n/m)$ is the integer resulting from the truncation of n/m. Thus, mod(44, 7) is 2, mod(3, 11) is 3, and $\text{mod}(n, m)$ is 0 whenever m divides n evenly. We also note that $x \equiv y$ modulo (z) means that $x - y$ is divisible by z.

Outlines of two other random-number generator algorithms follow:

■ **Algorithm**

Mother of All Pseudo-Random-Number Generators

Initialize x_0, x_1, x_2, x_3 and c to random values based on a value of the seed. Letting $s = 2111111111x_{n-4} + 1492x_{n-3} + 1776x_{n-2} + 5115x_{n-1} + c$, compute $x_n = s \bmod (2^{32})$ and $c = \lfloor s/2^{32} \rfloor$ for $n \geqq 4$.

Invented by George Marsaglia. (See `www.agner.org/random/`.)

■ **Algorithm**

`rand()` in Unix

Initialize the x_0 to a random value based on a value of the seed. Compute $x_{n+1} = (1103515245x_n + 12345)\ \text{mod}(2^{31})$ for $n \geqq 1$.

These algorithms are suitable for some applications, but they may not produce high-quality randomness and may not be suitable for applications requiring accurate statistics or in cryptographics. On the Internet, one can find new and improved pseudo-random-number generators, which are designed for the fast generations of high-quality random numbers with colossal periods and with special distributions. (See, for example, the GNU Scientific Library (GSL) at `www.gnu.org/software/gsl/`.)

Caution

A few words of caution about random-number generators in computing systems are needed. The fact that the sequences produced by these programs are not truly random has already been noted. In some simulations, the failure of randomness can lead to erroneous conclusions. Here are three specific points and examples to remember:

■ **Properties**

Characteristics of Random Number Algorithms

1. The algorithms of the type illustrated here by *Random* and those above produce **periodic** sequences; that is, the sequences eventually repeat themselves. (The period is of the order 2^{30} for *Random*, which is quite large.)
2. If a random-number generator is used to produce random points in n-dimensional space, these points lie on a relatively small number of planes or hyperplanes. (As Marsaglia [1968] reports, points obtained in this way in 3-space lie on a set of only 119086 planes for computers with integer storage of 48 bits. In 10-space they lie on a set of 126 planes, which is quite small.)
3. The individual digits that make up random numbers generated by routines such as *Random* are *not*, in general, independent random digits. (For example, it might happen that the digit 3 follows the digit 5 more (or less) often than would be expected.)

Examples

An example of a pseudocode to compute and print ten random numbers using procedure *Random* follows:

Test_Random **Pseudocode**

```
program Test_Random
real array (x_i)_{1:n};   integer n ← 10
call Random((x_i))
output (x_i)
end program Test_Random
```

The computer results from a typical run are as follows:

$$0.31852\,29,\ 0.53260\,59,\ 0.50676\,22,\ 0.15271\,48,\ 0.67687\,93,$$
$$0.31067\,89,\ 0.57963\,66,\ 0.95331\,68,\ 0.39584\,57,\ 0.97879\,35$$

Mathematical software systems such as MATLAB, Maple, and Mathematica have collections of random-number generators with various distributions. For example, one can generate uniformly distributed pseudo-random numbers in the interval (0, 1). Moreover, they are particularly useful for plotting and displaying random points generated within regions in one, two, and three dimensions.

As a coarse check on the random-number generator, let us compute a long sequence of random numbers and determine what proportion of them lie in the interval $\left(0, \frac{1}{2}\right]$. The computed answer should be approximately 50%. The results with different sequence lengths are tabulated. Here is the pseudocode to carry out this experiment:

```
program Coarse_Check
integer i, m;   real per;   real array (r_i)_{1:n}
integer n ← 10000
m ← 0
```

(Continued)

Coarse_Check Pseudocode

```
call Random((r_i))
for i = 1 to n
    if r_i ≦ 1/2 then m ← m + 1
    if mod(i, 1000) = 0 then
        per ← 100 real(m)/real(i)
        output i, per
    end if
end for
end program Coarse_Check
```

In this pseudocode, a sequence of 10000 random numbers is generated. Along the way, the current proportion of numbers less than $1/2$ is computed at the 1000th step and then at multiples of 1000. Some of the computer results of the experiment are 49.5, 50.2, 51.0, and 50.625.

The experiment described can also be interpreted as a computer simulation of the tossing of a coin. A single toss corresponds to the selection of a random number x in the interval $(0, 1)$. We arbitrarily associate heads with event $0 < x \leqq \frac{1}{2}$ and tails with event $\frac{1}{2} < x < 1$. One thousand tosses of the coin corresponds to 1000 choices of random numbers. The results show the proportion of heads that result from repeated tossing of the coin. Random integers can be used to simulate coin tossing as well.

Observe that (at least in this experiment) reasonable precision is attained with only a moderate number of random numbers (4000). Repeating the experiment 10000 times has only a marginal influence on the precision. Of course, theoretically, if the random numbers were truly random, the limiting value as the number of random numbers used increases without bound would be exactly 50%.

In this pseudocode and others in the chapter, all of the random numbers are generated initially, stored in an array, and used later in the program as needed. This is an efficient way to obtain these numbers because it minimizes the number of procedure calls but at the cost of storage space. If memory space is at a premium, the call to the random-number generator can be moved closer to its use (inside the loop(s)) so that it returns a single random number with each call.

Uniformly Distributed Random Points in:

Now we consider some basic questions about generating random points in various geometric configurations. Assume that procedure *Random* is used to obtain a random number r in the interval $[0, 1]$.

First, if uniformly distributed random points are needed on some interval (a, b), this statement accomplishes it:

(a, b)

$$x \leftarrow (b - a)r + a$$

Second, this pseudocode produces random integers in the set $\{0, 1, \ldots, n\}$:

$\{0, 1, \ldots, n\}$

$$i \leftarrow \text{integer}\,((n + 1)r)$$

Third, for random integers from j to k $(j \leqq k)$, use this assignment statement:

Integers j to k

$$i \leftarrow \text{integer}\,((k - j + 1)r + j)$$

Finally, the following statements can be used to obtain the first four digits in a random number:

First Four Digits in a Random Number Pseudocode

```
integer array (m_i)_{1:n};   integer i;   real r, x
integer n ← 4
call Random(r)
for i = 1 to n
    x ← 10r
    m_i ← integer(x)
    x ← x − real(m_i)
end for
output (m_i)
```

Uses of Pseudocode *Random*

We now illustrate both correct and incorrect uses of procedure *Random* for producing uniformly distributed points.

Consider the problem of generating 1000 random points uniformly distributed inside the ellipse $x^2 + 4y^2 = 4$.

One way to do so is to generate random points in the rectangle $-2 \leqq x \leqq 2$, $-1 \leqq y \leqq 1$, and discard those that do not lie in the ellipse (see Figure 10.2).

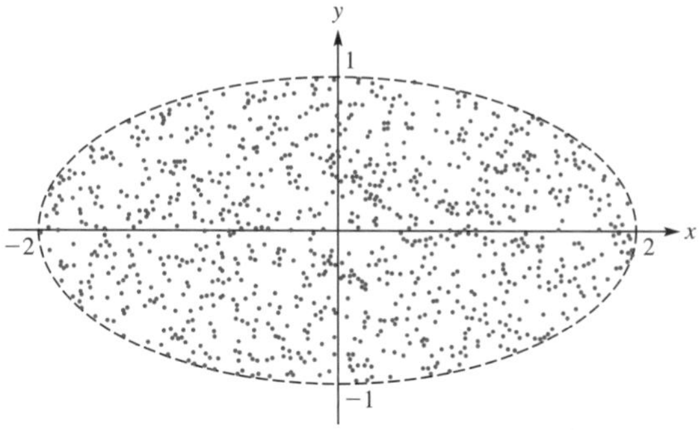

FIGURE 10.2 Uniformly distributed random points in ellipse $x^2 + 4y^2 = 4$

***Ellipse* Pseudocode**

```
program Ellipse
integer i, j;   real u, v;   real array (x_i)_{1:n}, (y_i)_{1:n}, (r_ij)_{1:npts×1:2}
integer n ← 1000, npts ← 2000
call Random((r_ij))
j ← 1
for i = 1 to npts
    u ← 4r_{i,1} − 2
    v ← 2r_{i,2} − 1
    if u^2 + 4v^2 ≦ 4 then
        x_j ← u
        y_j ← v
        j ← j + 1
    if j = n then exit loop i
    end if
end for
end program Ellipse
```

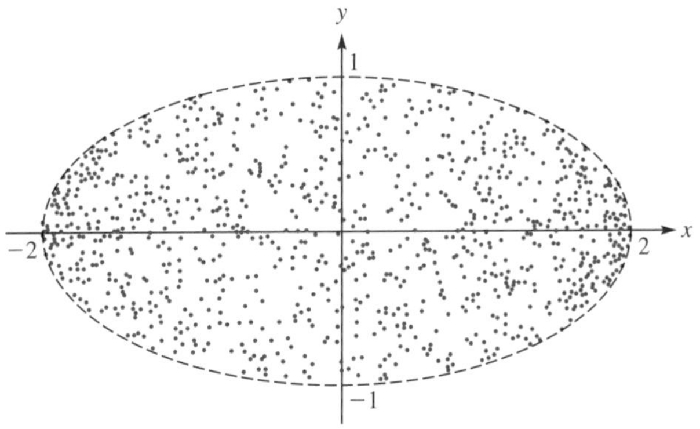

FIGURE 10.3 Nonuniformly distributed random points in the ellipse $x^2 + 4y^2 = 4$

To be less wasteful, we can *force* the $|y|$ value to be less than $\frac{1}{2}\sqrt{4 - x^2}$, as in the following pseudocode, *which produces erroneous results* (see Figure 10.3):

***Ellipse_Erroneous* Pseudocode**

```
program Ellipse_Erroneous
integer i;   real array (x_i)_{1:n}, (y_i)_{1:n}, (r_ij)_{1:n×1:2}
integer n ← 1000
call Random((r_ij))
for i = 1 to n
    x_i ← 4r_{i,1} − 2
    y_i ← [(2r_{i,2} − 1)/2] sqrt(4 − x_i^2)
end for
end program Ellipse_Erroneous
```

This pseudocode does *not* produce uniformly distributed points inside the ellipse. To be convinced of this, consider two vertical strips taken inside the ellipse (see Figure 10.4).

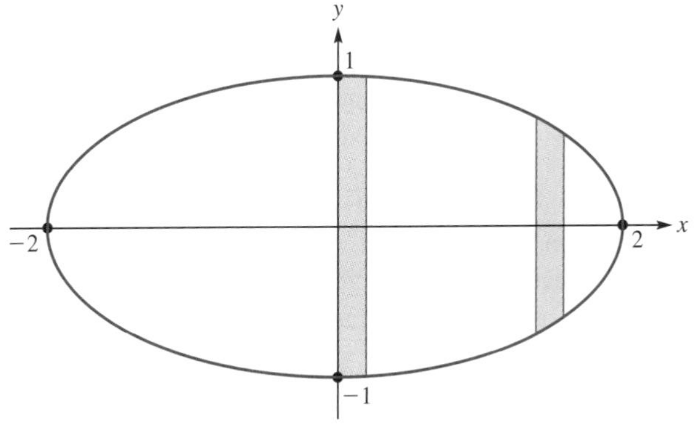

FIGURE 10.4 Vertical strips containing nonuniformly distributed points

If each strip is of width h, then approximately $1000(h/4)$ of the random points lie in each strip because the random variable x is uniformly distributed in $(-2, 2)$, and with each x, a corresponding y is generated by the program so that (x, y) is inside the ellipse. But the two strips shown should *not* contain approximately the same number of points because they do not have the same area. The points generated by the second program tend to be clustered at the left and right extremities of the ellipse in Figure 10.3.

For the same reasons, the following pseudocode does *not* produce uniformly distributed random points in the circle $x^2 + y^2 = 1$ (see Figure 10.5):

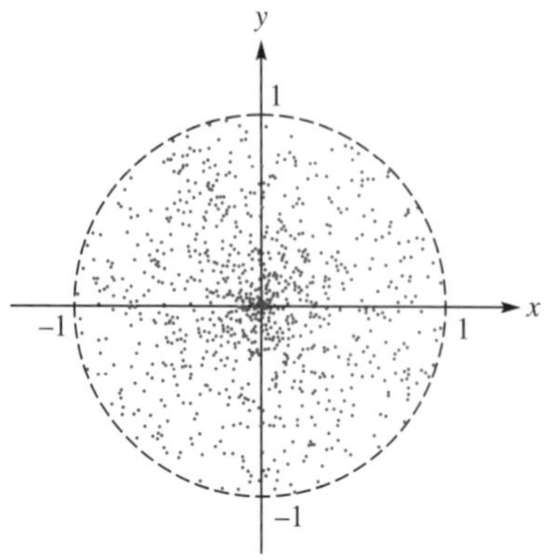

FIGURE 10.5 Nonuniformly distributed random points in the circle $x^2 + y^2 = 1$

Circle_Erroneous **Pseudocode**

program *Circle_Erroneous*
integer i; **real array** $(x_i)_{1:n}, (y_i)_{1:n}, (r_{ij})_{1:n\times 1:2}$
integer $n \leftarrow 1000$
call *Random*$((r_{ij}))$
for $i = 1$ **to** n
    $x_i \leftarrow r_{i,1}\cos(2\pi r_{i,2})$
    $y_i \leftarrow r_{i,1}\sin(2\pi r_{i,2})$
end for
end program *Circle_Erroneous*

In this pseudocode, $2\pi r_{i,2}$ is uniformly distributed in $(0, 2\pi)$, and $r_{i,1}$ is uniformly distributed in $(0, 1)$. However, in the transfer from polar to rectangular coordinates by the equations $x = r_{i,1}\cos(2\pi r_{i,2})$ and $y = r_{i,1}\sin(2\pi r_{i,2})$, the uniformity is lost. The random points are strongly clustered near the origin in Figure 10.5.

A random-number generator produces a sequence of numbers that are random in the sense that they are uniformly distributed over a certain interval such as $[0, 1)$, and it is not possible to predict the next number in the sequence from knowing the previous ones. One can increase the randomness of such a sequence by a suitable *shuffle* of them. The idea is to fill an array with the consecutive numbers from the random-number generator and then to use the generator again to choose at random which of the numbers in the array is to be selected as the next number in a new sequence. The hope is that the new sequence is more random than the original one. For example, a shuffle can remove any correlation between near successors of a number in a sequence. See Flowers [1995] for a shuffling procedure that can be used with a random-number generator based on a linear congruence. It is particularly useful on computers with a small word length.

There are statistical tests that can be performed on a sequence of random numbers. Although such tests do not certify the randomness of a sequence, they are particularly important in applications. For example, they are useful in choosing between different random-number generators, and it is comforting to know that the random-number generator being used has passed such tests. Situations exist when random-number generators are useful even though they do not pass rigid tests for true randomness. Thus, if one is producing random matrices for testing a linear algebra code, then strict randomness may not be important. On the other hand, strict randomness is *essential* in Monte Carlo integration and other applications. In these cases in which strict randomness is important, it is recommended that one use a machine with a large word size and a random-number generator with known statistical characteristics. (See Volume 2 of Knuth [1997] or Flowers [1995] for some tests of randomness.)

Quasi-Random

Quasi-random or low-discrepancy sequences are constructed to give a uniform coverage of an area or volume while maintaining a reasonably random appearance even though they are not in fact random.

Prime Number Brief History

A **prime number** is an integer greater than 1 whose only factors (divisors) are itself and 1. Prime numbers are some of the fundamental building blocks in mathematics. The search for large primes has a long and interesting history. In 1644, Mersenne (a French friar) conjectured that $2^n - 1$ was a prime number for $n = 17, 19, 31, 67, 127, 257$ and for no other n in the range $1 \leqq n \leqq 257$. In 1876, Lucas proved that $2^{127} - 1$ was prime. In 1937, however, Lehmer showed that $2^{257} - 1$ was *not* prime. Until 1952, $2^{127} - 1$ was the largest known prime. Then it was shown that $2^{521} - 1$ was prime. As a means of testing new computer systems, the search for ever-larger Mersenne primes continues. In fact, the search for ever larger primes has grown in importance for use in cryptology. In 1992, a Cray 2 supercomputer using the Lucas-Lehmer test determined after a 19-hour computation that the number $2^{756839} - 1$ was a prime. This number has 227,832 digits! The previous largest known Mersenne prime was identified in 1985 as $2^{216091} - 1$. In 2006, the largest known prime $2^{32582657} - 1$, with 9.8 million digits, was discovered using the Internet facility GIMPS (Great Internet Mersenne Prime Search). Thousands of individuals have used the GIMPS database to facilitate their search for large primes, and interaction with the database can be done automatically without human intervention. For more information on large primes and to find out the current record for the largest known prime, consult `http://www.mersenne.org/` or do a search on the World Wide Web (i.e., online).

Mersenne Prime

Summary 10.1

- An algorithm to generate an array (r_i) of **pseudo-random numbers** is

integer ℓ; **real array** $(x_i)_{1:n}$
$\ell \leftarrow$ an integer between 1 and $2^{31} - 1$
for $i = 1$ **to** n
  $\ell \leftarrow \text{mod}(7^5\ell, 2^{31} - 1)$
  $x_i \leftarrow \ell/(2^{31} - 1)$
end for

- If (r_i) is an array of random numbers, then use the following to generate random points in an interval (a, b)

$$x \leftarrow (b - a)r_i + a$$

to produce random integers in the set $\{0, 1, \dots, n\}$

$$i \leftarrow \text{integer}\,((n + 1)r_i)$$

and to obtain random integers from j to k $(j \leqq k)$

$$i \leftarrow \text{integer}\,((k - j + 1)r_i + j)$$

Exercises 10.1

[a]1. Taking the seed to be 123456, compute by hand the first three random numbers produced by procedure *Random*.

2. Show that if the seed ℓ is less than or equal to 12777, then the first random number produced by procedure *Random* is less than $1/10$.

3. Show that the numbers produced by procedure *Random* are not random because their products with $2^{31} - 1$ are integers.

4. Describe in what ways this algorithm for random numbers differs from procedure *Random*:

$$\begin{cases} x_0 \text{ arbitrary in } (0, 1) \\ x_i = \text{fractional part of } 7^5 x_{i-1} \qquad (i \geqq 1) \end{cases}$$

Computer Exercises 10.1

1. Write a program to generate 1000 random points uniformly distributed in the cardioid $r = 2 - \cos\theta$.

2. Using procedure *Random*, write code for procedure *Random_Trapezoid*(x, y), which generates a pseudorandom point (x, y) inside or on the trapezoid formed by the points $(1, 3)$, $(2, 5)$, $(4, 3)$, and $(3, 5)$.

3. Without using any procedures, write a program to generate and print 100 random numbers uniformly distributed in $(0, 1)$. Eight statements suffice.

4. Test some random-number generators found in mathematical software on the World Wide Web.

5. Test the random-number generator on your computer system in the following way: Generate 1000 random numbers $x_1, x_2, \ldots, x_{1000}$.

   a. In any small interval of width h, approximately $1000h$ of the x_i's should lie in that interval. Count the number of random numbers in each of ten intervals $[(n-1)/10, n/10]$, where $n = 1, 2, \ldots, 10$.

   b. The inequality $x_i < x_{i+1}$ should occur approximately 500 times. Count them in your sample.

6. Write a procedure to generate with each call a random vector of the form $\boldsymbol{x} = [x_1, x_2, \ldots, x_{20}]^T$, where each x_i is an integer from 1 to 100 and no two components of $\boldsymbol{x}$ are the same.

7. Write a program to generate $n = 1000$ random points uniformly distributed in the

   a. equilateral triangle in the following figure:

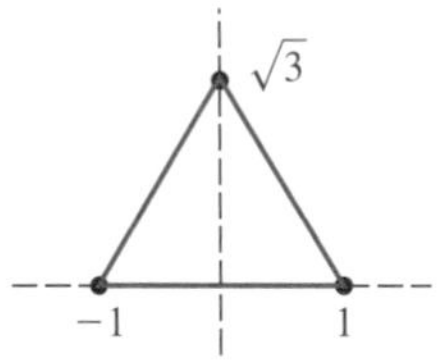

   b. diamond in the following figure:

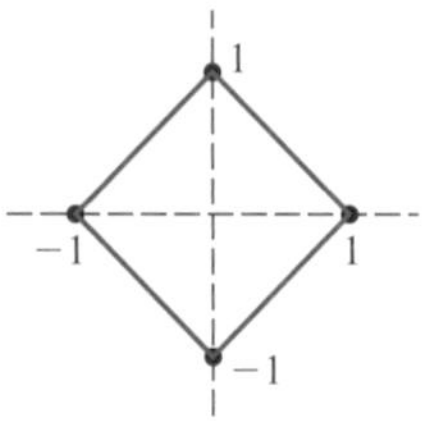

Store the random points (x_i, y_i) in arrays $(x_i)_{1:n}$ and $(y_i)_{1:n}$.

[a]8. If $x_1, x_2, \ldots$ is a random sequence of numbers uniformly distributed in the interval $(0, 1)$, what proportion would you expect to satisfy the inequality $40x^2 + 7 > 43x$? Write a program to test this on 1000 random numbers.

9. Write a program to generate and print 1000 points uniformly and randomly distributed in the circle $(x - 3)^2 + (y + 1)^2 \leqq 9$.

10. Generate 1000 random numbers x_i according to a uniform distribution in the interval $(0, 1)$. Define a function f on $(0, 1)$ as follows: $f(t)$ is the number of random numbers $x_1, x_2, \ldots, x_{1000}$ less than t. Compute $f(t)/1000$ for 200 points t uniformly distributed in $(0, 1)$. What do you expect $f(t)/1000$ to be? Is this expectation borne out by the experiment? If a plotter is available, plot $f(t)/1000$.

[a]11. Let n_i $(1 \leqq i \leqq 1000)$ be a sequence of integers that satisfies $0 \leqq n_i \leqq 9$. Write a program to test the given sequence for periodicity. (The sequence is **periodic** if there is an integer k such that $n_i = n_{i+k}$ for all i.)

12. Generate in the computer 1000 random numbers in the interval $(0, 1)$. Print and examine them for evidence of nonrandom behavior.

[a]**13.** Generate 1000 random numbers x_i $(1 \leqq i \leqq 1000)$ on your computer. Let n_i denote the eighth decimal digit in x_i. Count how many 0's, 1's, ..., 9's there are among the 1000 numbers n_i. How many of each would you expect? This code can be written with nine statements.

14. (Continuation) Using a random-number generator, generate 1000 random numbers, and count how many times the digit i occurs in the jth decimal place. Print a table of these values—that is, frequency of digit versus decimal place. By examining the table, determine which decimal place seems to produce the best uniform distribution of random digits.
Hint: Use Computer Exercise 1.1.7 (p. 17) to compute the arithmetic mean, variance, and standard deviations of the table entries.

[a]**15.** Using random integers, write a short program to simulate five people matching coin flips. Print the percentage of match-ups (five of a kind) after 125 flips.

[a]**16.** Write a program to generate 1600 random points uniformly distributed in the sphere defined by $x^2 + y^2 + z^2 \leqq 1$. Count the number of random points in the first octant.

17. Write a program to simulate 1000 simultaneous flips of three coins. Print the number of times that two of the three coins come up heads.

18. Compute 1000 triples of random numbers drawn from a uniform distribution. For each triple (x, y, z), compute the leading significant digit of the product xyz. (The leading significant digit is one of 1, 2, ..., 9.) Determine the frequencies with which the digits 1 through 9 occur among the 1000 cases. Try to account for the fact that these digits do not occur with the same frequency. (For example, 1 occurs approximately 7 times more often than 9.) If you are intrigued by this, you may wish to consult the articles by Flehinger [1966], Raimi [1969], and Turner [1982].

19. Run the example programs in this section and see whether similar results are obtained on your computer system.

20. Write a program to generate and plot 1000 pseudorandom points with the following **exponential** distribution inside the following figure: $x = -\ln(1-r)/\lambda$ for $r \in [0, 1)$ and $\lambda = 1/30$.

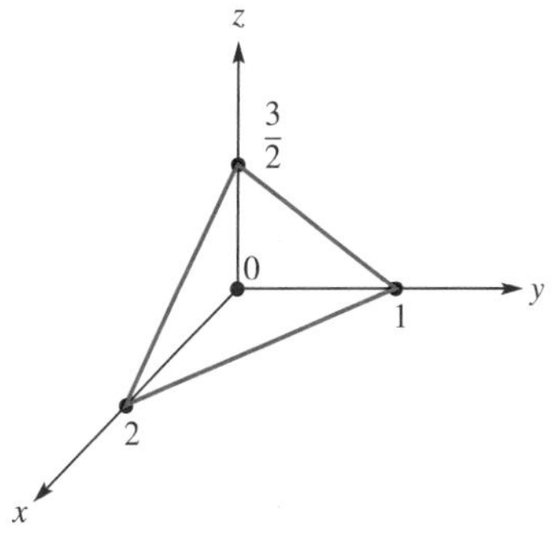

21. Improve the program *Coarse_Check* by using ten or a hundred *buckets* instead of two.

22. (**Student Research Project**) Investigate some of the latest developments on random-number generators and explore parallel random number generators. Random numbers are often needed for distributions other than the uniform distribution, so this has a statistical aspect.

10.2 Estimation of Areas and Volumes by Monte Carlo Techniques

Numerical Integration

Now we turn to applications, the first being the approximation of a definite integral by the Monte Carlo method. If we select the first n elements $x_1, x_2, \ldots, x_n$ from a random sequence in the interval $(0, 1)$, then

Average of n Numbers

$$\int_0^1 f(x)\,dx \approx \frac{1}{n}\sum_{i=1}^{n} f(x_i)$$

Here, the integral is approximated by the average of n numbers $f(x_1), f(x_2), \ldots, f(x_n)$. When this is actually carried out, the error is of order $1/\sqrt{n}$, which is not at all competitive with good algorithms, such as the Romberg method. However, in higher dimensions, the

Monte Carlo method can be quite attractive. For example, the triple integral is

Using Random Sequence of Points

$$\int_0^1\int_0^1\int_0^1 f(x,y,z)\,dx\,dy\,dz \approx \frac{1}{n}\sum_{i=1}^{n} f(x_i,y_i,z_i)$$

where (x_i, y_i, z_i) is a random sequence of n points in the unit cube $0 \leqq x \leqq 1$, $0 \leqq y \leqq 1$, and $0 \leqq z \leqq 1$. To obtain random points in the cube, we assume that we have a random sequence in (0, 1) denoted by $\xi_1, \xi_2, \xi_3, \xi_4, \xi_5, \xi_6, \ldots$ To get our first random point p_1 in the cube, just let $p_1 = (\xi_1, \xi_2, \xi_3)$. The second is, of course, $p_2 = (\xi_4, \xi_5, \xi_6)$, and so on.

If the interval (in a one-dimensional integral) is not of length 1 but, say, is the general case (a, b), then the average of f over n random points in (a, b) is not simply an approximation for the integral, but rather for

$$\frac{1}{b-a}\int_a^b f(x)\,dx$$

which agrees with our intention that the function $f(x) = 1$ have an average of 1. Similarly, in higher dimensions, the average of f over a region is obtained by integrating and dividing by the area, volume, or measure of that region. For instance,

$$\frac{1}{8}\int_1^3\int_{-1}^1\int_0^2 f(x,y,z)\,dx\,dy\,dz$$

is the average of f over the parallelepiped described by the following three inequalities: $0 \leqq x \leqq 2$, $-1 \leqq y \leqq 1$, $1 \leqq z \leqq 3$.

To keep the limits of integration straight, recall that

Order of Integration

$$\int_a^b\int_c^d f(x,y)\,dx\,dy = \int_a^b\left[\int_c^d f(x,y)\,dx\right]dy$$

and

$$\int_{a_1}^{a_2}\int_{b_1}^{b_2}\int_{c_1}^{c_2} f(x,y,z)\,dx\,dy\,dz = \int_{a_1}^{a_2}\left\{\int_{b_1}^{b_2}\left[\int_{c_1}^{c_2} f(x,y,z)\,dx\right]dy\right\}dz$$

So if (x_i, y_i) denote random points with appropriate uniform distribution, the following examples illustrate Monte Carlo techniques:

Example Integrals

$$\int_0^5 f(x)\,dx \approx \frac{5}{n}\sum_{i=1}^{n} f(x_i)$$

$$\int_2^5\int_1^6 f(x,y)\,dx\,dy \approx \frac{15}{n}\sum_{i=1}^{n} f(x_i,y_i)$$

In each case, the random points should be uniformly distributed in the regions involved.

In general, we have

General Case

$$\int_A f \approx (\text{measure of } A) \times (\text{average of } f \text{ over } n \text{ random points in } A)$$

Here, we are using the fact that the average of a function on a set is equal to the integral of the function over the set divided by the measure of the set.

Example and Pseudocode

Approximate
$\iint_\Omega f(x, y)\,dx\,dy$
Using Random Numbers

Let us consider the problem of obtaining the numerical value of the integral

$$\iint_\Omega \sin\sqrt{\ln(x+y+1)}\,dx\,dy = \iint_\Omega f(x, y)\,dx\,dy$$

over the disk in xy-space, defined by the inequality

$$\Omega = \left\{(x, y) : \left(x - \frac{1}{2}\right)^2 + \left(y - \frac{1}{2}\right)^2 \leqq \frac{1}{4}\right\}$$

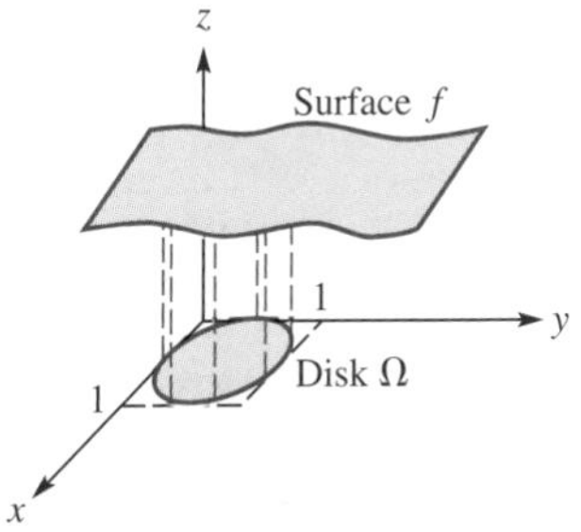

FIGURE 10.6
Sketch of surface $f(x, y)$ above disk Ω

A sketch of this domain, with a surface above it, is shown in Figure 10.6. We proceed by generating random points in the square and discarding those that do not lie in the disk. We take $n = 5000$ points in the disk. If the points are $p_i = (x_i, y_i)$, then the integral is estimated to be

General Case

$$\begin{aligned}\iint_\Omega f(x, y)\,dx\,dy &\approx (\text{area of disk } \Omega) \times \begin{pmatrix}\text{average height of } f\\ \text{over } n \text{ random points}\end{pmatrix}\\ &= (\pi r^2)\left[\frac{1}{n}\sum_{i=1}^{n} f(p_i)\right]\\ &= \frac{\pi}{4n}\sum_{i=1}^{n} f(p_i)\end{aligned}$$

The pseudocode for this example follows. Intermediate estimates of the integral are printed when n is a multiple of 1000. This gives us some idea of how the correct value is being approached by our averaging process.

***Double_Integral* Pseudocode**

```
program Double_Integral
integer i, j;   real sum, vol, x, y;   real array (r_ij)_{1:n×1:2}
integer n ← 5000, iprt ← 1000;   external function f
call Random((r_ij))
j ← 0;   sum ← 0
for i = 1 to n
    x = r_{i,1}; y = r_{i,2}
    if (x − 1/2)^2 + (y − 1/2)^2 ≦ 1/4 then
        j ← j + 1
        sum ← sum + f(x, y)
        if mod(j, iprt) = 0 then
            vol ← (π/4)sum/real(j)
            output j, vol
```

(Continued)

```
            end if
        end if
    end for
    vol ← (π/4)sum/real(j)
    output j, vol
    end program Double_Integral

    real function f(x, y)
    real x, y
    f ← sin(√ln(x + y + 1))
    end function
```

We obtain an approximate value of 0.57 for the integral.

Computing Volumes

The volume of a complicated region in 3-space can be computed by a Monte Carlo technique. Taking a simple case, let us determine the volume of the region whose points satisfy the inequalities

Complicated Region in 3D

$$\begin{cases} 0 \leqq x \leqq 1, \quad 0 \leqq y \leqq 1, \quad 0 \leqq z \leqq 1 \\ x^2 + \sin y \leqq z \\ x - z + e^y \leqq 1 \end{cases}$$

The first line defines a cube whose volume is 1. The region defined by *all* the given inequalities is therefore a subset of this cube. If we generate n random points in the cube and determine that m of them satisfy the last two inequalities, then the volume of the desired region is approximately m/n. Here is a pseudocode that carries out this procedure:

***Volume_Region* Pseudocode**

```
program Volume_Region
integer i, m;   real array (r_ij)_1:n×1:3;   real vol, x, y, z
integer n ← 5000, iprt ← 1000
call Random((r_ij))
for i = 1 to n
    x ← r_i,1
    y ← r_i,2
    z ← r_i,3
    if x² + sin y ≦ z, x − z + e^y ≦ 1 then m ← m + 1
    if mod(i, iprt) = 0 then
        vol ← real(m)/real(i)
        output i, vol
    end if
end for
end program Volume_Region
```

Observe that intermediate estimates are printed out when we reach 1000, 2000, . . . , 5000 points. An approximate value of 0.14 is determined for the volume of the region.

Ice Cream Cone Example

Consider the problem of finding the volume above the cone $z^2 = x^2 + y^2$ and inside the sphere $x^2 + y^2 + (z - 1)^2 = 1$ as shown in Figure 10.7.

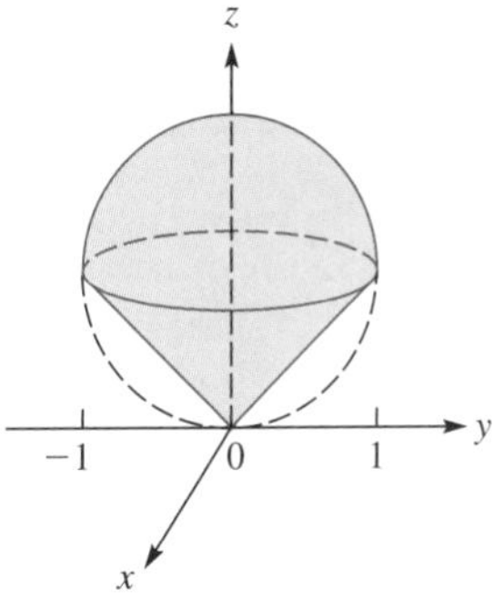

FIGURE 10.7 Ice cream cone region

The volume is contained in the box bounded by $-1 \leqq x \leqq 1$, $-1 \leqq y \leqq 1$, and $0 \leqq z \leqq 2$, which has volume 8. Thus, we want to generate random points inside this box and multiply by 8 the ratio of those inside the desired volume to the total number generated. A pseudocode for doing this follows:

Cone **Pseudocode**

```
program Cone
integer i, m;   real vol, x, y, z;   real array (r_ij)_1:n×1:3
integer n ← 5000, iprt ← 1000;   m ← 0
call Random((r_ij))
for i = 1 to n
    x ← 2r_i,1 − 1; y ← 2r_i,2 − 1; z ← 2r_i,3
    if x^2 + y^2 ≦ z^2, x^2 + y^2 ≦ z(2 − z) then m ← m + 1
    if mod(i, iprt) = 0 then
        vol ← 8 real(m)/real(i)
        output i, vol
    end if
end for
end program Cone
```

The volume of the cone is approximately 3.3.

Summary 10.2

- We can approximate integrals by using the **Monte Carlo method to estimate areas and volumes**:

$$\int_0^1 f(x)\,dx \approx \frac{1}{n}\sum_{i=1}^{n} f(x_i)$$

$$\int_0^1\int_0^1\int_0^1 f(x, y, z)\,dx\,dy\,dz \approx \frac{1}{n}\sum_{i=1}^{n} f(x_i, y_i, z_i)$$

where $\{x_i\}$ is a sequence of random numbers in the unit interval and (x_i, y_i, z_i) is a random sequence of n points in the unit cube.

- In general, we have

$$\int_A f \approx (\text{measure of } A) \times (\text{average of } f \text{ over } n \text{ random points in } A)$$

Exercises 10.2

[a]1. It is proposed to calculate π by using the Monte Carlo method. A circle of radius 1 is inside a square of side 2. We count how many of m random points in the square happen to lie in the circle. Assume that the error is $1/\sqrt{m}$. How many points must be taken to obtain π with three accurate figures (i.e., 3.142)?

2. The **Mean-Value Theorem for Integral** says that there exists a number c with $a < c < b$ such that

$$\int_a^b f(x)dx = (b-a)f(c)$$

assuming $f(x)$ is continuous over the interval $[a, b]$. Consequently, this computation shows that the area under the curve is the base width $b - a$ times the average height $f(c)$.

Let $f(x) = \sin x + \frac{1}{3}\sin 3x$. Find c so that

$$\int_0^{\pi} \left[\sin x + \frac{1}{3}\sin 3x\right] dx = \pi f(c)$$

3. An intuitive method of finding the area under a curve is to approximate that area with a series of rectangles that lie above the subintervals $[a = x_0, x_1]$, $[x_1, x_2], \ldots, [x_{n-1}, x_n = b]$. Here the interval $[a, b]$ is subdivided into n subinterval of equal width $h = (b - a)/n$. We form the equally spaced nodes $c_k = a+(k-t)h$ for $k = 1, 2, \ldots, n$.

The **Composite Midpoint Rule** for n subintervals is

$$\int_a^b f(x)\,dx = h\sum_{k=1}^{n} f(c_k)$$

Show that we can approximate the integral

$$\int_a^b f(x)\,dx \approx (b-a)\hat{f}$$

where

$$\hat{f} = \frac{1}{n}\sum_{k=1}^{n} f(c_k)$$

by sampling $f(x)$ at the n equally spaced points $c_k = a + \left(k - \frac{1}{2}\right)h$ for $h = 1, 2, \ldots, n$ where $h = (b-a)/n$.

Computer Exercises 10.2

1. Run the codes given in this section on your computer system and verify that they produce reasonable answers.

[a]2. Write and test a program to evaluate the integral $\int_0^1 e^x\,dx$ by the Monte Carlo method, using $n = 25, 50, 100, 200, 400, 800, 16000$, and 32000. Observe that 32,000 random numbers are needed and that the work in each case can be used in the next case. Print the exact answer. Plot the results using a logarithmic scale to show the rate of growth.

3. Write a program to verify numerically that

$$\pi = \int_0^2 (4 - x^2)^{1/2}\,dx$$

Use the Monte Carlo method and 2500 random numbers.

[a]4. Use the Monte Carlo method to approximate the integral

$$\int_{-1}^{1}\int_{-1}^{1}\int_{-1}^{1} (x^2 + y^2 + z^2)\,dx\,dy\,dz$$

Compare with the correct answer.

[a]5. Write a program to estimate

$$\int_0^2\int_3^6\int_{-1}^1 (yx^2 + z\log y + e^x)\,dx\,dy\,dz$$

6. Using the Monte Carlo technique, write a pseudocode to approximate the integral

$$\iiint_{\Omega} (e^x \sin y \log z)\,dx\,dy\,dz$$

where Ω is the circular cylinder that has height 3 and circular base $x^2 + y^2 \leq 4$.

[a]7. Estimate the area under the curve $y = e^{-(x+1)^2}$ and inside the triangle that has vertices $(1, 0)$, $(0, 1)$, and $(-1, 0)$ by writing and testing a short program.

8. Using the Monte Carlo approach, find the area of the irregular figure defined by

$$\begin{cases} 1 \leqq x \leqq 3, \quad -1 \leqq y \leqq 4 \\ x^3 + y^3 \leqq 29 \\ y \geqq e^x - 2 \end{cases}$$

[a]**9.** Use the Monte Carlo method to estimate the volume of the solid whose points (x, y, z) satisfy

$$\begin{cases} 0 \leqq x \leqq y, \quad 1 \leqq y \leqq 2, \quad -1 \leqq z \leqq 3 \\ e^x \leqq y \\ (\sin z) y \geqq 0 \end{cases}$$

[a]**10.** Using a Monte Carlo technique, estimate the area of the region determined by the inequalities $0 \leqq x \leqq 1$, $10 \leqq y \leqq 13$, $y \geqq 12 \cos x$, and $y \geqq 10 + x^3$. Print intermediate answers.

11. Use the Monte Carlo method to approximate the following integrals.

a. $\displaystyle\int_{-1}^{1}\int_{-1}^{1}\int_{-1}^{1} (x^2 - y^2 - z^2)\,dx\,dy\,dz$

b. $\displaystyle\int_{1}^{4}\int_{2}^{5} (x^2 - y^2 + xy - 3)\,dx\,dy$

c. $\displaystyle\int_{2}^{3}\int_{1+y}^{\sqrt{y}} (x^2 y + xy^2)\,dx\,dy$

d. $\displaystyle\int_{0}^{1}\int_{y^2}^{\sqrt{y}}\int_{0}^{y+z} xy\,dx\,dy\,dz$

12. The value of the integral

$$\int_0^{\pi/4}\int_0^{2\cos\phi}\int_0^{2\pi} \rho^2 \sin\phi\,d\theta\,d\rho\,d\phi$$

using spherical coordinates is the volume above the cone $z^2 = x^2 + y^2$ and inside the sphere $x^2 + y^2 + (z-1)^2 = 1$. Use the Monte Carlo method to approximate this integral and compare the results with that from the example in the text.

13. Let R denote the region in the xy-plane defined by the inequalities

$$\begin{cases} \frac{1}{3} \leqq 3x \leqq 9 - y \\ \sqrt{x} \leqq y \leqq 3 \end{cases}$$

Estimate the integral

$$\iint_R (e^x + \cos xy)\,dx\,dy$$

[a]**14.** Using a Monte Carlo technique, estimate the area of the region defined by the inequalities $4x^2 + 9y^2 \leqq 36$ and $y \leqq \arctan(x + 1)$.

15. Write a program to estimate the area of the region defined by the inequalities

$$\begin{cases} x^2 + y^2 \leqq 4 \\ |y| \leqq e^x \end{cases}$$

16. An integral can be estimated by the formula

$$\int_0^1 f(x)\,dx \approx \frac{1}{n}\sum_{i=1}^{n} f(x_i)$$

even if the x_i's are not random numbers; in fact, some nonrandom sequences may be better. Use the sequence $x_i = \left(\text{fractional part of } i\sqrt{2}\right)$ and test the corresponding numerical integration scheme. Test whether the estimates converge at the rate $1/n$ or $1/\sqrt{n}$ by using some simple examples, such as $\int_0^1 e^x\,dx$ and $\int_0^1 (1 + x^2)^{-1}\,dx$.

17. Consider the ellipsoid

$$\frac{x^2}{4} + \frac{y^2}{16} + \frac{z^2}{4} = 1$$

a. Write a program to generate and store 5000 random points uniformly distributed in the first octant of this ellipsoid.

[a]**b.** Write a program to estimate the volume of this ellipsoid in the first octant.

18. A Monte Carlo method for estimating $\int_a^b f(x)\,dx$ if $f(x) \geqq 0$ is as follows: Let $c \geqq \max_{a \leqq x \leqq b} f(x)$. Then generate n random points (x, y) in the rectangle $a \leqq x \leqq b$, $0 \leqq y \leqq c$. Count the number k of these random points (x, y) that satisfy $y \leqq f(x)$. Then

$$\int_a^b f(x)\,dx \approx kc(b - a)/n$$

Verify this and test the method on $\int_1^2 x^2\,dx$, $\int_0^1 (2x^2 - x + 1)\,dx$, and $\int_0^1 (x^2 + \sin 2x)\,dx$.

19. (Continuation) Use the method to estimate the value of

$$\pi = 4\int_0^1 \sqrt{1 - x^2}\,dx$$

Generate random points in $0 \leqq x \leqq 1$, $0 \leqq y \leqq 1$. Use $n = 1000, 2000, \ldots, 10000$ and try to determine whether the error is behaving like $1/\sqrt{n}$.

20. (Continuation) Modify the method to handle the case when f takes positive and negative values on $[a, b]$. Test the method on $\int_{-1}^1 x^3\,dx$.

21. Another Monte Carlo method for evaluating $\int_a^b f(x)\,dx$ is as follows: Generate an odd number of random numbers in (a, b). Reorder these points so that $a < x_1 < x_2 < \cdots < x_n < b$. Now compute

$$f(x_1)(x_2 - a) + f(x_3)(x_4 - x_2) + f(x_5)(x_6 - x_4) + \cdots + f(x_n)(b - x_{n-1})$$

Test this method on

$$\int_0^1 (1+x^2)^{-1}\,dx, \quad \int_0^1 (1-x^2)^{-1/2}\,dx, \quad \int_0^1 x^{-1}\sin x\,dx$$

22. What is the expected value of the volume of a tetrahedron formed by four points chosen randomly inside the tetrahedron whose vertices are (0, 0, 0), (0, 1, 0), (0, 0, 1), and (1, 0, 0)? (The precise answer is unknown!)

23. Write a program to compute the area under the curve $y = \sin x$ and above the curve $y = \ln(x + 2)$. Use the Monte Carlo method, and print intermediate results.

24. Estimate the integral

$$\int_{3.2}^{5.9} \left(\frac{e^{\sin x + x^2}}{\ln x}\right) dx$$

by the Monte Carlo method.

25. Test the random-number generator that is available to you in the following manner: Begin by creating a list of N random numbers r_k, uniformly distributed in the interval [0, 1]. Create a list of random integers n_k by extracting the integer part of $10r_k$ for $1 \le k \le N$. Compute the elements in a 10×10 matrix (m_{ij}), where m_{ij} is the number of times i is followed by j in the list (n_k). Compare these numbers to the values predicted by elementary probability theory. If possible, display the values of m_{ij} graphically.

26. (**Student Research Project**) Investigate some of the latest developments on Monte Carlo methods for multivariable integration.

27. Use the Monte Carlo method to calculate approximations to these integrals

a. $\displaystyle\int_0^4 \sqrt{x}\,dx$ **b.** $\displaystyle\int_0^1 \frac{4}{1+x^2}\,dx$

c. $\displaystyle\int_0^1 \sqrt{x+\sqrt{x}}\,dx$

d. $\displaystyle\int_0^{\frac{5}{4}}\int_0^{\frac{5}{4}} (4 - x^2 - y^2)\,dx\,dy$

e. $\displaystyle\int_0^{\frac{5}{4}}\int_0^{\frac{5}{4}} \sqrt{4 - x^2 - y^2}\,dx\,dy$

f. $\displaystyle\int_0^{0.9}\int_0^{1}\int_0^{1.1} (4 - x^2 - y^2 - z^2)\,dx\,dy\,dz$

g. $\displaystyle\int_0^{0.8}\int_0^{0.9}\int_0^{1}\int_0^{1.1} (5 - x^2 - y^2 - z^2 - u^2)\,dx\,dy\,dz\,du$

h. $\displaystyle\int_0^{0.7}\int_0^{0.8}\int_0^{0.9}\int_0^{1}\int_0^{1.1} \sqrt{6 - x^2 - y^2 - u^2 - w^2} \times dx\,dy\,dz\,du\,dw$

10.3 Simulation

Simulation

We next illustrate the idea of **simulation**. We consider a physical situation in which an element of chance is present and try to imitate the situation on the computer. Statistical conclusions can be drawn if the experiment is performed many times. Applications include the simulation of servers, clients, and queues as might occur in businesses such as banks or grocery stores.

Loaded Die Problem

In simulation problems, we must often produce random variables with a prescribed distribution. Suppose, for example, that we want to simulate the throw of a loaded die and that the probabilities of various outcomes have been determined as shown:

Example Loaded Die

Outcome	1	2	3	4	5	6
Probability	0.2	0.14	0.22	0.16	0.17	0.11

If the random variable x is uniformly distributed in the interval (0, 1), then by breaking this interval into six subintervals of lengths given by the table, we can simulate the throw of this loaded die. For example, we agree that if x is in (0, 0.2), the die shows 1; if x is in [0.2, 0.34), the die shows 2, and so on. A pseudocode to count the outcome of 5000 throws of this die and compute the probability might be written as follows:

Loaded_Die **Pseudocode**

```
program Loaded_Die
integer i, j;     real array (y_i)1:6, (m_i)1:6, (r_i)1:n
real n ← 5000
(y_i)6 ← (0.2, 0.34, 0.56, 0.72, 0.89, 1.0)
(m_i)6 ← (0.0, 0.0, 0.0, 0.0, 0.0, 0.0)
call Random((r_i))
for i = 1 to n
    for j = 1 to 6
        if r_i < y_j then
            m_j ← m_j + 1
            exit loop j
        end if
    end for
end for
output real(m_i)/real(n)
end program Loaded_Die
```

The results are 0.2024, 0.1344, 0.2252, 0.1600, 0.1734, and 0.1046, which are reasonable approximations to the probabilities in the table.

Birthday Problem

Birthday Problem

An interesting problem that can be solved by using simulation is the famous **birthday problem**. Suppose that in a room of n people, each of the 365 days of the year is equally likely to be someone's birthday. From probability theory, it can be shown that, contrary to intuition, only 23 people need be present for the chances to be better than fifty-fifty that at least two of them will have the same birthday! (It is always fun to try this experiment at a large party or in class to see it work in practice.)

Many people are curious about the theoretical reasoning behind this result, so we discuss it briefly before solving the simulation problem. After someone is asked his or her birthday, the chances that the next person asked will not have the same birthday are 364/365. The chances that the third person's birthday will not match those of the first two people are 363/365. The chances of two successive independent events occurring is the product of the probability of the separate events. (The sequential nature of the explanation does not imply that the events are dependent.) In general, the probability that the nth person asked will have a birthday different from that of anyone who has already been asked is

$$\left(\frac{365}{365}\right)\left(\frac{364}{365}\right)\left(\frac{363}{365}\right)\cdots\left(\frac{365-(n-1)}{365}\right)$$

The probability that the nth person asked will provide a match is 1 minus this value. A table of the quantity $1 - (365)(364)\cdots[365-(n-1)]/365^n$ shows that with 23 people, the chances are 50.7%; with 55 or more people, the chances are 98.6% or almost theoretically certain that at least two out of 55 people will have the same birthday. (See Table 10.1, p. 500.)

Without using probability theory, we can write a routine that uses the random-number generator to compute the approximate chances for groups of n people. Clearly, all that is needed is to select n random integers from the set $\{1, 2, 3, \ldots, 365\}$ and to examine them in some way to determine whether there is a match. By repeating this experiment a large number of times, we can compute the probability of at least one match in any gathering of n people.

TABLE 10.1 Birthday Problem

n	Theoretical	Simulation
5	0.027	0.028
10	0.117	0.110
15	0.253	0.255
20	0.411	0.412
22	0.476	0.462
23	0.507	0.520
25	0.569	0.553
30	0.706	0.692
35	0.814	0.819
40	0.891	0.885
45	0.941	0.936
50	0.970	0.977
55	0.986	0.987

One way of writing a routine for simulating the birthday problem follows. In it we use the approach of checking off days on a calendar to find out whether there is a match. Of course, there are many other ways of approaching this problem.

Function procedure *Probably* calculates the probability of repeated birthdays:

***Probably* Pseudocode**

```
real function Probably(n, npts)
integer i, npts;    logical Birthday;    real sum ← 0
for i = 1 to npts
    if Birthday (n) then sum ← sum + 1
end for
Probably ← sum/real(npts)
end function Probably
```

Logical function *Birthday* generates n random numbers and compares them. It returns a value of `true` if these numbers contain at least one repetition and `false` if all n numbers are different.

***Birthday* Pseudocode**

```
logical function Birthday(n)
integer i, n, number;    logical array (days_i)_{1:365}
real array (r_i)_{1:n}
call Random((r_i))
for i = 1 to 365
    days(i) ← false
end for
Birthday ← false
for i = 1 to n
    number ← integer (365r_i + 1)
    if days(number) then
        Birthday ← true
        exit loop i
```

(Continued)

```
        end if
        days(number) ← true
    end for
    end function Birthday
```

The results of the theoretical calculations and the simulation are given in Table 10.1.

Buffon's Needle Problem

Buffon's Needle Problem

The next example of a simulation is a very old problem known as **Buffon's needle problem**. Imagine that a needle of unit length is dropped onto a sheet of paper ruled by parallel lines 1 unit apart. What is the probability that the needle intersects one of the lines?

To make the problem precise, assume that the center of the needle lands between the lines at a random point. Assume further that the angular orientation of the needle is another random variable. Finally, assume that our random variables are drawn from a uniform distribution. Figure 10.8 shows the geometry of the situation.

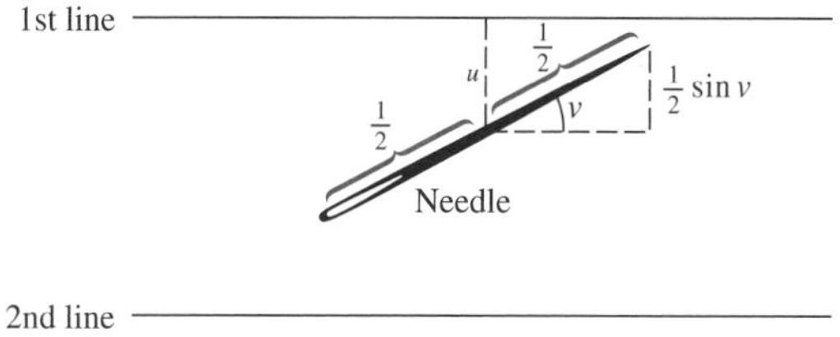

FIGURE 10.8 Buffon's needle problem

Let the distance of the center of the needle from the nearer of the two lines be u, and let the angle from the horizontal be v. Here, u and v are the two random variables. The needle intersects one of the lines if and only if $u \leqq \frac{1}{2}\sin v$. We perform the experiment many times, say, 5000. Because of the problem's symmetry, we select u from a uniform random distribution on the interval $\left(0, \frac{1}{2}\right)$ and v from a uniform random distribution on the interval $(0, \pi/2)$, and we determine the number of times that $2u \leqq \sin v$. We let $w = 2u$ and test $w \leqq \sin v$, where w is a random variable in $(0, 1)$. In this program, intermediate answers are printed out so that their progression can be observed. Also, the theoretical answer, $t = 2/\pi \approx 0.63662$, is printed for comparison.

***Needle* Pseudocode**

```
program Needle
integer i, m;   real prob, v, w;   real array (r_ij)_{1:n×1:2}
integer n ← 5000, iprt ← 1000
m ← 0
call Random((r_ij))
for i = 1 to n
    w ← r_i1
    v ← (π/2)r_i,2
    if w ≦ sin v then m ← m + 1
    if mod(i, iprt) = 0 then
        prob ← real(m)/real(i)
        output i, prob, (2/π)
    end if
end for
end program Needle
```

Two Dice Problem

Our next example also has an analytic solution. This is advantageous for us because we wish to compare the results of Monte Carlo simulations with theoretical solutions. Consider the experiment of tossing two dice. For an (unloaded) die, the numbers 1, 2, 3, 4, 5, and 6 are equally likely to occur. We ask: *What is the probability of throwing a* 12 *(i.e.,* 6 *appearing on each die) in* 24 *throws of the dice*?

Two Dice Problem

There are six possible outcomes from each die for a total of 36 possible combinations. Only one of these combinations is a double 6, so 35 out of the 36 combinations are not correct. With 24 throws, we have $(35/36)^{24}$ as the probability of a wrong outcome. Hence, $1 - (35/36)^{24} = 0.49140$ is the answer. Not all problems of this type can be analyzed like this, so we model the situation using a random-number generator.

If we simulate this process, a single experiment consists of throwing the dice 24 times, and this experiment must be repeated a large number of times, say, 1000. For the outcome of the throw of a single die, we need random integers that are uniformly distributed in the set $\{1, 2, 3, 4, 5, 6\}$. If x is a random variable in (0, 1), then $6x + 1$ is a random variable in $(1, 7)$, and the integer part is a random integer in $\{1, 2, 3, 4, 5, 6\}$. Here is a pseudocode:

***Two_Dice* Pseudocode**

```
program Two_Dice
integer i, j, i1, i2, m;   real prob;   real array (r_ijk)_{1:n×1:24×1:2}
integer n ← 5000, iprt ← 1000
call Random((r_ijk))
m ← 0
for i = 1 to n
    for j = 1 to 24
        i1 ← integer(6r_ij1 + 1)
        i2 ← integer(6r_ij2 + 1)
        if i1 + i2 = 12 then
            m ← m + 1
            exit loop j
        end if
    end for
    if mod(i, 1000) = 0 then
        prob ← real(m)/real(i)
        output i, prob
    end if
end for
end program Two_Dice
```

This program computes the probability of throwing a 12 in 24 throws of the dice at approximately *even money*—that is, 0.487.

Neutron Shielding Problem

Neutron Shielding Problem

Our final example concerns neutron shielding. We take a simple model of neutrons penetrating a lead wall. It is assumed that each neutron enters the lead wall at a right angle to the wall and travels a unit distance. Then it collides with a lead atom and rebounds in a random direction. Again, it travels a unit distance before colliding with another lead atom.

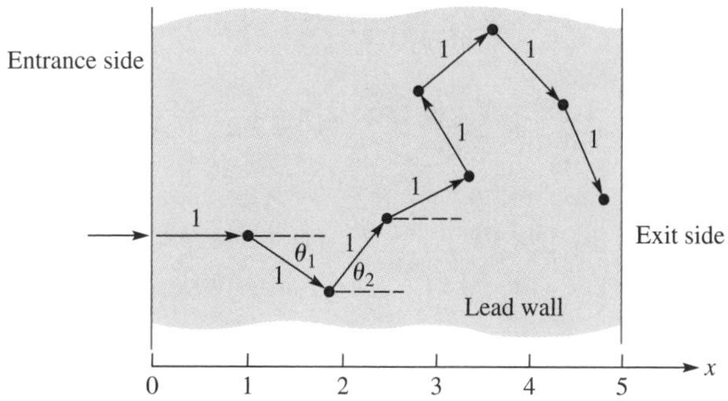

FIGURE 10.9
Neutron-shielding experiment

It rebounds in a random direction and so on. Assume that after eight collisions, all the neutron's energy is spent. Assume also that the lead wall is 5 units thick in the x direction and for all practical purposes infinitely thick in the y direction. The question is: *What percentage of neutrons can be expected to emerge from the other side of the lead wall?* (See Figure 10.9.)

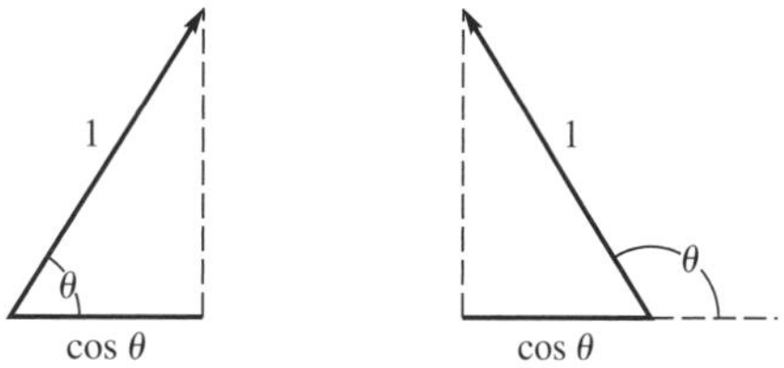

FIGURE 10.10
Right triangles with hypotenuse 1

Let x be the distance measured from the initial surface where the neutron enters. From trigonometry, we recall that in a right triangle with hypotenuse 1, one side is $\cos\theta$. Also note that $\cos\theta \leqq 0$ when $\pi/2 \leqq \theta \leqq \pi$ (see Figure 10.10). The first collision occurs at a point where $x = 1$. The second occurs at a point where $x = 1 + \cos\theta_1$. The third collision occurs at a point where $x = 1 + \cos\theta_1 + \cos\theta_2$, and so on. If $x \geqq 5$, the neutron has exited. If $x < 5$ for all eight collisions, the wall has shielded the area from that particular neutron. For a Monte Carlo simulation, we can use random angles θ_i in the interval $(0, \pi)$ because of symmetry. The simulation program then follows:

Shielding **Pseudocode**

```
program Shielding
integer i, j, m;   real x, per;   real array (r_ij)_{1:n×1:7}
integer n ← 5000, iprt ← 1000
m ← 0
call Random((r_ij))
for i = 1 to n
    x ← 1
    for j = 1 to 7
        x ← x + cos(π r_ij)
        if x ≦ 0 then exit loop j
        if x ≧ 5 then
            m ← m + 1
            exit loop j
        end if
    end for
```

(Continued)

```
        if mod(i, iprt) = 0 then
            per ← 100 real(m)/real(i)
            output i, per
        end if
    end for
end program Shielding
```

After running this program, we can say that approximately 1.85% of the neutrons can be expected to emerge from the lead wall.

Summary 10.3

- Random number generators are used in the **simulation** of a physical situation in which an element of chance is present. Statistical conclusions can be drawn if the numerical experiment is performed many times.

Computer Exercises 10.3

[a]**1.** A point (a, b) is chosen at random in a rectangle defined by inequalities $|a| \leqq 1$ and $|b| \leqq 2$. What is the probability that the resulting quadratic equation $ax^2 + bx + 1 = 0$ has *real* roots? Find the answer both analytically and by the Monte Carlo method.

[a]**2.** Compute the average distance between two points in the circle $x^2+y^2 = 1$. To solve this, generate N random pairs of points (x_i, y_i) and (v_i, w_i) in the circle, and compute

$$N^{-1} \sum_{i=1}^{N} \left[(x_i - v_i)^2 + (y_i - w_i)^2\right]^{1/2}$$

3. (**French Railroad System**) Define the distance between two points (x_1, y_1) and (x_2, y_2) in the plane to be

$$\sqrt{(x_1 - x_2)^2 + (y_1 - y_2)^2}$$

if the points are on a straight line through the origin but

$$\sqrt{x_1^2 + y_1^2} + \sqrt{x_2^2 + y_2^2}$$

in all other cases. Draw a picture to illustrate. Compute the average distance between two points randomly selected in a unit circle centered at the origin.

[a]**4.** Consider a circle of radius 1. A point is chosen at random inside the circle, and a chord that has the chosen point as midpoint is drawn. What is the probability that the chord will have length greater than $\frac{3}{2}$? Solve the problem analytically and by the Monte Carlo method.

5. Two points are selected at random on the circumference of a circle. What is the average distance from the center of the circle to the center of gravity of the two points?

[a]**6.** Consider the **cardioid** given by $(x^2 + y^2 + x)^2 = (x^2 + y^2)$. Write a program to find the average distance, *staying within the cardioid*, between two points randomly selected within the figure. Use 1000 points, and print intermediate estimates.

[a]**7.** Find the length of the **lemniscate** whose equation in polar coordinates is given by $r^2 = \cos 2\theta$.
Hint: In polar coordinates, $ds^2 = dr^2 + r^2\, d\theta^2$.

8. Suppose that a die is loaded so that the six faces are not equally likely to turn up when the die is rolled. The probabilities associated with the six faces are as follows:

Outcome	1	2	3	4	5	6
Probability	0.15	0.2	0.25	0.15	0.1	0.15

Write and run a program to simulate 1500 throws of such a die.

[a]**9.** Consider a pair of loaded dice as described in the text. By a Monte Carlo simulation, determine the probability of throwing a 12 in 25 throws of the dice.

10. Consider a neutron-shielding problem similar to the one in the text but modified as follows: Imagine the neutron beam impinging on the wall 1 unit above its base. The wall can be very high. Neutrons cannot escape from the top, but they can escape from the bottom as well as from the exit side. Find the percentage of escaping neutrons.

11. Rewrite the routine(s) for the birthday problem using some other scheme for determining whether or not there is a match.

[a]**12.** Write a program to estimate the probability that three random points on the edges of a square form an obtuse triangle (see the following figure).
Hint: Use the Law of Cosines: $\cos\theta = (b^2+c^2-a^2)/2bc$.

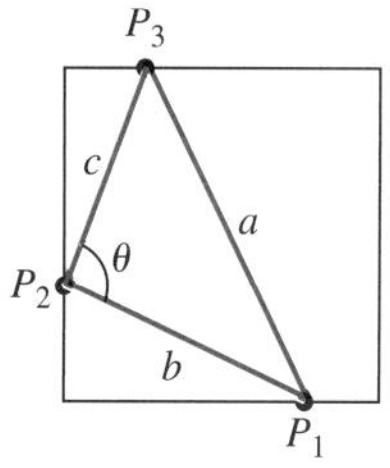

13. A **histogram** is a graphical device for displaying frequencies by means of rectangles whose heights are proportional to frequencies. For example, in throwing two dice 3600 times, the resulting sums 2, 3, ..., 12 should occur with frequencies close to those shown in the following histogram. By means of a Monte Carlo simulation, obtain a histogram for the frequency of digits 0, 1, ..., 9 that appear in 1000 random numbers.

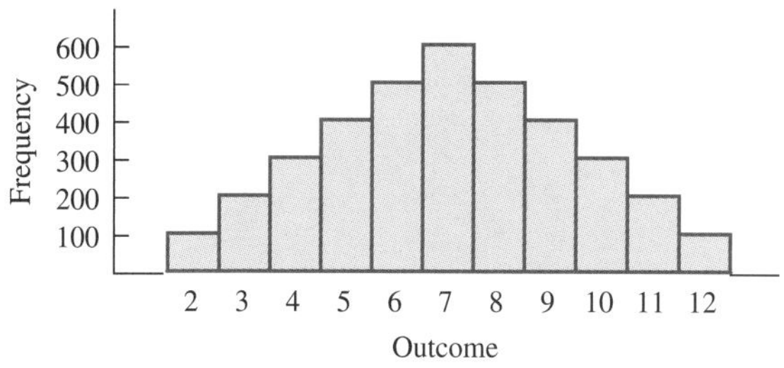

[a]**14.** Consider a circular city of diameter 20 kilometers (see the following figure). Radiating from the center are 36 straight roads, spaced 10° apart in angle. There are also 20 circular roads spaced 1 kilometer apart. What is the average distance, measured along the roads, between road intersection points in the city?

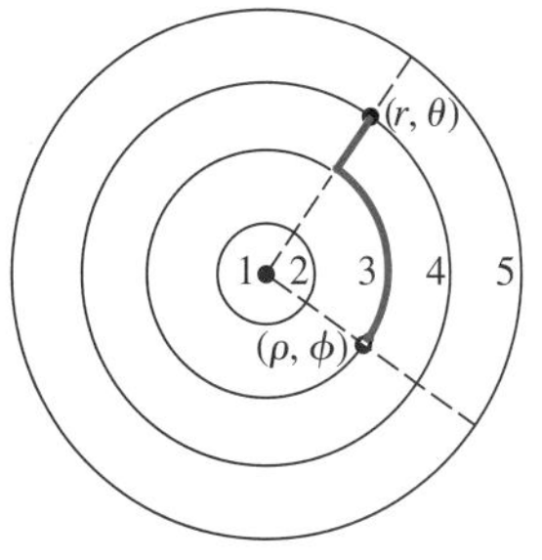

[a]**15.** A particle breaks off from a random point on a rotating flywheel. Referring to the following figure, determine the probability of the particle hitting the window. Perform a Monte Carlo simulation to compute the probability in an experimental way.

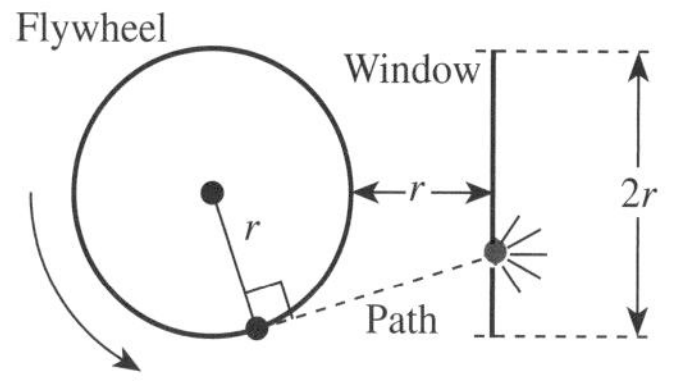

16. Write a program to simulate the following phenomenon: A particle is moving in the xy-plane under the effect of a random force. It starts at (0, 0). At the end of each second, it moves 1 unit in a random direction. We want to record in a table its position at the end of each second, taking altogether 1000 seconds.

[a]**17.** **(A Random Walk)** On a windy night, a drunkard begins walking at the origin of a two-dimensional coordinate system. His steps are 1 unit in length and are random in the following way: With probability $\frac{1}{6}$, he takes a step east; with probability $\frac{1}{4}$, he takes a step north; with probability $\frac{1}{4}$, he takes a step south; and with probability $\frac{1}{3}$, he takes a step west. What is the probability that after 50 steps, he will be more than 20 units distant from the origin? Write a program to simulate this problem.

18. **(Another Random Walk)** Consider the lattice points (points with integer coordinates) in the square $0 \leqq x \leqq 6$, $0 \leqq y \leqq 6$. A particle starts at the point (4, 4) and moves in the following way: At each step, it moves with equal probability to one of the four adjacent lattice points. What is the probability that when the particle first crosses the boundary of the square, it crosses the bottom side? Use Monte Carlo simulation.

19. What is the probability that within 20 generations, the Kzovck family name will die out? Use the following data: In the first generation, there is only one male Kzovck. In each succeeding generation, the probability that a male Kzovck will have exactly one male offspring is 4/11, the probability that he will have exactly two is 1/11, and the probability that he will have more than two is 0.

20. Write a program that simulates the random shuffle of a deck of 52 cards.

[a]**21.** A merry-go-round with a total of 24 horses allows children to jump on at three gates and jump off at only one gate while it continues to turn slowly. If the children get on and off randomly (at most one per gate), how many revolutions go by before someone must wait longer than

one revolution to ride? Assume a probability of $\frac{1}{2}$ that a child gets on or off.

22. Run the programs given in this section, and determine whether the results are reasonable.

^{a}**23.** In the unit cube $\{(x, y, z): 0 \leq x \leq 1,\ 0 \leq y \leq 1,\ 0 \leq z \leq 1\}$, if two points are randomly chosen, then what is the expected distance between them?

24. The lattice points in the plane are defined as those points whose coordinates are integers. A circle of diameter 1.5 is dropped on the plane in such a way that its center is a uniformly distributed random point in the square $0 \leq x \leq 1$, $0 \leq y \leq 1$. What is the probability that two or more lattice points lie inside the circle? Use the Monte Carlo simulation to compute an approximate answer.

25. Write a program to simulate a traffic flow problem similar to the one in the example that begins this chapter.

26. Can you modify and rerun the programs in this section so that large arrays are not used?

27. (**Student Research Project**) In their paper *Trailing the Dovetail Shuffle to Its Lair*, Bayer and Diaconis [1992] show that it takes seven **riffle shuffles** to randomize a deck of cards. Greenbaum [2002] uses this as an example of the application of polynomial numerical hulls of various degrees associated with the probability transition matrix. This is the cutoff phenomenon that is often observed in Markov processes.*

To find out more about these topics, do an online search for **rising sequences in card shuffling**.

Investigate some of the following questions:

- *How many times do we need to shuffle a deck of cards before the order of the cards is sufficiently random?*
- *Is there some minimum number of shuffles required to ensure the deck is not ordered or not predictable?*
- *Is there a point where continued shuffling no longer helps make the deck less predictable?*

---

*__Markov chains__ can be used to model the behavior of a system that depends only on its previous state. Markov chains involve a **transition matrix** $\boldsymbol{P} = (P_{ij})$, where the entries are the probability of going from state j to state i.

11 Boundary-Value Problems

In the design of pivots and bearings, the mechanical engineer encounters the following problem: The cross section of a pivot is determined by a curve $y = y(x)$ that must pass through two fixed points, $(0, 1)$ and $(1, a)$, as in Figure 11.1. Moreover, for optimal performance (principally low friction), the unknown function must minimize the value of a certain integral

$$\int_0^1 [y(y')^2 + b(x)y^2]\,dx$$

in which $b(x)$ is a known function. From this, it is possible to obtain a second-order differential equation (the so-called **Euler equation**) for y. The differential equation with its initial and terminal values is

$$\begin{cases} -(y')^2 - 2b(x)y + 2yy'' = 0 \\ y(0) = 1 \quad y(1) = a \end{cases}$$

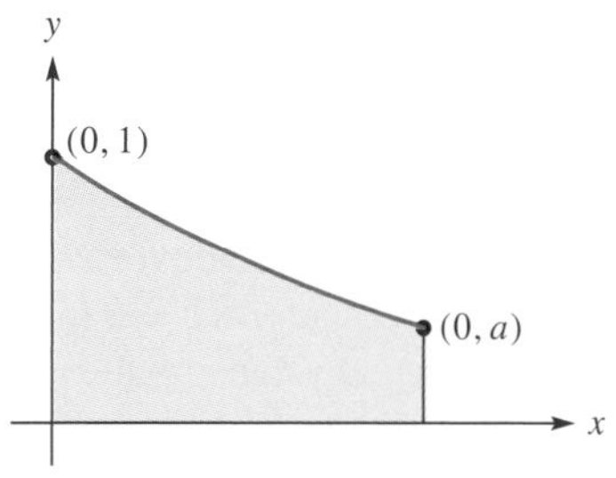

FIGURE 11.1
Pivot cross section

This is a nonlinear two-point boundary-value problem, and methods for solving it numerically are discussed in this chapter.

11.1 Shooting Method

Introduction

Boundary-Value Problem

In Chapter 7, we dealt with the initial-value problem for ordinary differential equations, but now we consider another type of numerical problem involving ordinary differential equations. A **boundary-value problem** is exemplified by a second-order ordinary differential equation whose solution function is prescribed at the endpoints of the interval of interest. An instance of such a problem is

$$\begin{cases} x'' = -x \\ x(0) = 1 \quad x\left(\frac{\pi}{2}\right) = -3 \end{cases}$$

Here, we have a differential equation whose general solution involves two arbitrary parameters. To specify a particular solution, two conditions must be given. If this were an initial-value problem, x and x' would be specified at some initial point. In this problem, however, we are given two points of the form $(t, x(t))$ through which the solution curve passes—namely, $(0, 1)$ and $(\pi/2, -3)$. The general solution of the differential equation is $x(t) = c_1 \sin(t) + c_2 \cos(t)$, and the two conditions (known as **boundary values**) enable us to determine that $c_1 = -3$ and $c_2 = 1$.

Now suppose that we have a similar problem in which we are unable to determine the general solution as above. We take as our model the **boundary-value problem (BVP)**

Model BVP

$$\begin{cases} x''(t) = f(t, x(t), x'(t)) \\ x(a) = \alpha, \quad x(b) = \beta \end{cases} \tag{1}$$

A step-by-step numerical solution of Problem (1) by the methods of Chapter 7 requires two initial conditions, but in Problem (1) only one condition is present at $t = a$. This fact makes a problem like (1) considerably more difficult than an initial-value problem. Several ways to attack it are considered in this chapter. Existence and uniqueness theorems for solutions of two-point boundary-value problems can be found in Keller [1976].

Shooting Method

One way to proceed in solving Problem (1) is to guess $x'(a)$, then carry out the solution of the resulting initial-value problem as far as b, and hope that the computed solution agrees with β; that is, $x(b) = \beta$. If it does not (which is quite likely), we can go back and change our guess for $x'(a)$. Thus, we repeat this procedure and learn something from the various trials until we hit the target β. There are systematic ways of utilizing this information, and the resulting method is known by the nickname **shooting**.

We observe that the final value $x(b)$ of the solution of our initial-value problem depends on the guess that was made for $x'(a)$. Everything else remains fixed in this problem. Thus, the differential equation $x'' = f(t, x, x')$ and the first initial value, $x(a) = \alpha$, do not change. If we assign a real value z to the missing initial condition,

$$x'(a) = z$$

$\varphi(z)$

then the initial-value problem can be solved numerically. The value of x at b is now a function of z, which we denote by $\varphi(z)$. In other words, for each choice of z, we obtain a new value for $x(b)$, and φ is the name of the function with this behavior. We know very little about $\varphi(z)$, but we can compute or evaluate it. It is, however, an *expensive* function to evaluate because each value of $\varphi(z)$ is obtained only after solving an initial-value problem!

It should be emphasized that the shooting method combines *any* algorithm for the initial-value problem with *any* algorithm for finding a zero of a function. The choice of these two algorithms should reflect the nature of the problem being solved.

The basic idea of the shooting method is illustrated in Figure 11.2. The solution curves are shown as well as two paths using different initial slopes. The goal is to keep adjusting the initial aim with each attempt.

Shooting Method Algorithm

To summarize, a function $\varphi(z)$ is computed as follows: Solve the **initial-value problem (IVP)**

Model IVP

$$\begin{cases} x'' = f(t, x(t), x'(t)) \\ x(a) = \alpha, \quad x'(a) = z \end{cases}$$

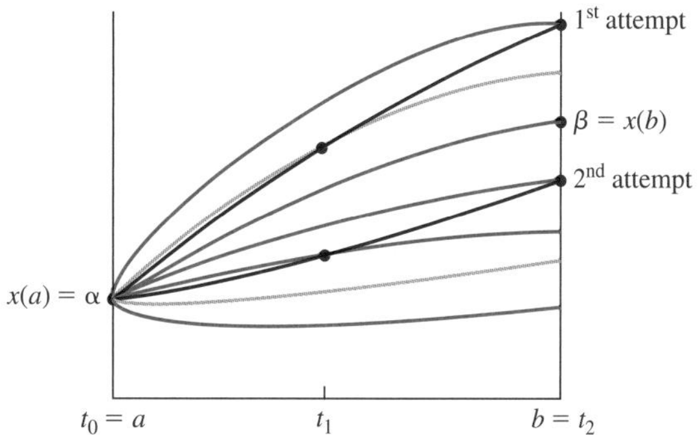

FIGURE 11.2 Shooting method illustrated

on the interval $[a, b]$. Let

$$\varphi(z) = x(b)$$

Our objective is to adjust z until we find a value for which

Linear Interpolation Strategy

$$\varphi(z) = \beta$$

One way to do so is to use linear interpolation between $\varphi(z_1)$ and $\varphi(z_2)$, where z_1 and z_2 are two guesses for the initial condition $x'(a)$. That is, given two values of φ, we pretend that φ is a **linear function** and determine an appropriate value of z based on this hypothesis. A sketch of the values of z versus $\varphi(z)$ might look like Figure 11.3. The strategy just outlined is the same as that used in secant method for finding a zero of $\varphi(z) - \beta$.

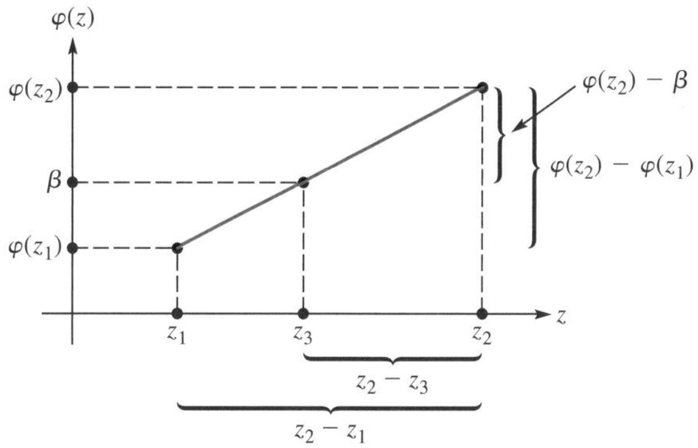

FIGURE 11.3 φ linear function

To obtain an estimating formula for the next value z_3, we compute $\varphi(z_1)$ and $\varphi(z_2)$ on the basis of values z_1 and z_2, respectively.

By considering similar triangles, we have

$$\frac{z_3 - z_2}{\beta - \varphi(z_2)} = \frac{z_2 - z_1}{\varphi(z_2) - \varphi(z_1)}$$

from which

$$z_3 = z_2 + [\beta - \varphi(z_2)]\left[\frac{z_2 - z_1}{\varphi(z_2) - \varphi(z_1)}\right]$$

We can repeat this process and generate the sequence

Recurrence z_{n+1} Equation

$$z_{n+1} = z_n + [\beta - \varphi(z_n)]\left[\frac{z_n - z_{n-1}}{\varphi(z_n) - \varphi(z_{n-1})}\right] \quad (n \geqq 2) \tag{2}$$

all based on two starting values z_1 and z_2.

This procedure for solving the **two-point boundary-value problem**

BVP

$$\begin{cases} x'' = f(t, x, x') \\ x(a) = \alpha, \quad x(b) = \beta \end{cases} \tag{3}$$

is then as follows: Solve the **initial-value problem**

IVP

$$\begin{cases} x'' = f(t, x, x') \\ x(a) = \alpha, \quad x'(a) = z \end{cases} \tag{4}$$

from $t = a$ to $t = b$, letting the value of the solution at b be denoted by $\varphi(z)$. Do this twice with two different values of z, say, z_1 and z_2, and compute $\varphi(z_1)$ and $\varphi(z_2)$. Now calculate a new z, called z_3, by Formula (2). Then compute $\varphi(z_3)$ by again solving (4). Obtain z_4 from z_2 and z_3 in the same way, and so on. Monitor

$$\varphi(z_{n+1}) - \beta$$

to see whether progress is being made. When it is satisfactorily small, stop. This process is called a **shooting method**. Note that the numerically obtained values $x(t_i)$ for $a \leqq t_i \leqq b$ must be saved until better ones are obtained (that is, one whose terminal value $x(b)$ is closer to β than the present one) because the objective in solving Problem (3) is to obtain values $x(t)$ for values of t between a and b.

Step Size *h*

The shooting method may be very time-consuming if each solution of the associated initial-value problem involves a small value for the step size h. Consequently, we use a relatively large value of h until $|\varphi(z_{n+1}) - \beta|$ is sufficiently small and then reduce h to obtain the required accuracy.

EXAMPLE 1 What is the function φ for this two-point boundary-value problem?

$$\begin{cases} x'' = x \\ x(0) = 1, \quad x(1) = 7 \end{cases}$$

Solution The general solution of the differential equation is $x(t) = c_1 e^t + c_2 e^{-t}$. The solution of the initial-value problem

$$\begin{cases} x'' = x \\ x(0) = 1, \quad x'(0) = z \end{cases}$$

is $x(t) = \frac{1}{2}(1 + z)e^t + \frac{1}{2}(1 - z)e^{-t}$. Therefore, we have

$$\varphi(z) = x(1) = \frac{1}{2}(1 + z)e + \frac{1}{2}(1 - z)e^{-1}$$

■

Modifications and Refinements

Many modifications and refinements are possible. For instance, when $\varphi(z_{n+1})$ is near β, one can use *higher-order interpolation* formulas to estimate successive values of z_i. Suppose, for example, that instead of utilizing two values $\varphi(z_1)$ and $\varphi(z_2)$ to obtain z_3, we utilize the four values

$$\varphi(z_1), \quad \varphi(z_2), \quad \varphi(z_3), \quad \varphi(z_4)$$

to estimate z_5. We could set up a cubic interpolating polynomial p_3 for the data

Cubic Interpolation

z_1	z_2	z_3	z_4
$\varphi(z_1)$	$\varphi(z_2)$	$\varphi(z_3)$	$\varphi(z_4)$

(5)

and solve

$$p_3(z_5) = \beta$$

for z_5. Since p_3 is a cubic, this would entail some additional work. A better way may be to set up a polynomial $\widehat{p}_3$ to interpolate the data

Inverse Interpolation

$\varphi(z_1)$	$\varphi(z_2)$	$\varphi(z_3)$	$\varphi(z_4)$
z_1	z_2	z_3	z_4

(6)

and then use $\widehat{p}_3(\beta)$ as the estimate for z_5. This procedure is known as **inverse interpolation**. (See Section 4.1, p. 169.)

Further remarks on the shooting method are made in Section 11.2 after the discussion of an alternative procedure.

Summary 11.1

- A generic **two-point boundary-value problem** on the interval $[a, b]$ is

$$\begin{cases} x'' = f(t, x, x') \\ x(a) = \alpha, \quad x(b) = \beta \end{cases}$$

There is a related **initial-value problem**

$$\begin{cases} x'' = f(t, x, x') \\ x(a) = \alpha, \quad x'(a) = z \end{cases}$$

We hope to find a value of z so that the computed solution to the initial-value problem will be the solution of the two-point boundary-value problem. We define a function $\varphi(z)$ whose value is the computed solution of the initial-value problem at $t = b$, namely, $\varphi(z) = x(b)$, where x solves the initial-value problem. We repeatedly adjust z until we find a value for which $\varphi(z) = \beta$. If z_1 and z_2 are two guesses for the initial condition $x'(a)$, we can use linear interpolation between $\varphi(z_1)$ and $\varphi(z_2)$ to find an improved value for z. This is done by solving the initial-value problem twice with z_1 and z_2 and thereby compute $\varphi(z_1)$ and $\varphi(z_2)$. We calculate a new z_3 using

$$z_{n+1} = z_n + [\beta - \varphi(z_n)]\left[\frac{z_n - z_{n-1}}{\varphi(z_n) - \varphi(z_{n-1})}\right] \qquad (n \geqq 2)$$

and compute $\varphi(z_{n+1})$ by again solving the initial-value problem. We monitor $\varphi(z_{n+1}) - \beta$ until it is satisfactorily small and then stop. This is called the **shooting method**.

- Improvements and refinements to the shooting method involve using **cubic polynomial interpolation** or **inverse interpolation**.

Exercises 11.1

1. Verify that $x = (2t + 1)e^t$ is the solution to each of the following problems:

$$\begin{cases} x'' = x + 4e^t \\ x(0) = 1, \quad x\left(\frac{1}{2}\right) = 2e^{1/2} \end{cases}$$

$$\begin{cases} x'' = x' + x - (2t - 1)e^t \\ x(1) = 3e, \quad x(2) = 5e^2 \end{cases}$$

[a]2. Verify that $x = c_1 e^t + c_2 e^{-t}$ solves the boundary-value problem

$$\begin{cases} x'' = x \\ x(0) = 1, \quad x(1) = 2 \end{cases}$$

if appropriate values of c_1 and c_2 are chosen.

3. Solve these boundary value problems by adjusting the general solution of the differential equation.

[a]**a.** $x'' = x, \quad x(0) = 0, \quad x(\pi) = 1$

[a]**b.** $x'' = t^2, \quad x(0) = 1, \quad x(1) = -1$

4.[a]**a.** Determine all pairs (α, β) for which the problem

$$\begin{cases} x'' = -x \\ x(0) = \alpha, \quad x\left(\frac{\pi}{2}\right) = \beta \end{cases}$$

has a solution.

[a]**b.** Repeat part **a** for $x(0) = \alpha$ and $x(\pi) = \beta$.

5. **a.** Verify the following algorithm for the inverse interpolation technique suggested in the text. Here we have set $\varphi_i = \varphi(z_i)$.

$$\begin{cases} u = \dfrac{z_2 - z_1}{\varphi_2 - \varphi_1}, \quad v = \dfrac{s - u}{\varphi_3 - \varphi_1}, \quad s = \dfrac{z_3 - z_2}{\varphi_3 - \varphi_2} \\ r = \dfrac{e - s}{\varphi_4 - \varphi_2}, \quad e = \dfrac{z_4 - z_3}{\varphi_4 - \varphi_3}, \quad w = \dfrac{r - v}{\varphi_4 - \varphi_1} \end{cases}$$

$$z_5 = z_1 + (\beta - \varphi_1)\{u + (\beta - \varphi_2)[v + w(\beta - \varphi_3)]\}$$

b. Find similar formulas for three points.

[a]6. Let $\varphi(z)$ denote $x(\pi/2)$, where x is the solution of the initial-value problem

$$\begin{cases} x'' = -x \\ x(0) = 0, \quad x'(0) = z \end{cases}$$

What is $\varphi(z)$?

[a]7. Determine the function φ explicitly in the case of this two-point boundary-value problem.

$$\begin{cases} x'' = -x \\ x(0) = 1, \quad x\left(\dfrac{\pi}{2}\right) = 3 \end{cases}$$

[a]8. (Continuation) Repeat the preceding problem for $x'' = -(x')^2/x$ with $x(1) = 3$ and $x(2) = 5$. Using your result, solve the boundary-value problem.
Hint: The general solution of the differential equation is $x(t) = c_1\sqrt{c_2 + t}$.

[a]9. Determine the function φ explicitly in the case of this two-point boundary-value problem:

$$\begin{cases} x'' = x \\ x(-1) = e, \quad x'(1) = \frac{1}{2}e \end{cases}$$

[a]10. Boundary-value problems may involve differential equations of order higher than 2. For example,

$$\begin{cases} x''' = f(t, x, x', x'') \\ x(a) = \alpha, \quad x'(a) = \gamma, \quad x(b) = \beta \end{cases}$$

Discuss the ways in which this problem can be solved using the shooting method.

[a]11. Solve analytically this three-point boundary-value problem:

$$\begin{cases} x''' = -e^t + 4(t + 1)^{-3} \\ x(0) = -1, \quad x(1) = 3 - e + 2\ln 2 \\ x(2) = 6 - e^2 + 2\ln 3 \end{cases}$$

12. Solve

$$\begin{cases} x'' = -x \\ x(0) = 2, \quad x(\pi) = 3 \end{cases}$$

analytically and analyze any difficulties.

13. Show that the following two problems are equivalent in the sense that a solution of one is easily obtained from a solution of the other:

$$\begin{cases} y'' = f(t, y) \\ y(0) = \alpha, \quad y(1) = \beta \end{cases}$$

$$\begin{cases} z'' = f(t, z + \alpha - \alpha t + \beta t) \\ z(0) = 0, \quad z(1) = 0 \end{cases}$$

14. Discuss in general terms the numerical solution of the following two-point boundary-value problems. Recommend specific methods for each, being sure to take advantage of any special structure.

[a]**a.** $\begin{cases} x'' = \sin t + \left(e^t\sqrt{t^2+1}\right)x + (\cos t)x' \\ x(0) = 0, \quad x(1) = 5 \end{cases}$

b. $\begin{cases} x_1' = x_1^2 + (t-3)x_1 + \sin t \\ x_2' = x_2^3 + \sqrt{t^2+1} + (\cos t)x_1 \\ x_1(0) = 1, \quad x_2(2) = 3 \end{cases}$

[a]**15.** What is $\varphi(z)$ in the case of this boundary-value problem?

$$\begin{cases} x'' = -x \\ x(0) = 1, \quad x(\pi) = 3 \end{cases}$$

Explain the implications.

16. Find the function φ explicitly for this two-point boundary-value problem:

$$\begin{cases} x'' = e^{-2t} - 4x - 4x' \\ x(0) = 1, \quad x(2) = 0 \end{cases}$$

What is the initial-value problem whose solution solves the boundary-value problem?
Hint: Find a solution of the form $x(t) = q(t)e^{-2t}$, where q is a quadratic polynomial.

Computer Exercises 11.1

[a]**1.** The *nonlinear* two-point boundary-value problem

$$\begin{cases} x'' = e^x \\ x(0) = \alpha, \quad x(1) = \beta \end{cases}$$

has the closed-form solution

$$x = \ln c_1 - 2\ln\left\{\cos\left[\left(\tfrac{1}{2}c_1\right)^{1/2} t + c_2\right]\right\}$$

where c_1 and c_2 are the solutions of

$$\begin{cases} \alpha = \ln c_1 - 2\ln\cos c_2 \\ \beta = \ln c_1 - 2\ln\left\{\cos\left[\left(\tfrac{1}{2}c_1\right)^{1/2} + c_2\right]\right\} \end{cases}$$

Use the shooting method to solve this problem with $\alpha = \beta = \ln 8\pi^2$. Start with $z_1 = -\frac{25}{2}$ and $z_2 = -\frac{23}{2}$. Determine c_1 and c_2 so that a comparison with the true solution can be made.
Remark: The corresponding discretization method, as discussed in the next section, involves a system of nonlinear equations with no closed-form solution.

2. Write a program to solve the example that begins this chapter (p. 507) for specific a and $b(x)$, such as $a = \frac{1}{4}$ and $b(x) = x^2$.

11.2 A Discretization Method

Finite-Difference Approximations

We turn now to a completely different approach to solving the two-point boundary-value problem—one based on a direct **discretization** of the differential equation. The problem that we want to solve is

Model BVP

$$\begin{cases} x'' = f(t, x, x') \\ x(a) = \alpha, \quad x(b) = \beta \end{cases} \tag{1}$$

Select a set of equally spaced points $t_0, t_1, \ldots, t_n$ on the interval $[a, b]$ by letting

$$t_i = a + ih, \quad \text{with} \quad h = \frac{b-a}{n} \quad (0 \leqq i \leqq n)$$

Approximate Derivatives Next, approximate the derivatives, using the standard central difference formulas (5) and (20) from Section 4.3:

Central Difference Formulas

$$x'(t) \approx \frac{1}{2h}[x(t+h) - x(t-h)]$$

$$x''(t) \approx \frac{1}{h^2}[x(t+h) - 2x(t) + x(t-h)] \tag{2}$$

The approximate value of $x(t_i)$ is denoted by x_i. Hence, the problem becomes

$$\begin{cases} x_0 = \alpha \\ \dfrac{1}{h^2}(x_{i-1} - 2x_i + x_{i+1}) = f\left(t_i, x_i, \dfrac{1}{2h}(x_{i+1} - x_{i-1})\right) & (1 \leqq i \leqq n-1) \\ x_n = \beta \end{cases} \tag{3}$$

Usually, this is a **nonlinear system of equations** in the $n - 1$ unknowns $x_1, x_2, \dots, x_{n-1}$, because f generally involves the x_i's in a nonlinear way. The solution of such a system is seldom easy, but could be approached by using the methods of Chapter 3.

The Linear Case

In some cases, System (3) *is* linear. This situation occurs exactly when f in Equation (1) has the form

$$f(t, x, x') = u(t) + v(t)x + w(t)x' \tag{4}$$

In this special case, the principal equation in System (3) looks like this:

$$\frac{1}{h^2}(x_{i-1} - 2x_i + x_{i+1}) = u(t_i) + v(t_i)x_i + w(t_i)\left[\frac{1}{2h}(x_{i+1} - x_{i-1})\right]$$

or, equivalently,

$$-\left(1 + \frac{h}{2}w_i\right)x_{i-1} + \left(2 + h^2 v_i\right)x_i - \left(1 - \frac{h}{2}w_i\right)x_{i+1} = -h^2 u_i \tag{5}$$

where $u_i = u(t_i)$, $v_i = v(t_i)$, and $w_i = w(t_i)$. Now let

$$\begin{cases} a_i = -\left(1 + \dfrac{h}{2}w_i\right) \\ d_i = 2 + h^2 v_i \\ c_i = -\left(1 - \dfrac{h}{2}w_i\right) \\ b_i = -h^2 u_i \end{cases} \qquad (1 \leqq i \leqq n-1)$$

Then the principal Equation (5) becomes

Typical Equation

$$a_i x_{i-1} + d_i x_i + c_i x_{i+1} = b_i$$

The equations corresponding to $i = 1$ and $i = n - 1$ are different because we know x_0 and x_n. The system can therefore be written as

$$\begin{cases} d_1 x_1 + c_1 x_2 = b_1 - a_1\alpha \\ a_i x_{i-1} + d_i x_i + c_i x_{i+1} = b_i & (2 \leqq i \leqq n-2) \\ a_{n-1}x_{n-2} + d_{n-1}x_{n-1} = b_{n-1} - c_{n-1}\beta \end{cases} \tag{6}$$

In matrix form, System (6) looks like this:

Tridiagonal System

$$\begin{bmatrix} d_1 & c_1 & & & & \\ a_2 & d_2 & c_2 & & & \\ & a_3 & d_3 & c_3 & & \\ & & \ddots & \ddots & \ddots & \\ & & & a_{n-2} & d_{n-2} & c_{n-2} \\ & & & & a_{n-1} & d_{n-1} \end{bmatrix} \begin{bmatrix} x_1 \\ x_2 \\ x_3 \\ \vdots \\ x_{n-2} \\ x_{n-1} \end{bmatrix} = \begin{bmatrix} b_1 - a_1\alpha \\ b_2 \\ b_3 \\ \vdots \\ b_{n-2} \\ b_{n-1} - c_{n-1}\beta \end{bmatrix}$$

Since this system is tridiagonal, we can attempt to solve it with the special procedure *Tri* for tridiagonal systems developed in Section 2.3. That procedure does not include pivoting and may fail in some cases; however, procedure *Gauss* with scaled partial pivoting should succeed. (See Exercise 11.2.5.)

Numerical Example and Pseudocode

The ideas just explained are now used to write a program for a specific test case. The problem is of the form (1) with f a *linear* function as in Equation (4):

Sample BVP

$$\begin{cases} x'' = e^t - 3\sin(t) + x' - x \\ x(1) = 1.09737\,491, \quad x(2) = 8.63749\,661 \end{cases} \tag{7}$$

The solution, known in advance to be $x(t) = e^t - 3\cos(t)$, can be used to check the computer solution. We use the discretization technique described earlier and procedure *Tri* for solving the resulting linear system.

First, we decide to use 100 points, including endpoints $a = 1$ and $b = 2$. Thus, $n = 99$, $h = \frac{1}{99}$, and $t_i = 1 + ih$ for $0 \leqq i \leqq 99$. Then we have $t_0 = 1$, $x_0 = x_0(t_0) = 1.09737\,491$, $t_{99} = 2$, and $x_{99} = x(t_{99}) = 8.63749\,661$. The unknowns in our problem are the remaining values of x_i, namely, $x_1, x_2, \ldots, x_{98}$. By the discretization of the derivatives using the central difference Formula (2), we obtain a linear system of type (3). Our principal equation is of the form (5) and is

$$-\left(1 + \frac{h}{2}\right)x_{i-1} + (2 - h^2)x_i - \left(1 - \frac{h}{2}\right)x_{i+1} = -h^2\left[e^{t_i} - 3\sin(t_i)\right]$$

since $u(t) = e^t - 3\sin t$, $v(t) = -1$, and $w(t) = 1$.

Pseudocode

We generalize the pseudocode so that with only a few changes, it can accommodate any two-point boundary value problem of type (1) with the right-hand side of form (4). Here, $u(x)$, $v(x)$, and $w(x)$ are statement functions.

```
program BVP1
integer i;  real error, h, t, u, v, w, x
real array (a_i)_1:n, (b_i)_1:n, (c_i)_1:n, (d_i)_1:n, (y_i)_1:n
integer n ← 99
real t_a ← 1, t_b ← 2, α ← 1.09737 491, β ← 8.63749 661
u(x) = e^x − 3 sin(x)
v(x) = −1
```

(Continued)

BVP1 Pseudocode

```
w(x) = 1
h ← (t_b − t_a)/n
for i = 1 to n − 1
    t ← t_a + ih
    a_i ← −[1 + (h/2)w(t)]
    d_i ← 2 + h^2 v(t)
    c_i ← −[1 − (h/2)w(t)]
    b_i ← −h^2 u(t)
end for
b_1 ← b_1 − a_1 α
b_{n−1} ← b_{n−1} − c_{n−1} β
for i = 1 to n − 1
    a_i ← a_{i+1}
end
call Tri(n − 1, (a_i), (d_i), (c_i), (b_i), (y_i))
error ← e^{t_a} − 3 cos(t_a) − α
output t_a, α, error
for i = 1 to n − 1 step 9
    t ← t_a + ih
    error ← e^t − 3 cos(t) − y_i
    output t, y_i, error
end for
error ← e^{t_b} − 3 cos(t_b) − β
output b, β, error
end program BVP1
```

The computer results are as follows:

Computer Results

t-Value	Solution	Error
1.00000 00	1.09737 49	0.00
1.09090 91	1.59203 02	-8.83×10^{-5}
1.18181 82	2.12274 17	-1.74×10^{-4}
1.27272 73	2.68980 86	-2.56×10^{-4}
1.36363 64	3.29367 04	-3.28×10^{-4}
1.45454 55	3.93494 53	-3.76×10^{-4}
1.54545 45	4.61449 10	-4.06×10^{-4}
1.63636 36	5.33343 17	-4.13×10^{-4}
1.72727 27	6.09319 59	-3.89×10^{-4}
1.81818 18	6.89557 22	-3.16×10^{-4}
1.90909 10	7.74277 78	-1.88×10^{-4}
2.00000 00	8.63749 69	0.00

Shooting Method in the Linear Case

We have just seen that this discretization method (also called a **finite-difference method**) is rather simple in the case of the linear two-point boundary-value problem:

Model BVP

$$\begin{cases} x'' = u(t) + v(t)x + w(t)x' \\ x(a) = \alpha, \quad x(b) = \beta \end{cases} \tag{8}$$

The shooting method is also especially simple in this case. Recall that the shooting method requires us to solve an initial-value problem:

Model IVP

$$\begin{cases} x'' = u(t) + v(t)x + w(t)x' \\ x(a) = \alpha, \quad x'(a) = z \end{cases} \tag{9}$$

and interpret the terminal value $x(b)$ as a function of z. We call that function φ and seek a value of z for which $\varphi(z) = \beta$. For the linear Problem (9), φ is a *linear* function of z, and so Figure 11.3 in Section 11.1 (p. 509) is actually realistic. Consequently, we need only solve Problem (9) with two values of z to determine the function precisely. To establish these facts, let us do a little more analysis.

Suppose that we have solved Problem (9) twice with particular values z_1 and z_2. Let the solutions that are so obtained be denoted by $x_1(t)$ and $x_2(t)$. Then we claim that the function

$$g(t) = \lambda x_1(t) + (1 - \lambda)x_2(t) \tag{10}$$

has properties

$$\begin{cases} g'' = u + vg + wg' \\ g(a) = \alpha \end{cases}$$

which are left to the reader to verify in Exercise 11.2.6. (The value of λ in this analysis is a constant but is completely arbitrary.)

The function g nearly solves the two-point boundary-value Problem (8), and g contains a parameter λ at our disposal. Imposing the condition $g(b) = \beta$, we obtain

$$\lambda x_1(b) + (1 - \lambda)x_2(b) = \beta$$

from which

$$\lambda = \frac{\beta - x_2(b)}{x_1(b) - x_2(b)}$$

Perhaps the simplest way to implement these ideas is to solve two initial-value problems

$$\begin{cases} x'' = u(t) + v(t)x + w(t)x' \\ x(a) = \alpha, \quad x'(a) = 0 \end{cases}$$

and

$$\begin{cases} y'' = u(t) + v(t)y + w(t)y' \\ y(a) = \alpha, \quad y'(a) = 1 \end{cases}$$

Then the solution to the original two-point boundary-value Problem (8) is

Solution

$$\lambda x(t) + (1 - \lambda)y(t), \quad \text{with} \quad \lambda = \frac{\beta - y(b)}{x(b) - y(b)} \tag{11}$$

In the computer realization of this procedure, we must save the entire solution curves x and y. They are stored in arrays (x_i) and (y_i).

Numerical Example Revisited

As an example of the shooting method, consider the problem of Equation (7). We solve the two initial-value problems

Two IVPs

$$\begin{cases} x'' = e^t - 3\sin(t) + x' - x \\ x(1) = 1.09737\,491 \\ x'(1) = 0 \end{cases} \qquad \begin{cases} y'' = e^t - 3\sin(t) + y' - y \\ y(1) = 1.09737\,491 \\ y'(1) = 1 \end{cases} \tag{12}$$

by using the fourth-order Runge-Kutta method. To do so, we introduce variables

$$x_0 = t, \quad x_1 = x, \quad x_2 = x'$$

Then the first initial-value problem is

$$\begin{bmatrix} x_0' \\ x_1' \\ x_2' \end{bmatrix} = \begin{bmatrix} 1 \\ x_2 \\ e^{x_0} - 3\sin(x_0) + x_2 - x_1 \end{bmatrix} \qquad \begin{bmatrix} x_0(1) \\ x_1(1) \\ x_2(1) \end{bmatrix} = \begin{bmatrix} 1 \\ 1.09737\,491 \\ 0 \end{bmatrix}$$

Now let

$$y_0 = t, \quad y_1 = y, \quad y_2 = y'$$

The second initial-value problem that we must solve is similar except that we modify the initial vector

$$\begin{bmatrix} y_0' \\ y_1' \\ y_2' \end{bmatrix} = \begin{bmatrix} 1 \\ y_2 \\ e^{y_0} - 3\sin(y_0) + y_2 - y_1 \end{bmatrix} \qquad \begin{bmatrix} y_0(1) \\ y_1(1) \\ y_2(1) \end{bmatrix} = \begin{bmatrix} 1 \\ 1.09737\,491 \\ 1 \end{bmatrix}$$

It is more efficient to solve these two problems together as a single system. Introducing

$$x_3 = y, \quad x_4 = y'$$

into the first system, we have

$$\begin{bmatrix} x_0' \\ x_1' \\ x_2' \\ x_3' \\ x_4' \end{bmatrix} = \begin{bmatrix} 1 \\ x_2 \\ e^{x_0} - 3\sin(x_0) + x_2 - x_1 \\ x_4 \\ e^{x_0} - 3\sin(x_0) + x_4 - x_3 \end{bmatrix} \qquad \begin{bmatrix} x_0(1) \\ x_1(1) \\ x_2(1) \\ x_3(1) \\ x_4(1) \end{bmatrix} = \begin{bmatrix} 1 \\ 1.09737\,491 \\ 0 \\ 1.09737\,491 \\ 1 \end{bmatrix}$$

Clearly, the $x_1(t)$ and $x_3(t)$ components of the solution vector at each t satisfy the first and second problems, respectively. Consequently, the solution is

$$\lambda x_1(t_i) + (1 - \lambda)x_3(t_i) \quad (1 \leqq i \leqq n - 1)$$

where

$$\lambda = \frac{8.63749\,661 - x_3(2)}{x_1(2) - x_3(2)}$$

We use 100 points as before, so $n = 99$.

Pseudocode

```
program BVP2
integer i;   real array (x_i)_{0:m}, (x1_i)_{0:n}, (x3_i)_{0:n};   real error, h, p, q, t
integer n ← 99, m ← 4
real a ← 1, b ← 2, α ← 1.09737 491, β ← 8.63749 661
x ← (1, α, 0, α, 1)
h ← (b − a)/n
for i = 1 to n
    call RK4_System2(m, h, (x_i), 1)
    (x1)_i ← x_1
    (x3)_i ← x_3
end for
```

(Continued)

BVP2 Pseudocode

$p \leftarrow [\beta - (x3)_n]/[(x1)_n - (x3)_n]$
$q \leftarrow 1 - p$
for $i = 1$ **to** n
  $(x1)_i \leftarrow p\,(x1)_i + q\,(x3)_i$
end for
$error \leftarrow e^a - 3\cos(a) - \alpha$
output $a, \alpha, error$
for $i = 9$ **to** n **step** 9
  $t \leftarrow a + ih$
  $error \leftarrow e^t - 3\cos(t) - (x1)_i$
  output $t, (x1)_i, error$
end for
end program *BVP2*

procedure *XP_System*$(m, (x_i), (f_i))$
real array $(x_i)_{0:m}, (f_i)_{0:m}$
$f_0 \leftarrow 1$
$f_1 \leftarrow x_2$
$f_2 \leftarrow e^{x_0} - 3\sin(x_0) + x_2 - x_1$
$f_3 \leftarrow x_4$
$f_4 \leftarrow e^{x_0} - 3\sin(x_0) + x_4 - x_3$
end procedure *XP_System*

Computer Results

The final computer results are as shown:

t-Value	Solution	Error
1.00000 00	1.09737 49	0.00
1.09090 91	1.59194 09	9.54×10^{-7}
1.18181 82	2.12256 57	1.91×10^{-6}
1.27272 73	2.68955 09	1.43×10^{-6}
1.36363 64	3.29334 26	2.38×10^{-7}
1.45454 55	3.93456 79	9.54×10^{-7}
1.54545 45	4.61408 57	-4.77×10^{-7}
1.63636 36	5.33301 78	4.77×10^{-7}
1.72727 27	6.09280 54	1.91×10^{-6}
1.81818 18	6.89525 56	9.54×10^{-7}
1.90909 10	7.74258 90	9.54×10^{-7}
2.00000 00	8.63749 69	0.00

Notice that the errors are smaller than those obtained in the discretization method for the same problem. (Why?)

Mathematical Software

By using mathematical software such as found in MATLAB, Maple, or Mathematica, this problem can be solved in various ways. In MATLAB and Mathematica, built-in routines can be used to obtain the numerical solution to this boundary-value problem and plot the solution curve. On the other hand, Maple can solve the two differential equations in (12) and combine the solutions as described earlier with an appropriate value for λ. Also, the code can evaluate the solution at 1, 1.5, and 2, for example. Note that this is an *analytic* solution. These mathematical software systems do not produce the solution instantaneously; there is a lot of calculation going on behind the scenes.

Existence of Solutions

In our brief discussion of two-point boundary-value problems, we have not touched upon the difficult question of the **existence of solutions**. Sometimes a boundary-value problem has no solution despite having smooth coefficients. An example was given in Exercise 11.1.4b. This behavior contrasts sharply with that of initial-value problems. These matters are beyond the scope of this book but are treated, for example, in Keller [1976] and Stoer and Bulirsch [1993].

Summary 11.2

- For the **two-point boundary-value problem**

$$\begin{cases} x''(t) = f(t, x(t), x'(t)) \\ x(a) = \alpha, \quad x(b) = \beta \end{cases}$$

we use finite differences over the interval $[a, b]$ with $n + 1$ points, namely, $t_i = a + ih$ with $0 \leqq i \leqq n$ and $h = (b - a)/n$. We obtain $x_0 = \alpha$, $x_n = \beta$, and

$$\frac{1}{h^2}(x_{i-1} - 2x_i + x_{i+1}) = f\left(t_i, x_i, \frac{1}{2h}(x_{i+1} - x_{i-1})\right) \qquad (1 \leqq i \leqq n - 1)$$

The **linear case** of this problem occurs when the right-hand side is

$$f(t, x, x') = u(t) + v(t)x + w(t)x'$$

In this case, the main equation becomes

$$\frac{1}{h^2}(x_{i-1} - 2x_i + x_{i+1}) = u(t_i) + v(t_i)x_i + w(t_i)\left[\frac{1}{2h}(x_{i+1} - x_{i-1})\right]$$

Then the computational form is

$$-\left(1 + \frac{h}{2}w_i\right)x_{i-1} + (2 + h^2 v_i)x_i - \left(1 - \frac{h}{2}w_i\right)x_{i+1} = -h^2 u_i$$

where $u_i = u(t_i)$, $v_i = v(t_i)$, and $w_i = w(t_i)$. This leads to a tridiagonal linear system to be solved.

- Consider the **linear two-point boundary-value problem**

$$\begin{cases} x'' = u(t) + v(t)x + w(t)x' \\ x(a) = \alpha, \quad x(b) = \beta \end{cases}$$

and the corresponding **initial-value problem**

$$\begin{cases} x'' = u(t) + v(t)x + w(t)x' \\ x(a) = \alpha, \quad x'(a) = z \end{cases}$$

Suppose that x_1 and x_2 are two solution curves to the initial-value problem with z_1 and z_2, respectively. The solution of the two-point boundary-value problem is

$$g(t) = \lambda x_1(t) + (1 - \lambda)x_2(t)$$

with

$$\lambda = \frac{\beta - x_2(b)}{x_1(b) - x_2(b)}$$

Then we find

$$\begin{cases} g'' = u + vg + wg' \\ g(a) = \alpha \quad g(b) = \lambda x_1(b) + (1 - \lambda)x_2(b) = \beta \end{cases}$$

A simple way to implement this is to solve two initial-value problems:

$$\begin{cases} x'' = u(t) + v(t)x + w(t)x' \\ x(a) = \alpha, \quad x'(a) = 0 \end{cases} \qquad \begin{cases} y'' = u(t) + v(t)y + w(t)y' \\ y(a) = \alpha, \quad y'(a) = 1 \end{cases}$$

Then the solution to the original two-point boundary-value problem is

$$\lambda x(t) + (1 - \lambda)y(t), \quad \text{with} \quad \lambda = \frac{\beta - y(b)}{x(b) - y(b)}$$

Exercises 11.2

[a]**1.** If standard finite-difference approximations to derivatives are used to solve a two-point boundary-value problem with $x'' = t + 2x - x'$, what is the typical equation in the resulting linear system of equations?

[a]**2.** Consider the two-point boundary-value problem

$$\begin{cases} x'' = -x \\ x(0) = 0, \quad x(1) = 1 \end{cases}$$

Set up and solve the tridiagonal system that arises from the finite-difference method when $h = \frac{1}{4}$. Explain any differences from the analytic solution at $x\left(\frac{1}{4}\right) \approx 0.29401$, $x\left(\frac{1}{2}\right) \approx 0.56975$, and $x\left(\frac{3}{4}\right) \approx 0.81006$.

3. Verify that Equation (11) gives the solution of boundary-value Problem (8).

[a]**4.** Consider the two-point boundary-value problem

$$\begin{cases} x'' = x^2 - t + tx \\ x(0) = 1, \quad x(1) = 3 \end{cases}$$

Suppose that we have solved two initial-value problems

$$\begin{cases} u'' = u^2 - t + tu \\ u(0) = 1, \quad u'(0) = 1 \end{cases} \qquad \begin{cases} v'' = v^2 - t + tv \\ v(0) = 1, \quad v'(0) = 2 \end{cases}$$

numerically and have found as terminal values $u(1) = 2$ and $v(1) = 3.5$. What is a reasonable initial-value problem to try *next* in attempting to solve the original two-point value problem?

5. Consider the tridiagonal System (6). Show that if $v_i > 0$, then some choice of h exists for which the matrix is diagonally dominant.

6. Establish the properties claimed for the function g in Equation (10).

7. Show that for the simple problem

$$\begin{cases} x'' = -x \\ x(a) = \alpha, \quad x(b) = \beta \end{cases}$$

the tridiagonal system to be solved can be written as

$$\begin{cases} (2 - h^2)x_1 - x_2 = \alpha \\ -x_{i-1} + (2 - h^2)x_i - x_{i+1} = 0 \quad (2 \le i \le n-2) \\ -x_{n-2} + (2 - h^2)x_{n-1} = \beta \end{cases}$$

[a]**8.** Write down the system of equations $\boldsymbol{Ax} = \boldsymbol{b}$ that results from using the usual second-order central difference approximation to solve

$$\begin{cases} x'' = (1 + t)x \\ x(0) = 0, \quad x(1) = 1 \end{cases}$$

[a]**9.** Let u be a solution of the initial-value problem

$$\begin{cases} u'' = e^t u + t^2 u' \\ u(1) = 0, \quad u'(1) = 1 \end{cases}$$

How do we solve the following two-point boundary-value problem by utilizing u?

$$\begin{cases} x'' = e^t x + t^2 x' \\ x(1) = 0, \quad x(2) = 7 \end{cases}$$

10. How would you solve the problem

$$\begin{cases} x' = f(t, x) \\ Ax(a) + Bx(b) = C \end{cases}$$

where a, b, A, B, and C are given real numbers? (Assume that A and B are not both zero.)

[a]**11.** Use the shooting method on this two-point boundary-value problem, and explain what happens:

$$\begin{cases} x'' = -x \\ x(0) = 3, \quad x(\pi) = 7 \end{cases}$$

This problem is to be solved analytically, not by computer or calculator.

Computer Exercises 11.2

1. Explain the main steps in setting up a program to solve this two-point boundary value problem by the finite-difference method.

$$\begin{cases} x'' = x\sin t + x'\cos t - e^t \\ x(0) = 0, \quad x(1) = 1 \end{cases}$$

Show any preliminary work that must be done before programming. Exploit the linearity of the differential equation. Program and compare the results when different values of n are used, say, $n = 10, 100$, and 1000.

2. Solve the following two-point boundary value problem numerically. For comparisons, the exact solutions are given.

[a] **a.** $$\begin{cases} x'' = \dfrac{(1-t)x + 1}{(1+t)^2} \\ x(0) = 1, \quad x(1) = 0.5 \end{cases}$$

[a] **b.** $$\begin{cases} x'' = \dfrac{1}{3}\left[(2-t)e^{2x} + (1+t)^{-1}\right] \\ x(0) = 0, \quad x(1) = -\log 2 \end{cases}$$

3. Solve the boundary-value problem

$$\begin{cases} x'' = -x + tx' - 2t\cos t + t \\ x(0) = 0, \quad x(\pi) = \pi \end{cases}$$

by discretization. Compare with the exact solution, which is $x(t) = t + 2\sin t$.

4. Repeat Computer Exercise 11.1.2 (p. 513), using a discretization method.

5. Write a computer program to implement

a. program *BVP1* **b. program** *BVP2*

6. (Continuation) Using built-in routines in mathematical software systems such as MATLAB, Maple, or Mathematica, solve and plot the solution curve for the boundary-value problem associated with

a. program *BVP1* **b. program** *BVP2*

7. Investigate the computation of numerical solutions to the following **challenging test problems**, which are nonlinear:

a. $$\begin{cases} x'' = e^x \\ x(0) = 0, \quad x(1) = 0 \end{cases}$$

b. $$\begin{cases} \varepsilon x'' + (x')^2 = 1 \\ x(0) = 0, \quad x(1) = 1 \end{cases}$$

Vary $\varepsilon = 10^{-1}, 10^{-2}, 10^{-3}, \ldots$. Compare to the true solution $x(t) = 1 + \varepsilon \ln\cosh((x - 0.745)/\varepsilon)$ which has a corner at $t = 0.745$.

c. Troesch's problem:

$$\begin{cases} x'' = \mu\sinh(\mu x) \\ x(0) = 0, x(1) = 1 \end{cases} \quad \text{using } \mu = 50.$$

d. Bratu's problem:

$$\begin{cases} x'' + \lambda e^x = 0 \\ x(0) = 0, x(1) = 0 \end{cases} \quad \text{using } \lambda = 3.55.$$

If we let $\lambda = 3.51383\ldots$, there are two solutions when $\lambda < \lambda^*$, one solution when $\lambda = \lambda^*$, and no solutions when $\lambda > \lambda^*$.

e. $$\begin{cases} \varepsilon x'' + tx' = 0 \\ x(-1) = 0, x(1) = 2 \end{cases} \quad \text{using } \varepsilon = 10^{-8}.$$

Compare to the true solution $x(t) = 1 + \text{erf}(t/\sqrt{2\varepsilon})/\text{erf}(1/\sqrt{2\varepsilon})$.

Cash [2003] uses these and other test problems in his research. For more information on them, see `www2.imperial.ac.uk/~jcash/`.

8. (**Bucking of a Circular Ring Project**) A model for a circular ring with compressibility c under hydrostatic pressure p from all directions is given by the following boundary-value problem involving a system of seven differential equations:

$$\begin{cases} y_1' = -1 - cy_5 + (c+1)y_7 \\ y_2' = [1 + c(y_5 - y_7)]\cos y_1 \\ y_3' = [1 + c(y_5 - y_7)]\sin y_1 \\ y_4' = 1 + c(y_5 - y_7) \\ y_5' = y_6[-1 - cy_5 + (c+1)y_7] \\ y_6' = y_5y_7 - [1 + c(y_5 - y_7)](y_5 + p) \\ y_7' = [1 + c(y_5 - y_7)]y_6 \end{cases}$$

where $y_1(0) = \pi/2$, $y_1(\pi/2) = 0$, $y_2(\pi/2) = 0$, $y_3(0) = 0$, $y_4(0) = 0$, and $y_6(0) = 0$, $y_6(\pi/2) = 0$. Various simplifications are useful in the study of the

buckling or collapse of the circular ring. Consider only a quarter-circle using symmetry. (See Sketch (a).) As the pressure increases, the radius of the circle decreases, and a **bifurcation or a change of state** can occur. (See Sketch (b).)

The shooting method together with more advanced numerical methods can be used to solve this problem. Explore some of them. See Huddleston [2000] and Sauer [2012].

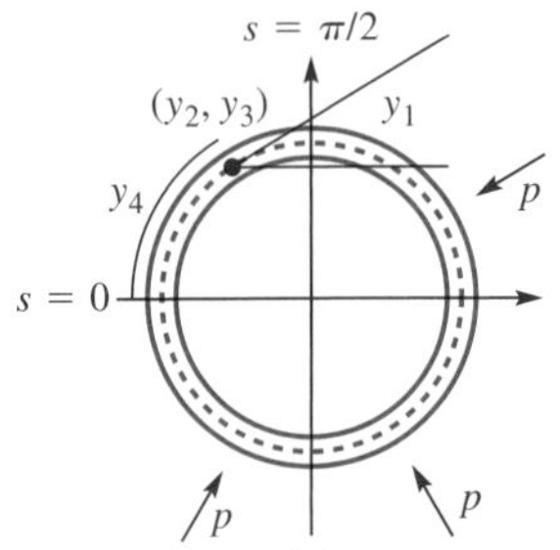

(a)

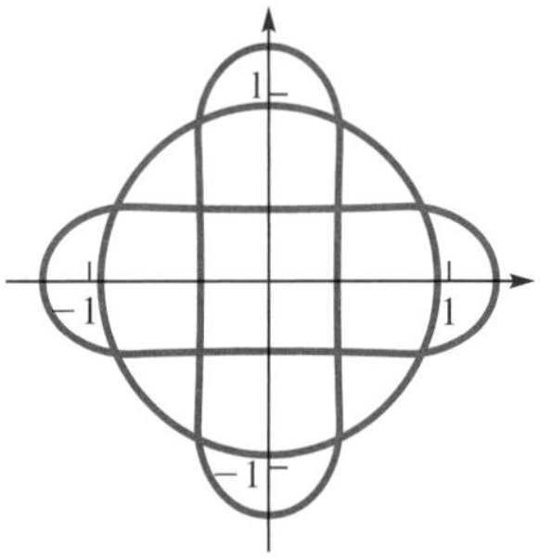

(b)

9. Use Conservation of Heat to determine the heat balance for a long thin rod. Assume the rod is not insulated along its length and the system is at a steady state.

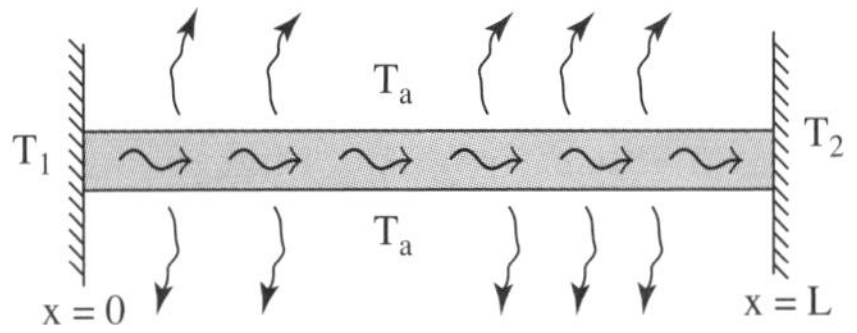

The figure shows a non-insulated uniform rod positioned between two bodies of constant temperature, but different values; say, $T_1 > T_2$ and $T_2 > T_a$. The resulting differential equation is

$$\frac{d^2T}{dx^2} + \alpha(T_a - T) = 0$$

Here T_a is the temperature of the surrounding air and α is the radiative heat loss coefficient for the rate of heat dissipation to the surrounding air. For a simple case, consider a 10 meter rod with the temperature held fixed at the end values $T(0) = T_1$ and $T(L) = T_2$. With these values $T_a = 20$, $T_1 = T(0) = 40$, $T_2 = T(10) = 200$, and $\alpha = 10^{-2}$, solve this problem using the following approaches.

a. Use calculus to determine the analytical solution.

b. Use the shooting method.

c. Use the finite-difference method.

Produce a table of computer results comparing the exact analytical solution with the shooting method and the finite-difference method at values of $x = 0$ to $x = 10$ in steps of 2. Show that the errors can be mitigated by decreasing the step size used in the numerical methods. (See Chapra [2012].)

12 Partial Differential Equations

In the theory of elasticity, it is shown that the stress in a cylindrical beam under torsion can be derived from a function $u(x, y)$ that satisfies the Poisson equation

$$\frac{\partial^2 u}{\partial x^2} + \frac{\partial^2 u}{\partial y^2} + 2 = 0$$

In the case of a beam whose cross section is the square defined by $|x| \leqq 1$, $|y| \leqq 1$, the function u must satisfy Poisson's equation *inside* the square and must be zero at each point on the *perimeter* of the square. By using the methods of this chapter, we can construct a table of approximate values of $u(x, y)$.

12.1 Parabolic Problems

Challenging Problems

Many physical phenomena can be modeled mathematically by differential equations. When the function that is being studied involves two or more independent variables, the differential equation is usually a *partial* differential equation. Since functions of several variables are intrinsically more complicated than those of one variable, partial differential equations (PDEs) can lead to some of the most challenging of numerical problems. In fact, their numerical solution is one type of scientific calculation in which the resources of the fastest and most expensive computing systems easily become taxed. We shall see later why this is so.

Some Partial Differential Equations from Applied Problems

Some important partial differential equations and the physical phenomena that they govern are listed next.

The **wave equation** in three spatial variables (x, y, z) and time t is

Wave Equation

$$\frac{\partial^2 u}{\partial t^2} = \frac{\partial^2 u}{\partial x^2} + \frac{\partial^2 u}{\partial y^2} + \frac{\partial^2 u}{\partial z^2}$$

The function u represents the displacement at time t of a particle whose position at rest is (x, y, z). With appropriate boundary conditions, this equation governs vibrations of a three-dimensional elastic body.

The **heat equation** is

Heat Equation

$$\frac{\partial u}{\partial t} = \frac{\partial^2 u}{\partial x^2} + \frac{\partial^2 u}{\partial y^2} + \frac{\partial^2 u}{\partial z^2}$$

The function u represents the temperature at time t in a physical body at the point that has coordinates (x, y, z).

Laplace's equation is

Laplace's Equation

$$\frac{\partial^2 u}{\partial x^2} + \frac{\partial^2 u}{\partial y^2} + \frac{\partial^2 u}{\partial z^2} = 0$$

It governs the steady-state distribution of heat in a body or the steady-state distribution of electrical charge in a body. Laplace's equation also governs gravitational, electric, and magnetic potentials and velocity potentials in irrotational flows of incompressible fluids. The form of Laplace's equation given above applies to rectangular coordinates. In cylindrical and spherical coordinates, it takes these respective forms:

$$\frac{\partial^2 u}{\partial r^2} + \frac{1}{r}\frac{\partial u}{\partial r} + \frac{1}{r^2}\frac{\partial^2 u}{\partial \phi^2} + \frac{\partial^2 u}{\partial z^2} = 0$$

$$\frac{1}{r}\frac{\partial^2}{\partial r^2}(ru) + \frac{1}{r^2 \sin\theta}\frac{\partial}{\partial \theta}\left(\sin\theta \frac{\partial u}{\partial \theta}\right) + \frac{1}{r^2 \sin^2\theta}\frac{\partial^2 u}{\partial \phi^2} = 0$$

The **biharmonic equation** is

Biharmonic Equation

$$\frac{\partial^4 u}{\partial x^4} + 2\frac{\partial^4 u}{\partial x^2\,\partial y^2} + \frac{\partial^4 u}{\partial y^4} = 0$$

It occurs in the study of elastic stress, and from its solution the shearing and normal stresses can be derived for an elastic body.

The **Navier-Stokes equations** are

Navier-Stokes Equations

$$\frac{\partial u}{\partial t} + u\frac{\partial u}{\partial x} + v\frac{\partial u}{\partial y} + \frac{\partial p}{\partial x} = \frac{\partial^2 u}{\partial x^2} + \frac{\partial^2 u}{\partial y^2}$$

$$\frac{\partial v}{\partial t} + u\frac{\partial v}{\partial x} + v\frac{\partial v}{\partial y} + \frac{\partial p}{\partial y} = \frac{\partial^2 v}{\partial x^2} + \frac{\partial^2 v}{\partial y^2}$$

Here, u and v are components of the velocity vector in a fluid flow. The function p is the pressure, and the fluid is assumed to be incompressible but viscous.

In three dimensions, the following operators are useful in writing many standard partial differential equations

Operators

$$\nabla = \frac{\partial}{\partial x} + \frac{\partial}{\partial y} + \frac{\partial}{\partial z}$$

$$\nabla^2 = \frac{\partial^2}{\partial x^2} + \frac{\partial^2}{\partial y^2} + \frac{\partial^2}{\partial z^2} \qquad \textbf{(Laplacian operator)}$$

For example, we have

Classical PDEs Using Operators

Heat equation	$\frac{1}{k}\frac{\partial u}{\partial t} = \nabla^2 u$
Diffusion equation	$\frac{\partial u}{\partial t} = \nabla(d\nabla u) + \rho$
Wave equation	$\frac{1}{v^2}\frac{\partial^2 u}{\partial t^2} = \nabla^2 u$
Laplace equation	$\nabla^2 u = 0$

Poisson equation	$\nabla^2 u = -4\pi\rho$
Helmholtz equation	$\nabla^2 u = -k^2 u$

The **diffusion equation** with diffusion constant d has the same structure as the heat equation because heat transfer is a diffusion process. Some authors use alternate notation such as $\Delta u = \mathtt{curl}(\mathtt{grad}(u)) = \nabla^2 u$.

Additional examples from quantum mechanics, electromagnetism, hydrodynamics, elasticity, and so on could also be given, but the five partial differential equations shown already exhibit a great diversity. The Navier-Stokes equation, in particular, illustrates a very complicated problem: a pair of nonlinear, simultaneous partial differential equations.

Boundary Conditions

To specify a unique solution to a partial differential equation, additional conditions must be imposed on the solution function. Typically, these conditions occur in the form of boundary values that are prescribed on all or part of the perimeter of the region in which the solution is sought. The nature of the boundary and the boundary values are usually the determining factors in setting up an appropriate numerical scheme for obtaining the approximate solution.

Mathematical Software

MATLAB includes a PDE Toolbox for partial differential equations. It contains many commands for such tasks as describing the domain of an equation, generating meshes, computing numerical solutions, and plotting. Within MATLAB, the command `pdetool` invokes a graphical user interface (GUI) that is a self-contained graphical environment for solving partial differential equations. One draws the domain and indicates the boundary, fills in menus with the problem and boundary specifications, and selects buttons to solve the problem and plot the results. Although this interface may provide a convenient working environment, there are situations in which command-line functions are needed for additional flexibility. A suite of demonstrations and help files is useful in finding one's way. For example, this software can handle PDEs of the following types:

Examples of Types of PDEs

Parabolic PDE	$b\dfrac{\partial u}{\partial t} - \nabla \cdot (c\nabla u) + au = f$
Hyperbolic PDE	$b\dfrac{\partial^2 u}{\partial t^2} - \nabla \cdot (c\nabla u) + au = f$
Elliptic PDE	$-\nabla \cdot (c\nabla u) + au = f$

for x and y on the two-dimensional domain Ω for the problem. On the boundaries of the domain, the following boundary conditions can be handled:

Boundary Conditions

Dirichlet	$hu = r$
Generalized Neumann	$\vec{n} \cdot (c\nabla u) + qu = g$
Mixed	combination of Dirichlet/Neumann

Unit Normal Derivative

Here, $\vec{n} = du/d\nu$ is the outward **unit length normal derivative**. While the PDE can be entered via a dialog box, both the boundary conditions and the PDE coefficients a, c, d can be entered in a variety of ways. One can construct the geometry of the domain by drawing solid objects (circle, polygon, rectangle, and ellipse) that may be overlapped, moved, and rotated. Also, Maple, Mathematica, and specialized software packages can be used to solve a wide variety of PDEs.

Heat Equation Model Problem

Now, we consider a model problem of modest scope to introduce some of the essential ideas. For technical reasons, the problem is said to be of the **parabolic** type. In it we have the heat equation in one spatial variable accompanied by boundary conditions appropriate to a certain physical phenomenon:

$$\begin{cases} \dfrac{\partial^2}{\partial x^2}u(x,t) = \dfrac{\partial}{\partial t}u(x,t) \\ u(0,t) = u(1,t) = 0 \\ u(x,0) = \sin \pi x \end{cases} \tag{1}$$

These equations govern the temperature $u(x,t)$ in a thin rod of length 1 when the ends are held at temperature 0, under the assumption that the initial temperature in the rod is given by the function $\sin \pi x$ (see Figure 12.1). In the xt-plane, the region in which the solution is sought is described by inequalities $0 \leqq x \leqq 1$ and $t \geqq 0$. On the boundary of this region (shaded in Figure 12.2), the values of u have been prescribed.

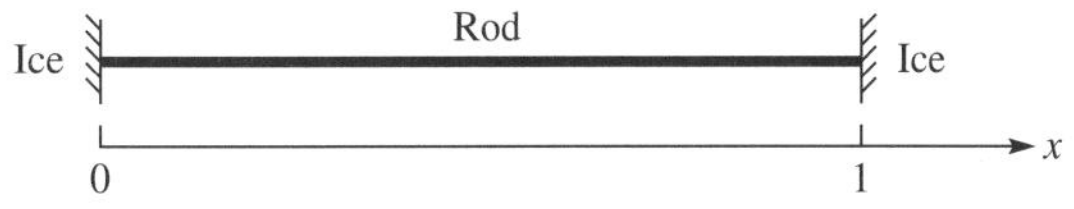

FIGURE 12.1
Heated rod

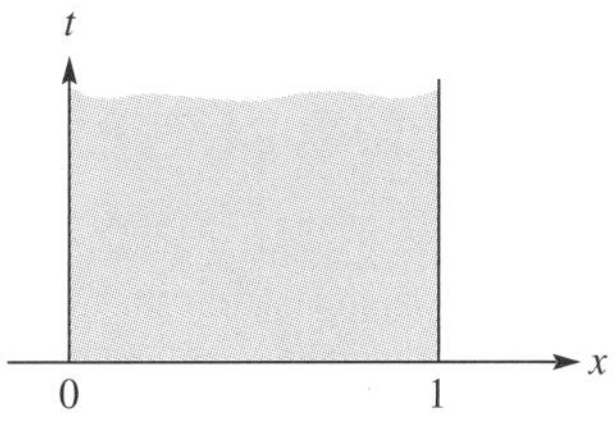

FIGURE 12.2
Heat equation: xt-plane

Finite-Difference Method

A principal approach to the numerical solution of such a problem is the **finite-difference method**. It proceeds by replacing the derivatives in the equation by finite differences. Two formulas from Section 4.3 are useful in this context:

f' and f'' Finite Difference Approximations

$$f'(x) \approx \frac{1}{h}[f(x+h) - f(x)]$$

$$f''(x) \approx \frac{1}{h^2}[f(x+h) - 2f(x) + f(x-h)]$$

If the formulas are used in the differential Equation (1), with possibly different step lengths h and k, the result is

$$\frac{1}{h^2}[u(x+h,t) - 2u(x,t) + u(x-h,t)] = \frac{1}{k}[u(x,t+k) - u(x,t)] \tag{2}$$

This equation is now interpreted as a means of advancing the solution step-by-step in the t variable. That is, if $u(x,t)$ is known for $0 \leqq x \leqq 1$ and $0 \leqq t \leqq t_0$, then Equation (2) allows us to evaluate the solution for $t = t_0 + k$.

Equation (2) can be rewritten in the form

Stencil (4-Point Explicit)

$$u(x, t+k) = \sigma u(x+h,t) + (1-2\sigma)u(x,t) + \sigma u(x-h,t) \tag{3}$$

where

$$\sigma = \frac{k}{h^2}$$

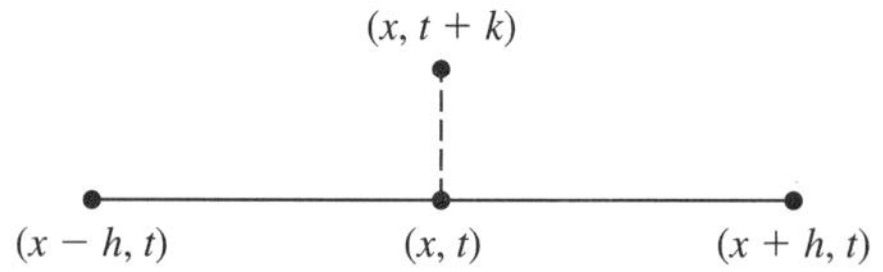

FIGURE 12.3 Heat equation: Explicit stencil

A sketch showing the location of the four points involved in this equation is given in Figure 12.3. Since the solution is known on the boundary of the region, it is possible to compute an approximate solution inside the region by systematically using Equation (3). It is, of course, an *approximate* solution because Equation (2) is only a finite-difference analog of Equation (1).

To obtain an approximate solution on a computer, we select values for h and k and use Equation (3). An analysis of this procedure, which is outside the scope of this text, shows that for *stability* of the computation, the coefficient $1-2\sigma$ in Equation (3) should be nonnegative. (If this condition is not met, errors made at one step may be magnified at subsequent steps, ultimately spoiling the solution.) The reader is referred to Kincaid and Cheney [2002] or Forsythe and Wasow [1960] for a discussion of stability. Using this algorithm, we can continue the solution indefinitely in the t-variable by computations involving only prior values of t. This is an example of a **marching problem** or **marching method**.

Marching Method

Pseudocode for Explicit Method

For utmost simplicity, we select $h = 0.1$ and $k = 0.005$. Coefficient σ is now 0.5. This choice makes the coefficient $1 - 2\sigma$ equal to zero. Our pseudocode first prints $u(ih, 0)$ for $0 \leqq i \leqq 10$ because they are known boundary values. Then it computes and prints $u(ih, k)$ for $0 \leqq i \leqq 10$ using Equation (3) and boundary values $u(0, t) = u(1, t) = 0$. This procedure is continued until t reaches the value 0.1. The single subscripted arrays (u_i) and (v_i) are used to store the values of the approximate solution at t and $t+k$, respectively. Since the analytic solution of the problem is

$$u(x, t) = e^{-\pi^2 t} \sin \pi x$$

(see Exercise 12.1.3), the error can be printed out at each step.

Explicit Method

The procedure described is an example of an **explicit method**. The approximate values of $u(x, t+k)$ are calculated explicitly in terms of $u(x, t)$. Not only is this situation atypical, but even in this problem the procedure is rather slow because considerations of stability force us to select

$$k \leqq \frac{1}{2} h^2$$

Since h must be rather small to represent the derivative accurately by the finite difference formula, the corresponding k must be extremely small. Values such as $h = 0.1$ and $k = 0.005$ are representative, as are $h = 0.01$ and $k = 0.00005$. With such small values of k, an inordinate amount of computation is necessary to make much progress in the t variable.

```
program Parabolic1
integer i, j;   real array (u_i)_{0:n}, (v_i)_{0:n}
integer n ← 10, m ← 20
real h ← 0.1, k ← 0.005
real u_0 ← 0, v_0 ← 0, u_n ← 0, v_n ← 0
```

(Continued)

*Parabolic*1 **Pseudocode**

```
for i = 1 to n − 1
    u_i ← sin(πih)
end for
output (u_i)
for j = 1 to m
    for i = 1 to n − 1
        v_i ← (u_{i−1} + u_{i+1})/2
    end for
    output (v_i)
    t ← jk
    for i = 1 to n − 1
        u_i ← e^{−π²t} sin(πih) − v_i
    end for
    output (u_i)
    for i = 1 to n − 1
        u_i ← v_i
    end for
end for
end program Parabolic1
```

Crank-Nicolson Method

An alternative procedure of the implicit type goes by the name of its inventors, John Crank and Phyllis Nicolson, and is based on a simple variant of Equation (2):

$$\frac{1}{h^2}[u(x+h,t) - 2u(x,t) + u(x-h,t)] = \frac{1}{k}[u(x,t) - u(x,t-k)] \tag{4}$$

If a numerical solution at grid points $x = ih$, $t = jk$ has been obtained up to a certain level in the t variable, Equation (4) governs the values of u on the next t level. Therefore, Equation (4) may be rewritten as

Stencil (4-Point Implicit)

$$-u(x-h,t) + ru(x,t) - u(x+h,t) = su(x,t-k) \tag{5}$$

in which

$$r = 2 + s \qquad \text{and} \qquad s = \frac{h^2}{k}$$

The locations of the four points in this equation are shown in Figure 12.4.

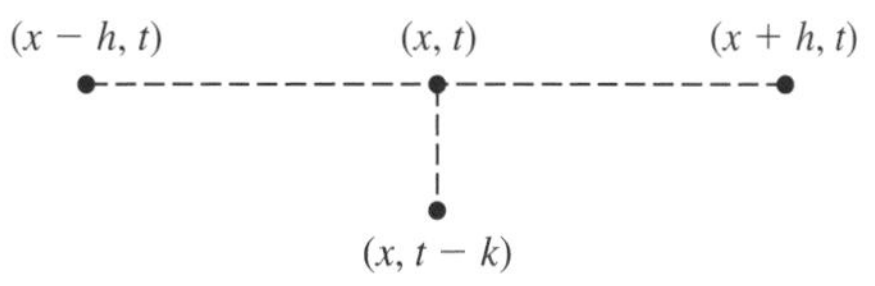

FIGURE 12.4 Crank-Nicolson method: Implicit stencil

On the t level, u is unknown, but on the $(t-k)$ level, u is known. So we can introduce unknowns $u_i = u(ih, t)$ and known quantities $b_i = su(ih, t-k)$ and write Equation (5) in matrix form:

Diagonally Dominant Tridiagonal System

$$\begin{bmatrix} r & -1 & & & & \\ -1 & r & -1 & & & \\ & -1 & r & -1 & & \\ & & \ddots & \ddots & \ddots & \\ & & & -1 & r & -1 \\ & & & & -1 & r \end{bmatrix} \begin{bmatrix} u_1 \\ u_2 \\ u_3 \\ \vdots \\ u_{n-2} \\ u_{n-1} \end{bmatrix} = \begin{bmatrix} b_1 \\ b_2 \\ b_3 \\ \vdots \\ b_{n-2} \\ b_{n-1} \end{bmatrix} \tag{6}$$

The simplifying assumption that $u(0, t) = u(1, t) = 0$ has been used here. Also, $h = 1/n$. The system of equations is **tridiagonal** and **diagonally dominant** because $|r| = 2+h^2/k > 2$. Hence, it can be solved by the efficient method of Section 2.3.

Stable

An elementary argument shows that this method is **stable**. We shall see that if the initial values $u(x, 0)$ lie in an interval $[\alpha, \beta]$, then values subsequently calculated by using Equation (5) will also lie in $[\alpha, \beta]$, thereby ruling out any unstable growth. Since the solution is built up line by line in a uniform way, we need only verify that the values on the first computed line, $u(x, k)$, lie in $[\alpha, \beta]$. Let j be the index of the largest u_i that occurs on this line $t = k$. Then we have

$$-u_{j-1} + ru_j - u_{j+1} = b_j$$

Since u_j is the largest of the u's, $u_{j-1} \leqq u_j$ and $u_{j+1} \leqq u_j$. Thus, we obtain

$$ru_j = b_j + u_{j-1} + u_{j+1} \leqq b_j + 2u_j$$

Since $r = 2 + s$ and $b_j = su(jh, 0)$, the previous inequality leads at once to

$$u_j \leqq u(jh, 0) \leqq \beta$$

Since u_j is the largest of the u_i, we have

$$u_i \leqq \beta \quad \text{for all } i$$

Similarly, we have

$$u_i \geqq \alpha \quad \text{for all } i$$

thus establishing our assertion.

Pseudocode for the Crank-Nicolson Method

Now we present a pseudocode for carrying out the Crank-Nicolson method on the model program. In it, $h = 0.1$, $k = h^2/2$, and the solution is continued until $t = 0.1$. The value of r is 4 and $s = 2$. It is easier to compute and print only the values of u at interior points on each horizontal line. At boundary points, we have $u(0, t) = u(1, t) = 0$. The program calls procedure *Tri* from Section 2.3.

```
program Parabolic2
integer i, j;   real array (c_i)_{1:n−1}, (d_i)_{1:n−1}, (u_i)_{1:n−1}, (v_i)_{1:n−1}
integer n ← 10, m ← 20
real h ← 0.1, k ← 0.005
real r, s, t
s ← h^2/k
r ← 2 + s
```

(Continued)

Parabolic2 Pseudocode

```
for i = 1 to n − 1
    d_i ← r
    c_i ← −1
    u_i ← sin(πih)
end for
output (u_i)
for j = 1 to m
    for i = 1 to n − 1
        d_i ← r
        v_i ← su_i
    end for
    call Tri(n − 1, (c_i), (d_i), (c_i), (v_i), (v_i))
    output (v_i)
    t ← jk
    for i = 1 to n − 1
        u_i ← e^(−π²t) sin(πih) − v_i
    end for
    output (u_i)
    for i = 1 to n − 1
        u_i ← v_i
    end for
end for
end program Parabolic2
```

We used the same values for h and k in the pseudocode for two methods (explicit and Crank-Nicolson), so a fair comparison can be made of the outputs. Because the Crank-Nicolson method is stable, a much larger k could have been used.

Alternative Version of the Crank-Nicolson Method

Another version of the Crank-Nicolson method can be obtained by using the central differences at $\left(x, t - \frac{1}{2}k\right)$ in Equation (4) to produce

$$\frac{1}{h^2}\left[u\left(x+h, t-\tfrac{1}{2}k\right) - 2u\left(x, t-\tfrac{1}{2}k\right) + u\left(x-h, t-\tfrac{1}{2}k\right)\right]$$

$$= \frac{1}{k}[u(x,t) - u(x, t-k)]$$

Since the u values are known only at integer multiples of k, terms such as $u\left(x, t - \frac{1}{2}k\right)$ are replaced by the average of u values at adjacent grid points; that is,

$$u\left(x, t - \tfrac{1}{2}x\right) \approx \frac{1}{2}[u(x,t) + u(x, t-k)]$$

So we have

$$\frac{1}{2h^2}[u(x+h,t) - 2u(x,t) + u(x-h,t) + u(x+h, t-k)$$

$$-2u(x, t-k) + u(x-h, t-k)] = \frac{1}{k}[u(x,t) - u(x,t-k)]$$

The computational form of this equation is

Stencil (6-Point Implicit)

$$\begin{aligned}&-u(x-h,t)+2(1+s)u(x,t)-u(x+h,t)\\&\quad=u(x-h,t-k)+2(s-1)u(x,t-k)+u(x+h,t-k)\end{aligned} \tag{7}$$

where

$$s=\frac{h^2}{k}\equiv\frac{1}{\sigma}$$

The six points in this equation are shown in Figure 12.5. This leads to a tridiagonal system of form (6) with $r=2(1+s)$ and

$$b_i=u((i-1)h,t-k)+2(s-1)u(ih,t-k)+u((i+1)h,t-k)$$

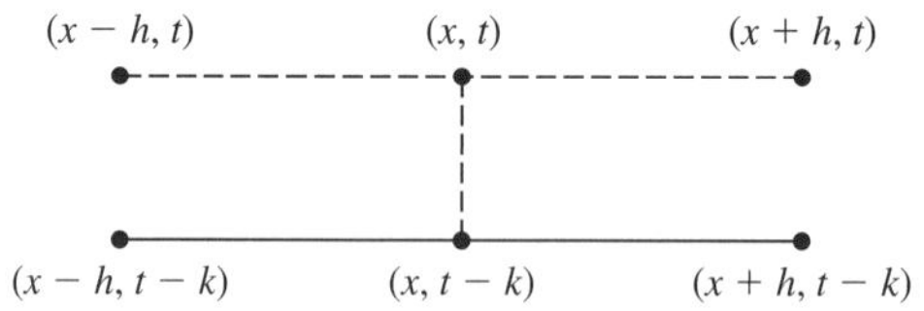

FIGURE 12.5 Crank-Nicolson method: Alternative stencil

Stability

At the heart of the explicit method is Equation (3), which shows how the values of u for $t+k$ depend on the values of u at the previous time step, t. If we introduce the values of u on the mesh by writing $u_{ij}=u(ih,jk)$, then we can assemble all the values for one t-level into a vector $\boldsymbol{v}^{(j)}$ as follows:

$$\boldsymbol{v}^{(j)}=[u_{0j},u_{1j},u_{2j},\ldots,u_{nj}]^T$$

Equation (3) can now be written in the form

$$u_{i,j+1}=\sigma u_{i+1,j}+(1-2\sigma)u_{ij}+\sigma u_{i-1,j}$$

This equation shows how $\boldsymbol{v}^{(j+1)}$ is obtained from $\boldsymbol{v}^{(j)}$. It is simply

Matrix Form (Explicit Method)

$$\boldsymbol{v}^{(j+1)}=\boldsymbol{A}\boldsymbol{v}^{(j)}$$

where $\boldsymbol{A}$ is the matrix whose elements are

$$\begin{bmatrix}1-2\sigma & \sigma & & & & \\ \sigma & 1-2\sigma & \sigma & & & \\ & \sigma & 1-2\sigma & \sigma & & \\ & & \ddots & \ddots & \ddots & \\ & & & \sigma & 1-2\sigma & \sigma \\ & & & & \sigma & 1-2\sigma\end{bmatrix}$$

Our equations tell us that

$$\boldsymbol{v}^{(j)}=\boldsymbol{A}\boldsymbol{v}^{(j-1)}=\boldsymbol{A}^2\boldsymbol{v}^{(j-2)}=\boldsymbol{A}^3\boldsymbol{v}^{(j-3)}=\cdots=\boldsymbol{A}^j\boldsymbol{v}^{(0)}$$

From physical considerations, the temperature in the bar should approach zero. After all, the heat is being lost through the ends of the rod, which are being kept at temperature 0. Hence, $\boldsymbol{A}^j\boldsymbol{v}^{(0)}$ should converge to 0 as $j\to\infty$.

At this juncture, we need a theorem in linear algebra that asserts (for any matrix $\boldsymbol{A}$) that $\boldsymbol{A}^j v \to \mathbf{0}$ for all vectors $\boldsymbol{v}$ if and only if all eigenvalues of $\boldsymbol{A}$ satisfy $|\lambda_i| < 1$. The eigenvalues of the matrix $\boldsymbol{A}$ in the present analysis are known to be

$$\lambda_i = 1 - 2\sigma(1 - \cos\theta_i) \quad \text{where} \quad \theta_i = \frac{i\pi}{n+1}$$

In our problem, we therefore must have

$$-1 < 1 - 2\sigma(1 - \cos\theta_i) < 1$$

This leads to $0 < \sigma \leqq \frac{1}{2}$, because θ_i can be arbitrarily close to π. This in turn leads to the **step-size condition**

Step-Size Condition

$$k \leqq \frac{1}{2}h^2$$

Mathematical Software

Mathematical software systems such as MATLAB, Maple, or Mathematica contain routines that solve partial differential equations. For example in Maple and Mathematica, we can invoke commands to verify the general analytical solution. (See Exercise 12.1.3.) In MATLAB, there is a sample program to numerically solve our model heat equation example. In Figure 12.6, we solve the heat equation, generate a three-dimensional plot of its solution surface, and produce a two-dimensional contour plot, which is displayed in color for indicating the various contours.

The PDE Toolbox within MATLAB produces solutions to partial differential equations using the finite-element formulation of the scalar PDE problem. (See Section 12.3

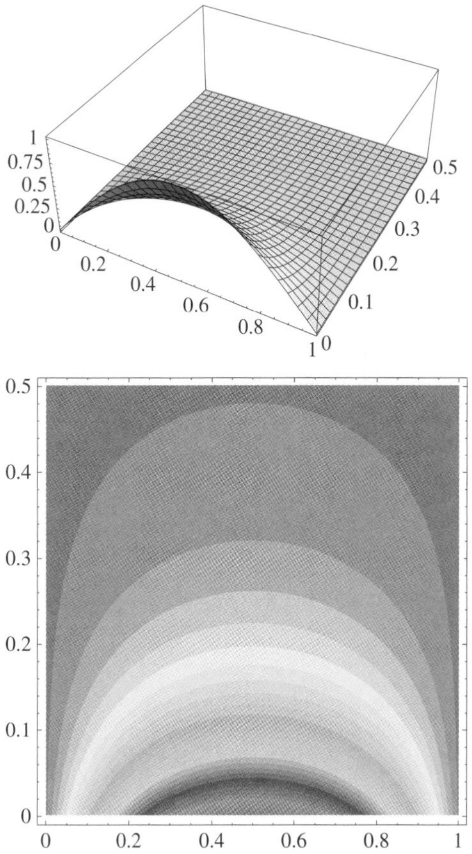

FIGURE 12.6
Heat equation:
(a) Solution surface;
(b) Contour plot

for additional discussion of the finite-element method.) This software library contains a graphical user interface with graphical tools for describing domains, generating triangular meshes on them, discretizing the PDEs on the mesh, building systems of equations, obtaining numerical approximations for their solution, and visualizing the results.

PDE Toolbox

In particular, MATLAB has the function `parabolic` for solving parabolic PDEs. As is found in the online documentation, one can solve the two-dimensional heat equation

$$\frac{\partial u}{\partial t} = \nabla^2 u$$

on the square $-1 \leqq x, y \leqq 1$. There are Dirichlet boundary conditions $u = 0$ and discontinuous initial conditions $u(0) = 1$ in the circle $x^2 + y^2 < \frac{2}{5}$ and $u(0) = 0$ otherwise. In fact, the demonstration continues with a movie of the solution curves.

Summary 12.1

- We consider a model problem involving the following parabolic partial differential equation

$$\frac{\partial^2}{\partial x^2} u(x, t) = \frac{\partial}{\partial t} u(x, t)$$

Using finite differences with step size h in the x-direction and k in the t-direction, we obtain

$$\frac{1}{h^2}[u(x+h, t) - 2u(x, t) + u(x-h, t)] = \frac{1}{k}[u(x, t+k) - u(x, t)]$$

The computational form is

$$u(x, t+k) = \sigma u(x+h, t) + (1 - 2\sigma)u(x, t) + \sigma u(x-h, t)$$

where $\sigma = k/h^2$. An alternative approach is the **Crank-Nicolson method** based on other finite differences for the right-hand side:

$$\frac{1}{h^2}[u(x+h, t) - 2u(x, t) + u(x-h, t)] = \frac{1}{k}[u(x, t) - u(x, t-k)]$$

Its computational form is

$$-u(x-h, t) + ru(x, t) - u(x+h, t) = su(x, t-k)$$

where $r = 2 + s$ and $s = h^2/k$. Yet another variant of the Crank-Nicolson method is based on these finite differences:

$$\frac{1}{h^2}\left[u\left(x+h, t-\tfrac{1}{2}k\right) - 2u\left(x, t-\tfrac{1}{2}k\right) + u\left(x-h, t-\tfrac{1}{2}k\right)\right]$$
$$= \frac{1}{k}[u(x, t) - u(x, t-k)]$$

Then by using

$$u\left(x, t-\tfrac{1}{2}k\right) \approx \frac{1}{2}[u(x, t) + u(x, t-k)]$$

the computational form is

$$-u(x-h, t) + 2(1+s)u(x, t) - u(x+h, t)$$
$$= u(x-h, t-k) + 2(s-1)u(x, t-k) + u(x+h, t-k)$$

where $s = h^2/k$. This results in a tridiagonal system of equations to be solved.

Exercises 12.1

1. A second-order linear differential equation with two variables has the form

$$A\frac{\partial^2 u}{\partial x^2} + B\frac{\partial^2 u}{\partial x\,\partial y} + C\frac{\partial^2 u}{\partial y^2} + \cdots = 0$$

Here, A, B, and C are functions of x and y, and the terms not written are of lower order. The equation is said to be **elliptic**, **parabolic**, or **hyperbolic** at a point (x, y), depending on whether $B^2 - 4AC$ is negative, zero, or positive, respectively. Classify each of these equations in this manner:

[a]**a.** $u_{xx} + u_{yy} + u_x + \sin x u_y - u = x^2 + y^2$

b. $u_{xx} - u_{yy} + 2u_x + 2u_y + e^x u = x - y$

[a]**c.** $u_{xx} = u_y + u - u_x + y$

d. $u_{xy} = u - u_x - u_y$

e. $3u_{xx} + u_{xy} + u_{yy} = e^{xy}$

[a]**f.** $e^x u_{xx} + \cos y u_{xy} - u_{yy} = 0$

g. $u_{xx} + 2u_{xy} + u_{yy} = 0$

h. $x u_{xx} + y u_{xy} + u_{yy} = 0$

[a]2. Derive the two-dimensional form of Laplace's equation in polar coordinates.

3. Show that the function

$$u(x, t) = \sum_{n=1}^{N} c_n e^{-(n\pi)^2 t} \sin n\pi x$$

is a solution of the heat conduction problem $u_{xx} = u_t$ and satisfies the boundary condition

$$u(0, t) = u(1, t) = 0$$
$$u(x, 0) = \textstyle\sum_{n=1}^{N} c_n \sin n\pi x \quad \text{for all } N \ge 1$$

[a]4. Refer to the model problem solved numerically in this section and show that if there is no roundoff, the approximate solution values obtained by using Equation (3) lie in the interval $[0, 1]$. (Assume $1 \ge 2k/h^2$.)

[a]5. Find a solution of Equation (3) that has the form $u(x, t) = a^t \sin \pi x$, where a is a constant.

[a]6. In using Equation (5), how must the linear System (6) be modified for $u(0, t) = c_0$ and $u(1, t) = c_n$ with $c_0 \neq 0$, $c_n \neq 0$? When using Equation (7)?

[a]7. Describe in detail how Equation (1) with boundary conditions $u(0, t) = q(t)$, $u(1, t) = g(t)$, and $u(x, 0) = f(x)$ can be solved numerically by using System (6). Here q, g, and f are known functions.

[a]8. What finite difference equation should be a suitable replacement for the equation $\partial^2 u/\partial x^2 = \partial u/\partial t + \partial u/\partial x$ in numerical work?

[a]9. Consider the partial differential equation $\partial u/\partial x + \partial u/\partial t = 0$ with $u = u(x, t)$ in the region $[0, 1] \times [0, \infty]$, subject to the boundary conditions $u(0, t) = 0$ and $u(x, 0)$ specified. For fixed t, we discretize only the first term using $(u_{i+1} - u_{i-1})/(2h)$ for $i = 1, 2, \ldots, n-1$ and $(u_n - u_{n-1})/h$, where $h = 1/n$. Here, $u_i = u(x_i, t)$ and $x_i = ih$ with fixed t. In this way, the original problem can be considered a first-order initial-value problem

$$\frac{d\boldsymbol{y}}{dx} + \frac{1}{2h}\boldsymbol{A}\boldsymbol{y} = 0$$

where

$$\boldsymbol{y} = [u_1, u_2, \ldots, u_n]^T, \quad \frac{d\boldsymbol{y}}{dx} = \left[u_1', u_2', \ldots, u_n'\right]^T$$
$$u_i' = \frac{\partial u_i}{\partial t}$$

Determine the $n \times n$ matrix $\boldsymbol{A}$.

10. Refer to the discussion of the stability of the Crank-Nicolson procedure, and establish the inequality $u_i \ge \alpha$.

11. What happens to System (6) when $k = h^2$?

12. (**Multiple Choice**) In solving the heat equation $u_{xx} = u_t$ on the domain $t \ge 1$ and $0 \le x \le 1$, one can use the **explicit method**. Suppose the approximate solution on one horizontal line is a vector $\boldsymbol{V}_j$. Then the whole process turns out to be described by

$$\boldsymbol{V}_{j+1} = \boldsymbol{A}\boldsymbol{V}_j$$

where $\boldsymbol{A}$ is a tridiagonal matrix, having $1 - 2\sigma$ on its diagonal and σ in the superdiagonal and subdiagonal positions. Here $\sigma = k/h^2$, where k is the time step and h is the x-step. For stability in the numerical solution, what should we require?

a. $\sigma = \frac{1}{2}$

b. All eigenvalues of $\boldsymbol{A}$ satisfy $|\lambda| < 1$.

c. $k \ge h^2/2$

d. $h = 0.01$ and $k = 5 \times 10^{-3}$ **e.** None of these.

13. (Continuation) The **fully implicit method** for solving the heat conduction problem requires at each step the solution of the equation

$$\boldsymbol{A}\boldsymbol{V}_{j-1} = \boldsymbol{V}_j$$

Here, $\boldsymbol{A}$ is not the same as in the preceding problem, but is similar: It has $1 + 2\sigma$ on the diagonal and $-\sigma$ on the

subdiagonal and superdiagonal. What do we know about the eigenvalues of this matrix, $\boldsymbol{A}$?
Hint: This question concerns eigenvalues of $\boldsymbol{A}$, not $\boldsymbol{A}^{-1}$.

a. They are all negative.

b. They are all in the open interval (0, 1).

c. They are greater than 1.

d. They are in the interval $(-1, 0)$.

e. None of these.

Computer Exercises 12.1

1. Solve the same heat conduction problem as in the text except use $h = 2^{-4}$, $k = 2^{-10}$, and $u(x, 0) = x(1 - x)$. Carry out the solution until $t = 0.0125$.

2. Modify the Crank-Nicolson code in the text so that it uses the alternative scheme (7). Compare the two programs on the same problems with the same spacing.

3. Recode and test the pseudocode in this section using a computer language that supports vector operations.

4. Run the Crank-Nicolson code with different choices of h and k, in particular, letting k be much larger. Try $k = h$, for example.

5. Try to take advantage of any special commands or procedures in mathematical software such as in MATLAB, Maple, or Mathematica to solve the numerical example (1).

6. (Continuation) Use the symbolic manipulation capabilities in MATLAB, Maple, or Mathematica to verify the general analytical solution of (1).
Hint: See Exercise 12.1.3.

12.2 Hyperbolic Problems

Wave Equation Model Problem

The **wave equation** with one space variable

Wave Equation

$$\frac{\partial^2 u}{\partial t^2} = \frac{\partial^2 u}{\partial x^2} \tag{1}$$

governs the vibration of a string (transverse vibration in a plane) or the vibration in a rod (longitudinal vibration). It is an example of a second-order linear differential equation of the hyperbolic type. If Equation (1) is used to model the vibrating string, then $u(x, t)$ represents the deflection at time t of a point on the string whose coordinate is x when the string is at rest.

To pose a definite model problem, we suppose that the points on the string have coordinates x in the interval $0 \leqq x \leqq 1$ (see Figure 12.7). Let's suppose that at time $t = 0$, the deflections satisfy equations $u(x, 0) = f(x)$ and $u_t(x, 0) = 0$. Assume also that the ends of the string remain fixed. Then $u(0, t) = u(1, t) = 0$. A fully defined **boundary-value problem (BVP)**, then, is

Model Problem

$$\begin{cases} u_{tt} - u_{xx} = 0 \\ u(x, 0) = f(x) \\ u_t(x, 0) = 0 \\ u(0, t) = u(1, t) = 0 \end{cases} \tag{2}$$

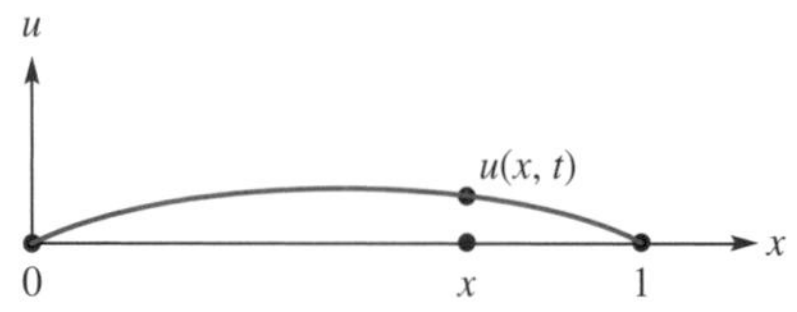

FIGURE 12.7
Vibrating string

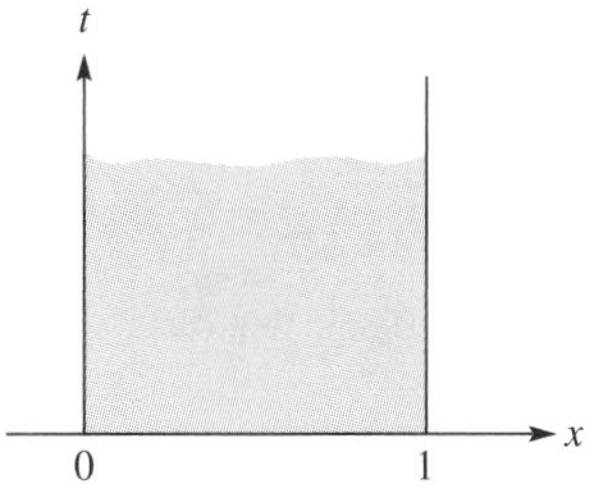

FIGURE 12.8 Wave equation: xt-plane

The region in the xt-plane where a solution is sought is the semi-infinite strip defined by inequalities $0 \leqq x \leqq 1$ and $t \geqq 0$. As in the heat conduction problem of Section 12.1, the values of the unknown function are prescribed on the boundary of the region shown (see Figure 12.8).

Analytic Solution

The model problem in (2) is so simple that it can be immediately solved. Indeed, the solution is

Solution

$$u(x, t) = \frac{1}{2}[f(x+t) + f(x-t)] \tag{3}$$

provided that f possesses two derivatives and has been extended to the whole real line by defining

Extension Conditions

$$f(-x) = -f(x), \qquad f(x+2) = f(x)$$

To verify that Equation (3) is a solution, we compute derivatives using the chain rule:

$$u_x = \frac{1}{2}[f'(x+t) + f'(x-t)], \qquad u_t = \frac{1}{2}[f'(x+t) - f'(x-t)]$$

$$u_{xx} = \frac{1}{2}[f''(x+t) + f''(x-t)], \qquad u_{tt} = \frac{1}{2}[f''(x+t) + f''(x-t)]$$

Obviously, we obtain

$$u_{tt} = u_{xx}$$

Also, we find

$$u(x, 0) = f(x)$$

Furthermore, we have

$$u_t(x, 0) = \frac{1}{2}[f'(x) - f'(x)] = 0$$

In checking endpoint conditions, we use the formulas by which f was extended:

$$u(0, t) = \frac{1}{2}[f(t) + f(-t)] = 0$$

$$\begin{aligned} u(1, t) &= \frac{1}{2}[f(1+t) + f(1-t)] \\ &= \frac{1}{2}[f(1+t) - f(t-1)] \\ &= \frac{1}{2}[f(1+t) - f(t-1+2)] = 0 \end{aligned}$$

The extension of f from its original domain to the entire real line makes it an **odd periodic** function of period 2. **Odd** means that

Odd Function

$$f(x) = -f(-x)$$

and the **periodicity** is expressed by

Periodic Function

$$f(x + 2) = f(x)$$

for all x. To compute $u(x, t)$, we need to know f at only two points on the x-axis, $x + t$ and $x - t$, as in Figure 12.9.

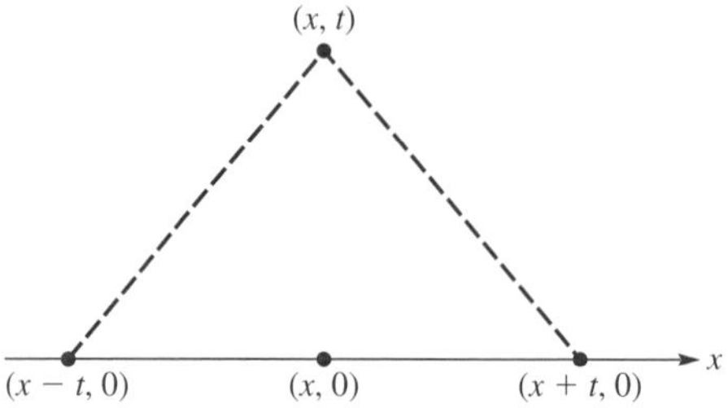

FIGURE 12.9 Wave equation: f stencil

Numerical Solution

The model problem is used next to illustrate again the principle of numerical solution. Choosing step sizes h and k for x and t, respectively, and using the familiar approximations for derivatives, we have from Equation (1)

$$\frac{1}{h^2}[u(x + h, t) - 2u(x, t) + u(x - h, t)]$$
$$= \frac{1}{k^2}[u(x, t + k) - 2u(x, t) + u(x, t - k)]$$

which can be rearranged as

Basic Scheme

$$u(x, t + k) = \rho u(x + h, t) + 2(1 - \rho)u(x, t) + \rho u(x - h, t) - u(x, t - k) \tag{4}$$

Here, we let

$$\rho = \frac{k^2}{h^2}$$

Figure 12.10 shows the point $(x, t + k)$ and the nearby points that enter into Equation (4).

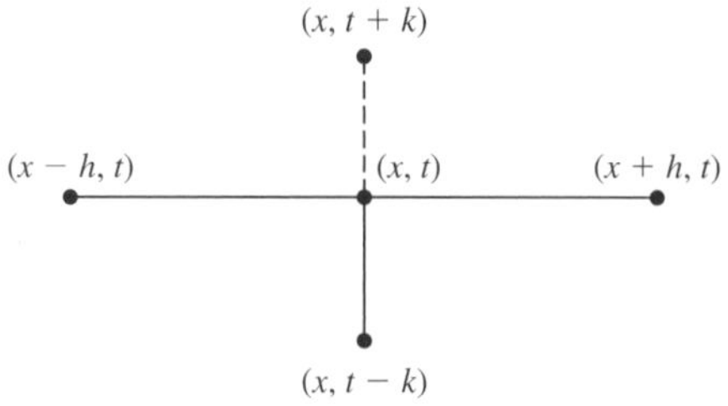

FIGURE 12.10 Wave equation: Explicit stencil

The boundary conditions in Problem (2) can be written as

Boundary Conditions

$$\begin{cases} u(x,0) = f(x) \\ \dfrac{1}{k}[u(x,k) - u(x,0)] = 0 \\ u(0,t) = u(1,t) = 0 \end{cases} \tag{5}$$

The problem defined by Equations (4) and (5) can be solved by beginning at the line $t = 0$, where u is known, and then progressing one line at a time with $t = k, t = 2k, t = 3k, \dots$. Note that because of (5), our approximate solution satisfies

$$u(x,k) = u(x,0) = f(x) \tag{6}$$

The use of the $\mathcal{O}(k)$ approximation for u_t leads to low accuracy in the computed solution to Model Problem (2). Suppose that there is a row of grid points $(x, -k)$. Letting $t = 0$ in Equation (4), we have

$$u(x,k) = \rho u(x+h,0) + 2(1-\rho)u(x,0) + \rho u(x-h,0) - u(x,-k)$$

Now in the equation

$$u_t(x,0) = 0$$

we use the central difference approximation to obtain

$$\frac{1}{2k}[u(x,k) - u(x,-k)] = 0$$

which eliminates the fictitious grid point $(x, -k)$. So instead of Equation (6), we set

Revised Scheme

$$u(x,k) = \frac{1}{2}\rho[f(x+h) + f(x-h)] + (1-\rho)f(x) \tag{7}$$

because $u(x,0) = f(x)$. Consequently, values of $u(x, nk)$, for $n \geqq 2$, can now be computed from Equation (4).

Pseudocode

A pseudocode to carry out this numerical process is given next. For simplicity, three one-dimensional arrays (u_i), (v_i), and (w_i) are used: (u_i) represents the solution being computed on the new t line; (v_i) and (w_i) represent solutions on the preceding two t lines.

```
program Hyperbolic
integer i, j;   real t, x, ρ;   real array (u_i)_{0:n}, (v_i)_{0:n}, (w_i)_{0:n}
integer n ← 10, m ← 20
real h ← 0.1, k ← 0.05
u_0 ← 0; v_0 ← 0; w_0 ← 0; u_n ← 0; v_n ← 0; w_n ← 0
ρ ← (k/h)^2
for i = 1 to n − 1
    x ← ih
    w_i ← f(x)
    v_i ← (1/2)ρ[f(x − h) + f(x + h)] + (1 − ρ)f(x)
end for
```

(*Continued*)

Hyperbolic Pseudocode

```
for j = 2 to m
    for i = 1 to n − 1
        u_i ← ρ(v_{i+1} + v_{i−1}) + 2(1 − ρ)v_i − w_i
    end for
    output j, (u_i)
    for i = 1 to n − 1
        w_i ← v_i
        v_i ← u_i
        t ← jk
        x ← ih
        u_i ← True_Solution(x, t) − v_i
    end for
    output j, (u_i)
end for
end program Hyperbolic

real function f(x)
real x
f ← sin(πx)
end function f

real function True_Solution(x, t)
real t, x
True_Solution ← sin(πx) cos(πt)
end function True_Solution
```

This pseudocode requires accompanying functions to compute values of $f(x)$ and the true solution. We chose $f(x) = \sin(\pi x)$ in our example. It is assumed that the x interval is $[0, 1]$, but when h or n is changed, the interval can be $[0, b]$; that is, $nh = b$. The numerical solution is printed on the t lines that correspond to $1k, 2k, \dots, mk$.

More advanced treatments show that the ratios

$$\rho = \frac{k^2}{h^2}$$

must not exceed 1 if the solution of the finite difference equations converges to a solution of the differential problem as $k \to 0$ and $h \to 0$. Furthermore, if $\rho > 1$, roundoff errors that occur at one stage of the computation would probably be magnified at later stages and thereby ruin the numerical solution.

Mathematical Software

In MATLAB, the PDE Toolbox has a function for producing the solution of hyperbolic problems using the finite element formulation of the scalar PDE problem. An example found in the online documentation finds the numerical solution of the two-dimensional wave propagation problem

$$\frac{\partial^2 u}{\partial t^2} = \nabla^2 u$$

on the square $-1 \leqq x, y \leqq 1$ with Dirichlet boundary conditions on the left and right boundaries, $u = 0$ for $x = \pm 1$, and zero values of the normal derivatives on the top and bottom boundaries. Further, there are Neumann boundary conditions $\partial u/\partial \nu = 0$ for $y = \pm 1$. The initial conditions $u(0) = \arctan\left(\cos\left(\frac{\pi}{2}x\right)\right)$ and $du(0)/dt = 3\sin(\pi x)\exp\left(\sin\left(\frac{\pi}{2}y\right)\right)$ are chosen to avoid putting too much energy into the higher vibration modes.

Advection Equation

We focus on the **advection equation**

Advection PDE

$$\frac{\partial u}{\partial t} = -c\frac{\partial u}{\partial x}$$

Here, $u = u(x, t)$ and $c = c(x, t)$ in which one can consider x as space and t as time. The advection equation is a hyperbolic partial differential equation that governs the motion of a conserved scalar as it is advected by a known velocity field. For example, the advection equation applies to the transport of dissolved salt in water. Even in one space dimension and constant velocity, the system remains difficult to solve. Since the advection equation is difficult to solve numerically, interest typically centers on discontinuous shock solutions, which are notoriously hard for numerical schemes to handle.

Using the forward difference approximation in time and the central-difference approximations in space, we have

$$\frac{1}{k}[u(x, t+k) - u(x, t)] = -c\frac{1}{2h}[u(x+h, t) - u(x-h, t)]$$

This gives

Central Difference Scheme

$$u(x, t+k) = u(x, t) - \frac{1}{2}\sigma\,[u(x+h, t) - u(x-h, t)]$$

where $\sigma = (k/h)c(x, t)$. All numerical solutions grow in magnitude for all time steps k. For all $\sigma > 0$, this scheme is *unstable* by Fourier stability analysis.

Lax Method

In the central-difference scheme above, replace the $u(x, t)$ term on the right-hand side by $\frac{1}{2}[u(x, t-k) + u(x, t+k)]$. Then we obtain

Lax Scheme

$$\begin{aligned} u(x, t+k) &= \frac{1}{2}[u(x, t-k) + u(x, t+k)] - \frac{1}{2}\sigma\,[u(x+h, t) - u(x-h, t)] \\ &= \frac{1}{2}(1+\sigma)u(x-h, t) + \frac{1}{2}(1+\sigma)u(x, t-k) \end{aligned}$$

This is the **Lax method**, and this simple change makes the method conditionally stable.

Upwind Method

Another way of obtaining a stable method is by using a one-sided approximation to u_x in the advection equation as long as the side is taken in the *upwind* direction. If $c > 0$, the transport is to the right. This can be interpreted as a wind of speed c blowing the solution from left to right. So the upwind direction is to the left for $c > 0$ and to the right for $c < 0$. Thus, the upwind difference approximation is

$$u_x(x, t) \approx \begin{cases} -c\,[u(x, t) - u(x-h, t)]/h & (c > 0) \\ -c\,[u(x+h, t) - u(x, t)]/h & (c < 0) \end{cases}$$

Then the upwind scheme for the advection equation is

Upwind Scheme

$$u(x, t+k) = u(x, t) - \sigma\begin{cases} -c\,[u(x, t) - u(x-h, t)]/h & (c > 0) \\ -c\,[u(x+h, t) - u(x, t)]/h & (c < 0) \end{cases}$$

Lax-Wendroff Method

The Lax-Wendroff scheme is second-order in space and time. The following is one of several possible forms of this method. We start with a Taylor series expansion over one time step:

Taylor Series Expansion

$$u(x, t+k) = u(x, t) + ku_t(x, t) + \frac{1}{2}k^2 u_{tt}(x, t) + \mathcal{O}(k^3)$$

Now use the advection equation to replace time derivatives on the right-hand side by space derivatives:

$$\begin{aligned} u_t &= -cu_x \\ u_{tt} &= (-cu_x)_t \\ &= -c_t u_x - c\,(u_x)_t \\ &= -c_t u_x - c\,(u_t)_x \\ &= -c_t u_x + c\,(cu_x)_t \end{aligned}$$

Here, we have let $c = c(x, t)$ and have *not* assumed c is a constant. Substituting for u_t and u_{xx} gives us

$$u(x, t+k) = u(x, t) - cku_x + \frac{1}{2}k^2\left[-c_t u_x + c\,(cu_x)_x\right] + \mathcal{O}(k^3)$$

where everything on the right-hand side is evaluated at (x, t). If we approximate the space derivative with second-order differences, we obtain a second-order scheme in space and time:

Second-Order Scheme

$$\begin{aligned} u(x, t+k) \approx u(x, t) &- ck\frac{1}{2h}\left[u(x+h, t) - u(x-h, t)\right] \\ &+ \frac{1}{2}k^2\left[-c_t\frac{1}{2h}\left[u(x+h, t) - u(x-h, t)\right] + c\,(cu_x)_x\right] \end{aligned}$$

The difficulty with this scheme arises when c depends on space and we must evaluate the last term in the expression above. In the case in which c is a constant, we obtain

$$\begin{aligned} c\,(cu_x)_x &= c^2 u_{xx} \\ &\approx \frac{1}{2h}[u(x+h, t) - 2u(x, t) + u(x-h, t)] \end{aligned}$$

The **Lax-Wendroff scheme** becomes

Lax-Wendroff Scheme

$$\begin{aligned} u(x, t+k) = u(x, t) &- \frac{1}{2}\sigma\left[u(x+h, t) - u(x-h, t)\right] \\ &+ \frac{1}{2}c\sigma^2\left[u(x+h, t) - 2u(x, t) + u(x-h, t)\right] \end{aligned}$$

where $\sigma = c(k/h)$. As does the Lax method, this method has numerical dissipation (lose of amplitude); however, it is relatively weak.

Summary 12.2

- We consider a model problem involving the following **hyperbolic partial differential equation**:

$$\frac{\partial^2 u}{\partial t^2} = \frac{\partial^2 u}{\partial x^2}$$

Using finite differences, we approximate it by

$$\frac{1}{h^2}[u(x+h,t) - 2u(x,t) + u(x-h,t)] = \frac{1}{k^2}[u(x,t+k) - 2u(x,t) + u(x,t-k)]$$

The computational form is

$$u(x,t+k) = \rho u(x+h,t) + 2(1-\rho)u(x,t) + \rho u(x-h,t) - u(x,t-k)$$

where $\rho = k^2/h^2 < 1$. At $t = 0$, we use

$$u(x,k) = \frac{1}{2}\rho[f(x+h) + f(x-h)] + (1-\rho)f(x)$$

Exercises 12.2

[a]**1.** What is the solution of the boundary-value problem

$$u_{tt} = u_{xx}, \qquad u(x,0) = x(1-x), \qquad u_t(x,0) = 0, \qquad u(0,t) = u(1,t) = 0$$

at the point where $x = 0.3$ and $t = 4$?

[a]**2.** Show that the function $u(x,t) = f(x+at) + g(x-at)$ satisfies the wave equation $u_{tt} = a^2u_{xx}$.

[a]**3.** (Continuation) Using the idea in the preceding problem, solve this boundary-value problem:

$$u_{tt} = u_{xx}, \qquad u(x,0) = F(x), \qquad u_t(x,0) = G(x), \qquad u(0,t) = u(1,t) = 0$$

4. Show that the boundary-value problem

$$u_{tt} = u_{xx}, \qquad u(x,0) = 2f(x), \qquad u_t(x,0) = 2g(x)$$

has the solution

$$u(x,t) = f(x+t) + f(x-t) + G(x+t) - G(x-t)$$

where G is an antiderivative (i.e., indefinite integral) of g. Here, we assume that $-\infty < x < \infty$ and $t \geqq 0$.

5. (Continuation) Solve the preceding problem on a finite x interval, for example, $0 \leqq x \leqq 1$, adding boundary condition $u(0,t) = u(1,t) = 0$. In this case, f and g are defined only for $0 \leqq x \leqq 1$.

Computer Exercises 12.2

[a]**1.** Given $f(x)$ defined on [0, 1], write and test a function for calculating the extended f that obeys the equations $f(-x) = -f(x)$ and $f(x+2) = f(x)$.

2. (Continuation) Write a program to compute the solution of $u(x,t)$ at any given point (x,t) for the boundary-value problem of Equation (2).

3. Compare the accuracy of the computed solution, using first Equation (6) and then Equation (7), in the computer program in the text.

4. Use the program in the text to solve boundary-value Problem (2) with

$$f(x) = \frac{1}{4}\left(\frac{1}{2} - \left|x - \frac{1}{2}\right|\right), \qquad h = \frac{1}{16}, \qquad k = \frac{1}{32}$$

5. Modify the code in the text to solve boundary-value Problem (2) when $u_t(x,0) = g(x)$.
Hint: Equations (5) and (7) will be slightly different (a fact that affects only the initial loop in the program).

6. (Continuation) Use the program that you wrote for the preceding computer problem to solve the following boundary-value problem:

$$\begin{cases} u_{tt} = u_{xx} & (0 \leqq x \leqq 1,\ t \geqq 0) \\ u(x,0) = \sin \pi x \\ u_t(x,0) = \dfrac{1}{4}\sin 2\pi x \\ u(0,t) = u(1,t) = 0 \end{cases}$$

7. Modify the pseudocode (p. 539–540) to solve the following boundary-value problem:

$$\begin{cases} u_{tt} = u_{xx} & (-1 \leqq x \leqq 1,\ t \geqq 0) \\ u(x,0) = |x| - 1 \\ u_t(x,0) = 0 \\ u(-1,t) = u(1,t) = 0 \end{cases}$$

8. Modify the code in the text to avoid storage of the (v_i) and (u_i) arrays.

9. Simplify the code in the text for the special case in which $\rho = 1$. Compare the numerical solution at the same grid points for a problem in which $\rho = 1$ and $\rho \neq 1$.

10. Use mathematical software such as in MATLAB, Maple, or Mathematica to solve the wave Equation (2) and plot both the solution surface and the contour plot.

11. Use the symbolic manipulation capabilities in Maple or Mathematica to verify that Equation (3) is the general analytical solution of the wave equation.

12.3 Elliptic Problems

One of the most important partial differential equations in mathematical physics and engineering is **Laplace's equation**, which has the following form in two variables:

Laplace's Equation

$$\nabla^2 u \equiv \frac{\partial^2 u}{\partial x^2} + \frac{\partial^2 u}{\partial y^2} = 0$$

Closely related to it is **Poisson's equation**:

Poisson's Equation

$$\nabla^2 u = g(x, y)$$

These are examples of **elliptic** equations. (Refer to Exercise 12.1.1 for the classification of equations.) The boundary conditions associated with elliptic equations generally differ from those for parabolic and hyperbolic equations. A model problem is considered here to illustrate the numerical procedures that are often used.

Helmholtz Equation

Suppose that a function $u = u(x, y)$ of two variables is the solution to a certain physical problem. This function is unknown but has some properties that, theoretically, determine it uniquely. We assume that on a given region R in the xy-plane,

Model Problem

$$\begin{cases} \nabla^2 u + f u = g \\ u(x, y) \quad \text{(known on the boundary of } R\text{)} \end{cases} \tag{1}$$

Here, $f = f(x, y)$ and $g = g(x, y)$ are given continuous functions defined in R. The boundary values could be given by a third function

$$u(x, y) = q(x, y)$$

on the perimeter of R. When f is a constant, this partial differential equation is called the **Helmholtz equation**. It arises in looking for oscillatory solutions of the wave equations.

Helmholtz Equation

Finite-Difference Method

As before, we find an approximate solution of such a problem by the finite-difference method. The first step is to select approximate formulas for the derivatives in our problem. In the present situation, we use the standard formula

f'' Central Difference Approximation

$$f''(x) \approx \frac{1}{h^2}[f(x+h) - 2f(x) + f(x-h)] \tag{2}$$

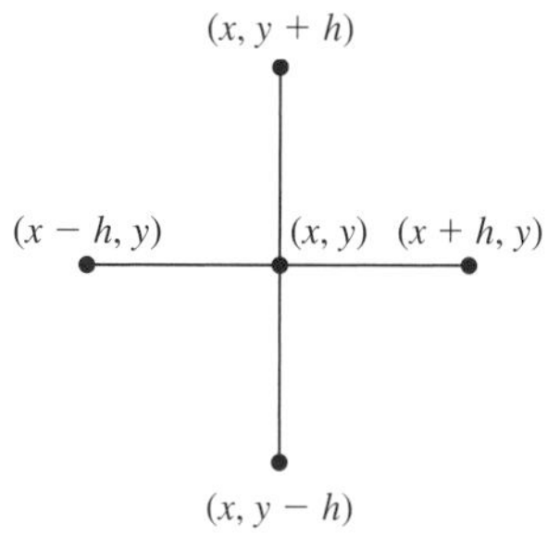

FIGURE 12.11
Laplace's equation:
Five-point stencil

derived in Section 4.3. If it is used on a function of two variables, we obtain the **five-point formula** approximation to Laplace's equation:

Five-Point Formula

$$\nabla^2 u \approx \frac{1}{h^2}[u(x+h, y) + u(x-h, y) + u(x, y+h) + u(x, y-h) - 4u(x, y)] \qquad (3)$$

This formula involves the five points displayed in Figure 12.11.

The local error inherent in the five-point formula is

Error Term

$$-\frac{h^2}{12}\left[\frac{\partial^4 u}{\partial x^4}(\xi, y) + \frac{\partial^4 u}{\partial y^4}(x, \eta)\right] \qquad (4)$$

and for this reason, Formula (3) is said to provide an approximation of order $\mathcal{O}(h^2)$. In other words, if grids are used with smaller and smaller spacing, $h \to 0$, then the error that is committed in replacing $\nabla^2 u$ by its finite-difference approximation goes to zero as rapidly as does h^2. Equation (3) is called the **five-point formula** because it involves values of u at (x, y) and at the four nearest grid points.

Problem Simulation

It should be emphasized that when the differential equation in (1) is replaced by the finite-difference analog, we have changed the problem. Even if the analogous finite-difference problem is solved with complete precision, the solution is that of a problem that only *simulates* the original one. This simulation of one problem by another becomes better and better as h is made to decrease to zero, but the computing cost inevitably increases.

We should also note that other representations of the derivatives can be used. For example, the **nine-point formula** is

Nine-Point Formula

$$\begin{aligned}\nabla^2 u \approx \frac{1}{6h^2}[&4u(x+h, y) + 4u(x-h, y) + 4u(x, y+h) + 4u(x, y-h)\\ &+ u(x+h, y+h) + u(x-h, y+h) + u(x+h, y-h)\\ &+ u(x-h, y-h) - 20u(x, y)]\end{aligned} \qquad (5)$$

This formula is of order $\mathcal{O}(h^2)$. In the special case that u is a **harmonic function** (which means it is a solution of Laplace's equation), the nine-point formula is of order $\mathcal{O}(h^6)$. For additional details, see Forsythe and Wasow [1960, pp. 194–195]. Hence, it is an extremely accurate approximation in using finite-difference methods and solving the Poisson equation $\nabla^2 u = g$, with g a harmonic function. For more general problems, the nine-point Formula (5) has the same order error term as the five-point Formula (3) [namely, $\mathcal{O}(h^2)$] and would not be an improvement over it.

If the mesh spacing is not regular (say, h_1, h_2, h_3, and h_4 are the left, bottom, right, and top spacing, respectively), then it is not difficult to show that at (x, y) the **irregular**

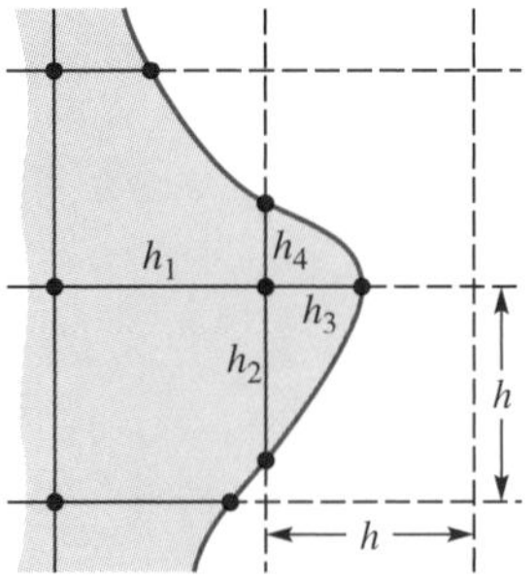

FIGURE 12.12 Boundary points: Irregular mesh spacing

five-point formula is

Irregular Five-Point Formula

$$\begin{aligned}\nabla^2 u \approx & \frac{1}{\frac{1}{2}h_1h_3(h_1+h_3)}[h_1u(x+h_3,y)+h_3u(x-h_1,y)] \\ & + \frac{1}{\frac{1}{2}h_2h_4(h_2+h_4)}[h_2u(x,y+h_4)+h_4u(x,y-h_2)] \\ & -2\left(\frac{1}{h_1h_3}+\frac{1}{h_2h_4}\right)u(x,y)\end{aligned} \tag{6}$$

which is only of order h when $h_1 = \alpha_i h$ for $0 < \alpha_i < 1$. This formula is usually used near boundary points, as in Figure 12.12. If the mesh is small, however, the boundary points can be moved over slightly to avoid the use of (6). This perturbation of the region R (in most cases for small h) produces an error no greater than that introduced by using the irregular Scheme (6).

Returning to the model Problem (1), we cover the region R by mesh points

$$x_i = ih, \qquad y_j = jh, \qquad (i, j \geqq 0) \tag{7}$$

At this time, it is convenient to introduce an abbreviated notation:

$$u_{ij} = u(x_i, y_i), \qquad f_{ij} = f(x_i, y_i), \qquad g_{ij} = g(x_i, y_j) \tag{8}$$

With it, the five-point formula takes on a simple form at the point (x_i, y_j):

$$(\nabla^2 u)_{ij} \approx \frac{1}{h^2}(u_{i+1,j} + u_{i-1,j} + u_{i,j+1} + u_{i,j-1} - 4u_{ij}) \tag{9}$$

If this approximation is made in the partial differential Equation (1), the result is (the reader should verify it)

Typical Linear Equation

$$-u_{i+1,j} - u_{i-1,j} - u_{i,j+1} - u_{i,j-1} + \left(4 - h^2 f_{ij}\right) u_{ij} = -h^2 g_{ij} \tag{10}$$

The coefficients of this equation can be illustrated by a five-point star in which each point corresponds to the coefficient of u in the grid (see Figure 12.13).

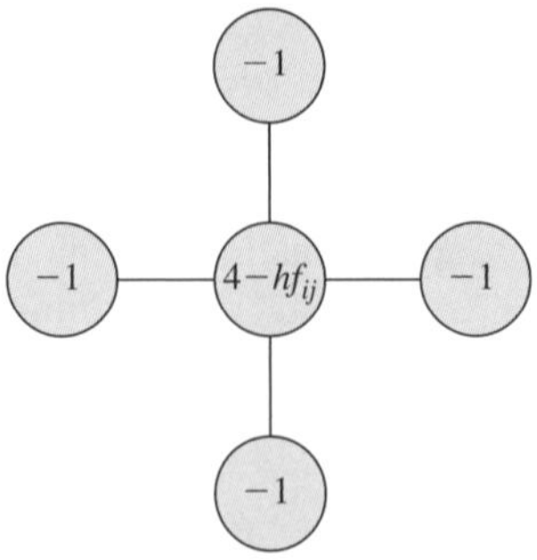

FIGURE 12.13 Five-point star

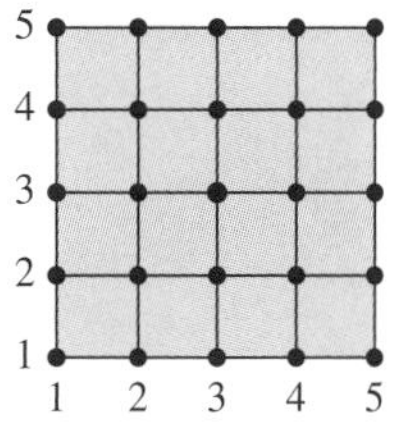

FIGURE 12.14
Uniform grid spacing

To be specific, we assume that the region R is a unit square and that the grid has spacing $h = \frac{1}{4}$ (see Figure 12.14). We obtain a single linear equation of the form (10) for each of the nine interior grid points. These nine linear equations are as follows:

$$
\begin{cases}
-u_{21} - u_{01} - u_{12} - u_{10} + (4 - h^2 f_{11})u_{11} = -h^2 g_{11} \\
-u_{31} - u_{11} - u_{22} - u_{20} + (4 - h^2 f_{21})u_{21} = -h^2 g_{21} \\
-u_{41} - u_{21} - u_{32} - u_{30} + (4 - h^2 f_{31})u_{31} = -h^2 g_{31} \\
-u_{22} - u_{02} - u_{13} - u_{11} + (4 - h^2 f_{12})u_{12} = -h^2 g_{12} \\
-u_{32} - u_{12} - u_{23} - u_{21} + (4 - h^2 f_{22})u_{22} = -h^2 g_{22} \\
-u_{42} - u_{22} - u_{33} - u_{31} + (4 - h^2 f_{32})u_{32} = -h^2 g_{32} \\
-u_{23} - u_{03} - u_{14} - u_{12} + (4 - h^2 f_{13})u_{13} = -h^2 g_{13} \\
-u_{33} - u_{13} - u_{24} - u_{22} + (4 - h^2 f_{23})u_{23} = -h^2 g_{23} \\
-u_{43} - u_{23} - u_{34} - u_{32} + (4 - h^2 f_{33})u_{33} = -h^2 g_{33}
\end{cases}
$$

This system of equations could be solved through Gaussian elimination, but let's examine them more closely. There are 45 coefficients. Since u is known at the boundary points, we move these 12 terms to the right-hand side, leaving only 33 nonzero entries out of 81 in our 9×9 system. The standard Gaussian elimination causes a great deal of fill-in, in the forward elimination phase—that is, zero entries are replaced by nonzero values. So we seek a method that retains the sparse structure of this system.

To illustrate how sparse this system of equations is, we write it in matrix notation:

$$\boldsymbol{A}\boldsymbol{u} = \boldsymbol{b} \tag{11}$$

Suppose that we order the unknowns from left to right and bottom to top:

$$\boldsymbol{u} = [u_{11}, u_{21}, u_{31}, u_{12}, u_{22}, u_{32}, u_{13}, u_{23}, u_{33}]^T \tag{12}$$

This is called the **natural ordering**. Now the coefficient matrix is

Sparse Matrix

$$
\boldsymbol{A} = \begin{bmatrix}
4-h^2 f_{11} & -1 & 0 & -1 & 0 & 0 & 0 & 0 & 0 \\
-1 & 4-h^2 f_{21} & -1 & 0 & -1 & 0 & 0 & 0 & 0 \\
0 & -1 & 4-h^2 f_{31} & 0 & 0 & -1 & 0 & 0 & 0 \\
-1 & 0 & 0 & 4-h^2 f_{12} & -1 & 0 & -1 & 0 & 0 \\
0 & -1 & 0 & -1 & 4-h^2 f_{22} & -1 & 0 & -1 & 0 \\
0 & 0 & -1 & 0 & -1 & 4-h^2 f_{32} & 0 & 0 & -1 \\
0 & 0 & 0 & -1 & 0 & 0 & 4-h^2 f_{13} & -1 & 0 \\
0 & 0 & 0 & 0 & -1 & 0 & -1 & 4-h^2 f_{23} & -1 \\
0 & 0 & 0 & 0 & 0 & -1 & 0 & -1 & 4-h^2 f_{33}
\end{bmatrix}
$$

and the right-hand side is

$$
\boldsymbol{b} = \begin{bmatrix} -h^2 g_{11} + u_{10} + u_{01} \\ -h^2 g_{21} + u_{20} \\ -h^2 g_{31} + u_{30} + u_{41} \\ -h^2 g_{12} + u_{02} \\ -h^2 g_{22} \\ -h^2 g_{32} + u_{42} \\ -h^2 g_{13} + u_{14} + u_{03} \\ -h^2 g_{23} + u_{24} \\ -h^2 g_{33} + u_{34} + u_{43} \end{bmatrix}
$$

Notice that if $f(x, y) < 0$, then $\boldsymbol{A}$ is a diagonally dominant matrix.

Gauss-Seidel Iterative Method

Since the equations are similar in form, iterative methods are often used to solve such sparse systems. Solving for the diagonal unknown, we have from Equation (10) the **Gauss-Seidel method** or **iteration** given by

Gauss-Seidel Iteration

$$
u_{ij}^{(k+1)} = \frac{1}{4 - h^2 f_{ij}} \left(u_{i+1,\,j}^{(k)} + u_{i-1,\,j}^{(k+1)} + u_{i,\,j+1}^{(k)} + u_{i,\,j-1}^{(k+1)} - h^2 g_{ij} \right)
$$

If we have approximate values of the unknowns at each grid point, this equation can be used to generate new values. We call $u^{(k)}$ the current values of the unknowns at iteration k and $u^{(k+1)}$ the value in the next iteration. Moreover, the new values are used in this equation as soon as they become available. The Gauss-Seidel method were other iterative methods were discussed in Section 8.4.

The pseudocode for this method on a rectangle is as follows:

Seidel **Pseudocode**

```
procedure Seidel(a_x, a_y, n_x, n_y, h, itmax, (u_ij))
integer i, j, k, n_x, n_y, itmax
real a_x, a_y, x, y;   real array (u_ij)_{0:n_x, 0:n_y}
for k = 1 to itmax
    for j = 1 to n_y − 1
        y ← a_y + jh
        for i = 1 to n_x − 1
            x ← a_x + ih
            v ← u_{i+1, j} + u_{i−1, j} + u_{i, j+1} + u_{i, j−1}
            u_ij ← (v − h^2 g(x, y))/(4 − h^2 f(x, y))
        end for
    end for
end for
end procedure Seidel
```

In using this procedure, one must decide on the number of iterative steps to be computed, *itmax*. The coordinates of the lower left-hand corner of the rectangle, (a_x, a_y), and the step size h are specified. The number of x grid points is n_x, and the number of y grid points is n_y.

Numerical Example and Pseudocode

Let's illustrate this procedure on the boundary-value problem

$$\begin{cases} \nabla^2 u - \dfrac{1}{25} u = 0 & \text{inside } R \text{ (unit square)} \\ u = q & \text{on the boundary of } R \end{cases} \tag{13}$$

where $q = \cosh\left(\frac{1}{5}x\right) + \cosh\left(\frac{1}{5}y\right)$. This problem has the known solution $u = q$. A driver pseudocode for the Gauss-Seidel procedure, starting with $u = 1$ and taking 20 iterations, is given next. Notice that, for $h = \frac{1}{8}$, only 81 words of storage are needed for the array in solving the 49×49 linear system iteratively.

***Elliptic* Pseudocode**

```
program Elliptic
integer i, j;   real h, x, y;   real array (u_ij)_{0:n_x, 0:n_y}
integer n_x ← 8, n_y ← 8, itmax ← 20
real a_x ← 0, b_x ← 1, a_y ← 0, b_y ← 1
h ← (b_x − a_x)/n_x
for j = 0 to n_y
    y ← a_y + jh
    u_0j ← Bndy(a_x, y)
    u_{n_x, j} ← Bndy(b_x, y)
end for
for i = 0 to n_x
    x ← a_x + ih
    u_i0 ← Bndy(x, a_y)
    u_{i,n_y} ← Bndy(x, b_y)
end for
for j = 1 to n_y − 1
    y ← a_y + jh
    for i = 1 to n_x − 1
        x ← a_x + ih
        u_ij ← Ustart(x, y)
    end for
end for
output 0, Norm((u_ij), n_x, n_y)
call Seidel(a_x, a_y, n_x, n_y, h, itmax, (u_ij))
output itmax, Norm((u_ij), n_x, n_y)
for j = 0 to n_y
    y ← a_y + jh
    for i = 0 to n_x
        x ← a_x + ih
        u_ij ← |True_Solution(x, y) − u_ij|
    end for
end for
output itmax, Norm((u_ij), n_x, n_y)
end program Elliptic
```

For this model problem, the accompanying functions are given next:

Accompanying Functions Pseudocodes

```
real function f(x, y)
real x, y
f ← −0.04
end function f

real function g(x, y)
real x, y
g ← 0
end function g

real function Bndy(x, y)
real x, y
Bndy ← True_Solution(x, y)
end function Bndy

real function Ustart(x, y)
real x, y
Ustart ← 1
end function Ustart

real function True_Solution(x, y)
real x, y
True_Solution ← cosh(0.2x) + cosh(0.2y)
end function True_Solution

real function Norm((u_ij), n_x, n_y)
real array (u_ij)_{0:n_x,0:n_y}
t ← 0
for i = 1 to n_x − 1
for j = 1 to n_y − 1
    t ← t + u_ij^2
    end for
end for
Norm ← √t
end function Norm
```

After 75 iterations, the computed values at the 49 interior grid points are as follows:

2.0000	2.0003	2.0013	2.0028	2.0050	2.0078	2.0113	2.0154	2.0201
2.0003	2.0006	2.0016	2.0031	2.0053	2.0081	2.0116	2.0157	2.0204
2.0013	2.0016	2.0025	2.0041	2.0062	2.0091	2.0125	2.0166	2.0213
2.0028	2.0031	2.0041	2.0056	2.0078	2.0106	2.0141	2.0182	2.0229
2.0050	2.0053	2.0062	2.0078	2.0100	2.0128	2.0163	2.0204	2.0251
2.0078	2.0081	2.0091	2.0106	2.0128	2.0156	2.0191	2.0232	2.0279
2.0113	2.0116	2.0125	2.0141	2.0163	2.0191	2.0225	2.0266	2.0313
2.0154	2.0157	2.0166	2.0182	2.0204	2.0232	2.0266	2.0307	2.0354
2.0201	2.0204	2.0213	2.0229	2.0251	2.0279	2.0313	2.0354	2.0401

The Euclidean norm $||u||_2^2 = \sum_{i=1}^{n_x-1}\sum_{j=1}^{n_y-1} u_{ij}^2$ of the difference between the computed values and the known solution of the boundary-value problem (13) is approximately 0.47×10^{-4}.

This example is a good illustration of the fact that the numerical problem being solved is the system of linear Equations (11), which is a discrete approximation to the continuous-boundary-value Problem (13). When comparing the true solution of (13) with the computed solution of the system, remember the discretization error involved in making the approximation. This error is $\mathcal{O}(h^2)$. With h as large as $h = \frac{1}{8}$, most of the errors in the computed solution are due to the discretization error! To obtain a better agreement between the discrete and continuous problems, select a much smaller mesh size. Of course, the resulting linear system will have a coefficient matrix that is extremely large and quite sparse. Iterative methods are ideal for solving such systems that arise from partial differential equations. For additional information, see the references listed at the end of this section.

For a range of engineering and science applications, MATLAB has a PDE Toolbox for the numerical solution of elliptic partial differential equations. It can accommodate two space variables and one time variable. After discretizing the equation over an unstructured mesh, the software applies finite elements to solve the PDE and offers a provision for visualizing the results. The first example is Poisson's equation

$$\nabla^2 u = -1$$

in the unit circle with $u = 0$ on the boundary. A comparison of the finite-element solution is made with the exact solution. We discuss finite element methods next.

Finite-Element Methods

The finite-element method has become one of the major strategies for solving partial differential equations. It provides an alternative to the finite-difference methods discussed up to now in this chapter.

As an illustration, we develop a version of the finite-element method for Poisson's equation

Poisson's Equation

$$\nabla^2 u \equiv u_{xx} + u_{yy} = r$$

where r is a constant function. The partial differential equation holds over a specified region R in a two-dimensional plane. Solving Poisson's equation is equivalent to minimizing the expression

$$J(u) = \int\int_R \left[\frac{1}{2}\left(u_x^2 + u_y^2\right) + ru\right]\,dx\,dy$$

This means that if the function u minimizes the expression above, then u obeys Poisson's equation. Suppose the region is subdivided into triangles using approximations as necessary. The function u is approximated by a function φ that is a composite of plane triangular elements, each defined over a triangular piece of R. Then consider the substitute problem of minimizing

$$\sum_e J_e\left(\varphi^{(e)}\right)$$

where each term in the summation is evaluated over its own base triangle T as described below. (By accepting this theory on faith, you should be able to grasp the general idea of the finite-element method.)

Plane Triangular Element

Assume that a base triangle has vertices (x_i, y_i), (x_j, y_j), and (x_k, y_k). The solution surface above the triangle is approximated by a **plane triangular element** denoted $\varphi^{(e)}(x, y)$, where the superscript indicated this element. Let z_i, z_j, and z_k be the distances up to the plane at the triangle corners called **nodes**. Let $L_i^{(e)}$ be 1 at node i and 0 at nodes j and k. Similarly, let $L_j^{(e)}$ be 1 at node j and 0 at nodes i and k, and let $L_k^{(e)}$ be 1 at node k and 0 at nodes i and j.

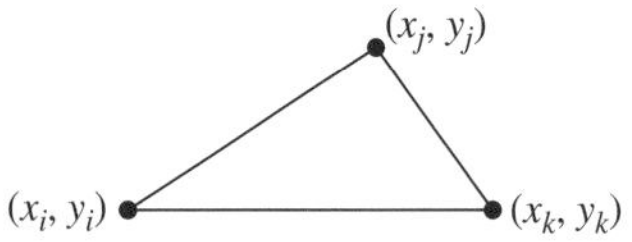

FIGURE 12.15
Base triangle

As is shown in Figure 12.15, the area of the base triangle, denoted Δ_e, is given by

Area of Base Triangle

$$\Delta_e = \frac{1}{2}\text{Det}\begin{bmatrix} 1 & x_i & y_i \\ 1 & x_j & y_j \\ 1 & x_k & y_k \end{bmatrix}$$

$$= x_j y_k + x_i y_j + x_k y_i - x_j y_i - x_i y_k - x_k y_j$$

Consequently, we obtain

$L_i^{(e)}$

$$L_i^{(e)} = \frac{1}{2}\Delta_e^{-1}\text{Det}\begin{bmatrix} 1 & x & y \\ 1 & x_j & y_j \\ 1 & x_k & y_k \end{bmatrix}$$

$$= \frac{1}{2}\Delta_e^{-1}[(x_j y_k - x_k y_j) + (y_j - y_k)x + (x_k - x_j)y]$$

$$\equiv \frac{1}{2}\Delta_e^{-1}\left(a_i^{(e)} + b_i^{(e)}x + c_i^{(e)}y\right)$$

We have defined the coefficients $a_i^{(e)}$, $b_i^{(e)}$, and $c_i^{(e)}$. Similarly, we find

$L_j^{(e)}$

$$L_j^{(e)} = \frac{1}{2}\Delta_e^{-1}\text{Det}\begin{bmatrix} 1 & x & y \\ 1 & x_k & y_k \\ 1 & x_i & y_i \end{bmatrix}$$

$$= \frac{1}{2}\Delta_e^{-1}[(x_k y_i - x_i y_k) + (y_k - y_i)x + (x_i - x_k)y]$$

$$\equiv \frac{1}{2}\Delta_e^{-1}\left(a_j^{(e)} + b_j^{(e)}x + c_j^{(e)}y\right)$$

and

$L_k^{(e)}$

$$L_k^{(e)} = \frac{1}{2}\Delta_e^{-1}\text{Det}\begin{bmatrix} 1 & x & y \\ 1 & x_i & y_i \\ 1 & x_j & y_j \end{bmatrix}$$

$$= \frac{1}{2}\Delta_e^{-1}[(x_i y_j - x_j y_i) + (y_i - y_j)x + (x_j - x_i)y]$$

$$\equiv \frac{1}{2}\Delta_e^{-1}\left(a_k^{(e)} + b_k^{(e)}x + c_k^{(e)}y\right)$$

Finally, we obtain

$$\varphi^{(e)} = L_i^{(e)}z_i + L_j^{(e)}z_j + L_k^{(e)}z_k$$

We have

$$J_e\left(\varphi^{(e)}\right) = \int\!\!\int_T \left[\frac{1}{2}\left(\left(\varphi_x^{(e)}\right)^2 + \left(\varphi_y^{(e)}\right)^2\right) + r\varphi^{(e)}\right] dx\,dy \equiv F(z_i, z_j, z_k)$$

To solve the minimization problem, we set the appropriate derivatives to zero, which requires derivatives of the components. Notice that

$$\varphi_x^{(e)} = \frac{1}{2}\Delta_e^{-1}\left(b_i^{(e)}z_i + b_j^{(e)}z_j + b_k^{(e)}z_k\right)$$

and

$$\varphi_y^{(e)} = \frac{1}{2}\Delta_e^{-1}\left(c_i^{(e)} z_i + c_j^{(e)} z_j + c_k^{(e)} z_k\right)$$

We carry out the differentiations

$$\begin{aligned}
\partial F/\partial z_i &= \int\!\!\int_T \left(\varphi_x^{(e)}\varphi_{xz_i}^{(e)} + \varphi_y^{(e)}\varphi_{yz_i}^{(e)} + r\varphi_{z_i}^{(e)}\right) dx\,dy \\
&= \int\!\!\int_T \left(\varphi_x^{(e)}\frac{1}{2}\Delta_e^{-1} b_i^{(e)} + \varphi_y^{(e)}\frac{1}{2}\Delta_e^{-1} c_i^{(e)} + rL_i^{(e)}\right) dx\,dy \\
&= \frac{1}{4}\Delta_e^{-1}\Big[\left(\left(b_i^{(e)}\right)^2 + \left(c_i^{(e)}\right)^2\right) z_i + \left(b_i^{(e)} b_j^{(e)} + c_i^{(e)} c_j^{(e)}\right) z_j \\
&\qquad + \left(b_i^{(e)} b_k^{(e)} + c_i^{(e)} c_k^{(e)}\right) z_k\Big] + r\frac{1}{3}\Delta_e
\end{aligned}$$

Here, the integrations are straightforward by elementary calculus. Moreover, it can be shown that

$$\int\!\!\int_T L_i^{(e)}\,dx\,dy = \int\!\!\int_T L_j^{(e)}\,dx\,dy = \int\!\!\int_T L_k^{(e)}\,dx\,dy = \frac{1}{3}\Delta_e$$

where Δ_e is the area of each triangle T. Similar results are obtained for $\partial F/\partial z_j$ and $\partial F/\partial z_k$. Consequently, we set

$$\begin{bmatrix} \partial F/\partial z_i \\ \partial F/\partial z_j \\ \partial F/\partial z_k \end{bmatrix} = \begin{bmatrix} 0 \\ 0 \\ 0 \end{bmatrix}$$

and we obtain

$$\begin{bmatrix}
\left(b_i^{(e)}\right)^2 + \left(c_i^{(e)}\right)^2 & b_i^{(e)} b_j^{(e)} + c_i^{(e)} c_j^{(e)} & b_i^{(e)} b_k^{(e)} + c_i^{(e)} c_k^{(e)} \\
b_i^{(e)} b_j^{(e)} + c_i^{(e)} c_j^{(e)} & \left(b_j^{(e)}\right)^2 + \left(c_j^{(e)}\right)^2 & b_j^{(e)} b_k^{(e)} + c_j^{(e)} c_k^{(e)} \\
b_i^{(e)} b_k^{(e)} + c_i^{(e)} c_k^{(e)} & b_j^{(e)} b_k^{(e)} + c_j^{(e)} c_k^{(e)} & \left(b_k^{(e)}\right)^2 + \left(c_k^{(e)}\right)^2
\end{bmatrix}
\begin{bmatrix} z_1 \\ z_2 \\ z_3 \end{bmatrix} = -\frac{4}{3} r \Delta_e^2 \begin{bmatrix} 1 \\ 1 \\ 1 \end{bmatrix}$$

This matrix equation contains all the ingredients we need to assemble the partial derivatives. In a particular application, we need to do the proper assembling. For each element $\varphi^{(e)}$, the active nodes i, j, and k are those that contribute nonzero values. These contributions are recorded for derivatives relative to the corresponding variables among the z_i, z_j, z_k, and so on.

EXAMPLE 1 Apply the finite-element method to solve Poisson's equation $u_{xx} + u_{yy} = 4$ over the unit square with the triangularizations shown in Figure 12.16 and using boundary values corresponding to the exact solution $u(x, y) = x^2 + y^2$.

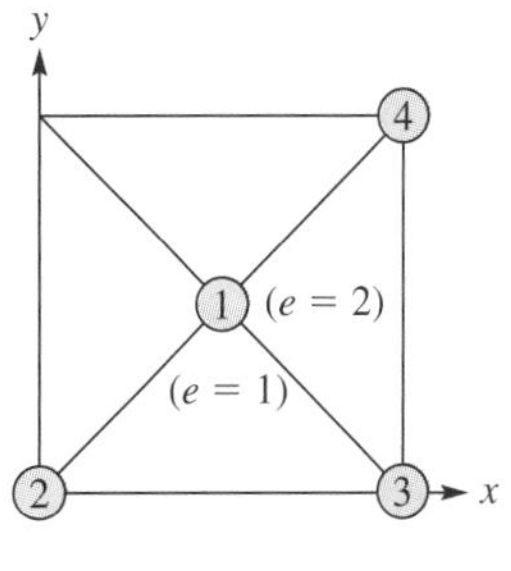

FIGURE 12.16
Triangularization

Solution By symmetry, we need to consider only the bottom right-hand part of the square, which has been split into two triangles. The input ingredients are nodes 1 to 4, where the coordinates (x, y) are as follows: node 1: $\left(\frac{1}{2}, \frac{1}{2}\right)$, node 2: (0, 0), node 3: (1, 0), and node 4: (1, 1). The elements are two triangles with node numbers indicated: $e = 1$: 1, 2, 3 and $e = 2$: 1, 3, 4. The astute reader will notice that the z coordinates need to be determined only for node 1, since they are boundary values for nodes 2, 3, 4! However, we will ignore this fact for the moment to illustrate the assembly process in the finite-element method. Notice that the areas of the triangular elements are $\Delta_1 = \Delta_2 = \frac{1}{4}$ and $r = 4$. First, we compute the $a^{(e)}, b^{(e)}, c^{(e)}$ coefficients from this basic information. In the following table, each column corresponds to a node (i, j, k):

$$
\begin{array}{c|ccc|ccc}
 & & e=1 & & & e=2 & \\
\hline
a^{(e)} & 0 & \frac{1}{2} & 0 & 1 & 0 & -\frac{1}{2} \\
b^{(e)} & 0 & -\frac{1}{2} & \frac{1}{2} & -1 & \frac{1}{2} & \frac{1}{2} \\
c^{(e)} & 1 & -\frac{1}{2} & -\frac{1}{2} & 0 & -\frac{1}{2} & \frac{1}{2}
\end{array}
$$

One can verify that the columns do produce the desired $L_i^{(e)}$, $L_j^{(e)}$, and $L_k^{(e)}$ functions. For example, the first column gives $L_i^{(1)} = \frac{1}{2}\Delta_1^{-1}[0 + 0 \cdot x + 1 \cdot y] = 2y$. At node 1, this gives the value of 1, while at nodes 2 and 3, it gives the value 0. Similarly, the other columns produce the desired results.

Next, we obtain the matrix equation for element $e = 1$:

$$
\begin{bmatrix} 1 & -\frac{1}{2} & -\frac{1}{2} \\ -\frac{1}{2} & \frac{1}{2} & 0 \\ -\frac{1}{2} & 0 & \frac{1}{2} \end{bmatrix}
\begin{bmatrix} z_1 \\ z_2 \\ z_3 \end{bmatrix}
=
\begin{bmatrix} -\frac{1}{3} \\ -\frac{1}{3} \\ -\frac{1}{3} \end{bmatrix}
$$

and the matrix equation for element $e = 2$:

$$
\begin{bmatrix} 1 & -\frac{1}{2} & -\frac{1}{2} \\ -\frac{1}{2} & \frac{1}{2} & 0 \\ -\frac{1}{2} & 0 & -\frac{1}{2} \end{bmatrix}
\begin{bmatrix} z_1 \\ z_3 \\ z_4 \end{bmatrix}
=
\begin{bmatrix} -\frac{1}{3} \\ -\frac{1}{3} \\ -\frac{1}{3} \end{bmatrix}
$$

Then we assemble the two matrices, obtaining

$$
\begin{bmatrix} 2 & -\frac{1}{2} & -1 & -\frac{1}{2} \\ -\frac{1}{2} & \frac{1}{2} & 0 & 0 \\ -1 & 0 & 1 & 0 \\ -\frac{1}{2} & 0 & 0 & -\frac{1}{2} \end{bmatrix}
\begin{bmatrix} z_1 \\ z_2 \\ z_3 \\ z_4 \end{bmatrix}
=
\begin{bmatrix} -\frac{2}{3} \\ -\frac{1}{3} \\ -\frac{2}{3} \\ -\frac{1}{3} \end{bmatrix}
$$

Now that we have illustrated the process of assembling the elements, we can quickly find the solution using the fact that $z_2 = 0$, $z_3 = 1$, and $z_4 = 2$, since they are boundary values. Using these values in the last matrix equation shown, we immediately find that $z_1 = \frac{2}{3}$. This is a rough approximation, since the true value is $\frac{1}{2}$. Remember that $u(x, y) = x^2 + y^2$ is the exact solution. ■

We can obtain more accurate approximations by adding more elements and writing a computer program to handle the computations. (See Computer Exercise 12.3.14.) For additional details, see Scheid [1990] and Sauer [2012].

More on Finite Elements

At first, we take a very general approach to this topic, supposing that we have a linear transformation $\boldsymbol{A}$ and want to solve the equation

$$\boldsymbol{A}\boldsymbol{u} = \boldsymbol{b}$$

for $\boldsymbol{u}$, when $\boldsymbol{b}$ is given. This obviously includes the case when $\boldsymbol{A}$ is an $m \times n$ matrix and $\boldsymbol{b}$ is a vector of m components. But there are many complicated problems that fit this same mold.

For example, $\boldsymbol{A}$ can be a linear differential operator, and we may wish to solve a two-point boundary-value problem involving it, such as

$$\begin{cases} u''(t) + 2u(t) = t^2 & (0 \leqq t \leqq 1) \\ u(0) = u(1) = 0 \end{cases}$$

Here, $\boldsymbol{A}$ operates on functions and is defined by the equation $\boldsymbol{A}\boldsymbol{u} = \boldsymbol{u}'' + 2\boldsymbol{u}$.

Another example of great importance is the model problem Equation (1). In this case, $\boldsymbol{A}$ would be the Laplacian differential operator. This problem is discussed in Chapter 13 as well.

Basic Functions

The basic strategy of the finite-element method for solving the equation $\boldsymbol{A}\boldsymbol{u} = \boldsymbol{b}$ is to select *basic functions* $\boldsymbol{v}_1, \boldsymbol{v}_2, \dots, \boldsymbol{v}_n$ and try to solve the equation with a linear combination of these basic functions. Since $\boldsymbol{A}$ is assumed to be a linear transformation, we obtain

$$\boldsymbol{A}\boldsymbol{u} = \boldsymbol{A} \sum_{j=1}^{n} c_j \boldsymbol{v}_j = \sum_{j=1}^{n} c_j \left(\boldsymbol{A}\boldsymbol{v}_j\right) = \boldsymbol{b}$$

Now the unknowns in the problem are the coefficients c_j. Typically, the equation just displayed is inconsistent because $\boldsymbol{b}$ is not in the linear span of the set of functions $\{\boldsymbol{A}\boldsymbol{v}_1, \boldsymbol{A}\boldsymbol{v}_2, \dots, \boldsymbol{A}\boldsymbol{v}_n\}$. In this case, one must compromise and accept an approximate solution to the set of equations. Many different tactics can be used to arrive at an approximate solution to the problem. For example, a least-squares approach can be used if the linear space involved has an inner product, $\langle \cdot, \cdot \rangle$. The coefficients c_j would then be chosen so that the orthogonality condition was fulfilled; that is,

$$\sum_{j=1}^{n} c_j \boldsymbol{A}\boldsymbol{v}_j - \boldsymbol{b} \quad \perp \quad \operatorname{Span}\{\boldsymbol{v}_1, \boldsymbol{v}_2, \dots, \boldsymbol{v}_n\}$$

This leads to the **normal equations**

Normal Equations

$$\sum_{j=1}^{n} \langle \boldsymbol{A}\boldsymbol{v}_j, \boldsymbol{v}_i \rangle c_j = \langle \boldsymbol{b}, \boldsymbol{v}_i \rangle \qquad (1 \leqq i \leqq n)$$

These equations for the unknown coefficients c_j are also known (in this context) as the **Galerkin equations**. They form a system of n linear equations in n unknowns.

We shall illustrate this process with a two-point boundary-value problem involving a second-order ordinary differential equation:

$$\begin{cases} u''(t) + g(t)u(t) = f(t) \\ u(0) = a, \qquad u(1) = b \end{cases}$$

The finite-element method usually uses *local* functions as the basic functions in the previous discussion. This means that each basic function should be zero except on a short interval. B splines have this property and are therefore often used in the finite-element method. In the present problem, we shall want to use B splines having two continuous derivatives because the operator $\boldsymbol{A}$ will be defined by

$$\boldsymbol{A}\boldsymbol{u} = \boldsymbol{u}'' + \boldsymbol{g}\boldsymbol{u}$$

Hence, cubic splines would suggest themselves. Define knots $t_i = ih$, where h is a chosen step size. (Its reciprocal should be an integer in this example.) Let $\boldsymbol{B}_j^3$ be the cubic B splines corresponding to the given knots. This is an infinite list of B splines, as discussed in Chapter 6. All but a finite number are zero on the interval [0, 1]. The ones that are not identically zero on the interval [0, 1] can be relabeled as $\boldsymbol{v}_1, \boldsymbol{v}_2, \ldots, \boldsymbol{v}_n$. These are our *test functions*. Proceeding as before, we arrive at a set of n linear equations in n unknowns. The details require one to find the functions $\boldsymbol{A}\boldsymbol{v}_j$ by using the B spline formulas in Chapter 6. This is tedious and not very instructive.

Similar considerations can be applied to Laplace's equation on a given domain. To illustrate, we take the domain to be a square of side 2, where $0 \leqq x, y \leqq 2$. On the boundary of the square, we require $u(x, y) = \sin(xy)$. Such a problem is called a **Dirichlet problem**. For base functions, we use functions v_j that already satisfy the homogeneous part of the problem. That is, we want each v_j to satisfy Laplace's equation inside the square domain. Functions that satisfy Laplace's equation are said to be **harmonic**. We can exploit the fact that the real and imaginary parts of an analytic function are harmonic. Thus, if we set $z = x + iy$ and compute z^k, we will be able to extract harmonic functions that are polynomials. Here are a few harmonic polynomials, v_j for $0 \leqq j \leqq 6$:

Harmonic Polynomials

$$\begin{array}{lll} z = 1, & v_0(x, y) = 1 & \\ z = x + iy, & v_1(x, y) = x, & v_2(x, y) = y \\ z^2 = (x + iy)^2, & v_3(x, y) = x^2 - y^2, & v_4(x, y) = 2xy \\ z^3 = (x + iy)^3, & v_5(x, y) = x^3 - 3xy^2, & v_6(x, y) = 3x^2y - y^3 \end{array}$$

Using these seven functions, we form $u = \sum_{j=0}^{6} c_j v_j$. This satisfies Laplace's equation, and we can concentrate on making u close to the specified boundary value $x^3 - y^2$ on the perimeter of the square. There are many ways to proceed, and we choose first to use a method called **collocation**. In this process, we select a number of points on the boundary and write down an equation at each point that says the value of $\sum_{j=0}^{6} c_j v_j$ equals the prescribed value. If the number of points equals the number of basic functions, we have the classical collocation method. Here, we took eight points, whereas there are only seven functions and seven coefficients. Hence, we ask for a least-squares solution. We took the so-called **collocation points** to be (0, 2), (1, 2), (2, 2), (2, 1), (2, 0), (1, 0), (0, 0) and (0, 1). This led to the following system of eight equations:

$$\begin{bmatrix} 1 & 0 & 2 & -4 & 0 & 0 & -8 \\ 1 & 1 & 2 & -3 & 4 & -11 & -2 \\ 1 & 2 & 2 & 0 & 8 & -16 & 16 \\ 1 & 2 & 1 & 3 & 4 & 2 & 11 \\ 1 & 2 & 0 & 4 & 0 & 8 & 0 \\ 1 & 1 & 0 & 1 & 0 & 1 & 0 \\ 1 & 0 & 0 & 0 & 0 & 0 & 0 \\ 1 & 0 & 1 & -1 & 0 & 0 & -1 \end{bmatrix} \begin{bmatrix} c_0 \\ c_1 \\ c_2 \\ c_3 \\ c_4 \\ c_5 \\ c_6 \\ c_7 \end{bmatrix} = \begin{bmatrix} 0 \\ \sin(2) \\ \sin(4) \\ \sin(2) \\ 0 \\ 0 \\ 0 \\ 0 \end{bmatrix}$$

The least-squares solution is a $\boldsymbol{c}$-vector having components

$$\boldsymbol{c} = [0.3219, \ -0.8585, \ -0.8585, \ 0, \ 1.1931, \ 0.2146, \ -0.2146]^T$$

The **residual function** is $\sum_{j=0}^{6} c_j \boldsymbol{v}_j - \boldsymbol{b}$, where $\boldsymbol{b}_i(x, y) = \sin(xy)$. Its absolute value is 0.3219 at each of the eight collocation points. To improve the accuracy, one must employ more basic functions and more collocation points.

Another technique that is often used in the finite-element method is the replacement of a differential equation by an optimization problem. This can be illustrated by a two-point boundary-value problem such as

$$\begin{cases} (hu')' - gu = f \\ u(a) = \alpha, \quad u(b) = \beta \end{cases}$$

Here, u is the unknown function, and h, g, and f are prescribed functions, all defined on the interval $[a, b]$. This problem is called a **Sturm-Liouville problem**. There is an accompanying functional, defined by

$$\Phi(u) = \int_a^b \left[(u')^2 h + u^2 g + 2uf\right] dx$$

The functional and the two-point boundary-value problem are related by several theorems. One of these states roughly that if we find the function u that minimizes the functional $\Phi(u)$ subject to the side conditions $u(a) = \alpha$ and $u(b) = \beta$, then we obtain the solution of the boundary-value problem. It is possible to exploit the fact that $\Phi(u)$ is defined as long as u has a derivative, whereas in the differential equation, we require a function possessing two derivatives. In fact, for the functional, we require only that u be piecewise differentiable, a property that spline functions of degrees 0 and 1 possess. These ideas extend to functions of two or more variables and allow one to use spline functions of low degree in two or more variables to approximate the solution to a differential equation. These are the principal features of the finite-element method. For the mathematical theory of finite-element methods, see the books by Brenner and Scott [2002], Strang [2006], as well as others.

Summary 12.3

- We study a model problem involving the following elliptic partial differential equation

  $$\nabla^2 u + fu = g$$

  over a region, with the value of u given on the boundary. The first term involves the **Laplace operator** ∇^2, which is

  $$\nabla^2 u \equiv \frac{\partial^2 u}{\partial x^2} + \frac{\partial^2 u}{\partial y^2}$$

  By placing a grid over the region with uniform spacing h in both directions the Laplacian term can be approximated by using the **five-point finite differences**

  $$\nabla^2 u \approx \frac{1}{h^2}[u(x+h, y) + u(x-h, y) + u(x, y+h) + u(x, y-h) - 4u(x, y)]$$

  At each interior grid point, we write $u_{ij} = u(x_i, y_j) = u(ih, jh)$, and we obtain the following **equation for our model problem**:

  $$-u_{i+1,j} - u_{i-1,j} - u_{i,j+1} - u_{i,j-1} + \left(4 - h^2 f_{ij}\right) u_{ij} = -h^2 g_{ij}$$

  Usually, the resulting linear system of equations is large and sparse, and iterative methods can be used to solve it.

- For example, the **Gauss-Seidel iterative method** for our linear system is

$$u_{ij}^{(k+1)} = \frac{1}{4 - h^2 f_{ij}} \left(u_{i+1,j}^{(k)} + u_{i-1,j}^{(k+1)} + u_{i,j+1}^{(k)} + u_{i,j-1}^{(k+1)} - h^2 g_{ij} \right)$$

The grid points can be ordered in different ways, such as the natural ordering or the red-black ordering, which affects the rate of convergence of the iterative procedures.

- The distinguishing feature of the finite-element method is that we solve an equation $\boldsymbol{Ax} = \boldsymbol{b}$ approximately by setting $\boldsymbol{u} = \sum_{j=1}^{n} c_j \boldsymbol{v}_j$, where $\boldsymbol{v}_1, \boldsymbol{v}_2, \ldots, \boldsymbol{v}_n$ are chosen by the user. The unknown coefficients c_j are computed so that $\sum_{j=1}^{n} c_j \boldsymbol{A}\boldsymbol{v}_j$ is as close as possible to $\boldsymbol{b}$. Typically, in partial differential equations, the functions $\boldsymbol{v}_j$ will be multidimensional spline functions.

Exercises 12.3

1. Establish the formula for the error in the
   a. five-point formula, Equation (3).
   b. nine-point formula, Equation (5).

2. Establish the irregular five-point Formula (6) and its error term.

3. Write the matrices that occur in Equation (11) when the unknowns are ordered according to the vector $\boldsymbol{u} = [u_{11}, u_{31}, u_{22}, u_{13}, u_{33}, u_{21}, u_{12}, u_{32}, u_{23}]^T$. This is **red-black** or **checkerboard ordering**.

4. a. Verify Equation (10).
   b. Verify that the solution of Equation (13) is as given in the text.

[a]5. Consider the problem of solving the partial differential equation

$$20u_{xx} - 30u_{yy} + \frac{5}{x+y}u_x + \frac{1}{y}u_y = 69$$

in a region R with u prescribed on the boundary. Derive a five-point finite difference equation of order $\mathcal{O}(h^2)$ that corresponds to this equation at some interior point (x_i, y_j).

[a]6. Solve this boundary-value problem to estimate $u\left(\frac{1}{2}, \frac{1}{2}\right)$ and $u\left(0, \frac{1}{2}\right)$:

$$\begin{cases} \nabla^2 u = 0, & (x, y) \in R \\ u = x, & (x, y) \in \partial R \end{cases}$$

The region R with boundary ∂R is shown in the following figure (the arc is circular). Use $h = \frac{1}{2}$.
Note: This problem (and many others in this text) can be stated in physical terms also. For example, in this problem, we are finding the steady-state temperature in a beam of cross section R if the surface of the beam is held at temperature $u(x, y) = x$.

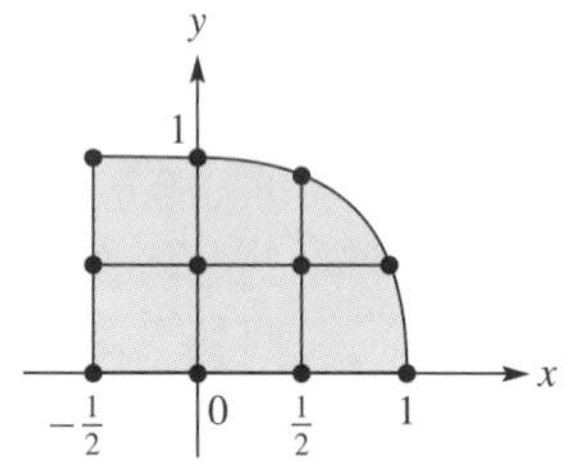

[a]7. Consider the boundary-value problem

$$\begin{cases} \nabla^2 u = 9(x^2 + y^2), & (x, y) \in R \\ u = x - y, & (x, y) \in \partial R_1 \end{cases}$$

for the region in the unit square with $h = \frac{1}{3}$ in the figure below. Here, ∂R is the boundary of R, $\partial R_2 = \left\{(x, y) \in \partial R\colon \frac{2}{3} \leqq x < 1,\ \frac{2}{3} \leqq y < 1\right\}$, and $\partial R_1 = \partial R - \partial R_2$. At the mesh points, determine the system of linear equations that yields an approximate value for $u(x, y)$. Write the system in the form $\boldsymbol{Au} = \boldsymbol{b}$.

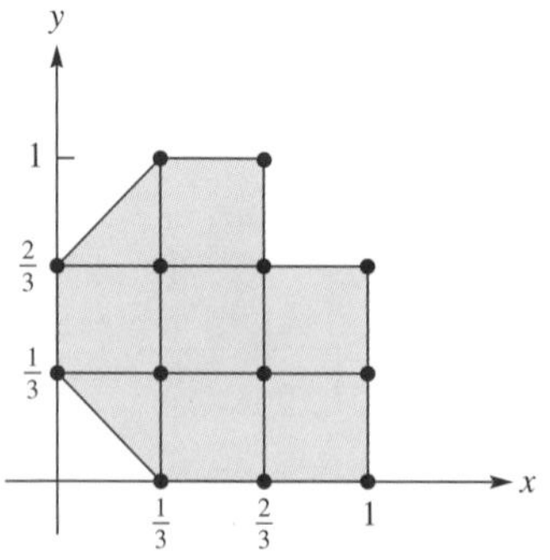

8. Determine the linear system to be solved if the nine-point Formula (5) is used as the approximation in the problem of Equation (1). Notice the pattern in the coefficient matrix with both the five-point and nine-point formulas

when unknowns in each row are grouped together. (Draw dotted lines through A to form 3×3 submatrices.)

9. In Equation (11), show that A is diagonally dominant when $f(x, y) \leq 0$.

10. What is the linear system if an alternative **nine-point formula**

$$\begin{aligned}\nabla^2 u \approx \frac{1}{12h^2}[&16u(x+h, y) + 16u(x-h, y)\\ &+ 16u(x, y+h) + 16u(x, y-h)\\ &- u(x+2h, y) - u(x-2h, y)\\ &- u(x, y+2h) - u(x, y-2h)\\ &- 60u(x, y)]\end{aligned}$$

is used? What are the advantages and disadvantages of using it?
Hint: It has accuracy $\mathcal{O}(h^4)$.

11. (**Multiple Choice**) What is Laplace's equation in three variables?

a. $u - x + u_y + u_z = 0$ **b.** $u_{xx} + u_{yy} = 0$
c. $u_{xx} + u_{yy} + y_{zz} = 0$ **d.** $u_{xx} + u_{yy} = yu_t$
e. None of these.

12. (**Multiple Choice**) Which of these is *not* a harmonic function of (x, y)?

a. $x^2 - y^2$ **b.** $2xy$ **c.** $x^3y - xy^3$
d. $x^3 - xy^3$ **e.** None of these.

13. (**Multiple Choice**) In solving the Dirichlet problem on the unit square, where $0 < x < 1$ and $0 < y < 1$, suppose that we have chosen step size $h = 1/100$. How many unknown function values $u(x, y)$ will there be in this discrete version of the problem? Take into account that $x_i = ih$ for $0 \leq i \leq n+1$, and similarly for y_i. Also, $x_0 = 0$ and $x_{n+1} = 1$, and similarly for y.
Hint: Boundary values on the perimeter of the square are given and are *not* unknowns.

a. $9801 = 99^2$ **b.** $10{,}000 = 100^2$
c. $10{,}404 = 102^2$ **d.** $10{,}201 = 101^2$
e. None of these.

14. Let $z^n = u_n + iv_n$. Verify that u_n and v_n can be determined by the algorithm $u_0 = 1$, $v_0 = 0$, $u_{n+1} = xu_n - yv_n$, and $v_{n+1} = xv_n + yu_n$.

Computer Exercises 12.3

1. Print the system of linear equations for solving Equation (13) with $h = \frac{1}{4}$ and $\frac{1}{8}$. Solve these systems using procedures *Gauss* and *Solve* of Chapter 2.

2. Try the Gauss-Seidel routine on the problem

$$\begin{cases} \nabla^2 u = 2e^{x+y}, & (x, y) \in R \\ u = e^{x+y}, & (x, y) \in \partial R \end{cases}$$

R is the rectangle shown in the following figure.

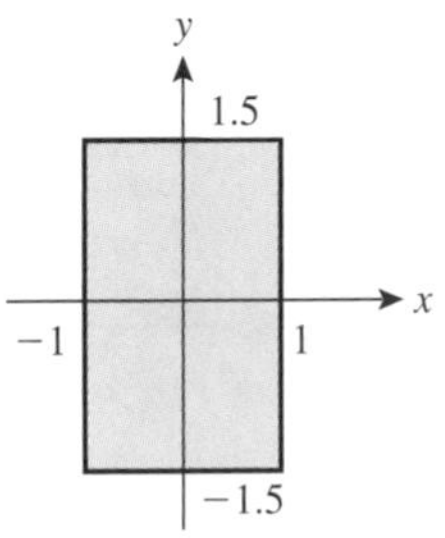

Starting values and mesh sizes are in the following table. Compare your numerical solutions with the exact solutions after *itmax* iterations.

Starting Values	h	itmax
$u = xy$	0.1	15
$u = 0$	0.2	20
$u = (1+x)(1+y)$	0.25	40
$u = \left(1 + x + \frac{1}{2}x^2\right)\left(1 + y + \frac{1}{2}y^2\right)$	0.05	100
$u = 1 + xy$	0.25	200

3. Modify the Gauss-Seidel procedure to handle the red-black ordering. Redo the preceding computer problem with this ordering. Does the ordering make any difference? (See Exercise 12.3.3.)

4. Rewrite the Gauss-Seidel pseudocode so that it can handle any ordering; that is, introduce an ordering array (ℓ_i). Try several different orderings—natural, red-black, spiral, and diagonal.

[a]5. Consider the heat transfer problem on the irregular region shown in the following figure (p. 560).

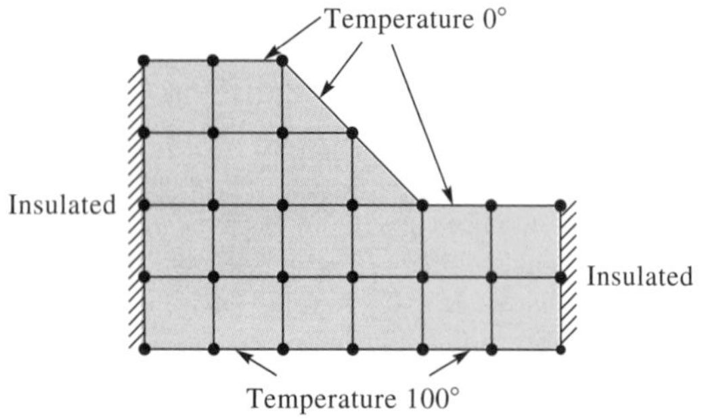

The mathematical statement of this problem is as follows:

$$\begin{cases} \dfrac{\partial^2 u}{\partial x^2} + \dfrac{\partial^2 u}{\partial y^2} = 0, & \text{inside} \\ \dfrac{\partial u}{\partial x} = 0, & \text{sides} \\ u = 0, & \text{top} \\ u = 100, & \text{bottom} \end{cases}$$

Here, the partial derivative $\partial u/\partial x$ can be approximated by a divided-difference formula. Establish that the insulated boundaries act like mirrors so that we can assume that the temperature is the same as at an adjacent interior grid point. Determine the associated linear system, and solve for the temperature u_i with $1 \leq i \leq 10$.

6. Modify procedure *Seidel* so that it uses the nine-point Formula (5). Re-solve model Problem (13) and compare results.

7. Solve the example that begins this chapter with $h = \frac{1}{9}$.

8. Solve the boundary-value problem

$$\begin{cases} \nabla^2 u + 2u = g, & \text{inside } R \\ u = 0, & \text{on boundary of } R \end{cases}$$

where $g(x, y) = (xy + 1)(xy - x - y) + x^2 + y^2$ and R is the unit square. This problem has the known solution $u = \frac{1}{2}xy(x-1)(y-1)$. Use the Gauss-Seidel procedure *Seidel* starting with $u = xy$ and take 30 iterations.

9. (Continuation) Using the modified procedure *Seidel* of Computer Exercise 12.3.6, in which the nine-point Formula (5) is used, re-solve this problem. Compare results and explain the difference.

10. For the elliptic PDE problem (13), use Maple, Mathematica, or MATLAB to find the numerical solution of the linear system (11), where $h = \frac{1}{4}$, $f_{ij} = \frac{1}{25}$, and $g_{ij} = 0$ in the 7×7 coefficient matrix and the 1×7 right-hand side. Compare it with the exact solution of the boundary-value problem, which is

$$u_{ij} = \cosh\left(\tfrac{1}{5}ih\right) + \cosh\left(\tfrac{1}{5}jh\right)$$

Also, compare these results with those obtained in the example in text when $h = \frac{1}{8}$ and the Gauss-Seidel method was used. What conclusions can you draw?

11. Find, approximately, a harmonic function on the circular domain $x^2 + y^2 < 1$ that takes the values $\sin 3\theta$ on the boundary circle. Here, θ is the angular coordinate of the point in polar coordinates. Use the seven basic harmonic polynomials employed in the example of this section. Choose 100 equally spaced points on the circumference, and use the (extended) collocation method, in which a least-squares solution to the system of linear equations is computed.

12. In the collocation example in the text, solve the Dirichlet problem but substitute the boundary values $x^3 - x^2$.

13. Take advantage of any special commands or procedures in mathematical software systems such as MATLAB, Maple, or Mathematica to solve the numerical example (13).

14. (Continuation) Use the symbolic manipulation capabilities in mathematical software such as in Maple or Mathematica to verify the general solution of (13).

15. Write a computer program using the finite-element method to solve Poisson's equation $u_{xx} + u_{yy} = 4$ with boundary conditions $u(x, y) = x^2 + y^2$ using nine nodes in the finer triangularization shown. See Scheid (1988) for additional details.

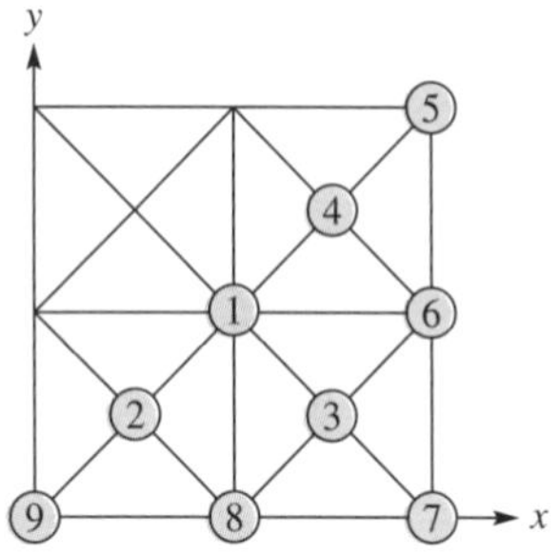

13

Minimization of Functions

An engineering design problem leads to a function

$$F(x, y) = \cos(x^2) + e^{(y-6)^2} + 3(x + y)^4$$

in which x and y are parameters to be selected and $F(x, y)$ is a function related to the cost of manufacturing and is to be minimized. Methods for locating optimal points (x, y) in such problems are developed in this chapter.

13.1 One-Variable Case

An important application of calculus is the problem of finding the local minima of a function. Problems of maximization are covered by the theory of minimization because the maxima of F occur at points where $-F$ has its minima. In calculus, the principal technique for minimization is to differentiate the function whose minimum is sought, set the derivative equal to zero, and locate the points that satisfy the resulting equation.

This technique can be used on functions of one or several variables. For example, if a minimum value of $F(x_1, x_2, x_3)$ is sought, we look for the points where all three partial derivatives are simultaneously zero:

Calculus: Set (Partial) Derivative(s) to 0 and Solve

$$\frac{\partial F}{\partial x_1} = \frac{\partial F}{\partial x_2} = \frac{\partial F}{\partial x_3} = 0$$

This procedure cannot be readily accepted as a *general-purpose* numerical method because it requires differentiation followed by the solution of one or more equations in one or more variables using methods from Chapter 3. This task may be as difficult to carry out as a direct frontal attack on the original problem.

Unconstrained and Constrained Minimization Problems

The minimization problem has two forms: the *unconstrained* and the *constrained*. In an **unconstrained** minimization problem, a function F is defined from the n-dimensional space $\mathbb{R}^n$ into the real line $\mathbb{R}$, and a point $z \in \mathbb{R}^n$ is sought with the property that

Unconstrained Minimization

$$F(\boldsymbol{z}) \leqq F(\boldsymbol{x}) \quad \text{for all } \boldsymbol{x} \in \mathbb{R}^n$$

It is convenient to write points in $\mathbb{R}^n$ simply as $\boldsymbol{x}$, $\boldsymbol{y}$, $\boldsymbol{z}$, and so on. If it is necessary to display the components of a point, we write $\boldsymbol{x} = [x_1, x_2, \ldots, x_n]^T$. In a **constrained** minimization problem, a subset $\boldsymbol{K}$ in $\mathbb{R}^n$ is prescribed, and a point $\boldsymbol{z} \in \boldsymbol{K}$ is sought for which

Constrained Minimization

$$F(\boldsymbol{z}) \leqq F(\boldsymbol{x}) \quad \text{for all } \boldsymbol{x} \in \boldsymbol{K}$$

Such problems are more difficult because of the need to keep the points within the set $\boldsymbol{K}$. Sometimes the set $\boldsymbol{K}$ is defined in a complicated way.

Consider the elliptic paraboloid $F(x_1, x_2) = x_1^2 + x_2^2 - 2x_1 - 2x_2 + 4$, which is sketched in Figure 13.1. The unconstrained minimum occurs at (1, 1) because $F(x_1, x_2) = (x_1 - 1)^2 + (x_2 - 1)^2 + 2$. If $\boldsymbol{K} = \{(x_1, x_2) : x_1 \leqq 0,\ x_2 \leqq 0\}$, the constrained minimum is 4 at (0, 0).

Elliptic Paraboloid

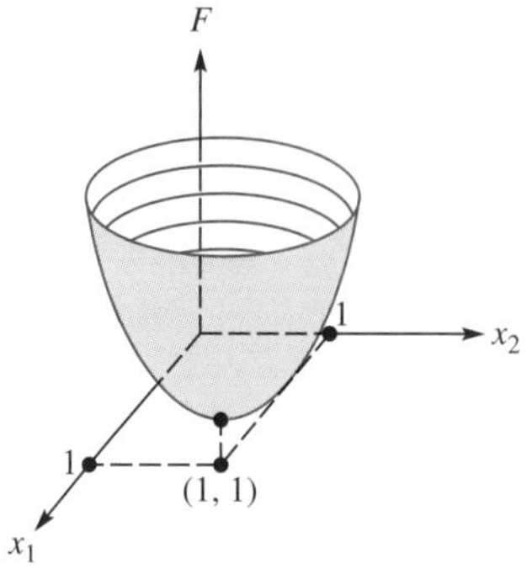

FIGURE 13.1
Elliptic paraboloid

Mathematical Software

Mathematical software systems such as MATLAB, Maple, and Mathematica contain commands for the optimization of general linear and nonlinear functions. For example, we can solve the minimization problem corresponding to the elliptic paraboloid shown in Figure 13.1. First, we define the function, find the minimum value close to the point $\left(\frac{1}{2}, \frac{1}{2}\right)$, and plot this function. We obtain the minimum point as (1, 1) and the value of the function at this point as 2.

One-Variable Case

The special case in which a function F is defined on $\mathbb{R}$ is considered first because the more general problem with n variables is often solved by a sequence of one-variable problems.

Suppose that $F\colon \mathbb{R} \to \mathbb{R}$ and that we seek a point $z \in \mathbb{R}$ with the property that $F(z) \leqq F(x)$ for all $x \in \mathbb{R}$. Note that if no assumptions are made about F, this problem is insoluble in its general form. For instance, the function

$$f(x) = \frac{1}{1 + x^2}$$

has no minimum point. Even for relatively well-behaved functions, such as

$$F(x) = x^2 + \sin(53x)$$

Local/Global Minimum

numerical methods may encounter some difficulties because of the large number of purely local minima. See Figure 13.2. Recall that a point z is a **local minimum** point of a function F if there is some neighborhood of z in which all points satisfy $F(z) \leqq F(x)$. We can use mathematical software such as MATLAB and Mathematica to find local minimum values for the function $F(x) = x^2 + \sin(53x)$. First, we define the function, find a local minimum value

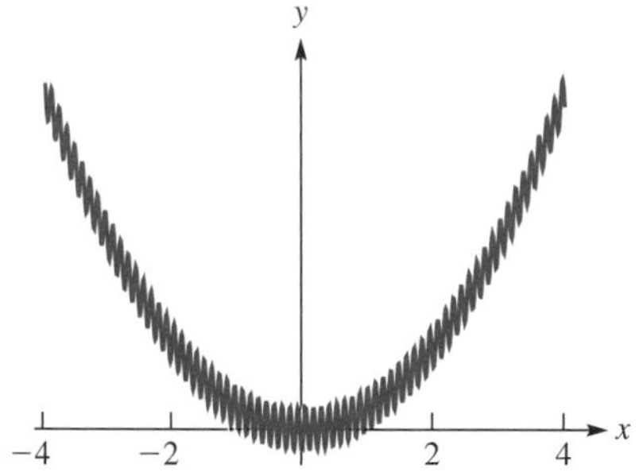

FIGURE 13.2
$F(x) = x^2 + \sin(53x)$

in the interval $\left[-\frac{1}{2}, \frac{1}{2}\right]$, and plot the curve. The point that is computed may not be a global minimum point! To try to find the **global minimum** point, we could use various starting values to find local minimum values and then find the minimum of them. (See Computer Exercise 13.1.6.) In fact, we find a local minimum -0.999122 at $t = -0.0296166$, which is the global minimum for this function.

Unimodal Functions *F*

F Unimodal: A Single Minimum

In attacking a minimization problem, one reasonable assumption is that on some interval $[a, b]$ given to us in advance, F has only a single local minimum. This property is often expressed by saying that F is **unimodal** on $[a, b]$. Some unimodal functions are sketched in Figure 13.3.

Caution. In statistics, *unimodal* refers to a single local maximum.

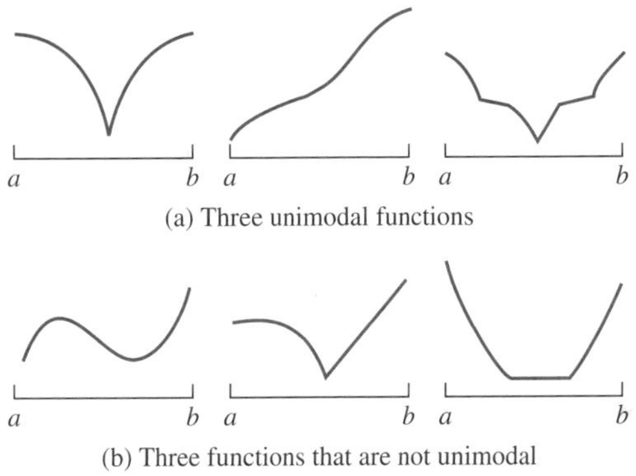

FIGURE 13.3 Examples of unimodal and nonunimodal functions

Continuous Unimodal Function Property

An important property of a continuous unimodal function, which might be surmised from Figure 13.3, is that it is strictly decreasing up to the minimum point and strictly increasing thereafter.

To be convinced of this, let x^* be the minimum point of F on $[a, b]$ and suppose, for instance, that F is not strictly decreasing on the interval $[a, x^*]$. Then points x_1 and x_2 that satisfy $a \leqq x_1 < x_2 \leqq x^*$ and $F(x_1) \leqq F(x_2)$ must exist. Now let x^{**} be a minimum point of F on the interval $[a, x_2]$. (Recall that a continuous function on a closed finite interval attains its minimum value.) We can assume that $x^{**} \neq x_2$ because if x^{**} were initially chosen as x_2, it could be replaced by x_1 inasmuch as $F(x_1) \leqq F(x_2)$. But now we see that x^{**} is a local minimum point of F in the interval $[a, b]$ because it is a minimum point of F on $[a, x_2]$, but it is not x_2 itself. The presence of two local minimum points, of course, contradicts the unimodality of F.

Fibonacci Search Algorithm

Now we pose a problem concerning the search for a minimum point x^* of a continuous unimodal function F on a given interval $[a, b]$. *How accurately can the true minimum point x^* be computed with only n evaluations of F?* With no evaluations of F, the best that can be said is that $x^* \in [a, b]$; taking the midpoint $\widehat{x} = \frac{1}{2}(b + a)$ as the best estimate gives an error of $\left|x^* - \widehat{x}\right| \leqq \frac{1}{2}(b - a)$. One evaluation by itself does not improve this situation, so the best estimate and the error remain the same as in the previous case. Consequently, we need at least two function evaluations to obtain a better estimate.

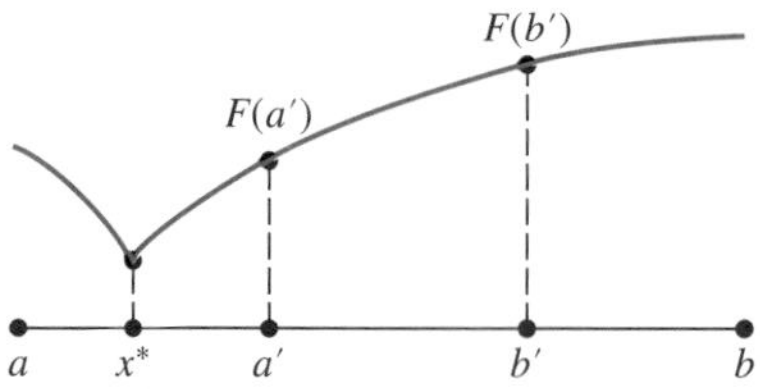

FIGURE 13.4 Fibonacci search algorithm: F evaluated at a' and b'

Suppose that F is evaluated at a' and b' with the results shown in Figure 13.4. If $F(a') < F(b')$, then because F is increasing to the right of x^*, we can be sure that $x^* \in [a, b']$. On the other hand, similar reasoning for the case $F(a') \geqq F(b')$ shows that $x^* \in [a', b]$. To make both intervals of uncertainty as small as possible, we move b' to the left and a' to the right. Thus, F should be evaluated at two nearby points on either side of the midpoint, as shown in Figure 13.5.

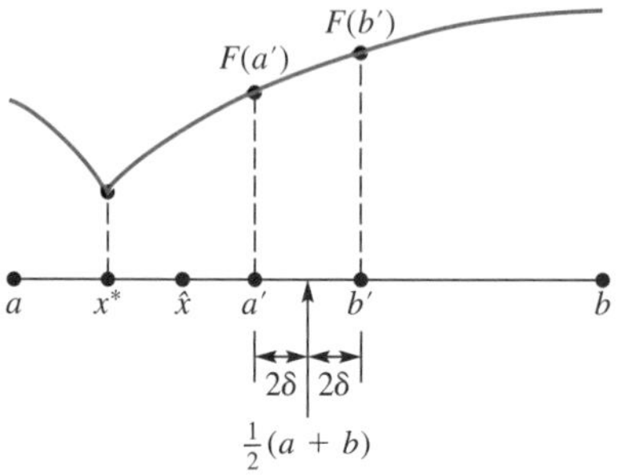

FIGURE 13.5 Fibonacci search algorithm: F evaluated on either side of the midpoint

Suppose that

$$a' = \frac{1}{2}(a+b) - 2\delta, \qquad b' = \frac{1}{2}(a+b) + 2\delta$$

Taking the midpoint of the appropriate subinterval $[a, b']$ or $[a', b]$ as the best estimate $\widehat{x}$ of x^*, we find that the error does not exceed $\frac{1}{4}(b-a) + \delta$. The reader can easily verify this.

For $n = 3$, two evaluations are first made at the $\frac{1}{3}$ and $\frac{2}{3}$ points of the initial interval $[a, b]$; that is,

$$a' = a + \frac{1}{3}(b-a), \qquad b' = a + \frac{2}{3}(b-a)$$

From the two values $F(a')$ and $F(b')$, it can be determined whether $x^* \in [a, b']$ or $x^* \in [a', b]$. The two cases are, of course, similar. Let us suppose that $F(a') \geqq F(b')$, so that our minimum point x^* must be in $[a', b]$, as shown in Figure 13.6. The third (final) evaluation is made close to b', for example, at $b' + \delta$ (where $\delta > 0$). If $F(b') \geqq F(b' + \delta)$, then $x^* \in [b', b]$. Taking the midpoint of this interval, we obtain $\widehat{x} = \frac{1}{2}(b' + b)$ as our

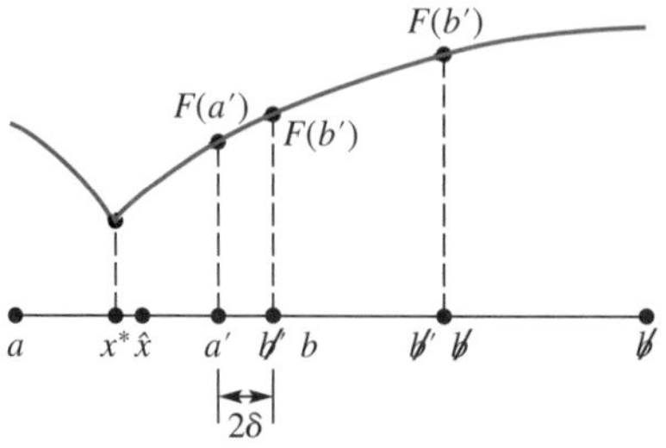

FIGURE 13.6 Fibonacci search algorithm: Reset $b = b'$

estimate of x^* and find that $|\widehat{x} - x^*| \leqq \frac{1}{6}(b-a)$. On the other hand, if $F(b') < F(b'+\delta)$, then $x^* \in [a', b'+\delta]$. Again we take the midpoint, $\widehat{x} = \frac{1}{2}(a'+b'+\delta)$, and find that $|\widehat{x} - x^*| \leqq \frac{1}{6}(b-a) + \frac{1}{2}\delta$. So if we ignore the small quantity $\delta/2$, our accuracy is $\frac{1}{6}(b-a)$ in using three evaluations of F.

By continuing the search pattern outlined, we find an estimate $\widehat{x}$ of x^* with only n evaluations of F and with an error not exceeding

$$\frac{1}{2}\left(\frac{b-a}{\lambda_n}\right) \tag{1}$$

where λ_n is the $(n+1)$st member of the **Fibonacci sequence**:

Fibonacci Sequence

$$\begin{cases} \lambda_1 = 1, \quad \lambda_2 = 1 \\ \lambda_k = \lambda_{k-1} + \lambda_{k-2} \qquad (k \geqq 3) \end{cases} \tag{2}$$

For example, elements λ_1 through λ_8 are 1, 1, 2, 3, 5, 8, 13, and 21.

In the **Fibonacci search algorithm**, we initially determine the number of steps N for a desired accuracy $\epsilon > \delta$ by selecting N to be the subscript of the smallest Fibonacci number greater than $\frac{1}{2}(b-a)/\epsilon$. We define a sequence of intervals, starting with the given interval $[a, b]$ of length $\ell = b - a$, and, for $k = N, N-1, \ldots, 3$, use these formulas for updating:

Fibonacci Search Algorithm

$$\Delta = \left(\frac{\lambda_{k-2}}{\lambda_k}\right)(b-a) \tag{3}$$

$$a' = a + \Delta \qquad b' = b - \Delta$$

$$\begin{cases} a = a', & \text{if } F(a') \geqq F(b') \\ b = b', & \text{if } F(a') < F(b') \end{cases}$$

At the Step $k = 2$, we set

$$a' = \frac{1}{2}(a+b) - 2\delta, \qquad b' = \frac{1}{2}(a+b) + 2\delta$$

$$\begin{cases} a = a', & \text{if } F(a') \geqq F(b') \\ b = b', & \text{if } F(a') < F(b') \end{cases}$$

and we have the final interval $[a, b]$, from which we compute $\widehat{x} = \frac{1}{2}(a+b)$. This algorithm requires only one function evaluation per step after the initial step.

To verify the algorithm, consider the situation shown in Figure 13.7. Since $\lambda_k = \lambda_{k-1} + \lambda_{k-2}$, we have

$$\ell' = \ell - \Delta = \ell - \left(\frac{\lambda_{k-2}}{\lambda_k}\right)\ell = \left(\frac{\lambda_{k-1}}{\lambda_k}\right)\ell \tag{4}$$

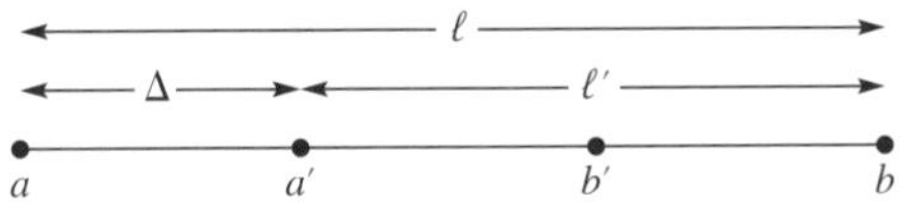

FIGURE 13.7 Fibonacci search algorithm: Verify using a typical situation

and the length of the interval of uncertainty has been reduced by the factor $(\lambda_{k-1}/\lambda_k)$. The next step yields

$$\Delta' = \left(\frac{\lambda_{k-3}}{\lambda_{k-1}}\right)\ell' \tag{5}$$

and Δ' is actually the distance between a' and b'. Therefore, one of the preceding points at which the function was evaluated is at one end or the other of $[a, b]$; that is,

$$\begin{aligned} b' - a' = \ell = 2\Delta &= \left(\frac{\lambda_k - 2\lambda_{k-2}}{\lambda_k}\right)\ell \\ &= \left(\frac{\lambda_{k-1} - \lambda_{k-2}}{\lambda_k}\right)\ell = \left(\frac{\lambda_{k-3}}{\lambda_k}\right)\ell \\ &= \left(\frac{\lambda_{k-3}}{\lambda_{k-1}}\right)\ell' = \Delta' \end{aligned}$$

by Equations (2), (4), and (5).

It is clear by Equation (4) that after $N-1$ function evaluations, the next-to-last interval has length $(1/\lambda_N)$ times the length of the initial interval $[a, b]$. So the final interval is $(b-a)(1/\lambda_N)$ wide, and the maximum error (1) is established. The final step is similar to that outlined, and F is evaluated at a point 2δ away from the midpoint of the next-to-last interval. Finally, set $\widehat{x} = \frac{1}{2}(b+a)$ from the last interval $[a, b]$.

One disadvantage of the Fibonacci search is that the algorithm is rather complicated. Also, the desired precision must be given in advance, and the number of steps to be computed for this precision must be determined before beginning the computation. Thus, the initial evaluation points for the function F depend on N, the number of steps.

Golden Section Search Algorithm

A similar algorithm that is free of these drawbacks is described next. It has been termed the **golden section search** because it depends on a ratio ρ known to the early Greeks as the **golden section ratio**:

Golden Section Ratio

$$\rho = \frac{1}{2}\left(1 + \sqrt{5}\right) \approx 1.61803\ 39887$$

The mathematical history of this number can be found in Roger [1998], and ρ satisfies the equation $\rho^2 = \rho + 1$, which has roots $\frac{1}{2}\left(1 + \sqrt{5}\right) \approx 1.61803\ldots$ and $\frac{1}{2}\left(1 - \sqrt{5}\right) \approx -0.61803\ldots$.

In each step of this iterative algorithm, an interval $[a, b]$ is available from the previous work. It is an interval that is known to contain the minimum point x^*, and our objective is to replace it by a smaller interval that is also known to contain x^*. In each step, two values of F are needed:

Outline of Golden Section Search Algorithm

$$\begin{cases} x = a + r(b-a), & u = F(x) \\ y = a + r^2(b-a), & v = F(y) \end{cases} \tag{6}$$

where $r = 1/\rho$ and $r^2 + r = 1$, which has roots $\frac{1}{2}\left(-1 + \sqrt{5}\right) \approx 0.61803\ldots$ and $\frac{1}{2}\left(-1 - \sqrt{5}\right) \approx -1.61803\ldots$. There are two cases to consider: Either $u > v$ or $u \leqq v$.

Case $u > v$

Let us take the first. Figure 13.8 depicts this situation. Since F is assumed continuous and unimodal, the minimum of F must be in the interval $[a, x]$. This interval is the input interval at the beginning of the next step. Observe now that within the interval $[a, x]$, one

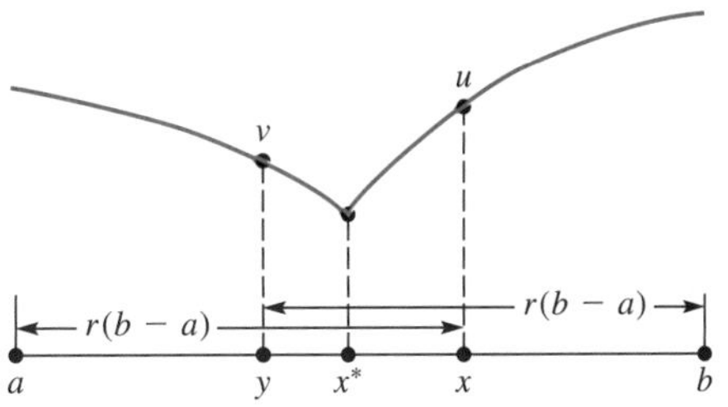

FIGURE 13.8
Golden section search algorithm: $u > v$

evaluation of F is already available, at y. Also note that

$$a + r(x - a) = y$$

because $x - a = r(b - a)$. In the next step, therefore, y will play the role of x, and we shall need the value of $\boldsymbol{F}$ at the point at $a + r^2(x - a)$. In this step we must carry out the following replacements *in order*:

$$\begin{aligned} b &\leftarrow x \\ x &\leftarrow y \\ u &\leftarrow v \\ y &\leftarrow a + r^2(b - a) \\ v &\leftarrow F(y) \end{aligned}$$

The other case is similar. If $u \leqq v$, the picture might be as in Figure 13.9. In this case, the minimum point must lie in $[y, b]$. Within this interval, one value of F is available, at x. Observe that

Case $u \geqq v$

$$y + r^2(b - y) = x$$

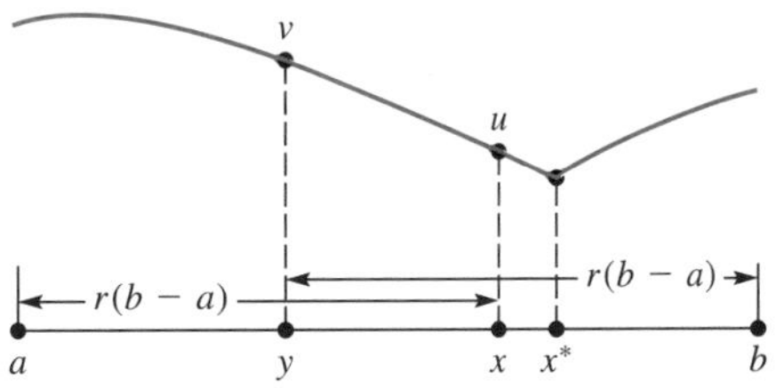

FIGURE 13.9
Golden section search algorithm: $u \leqq v$

(See Exercise 13.1.9.) Thus, x should now be given the role of y, and the value of F is to be computed at $y + r(b - y)$. The following ordered replacements accomplish this:

$$\begin{aligned} a &\leftarrow y \\ y &\leftarrow x \\ v &\leftarrow u \\ x &\leftarrow a + r(b - a) \\ u &\leftarrow F(x) \end{aligned}$$

Exercises 13.1.10–13.1.11 hint at a shortcoming of this procedure: It is quite slow. Slowness in this context refers to the large number of function evaluations that are needed to achieve reasonable precision. This slowness is attributable to the extreme generality of the algorithm. No advantage has been taken of any smoothness that the function F may possess.

Golden Section Search versus Fibonacci Search

If $[a, b]$ is the starting interval in the search for a minimum of F, then at the beginning, with one evaluation of F, we can be sure only that the minimum point, x^*, is in an interval of width $b - a$. In the golden section search, the corresponding lengths in successive steps are $r(b-a)$ for two evaluations of F, $r^2(b-a)$ for three evaluations of F, and so on. After n steps, the minimum point has been pinned down to an interval of length $r^{n-1}(b-a)$.

How does this compare with the Fibonacci search algorithm using n evaluations? The corresponding width of interval, at the last step of this algorithm, is $\lambda_n^{-1}(b-a)$. Now, the Fibonacci algorithm should be better, because it is designed to do as well as possible with a prescribed number of steps. So we expect the ratio r^{n-1}/λ_n^{-1} to be greater than 1. But it approaches 1.17 as $n \to \infty$. (See Exercise 13.1.8.) Thus, one may conclude that the extra complexity of the Fibonacci algorithm, together with the disadvantage of having the algorithm itself depend on the number of evaluations permitted, mitigates against its use in general.

How to Determine the Correct Ratio

In the golden section search algorithm, how is the correct ratio r determined? Remember that when we pass from one interval to the next in the algorithm, one of the points x or y is to be retained in the next step. Here, we present first a sketch of the first interval in which we let $x = a + r(b-a)$ and $y = b + r(a-b)$. It is followed by a sketch of the next interval.

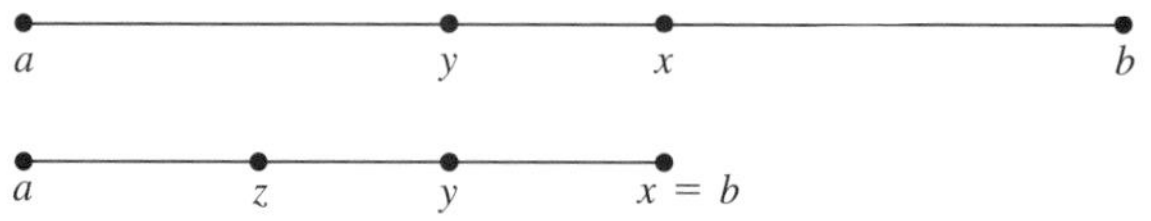

In this new interval, the same ratios should hold, so we have $y = a + r(x - a)$. Since $x - a = r(b-a)$, we can write $y = a + r[r(b-a)]$. Setting the two formulas for y equal to each other gives us

$$a + r^2(b-a) = b + r(a-b)$$

whence

$$a - b + r^2(b-a) = r(a-b)$$

Dividing by $(a-b)$ gives

$$r^2 + r - 1 = 0$$

The roots of this quadratic equation are as given previously.

Quadratic Interpolation Algorithm

Suppose that F is represented by a Taylor series in the vicinity of the point x^*. Then

$$F(x) = F(x^*) + (x - x^*)F'(x^*) + \frac{1}{2}(x - x^*)^2 F''(x^*) + \cdots$$

Since x^* is a minimum point of F, we have $F'(x^*) = 0$. Thus,

$$F(x) \approx F(x^*) + \frac{1}{2}(x - x^*)^2 F''(x^*)$$

This tells us that, in the neighborhood of x^*, $F(x)$ is approximated by a quadratic function whose minimum is also at x^*. Since we do not know x^* and do not want to involve derivatives in our algorithms, a natural stratagem is to interpolate F by a quadratic polynomial. Any three values $(x_i, F(x_i))$, $i = 1, 2, 3$, can be used for this purpose. The minimum point of the resulting quadratic function may be a better approximation to x^* than is x_1, x_2, or x_3. Writing an algorithm that carries out this idea iteratively is not trivial, and many unpleasant cases must be handled. *What should be done if the quadratic interpolant has a maximum instead of a minimum, for example?* There is also the possibility that $F''(x^*) = 0$, in which case higher-order terms of the Taylor series determine the nature of F near x^*.

Here is the outline of an algorithm for this procedure. At the beginning, we have a function F whose minimum is sought. Two starting points x and y are given, as well as two control numbers δ and ε. Computing begins by evaluating the two numbers

Outline of Quadratic Interpolation Algorithm

$$\begin{cases} u = F(x) \\ v = F(y) \end{cases}$$

Now let

$$z = \begin{cases} 2x - y, & \text{if } u < v \\ 2y - x, & \text{if } u \geqq v \end{cases}$$

In either case, the number

$$w = F(z)$$

is to be computed.

At this stage, we have three points x, y, and z together with corresponding function values u, v, and w. In the main iteration step of the algorithm, one of these points and its accompanying function value are replaced by a new point and new function value. The process is repeated until a success or failure is reached.

In the main calculation, a quadratic polynomial q is determined to interpolate F at the three current points x, y, and z. The formulas are discussed below. Next, the point t where $q'(t) = 0$ is determined. Under ideal circumstances, t is a *minimum* point of q and an *approximate minimum* point of F. So one of the x, y, or z should be replaced by t. We are interested in examining $q''(t)$ to determine the shape of the curve q near t.

For the complete description of this algorithm, the formulas for t and $q''(t)$ must be given. They are obtained as follows:

$$\begin{cases} a = \dfrac{v - u}{y - x} \\ b = \dfrac{w - v}{z - y} \\ c = \dfrac{b - a}{z - x} \\ t = \dfrac{1}{2}\left[x + y - \dfrac{a}{c}\right] \\ q''(t) = 2c \end{cases}$$

Their derivation is outlined in Exercise 13.1.12.

The **solution case** occurs if

Solution Case

$$q''(t) > 0, \qquad \max\{|t-x|, |t-y|, |t-z|\} < \varepsilon$$

The condition $q''(t) > 0$ indicates, of course, that q' is *increasing* in the vicinity of t, so t is indeed a minimum point of q. The second condition indicates that this estimate, t, of the minimum point of F is within distance ε of each of the three points x, y, and z. In this case, t is accepted as a solution.

The **usual case** occurs if

Usual Case

$$q''(t) > 0, \qquad \delta \geqq \max\{|t-x|, |t-y|, |t-z|\} \geqq \varepsilon$$

These inequalities indicate that t is a minimum point of q but not near enough to the three initial points to be accepted as a solution. Also, t is not farther than δ units from each of x, y, and z and can thus be accepted as a reasonable new point. The old point that has the greatest function value is now replaced by t and its function value by $F(t)$.

The **first bad case** occurs if

First Bad Case

$$q''(t) > 0, \qquad \max\{|t-x|, |t-y|, |t-z|\} > \delta$$

Here, t is a minimum point of q but is so remote that there is some danger in using it as a new point. We identify one of the original three points that is farthest from t, for example, x, and also we identify the point closest to t, say z. Then we replace x by $z + \delta\ \text{sign}(t - z)$ and u by $F(x)$. Figure 13.10 shows this case. The curve is the graph of q.

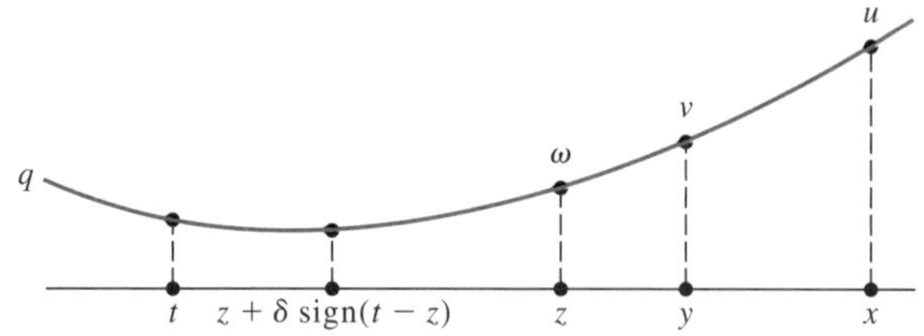

FIGURE 13.10 Taylor series algorithm: First bad case

The **second bad case** occurs if

Second Bad Case

$$q''(t) < 0$$

thus indicating that t is a maximum point of q. In this case, identify the greatest and the least among u, v, and w. Suppose, for example, that $u \geqq v \geqq w$. Then replace x by $z + \delta$ sign$(z - x)$. An example is shown in Figure 13.11.

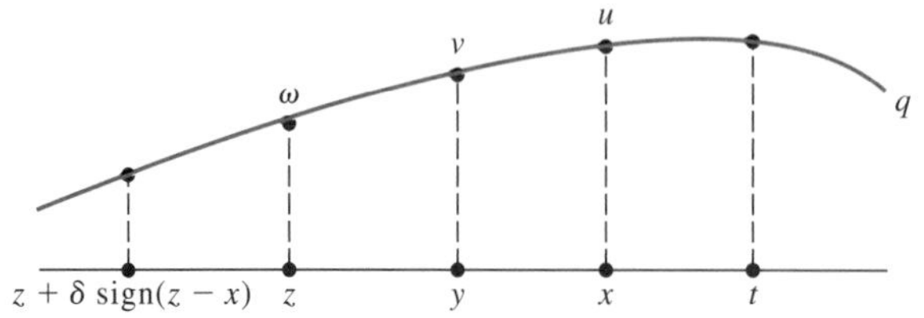

FIGURE 13.11 Taylor series algorithm: Second bad case

Summary 13.1

- We consider the problem of finding the **local minimum** of a unimodal function of a one-variable.
- Algorithms discussed are **Fibonacci search**, **golden section search**, and **quadratic interpolation**.

Exercises 13.1

[a]**1.** For the function $F(x_1, x_2, x_3) = x_1^2 + 3x_2^2 + 2x_3^2 - 4x_1 - 6x_2 + 8x_3$, find the unconstrained minimum point. Then find the constrained minimum over the set $\boldsymbol{K}$ defined by inequalities $x_1 \leqq 0$, $x_2 \leqq 0$, and $x_3 \leqq 0$. Next, solve the same problem when $\boldsymbol{K}$ is defined by $x_1 \leqq 2$, $x_2 \leqq 0$, and $x_3 \leqq -2$.

[a]**2.** For the function $F(x, y) = 13x^2 + 13y^2 - 10xy - 18x - 18y$, find the unconstrained minimum.
Hint: Try substituting $x = u + v$ and $y = u - v$.

3. If F is unimodal and continuous on the interval $[a, b]$, how many local maxima may F have on $[a, b]$?

[a]**4.** For the Fibonacci search algorithm, write expressions for $\widehat{x}$ in the two cases $n = 2, 3$.

5. Carry out four steps of the Fibonacci search algorithm using $\epsilon = \frac{1}{4}$ to determine the following:

[a]**a.** Minimum of $F(x) = x^2 - 6x + 2$ on $[0, 10]$

b. Minimum of $F(x) = 2x^3 - 9x^2 + 12x + 2$ on $[0, 3]$

c. Maximum of $F(x) = 2x^3 - 9x^2 + 12x$ on $[0, 2]$

6. Let F be a continuous unimodal function defined on the interval $[a, b]$. Suppose that the values of F are known at n points, namely, $a = t_1 < t_2 < \cdots < t_n = b$. How accurately can one estimate the minimum point x^* from only the values of t_i and $F(t_i)$?

[a]**7.** The equation satisfied by Fibonacci numbers, namely, $\lambda_n - \lambda_{n-1} - \lambda_{n-2} = 0$, is an example of a linear difference equation with constant coefficients. Solve it by postulating that $\lambda_n = \lambda^n$ and finding that $\alpha = \frac{1}{2}\left(1+\sqrt{5}\right)$ or $\beta = \frac{1}{2}\left(1 - \sqrt{5}\right)$ will serve for λ. Initial conditions $\lambda_1 = \lambda_2 = 1$ can be met by a solution of the form $\lambda_n = A\alpha^n + B\beta^n$. Find A and B. Establish that

$$\lim_{n\to\infty} \left(\frac{\lambda_n}{\lambda_{n-1}}\right) = \alpha = \frac{1}{2}\left(1 + \sqrt{5}\right)$$

Show that this agrees with Equations (10) and (11) of Section 3.3.

8. (Continuation) Refer to the golden section search algorithm and to the preceding problem. Prove that $\alpha\beta = -1$ and $\alpha + \beta = 1$ so that $\alpha = 1/r$ and $\beta = -r$. Then establish that $r^n\lambda_n$ converges to $1/\sqrt{5}$ as $n \to \infty$.

[a]**9.** Verify that $y + r^2(b - y) = x$ in the golden section algorithm.
Hint: Use $r^2 + r = 1$.

[a]**10.** If F is unimodal on an interval of length ℓ, how many evaluations are necessary in the golden section algorithm to estimate the minimum point with an error of at most 10^{-k}?

[a]**11.** (Continuation) In the preceding problem, how large must n be if $\ell = 1$ and $k = 10$?

12. Using the divided-difference algorithm on the table

x	y	z
u	v	w

show that the quadratic interpolant in Newton form is

$$q(t) = u + a(t - x) + c(t - x)(t - y)$$

with a, b, and c given by Equation (7). Then verify the formulas for t and $q''(t)$ given in (7).

[a]**13.** If routines can be written easily for F, F', and F'', how can Newton's method be used to locate the minimum point of F? Write down the formula that defines the iterative process. Does it involve F?

[a]**14.** If routines are available for F and F', how can the secant method be used to minimize F?

15. The **golden section ratio**, $\rho = \frac{1}{2}\left(1 + \sqrt{5}\right)$, has many mystical properties; for example,

a. $\rho = 1 + \cfrac{1}{1 + \cfrac{1}{1 + \cfrac{1}{1 + \cfrac{1}{1 + \cdots}}}}$

[a]**b.** $\rho = \sqrt{1 + \sqrt{1 + \sqrt{1 + \sqrt{1 + \cdots}}}}$

c. $\rho^n = \rho^{n-1} + \rho^{n-2}$

[a]**d.** $\rho = \rho^{-1} + \rho^{-2} + \rho^{-3} + \cdots$

Establish these properties.

16. (**Multiple Choice**) In the golden section search algorithm, we use a number $r = 0.618\ldots$, which is the larger of the two roots of the quadratic equation $r^2+r = 1$. Let f be a unimodal function on the interval $[a, b]$. Thus, f has a single local minimum in $[a, b]$, where here we assume that $a < b$. Let $x = a+r(b-a)$ and $y = a+r^2(b-a)$. Also, let $u = f(x)$ and $v = f(y)$, where we suppose that $u < v$. What interval must contain the minimum point of f?

a. $[y, b]$ **b.** $[a, x]$ **c.** $[a, y]$ **d.** $[y, x]$

e. None of these.

Computer Exercises 13.1

1. Write a routine to carry out the golden section algorithm for a given function and interval. The search should continue until a preassigned error bound is reached but not beyond 100 steps in any case.

2. (Continuation) Test the routine of the preceding computer problem on these examples or use a routine from a package such as MATLAB, Maple, or Mathematica:

a. $F(x) = \sin x$, on $[0, \pi/2]$

b. $F(x) = (\arctan x)^2$, on $[-1, 1]$

c. $F(x) = |\ln x|$, on $\left[\frac{1}{2}, 4\right]$

d. $F(x) = |x|$, on $[-1, 1]$

3. Code and test the following algorithm for approximating the minima of a function F of one variable over an interval $[a, b]$: The algorithm defines a sequence of quadruples $a < a' < b' < b$ by initially setting $a' = \frac{2}{3}a + \frac{1}{3}b$ and $b' = \frac{1}{3}a + \frac{2}{3}b$ and repeatedly updating by $a = a'$, $a' = b'$, and $b' = \frac{1}{2}(b + b')$ if $F(a') > F(b')$; $b = b'$, $a' = \frac{1}{2}(a + a')$, and $b' = a$ if $F(a') < F(b')$; $a = a'$, $b = b'$, $a' = \frac{2}{3}a + \frac{1}{3}b$, and $b' = \frac{1}{3}a + \frac{2}{3}b$ if $F(a') = F(b')$. *Note*: The construction ensures that $a < a' < b' < b$, and the minimum of F always occurs between a and b. Furthermore, only one new function value needs to be computed at each stage of the calculation after the first unless the case $F(a') = F(b')$ is obtained. The values of a, a', b', and b tend to have the same limit, which is a minimum point of F. Notice the similarity to the method of bisection of Section 3.1.

4. Write and test a routine for the Fibonacci search algorithm. Verify that a partial algorithm for the Fibonacci search is as follows: Initially, set

$$\begin{cases} \Delta = \left(\dfrac{\lambda_{N-2}}{\lambda_N}\right)(b-a) \\ a' = a + \Delta \\ b' = b - \Delta \\ u = F(a') \\ v = F(b') \end{cases}$$

Then *loop* on k from $N - 1$ downward to 3, updating as follows:

If $u \geqq v$:

$$\begin{cases} a \leftarrow a' \\ a' \leftarrow b' \\ u \leftarrow v \\ \Delta \leftarrow \left(\dfrac{\lambda_{k-2}}{\lambda_k}\right)(b-a) \\ b' \leftarrow b - \Delta \\ v \leftarrow F(b') \end{cases}$$

If $v > u$:

$$\begin{cases} b \leftarrow b' \\ b' \leftarrow a' \\ v \leftarrow u \\ \Delta \leftarrow \left(\dfrac{\lambda_{k-2}}{\lambda_k}\right)(b-a) \\ a' \leftarrow a + \Delta \\ u \leftarrow F(a') \end{cases}$$

Add steps for $k = 2$.

5. (**Berman Algorithm**) Suppose that F is unimodal on $[a, b]$. Then if x_1 and x_2 are any two points such that $a \leqq x_1 < x_2 \leqq b$, we have

$$\begin{aligned} F(x_1) > F(x_2) &\quad\Rightarrow\quad x^* \in (x_1, b] \\ F(x_1) = F(x_2) &\quad\Rightarrow\quad x^* \in [x_1, x_2] \\ F(x_1) < F(x_2) &\quad\Rightarrow\quad x^* \in [a, x_2) \end{aligned}$$

So by evaluating F at x_1 and x_2 and comparing function values, we are able to reduce the size of the interval that is known to contain x^*. The simplest approach is to start at the midpoint $x_0 = \frac{1}{2}(a+b)$ and if F is, say, decreasing for $x > x_0$, we test F at $x_0 + ih$, $i = 1, 2, \ldots, q$, with $h = (b-a)/2q$, until we find a point x_1 from which F begins to increase again (or until we reach b). Then we repeat this procedure starting at x_1 and using a smaller step length h/q. Here, q is the maximal number of evaluations at each step, say, 4.

Write a subroutine to perform the Berman algorithm and test it for evaluating the approximate minimization of one-dimensional functions.

Note: The total number of evaluations of F needed for executing this algorithm up to some iterative step k depends on the location of x^*. If, for example, $x^* = b$, then clearly we need q evaluations at each iteration and hence

kq evaluations. This number will decrease the closer x^* is to x_0, and it can be shown that with $q = 4$, the *expected* number of evaluations is three per step. It is interesting to compare the efficiency of the Berman algorithm ($q = 4$) with that of the Fibonacci search algorithm. The expected number of evaluations per step is three, and the uncertainty interval decreases by a factor $4^{-1/3} \approx 0.63$ per evaluation. In comparison, the Fibonacci search algorithm has a reduction factor of $\frac{1}{2}\left(1+\sqrt{5}\right) \approx 0.62$. Of course, the factor 0.63 in the Berman algorithm represents only an average and can be considerably lower but also as high as $4^{-1/4} \approx 0.87$.

6. Select a routine from your program library or from a package such as MATLAB, Maple, or Mathematica for finding the minimum point of a function of one variable. Experiment with the function $F(x) = x^4 + \sin(23x)$ to determine whether this routine encounters any difficulties in finding a global minimum point. Use starting values both near to and far from the global minimum point. (See Figure 13.2.)

7. **(Student Research Project)** The Greek mathematician Euclid of Alexandria (325–265 B.C.E.) wrote a collection of 13 books on mathematics and geometry. In book six, Proposition 30 shows how to divide a line into its mean and extreme mean, which is finding the golden section point on a line. This states that the ratio of the smaller part of a line segment to the larger part is the same as the ratio of the larger part to the whole line segment. For a line segment of length 1, denote the larger part by r and the smaller part by $1 - r$ as shown here:

Hence, we have the ratios

$$\frac{1-r}{r} = \frac{r}{1}$$

and we obtain the quadratic equation

$$r^2 = 1 - r$$

This equation has two roots, one positive and one negative. The reciprocal of the positive root is the **golden ratio** $\frac{1}{2}\left(1+\sqrt{5}\right)$, which was of interest to Pythagoras (580–500 B.C.E.). It was also used in the construction of the Great Pyramid of Gizah. Mathematical software systems such as MATLAB, Maple, or Mathematica contain the golden ratio constant. In fact, the default width-to-height ratio for the plot function is the golden ratio. Investigate the golden section ratio and its use in scientific computing.

8. Using a mathematical software system such as MATLAB, Maple, or Mathematica, write a computer program to reproduce

a. Figure 13.1.

b. Figure 13.2. Also, find the global minimum of the function as well as several local minimum points near the origin.

13.2 Multivariate Case

Now we consider a real-valued function of n real variables $F: \mathbb{R}^n \to \mathbb{R}$. As before, a point $\boldsymbol{x}^*$ is sought such that

$$F(\boldsymbol{x}^*) \leq F(\boldsymbol{x}) \quad \text{for all } \boldsymbol{x} \in \mathbb{R}^n$$

Some of the theory of multivariate functions must be developed to understand the rather sophisticated minimization algorithms in current use.

Taylor Series for F: Gradient Vector and Hessian Matrix

If the function F possesses partial derivatives of certain low orders (which is usually assumed in the development of these algorithms), then at any given point $\boldsymbol{x}$, a **gradient vector** $\boldsymbol{G}(\boldsymbol{x}) = (G_i)_n$ is defined with components

Gradient Vector

$$G_i = G_i(\boldsymbol{x}) = \frac{\partial F(\boldsymbol{x})}{\partial x_i} \qquad (1 \leq i \leq n) \tag{1}$$

and a **Hessian matrix** $\boldsymbol{H}(\boldsymbol{x}) = (H_{ij})_{n\times n}$ is defined with components

Hessian Matrix

$$H_{ij} = H_{ij}(\boldsymbol{x}) = \frac{\partial^2 F(\boldsymbol{x})}{\partial x_i\,\partial x_j} \qquad (1 \leq i,\ j \leq n) \tag{2}$$

We interpret $\boldsymbol{G}(\boldsymbol{x})$ as an n-component vector and $\boldsymbol{H}(\boldsymbol{x})$ as an $n \times n$ matrix, both depending on $\boldsymbol{x}$.

Using the gradient and Hessian, we can write the first few terms of the Taylor series for F as

Taylor Series for F

$$F(\boldsymbol{x} + \boldsymbol{h}) = F(\boldsymbol{x}) + \sum_{i=1}^{n} G_i(\boldsymbol{x})h_i + \frac{1}{2}\sum_{i=1}^{n}\sum_{j=1}^{n} h_i H_{ij}(\boldsymbol{x})h_j + \cdots \tag{3}$$

Equation (3) can also be written in an elegant matrix-vector form:

Matrix-Vector Form

$$F(\boldsymbol{x} + \boldsymbol{h}) = F(\boldsymbol{x}) + \boldsymbol{G}(\boldsymbol{x})^T\boldsymbol{h} + \frac{1}{2}\boldsymbol{h}^T\boldsymbol{H}(\boldsymbol{x})\boldsymbol{h} + \cdots \tag{4}$$

Here, $\boldsymbol{x}$ is the fixed point of expansion in $\mathbb{R}^n$, and $\boldsymbol{h}$ is the variable in $\mathbb{R}^n$ with components $h_1, h_2, \ldots, h_n$. The three dots indicate higher-order terms in $\boldsymbol{h}$ that are not needed in this discussion.

A result in calculus states that the *order* in which partial derivatives are taken is immaterial if all partial derivatives that occur are continuous. In the special case of the Hessian matrix, if the second partial derivatives of F are all continuous, then $\boldsymbol{H}$ is a **symmetric matrix**; that is, $\boldsymbol{H} = \boldsymbol{H}^T$ because

***H* Symmetric Matrix**

$$H_{ij} = \frac{\partial^2 F}{\partial x_i\,\partial x_j} = \frac{\partial^2 F}{\partial x_j\,\partial x_i} = H_{ji}$$

EXAMPLE 1 To illustrate Formula (4), let us compute the first three terms in the Taylor series for the function

$$F(x_1, x_2) = \cos(\pi x_1) + \sin(\pi x_2) + e^{x_1 x_2}$$

taking $(1, 1)$ as the point of expansion.

Solution Partial derivatives are

$$\frac{\partial F}{\partial x_1} = -\pi\sin(\pi x_1) + x_2 e^{x_1x_2}, \qquad \frac{\partial F}{\partial x_2} = \pi\cos(\pi x_2) + x_1 e^{x_1x_2}$$

$$\frac{\partial^2 F}{\partial x_1^2} = -\pi^2\cos(\pi x_1) + x_2^2 e^{x_1x_2}, \qquad \frac{\partial^2 F}{\partial x_2\,\partial x_1} = (x_1x_2 + 1)e^{x_1x_2}$$

$$\frac{\partial^2 F}{\partial x_1\,\partial x_2} = (x_1x_2 + 1)e^{x_1x_2}, \qquad \frac{\partial^2 F}{\partial x_2^2} = -\pi^2\sin(\pi x_2) + x_1^2 e^{x_1x_2}$$

Note the equality of cross derivatives; that is, $\partial^2 F/\partial x_1\,\partial x_2 = \partial^2 F/\partial x_2\,\partial x_1$. At the particular point $\boldsymbol{x} = [1, 1]^T$, we have

$$F(\boldsymbol{x}) = -1 + e, \qquad \boldsymbol{G}(\boldsymbol{x}) = \begin{bmatrix} e \\ -\pi + e \end{bmatrix}, \qquad \boldsymbol{H}(\boldsymbol{x}) = \begin{bmatrix} \pi^2 + e & 2e \\ 2e & e \end{bmatrix}$$

So by Equation (4),

$$F(1+h_1, 1+h_2) = -1 + e + [e, -\pi + e]\begin{bmatrix} h_1 \\ h_2 \end{bmatrix} + \frac{1}{2}[h_1, h_2]\begin{bmatrix} \pi^2 + e & 2e \\ 2e & e \end{bmatrix}\begin{bmatrix} h_1 \\ h_2 \end{bmatrix} + \cdots$$

or equivalently, by Equation (3),

$$F(1+h_1, 1+h_2) = -1 + e + eh_1 + (-\pi + e)h_2 + \frac{1}{2}\left[(\pi^2 + e)h_1^2 + (2e)h_1h_2 + (2e)h_2h_1 + eh_2^2\right] + \cdots$$

■

Mathematical Software

In mathematical software systems such Maple or Mathematica, we can verify these calculations using built-in routines for the gradient and Hessian. Also, we can obtain two terms in the Taylor series in two variables expanded about the point (1, 1) and then carry out a change of variables to obtain similar results as shown.

Alternative Form of Taylor Series

Another form of the Taylor series is useful. First let z be the point of expansion, and then let $\boldsymbol{h} = \boldsymbol{x} - z$. Now from Equation (4),

$$F(\boldsymbol{x}) = F(z) + \boldsymbol{G}(z)^T(\boldsymbol{x} - z) + \frac{1}{2}(\boldsymbol{x} - z)^T \boldsymbol{H}(z)(\boldsymbol{x} - z) + \cdots \tag{5}$$

We illustrate with two special types of functions.

First, the **linear function** has the form

Linear Function

$$F(\boldsymbol{x}) = c + \sum_{i=1}^{n} b_i x_i = c + \boldsymbol{b}^T\boldsymbol{x}$$

for appropriate coefficients $c, b_1, b_2, \ldots, b_n$. Clearly, the gradient and Hessian are $G_i(z) = b_i$ and $H_{ij}(z) = 0$, so Equation (5) yields

Case 2 Variables

$$F(\boldsymbol{x}) = F(z) + \sum_{i=1}^{n} b_i(x_i - z_i) = F(z) + \boldsymbol{b}^T(\boldsymbol{x} - z)$$

Second, consider a general **quadratic function**. For simplicity, we take only two variables. The form of the function is

Quadratic Function

$$F(x_1, x_2) = c + (b_1x_1 + b_2x_2) + \frac{1}{2}\left(a_{11}x_1^2 + 2a_{12}x_1x_2 + a_{22}x_2^2\right) \tag{6}$$

which can be interpreted as the Taylor series for F when the point of expansion is (0, 0).

To verify this assertion, the partial derivatives must be computed and evaluated at (0, 0):

$$\frac{\partial F}{\partial x_1} = b_1 + a_{11}x_1 + a_{12}x_2, \qquad \frac{\partial F}{\partial x_2} = b_2 + a_{22}x_2 + a_{12}x_1$$

$$\frac{\partial^2 F}{\partial x_1^2} = a_{11}, \qquad \frac{\partial^2 F}{\partial x_1\,\partial x_2} = a_{12}$$

$$\frac{\partial^2 F}{\partial x_2\,\partial x_1} = a_{12}, \qquad \frac{\partial^2 F}{\partial x_2^2} = a_{22}$$

Letting $z = [0, 0]^T$, we obtain from Equation (5)

Matrix Form

$$F(\boldsymbol{x}) = c + [b_1, b_2]\begin{bmatrix} x_1 \\ x_2 \end{bmatrix} + \frac{1}{2}[x_1, x_2]\begin{bmatrix} a_{11} & a_{12} \\ a_{12} & a_{22} \end{bmatrix}\begin{bmatrix} x_1 \\ x_2 \end{bmatrix}$$

This is the matrix form of the original quadratic function of two variables. It can also be written as

$$F(\boldsymbol{x}) = c + \boldsymbol{b}^T\boldsymbol{x} + \frac{1}{2}\boldsymbol{x}^T\boldsymbol{A}\boldsymbol{x} \tag{7}$$

where c is a scalar, $\boldsymbol{b}$ a vector, and $\boldsymbol{A}$ a matrix. Equation (7) holds for a general quadratic function of n variables, with $\boldsymbol{b}$ an n-component vector and $\boldsymbol{A}$ an $n \times n$ matrix.

Returning to Equation (3), we now write out the complicated double sum in complete detail to assist in understanding it:

$\boldsymbol{x}^T\boldsymbol{H}\boldsymbol{x}$ Term in Detail

$$\boldsymbol{x}^T\boldsymbol{H}\boldsymbol{x} = \sum_{i=1}^{n}\sum_{j=1}^{n} x_i H_{ij} x_j = \left\{\begin{array}{l} \sum_{j=1}^{n} x_1 H_{1j} x_j \\ + \sum_{j=1}^{n} x_2 H_{2j} x_j \\ + \cdots \\ + \cdots \\ + \sum_{j=1}^{n} x_n H_{nj} x_j \end{array}\right\}$$

$$= \left\{\begin{array}{llll} x_1 H_{11} x_1 & + x_1 H_{12} x_2 & + \cdots & + x_1 H_{1n} x_n \\ + x_2 H_{21} x_1 & + x_2 H_{22} x_2 & + \cdots & + x_2 H_{2n} x_n \\ + \cdots & & + \cdots & \\ + \cdots & & + \cdots & \\ + x_n H_{n1} x_1 & + x_n H_{n2} x_2 & + \cdots & + x_n H_{nn} x_n \end{array}\right\}$$

Thus, $\boldsymbol{x}^T\boldsymbol{H}\boldsymbol{x}$ can be interpreted as the sum of all n^2 terms in a square array of which the (i, j) element is $x_i H_{ij} x_j$.

Steepest Descent Procedure

A crucial property of the gradient vector $\boldsymbol{G}(\boldsymbol{x})$ is that it points in the direction of the most rapid increase in the function F, which is the direction of **steepest ascent**. Conversely, $-\boldsymbol{G}(\boldsymbol{x})$ points in the direction of the **steepest descent**. This fact is so important that it is worth a few words of justification. Suppose that $\boldsymbol{h}$ is a unit vector, $\sum_{i=1}^{n} h_i^2 = 1$. The **rate of change of F (at $\boldsymbol{x}$)** in the direction $\boldsymbol{h}$ is defined naturally by

Rate of Change of F at x in Direction h

$$\left.\frac{d}{dt}F(\boldsymbol{x} + t\boldsymbol{h})\right|_{t=0}$$

This rate of change can be evaluated by using Equation (4). From that equation, it follows that

$$F(\boldsymbol{x} + t\boldsymbol{h}) = F(\boldsymbol{x}) + t\boldsymbol{G}(\boldsymbol{x})^T\boldsymbol{h} + \frac{1}{2}t^2\boldsymbol{h}^T\boldsymbol{H}(\boldsymbol{x})\boldsymbol{h} + \cdots \tag{8}$$

Differentiation with respect to t leads to

$$\frac{d}{dt}F(\boldsymbol{x} + t\boldsymbol{h}) = \boldsymbol{G}(\boldsymbol{x})^T\boldsymbol{h} + t\boldsymbol{h}^T\boldsymbol{H}(\boldsymbol{x})\boldsymbol{h} + \cdots \tag{9}$$

By letting $t = 0$ here, we see that the rate of change of F in the direction $\boldsymbol{h}$ is nothing else than

Key Results

$$\boldsymbol{G}(\boldsymbol{x})^T\boldsymbol{h}$$

Now we ask: *For what unit vector $\boldsymbol{h}$ is the rate of change a maximum*? The simplest path to the answer is to invoke the powerful **Cauchy-Schwarz inequality**:

Cauchy-Schwarz Inequality

$$\sum_{i=1}^{n} u_i v_i \leqq \left(\sum_{i=1}^{n} u_i^2\right)^{1/2} \left(\sum_{i=1}^{n} v_i^2\right)^{1/2} \tag{10}$$

where equality holds only if one of the vectors $\boldsymbol{u}$ or $\boldsymbol{v}$ is a nonnegative multiple of the other. Applying this to

$$\boldsymbol{G}(\boldsymbol{x})^T \boldsymbol{h} = \sum_{i=1}^{n} G_i(\boldsymbol{x}) h_i$$

and remembering that $\sum_{i=1}^{n} h_i^2 = 1$, we conclude that the maximum occurs when $\boldsymbol{h}$ is a positive multiple of $\boldsymbol{G}(\boldsymbol{x})$, that is, when $\boldsymbol{h}$ points in the direction of $\boldsymbol{G}$.

Best Step Steepest Descent

On the basis of the foregoing discussion, a minimization procedure called **best-step steepest descent** can be described. At any given point $\boldsymbol{x}$, the gradient vector $\boldsymbol{G}(\boldsymbol{x})$ is calculated. Then a one-dimensional minimization problem is solved by determining the value t^* for which the function

$$\phi(t) = F(\boldsymbol{x} + t\boldsymbol{G}(\boldsymbol{x}))$$

is a minimum. Then we replace $\boldsymbol{x}$ by $\boldsymbol{x} + t^*\boldsymbol{G}(\boldsymbol{x})$ and begin anew.

The general method of steepest descent takes a step of any size in the direction of the negative gradient. It is not usually competitive with other methods, but it has the advantage of simplicity. One way of speeding it up is described in Computer Exercise 13.2.2.

Contour Diagrams

In understanding how these methods work on functions of two variables, it is often helpful to draw contour diagrams. A **contour** of a function F is a set of the form

Contours (Level Sets) of F

$$\{\boldsymbol{x} : F(\boldsymbol{x}) = c\}$$

where c is a given constant. For example, the contours of function

$$F(\boldsymbol{x}) = 25x_1^2 + x_2^2$$

are ellipses, as shown in Figure 13.12 (p. 578). Contours are also called **level sets** by some authors. At any point on a contour, the gradient of F is perpendicular to the curve. So, in general, the path of steepest descent may look like Figure 13.13 (p. 578).

More Advanced Algorithms

F General Real-Valued Function

To explain more advanced algorithms, we consider a general real-valued function F of n variables. Suppose that we have obtained the first three terms in the Taylor series of F in the vicinity of a point z. *How can they be used to guess the minimum point of F?* Obviously, we could ignore all terms beyond the quadratic terms and find the minimum of the resulting quadratic function:

Quadratic Function

$$F(\boldsymbol{x} + z) = F(z) + \boldsymbol{G}(z)^T \boldsymbol{x} + \frac{1}{2}\boldsymbol{x}^T \boldsymbol{H}(z)\boldsymbol{x} + \cdots \tag{11}$$

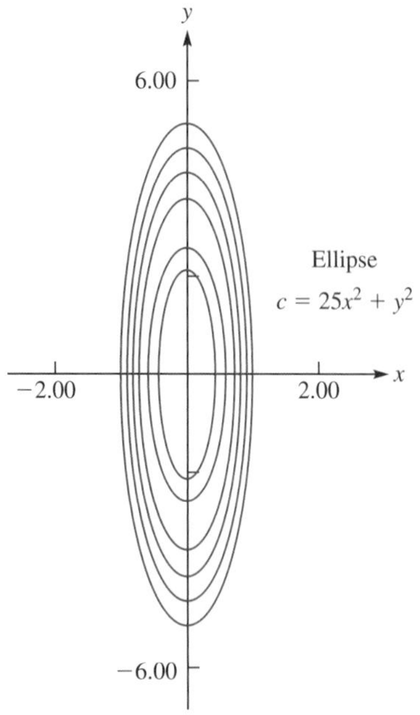

FIGURE 13.12 Contours of $F(x) = 25x_1^2 + x_2^2$

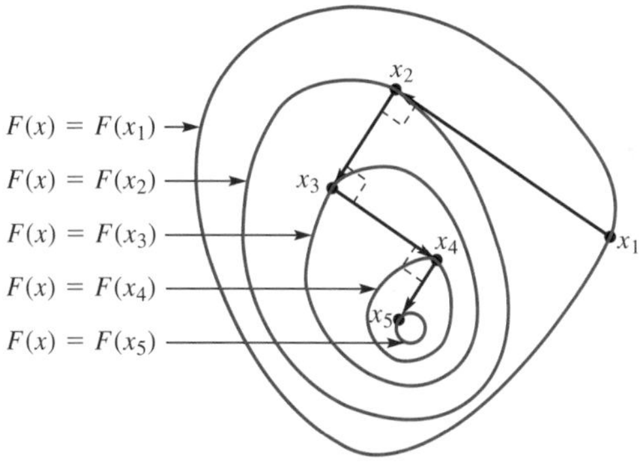

FIGURE 13.13 Path of steepest descent

Here, $\boldsymbol{z}$ is fixed and $\boldsymbol{x}$ is the variable. To find the minimum of this quadratic function of $\boldsymbol{x}$, we must compute the first partial derivatives and set them equal to zero. Denoting this quadratic function by Q and simplifying the notation slightly, we have

$$Q(\boldsymbol{x}) = F(\boldsymbol{z}) + \sum_{i=1}^{n} G_i x_i + \frac{1}{2}\sum_{i=1}^{n}\sum_{j=1}^{n} x_i H_{ij} x_j \tag{12}$$

from which it follows that

$$\frac{\partial Q}{\partial x_k} = G_k + \sum_{j=1}^{n} H_{kj} x_j \qquad (1 \leqq k \leqq n) \tag{13}$$

(See Exercise 13.2.13.) The point $\boldsymbol{x}$ that is sought is thus a solution of the system of n equations

$$\sum_{j=1}^{n} H_{kj} x_j = -G_k \qquad (1 \leqq k \leqq n)$$

or, equivalently,

$$H(z)x = -G(z) \tag{14}$$

Iterative Algorithm

The preceding analysis suggests the following iterative algorithm for locating a minimum point of a function F: Start with a point z that is a current estimate of the minimum point. Compute the gradient and Hessian of F at the point z. They can be denoted by $\boldsymbol{G}$ and $\boldsymbol{H}$, respectively. Of course, $\boldsymbol{G}$ is an n-component vector of numbers and $\boldsymbol{H}$ is an $n \times n$ matrix of numbers. Then solve the matrix equation

Matrix Equation: *G* Gradient, *H* Hessian

$$Hx = -G$$

obtaining an n-component vector x. Replace z by $z + x$ and return to the beginning of the algorithm.

Minimum, Maximum, and Saddle Points

There are many reasons for expecting trouble from the iterative procedure just outlined. One especially noisome aspect is that we can expect to find a point only where the first partial derivatives of F vanish; it need not be a minimum point. It is what we call a **stationary point**. Such points can be classified into three types: **minimum point**, **maximum point**, and **saddle point**. They can be illustrated by simple quadratic surfaces familiar from analytic geometry:

Stationary Points: Minimum Point, Maximum Point, Saddle Point

- Minimum of $F(x, y) = x^2 + y^2$ at $(0, 0)$ (See Figure 13.14(a).)
- Maximum of $F(x, y) = 1 - x^2 - y^2$ at $(0, 0)$ (See Figure 13.14(b).)
- Saddle point of $F(x, y) = x^2 - y^2$ at $(0, 0)$ (See Figure 13.14(c).)

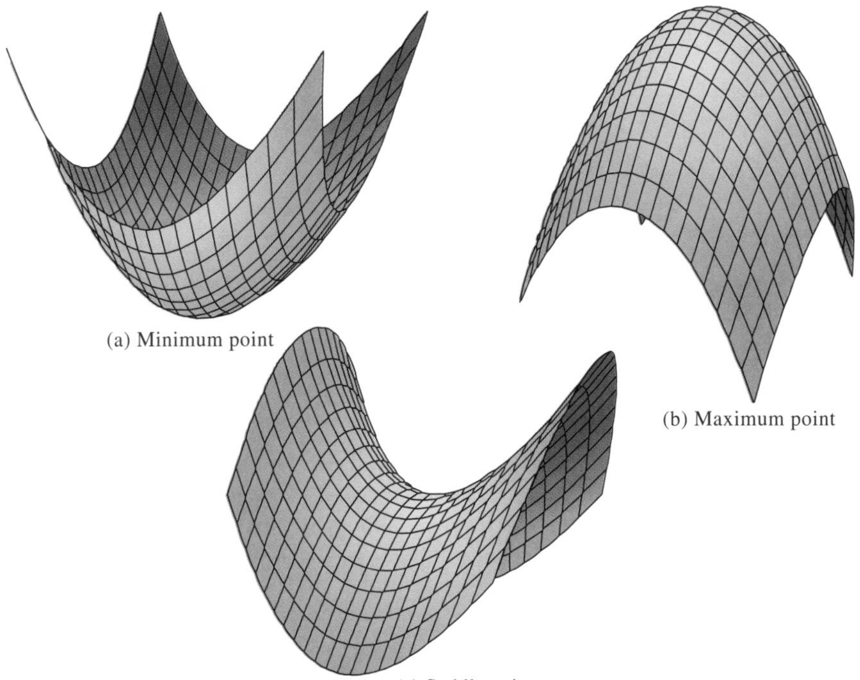

(a) Minimum point

(b) Maximum point

(c) Saddle point

FIGURE 13.14
Simple quadratic surfaces

Positive Definite Matrix

If z is a stationary point of F, then

$$G(z) = 0$$

Moreover, a criterion ensuring that Q, as defined in Equation (12), has a minimum point is as follows:

■ **Theorem 1**

Quadratic Function Theorem

If the matrix $\boldsymbol{H}$ has the property that $\boldsymbol{x}^T\boldsymbol{H}\boldsymbol{x} > 0$ for every nonzero vector $\boldsymbol{x}$, then the quadratic function Q has a minimum point.

Matrix *H* Positive Definite

(See Exercise 13.2.15.) A matrix that has this property is said to be **positive definite**. Notice that this theorem involves only second-degree terms in the quadratic function Q.

As examples of quadratic functions that do not have minima, consider the following:

Sample Quadratic Functions

$$-x_1^2 - x_2^2 + 13x_1 + 6x_2 + 12, \qquad x_1^2 - x_2^2 + 3x_1 + 5x_2 + 7$$
$$x_1^2 - 2x_1x_2 + x_1 + 2x_2 + 3, \qquad 2x_1 + 4x_2 + 6$$

In the first two examples, let $x_1 = 0$ and $x_2 \to \infty$. In the third, let $x_1 = x_2 \to \infty$. In the last, let $x_1 = 0$ and $x_2 \to -\infty$. In each case, the function values approach $-\infty$, and no global minimum can exist.

Quasi-Newton Methods

Algorithms: Quasi-Newton, Variable Metric

Algorithms that converge faster than steepest descent in general and that are currently recommended for minimization are of a type called **quasi-Newton**. The principal example is an algorithm introduced in 1959 by Davidon, called the **variable metric algorithm**. Subsequently, important modifications and improvements were made by others, such as R. Fletcher, M. J. D. Powell, C. G. Broyden, P. E. Gill, and W. Murray. These algorithms proceed iteratively, assuming in each step that a local quadratic approximation is known for the function F whose minimum is sought. The minimum of this quadratic function either provides the new point directly or is used to determine a line along which a one-dimensional search can be carried out. In implementation of the algorithm, the gradient can be either provided in the form of a procedure or computed numerically by finite differences. The Hessian $\boldsymbol{H}$ is *not* computed, but an estimate of its $\boldsymbol{LU}$ factorization is kept up to date as the process continues.

Nelder-Mead Algorithm

Nelder-Mead Algorithm

For minimizing a function $F\colon \mathbb{R}^n \to \mathbb{R}$, another method, called the **Nelder-Mead algorithm**, is available. It is a method of *direct search* and proceeds without involving any derivatives of the function F and without any *line searches*.

Before beginning the calculations, the user assigns values to three parameters: α, β, and γ. The default values are 1, $\frac{1}{2}$, and 1, respectively. In each step of the algorithm, a set of $n + 1$ points in $\mathbb{R}^n$ is given: $\{x_0, x_1, \ldots, x_n\}$. This set is in *general position* in $\mathbb{R}^n$. This means that the set of n points $x_i - x_0$, with $1 \leq i \leq n$, is linearly independent. A consequence of this assumption is that the convex hull of the original set $\{x_0, x_1, \ldots, x_n\}$ is an $\boldsymbol{n}$**-simplex**.

Convex Hull *n*-Simplex

2-Simplex: Triangle
3-Simplex: Tetrahedron

For example, a **2-simplex** is a triangle in $\mathbb{R}^2$, and a **3-simplex** is a tetrahedron in $\mathbb{R}^3$. To make the description of the algorithm as simple as possible, we assume that the points have been relabeled (if necessary) so that $F(x_0) \geqq F(x_1) \geqq \cdots \geqq F(x_n)$. Since we are trying to minimize the function F, the point x_0 is the worst of the current set, because it produces the highest value of F.

We compute the point

Centroid

$$u = \frac{1}{n}\sum_{i=1}^{n} x_i$$

Reflected Point

This is the **centroid** of the face of the current simplex opposite the worst vertex, x_0. Next, we compute a **reflected point** $v = (1+\alpha)u - \alpha x_0$.

Expended Reflected Point

If $F(v)$ is less than $F(x_n)$, then this is a favorable situation, and one is tempted to replace x_0 by v and begin anew. However, we first compute an **expanded reflected point** $w = (1+\gamma)v - \gamma u$ and test to see whether $F(w)$ is less than $F(x_n)$. If so, we replace x_0 by w and begin anew. Otherwise, we replace x_0 by v as originally suggested and begin with the new simplex.

Assume now that $F(v)$ is not less than $F(x_n)$. If $F(v) \leqq F(x_1)$, then replace x_0 by v and begin again. Having disposed of all cases when $F(v) \leqq F(x_1)$, we now consider two further cases. First, if $F(v) \leqq F(x_0)$, then define $w = u + \beta(v - u)$. If $F(v) > F(x_0)$, compute $w = u + \beta(x_0 - u)$. With w now defined, test whether $F(w) < F(x_0)$. If this is true, replace x_0 by w and begin anew. However, if $F(w) \geqq F(x_0)$, *shrink* the simplex by using $x_i \leftarrow \frac{1}{2}(x_i + x_n)$ for $0 \leqq i \leqq n-1$. Then begin anew.

The algorithm needs a stopping test in each major step. One such test is whether the relative **flatness** is small. That is the quantity

Flatness Test

$$\frac{F(x_0) - F(x_n)}{|F(x_0)| + |F(x_n)|}$$

Other tests to make sure progress is being made can be added. In programming the algorithm, one keeps the number of evaluations of f to a minimum. In fact, only three indices are needed: the indices of the greatest $F(x_i)$, the next greatest, and the least.

In addition to the original paper of Nelder and Mead [1965], one can consult Dennis and Woods [1987], Dixon [1974], and Torczon [1997]. Different authors give slightly different versions of the algorithm. We have followed the original description by Nelder and Mead.

Method of Simulated Annealing

Minimizing Difficult Functions

This method has been proposed and found to be effective for the *minimization* of *difficult* functions, especially if they have many purely local minimum points. It involves *no* derivatives or *line searches*; indeed, it has found great success in minimizing discrete functions, such as arise in the *traveling salesman problem*.

Suppose we are given a real-valued function of n real variables; $F\colon \mathbb{R}^n \to \mathbb{R}$. We must be able to compute the values $F(x)$ for any x in $\mathbb{R}^n$. It is desired to locate a **global minimum point of F**, which is a point x^* such that

Global Minimum Point of F

$$F(x^*) \leqq F(x) \text{ for all } x \text{ in } \mathbb{R}^n$$

In other words, $F(x^*)$ is equal to $\inf_{x\in\mathbb{R}^n} F(x)$. The algorithm generates a sequence of points $x_1, x_2, x_3, \ldots,$ and one hopes that $\min_{j\leqq k} F(x_j)$ converges to $\inf F(x)$ as $k \to \infty$.

In describing the computation that leads to x_{k+1}, assuming that x_k has been computed, we begin by generating a modest number of random points $u_1, u_2, \ldots, u_m$ in a large neighborhood of x_k. For each of these points, the value of F must be computed. The next point, x_{k+1}, in our sequence is chosen to be one of the points $u_1, u_2, \ldots, u_m$. This choice is made as follows. Select an index j such that

$$F(u_j) = \min\{F(u_1), F(u_2), \ldots, F(u_m)\}$$

If $F(u_j) < F(x_k)$, then set $x_{k+1} = u_j$. In the other case, for each i, we assign a probability p_i to u_i by the formula

$$p_i = e^{\alpha[F(x_k) - F(u_i)]} \qquad (1 \leqq i \leqq m)$$

Here, α is a positive parameter chosen by the user of the code. We normalize the probabilities by dividing each by their sum. That is, we compute

$$S = \sum_{i=1}^{m} p_i$$

and then carry out a replacement

$$p_i \leftarrow p_i / S$$

Finally, a **random choice** is made among the points $u_1, u_2, \ldots, u_m$, taking account of the probabilities p_i that have been assigned to them. This randomly chosen u_i becomes x_{k+1}.

Simple Way to Make Random Choice

The simplest way to make this random choice is to employ a random number generator to get a random point ξ in the interval (0, 1). Select i to be the first integer such that

$$\xi \leqq p_1 + p_2 + \cdots + p_i$$

Thus, if $\xi \leqq p_1$, let $i = 1$ (and $x_{n+1} = u_1$). If $p_1 < \xi \leqq p_1 + p_2$, then let $i = 2$ (and $x_{n+1} = u_2$), and so on.

The formula for the probabilities p_i is taken from the theory of thermodynamics. The interested reader can consult the original articles by Metropolis et al. [1953] or Otten and van Ginneken [1989]. Presumably, other functions can serve in this role as well.

Purpose of Choice for x_{k+1}

What is the purpose of the complicated choice for x_{k+1}? Because of the possibility of encountering local minima, the algorithm must occasionally choose a point that is *uphill* from the current point. Then there is a chance that subsequent points might begin to move toward a different local minimum. An element of randomness is introduced to make this possible.

With minor modifications, the algorithm can be used for functions $f\colon X \to \mathbb{R}$, where X is any set. For example, in the *traveling salesman problem*, X is the set of all permutations of a set of integers $\{1, 2, 3, \ldots, N\}$. All that is required is a procedure for generating random permutations and, of course, a code for evaluating the function f.

Computer programs for a variety of algorithms can be found online at the websites `http://www.netlib.org/`. A collection of papers on simulated annealing, emphasizing parallel computation, is Azencott [1992].

Summary 13.2

- In a typical **minimization problem**, we seek a point $\boldsymbol{x}^*$ such that

$$F(\boldsymbol{x}^*) \leqq F(\boldsymbol{x}) \quad \text{for all } \boldsymbol{x} \in \mathbb{R}^n$$

where F is a real-valued multivariate function.

- A **gradient vector** $\boldsymbol{G}(\boldsymbol{x})$ has components

$$G_i = G_i(\boldsymbol{x}) = \frac{\partial F(\boldsymbol{x})}{\partial x_i} \qquad (1 \leqq i \leqq n)$$

and a **Hessian matrix** $\boldsymbol{H}(\boldsymbol{x})$ has components

$$H_{ij} = H_{ij}(\boldsymbol{x}) = \frac{\partial^2 F(\boldsymbol{x})}{\partial x_i\, \partial x_j} \qquad (1 \leqq i,\ j \leqq n)$$

It is a symmetric matrix if the second-order derivatives are continuous.

- The **Taylor series** for F is

$$F(\boldsymbol{x}+\boldsymbol{h}) = F(\boldsymbol{x}) + \boldsymbol{G}(\boldsymbol{x})^T\boldsymbol{h} + \frac{1}{2}\boldsymbol{h}^T\boldsymbol{H}(\boldsymbol{x})\boldsymbol{h} + \cdots$$

Here, $\boldsymbol{x}$ is the fixed point of expansion in $\mathbb{R}^n$ and $\boldsymbol{h}$ is the variable in $\mathbb{R}^n$ with components $h_1, h_2, \ldots, h_n$. The three dots indicate higher-order terms in $\boldsymbol{h}$ that are not needed in this discussion.

- An alternative form of the Taylor series is

$$F(\boldsymbol{x}) = F(\boldsymbol{z}) + \boldsymbol{G}(\boldsymbol{z})^T(\boldsymbol{x}-\boldsymbol{z}) + \frac{1}{2}(\boldsymbol{x}-\boldsymbol{z})^T\boldsymbol{H}(\boldsymbol{z})(\boldsymbol{x}-\boldsymbol{z}) + \cdots$$

For example, a **linear function** $F(\boldsymbol{x}) = c + \boldsymbol{b}^T\boldsymbol{x}$ has the Taylor series

$$F(\boldsymbol{x}) = F(\boldsymbol{z}) + \boldsymbol{b}^T(\boldsymbol{x}-\boldsymbol{z})$$

A **quadratic function** is

$$F(\boldsymbol{x}) = c + \boldsymbol{b}^T\boldsymbol{x} + \frac{1}{2}\boldsymbol{x}^T\boldsymbol{A}\boldsymbol{x}$$

- An iterative procedure for locating a minimum point of a function F is to start with a point $\boldsymbol{z}$ that is a current estimate of the minimum point, compute the gradient $\boldsymbol{G}$ and Hessian $\boldsymbol{H}$ of F at the point $\boldsymbol{z}$, and solve the matrix equation

$$\boldsymbol{H}\boldsymbol{x} = -\boldsymbol{G}$$

for $\boldsymbol{x}$. Then replace $\boldsymbol{z}$ by $\boldsymbol{z}+\boldsymbol{x}$ and repeat.

- If the matrix $\boldsymbol{H}$ has the property that $\boldsymbol{x}^T\boldsymbol{H}\boldsymbol{x} > 0$ for every nonzero vector $\boldsymbol{x}$, then the quadratic function Q has a unique minimum point.
- Algorithms that are discussed are **steepest descent**, **Nelder-Mead**, and **simulated annealing**.

Exercises 13.2

1. Determine whether these functions have minimum values in $\mathbb{R}^2$:

   [a]**a.** $x_1^2 - x_1x_2 + x_2^2 + 3x_1 + 6x_2 - 4$

   [a]**b.** $x_1^2 - 3x_1x_2 + x_2^2 + 7x_1 + 3x_2 + 5$

   c. $2x_1^2 - 3x_1x_2 + x_2^2 + 4x_1 - x_2 + 6$

   d. $ax_1^2 - 2bx_1x_2 + cx_2^2 + dx_1 + ex_2 + f$

   Hint: Use the method of completing the square.

[a]2. Locate the minimum point of $3x^2 - 2xy + y^2 + 3x - 46 + 7$ by finding the gradient and Hessian and solving the appropriate linear equations.

[a]3. Using $(0, 0)$ as the point of expansion, write the first three terms of the Taylor series for $F(x, y) = e^x \cos y - y\ln(x+1)$.

4. Using $(1, 1)$ as the point of expansion, write the first three terms of the Taylor series for $F(x, y) = 2x^2 - 4xy + 7y^2 - 3x + 5y$.

5. The Taylor series expansion about zero can be written as

$$F(\boldsymbol{x}) = F(0) + \boldsymbol{G}(0)^T\boldsymbol{x} + \frac{1}{2}\boldsymbol{x}^T\boldsymbol{H}(0)\boldsymbol{x} + \cdots$$

Show that the Taylor series about $\boldsymbol{z}$ can be written in a similar form by using matrix-vector notation; that is,

$$F(\boldsymbol{x}) = F(\boldsymbol{z}) + \mathcal{G}(\boldsymbol{z})^T\mathcal{X} + \frac{1}{2}\mathcal{X}^T\mathcal{H}(\boldsymbol{z})\mathcal{X} + \cdots$$

where

$$\mathcal{X} = \begin{bmatrix} \boldsymbol{x} \\ \boldsymbol{z} \end{bmatrix}, \quad \mathcal{G}(\boldsymbol{z}) = \begin{bmatrix} \boldsymbol{G}(\boldsymbol{z}) \\ -\boldsymbol{G}(\boldsymbol{z}) \end{bmatrix}$$

$$\mathcal{H}(\boldsymbol{z}) = \begin{bmatrix} \boldsymbol{H}(\boldsymbol{z}) & -\boldsymbol{H}(\boldsymbol{z}) \\ -\boldsymbol{H}(\boldsymbol{z}) & \boldsymbol{H}(\boldsymbol{z}) \end{bmatrix}$$

[a]**6.** Show that the gradient of $F(x, y)$ is perpendicular to the contour.
Hint: Interpret the equation $F(x, y) = c$ as defining y as a function of x. Then by the chain rule,

$$\frac{\partial F}{\partial x} + \frac{\partial F}{\partial y}\frac{dy}{dx} = 0$$

From it, obtain the slope of the tangent to the contour.

7. Consider the function

$$F(x_1, x_2, x_3) = 3e^{x_1x_2} - x_3 \cos x_1 + x_2 \ln x_3$$

a. Determine the gradient vector and Hessian matrix.

[a]**b.** Derive the first three terms of the Taylor series expansion about $(0, 1, 1)$.

c. What linear system should be solved for a reasonable guess as to the minimum point for F? What is the value of F at this point?

8. It is asserted that the Hessian of an unknown function F at a certain point is

$$\begin{bmatrix} 3 & 2 \\ 1 & 4 \end{bmatrix}$$

What conclusion can be drawn about F?

9. What are the gradients of the following functions at the points indicated?

[a]**a.** $F(x, y) = x^2y - 2x + y$ at $(1, 0)$

[a]**b.** $F(x, y, z) = xy + yz^2 + x^2z$ at $(1, 2, 1)$

[a]**10.** Consider $F(x, y, z) = y^2z^2(1+\sin^2 x)+(y+1)^2(z+3)^2$. We want to find the minimum of the function. The program to be used requires the gradient of the function. What formulas must we program for the gradient?

11. Let F be a function of two variables whose gradient at $(0, 0)$ is $[-5, 1]^T$ and whose Hessian is

$$\begin{bmatrix} 6 & -1 \\ -1 & 2 \end{bmatrix}$$

Make a reasonable guess as to the minimum point of F. Explain.

[a]**12.** Write the function $F(x_1, x_2) = 3x_1^2 + 6x_1x_2 - 2x_2^2 + 5x_1 + 3x_2 + 7$ in the form of Equation (7) with appropriate $\boldsymbol{A}$, $\boldsymbol{b}$, and c. Show in matrix form the linear equations that must be solved in order to find a point where the first partial derivatives of F vanish. Finally, solve these equations to locate this point numerically.

13. Verify Equation (13). In differentiating the double sum in Equation (12), first write all terms that contain x_k. Then differentiate and use the symmetry of the matrix $\boldsymbol{H}$.

14. Consider the quadratic function Q in Equation (12). Show that if $\boldsymbol{H}$ is positive definite, then the stationary point is a minimum point.

15. (**General Quadratic Function**) Generalize Equation (6) to n variables. Show that a general quadratic function $Q(\boldsymbol{x})$ of n variables can be written in the matrix-vector form of Equation (7), where $\boldsymbol{A}$ is an $n \times n$ symmetric matrix, $\boldsymbol{b}$ a vector of length n, and c a scalar. Establish that the gradient and Hessian are

$$\boldsymbol{G}(\boldsymbol{x}) = \boldsymbol{A}\boldsymbol{x} + \boldsymbol{b}, \qquad \boldsymbol{H}(\boldsymbol{x}) = \boldsymbol{A}$$

respectively.

16. Let $\boldsymbol{A}$ be an $n \times n$ symmetric matrix and define an upper triangular matrix $\boldsymbol{U} = (u_{ij})$ by putting

$$u_{ij} = \begin{cases} a_{ij}, & i = j \\ 2a_{ij}, & i < j \\ 0, & i > j \end{cases}$$

Show that $\boldsymbol{x}^T\boldsymbol{U}\boldsymbol{x} = \boldsymbol{x}^T\boldsymbol{A}\boldsymbol{x}$ for all vectors $\boldsymbol{x}$.

17. Show that the general quadratic function $Q(\boldsymbol{x})$ of n variables can be written

$$Q(\boldsymbol{x}) = c + \boldsymbol{b}^T\boldsymbol{x} + \frac{1}{2}\boldsymbol{x}^T\boldsymbol{U}\boldsymbol{x}$$

where $\boldsymbol{U}$ is an upper triangular matrix. Can this simplify the work of finding the stationary point of Q?

18. Show that the gradient and Hessian satisfy the equation

$$\boldsymbol{H}(\boldsymbol{z})(\boldsymbol{x} - \boldsymbol{z}) = \boldsymbol{G}(\boldsymbol{x}) - \boldsymbol{G}(\boldsymbol{z})$$

for a general quadratic function of n variables.

19. Using Taylor series, show that a general quadratic function of n variables can be written in block form

$$Q(x) = \frac{1}{2}\mathcal{X}^T \mathcal{A} \mathcal{X} + \mathcal{B}^T \mathcal{X} + c$$

where

$$\mathcal{X} = \begin{bmatrix} x \\ z \end{bmatrix}, \quad \mathcal{A} = \begin{bmatrix} A & -A \\ -A & A \end{bmatrix}, \quad \mathcal{B} = \begin{bmatrix} b \\ -b \end{bmatrix}$$

Here z is the point of expansion.

20. (**Least-Squares Problem**) Consider the function

$$F(\boldsymbol{x}) = (\boldsymbol{b} - \boldsymbol{A}\boldsymbol{x})^T(\boldsymbol{b} - \boldsymbol{A}\boldsymbol{x}) + \alpha \boldsymbol{x}^T\boldsymbol{x}$$

where $\boldsymbol{A}$ is a real $m \times n$ matrix, $\boldsymbol{b}$ is a real column vector of order m, and α is a positive real number. We want the minimum point of F for given $\boldsymbol{A}$, $\boldsymbol{b}$, and α. Show that

$$F(\boldsymbol{x} + \boldsymbol{h}) - F(\boldsymbol{x}) = (\boldsymbol{A}\boldsymbol{h})^T(\boldsymbol{A}\boldsymbol{h}) + \alpha \boldsymbol{h}^T\boldsymbol{h} \geq 0$$

for $\boldsymbol{h}$ a vector of order n, provided that

$$(\boldsymbol{A}^T\boldsymbol{A} + \alpha \boldsymbol{I})\boldsymbol{x} = \boldsymbol{A}^T\boldsymbol{b}$$

This means that any solution of this linear system minimizes $F(\boldsymbol{x})$; hence, this is the normal equation.

21. (**Multiple Choice**) What is the gradient of the function $f(\boldsymbol{x}) = 3x_1^2 - \sin(x_1x_2)$ at the point $(3, 0)$?

a. $(6, -3)$ **b.** $(3, -1)$ **c.** $(18, 0)$ **d.** $(18, -3)$ **e.** None of these.

22. (**Multiple Choice**, continuation) The directional derivative of the function f at the point $\boldsymbol{x}$ in the direction $\boldsymbol{u}$ is given by the expression

$$\frac{d}{dt}f(\boldsymbol{x} + t\boldsymbol{u})|_{t=0}$$

In this description, $\boldsymbol{u}$ should be a unit vector. What is the numerical value of the directional derivative where $f(\boldsymbol{x})$ is the function defined in the preceding problem, $\boldsymbol{x} = (1, \pi/2)$, and $\boldsymbol{u} = (1, 1)/\sqrt{2}$.

a. $6/\sqrt{2}$ **b.** 6 **c.** 18 **d.** 3 **e.** None of these.

23. (**Multiple Choice**, continuation) If f is a real-valued function of n variables, the Hessian $\boldsymbol{H} = (H_{ij})$ is given by $H_{ij} = \partial^2 f/\partial x_i \partial x_j$, all terms being evaluated at a specific point $\boldsymbol{x}$. What is the entry H_{22} in this matrix in the case of f as given in the previous problem and $\boldsymbol{x} = (1, \pi/2)$?

a. 6 **b.** $6/\sqrt{2}$ **c.** 1 **d.** $\pi^2/2$ **e.** None of these.

24. (**Multiple Choice**) Let f be a real-valued function of n real variables. Let $\boldsymbol{x}$ and $\boldsymbol{u}$ be given as numerical vectors, and $\boldsymbol{u} \neq 0$. Then the expression $f(\boldsymbol{x} + t\boldsymbol{u})$ defines a function of t. Suppose that the minimum of $f(\boldsymbol{x} + t\boldsymbol{u})$ occurs at $t = 0$. What conclusion can be drawn?

a. The gradient of f at $\boldsymbol{x}$, denoted by $\boldsymbol{G}(\boldsymbol{x})$, is 0.

b. $\boldsymbol{u}$ is perpendicular to the gradient of f at $\boldsymbol{x}$.

c. $\boldsymbol{u} = \boldsymbol{G}(\boldsymbol{x})$, where $\boldsymbol{G}(\boldsymbol{x})$ denotes the gradient of f at $\boldsymbol{x}$.

d. $\boldsymbol{G}(\boldsymbol{x})$ is perpendicular to $\boldsymbol{x}$.

e. None of these.

25. (**Multiple Choice**) If f is a (real-valued) quadratic function of n real variables, we can write it in the form $f\boldsymbol{x}) = c - \boldsymbol{b}^T\boldsymbol{x} + \frac{1}{2}\boldsymbol{x}^T\boldsymbol{A}\boldsymbol{x}$. The gradient of f is then:

a. $\boldsymbol{A}\boldsymbol{x}$ **b.** $\boldsymbol{b} - \boldsymbol{A}\boldsymbol{x}$ **c.** $\boldsymbol{A}\boldsymbol{x} - \boldsymbol{b}$ **d.** $\frac{1}{2}\boldsymbol{A}\boldsymbol{x} - \boldsymbol{b}$ **e.** None of these.

Computer Exercises 13.2

1. Select a routine from your program library or from a package such as MATLAB, Maple, or Mathematica for minimizing a function of many variables without the need to program derivatives. Test it on one or more of the following well-known functions. The ordering of our variables is (x, y, z, w).

[a]**a.** **Rosenbrock**: $100(y - x^2)^2 + (1 - x)^2$. Start at $(-1.2, 1.0)$.

b. **Powell 1**: $(x + 10y)^2 + 5(z - w)^2 + (y - 2z)^4 + 10(x - w)^4$. Start at $(3, -1, 0, 1)$.

[a]**c.** **Powell 2**: $x^2 + 2y^2 + 3z^2 + 4w^2 + (x + y + z + w)^4$. Start at $(1, -1, 1, 1)$.

d. **Fletcher and Powell**: $100(z - 10\phi)^2 + \left(\sqrt{x^2 + y^2} - 1\right)^2 + z^2$ in which ϕ is an angle determined from (x, y) by

$$\cos 2\pi\phi = \frac{x}{\sqrt{x^2 + y^2}}, \qquad \sin 2\pi\phi = \frac{y}{\sqrt{x^2 + y^2}}$$

where $-\pi/2 < 2\pi\phi \leq 3\pi/2$. Start at $(1, 1, 1)$.

[a]**e.** **Woods**: $100(x^2 - y)^2 + (1 - x)^2 + 90(z^2 - w)^2 + (1 - z)^2 + 10(y - 1)^2 + (w - 1)^2 + 19.8(y - 1)(w - 1)$. Start at $(-3, -1, -3, -1)$.

2. (**Accelerated Steepest Descent**) This version of steepest descent is superior to the basic one. A sequence of points $\boldsymbol{x}_1, \boldsymbol{x}_2, \ldots$ is generated as follows: Point $\boldsymbol{x}_1$ is specified as the starting point. Then $\boldsymbol{x}_2$ is obtained by one step of steepest descent from $\boldsymbol{x}_1$. In the general step, if $\boldsymbol{x}_1, \boldsymbol{x}_2, \ldots, \boldsymbol{x}_m$ have been obtained, we find a point z by steepest descent from $\boldsymbol{x}_m$. Then $\boldsymbol{x}_{m+1}$ is taken as the minimum point on the line $\boldsymbol{x}_{m-1} + t(z - z_{m-1})$. Program and test this algorithm on one of the examples in Computer Exercise 13.2.1.

3. Using a routine in your program library or in MATLAB, Maple, or Mathematica, do the following:

a. Solve the minimization problem that begins this chapter.

b. Plot and solve for the minimum point, the maximum point, and the saddle point of these functions, respectively: $x^2 + y^2$, $1 - x^2 - y^2$, $x^2 - y^2$.

c. Plot and numerically experiment with these functions that do not have minima: $-x^2 - y^2 + 13x + 6y + 12$, $x^2 - y^2 + 3x + 5y + 7$, $x^2 - 2xy + x + 2y + 3$, $2x + 4y + 6$.

4. We want to find the minimum of $F(x, y, z) = z^2 \cos x + x^2y^2 + x^2e^z$ using a computer program that requires procedures for the gradient of F together with F. Write the necessary procedures. Find the minimum using a preprogrammed code that uses the gradient.

5. Assume that

procedure *Xmin*$(f, (grad_i), n, (xi), (g_{ij}))$

is available to compute the minimum value of a function of two variables. Suppose that this routine requires not only the function but also its gradient. If we are going to use this routine with the function $F(x, y) = e^x \cos^2(xy)$, what procedure will be needed? Write the appropriate code. Find the minimum using a preprogrammed code that uses the gradient.

6. Program and test the Nelder-Mead algorithm.

7. Program and test the simulated annealing algorithm.

8. (**Student Research Project**) Explore one of the newer methods for minimization such as generic algorithms, methods of simulated annealing, or the Nelder-Mead algorithm. Use some of the software that is available for them.

9. Use built-in routines in mathematical software systems such as Maple or Mathematica to verify the calculations in Example 1.
Hint: In Maple, use `grad` and `Hessian`, and in Mathematica, use `Series`. For example, obtain two terms in the Taylor series in two variables expanded about the point $(1, 1)$, and then carry out a change of variables.

10. (**Molecular Conformation: Protein Folding Project**) Forces that govern folding of amino acids into proteins are due to bonds between individual atoms and to weaker interactions between unbound atoms such as electrostatic and Van der Waals forces. The Van der Waals forces are modeled by the Lennard-Jones potential

$$U(r) = \frac{1}{r^{12}} - \frac{2}{r^6}$$

where r is the distance between atoms.

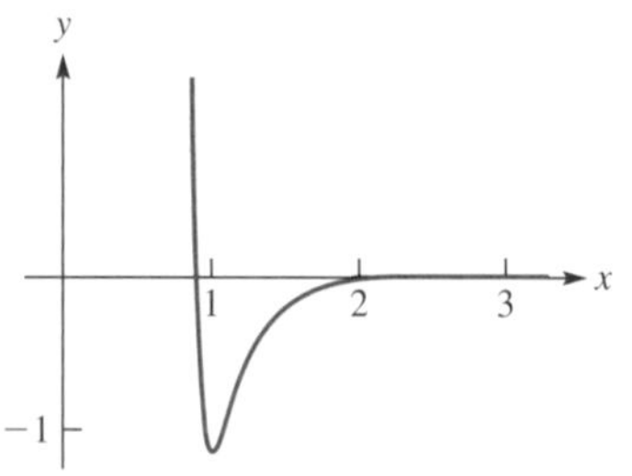

In the figure, the energy minimum is -1, and it is achieved at $r = 1$. Explore this subject and the numerical methods used. One approach is to predict the conformation of the proteins in finding the minimum potential energy of the total configuration of amino acids. For a cluster of atoms with positions (x_1, y_1, z_1) to (x_n, y_n, z_n), the objective function to be minimized is

$$U = \sum_{i<j} \frac{1}{r_{ij}^{12}} - \frac{2}{r_{ij}^{6}}$$

over all pairs of atoms. Here, $r_{ij} = \left[(x_i - x_j)^2 + (y_i - y_j)^2 + (z_i - z_j)^2\right]^{2}$ is the distance between atoms i and j. This optimization problem finds the rectangular coordinates of the atoms. See Sauer [2012] for additional details.

14

Linear Programming Problems

In the study of how the U.S. economy is affected by changes in the supply and cost of energy, it is appropriate to use a linear programming model. This is a large system of linear inequalities that govern the variables in the model, together with a linear function of these variables to be maximized. Typically, the variables are the activity levels of various processes in the economy, such as the number of barrels of oil pumped per day or the number of men's shirts produced per day. A model that contains reasonable detail could easily involve thousands of variables and thousands of linear inequalities. Such problems are discussed in this chapter, and some guidance is offered on how to use existing software.

14.1 Standard Forms and Duality

First Primal Form

Linear programming is a branch of mathematics that deals with finding extreme values of linear functions when the variables are constrained by linear inequalities. Any problem of this type can be put into a standard form known as *first primal form* by simple manipulations (to be discussed later in this chapter).

In matrix notation, the linear programming problem in first primal form looks like this:

First Primal Form

$$\begin{cases} \text{Maximize:} & \boldsymbol{c}^T \boldsymbol{x} \\ \text{Constraints:} & \begin{cases} \boldsymbol{A}\boldsymbol{x} \leqq \boldsymbol{b} \\ \boldsymbol{x} \geqq 0 \end{cases} \end{cases} \tag{1}$$

■ **Theorem 1** **First Primal Form**

Given data c_j, a_{ij}, b_i (for $1 \leqq j \leqq n$, $1 \leqq i \leqq m$), we wish to determine the x_j's ($1 \leqq j \leqq n$) that maximize the linear function

Linear Objective Function

$$\sum_{j=1}^{n} c_j x_j$$

subject to the constraints

Constraints

$$\begin{cases} \displaystyle\sum_{j=1}^{n} a_{ij} x_j \leqq b_i & (1 \leqq i \leqq m) \\ x_j \geqq 0 & (1 \leqq j \leqq n) \end{cases}$$

Here, $\boldsymbol{c}$ and $\boldsymbol{x}$ are n-component vectors, $\boldsymbol{b}$ is an m-component vector, and $\boldsymbol{A}$ is an $m \times n$ matrix. A **vector inequality** $\boldsymbol{u} \leqq \boldsymbol{v}$ means that $\boldsymbol{u}$ and $\boldsymbol{v}$ are vectors with the same number of components and that *all* the individual components satisfy the inequality $u_i \leqq v_i$. The linear function $\boldsymbol{c}^T\boldsymbol{x}$ is called the **objective function**.

Objective Function

In a linear programming problem, the set of all vectors that satisfy the constraints is called the **feasible set**, and its elements are the **feasible points**. So in the preceding notation, the feasible set is

Feasible Set

$$\boldsymbol{K} = \{\boldsymbol{x} \in \mathbb{R}^n\colon \boldsymbol{x} \geqq 0 \quad \text{and} \quad \boldsymbol{A}\boldsymbol{x} \leqq \boldsymbol{b}\}$$

A more precise (and concise) statement of the linear programming problem, then, is as follows: Determine $\boldsymbol{x}^* \in \boldsymbol{K}$ such that $\boldsymbol{c}^T\boldsymbol{x}^* \geqq \boldsymbol{c}^T\boldsymbol{x}$ for all $\boldsymbol{x} \in \boldsymbol{K}$.

Numerical Example

To get an idea of the type of practical problem that can be solved by linear programming, consider a simple example of optimization. Suppose that a certain factory uses two raw materials to produce two products. Suppose also that the following are true:

Sample LPP

1. Each unit of the first product requires 5 units of the first raw material and 3 of the second.
2. Each unit of the second product requires 3 units of the first raw material and 6 of the second.
3. On hand are 15 units of the first raw material and 18 units of the second.
4. The profits on sales of the products are 2 per unit for the first product and 3 per unit for the second product.

How should the raw materials be used to realize a maximum profit? To answer this question, variables x_1 and x_2 are introduced to represent the number of units of the two products to be manufactured. In terms of these variables, the profit is

$$2x_1 + 3x_2$$

The process uses up $5x_1 + 3x_2$ units of the first raw material and $3x_1 + 6x_2$ units of the second. The limitations in the third fact are expressed by these inequalities:

$$\begin{cases} 5x_1 + 3x_2 \leqq 15 \\ 3x_1 + 6x_2 \leqq 18 \end{cases}$$

Of course, $x_1 \geqq 0$ and $x_2 \geqq 0$. Thus, the solution to the problem is a vector $\boldsymbol{x} \geqq 0$ that maximizes the objective function $2x_1 + 3x_2$ while satisfying the constraints above. So the linear programming problem is

Linear Programming Problem (LPP)

$$\begin{cases} \text{Maximize:} \quad 2x_1 + 3x_2 \\ \text{Constraints:} \quad \begin{cases} 5x_1 + 3x_2 \leqq 15 \\ 3x_1 + 6x_2 \leqq 18 \\ x_1 \geqq 0, \quad x_2 \geqq 0 \end{cases} \end{cases} \tag{2}$$

More precisely, among all vectors $\boldsymbol{x}$ in the set

$$\boldsymbol{K} = \{\boldsymbol{x}\colon \boldsymbol{x} \geqq 0, \quad 5x_1 + 3x_2 \leqq 15, \quad 3x_1 + 6x_2 \leqq 18\}$$

we want the one that makes $2x_1 + 3x_2$ as large as possible.

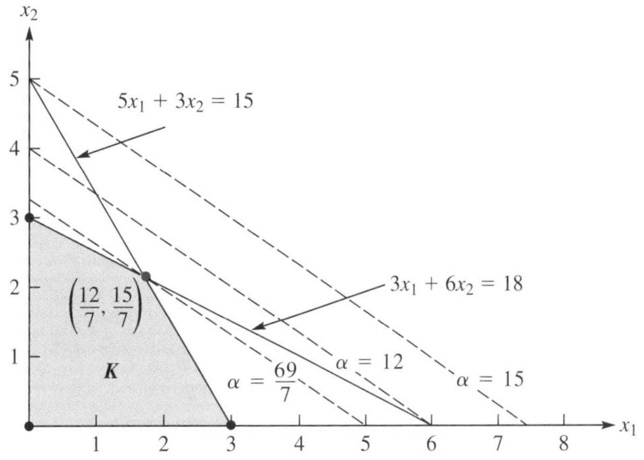

FIGURE 14.1
Graphical solution method

Because the number of variables in this example is only two, the problem can be solved graphically. To locate the solution, we begin by graphing the set $\boldsymbol{K}$. This is the shaded region in Figure 14.1. Then we draw some of the dashed lines $2x_1 + 3x_2 = \alpha$, where α is given various values. These lines are dashed in the figure and labeled with the values of α. Finally, we select one of these lines with a maximum α that intersects $\boldsymbol{K}$. That intersection is the solution point and a vertex of $\boldsymbol{K}$. It is obtained numerically by solving simultaneously the equations $5x_1 + 3x_2 = 15$ and $3x_1 + 6x_2 = 18$. Thus, we obtain $\boldsymbol{x} = \left[\frac{12}{7}, \frac{15}{7}\right]^T$, and the corresponding profit from Equation (2) is $2\left(\frac{12}{7}\right) + 3\left(\frac{15}{7}\right) = \frac{69}{7}$.

Mathematical Software

We can use mathematical software systems such as MATLAB, Maple, or Mathematica to solve this linear programming problem. For example, we obtain the solution $x = \frac{12}{7}$ and $y = \frac{15}{7}$ with objective function value $\frac{69}{7}$ using one system, and we obtain the solution $x = 1.7143$ and $y = 2.1429$ with the value of the objective function used as -9.8571 on another. (Why?)

Some of these mathematical systems contain large collections of commands for the optimization of general linear and nonlinear functions. For nonlinear optimization, these functions can handle unconstrained and constrained minimization as well as a large number of other tasks. If the program performs minimization of the objective function and we wish to maximize the objective function, we need to minimize the negative of the objective function. Also, it may allow for additional equality constraints, and since we do not have any, we set them to null entries.

Integer Constraints

Note in this example that the units that are used—whether dollars, pesos, pounds, or kilograms—do not matter for the mathematical method as long as they are used consistently. Notice also that x_1 and x_2 are permitted to be arbitrary real numbers. The problem would be quite different if only integer values were acceptable as a solution. This situation occurs if the products being produced consisted of indivisible units, such as a manufactured article. If the integer constraint is imposed, only points with integer coordinates inside $\boldsymbol{K}$ are acceptable. So (0, 3) is the best of them. Observe particularly that we *cannot* simply round off the solution (1.71, 2.14) to the nearest integers to solve the problem with integer constraints. The point (2, 2) lies just outside $\boldsymbol{K}$. However, if the company could alter the constraints slightly by increasing the amount of the first raw material to 16, the integer solution (2, 2) would be allowable. Special programs for integer linear programming are available but are outside the scope of this book.

Observe how the solution would be altered if our profit or objective function were $2x_1 + x_2$. In this case, the dashed lines in the figure would have a different slope (namely, -2)

and a different vertex of the shaded region would occur as the solution—namely, (3, 0). A characteristic feature of linear programming problems is that the solutions (if any exist) can always be found among the vertices.

Transforming Problems into First Primal Form

A linear programming problem that is not already in the first primal form can be put into that form by some standard techniques:

Putting LPP in First Primal Form

1. If the original problem calls for the minimization of the linear function $\boldsymbol{c}^T\boldsymbol{x}$, this is the same as maximizing $(-\boldsymbol{c})^T\boldsymbol{x}$.
2. If the original problem contains a constraint like $\boldsymbol{a}^T\boldsymbol{x} \geqq \beta$, it can be replaced by the constraint $(-\boldsymbol{a})^T\boldsymbol{x} \leqq -\beta$.
3. If the objective function contains a constant, this fact has no effect on the solution. For example, the maximum of $\boldsymbol{c}^T\boldsymbol{x} + \lambda$ occurs for the same $\boldsymbol{x}$ as the maximum of $\boldsymbol{c}^T\boldsymbol{x}$.
4. If the original problem contains equality constraints, each can be replaced by two inequality constraints. Thus, the equation $\boldsymbol{a}^T\boldsymbol{x} = \beta$ is equivalent to $\boldsymbol{a}^T\boldsymbol{x} \leqq \beta$ and $\boldsymbol{a}^T\boldsymbol{x} \geqq \beta$.
5. If the original problem does not require a variable (say, x_i) to be nonnegative, we can replace x_i by the difference of two nonnegative variables, say, $x_i = u_i - v_i$, where $u_i \geqq 0$ and $v_i \geqq 0$.

Here is an example that illustrates all five techniques. Consider the linear programming problem

Sample LPP

$$\begin{cases} \text{Minimize:} & 2x_1 + 3x_2 - x_3 + 4 \\ \text{Constraints:} & \begin{cases} x_1 - x_2 + 4x_3 \geqq 2 \\ x_1 + x_2 + x_3 = 15 \\ x_2 \geqq 0 \geqq x_3 \end{cases} \end{cases} \tag{3}$$

It is equivalent to the following problem in first primal form:

Equivalent to First Primal LPP

$$\begin{cases} \text{Maximize:} & -2u + 2v - 3z - w \\ \text{Constraints:} & \begin{cases} -u + v + z + 4w \leqq -2 \\ u - v + z - w \leqq 15 \\ -u + v - z + w \leqq -15 \\ u \geqq 0, \quad v \geqq 0, \quad z \geqq 0, \quad w \geqq 0 \end{cases} \end{cases}$$

Dual Problem

Corresponding to a given linear programming problem in first primal form is another problem, called its **dual**. It is obtained from the original primal problem

First Primal Form LPP

$$\textbf{(P)} \begin{cases} \text{Maximize:} & \boldsymbol{c}^T\boldsymbol{x} \\ \text{Constraints:} & \begin{cases} \boldsymbol{A}\boldsymbol{x} \leqq \boldsymbol{b} \\ \boldsymbol{x} \geqq 0 \end{cases} \end{cases}$$

by defining the dual to be the problem

Dual LPP

$$
\textbf{(D)} \begin{cases} \text{Minimize:} \quad \boldsymbol{b}^T\boldsymbol{y} \\ \text{Constraints:} \begin{cases} \boldsymbol{A}^T\boldsymbol{y} \geqq \boldsymbol{c} \\ \boldsymbol{y} \geqq 0 \end{cases} \end{cases}
$$

For example, we consider

Another Sample LPP

$$
\begin{cases} \text{Maximize:} \quad 2x_1 + 3x_2 \\ \text{Constraints:} \begin{cases} 4x_1 + 5x_2 \leqq 6 \\ 7x_1 + 8x_2 \leqq 9 \\ 10x_1 + 11x_2 \leqq 12 \\ x_1 \geqq 0, \quad x_2 \geqq 0 \end{cases} \end{cases} \tag{4}
$$

The dual of the problem is this linear programming problem

Dual LPP

$$
\begin{cases} \text{Minimize:} \quad 6y_1 + 9y_2 + 12y_3 \\ \text{Constraints:} \begin{cases} 4y_1 + 7y_2 + 10y_3 \geqq 2 \\ 5y_1 + 8y_2 + 11y_3 \geqq 3 \\ y_1 \geqq 0, \quad y_2 \geqq 0, \quad y_3 \geqq 0 \end{cases} \end{cases}
$$

Note that, in general, the dual problem has different dimensions from those of the original problem. Thus, the number of *inequalities* in the original problem becomes the number of *variables* in the dual problem.

An elementary relationship between the original primal problem and its dual is as follows:

■ **Theorem 2**

Theorem on Primal and Dual Problems

If $\boldsymbol{x}$ satisfies the constraints of the primal problem and $\boldsymbol{y}$ satisfies the constraints of its dual, then $\boldsymbol{c}^T\boldsymbol{x} \leqq \boldsymbol{b}^T\boldsymbol{y}$. Consequently, if $\boldsymbol{c}^T\boldsymbol{x} = \boldsymbol{b}^T\boldsymbol{y}$, then $\boldsymbol{x}$ and $\boldsymbol{y}$ are solutions of the primal problem and the dual problem, respectively.

Proof By the assumptions made, $\boldsymbol{x} \geqq 0$, $\boldsymbol{A}\boldsymbol{x} \leqq \boldsymbol{b}$, $\boldsymbol{y} \geqq 0$, and $\boldsymbol{A}^T\boldsymbol{y} \geqq \boldsymbol{c}$. Consequently, ■

$$
\boldsymbol{c}^T\boldsymbol{x} \leqq \left(\boldsymbol{A}^T\boldsymbol{y}\right)^T\boldsymbol{x} = \boldsymbol{y}^T\boldsymbol{A}\boldsymbol{x} \leqq \boldsymbol{y}^T\boldsymbol{b} = \boldsymbol{b}^T\boldsymbol{y}
$$

This relationship can be used to estimate the number

Value of LPP

$$
\lambda = \max\left\{\boldsymbol{c}^T\boldsymbol{x} : \boldsymbol{x} \geqq 0 \text{ and } \boldsymbol{A}\boldsymbol{x} \leqq \boldsymbol{b}\right\}
$$

(This number is often termed the **value** of the linear programming problem.) To estimate λ, take any $\boldsymbol{x}$ and $\boldsymbol{y}$ that satisfy $\boldsymbol{x} \geqq 0$, $\boldsymbol{y} \geqq 0$, $\boldsymbol{A}\boldsymbol{x} \leqq \boldsymbol{b}$, and $\boldsymbol{A}^T\boldsymbol{y} \geqq \boldsymbol{c}$. Then $\boldsymbol{c}^T\boldsymbol{x} \leqq \lambda \leqq \boldsymbol{b}^T\boldsymbol{y}$. The importance of the dual problem stems from the fact that the extreme values in the primal and dual problems are the same. Formally stated, we have the following:

■ **Theorem 3**

Duality Theorem

If the original problem has a solution $\boldsymbol{x}^*$, then the dual problem has a solution $\boldsymbol{y}^*$; furthermore, $\boldsymbol{c}^T\boldsymbol{x}^* = \boldsymbol{b}^T\boldsymbol{y}^*$.

This result is nicely illustrated by using the linear programming Problem (2). The dual to that problem is

Dual LPP

$$\begin{cases} \text{Minimize:} \quad 15y_1 + 18y_2 \\ \text{Constraints:} \begin{cases} 5y_1 + 3y_2 \geqq 2 \\ 3y_1 + 6y_2 \geqq 3 \\ y_1 \geqq 0, \quad y_2 \geqq 0 \end{cases} \end{cases} \tag{5}$$

The graph of this problem is given in Figure 14.2. Moving the line $15y_1 + 18y_2 = \alpha$, we see that the vertex $\left(\frac{1}{7}, \frac{3}{7}\right)$ is the minimum point. The values of the objective functions are indeed identical because $15\left(\frac{1}{7}\right) + 18\left(\frac{3}{7}\right) = \frac{69}{7}$. Moreover, the solutions $\boldsymbol{x} = \left[\frac{12}{7}, \frac{15}{7}\right]^T$ and $\boldsymbol{y} = \left[\frac{1}{7}, \frac{3}{7}\right]^T$ can be related, but we will not discuss this.

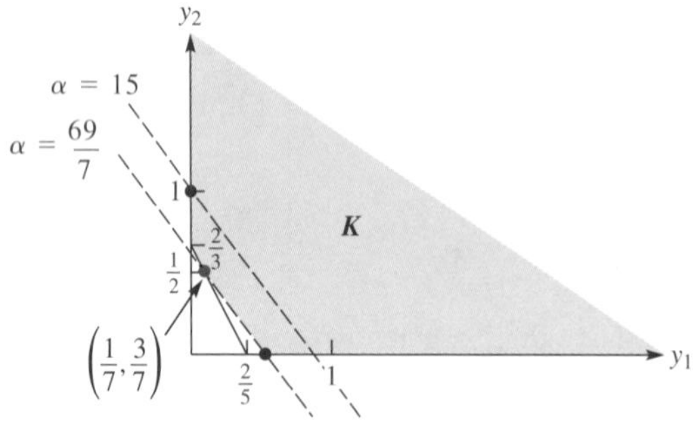

FIGURE 14.2 Graphical method for the dual problem

Mathematical Software

We can use mathematical software systems such as MATLAB, Maple, or Mathematica to solve this linear programming problem. For example, we obtain $x = 0.1429$ and $y = 0.4286$ with $f(x, y) = 9.8571$.

Second Primal Form

Slack Variables

Returning to the general problem in the first primal form, we introduce additional nonnegative variables $x_{n+1}, x_{n+2}, \dots, x_{n+m}$, known as **slack variables**, so that some of the inequalities can be written as equalities. Using this device, we can put the original problem into the following standard form:

■ **Theorem 4** **Second Primal Form**

Second Primal Form

Maximize the linear function

$$\sum_{j=1}^{n} c_j x_j$$

subject to the constraints

$$\begin{cases} \displaystyle\sum_{j=1}^{n} a_{ij} x_j + x_{n+i} = b_i & (1 \leqq i \leqq m) \\ x_j \geqq 0 & (1 \leqq j \leqq m + n) \end{cases}$$

Using matrix notation, we have

Matrix Form LPP

$$\begin{cases} \text{Maximize:} \quad \boldsymbol{c}^T\boldsymbol{x} \\ \text{Constraints:} \quad \begin{cases} \boldsymbol{A}\boldsymbol{x} = \boldsymbol{b} \\ \boldsymbol{x} \geqq 0 \end{cases} \end{cases}$$

Here, it is assumed that the $m \times n$ matrix $\boldsymbol{A}$ contains an $m \times m$ identity matrix in its last m columns and that the last m entries of $\boldsymbol{c}$ are 0. Also, note that when a problem in first primal form is changed to second primal form, we increase the number of variables and thus alter the quantities n, $\boldsymbol{x}$, $\boldsymbol{c}$, and $\boldsymbol{A}$. That is, a problem in the first primal form with n variables would contain $n + m$ variables in the second form.

To illustrate the transformation from first to second primal form, consider again linear programming Problem (2):

First Primal Form

$$\begin{cases} \text{Maximize:} \quad 2x_1 + 3x_2 \\ \text{Constraints:} \quad \begin{cases} 5x_1 + 3x_2 \leqq 15 \\ 3x_1 + 6x_2 \leqq 18 \\ x_1 \geqq 0, \quad x_2 \geqq 0 \end{cases} \end{cases} \tag{6}$$

Two slack variables x_3 and x_4 are introduced to take up the slack in two of the inequalities. The new problem in second primal form is then

Second Primal Form

$$\begin{cases} \text{Maximize:} \quad 2x_1 + 3x_2 + 0x_3 + 0x_4 \\ \text{Constraints:} \quad \begin{cases} 5x_1 + 3x_2 + x_3 = 15 \\ 3x_1 + 6x_2 + x_4 = 18 \\ x_1 \geqq 0, \quad x_2 \geqq 0, \quad x_3 \geqq 0, \quad x_4 \geqq 0 \end{cases} \end{cases}$$

Problems involving absolute values of the variables or absolute values of linear expressions can often be turned into linear programming problems. To illustrate, consider the problem of minimizing $|x - y|$ subject to linear constraints on x and y. We can introduce a new variable $z \geqq 0$ and then impose constraints $x - y \leqq z$ and $-x + y \leqq z$. Then we seek to minimize the linear form $0x + 0y + 1z$.

Summary 14.1

- The linear programming problem in **first primal form** is

$$\begin{cases} \text{Maximize:} \quad \boldsymbol{c}^T\boldsymbol{x} \\ \text{Constraints:} \quad \begin{cases} \boldsymbol{A}\boldsymbol{x} \leqq \boldsymbol{b} \\ \boldsymbol{x} \geqq 0 \end{cases} \end{cases}$$

- Its **dual problem** is

$$\begin{cases} \text{Minimize:} \quad \boldsymbol{b}^T\boldsymbol{y} \\ \text{Constraints:} \quad \begin{cases} \boldsymbol{A}^T\boldsymbol{y} \geqq \boldsymbol{c} \\ \boldsymbol{y} \geqq 0 \end{cases} \end{cases}$$

- The **second primal form** is

$$\begin{cases} \text{Maximize:} \quad \boldsymbol{c}^T\boldsymbol{x} \\ \text{Constraints:} \begin{cases} \boldsymbol{A}\boldsymbol{x} = \boldsymbol{b} \\ \boldsymbol{x} \geqq 0 \end{cases} \end{cases}$$

where the $m \times n$ matrix $\boldsymbol{A}$ contains an $m \times m$ identity matrix in its last m columns and where the last m entries of $\boldsymbol{c}$ are 0.

- If $\boldsymbol{x}$ satisfies the constraints of the primal problem and $\boldsymbol{y}$ satisfies the constraints of its dual, then $\boldsymbol{c}^T\boldsymbol{x} \leqq \boldsymbol{b}^T\boldsymbol{y}$. Consequently, if $\boldsymbol{c}^T\boldsymbol{x} = \boldsymbol{b}^T\boldsymbol{y}$, then $\boldsymbol{x}$ and $\boldsymbol{y}$ are solutions of the primal problem and the dual problem, respectively.
- The extreme values in the primal and dual problems are the same.

Exercises 14.1

1. Put the following problem into first primal form:

$$\begin{cases} \text{Minimize:} \quad |x_1 + 2x_2 - x_3| \\ \text{Constraints:} \begin{cases} x_1 + 3x_2 - x_3 \leqq 8 \\ 2x_1 - 4x_2 - x_3 \geqq 1 \\ |4x_1 + 5x_2 + 6x_3| \leqq 12 \\ x_1 \geqq 0, \quad x_2 \geqq 0, \quad x_3 \geqq 0 \end{cases} \end{cases}$$

Hint: $|\alpha| \leqq \beta$ can be written as $-\beta \leqq \alpha \leqq \beta$.

[a]**2.** A program is available for solving linear programming problems in first primal form. Put the following problem into that form:

$$\begin{cases} \text{Minimize:} \quad 5x_1 + 6x_2 - 2x_3 + 8 \\ \text{Constraints:} \begin{cases} 2x_1 - 3x_2 \geqq 5 \\ x_1 + x_2 \leqq 15 \\ 2x_1 - x_2 + x_3 \leqq 25 \\ x_1 + x_2 - x_3 \geqq 1 \\ x_1 \geqq 0, \quad x_2 \geqq 0, \quad x_3 \geqq 0 \end{cases} \end{cases}$$

3. Consider the following linear programming problems:

a.
$$\begin{cases} \text{Maximize:} \quad 2x_1 + 3x_2 \\ \text{Constraints:} \begin{cases} x_1 + 2x_2 \geqq -6 \\ -x_1 + 3x_2 \leqq 3 \\ |2x_1 - 5x_2| \leqq 5 \\ x_1 \geqq 0, \quad x_2 \geqq 0 \end{cases} \end{cases}$$

b.
$$\begin{cases} \text{Minimize:} \quad 7x_1 + x_2 - x_3 + 4 \\ \text{Constraints:} \begin{cases} x_1 - x_2 + x_3 \geqq 2 \\ x_1 + x_2 + x_3 \leqq 10 \\ -2x_1 - x_2 \leqq -4 \\ x_1 \geqq 0, \quad x_2 \geqq 0 \end{cases} \end{cases}$$

Rewrite each problem in first primal form and give the dual problem.

4. Sketch the feasible region for the following constraints:

$$\begin{cases} x - y \leqq 2 \\ x + y \leqq 3 \\ 2x + y \geqq 3 \\ x \geqq 0, \quad y \geqq 0 \end{cases}$$

[a]**a.** By substituting the vertices into the objective function

$$z(x, y) = x + 2y$$

determine the minimum value of this function on the feasible region.

b. Let

$$z(x, y) = \left(x - \frac{1}{2}\right)^2 + \left(y - \frac{1}{2}\right)^2$$

Show that the minimum value of z over the feasible region does not occur at a vertex.

5. Put the following linear programming problems into first primal form. What is the dual of each?

a. $$\begin{cases} \text{Minimize:} & 2x + y - 3z + 1 \\ \text{Constraints:} & \begin{cases} x - y \geqq 3 \\ |x - z| \leqq 2 \\ x \geqq 0, \quad y \geqq 0 \end{cases} \end{cases}$$

[a]**b.** $$\begin{cases} \text{Minimize:} & 3x - 2y + 5z + 3 \\ \text{Constraints:} & \begin{cases} x + y + z \geqq 4 \\ x - y - z = 2 \\ x \geqq 0, \quad y \geqq 0, \quad z \geqq 0 \end{cases} \end{cases}$$

c. $$\begin{cases} \text{Maximize:} & 3x + 2y \\ \text{Constraints:} & \begin{cases} 6x + 5y \leqq 17 \\ 2x + 11y \leqq 23 \\ x \leqq 0 \end{cases} \end{cases}$$

6. Consider the following linear programming problem:

$$\begin{cases} \text{Maximize:} & 2x_1 + 2x_2 - 6x_3 - x_4 \\ \text{Constraints:} & \begin{cases} 3x_1 + x_4 = 25 \\ x_1 + x_2 + x_3 + x_4 = 20 \\ 4x_1 + 6x_3 \geqq 5 \\ 2x_1 + 3x_3 + 2x_4 \geqq 0 \\ x_1 \geqq 0, \quad x_2 \geqq 0, \quad x_3 \geqq 0, \quad x_4 \geqq 0 \end{cases} \end{cases}$$

[a]**a.** Reformulate this problem in second primal form.

[a]**b.** Formulate the dual problem.

[a]**7.** Solve the following linear programming problem graphically:

$$\begin{cases} \text{Maximize:} & 3x_1 + 5x_2 \\ \text{Constraints:} & \begin{cases} x_1 \leqq 4 \\ x_2 \leqq 6 \\ 3x_1 + 2x_2 \leqq 18 \\ x_1 \geqq 0, \quad x_2 \geqq 0 \end{cases} \end{cases}$$

[a]**8.** (Continuation) Solve the dual problem of the preceding problem.

9. Show that the dual problem may be written as

$$\begin{cases} \text{Maximize:} & \boldsymbol{b}^T\boldsymbol{y} \\ \text{Constraints:} & \begin{cases} \boldsymbol{y}^T A \geqq \boldsymbol{c}^T \\ \boldsymbol{y} \geqq 0 \end{cases} \end{cases}$$

10. Describe how $\max\{|x - y - 3|, |2x + y + 4|, |x + 2y - 7|\}$ can be minimized by using a linear programming code.

[a]**11.** Show how this problem can be solved by linear programming:

$$\begin{cases} \text{Minimize:} & |x - y| \\ \text{Constraints:} & \begin{cases} x \leqq 3y \\ x \geqq y \\ y \leqq x - 2 \end{cases} \end{cases}$$

12. Consider the linear programming problem

$$\begin{cases} \text{Minimize:} & x_1 + x_4 + 25 \\ \text{Constraints:} & \begin{cases} 2x_1 + 2x_2 + x_3 < 7 \\ 2x_1 - 3x_2 + x_4 = 4 \\ x_2 - x_4 > 1 \\ 3x_2 - 8x_3 + x_4 = 5 \\ x_1, x_2, x_3, x_4 \geqq 0 \end{cases} \end{cases}$$

Write in matrix-vector form the dual problem and the second primal problem.

13. Solve each of the linear programming problems by the graphical method. Determine $\boldsymbol{x}$ to

$$\begin{cases} \text{Maximize:} & \boldsymbol{c}^T\boldsymbol{x} \\ \text{Constraints:} & \begin{cases} A\boldsymbol{x} \leqq \boldsymbol{b} \\ \boldsymbol{x} \geqq 0 \end{cases} \end{cases}$$

Here, nonunique and unbounded "solutions" may be obtained.

[a]**a.** $\boldsymbol{c} = [2, \ -4]^T \quad A = \begin{bmatrix} -3 & -5 \\ 4 & 9 \end{bmatrix} \quad \boldsymbol{b} = [-15, \ 36]$

b. $\boldsymbol{c} = \left[2, \ \frac{1}{2}\right]^T \quad A = \begin{bmatrix} 6 & 5 \\ 4 & 1 \end{bmatrix} \quad \boldsymbol{b} = [30, \ 12]^T$

[a]**c.** $\boldsymbol{c} = [3, \ 2]^T \quad A = \begin{bmatrix} -3 & 2 \\ -4 & 9 \end{bmatrix} \quad \boldsymbol{b} = [6, \ 36]^T$

d. $\boldsymbol{c} = [2, \ -3]^T \quad A = \begin{bmatrix} -1 & 1 \\ 0 & 1 \end{bmatrix} \quad \boldsymbol{b} = [0, \ 5]^T$

e. $\boldsymbol{c} = [-4, \ 11]^T \quad A = \begin{bmatrix} -3 & 4 \\ -4 & 11 \end{bmatrix} \quad \boldsymbol{b} = [12, \ 44]^T$

[a]**f.** $\boldsymbol{c} = [-3, \ 4]^T \quad A = \begin{bmatrix} 2 & 3 \\ -4 & -5 \end{bmatrix} \quad \boldsymbol{b} = [6, \ -20]^T$

g. $\boldsymbol{c} = [2, \ 1]^T \quad A = \begin{bmatrix} 1 & 1 \\ 1 & 2 \end{bmatrix} \quad \boldsymbol{b} = [0, \ -2]^T$

[a]**h.** $\boldsymbol{c} = [3, \ 1]^T \quad A = \begin{bmatrix} 2 & 4 \\ 5 & 3 \end{bmatrix} \quad \boldsymbol{b} = [21, \ 18]^T$

[a]**14.** Solve the following linear programming problem by hand, using a graph for help:

$$\begin{cases} \text{Maximize:} & 4x + 4y + z \\ \text{Constraints:} & \begin{cases} 3x + 2y + \ z = \ 12 \\ 7x + 7y + 2z \leqq 144 \\ 7x + 5y + 2z \leqq \ 80 \\ 11x + 7y + 3z \leqq 132 \\ x \geqq 0, \quad y \geqq 0 \end{cases} \end{cases}$$

Hint: Use the equation to eliminate z from all other expressions. Solve the resulting two-dimensional problem.

15. Put this linear programming problem into second primal form. You may want to make changes of variables. If so, include a *dictionary* relating new and old variables.

$$\begin{cases} \text{Minimize:} & \varepsilon_1 + \varepsilon_2 + \varepsilon_3 \\ \text{Constraints:} & \begin{cases} |3x + 4y + 6| \ \leqq \varepsilon_1 \\ |2x - 8y - 4| \ \leqq \varepsilon_2 \\ |-x - 3y + 5| \leqq \varepsilon_3 \\ \varepsilon_1 > 0, \ \ \varepsilon_2 > 0, \ \ \varepsilon_3 > 0, \ \ x > 0, \ \ y > 0 \end{cases} \end{cases}$$

Solve the resulting problem.

16. Consider the following linear programming problem:

$$\begin{cases} \text{Maximize:} & c_1x_1 + c_2x_2 \\ \text{Constraints:} & \begin{cases} a_1x_1 + a_2x_2 \leqq b \\ x_1 \geqq 0, \quad x_2 \geqq 0 \end{cases} \end{cases}$$

In the special case in which all data are positive, show that the dual problem has the same extreme value as the original problem.

[a]**17.** Suppose that a linear programming problem in first primal form has the property that $c^T x$ is not bounded on the feasible set. What conclusion can be drawn about the dual problem?

18. (**Multiple Choice**) Which of these problems is formulated in the first primal form for a linear programming problem?

a. Maximize $\boldsymbol{c}^T\boldsymbol{x}$ subject to $\boldsymbol{Ax} \leqq \boldsymbol{b}$
b. Minimize $\boldsymbol{c}^T\boldsymbol{x}$ subject to $\boldsymbol{Ax} \leqq \boldsymbol{b}, \boldsymbol{x} \geqq 0$
c. Maximize $\boldsymbol{c}^T\boldsymbol{x}$ subject to $\boldsymbol{Ax} = \boldsymbol{b}, \boldsymbol{x} \geqq 0$
d. Maximize $\boldsymbol{c}^T\boldsymbol{x}$ subject to $\boldsymbol{Ax} \leqq \boldsymbol{b}, \boldsymbol{x} \geqq 0$
e. None of these.

Computer Exercises 14.1

[a]**1.** A western shop wishes to purchase 300 felt and 200 straw cowboy hats. Bids have been received from three wholesalers. Texas Hatters has agreed to supply not more than 200 hats, Lone Star Hatters not more than 250, and Lariat Ranch Wear not more than 150. The owner of the shop has estimated that his profit per hat sold from Texas Hatters would be \$3/felt and \$4/straw, from Lone Star Hatters \$3.80/felt and \$3.50/straw, and from Lariat Ranch Wear \$4/felt and \$3.60/straw. Set up a linear programming problem to maximize the owner's profits. Solve by using mathematical software.

2. The ABC Drug Company makes two types of liquid painkiller that have brand names Relieve (R) and Ease (E) and contain different mixtures of three basic drugs, A, B, and C, produced by the company. Each bottle of R requires $\frac{7}{9}$ unit of drug A, $\frac{1}{2}$ unit of drug B, and $\frac{3}{4}$ unit of drug C. Each bottle of E requires $\frac{4}{9}$ unit of drug A, $\frac{5}{2}$ unit of drug B, and $\frac{1}{4}$ unit of drug C. The company is able to produce each day only 5 units of drug A, 7 units of drug B, and 9 units of C. Moreover, Food and Drug Administration regulations stipulate that the number of bottles of R manufactured cannot exceed twice the number of bottles of E. The profit margin for each bottle of E and R is \$7 and \$3, respectively. Set up the linear programming problem in first primal form to determine the number of bottles of the two painkillers that the company should produce each day so as to maximize their profits. Solve by using mathematical software.

[a]**3.** Suppose that the university student government wishes to charter planes to transport at least 750 students to the bowl game. Two airlines, α and β, agree to supply aircraft for the trip. Airline α has five aircraft available carrying 75 passengers each, and airline β has three aircraft available carrying 250 passengers each. The cost per aircraft is \$900 and \$3250 for the trip from airlines α and β, respectively. The student government wants to charter at most six aircraft. How many of each type should be chartered to minimize the cost of the airlift? How much should the student government charge to make 50¢ profit per student? Solve by the graphical method, and verify by using mathematical software.

4. (Continuation) Rework the preceding computer problem in the following two possibly different ways:

a. The number of students going on the airlift is maximized.
b. The cost per student is minimized.

[a]5. **(Diet Problem)** A university dining hall wishes to provide at least 5 units of vitamin C and 3 units of vitamin E per serving. Three foods are available containing these vitamins. Food f_1 contains 2.5 and 1.25 units per ounce of vitamins C and E, respectively, whereas food f_2 contains just the opposite amounts. The third food f_3 contains an equal amount of each vitamin at 1 unit per ounce. Food f_1 costs 25¢ per ounce, food f_2 costs 56¢ per ounce, and food f_3 costs 10¢ per ounce. The dietitian wishes to provide the meal at a minimum cost per serving that satisfies the minimum vitamin requirements. Set up this linear programming problem in second primal form. Solve with the aid of mathematical software.

6. Use built-in routines in mathematical software systems such as MATLAB, Maple, or Mathematica to solve each of these linear programming problems in first primal form, in second primal form, and in dual form:

   a. LPP (2) **b.** LPP (3) **c.** LPP (4)
   d. LPP (5) **e.** LPP (6)

14.2 Simplex Method

The principal algorithm that is used in solving linear programming problems is the *simplex method*. Here, enough of the background of this method is described that the reader can use available computer programs that incorporate it.

Consider a linear programming problem in second primal form:

Second Primal Form LPP

$$\begin{cases} \text{Maximize:} & \boldsymbol{c}^T\boldsymbol{x} \\ \text{Constraints:} & \begin{cases} \boldsymbol{A}\boldsymbol{x} = \boldsymbol{b} \\ \boldsymbol{x} \geqq 0 \end{cases} \end{cases}$$

It is assumed that $\boldsymbol{c}$ and $\boldsymbol{x}$ are n-component vectors, $\boldsymbol{b}$ is an m-component vector, and $\boldsymbol{A}$ is an $m \times n$ matrix. Also, it is assumed that $\boldsymbol{b} \geqq 0$ and that $\boldsymbol{A}$ contains an $m \times m$ identity matrix in its last m columns. As before, we define the set of feasible points as

$$K = \{\boldsymbol{x} \in \mathbb{R}^n : \boldsymbol{A}\boldsymbol{x} = \boldsymbol{b}, \boldsymbol{x} \geqq 0\}$$

The points of K are exactly the points that are competing to maximize $\boldsymbol{c}^T\boldsymbol{x}$.

Vertices in *K* and Linearly Independent Columns of *A*

Polyhedral Set

The set K is a **polyhedral** set in $\mathbb{R}^n$, and the algorithm to be described proceeds from vertex to vertex in K, always increasing the value of $\boldsymbol{c}^T\boldsymbol{x}$ as it goes from one to another. Let us give a precise definition of *vertex*. A point $\boldsymbol{x}$ in K is called a **vertex** if it is impossible to express it as $\boldsymbol{x} = \frac{1}{2}(\boldsymbol{u} + \boldsymbol{v})$, with both $\boldsymbol{u}$ and $\boldsymbol{v}$ in K and $\boldsymbol{u} \neq \boldsymbol{v}$. In other words, $\boldsymbol{x}$ is not the midpoint of any line segment whose endpoints lie in K.

We denote by $\boldsymbol{a}^{(1)}, \boldsymbol{a}^{(2)}, \ldots, \boldsymbol{a}^{(n)}$ the column vectors constituting the matrix $\boldsymbol{A}$. The following theorem relates the columns of $\boldsymbol{A}$ to the vertices of K:

■ **Theorem 1** **Theorem on Vertices and Column Vectors**

Let $\boldsymbol{x} \in K$ and define $\mathcal{I}(\boldsymbol{x}) = \{i : x_i > 0\}$. Then the following are equivalent:

1. $\boldsymbol{x}$ is a vertex of K.
2. The set $\{\boldsymbol{a}^{(i)} : i \in \mathcal{I}(\boldsymbol{x})\}$ is linearly independent.

Proof If Statement 1 is false, then we can write $\boldsymbol{x} = \frac{1}{2}(\boldsymbol{u}+\boldsymbol{v})$, with $\boldsymbol{u} \in \boldsymbol{K}, \boldsymbol{v} \in \boldsymbol{K}$, and $\boldsymbol{u} \neq \boldsymbol{v}$. For every index i that is not in the set $\mathcal{I}(\boldsymbol{x})$, we have $x_i = 0$, $u_i \geqq 0$, $v_i \geqq 0$, and $x_i = \frac{1}{2}(u_i + v_i)$. This forces u_i and v_i to be zero. Thus, all the nonzero components of $\boldsymbol{u}$ and $\boldsymbol{v}$ correspond to indices i in $\mathcal{I}(\boldsymbol{x})$. Since $\boldsymbol{u}$ and $\boldsymbol{v}$ belong to the set $\boldsymbol{K}$, we have

$$\boldsymbol{b} = \boldsymbol{A}\boldsymbol{u} = \sum_{i=1}^{n} u_i \boldsymbol{a}^{(i)} = \sum_{i \in \mathcal{I}(\boldsymbol{x})} u_i \boldsymbol{a}^{(i)}$$

and

$$\boldsymbol{b} = \boldsymbol{A}\boldsymbol{v} = \sum_{i=1}^{n} v_i \boldsymbol{a}^{(i)} = \sum_{i \in \mathcal{I}(\boldsymbol{x})} v_i \boldsymbol{a}^{(i)}$$

Hence, we obtain

$$\sum_{i \in \mathcal{I}(\boldsymbol{x})} (u_i - v_i)\, \boldsymbol{a}^{(i)} = 0$$

showing the linear dependence of the set $\{\boldsymbol{a}^{(i)} : i \in \mathcal{I}(\boldsymbol{x})\}$. Thus, Statement 2 is false. Consequently, Statement 2 implies Statement 1.

For the converse, assume that Statement 2 is false. From the linear dependence of column vectors $\boldsymbol{a}^{(i)}$ for $i \in \mathcal{I}(\boldsymbol{x})$, we have

$$\sum_{i \in \mathcal{I}(\boldsymbol{x})} y_i\, \boldsymbol{a}^{(i)} = 0 \qquad \text{with} \qquad \sum_{i \in \mathcal{I}(\boldsymbol{x})} |y_i| \neq 0$$

for appropriate coefficients y_i. For each $i \notin \mathcal{I}(\boldsymbol{x})$, let $y_i = 0$. Form the vector $\boldsymbol{y}$ with components y_i for $i = 1, 2, \dots, n$. Then, for any λ, we see that because $\boldsymbol{x} \in \boldsymbol{K}$,

$$\boldsymbol{A}(\boldsymbol{x} \pm \lambda \boldsymbol{y}) = \sum_{i=1}^{n} (x_i \pm \lambda y_i)\, \boldsymbol{a}^{(i)} = \sum_{i=1}^{n} x_i\, \boldsymbol{a}^{(i)} \pm \lambda \sum_{i \in \mathcal{I}(\boldsymbol{x})} y_i\, \boldsymbol{a}^{(i)} = \boldsymbol{A}\boldsymbol{x} = \boldsymbol{b}$$

Now select the real number λ positive but so small that $\boldsymbol{x} + \lambda \boldsymbol{y} \geqq 0$ and $\boldsymbol{x} - \lambda \boldsymbol{y} \geqq 0$. (To see that it is possible, consider separately the components for $i \in \mathcal{I}(\boldsymbol{x})$ and $i \notin \mathcal{I}(\boldsymbol{x})$.) The resulting vectors, $\boldsymbol{u} = \boldsymbol{x} + \lambda \boldsymbol{y}$ and $\boldsymbol{v} = \boldsymbol{x} - \lambda \boldsymbol{y}$, belong to $\boldsymbol{K}$. They differ, and obviously, $\boldsymbol{x} = \frac{1}{2}(\boldsymbol{u} + \boldsymbol{v})$. Thus, $\boldsymbol{x}$ is not a vertex of $\boldsymbol{K}$; that is, Statement 1 is false. So Statement 1 implies Statement 2. ■

Given a linear programming problem, there are three possibilities:

1. There are no feasible points; that is, the set $\boldsymbol{K}$ is empty.
2. $\boldsymbol{K}$ is not empty, and $\boldsymbol{c}^T\boldsymbol{x}$ is not bounded on $\boldsymbol{K}$.
3. $\boldsymbol{K}$ is not empty, and $\boldsymbol{c}^T\boldsymbol{x}$ is bounded on $\boldsymbol{K}$.

It is true (but not obvious) that in the third case, there is a point $\boldsymbol{x}$ in $\boldsymbol{K}$ such that $\boldsymbol{c}^T\boldsymbol{x} \geqq \boldsymbol{c}^T\boldsymbol{y}$ for all $\boldsymbol{y}$ in $\boldsymbol{K}$. We have assumed that our problem is in the second primal form so that possibility 1 cannot occur. Indeed, $\boldsymbol{A}$ contains an $m \times m$ identity matrix and so has the form

Augmented Matrix

$$\boldsymbol{A} = \begin{bmatrix} a_{11} & a_{12} & \cdots & a_{1k} & 1 & 0 & \cdots & 0 \\ a_{21} & a_{22} & \cdots & a_{2k} & 0 & 1 & \cdots & 0 \\ \vdots & \vdots & \ddots & \vdots & \vdots & \vdots & \ddots & \vdots \\ a_{m1} & a_{m2} & \cdots & a_{mk} & 0 & 0 & \cdots & 1 \end{bmatrix}$$

where $k = n - m$. Consequently, we can *construct* a feasible point $\boldsymbol{x}$ easily by setting $x_1 = x_2 = \cdots = x_k = 0$ and $x_{k+1} = b_1$, $x_{k+2} = b_2$, and so on. It is then clear that $\boldsymbol{Ax} = \boldsymbol{b}$. The inequality $\boldsymbol{x} \geqq 0$ follows from our initial assumption that $\boldsymbol{b} \geqq 0$.

Simplex Method

Next we present a brief outline of the simplex method for solving linear programming problems. It involves a sequence of exchanges so that the trial solution proceeds systematically from one vertex to another in $\boldsymbol{K}$. This procedure is stopped when the value of $\boldsymbol{c}^T\boldsymbol{x}$ is no longer increased as a result of the exchange.

The following is an outline of the **simplex algorithm**.

■ **Algorithm**

Simplex Algorithm

Simplex Method

Select a small positive value for ε. In each step, we have a set of m indices $\{k_1, k_2, \dots, k_m\}$.

1. Put columns $\boldsymbol{a}^{(k_1)}, \boldsymbol{a}^{(k_2)}, \dots, \boldsymbol{a}^{(k_m)}$ into $\boldsymbol{B}$, and solve $\boldsymbol{Bx} = \boldsymbol{b}$.
2. If $x_i > 0$ for $1 \leqq i \leqq m$, continue. Otherwise, exit because the algorithm has failed.
3. Set $\boldsymbol{e} = [c_{k_1}, c_{k_2}, \dots, c_{k_m}]^T$, and solve $\boldsymbol{B}^T\boldsymbol{y} = \boldsymbol{e}$.
4. Choose any s in $\{1, 2, \dots, n\}$ but not in $\{k_1, k_2, \dots, k_m\}$ for which $c_s - \boldsymbol{y}^T\boldsymbol{a}^{(s)}$ is greatest.
5. If $c_s - \boldsymbol{y}^T\boldsymbol{a}^{(s)} < \varepsilon$, exit because $\boldsymbol{x}$ is the solution.
6. Solve $\boldsymbol{Bz} = \boldsymbol{a}^{(s)}$.
7. If $z_i \leqq \varepsilon$ for $1 \leqq i \leqq m$, then exit because the objective function is unbounded on $\boldsymbol{K}$.
8. Among the ratios x_i/z_i that have $z_i > 0$ for $1 \leqq i \leqq m$, let x_r/z_r be the smallest. In case of a tie, let r be the first occurrence.
9. Replace k_r by s, and go to Step 1.

A few remarks on this algorithm are in order. In the beginning, select the indices $k_1, k_2, \dots, k_m$ such that $\boldsymbol{a}^{(k_1)}, \boldsymbol{a}^{(k_2)}, \dots, \boldsymbol{a}^{(k_m)}$ form an $m \times m$ identity matrix. At Step 5, where we say that $\boldsymbol{x}$ *is a solution*, we mean that the vector $\boldsymbol{v} = (v_i)$ given by $v_{k_i} = x_i$ for $1 \leqq i \leqq n$ and $v_i = 0$ for $i \notin \{k_1, k_2, \dots, k_m\}$ is the solution. A convenient choice for the tolerance ε that occurs in Steps 5 and 7 might be 10^{-6}.

In any reasonable implementation of the simplex method, advantage must be taken of the fact that succeeding occurrences of Step 1 are very similar. In fact, only one column of $\boldsymbol{B}$ changes at a time. Similar remarks hold for Steps 3 and 6.

We do not recommend that the reader attempt to program the simplex algorithm. Efficient codes, refined over many years of experience, are usually available in software libraries. Many of them can provide solutions to a given problem *and* to its dual with very little additional computing. Sometimes this feature can be exploited to decrease the execution time of a problem. To see why, consider a linear programming problem in **first**

primal form:

First Primal Form LPP

$$(\mathbf{P}) \begin{cases} \text{Maximize: } \boldsymbol{c}^T\boldsymbol{x} \\ \text{Constraints: } \begin{cases} \boldsymbol{A}\boldsymbol{x} \leqq \boldsymbol{b} \\ \boldsymbol{x} \geqq 0 \end{cases} \end{cases}$$

As usual, we assume that $\boldsymbol{x}$ is an n vector and that $\boldsymbol{A}$ is an $m \times n$ matrix. When the simplex algorithm is applied to this problem, it performs an iterative process on an $m \times m$ matrix denoted by $\boldsymbol{B}$ in the preceding description. If the number of inequality constraints m is very large relative to n, then the dual problem may be easier to solve, since the $\boldsymbol{B}$ matrices for it will be of dimension $n \times n$. Indeed, the **dual problem** is

Dual Form LPP

$$(\mathbf{D}) \begin{cases} \text{Minimize: } \boldsymbol{b}^T\boldsymbol{y} \\ \text{Constraints: } \begin{cases} \boldsymbol{A}^T\boldsymbol{y} \geqq \boldsymbol{c} \\ \boldsymbol{y} \geqq 0 \end{cases} \end{cases}$$

and the number of inequality constraints here is n. An example of this technique appears in the next section.

Summary 14.2

- For the second primal form, the set of **feasible points** is

$$\boldsymbol{K} = \{\boldsymbol{x} \in \mathbb{R}^n \colon \boldsymbol{A}\boldsymbol{x} = \boldsymbol{b}, \boldsymbol{x} \geqq 0\}$$

  which are the points of $\boldsymbol{K}$ competing to maximize $\boldsymbol{c}^T\boldsymbol{x}$.
- For a linear programming problem, there are these possibilities: There are no feasible points, that is, the set $\boldsymbol{K}$ is empty; $\boldsymbol{K}$ is not empty, and $\boldsymbol{c}^T\boldsymbol{x}$ is not bounded on $\boldsymbol{K}$; $\boldsymbol{K}$ is not empty, and $\boldsymbol{c}^T\boldsymbol{x}$ is bounded on $\boldsymbol{K}$.
- Denote by $\boldsymbol{a}^{(1)}, \boldsymbol{a}^{(2)}, \ldots, \boldsymbol{a}^{(n)}$ the column vectors constituting the matrix $\boldsymbol{A}$. Let $\boldsymbol{x} \in \boldsymbol{K}$ and define $\mathcal{I}(\boldsymbol{x}) = \{i \colon x_i > 0\}$. Then $\boldsymbol{x}$ is a vertex of $\boldsymbol{K}$ if and only if the set $\{\boldsymbol{a}^{(i)} \colon i \in \mathcal{I}(\boldsymbol{x})\}$ is linearly independent.
- The **simplex method** involves a sequence of exchanges so that the trial solution proceeds systematically from one vertex to another in the set of feasible points $\boldsymbol{K}$. This procedure is stopped when the value of $\boldsymbol{c}^T\boldsymbol{x}$ is no longer increased as a result of exchanges.

Exercises 14.2

[a]**1.** Put this linear programming problem into first primal form by increasing the number of variables by just one:

$$\begin{cases} \text{Maximize: } \boldsymbol{c}^T\boldsymbol{x} \\ \text{Constraints: } \boldsymbol{A}\boldsymbol{x} \leqq \boldsymbol{b} \end{cases}$$

Hint: Replace x_j by $y_j - y_0$.

[a]**2.** Show that the set $\boldsymbol{K}$ can have only a finite number of vertices.

3. Suppose that $\boldsymbol{u}$ and $\boldsymbol{v}$ are solution points for a linear programming problem and that $\boldsymbol{x} = \frac{1}{2}(\boldsymbol{u} + \boldsymbol{v})$. Show that $\boldsymbol{x}$ is also a solution.

4. Using the simplex method as described, solve the numerical example in the text.

[a]**5.** Using standard manipulations, put the dual problem (D) into first and second primal forms.

[a]**6.** Show how a code for solving a linear programming problem in first primal form can be used to solve a system of n linear equations in n variables.

7. Using standard techniques, put the dual problem (D) into first primal form (P); then take the dual of it. What is the result?

Computer Exercises 14.2

1. Select a linear programming code from a mathematical software package or library and use it to solve these linear programming problems:

a. $$\begin{cases} \text{Minimize:} & 8x_1 + 6x_2 + 6x_3 + 9x_4 \\ \text{Constraints:} & \begin{cases} x_1 + 2x_2 \qquad + x_4 \geqq 2 \\ 3x_1 + x_2 \qquad + x_4 \geqq 4 \\ \qquad\qquad x_3 + x_4 \geqq 1 \\ x_1 \qquad + x_3 \qquad \geqq 1 \\ x_1 \geqq 0, \quad x_2 \geqq 0, \quad x_3 \geqq 0, \quad x_4 \geqq 0 \end{cases} \end{cases}$$

[a]b. $$\begin{cases} \text{Minimize:} & 10x_1 - 5x_2 - 4x_3 + 7x_4 + x_5 \\ \text{Constraints:} & \begin{cases} 4x_1 - 3x_2 - x_3 + 4x_4 + x_5 = 1 \\ -x_1 + 2x_2 + 2x_3 + x_4 + 3x_5 = 4 \\ x_1 \geqq 0, \quad x_2 \geqq 0, \quad x_3 \geqq 0, \quad x_4 \geqq 0, \quad x_5 \geqq 0 \end{cases} \end{cases}$$

[a]c. $$\begin{cases} \text{Maximize:} & 2x_1 + 4x_2 + 3x_3 \\ \text{Constraints:} & \begin{cases} 4x_1 + 2x_2 + 3x_3 \leqq 15 \\ 3x_1 + 2x_2 + x_3 \leqq 7 \\ x_1 + x_2 + 2x_3 \leqq 6 \\ x_1 \geqq 0, \quad x_2 \geqq 0, \quad x_3 \geqq 0 \end{cases} \end{cases}$$

2. (**Student Research Project**) Investigate recent developments in computational linear programming algorithms, especially by interior-point methods.

14.3 Inconsistent Linear Systems

Linear programming can be used for the approximate solution of systems of linear equations that are inconsistent. An $m \times n$ system of equations

$$\sum_{j=1}^{n} a_{ij} x_j = b_i \qquad (1 \leqq i \leqq m)$$

is said to be **inconsistent** if there is no vector $\boldsymbol{x} = [x_1, x_2, \ldots, x_n]^T$ that simultaneously satisfies all m equations in the system. For instance, the system

Sample Inconsistent System

$$\begin{cases} 2x_1 + 3x_2 = 4 \\ x_1 - x_2 = 2 \\ x_1 + 2x_2 = 7 \end{cases} \tag{1}$$

is inconsistent, as can be seen by attempting to carry out the Gaussian elimination process.

ℓ_1 Problem

Since no vector $\boldsymbol{x}$ can solve an inconsistent system of equations, the **residuals**

$$r_i = \sum_{j=1}^{n} a_{ij} x_j - b_i \qquad (1 \leqq i \leqq m)$$

cannot be made to vanish simultaneously. Hence, we have $\sum_{i=1}^{m} |r_i| > 0$. Now it is natural to ask for an $\boldsymbol{x}$ vector that renders the expression $\sum_{i=1}^{m} |r_i|$ as small as possible. This problem is called the **ℓ_1 problem** for this system of equations. Other criteria, leading to different approximate solutions, might be to minimize $\sum_{i=1}^{m} r_i^2$ or $\max_{1 \leqq i \leqq m} |r_i|$. Chapter 9 discusses in detail the problem of minimizing $\sum_{i=1}^{m} r_i^2$.

The minimization of $\sum_{i=1}^{n} |r_i|$ by appropriate choice of the $\boldsymbol{x}$ vector is a problem for which special algorithms have been designed (see Barrodale and Roberts [1974]). However, if one of these special programs is not available or if the problem is small in scope, linear programming can be used.

A simple, direct restatement of the problem is

Direct Restatement LPP

$$\begin{cases} \text{Minimize:} \quad \displaystyle\sum_{i=1}^{m} \varepsilon_i \\ \text{Constraints:} \begin{cases} \displaystyle\sum_{j=1}^{n} a_{ij}x_j - b_i \leqq \varepsilon_i & (1 \leqq i \leqq m) \\ \displaystyle -\sum_{j=1}^{n} a_{ij}x_j + b_i \leqq \varepsilon_i & (1 \leqq i \leqq m) \end{cases} \end{cases} \tag{2}$$

If a linear programming code is at hand in which the variables are not required to be nonnegative, then it can be used on Problem (2). If the variables must be nonnegative, the following technique can be applied. Introduce a variable y_{n+1}, and write $x_j = y_j - y_{n+1}$. Then define $a_{i,n+1} = -\sum_{j=1}^{n} a_{ij}$. This step creates an additional column in the matrix $\boldsymbol{A}$. Now consider the linear programming problem

First Primal Form LPP (Nonnegative Variables)

$$\begin{cases} \text{Maximize:} \quad \displaystyle -\sum_{i=1}^{m} \varepsilon_i \\ \text{Constraints:} \begin{cases} \displaystyle\sum_{j=1}^{n+1} a_{ij}y_j - \varepsilon_i \leqq b_i & (1 \leqq i \leqq m) \\ \displaystyle -\sum_{j=1}^{n+1} a_{ij}y_j - \varepsilon_i \leqq -b_i & (1 \leqq i \leqq m) \\ y \geqq 0, \quad \varepsilon \geqq 0 \end{cases} \end{cases} \tag{3}$$

which is in first primal form with $m + n + 1$ variables and $2m$ inequality constraints.

It is not hard to verify that Problem (3) is equivalent to Problem (2). The main point is that

$$\begin{aligned} \sum_{j=1}^{n+1} a_{ij}y_j &= \sum_{j=1}^{n} a_{ij}(x_j + y_{n+1}) + a_{i,n+1}y_{n+1} \\ &= \sum_{j=1}^{n} a_{ij}x_j + y_{n+1}\sum_{j=1}^{n} a_{ij} + y_{n+1}\left(-\sum_{j=1}^{n} a_{ij}\right) \\ &= \sum_{j=1}^{n} a_{ij}x_j \end{aligned}$$

Another technique can be used to replace the $2m$ inequality constraints in Problem (3) by a set of m equality constraints. We write

$$\varepsilon_i = |r_i| = u_i + v_i$$

where $u_i = r_i$ and $v_i = 0$ if $r_i \geqq 0$ but $v_i = -r_i$ and $u_i = 0$ if $r_i < 0$. The resulting linear programming problem is

LPP Inequality Constraints

$$\begin{cases} \text{Maximize:} & -\sum_{i=1}^{m} u_i - \sum_{i=1}^{m} v_i \\ \text{Constraints:} & \begin{cases} \sum_{j=1}^{n+1} a_{ij} y_j - u_i + v_i = b_i & (1 \leqq i \leqq m) \\ u \geqq 0, \quad v \geqq 0, \quad y \geqq 0 \end{cases} \end{cases}$$

Using the preceding formulas, we have

$$\begin{aligned} r_i &= \sum_{j=1}^{n} a_{ij} x_j - b_i = \sum_{j=1}^{n} a_{ij}(y_j - y_{n+1}) - b_i \\ &= \sum_{j=1}^{n} a_{ij} y_j - y_{n+1} \sum_{j=1}^{n} a_{ij} - b_i \\ &= \sum_{j=1}^{n+1} a_{ij} y_j - b_i = u_i - v_i \end{aligned}$$

From it, we conclude that $r_i + v_i = u_i \geqq 0$. Now v_i and u_i should be as small as possible, consistent with this restriction, because we are attempting to minimize $\sum_{i=1}^{m}(u_i + v_i)$. So if $r_i \geqq 0$, we take $v_i \geqq 0$ and $u_i = r_i$, whereas if $r_i < 0$, we take $v_i = -r_i$ and $u_i = 0$. In either case, $|r_i| = u_i + v_i$. Thus, minimizing $\sum_{i=1}^{m}(u_i + v_i)$ is the same as minimizing $\sum_{i=1}^{m} |r_i|$.

The example of the inconsistent linear system given by System (1) could be solved in the ℓ_1 sense by solving the linear programming problem

ℓ_1 Sense LPP

$$\begin{cases} \text{Minimize:} & u_1 + v_1 + u_2 + v_2 + u_3 + v_3 \\ \text{Constraints:} & \begin{cases} 2y_1 + 3y_2 - 5y_3 - u_1 + v_1 = 4 \\ y_1 - y_2 - u_2 + v_2 = 2 \\ y_1 + 2y_2 - 3y_3 - u_3 + v_3 = 7 \\ y_1, y_2, y_3 \geqq 0, \quad u_1, u_2, u_3 \geqq 0, \quad v_1, v_2, v_3 \geqq 0 \end{cases} \end{cases} \tag{4}$$

The solution is

$$\begin{array}{lll} u_1 = 0, & u_2 = 0, & u_3 = 0 \\ v_1 = 0, & v_2 = 0, & v_3 = 5 \\ y_1 = 2, & y_2 = 0, & y_3 = 0 \end{array}$$

From it, we recover the ℓ_1 solution of System (1) in the form

$$\begin{array}{ll} x_1 = y_1 - y_3 = 2, & r_1 = u_1 - v_1 = 0 \\ x_2 = y_2 - y_3 = 0, & r_2 = u_2 - v_2 = 0 \\ & r_3 = u_3 - v_3 = -5 \end{array}$$

Mathematical Software

We can use mathematical software systems such as MATLAB, Maple, or Mathematica to solve this linear programming problem. For example, we obtain $u_1 = v_1 = u_2 = v_2 = u_3 = y_2 = y_3 = 0$, $v_3 = 5$, and $y_1 = 2$, with 5 as the value of the objective function. For another system, we need to set the equality constraints. We obtain the solution corresponding to $y_1 = y_2 = y_3 = 684.2887$, $u_1 = u_2 = u_3 = v_1 = v_2 = 0$, and $v_3 = 5$ with 5 as the value of the objective function. The x vector is $x_1 = 2$ and $x_2 = 3.1494 \times 10^{-11}$. This solution is slightly different from the one previously obtained, owing to roundoff errors, but the minimum value for the objective function is the same and all the constraints are satisfied.

ℓ_∞ Problem

Consider again a system of m linear equations in n unknowns:

$$\sum_{j=1}^{n} a_{ij}x_j = b_i \qquad (1 \leqq i \leqq m)$$

If the system is inconsistent, we know that the residuals $r_i = \sum_{j=1}^{n} a_{ij}x_j - b_i$ cannot all be zero for any $\boldsymbol{x}$ vector. So the quantity $\varepsilon = \max_{1 \leqq i \leqq m} |r_i|$ is positive. The problem of making ε a minimum is called the **ℓ_∞ problem** for the system of equations. An equivalent linear programming problem is

ℓ_∞ Sense LPP

$$\begin{cases} \text{Minimize:} \quad \varepsilon \\ \text{Constraints:} \begin{cases} \displaystyle\sum_{j=1}^{n} a_{ij}x_j - \varepsilon \leqq b_i & (1 \leqq i \leqq m) \\ \displaystyle -\sum_{j=1}^{n} a_{ij}x_j - \varepsilon \leqq -b_i & (1 \leqq i \leqq m) \end{cases} \end{cases}$$

If a linear programming code is available in which the variables need not be greater than or equal to zero, then it can be used to solve the ℓ_∞ problem as formulated above. If the variables must be nonnegative, we first introduce a variable y_{n+1} so large that the quantities $y_j = x_j + y_{n+1}$ are positive. Next, we solve the linear programming problem

ℓ_∞ Sense LPP (Nonnegative Variables)

$$\begin{cases} \text{Minimize:} \quad \varepsilon \\ \text{Constraints:} \begin{cases} \displaystyle\sum_{j=1}^{n+1} a_{ij}y_j - \varepsilon \leqq b_i & (1 \leqq i \leqq m) \\ \displaystyle -\sum_{j=1}^{n+1} a_{ij}y_j - \varepsilon \leqq -b_i & (1 \leqq i \leqq m) \\ \varepsilon \geqq 0, \quad y_j \geqq 0 & (1 \leqq j \leqq n+1) \end{cases} \end{cases} \tag{5}$$

Here, we have again defined $a_{i,n+1} = -\sum_{j=1}^{n} a_{ij}$.

For our System (1), the solution that minimizes the quantity

$$\max\{|2x_1 + 3x_2 - 4|, |x_1 - x_2 - 2|, |x_1 + 2x_2 - 7|\}$$

is obtained from the linear programming problem

System (1) LPP

$$\begin{cases} \text{Minimize:} \quad \varepsilon \\ \text{Constraints:} \begin{cases} 2y_1 + 3y_2 - 5y_3 - \varepsilon \leqq 4 \\ y_1 - y_2 \qquad\quad - \varepsilon \leqq 2 \\ y_1 + 2y_2 - 3y_3 - \varepsilon \leqq 7 \\ -2y_1 - 3y_2 + 5y_3 - \varepsilon \leqq -4 \\ -y_1 + y_2 \qquad\quad - \varepsilon \leqq -2 \\ -y_1 - 2y_2 + 3y_3 - \varepsilon \leqq -7 \\ y_1, y_2, y_3 \geqq 0, \quad \varepsilon \geqq 0 \end{cases} \end{cases} \tag{6}$$

The solution is

$$y_1 = \frac{8}{9}, \qquad y_2 = \frac{5}{3}, \qquad y_3 = 0, \qquad \varepsilon = \frac{25}{9}$$

From it, the ℓ_∞ solution of System (1) is recovered as follows:

$$x_1 = y_1 - y_3 = \frac{8}{9}, \qquad x_2 = y_2 - y_3 - \frac{5}{3}$$

Mathematical Software

We can use mathematical software systems such as MATLAB, Maple, or Mathematica to solve the linear programming Problem (6). For example, we obtain the solution $y_1 = \frac{8}{9}$, $y_2 = \frac{5}{3}$, $y_3 = 0$, and $\varepsilon = \frac{25}{9}$ from two of these systems. But for one of the mathematical systems, we obtain the solution corresponding to $y_1 = 1.0423 \times 10^3$, $y_2 = 1.0431 \times 10^3$, $y_3 = 1.0414 \times 10^3$, and $\varepsilon = 2.778$. We do obtain the same results as before $(0.8889, 1.6667) \approx \left(\frac{8}{9}, \frac{5}{3}\right)$.

In problems like (6), m is often much larger than n. Thus, in accordance with remarks made in Section 14.2, it may be preferable to solve the dual problem because it would have $2m$ variables but only $n + 2$ inequality constraints. To illustrate, the dual of Problem (6) is

System (1) Dual LPP

$$\begin{cases} \text{Maximize:} \quad 4u_1 + 2u_2 + 7u_3 - 4u_4 - 2u_5 - 7u_6 \\ \text{Constraints:} \begin{cases} 2u_1 + u_2 + u_3 - 2u_4 - u_5 - u_6 \geqq 0 \\ 3u_1 - u_2 + 2u_3 - 3u_4 + u_5 - 2u_6 \geqq 0 \\ -5u_1 - 3u_3 + 5u_4 + 3u_6 \geqq 0 \\ -u_1 - u_2 - u_3 - u_4 - u_5 - u_6 \geqq -1 \\ u_i \geqq 0 \quad (1 \leqq i \leqq 6) \end{cases} \end{cases}$$

The three types of approximate solution that have been discussed (for an overdetermined system of linear equations) are useful in different situations. Broadly speaking, an ℓ_∞ solution is preferred when the data are known to be accurate. An ℓ_2 solution is preferred when the data are contaminated with errors that are believed to conform to the normal probability distribution. The ℓ_1 solution is often used when data are suspected of containing *wild* points—points that result from gross errors, such as the incorrect placement of a decimal point. Additional information can be found in Rice and White [1964]. The ℓ_2 problem is discussed in Chapter 9 also.

Summary 14.3

- We consider an **inconsistent system** of m linear equations in n unknowns

$$\sum_{j=1}^{n} a_{ij}x_j = b_i \qquad (1 \leqq i \leqq m)$$

For the residuals $r_i = \sum_{j=1}^{n} a_{ij}x_j - b_i$, the **$\ell_1$ problem** for this system is to minimize the expression $\sum_{i=1}^{m} |r_i|$. A direct restatement of the problem is

$$\begin{cases} \text{Minimize:} & \displaystyle\sum_{i=1}^{m} \varepsilon_i \\ \text{Constraints:} & \begin{cases} \displaystyle\sum_{j=1}^{n} a_{ij}x_j - b_i \leqq \varepsilon_i & (1 \leqq i \leqq m) \\ \displaystyle -\sum_{j=1}^{n} a_{ij}x_j + b_i \leqq \varepsilon_i & (1 \leqq i \leqq m) \end{cases} \end{cases}$$

where $\varepsilon_i = |r_i|$. If the variables must be nonnegative, we introduce a variable y_{n+1} and write $x_j = y_j - y_{n+1}$. Define $a_{i,n+1} = -\sum_{j=1}^{n} a_{ij}$; an equivalent linear programming problem is

$$\begin{cases} \text{Maximize:} & \displaystyle -\sum_{i=1}^{m} \varepsilon_i \\ \text{Constraints:} & \begin{cases} \displaystyle\sum_{j=1}^{n+1} a_{ij}y_j - \varepsilon_i \leqq b_i & (1 \leqq i \leqq m) \\ \displaystyle -\sum_{j=1}^{n+1} a_{ij}y_j - \varepsilon_i \leqq -b_i & (1 \leqq i \leqq m) \\ y \geqq 0, \quad \varepsilon \geqq 0 \end{cases} \end{cases}$$

which is in first primal form with $m + n + 1$ variables and $2m$ inequality constraints.

- Another technique is to replace the $2m$ inequality constraints by a set of m equality constraints. We write $\varepsilon_i = |r_i| = u_i + v_i$, where $u_i = r_i$ and $v_i = 0$ if $r_i \geqq 0$ but $v_i = -r_i$ and $u_i = 0$ if $r_i < 0$. The resulting linear programming problem is

$$\begin{cases} \text{Maximize:} & \displaystyle -\sum_{i=1}^{m} u_i - \sum_{i=1}^{m} v_i \\ \text{Constraints:} & \begin{cases} \displaystyle\sum_{j=1}^{n+1} a_{ij}y_j - u_i + v_i = b_i & (1 \leqq i \leqq m) \\ u \geqq 0, \quad v \geqq 0, \quad y \geqq 0 \end{cases} \end{cases}$$

- For an inconsistent system, the problem of making $\varepsilon = \max_{1 \leqq i \leqq m} |r_i|$ a minimum is the **ℓ_∞ problem** for the system. An equivalent linear programming problem is

$$\begin{cases} \text{Minimize:} & \varepsilon \\ \text{Constraints:} & \begin{cases} \displaystyle\sum_{j=1}^{n} a_{ij}x_j - \varepsilon \leqq b_i & (1 \leqq i \leqq m) \\ \displaystyle -\sum_{j=1}^{n} a_{ij}x_j - \varepsilon \leqq -b_i & (1 \leqq i \leqq m) \end{cases} \end{cases}$$

If the variables must be nonnegative, we introduce a large variable y_{n+1} so that the quantities $y_j = x_j + y_{n+1}$ are positive and we have an equivalent linear programming problem:

$$\begin{cases} \text{Minimize:} & \varepsilon \\ \text{Constraints:} & \begin{cases} \displaystyle\sum_{j=1}^{n+1} a_{ij} y_j - \varepsilon \leqq b_i & (1 \leqq i \leqq m) \\ \displaystyle -\sum_{j=1}^{n+1} a_{ij} y_j - \varepsilon \leqq -b_i & (1 \leqq i \leqq m) \\ \varepsilon \geqq 0, \quad y_j \geqq 0 & (1 \leqq j \leqq n+1) \end{cases} \end{cases}$$

where we defined $a_{i,n+1} = -\sum_{j=1}^{n} a_{ij}$.

Exercises 14.3

1. Consider the inconsistent linear system

$$\begin{cases} 5x_1 + 2x_2 = 6 \\ x_1 + x_2 + x_3 = 2 \\ 7x_2 - 5x_3 = 11 \\ 6x_1 + 9x_3 = 9 \end{cases}$$

Write the following with nonnegative variables:

[a]**a.** The equivalent linear programming problem for solving the system in the ℓ_1 sense.

[a]**b.** The equivalent linear programming problem for solving the system in the ℓ_∞ sense.

2. (Continuation) Repeat the preceding exercise for the system

$$\begin{cases} 3x + y = 7 \\ x - y = 11 \\ x + 6y = 13 \\ -x + 3y = -12 \end{cases}$$

[a]3. We want to find a polynomial p of degree n that approximates a function f as well as possible *from below*; that is, we want $0 \leqq f - p \leqq \varepsilon$ for minimum ε. Show how p could be obtained with reasonable precision by solving a linear programming problem.

[a]4. To solve the ℓ_1 problem for the system of equations

$$\begin{cases} x - y = 4 \\ 2x - 3y = 7 \\ x + y = 2 \end{cases}$$

we can solve a linear programming problem. What is it?

Computer Exercises 14.3

[a]1. Obtain numerical answers for Parts **a** and **b** of Exercise 14.3.1.

2. (Continuation) Repeat for Exercise 14.3.2.

[a]3. Find a polynomial of degree 4 that represents the function e^x in the following sense: Select 20 equally spaced points x_i in interval [0, 1] and require the polynomial to minimize the expression $\max_{1 \leqq i \leqq 20} |e^{x_i} - p(x_i)|$.
Hint: This is the same as solving 20 equations in five variables in the ℓ_∞ sense. The ith equation is $A + Bx_i + Cx_i^2 + Dx_i^3 + Ex_i^4 = e^{x_i}$, and the unknowns are A, B, C, D, and E.

4. Use built-in routines in mathematical software systems such as MATLAB, Maple, or Mathematica to solve each of these linear programming problems in first primal form, in second primal form, and in dual form:

a. LPP (4) **b.** LPP (6)

A

Advice on Good Programming Practices

Because the programming of numerical schemes is essential to understanding them, we offer here a few words of advice on good programming practices.

A.1 Programming Suggestions

The suggestions and techniques given here should be considered in context. They are not intended to be complete, and some good programming suggestions have been omitted to keep the discussion brief. Our purpose is to encourage the reader to be attentive to considerations of efficiency, economy, readability, and roundoff errors. Of course, some of these suggestions and admonitions may vary depending on the particular mathematical software system or programming language that is being used.

Be Careful and Be Correct Strive to write programs carefully and correctly. This is of utmost importance.

Use Pseudocode Before beginning the coding, write out in complete detail the mathematical algorithm to be used in *pseudocode* such as that used in this text. The pseudocode serves as a bridge between the mathematics and the computer program. It need not be defined in a formal way, as is done for a computer language, but it should contain sufficient detail that the implementation is straightforward. When writing the pseudocode, use a style that is easy to read and understand. For maintainability, it should be easy for a person who is unfamiliar with the code to read it and understand what it does.

Check and Double-Check Check the code thoroughly for errors and omissions before beginning to edit on a computer terminal. Spend time checking the code before running it to avoid executing the program, showing the output, discovering an error, correcting the error, and repeating the process ad nauseam.*

Modern computing environments may allow the user to accomplish this process in only a few seconds, but this advice is still valid if for no other reason than that it is dangerously easy to write programs that may work on a simple test but not on a more complicated one. No function key or mouse can tell you what is wrong!

---

*In 1962, the rocket carrying the Mariner I space probe to Venus went off course after only five minutes of flight and was destroyed. An investigation revealed that a single line of faulty Fortran code caused the disaster. A period was typed in the code `DO 5 I=1,3` instead of the comma, resulting in the loop being executed once instead of three times. It has been estimated that this single typographical error cost the U.S. National Aeronautics and Space Administration $18.5 million dollars! For additional details, see material available online: `history.nasa.gov/mariner.html`

Use Test Cases After writing the pseudocode, check and trace through it using pencil-and-paper calculations on a typical yet simple example. Checking boundary cases, such as the values of the first and second iterations in a loop and the processing of the first and last elements in a data structure, may reveal embarrassing errors. These same sample cases can be used as the first set of test cases on the computer.

Modularize Code Build a program in steps by writing and testing a series of segments (subprograms, procedures, or functions); that is, write self-contained subtasks as separate routines. Try to keep these program segments reasonably small, less than a page or computer screen whenever possible, to make reading and debugging easier.

Generalize Slightly If the code can be written to handle a slightly more general situation, then in many cases it is worth the extra effort to do so. A program that was written for only a particular set of numbers may have to be completely rewritten for another set. For example, only a few additional statements are required to write a program with an arbitrary step size compared with a program in which the step size is fixed numerically. However, be careful not to introduce too much generality into the code because it can make a simple programming task overly complicated.

Show Intermediate Results Print out or display intermediate results and diagnostic messages to assist in debugging and understanding the program's operation. Always echo-print or display the input data unless it is impractical to do so, such as with a large amount of data. Using simple read and print commands frees the programmer from errors associated with misalignment of data. Fancy output formats are not necessary, but some simple labeling of the output is recommended.

Include Warning Messages A robust program always warns the user of a situation that it is not designed to handle. In general, write programs so that they are easy to debug when the inevitable bug appears.

Use Meaningful Variable Names It is often helpful to assign meaningful names to the variables because they may have greater mnemonic value. There is perennial confusion between the characters `O` (letter "oh") and `0` (number zero) and between `l` (letter "ell") and `1` (number one).

Declare All Variables All variables should be listed in the type declarations in each program or program segment. Implicit type assignments can be ignored when declaration statements include all variables used. Historically, in Fortran, variables beginning with `I/i`, `J/j`, `K/k`, `L/l`, `M/m`, and `N/n` are integer variables, and ones beginning with other letters are floating-point real variables. It may be a good idea to adhere to this scheme to immediately recognize the type of a variable without looking it up in the type declarations. Nevertheless, algorithms written in pseudocode do not always need to follow this advice.

Include Comments Comments within a routine are helpful for revealing at some later time what the program does. Extensive comments are not necessary, but we recommend that you include a preface to each program or subprogram explaining the purpose, the input and output variables, and the algorithm used as well as providing a few comments between major segments of the code. Indent each block of code a consistent number of spaces to improve readability. Inserting blank comment lines and blank spaces can greatly improve the readability of the code as well. To save space, we have not included any comments in the pseudocode.

Use Clean Loops Never put unnecessary statements within loops. Move expressions and variables outside a loop from inside a loop if they do not depend on the loop or do not change. Also, indenting loops can add to the readability of the code, particularly for nested loops. Use a nonexecutable statement as the terminator of a loop so that the code may be altered easily.

Declare Nonchanging Constants Use a parameter statement for assigning values to key constants. Parameter values correspond to constants that do not change throughout the routine. Such parameter statements are easy to change when one needs to rerun the program with different values. Also, they clarify the role key constants play in the code and make the routines more readable and easier to understand.

Use Appropriate Data Structures Use data structures that are natural to the problem at hand. If the problem adapts more easily to a three-dimensional array than to several one-dimensional arrays, then by all means, use a three-dimensional array.

Use Arrays of All Types The elements of arrays, whether one-, two-, or higher-dimensional, are usually stored in consecutive words of memory. Since the compiler may map the value of an index for two- and higher-subscripted arrays into a single subscript value that is used as a pointer to determine the location of elements in storage, the use of two- and higher-dimensional arrays can be considered a notational convenience for the user. However, any advantage in using only a one-dimensional array and performing complicated subscript calculation is slight. Such matters are best left to the compiler.

Use Built-in Functions In scientific programming languages, many built-in mathematical functions are available for common functions such as *sin*, *log*, *exp*, *arcsin*, and so on. Also, numeric functions such as *integer*, *real*, *complex*, and *imaginary* are usually available for type conversion. One should utilize these and others as much as possible. Some of these intrinsic functions accept arguments of more than one type and return a result whose type may vary depending on the type of the argument used. Such functions are called **generic functions**, for they represent an entire family of related functions. Of course, care should be taken not to use the wrong argument type.

Use Program Libraries In preference to one that you might write yourself for a programming project, a *preprogrammed* routine from a mathematical software system or a computer program library should be used when applicable. Such routines can be expected to be state-of-the-art software, well tested, and, of course, completely debugged.

Do Not Overoptimize Students should be primarily concerned with writing readable code that correctly computes the desired results. There are any number of tricks of the trade for making code run faster or more efficiently. Save them for use later on in your programming life. We are primarily concerned with understanding and testing various numerical methods. Do not sacrifice the clarity of a program in an effort to make the code run faster. Clarity of code may be preferable to optimization of code when the two criteria conflict.

Case Studies

We present some case studies that may be helpful.

Computing Sums When a long list of floating-point numbers is added in the computer, there may be less roundoff error if the numbers are added in order of increasing magnitude. (Roundoff errors are discussed in detail in Section 1.4.)

Mathematical Constants Some students are surprised to learn that in many programming languages, the computer does not automatically know the values of common mathematical constants such as π and e and must be explicitly told their values! It is easy to mistype a long sequence of digits in a mathematical constant, such as in the real number π coded as

$$pi \leftarrow 3.14159\,26535\,89793$$

We recommend using simple calculations involving mathematical functions. For example, the real numbers π and e can be easily and safely entered with nearly full machine precision by using standard intrinsic functions such as

$$pi \leftarrow 4.0 \arctan(1.0)$$
$$e \leftarrow \exp(1.0)$$

Another reason for this advice is to avoid the problem that arises if you use a short approximation such as $pi \leftarrow 3.14159$ on a computer with limited precision but later move the code to another computer that has more precision. If you overlook changing this assignment statement, then all results that depend on this value may be less accurate than they should be.

Exponents In coding for the computer, exercise care in writing statements that involve exponents. The general function x^y may be calculated on many computers as $\exp(y \ln x)$ whenever y is not an integer. Sometimes this is unnecessarily complicated and may contribute to roundoff errors. For example, it may be preferable to write code with integer exponents such as 5 rather than 5.0. Similarly, using exponents such as $\frac{1}{2}$ or 0.5 is not recommended because the built-in function *sqrt* may be used.

There is rarely any need for a calculation such as $j \leftarrow (-1)^k$ because there are better ways of obtaining the same result. For example, in a loop, we can write $j \leftarrow 1$ before the loop and $j \leftarrow -j$ inside the loop.

Avoid Mixed Mode In general, one should avoid mixing real and integer expressions in the computer code. *Mixed expressions* are formulas in which variables and constants of different types appear together. If the floating-point form of an integer variable is needed, use a function such as *real*. Similarly, a function such as *integer* is generally available for obtaining the integer part of a real variable. In other words, use the intrinsic type conversion functions whenever converting from complex to real, real to integer, or vice versa. For example, in floating-point calculations, m/n should be coded as $\text{real}(m)/\text{real}(n)$ when m and n are integer variables so that it computes the correct real value of m/n. Similarly, $1/m$ should be coded as $1.0/\text{real}(m)$ and $1/2$ as 0.5 and so on.

Precision In the usual mode of representing numbers in a computer, one word of storage is used for each number. This mode of representation is called **single precision**. In calculations that require greater precision (called **double precision** or **extended precision**), two or more words of storage are alloted to each number. On a 32-bit computer, one can obtain approximately 6 decimal places of precision in single precision, and approximately 15 decimal places of precision in double precision. If more accuracy is needed than single precision can provide, then double or extended precision should be used.* This is particularly true on computers with limited precision, such as a 32-bit computer, on which roundoff errors can quickly accumulate in long computations and reduce the accuracy to only three or four decimal places! (This topic is discussed in Section 1.3.)

Usually, two words of memory are used to store the real and imaginary parts of a complex number. Complex variables and arrays must be explicitly declared as being of complex type. Expressions involving variables and constants of complex type are evaluated according to the normal rules of complex arithmetic. Intrinsic functions such as *complex*, *real*, and *imaginary* should be used to convert between real and complex types.

Memory Fetches When using loops, write the code so that fetches are made from *adjacent* words in memory. To illustrate, suppose we want to store values in a two-dimensional array (a_{ij}) in which the elements of each column are stored in consecutive memory locations. Using i and j loops with the ith loop as the innermost one would process elements down the columns. For some mathematical software systems and computer programming languages, this detail may be of only secondary concern. However, some computers have immediate access to only a portion or a few *pages* of memory at a time. In this case, it is advantageous to process the elements of an array so that they are taken from, or stored in, adjacent or nearby memory locations.

When to Avoid Arrays Although the mathematical description of an algorithm may indicate that a sequence of values is computed, thus seeming to imply the need for an array, it is often possible to avoid arrays. (This is especially true if only the final value of a sequence is required.) For example, the theoretical description of Newton's method (Section 3.2) reads

$$x_{n+1} = x_n - \frac{f(x_n)}{f'(x_n)}$$

but the pseudocode can be written within a loop simply as

for $n = 1$ **to** 10
  $x \leftarrow x - f(x)/f'(x)$
end for

where x is a real variable and function procedures for f and f' have been written. Such an assignment statement automatically effects the replacement of the value of the *old* x with the *new* numerical value of $x - f(x)/f'(x)$.

Limit Iterations In a repetitive algorithm, one should always limit the number of permissible steps by the use of a loop with a control variable. This prevents endless cycling due to unforeseen problems (e.g., programming errors and roundoff errors). For example, in

---

*With the proliferation of 64-bit microprocessor(s), 64-bit computing is becoming more commonplace with the resulting improvement in performance and increase in precision. The meanings of single precision and double precision are changing!

Newton's pseudocode, one might replace `x ← x - f(x)/f'(x)` with

```
d ← f(x)/f'(x)
while |d| > 1/2 × 10^-6
    x ← x − d
    output x
    d ← f(x)/f'(x)
end while
```

If the function involves some erratic behavior, there is a danger here in not limiting the number of repetitions. It is better to use a loop with a control variable:

```
for n = 1 to n_max
    d ← f(x)/f'(x)
    x ← x − d
    output n, x
    if |d| ≤ 1/2 × 10^-6 then exit loop
end for
```

where n and n_max are integer variables and the value of n_max is an upper bound on the number of desired repetitions. All others are real variables.

Floating-Point Equality The sequence of steps in a routine should not depend on whether two floating-point numbers are equal. Instead, reasonable tolerances should be permitted to allow for floating-point arithmetic roundoff errors. For example, a suitable branching statement for n decimal digits of accuracy might be

```
if |x − y| < ε then ... end if
```

provided that it is known that x and y have magnitude comparable to 1. Here, x, y, and ε are real variables with $\varepsilon = \frac{1}{2} \times 10^{-n}$. This corresponds to requiring that the *absolute error* between x and y be less than ε. However, if x and y have very large or small orders of magnitude, then the *relative error* between x and y would be needed, as in the branching statement

```
if |x − y| < ε max{|x|, |y|} then ... end if
```

Equal Floating-Point Steps In some situations, notably in solving differential equations (see Chapter 7), a variable t assumes a succession of values equally spaced a distance of h apart along the real line. One way of coding this is

```
t ← t_0
output 0, t
for i = 1 to n
    ⋮
    t ← t + h
    output i, t
end for
```

Here, i and n are integer variables, and t_0, t, and h are real variables. An alternative way is

for $i = 0$ **to** n
  ⋮
  $t \leftarrow t_0 + \text{real}(i)h$
  output i, t
end for

In the first pseudocode, n additions occur, each with possible roundoff error. In the second, this situation is avoided, but at the added cost of n multiplications. Which is better depends on the particular situation at hand.

Function Evaluations When values of a function at arbitrary points are needed in a program, several ways of coding this are available. For example, suppose values of the function

$$f(x) = 2x + \ln x - \sin x$$

are needed. A simple approach is to use an assignment statement such as

$$y \leftarrow 2x + \ln(x) - \sin(x)$$

at appropriate places within the program. Here, x and y are real variables. Equivalently, by using an *internal* function procedure corresponding to the pseudocode

$$f(x) \leftarrow 2x + \ln(x) - \sin(x)$$

it could be evaluated at 2.5 by

$$y \leftarrow f(2.5)$$

or whatever value of x is desired. Finally, a function subprogram can be used such as in the following pseudocode:

real function $f(x)$
real x
  $f \leftarrow 2x + \ln(x) - \sin(x)$
end function f

Which implementation is best?

It depends on the situation at hand. The assignment statement is simple and safe. An internal or external function procedure can be used to avoid duplicating code. A separate external function subprogram is the best way to avoid difficulties that inadvertently occur when someone must insert code into another's program. In using program library routines, the user may be required to furnish an external function procedure to communicate function values to the library routine. If the external function procedure f is passed as an argument in another procedure, then a special *interface* must be used to designate it as an external function.

On Developing Mathematical Software

Fred Krogh [2003] has written a paper listing some of the things he has learned from a career at the Jet Propulsion Laboratory involving the development and writing of mathematical software used in application packages. Some of his helpful hints and random thoughts to remember in code development are as follows:

- Include internal output in order to see what your algorithm is doing
- Support debugging by including output at the interfaces
- Provide detailed error messages
- Fine-tune your code
- Provide understandable test cases
- Verify results with care
- Take advantage of your mistakes
- Keep units consistent
- Test the extremes
- The algorithm matters
- Work on what does work
- Toss out what does not work
- Do not give up too soon on ideas for improving or debugging your code
- Your subconscious is a powerful tool, so learn to use it
- Test your assumptions
- In the comments, keep a dictionary of variables in alphabetical order because it is quite helpful when looking at a code years after it was written
- Write the user documentation first
- Know what performance you should expect to get
- Do not pay too much, but just enough, attention to others
- See setbacks as learning opportunities and as the staircase for keeping one's spirits up
- When comparing codes, do not change their features or capabilities in order to make the comparison fair, since you may not fully understand the other person's code
- Keep action lists
- Categorize code features
- Organize things into groups
- The organization of the code may be one of the most important decisions the developer makes
- Isolate the linear algebra parts of the code in an application package so that the user may make modifications to them
- *Reverse communication* is a helpful feature that allows users to leave the code and carry out matrix-vector operations using their own data structures
- Save and restore variables when the user is allowed to leave the code and return
- Portability is more important than efficiency

These are just a random sampling of some of his insights. Use them as you see fit!

B Representation of Numbers in Different Bases

In this appendix, we review some basic concepts on number representation in different bases.

B.1 Representation of Numbers in Different Bases

We begin with a discussion of general number representation, but move quickly to bases 2, 8, and 16, as they are the bases primarily used in computer arithmetic.

The familiar decimal notation for numbers uses the digits 0, 1, 2, 3, 4, 5, 6, 7, 8, and 9. When we write a whole number such as 37294, the individual digits represent coefficients of powers of 10 as follows:

$$\begin{aligned} 37294 &= 4 + 90 + 200 + 7000 + 30000 \\ &= 4 \times 10^0 + 9 \times 10^1 + 2 \times 10^2 + 7 \times 10^3 + 3 \times 10^4 \end{aligned}$$

Thus, in general, a string of digits represents a number according to the formula

Base 10 Integer Part

$$(a_n a_{n-1} \ldots a_2 a_1 a_0)_{10} = a_0 \times 10^0 + a_1 \times 10^1 + \cdots + a_{n-1} \times 10^{n-1} + a_n \times 10^n$$

This takes care of only the positive whole numbers. A number between 0 and 1 is represented by a string of digits to the right of a decimal point. For example, we see that

$$\begin{aligned} 0.7215 &= \frac{7}{10} + \frac{2}{100} + \frac{1}{1000} + \frac{5}{10000} \\ &= 7 \times 10^{-1} + 2 \times 10^{-2} + 1 \times 10^{-3} + 5 \times 10^{-4} \end{aligned}$$

In general, we have the formula

Base 10 Fractional Part

$$(0.b_1 b_2 b_3 \ldots)_{10} = b_1 \times 10^{-1} + b_2 \times 10^{-2} + b_3 \times 10^{-3} + \cdots$$

Note that there can be an infinite string of digits to the right of the decimal point; indeed, there *must* be an infinite string to represent some numbers. For example, we note that

Infinite String of Digits

$$\begin{aligned} \sqrt{2} &= 1.41421\,35623\,73095\,04880\,16887\,24209\,69\ldots \\ e &= 2.71828\,18284\,59045\,23536\,02874\,71352\,66\ldots \\ \pi &= 3.14159\,26535\,89793\,23846\,26433\,83279\,50\ldots \\ \ln 2 &= 0.69314\,71805\,59945\,30941\,72321\,21458\,17\ldots \\ \tfrac{1}{3} &= 0.33333\,33333\,33333\,33333\,33333\,33333\,33\ldots \end{aligned}$$

For a real number of the form

Base 10 General Expression

$$(a_n a_{n-1} \ldots a_1 a_0 . b_1 b_2 b_3 \ldots)_{10} = \sum_{k=0}^{n} a_k 10^k + \sum_{k=1}^{\infty} b_k 10^{-k}$$

the **integer part** is the first summation in the expansion and the **fractional part** is the second summation. If ambiguity can arise, a number represented in base β is signified by enclosing it in parentheses and adding a subscript β.

Base β Numbers

The foregoing discussion pertains to the usual representation of numbers with base 10. Other bases are also used, especially in computers. For example, the **binary** system uses 2 as the base, the **octal** system uses 8, and the **hexadecimal** system uses 16.

In the octal representation of a number, the digits that are used are 0, 1, 2, 3, 4, 5, 6, and 7. Thus, we see that

Integer Base 8 Example

$$\begin{aligned}(21467)_8 &= 7 + 6 \times 8 + 4 \times 8^2 + 1 \times 8^3 + 2 \times 8^4 \\ &= 7 + 8(6 + 8(4 + 8(1 + 8(2)))) \\ &= 9015\end{aligned}$$

A number between 0 and 1, expressed in octal, is represented with combinations of 8^{-1}, 8^{-2}, and so on. For example, we have

Fraction Base 8 Example

$$\begin{aligned}(0.36207)_8 &= 3 \times 8^{-1} + 6 \times 8^{-2} + 2 \times 8^{-3} + 0 \times 8^{-4} + 7 \times 8^{-5} \\ &= 8^{-5}(3 \times 8^4 + 6 \times 8^3 + 2 \times 8^2 + 7) \\ &= 8^{-5}(7 + 8^2(2 + 8(6 + 8(3)))) \\ &= \frac{15495}{32768} \\ &= 0.47286\,987\ldots\end{aligned}$$

We shall see presently how to convert easily to decimal form without having to find a common denominator.

If we use another base, say β, then numbers represented in the β-system look like this:

Base β General Expression

$$(a_n a_{n-1} \ldots a_1 a_0 . b_1 b_2 b_3 \ldots)_\beta = \sum_{k=0}^{n} a_k \beta^k + \sum_{k=1}^{\infty} b_k \beta^{-k}$$

The digits are $0, 1, \ldots, \beta - 2$, and $\beta - 1$ in this representation. If $\beta > 10$, it is necessary to introduce symbols for $10, 11, \ldots, \beta - 1$. The separator between the integer and fractional part is called the **radix point**, since **decimal point** is reserved for base-10 numbers.

Conversion of Integer Parts

We now formalize the process of converting a number from one base to another. It is advisable to consider separately the integer and fractional parts of a number. Consider, then, a positive integer N in the number system with base γ:

Integer Base γ to Base β

$$N = (a_n a_{n-1} \ldots a_1 a_0)_\gamma = \sum_{k=0}^{n} a_k \gamma^k$$

Suppose that we wish to convert this to the number system with base β and that the calculations are to be performed in arithmetic with base β. Write N in its nested form:

$$N = a_0 + \gamma(a_1 + \gamma(a_2 + \cdots + \gamma(a_{n-1} + \gamma(a_n))\cdots))$$

Then replace each of the numbers on the right by its representation in base β. Next, carry out the calculations in β-arithmetic. The replacement of the a_k's and γ by equivalent base-β numbers requires a table showing how each of the numbers $0, 1, \ldots, \gamma - 1$ appears in the β-system. Moreover, a base-β multiplication table may be required.

To illustrate this procedure, consider the conversion of the decimal number 3781 to binary form. Using the decimal binary equivalences and longhand multiplication in base 2, we have

Integer Base 10 to Base 2

$$\begin{aligned}(3781)_{10} &= 1 + 10(8 + 10(7 + 10(3)))\\ &= (1)_2 + (1\,010)_2\,((1\,000)_2 + (1\,010)_2\,((111)_2 + (1\,010)_2(11)_2))\\ &= (111\,011\,000\,101)_2\end{aligned}$$

This arithmetic calculation in binary is easy for a computer that operates in binary but tedious for humans!

Another procedure should be used for hand calculations. Write down an equation containing the digits $c_0, c_1, \ldots, c_m$ that we seek:

$$N = (c_m c_{m-1} \ldots c_1 c_0)_\beta = c_0 + \beta(c_1 + \beta(c_2 + \cdots + \beta(c_m)\cdots))$$

Next, observe that if N is divided by β, then the **remainder** in this division is c_0, and the **quotient** is

$$c_1 + \beta(c_2 + \cdots + \beta(c_m)\cdots)$$

If *this* number is divided by β, the remainder is c_1, and so on. Thus, we divide repeatedly by β, saving remainders $c_0, c_1, \ldots, c_m$ and quotients.

EXAMPLE 1 Convert the decimal number 3781 to binary form using the division algorithm.

Solution As was indicated above, we divide repeatedly by 2, saving the remainders along the way. Here is the work:

Quotients	Remainders	
2) 3781		
2) 1890	$1 = c_0$	↓
2) 945	$0 = c_1$	
2) 472	$1 = c_2$	
2) 236	$0 = c_3$	
2) 118	$0 = c_4$	
2) 59	$0 = c_5$	
2) 29	$1 = c_6$	
2) 14	$1 = c_7$	
2) 7	$0 = c_8$	
2) 3	$1 = c_9$	
2) 1	$1 = c_{10}$	
0	$1 = c_{11}$	

Here, the symbol ↓ is used to remind us that the digits c_i are obtained beginning with the digit next to the binary point. Thus, we have

$$(3781.)_{10} = (111\,011\,000\,101.)_2$$

and not the other way around: $(101\,000\,110\,111.)_2 = (2615)_{10}$. ■

EXAMPLE 2 Convert the number $N = (111\,011\,000\,101)_2$ to decimal form by nested multiplication.

Solution

$$\begin{aligned}
N &= 1 \times 2^0 + 0 \times 2^1 + 1 \times 2^2 + 0 \times 2^3 + 0 \times 2^4 + 0 \times 2^5 \\
&\qquad + 1 \times 2^6 + 1 \times 2^7 + 0 \times 2^8 + 1 \times 2^9 + 1 \times 2^{10} + 1 \times 2^{11} \\
&= 1 + 2(0 + 2(1 + 2(0 + 2(0 + 2(0 + 2(1 + 2(1 + 2(0 \\
&\qquad + 2(1 + 2(1 + 2(1)))))))))))\\
&= 3781
\end{aligned}$$

The **nested multiplication** with repeated multiplication and addition can be carried out on a hand-held calculator more easily than can the previous form with exponentiation. ■

Another conversion problem exists in going from an integer in base γ to an integer in base β when using calculations in base γ. As before, the unknown coefficients in the equation

$$N = c_0 + c_1\beta + c_2\beta^2 + \cdots + c_m\beta^m$$

are determined by a process of successive division, and this arithmetic is carried out in the γ-system. At the end, the numbers c_k are in base γ, and a table of γ-β equivalents is used.

For example, we can convert a binary integer into decimal form by repeated division by $(1\,010)_2$ (which equals $(10)_{10}$), carrying out the operations in binary. A table of binary-decimal equivalents is used at the final step. However, since binary division is easy only for computers, we shall develop alternative procedures presently.

Conversion of Fractional Parts

We can convert a fractional number such as $(0.372)_{10}$ to binary by using a direct yet naive approach as follows:

Fraction Base 10 to Base 2

$$\begin{aligned}
(0.372)_{10} &= 3 \times 10^{-1} + 7 \times 10^{-2} + 2 \times 10^{-3} \\
&= \frac{1}{10}\left(3 + \frac{1}{10}\left(7 + \frac{1}{10}(2)\right)\right) \\
&= \frac{1}{(1\,010)_2}\left((011)_2 + \frac{1}{(1\,010)_2}\left((111)_2 + \frac{1}{(1\,010)_2}(010)_2\right)\right)
\end{aligned}$$

Dividing in binary arithmetic is *not* straightforward, so we look for easier ways of doing this conversion!

Suppose that x is in the range $0 < x < 1$ and that the digits c_k in the representation

$$x = \sum_{k=1}^{\infty} c_k\beta^{-k} = (0.c_1c_2c_3\ldots)_\beta$$

are to be determined. Observe that

$$\beta x = (c_1.c_2c_3c_4\ldots)_\beta$$

because it is necessary to shift the radix point only when multiplying by base β.

Thus, the unknown digit c_1 can be described as the **integer part** of βx. It is denoted by $\mathcal{I}(\beta x)$. The **fractional part,** $(0.c_2c_3c_4\ldots)_\beta$, is denoted by $\mathcal{F}(\beta x)$. The process is repeated in the following pattern:

$$\begin{array}{lll} d_0 = x & & \\ d_1 = \mathcal{F}(\beta d_0) & c_1 = \mathcal{I}(\beta d_0) & \downarrow \\ d_2 = \mathcal{F}(\beta d_1) & c_2 = \mathcal{I}(\beta d_1) & \\ & \text{etc.} & \end{array}$$

In this algorithm, the arithmetic is carried out in the decimal system.

EXAMPLE 3 Use the preceding algorithm to convert the decimal number $x = (0.372)_{10}$ to binary form.

Solution The algorithm consists in repeatedly multiplying by 2 and removing the integer parts. Here is the work:

$$\begin{array}{lr} & 0.372 \\ & \underline{\quad 2} \\ \downarrow \quad c_1 = & \boxed{0}.744 \\ & \underline{\quad 2} \\ c_2 = & \boxed{1}.488 \\ & \underline{\quad 2} \\ c_3 = & \boxed{0}.976 \\ & \underline{\quad 2} \\ c_4 = & \boxed{1}.952 \\ & \underline{\quad 2} \\ c_5 = & \boxed{1}.904 \\ & \underline{\quad 2} \\ c_6 = & \boxed{1}.808 \\ & \\ & \text{etc.} \end{array}$$

Thus, we have $(0.372)_{10} = (0.010\,111\ \ldots)_2$. ■

Base Conversion 10 $\leftrightarrow$ 8 $\leftrightarrow$ 2

Most computers use the binary system (base 2) for their internal representation of numbers. The octal system (base 8) is particularly useful in converting from the decimal system (base 10) to the binary system and vice versa. With base 8, the positional values of the numbers are $8^0 = 1$, $8^1 = 8$, $8^2 = 64$, $8^3 = 512$, $8^4 = 4096$, and so on. Thus, for example, we have

Examples Base 8 to Base 10

$$\begin{aligned} (26031)_8 &= 2 \times 8^4 + 6 \times 8^3 + 0 \times 8^2 + 3 \times 8 + 1 \\ &= ((((2)8 + 6)8 + 0)8 + 3)8 + 1 \\ &= 11289 \end{aligned}$$

and

$$\begin{aligned} (7152.46)_8 &= 7 \times 8^3 + 1 \times 8^2 + 5 \times 8 + 2 + 4 \times 8^{-1} + 6 \times 8^{-2} \\ &= (((7)8 + 1)8 + 5)8 + 2 + 8^{-2}[(4)8 + 6] \\ &= 3690 + \tfrac{38}{64} \\ &= 3690.59375 \end{aligned}$$

When numbers are converted between decimal and binary form by hand, it is convenient to use octal representation as an intermediate step. In the octal system, the base is 8, and, of course, the digits 8 and 9 are not used. Conversion between octal and decimal proceeds according to the principles already stated. Conversion between octal and binary is especially simple. Groups of three binary digits can be translated directly to octal according to the following table:

Binary-Odd Table

Binary	000	001	010	011	100	101	110	111
Octal	0	1	2	3	4	5	6	7

This grouping starts at the binary point and proceeds in both directions. Thus, we have

$$(101\,101\,001.110\,010\,100)_2 = (551.624)_8$$

To justify this convenient sleight of hand, we consider, for instance, a fraction expressed in binary form:

$$\begin{aligned} x &= (0.b_1 b_2 b_3 b_4 b_5 b_6 \ldots)_2 \\ &= b_1 2^{-1} + b_2 2^{-2} + b_3 2^{-3} + b_4 2^{-4} + b_5 2^{-5} + b_6 2^{-6} + \cdots \\ &= (4b_1 + 2b_2 + b_3)8^{-1} + (4b_4 + 2b_5 + b_6)8^{-2} + \cdots \end{aligned}$$

In the last line of this equation, the parentheses enclose numbers from the set {0, 1, 2, 3, 4, 5, 6, 7} because the b_i's are either 0 or 1. Hence, this must be the octal representation of x.

Conversion of an octal number to binary can be done in a similar manner but in reverse order. It is easy! Just replace each octal digit with the corresponding three binary digits. Thus, for example, we obtain

$$(5362.74)_8 = (101\,011\,110\,010.111\,100)_2$$

EXAMPLE 4 What is $(2576.35546\,875)_{10}$ in octal and binary forms?

Solution We convert the original decimal number first to octal and then to binary. For the integer part, we repeatedly divide by 8:

$$\begin{array}{rl} 8\,\underline{)\,2576} & \\ 8\,\underline{)\,322}\; 0 & \downarrow \\ 8\,\underline{)\,40}\; 2 & \\ 8\,\underline{)\,5}\; 0 & \\ 0\; 5 & \end{array}$$

Thus, we have

$$2576. = (5020.)_8 = (101\,000\,010\,000.)_2$$

using the rules for grouping binary digits. For the fractional part, we repeatedly multiply by 8

$$\begin{array}{rr} & 0.35546875 \\ & \underline{8} \\ \downarrow & \boxed{2}.84375000 \\ & \underline{8} \\ & \boxed{6}.75000000 \\ & \underline{8} \\ & \boxed{6}.00000000 \end{array}$$

so that

$$0.35546\,875 = (0.266)_8 = (0.010\,110\,110)_2$$

Finally, we obtain the result

$$2576.35546\,875 = (101\,000\,010\,000.010\,110\,110)_2$$

Although this approach is longer for this example, we feel that it is easier, in general, and less likely to lead to errors because one is working with single-digit numbers most of the time. ■

Base 16

Some computers whose word lengths are multiples of 4 use the **hexadecimal** system (base 16) in which A, B, C, D, E, and F represent 10, 11, 12, 13, 14, and 15, respectively, as given in the following table of equivalences:

Hexadecimal-Binary Table

Hexadecimal	0	1	2	3	4	5	6	7
Binary	0000	0001	0010	0011	0100	0101	0110	0111

8	9	A	B	C	D	E	F
1000	1001	1010	1011	1100	1101	1110	1111

Conversion between binary numbers and hexadecimal numbers is particularly easy. We need only regroup the binary digits from groups of three to groups of four. For example, we have

$$(010\,101\,110\,101\,101)_2 = (0010\,1011\,1010\,1101)_2 = (2BAD)_{16}$$

and

$$\begin{aligned}(111\,101\,011\,110\,010.110\,010\,011\,110)_2 &= (1010\,1111\,0010.1100\,1001\,1110)_2 \\ &= (7AF2.C9E)_{16}\end{aligned}$$

More Examples

Continuing with more examples, let us convert $(0.276)_8$, $(0.C8)_{16}$, and $(492)_{10}$ into different number systems. We show one way for each number and invite the reader to work out the details for other ways and to verify the answers by converting them back into the original base.

$$\begin{aligned}(0.276)_8 &= 2 \times 8^{-1} + 7 \times 8^{-2} + 6 \times 8^{-3} \\ &= 8^{-3}[((2)8 + 7)8 + 6] \\ &= (0.37109\,375)_{10}\end{aligned}$$

More Example

$$\begin{aligned}(0.C8)_{16} &= (0.110\,010)_2 \\ &= (0.62)_8 \\ &= 6 \times 8^{-1} + 2 \times 8^{-2} \\ &= 8^{-2}[(6)8 + 2] \\ &= (0.78125)_{10}\end{aligned}$$

$$\begin{aligned}(492)_{10} &= (754)_8 \\ &= (111\,101\,100)_2 \\ &= (1EC)_{16}\end{aligned}$$

because

$$\begin{array}{rl} 8\overline{)\,492} & \\ 8\overline{)\,61}\ 4 & \downarrow \\ 8\overline{)\,7}\ 5 & \\ 0\ 7 & \end{array}$$

Summary B.1

It might seem that there are several different procedures for converting between number systems. Actually, there are only *two* basic techniques. The first procedure for converting the number $(N)_\gamma$ to base β can be outlined as follows:

- Express $(N)_\gamma$ in nested form using powers of γ.
- Replace each digit by the corresponding base-β numbers.
- Carry out the indicated arithmetic in base β.

This outline holds whether N is an integer or a fraction. The second procedure is either the divide-by-β and **remainder-quotient-split** process for N an integer or the multiply-by-β and **integer-fraction-split** process for N a fraction. The first procedure is preferred when $\gamma < \beta$ and the second when $\gamma > \beta$. Of course, the $10 \leftrightarrow 8 \leftrightarrow 2 \leftrightarrow 16$ base conversion procedure should be used whenever possible because it is the easiest way to convert numbers between the decimal, octal, binary, or hexadecimal systems.

Exercises B.1

1. Find the binary representation and check by reconverting to decimal representation.

[a]**a.** $e \approx (2.718)_{10}$ **b.** $\frac{7}{8}$ **c.** $(592)_{10}$

2. Convert the following decimal numbers to octal numbers.

a. 27.1 **b.** 12.34 **c.** 3.14 [a]**d.** 23.58

[a]**e.** 75.232 [a]**f.** 57.321

3. Convert to hexadecimal, to octal, and then to decimal.

[a]**a.** $(110\,111\,001.101\,011\,101)_2$ [a]**b.** $(1\,001\,100\,101.011\,01)_2$

4. Convert the following numbers:

a. $(100\,101\,101)_2 = (\quad)_8 = (\quad)_{10}$

b. $(0.782)_{10} = (\quad)_8 = (\quad)_2$

[a]**c.** $(47)_{10} = (\quad)_8 = (\quad)_2$

d. $(0.47)_{10} = (\quad)_8 = (\quad)_2$

[a]**e.** $(51)_{10} = (\quad)_8 = (\quad)_2$

f. $(0.694)_{10} = (\quad)_8 = (\quad)_2$

[a]**g.** $(110011.111\,010\,110\,110\,1)_2 = (\quad)_8 = (\quad)_{10}$

h. $(361.4)_8 = (\quad)_2 = (\quad)_{10}$

5. Convert $(45653.127664)_8$ to binary and to decimal.

[a]**6.** Convert $(0.4)_{10}$ first to octal and then to binary. *Check:* Convert directly to binary.

7. Prove that the decimal number $\frac{1}{5}$ cannot be represented by a finite expansion in the binary system.

8. Do you expect your computer to calculate $3 \times \frac{1}{3}$ with infinite precision? What about $2 \times \frac{1}{2}$ or $10 \times \frac{1}{10}$?

[a]**9.** Explain the algorithm for converting an integer in base 10 to one in base 2, assuming that the calculations are performed in binary arithmetic. Illustrate by converting $(479)_{10}$ to binary.

10. Justify mathematically the conversion between binary and hexadecimal numbers by regrouping.

11. Justify for integers the rule given for the conversion between octal and binary numbers.

[a]**12.** Prove that a real number has a finite representation in the binary number system if and only if it is of the form $\pm m/2^n$, where n and m are positive integers.

13. Prove that any number that has a finite representation in the binary system must have a finite representation in the decimal system.

14. Some countries measure temperature in Fahrenheit (F), while other countries use Celsius (C). Similarly, for distance, some use miles and others use kilometers. As a frequent traveler, you may be in need of a quick approximate conversion scheme that you can do in your head.

 a. Fahrenheit and Celsius are related by the equation $F = 32+(9/5)C$. Verify the following simple conversion scheme for going from Celsius to Fahrenheit: A rough approximation is to double the Celsius temperature and add 32. To refine your approximation, shift the decimal place to the left in the doubled number ($2C$) and subtract it from the approximation obtained previously:

$$F = [(2C)+32]-(2C)/10$$

 b. Determine a simple scheme to convert from Fahrenheit to Celsius.

 c. Determine a simple scheme to convert from miles to kilometers.

 d. Determine a simple scheme to convert from kilometers to miles.

15. Convert fractions such as $\frac{1}{3}$ and $\frac{1}{11}$ into their binary represention.

16. (**Mayan Arithmetic**) The Maya civilization of Central America (2600 B.C. to 1200 A.D.) understood the concept of zero hundreds of years before many other civilizations. For their calculations, the vigesimal (base 20) system was used, not the decimal (base 10) system. So instead of 1, 10, 100, 1000, 10000, they used 1, 20, 400, 8000, 16000. They used a dot for 1 and a bar for 5, and 0 was represented by the shell symbol. For example, the calculations

$$11131 + 7520 = 18651, \qquad 11131 - 7520 = 3611$$

was as follows:

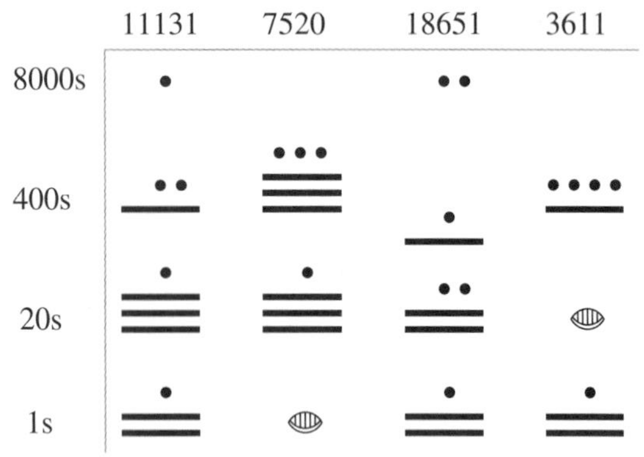

In the table above, some decimal numbers are included as an aid: on the left are the powers used, and across the top are the numbers represented by the columns.

Do these calculations using Mayan symbols and arithmetic:

 a. $92819 + 56313 = 149132$,
 $92819 - 56313 = 36506$

 b. $3296 + 853 = 4149, \qquad 3296 - 853 = 2443$

 c. $2273 + 729 = 1544, \qquad 2273 - 729 = 1544$

 d. Investigate how the Mayans might have done multiplication and division in their number system. Work out some simple examples.

17. (**Babylonian Arithmetic**) Babylonians of ancient Mesopotania (now Iraq) used a sexagesimal (base 60) positional number system with a decimal (base 10) system within it. The Babylonians based their number system on only two symbols! The influence of Babylonian arithmetic is still with us today. An hour consists of 60 minutes and is divided into 60 seconds, and a circle is measured in divisions of 360 degrees. Numbers are frequently called *digits*, from the Latin word for "finger." The base-10 and base-20 systems most likely arose from the fact that 10 fingers and 10 toes could be used in counting. Investigate the early history of numbers and doing aritmetic calculations in different number systems.

Computer Exercises B.1

1. Read into your computer $x = 1.1$ (base 10), and print it out using several different formats. Explain the results.

2. Show that $e^{\pi\sqrt{163}}$ is incredibly close to being the 18-digit integer

$$262\ 53741\ 26407\ 68744$$

Hint: More than 30 decimal digits will be needed to see any difference.

3. Write and test a routine for converting integers into octal and binary forms.

4. (Continuation) Write and test a routine for converting decimal fractions into octal and binary forms.

5. (Continuation) Using the two routines of the preceding problems, write and test a program that reads in decimal numbers and prints out the decimal, octal, and binary representations of these numbers.

6. See how many binary digits your computer has for $(0.1)_{10}$.

7. Some mathematical software systems have commands for converting numbers between binary, decimal, hex, octal, and vice versa. Explore these commands using various numerical values. Also, see whether there are commands for determining the *precision* (the number of significant decimal digits in a number) and the *accuracy* (the number of significant decimal digits to the right of the decimal point in a number).

8. Write a computer program to verify the conclusions in evaluating

$$f(x) = x - \sin x$$

for various values of x near 1.9, say, over the interval $[0.1, 2.5]$ with increments of 0.1. For these values, compute the approximate value of f, the true calculated value, and the absolute error between them. Single-precision and double-precision computations may be necessary.

C Additional Details on IEEE Floating-Point Arithmetic

In this appendix, we summarize some additional features in IEEE standard floating-point arithmetic. (See Overton [2001] for additional details.)

C.1 More on IEEE Standard Floating-Point Arithmetic

In the early 1980s, a working committee of the Institute for Electrical and Electronics Engineers (IEEE) established a standard floating-point arithmetic system for computers that is now known as the **IEEE floating-point standard** (IEEE-754). Previously, manufacturers of different computers each developed their own internal floating-point number systems. This led to inconsistencies in numerical results in moving code from machine to machine, for example, in porting source code from an IBM computer to a Cray machine. Some important requirements for all machines adopting the IEEE floating-point standard include the following:

- Correctly rounded arithmetic
- Consistent representation of floating-point numbers across machines
- Consistent and sensible treatment of exceptional situations

Suppose that we are using a 32-bit computer with IEEE standard floating-point arithmetic. There are exactly 23 bits of precision in the fraction field in a single-precision normalized number. By counting the hidden bit, this means that there are 24 bits in the significand and the **unit roundoff error** is $u = 2^{-24}$. In single precision, the **machine epsilon** is $\varepsilon_{\text{single}} = 2^{-23}$ because $1 + 2^{-23}$ is the *first* such single-precision number larger than 1. Since $2^{-23} \approx 1.19 \times 10^{-7}$, we can expect only approximately 6 accurate decimal digits in the output. This accuracy may be reduced further by errors of various types, such as roundoff errors in the arithmetic, truncation errors in the formulas used, and so on.

For example, when computing the single-precision approximation to π, we obtain 6 accurate digits: 3.14159. Converting and printing the 24-bit binary number results in an actual decimal number with more than six nonzero digits, but only the first six digits are considered accurate approximations to π.

The *first* double-precision number larger than 1 is $1 + 2^{-52}$. So the double-precision **machine epsilon** is $\varepsilon_{\text{double}} = 2^{-52}$. Since $2^{-52} \approx 2.22 \times 10^{-16}$, there are only approximately 15 accurate decimal digits in the output in the absence of errors. The fraction field has exactly 52 bits of precision, and this results in 53 bits in the significand when the hidden bit is counted.

For example, when approximating π in double precision, we obtain 15 accurate digits: 3.14159 26535 8979. As in the case with single precision, converting and printing the 54-bit binary significand results in more than 15 digits, but only the first 15 digits are accurate approximations to π.

There are some useful special numbers in the IEEE standard. Instead of terminating with an overflow when dividing a nonzero number by 0, the machine representation for ∞ is stored, which is the mathematically sensible thing to do. Because of the hidden bit representation, a special technique for storing zero is necessary. Note that all zeros in the fraction field (mantissa) represent the significand 1.0 rather than 0.0. Moreover, there are two different representations for the same number zero, namely, $+0$ and -0. On the other hand, there are two different representations for infinity that correspond to two quite different numbers, $+\infty$ and $-\infty$. **NaN** stands for **Not a Number** and is an error pattern rather than a number.

Is it possible to represent numbers smaller than the smallest normalized floating-point number 2^{-126} *in IEEE standard floating-point format?*

Yes! If the exponent field contains a bit string of all zeros and the fraction field contains a nonzero bit string, then this representation is called a **subnormal number**. Subnormal numbers cannot be normalized because this would result in an exponent that does not fit into the exponent field. These subnormal numbers are less accurate than normal numbers because they have less room in the fraction field for nonzero bits.

By using various system inquiry functions (such as those in Table C.1 from Fortran), we can determine some of the characteristics of the floating-point number system on a typical PC with 32-bit IEEE standard floating-point arithmetic. Table C.2 contains the results. In most cases, simple programs can also be written to determine these values.

TABLE C.1 Some Numeric Inquiry Functions in Fortran

`EPSILON(X)`	Machine epsilon (number almost negligible compared to 1)
`TINY(X)`	Smallest positive number
`HUGE(X)`	Largest number
`PRECISION(X)`	Decimal precision (number of significant decimal digits in output)

TABLE C.2 Results with IEEE Standard Floating-Point on 32-Bit Machine

	`X` **Single Precision**	`X` **Double Precision**
`EPSILON(X)`	$1.192 \times 10^{-7} \approx 2^{-23}$	$2.220 \times 10^{-16} \approx 2^{-52}$
`TINY(X)`	$1.175 \times 10^{-38} \approx 2^{-126}$	$2.225 \times 10^{-308} \approx (2 - 2^{-23}) \times 2^{127}$
`HUGE(X)`	$3.403 \times 10^{38} \approx 2^{128}$	$1.798 \times 10^{308} \approx 2^{1024}$
`PRECISION(X)`	6	15

In Table C.3 (p. 628), we show the relationship between the exponent field and the possible single-precision 32-bit floating-points numbers corresponding to it. In this table, all lines except the first and the last are normalized floating-point numbers. The first line shows that zero is represented by $+0$ when all bits $b_i = 0$, and by -0 when all bits are zero except $b_1 = 1$. The last line shows that $+\infty$ and $-\infty$ have bit strings of all ones in the exponent field except for possibly the sign bit together with all zeros in the mantissa field.

In the IEEE floating-point standard, the **round to nearest** or **correctly rounded** value of the real number x, denoted round(x), is defined as follows. First, let x_+ be the closest floating-point number greater than x, and let x_- be the closest one less than x. If x is a

TABLE C.3 Single-Precision 32-Bit Word $\boxed{b_1\ b_2 b_3 b_4 \cdots b_9\ b_{10} b_{11} \cdots b_{32}}$ with Sign Bit $b_1 = 0$ for $+$ and $b_1 = 1$ for $-$.

$(b_2 b_3 \ldots b_9)_2$ Exponent Field	Numerical Representation
$(00000000)_2 = (0)_{10}$	$\begin{cases} \pm 0, & \text{if } b_{10} = b_{11} = \cdots = b_{32} = 0 \\ \text{subnormal}, & \text{otherwise} \end{cases}$
$(00000001)_2 = (1)_{10}$	$\pm(1.b_{10}b_{11}b_{12}\cdots b_{32})_2 \times 2^{-126}$
$(00000010)_2 = (2)_{10}$	$\pm(1.b_{10}b_{11}b_{12}\cdots b_{32})_2 \times 2^{-125}$
$(00000011)_2 = (3)_{10}$	$\pm(1.b_{10}b_{11}b_{12}\cdots b_{32})_2 \times 2^{-124}$
$(00000100)_2 = (4)_{10}$	$\pm(1.b_{10}b_{11}b_{12}\cdots b_{32})_2 \times 2^{-123}$
$\vdots$	$\vdots$
$(01111101)_2 = (125)_{10}$	$\pm(1.b_{10}b_{11}b_{12}\cdots b_{32})_2 \times 2^{-2}$
$(01111110)_2 = (126)_{10}$	$\pm(1.b_{10}b_{11}b_{12}\cdots b_{32})_2 \times 2^{-1}$
$(01111111)_2 = (127)_{10}$	$\pm(1.b_{10}b_{11}b_{12}\cdots b_{32})_2 \times 2^{0}$
$(10000000)_2 = (128)_{10}$	$\pm(1.b_{10}b_{11}b_{12}\cdots b_{32})_2 \times 2^{1}$
$(10000001)_2 = (129)_{10}$	$\pm(1.b_{10}b_{11}b_{12}\cdots b_{32})_2 \times 2^{2}$
$\vdots$	$\vdots$
$(11111011)_2 = (251)_{10}$	$\pm(1.b_{10}b_{11}b_{12}\cdots b_{32})_2 \times 2^{124}$
$(11111100)_2 = (252)_{10}$	$\pm(1.b_{10}b_{11}b_{12}\cdots b_{32})_2 \times 2^{125}$
$(11111101)_2 = (253)_{10}$	$\pm(1.b_{10}b_{11}b_{12}\cdots b_{32})_2 \times 2^{126}$
$(11111110)_2 = (254)_{10}$	$\pm(1.b_{10}b_{11}b_{12}\cdots b_{32})_2 \times 2^{127}$
$(11111111)_2 = (255)_{10}$	$\begin{cases} \pm\infty, & \text{if } b_{10} = b_{11} = \cdots = b_{32} = 0 \\ \texttt{NaN}, & \text{otherwise} \end{cases}$

floating-point number, then $\text{round}(x) = x$. Otherwise, the value of $\text{round}(x)$ depends on the **rounding mode** selected:

Rounding Modes

- **Round to nearest:** $\text{round}(x)$ is either x_- or x_+, whichever is nearer to x. (If there is a tie, choose the one with the least significant bit equal to 0.)
- **Round toward 0:** $\text{round}(x)$ is either x_- or x_+, whichever is between 0 and x.
- **Round toward $-\infty$/round down:** $\text{round}(x) = x_-$.
- **Round toward $+\infty$/round up:** $\text{round}(x) = x_+$.

Round to nearest is almost always used, since it is the most useful and gives the floating-point number closest to x.

D

Linear Algebra Concepts and Notation

In this appendix, we review some basic concepts and standard notation used in linear algebra.

D.1 Elementary Concepts

The two concepts from linear algebra that we are most concerned with are *vectors* and *matrices* because of their usefulness in compressing complicated expressions into a compact notation. The vectors and matrices in this text are most often *real*, since they consist of real numbers. These concepts easily generalize to *complex* vectors and matrices.

Vectors

A **vector** $\boldsymbol{x} \in \mathbb{R}^n$ can be thought of as a one-dimensional array of numbers and is written as

Column Vector

$$\boldsymbol{x} = \begin{bmatrix} x_1 \\ x_2 \\ \vdots \\ x_n \end{bmatrix}$$

where x_i is called the i-**th element**, **entry**, or **component**. An alternative notation that is useful in pseudocodes is $\boldsymbol{x} = (x_i)_n$. Sometimes the vector $\boldsymbol{x}$ displayed above is said to be a **column vector** to distinguish it from a **row vector** $\boldsymbol{y}$ written as

Row Vector

$$\boldsymbol{y} = [y_1, y_2, \ldots, y_n]$$

For example, here are some vectors:

$$\begin{bmatrix} \frac{1}{5} \\ 3 \\ -\frac{5}{6} \\ \frac{2}{7} \end{bmatrix}, \qquad [\pi, \quad e, \quad 5, \quad -4], \qquad \begin{bmatrix} \frac{1}{2} \\ \frac{1}{3} \end{bmatrix}$$

To save space, a column vector $\boldsymbol{x}$ can be written as a row vector such as

Transpose

$$\boldsymbol{x} = [x_1, x_2, \ldots, x_n]^T \qquad \text{or} \qquad \boldsymbol{x}^T = [x_1, x_2, \ldots, x_n]$$

by adding a T (for **transpose**) to indicate that we are interchanging or transposing a row or column vector. As an example, we have

$$[1 \quad 2 \quad 3 \quad 4]^T = \begin{bmatrix} 1 \\ 2 \\ 3 \\ 4 \end{bmatrix}$$

Many operations involving vectors are **component-by-component operations**. For vectors $\boldsymbol{x}$ and $\boldsymbol{y}$

Column Vectors

$$\boldsymbol{x} = \begin{bmatrix} x_1 \\ x_2 \\ \vdots \\ x_n \end{bmatrix}, \qquad \boldsymbol{y} = \begin{bmatrix} y_1 \\ y_2 \\ \vdots \\ y_n \end{bmatrix}$$

the following definitions apply:

Equality $\boldsymbol{x} = \boldsymbol{y}$ if and only if $x_i = y_i$ for all i $(1 \leqq i \leqq n)$

Inequality $\boldsymbol{x} < \boldsymbol{y}$ if and only if $x_i < y_i$ for all i $(1 \leqq i \leqq n)$

Addition/Subtraction

$$\boldsymbol{x} \pm \boldsymbol{y} = \begin{bmatrix} x_1 \pm y_1 \\ x_2 \pm y_2 \\ \vdots \\ x_n \pm y_n \end{bmatrix}$$

Scalar Product

$$\alpha\boldsymbol{x} = \begin{bmatrix} \alpha x_1 \\ \alpha x_2 \\ \vdots \\ \alpha x_n \end{bmatrix}, \quad \text{for } \alpha \text{ a constant or scalar}$$

Here is an example:

$\boldsymbol{y} = a\boldsymbol{x} + \boldsymbol{b}$

$$\begin{bmatrix} 2 \\ 4 \\ 6 \\ 8 \end{bmatrix} = 2\begin{bmatrix} 0 \\ 2 \\ 0 \\ 4 \end{bmatrix} + \begin{bmatrix} 2 \\ 0 \\ 6 \\ 0 \end{bmatrix}.$$

For m vectors $\boldsymbol{x}^{(1)}, \boldsymbol{x}^{(2)}, \ldots, \boldsymbol{x}^{(m)}$ and m scalars $\alpha_1, \alpha_2, \ldots, \alpha_m$, we define a **linear combination** as

Linear Combination

$$\sum_{i=1}^{m} \alpha_i \boldsymbol{x}^{(i)} = \alpha_1\boldsymbol{x}^{(1)} + \alpha_2\boldsymbol{x}^{(2)} + \cdots + \alpha_m\boldsymbol{x}^{(m)} = \begin{bmatrix} \sum_{i=1}^{m} \alpha_i x_1^{(i)} \\ \sum_{i=1}^{m} \alpha_i x_2^{(i)} \\ \vdots \\ \sum_{i=1}^{m} \alpha_i x_n^{(i)} \end{bmatrix}$$

Special vectors are the standard **unit vectors**:

Unit Vectors

$$\boldsymbol{e}^{(1)} = \begin{bmatrix} 1 \\ 0 \\ 0 \\ \vdots \\ 0 \end{bmatrix}, \quad \boldsymbol{e}^{(2)} = \begin{bmatrix} 0 \\ 1 \\ 0 \\ \vdots \\ 0 \end{bmatrix}, \quad \ldots, \quad \boldsymbol{e}^{(n)} = \begin{bmatrix} 0 \\ 0 \\ 0 \\ \vdots \\ 1 \end{bmatrix}$$

Clearly, we obtain

Using Unit Vectors

$$\sum_{i=1}^{n} \alpha_i \boldsymbol{e}^{(i)} = \begin{bmatrix} \alpha_1 \\ \alpha_2 \\ \vdots \\ \alpha_n \end{bmatrix}$$

Hence, any vector $\boldsymbol{x}$ can be written as a linear combination of the standard unit vectors

$$\boldsymbol{x} = x_1\boldsymbol{e}^{(1)} + x_2\boldsymbol{e}^{(2)} + \cdots + x_n\boldsymbol{e}^{(n)} = \sum_{i=1}^{n} x_i \boldsymbol{e}^{(i)}$$

As an example, notice that

$$\begin{bmatrix} 1 \\ 2 \\ 3 \\ 4 \end{bmatrix} = \begin{bmatrix} 1 \\ 0 \\ 0 \\ 0 \end{bmatrix} + 2\begin{bmatrix} 0 \\ 1 \\ 0 \\ 0 \end{bmatrix} + 3\begin{bmatrix} 0 \\ 0 \\ 1 \\ 0 \end{bmatrix} + 4\begin{bmatrix} 0 \\ 0 \\ 0 \\ 1 \end{bmatrix}$$

The **dot product** or **inner product** of vectors $\boldsymbol{x}$ and $\boldsymbol{y}$ is the number

Dot Product

$$\boldsymbol{x}^T\boldsymbol{y} = [x_1, \quad x_2, \quad \ldots, \quad x_n] \begin{bmatrix} y_1 \\ y_2 \\ \vdots \\ y_n \end{bmatrix} = \sum_{i=1}^{n} x_i y_i$$

As an example, we see that

$$[1, \quad 1, \quad 1, \quad 1] \begin{bmatrix} 1 \\ 1 \\ 1 \\ 1 \end{bmatrix} = 4$$

Matrices

A **matrix** is a two-dimensional array of numbers that can be written as

Matrix $m \times n$

$$\boldsymbol{A} = \begin{bmatrix} a_{11} & a_{12} & \cdots & a_{1m} \\ a_{21} & a_{22} & \cdots & a_{2m} \\ \vdots & \vdots & \ddots & \vdots \\ a_{n1} & a_{n2} & \cdots & a_{nm} \end{bmatrix}$$

where a_{ij} is called the **element** or **entry** in the ith row and jth column. An alternative notation is $\boldsymbol{A} = (a_{ij})_{n\times m}$. A column vector is also an $n \times 1$ matrix and a row vector is also a $1 \times m$ matrix. For example, here are three matrices:

$$\begin{bmatrix} \frac{1}{5} & \frac{2}{7} & -1 \\ 3 & 2 & \frac{1}{8} \\ -\frac{5}{6} & \frac{2}{5} & 3 \end{bmatrix}, \qquad \begin{bmatrix} 1 & 6 & \frac{9}{8} & -5 \end{bmatrix}, \qquad \begin{bmatrix} \frac{11}{2} & \frac{4}{9} \\ \frac{2}{3} & -\frac{7}{8} \\ \pi & e \\ \frac{1}{\pi} & \frac{1}{e} \end{bmatrix}$$

The entries in $\boldsymbol{A}$ can be grouped into column vectors:

Matrix by Columns

$$\boldsymbol{A} = \begin{bmatrix} \begin{bmatrix} a_{11} \\ a_{21} \\ \vdots \\ a_{n1} \end{bmatrix} & \begin{bmatrix} a_{12} \\ a_{22} \\ \vdots \\ a_{n2} \end{bmatrix} & \cdots & \begin{bmatrix} a_{1m} \\ a_{2m} \\ \vdots \\ a_{nm} \end{bmatrix} \end{bmatrix} = \begin{bmatrix} \boldsymbol{a}^{(1)} & \boldsymbol{a}^{(2)} & \cdots & \boldsymbol{a}^{(m)} \end{bmatrix}$$

where $\boldsymbol{a}^{(j)}$ is the jth column vector. Also, $\boldsymbol{A}$ can be grouped into row vectors:

Matrix by Rows

$$\boldsymbol{A} = \begin{bmatrix} [a_{11} & a_{12} & \cdots & a_{1m}] \\ [a_{21} & a_{22} & \cdots & a_{2m}] \\ & & \vdots & \\ [a_{n1} & a_{n2} & \cdots & a_{nm}] \end{bmatrix} = \begin{bmatrix} \boldsymbol{A}^{(1)} \\ \boldsymbol{A}^{(2)} \\ \vdots \\ \boldsymbol{A}^{(n)} \end{bmatrix}$$

where $\boldsymbol{A}^{(i)}$ is the ith row vector. Notice that

$$\begin{bmatrix} 1 & 5 & 9 & 13 \\ 2 & 6 & 10 & 14 \\ 3 & 7 & 11 & 15 \\ 4 & 8 & 12 & 16 \end{bmatrix} = \begin{bmatrix} \begin{bmatrix} 1 \\ 2 \\ 3 \\ 4 \end{bmatrix} & \begin{bmatrix} 5 \\ 6 \\ 7 \\ 8 \end{bmatrix} & \begin{bmatrix} 9 \\ 10 \\ 11 \\ 12 \end{bmatrix} & \begin{bmatrix} 13 \\ 14 \\ 15 \\ 16 \end{bmatrix} \end{bmatrix} = \begin{bmatrix} [1 & 5 & 9 & 13] \\ [2 & 6 & 10 & 14] \\ [3 & 7 & 11 & 15] \\ [4 & 8 & 12 & 16] \end{bmatrix}$$

An $n \times n$ matrix of special importance is the **identity** matrix, denoted by $\boldsymbol{I}$, composed of all 0's, except that the main diagonal consists of 1's:

Identity Matrix

$$\boldsymbol{I} = \begin{bmatrix} 1 & 0 & \cdots & 0 \\ 0 & 1 & \cdots & 0 \\ \vdots & \vdots & \ddots & \vdots \\ 0 & 0 & \cdots & 1 \end{bmatrix} = \begin{bmatrix} \boldsymbol{e}^{(1)} & \boldsymbol{e}^{(2)} & \cdots & \boldsymbol{e}^{(n)} \end{bmatrix}$$

A matrix of this same general form with entries d_i on the main diagonal is called a **diagonal** matrix and is written as

Diagonal Matrix

$$\boldsymbol{D} = \begin{bmatrix} d_1 & & & \\ & d_2 & & \\ & & \ddots & \\ & & & d_n \end{bmatrix} = \text{Diag}(d_1, d_2, \ldots, d_n)$$

where the blank space indicates 0 entries. A **tridiagonal** matrix is a square matrix of the form

Tridiagonal Matrix

$$\boldsymbol{T} = \begin{bmatrix} d_1 & c_1 & & & & \\ a_1 & d_2 & c_2 & & & \\ & a_2 & d_3 & c_3 & & \\ & & \ddots & \ddots & \ddots & \\ & & & a_{n-2} & d_{n-1} & c_{n-1} \\ & & & & a_{n-1} & d_n \end{bmatrix}$$

where the diagonal entries $\{a_i\}$, $\{d_i\}$, and $\{c_i\}$ are called the **subdiagonal**, **main diagonal**, and **superdiagonal**, respectively.

For the general $n \times n$ matrix $\boldsymbol{A} = (a_{ij})$, $\boldsymbol{A}$ is a diagonal matrix if $a_{ij} = 0$ whenever $i \neq j$, and $\boldsymbol{A}$ is a tridiagonal matrix if $a_{ij} = 0$ whenever $|i - j| \geq 2$. The matrix $\boldsymbol{A}$ is a **lower triangular matrix** whenever $a_{ij} = 0$ for all $i < j$ and is an **upper triangular matrix** whenever $a_{ij} = 0$ for all $i > j$. Examples of identity, diagonal, tridiagonal, lower triangular, and upper triangular matrices, respectively, are as follows:

$$\begin{bmatrix} 1 & 0 & 0 & 0 \\ 0 & 1 & 0 & 0 \\ 0 & 0 & 1 & 0 \\ 0 & 0 & 0 & 1 \end{bmatrix}, \quad \begin{bmatrix} 3 & 0 & 0 & 0 \\ 0 & 5 & 0 & 0 \\ 0 & 0 & 7 & 0 \\ 0 & 0 & 0 & 9 \end{bmatrix}, \quad \begin{bmatrix} 5 & 3 & 0 & 0 & 0 \\ 2 & 5 & 3 & 0 & 0 \\ 0 & 2 & 9 & 2 & 0 \\ 0 & 0 & 3 & 7 & 2 \\ 0 & 0 & 0 & 3 & 7 \end{bmatrix}$$

$$\begin{bmatrix} 6 & 0 & 0 & 0 \\ 3 & 6 & 0 & 0 \\ 4 & -2 & 7 & 0 \\ 5 & -3 & 9 & 21 \end{bmatrix}, \quad \begin{bmatrix} 1 & -1 & 2 & 1 \\ 0 & 5 & -5 & 1 \\ 0 & 0 & 9 & -3 \\ 0 & 0 & 0 & 2 \end{bmatrix}$$

As with vectors, many operations involving matrices correspond to component operations. For matrices $\boldsymbol{A}$ and $\boldsymbol{B}$, we write

$$\boldsymbol{A} = \begin{bmatrix} a_{11} & a_{12} & \cdots & a_{1m} \\ a_{21} & a_{22} & \cdots & a_{2m} \\ \vdots & \vdots & \ddots & \vdots \\ a_{n1} & a_{n2} & \cdots & a_{nm} \end{bmatrix}, \qquad \boldsymbol{B} = \begin{bmatrix} b_{11} & b_{12} & \cdots & b_{1m} \\ b_{21} & b_{22} & \cdots & b_{2m} \\ \vdots & \vdots & \ddots & \vdots \\ b_{n1} & b_{n2} & \cdots & b_{nm} \end{bmatrix}$$

the following definitions apply:

Equality $\quad \boldsymbol{A} = \boldsymbol{B}$ if and only if $a_{ij} = b_{ij}$ for all $i\,(1 \leqq i \leqq n)$ and all $j\ (1 \leqq j \leqq m)$

Inequality $\quad \boldsymbol{A} < \boldsymbol{B}$ if and only if $a_{ij} < b_{ij}$ for all $i\,(1 \leqq i \leqq n)$ and all $j\ (1 \leqq j \leqq m)$

Addition/Subtraction

$$\boldsymbol{A} \pm \boldsymbol{B} = \begin{bmatrix} a_{11} \pm b_{11} & a_{12} \pm b_{12} & \cdots & a_{1m} \pm b_{1m} \\ a_{21} \pm b_{21} & a_{22} \pm b_{22} & \cdots & a_{2m} \pm b_{2m} \\ \vdots & \vdots & \ddots & \vdots \\ a_{n1} \pm b_{n1} & a_{n2} \pm b_{n2} & \cdots & a_{nm} \pm b_{nm} \end{bmatrix}$$

Scalar Product

$$\alpha \boldsymbol{A} = \begin{bmatrix} \alpha a_{11} & \alpha a_{12} & \cdots & \alpha a_{1m} \\ \alpha a_{21} & \alpha a_{22} & \cdots & \alpha a_{2m} \\ \vdots & \vdots & \ddots & \vdots \\ \alpha a_{n1} & \alpha a_{n2} & \cdots & \alpha a_{nm} \end{bmatrix} \quad \text{for } \alpha \text{ a constant}$$

As an example, we have

$$\begin{bmatrix} \frac{1}{5} & \frac{7}{5} & -1 \\ -3 & 2 & -8 \\ \frac{6}{5} & \frac{2}{5} & -3 \end{bmatrix} = \frac{1}{5}\begin{bmatrix} 1 & 7 & 0 \\ 0 & 10 & 0 \\ 6 & 2 & 0 \end{bmatrix} - \begin{bmatrix} 0 & 0 & 1 \\ 3 & 0 & 8 \\ 0 & 0 & 3 \end{bmatrix}$$

Matrix-Vector Product

The product of an $n \times m$ matrix $\boldsymbol{A}$ and an $m \times 1$ vector $\boldsymbol{b}$ is of special interest. Considering the matrix $\boldsymbol{A}$ in terms of its columns, we have

Matrix-Vector Product by Columns of *A*

$$\boldsymbol{Ab} = \begin{bmatrix} \boldsymbol{a}^{(1)} & \boldsymbol{a}^{(2)} & \cdots & \boldsymbol{a}^{(m)} \end{bmatrix} \begin{bmatrix} b_1 \\ b_2 \\ \vdots \\ b_m \end{bmatrix}$$

$$= b_1\boldsymbol{a}^{(1)} + b_2\boldsymbol{a}^{(2)} + \cdots + b_m\boldsymbol{a}^{(m)}$$

$$= \sum_{i=1}^{m} b_i \boldsymbol{a}^{(i)}$$

Thus, $\boldsymbol{Ab}$ is a vector and can be thought of as a linear combination of the columns of $\boldsymbol{A}$ with coefficients the entries of $\boldsymbol{b}$. Considering matrix $\boldsymbol{A}$ in terms of its rows, we have

Matrix-Vector Product by Rows of *A*

$$\boldsymbol{Ab} = \begin{bmatrix} \boldsymbol{A}^{(1)} \\ \boldsymbol{A}^{(2)} \\ \vdots \\ \boldsymbol{A}^{(n)} \end{bmatrix} \boldsymbol{b} = \begin{bmatrix} \boldsymbol{A}^{(1)}\boldsymbol{b} \\ \boldsymbol{A}^{(2)}\boldsymbol{b} \\ \vdots \\ \boldsymbol{A}^{(n)}\boldsymbol{b} \end{bmatrix}$$

Thus, the jth element of $\boldsymbol{Ab}$ can be viewed as the **dot product** of the jth row of $\boldsymbol{A}$ and the vector $\boldsymbol{b}$.

Matrix Product

The product of the matrix $\boldsymbol{A} = (a_{ij})_{n\times m}$ and the matrix $\boldsymbol{B} = (b_{ij})_{m\times r}$ is the matrix $\boldsymbol{C} = (c_{ij})_{n\times r}$ such that

Matrix-Matrix Product

$$\boldsymbol{AB} = \boldsymbol{C}$$

where

$$c_{ij} = a_{i1}b_{1j} + a_{i2}b_{2j} + \cdots + a_{im}b_{mj} = \sum_{k=1}^{m} a_{ik}b_{kj} \qquad (1 \leqq i \leqq n,\ 1 \leqq j \leqq r)$$

The element c_{ij} is the **dot product** of the ith row vector of $\boldsymbol{A}$

Row Vector

$$\boldsymbol{A}^{(i)} = [a_{i1}, a_{i2}, \ldots, a_{im}]$$

and the jth column vector of $\boldsymbol{B}$

Column Vector

$$\boldsymbol{b}^{(j)} = \begin{bmatrix} b_{1j} \\ b_{2j} \\ \vdots \\ b_{mj} \end{bmatrix}$$

that is,

$$c_{ij} = \boldsymbol{A}^{(i)}\boldsymbol{b}^{(j)}$$

Similarly, the matrix product $\boldsymbol{AB}$ can be thought of in two different ways. We can write either

Matrix-Matrix Product Post-Multiplication

$$\begin{aligned} \boldsymbol{AB} &= \boldsymbol{A}\left[\boldsymbol{b}^{(1)} \quad \boldsymbol{b}^{(2)} \quad \cdots \quad \boldsymbol{b}^{(r)}\right] \\ &= \left[\boldsymbol{Ab}^{(1)} \quad \boldsymbol{Ab}^{(2)} \quad \cdots \quad \boldsymbol{Ab}^{(r)}\right] \\ &= \boldsymbol{C} \end{aligned} \tag{1}$$

or

Matrix-Matrix Product Pre-Multiplication

$$\boldsymbol{AB} = \begin{bmatrix} \boldsymbol{A}^{(1)} \\ \boldsymbol{A}^{(2)} \\ \vdots \\ \boldsymbol{A}^{(n)} \end{bmatrix} \boldsymbol{B} = \begin{bmatrix} \boldsymbol{A}^{(1)}\boldsymbol{B} \\ \boldsymbol{A}^{(2)}\boldsymbol{B} \\ \vdots \\ \boldsymbol{A}^{(n)}\boldsymbol{B} \end{bmatrix} = \boldsymbol{C} \tag{2}$$

Equation (1) implies that the jth column of $\boldsymbol{C} = \boldsymbol{AB}$ is

Post-Multiplication

$$\boldsymbol{c}^{(j)} = \boldsymbol{Ab}^{(j)}$$

That is, each column of $\boldsymbol{C}$ is the result of **postmultiplying** $\boldsymbol{A}$ by the jth column of $\boldsymbol{B}$. In other words, each column of $\boldsymbol{C}$ can be obtained by taking inner products of a column of $\boldsymbol{B}$ with all rows of $\boldsymbol{A}$:

$$\boldsymbol{c}^{(j)} = \boldsymbol{Ab}^{(j)} = \begin{bmatrix} \longleftarrow \\ \longleftarrow \\ \vdots \\ \longleftarrow \end{bmatrix} \begin{bmatrix} b_{1j} \\ b_{2j} \\ \vdots \\ b_{mj} \end{bmatrix} = \begin{bmatrix} c_{1j} \\ c_{2j} \\ \vdots \\ c_{nj} \end{bmatrix}$$

The long left-arrow means an inner product is formed across the elements in the row; that is, $c_{ij} = \sum_{k=1}^{n} a_{ik} b_{kj}$. Equation (2) implies that the ith row of the result $\boldsymbol{C}$ of multiplying $\boldsymbol{A}$ times $\boldsymbol{B}$ is

Pre-Multiplication

$$\boldsymbol{C}^{(i)} = \boldsymbol{A}^{(i)}\boldsymbol{B}$$

That is, each row of $\boldsymbol{C}$ is the result of **premultiplying** $\boldsymbol{B}$ by the ith row of $\boldsymbol{A}$. In other words, each row of $\boldsymbol{C}$ can be obtained by taking inner products of a row of $\boldsymbol{A}$ with all columns of $\boldsymbol{B}$:

$[\uparrow \;\; \uparrow \;\; \cdots \;\; \uparrow]$

$$\begin{aligned} \boldsymbol{C}^{(i)} = \boldsymbol{A}^{(i)}\boldsymbol{B} &= [a_{i1} \quad a_{i2} \quad \cdots \quad a_{im}]\,[\uparrow \quad \uparrow \quad \cdots \quad \uparrow] \\ &= [c_{i1} \quad c_{i2} \quad \cdots \quad c_{ir}] \end{aligned}$$

The long up-arrow means an inner product is formed from the elements in the column; that is, $c_{ij} = \sum_{k=1}^{n} a_{ik} b_{kj}$.

As an example, we can determine the matrix product **columnwise** as

Example Columnwise

$$\begin{bmatrix} 3 & 1 & 7 \\ 2 & 4 & -5 \\ 1 & -3 & 2 \end{bmatrix} \begin{bmatrix} -1 & -3 & 2 \\ 1 & 1 & 1 \\ -3 & -2 & 1 \end{bmatrix} = \left[\boldsymbol{c}^{(1)} \quad \boldsymbol{c}^{(2)} \quad \boldsymbol{c}^{(3)}\right]$$

where

$$\boldsymbol{c}^{(1)} = \begin{bmatrix} 3 & 1 & 7 \\ 2 & 4 & -5 \\ 1 & -3 & 2 \end{bmatrix} \begin{bmatrix} -1 \\ 1 \\ -3 \end{bmatrix} = \begin{bmatrix} -23 \\ 17 \\ -10 \end{bmatrix}$$

$$c^{(2)} = \begin{bmatrix} 3 & 1 & 7 \\ 2 & 4 & -5 \\ 1 & -3 & 2 \end{bmatrix} \begin{bmatrix} -3 \\ 1 \\ -2 \end{bmatrix} = \begin{bmatrix} -22 \\ 8 \\ -10 \end{bmatrix}$$

$$c^{(3)} = \begin{bmatrix} 3 & 1 & 7 \\ 2 & 4 & -5 \\ 1 & -3 & 2 \end{bmatrix} \begin{bmatrix} 2 \\ 1 \\ 1 \end{bmatrix} = \begin{bmatrix} 14 \\ 3 \\ 1 \end{bmatrix}$$

or we can determine it **rowwise** as

Example Rowwise

$$\begin{bmatrix} 3 & 1 & 7 \\ 2 & 4 & -5 \\ 1 & -3 & 2 \end{bmatrix} \begin{bmatrix} -1 & -3 & 2 \\ 1 & 1 & 1 \\ -3 & -2 & 1 \end{bmatrix} = \begin{bmatrix} \boldsymbol{C}^{(1)} \\ \boldsymbol{C}^{(2)} \\ \boldsymbol{C}^{(3)} \end{bmatrix}$$

where

$$\boldsymbol{C}^{(1)} = [3 \quad 1 \quad 7] \begin{bmatrix} -1 & -3 & 2 \\ 1 & 1 & 1 \\ -3 & -2 & 1 \end{bmatrix} = [-23 \quad -22 \quad 14]$$

$$\boldsymbol{C}^{(2)} = [2 \quad 4 \quad -5] \begin{bmatrix} -1 & -3 & 2 \\ 1 & 1 & 1 \\ -3 & -2 & 1 \end{bmatrix} = [17 \quad 8 \quad 3]$$

$$\boldsymbol{C}^{(3)} = [1 \quad -3 \quad 2] \begin{bmatrix} -1 & -3 & 2 \\ 1 & 1 & 1 \\ -3 & -2 & 1 \end{bmatrix} = [-10 \quad -10 \quad 1]$$

Other Concepts

The **transpose** of the $n \times m$ matrix $\boldsymbol{A}$, denoted $\boldsymbol{A}^T$, is obtained by interchanging the rows and columns of $\boldsymbol{A} = (a_{ij})_{n\times m}$:

Matrix Transpose by Rows

$$\boldsymbol{A}^T = \begin{bmatrix} \boldsymbol{A}^{(1)} \\ \boldsymbol{A}^{(2)} \\ \vdots \\ \boldsymbol{A}^{(n)} \end{bmatrix}^T = \begin{bmatrix} \boldsymbol{A}^{(1)^T} & \boldsymbol{A}^{(2)^T} & \cdots & \boldsymbol{A}^{(n)^T} \end{bmatrix}$$

or

Matrix Transpose by Columns

$$\boldsymbol{A}^T = \begin{bmatrix} \boldsymbol{a}^{(1)} & \boldsymbol{a}^{(2)} & \cdots & \boldsymbol{a}^{(m)} \end{bmatrix}^T = \begin{bmatrix} \boldsymbol{a}^{(1)^T} \\ \boldsymbol{a}^{(2)^T} \\ \vdots \\ \boldsymbol{a}^{(m)^T} \end{bmatrix}$$

Hence, $\boldsymbol{A}^T$ is the $m \times n$ matrix:

$$\boldsymbol{A}^T = \begin{bmatrix} a_{11} & a_{21} & \cdots & a_{n1} \\ a_{12} & a_{22} & \cdots & a_{n2} \\ \vdots & \vdots & & \vdots \\ a_{1m} & a_{2m} & \cdots & a_{nm} \end{bmatrix} = (a_{ji})_{m\times n}$$

As an example, we have

$$\begin{bmatrix} 2 & 4 & 9 \\ 5 & 7 & 3 \\ 10 & 6 & 2 \end{bmatrix}^T = \begin{bmatrix} 2 & 5 & 10 \\ 4 & 7 & 6 \\ 9 & 3 & 2 \end{bmatrix}$$

An $n \times n$ matrix $\boldsymbol{A}$ is **symmetric** if $a_{ij} = a_{ji}$ for all i $(1 \leqq i \leqq n)$ and all j $(1 \leqq j \leqq n)$. In other words, $\boldsymbol{A}$ is symmetric if $\boldsymbol{A} = \boldsymbol{A}^T$.

Some useful properties for matrices of compatible sizes are as follows:

■ **Properties**

Elementary Consequences of the Definitions

1. $\boldsymbol{AB} \neq \boldsymbol{BA}$ (in general)
2. $\boldsymbol{AI} = \boldsymbol{IA} = \boldsymbol{A}$
3. $\boldsymbol{A0} = \boldsymbol{0A} = \boldsymbol{0}$
4. $\left(\boldsymbol{A}^T\right)^T = \boldsymbol{A}$
5. $(\boldsymbol{A} + \boldsymbol{B})^T = \boldsymbol{A}^T + \boldsymbol{B}^T$
6. $(\boldsymbol{AB})^T = \boldsymbol{B}^T \boldsymbol{A}^T$

If $\boldsymbol{A}$ and $\boldsymbol{B}$ are square matrices that satisfy

Matrix Inverse

$$\boldsymbol{AB} = \boldsymbol{BA} = \boldsymbol{I}$$

then $\boldsymbol{B}$ is said to be the **inverse** of $\boldsymbol{A}$, which is denoted $\boldsymbol{A}^{-1}$.

To illustrate Property 1, we can form the following products to observe that matrix multiplication is *not* commutative:

Example Matrice Multiplication not Commutative

$$\mathbf{AB} = \begin{bmatrix} 3 & 1 & 7 \\ 2 & 4 & -5 \\ 1 & -3 & 2 \end{bmatrix} \begin{bmatrix} -1 & -3 & 2 \\ 1 & 1 & 1 \\ -3 & -2 & 1 \end{bmatrix} = \begin{bmatrix} -23 & -22 & 14 \\ 17 & 8 & 3 \\ -10 & -10 & 1 \end{bmatrix}$$

$$\mathbf{BA} = \begin{bmatrix} -1 & -3 & 2 \\ 1 & 1 & 1 \\ -3 & -2 & 1 \end{bmatrix} \begin{bmatrix} 3 & 1 & 7 \\ 2 & 4 & -5 \\ 1 & -3 & 2 \end{bmatrix} = \begin{bmatrix} -7 & -19 & 12 \\ 6 & 2 & 4 \\ -12 & -14 & -9 \end{bmatrix}$$

Also, verify that $\boldsymbol{AA}^{-1} = \boldsymbol{A}^{-1}\boldsymbol{A} = \boldsymbol{I}$ for

$$\boldsymbol{A} = \begin{bmatrix} 1 & 1 & 1 \\ -1 & 3 & 2 \\ 2 & 1 & 1 \end{bmatrix}$$

and

Example Matrix Inverse

$$\boldsymbol{A}^{-1} = \begin{bmatrix} -1 & 0 & 1 \\ -5 & 1 & 3 \\ 7 & -1 & -4 \end{bmatrix}$$

As our final set of examples, we take the product of a matrix times a vector and the product of two matrices:

$$\boldsymbol{Ax} = \begin{bmatrix} 3 & 2 & -1 \\ 5 & 3 & 2 \\ -1 & 1 & -3 \end{bmatrix} \begin{bmatrix} x_1 \\ x_2 \\ x_3 \end{bmatrix} = \begin{bmatrix} 3x_1 & + & 2x_2 & - & x_3 \\ 5x_1 & + & 3x_2 & + & 2x_3 \\ -x_1 & + & x_2 & - & 3x_3 \end{bmatrix}$$

$$\boldsymbol{L}^{-1}\boldsymbol{A} = \begin{bmatrix} 1 & 0 & 0 \\ -\frac{5}{3} & 1 & 0 \\ -8 & 5 & 1 \end{bmatrix} \begin{bmatrix} 3 & 2 & -1 \\ 5 & 3 & 2 \\ -1 & 1 & -3 \end{bmatrix} = \begin{bmatrix} 3 & 2 & -1 \\ 0 & -\frac{1}{3} & \frac{11}{3} \\ 0 & 0 & 15 \end{bmatrix} = \boldsymbol{U}$$

The reader should show how these relate to solving a linear system using naive Gaussian elimination (see Sections 2.1 and 8.1):

$$\begin{cases} 3x_1 + 2x_2 - x_3 = 7 \\ 5x_1 + 3x_2 + 2x_3 = 4 \\ -x_1 + x_2 - 3x_3 = -1 \end{cases} \qquad (\boldsymbol{Ax} = \boldsymbol{b})$$

Next, compute the products shown and relate them to this problem:

$$\boldsymbol{L}^{-1}\boldsymbol{b} = \begin{bmatrix} 1 & 0 & 0 \\ -\frac{5}{3} & 1 & 0 \\ -8 & 5 & 1 \end{bmatrix} \begin{bmatrix} 7 \\ 4 \\ -1 \end{bmatrix} = \boldsymbol{y}$$

$$\boldsymbol{U}^{-1}\boldsymbol{U} = \begin{bmatrix} \frac{1}{3} & 2 & -\frac{7}{15} \\ 0 & -3 & \frac{11}{15} \\ 0 & 0 & \frac{1}{15} \end{bmatrix} \begin{bmatrix} 3 & 2 & -1 \\ 0 & -\frac{1}{3} & \frac{11}{3} \\ 0 & 0 & 15 \end{bmatrix} = \boldsymbol{I}$$

$$\boldsymbol{U}^{-1}\boldsymbol{y} = \begin{bmatrix} \frac{1}{3} & 2 & -\frac{7}{15} \\ 0 & -3 & \frac{11}{15} \\ 0 & 0 & \frac{1}{15} \end{bmatrix} \begin{bmatrix} 7 \\ -\frac{23}{3} \\ -37 \end{bmatrix} = \boldsymbol{x}$$

Cramer's Rule

The solution of a 2×2 linear system of the form

Solving 2×2 System

$$\begin{bmatrix} a & c \\ b & d \end{bmatrix} \begin{bmatrix} x \\ y \end{bmatrix} = \begin{bmatrix} f \\ g \end{bmatrix}$$

is given by

$$x = \frac{1}{D}\text{Det}\begin{bmatrix} f & c \\ g & d \end{bmatrix} = \frac{1}{D}(fd - gc)$$

$$y = \frac{1}{D}\text{Det}\begin{bmatrix} a & f \\ b & g \end{bmatrix} = \frac{1}{D}(ag - bf)$$

where

$$D = \text{Det}\begin{bmatrix} a & c \\ b & d \end{bmatrix} = ad - bc \neq 0$$

As an example, consider

Example 2×2 System

$$\begin{bmatrix} -1 & 1 \\ 2 & 1 \end{bmatrix} \begin{bmatrix} x \\ y \end{bmatrix} = \begin{bmatrix} 1 \\ 4 \end{bmatrix}$$

Using Cramer's rule, we have

$$D = \text{Det}\begin{bmatrix} -1 & 1 \\ 2 & 1 \end{bmatrix} = (-1)(1) - (2)(1) = -3$$

and

$$x = -\frac{1}{3}\text{Det}\begin{bmatrix} 1 & 1 \\ 4 & 1 \end{bmatrix} = -\frac{1}{3}[(1)(1) - (4)(1)] = 1$$

$$y = -\frac{1}{3}\text{Det}\begin{bmatrix} -1 & 1 \\ 2 & 4 \end{bmatrix} = -\frac{1}{3}[(-1)(4) - (2)(1)] = 2$$

Answers for Selected Exercises*

Exercises 1.1

2. $x = \dfrac{6032}{9990}$; $x = \dfrac{6032}{10010}$ **3.** 6×10^{-5}

4. Two other ways:

$pi \leftarrow 2.0 \arcsin(1.0)$ or $pi \leftarrow 2.0 \arccos(0.0)$

5a.
$sum \leftarrow 0$
for $i = 1$ **to** n
  for $j = 1$ **to** n
    $sum \leftarrow sum + a_{ij}$
  end for
end for

5d.
$sum \leftarrow 0.0$
for $i = 1$ **to** n
  $sum \leftarrow sum + a_{ii}$
end for
for $j = 2$ **to** n
  for $i = j$ **to** n
    $sum \leftarrow sum + a_{i,i-j+1} + a_{i-j+1,i}$
  end for
end for

6. n multiplications and n additions/subtractions

8a.
for $i = 1$ **to** 5
  $x \leftarrow x \cdot x$
end for
$p \leftarrow x$

8c.
$z \leftarrow x + 2$
$p \leftarrow z^3(6 + z^4(9 + z^8(3 - z^{16})))$

10.
$z \leftarrow a_n/b_n$
for $i = 1$ **to** $n - 1$
  $z \leftarrow a_{n-i}(z + 1/b_{n-i})$
end for

11b.
$z \leftarrow 1$
$v \leftarrow 1$
for $i = 1$ **to** $n - 1$
  $v \leftarrow vx$
  $z \leftarrow vz + 1$
end for
$z \leftarrow vxz$

12b. $v = \displaystyle\sum_{i=0}^{n} a_i x^i$

12e. $v = a_n x^n + x \displaystyle\sum_{i=1}^{n} a_{n-i} x^{n-i}$

13. $z = 1 + \displaystyle\sum_{i=2}^{n} \prod_{j=2}^{i} b_j$ **14.** $n(n + 1)/2$

15b.
for $j = 1$ **to** n
  for $i = 1$ **to** n
    $a_{ij} \leftarrow 1.0/\text{real}(i + j - 1)$
  end for
end for

Computer Exercises 1.1

4. $\exp(1.0) \approx 2.71828\,18284\,6$

9. Computation deviates from theory; for example, when $a_1 = 10^{-12}, 10^{-8}, 10^{-4}, 10^{20}$.

10. x may underflow and be set to zero.

12. 40 different spellings

20a. The computation m/n may result in truncation so that $x \neq y$.

Exercises 1.2

4a. First derivative $+\infty$ at 0.

4b. First derivative not continuous.

*Answers to exercises marked in the text with the symbol ^{a} are given here and in the *Student Solution Manual* with more details.

4e. Function $-\infty$ at 0.

5. $\cosh x = \sum_{k=0}^{\infty} \frac{x^{2k}}{(2k)!}$; $\cosh 0.7 \approx 1.25517$

6a. $e^{\cos x} = e\left(1 - \frac{x^2}{2} + \cdots\right)$

6b. $\sin(\cos x) = (\sin 1) - (\cos 1)\left(\frac{x^2}{2}\right) + \cdots$

7. $m = 2$ **8.** At least 18 terms

9. Yes. By using this formula, we avoid the series for e^{-x} and use the series for e^x.

11. $\ln(1 - x) = -\sum_{k=1}^{\infty} \frac{x^k}{k}$;

$$\ln\left(\frac{1+x}{1-x}\right) = 2\sum_{k=1}^{\infty} \frac{x^{2k-1}}{(2k-1)}$$

12. $x = \frac{1}{3}$; $\ln 2 = 0.69313$ (four terms). At least 10 terms.

15a. $\sin x + \cos x = 1 + x - \frac{x^2}{2} - \frac{x^3}{6} + \cdots$;
$\sin(0.001) + \cos(0.001) \approx 1.00099\,94998\,3$

15b. $(\sin x)(\cos x) = x - \frac{2}{3}x^3 + \frac{2}{15}x^5 - \frac{4}{315}x^7 + \cdots$;
$\sin(0.0006)\cos(0.0006) \approx 0.00059\,99998\,57$

16. $\ln(e + x) = 1 + \sum_{n=1}^{\infty} (-1)^{n-1}\frac{1}{n}\left(\frac{x}{e}\right)^n$

17. At least seven terms.

18. At least 100 terms. **20.** $-\frac{5}{8}h^4$

23. $\frac{1}{8}\left(x - \frac{17}{4}\right)$

24.
```
s ← 0
for i = 2 to n
    s ← s + log(i)
    output i, s
end for
```

28. $\left|\cos x - \left(1 - \frac{x^2}{2}\right)\right| < \frac{1}{16 \times 24} = \frac{1}{384}$

32. Maclaurin series:
$f(x) = 3 + 7x - 1.33x^2 + 19.2x^4$;

$$f(x) = 318.88 + (x-2)616.08 + \frac{(x-2)^2}{2!}918.94 + \frac{(x-2)^3}{3!}921.6 + \frac{(x-2)^4}{4!}460.8$$

35. 400 terms.

38. $$\cos\left(\frac{\pi}{3} + h\right) = \frac{1}{2}\sum_{k=0}^{\infty}(-1)^k\frac{h^{2k}}{(2k)!} + \frac{\sqrt{3}}{2}\sum_{k=1}^{\infty}(-1)^k\frac{h^{2k-1}}{(2k-1)!};$$
$\cos(60.001^\circ) \approx 0.49998\,488$

39. $\sin(45.0005^\circ) \approx 0.70711\,295$

42. $$f(x - h) = (x - h)^m = x^m - mhx^{m-1} + m(m-1)\frac{h^2}{2!}x^{m-2} + \cdots$$

47. $n = 16$ or $n = 17$

50b. $\lim_{x\to 0} \frac{\arctan x}{x} = 1$ **50c.** $\lim_{x\to\pi} \frac{\cos x + 1}{\sin x} = 0$

51. At least 38 terms.

52. $\text{erf}(x) = \frac{2}{\sqrt{\pi}}\left[x - \frac{x^3}{3} + \frac{x^5}{5(2!)} - \frac{x^7}{7(3!)} + \cdots\right]$;
$\text{erf}(1) \approx 0.8382$

53. 10^{10} **54.** 10^5

Computer Exercises 1.2

1.

	$c = 1$	$c = 10^8$
x_1	0	-1
x_2	-10^8	-10^8

14. g converges faster (in five iterations)

16. $\lambda_{50} = 1\,25862\,69025$ **17.** $\alpha_{50} = 2\,81437\,53123$

Exercises 1.3

1c. $[B5\,000000]_{16}$

2d. $[3FA\,0000000000000]_{16}$; $[BFA\,0000000000000]_{16}$

4d. $[3E7\,00000]_{16}$, **4e.** $[3FCE\,0000000000000]_{16}$

5d. $-\infty$ **8a.** -3.131968×10^6

8d. 9.992892×10^6 **8g.** -3.39×10^3

11c. $m = -1, 0, 1$. Nonnegative machine numbers:
$0, \frac{1}{8}, \frac{1}{4}, \frac{3}{8}, \frac{1}{2}, \frac{3}{4}, 1, \frac{3}{2}$

15. 1 **17.** 1.00005; 1.0 **18.** $|x| < 5 \times 10^{-5}$

19. β^{1-n} **21.** $\approx 3 \times 2^{-25}$ **25.** $\approx 3 \times 2^{-24}$

26. $\approx 2^{-22}$ **30.** $\approx n \times 2^{-24}$; $n = 1000, \approx 2^{-14}$

37. $\frac{1}{2} \times 10^{-12}$ rounding; 10^{-12} chopping **38.** 9%

39. The relative error cannot exceed 5×2^{-24}.

42. $\left((q - 2^{-25})2^m,\ (q + 2^{-25})2^m\right)$

Computer Exercises 1.3

3. 1.41423 56237 30922 58894 75783 33318 23348 99902 34375

7. On 32-bit computer, $s \leqq 33.2710$

8. Two limits are equal.

Exercises 1.4

4. $y = \dfrac{\cos^2 x}{1 + \sin x}$ **6.** $f(x) = -\frac{1}{2}x^3 - \frac{1}{2}x^4$; $f(0.0125) \approx -9.888 \times 10^{-7}$

8. $f(x) = \dfrac{1}{\sqrt{1+x^2}+1} + 3 - 1.7x^2$; $f(0) = 3.5$

10. $f(x) = \dfrac{1}{\sqrt{x^2+1}+x}$

11. $f(x) = \begin{cases} \ln\left(x + \sqrt{x^2+1}\right), & x > 0 \\ 0, & x = 0 \\ -\ln\left(-x + \sqrt{x^2+1}\right), & x < 0 \end{cases}$

13. $z = \dfrac{x^4}{\sqrt{x^4+4}+2}$

16. $f(x) \approx 1 - x + \dfrac{x^2}{3} - \dfrac{x^3}{6}$; $f(0.008) \approx 0.99202\,0915$

20. $\arctan x - x \approx x^3\left(-\frac{1}{3} + x^2\left(\frac{1}{5} + x^2\left(-\frac{1}{7}\right)\right)\right)$

22. $\left(e^{2x} - 1\right)/2x \approx 1 + x(1 + (x/3)(2 + x))$

24a. Near $\pi/2$, sine curve is relatively flat.

26b. $\ln x - 1 = \ln(x/e)$

26d. $x^{-2}(\sin x - e^x + 1) \approx -\dfrac{1}{2} - \dfrac{x}{3}$ when $x \to 0$

28. $|x| < \sqrt{6\varepsilon}$, where ε machine precision

29. $x_1 \approx 10^5$, $x_2 \approx 10^{-5}$

30. Not much. Expect to compute $b^2 - 4ac$ in double precision.

Computer Exercises 1.4

1. No solution; (0, 0); (0, 0);
Any solution; (−1., 0.);
(−0.10208 42383, −4.89791 57617);
(4.00000 00001, 4.0009 99999);
(−0.10208 42383, −4.89791 57617);
(1.0000 00000, 1.00000 0000E34);
(1.99683 77223, 2.00316 22777)

10.

x	Series	n
0	1.0	1
1	2.71828 18285	10
−1	0.36787 94412	10
0.5	1.64872 12707	8
−0.123	0.88426 36626	5
−25.5	$8.42346\,37545 \times 10^{-12}$	25
−1776	0	25
3.14159	23.14063 12270	17

14. $|x| < 10^{-15}$ **15.** $\rho_{50} = 2.85987$

Exercises 2.1

1. Homogeneous: $\alpha = 0$, zero solution; $\alpha = \pm 1$, infinite number of solutions

2. For $\alpha \approx 1$, erroneous answer is produced.

3a. No solution

3b. Infinite number of solutions

4. $\begin{cases} x_1 = -697.3 \\ x_2 = 343.9 \end{cases}$ $\begin{cases} x_1 = -720.79976 \\ x_2 = 356.28760 \end{cases}$

5. $\mathbf{r} = \begin{bmatrix} -0.001343 \\ -0.001572 \end{bmatrix}$, $\widehat{\mathbf{r}} = \begin{bmatrix} -0.0000001 \\ 0.0000000 \end{bmatrix}$,

$\mathbf{e} = \begin{bmatrix} -0.001 \\ -0.001 \end{bmatrix}$, $\widehat{\mathbf{e}} = \begin{bmatrix} -0.659 \\ 0.913 \end{bmatrix}$

6a. $x_2 = 1,\ x_1 = 0$ **6b.** $x_2 = 1,\ x_1 = 1$

6c. Let $b_1 = b_2 = 1$. Then $x_2 = 1,\ x_1 = 0$, which is exact.

7a. $x_1 = -0.2752 + 0.9174i$,
$x_2 = 5.4312 - 0.7706i$,
$x_3 = -3.3394 - 0.018i$

7b. $x_1 = 1.1927 - 0.6422i$,
$x_2 = -0.2018 + 3.3394i$,
$x_3 = 1.1376 - 2.4587i$

7c. $x_1 \approx -7.233, \quad x_2 \approx 1.133,$
$x_3 \approx 2.433, \quad x_4 = 4.5$

Computer Exercises 2.1

6. $z = [2i, i, i, i]^T, \lambda = 1 + 5i;$
$z = [1, 2, 1, 1]^T, \lambda = 2 + 6i;$
$z = [-i, -i, 0, -i]^T, \lambda = -3 - 7i;$
$z = [1, 1, 1, 0]^T, \lambda = -4 - 8i$

7a. $(3.75, 90°); \quad (3.27, -65.7°); \quad (0.775, 172.9°)$

7b. $(2.5, -90°); \quad (2.08, 56.3°); \quad (1.55, -60.2°)$

Exercises 2.2

1. $$\begin{bmatrix} 1/2 & 5/2 & -4 & -1 \\ 1/4 & -1/2 & -5/19 & -62/19 \\ 3/4 & 9/10 & 38/5 & 9/10 \\ 4 & 1 & 0 & 4 \end{bmatrix}$$

2. $\mathbf{x} = [1/3, 3, 1/3]^T$

3. $$\begin{bmatrix} 1 & 0 & 3 & 0 \\ 0 & 1 & 3 & -1 \\ 3 & -3 & 0 & 6 \\ 0 & 2 & 4 & -6 \end{bmatrix} \Rightarrow \begin{bmatrix} 0 & 1 & 3 & -2 \\ 0 & 1 & 3 & -1 \\ 3 & -3 & 0 & 6 \\ 0 & 2 & 4 & -6 \end{bmatrix}$$
$$\Rightarrow \begin{bmatrix} 0 & 1 & 3 & -2 \\ 0 & 0 & 0 & 1 \\ 3 & -3 & 0 & 6 \\ 0 & 0 & -2 & -2 \end{bmatrix}$$

5. $$\begin{bmatrix} 1/4 & 5/2 & 7/4 & 1/2 \\ \underline{4} & 2 & 1 & 2 \\ \underline{1/2} & \underline{0} & 5/9 & 17/9 \\ \underline{1/4} & \underline{3/5} & \underline{27/10} & 1/5 \end{bmatrix}$$

6. $\ell = (1, 3, 2)$, the second pivot row is the third row.

8. $x_3 = -1, \quad x_2 = 1, \quad x_1 = 0$

10. $x_4 = -1, \quad x_3 = 0, \quad x_2 = 2, \quad x_1 = 1$

13b. $x_3 = 1, \quad x_2 = 1, \quad x_1 = 1$

13d. $x_1 \approx 4.267, \quad x_2 \approx -4.133, \quad x_3 \approx -2.467$

17. $n(n+1)$

18. $\left[\frac{29}{10}(n^2 - 1) + \frac{7}{30}n(n-1)(2n-1)\right] 10^{-6}$ seconds

19.

n	10	10^2	10^3	10^4
Time	$\frac{1}{3} \times 10^{-3}$ sec.	$\frac{1}{3}$ sec.	5.56 min.	3.86 days
Cost	0.005 cents	5 cents	\$46.30	\$46,296.30

21. Solve these: $\mathbf{U}^T\mathbf{y} = \mathbf{b}, \quad \mathbf{L}^T\mathbf{x} = \mathbf{y}$

23a. $x_1 = \frac{5}{9}, \quad x_2 = \frac{2}{9}, \quad x_3 = \frac{1}{9} \times 10^{-9}$

Computer Exercises 2.2

2. $[3.4606, 1.5610, -2.9342, -0.4301]^T$

3. $[6.7831, 3.5914, -6.4451, -1.5179]^T$

4. $2 \leq n \leq 10$, $x_i \approx 1$ for all i; for large n, many $x_i \neq 1$

5. $b_i = n^2 + 2(i-1)$ **6.** $x_2 = 1, \quad x_i = 0$ for $i \neq 2$

Exercises 2.3

2a. $5n - 4$ **3.** $n + 2nk - k(k+1)$ **6.** Yes, it does.

7. $D^{-1}AD = \text{Tridiagonal}\left[\pm\sqrt{a_{i-1}c_{i-1}},\ d_i,\ \pm\sqrt{a_i c_i}\right]$

Computer Exercises 2.3

3. $$\begin{cases} d_i \leftarrow d_i - 1/d_{i-1} \\ b_i \leftarrow b_i - b_{i-1}/d_{i-1} \quad (2 \leq i \leq n) \end{cases}$$
$$\begin{cases} x_n \leftarrow b_n \\ x_i \leftarrow (b_i - x_{i+1})/x_i \quad (n-1 \geq j \geq 1) \end{cases}$$

4. $$\begin{cases} x_1 = 1 \\ x_i = 1 - (4x_{i-1})^{-1} \quad (2 \leq i \leq 100) \end{cases}$$

11a. $$\begin{cases} x_1 \leftarrow b_1/a_{11} \\ x_i \leftarrow \left(b_i - \sum_{j=1}^{n-1} a_{ij}x_j\right) \Big/ a_{ii} \quad (2 \leq i \leq n) \end{cases}$$

12. $$\begin{cases} c_i \leftarrow c_i/d_i \\ b_i \leftarrow b_i/d_i \\ d_{i+1} \leftarrow d_{i+1} - a_{i+1}c_i \\ b_{i+1} \leftarrow b_{i+1} - a_{i+1}b_i \quad (1 \leq i \leq n-1) \end{cases}$$
$$\begin{cases} b_n \leftarrow b_n/d_n \\ b_i \leftarrow b_i - c_i b_{i+1} \quad (1 = n-1, \ldots, 1) \end{cases}$$

Exercises 3.1

1. 0.61906; 1.51213

4. $\{-\frac{\pi}{4} - \delta, 0, \frac{\pi}{4} + \varepsilon, \frac{3\pi}{4} + \varepsilon, \frac{5\pi}{4} + \varepsilon, \ldots\}$,
where $\delta \approx 0.2$ and ε starts at approximately 0.4 and decreases.

9. $\left\{0, \pm\frac{\pi}{2}, \pm\pi, \pm\frac{3\pi}{2}, \pm 2\pi, \ldots\right\}$ **10.** $x = 0$

12. If the original interval is of width h, then after, say, k steps, we have reduced the interval containing the root to width $h2^{-k}$. From then on, we add one bit at each step. About three steps are needed for each decimal digit.

17. 20 steps

18b. False, because if r is close to b_n, then $r - a_n \approx b_n - a_n = 2^{-n}(b_0 - a_0)$.

18d. True, because $0 \leqq r - a_n$ (obvious) and $r - a_n \leqq b_n - a_n = 2^{-n}(b_0 - a_0)$.

19a. False, in some cases. **19e.** True.

21. $n \geqq 24 - m$. **23.** No; No.

Computer Exercises 3.1

10. $1, 2, 3, 3 - 2i, 3 + 2i, 5 + 5i, 5 - 5i, 16$

11. 2.365

Exercises 3.2

3. $x_{n+1} = \frac{1}{2}[x_n + 1/(Rx_n)]$ **4.** 0.79; 1.6

7. $y = \frac{\sqrt{2}}{2}x + \frac{\sqrt{2}}{2}\left(1 - \frac{\pi}{4}\right)$ **9.** π

11. $x_{n+1} = 2x_n / \left(x_n^2 R + 1\right)$; -0.49985

12a. Yes, $-\sqrt[3]{R}$. **13a.** $x_{n+1} = \frac{1}{3}\left(2x_n + R/x_n^2\right)$

13c. $x_{n+1} = x_n\left(x_n^3 + 2R\right)/\left(2x_n^3 + R\right)$

13e. $x_{n+1} = \frac{x_n}{3R}\left(4R - x_n^3\right)$

13g. $x_{n+1} = \frac{R}{x_n^2}\left(2x_n^6 + 1\right)/\left(2Rx_n^3 + 1\right)$

15. $x_1 = \frac{1}{2}$ **17.** $x_{n+1} = -\frac{1}{2}$

19. $|x_0| < \sqrt{3}$

21. Newton's method cycles if $x_0 \neq 0$.

22.
```
x ← R
for n = 1 to n_max
    x ← (2x + Rx^2)/3
end for
```

27. $x_{n+1} = \left[(m-1)x_n^m + R\right]/\left(mx_n^{m-1}\right)$;

$x_{n+1} = x_n\left[(m+1)R - x_n^m\right]/(mR)$

29. Diverges.

31. $x_{n+1} = x_n - \dfrac{f(x_n)f'(x_n)}{[f'(x_n)]^2 - f(x_n)f''(x_n)}$

32. $x_{n+1} = x_n - \dfrac{f'(x_n)}{f''(x_n)} + \dfrac{\sqrt{[f'(x_n)]^2 - 2f(x_n)f''(x_n)}}{f''(x_n)}$

35. $$e_{n+1} = e_n^2\left[\frac{\dfrac{f^{(m+1)}(\eta_n)}{m!} - \dfrac{f^{(m+1)}(\xi_n)}{(m+1)(m-1)!}}{\dfrac{f^{(m)}(r)}{(m-1)!} + \dfrac{e_n f^{(m+1)}(\eta_n)}{m!}}\right]$$

36. $e_{n+1} = \frac{1}{2}e_n^2\frac{f''}{g}$

37. $|g'(r)| < 1$ if $0 < \omega < 2$ **41.** 4th order

Computer Exercises 3.2

4. 0.32796 77853 31818 36223 77546

5. 2.09455 14815 42326 59148 23865 40579

8. 1.83928 67552 **9.** 0.47033 169

10a. 1.89549 42670 340

10b. 1.99266 68631 307

10c. 0.51097 34293 8857

10d. 2.58280 14730 552

14. 3.13108; 3.15145 (two nearby roots)

Exercises 3.3

1. 2.7385 **3.** $-\frac{3}{2}$ **4.** $\ln 2$

9. $x_{n+1} = x_n - \dfrac{x_n^2 - R}{x_n + x_{n-1}}$

12. $e_{n+1} = \left[1 - \left(\dfrac{x_n - x_0}{f(x_n) - f(x_0)}\right)f'(\xi_n)\right]e_n$

13a. Linear convergence

13c. Quadratic convergence

15. Show $|\xi - x_{n+1}| \leqq c|\xi - x_n|$. **16.** $\sqrt{2}$

17. $x = 4.510187$

Computer Exercises 3.3

1. −0.45896; 3.73308 **6a.** 1.53209

6b. 1.23618 **7.** 1.3880 81078 21373

9. 20.80485 4

Exercises 4.1

1. $p_3(x) = 7 - 2x + x^3$

3. $\ell_2(x) = -(x-4)(x^2-1)/8$

7a. $p_3(x) = 2 + (x+1)\left(-3 + (x-1)\left(2-(x-3)\dfrac{11}{24}\right)\right)$

8. $p_4(x) = -1 + (x-1)\left(\dfrac{2}{3} + (x-2)\left(\dfrac{1}{8} + (x-2.5)\left(\dfrac{3}{4} + (x-3)\dfrac{11}{6}\right)\right)\right)$

9a.

0	1	8	3	$\boxed{1}$	0
1	9	14	$\boxed{7}$	1	
2	23	$\boxed{35}$	12		
4	$\boxed{93}$	83			
6	259				

9b. $f(4.2) = 104.488$

12. $q(x) = x^4 - x^3 + x^2 - x + 1 - \dfrac{31}{120}(x+2)(x+1)(x)(x-1)(x-2)$

13a. $x^3 - 3x^2 + 2x - 1$ **14.** $p(x) = x - 2.5$

16. $\alpha_0 = \dfrac{1}{2}$ **18.** $2 + x\big(-1 + (x-1)\big(1-(x-3)x\big)\big)$

19. $p_4(x) = -1 + 2(x+2) - (x+2)(x+1) + (x+2)(x+1)x$; $p_2(x) = 1 + 2(x+1)x$

22. $p(x) = 0.76(x-1.73)(x-1.82)(x-5.22)(x-8.26)$

25. 1.5727; No advantage

27. $p(x) = -\dfrac{3}{5}x^3 - \dfrac{2}{5}x^2 + 1$ **28.** 0.38099; 0.077848

39. 0.85527; 0.87006

40. Divisions: $\dfrac{1}{2^n}(n-1)$;
Additions/subtractions: $n(n-1)$

42. 0

45. False, only unique for polynomial p of degree $\leqq n-1$.

Computer Exercises 4.1

1. $p(x) = 2 + 46(x-1) + 89(x-1)(x-2) + 6(x-1)(x-2)(x-3) + 4(x-1)(x-2)(x-3)(x+4)$

Exercises 4.2

1. $f[x_0, x_1, x_2, x_3, x_4] = 0$ **6.** 1.25×10^{-5}

7. Errors: 8.1×10^{-6}, 6.1×10^{-6}

8. 497 table entries

9. 4.105×10^{-14} (Thm 1); 1.1905×10^{-23} (Thm 2)

10. 2.6×10^{-6} **13.** $n \geqq 7$

14. $\displaystyle\prod_{i=0}^{n-1} |x - x_i| \leqq \frac{h^n(2n)!}{2^{2n}n!}$ **16.** Yes.

Computer Exercises 4.2

10. On $[-1, 1]$, the interpolation error
$f(x) - p(x) = \dfrac{f^{(n)}(\xi)}{n!}w(x)$
where $w(x) = (x-x_0)(x-x_1)\cdots(x-x_n)$
the infinity norm of w on $[-1, 1]$
with the Chebyshev nodes minimize:
$||w||_\infty = \max_{-1 \leqq x \leqq 1} |w(x)|$.
So they make the error formula most favorable.

Exercises 4.3

1. $-hf''(\xi)$ **2.** Error term $= -hf''(\xi)$ for $\xi \in (0, 2h)$

4. No such formula exists.

6. The point ξ for the first Taylor series is such that $\xi \in (x, x+h)$, while the second is $\xi \in (x-h, x)$. They are *not* the same.

8a. $-\dfrac{2}{3}h^2 f'''(\xi)$ **9a.** $-\dfrac{h^2}{4}f^{(5)}(\xi)$ **9b.** $-\dfrac{h^2}{6}f^{(6)}(\xi)$

11. $\alpha = 1$, error term $= -\dfrac{h^2}{6} f'''(\xi)$;

$\alpha \neq 1$, error term $= -(\alpha - 1)\dfrac{h}{2} f''(\xi)$

12. Error term $= -\dfrac{h^2}{6}\left[f'''(\xi_1) + \dfrac{1}{2} f^{(4)}(\xi_2)\right]$

for some $\xi_i \in (x - h, x + h)$.

13. $p'\left(\dfrac{x_0 + x_1}{2}\right) = \dfrac{f(x_1) - f(x_0)}{x_1 - x_0}$

16. $L \approx 2\varphi\left(\dfrac{h}{2}\right) - \varphi(h)$

20. $$L \approx \left\{\left[\varphi\left(\frac{h}{2}\right)\right]^2 - \varphi(h)\varphi\left(\frac{h}{3}\right)\right\} \Big/ \left\{2\varphi\left(\frac{h}{2}\right) - \varphi(h) - \varphi\left(\frac{h}{3}\right)\right\}$$

Computer Exercises 4.3

3. 0.20211 58503

Exercises 5.1

1. $\dfrac{31}{72}$ **3.** 0 00010 00025 0006 **8.** ≈ 0.70833

9. $T(f; P) = 0.775$; $\displaystyle\int_0^1 \frac{dx}{x^2 + 1} \approx 0.7854$;

Error $= 0.0104$

13.

h	2	1	1/2	1/4
T	2	1	1	1

14. $n \geqq 16\,07439$; Too small.

15. $T = \dfrac{1}{n^3}\left[\dfrac{1}{6}(n - 1)(2n - 1)n\right] + \dfrac{1}{2n}$

19. 0.000025 **20.** $T(f; P) \approx 4.37132$

21. $T(f; P) \approx 0.43056$

22. $|$ error term $| \leqq 0.3104$

23. $T(f; P) = 7.125$; No, they cannot be computed from the given data.

24a. $-\dfrac{1}{2}(b - a)hf'(\xi)$ for some $\xi \in (a, b)$.

24b. $-\dfrac{1}{6}(b - a)h^2 f''(\xi)$ for some $\xi \in (a, b)$.

25a. $\dfrac{1}{24}h^3 f''(\xi)$ **25b.** $\dfrac{1}{24}\sum_{i=1}^{n} h_i^3 f''(\xi_i)$

25c. $\dfrac{1}{24}(b - a)h^2 f''(\xi)$

29. $f(x) = x^n$ $(n > 3)$ on $[0, 1]$, with partition $\{0, 1\}$

30. $\displaystyle\int_a^b f(x)\,dx \leqq T$ **31.** $n \geqq 1155$

34. $-\dfrac{1}{2}(b - a)hf'(\xi)$

35. $\displaystyle\int_a^b f(x)\,dx = h\sum_{i=0}^{2^n} f(a + ih) + E$

where $E = \dfrac{1}{2}(b - a)hf'(\xi)$ for $\xi \in (a, b)$

Computer Exercises 5.1

2a. 2 **2b.** 1.71828 **2c.** 0.43882

9. 0.94598 385; 0.94723 395 **11a.** 4.84422

Exercises 5.2

1. 13 **3.** $-\dfrac{136}{15}$ **5.** 4.267 **7.** Not well.

8. $R(1, 1) = \dfrac{h}{3}\{f(-h) + 4f(0) + f(h)\}$ is Simpson's rule.

10. $1 + 2^{m-1}$

13. $$R(2, 2) = \frac{2h}{45}\big[7f(a) + 32f(a + h) + 12f(a + 2h) + 32f(a + 3h) + 7f(b)\big]$$

14. $X = (27v - u)/26$

15. $$Z = \frac{4096}{2835}f\left(\frac{h}{8}\right) - \frac{1344}{2835}f\left(\frac{h}{4}\right) + \frac{84}{2835}f\left(\frac{h}{2}\right) - \frac{1}{2835}f(h)$$

17. $x_{n+1} + n^3(x_{n+1} - x_n)/(3n^2 + 3n + 1)$

18. $|I - R(n, m)| = \mathcal{O}(h^{2m})$ as $h \to 0$

22. $$R(n + 1, m + 1) = R(n + 1, m) + [R(n + 1, m) - R(n, m)]/(8^m - 1)$$

23. Show $\displaystyle\int_a^b f(x)\,dx - R(n, 0) \approx c4^{-(n+1)}$.

24. Let $m = 1$ and let $n \to \infty$ in Formula (2).

27. $$E = A_{2m}(2\pi)\left(\frac{2\pi}{4}\right)^{2m}[\pm 4^{2m}\cos(4\xi)] \pm (2\pi)^{2m+1}4^{2m+1}A_{2m}\cos(4\xi)$$

Computer Exercises 5.2

1. $R(7,7) = 0.49996\,9819$

5. $R(5,0) = 1.81379\,9364$ **6.** $\dfrac{2}{9} = 0.22222\ldots$

7. $0.62135\,732$ **9.** 0.61748

11. $R(7,7) = 0.76519\,7687$

Exercises 5.3

1. $\dfrac{\pi}{4}$ **2a.** $h < 0.03$ or $n > 33.97$.

2b. $h < 0.15$ or $n > 7.5$.

3a. 7.1667 **3b.** 7.0833 **3c.** 7.0777

4. $\displaystyle\int_1^2 \frac{dx}{x} = 0.6933$; Bound is 5.2×10^{-4}.

7. $\displaystyle\int_a^b f(x)\,dx = \frac{16}{15}S_{2(n-1)} - \frac{1}{15}S_{n-1}$

8. $-\dfrac{3}{80}h^5 f^{(4)}(\xi)$

Computer Exercises 5.3

1. 3.1416; 3.1416

2. 0.1996

Exercises 5.4

1. ≈ 0.91949 **4a.** $x = \pm\sqrt{\dfrac{1}{3}}$

4b. $x = \pm 0.861136, \quad \pm 0.339981$

5. $\alpha = \gamma = \dfrac{4}{3}, \quad \beta = -\dfrac{2}{3}$

6. $A = (b-a), \quad B = \dfrac{1}{2}(b-a)^2$

7. $\dfrac{5h}{12}f(a) + \dfrac{2h}{3}f(a+h) - \dfrac{h}{12}f(a+2h)$

9. $\alpha = \sqrt{\dfrac{5}{7}}, \quad a = c = \dfrac{7}{25}, \quad b = \dfrac{8}{75}$

10. $w_1 = w_2 = \dfrac{h}{2}, \quad w_3 = w_4 = -\dfrac{h^3}{24}$

11. $A = 2h, \quad B = 0, \quad C = \dfrac{h^3}{3}$

12. $A = \dfrac{8}{3}, \quad B = -\dfrac{4}{3}, \quad C = \dfrac{8}{3}$
Yes. Exact for polynomials of degree $\leqq 3$.

13. $A = \dfrac{h}{3}, \quad D = 0, \quad C = \dfrac{h}{3}, \quad B = \dfrac{4}{3}h$

14. True for $n \leqq 3$

Computer Exercises 5.4

2a. 1.4183 **8a.** 2.03480 53185 77

8b. 0.89297 95115 69 **8c.** 0.43398 771

Exercises 6.1

1. Yes

6. In Exercise 6.1.5, the bracketed expression is $f'(\xi_1) - f'(\xi_2)$ and in magnitude does not exceed $2C$.

9. Knots $\approx 1.57 \times 10^{10}$.

10. $\sum_{i=1}^{n} f(t_i)S_i$ is a linear combination of 1st-degree spline functions having knots $t_0, t_1, \ldots, t_n$. Hence, it is also such a function. Its value at t_j is $\sum_{i=1}^{n} f(t_i)S_i(t_j) = f(t_j)$.
$S_i(x) = 0$ if $x < t_{i-1}$ or $x > t_{i+1}$.
On (t_{i-1}, t_i), $S_i(x)$ is given by $(x - t_{i-1})/(t_i - t_{i-1})$.
On (t_i, t_{i+1}), $S_i(x)$ is given by $(x - t_{i+1})/(t_i - t_{i+1})$.
S_0 and S_n are slightly different.

12. If S is piecewise quadratic, then clearly S' is piecewise linear. If S is a quadratic spline then $S \in C^1$. Hence, $S' \in C$. Hence, S' is piecewise linear and continuous.

17. $$\begin{cases} Q_0(x) = -(x+1)^2 + 2 \\ Q_1(x) = -2x + 1 \\ Q_2(x) = 8\left(x - \frac{1}{2}\right)^2 - 2\left(x - \frac{1}{2}\right) \\ Q_3(x) = -5(x-1)^2 + 6(x-1) + 1 \\ Q_4(x) = 12(x-2)^2 - 4(x-2) + 2 \end{cases}$$

19. The answer is given by Equation (8).

20a. Yes **20b.** No **20c.** No **21.** Yes

Exercises 6.2

1. No **2.** No

4. $a = -4, \quad b = -6, \quad c = -3, \quad d = -1, \quad e = -3$

5. $a = -5, \quad b = -26, \quad c = -27, \quad d = \dfrac{27}{2}$

6. No

7a. $S(x)$ is not continuous at $x = -1$. $S''(x)$ is not continuous at $x = -1, 1$.

8a. $(m+1)n$ **8b.** $2n$

8c. $(m-1)(n-1)$ **8d.** $m - 1$

10. $S = \begin{cases} x^2, & [0, 1] \\ 1 + 2(x-1) + (x-1)^2 + (x-1)^3, & [1, 2] \\ 5 + 7(x-2) + 4(x-2)^2, & [2, 3] \end{cases}$

12. $a = 3, \quad b = 3, \quad c = 1$ **13.** No

15. $a = -1, \quad b = 3, \quad c = -2, \quad d = 2$

17. $n + 3$ **19.** f is not a cubic spline

22. $p_3(x) = x - 0.0175x^2 + 0.1927x^3$; No

26. S is linear.

32. $S_0(x) = -\frac{5}{7}(x-1)^3 + \frac{12}{7}(x-1)$,
$S_1(x) = \frac{6}{7}(x-2)^3 - \frac{5}{7}(3-x)^3 - \frac{6}{7}(x-2) + \frac{12}{7}(3-x)$,
$S_2(x) = -\frac{5}{7}(x-3)^3 + \frac{6}{7}(4-x)^3 + \frac{12}{7}(x-3) - \frac{6}{7}(4-x)$,
$S_3(x) = -\frac{5}{7}(5-x)^3 + \frac{12}{7}(5-x)$

33. The conditions on S make it an even function. If $S(x) = S_0(x)$ in $[-1, 0]$ and $S(x) = S_1(x)$ in $[0, 1]$, then $S_1(0) = 1$, $S_1'(0) = 0$, $S_1''(1) = 0$, and $S_1(1) = 0$. An easy calculation yields $S_1(x) = 1 - \frac{3}{2}x^2 + \frac{1}{2}x^3$.

38. $5n, n + 4$ **39.** Yes

Exercises 6.3

2. Chebyshev polynomials recurrence relation. See Section 10.2.

3. $B_i^2(x) = \begin{cases} \dfrac{(x-t_i)^2}{(t_{i+2}-t_i)(t_{i+1}-t_i)}, & [t_i, t_{i+1}] \\ \dfrac{(x-t_i)(t_{i+2}-x)}{(t_{i+2}-t_i)(t_{i+2}-t_{i+1})} + \dfrac{(t_{i+3}-x)(x-t_{i+1})}{(t_{i+3}-t_{i+1})(t_{i+2}-t_{i+1})}, & [t_{i+1}, t_{i+2}] \\ \dfrac{(t_{i+3}-x)^2}{(t_{i+3}-t_{i+1})(t_{i+3}-t_{i+2})}, & [t_{i+2}, t_{i+3}] \\ 0, \quad \text{elsewhere} \end{cases}$

5. $\sum_{i=-\infty}^{\infty} f(t_i) B_i^0(x)$ **14.** $n - k \leqq i \leqq m - 1$

15. Use induction on k and $B_{i+i}^{k+i}(x) = 0$ on $[t_i, t_{i+1}]$.

16. No **17.** No

19. $\sum_{i=-\infty}^{\infty} t_{i+1} B_i^1(x)$

20. In Equation (9), take all $c_i = 1$. Then $d_i = 0$. Hence, $\frac{d}{dx}\sum_{i=1}^{n} B_i^k(x) = 0$ and $\sum_{i=1}^{n} B_i^k(x)$ are constants.

24. Use Equation (14) with all A's zero except $A_j = 1$. Next, take all A's zero except $A_{j+1} = 1$.

28. No **30.** Let $C_i^2 = t_{i+1}t_{i+2}$, then $C_i^1 = xt_{i+1}$, and $C_i^0 = x^2$.

32. $B_i^k(t_j) = 0$ iff $t_j \geqq t_{i+k+1}$ or $t_j \leqq t_i$

33. $x = (t_{i+3}t_{i+2} - t_i t_{i+1})/(t_{i+3} + t_{i+2} - t_{i+1} - t_i)$

Computer Exercises 6.3

7. 47040

Exercises 7.1

1a. $x = \frac{1}{4}t^4 + \frac{7}{3}t^3 - \frac{2}{3}t^{3/2} + c$ **1b.** $x = ce^t$

1e. $x = c_1 e^t + c_2 e^{-t}$ or $x = c_1 \cosh t + c_2 \sinh t$

2a. $x = \frac{1}{3}t^3 + \frac{3}{4}t^{4/3} + 7$

3c. $x = \sum_{n=0}^{\infty} (-1)^n \dfrac{t^{2n+1}}{(2n+1)(2n+1)!} + c$

3d. $x = e^{-t^2/2}\left[\int t^2 e^{t^2/2}\,dt + c\right]$

4. $x = a_0 + a_0 \sum_{n=1}^{\infty} (-1)^n \left(\dfrac{(2n-1)!}{2^{n-1}(2n)!}\right) t^{2n} + \sum_{n=1}^{\infty} (-1)^{n-1} \left(\dfrac{n!2^n}{(2n+1)!}\right) t^{2n+1}$

6. Let $p(t) = a_0 + a_1 t + a_2 t^2 + \cdots$ and determine a_i.

9. $t = 10$, Error $= 2.2 \times 10^4 \varepsilon$; $t = 20$, Error $= 4.8 \times 10^8 \varepsilon$

10. $x^{(iv)} = 18xx'x'' + 6(x')^3 + 3x^2 x'''$

11a. $x' = x + e^x$;
$x'' = (1 + e^x)x'$;
$x''' = (1 + e^x)x'' + e^x(x')^2$;
$x^{(iv)} = (1 + e^x)x''' + 3e^x x' x'' + e^x (x')^3$.

12. $x(0.1) = 1.21633$

14.
```
n ← 20
s ← x^(n)
for i = 1 to n − 1
    s ← x^(n−i) + [h/real(n + 1 − i)]s
end for
s ← x + h[s]
```

Computer Exercises 7.1

1. $x(2.77) = 385.79118$

2b. $x(1.75) = 0.63299\,9983$

2c. $x(5) = -0.20873\,51554$

3. $x(10) = 22026.47$

4a. Error at $t = 1$ is 1.8×10^{-10}.

5. $x(0) = 0.03245\,34427$

7. $x(1) = 1.64872\,12691$

9. $x(0) = 1.67984\,09205 \times 10^{-3}$

10. $x(0) = -3.75940\,73450$

Exercises 7.2

2c. $f(t, x) = +\sqrt{x/\left(1-t^2\right)}$ **3.** $x(-0.2) = 1.92$

5a.
```
real function f(t, x)
real t, x
   f ← t²/(1 − t + 2x)
end function f
```

8. Solve $\dfrac{df}{dx} = e^{-x^2}, \quad f(0) = 0.$

10. $h^3\left(\dfrac{1}{6} - \dfrac{\alpha}{4}\right)D^2 f + \dfrac{h^3}{6} f_x Df$ where

$$D = \frac{\partial}{\partial t} + f\frac{\partial}{\partial x}$$

$$D^2 = \frac{\partial^2}{\partial t^2} + 2f\frac{\partial^2}{\partial x\,\partial t} + f^2\frac{\partial^2}{\partial x^2}$$

11. $h = 1/1024$

12. Let's make local truncation error $\leqq 10^{-13}$.
Thus, $100h^5 \leqq 10^{-13}$ or $h \leqq 10^{-3}$.
So take $h = 10^{-3}$ and hope that the three extra digits are enough to preserve 10-digit precision.

14b. $x^{(iv)} = D^3 f + f_x D^2 f + 3Df_x\, Df + f_x^2\, Df$ where

$$D^3 = \frac{\partial^3}{\partial t^3} + 3f\frac{\partial^3}{\partial x\,\partial t^2} + 3f^2\frac{\partial^3}{\partial t\,\partial x^2} + f^3\frac{\partial^3}{\partial x^2}$$

15.
$$\begin{aligned} f(x+th, y+tk) &= f(x,y) + t[f_1(x,y)h + f_2(x,y)k] \\ &\quad + \frac{1}{2}t^2\left[f_{11}(x,y)h^2\right. \\ &\quad \left. + 2f_{12}(x,y)hk + f_{22}(x,y)k^2\right] \\ &\quad + \cdots \end{aligned}$$

Now let $t = 1$ to get the usual form of Taylor's series in two variables.

17. Taylor series of $f(x, y) = g(x) + h(y)$ about (a, b) is equal to the Taylor series of $g(x)$ about a plus that of $h(y)$ about b.

18. $f(1+h, k) \approx -3h + \dfrac{3}{2}h^2 + k^2$

19. $e^{1-xy} \approx 3 - x - y$

20. $A = 1 + k + \dfrac{1}{2}k^2, \quad B = h(1+k)$

21. $A = 1, \quad B = h - k, \quad C = (h-k)^2$

22. $f(x+h, y+k) \approx$
$$\left(1 + 2xh + k + (1+2x^2)h^2 + 2hkx + \tfrac{1}{2}k^2\right) f;$$
$f(0.001, 0.998) \approx 2.71285\,34$

Computer Exercises 7.2

2. $x(1) = 1.5708$

3b. $n = 7;\quad x(2) = 0.82356\,78972$ (RK),
$x(2) = 0.82356\,78970$ (TS)

3c. $n = 7;\quad x(2) = -0.49999\,99998$ (RK),
$x(2) = -0.50000\,00012$ (TS)

4. $x(1) = 0.60653 = x(3)$ **5.** $x(3) = 1.5$

6. $x(0) = 1.0 = x(1.6)$ **8.** $x(1) = 3.95249$

9. $x(10) = 1.344 \times 10^{43}$

Exercises 7.3

1. Let $h = 1/n$. Then $x(1) = e^{-1}$ (True Soln.) and
$x_n = \{[1 - 1/(2n)]/[1 + 1/(2n)]\}^n$ (Approx. Soln.)

2.
$$x(t+h) = x(t-h) + \frac{h}{3}[f(t-h, x(t-h)) + 4f(t, x(t)) + f(t+h, x(t+h))]$$

4. $a = \dfrac{24}{13}, \quad b = -\dfrac{11}{13}, \quad c = \dfrac{2}{13},$
$d = \dfrac{10}{13}, \quad e = -\dfrac{2}{39}h^2$

5. $a = 1, b = c = \dfrac{h}{2}$; Error term is $\mathcal{O}(h^3)$.

8. $\dfrac{\partial}{\partial s}x(9, s) = e^{252} \approx 10^{109}$ **9a.** All t.

9c. Positive t. **9e.** No t. **11.** Divergent for all t.

Computer Exercises 7.3

5. $x(\frac{1}{2}) = 2.25$ **6.** $x(-\frac{1}{2}) = -4.5$

9. $y(e) = -6.38905\,60989$ where $y(x) = [1 - \ln v(x)]v(x)$

12. $0.21938\,39244$ **13.** $0.99530\,87432$

15. $\mathrm{Si}(1) = 0.94608\,30703$

Exercises 7.4

1. $$x(t+h) = x\left(1 + \frac{1}{2}h^2 + \frac{1}{24}h^4\right) + y\left(h + \frac{1}{6}h^3 + \frac{1}{120}h^5\right)$$
$$y(t+h) = y\left(1 + \frac{1}{2}h^2 + \frac{1}{24}h^4\right) + x\left(h + \frac{1}{6}h^3 + \frac{1}{120}h^5\right)$$

2. Since system is not coupled, solve two separate problems.

3. System is not coupled so each differential equation can be solved individually by the program.

4. $$\mathbf{X}' = \begin{bmatrix} 1 \\ x_1^2 + \log x_2 + x_0^2 \\ e^{x_2} - \cos x_1 + \sin(x_0 x_1) - (x_1 x_2)^7 \end{bmatrix},$$
$\mathbf{X}(0) = [0, 1, 3]^T$

5. $$\mathbf{X}' = \begin{bmatrix} x_2 \\ x_3 \\ 2x_2 + \log x_3 + \cos x_1 \end{bmatrix}, \quad \mathbf{X}(0) = [1, -3, 5]^T$$

7. Solve each equation separately since they are not coupled.

8. $$\mathbf{X}' = \begin{bmatrix} x_2 \\ -x_1\left(x_1^2 + x_3^2\right)^{-3/2} \\ x_4 \\ -x_3\left(x_1^2 + x_3^2\right)^{-3/2} \end{bmatrix}, \quad \mathbf{X}(0) = \begin{bmatrix} 0.5 \\ 0.75 \\ 0.25 \\ 1.0 \end{bmatrix}$$

9. $$\mathbf{X}' = \begin{bmatrix} 1 \\ x_2 \\ x_3 \\ x_4 \\ x_4^2 + \cos(x_2 x_3) - \sin(x_0 x_1) + \log(x_1/x_0) \end{bmatrix},$$
$\mathbf{X}(0) = [0, 1, 3, 4, 5]^T$

10. $$\mathbf{X}' = \begin{bmatrix} x_4 \\ x_5 \\ x_6 \\ 2x_1 x_3 x_4 + 3x_1^2 x_2 t^2 \\ e^{x_2} x_5 + 4x_1 t^2 x_3 \\ 2t x_6 + 2t e^{x_1 x_3} \end{bmatrix}, \quad \mathbf{X}(1) = \begin{bmatrix} 3 \\ 3 \\ 2 \\ -79/12 \\ 2 \\ 3 \end{bmatrix}$$

11a. Let $x_1 = x$, $x_2 = x'$, $x_3 = x''$.

Then $$\mathbf{X}' = \begin{bmatrix} x_2 \\ x_3 \\ -x_3 \sin x_1 - t x_2 - x_3 \end{bmatrix}$$

12. $$\mathbf{X}' = \begin{bmatrix} x_2 \\ x_2 - x_1 \end{bmatrix}, \quad \mathbf{X}(0) = [0, 1]^T$$

13. Let $x_0 = t$, $x_1 = x$, $x_2 = y$, $x_3 = x'$, $x_4 = y'$.

Then $$\mathbf{X}' = \begin{bmatrix} 1 \\ x_3 \\ x_4 \\ x_1 + x_2 - 2x_3 + 3x_4 + \log x_0 \\ 2x_1 - 3x_2 + 5x_3 + x_0 x_2 - \sin x_0 \end{bmatrix},$$
$\mathbf{X}(0) = [0, 1, 3, 2, 4]^T$

Computer Exercises 7.4

1. $x(1) = 2.46869\,39399$, $y(1) = 1.28735\,52872$

2. $x(0.38) = 1.90723 \times 10^{12}$, $y(0.38) = -8.28807 \times 10^4$

4. $x(-1) = 3.36788$, $y(-1) = 2.36788$

5. $x_1\left(\frac{\pi}{2}\right) = x_4\left(\frac{\pi}{2}\right) = 0$, $x_2\left(\frac{\pi}{2}\right) = 1$, $x_3\left(\frac{\pi}{2}\right) = -1$

7. $x(6) = 4.39411$, $y(6) = 3.10378$

Exercises 7.5

1. $x_j(t) = e^{\lambda_j t} x_j(0)$

Exercises 8.1

1a. $$\mathbf{L} = \begin{bmatrix} 1 & 0 & 0 \\ 0 & 1 & 0 \\ 1/3 & -3 & 1 \end{bmatrix}, \quad \mathbf{U} = \begin{bmatrix} 3 & 0 & 3 \\ 0 & -1 & 3 \\ 0 & 0 & 8 \end{bmatrix}$$

2a. $$\mathbf{M} = \begin{bmatrix} 1 & 0 & 0 & 0 \\ 0 & 1 & 0 & 0 \\ 0 & -3 & 1 & 0 \\ -5 & 6 & -2 & 1 \end{bmatrix}, \quad \mathbf{U} = \begin{bmatrix} 1 & 0 & 0 & 2 \\ 0 & 3 & 0 & 0 \\ 0 & 0 & 4 & 0 \\ 0 & 0 & 0 & 0 \end{bmatrix}$$

3a. $$\mathbf{M} = \begin{bmatrix} 1 & 0 & 0 & 0 & 0 \\ 0 & 1 & 0 & 0 & 0 \\ 0 & -2 & 1 & 0 & 0 \\ 0 & 0 & -2 & 1 & 0 \\ -4 & 0 & 0 & 0 & 1 \end{bmatrix},$$
$$\mathbf{U} = \begin{bmatrix} 25 & 0 & 0 & 0 & 1 \\ 0 & 27 & 4 & 3 & 2 \\ 0 & 0 & 50 & -6 & -4 \\ 0 & 0 & 0 & 0 & 0 \\ 0 & 0 & 0 & 0 & 20 \end{bmatrix}$$

4b. $$\mathbf{A} = \begin{bmatrix} 3 & 2 & 1 \\ 2 & 2 & 1 \\ 1 & 1 & 1 \end{bmatrix}$$

5a. $\mathbf{M} = \begin{bmatrix} 1 & 0 & 0 & 0 \\ 0 & 1 & 0 & 0 \\ 0 & -x/b & 1 & 0 \\ -w/a & (xy)/(bc) & -y/c & 1 \end{bmatrix}$,

$\mathbf{U} = \begin{bmatrix} a & 0 & 0 & z \\ 0 & b & 0 & 0 \\ 0 & 0 & c & 0 \\ 0 & 0 & 0 & d-(wz)/a \end{bmatrix}$

5b. $\mathbf{L}' = \begin{bmatrix} a & 0 & 0 & 0 \\ 0 & b & 0 & 0 \\ 0 & x & c & 0 \\ 0 & 0 & y & d-(wz)/a \end{bmatrix}$,

$\mathbf{U}' = \begin{bmatrix} 1 & 0 & 0 & z/a \\ 0 & 1 & 0 & 0 \\ 0 & 0 & 1 & 0 \\ 0 & 0 & 0 & 1 \end{bmatrix}$

6a. $\mathbf{L} = \begin{bmatrix} 1 & 0 & 0 & 0 \\ -1/4 & 1 & 0 & 0 \\ -1/4 & -1/15 & 1 & 0 \\ 0 & -4/15 & -2/7 & 1 \end{bmatrix}$,

$\mathbf{U} = \begin{bmatrix} 4 & -1 & -1 & 0 \\ 0 & 15/4 & -1/4 & -1 \\ 0 & 0 & 56/15 & -16/15 \\ 0 & 0 & 0 & 24/7 \end{bmatrix}$

6b. $\mathbf{D} = \begin{bmatrix} 4 & 0 & 0 & 0 \\ 0 & 15/4 & 0 & 0 \\ 0 & 0 & 56/15 & 0 \\ 0 & 0 & 0 & 24/7 \end{bmatrix}$,

$\mathbf{U}' = \begin{bmatrix} 1 & -1/4 & -1/4 & 0 \\ 0 & 1 & -1/15 & -4/15 \\ 0 & 0 & 1 & -2/7 \\ 0 & 0 & 0 & 1 \end{bmatrix}$

6c. $\mathbf{L}' = \begin{bmatrix} 4 & 0 & 0 & 0 \\ -1 & 15/4 & 0 & 0 \\ -1 & -1/4 & 56/15 & 0 \\ 0 & -1 & -16/15 & 24/7 \end{bmatrix}$

6d. $\mathbf{L}'' = \begin{bmatrix} 2 & 0 & 0 & 0 \\ -1/2 & (1/2)\sqrt{15} & 0 & 0 \\ -1/2 & -1/\left(2\sqrt{15}\right) & 2\sqrt{14/15} & 0 \\ 0 & -2/\left(\sqrt{15}\right) & -(4/7)\sqrt{14/15} & 2\sqrt{6/7} \end{bmatrix}$

6e. 192

8. $\mathbf{U} = \begin{bmatrix} 1 & 0 & 0 & 1 \\ 0 & 1 & 0 & -2 \\ 0 & 0 & 1 & 4 \\ 0 & 0 & 0 & -8 \end{bmatrix}$,

$\mathbf{L} = \begin{bmatrix} 1 & 0 & 0 & 0 \\ 1 & 1 & 0 & 0 \\ -1 & 1 & 1 & 0 \\ 1 & -1 & 1 & 1 \end{bmatrix}$

9a. $\mathbf{L} = \begin{bmatrix} 1 & 0 & 0 \\ 1 & 1 & 0 \\ 3 & -1 & 1 \end{bmatrix}$, $\mathbf{D} = \begin{bmatrix} 2 & 0 & 0 \\ 0 & -2 & 0 \\ 0 & 0 & 3 \end{bmatrix}$,

$\mathbf{U}' = \begin{bmatrix} 1 & -1/2 & 1 \\ 0 & 1 & -1/2 \\ 0 & 0 & 1 \end{bmatrix}$

9b. $\mathbf{x} = [-1, 2, 1]^T$

10a. $\mathbf{L} = \begin{bmatrix} 1 & 0 & 0 \\ 2 & 1 & 0 \\ -1 & 3 & 1 \end{bmatrix}$, $\mathbf{D} = \begin{bmatrix} -2 & 0 & 0 \\ 0 & 1 & 0 \\ 0 & 0 & -1 \end{bmatrix}$,

$\mathbf{U}' = \begin{bmatrix} 1 & -1/2 & 1 \\ 0 & 1 & 1 \\ 0 & 0 & 1 \end{bmatrix}$

10b. $\mathbf{x} = [-1, 1, 1]^T$

12. $\mathbf{A}^{-1} = \dfrac{1}{15}\begin{bmatrix} 11 & -5 & -7 \\ -13 & 10 & 11 \\ -8 & 5 & 1 \end{bmatrix}$

14a. $\begin{bmatrix} \ell_{11} & \ell_{11}u_{12} & 0 & 0 \\ \ell_{21} & \ell_{21}u_{12}+\ell_{22} & \ell_{22}u_{23} & 0 \\ 0 & \ell_{32} & \ell_{32}u_{23}+\ell_{33} & \ell_{33}u_{34} \\ 0 & 0 & \ell_{43} & \ell_{43}u_{34}+\ell_{44} \end{bmatrix}$

16a. $\mathcal{X}^{-1} = \begin{bmatrix} 1 & 0 & 0 & -1 \\ 1 & 1 & -1 & 1 \\ -1 & 0 & 1 & -1 \\ 0 & 0 & 0 & 1 \end{bmatrix}$

16b. $\mathcal{X}^{-1} = \begin{bmatrix} 0 & -1 & -1 & 1 \\ -1 & 0 & -1 & 1 \\ -1 & -1 & 0 & 1 \\ 1 & 1 & 1 & -1 \end{bmatrix}$

Computer Exercises 8.1

3. **Case 4**:

$p_5(A) = \begin{bmatrix} 536 & -668 & 458 & -186 \\ -668 & 994 & -854 & 458 \\ 458 & -854 & 994 & -668 \\ -186 & 458 & -668 & 536 \end{bmatrix}$

Exercises 8.2

1. Yes.
3. Bases vectors mapping: $(1, 0) \to (\cos\theta, \sin\theta,)$. Counterclockwise rotations through the angle θ.
7. Eigenvalues: **a.** $9.8393, 3.0804 \pm 1.3763i$
   c. $0.4660 - 1.4971i, 2.6112 + 0.3313i, -0.0773 + 2.1658i$
9. c. **11.** d. **13.** True. **15.** True.

Computer Exercises 8.2

1a. Eigenvalues: $-2 \pm \sqrt{23} = -6.79583, 2.79583$.
   Eigenvectors: $[1, 9 - (3 \pm \sqrt{23})/7]^T$ or $[1, -1.11369]^T$ and $[1, 0.25655]^T$.
1d. $n = 5$: Eigenvalues: 0.26795, 1, 2, 3, 3.73205.
7. Eigenvalues/Eigenvectors:
   a. 5.2426, (0.6656, 0.7463); −3.2426, (−0.3051, 0.9523)
   b. $3.9893 \pm 5.5601i$, $(0.7267, -0.0680 \pm 0.4533i, -0.3395 \mp 0.3829i)$; 0.0214, (0.7916, 0.5137, 0.3308)
11. Eigenvalues/Eigenvectors: 1, (−1, 1, 0, 0); 2, (0, 0, −1, 1); 5, (−1, 1, 2, 2)

Exercises 8.3

1. a. **2.** b.

Computer Exercises 8.3

4. Eigenvalues are −5, 7, and 3.

Exercises 8.4

1. Jacobi and Gauss-Seidel converge because **A** is diagonally dominant, SOR converges because **A** is symmetric and postive definite.
3. d. **5.** e. **9.** b.

Computer Exercises 8.4

3. Iterations: Jacobi 77, Gauss-Seidel 38, and SOR 12.
7. Both converge to approximately the true solution $x = -4/11$ and $y = 6/11$.

Exercises 9.1

1. $y(x) = 1$
2. $$f(x) = \frac{1}{m+1}\sum_{k=0}^{m} y_k = (y_0 + \cdots + y_m)/(m+1),$$
   the *average* of the y values which does not involve any x_i.
3. $a = (1 + 2e)/(1 + 2e^2), \quad b = 1$
5. $a = 2.1, \quad b = 0.9$
7. $$c = \left[\sum_{k=0}^{m} y_k \log x_k\right] \Big/ \left[\sum_{k=0}^{m} (\log x_k)^2\right]$$
11. φ involves the sum of $m + 1$ polynomials of degree two in c which is either concave upward or a constant. Thus, no maxima exists—only a minima.
12. $c = 10^\alpha$ where $\alpha = \left[(m+1)^{-1}\sum_{k=0}^{m}(y_k - \log x_k)\right]$.
13. $y = (6x - 5)/10$
16. $a \approx 2.5929, \quad b \approx -0.32583, \quad c \approx 0.022738$
18. $a = 1, \quad b = \frac{1}{3}$ **19.** $y(x) = \frac{2}{7}x^2 + \frac{29}{35}$
20. $y = x + 1$ **21.** $c = \left[\sum_{k=0}^{m} e^{x_k} f(x_k)\right] \Big/ \left[\sum_{k=0}^{m} e^{2x_k}\right]$

Exercises 9.2

2. $$\begin{cases} w_{n+2} = w_{n+1} = 0 \\ w_k = c_k + 3x w_{k+1} + 2w_{k+2} \quad (k = n, n-1, \ldots, 0) \\ f(x) = w_0 - (1 + 2x)w_1 \end{cases}$$
3. Since $\cos(n-2)\theta = \cos[(n-1)\theta - \theta] =$
   $\cos(n-1)\theta\cos\theta + \sin(n-1)\theta\sin\theta$,
   we have $2\cos\theta\cos(n-1)\theta - \cos(n-2)\theta =$
   $\cos(n-1)\theta\cos\theta - \sin(n-1)\theta\sin\theta = \cos(n\theta)$. Note:
   If $g_n(\theta) = \cos n\theta$, then $g_n(\theta) = 2\cos\theta g_{n-1}(\theta) - g_{n-2}(\theta)$.
5. By the previous problem, the recursive relation is the same as (2) so that $T_n(x) = f_n(x) = \cos(n \arccos x)$.
6. $$T_n(T_m(x)) = \cos(n\arccos(\cos(m\arccos x))) = \cos(nm\arccos x) = T_{nm}(x)$$
7. $|T_n(x)| = |\cos(n\arccos x)| \leqq 1$ for all $x \in [-1, 1]$ since $|\cos y| \leqq 1$ and for $\arccos x$ to exist x must be $|x| \leqq 1$.
8. $$\begin{cases} g_0(x) = 1 \\ g_1(x) = (x+1)/2 \\ g_j(x) = (x+1)g_{j-1}(x) - g_{j-2}(x) \quad (j \geqq 2) \end{cases}$$
10. $n + 2$ multiplications, $2n + 1$ additions/subtractions if $2x$ is computed as $x + x$
12. n multiplications, $2n$ additions/subtractions
13. $T_6(x) = 32x^6 - 48x^4 + 18x^2 - 1$
17. $$\alpha = \frac{y_1 x_2^{13} - y_2 x_1^{13}}{x_1^{12} x_2^{12}(x_2 - x_1)}.$$
   α is very sensitive to perturbations in y_1.

Computer Exercises 9.2

7. $a_{ij} = \begin{cases} 0 & (i \neq j) \\ (m+1) & (i = j = 1) \\ (m+1)/2 & (i = j > 1) \end{cases}$

Exercises 9.3

2. Coefficient matrix for the normal equations has elements $a_{ij} = 1/(i + j - 1)$ by (5).

3. $c = 0$ **4.** $y = b^x$ **6.** $c = \ln 2$

8. $x = -1, \quad y = \dfrac{20}{13}$ **9a.** $c = \dfrac{24}{\pi^3}$ **9b.** $c = 3$

14. No. **15.** $y \approx \dfrac{1}{a + bx}$. Change to $\dfrac{1}{y} \approx a + bx$.

16. $\begin{bmatrix} \pi & 0 & 2 \\ 0 & \pi/2 & 0 \\ 2 & 0 & \pi/2 \end{bmatrix} \begin{bmatrix} a \\ b \\ c \end{bmatrix} = \begin{bmatrix} (e^{2\pi} - 1)/2 \\ -2(e^{2\pi} + 1)/5 \\ (e^{2\pi} + 1)/5 \end{bmatrix}$

17. $c = 3$ **20.** $c = \left[\sum_{i=1}^{n} y_i \sin x_i \right] \Big/ \left[\sum_{i=1}^{n} (\sin x_i)^2 \right]$

Computer Exercises 9.3

1. $a = 2, \quad b = 3$

Exercises 9.4

5. $x^2 = \dfrac{1}{3} + \sum_{n=1}^{\infty} (-1)^n \dfrac{4}{n^2 \pi^2} \cos n\pi x$

7d. $f(x) = \dfrac{4}{\pi} \sum_{n=1,3,5,\ldots}^{\infty} \dfrac{1}{n} \sin nx$

11a. $\sqrt[n]{a} = \sqrt[n]{r} \left[\cos\left(2\pi k/n\right) + i \sin(2\pi k/n) \right]$

11b. $\sqrt[4]{2}, i\sqrt[4]{2}, -i\sqrt[4]{2}, -\sqrt[4]{2}$

17. Can reduce additions from 12 to 6 and multiplications from 16 to 2.

Computer Exercises 9.4

6. $f(x) = \dfrac{4A}{\pi} \sum_{n=1,3,5,\ldots}^{\infty} \dfrac{1}{n} \sin n\omega_0 x$

8a. $f(x) = 2 - \sum_{n=1}^{\infty} \dfrac{4}{n\pi} \sin \dfrac{1}{2} n\pi x$

Exercises 10.1

1. $\ell_0 = 123456; \quad x_1 = .96621\,2243;$
$x_2 = .12917\,3003; \quad x_3 = .01065\,6910$

Computer Exercises 10.1

8. 32.5% **11.** Sequence not periodic.

13.

0	1	2	3	4	5	6	7	8	9
97	93	97	107	90	115	88	101	113	99

15. 5.6% **16.** 200

Exercises 10.2

1. $m > 4$ million

Computer Exercises 10.2

2. 1.71828 **4.** 8 **5.** 49.9 **7.** 0.518

9. 1.11 **10.** 2.00034 6869 **14.** 0.635 **17b.** 8.3

Computer Exercises 10.3

1. $\dfrac{2}{3}$ **2.** 0.898 **4.** $\dfrac{7}{16}$ **6.** 1.05

7. 5.24 **9.** 0.996 **12.** 0.6394

14. 11.6 kilometers **15.** 0.14758

17. 0.009 **21.** 24.2 revolutions **23.** 0.6617

Exercises 11.1

2. $c_1 = \left(1 - 2e\right) / \left(1 - e^2\right),$
$c_2 = \left(2e - e^2\right) / \left(1 - e^2\right)$

3a. $x(t) = \left(e^{\pi+t} - e^{\pi-t}\right) / \left(e^{2\pi} - 1\right)$

3b. $x(t) = \left(t^4 - 25t + 12\right) / 12$

4a. $x(t) = \beta \sin t + \alpha \cos t$ for all (α, β)

4b. $x(t) = c_1 \sin t + \alpha \cos t$ for all $\alpha + \beta = 0$ with c_1 arbitrary

6. $\varphi(t) = z$ **7.** $\varphi(z) = z$ **8.** $\varphi(z) = \sqrt{9 + 6z}$

9. $\varphi(z) = \left(e^5 + e + ze^4 - z\right) / \left(2e^2\right)$

10. Two ways: Use $x''(a) = z$ or $x'(b) = z$, $x''(b) = w$.

11. $x(t) = -e^t + 2\ln(t + 1) + 3t$

14a. This is a linear problem. So two initial-value problems can be solved to obtain the solution. The two sets of initial values would be

$$\begin{cases} x(0) = 0 \\ x'(0) = 1 \end{cases} \quad \text{and} \quad \begin{cases} x(0) = 1 \\ x'(0) = 0 \end{cases}.$$

15. Solution of $x'' = -x, x(0) = 1, x'(0) = z$ is $x(t) = \cos t + z \sin t$. So $\varphi(z) = x(\pi) = -1$. Since φ is constant, we cannot get $\varphi(z) = 3$ by *any* choice of z!

Computer Exercises 11.1

1.

t	x
0.00000	1.00000 00000 000
0.11111	1.04596 43628 148
0.22222	1.08421 37270 667
0.33333	1.11366 39070 190
0.44444	1.13294 09456 874
0.55556	1.14031 38573 989
0.66667	1.13362 66672 380
0.77778	1.11023 44247 917
0.88889	1.06694 88627 917
1.00000	1.00000 00000 000

Exercises 11.2

1. $-\left(1-\dfrac{h}{2}\right)x_{i-1} + 2(1+h^2)x_1 - \left(1-\dfrac{h}{2}\right)x_{i+1} = -h^2 t$

2. $x_1 \approx 0.29427, \quad x_2 \approx 0.57016, \quad x_3 \approx 0.81040$

4. $x'(0) = \dfrac{5}{3}$ **8.** $-x_{i-1} + \left[2 + (1+t_i)h^2\right] x_i - x_{i+1} = 0$

9. $x(t) = [7/u(2)]u(t)$

11. $x_1'' = -x_1, x_1(0) = 3, x_1'(0) = z_1$ implies
$x = A\cos t + B\sin t, 3 = x(0) = A$,
$x' = -A\sin t + B\cos t$. Let $z_1 = x'(0) = B$.
So $x_1 = 3\cos t + z_1 \sin t, x_2 = 3\cos t + z_2 \sin t$.
By Equation (10), $x = \lambda x_1 + (1-\lambda)x_2$ and
$\lambda = [\beta - x_2(b)]/[x_1(b) - x_2(b)]$
$= x[7-(-3)]/[(-3)-(-3)] = 10/0$

Computer Exercises 11.2

2a. $x = 1/(1+t)$ **2b.** $x = -\log(1+t)$

Exercises 12.1

1a. Elliptic. **1c.** Parabolic. **1f.** Hyperbolic.

2. $\dfrac{1}{r}\dfrac{\partial}{\partial r}\left(r\dfrac{\partial u}{\partial r}\right) + \dfrac{1}{r^2}\dfrac{\partial^2 u}{\partial \theta^2} = 0$

4. Equation (3) shows that $u(x, t+k)$ is a convex combination of values of $u(x,t)$ in the interval [0, 1]. So it remains in the interval.

5. $a = [1 + 2kh^{-2}(\cos \pi h - 1)]^{1/k}$

6. The right-hand side is changed by $b_1 + c_0$ in place of b_1 and $b_{n-1} + c_n$ replacing b_{n-1} for both (5) and (7).

7. In (6), b_1 is replaced by $b_1 + g(t)$, b_{n-1} by $b_{n-1} + g(t)$. At the level zero, $b_i = f(ih)$ for $1 \leqq i \leqq n-1$.

8. $$u(x, t+k) = \frac{k}{h^2}(1-h)u(x+h,t) + \frac{k}{h^2}\left(\frac{h^2}{k} + h - 2\right)u(x,t) + \frac{k}{h^2}u(x-h,t)$$

9. $$A = \begin{bmatrix} 0 & 1 & & & \\ -1 & 0 & 1 & & \\ & \ddots & \ddots & \ddots & \\ & & -1 & 0 & 1 \\ & & & -2 & 2 \end{bmatrix}$$

Exercises 12.2

1. -0.21

2. $u_{xx} = f''(x+at) + g''(x-at)$,
$u_{tt} = a^2 f''(x+at) + a^2 g''(x-at) = a^2 u_{xx}$

3. $$u(x,t) = \frac{1}{2}[F(x+t) - F(-x+t)] + \frac{1}{2}\left[\overline{G}(x+t) - \overline{G}(-x+t)\right]$$
where $\overline{G}$ is the antiderivative of G

Computer Exercises 12.2

1.
```
real function fbar(x)
real x, xbar
xbar ← x + 2 real(integer(−(1 + x)/2))
if xbar < 0 then
    fbar ← −f(−xbar)
else
    fbar ← f(xbar)
end if
end function fbar
```

Exercises 12.3

5. $$\left(20 + \frac{2.5h}{x_i + y_j}\right)u_{i+1,\,j} + \left(20 - \frac{2.5h}{x_i + y_j}\right)u_{i-1,\,j} + \left(-30 + \frac{0.5h}{y_j}\right)u_{i,\,j+1} + \left(-30 + \frac{0.5h}{y_j}\right)u_{i,\,j-1} + 20u_{ij} = 69h^2$$

6. $u\left(0, \frac{1}{2}\right) \approx -8.932 \times 10^{-3}$; $u\left(\frac{1}{2}, \frac{1}{2}\right) \approx 4.643 \times 10^{-1}$

7. $A = \begin{bmatrix} -4 & 1 & 1 & 0 \\ 1 & -4 & 0 & 1 \\ 1 & 0 & -4 & 1 \\ 0 & 1 & 1 & -4 \end{bmatrix}$

Computer Exercises 12.3

5. 18.41° 13.75°
41.47° 36.60° 24.41°
69.41° 66.77° 61.05° 53.01° 51.00°

Exercises 13.1

1. $F(2, 1, -2) = -15$; $F(0, 0, -2) = -8$;
$F(2, 0, -2) = -12$

2. $F\left(\frac{9}{8}, \frac{9}{8}\right) = -20.25$

4. Case $n = 2$:
$$\begin{cases} \widehat{x} = (3a + b)/4 + \delta, & a \leqq x^* \leqq b' \\ \widehat{x} = (a + 3b)/4 - \delta, & a' \leqq x^* \leqq b \end{cases}$$

5a. Exact solution $F(3) = -7$.

7. $A = \alpha/\sqrt{5}$, $A = -\beta/\sqrt{5}$

9. By (6), $y + rb = a + r^2(b - a) + rb = ar + b$ since $r^2 + r = 1$. Moreover, we have $r(y + rb) = a + r(b - a) = x$. Thus, we obtain $yr + r^2 b = x$ or $y + r^2(b - y) = x$.

10. $n \geqq 1 + (k + \log \ell - \log 2)/|\log r|$ **11.** $n \geqq 48$

13. Minimum point of F is a root of F'.
Newton's method to find a root of F':
$$x_{n+1} = x_n - \frac{F'(x_n)}{F''(x_n)}$$
Formula does *not* involve F itself.

14. To find minimum of F, look for root of F'.
Secant method to find a root of F' is:
$$x_{n+1} = x_n - F'(x_n)\left[\frac{x_n - x_{n-1}}{F'(x_n) - F'(x_{n-1})}\right]$$
Formula does *not* involve F.

15b. Square both sides to obtain
$$r^2 = 1 + \sqrt{1 + \sqrt{1 + \cdots}} = 1 + r.$$

15d. By series expansion, we have
$$1 + r^{-1} + r^{-2} + \cdots = (1 - r^{-1})^{-1}$$
Hence, we have $r = (1 - r^{-1})^{-1} - 1 = \dfrac{1}{r - 1}$
or $r^2 = r + 1$.

Exercises 13.2

1a. Yes **1b.** No **2.** $\left(\frac{1}{4}, \frac{9}{4}\right)$

3. $F(x, y) = 1 + x - xy + \frac{1}{2}x^2 - \frac{1}{2}y^2 + \cdots$

6. The slope of the tangent is $\dfrac{dy}{dx} = -\dfrac{F_x}{F_y} \equiv m_1$.
The gradient has direction numbers F_x and F_y, and its slope is $\dfrac{F_y}{F_x} \equiv m_2$. The condition of perpendicularity $m_1 m_2 = -1$ is met.

7b. $F(x) = \frac{3}{2} - \frac{1}{2}x_2 + 3x_1x_2 + x_2x_3 + 2x_1^2 - \frac{1}{2}x_3^2 + \cdots$

9a. $G(1, 0) = \begin{bmatrix} -2 \\ 2 \end{bmatrix}$ **9b.** $G(1, 2, 1) = \begin{bmatrix} 5 \\ 2 \\ 5 \end{bmatrix}$

10. $G = \begin{bmatrix} 2y^2z^2 \sin x \cos x \\ 2yz^2(1 + \sin^2 x) + 2(y + 1)(z + 3)^2 \\ 2y^2z(1 + \sin^2 x) + 2(y + 1)^2(z + 3) \end{bmatrix}$

12. $\left(-\frac{19}{30}, -\frac{1}{5}\right)$

Computer Exercises 13.2

1a. (1, 1) **1c.** (0, 0, 0, 0) **1e.** (1, 1, 1, 1)

Exercises 14.1

2.
$$\begin{cases} \text{Maximize:} & -5x_1 - 6x_2 + 2x_3 \\ \text{Constraints:} & \begin{cases} -2x_1 + 3x_2 \leqq -5 \\ x_1 + x_2 \leqq 15 \\ 2x_1 - x_2 + x_3 \leqq 25 \\ -x_1 - x_2 + x_3 \leqq -1 \\ x_1 \geqq 0,\ x_2 \geqq 0,\ x_3 \geqq 0 \end{cases} \end{cases}$$

4a. Minimum value 1.5 at (1.5, 0).

5b.
$$\begin{cases} \text{Maximize:} & -3x + 2y - 5z \\ \text{Constraints:} & \begin{cases} -x - y - z \leqq -4 \\ x - y - z \leqq 2 \\ -x + y + z \leqq -2 \\ x \geqq 0,\ y \geqq 0,\ z \geqq 0 \end{cases} \end{cases}$$

6a. $$\begin{cases} \text{Maximize:} & 2x_1 + 2x_2 - 6x_3 - x_4 \\ \text{Constraints:} & \begin{cases} 3x_1 + x_4 = 25 \\ x_1 + x_2 + x_3 + x_4 = 20 \\ -4x_1 - 6x_3 + x_5 = -5 \\ -2x_1 - 3x_3 - 2x_4 + x_6 = 0 \\ x_1,\ x_2,\ x_3,\ x_4,\ x_5,\ x_6 \geqq 0 \end{cases} \end{cases}$$

6b. $$\begin{cases} \text{Minimize:} & 25y_1 + 20y_2 - 5y_3 \\ \text{Constraints:} & \begin{cases} 3y_1 + y_2 - 4y_3 - 2y_4 \leqq 2 \\ y_2 \leqq 2 \\ y_2 - 6y_3 - 3y_4 \leqq -6 \\ y_1 + y_2 - 2y_4 \leqq -1 \\ y_1,\ y_2,\ y_3,\ y_4 \geqq 0 \end{cases} \end{cases}$$

7. Maximum of 36 at (2, 6)

8. Minimum of 36 at (0, 3, 1)

11. Minimum 2 for $(x, x-2)$ where $x \geqq 3$

13a. Maximum of 18 at (9, 0)

13c. Unbounded solution

13f. No solution

13h. Maximum of $\frac{54}{5}$ at $\left(\frac{18}{5}, 0\right)$

14. Maximum of 100 at (24, 32, −124)

17. Its feasible set is empty.

Computer Exercises 14.1

1.

	Felt	Straw
Texas Hatters	0	200
Lone Star Hatters	150	0
Lariat Ranch Wear	150	0

3. \$13.50

5. Cost 50¢ for 1.6 ounces of food f_1, 1 ounce of food f_3, and none of food f_2.

Exercises 14.2

1. $$\begin{cases} \text{Maximize:} & \sum_{j=0}^{n} c_j y_j \\ \text{Constraints:} & \begin{cases} \sum_{j=0}^{n} a_{ij} y_j \leqq b_i \\ y_i \geqq 0 \quad (0 \leqq i \leqq n) \end{cases} \end{cases}$$

Here $c_0 = -\sum_{j=1}^{n} c_j$ and $a_{i0} = -\sum_{j=1}^{n} a_{ij}$.

2. At most 2^n.

5. First Primal form:

$$\begin{cases} \text{Maximize:} & -\mathbf{b}^T\mathbf{y} \\ \text{Constraints:} & \begin{cases} -\mathbf{A}^T\mathbf{y} \leqq -\mathbf{c} \\ \mathbf{y} \geqq \mathbf{0} \end{cases} \end{cases}$$

6. Given $\mathbf{A}\mathbf{x} = \mathbf{b}$. Let $y_j = x_j + y_{n+1}$.

Now $\sum_{j=1}^{n} a_{ij}x_j - b_i = \sum_{j=1}^{n} a_{ij}y_j - y_{n+1}\sum_{j=1}^{n} a_{ij} - b_i$.

$$\begin{cases} \text{Minimize:} & y_{n+1} \\ \text{Constraints:} & \begin{cases} \sum_{j=1}^{n} a_{ij}y_j + \left(-\sum_{j=1}^{n} a_{ij}\right) y_{n+1} = b_i \quad (1 \leqq i \leqq n+1) \\ \mathbf{y} \geqq \mathbf{0} \end{cases} \end{cases}$$

Computer Exercises 14.2

1b. $\mathbf{x} = \left[0, 0, \frac{5}{3}, \frac{2}{3}, 0\right]^T$ **1c.** $\mathbf{x} = \left[0, \frac{8}{3}, \frac{5}{3}\right]^T$

Exercises 14.3

1a. $$\begin{cases} \text{Maximize:} & -\sum_{i=1}^{4} (u_i + v_i) \\ \text{Constraints:} & \begin{cases} 5y_1 + 2y_2 - 7y_4 - u_1 + v_1 = 6 \\ y_1 + y_2 + y_3 - 3y_4 - u_2 + v_2 = 2 \\ 7y_2 - 5y_3 - 2y_4 - u_3 + v_3 = 11 \\ 6y_1 + 9y_3 - 15y_4 - u_4 + v_4 = 9 \\ \mathbf{u} \geqq \mathbf{0},\quad \mathbf{v} \geqq \mathbf{0},\quad \mathbf{y} \geqq \mathbf{0} \end{cases} \end{cases}$$

1b. $$\begin{cases} \text{Minimize:} & \varepsilon \\ \text{Constraints:} & \begin{cases} 5y_1 + 2y_2 - 7y_4 - \varepsilon \leqq 6 \\ y_1 + y_2 + y_3 - 3y_4 - \varepsilon \leqq 2 \\ 7y_2 - 5y_3 - 2y_4 - \varepsilon \leqq 11 \\ 6y_1 + 9y_3 - 15y_4 - \varepsilon \leqq 9 \\ -5y_1 - 2y_2 + 7y_4 - \varepsilon \leqq -6 \\ -y_1 - y_2 - y_3 + 3y_4 - \varepsilon \leqq -2 \\ -7y_2 + 5y_3 + 2y_4 - \varepsilon \leqq -11 \\ -6y_1 - 9y_3 + 15y_4 - \varepsilon \leqq -9 \\ \varepsilon \geqq 0,\quad y_j \geqq 0 \quad (1 \leqq i \leqq 4) \end{cases} \end{cases}$$

3. Take m points x_i $(i = 1, 2, \dots, m)$.

Let $p(x) = \sum_{j=0}^{n} a_j x^j$.

$$\begin{cases} \text{Minimize:} \quad \varepsilon \\ \text{Constraints:} \begin{cases} \sum_{j=0}^{n} a_j x_i^j \leqq f(x_i) & (1 \leqq i \leqq m) \\ \sum_{j=0}^{n} a_j x_i^j + \varepsilon \geqq f(x_i) & (1 \leqq i \leqq m) \\ \varepsilon \geqq 0 \end{cases} \end{cases}$$

4.
$$\begin{cases} \text{Minimize:} \quad u_1 + v_1 + u_2 + v_2 + u_3 + v_3 \\ \text{Constraints:} \begin{cases} y_1 - y_2 - u_1 + v_1 = 4 \\ 2y_1 - 3y_2 + y_3 - u_2 + v_2 = 7 \\ y_1 + y_2 - 2y_3 - u_3 + v_3 = 2 \\ y_1, y_2, y_3 \geqq 0; u_1, u_2, u_3 \geqq 0; v_1, v_2, v_3 \geqq 0 \end{cases} \end{cases}$$

Computer Exercises 14.3

1a. $x_1 = 0.353, \quad x_2 = 2.118, \quad x_3 = 0.765$

1b. $x_1 = 0.671, \quad x_2 = 1.768, \quad x_3 = 0.453$

3. $p(x) = 1.0001 + 0.9978x + 0.51307x^2 + 0.13592x^3 + 0.071344x^4$

Exercises Appendix B.1

1a. $e \approx (2.718)_{10} = (010.101\,101\,111\,100\,111\dots)_2$

2d. $(27.45075\,341\dots)_8$ **2e.** $(113.16662\,13\dots)_8$

2f. $(71.24426\,416\dots)_8$

3a. $(441.68164\,0625)_{10}$ **3b.** $(613.40625)_{10}$

4c. $(101\,111)_2$ **4e.** $(110\,011)_2$ **4g.** $(63.72664)_8$

6. $(0.3146\,3146\dots)_8$ **9.** $(479)_{10} = (111\,011\,111)_2$

12. A real number R has a finite representation in the binary system.
$\Leftrightarrow R = (a_m a_{m-1} \dots a_1 a_0 . b_1 b_2 \dots b_n)_2$.
$\Leftrightarrow R = (a_m \dots a_1 a_0 b_1 b_2 \dots b_n)_2 \times 2^{-n} = m \times 2^{-n}$
where $m = (a_m a_{m-1} \dots a_1 a_0 b_1 b_2 \dots b_n)_2$.

Bibliography

Abell, M. L., and J. P. Braselton. 1993. *The Mathematical Handbook*. New York: Academic Press.

Abramowitz, M., and I. A. Stegun (eds.). 1964. *Handbook of Mathematical Functions with Formulas, Graphs, and Mathematical Tables*. National Bureau of Standards. New York: Dover, 1965 (reprint).

Acton, F. S. 1959. *Analysis of Straight-Line Data*. New York: Wiley. New York: Dover, 1966 (reprint).

Acton, F. S. 1990. *Numerical Methods That (Usually) Work*. Washington, D.C.: Mathematical Association of America.

Acton, F. S. 1996. *Real Computing Made Real: Preventing Errors in Scientific and Engineering Calculations*. Princeton, New Jersey: Princeton University Press.

Ahlberg, J. H., E. N. Nilson, and J. L. Walsh. 1967. *The Theory of Splines and Their Applications*. New York: Academic Press.

Aiken, R. C., ed. 1985. *Stiff Computation*. New York: Oxford University Press.

Ames, W. F. 1992. *Numerical Methods for Partial Differential Equations*, 3rd Ed. New York: Academic Press.

Ammar, G. S., D. Calvetti, and L. Reichel, 1999. "Computation of Gauss-Kronrod quadrature rules with non-positive weights," *Electronic Transactions on Numerical Analysis* **9**, 26–38. `http://etna.mcs.kent.edu`

Anderson, E., Z. Bai, C. Bischof, S. Blackford, J. Demmel, J. Dongarra, J. Du Croz, A. Greenbaum, S. Hammarling, A. McKenney, and D. Sorensen. 1999. *LAPACK User's Guide*, 3rd Ed.. Philadelphia: SIAM.

Armstrong, R. D., and J. Godfrey. 1979. "Two linear programming algorithms for the linear discrete ℓ_1 norm problem." *Mathematics of Computation* **33**, 289–300.

Ascher, U. M., R. M. M. Mattheij, and R. D. Russell. 1995. *Numerical Solution of Boundary Value Problems for Ordinary Differential Equations*. Philadelphia: SIAM.

Ascher, U. M., and L. R. Petzold. 1998. *Computer Methods for Ordinary Differential Equations and Differential Algebraic Equations*. Philadelphia: SIAM.

Atkinson, K. 1993. *Elementary Numerical Analysis*. New York: Wiley.

Atkinson, K. A. 1988. *An Introduction to Numerical Analysis*, 2nd Ed. New York: Wiley.

Axelsson, O. 1994. *Iterative Solution Methods*. New York: Cambridge University Press.

Axelsson, O., and V. A. Barker. 2001. *Finite Element Solution of Boundary Value Problems: Theory and Computations*. Philadelphia: SIAM.

Azencott, R., ed. 1992. *Simulated Annealing: Parallelization Techniques*. New York: Wiley.

Bai, Z., J. Demmel, J. Dongarra, A. Ruhe, and H. van der Vorst. 2000. *Templates for the Solution of Algebraic Eigenvalue Problems: A Practical Guide*. Philadelphia: SIAM.

Baldick, R. 2006. *Applied Optimization*. New York, Cambridge University Press.

Barnsley, M. F. 2006. *SuperFractals*. New York, Cambridge University Press.

Barrett, R., M. Berry, T. F. Chan, J. Demmel, J. Donato, J. Dongarra, V. Eijkhout, R. Pozo, C. Romine, and H. van der Vorst. 1994. *Templates for the Solution of Linear Systems: Building Blocks for Iterative Methods* Philadelphia: SIAM.

Barrodale, I., and C. Phillips. 1975. "Solution of an overdetermined system of linear equations in the Chebyshev norm." *Association for Computing Machinery Transactions on Mathematical Software* **1**, 264–270.

Barrodale, I., and F. D. K. Roberts. 1974. "Solution of an overdetermined system of equations in the ℓ_1 norm." *Communications of the Association for Computing Machinery* **17**, 319–320.

Barrodale, I., F. D. K. Roberts, and B. L. Ehle. 1971. *Elementary Computer Applications*. New York: Wiley.

Bartels, R. H. 1971. "A stabilization of the simplex method." *Numerische Mathematik* **16**, 414–434.

Bartels, R., J. Beatty, and B. Barskey. 1987. *An Introduction to Splines for Use in Computer Graphics and Geometric Modelling*. San Francisco: Morgan Kaufmann.

Bassien, S. 1998. "The dynamics of a family of one-dimensional maps." *American Mathematical Monthly* **105**, 118–130.

Bayer, D., and P. Diaconis. 1992. "Trailing the dovetail shuffle to its lair." *Annals of Applied Probability*, **2**, 294–313.

Beale, E. M. L. 1988. *Introduction to Optimization*. New York: Wiley.

Berry, M. W., and M. Browne 2005. *Understanding Search Engineerings in Mathematical Modeling and Text Retrieval*, 2nd Ed. Philadelphia: SIAM.

Björck, Å. 1996. *Numerical Methods for Least Squares Problems*. Philadelphia: SIAM.

Bloomfield, P., and W. Steiger. 1983. *Least Absolute Deviations, Theory, Applications, and Algorithms*. Boston: Birkhäuser.

Bornemann, F., D. Laurie, S. Wagon, and J. Waldvogel. 2004. *The SIAM 100-Digit Challenge: A Study in High-Accuracy Numerical Computing*. Philadelphia: SIAM.

Borwein, J. M., and P. B. Borwein. 1984. "The arithmetic-geometric mean and fast computation of elementary functions." *Society for Industrial and Applied Mathematics Review* **26**, 351–366.

Borwein, J. M., and P. B. Borwein. 1987. *Pi and the AGM: A Study in Analytic Number Theory and Computational Complexity*. New York: Wiley.

Boyce, W. E., and R. C. DiPrima. 2008. *Elementary Differential Equations and Boundary Value Problems*, 8th Ed. New York: Wiley.

Branham, R. 1990. *Scientific Data Analysis: An Introduction to Overdetermined Systems*. New York: Springer-Verlag.

Brenner, S., and R. Scott. 2002. *The Mathematical Theory of Finite Element Methods*. New York: Springer-Verlag.

Brent, R. P. 1976. "Fast multiple precision evaluation of elementary functions." *Journal of the Association for Computing Machinery* **23**, 242–251.

Briggs, W. 2004. *Ants, Bikes, and Clocks: Problems Solving for Undergraduates*. Philadelphia: SIAM.

Buchanan, J. L., and P. R. Turner. 1992. *Numerical Methods and Analysis*. New York: McGraw-Hill.

Burden, R. L., and J. D. Faires. 2011. *Numerical Analysis*, 9th Ed. Boston: Brooks/Cole Cengage Learning.

Bus, J. C. P., and T. J. Dekker. 1975. "Two efficient algorithms with guaranteed convergence for finding a zero of a function." *Association for Computing Machinery Transactions on Mathematical Software* **1**, 330–345.

Butcher, J. C. 1987. *The Numerical Analysis of Ordinary Differential Equations: Runge-Kutta and General Linear Methods*. New York: Wiley.

Calvetti, D., G. H. Golub, W. B. Gragg, and L. Reichel. 2000. "Computation of Gauss-Kronrod quadrature rules." *Mathematics of Computation* **69**, 1035–1052.

Carrier, G., and C. Pearson. 1991. *Ordinary Differential Equations*. Philadelphia: SIAM.

Cärtner, B. 2006. *Understanding and Using Linear Programming*. New York: Springer.

Cash, J. "Mesh selection for nonlinear two-point boundary-value problems." *Journal of Computational Methods in Science and Engineering*, 2003.

Chaitlin, G. J. 1975. "Randomness and mathematical proof." *Scientific American* May, 47–52.

Chapman, S. J. 2000. *MATLAB Programming for Engineering*, Pacific Grove, California: Brooks/Cole.

Chapra, S. L. 2012. *Applied Numerical Methods for Engineers and Scientist* 3rd Ed. New York: McGraw-Hill.

Cheney, E. W. 1982. *Introduction to Approximation Theory*, 2nd Ed. Washington, D.C.: AMS.

Cheney, E. W. 2001. *Analysis for Applied Mathematics*, New York: Springer.

Cheney, W., and D. Kincaid. 2012. *Linear Algebra: Theory and Applications*, 2nd Ed., Sudbury, Massachusetts, Jones & Bartlett Learning.

Chicone, C. 2006. *Ordinary Differential Equations with Applications*. 2nd Ed. New York: Springer.

Clenshaw, C. W., and A. R. Curtis. 1960. "A method for numerical integration on an automatic computer." *Numerische Mathematik* **2**, 197–205.

Colerman, T. F., and C. Van Loan. 1988. *Handbook for Matrix Computations*. Philadelphia: SIAM.

Collatz, L. 1966. *The Numerical Treatment of Differential Equations*, 3rd Ed. Berlin: Springer-Verlag.

Conte, S. D., and C. de Boor. 1980. *Elementary Numerical Analysis*, 3rd Ed. New York: McGraw-Hill.

Cooper, L., and D. Steinberg. 1974. *Methods and Applications of Linear Programming*. Philadelphia: Saunders.

Crilly, A. J., R. A. Earnshaw, H. Jones, eds. 1991. *Fractals and Chaos*. New York: Springer-Verlag.

Cvijovic, D., and J. Klinowski. 1995. "Taboo search: An approach to the multiple minima problem." *Science* **267**, 664–666.

Dahlquist, G., and A. Björck. 1974. *Numerical Methods*. Englewood Cliffs, New Jersey: Prentice-Hall.

Dantzi, G. B., A. Orden, and P. Wolfe. 1963. "Generalized simplex method for minimizing a linear from under linear inequality constraints." *Pacific Journal of Mathematics* **5**, 183–195.

Davis, P. J., and P. Rabinowitz. 1984. *Methods of Numerical Integration*, 2nd Ed. New York: Academic Press.

Davis, T. 2006. *Direct Methods for Sparse Linear Systems*. Philadelphia: SIAM.

de Boor, C. 1971. "CADRE: An algorithm for numerical quadrature." In *Mathematical Software*, edited by J. R. Rice, 417–449. New York: Academic Press.

de Boor, C. 1984. *A Practical Guide to Splines*. 2nd Ed. New York: Springer-Verlag.

Dekker, T. J. 1969. "Finding a zero by means of successive linear interpolation." In *Constructive Aspects of the Fundamental Theorem of Algebra,* edited by B. Dejon and P. Henrici. New York: Wiley-Interscience.

Dekker, T. J., and W. Hoffmann. 1989. "Rehabilitation of the Gauss-Jordan algorithm." *Numerische Mathematik* **54**, 591–599.

Dekker, T. J., W. Hoffmann, and K. Potma. 1997. "Stability of the Gauss-Huard algorithm with partial pivoting." *Computing* **58**, 225–244.

Dekker, K., and J. G. Verwer. 1984. "Stability of Runge-Kutta methods for stiff nonlinear differential equations." CWI Monographs **2**. Amsterdam: Elsevier Science.

Demmel, J. W., 1997. *Applied Numerical Linear Algebra.* Philadelphia: SIAM.

Dennis, J. E., and R. Schnabel. 1983. *Quasi-Newton Methods for Nonlinear Problems.* Englewood Cliffs, New Jersey: Prentice-Hall.

Dennis, J. E., and R. B. Schnabel. 1996. *Numerical Methods for Unconstrained Optimization and Nonlinear Equations.* Philadelphia: SIAM.

Dennis, J. E., and D. J. Woods. 1987. "Optimization on microcomputers: The Nelder-Mead simplex algorithm." In *New Computing Environments*, edited by A. Wouk. Philadelphia: SIAM.

de Temple, D. W. 1993. "A quicker convergence to Euler's Constant." *American Mathematical Monthly* **100**, 468–470.

Devitt, J. S. 1993. *Calculus with Maple V.* Pacific Grove, California: Brooks/Cole.

Dixon, V. A. 1974. "Numerical quadrature: a survey of the available algorithms." In *Software for Numerical Mathematics*, edited by D. J. Evans. New York: Academic Press.

Dongarra, J. J., I. S. Duff, D. C. Sorenson, and H. van der Vorst. 1990. *Solving Linear Systems on Vector and Shared Memory Computers.* Philadelphia: SIAM.

Dorn, W. S., and D. D. McCracken. 1972. *Numerical Methods with FORTRAN IV Case Studies.* New York: Wiley.

Edwards, C., and D. Penny. 2004. *Differential Equations and Boundary Value Problems*, 5th Ed. Upper Saddle River, New Jersey: Prentice-Hall.

Ellis, W., Jr., E. W. Johnson, E. Lodi, and D. Schwalbe. 1997. *Maple V Flight Manual: Tutorials for Calculus, Linear Algebra, and Differential Equations.* Pacific Grove, California: Brooks/Cole.

Ellis, W., Jr., and E. Lodi. 1991. *A Tutorial Introduction to Mathematica.* Pacific Grove, California: Brooks/Cole.

Elman, H., D. J. Silvester, and A. Wathen. 2004. *Finite Element and Fast Iterative Solvers.* New York: Oxford University Press.

England, R. 1969. "Error estimates for Runge-Kutta type solutions of ordinary differential equations." *Computer Journal* **12**, 166–170.

Enright, W. H. 2006. "Verifying approximate solutions to differential equations." *Journal of Computational and Applied Mathematics* **185**, 203–311.

Epureanu, B. I., and H. S. Greenside. 1998. "Fractal basins of attraction associated with a damped Newton's method." *SIAM Review* **40**, 102–109.

Evans, G., J. Blackledge, and P. Yardlay. 2000. *Numerical Methods for Partial Differential Equations.* New York: Springer-Verlag.

Evans, G. W., G. F. Wallace, and G. L. Sutherland. 1967. *Simulation Using Digital Computers.* Englewood Cliffs, New Jersey: Prentice-Hall.

Farin, G. 1990. *Curves and Surfaces for Computer Aided Geometric Design: A Practical Guide*, 2nd Ed. New York: Academic Press.

Fauvel, J., R. Flood, M. Shortland, and R. Wilson (eds.). 1988. *Let Newton Be!* London: Oxford University Press.

Feder, J. 1988. *Fractals.* New York: Plenum Press.

Fehlberg, E. 1969. "Klassische Runge-Kutta formeln fünfter und siebenter ordnung mit schrittweitenkontrolle." *Computing* **4**, 93–106.

Ferris, M. C., O. L. Mangasarian, and S. J. Wright 2007. *Linear Programming with MATLAB.* Philadelphia: SIAM.

Flehinger, B. J. 1966. "On the probability that a random integer has initial digit A." *American Mathematical Monthly* **73**, 1056–1061.

Fletcher, R. 1976. *Practical Methods of Optimization.* New York: Wiley.

Floudas, C. A., and P. M. Pardalos (eds.). 1992. *Recent Advances in Global Optimization.* Princeton, New Jersey: Princeton University Press.

Flowers, B. H. 1995. *An Introduction to Numerical Methods in C++.* New York: Oxford University Press.

Ford, J. A. 1995. "Improved Algorithms of Ilinois-Type for the Numerical Solution of Nonlinear Equations." Technical Report, Department of Computer Science, University of Essex, Colchester, Essex, UK.

Forsythe, G. E. 1957. "Generation and use of orthogonal polynomials for data-fitting with a digital computer." *Society for Industrial and Applied Mathematics Journal* **5**, 74–88.

Forsythe, G. E. 1970. "Pitfalls in computation, or why a math book isn't enough," *American Mathematical Monthly* **77**, 931–956.

Forsythe, G. E., M. A. Malcolm, and C. B. Moler. 1977. *Computer Methods for Mathematical Computations.* Englewood Cliffs, New Jersey: Prentice- Hall.

Forsythe, G. E., and C. B. Moler. 1967. *Computer Solution of Linear Algebraic Systems.* Englewood Cliffs, New Jersey: Prentice-Hall.

Forsythe, G. E., and W. R. Wasow. 1960. *Finite Difference Methods for Partial Differential Equations.* New York: Wiley.

Fox, L. 1957. *The Numerical Solution of Two-Point Boundary Problems in Ordinary Differential Equations.* Oxford: Clarendon Press.

Fox, L. 1964. *An Introduction to Numerical Linear Algebra, Monograph on Numerical Analysis.* Oxford: Clarendon Press. Reprinted 1974. New York: Oxford University Press.

Fox, L., D. Juskey, and J. H. Wilkinson, 1948. "Notes on the solution of algebraic linear simultaneous equations," *Quarterly Journal of Mechanics and Applied Mathematics.* 1, 149–173.

Frank, W. 1958. "Computing eigenvalues of complex matrices by determinant evaluation and by methods of Danilewski and Wielandt." *Journal of SIAM* **6**, 37–49.

Fraser, W., and M. W. Wilson. 1966. "Remarks on the Clenshaw-Curtis quadrature scheme." *SIAM Review* **8**, 322–327.

Friedman, A., and N. Littman. 1994. *Industrial Mathematics: A Course in Solving Real-World Problems*. Philadelphia: SIAM.

Fröberg, C.-E. 1969. *Introduction to Numerical Analysis*. Reading, Massachusetts: Addison-Wesley.

Gallivan, K. A., M. Heath, E. Ng, B. Peyton, R. Plemmons, J. Ortega, C. Romine, A. Sameh, and R. Voigt. 1990. *Parallel Algorithms for Matrix Computations*. Philadelphia: SIAM.

Gander, W., and W. Gautschi. 2000. "Adaptive quadrature—revisited." *BIT* **40**, 84–101.

Garvan, F. 2002. *The Maple Book*. Boca Raton, Florida: Chapman & Hall/CRC.

Gautschi, W. 1990. "How (un)stable are Vandermonde systems?" in *Asymptotic and Computational Analysis*, 193–210, Lecture Notes in Pure and Applied Mathematics, 124. New York: Dekker.

Gautschi, W. 1997. *Numerical Analysis: An Introduction*. Boston: Birkhäuser.

Gear, C. W. 1971. *Numerical Initial Value Problems in Ordinary Differential Equations*. Englewood Cliffs, New Jersey: Prentice-Hall.

Gentle, J. E. 2003. *Random Number Generation and Monte Carlo Methods*, 2nd Ed. New York: Springer-Verlag.

Gentleman, W. M. 1972. "Implementing Clenshaw-Curtis quadrature." *Communications of the ACM* **15**, 337–346, 353.

Gerald, C. F., and P. O. Wheatley 2003. *Applied Numerical Analysis*, 7th Ed. Reading, Massachusetts: Addison-Wesley Longman.

Ghizetti, A., and A. Ossiccini. 1970. *Quadrature Formulae*. New York: Academic Press.

Gill, P. E., W. Murray, and M. H. Wright. 1981. *Practical Optimization*. New York: Academic Press.

Gleick, J. 1992. *Genius: The Life and Science of Richard Feynman*. New York: Pantheon.

Gockenbach, M. S. 2002. *Partial Differential Equations: Analytical and Numerical Methods*. Philadelphia: SIAM.

Goldberg, D. 1991. "What every computer scientist should know about floating-point arithmetic." *ACM Computing Surveys* **23**, 5–48.

Goldstine, H. H. 1977. *A History of Numerical Analysis from the 16th to the 19th Century*. New York: Springer-Verlag.

Golub, G. H., and J. M. Ortega. 1992. *Scientific Computing and Differential Equations*. New York: Harcourt Brace Jovanovich.

Golub, G. H., and J. M. Ortega. 1993. *An Introduction with Parallel Scientific Computing*. New York: Academic Press.

Golub, G. H., and C. F. Van Loan. 1996. *Matrix Computations*, 3rd Ed. Baltimore: Johns Hopkins University Press.

Good, I. J. 1972. "What is the most amazing approximate integer in the universe?" *Pi Mu Epsilon Journal* **5**, 314–315.

Greenbaum, A. 1997. *Iterative Methods for Solving Linear Systems*. Philadelphia: SIAM.

Greenbaum, A. 2002. "Card Shuffling and the Polynomial Numerical Hull of Degree *k*," Mathematics Department, University of Washington, Seattle, Washington.

Gregory, R. T., and D. Karney, 1969. *A Collection of Matrices for Testing Computational Algorithms*. New York: Wiley.

Griewark, A. 2000. *Evaluating Derivatives: Principles and Techniques of Algorithmic Differentiation*. Philadelphia: SIAM.

Groetsch, C. W. 1998. "Lanczos' generalized derivative." *American Mathematical Monthly* **105**, 320–326.

Haberman, R. 2004. *Applied Partial Differential Equations with Fourier Series and Boundary Value Problems*. Upper Saddle River, New Jersey: Prentice-Hall.

Hageman, L. A., and D. M. Young. 1981. *Applied Iterative Methods*. New York: Academic Press; Dover 2004 (reprint).

Hämmerlin, G., and K.-H. Hoffmann. 1991. *Numerical Mathematics*. New York: Springer-Verlag.

Hammersley, J. M., and D. C. Handscomb. 1964. *Monte Carlo Methods*. London: Methuen.

Hansen, T., G. L. Mullen, and H. Niederreiter. 1993. "Good parameters for a class of node sets in quasi-Monte Carlo integration." *Mathematics of Computation* **61**, 225–234.

Haruki, H., and S. Haruki. 1983. "Euler's Integrals." *American Mathematical Monthly* **7**, 465.

Hastings, H. M., and G. Sugihara. 1993. *Fractals: A User's Guide for the Natural Sciences*. New York: Oxford University Press.

Havie, T. 1969. "On a modification of the Clenshaw-Curtis quadrature formula." *BIT* **9**, 338–350.

Heath, J. M. 2002. *Scientific Computing: An Introductory Survey*, 2nd Ed. New York: McGraw-Hill.

Henrici, P. 1962. *Discrete Variable Methods in Ordinary Differential Equations*. New York: Wiley.

Heroux, M., P. Raghavan, and H. Simon. 2006. *Parallel Processing for Scientific Computing*. Philadelphia: SIAM.

Herz-Fischler, 1998. R. *A Mathematical History of the Golden Number*. New York: Dover

Hestenes, M. R., and E. Stiefel. 1952. "Methods of conjugate gradient for solving linear systems." *Journal Research National Bureau of Standards* **49**, 409–436.

Higham, N. J. 2002. *Accuracy and Stability of Numerical Algorithms*, 2nd Ed. Philadelphia: SIAM.

Hildebrand, F. B. 1974. *Introduction to Numerical Analysis*. New York: McGraw-Hill.

Hodges, A. 1983. *Alan Turing: The Enigma*. New York: Simon & Schuster.

Hoffmann, W. 1989. "A fast variant of the Gauss-Jordan algorithm with partial pivoting. Basic transformations in linear algebra for vector computing." Doctoral dissertation, University of Amsterdam, The Netherlands.

Hofmann-Wellenhof, B., H. Lichtenegger, and J. Collins. 2001. *Global Positioning System: Theory and Practice*, 5th Ed. New York: Springer-Verlag.

Horst, R., P. M. Pardalos, and N. V. Thoai. 2000. *Introduction to Global Optimization*, 2nd Ed. Boston: Kluwer.

Householder, A. S. 1970. *The Numerical Treatment of a Single Nonlinear Equation*. New York: McGraw-Hill.

Huard, P. 1979. "La méthode du simplexe sans inverse explicite." *Bull. E.D.F. Série C* **2**.

Huddleston, J. V. 2000. *Extensibility and Compressibility in One-Dimensional Structures*. 2nd Ed. Buffalo, NY: ECS Publ.

Hull, T. E., and A. R. Dobell. 1962. "Random number generators." *Society for Industrial and Applied Mathematics Review* **4**, 230–254.

Hull, T. E., W. H. Enright, B. M. Fellen, and A. E. Sedgwick. 1972. "Comparing numerical methods for ordinary differential equations." *Society for Industrial and Applied Mathematics Journal on Numerical Analysis* **9**, 603–637.

Hundsdorfer, W. H. 1985. "The numerical solution of nonlinear stiff initial value problems: an analysis of one step methods." CWI Tract, 12. Amsterdam: Stichting Mathematisch Centrum, Centrum voor Wiskunde en Informatica.

Isaacson, E., and H. B. Keller. 1966. *Analysis of Numerical Methods*. New York: Wiley.

Jeffrey, A. 2000. *Handbook of Mathematical Formulas and Integrals*. Boston: Academic Press.

Jennings, A. 1977. *Matrix Computation for Engineers and Scientists*. New York: Wiley.

Johnson, L. W., R. D. Riess, and J. T. Arnold. 1997. *Introduction to Linear Algebra*. New York: Addison-Wesley.

Kahaner, D. K. 1971. "Comparison of numerical quadrature formulas." In *Mathematical Software*, edited by J. R. Rice. New York: Academic Press.

Kahaner, D., C. Moler, and S. Nash. 1989. *Numerical Methods and Software*. Englewood Cliffs, New Jersey: Prentice-Hall.

Keller, H. B. 1968. *Numerical Methods for Two-Point Boundary-Value Problems*. Toronto: Blaisdell.

Keller, H. B. 1976. *Numerical Solution of Two-Point Boundary Value Problems*. Philadelphia: SIAM.

Kelley, C. T. 1995. *Iterative Methods for Linear and Nonlinear Equations*. Philadelphia: SIAM.

Kelley, C. T. 2003. *Solving Nonlinear Equations with Newton's Method*. Philadelphia: SIAM.

Kincaid, D., and W. Cheney. 2002. *Numerical Analysis: Mathematics of Scientific Computing*, 3rd Ed. Providence, Rhode Island: American Mathematical Society.

Kincaid, D. R., and D. M. Young. 1979. "Survey of iterative methods." In *Encyclopedia of Computer Science and Technology*, edited by J. Belzer, A. G. Holzman, and A. Kent. New York: Dekker.

Kincaid, D. R., and D. M. Young. 2000. "Partial differential equations." In *Encyclopedia of Computer Science*, 4th Ed., edited by A. Ralston, E. D. Reilly, D. Hemmendinger. New York: Grove's Dictionaries.

Kinderman, A. J., and J. F. Monahan. 1977. "Computer generation of random variables using the ratio of uniform deviates." *Association of Computing Machinery Transactions on Mathematical Software* **3**, 257–260.

Kirkpatrick, S., C. D. Gelatt, Jr., and M. P. Vecchi. 1983. "Optimization by simulated annealing." *Science* **220**, 671–680.

Knuth, D. E. 1997. *The Art of Computer Programming*, 3rd Ed. Vol. 2, *Seminumerical Algorithms*. New York: Addison-Wesley.

Krogh, F. T. 2003. "On developing mathematical software." *Journal of Computational and Applied Mathematics* **185**, 196–202.

Kronrod, A. S. 1964. "Nodes and Weights of Quadrature Rules." *Doklady Akad. Nauk SSSR*, **154**, 283–286. [Russian] (1965. New York: Consultants Bureau.)

Krylov, V. I. 1962. *Approximate Calculation of Integrals*, translated by A. Stroud. New York: Macmillan.

Lambert, J. D. 1973. *Computational Methods in Ordinary Differential Equations*. New York: Wiley.

Lambert, J. D. 1991. *Numerical Methods for Ordinary Differential Equations*. New York: Wiley.

Lapidus, L., and J. H. Seinfeld. 1971. *Numerical Solution of Ordinary Differential Equations*. New York: Academic Press.

Laurie, D. P. 1997. "Calculation of Gauss-Kronrod quadrature formulae." *Mathematics of Computation*, 1133–1145.

Lawson, C. L., and R. J. Hanson. 1995. *Solving Least-Squares Problems*. Philadelphia: SIAM.

Leva, J. L. 1992. "A fast normal random number generator." *Association of Computing Machinery Transactions on Mathematical Software* **18**, 449–455.

Lindfield, G., and J. Penny. 2000. *Numerical Methods Using MATLAB*, 2nd Ed. Upper Saddle River: New Jersey: Prentice-Hall.

Lootsam, F. A., ed. 1972. *Numerical Methods for Nonlinear Optimization*. New York: Academic Press.

Lozier, D. W., and F. W. J. Olver. 1994. "Numerical evaluation of special functions." In *Mathematics of Computation 1943–1993: A Half-Century of Computational Mathematics* **48**, 79–125. Providence, Rhode Island: AMS.

Lynch, S. 2004. *Dynamical Systems with Applications*. Boston: Birkhäuser.

MacLeod, M. A. 1973. "Improved computation of cubic natural splines with equi-spaced knots." *Mathematics of Computation* **27**, 107–109.

Maron, M. J. 1991. *Numerical Analysis: A Practical Approach*. Boston: PWS Publishers.

Marsaglia, G. 1968. "Random numbers fall mainly in the planes." *Proceedings of the National Academy of Sciences* **61**, 25–28.

Marsaglia, G., and W. W. Tsang. 2000. "The Ziggurat Method for generating random variables." *Journal of Statistical Software* **5**, 1–7.

Mattheij, R. M. M., S. W. Rienstra, and J. H. M. ten Thije Boonkkamp. 2005. *Partial Differential Equations: Modeling, Analysis, Computation*. Philadelphia: SIAM.

McCartin, B. J. 1998. "Seven deadly sins of numerical computations," *American Mathematical Monthly* **105**, No. 10, 929–941.

McKenna, P. J., and C. Tuama. 2001. "Large torsional oscillations in suspension bridges visited again: Vertical forcing creates torsional response." *American Mathematical Monthly* **108**, 738–745.

Mehrotra, S. 1992. "On the implementation of a primal-dual interior point method." *SIAM Journal on Optimization* **2**, 575–601.

Metropolis, N., A. W. Rosenbluth, M. N. Rosenbluth, A. H. Teller, and E. Teller. 1953. "Equation of state calculations by fast computing machines." *Journal of Physical Chemistry* **21**, 1087–1092.

Meurant, G. 2006. *The Lanczos and Conjugate Gradient Algorithms: From Theory to Finite Precision Computations*. Philadelphia: SIAM.

Meyer, C. D. 2000. *Matrix Analysis and Applied Linear Algebra*. Philadelphia: SIAM.

Miranker, W. L. 1981. "Numerical methods for stiff equations and singular perturbation problems." In *Mathematics and its Applications, Vol. 5*. Dordrecht-Boston, Massachusetts: D. Reidel.

Moler, C. B., 2008. *Numerical Computing with MATLAB*, Revised Reprint. Philadelphia: SIAM.

Moler, C. 2011. *Clever's Corner – Computing* π, Mathworks News & Notes, `www.mathworks.com`

Moré, J. J., and S. J. Wright. 1993. *Optimization Software Guide*. Philadelphia: SIAM.

Moulton, F. R. 1930. *Differential Equations*. New York: Macmillan.

Nelder, J. A., and R. Mead. 1965. "A simplex method for function minimization." *Computer Journal* **7**, 308–313.

Nerinckx, D., and A. Haegemans. 1976. "A comparison of nonlinear equation solvers." *Journal of Computational and Applied Mathematics* **2**, 145–148.

Nering, E. D., and A. W. Tucker. 1992. *Linear Programs and Related Problems*. New York: Academic Press.

Niederreiter, H. 1978. "Quasi-Monte Carlo methods." *Bulletin of the American Mathematical Society* **84**, 957–1041.

Niederreiter, H. 1992. *Random Number Generation and Quasi-Monte Carlo Methods*. Philadelphia: SIAM.

Nievergelt, J., J. G. Farrar, and E. M. Reingold. 1974. *Computer Approaches to Mathematical Problems*. Englewood Cliffs, New Jersey: Prentice-Hall.

Noble, B., and J. W. Daniel. 1988. *Applied Linear Algebra*, 3rd Ed. Englewood Cliffs, New Jersey: Prentice-Hall.

Nocedal, J., and S. Wright. 2006. *Numerical Optimization*. 2nd Ed. New York: Springer.

Novak, E., K. Ritter, and H. Woźniakowski. 1995. "Average-case optimality of a hybrid secant-bisection method." *Mathematics of Computation* **64**, 1517–1540.

Novak, M., ed. 1998. *Fractals and Beyond: Complexities in the Sciences*. River Edge, New Jersey: World Scientific.

O'Hara, H., and F. J. Smith. 1968. "Error estimation in Clenshaw-Curtis quadrature formula." *Computer Journal* **11**, 213–219.

O'Leary, D. P. 2009. *Scientific Computing with Case Studies*. Philadelphia: SIAM.

Oliveira, S., and D. E. Stewart. 2006. *Writing Scientific Software: A Guide to Good Style*. New York: Cambridge University Press.

Orchard-Hays, W. 1968. *Advanced Linear Programming Computing Techniques*. New York: McGraw-Hill.

Ortega, J., and R. G. Voigt. 1985. *Solution of Partial Differential Equations on Vector and Parallel Computers*. Philadelphia: SIAM.

Ortega, J. M. 1990a. *Numerical Analysis: A Second Course*. Philadelphia: SIAM.

Ortega, J. M. 1990b. *Introduction to Parallel and Vector Solution of Linear Systems*. New York: Plenum.

Ortega, J. M., and W. C. Rheinboldt. 1970. *Iterative Solution of Nonlinear Equations in Several Variables*. New York: Academic Press. (2000. Reprint. Philadelphia: SIAM.)

Ostrowski, A. M. 1966. *Solution of Equations and Systems of Equations*, 2nd Ed. New York: Academic Press.

Overton, M. L. 2001. *Numerical Computing with IEEE Floating Point Arithmetic*. Philadelphia: SIAM.

Otten, R. H. J. M., and L. P. P. van Ginneken. 1989. *The Annealing Algorithm*. Dordrecht, Germany: Kluwer.

Pacheco, P. 1997. *Parallel Programming with MPI*. San Francisco: Morgan Kaufmann.

Patterson, T. N. L. 1968. "The optimum addition of points to quadrature formulae." *Mathematics of Computations* **22**, 847–856, and in 1969 *Mathematics of Computations* **23**, 892.

Parlett, B. N. 1997. *The Symmetric Eigenvalue Problem*. Philadelphia: SIAM.

Parlett, B. 2000. "The QR Algorithm," *Computing in Science and Engineering* **2**, 38–42.

Pessens, R., E. de Doncker, C. W. Uberhuber, and D. K. Kahaner, 1983. *QUADPACK: A Subroutine Package for Automatic Integration*. New York: Springer-Verlag.

Peterson, I. 1997. *The Jungles of Randomness: A Mathematical Safari*. New York: Wiley.

Phillips, G. M., and P. J. Taylor. 1973. *Theory and Applications of Numerical Analysis*. New York: Academic Press.

Press, W. H., S. A. Teukolsky, W. T. Vetterling, and B. P. Flannery. 2007. *Numerical Recipes in C++*, 3rd Ed. New York: Cambridge University Press.

Quinn, M. J. 1994. *Parallel Computing: Theory and Practice*. New York: McGraw-Hill.

Rabinowitz, P. 1968. "Applications of linear programming to numerical analysis." *Society for Industrial and Applied Mathematics Review* **10**, 121–159.

Rabinowitz, P. 1970. *Numerical Methods for Nonlinear Algebraic Equations*. London: Gordon & Breach.

Raimi, R. A. 1969. "On the distribution of first significant figures." *American Mathematical Monthly* **76**, 342–347.

Ralston, A. 1965. *A First Course in Numerical Analysis*. New York: McGraw-Hill.

Ralston, A., and C. L. Meek (eds.) 1976. *Encyclopedia of Computer Science*. New York: Petrocelli/Charter.

Ralston, A., and P. Rabinowitz 2001. *A First Course in Numerical Analysis*, 2nd Ed. New York: Dover.

Reid, J. 1971. "On the method of conjugate gradient for the solution of large sparse systems of linear equations." In *Large Sparse Sets of Linear Equations*, J. Reid (ed.), London: Academic Press.

Rheinboldt, 1998. *Methods for Solving Systems of Nonlinear Equations*, 2nd Ed. Philadelphia: SIAM.

Rice, J. R. 1971. "SQUARS: An algorithm for least squares approximation." In *Mathematical Software*, edited by J. R. Rice. New York: Academic Press.

Rice, J. R. 1983. *Numerical Methods, Software, and Analysis*. New York: McGraw-Hill.

Rice, J. R., and R. F. Boisvert. 1984. *Solving Elliptic Problems Using ELLPACK*. New York: Springer-Verlag.

Rice, J. R., and J. S. White. 1964. "Norms for smoothing and estimation." *Society for Industrial and Applied Mathematics Review* **6**, 243–256.

Rivlin, T. J. 1990. *The Chebyshev Polynomials*, 2nd Ed. New York: Wiley.

Roger, H.-F. 1998. *A Mathematical History of the Golden Number*. New York: Dover.

Roos, C., T. Terlaky, and J.-Ph. Vial. 1997. *Theory and Algorithms for Linear Optimization: An Interior Point Approach*. New York: Wiley.

Saad, Y. 2011. *Numerical Methods for Large Eigenvalues Problems*, 2nd Ed. Philadelphia: SIAM.

Saad, Y., 2003. *Iterative Methods for Sparse Linear Systems*, 3rd Ed. Philadelphia: SIAM.

Salamin, E. 1976. "Computation of π using arithmetic-geometric mean." *Mathematics of Computation* **30**, 565–570.

Sauer, T. 2012. *Numerical Analysis*. New York: Pearson, 2nd Ed., Addison-Wesley.

Scheid, F. 1968. *Theory and Problems of Numerical Analysis*. New York: McGraw-Hill.

Scheid, F. 1990. *2000 Solved Problems in Numerical Analysis*. Schaum's Solved Problem Series. New York: McGraw-Hill.

Schilling, R. J., and S. L. Harris. 2000. *Applied Numerical Methods for Engineering Using MATLAB and C*. Pacific Grove, California: Brooks/Cole.

Schmidt 1908. Title unknown. Rendiconti del Circolo Matematico di Palermo **25**, 53–77.

Schoenberg, I. J. 1946. "Contributions to the problem of approximation of equidistant data by analytic functions." *Quarterly of Applied Mathematics* **4**, 45–99, 112–141.

Schoenberg, I. J. 1967. "On spline functions." In *Inequalities*, edited by O. Shisha, 255–291. New York: Academic Press.

Schrage, L. 1979. "A more portable Fortran random number generator." *Association for Computing Machinery Transactions on Mathematical Software* **5**, 132–138.

Schrijver, A. 1986. *Theory of Linear and Integer Programming*. Somerset, New Jersey: Wiley.

Schultz, M. H. 1973. *Spline Analysis*. Englewood Cliffs, New Jersey: Prentice-Hall.

Schumaker, L. L. 1981. *Spine Function: Basic Theory*. New York: Wiley.

Shampine, J. D. 1994. *Numerical Solutions of Ordinary Differential Equations*. London: Chapman and Hall.

Shampine, L. F., R. C. Allen, and S. Pruess. 1997. *Fundamentals of Numerical Computing*. New York: Wiley.

Shampine, L. F., and M. K. Gordon. 1975. *Computer Solution of Ordinary Differential Equations*. San Francisco: W. H. Freeman.

Shewchuk, J. R. 1994. "An introduction to the conjugate gradient method without the agonizing pain," `www.wikipedia.com`.

Skeel, R. D., and J. B. Keiper. 1992. *Elementary Numerical Computing with Mathematica*. New York: McGraw-Hill.

Smith, G. D. 1965. *Solution of Partial Differential Equations*. New York: Oxford University Press.

Sobol, I. M. 1994. *A Primer for the Monte Carlo Method*. Boca Raton, Florida: CRC Press.

Southwell, R. V. 1946. *Relaxation Methods in Theoretical Physics*. Oxford: Clarendon Press.

Späth, H. 1992. *Mathematical Algorithms for Linear Regression*. New York: Academic Press.

Stakgold, I., 2000. *Boundary Value Problems of Mathematical Physics*. Philadelphia: SIAM.

Steele, J. M., 1997. *Random Number Generation and Quasi-Monte Carlo Methods*. Philadelphia: SIAM.

Stetter, H. J. 1973. *Analysis of Discretization Methods for Ordinary Differential Equations*. Berlin: Springer-Verlag.

Stewart, G. W. 1973. *Introduction to Matrix Computations*. New York: Academic Press.

Stewart, G. W. 1996. *Afternotes on Numerical Analysis*. Philadelphia: SIAM.

Stewart, G. W. 1998a. *Afternotes Goes to Graduate School*. Philadelphia: SIAM.

Stewart, G. W. 1998b. *Matrix Algorithms: Basic Decompositions*, Vol. 1. Philadelphia: SIAM.

Stewart, G. W. 2001. *Matrix Algorithms: Eigensystems*, Vol. 2. Philadelphia: SIAM.

Stoer, J., and R. Bulirsch. 1993. *Introduction to Numerical Analysis*, 2nd Ed. New York: Springer-Verlag.

Strang, G. 2006. *Linear Algebra and Its Applications*, 4th Ed. Belmont, California: Thomson Brooks/Cole.

Strang, G., and K. Borre. 1997. *Linear Algebra, Geodesy, and GPS*. Cambridge, Massachusetts: Wellesley Cambridge Press.

Street, R. L. 1973. *The Analysis and Solution of Partial Differential Equations*. Pacific Grove, California: Brooks/Cole.

Stroud, A. H. 1974. *Numerical Quadrature and Solution of Ordinary Differential Equations*. New York: Springer-Verlag.

Stroud, A. H., and D. Secrest. 1966. *Gaussian Quadrature Formulas*. Englewood Cliffs, New Jersey: Prentice-Hall.

Subbotin, Y. N. 1967. "On piecewise-polynomial approximation." *Matematicheskie Zametcki* **1**, 63–70. (Translation: 1967. *Math. Notes* **1**, 41–46.)

Szabo, F. 2002. *Linear Algebra: An Introduction Using MAPLE*. San Diego, California: Harcourt/Academic Press.

Torczon, V. 1997. "On the convergence of pattern search methods." *Society for Industrial and Applied Mathematics Journal on Optimization* **7**, 1–25.

Törn, A., and A. Zilinskas. 1989. *Global Optimization*. Lecture Notes in Computer Science 350. Berlin: Springer-Verlag.

Traub, J. F. 1964. *Iterative Methods for the Solution of Equations*. Englewood Cliffs, New Jersey: Prentice-Hall.

Trefethen, L. N., and D. Bau. 1997. *Numerical Linear Algebra*. Philadelphia: SIAM.

Turner, P. R. 1982. "The distribution of leading significant digits." *Journal of the Institute of Mathematics and Its Applications* **2**, 407–412.

van Huffel, S., and J. Vandewalle. 1991. *The Total Least Squares Problem: Computational Aspects and Analsyis*. Philadelphia: SIAM.

Van Loan, C. F. 1997. *Introduction to Computational Science and Mathematics*. Sudbury, Massachusetts: Jones and Bartlett.

Van Loan, C. F. 2000. *Introduction to Scientific Computing*, 2nd Ed. Upper Saddle River, New Jersey: Prentice-Hall.

Van der Vorst, H. A. 2003. *Iterative Krylov Methods for Large Linear Systems*. New York: Cambridge University Press.

Varga, R. S. 2004. *Geršgorin and His Circles*, New York: Springer.

Varga, R. S. 1962. *Matrix Iterative Analysis*. Englewood Cliffs: New Jersey: Prentice-Hall. (2000. *Matrix Iterative Analysis: Second Revised and Expanded Edition*. New York: Springer-Verlag.)

Wachspress, E. L. 1966. *Iterative Solutions to Elliptic Systems*. Englewood Cliffs, New Jersey: Prentice-Hall.

Watkins, D. S. 1991. *Fundamentals of Matrix Computation*. New York: Wiley.

Westfall, R. 1995. *Never at Rest: A Biography of Isaac Newton*, 2nd Ed. London: Cambridge University Press.

Whittaker, E., and G. Robinson. 1944. *The Calculus of Observation*, 4th Ed. London: Blackie. New York: Dover, 1967 (reprint).

Wilkinson, J. H. 1965. *The Algebraic Eigenvalue Problem*. Oxford: Clarendon Press. Reprinted 1988. New York: Oxford University Press.

Wilkinson, J. H. 1963. *Rounding Errors in Algebraic Processes*. Englewood Cliffs, New Jersey: Prentice-Hall. New York: Dover 1994 (reprint).

Wood, A. 1999. *Introduction to Numerical Analysis*. New York: Addison-Wesley.

Wright, S. J. 1997. *Primal-Dual Interior-Point Methods*. Philadelphia: SIAM.

Yamaguchi, F. 1988. *Curves and Surfaces in Computer Aided Geometric Design*. New York: Springer-Verlag.

Ye, Yinyu. 1997. *Interior Point Algorithms*. New York: Wiley.

Young, D. M. 1950. Iterative methods for solving partial difference equations of elliptic type. Ph.D. thesis. Cambridge, MA: Harvard University. See `www.sccm.stanford.edu/pub/sccm/david_young_thesis.ps.gz`

Young, D. M., 1971. *Iterative Solution of Large Linear Systems*. New York: Academic Press: Dover 2003 (reprint).

Young, D. M., and R. T. Gregory. 1972. *A Survey of Numerical Mathematics*, Vols. 1–2. Reading, Massachusetts: Addison-Wesley. New York: Dover 1988 (reprint).

Ypma, T. J. 1995, "Historical development of the Newton-Raphson method." *Society for Industrial and Applied Mathematics Review* **37**, 531–551.

Zhang, Y. 1995. "Solving large-scale linear programs by interior-point methods under the MATLAB environment." Technical Report TR96–01, Department of Mathematics and Statistics, University of Maryland, Baltimore County, Baltimore, MD.

Index